CHILTON'S IMPORT EMISSION DIAGNOSIS and SERVICE MANUAL with VACUUM CIRCUITS

Managing Editor John H. Weise, S.A.E. □ **Assistant Managing Editor** David H. Lee, A.S.E., S.A.E

Service Editors Dennis J. Carroll, Nick D'Andrea, Martin Gunther, Robert McAnally, Jack T. Kaufmann, Lawrence C. Braun, A.S.C.
Editorial Consultants Edward K. Shea, S.A.E., Stan Stephenson

Production Manager John Cantwell
Manager Editing & Design Dean F. Morgantini
Art & Production Coordinator Robin S. Miller
Mechanical Artists Margaret A. Stoner, Bill Gaskins, Cynthia Fiore

National Sales Manager Albert M. Kushnerick □ **Assistant** Jacquelyn T. Powers
Regional Managers Joseph Andrews, Jr., James O. Callahan, David Flaherty

OFFICERS
President Lawrence A. Fornasieri
Vice President & General Manager John P. Kushnerick

CHILTON BOOK COMPANY Chilton Way, Radnor, Pa. 19089

 ISBN 0-8019-7652-9
Library of Congress Catalog No. 83-45334
1234567890 5432109876

CONTENTS

CARS

PICK-UP TRUCKS
(SPECIFICATIONS)

SAFETY NOTICE

Proper service and repair procedures are vital to the safe, reliable operation of all motor vehicles, as well as the personal safety of those performing repairs. This manual outlines procedures for servicing and repairing vehicles using safe effective methods. The procedures contain many NOTES, CAUTIONS and WARNINGS which should be followed along with standard safety procedures to eliminate the possibility of personal injury or improper service which could damage the vehicle or compromise its safety.

It is important to note that repair procedures and techniques, tools and parts for servicing motor vehicles, as well as the skill and experience of the individual performing the work vary widely. It is not possible to anticipate all of the conceivable ways or conditions under which vehicles may be serviced, or to provide cautions as to all of the possible hazards that may result. Standard and accepted safety precautions and equipment should be used when handling toxic or flammable fluids, and safety goggles or other protection should be used during cutting, grinding, chiseling, prying, or any other process that can cause material removal or projectiles.

Some procedures require the use of tools specially designed for a specific purpose. Before substituting another tool or procedure, you must be completely satisfied that neither your personal safety, nor the performance of the vehicle will be endangered.

Although information in this manual is based on industry sources and is as complete as possible at the time of publication, the possibility exists that some car manufacturers made later changes which could not be included here. While striving for total accuracy, Chilton Book Company cannot assume responsibility for any errors, changes, or omissions that may occur in the compilation of this data.

Audi

INDEX

VEHICLE IDENTIFICATION NUMBER AND ENGINE IDENTIFICATION

Vehicle Identification Number (VIN)

4000 MODELS

The vehicle identification number (VIN) is located on the left windshield pillar and in the engine compartment on a plate attached to a front brace.

5000 MODELS

The vehicle identification number (VIN) is on a plate mounted on the left side of the instrument panel and clearly visible through the windshield. It is also stamped into the upper right corner of the firewall. The vehicle identification plate is mounted on the right wheel housing.

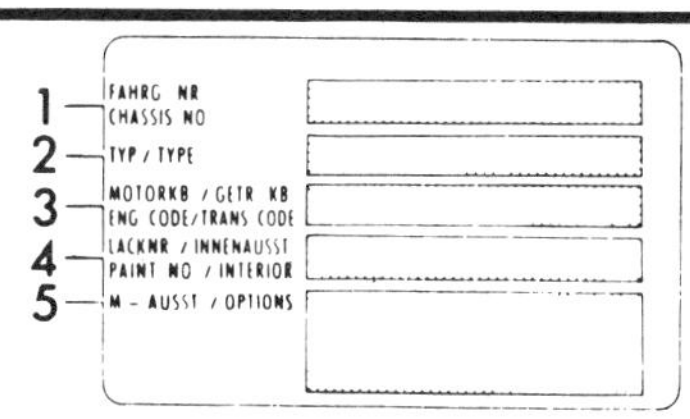

1. Vehicle Identification Number
2. Vehicle Code
3. Engine and Transmission Code
4. Paint and Interior Code
5. Option Codes

Typical vehicle identification label showing serial number location

Engine Identification

4000 MODELS

The engine identification number is located on the left side of the cylinder block, below the cylinder head and next to the distributor.

5000 MODELS

The engine number is stamped on the left side of the engine block (clutch housing). In addition to the engine number, a code number is stamped on the starter end of the cylinder block, just below the cylinder head. This number indicates the exact cylinder bore of the particular engine.

ENGINE IDENTIFICATION

Year	Model	Engine Code	Displacement (cu. in.)	(cc)	(liter)
'85–'86	4000	JN	109	1780	1.8
	4000 Coupe, Quattro	KX, KZ	136	2226	2.2
	5000, Wagon	KX, KZ	136	2226	2.2
	5000 Turbo	KH	131	2144	2.1
	Quattro	WX	131	2144	2.1

GASOLINE ENGINE TUNE-UP SPECIFICATIONS

Year	Model	Engine Displacement Cu. In. (cc)	Spark Plugs Type	Spark Plugs Gap (in.)	Distributor Point Dwell (deg)	Distributor Point Gap (in)	Ignition Timing (deg)	Fuel Pump Pressure (psi)	Idle Speed (rpm)	Valve Clear (in.) In	Valve Clear (in.) Ex
'85–'86	4000	109 (1780)	N8GY N281BY	.028	Electronic		6B @ Idle ②	64–74	850–1000	③	③
	5000 Turbo	131 (2144)	N8GY	.028	Electronic		21B @ 3,000 ①	72–82	750–850	③	③
	Quattro	131 (2144)	N6GY	.028	Electronic		①	75–85	790–910	③	③
	4000, Coupe	136 (2226)	N8GY	.028	Electronic		6B @ Idle	—	750–850	③	③
	4000 Quattro	136 (2226)	N8GY	.028	Electronic		6B @ Idle	—	750–850	③	③
	5000, Wagon	136 (2226)	N8GY	.028	Electronic		6B @ Idle	—	750–850	③	③

NOTE: The underhood specifications sticker often reflects tune-up specification changes made in production. Sticker figures must be used if they disagree with those in this chart. Spark plug torque is 14 ft. lbs. (20 Nm).

IN Intake
EX Exhaust
A After top dead center
B Before top dead center
① Distributor housing aligned with mark on Quattro and 5000 Turbo
② Supply hose to idle by-pass pinched shut
③ Models after Jan '84 have hydraulic lash adjusters—no adjustment is necessary.

FIRING ORDERS

NOTE: To avoid confusion, always replace spark plug wires one at a time.

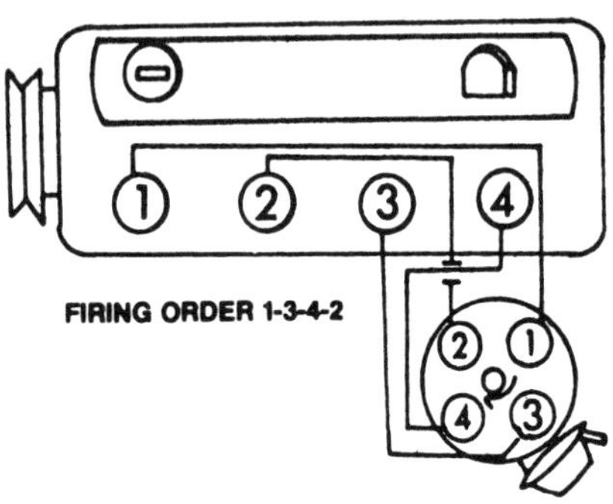

4 cylinder firing order

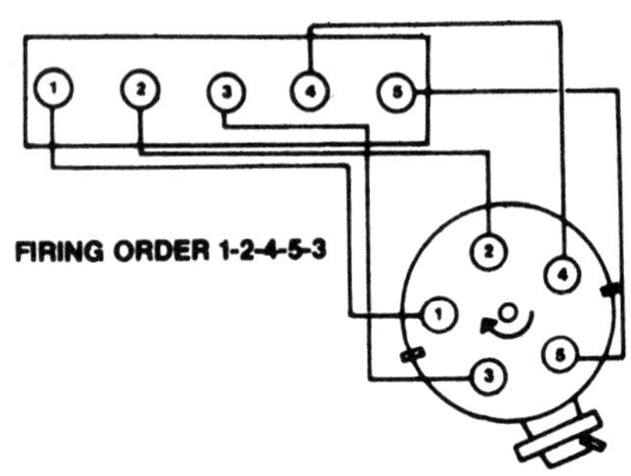

5 cylinder firing order

EMISSION EQUIPMENT USED

1985–86—ALL MODELS

Bosch Constant Injection System (CIS-E) Fuel Injection
Catalytic Converter (CAT)
Fuel Evaporation System (FES)
Positive Crankcase Ventilation (PCV)
Oxygen Sensor System (OXS)
Thermostatic Air Cleaner (TAC)
Idle Speed Stabilizer (ISS)
Exhaust Gas Recirculation (EGR)—Canada Only

REQUIRED EMISSION CONTROL MAINTENANCE

POSITIVE CRANKCASE VENTILATION (PCV)

Every 30,000 miles, all hoses and connections should be checked for leaks or deterioration. Remove and flush all lines with solvent, then blow dry using compressed air to clear any blockage. Check the PCV valve and, if found to be clogged, replace with a new one. Never attempt to clean an old PCV valve.

FUEL EVAPORATION SYSTEM (FES)

Every 30,000 miles, visually inspect all hoses, connections and components for damage, deterioration or leaks. Check all system components for proper operation and check charcoal canister for liquid fuel contamination.

THERMOSTATIC AIR CLEANER (TAC)

Replace the air cleaner element every 15,000 miles (30,000 miles on California models). The replacement interval should be cut in half if the vehicle is driven under extremely dusty conditions or in heavy traffic regularly. Check all system components for proper operation, damage or loose air duct connections.

EXHAUST GAS RECIRCULATION (EGR) SYSTEM

Inspect the system for leaks and deterioration every 30,000 miles. Clean the EGR valve and fittings and inspect manifolds for carbon deposits. Check the system for proper operation.

OXYGEN SENSOR

Replace the oxygen sensor every 30,000 miles and reset the service reminder light. Check the system for proper operation.

CATALYTIC CONVERTER

Check the exhaust system for damage, deterioration or leaks and correct as required. Check the converter body for external damage such as dents or cracks. If severe damage is noted, the converter should be removed and the catalyst visually inspected for internal damage.

TURBOCHARGER SYSTEM

Check the torque of all mounting bolts every 15,000 miles. Road test the vehicle and check for proper turbocharger and wastegate operation. Check all turbocharger oil lines for leaks and correct as necessary.

NOTE: Turbocharged models require more frequent oil changes to assure proper operation.

UNDERHOOD RUBBER AND PLASTIC COMPONENTS (EMISSION HOSES)

Inspect all hose surfaces for evidence of heat or mechanical damage. Hard or brittle rubber, cracking, tears, cuts, abrasions and excessive swelling indicate deterioration of the rubber. Inspect all hose connections and hoses and replace defective parts as required.

IDLE SPEED AND MIXTURE ADJUSTMENTS

NOTE: The following information is being published from the latest information available at the time of publication. If the information differs from the values given on the underhood emission control label, use the data on the label.

Adjustment

1985–86—ALL MODELS

NOTE: On all models, the idle speed, mixture adjustment and ignition timing must be checked together. Exhaust gas mixture must be checked and adjusted by measuring milliamps (mA), with a special meter and with the oxygen sensor connected. Refer to the underhood vacuum diagram for all hose locations.

4000 MODELS (4 CYL)

1. Start the engine and allow it to reach normal operating temperature. The radiator fan must come on at least once.
2. Switch OFF all accessories. The radiator fan and air conditioner must not be running during the test.
3. Pinch the hose to the idle speed boost valve.
4. Pull the crankcase breather hose off the valve cover and allow it to vent to the atmosphere.
5. Remove the cap from the T-piece.
6. Connect SIEMENS 451 tester (or equivalent) according to the manufacturer's instructions.
7. Remove the cap from the CO probe receptacle and insert the CO probe. The hose must fit tightly so there is no exhaust leak.
8. To measure milliamps (mA), connect a digital multimeter to the differential pressure regulator. Remove the connector from the differential pressure regulator and install adaptor VW 1515 A/1 or equivalent. Connect the multimeter to the adaptor and set the selector switch to DCA 20 mA.

NOTE: If the engine does not run after the adaptor is connected, the connections may be improper. Reverse the plug and try again.

9. Check the idle speed and adjust to specifications if necessary.
10. Check the ignition timing and adjust to specifications if necessary.

IDLE SPEED

Model	Engine Displacement Cu. In. (cc)	Idle Speed (rpm)
4000	109 (1780)	850–1000
5000 Turbo	131 (2144)	750–850
Quattro	131 (2144)	790–910
4000, Coupe	136 (2226)	750–850
4000 Quattro	136 (2226)	750–850
5000, Wagon	136 (2226)	750–850

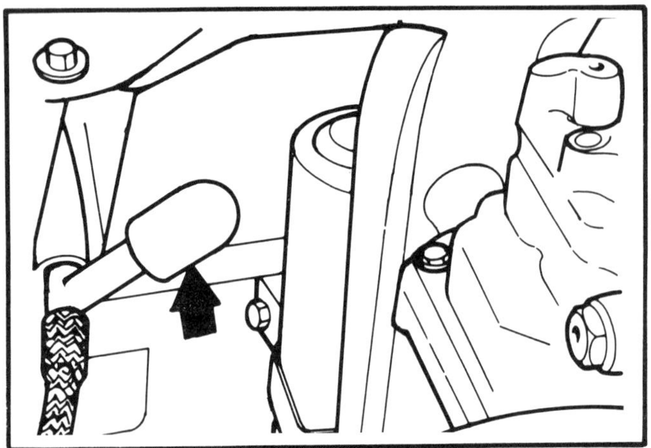

Location of cap (arrow) and T-connection at activated charcoal canister hose

11. Read the milliamp and CO values on the testers. The multimeter should read between 4–16 mA (reading fluctuates) and the CO value should be 0.3–1.2%.

12. If the current reading is less than 4 mA or more than 16 mA, remove the CO adjustment plug as follows:

a. Stop the engine and remove the boot on the mixture control unit.

b. Centerpunch the plug in the CO adjusting hole.

c. Drill a $^{3}/_{32}$ in. (2.5mm) hole in the center of the plug $^{9}/_{64}$–$^{5}/_{32}$ in. (3.5–4mm) deep.

d. Carefully clean up any metal shavings and insert a $^{1}/_{8}$ in. sheet metal screw into the drilled hole, then remove the screw and plug with pliers.

13. Start the engine and allow it to idle.

14. Adjust the current reading on the multimeter by turning the CO adjusting screw with tool P377 or equivalent to obtain 8–12 mA (reading fluctuates). Turn the adjusting screw clockwise to lower the current reading (CO higher) and counterclockwise to raise the current reading (CO lower).

CAUTION

Do not push adjustment tool down when making CO adjustment and do not accelerate the engine with the tool in place. Remove the tool after each adjustment and accelerate the engine briefly before reading multimeter. Always adjust in direction from high to low.

15. After CO adjustment is complete, recheck the idle speed and adjust if necessary. Reconnect the crankcase breather hose. If, after reconnecting the breather hose, the current reading drops below specifications, an oil change may be necessary.

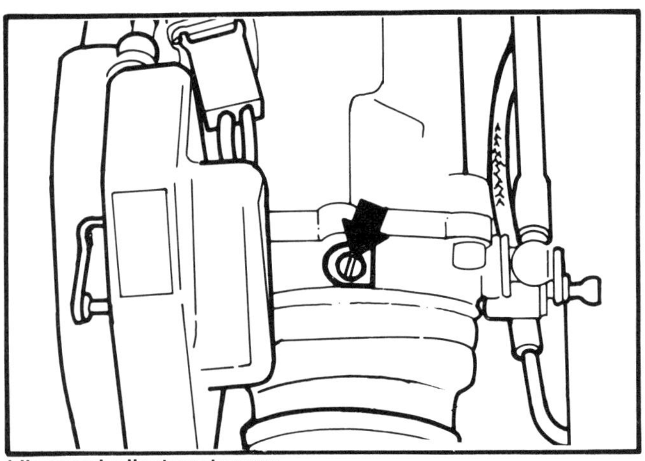

Idle speed adjustment screw

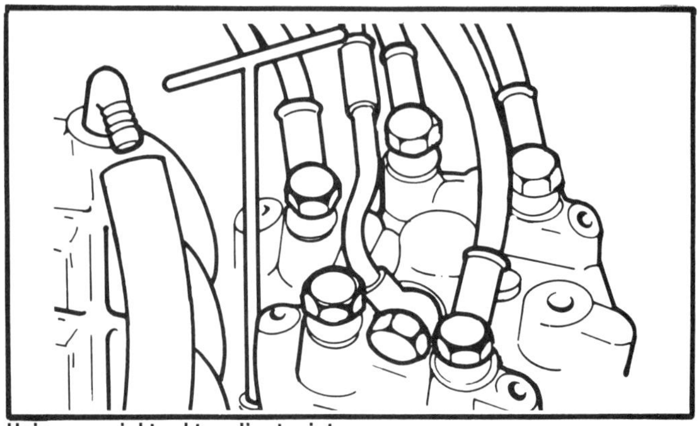

Using special tool to adjust mixture

16. Turn the ignition OFF and drive in a new mixture plug, flush with the mixture control unit.

17. Remove all test equipment, attach CO probe cap and remove the device used to pinch the idle speed boost valve hose.

4000 AND 5000 MODELS (5 CYL)

NOTE: Idle speed, ignition timing and oxygen sensor duty cycle (mixture) must be checked and adjusted together.

1. Check that there are no leaks in the exhaust system and connect Siemens 451 tester or equivalent according to the manufacturer's instructions. Make sure the TDC sending unit is installed snugly into the transmission housing.

2. Remove the cap from the CO probe receptacle and install CO test probe. Make sure the hose fits snugly so there is no exhaust leak.

3. Disconnect the crankcase breather hose at the cylinder head cover and plug the hose.

4. Disconnect both plugs at the idle stabilizer and plug the connectors together to bypass the unit. Make sure the connectors are tight.

5. Turn OFF all accessories. If any fuel lines were disconnected or replaced, start the engine and run it to 3000 rpm several times, then let idle for at least two minutes.

6. Start the engine and allow it to reach normal operating temperature. The radiator fan must come on at least once.

NOTE: The radiator fan must not be running during all tests and adjustments.

7. Check and adjust the idle speed if necessary.

8. Check and adjust the ignition timing if necessary. Turn ignition OFF.

9. Connect the dwell lead on the tester to the oxygen sensor blue/white test connection. Remove the cap from the charcoal canister purge line.

10. Remove the test lead from No. 1 ignition wire and TDC sending unit. Make sure the tester is on the 4 cylinder scale and press "%" button. Start the engine and check the oxygen sensor duty cycle (dwell) and CO at idle speed.

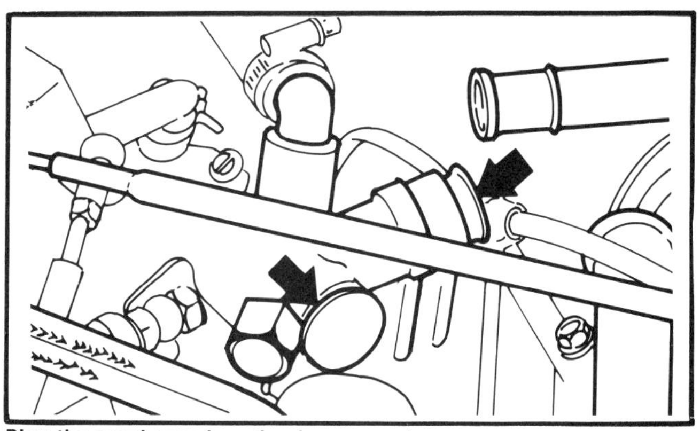

Plug the crankcase breather hoses as shown

11. The OXS duty cycle should be 25–65% (reading should fluctuate) and the CO should be between 0.3–1.2%. If the CO is more than 1.2%, but the dwell range is correct, check for leaks in the intake or exhaust system, malfunctioning fuel distributor, or a faulty injector spray pattern.

12. If the OXS duty cycle is less than 25% or more than 65%, remove the tamper-proof plug as described in Step 12 of the 4 cylinder procedure, above.

13. Using adjusting tool P377 or equivalent, adjust the OXS duty cycle (mixture) by turning the adjusting screw clockwise to lower the meter reading, or counterclockwise to increase the reading. Adjust to 44–56% (reading should fluctuate). The radiator fan must not be running during adjustment.

CAUTION

Do not push adjustment tool down while adjusting CO level. Never accelerate the engine with the tool in place and remove the tool and accelerate the engine briefly after each adjustment.

14. Readjust the idle speed, if necessary, then turn ignition OFF and remove the test equipment. Be sure to replace the CO probe cap to prevent exhaust leaks. Reconnect the crankcase breather hose and idle stabilizer. Install the cap to the charcoal canister purge line and replace the tamper-proof adjustment plug.

TURBOCHARGED MODELS

NOTE: Idle speed, ignition timing and oxygen sensor duty cycle (mixture) must be checked and adjusted together with the engine at normal operating temperature, throttle valve in the idle position and all electrical accessories OFF. If injector lines were disconnected or replaced, start and run the engine to 3000 rpm several times, then let idle for two minutes prior to test.

1. Connect Siemens 451, VW 1367, or equivalent tester according to manufacturer's instructions. Make sure the TDC sensor is firmly seated into the transmission housing.
2. Disconnect the green wire from terminal #1 of the ignition coil.
3. Remove the cap from the CO test tube and connect the hose from the CO tester. Make sure the hose connection is tight to eliminate any exhaust leak.
4. Disconect the crankcase ventilation hoses at the valve cover and seal the ends.
5. Remove the cap from the T-connection at the charcoal canister hose.
6. Start the engine and allow it to reach normal operating temperature. The cooling fan should come on at least once. Adjust the idle speed to specifications if necessary.

NOTE: The cooling fan must not be running during any tests or adjustments.

7. Switch ignition OFF, then disconnect the test cable from terminal #1 of the ignition coil and connect to the blue/white wire of the oxygen sensor. Disconnect the test connections from No. 1 cylinder ignition cable and TDC sensor. Push "Dwell Angle %" button on the tester.
8. Start the engine and allow it to idle. Raise the engine speed above 2000 rpm for five seconds, then check the oxygen sensor (OXS) duty cycle. Duty cycle should read 25–65% (reading should fluctuate) and the CO value should be 0.3–1.2% with the oxygen sensor connected.
9. If the CO value exceeds 1.2% with the duty cycle within 25–65%, check for a fault in the ignition system, leaks in the exhaust system or a problem with the fuel distributor. If the duty cycle is less than 25% or exceeds 65%, a CO adjustment is necessary.
10. Switch the ignition OFF and remove the CO adjustment plug as outlined in Step 12 of the 4 cylinder adjustment procedure, above.
11. Using CO adjustment tool P377 or equivalent, turn the adjustment screw clockwise to lower the OXS duty cycle, or counterclockwise to raise it to obtain a reading of 42–58% (reading should fluctuate). Do not lift or press down on the adjustment tool when making adjustments and remove the tool and briefly accelerate the engine after each adjustment. Never accelerate the engine with the tool in place.
12. Once the correct duty cycle reading is obtained, reset the idle speed if necessary.
13. Switch the ignition OFF and disconnect the CO tester. Install the cap on the CO test tube, making sure it is tight to prevent exhaust leaks. Disconnect all test equipment and restore all disconnected hoses. Install the cap on the charcoal canister T-connection and drive in a new tamper-proof mixture plug.

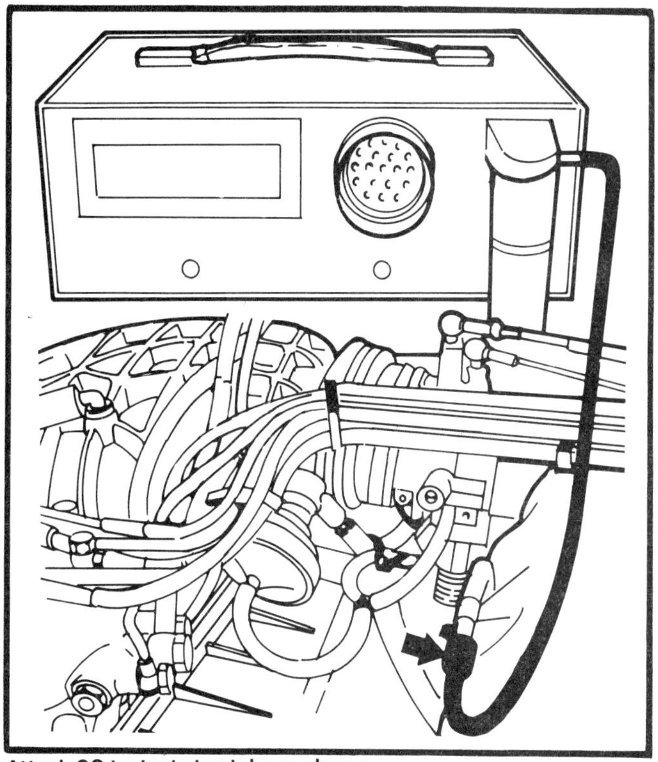

Attach CO tester to tap tube as shown

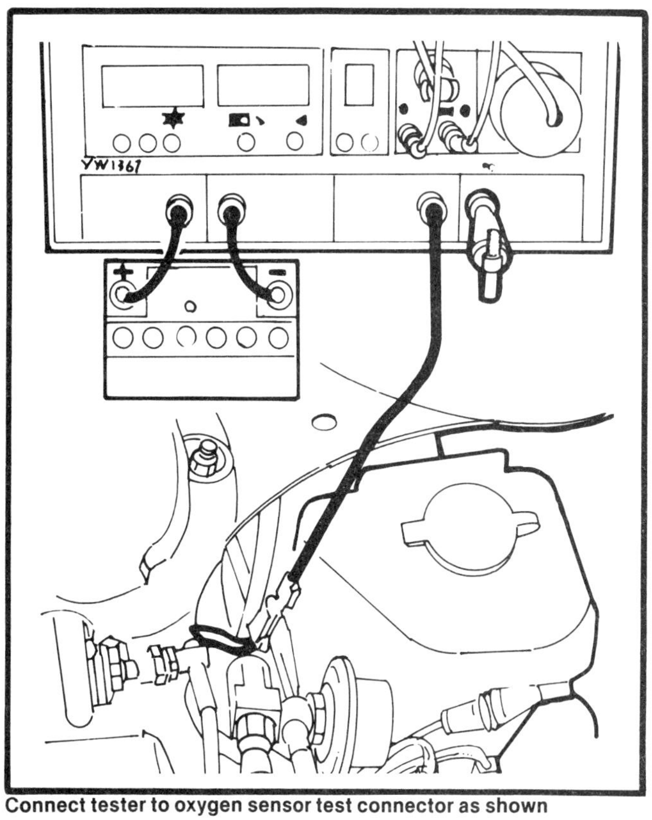

Connect tester to oxygen sensor test connector as shown

INITIAL TIMING SETTINGS

STATIC TIMING

A basic timing adjustment may be made by turning the engine in the normal direction of rotation until the timing mark is aligned with the timing pointer and the distributor rotor points toward the No. 1 cylinder mark on the rim of the distributor body (timing marks are located on the flywheel on all engines). This will put the No. 1 cylinder at top dead center (TDC) on the compression stroke. Connect a 12 volt test lamp between the ignition coil terminal connected to the distributor and a ground. Rotate the distributor clockwise until the lamp goes out. Turn the distributor counterclockwise until the lamp just lights, then tighten the distributor hold-down at that point. The distributor timing will now be approximately set. Connect a suitable timing light and set dynamic timing to specifications as outlined below.

DYNAMIC TIMING

Model	Engine Displacement Cu. In. (cc)	Ignition Timing (deg)
4000	109 (1780)	6B @ Idle ②
5000 Turbo	131 (2144)	21B @ 3,000 ①
Quattro	131 (2144)	①
4000, Coupe	136 (2226)	6B @ Idle
4000 Quattro	136 (2226)	6B @ Idle
5000, Wagon	136 (2226)	6B @ Idle

① Distributor housing aligned with mark on Quattro and 5000 Turbo.
② Supply hose to idle bypass pinched shut.

SPARK PLUGS

NOTE: The following information is being published from the latest available information at the time of publication. If the information differs from the information on the underhood emission control label, use the data on the label. Audi Coupe and 5000S models require spark plug replacement every 15,000 miles for 49 State vehicles and every 30,000 miles for California vehicles.

SPARK PLUGS

Model	Engine Displacement Cu. In. (cc)	Spark Plugs Type	Spark Plugs Gap (in.)
4000	109 (1780)	N8GY N281BY	.028
5000 Turbo	131 (2144)	N8GY	.028
Quattro	131 (2144)	N6GY	.028
4000, Coupe	136 (2226)	N8GY	.028
4000 Quattro	136 (2226)	N8GY	.028
5000, Wagon	136 (2226)	N8GY	.028

NOTE: Spark plug torque is 14 ft. lbs.

EMISSION CONTROL SYSTEMS

CRANKCASE EMISSION CONTROL SYSTEM

All Audi engines use a closed, positive crankcase ventilation system which cycles crankcase vapors back into the intake manifold for combustion with the fuel/air mixture. The oil filler cap is sealed and crankcase vapors are drawn from the top of the valve cover into the intake manifold through a reduction piece which is installed in the breather hose between the intake manifold and valve cover.

Maintenance

The positive crankcase ventilation (PCV) system should be checked for proper operation every 30,000 miles. Check the condition of all hoses and connectors and check for leaks in the system. Disconnect all hoses completely, wash with solvent, then blow clean with compressed air. If extreme blockage is encountered, the hoses should be replaced.

EVAPORATIVE EMISSION SYSTEM

A sealed filler cap allows vacuum to draw ambient air into the tank to equalize the pressure as fuel is used, but does not allow fuel vapors to escape into the atmosphere. Fuel evaporative system (FES) components include an expansion chamber integrated into the fuel tank, filler cap, charcoal canister, charcoal filter valve, T-piece with restrictor and the various vapor and vacuum lines which tie the system together. The charcoal canister stores the fuel vapors until the engine is started, when engine vacuum draws the vapors through the top of the canister into the intake bellows where they are drawn into the cylinders and burned with the normal fuel charge.

Maintenance

The fuel evaporative system should be inspected every 30,000 miles. Check all hoses and connections for leaks, cracks and deterioration. With the engine OFF, remove the hose connector from the top of the charcoal filter valve, then blow into the valve port. No air should pass through the valve, if it does, replace the valve.

EXHAUST EMISSION CONTROL SYSTEM

Thermostatic Air Cleaner (TAC)

To allow more efficient combustion, the intake air is maintained at 86°F by the thermostatic air cleaner. A thermostat mounted in an air mixture housing in line with the air intake ducts opens the control flap to draw air from the hot air duct. Above 86°F, the flap closes and allows cooler air into the engine intake system.

Maintenance

The air filter element should be changed every 15,000 miles, more frequently if the vehicle is driven under extremely dusty conditions or in heavy traffic regularly. Aside from periodically checking for free movement of the air flap, the TAC system requires no routine maintenance. California vehicles require air filter element replacement at 30,000 miles.

Exhaust Gas Recirculation (EGR) System

NOTE: The EGR system is used on Canadian models only.

Exhaust gas recirculation is used to lower the combustion chamber temperature, thereby limiting the amount of oxides of nitrogen (NOx) produced in the exhaust gas. EGR is not needed at idle and is cut off at full throttle. Audi uses a vacuum operated EGR valve controlled by a connection near the throttle plate. The EGR system includes a temperature valve, vacuum amplifier and deceleration valve.

Maintenance

If equipped with a service reminder, a dash light will illuminate at 15,000 or 30,000 mile intervals to indicate the need for EGR service. Replace the EGR valve and reset the service reminder by depressing the button marked "EGR" on the mileage counter located in the engine compartment or under the rear seat. Clean any carbon deposits from all components. If excessive carbon buildup is noted on the intake ports, the intake manifold should be removed and cleaned. Never attempt to clean the intake manifold while it is attached to the engine.

EGR Valve Test

The EGR operation can be tested by switching the vacuum lines temporarily so that exhaust gas is circulated at idle. If the EGR valve is working properly, the engine will run roughly or stall. To test an EGR valve with a single vacuum connection, pull a vacuum line from the distributor vacuum unit retard side and connect it to the EGR valve. The engine should stall or run roughly.

EGR SYSTEM DIAGNOSIS

Symptom	Possible Cause	Repair
EGR valve does not move during test.	Leaking or clogged vacuum line or hose connection. Defective EGR valve.	Repair vacuum line or replace valve.
Engine won't idle or dies on return to idle.	EGR control system blocked or inoperative.	Check all vacuum connections and control valve. Replace defective control valve.
EGR valve leaks in closed position. Poor wide-open throttle performance.	EGR valve defective or clogged with carbon. Defective EGR control valve.	Clean or replace EGR Valve. Replace control valve.
EGR warning light on.	Normal service due.	Service EGR system and reset counter.
Excessive HC and CO levels in exhaust.	Air injection system inoperative.	Check all components for proper operation. Check all hoses for unrestricted flow.

Temperature Valve Test

The temperature valve used is a bimetal snap-switch, installed in a cooling system fitting so that it will open and close according to the temperature of the coolant. The temperature valve controls the vacuum which operates the EGR valve. It can be tested as follows:

1. With the engine cold, disconnect the vacuum line at the EGR valve and replace with an extra length of hose. Blow through the hose; the valve should be closed.
2. Start the engine and allow it to reach normal operating temperature.
3. With the engine warm, again blow through the hose; the valve should now be open. If not, replace the valve.

Vacuum Amplifier Test

The vacuum amplifier is used to control the EGR vacuum signal. At idle, there is not enough control vacuum to operate the amplifier. At part throttle, control vacuum increases and trips the amplifier which then allows engine vacuum to operate the EGR valve. At full throttle, the engine vacuum has dropped to the point where it is not strong enough to allow EGR operation, thereby cutting off exhaust gas recirculation under load. To test the vacuum amplifier operation:

1. Start the engine and allow it to reach normal operating temperature.
2. Connect a vacuum gauge in-line between the vacuum amplifier and the throttle valve port. The reading should be 0.2–0.3 in. Hg for 4000 models and 0.43 in. Hg for 5000 models.
3. Connect a vacuum gauge in-line between the vacuum amplifier and the temperature valve. The reading should be 2–4 in. Hg for 4000 models and 2.5–2.8 in. Hg for 5000 models.

Deceleration Valve Test

During deceleration, engine vacuum suddenly increases causing tiny droplets of condensed fuel in the intake manifold to evaporate, resulting in an overly rich fuel mixture causing high emissions or backfiring. The deceleration valve opens to allow more air to flow into the intake manifold to lean out this rich mixture. To test deceleration valve operation:

1. Start the engine and allow it to reach normal operating temperature.
2. Remove the vacuum hose on the intake side of the deceleration valve and plug the connection. If the valve is working properly, the idle speed should not change.
3. Remove the plug and place a finger over the connection. Increase the engine speed and then release the throttle. If the valve is operating properly, you should feel vacuum; if not, replace the deceleration valve.

NOTE: If the deceleration valve is operating properly, but the engine does not return to the correct idle speed after deceleration, the valve may be adjusted. Loosen the lock nut and turn the adjusting rod until the correct idle speed is obtained. Do not turn the adjusting rod more than two complete turns.

FUEL INJECTION

CIS-E Electronic Control System

The CIS-E is an electronically controlled continuous fuel injection system. This system uses the basic CIS mechanical system for injection, with electronically controlled correction functions. The electronic portion of the system consists of an air flow sensor position indicator, electro-hydraulic actuator, thermo time switch, coolant temperature sensor, electronic control unit (ECU), transistorized ignition switching unit, microswitch, throttle valve switch, altitude correction indicator, lambda control and oxygen sensor. The mechanical portion of the CIS-E system consists of a mixture control unit, control pressure regulator, auxiliary air valve, cold start valve, injector nozzles, fuel pump and fuel filter.

ELECTRO-HYDRAULIC ACTUATOR (EHA)

This actuator is flanged onto the fuel distributor and acts as a pressure regulator which operates as a plate valve. The position of the plate valve can be varied causing a differential pressure change in the actuator and lower chamber, and this will cause a mixture correction. When the engine is running a constant system pressure of 78 psi is applied to the fuel inlet. The plate valve is adjusted depending on current intensity and therby determines the flow rate, in combination with the fixed orfice. The corresponding pressure change in the lower chamber cause movement of the diaphragm and influneces the volume of fuel flowing to the fuel injector valve.

COLD START

Electrical current at the actuator is 120 mA to 8 mA or less. The plate valve is positioned in the direction of the intake port. Differential pressure drop in the lower chamber, is 5.8–22 psi. As the coolant temperature increases, the current at the actuator drops to 8 mA and at the same rate, the differential pressure drops down to 5.8 psi.

ACCELERATION ENRICHMENT

Electrical current to the actuator depends on the coolant temperature and on the speed at which the air flow meter plate is deflected. When the plate valve is moved closer to the intake, the differential pressure will decrease by 22 psi. When the coolant temperature reaches 168°F., the acceleration enrichment is cancelled and the ECU will provide acceleration enrichment as an impulse, which will increase the instantaneous current value.

NOTE: During acceleration enrichment, the lambda control is influenced by the ECU.

VOLTAGE PROTECTION RELAY

This relay is located in the engine compartment on the right side at the electrical center. The function of this relay is to protect the electronic components of the CIS-E system. There is a ten (10) amp fuse located on top of the relay. When the ignition switch is turned to the ON position, the relay closes and directs battery voltage to the ECU.

NOTE: The locations of the various components in this system may differ from model to model.

ELECTRONIC CONTROL UNIT (ECU)

This unit is located on the main firewall in the engine compartment on the right side. There is a vent system installed for dissipating heat. The ECU acts as the brain of the system and is supplied with a voltage correction circuit (operating voltage—8 volts) to prevent voltage fluctuations when different vehicle components are switched on. The ECU uses various input signals for control of fuel delivery and exhaust emissions. The input signals are then converted into corresponding current values and sent to the electro-hydraulic actuator and to the idle speed air valve. During cranking, an enrichment signal is provided through terminal #50 and the amount of enrichment depends on the coolant temperature. A timing element requlates the enrichment after one second to the warm-up and after starting value. The value will remain constant, as long as the enrichment is cranked. After-start enrichment is used to provide a smooth runing engine after starting and

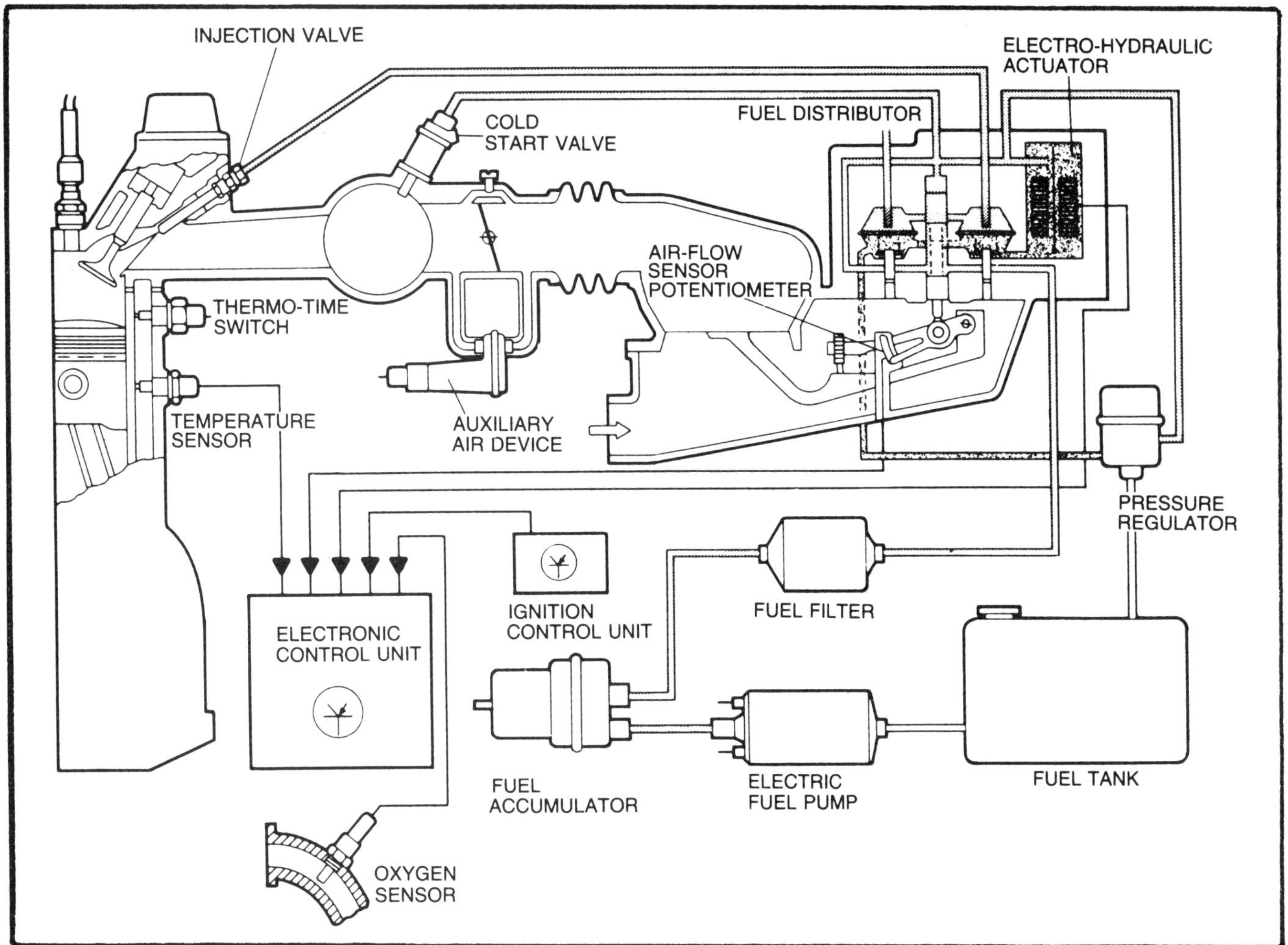

Typical schematic of Bosch KE-Jetronic fuel system

the amount and duration of the after start enrichment depends on the coolant temperature. Warmup enrichment again depends on the coolant temperature, the lower the coolant temperture, the higher the current at the actuator and the greater the fuel enrichment.

The maximum engine speed is limited by the ECU by sending the necessary current to the electro-hydraulic actuator. The lower chamber pressure is increased to the system pressure and the fuel supply to the injection valves is interrupted. Depending on the altitude, the amount of fuel is changed based on a signal from the altitude correction capsule. Once the ignition switch is turned on, the altitude correction indicator will receive a constant voltage signal of 8 volts from the ECU.

The lambda (oxygen sensor) control monitors the input signals from the sensors and amplifies them and calculates the output signals for the electro-hydraulic actuator. The electronic idle speed control is also incorporated into the ECU and the signals are received by the idle speed air valve, which is located in a hose which by-passes the throttle valve. When the ignition switch is turned on, the ECU is energized and the electronic control system will generate a basic frequency of 100 Hz. The idle speed air valve is opened to the maximum position by a set of return springs, when the ignition switch is in the off position. Once the ignition switch is turned on the valve is activated by a specific voltage, providing an air valve opening (dependent on the coolant temperature). The nominal speed is controlled, depending on the coolant temperature, 1000 rpm at 0°F to 720 rpm at 68°F.

The microswitch is energized by the ECU with a constant voltage signal of 8 volts. During decel operation, the circuit to the ECU is closed by the microswitch. The setting-in rate of speed of decel shutoff depends on the coolant temperature.

FUEL SYSTEM

Testing & Adjustments

1. Remove the air cleaner and may a visible check for any fuel leaks. Pull off the fuel pump relay and place a jumper wire between terminals 7 and 8 for a minute or two, so as to establish fuel pressure.
2. Push the air flow sensor plate down by hand, take note that a equal resistance should be felt across the entire path and there should be no resistance felt during the fast upward movement. If upward movement is slow, check the control piston.
3. Check the control piston in the fuel distributor for any fuel leaks.
4. Push the air flow sensor plate as far down as it can go for a minute or less, if a small amount of fuel leaks it is in proper order, and if there is no signs of large fuel leaks go on to the next test.

Fuel Pressure Check

1. Using fuel pressure gauge No. 100-589-13-21-00 or equivalent, connect the gauge to the fuel line on fuel distributor lower chamber and connect the other hose to the upper chamber of the fuel distributor.
2. Disconnect the fuel pump relay and with a jumper wire connect the number 7 and 8 terminals and check the lower chamber fuel pressure with the engine off.
3. Start the engine and bring up to operating temperature, pull off the electrical connection on the electro-hydraulic controlling element. Be sure to note the position of the pressure gauge control valve.

4. Take the pressure reading on the lower chamber, the pressure at normal operating temperature should be 6 psi below system pressure. When plugging in the electrical connections, there should be no changes in the fuel pressure.

5. Disconnect the plug from the coolant temperature sensor and using a multi-meter or equivalent connect the leads to the mA scale. Turn the ignition switch to the ON position and bypass the fuel pump relay, by inserting a jumper wire into the relay terminals 7 and 8. Read The lower chamber pressure and amperage.

NOTE: Whenever switching from amps to volts, pull off connecting lines from the meter.

6. The pressure reading in the lower chamber should be 78 psi or 5.8 psi below system pressure. Amperage should be 75 mA.

7. If the pressure is not within specifications, check the following:
   a. Check the coolant temperature sensor and the coolant level.
   b. Check the ECU and the electro-hydraulic controlling element.

8. If the pressure in the lower chamber is higher than specifications, check the throttle orifce in the fuel distributor for a blocked passage.

Decel Shut-Off Test

1. With the fuel pump relay installed and the engine running at normal operating temperature, be sure that the control valve on the fuel gauge is open.

2. Increase the engine speed for a minute or two to 2500 rpm, when the engine speed drops, the fuel pressure should increase by 6.0 psi. Combustion should start again at around 900 rpm.

3. If the pressure can not be attained, check the microswitch, control current on the controlling element, the TDC signal and The ECU.

Full-Load Enrichment Test

1. Seperate the coupling between the throttle valve switch and the ECU. Check the resistance from the throttle valve switch with a ohmmeter.

2. In the idle speed position the resistance from the throttle valve switch should read to the maximum resistance of the scale, and under a full load the resistance should read 0 ohms.

3. If the throttle valve switch does not pass the test, remove and replace it.

4. Connect test cable No. 102-589-04-63-00 or equivalent to the controlling element and set the multi-meter to the mA scale. Use a jumper wire to connect the throttle valve switch to the ECU.

5. Turn the ignition switch ON, the reading on the meter should be 7–9 mA. Check the circuit between the ECU and controlling element for passage, there should be 0 ohms there. If the readings check out all right, replace the ECU. If the readings do not check out all right, repair the short in the circuit.

Electronic Idle Speed Test

1. Connect the test cable of the Bosch tester No. KDJE-P-600 or equivalent to the idle speed air valve. Push the IR 100% button, with the engine idling and at operating temperature the reading should be 27–29% at 750 rpm. If the reading is within specifications the test is over, if the reading is out of specifications go on to the next step.

2. If the reading is higher or lower than specifications, adjust the nominal value or test the microswitch. If the reading is 0, test the voltage on the plug for the idle speed air valve.

3. Use pin socket number 2 to ground, the reading should be 12 volts, if the reading checks out, go on to the next step. If the reading does not check out, repair the short in the circuit.

4. Test the resistance on the idle speed air valve terminals 2 and 3, it should be 12 ohms. Terminal 2 and 1 should be 12 ohms, if the reading are out of range replace the idle speed air valve.

Throttle Valve Switch (TVS) Test

1. Using a multimeter, set the ohm scale to zero and check the full throttle stop.

2. Push the TVS against the full throttle stop, the reading should be 0, turn the TVS slightly in the direction of the idle, the reading should go to the maximum resistance of the scale.

3. If the readings do not meet the specifications, replace the throttle valve switch.

OXYGEN SENSOR

Removal and Installation

The oxygen sensor must be replaced every 30,000 miles. A service reminder light located on the dash board will illuminate at 30,000 mile intervals to indicate the need for oxygen sensor service. Once the sensor is replaced, the service reminder must be reset to extinguish the dash indicator light.

The oxygen sensor is located in the exhaust manifold. The exhaust system should be slightly warm to make removal easier. When installing a new oxygen sensor, coat the threads with an anti-seize compound but do not allow any anti-seize to contaminate the sensor body or it will ruin the probe. The service reminder is located either behind the dash panel or under the rear seat and is reset by depressing a button on the unit.

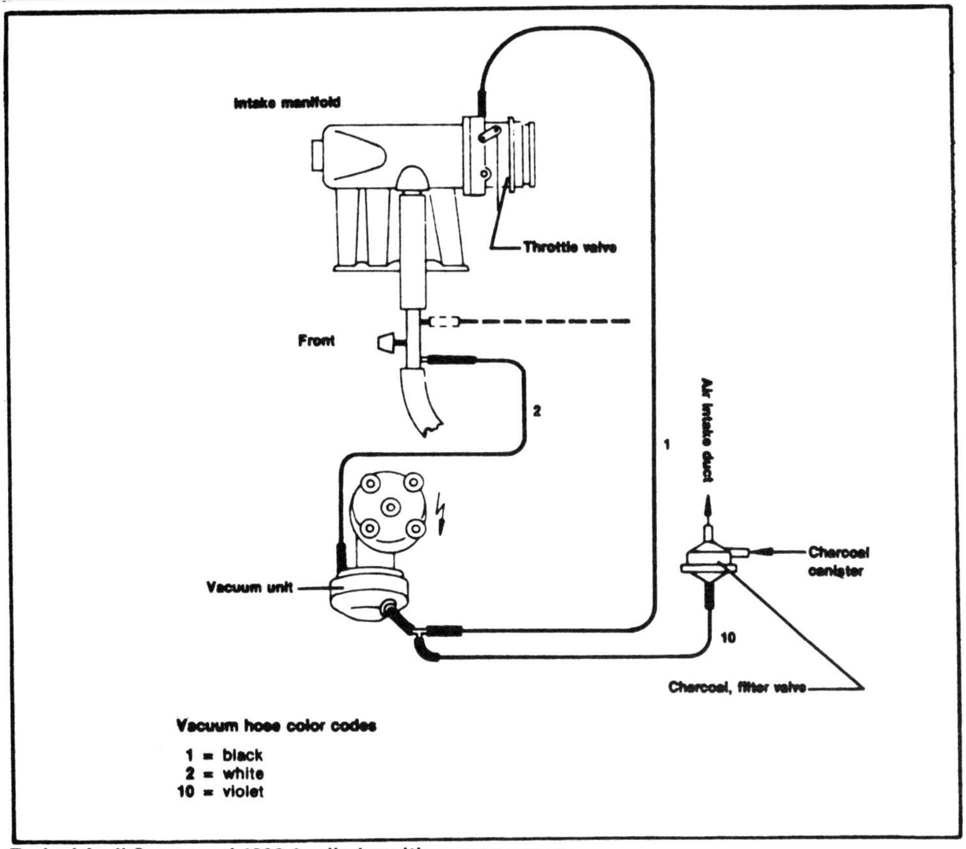

Typical Audi Coupe and 4000 4 cylinder with oxygen sensor

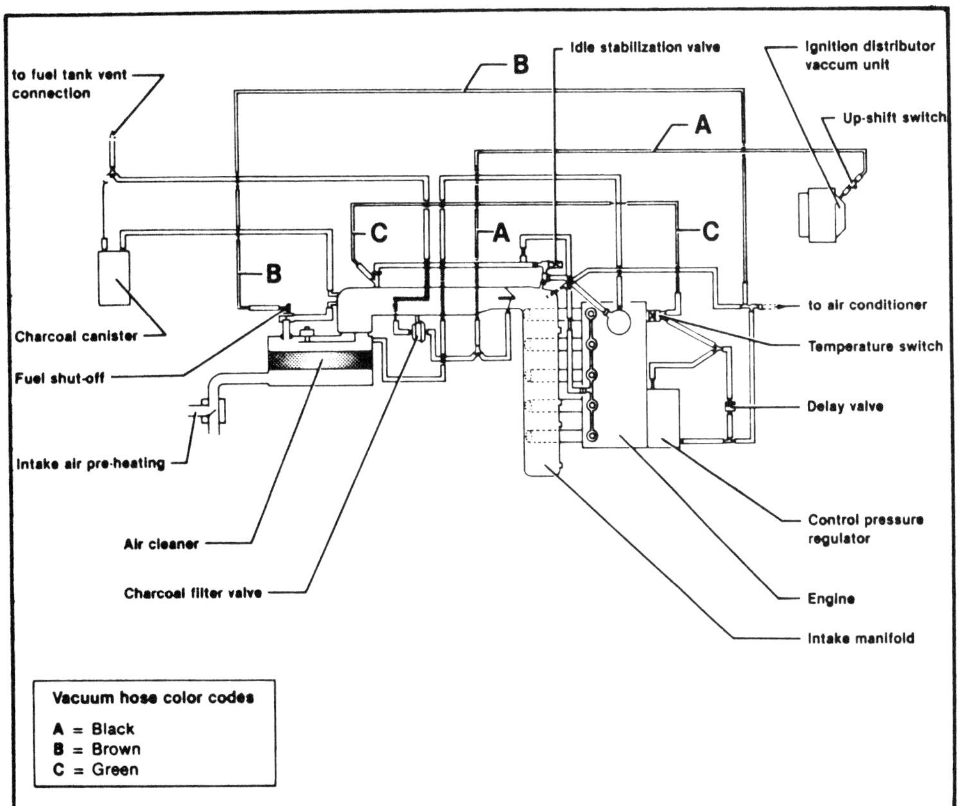

Audi 5000 USA models with manual transmission

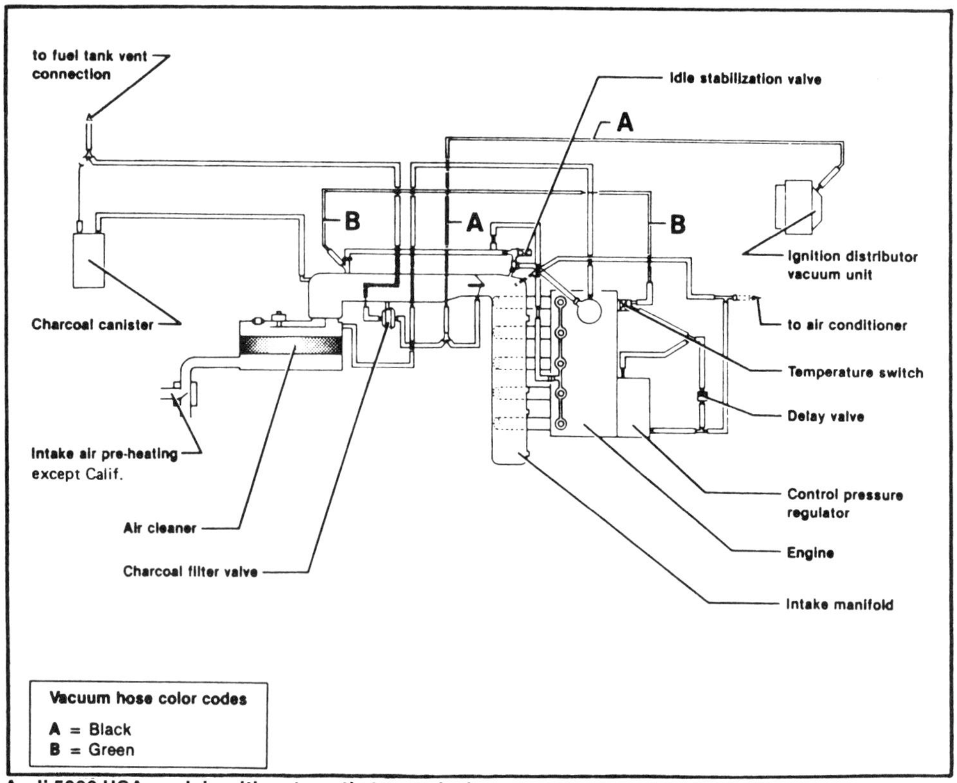

Audi 5000 USA models with automatic transmission

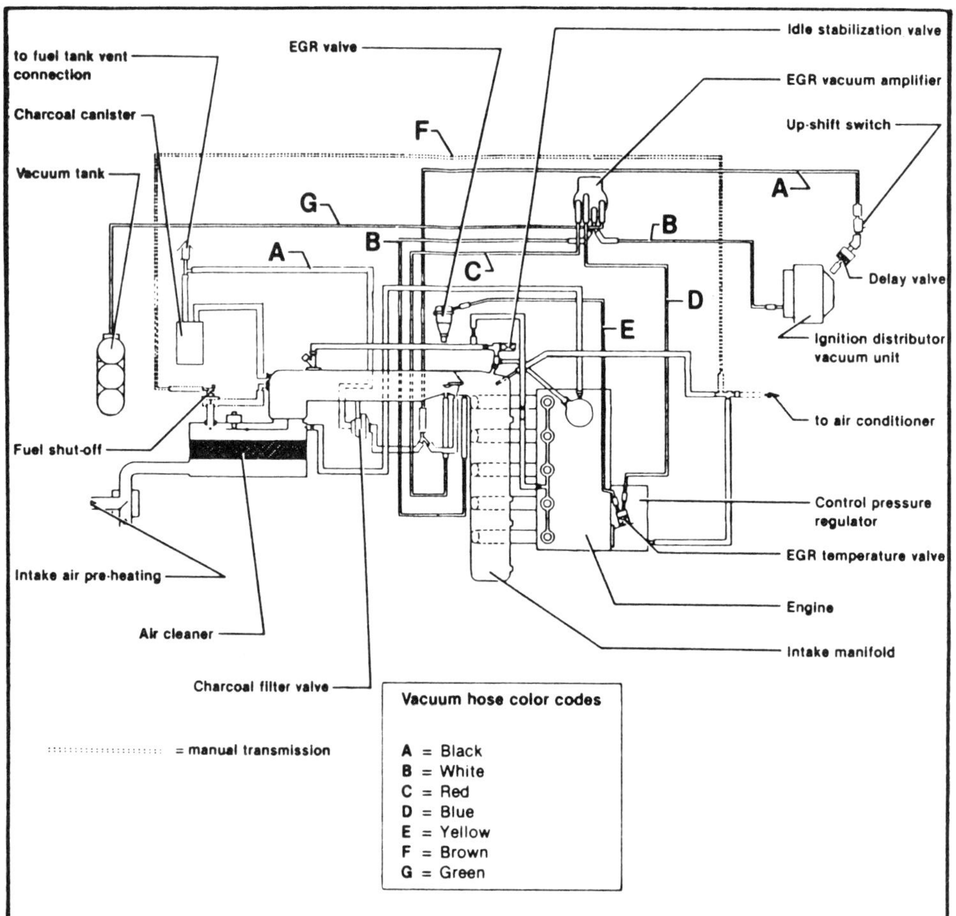

Audi 5000 Canadian models

BMW

INDEX

VEHICLE IDENTIFICATION NUMBER AND ENGINE IDENTIFICATION

Vehicle Identification Number (VIN)

The vehicle identification number (VIN) is located on a plate on the upper left of the instrument panel, visible through the windshield. The manufacturer's plate is located in the engine compartment on the right side inner fender panel or support, or on the right side of the firewall. The chassis number can be found in the engine compartment on the right inner fender support or facing forward on the right side of the heater bulkhead. A label is also attached to the upper steering column cover inside the vehicle.

Underhood serial number locations—typical

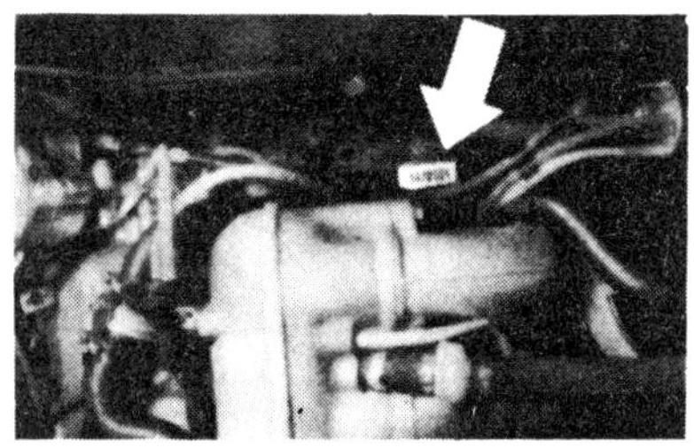

Engine serial number locations

Engine Identification

The engine identification number is located on the left rear side of the engine, above the starter motor.

TUNE-UP SPECIFICATIONS

Year	Model	Spark Plugs Type	Spark Plugs Gap (in.)	Distributor Dwell (deg.)	Distributor Point Gap (in.)	Ignition Timing (deg.) ① MT	Ignition Timing (deg.) ① AT	Intake Valve Opens (deg.) ●	Fuel Pump Pressure (psi)	Idle Speed (rpm) MT	Idle Speed (rpm) AT	Cold Valve Clearance (in.)
'85–'86	318i	WR9DS	.027–.031	Electronic		①	①	NA	43	①	①	.008
'85–'86	325e, 528e	WR9LS	.031–.035	Electronic		①	①	NA	43	①	①	.010

TUNE-UP SPECIFICATIONS

Year	Model	Spark Plugs Type	Spark Plugs Gap (in.)	Distributor Dwell (deg.)	Distributor Point Gap (in.)	Ignition Timing (deg.) ① MT	Ignition Timing (deg.) ① AT	Intake Valve Opens (deg.) ●	Fuel Pump Pressure (psi)	Idle Speed (rpm) MT	Idle Speed (rpm) AT	Cold Valve Clearance (in.)
'85–'86	535i 635CSi 735i	Bosch WR9LS	.027–.031	Electronic		①	①	NA	NA	①	①	.012

① Motronic injection system; controlled by computer, please refer to the underhood sticker for specifications.

FIRING ORDERS

NOTE: To avoid confusion, always replace spark plug wires one at a time.

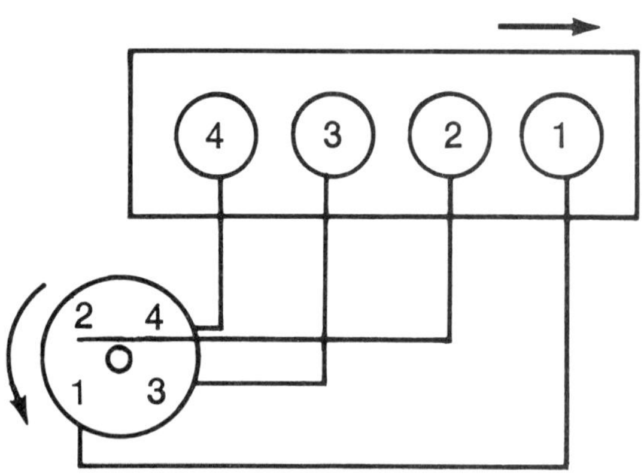

1985–86 four cylinder firing order: 1-3-4-2

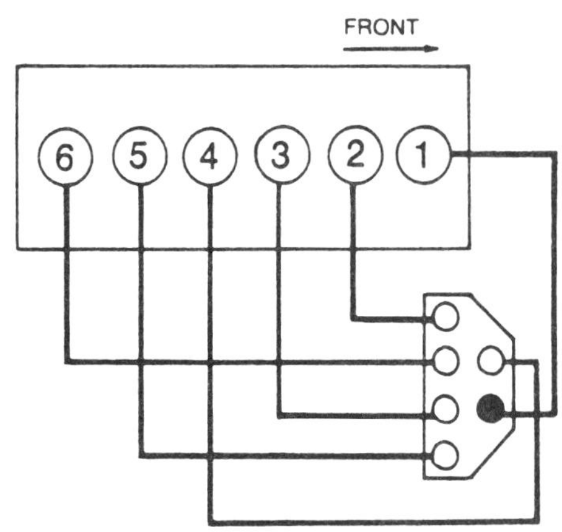

1985–86 six cylinder firing order: 1-5-3-6-2-4

EMISSION EQUIPMENT USED

1985–86

Catalytic converter (CAT).
Fuel evaporation system (FES).
Oxygen sensor (OS).
Positive crankcase vent (PCV).
Air flow controlled fuel injection (AFC).
Bosch Motronic System

REQUIRED EMISSION CONTROL MAINTENANCE

EGR VALVE

Every 15,000 miles (24000 km) the EGR flow passages should be checked for deposits.

SPARK PLUGS

On these vehicles the spark plugs should last for 30,000 miles, or (48000 km). If there is any indication of engine misfire the plugs should be replaced immediately.

UNDERHOOD RUBBER AND PLASTIC COMPONENTS (EMISSION HOSES)

Inspect hose surfaces for evidence of heat or mechanical damage. Hard or brittle rubber, cracking, checking, tears, cuts, abrasion, and excessive swelling indicate deterioration of the rubber. Inspect all hose connections, couplings and clamps for leakage.

CRANKCASE INLET AIR CLEANER

The crankcase inlet air filter must be kept clean and lubricated. The inlet air cleaner should be washed in kerosene or a similar solvent.

Lubricate the filter by inverting it and filling it with SAE 30 engine oil. Position the air cleaner to allow excess oil to drain through the vent nipple located in the top of the air cleaner. More frequent service may be required on vehicles that are operated extensively on short run, stop and go, or extended idle service.

FUEL VAPOR STORAGE CANISTER

Replace the filter element in the base of the canister with a new element every 30,000 miles or (48000 km). The filter should be replaced more often if the vehicle is driven under dusty or sandy conditions.

POSITIVE CRANKCASE VENT (PCV) VALVE

Every 15,000 miles (24000 km), the valve should be checked. If the valve is found to be clogged, replace it with a new valve. Never attempt to clean an old PCV valve.

AIR FILTER

The filter in the carburetor air cleaner should be replaced every 30,000 miles (48000 km). Replace the filter more often if the car is driven in dusty or sandy area.

IDLE SPEED AND MIXTURE ADJUSTMENTS

NOTE: Before any adjustments are made be sure that the air cleaner is good, the ignition timing is correct, the valve clearance is correct and the engine is at normal operating temperature.

IDLE MIXTURE

Adjustment

318i MODELS

1. The engine must be run until it is at operating temperature. Ignition timing and valve clearance must be correct. Connect the BMW digital mixture measurement unit 12 6 400 or equivalent according to the instrument instructions. Disconnect the hose going to the active carbon filter on the throttle housing and do not plug the open connections.

2. Operate the engine to at least 3,000 rpm for at least 30 seconds to ensure that the oxygen sensor is at operating temperature.

3. Disconnect the oxygen sensor wire, and fasten it where it cannot touch a ground. The nominal value you will be looking for will now appear in the test unit's display. Make note of it and then reconnect the oxygen sensor into the test unit. The actual value will appear in the display. If the actual value is within plus or minus .3 volt of the nominal value, CO is within tolerance. If not, proceed as follows.

4. Drill a hole through the tamper plug with a special tool 13 1 092 or equivalent. Then screw special tool 13 1 094 or equivalent into the plug, and use the slide hammer on the tool to draw the plug out.

5. Use a special tool 13 1 060 to turn the adjusting screw to bring the actual value to within 0.3 volts of the nominal value plus or minus. Turn off the engine, disconnect the test unit and reconnect the oxygen sensor wire to the oxygen sensor. Replace the anti-tamper plug with a new one.

325e MODELS

1. Disconnect the hose, leading from the throttle housing, that goes to the carbon canister. Do not plug the openings. Remove the bolts.

2. Remove the plug in the exhaust manifold, install the test nipple, BMW part No. 13 0 100 or equivalent and connect the CO tester 133 0 070 or equivalent, into the open nipple.

3. With the engine valve clearances correctly adjusted, ignition timing correct and the engine at operating temperature, measure the CO percentage at idle speed. CO nominal value is 0.2–1.2%.

4. CO level at idle speed should not be changed by pulling off oxygen sensor. To adjust CO level, drill a hole in anti-tamper lock on air flow sensor using removal tool (13 11 092).

5. Screw removal tool (13 1 094) into anti-tamper lock. Knock tool and anti-tamper lock out of air flow sensor with sharp impact. Insert adjuster tool 13 1 060) into anti-tamper lock hole and adjust engine speed by turning control screw. After completion, install new anti-tamper lock.

4. If the CO level is within the specified range, disconnect the test unit, replace the plug in the exhaust manifold, and conclude the test. If not, adjust the CO as described below.

5. Turn off the engine and then unplug the oxygen sensor plug. Drill a hole in the anti-tamper plug in the throttle body with special tool No 13 1 092 or equivalent. Then screw the special extractor tool No. 13 1 094 or equivalent into the hole drilled into the plug and draw the plug out with the impact mass. Finally, use an adjustment tool 13 1 060 or 13 1 100 or equivalent to turn the adjustment, with the engine running, until the CO meets nominal values.

6. When the adjustment is complete, install a new anti-tamper plug, and reconnect the oxygen sensor plug and the carbon canister hose. Also, remove the nipple in the exhaust manifold and replace the plug. Reinstall the exhaust manifold bolts.

IDLE SPEED

Adjustment

528e, 535i, 635CSi AND 735i MODELS

1. Idle speed is controlled by an idle speed control unit. Connect dwell/tachometer to engine. If idle speed is not as specified, check idle speed control unit. Replace if required.

2. On 318i, also check the idle valve, located on the intake manifold. To test valve, engine must be at normal operating temperature. Unplug wire connector from idle valve. Idle speed should rise to 2,000 rpm. Connect wire and rpm should drop to below idle speed. Replace valve if it does not perform as indicated.

3. Remove hose from collector to carbon canister. Connect special tool (13 0 070) with adapter (13 0 100) on exhaust manifold. Connect CO tester. Check CO value.

IDLE SPEED AND MIXTURE SPECIFICATIONS
1985 and Later

Vehicle	Engine RPM	Percent CO
318i (Distributor)	700-800	(See Text)
325e (Motronic)	650–750	0.2–1.2
528e (Motronic)	650–750	0.2–1.2
535i (Motronic)	650–750	0.2–1.2
635CSi (Motronic)	650–750	0.2–1.2
735i (Motronic)	650–750	0.2–1.2

INITIAL TIMING SETTINGS

NOTE: Refer to the vehicle emission control label located in the engine compartment before making any adjustments. If the information on the label differs with the information given, use the information on the label.

On cars with the Motronic control unit, the timing can be checked in a fairly straight-forward manner; however, timing cannot be adjusted. The only cure for improper timing is to replace the control unit. Also, timing must be within a specified range, as the computer changes the timing slightly to allow for various changes in operating conditions. In other words, the timing does not have to be right on, but anywhere within the specified range.

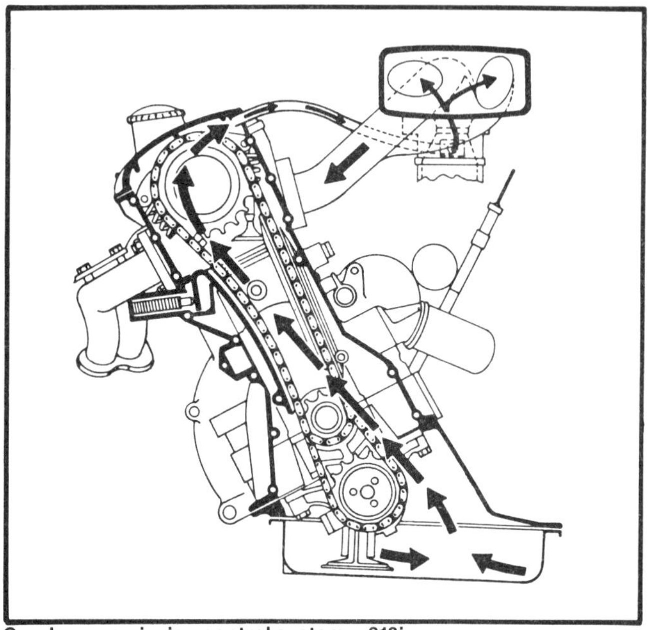

Crankcase emission control system—318i

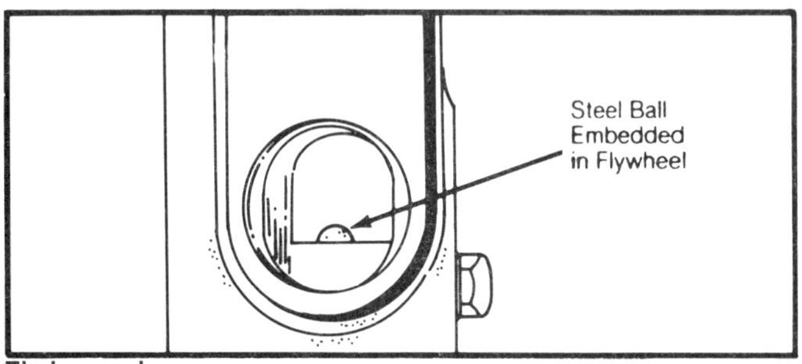

Timing mark

The engine should be at normal operating temperature and the operation should be performed at normal room temperatures. The engine rpm should be within the specified range under the control of the computer.

Look up the control unit number on the unit itself. On 3, 5 and 6 Series, it is in the right side speaker cutout. Find the control unit number on the underside of unit and then reference that number on the Computer Controlled Ignition Timing Chart.

Connect a tachometer and a timing light to the engine (the latter to the No. 1 cylinder). Start the engine and check the rpm. If it is not correct, see the appropriate checks under Idle Speed and Mixture Adjustment. Then, operate the timing light to see if timing is within the range specified on the chart. If it is significantly outside the range, the Motronic control unit must be replaced.

NOTE: The flywheel mark is either a pressed-in steel ball or a long tapered peg on the side of the starter ring gear.

MAINTENANCE INDICATOR LIGHT

ALL MODELS EXCEPT 735i

The oxygen sensor light on the dash will light the first time the mileage reaches 30,000. Replace the oxygen sensor and remove the light bulb from the dash. The bulb lights the first time only. However, the sensor must be changed every 30,000 miles.

1. Disconnect the oxygen sensor wire connector and remove the wires from the clip.
2. Pull off the sensor protective plate.
3. Unscrew the oxygen sensor.
4. Before installation, coat the threads of the new sensor with CRC® copper paste.
5. Install the sensor unit.
6. Remove the call unit from the dash by unscrewing the bolt, push the unit to the right and remove the oxygen sensor display bulb. On some models, the bulb must be broken to remove it.

735i MODELS

This heated type oxygen sensor needs to be replaced only at 50,000 mile intervals. The sensor is located on the engine exhaust pipe, just in front of the catalytic convertor.

To replace it, unscrew the protective plate and disconnect the plug in wire leading to the sensor. Then, unscrew the oxygen sensor. Replace in reverse order, coating the threads of the new sensor with an anti-seize compound.

NOTE: A special tool, only available through BMW sources, must be used to reset certain electronic type emission or oxygen sensor indicator light controls.

VACUUM ADVANCE SPECIFICATIONS
Except Motronic Control Unit

Part Number (Distributor)	Degrees	In.Hg.	Degrees	In.Hg.
0237-002-049	zero	3.0	10	11.0
0237-304-015	zero	2.6	17	10.1

SPARK PLUG APPLICATION CHART

Vehicle	Spark Plug	Gap (inches)	Fuel System
318i	Bosch—WRDS	0.024	Fuel Injection—Bosch—L-Jetronic
325e, 528e	Bosch—WR9LS	0.027	Fuel Injection—Bosch—Motronic
535i, 635CSi, 735i	Bosch—WR9LS	0.029	Fuel Injection—Bosch—Motronic

EMISSION CONTROL SYSTEMS
CRANKCASE EMISSION CONTROLS

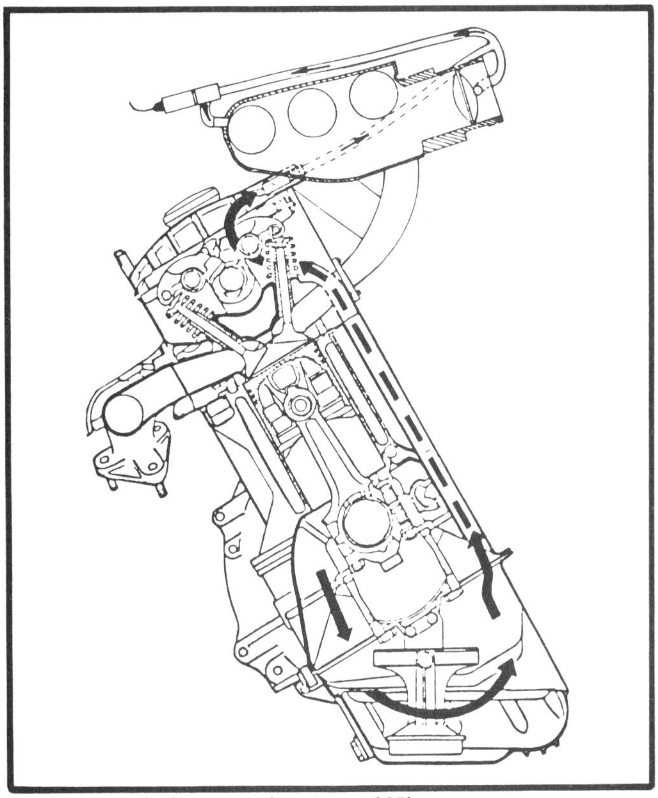

Crankcase emission control system—325i

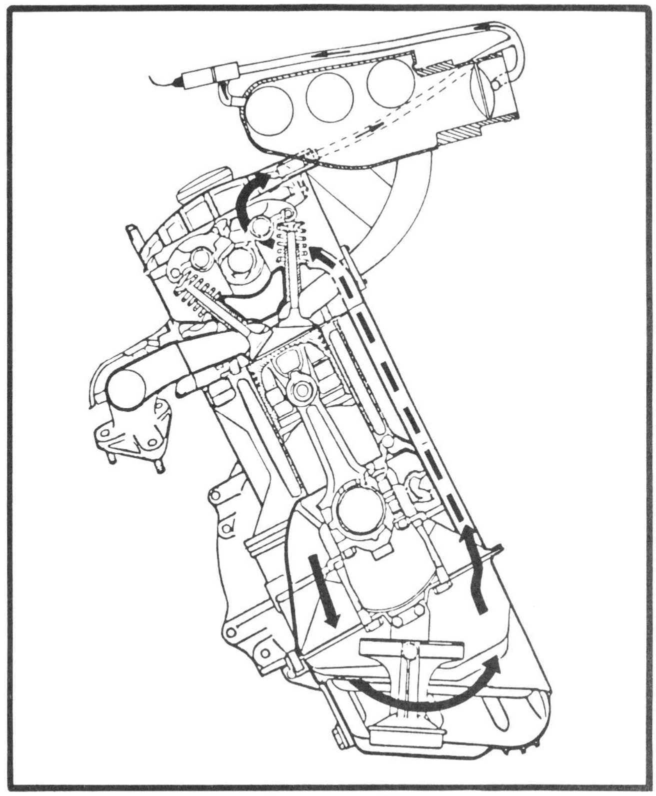

Crankcase emission control system—528e

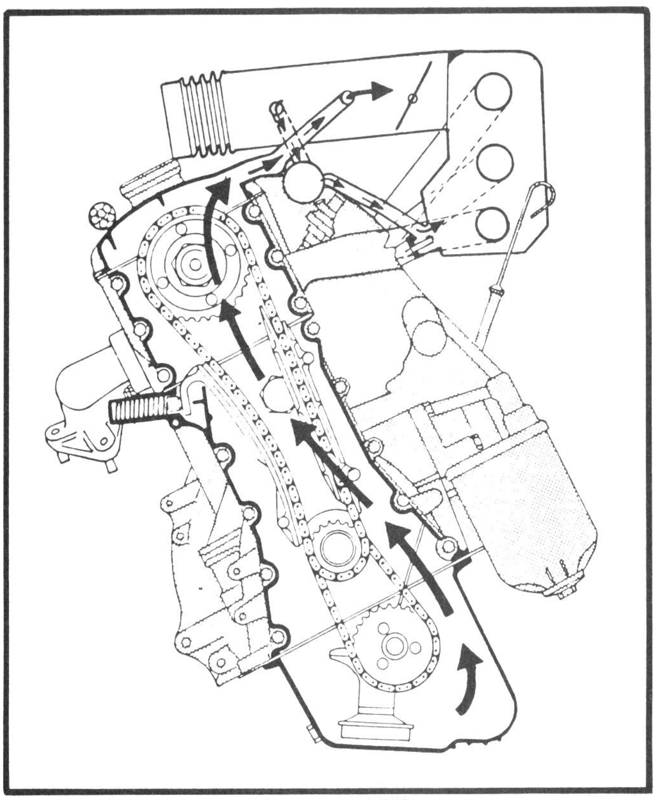

Crankcase emission control system—535i

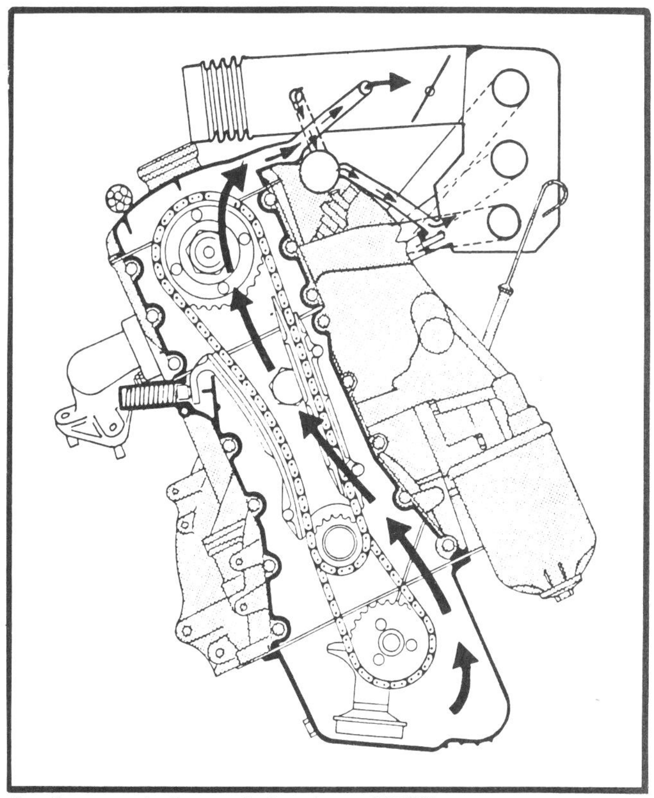

Crankcase emission control system—635CSi

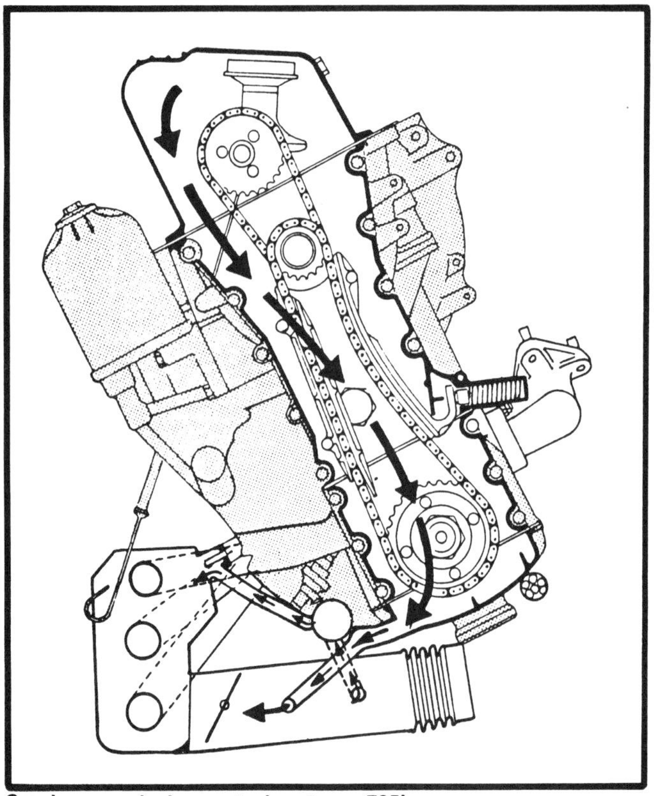

Crankcase emission control system—735i

POSITIVE CRANKCASE VENTILATION SYSTEM TROUBLE DIAGNOSIS

Symptom	Possible Cause	Repair
Rough engine idle	Defective canister, clogged or leaking vacuum hose.	Replace canister, clean or replace hose.
No PCV flow	Clogged lines or flame trap.	Clean or replace lines and flame trap.
Engine stalls at idle	Saturated carbon canister, clogged breather filter.	Replace canister or filter.
Surging at road speed	Improper PCV flow.	Check PCV system for proper function. Clean all lines and component.

FUEL EVAPORATION EMISSION CONTROLS

FUEL EVAPORATIVE SYSTEM DIAGNOSIS

Symptom	Possible Cause	Repair
Noticeable fuel odor or leaks	Damaged or loose lines, defective or saturated canister, broken or leaking vent valve.	Repair lines, replace canister or vent valve
Fuel tank deformed	Canister clogged, tank cap defective, hose clogged.	Replace defective canister or cap, clear vapor lines
Insufficient fuel delivery or vapor lock	Clogged or collapsed vapor lines, vent valve or canister. Clogged fuel feed lines.	Repair as necessary
Tank won't take fuel	Clogged or defective vent valve.	Replace vent valve

EXHAUST EMISSION CONTROL SYSTEMS

Motronic Emission Control System

The Motronic Emission Control system is an electronically controlled, computerized engine system which controls fuel injection and ignition timing as well as air/fuel ratio.

The system uses this information to determine engine operating.
The system uses this information to determine engine operating conditions, and adjusts timing and fuel ratio accordingly. The motronic control unit is located behind the speaker in the right kick panel of 635CSi and 735i models and in the glove compartment of the 325e, 528e and 535i models.

The motronic control unit is the brain of the system. Various engine sensors supply the unit with operating information including air flow, air temperature, throttle position, coolant temperature, engine speed, piston position and oxygen content of exhaust gases.

The system receives electronic input signals from several engine sensors. Information supplied by these sensors is used to determine optimum ignition and fuel injection timing under various engine operating conditions.

An ideal air/fuel ratio of 14:1 is maintained under most driving conditions. This is the ratio at which the catalytic converter operates most efficiently to reduce exhaust emissions.

The main components that make up the motronic control system are: oxygen sensor, air flow sensor, three coolant temperature sensors, reference point pick up, engine speed sensor and the throttle position sensor.

COOLANT TEMPERATURE SENSORS

There are three components which supply coolant temperature information to the motronic control unit. They are the coolant temperature switch, coolant temperature sensor and the thermo timer. All three devices are located in the water jacket of the engine block. They supply coolant temperature information to the motronic control unit in the form of electronic signals.

The system interprets these signals as cold or normal operating temperatures. During cold operating conditions, the air/fuel mixture is enriched by the cold start valve. This valve is located in the intake manifold, downstream from the butterfly valve. It supplies additional fuel to the inlet charge when signaled by the control unit. Extra rich conditions are maintained until normal operating temperature is attained.

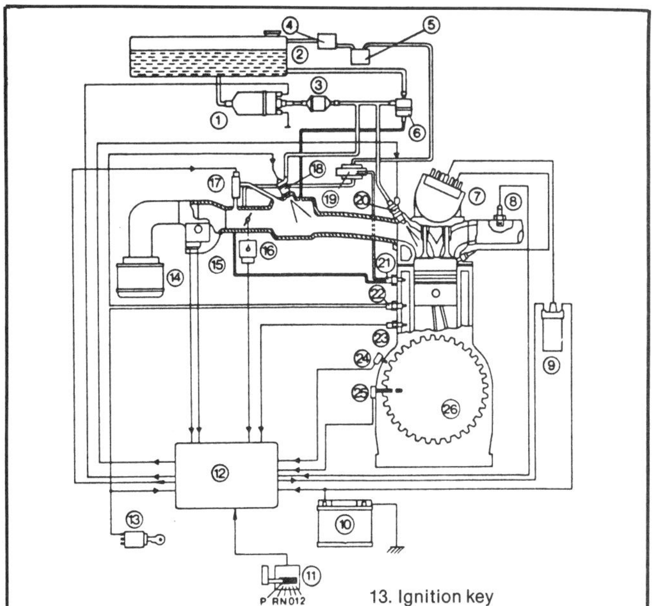

1. Fuel pump
2. Fuel tank
3. Fuel filter
4. Vapor liquid separator
5. Carbon filter
6. Fuel pressure regulator
7. High voltage distributor
8. Oxygen sensor
9. Ignition coil
10. Battery
11. Gearshift lever position automatic only
12. Control unit
13. Ignition key
14. Air filter
15. Airflow meter
16. Throttle switch
17. Idle control unit
18. Cold start valve
19. Purge valve
20. Injection valve
21. Thermo switch
22. Thermo-time switch
23. Temp. sensor water
24. RPM-pickup
25. Reference-point pickup
26. Starter ring gear

Typical digital motor electronic fuel injection system

THROTTLE POSITION SENSOR

The throttle position sensor is located in the throttle linkage at the intake butterfly valve where it detects position of the throttle valve. this data is converted into an electrical impulse and sent to the control unit. The control unit interprets the signal as either full throttle, idle or normal operating condition and makes adjustments accordingly.

ENGINE SPEED SENSOR

The engine speed sensor is located on the bellhousing, next to the starter ring gear. A steel ball, embedded in the ring gear, causes an electronic

1. Fuel tank
2. Fuel delivery pump
3. Pressure reservoir
4. Fuel filter
5. Fuel distributor
6. System pressure regulator
7. Warm-up regulator
8. Safety switch
9. Sensor plate
10. Throttle valve
11. Idle adjusting screw
12. Auxiliary air regulator
13. Start valve
14. Thermo timing valve
15. Injection valves
16. Air inlet
17. Vacuum regulator
18. Auxiliary air valve

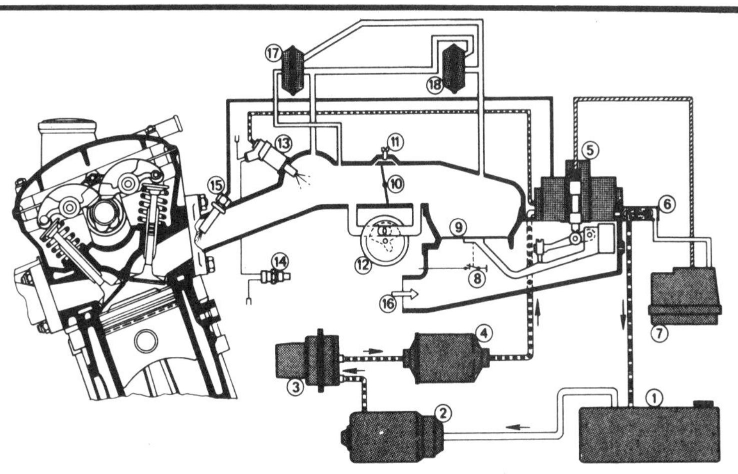

Injection press.
System pressure
Return
Control pressure

Bosch K-Jetronic fuel injection system

pulse in the speed sensor with each engine revolution. These pulses are converted into an electrical signal which is sent to the motronic control unit. The motronic control unit uses this information to determine rpm.

REFERENCE-POINT PICKUP

This sensor is located in the bellhousing, next to the engine speed sensor. It supplies the control unit with piston position data. When the control unit has determined optimum ignition timing data the reference-point pickup is used to signal ignition firing.

OXYGEN SENSOR

Oxygen content of exhaust gases is detected by the oxygen sensor which is located in the exhaust manifold. This sensor converts the amount of oxygen present in exhaust gases into an electrical signal which is transmitted to the motronic control unit. Motronic control unit uses this information to determine air/fuel ratio and adjusts injection pulse width to obtain the desired 14:1 ratio.

AIR FLOW SENSOR

Intake air flow is detected by the air flow sensor. It is located in the intake passage between the air filter and the intake manifold and informs the control unit of the rate of air intake. Incorporated into the air flow sensor is the air temperature sensor. This sensor informs the control unit of ambient temperature of incoming air.

Motronic Emission Control System Diagnosis

Before suspecting the motronic control system to be at fault, be sure that all other systems are in proper working order. Any engine system that would normally be checked in a vehicle not equipped with the motronic control system, should be checked first.

If the motronic control unit has been found to be causing the problem, determine which component or area is the most probable source of performance difficulty and begin testing there. Many component failures may be traced to faults in the wiring circuit. Before beginning other diagnostic procedures, check the appropriate circuit for breaks or shorts and be sure that all electrical connections are clean and tight.

REQUIRED TESTING EQUIPMENT

In order to properly diagnosis and repair any defects in the motronic control system the following test equipment will be necessary. The BMW service test kit, Bosch L-Jetronic fuel injection test kit and service procedures, BMW test meter 22-13-100 and a standard volt/ohm meter. Failure to use the proper test equipment may result in unnecessary replacement of good components or damage to the system.

EGR SYSTEM DIAGNOSIS

Symptom	Possible Cause	Repair
EGR valve does not move during test.	Leaking or clogged vacuum line or hose connection. Defective EGR valve.	Repair vacuum line or replace valve.
Engine won't idle or dies on return to idle.	EGR control system blocked or inoperative.	Check all vacuum connections and control valve. Replace defective control valve.
EGR valve leaks in closed position. Poor wide-open throttle performance.	EGR valve defective or clogged with carbon. Defective EGR control valve.	Clean or replace EGR valve. Replace control valve.
EGR warning light on.	Normal service due.	Service EGR system and reset counter.
Excessive HC and CO levels in exhaust.	Air injection system inoperative.	Check all components for proper operation. Check all hoses for unrestricted flow.

CATALYTIC CONVERTER DIAGNOSIS

Symptom	Possible Cause	Repair
Leaking exhaust gases	Leaks at pipe joints, damaged gaskets or rusted exhaust pipes	Tighten clamps, repair exhaust system as necessary
Loss of engine power, internal rattles in exhaust system	Dislodged baffles in muffler, broken ceramic insert in converter	Replace muffler or converter
Excessive CO	Contaminated catalyst	Replace converter
Excessive catalyst temperature	Mixture set too rich, oxygen sensor malfunction	Reset mixture, check oxygen sensor system
Catalyst warning light on	Normal service due	Replace converter and reset warning light

FUEL PRESSURE DIAGNOSIS

Testing

NO FUEL PRESSURE

1. Check fuel pump fuse and replace if defective. If fuse is okay, pull off electrical connector on fuel pump. Connect voltmeter between the two wires in connector and start engine. Voltmeter should read battery voltage. If voltage is correct, replace fuel pump.

2. If voltage reading is incorrect, check ground (Brown) wire: Connect ohmmeter between wire and ground. Resistance should be zero. If not, repair wire. If resistance is correct, check power (Green/Violet) wire: Disconnect relay 1 and connect ohmmeter between power wire and connector 87 of relay socket. Resistance should be zero. If value is incorrect, repair wire.

3. If resistance is correct, check power supply from relay 1: Connect voltmeter between wire 30 (Green/Yellow) on relay socket and ground. Turn ignition on. Meter should read battery voltage. If not, repair wire.

4. If voltage is correct, check power supply to relay 1: Connect voltmeter between wire 86 (Red/White) of plug and ground. Start engine. Voltmeter should show battery voltage. If not, connect a voltmeter between wire 87 (Red/White) of relay 2 and ground (relay still connected). Turn on ignition. Voltmeter should read battery voltage. If reading is incorrect, repair wire.

5. If power supply to relay is correct, check relay 1 ground: Connect ohmmeter between wire 85 (Brown/Green) on socket and ground. Start engine. Resistance should be about zero. If resistance is correct, replace relay 1.

6. If voltage in last part of step is correct, or if resistance is incorrect in step, check resistance of Brown/Green wire between connector 85 on Relay 1 plug and pin 20 of motronic control unit connector. Resistance should be zero. If not, repair wire. If resistance is correct, test the motronic control unit.

FUEL PRESSURE TOO HIGH

1. Check the vacuum connection of the fuel pressure regulator and vacuum hoses for leaks or kinks. Repair or replace s needed.

2. Check for defective pressure regulator and replace if faulty. Check the fuel return line for bends, pinches or clogs. Repair or replace as needed.

FUEL PRESSURE TOO LOW

1. Fuel line is probably restricted. Check for kinked, bent or clogged fuel line. Repair or re-route as needed. Check fuel filter for restriction and replace if dirty. Clean filter screen in fuel intake. Check pressure regulator operation and replace if defective.

FUEL INJECTION DIAGNOSIS

Testing

1. Check for fuel delivery at injectors. If one or more injectors are not operating correctly, start the engine and check for movement of needles in injectors. Movement can be felt with finger.

2. If no movement is detected, check power lines and coil of fuel injectors according to procedures in BMW service test, L-Jetronic step 5. If test value is incorrect, check wiring for shorts, breaks or poor connections and repair as needed. If wiring is okay replace defective fuel injectors.

3. If test value is correct, check power supply to injectors: Connect a voltmeter between wire 87 (Red/Blue) on relay 2 an ground (relay plugged in). Meter should read battery voltage. If not, test the motronic control unit.

4. If voltage is incorrect, check activation of injectors BMW service test, L-Jetronic Step 2. If test value is incorrect, replace control unit.

MOTRONIC CONTROL UNIT DIAGNOSIS

Testing

1. Check power supply to motronic control unit. Connect BMW service test unit an perform L-Jetronic Test Step 1. If values are correct, replace control unit.

2. If voltage value between wire 18/35 and 5, or 4 and 5 is insufficient or incorrect, connect voltmeter between wire 4 on disconnected control unit plug ground. Start engine. If voltmeter does not show battery voltage, trace circuit and repair wiring.

3. Check wire 18 and 35 for breaks or poor connections. Connect ohmmeter between wire 5 on control unit plug and ground. Resistance should be zero. Connect ohmmeter between wire 18 or 35 on control unit plug, and wire 87 on plug of relay 2. Resistance should be zero. If either test shows resistance, repair wiring.

4. If resistance values are correct, check power supply to relay 2; Disconnect relay 1 and connect voltmeter between terminal 86 on relay socket and ground. Turn ignition on. Voltmeter should read battery voltage. If not, trace circuit and repair wiring.

5. If voltage is correct, check ground of relay 2; pull off relay and check resistance between connector 85 and ground. Resistance should be zero. If not, trace circuit and repair wiring.

6. If resistance is correct, check power supply of relay 2: Pull of relay and connect voltmeter between connector 30 of relay socket and ground. Voltmeter should read battery voltage. If not, trace circuit and repair wiring. If voltage reading is correct, replace relay 2.

COLD START VALVE DIAGNOSIS

Testing

COLD START VALVE DOES NOT OPEN

1. Remove valve, leaving fuel lines connected. Supply battery voltage to valve with jumper wire and be sure valve is properly grounded. Pull off relay 1. Apply battery voltage to connector 87 in relay plug and check that fuel pump runs. Cold start valve should deliver fuel. If not, replace valve.

2. If valve functions properly, check power supply to valve: Pull plug off valve and connect voltmeter between wires of plug. Start engine. Meter should read battery voltage while cranking engine. If not, trace circuit and repair wiring.

3. Check thermo timer and replace if resistance values are not correct.

COLD START VALVE LEAKS

Check valve operation as in first cold start valve test. If valve operates properly (fuel is delivered), remove jumper wire to battery voltage and check that fuel delivery stops. If fuel is still delivered, or leaks, or seeps out, replace valve.

IDLE CONTROL VALVE DIAGNOSIS

Testing

1. Valve should be open when vehicle is at rest (no voltage to valve). When voltage is applied to valve (engine on), valve should close. Remove 2 valve hoses and observe valve operation. If valve does not operate as described, replace valve.

2. If valve operates properly, pull off connector plug and connect voltmeter between the 2 wires in plug. Start engine and turn A/C on. Voltmeter should read battery voltage. If it does not, see idle control diagnosis test.

IDLE CONTROL UNIT DIAGNOSIS

Testing

1. Check power supply to idle control unit: Pull connector plug off of the idle control unit and connect voltmeter between terminal 2 of plug and ground. Start engine. Voltmeter should read battery voltage. If not, repair circuit.

2. If voltage is correct, check ground connection of the idle control unit. Connect ohmmeter between terminal 4 of plug an ground. Ohmmeter should read zero. If not, trace circuit and repair wiring.

3. If resistance is correct, check speed signal to the idle control unit at terminal 3 of plug with BMW test meter 22-13-100. If signal is not correct, trace circuit and repair wiring.

4. If signal is correct, check for ground at terminal 6 of the idle control unit; connect ohmmeter between terminal 6 of the idle control

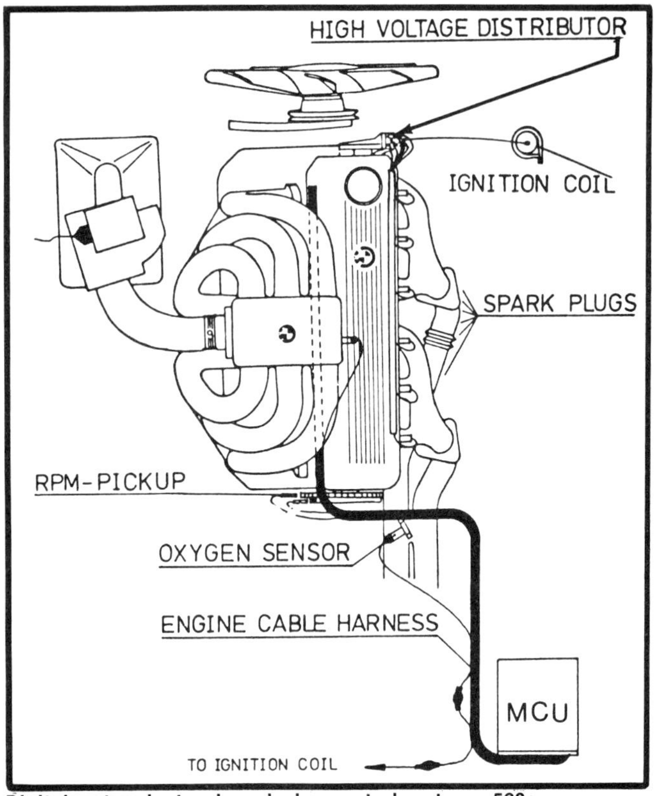

Digital motor electronic emission control system—528e

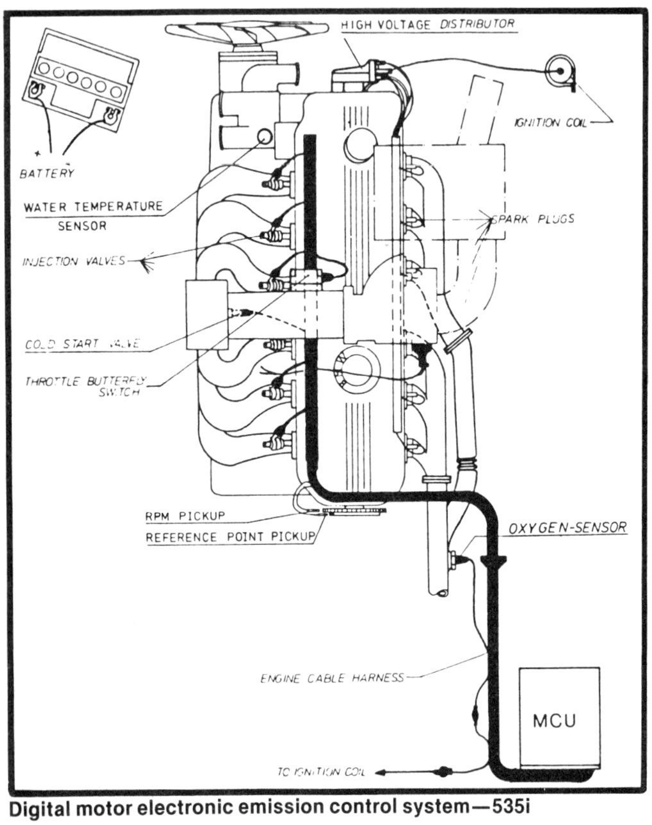

Digital motor electronic emission control system—535i

unit plug and ground. With coolant temperatures below 106°F, resistance should be zero. With coolant temperature above 117°F, resistance should be infinite. If values are incorrect, check coolant temperature switch. If switch is good, trace idle control unit circuit and repair wiring. If resistance values are correct, go to the next step.

5. On automatic transmission equipped vehicles, check for battery voltage at terminal 7 of lug with ignition on and transmission shift lever in "N". If voltage is absent, trace circuit and repair wiring. If voltage is present at terminal 7, check for battery voltage at terminal 8 with shift lever in "P". If voltage is absent, trace circuit and repair wiring. If reading is correct, go to next step.

6. On vehicles with air conditioning, check for battery voltage at terminal 9 of plug with air conditioner "ON". Trace circuit and repair wiring if voltage is incorrect. If reading is correct, go to next step.

7. On all models, check for battery voltage at terminal 10 of the idle control unit plug with ignition on. Voltage should be present with air temperature below 41°F, and absent at higher temperatures. If values are incorrect, check air temperature sensor. Replace if faulty. If sensor is okay, repair wiring.

8. If voltage readings are correct, check for ground at terminal 12 of the idle control unit. Connect ohmmeter between terminal 12 of plug of the idle control unit. Connect ohmmeter between terminal 12 of plug and ground. Resistance should be zero with throttle closed, and infinite with throttle open. If values are incorrect, check throttle position sensor. Replace or adjust as needed. If switch is good, repair wiring.

9. If resistance values are correct, check ground connection at terminal 11 of the idle control unit. Connect ohmmeter between terminal 11 and ground. Resistance value should be zero. If not, check coolant temperature sensor. Replace sensor if faulty. If sensor is good, repair wiring.

10. If resistance values are correct, and idle control valve is good, replace idle control unit.

ENGINE SPEED SENSOR AND REFERENCE POINT PICKUP

Diagnosis Testing

1. Check sensor and pickup general condition. Ensure that they are installed in the proper position and firmly seated. Electrical contacts must be clean and tight. Check that sensor plugs are not reserved. Plugs are color coded for identification.

2. Install BMW test meter 22-13-100. Connect terminals 8 and 27 and check for speed signal. Connect terminals 25 and 26 an check for reference signal. If signals are not correct, repair wiring. Remove terminal connections and recheck signals. If either signal is absent, replace sensor and/or pickup as needed.

AIR FLOW SENSOR

Diagnosis Testing

1. Check that sensor is properly installed and firmly seated. Ensure that sensor plate moves freely.

2. Run L-Jetronic test 4 to check air flow value. If value is incorrect, repair wiring. Check resistance at terminals 7 and 9 of sensor. Resistance should be checked with sensor plate in several different positions.

THROTTLE POSITION SENSOR

Diagnosis Testing

Check that sensor is properly installed and firmly seated. Run L-Jetronic test 3 to determine throttle sensor values. If values re incorrect, adjust or replace sensor as needed.

COOLANT TEMPERATURE SENSOR

Diagnosis Testing

Check that sensor is properly installed and firmly seated. Check that cooling system is full. Bleed system. Check resistance between switch connections. If resistance is incorrect, replace sensor. If resistance is correct, trace sensor circuit and repair wiring.

COOLANT TEMPERATURE SWITCH

Diagnosis Testing

Switch must be tightly installed. Check that cooling system is full. Bleed system. Check resistance between switch contacts. Resistance below 106°F (41°C) should be zero. At higher temperatures, resistance should be infinite. If values are correct, trace circuit and repair wiring. If values are incorrect, replace switch.

THERMO TIMER

Diagnosis Testing

1. Check that timer is properly installed and firmly seated. Check radiator for correct coolant level. Bleed cooling system.

2. Disconnect timer and check resistance values between plug terminals G and W, G and ground, and W and ground. If values are correct, trace timer circuit and repair wiring. If values are incorrect, replace timer.

AIR TEMPERATURE SENSOR

Diagnosis Testing

1. Check electrical connections on sensor. Use L-Jetronic test 4 to check air temperature sensor values. If values are incorrect, remove connector from sensor and attach ohmmeter. Check resistance between terminals 22 and 6. The resistance should be 2280–2720 ohms at 66–70°F and 760–910 ohms at 120–124°F.

2. If resistance values are correct, trace sensor circuit and repair wiring. If values are incorrect, replace air flow sensor.

VACUUM CIRCUITS

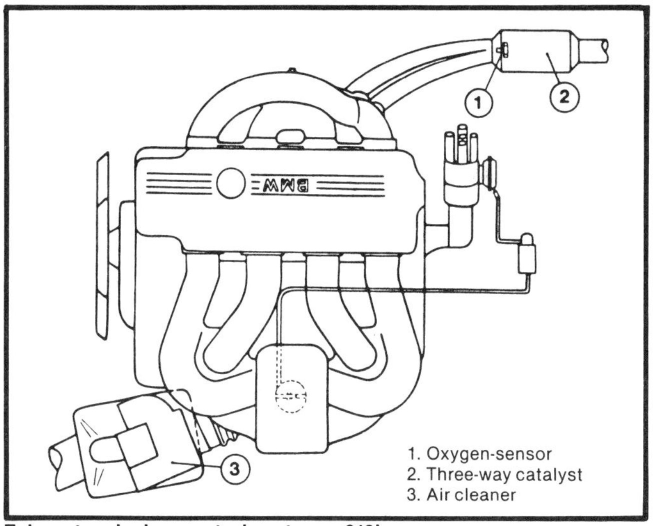

1. Oxygen-sensor
2. Three-way catalyst
3. Air cleaner

Exhaust emission control system—318i

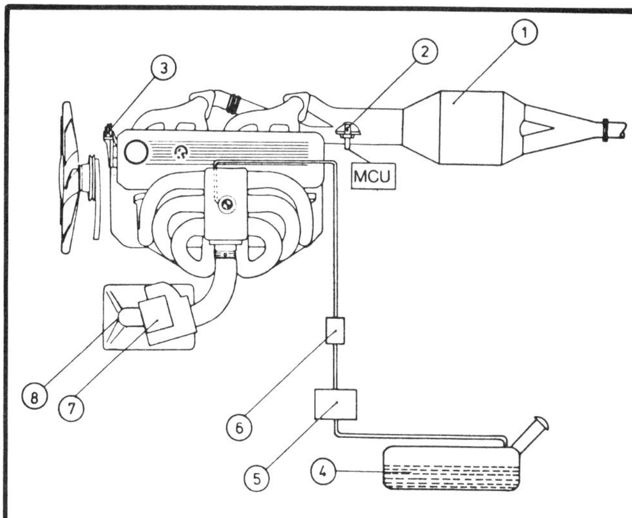

1. Three-way catalyst
2. Oxygen sensor
3. High voltage distributor
4. Fuel tank
5. Vapor liquid separator
6. Charcoal canister
7. Air flow meter
8. Air filter

Exhaust emission control system—528e

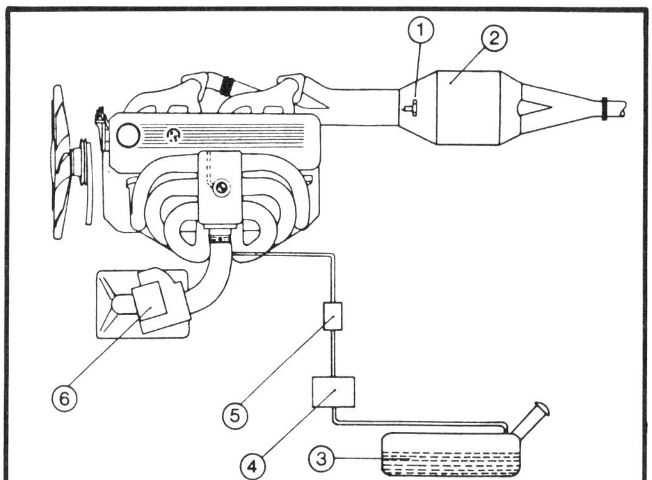

1. Oxygen-sensor
2. Three-way catalyst
3. Fuel tank
4. Vapor liquid separator
5. Charcoal canister
6. Air flow meter

Exhaust emission control system—325e

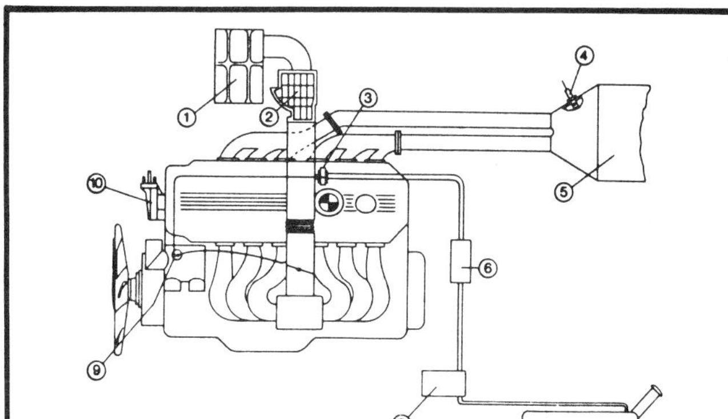

1. Air filter
2. Airflow meter
3. Purge valve
4. Oxygen sensor
5. Three-way catalyst
6. Activated carbon filter
7. Fuel tank
8. Liquid-Vapor separator
9. Thermo switch
10. High voltage distributor

Exhaust emission control system—635CSi 735i 535i

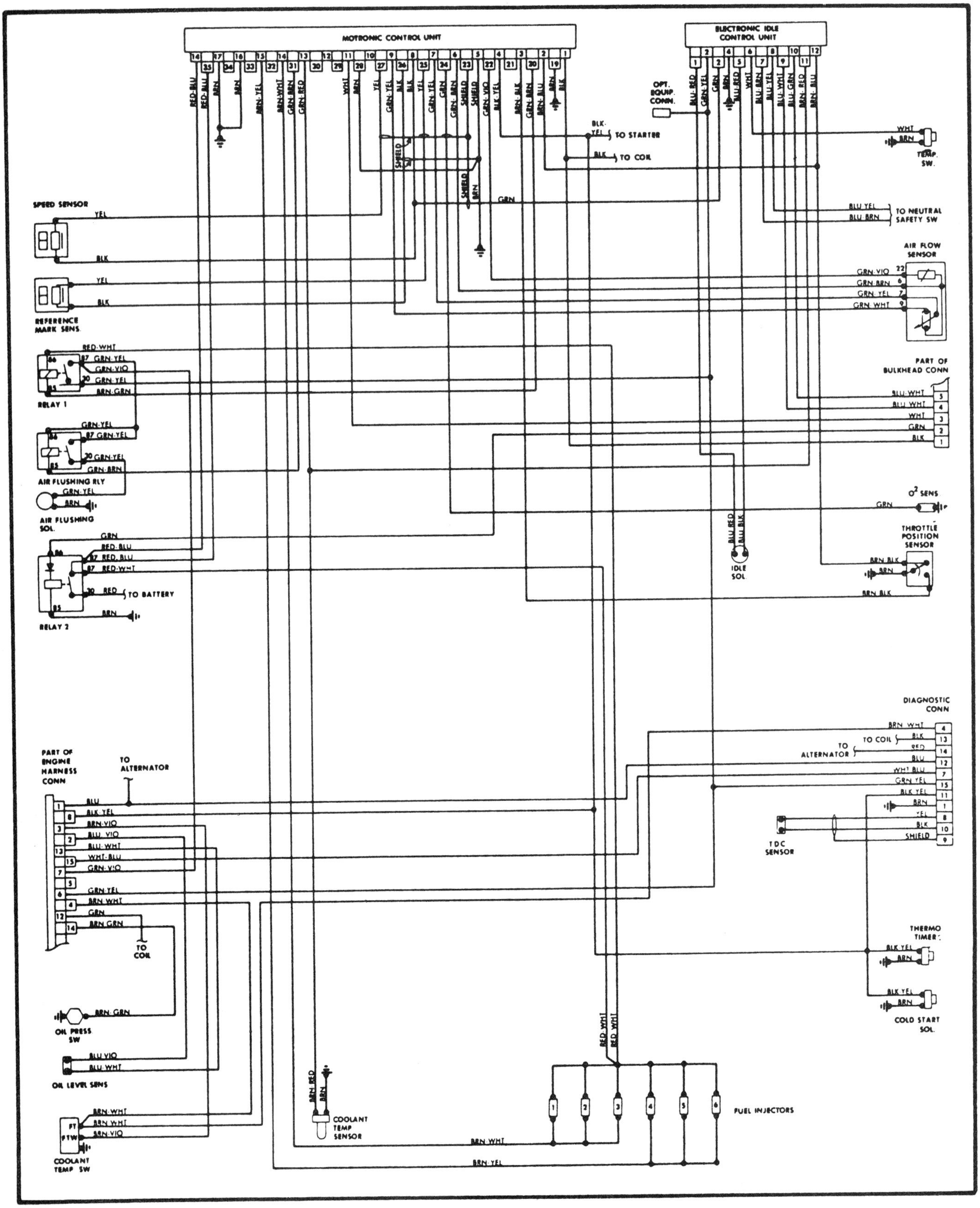

Motronic emission control system wiring schematic

Colt, Conquest (Chrysler)

INDEX

VEHICLE IDENTIFICATION NUMBER AND ENGINE IDENTIFICATION

Vehicle Identification Number (VIN)

The vehicle identification number (VIN) plate is mounted on the instrument panel, adjacent to the lower corner of the windshield on the left side of the vehicle, and is visible through the windshield. The VIN number is composed of seventeen (17) digits. The tenth (10) digit of the VIN number represents the year of the vehicle, with 'F' representing 1985 and 'G' representing 1986.

Engine Identification

The eighth (8) digit of the VIN number represents the engine model. The engine Identification number for all engine models is stamped at the right front side, on the top edge of the cylinder block.

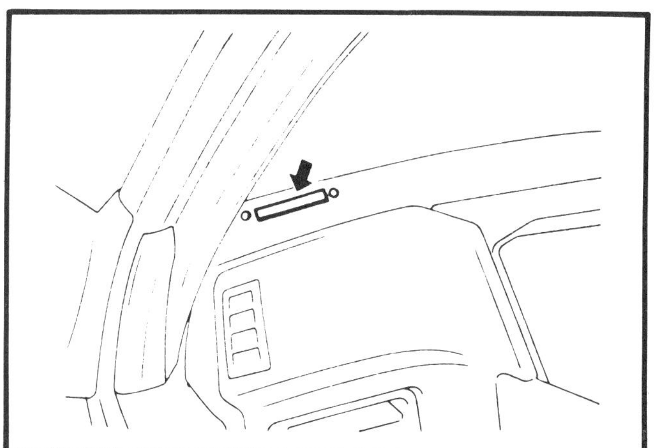

Typical serial number location (© Chrysler Corp.)

SERIAL NUMBER & ENGINE IDENTIFICATION

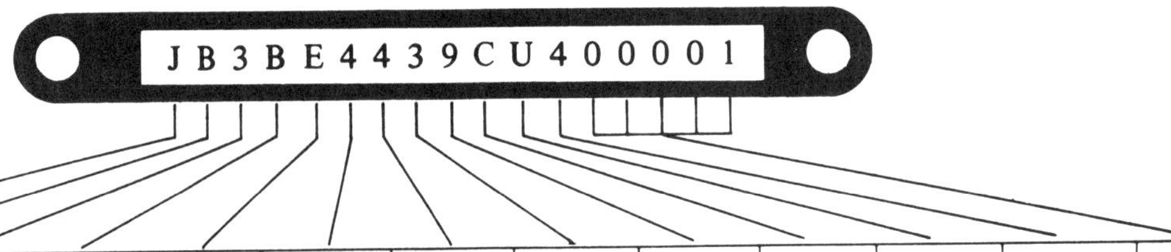

1st Digit	2nd Digit	3rd Digit	4th Digit	5th Digit	6th Digit	7th Digit	8th Digit	9th Digit	10th Digit	11th Digit	12th Digit	13th to 17th Digits
Manufacturing country	Sales channel	Vehicle type	Other	Vehicle line	Trim code	Body type	Engine displacement	•Check digit	Model year	Assembly plant	Transmission code	Sequence number
J–Japan	B–Dodge P–Plymouth	3–Passenger car	B–Manual seat belt	E–Colt/ Champ	2–Low 3–Medium 4–High, High (RS, LS)	4–2 door hatchback	2–1.4 litre (86.0 CID) 3–1.6 litre (97.5 CID)	1 2 · · · 9 X	C–1982	U–MIZUSHIMA Plant	1–4 speed M/T Federal 2–4 speed M/T California 3–4 speed M/T Canada 4–4 x 2 speed M/T Federal 5–4 x 2 speed M/T California 6–4 x 2 speed M/T Canada 7–3 speed A/T Federal 8–3 speed A/T California 9–3 speed A/T Canada	00001 to 99999

NOTE: •"Check Digit" means a single number or letter X used to verify the accuracy of transcription of vehicle identification number.
M/T is an abbreviation for manual transaxle.
A/T is an abbreviation for automatic transaxle.
RS is an abbreviation for "Rally Sports".
LS is an abbreviation for "Luxury Sports"

Typical 1985 and later 17 digit vehicle identification number (© Chrysler Corp.)

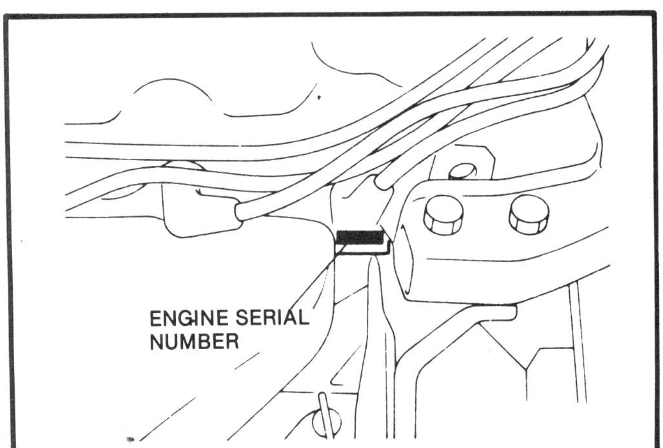

Colt engine serial number location (© Chrysler Corp.)

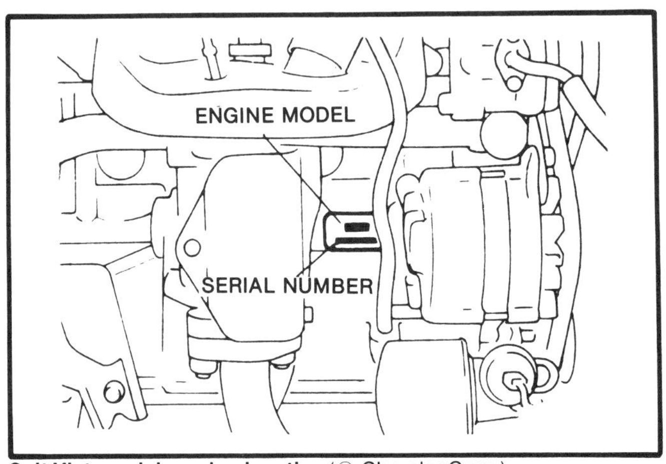

Colt Vista serial number location (© Chrysler Corp.)

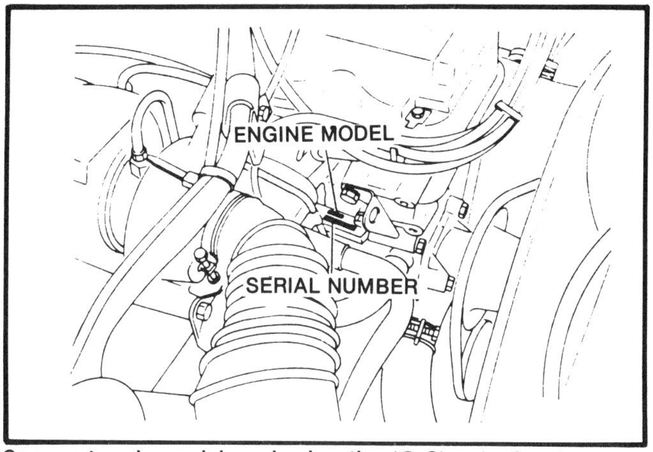

Conquest engine serial number location (© Chrysler Corp.)

TUNE-UP SPECIFICATIONS

(When analyzing compression test results, look for uniformity among cylinders, rather than specific pressures)

Year	Engine Displace. cu. in. (cc)	Spark Plugs Type	Spark Plugs Gap (in.)	Distributor Point Dwell (deg)	Distributor Point Gap (in.)	Ignition Timing (deg) MT	Ignition Timing (deg) AT	Intake Valve Opens (deg) BTDC	Fuel Pump Pressure (psi)	Idle Speed (rpm)	Valve Clear (in)• In.	Valve Clear (in)• Ex.
'85–'86	91.5 (1500)	W20EPR-S11 ①	0.035 0.039	Electronic		8B	8B ②	18	2.7–3.7	④	0.006 ③	0.010
	97.5 (1600)	W20EPR-V10 ①	0.035 0.039	Electronic		8B	8B ②	20	2.4–3.4	⑤	0.006 ③	0.010
	121.7 (2000)	W20EPR-S11 ①	0.035 0.039	Electronic		5B	5B ②	19	N/A	⑥	0.006 ③	0.010
	158.6 (2600)	W20EP-V10 ①	0.039 0.043	Electronic		10B	10B ②	N/A	35.6	⑦	0.006 ③	0.010

NOTE: The underhood sticker often reflects tune-up specifications changes made in production. Sticker figures must be used if they disagree with those in this chart.

B Before top dead center
N/A Not available at the time of publication
• All clearances set hot
① See text
② ± 2° at curb idle
③ Jet valve clearance: 0.010
④ Man. Trans: 700 ± 100 rpm
Auto. Trans: 750 ± 100 rpm
⑤ 700 ± 100 rpm
⑥ 700 ± 100 rpm
⑦ 850 ± 100 rpm

NOTE: This spark plug gap on the Canadian models is 0.028 to 0.031 in.

FIRING ORDERS

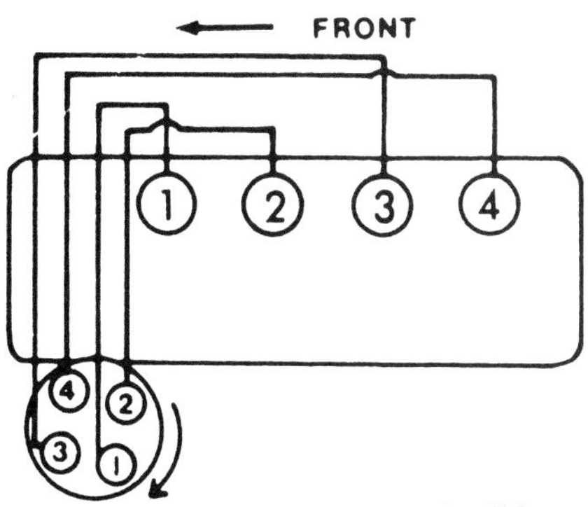

1500cc, 1600cc and 2000cc engine firing order (© Chrysler Corp.)

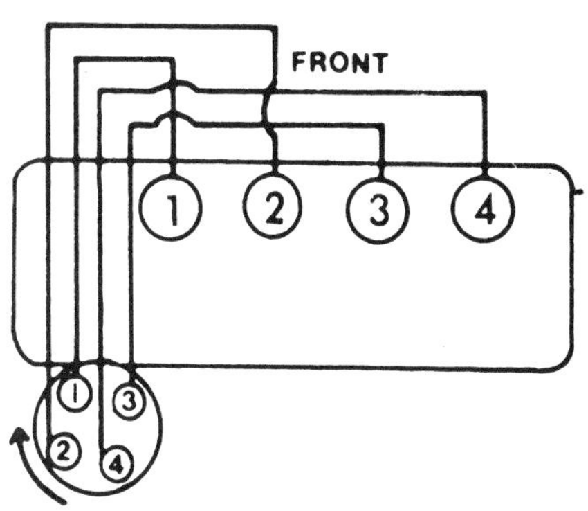

2600cc engine firing order (© Chrysler Corp.)

EMISSION EQUIPMENT USED

1985–86 COLT AND COLT VISTA

Crankcase Emission Control
- Closed system
- Positive crankcase vent valve

Evaporative Emission Control
- Canister-Single
- Bowl vent valve
- Purge control valve
- Fuel filler cap-with relief valve
- Overfill-limiter-two way valve
- Fuel check valve

Exhaust Emission Control
- Jet valve
- Catalytic converter-dual three way type (single Canada)
- Secondary air supply system-reed valve
- Exhaust gas recirculation (EGR) system
- EGR valve-single + sub (single Canada)
- Thermo valve-Single wax pellet type
- Fuel control system-Electronic Controlled Injection System (ECI)
- Feedback carburetor system
- Heated air intake system-vacuum type
- Deceleration device
- Dash pot
- Deceleration spark advance
- High altitude compensation
- Throttle opener
- Tamper-proof (mixture, choke-ECI system not available)

1985–86 CONQUEST

Crankcase Emission Control
- Closed system
- Positive crankcase vent valve
- Oil separator

Evaporative Emission Control
- Canister-Single
- Carbon element
- Purge control valve
- Fuel filler cap-with relief valve
- Vapor separator tank-Vapor-liquid
- Overfill-limiter-two way valve
- Fuel check valve
- Thermo Valve

Exhaust Emission Control
- Jet valve
- Catalytic converter-dual three way type
- Secondary air supply system-reed valve
- Exhaust gas recirculation (EGR) system
- EGR valve-single + sub
- Fuel control system-Electronic Controlled Injection System (ECI)
- Feedback carburetor system
- Idle speed control system (for A/C unit only)
- High altitude compensation

REQUIRED EMISSION CONTROL MAINTENANCE
1985-86 All Models

Emission Control System Maintenance	Service Intervals	Mileage in Thousands	7.5	15	22.5	30	37.5	45	50
		Kilometers in Thousands	12	24	36	48	60	72	80
Change engine oil every 12 months		OR	X	X	X	X	X	X	
Replace engine oil filter every 12 months		OR		X		X		X	
Change engine oil every 6 months*1		OR	Every 4,800 km (3,000 miles)						
Replace engine oil filter every 12 months*1		OR	Every 9,600 km (6,000 miles)						
Check drive belt (for water pump and alternator) for condition; adjust tension as required		AT		X				X	
Replace drive belt (for water pump and alternator)		AT				X			
Check valve clearance; adjust as required		AT		X		X		X	
Check ignition timing; adjust as required every 5 years		OR							X
Check engine idle speed; adjust as required		AT		X		X		X	
Clean carburetor choke mechanism and linkage		AT				X			
Check throttle position system; adjust as required	Initial Check	AT		X					
	After initial check at 50,000 miles (80,000 km), thereafter every 50,000 miles (80,000 km)								
Replace fuel filter every 5 years		OR							X
Check fuel system (tank, line and connections) for leaks every 5 years		OR							X

REQUIRED EMISSION CONTROL MAINTENANCE

Emission Control System Maintenance	Service Intervals	Mileage in Thousands	7.5	15	22.5	30	37.5	45	50
		Kilometers in Thousands	12	24	36	48	60	72	80
Replace vacuum hoses, secondary air hoses and crankcase ventilation hoses every 5 years	OR								X
Replace fuel hoses, water hoses and fuel filler cap every 5 years	OR								X
Replace turbocharger air intake hoses every 5 years	OR								X
Replace Turbocharger oil hoses every 5 years	OR								X
Replace air cleaner filter	AT					X			
Clean crankcase emission control system (PCV valve) every 5 years	OR								X
Check evaporative emission control system (except canister) for leaks and clogging every 5 years	OR								X
Replace canister	AT								X
Replace spark plugs	AT					X			
Replace ignition cables every 5 years	OR								X
Replace oxygen sensor*[2]	AT								X

NOTE:
*[1]Vehicles with a turbocharger
*[2]Except vehicles without a turbocharger for Canada

IDLE SPEED AND MIXTURE ADJUSTMENTS

NOTE: The following specifications are published from the latest information available. if the published information differs from the information on the vehicle emission label, use the data on the emission label.

IDLE SPEED SPECIFICATIONS

IDLE SPEED

Adjustment

1985–86 COLT AND COLT VISTA MODELS

NOTE: The improper setting (throttle valve opening) will increase the exhaust gas temperature and deceleration, which in turn will reduce the life of the catalyst greatly and deteriorate the exhaust gas cleaning performance. It will also effect the fuel consumption and the engine braking.

1. With the vehicle in park, the drive wheels blocked and all the accessories off, run the engine until it reaches normal operating temperature.
2. Bring the engine speed up to 2000–3000 rpm for about ten seconds, then let the engine idle for a least two minutes.
3. Connect a tachometer to the engine and check the idling speed. If it does not meet specifications, readjust the idle speed to the nominal specification using the idle speed adjusting screw which is located closest to the primary throttle valve shaft.

1985–86 CONQUEST

Idle mixture adjustment cannot be accomplished with Electronically

IDLE SPEED SPECIFICATIONS

Engine	Transaxle	Curb Idle Speed (RPM)
1985-86 Colt		
1.5L	M/T	700 ± 100
1.5L	A/T	750 ± 100
1.6L—T/C	M/T & A/T	700 ± 100 ①
1985-86 Colt Vista		
2.0L	M/T	700 ± 100
2.0L	A/T	700 ± 100 ①
1985-86 Conquest		
2.6L EFI	M/T & A/T	850 ± 100

M/T—Manual Transaxle
A/T—Automatic Transaxle
T/C—Turbocharger
EFI—Electronic Fuel Injection
① With Idle Speed Control (ISC)

Controlled Injection system. The fuel delivery is controlled by the Electronic Control Unit (ECU), reacting to varied operating conditions.

IDLE-UP ADJUSTMENT FOR CARBURETOR EQUIPPED ENGINES

Adjustment

1985–86 COLT AND COLT VISTA MODELS WITHOUT A/C

NOTE: Adjustment condition: lights, electric cooling fan and all accessories are off, and transaxle is in neutral.

1. Make sure the curb idle speed is within specifications and adjust if necessary.
2. By using the auxilary lead wire, activate the idle up solenoid valve, apply the intake manifold vacuum to the throttle opener and activate the throttle opener.
3. Open the throttle slightly (to engine speed of about 2000 rpm) and then slowly close it.
4. Adjust the engine speed to the specifications with the idle-up adjusting screw.
5. After repeating Step 3, check the engine speed.
6. Remove the auxilary lead wire used in Step 2, and reconnect the idle-up solenoid valve wiring.

Adjustment

1985–86 COLT AND COLT VISTA WITH A/C

1. With the vehicle in park, the drive wheels blocked and all the accessories off, run the engine until it reaches normal operating temperature.
2. Disconnect the electric cooling fan connector and on vehicles with power steering, set the tires in the straight ahead position to prevent the pump from being loaded. Set the steering wheel in the stationary position.
3. Be sure that the curb idle speed is within the specifications and adjust if necessary.
4. With the air conditioner on, adjust the engine speed to the specified speed with the throttle opener setting screw (idle-up adjusting screw).
5. Reconnect the electric cooling fan connector and turn the A/C on and off several times to check the operation of the throttle opener.

IDLE SPEED CONTROL SERVO AND THROTTLE POSITION SENSOR

Adjustment

1985–86 CONQUEST

NOTE: This adjustment is very important, since the vehicle driveability is dependent upon it.

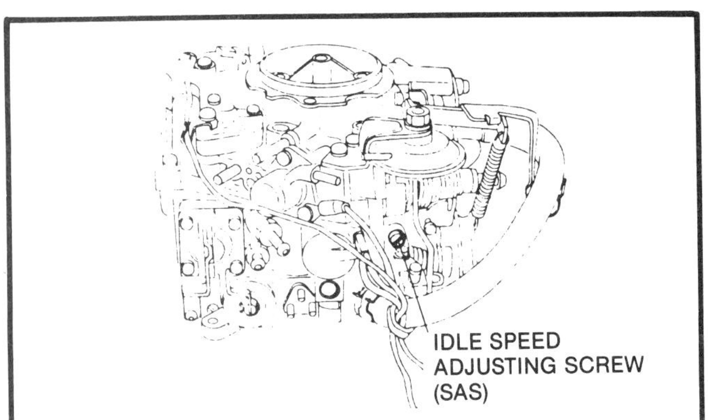

Location of speed adjusting screw (© Chrysler Corp.)

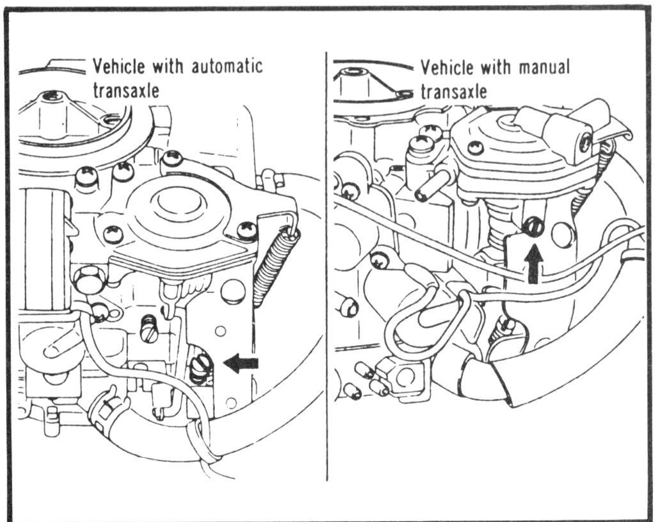

Location of speed adjusting screw with A/C (© Chrysler Corp.)

If the ISC servo, throttle position sensor, mixing body, or throttle body has been replaced or removed for any reason, use the following procedure to make the adjustments.

1. With the vehicle in park, the drive wheels blocked and all the accessories off, run the engine until it reaches normal operating temperature.
2. Stop the engine and loosen the two throttle position sensor mounting screws, turn the throttle position sensor clockwise as far as it will go, and then temporarily tighten the screws.
3. Turn the ignition switch to the on position for 15 seconds and then turn it off. This will set the ISC servo to the specified position.
4. Disconnect the ISC servo harness connector, start the engine and check the engine speed and adjust to specifications.
5. Stop the engine and disconnect the throttle position sensor harness connector.
6. Connect and adapter and a digital voltmeter between the throttle position sensor connector. Place the ignition switch to the on position, but do not start the engine.
7. Read the throttle position sensor output voltage.
8. If the measurement of the output voltage does not agree with the 0.48 ± 0.03 volts, loosen the throttle position sensor screws and turn the sensor left or right to bring the sensor into specifications. After applying sealant to the sensor, tighten the mounting screws.
9. Fully open the throttle valve and confirm that the output voltage is correct when it is returned to the closed position.
10. Remove the adapter and the digital voltmeter, reconnect the ISC servo harness connector and confirm that the curb idle is correct.

IDLE SPEED CONTROL (ISC)

Adjustment

1985–86 CONQUEST

NOTE: When replacing the ISC servo, the engine speed should be adjusted.

1. With the vehicle in park, the drive wheels blocked and all the accessories off, run the engine until it reaches normal operating temperature.
2. Stop the engine, place the ignition in the on position for over 15 seconds and then turn it off. This will set the ISC servo to the specified position.
3. Disconnect the ISC servo harness connector, start the engine and check the engine speed, adjust if necessary.
4. Reconnect the ISC servo harness connector.

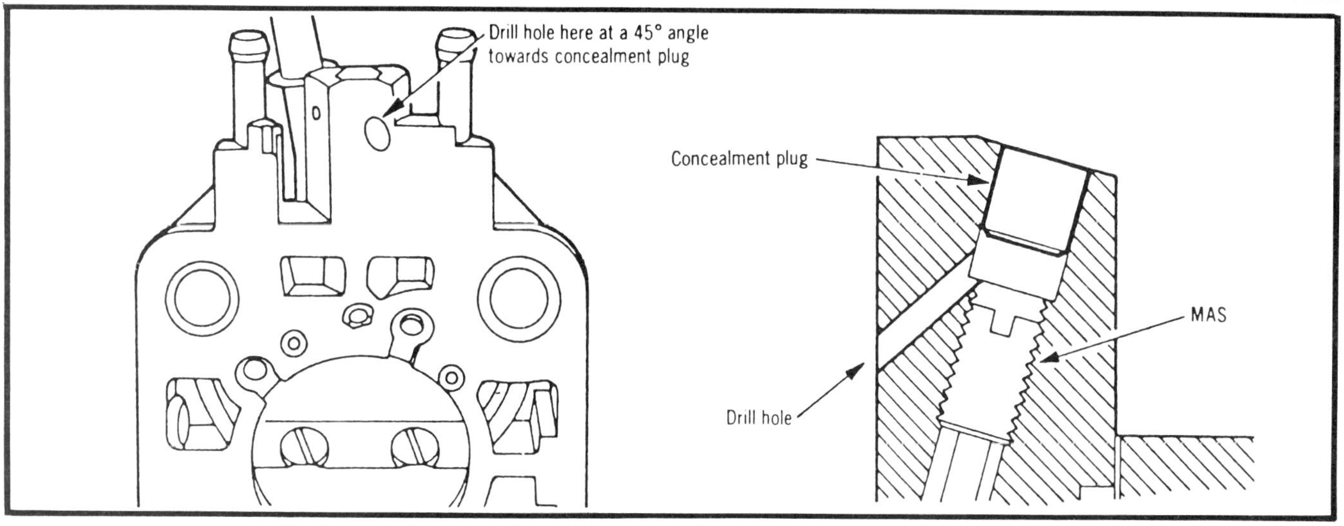

Removal of concealment plugs (© Chrysler Corp.)

Idle Mixture

Adjustment

1985–86 COLT, COLT VISTA WITH FEEDBACK CARBURETOR

1. Remove the carburetor from the engine and place the carburetor in a suitable fixture in order to remove the concealment plug.
2. Reinstall the carburetor on the engine without the concealment plug.
3. With the vehicle in park, the drive wheels blocked and all the accessories off, run the engine until it reaches normal operating temperature.
4. Turn off the engine and disconnect the negative battery cable for about three seconds, and reconnect the cable.
5. Disconnect the connector of the exhaust oxygen sensor, run the engine for five minutes at an rpm speed equivalent to 30 mph or run the engine for more than five seconds at the engine speed of 2,000–3,000 rpm.
6. Run the engine at idle for two minutes and set the idle CO and the engine speed to specifications (the idle CO: 0.1–0.3% at nominal curb idle speed).
7. Reconnect the oxygen sensor connector, readjust the engine speed, if necessary and install the concealment plug into the hole to seal the idle mixture adjusting screw.

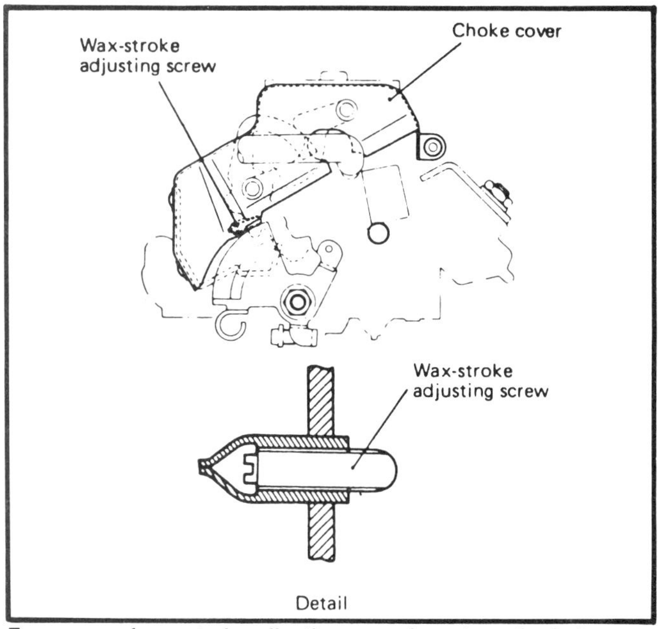

Tamper-proof wax-stroke adjusting screw for automatic choke (© Chrysler Corp.)

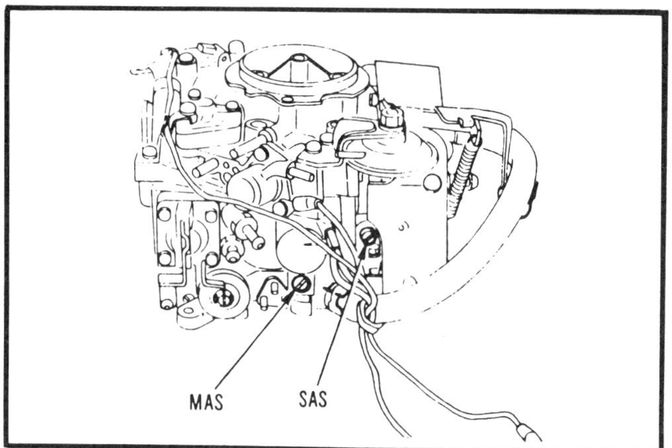

Location of mixture adjusting screw (MAS) and speed adjusting screw (SAS)—typical (© Chrysler Corp.)

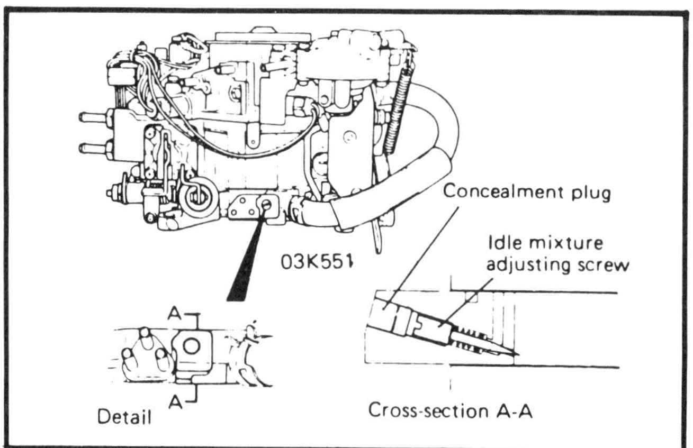

Removal of concealment plug—Feedback Carburetor (FBC) system (© Chrysler Corp.)

1985–86 CONQUEST

Idle mixture adjustment cannot be accomplished with Electronically Controlled Injection system. The fuel delivery is controlled by the Electronic Control Unit (ECU), reacting to varied operating conditions.

Tamper-Proof Automatic Choke

All carburetors have a tamper-proof choke. The choke related parts are factory adjusted. Neither removal of the choke cover or tampering with the W.A.S. wax-stroke adjustment screw is required in service, except when major carburetor overhaul or adjustment of choke-calibration-related parts is required by state or local inspections.

INITIAL TIMING

Procedure

ALL MODELS

1. Connect a suitable timing light, according to the manufacturers instructions, to number one cylinder spark plug wire.
2. Connect the positive lead of the test tachometer to the negative primary terminal of the ignition coil. Connect the negative lead of the test tachometer to a good ground.
3. Start the engine and run until operating temperature has been reached.
4. With the engine at operating temperature, fast idle off, momentarily open the throttle and release it to make sure that there is no bind in the linkage and that the idle speed screw is against its stop.
5. Disconnect and plug the vacuum line at the distributor.
6. Adjust the curb idle speed to specification.
7. Aim the timing light and check the engine timing. Advance or retard the ignition timing as required.
8. Once the timing has been set, recheck the curb idle specification and adjust as required. If the curb idle speed needs adjusting be sure to recheck the ignition timing again.

IGNITION TIMING SPECIFICATIONS

Engine	Transaxle	Ignition Timing
1985-86 Colt		
1.5L	M/T & A/T	8 BTDC ± 2 at curb idle
1.6L—T/C	M/T & A/T	8 BTDC ± 2 at curb idle
1985-86 Colt Vista		
2.0L	M/T & A/T	5 BTDC ± 2 at curb idle
1985-86 Conquest		
2.6L EFI	M/T & A/T	10 BTDC ± 2 at curb idle

M/T—Manual Transaxle
A/T—Automatic Transaxle
T/C—Turbocharger
EFI—Electronic Fuel Injection

DISTRIBUTOR SPECIFICATIONS 1985-86 Colt Vista

Identification Model No./Part No.	T4T63473/MD060266 T4T62182/MD060263 T4T62181/MD013880
Type	Breaker pointless type
Rotation	Clockwise

DISTRIBUTOR SPECIFICATIONS 1985-86 Colt Vista

Identification Model No./Part No.	T4T63473/MD060266 T4T62182/MD060263 T4T62181/MD013880
Firing order	1—3—4—2
Basic timing	5° BTDC ± 2°
Advance-Centrifugal (Distributor degrees at distributor rpm)	T4T63473, T4T62182 0° at 500 rpm 7° at 1,400 rpm 10° at 2,500 rpm
Advance-Vacuum [Distributor degrees at mm (in.) mercury]	0° at 60 mm 8.5° at 150 MM 14° at 240 mm (T4T62181—12.5° at 210 mm)
Idle advance [Distributor degrees at mm (in.) mercury]	0° at 150 mm 5° at 300 mm
Signal generator coil resistance	920—1,120Ω at 20°C (68°F)
Signal rotor gap	0.2 mm (.008 in.) or more

VACUUM ADVANCE

1985–86 COLT EQUIPPED WITH MANUAL FOUR SPEED TRANSAXLE (FEDERAL)

The distributor is equipped with a dual diaphragm type vacuum advance mechanism, each being independent of each other. The main diaphragm utilizes carburetor ported vacuum to advance the ignition timing, and the sub diaphragm is actuated by manifold vacuum below the throttle valve to provide additional ignition advance during periods of closed throttle deceleration and idle.

1985–86 COLT (CALIFORNIA)

The vacuum advance is operated by ported vacuum during periods of acceleration and cruising. During deceleration, a switching solenoid and speed sensor are used to change the vacuum source from ported to manifold vacuum, at predetermined speeds, preventing excessive HC emissions. However, when engine speeds drop to or below below a specified level, the vacuum source is changed from manifold vacuum to ported vacuum, in order to maintain a smooth engine operation.

1985–86 COLT VISTA AND CONQUEST

The distributor is equipped with a vacuum advance unit, which is controlled by the Electronic Spark Controller unit, to allow advance and retard at pre-determined speeds and engine operating conditions.

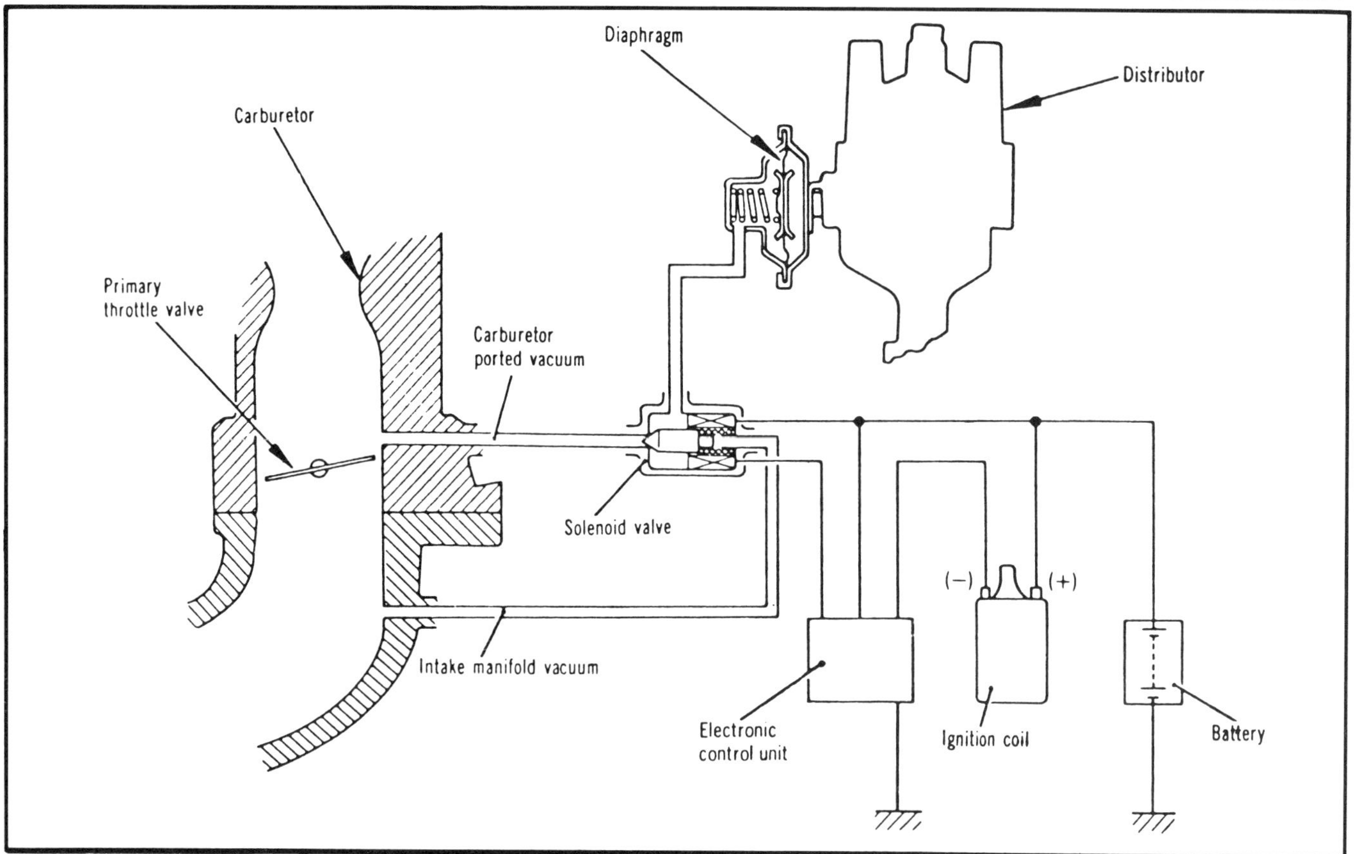

Vacuum advance system used with Electronic Control Injection (ECI) system (© Chrysler Corp.)

DISTRIBUTOR SPECIFICATIONS
1985-86 Colt

Description	Federal (not available in California) 1.5L engine with 4 speed M/T	Federal (not available in California) 1.5L engine with 4x2 speed M/T and 1.6L engine	California 1.5L and 1.6L engines	U.S.A. 1.6L turbo engine
Model No.	T3T62172	T3T61872		T3T63175
Type	Contact pointless type with electronic control unit (Electronic ignition system)			
Advance Mechanism	Centrifugal type and vacuum type with idle advance	Centrifugal type and vacuum type		
Turning directing		Clockwise		
Firing order		1—3—4—2		
Centrifugal advance—Crankshaft degrees at crankshaft speed				
Initial	0°/1,000 rpm	0°/1,000 rpm		0°/1,000 rpm
Final	20°/4,400 rpm	20°/4,400 rpm		26°/5,300 rpm
Vacuum advance—Crankshaft degrees at mmHg (in.Hg)				
Initial	0°/80 (3.15)	0°/80 (3.15)		0°/80 (3.15)
Final	23°/280 (11.02)	20°/360 (14.17)		15°/260 (10.24)
Idle advance—Crankshaft degrees at mmHg (in.Hg)	5°/245 (9.65)	—		—

DISTRIBUTOR SPECIFICATIONS
1985-86 Conquest

Identification model No./part No.	**T4T63371/MD061593**
Type	Breaker pointless type
Rotation	Clockwise
Firing order	1—3—4—2
Basic timing	10° ± 2° BTDC
Curb idle speed rpm	750^{+150}_{-100} [First 500 km (300 miles)] 800 ± 100 [After 500 km (300 miles)]
Advance—Centrifugal (Distributor degrees at distributor rpm)	0° at 600 rpm 6.5° at 1,150 rpm 12.5° at 2,500 rpm
Advance—Vacuum [Distributor degrees at mm (in.) of mercury]	0°/80 (3.1) 6°/150 (5.9) 11.5°/280 (11.0)
Retard—Pressure [Distributor degrees at mm (in.) of mercury]	0°/100 (3.9) 7°/450 (17.7)
Signal generator coil resistance Ω	920—1,120 at 20°C (68°F)
Signal rotor gap mm (in.)	0.2 (.008) or more

SPARK PLUGS

ORIGINAL EQUIPMENT

1985–86 COLT AND COLT VISTA

Nippondenso
W20EPR—U10 (1.6L only)
W20EPR—S11
W20EPR—Canada
W22EPR—S11

NGK
BPR6ES–11
BPR6EP–11
BPR6ES—Canada

Electrode Gap
0.035–0.039 in. (0.9–1.Omm) USA
0.028–0.031 in. (0.7–0.8mm) Canada
0.039–0.043 in. (1.0–1 1mm) NGK

ALTERNATE SPARK PLUGS
NGK BP6ES–11
Champion N–9Y
RN–9Y (USA and Canada)
Nippondenso W20EPR–U10

ORIGNIAL EQUIPMENT

1985–86 CONQUEST
Nippodenso
W20EP–U10
W20EPR–U10

Champion
N–9Y
RN–9Y

NGK
BP6ES–11
BPR6ES–11

Electrode Gap
0.039–0.043 in. (1.0–1.1mm)

CARBURETOR SPECIFICATIONS

1985–86 COLT

Type — Down-draft, 2 barrel, feedback typeModel No.:
28–32DIDTF–407 M/T 1.5L Federal
28–32DIDTF–408 A/T 1.5L Federal
28–32DIDTF–405 M/T 1.5L California
28–32DIDTF–406 A/T 1.5L California

Throttle bore:
Primary — 1.102 in. (28mm)
Secondary — 1.260 in. (32mm)Fast idle opening at 73 F. (23 C.):
0.019 in. (0.47mm) M/T
0.021 in. (0.54mm) A/T & CanadaAdditional:
Deceleration solenoid valve (DSV)
Jet mixture solenoid valve (JSV)
Enrichment solenoid valve (ESV)
Throttle position sensor (TPS)
Bowl vent valve (BVV)
Dash pot
Fuel cut-off solenoidChoke type — Full-automatic type

1985–86 COLT VISTA

Type — Down-draft, 2 barrel, feedback typeModel No.:

32–35DIDTF–200 M/T 2.0L Federal
32–35DIDTF–201 A/T 2.0L Federal
32–35DIDTF–195 M/T 2.0L California
32–35DIDTF–196 A/T 2.0L CaliforniaThrottle bore:
Primary — 1.260 in. (32mm)
Secondary — 1.378 in. (35mm)Fast idle opening at 73 F. (23 C.):
0.025 in. (0.63mm) M/T
0.028 in. (0.71mm) A/TAdditional:
Deceleration solenoid valve (DSV)
Jet mixture solenoid valve (JSV)
Enrichment solenoid valve (ESV)
Throttle position sensor (TPS)
Bowl vent valve (BVV)
Dash potChoke type — Full-automatic type

ELECTRONICALLY CONTROLLED INJECTION SYSTEM

1985–86 CONQUEST

Injection mixer:
Model No — 46EID–603
Injector:
Type — Electromagnetic type
No. of injectors — 2
Coil resistance — 2 ohms
Injector identification mark — H
Throttle bore — 1.811 in. (46mm)
Fuel pressure regulator:
Regulated pressure — 35.6 psi (245 kPa)
Throttle position sensor (TPS):
Type — Variable resistor type
Resistance — 5 kilo-ohms
Idle position switch:
Type — Contact point type
Idle speed control (ISC)
Electronic control unit (ECU):
Model No — E2T13481 M/T, E2T13482 A/T
Air Flow Sensor:
Type — Karman vortex type with intake air temperature sensor
Intake air temperature sensor:
Type — Thermistor
Resistance — 2.2 kilo-ohms at 77 F. (25 C.)
Resistor:
Model No — E8T00271
Resistance — 6 ohms
Oxygen sensor
Type — Zirconia sensor
Coolant temperature sensor
Type — Thermistor
Resistance — 16,200 ohms at -4 F. (-20 C.):
2,450 ohms at 68 F. (20 C.)
296 ohms at 176 F. (80 C.)
Pressure sensor
Model No — E1T15271
Control relay
Model No — E8T00571
Accelerator control method — Cable type

1985–86 COLT WITH 1.6L ENGINE

Injection mixer
Model No — 46EID-602 M/T, 46EID-652 A/T
Injector
Type — Electromagnetic type
No. of injectors — 2
Coil resistance — 2 ohms
Injector identification mark — B
Throttle bore — 1.811 in. (46mm)
Fuel pressure regulator
Regulated pressure — 35.6 psi (245 kPa)
Throttle position sensor (TPS)
Type — Variable resistor type
Resistance — 5 kilo-ohms
Idle position switch
Type — Contact point type
Idle speed control (ISC)
Electronic control unit (ECU)
Model No — E2T13271 M/T, E2T13272 A/T
Air Flow Sensor
Type — Karman vortex type with air intake temperature sensor
Intake air temperature sensor
Type — Thermistor
Resistance — 2.2 kilo-ohms at 77 F. (25 C.)
Resistor
Model No — E8T00271
Resistance — 6 ohms
Oxygen sensor
Type — Zirconia sensor
Coolant temperature sensor
Type — Thermistor
Resistance — 16,200 ohms at -4 F. (-20 C.)
2,450 ohms at 68 F. (20 C.)
296 ohms at 176 F. (80 C.)
Pressure sensor
Model No — E1T15271
Control relay
Model No — E8T00171
Accelerator control method—Cable type

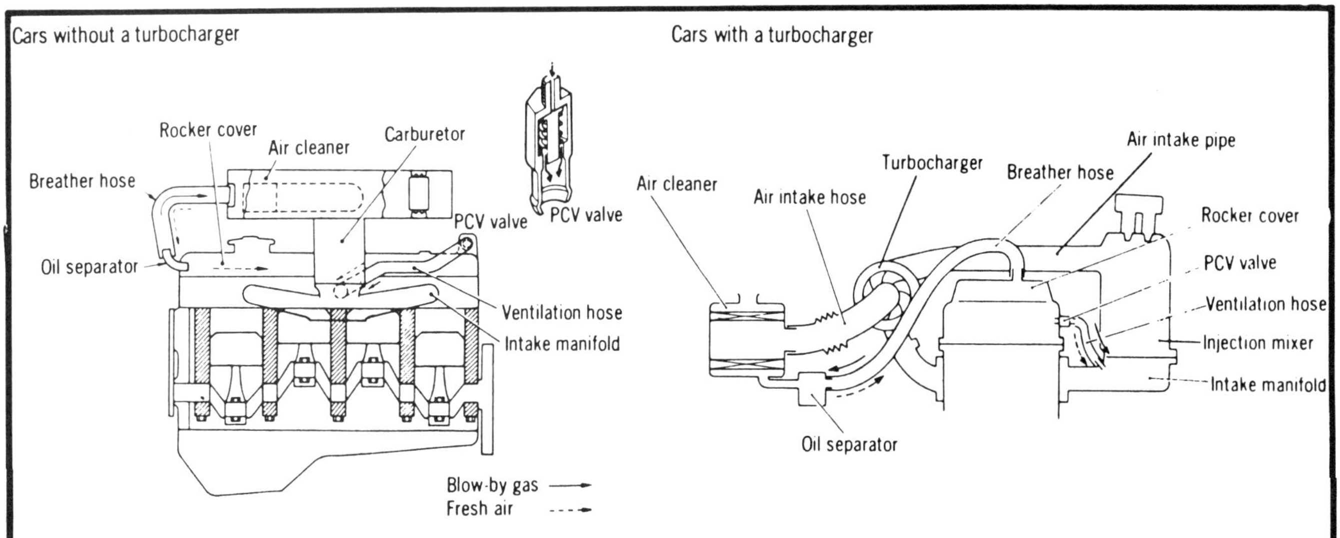

Crankcase Emission Control System—with and without a turbocharger (© Chrysler Corp.)

EMISSION CONTROL SYSTEM

CRANKCASE EMISSION CONTROL SYSTEM

PCV System

All vehicles are equipped with a closed crankcase ventilation system. The system consists of a PCV valve mounted on the cylinder head cover, with a hose extending from the valve to the base of the carburetor. A closed engine oil inlet air cleaner with a hose connecting it to the carburetor air cleaner housing provides a source of air to the system. The ventilation system operates by manifold vacuum, which is air drawn from the carburetor air cleaner through the crankcase air cleaner and into the engine crankcase. It is then circulated through the engine and drawn out through the PVC valve hose and into a passage at the base of the carburetor where it becomes part of the air/fuel mixture. It is then drawn into the combustion chamber where it is burned and expelled with the exhaust gases.

This system is essentially the same as the system used on vehicles equipped with electronic fuel injection. If the vehicle is equipped with electronic fuel injection, the engine, when operating, generates crankcase pressure the is used to purge crankcase vapors through the PCV valve to a port in the throttle body. When the engine is shut down the cooling gases in the crankcase causes a partial vacuum which is relieved by a breather hose that draws filtered air from the air cleaner assembly to the right valve cover. However, all air entering the intake manifold must be measured by the air flow meter to insure proper air/fuel ratio.

It is very important that the PCV system on a vehicle equipped with electronic fuel injection remain a closed system and be free from any disconnects or breaks in the hoses which may allow unmetered air to enter the intake system. Such conditions would cause poor driveability, fuel economy and performance.

System Service

NOTE: The PCV valve should be checked every 15,000 miles and replaced as required. If the valve is found to be defective, do not attempt to clean it. Replace it.

1. With the engine running remove the valve from its mounting. A hissing sound should be heard and vacuum should be felt from the inlet side of the valve.
2. Reinstall the valve. Remove the crankcase inlet air cleaner. Loosely hold a piece of stiff paper over the opening in the rocker cover. Allow one minute for the crankcase pressure to reduce itself. The paper should then suck itself against the rocker cover with noticeable force. Replace the inlet air cleaner in the rocker cover.
3. With the engine, stopped remove the PCV valve. Shake the valve, a clicking sound should be heard indicating that the valve is not stuck.
4. If the valve fails any of the above tests, it should be replaced.

FUEL EVAPORATIVE EMISSION CONTROL SYSTEM

Evaporation Control System

The function of the evaporation control system is to prevent the emission of gas vapors from the fuel tank and carburetor to be expelled into the atmosphere. When fuel evaporates in the gas tank or the carburetor float chamber the vapors pass through the vent hoses to the charcoal canister where they are stored until they can be drawn into the intake manifold when the engine is running. All vehicles are equipped with the charcoal canister for the storage of fuel vapors. The fuel bowls of all carburetors are vented internally and on some applications do not require venting the canister. If this is the case the bowl vent port on the canister will be capped. Most carburetors are also externally vented to the charcoal canister.

System Service

The charcoal canister is a non-serviceable component. The only service required for the system is to replace the filter pad on the bottom of the canister. The filter requires replacement every 12 months or 12,000 miles. If the vehicle is driven under severe conditions, the filter should be replaced more often.

BOWL VENT VALVE

The bowl vent valve is connected to the carburetor fuel bowl, the charcoal canister and the air pump discharge. When the engine is not running there is no air pump pressure applied there is a direct connection

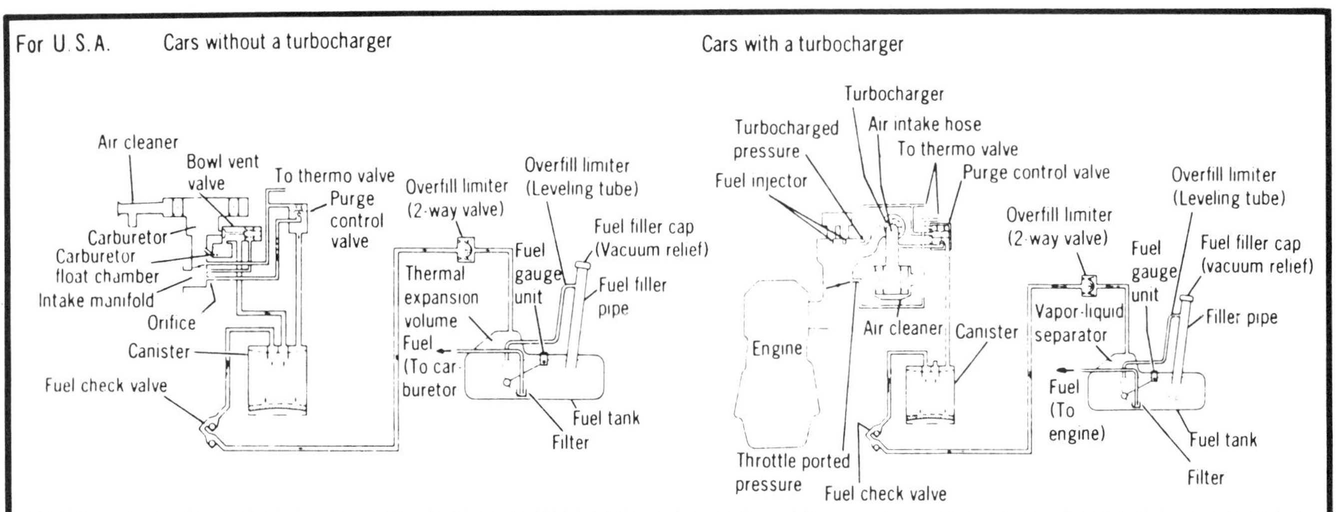

Evaporative Emission Control system—with or without a turbocharger (© Chrysler Corp.)

between the fuel bowl and the canister. When the engine is shut down, the valve air pressure bleeds down and the fuel bowl is allowed to vent into the canister.

OVERFILL LIMITER (TWO WAY VALVE)

The overfill limiter consists of a pressure valve and a vacuum valve. The pressure valve is designed to open when the fuel tank internal pressure has increased over the normal pressure and the vacuum valve opens when a vacuum has been produced in the tank.

PURGE CONTROL VALVE

CARBURETED MODELS

The purge control valve is kept closed during idling in order to prevent vaporized fuel from entering into the intake manifold for positive control of of high idle CO emission, which is a particular problem under high ambient temperatures. When the carburetor vacuum working on the diaphragm of the valve exceeds the pre-set value, the purge control valve is opened.

ECI AND TURBOCHARGER EQUIPPED ENGINES

The purge control valve is closed at idling to prevent vaporized fuel from entering into the air intake hose for positive control of high idle CO emissions which is particularly under high ambient temperature. Once the pressure difference between the turbocharger and the throttle ported pressures exceeds the pre-set value, the purge control valve is opened.

THERMO VALVE

A thermo valve, incorporated in the EGR system for sensing the coolant temperature at the intake manifold, closes the purge control valve when

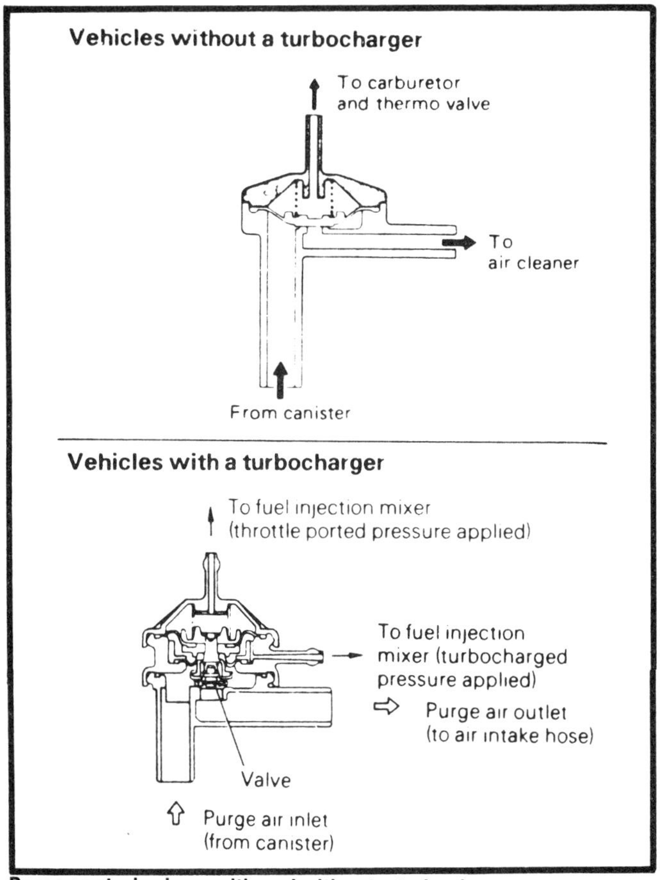

Purge control valve—with and without a turbocharger (© Chrysler Corp.)

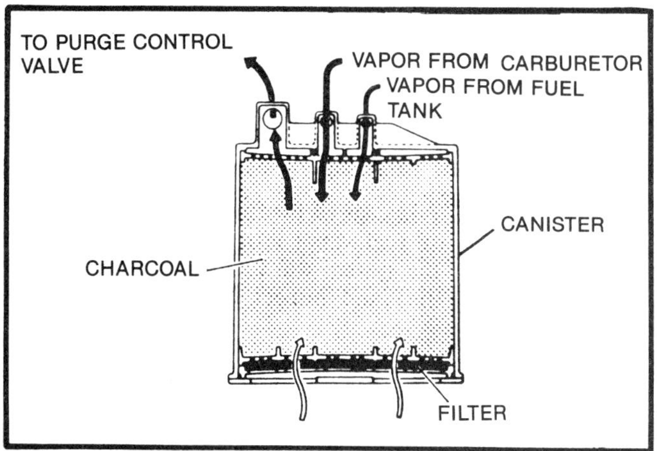

Typical charcoal canister (© Chrysler Corp.)

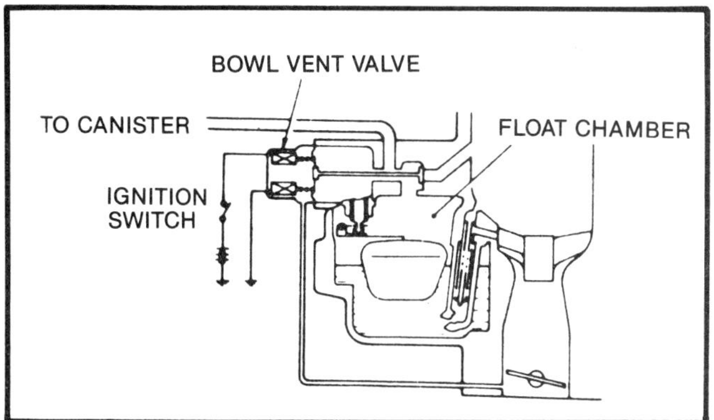

Bowlvent system (© Chrysler Corp.)

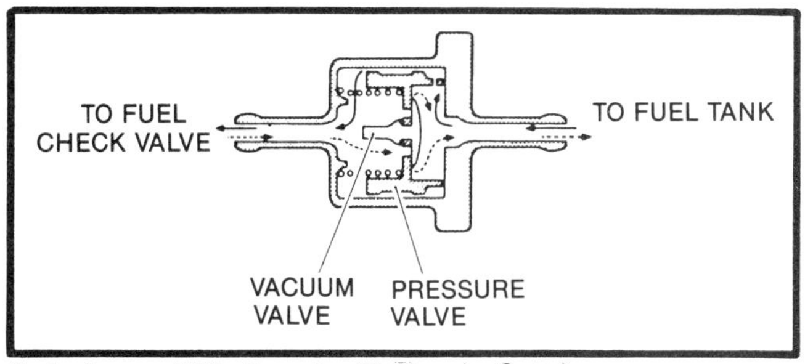

Typical—overfill limiter valve (© Chrysler Corp.)

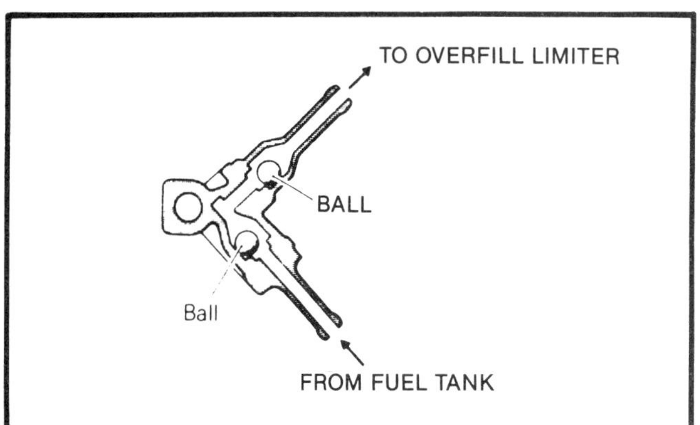

Cross-section of fuel check valve (© Chrysler Corp.)

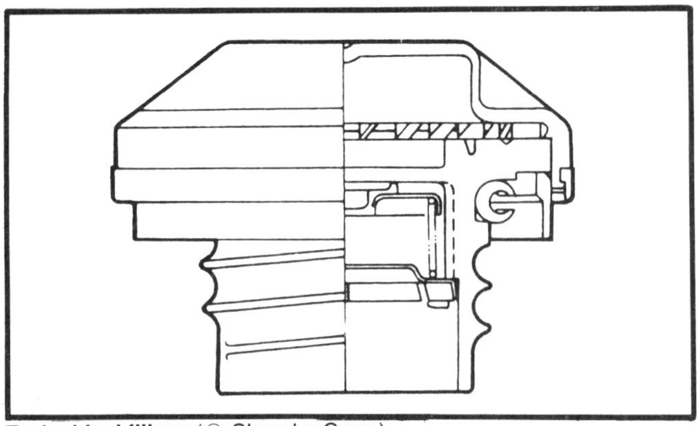

Typical fuel fillcap (© Chrysler Corp.)

the coolant temperature is lower than a pre-set value, in order to reduce CO and HC emissions under warm-up conditions, and opens the purge control valve when the coolant temperature rises above the preset temperature.

FUEL CHECK VALVE

The fuel check valve is used to prevent fuel leakages, should the vehicle suddenly roll over. The valve is connected in the fuel vapor line (between canister and overfill limiter) and is installed on the firewall. The fuel check valve contains two balls and under normal conditions, the gasoline vapor passage in the valve is opened, but if a vehicle roll-over occurs, one of the balls closes the fuel passage, thus preventing fuel leaks.

FUEL FILLER CAP

The fuel filler cap is equipped with relief valve to prevent the escape of fuel vapor into the atmosphere.

EXHAUST EMISSION CONTROL SYSTEM

CARBURETED ENGINES

Exhaust emissions, composed of carbon monoxide (CO), hydro carbons (HC) and oxides of nitrogen (NOx), are controlled by a combination of engine modifications and the addition of special control components. Modifications to the combustion chamber, intake manifold, camshaft, carburetor and ignition system form a basic control system. Additional control devices include an exhaust gas recirculation system (EGR valves and thermo valves), jet air control valve, deceleration spark-advance system, coasting air valve, air switching valve, dashpot, dual catalytic converters, secondary air supply system (pulse air feeder) and heated air intake system. These systems have been integrated into a highly effective system which controls exhaust emissions while maintaining good vehicle performance.

ECI TURBOCHARGED ENGINES

Exhaust emission, composed of carbon monoxide (CO), hydrocarbons (HC) and oxides of nitrogen (NOx), are controlled by a combination of engine modifications and the addition of special control components. Modifications to the combustion chamber, intake manifold and camshaft. Additional control devices include an exhaust gas recirculation system (EGR), dual catalytic converters, secondary air supply system and fuel control system. These systems have been integrated into a highly effective system which controls exhaust emissions while maintaining good vehicle performance.

Jet Air System

In addition to the intake valve and the exhaust valve, a jet valve has been added to draw jet air (super lean mixture or air) into the combustion chamber. The jet valve assembly consists of a jet valve, jet body and spring, which is all screwed into the jet piece and press fitted into the cylinder head with the jet opening toward the spark plug. The jet valve draws its air from an air passage in the carburetor, intake manifold and cylinder head. As the air flows through the two intake openings, the air goes through the carburetor passage, into the intake manifold and cylinder head and flows through the jet valve and opening into the combustion chamber. The jet valve is operated by the same cam as the intake valve and are joined by a common rocker arm so that the jet valve and intake valve open and close at the same time.

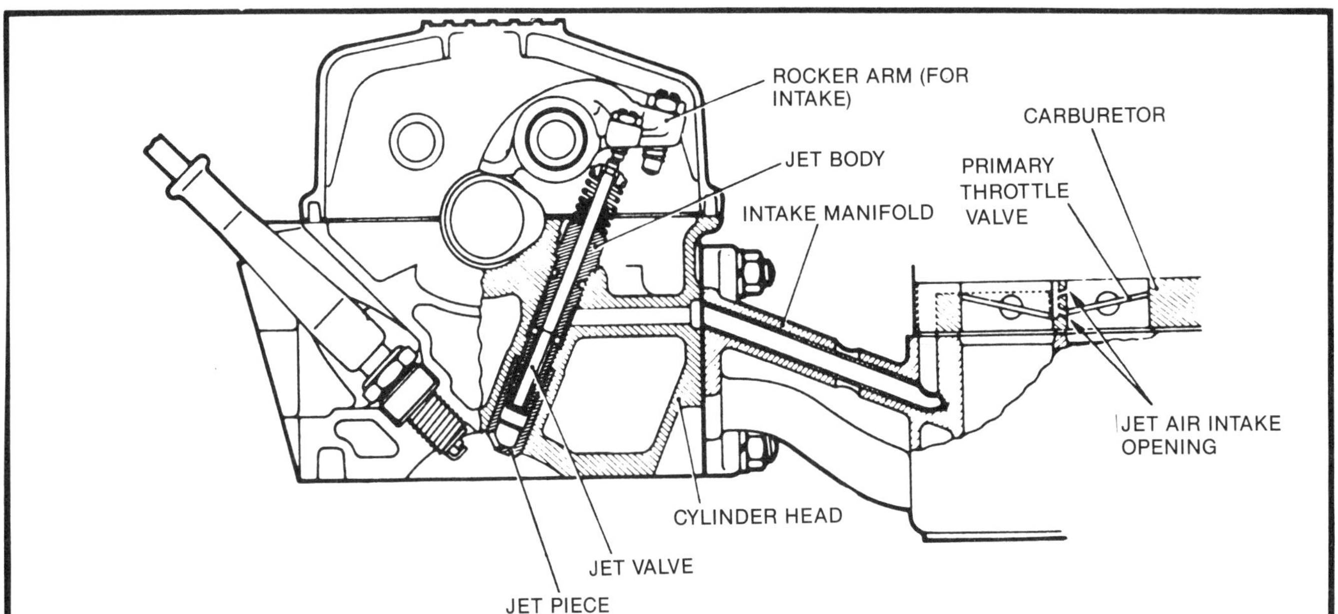

Exploded view of the jet valve system (© Chrysler Corp.)

On the intake stroke, the air-fuel mixture flows through the intake valve port into the combustion chamber. At the same time, the jet air is forced into the combustion chamber and the jet air, running out of the jet opening, scavanges the residual gases around the spark plug and creates a good ignition condition. It also produces a strong swirl in the combustion chamber, which continues through out the compression stroke and improves the flame spread after ignition, assuring a high combustion efficiency. The jet air swirl dwindles with the increase of the throttle valve opening, but the intensified inflow of normal intake air-mixture can satisfactorily promote combustion.

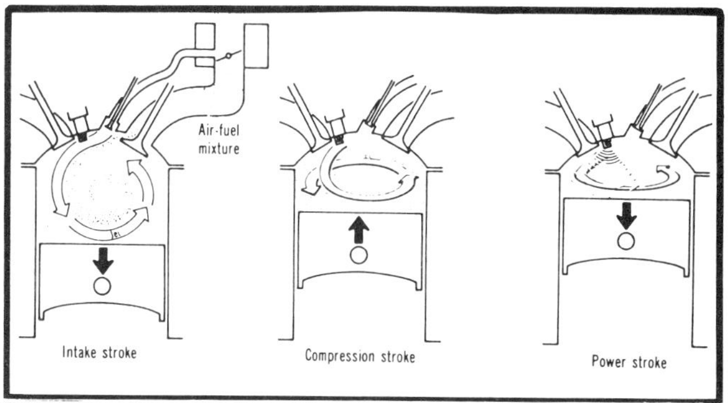

Operation of the jet valve system (© Chrysler Corp.)

JET VALVE

Clearance Adjustment

ALL MODELS EQUIPPED

NOTE: An incorrect jet valve clearance would affect the emission levels and could also cause engine troubles. So the jet valve clearance must be correctly adjusted. Adjust the jet valve clearance before adjusting the intake valve clearance. Furthermore, the cylinder head bolts should be retightened before making this adjustment. The jet valve clearance should be adjusted with the adjusting screw on the intake valve side fully loosened.

1. Start the engine and let it run at idle until it reaches normal operating temperature.
2. With the piston in the cylinder positioned at top dead center on the compression stroke, loosen the adjusting screw for the intake valve two or more turns.
3. Loosen the lock-nut on the adjusting screw for the jet valve, turn the adjusting screw counter-clockwise and insert a 0.010 in. (0.25mm) feeler gauge between the jet valve stem and the adjusting screw.
4. Tighten the adjusting screw until it touches the feeler gauge. Turn the lock-nut to secure it while holding the rocker arm adjusting screw with a suitable tool to keep it from turning.
5. Be sure that the 0.010 in. (0.25mm) feeler gauge can be easily inserted and then adjust the intake valve clearance 0.006 in. (0.15mm).

NOTE: The exhaust valve clearance is 0.010 in. (0.25mm) and all the valve clearance's are to bet set with the engine hot.

6. Check the idle speed and CO, and adjust if necessary.

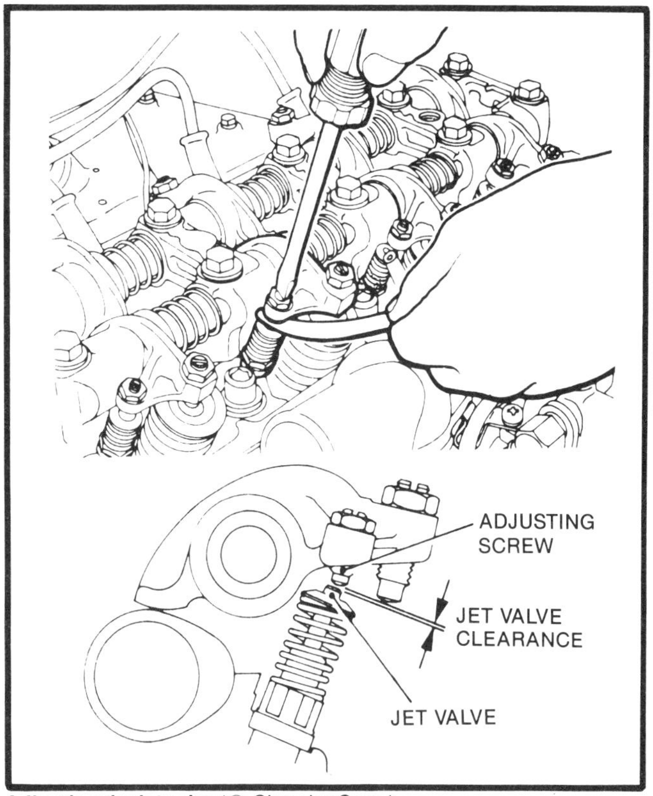

Adjusting the jet valve (© Chrysler Corp.)

JET VALVE

Removal

1. Disconnect the negative battery cable and remove the rocker arm cover, rocker arm and the rocker shaft, using the special jet valve socket wrench (MD998310), or its equivalent and remove the jet valve.

NOTE: When the jet valve socket wrench is used, be sure that the wrench is not tilted with respect for the center of the jet valve. If the tool is tilted, the valve stem might be bent by the force exerted on the valve spring retainer, resulting in defective jet valve operation.

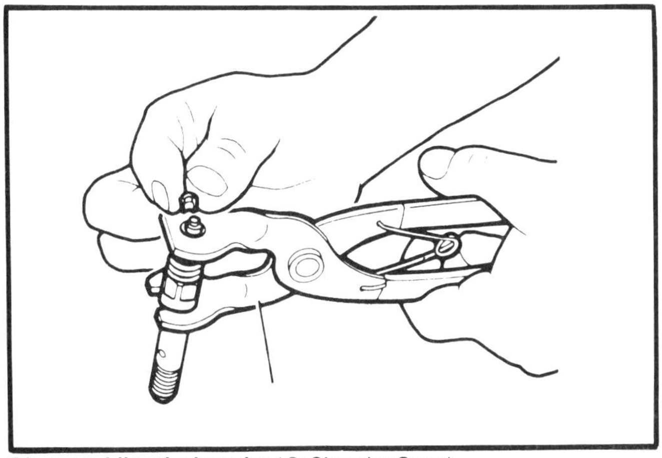

Disassembling the jet valve (© Chrysler Corp.)

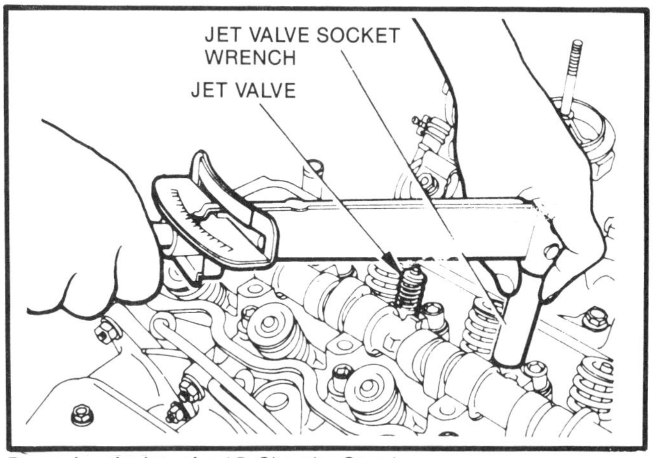

Removing the jet valve (© Chrysler Corp.)

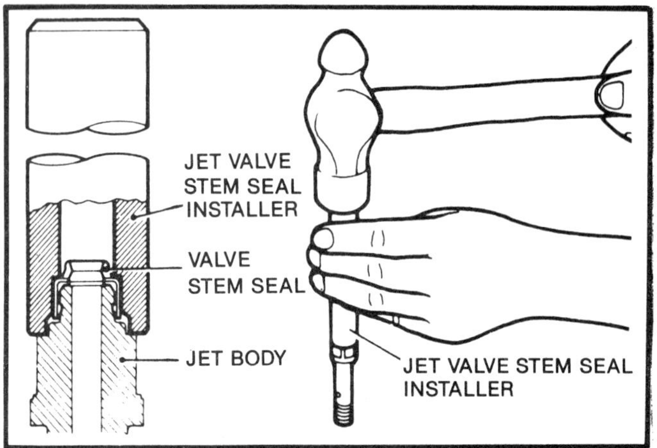

Installing the jet valve (© Chrysler Corp.)

2. Using a hand held spring compressor, remove the jet valve spring retainer lock, valve spring retainer and the valve spring.
3. Check the jet valve assembly for deterioration, cracks or damage, and replace if defective.

Installation

1. Using a seal installer, install a new seal jet valve stem seal. Do not reuse the old jet valve stem seal.
2. Apply engine oil to the stem of the jet valve and insert the jet valve into the jet body. Be sure that the valve slides smoothly.
3. Using the spring compressor, compress the jet valve spring and install the jet valve spring retainer. Be careful not to damage the valve stem with the spring compressor.
4. Install a new O-ring in the groove around the jet body and apply engine oil to the O-ring.
5. Screw the jet valve assembly into the cylinder head, using special tool MD 998310 jet valve socket wrench or equivalent. Torque the jet valve assembly to 13–15 ft.lbs. (18–21 Nm).
6. Install the rocker shaft, rocker arm, rocker arm cover and install the negative battery cable.

Catalytic Converter

Catalytic converters require the use of unleaded fuel only. Leaded fuel will destroy the effectivness of the catalysts as an emisssion control device. Under normal operating conditions, the converter will not require any maintenance. However, it is important to keep the engine properly tuned. If the engine is not kept properly tuned engine misfiring may cause overheating of the cataylsts. This may cause heat damage to the converters or vehicle components. This situation can also occur during diagnostic testing if any spark plug cables are removed and the engine is allowed to idle for a long period of time.

FRONT CATALYTIC CONVERTER

Removal

COLT AND COLT VISTA

NOTE: Be sure that the exhaust system is cooled down before attempting to remove the converter.

1. Remove the air cleaner, air duct and heat cowl.
2. Disconnect the front exhaust pipe at the exhaust manifold and the secondary air supply pipe.
3. Remove the stud nuts attaching the exhaust manifold to the cylinder head. Slide the manifold off the studs and away from the cylinder head.
4. Remove the bolts tightening the exhaust manifold to the catalyst case assembly.

Installation

NOTE: Replace the interior parts of the front catalytic converter, by placing a new cushion on the catalayst. Be sure the cushion is not deformed and place a new stainless steel gasket on the catalyst case. Be sure the gasket fits the inside diameter of the cushion.

1. Put the exhaust manifold and the catalyst case assembly together and torque the bolts evenly to 22–25 ft.lbs. (30–34 Nm).
2. After installing the exhaust manifold gasket to the cylinder head, install the exhaust manifold assembly and torque the nuts to 11–14 ft.lbs. (15–19 Nm).

NOTE: If the gasket has to be reused, check both sides of the gasket for damage. The gasket should not be reused if there are any indications of damage to the gasket.

3. Install the heat cowl, air duct and air cleaner.

Removal and Installation

CONQUEST

1. Remove the heat cowl and the heat protector.
2. Remove the oxygen sensor and the secondary airpipe from the front catalytic converter.
3. Disconnect the front exhaust pipe from the front converter and remove the stud nuts attaching the front converter to the turbocharger.
4. Slide the catalytic converter off the studs and away from the turbocharger.
5. Installation is the reverse order of the removal procedure, be sure to use new gaskets where ever needed.

REAR CATALYTIC CONVERTER

Removal and Installation

COLT AND COLT VISTA

1. Disconnect the main muffler from the rear catalytic converter and raise the front exhaust pipe.
2. Remove the rubber O-rings from the hanger and disconnect the rear catalytic converter from the front exhaust pipe.
3. Installation is the reverse order of the removal procedure.

Removal and Installation

CONQUEST

1. Disconnect the rear catalytic converter from the center of the exhaust pipe at the front catalytic converter.
2. Remove the rubber O-rings and take out the rear catalytic converter.
3. Installation is the reverse order of the removal procedure.

Secondary Air Supply System

REED VALVE

VEHICLES WITHOUT ELECTRONICALLY CONTROLLED INJECTION (ECI) SYSTEM

The air injection system consists of a reed valve with a secondary air control valve, solenoid valve, check valve, ECU and sensors. The reed valve supplies secondary air into the front catalytic converter for the purpose of promoting oxidation of exhaust emissions during the engine warm-up operation, deceleration and heavy engine load operation zone.

The reed valve is actuated by exhaust vacuum being generated from negative and positive pulsation in the exhaust manifold. Extra air is supplied into the exhaust manifold through the secondary air control valve. The secondary control valve is opened by the intake manifold pressure when the solenoid valve is energized by the ECU, based on the information on coolant temperature, engine speed, time, throttle position and idle positon.

VEHICLES EQUIPPED WITH ELECTRONICALLY CONTROLLED INJECTION (ECI) SYSTEM

The air injection system consists of a reed valve with a secondary air

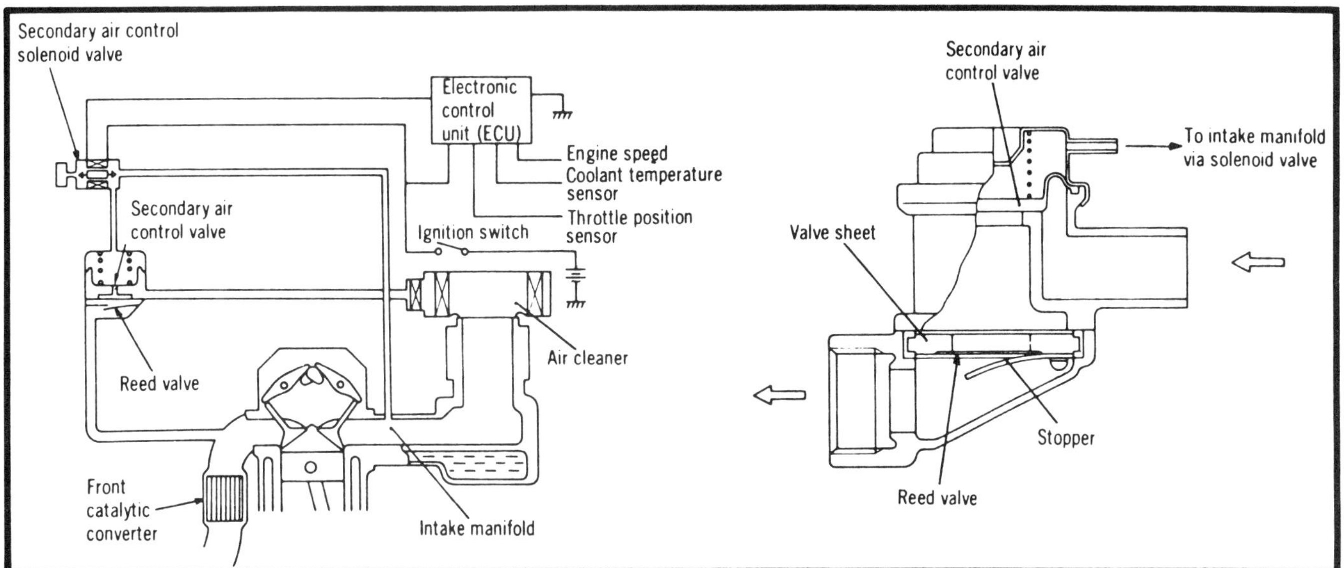

Secondary air supply system and reed valve assembly (© Chrysler Corp.)

control valve, and a solenoid valve. The reed valve supplies secondary air into the front catalytic converter for the purpose of promoting oxidation of exhaust emissions during the engine warm-up operation, the engine hot-start operation and the vehicle deceleration. The reed valve is actuated by exhaust vacuum being generated from pulsation in the exhaust manifold, and extra air is supplied into the exhaust manifold through the secondary air control valve. The secondary control valve is opened by the intake manifold pressure when the solenoid valve is energized by the ECU based on the information on coolant temperature, intake air flow, engine speed, ignition switch start and idle position.

Exhaust Gas Recirculation System (EGR)

COLT

In this exhaust gas recirculation system, the exhaust gas is partially recirculated from an exhaust port of the cylinder head, into a port located at the intake manifold below the carburetor, while the exhaust gas recirculation flow is controlled the the EGR control valve, a sub EGR valve and the thermo valve. Stringent oxides of nitrogen standard requires increased exhaust gas recirculation flow, which in turn affects driveability. To stop this problem, the EGR flow is increased to attain effective oxides of nitrogen reduction during high load vehicle operation.

COLT VISTA

In this exhaust gas recirculation system, the exhaust gas is partially recirculated from an exhaust port of the cylinder head into a port located

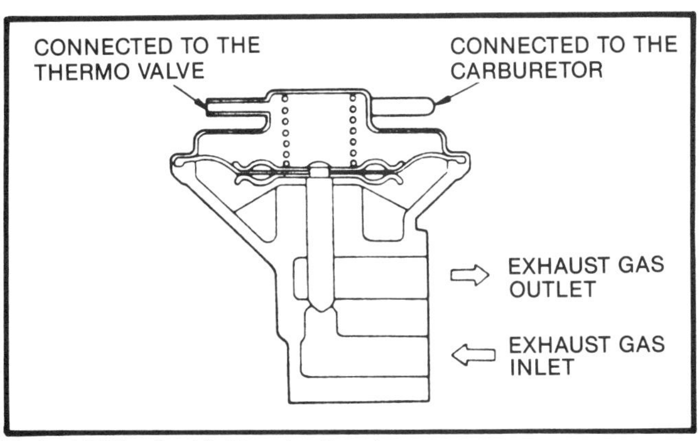

Typical EGR valve (© Chrysler Corp.)

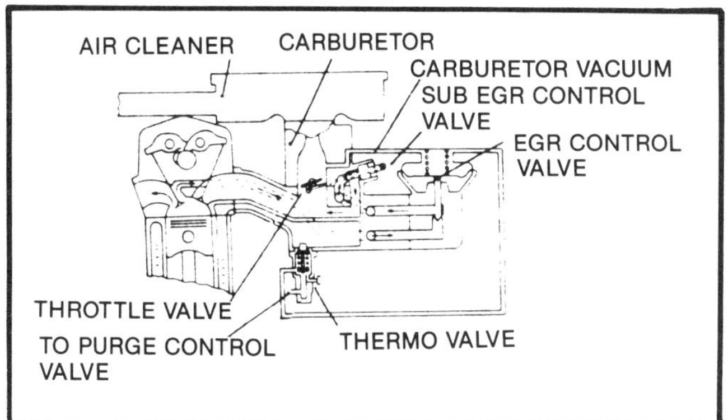

EGR—Gas Recirculation System (© Chrysler Corp.)

at the intake manifold below the carburetor, while the exhaust gas recirculation flow is controlled the the EGR control valve and the thermo valve. Stringent oxides of nitrogen standard requires increased exhaust gas recirculation flow which in turn affects driveability. To stop this problem, the EGR flow is increased to attain effective oxides of nitrogen reduction during high load vehicle operation and decreased to attain effective oxides of nitrogen during poor driveability under low load vehicle operation. The EGR flow is controlled by a combination of a dual EGR control valve and a sub EGR control valve. The dual EGR control valve consists of a primary and secondary set of valves which are controlled by different carburetor vacuums in response to the throttle valve openings, while the EGR flow is suspended at idle and wide open throttle operation. The primary valve controls the EGR flow for vehicle operation with relatively narrow throttle valve openings. The secondary control valve allows EGR flow to travel into the intake mixture when the throttle valve is opened wider. The applied vacuum on the dual EGR valve is controlled by a thermo valve. The sub EGR control valve is directly opened and closed with the motion of the throttle valve through linkage. A linkage control is used in order to closely modulate the EGR flow that is controlled by the EGR control valve, in response to throttle valve opening.

CONQUEST

In this exhaust gas recirculation system, the exhaust gas is partially recirculated from an exhaust port of the cylinder head into a port located at the intake manifold below the fuel injection mixer, while the exhaust gas recirculation flow is controlled the the EGR control valve and a solenoid valve. The pressure applied to the EGR control valve is controlled by the solenoid valve motion, which mixes the throttle ported pressure with the turbo charged pressure. The solenoid valve motion is controlled by the ECU based on the information from the coolant temperature and engine speed.

System Service

The valve should be inspected every 15,000 miles for deposits, particular attention should be given to the poppet and seat area of the valve. If deposits are present, the valve should be cleaned. Manifold heat control valve solvent or equivalent should be used when cleaning the valve, this compound will allow the deposits on the poppet and seat area to soften. The deposits may be removed by opening the valve, using an external vacuum source and scrapping the deposits from the poppet and seat with a sharp suitable tool. If wear to the stem or other moving components is noticed, the EGR valve should be replaced. Care should be used when cleaning the valve, so as not to damage the diaphragm. Once the diaphragm has be damaged, the valve will have to be replaced.

System Testing

COLT AND CONQUEST

1. Inspect all hose connections between the carburetor, intake manifold, and the EGR control valve. Replace any and all hoses that are found to be defective.
2. With the engine warmed up and running, position the selector lever in the neutral position and accelerate the engine to about 2000 rpm, but not over 3,000 rpm.
3. Visable movement in the EGR valve stem should be seen. During this operation, a change in the relative position of the groove on the EGR valve stem will occur.
4. This operation can be repeated several times to confirm EGR operation. Movement of the stem indicates the the valve is functioning properly.
5. If the stem does not move accordingly, the EGR valve is not functioning properly and the valve should be checked.

COLT VISTA

1. Check all vacuum hoses for proper routing, defects and installation.
2. Cold start and run engine at idle speed.
3. Check to be sure that increasing the engine idle speed to 2500 rpm does not cause the secondary EGR valve to operate. If the secondary EGR valve operates, replace the thermo valve.
4. Warm the engine until the coolant temperature exceeds 149 degrees F.
5. Check to be sure that increasing the engine idle speed to 2500 rpm operates the secondary EGR valve. If the valve does not operate, inspect the EGR control valve or the thermo valve.
6. Disconnect the green stripe hose from the nipple of the carburetor. Connect a vacuum pump to the green stripe hose.
7. Open the sub EGR valve by pulling it by hand. Apply over six inches of vacuum, using the pump.
8. If the engine idling speed becomes unstable, the secondary valve is operating properly. If the engine idling speed remains unchanged, the secondary valve or the thermo valve is not operating properly. Replace the defective component.
9. Disconnect the vacuum pump and connect the green stripe hose to the carburetor. Connect a vacuum pump to the yellow stripe hose.

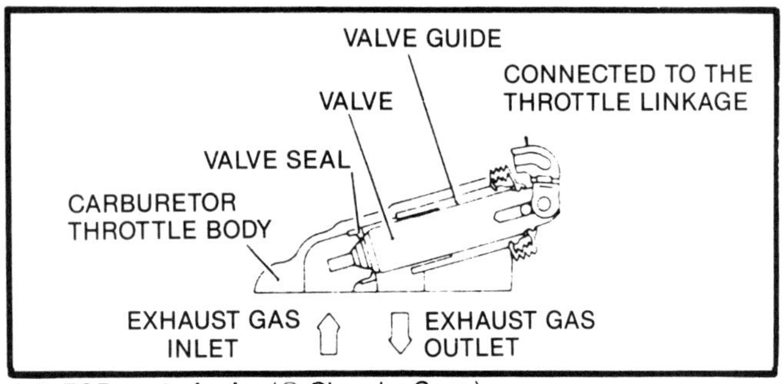

Sub-EGR control valve (© Chrysler Corp.)

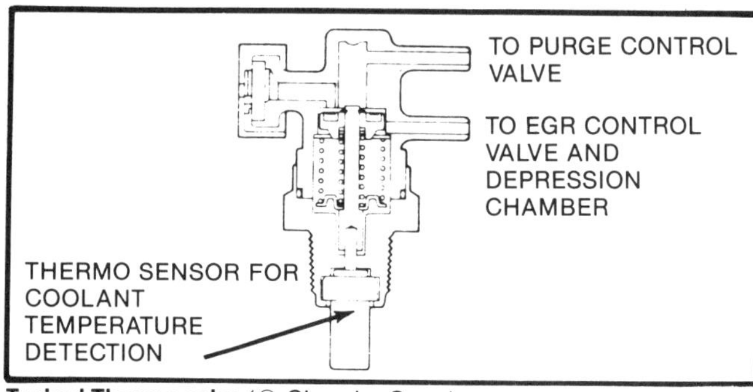

Typical Thermo-valve (© Chrysler Corp.)

10. Open the sub EGR valve by pulling it by hand, apply over six inches of vacuum using the pump.
11. If the engine idle speed becomes unstable, the primary valve is operating properly.
12. If the engine idling speed remains unchanged, the primary valve or the thermo valve is not operating. Replace the defective component.

SUB EGR CONTROL VALVE

COLT AND COLT VISTA

The Sub EGR control valve is directly opened and closed with the motion of the throttle valve through a linkage set up. In order to closely modulate the EGR flow controlled by the EGR control valve, in response to the throttle valve opening.

THERMO VALVE

COLT AND COLT VISTA

The Thermo valve switch in the EGR vacuum supply line is sensitive to engine temperature. The thermo valve eliminates the vacuum signal to the EGR valve during warm-up, when less oxides of nitrogen NOx is generated and the less EGR helps to maintain acceptable driveability.

SOLENOID VALVE

CONQUEST

When the solenoid valve is incorporated in the EGR system and the system is energized by the electronic control unit (ECU), the EGR control valve is closed, due to the throttle ported pressure mixed with the turbocharged pressure, and is opened with the aid of the throttle ported vacuum, applied to its diaphragm when the solenoid valve is de-energized by the ECU.

Solenoid Valve Test

1. Start the engine and let it idle until it reaches normal operating temperature.
2. Remove the green stripe hose from the injector mixer and connect a hand held vacuum pump or equivalent, to the end of the hose.
3. Disconnect the connector of the solenoid valve and apply a vacuum of 9.8 in. Hg. If a unstable idling occurs or engine stalls at this time, the EGR system is good.
4. Reconnect the connector of the solenoid valve and perform the test in step four again. This time, the engine should remain unaffected, as the negative pressure chamber of the EGR valve is opened to the atmosphere.
5. If an unstable idle occurs, it means that the solenoid valve is not operating properly and it is staying in the close position. Therefore, the valve should be removed and replaced.

Heated Air Intake System

COLT AND COLT VISTA (EXCEPT VEHICLES WITH TURBOCHARGER)

These models are equipped with a temperature regulated air cleaner which improves engine warm up characteristics and helps to minimize carburetor icing. This air cleaner has a door inside the snorkel to modulate the temperature of the air entering the carburetor. The door is controlled by a vacuum motor and a bi-metal temperature sensor combination system which reacts to intake manifold vacuum and the temperature inside the air horn. When the bi-metal senses the inside air horn temperature at or below 86 degrees F., the air bleed valve portion of the temperature sensor remains closed. Intake manifold vacuum is applied to the vacuum motor which opens the air control door to let the pre-heated air flow through the heat cowl and air duct and into the air cleaner. When the bi-metal sensor senses the air horn inside temperature at or above 113 degrees F., the air bleed valve portion of the temperature sensor is fully open, allowing the intake air to go directly to the carburetor through the outside air duct and closing the heated air duct, regardless of the intake manifold vacuum. At intermediate temperatures, the air entering the carburetor is a blend of outside air and pre-heated air as regulated by the bi-metal sensor controlling the degree of opening on the air control door.

System Service

1. Make sure that all vacuum hoses and the stove to the air cleaner flexible connector are properly installed and in good condition.
2. With a cold engine and ambient temperature less than 84 degrees F., the heat control door or valve should be in the up or heat-on position.
3. With the engine warm and running, check the air temperature entering the snorkel or at the sensor. When the air temperature entering the outer end of the snorkel is 113 degrees F. or higher, the door should be in the down or heat-off position.
4. Remove the air cleaner from the engine and allow it to cool down to 84 degrees F. With 15 in. Hg. of vacuum applied to the sensor, the door or valve should be in the up or heat-on position. Should the door not rise to the heat-on position, check the diaphragm or motor for proper operation.
5. To test the diaphragm or motor, apply 10 in. Hg. of vacuum with vacuum pump, tool number C-4207 or equivalent.
6. The valve should be in the full up position. Should this not be the case, replace the air cleaner and body assembly.

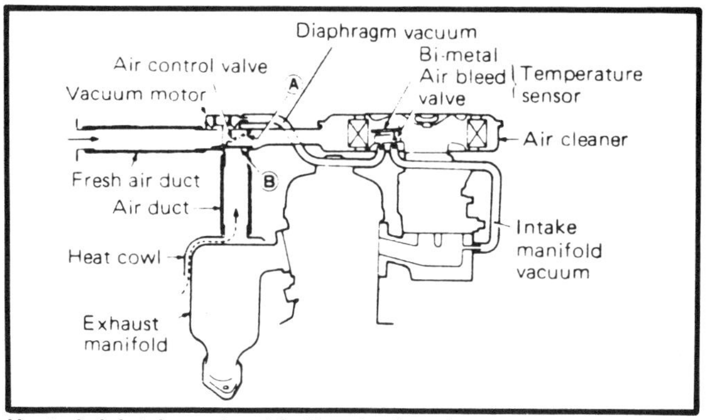

Heated air intake system (© Chrysler Corp.)

Deceleration Devices

COLT AND COLT VISTA

The deceleration devices are used to decrease hydrocarbon (HC) emissions during vehicle deceleration.

DECELERATION SPARK ADVANCE SYSTEM

COLT AND COLT VISTA (EXCEPT WITH TURBOCHARGER)

In order to decrease the hydro-carbon (HC) emissions emitted during vehicle deceleration, ignition timing is advanced by the solenoid operated vacuum valve on the distributor, changing the vacuum supplied to the valve from carburetor ported vacuum to intake manifold vacuum. The solenoid valve is controlled by the electronic control unit, based on engine speed.

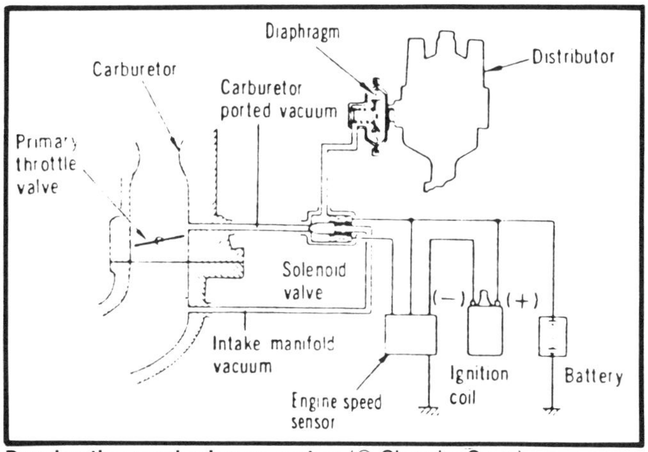

Deceleration spark advance system (© Chrysler Corp.)

DECELERATION FUEL CUT-OFF

CONQUEST

To decrease the hydro carbon emissions emitted during vehicle deceleration, the fuel delivery time is dramatically decreased by the electronic control unit changing the injection interval which is determined by the specific pulses from the air flow sensor, with specified duration of injection. During the vehicle deceleration, while the idle position switch is not yet in the on position, the fuel delivery time is determined by the electronic control unit responding to the throttle valve closing speeds.

DASH POT

COLT AND COLT VISTA

The carburetor is equipped with a dash pot, which delays the throttle valve closure to its normal idling position, thereby reducing the amount of hydrocarbon (HC) emissions emitted.

HIGH ALTITUDE COMPENSATION SYSTEM

FEDERAL CARBURETED MODELS

So that all Federal models will meet the federal requirements at high altitude, all the carbureted vehicles are equipped with a high altitude

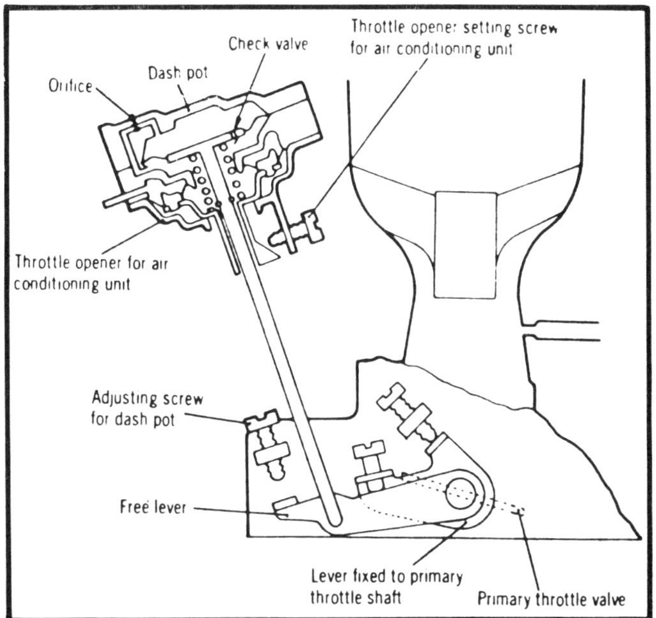

Cross section of dashpot assembly (© Chrysler Corp.)

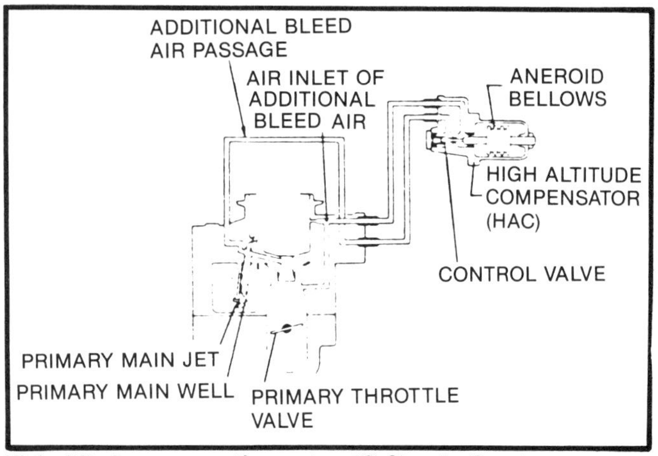

High altitude compensation system (© Chrysler Corp.)

compensation system, in addition to the feedback carburetor system. The high altitude compensation system is made up of the following components; a high altitude compensator (HAC), a vacuum switching valve and a distributor equipped with a high altitude advance system. Air/fuel ratio at a high altitude is controlled by the HAC to be approximately the same value as the one at sea level, by supplying additional bleed air into the primary and secondary main wells through HAC and vacuum switching valves, controlled by the high altitude compensator. In order to reduce hydrocarbon and carbon monoxide emissions and to get better driveability at high altitude, the igniton timing is advance by specified degrees at high altitude. The spark advance signal is sent to the sub-diaphragm chamber of the distributor, through the high altitude compensator. On California models, instead of using the vacuum switching valve, a check valve is used in its place.

Electronically Controlled Injection Models

The fuel injected vehicles are equipped with a pressure sensor and a solenoid valve which are utilized as a high altitude compensator. When the ignition switch is in the on position the solenoid valve is energized by the ECU for the specific time, which in turn allows the pressure sensor to detect the ambient atmospheric pressure. The fuel delivery time is then determined by the ECU in response to the ambient atmospheric pressure so that the air/fuel mixture at high altitudes can be maintained at the same ratio as at sea level.

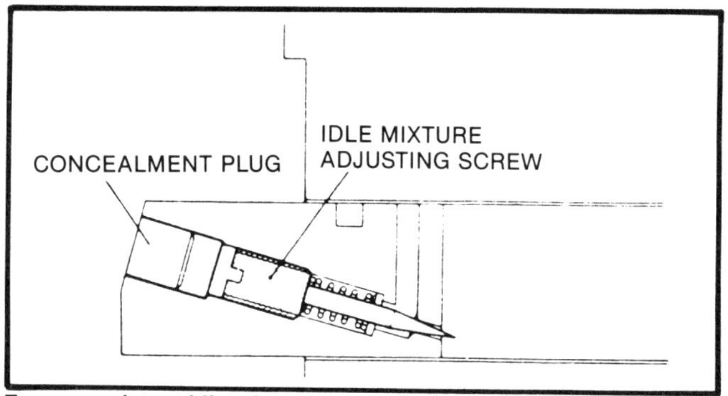

Tamper resistant idle mixture screw (© Chrysler Corp.)

THROTTLE OPENER SYSTEM

COLT AND COLT VISTA (WITH A/C)

This system contains the following components; a solenoid valve, an engine speed sensor and the compressor switch of the A/C unit. When the A/C compressor is turned on, the engine speed will fall below specified value. When this occurs, the solenoid valve will open and in turn, transfer the intake manifold vacuum to the throttle opener. The result is the engine running at a higher speed due to the increase setting of the new throttle valve opening against the compressor's load. When the A/C compressor is shut off, the throttle opener system stops working, and the engine runs at the original engine speed.

TAMPER RESISTANT IDLE MIXTURE

All carburetors have a tamper resistance idle mixture adjustment. The CO setting has been done as a factory adjustment. Neither removal of the plug nor adjustment of the mixture screw is required in service unless a major carburetor overhaul, throttle body replacement, or high CO adjustments are required by state or local inspections.

TAMPER PROOF AUTOMATIC CHOKE

All carburetors have tamper-proof choke mechanism. The choke-related parts are factory adjusted. Neither removal of the choke cover nor tampering with wax-stroke adjusting screw (WAS) is required in service except when major carburetor overhaul or adjustment of choke/calibration related parts are required by state or local inspection.

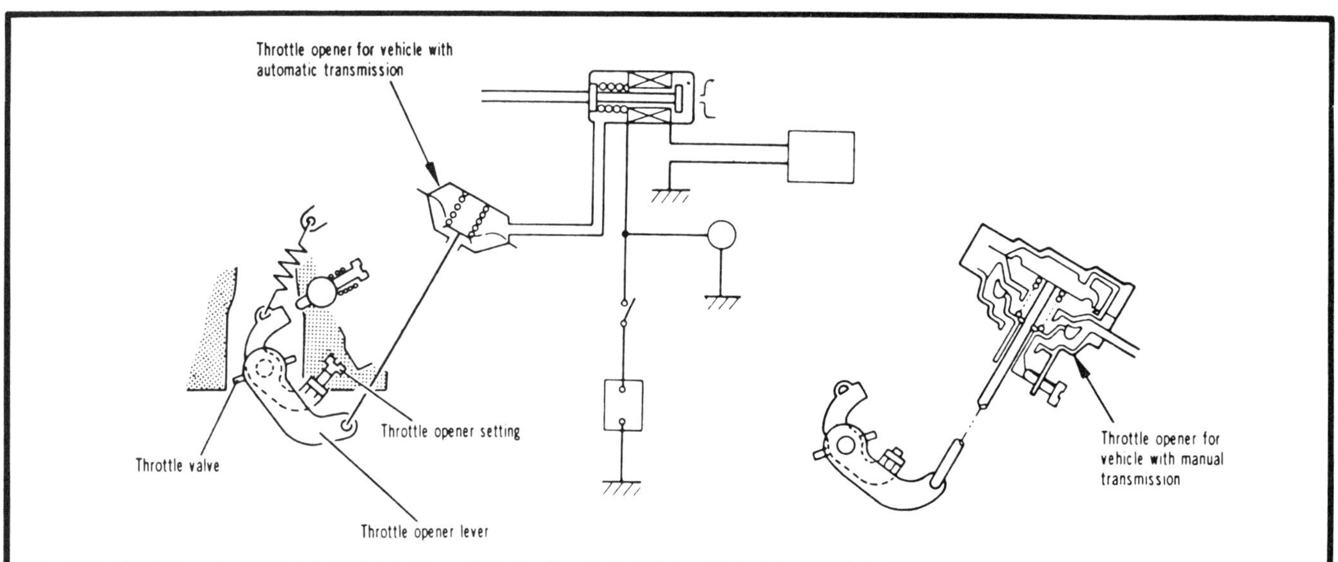

Throttle opener system used on A/C equipped vehicles (© Chrysler Corp.)

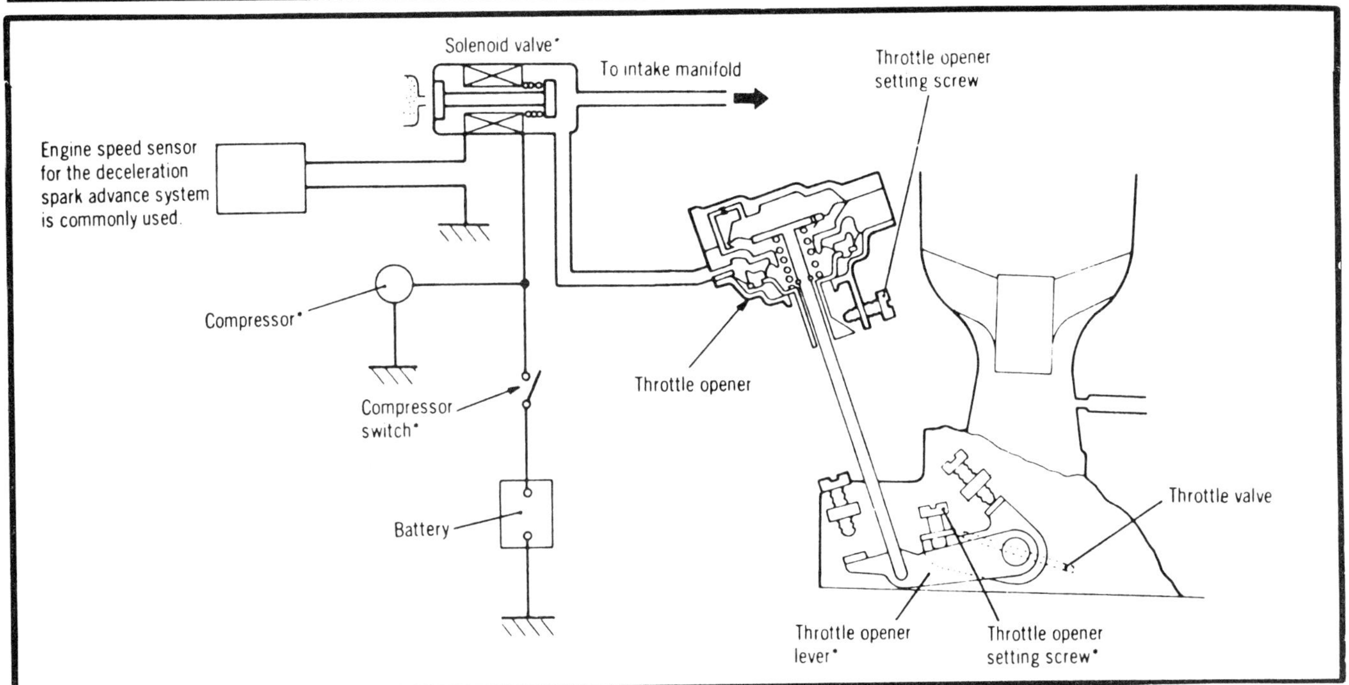

Throttle opener system used on most vehicles (© Chrysler Corp.)

FUEL CUT-OFF SYSTEM

When the ignition key is turned off, the deceleration solenoid valve cuts off the fuel flow to prevent engine run-on. During certain deceleration modes, the deceleration solenoid valve reduces the fuel flow in order to decrease hydrocarbon emissions and improve fuel economy. Under normal engine operation, the needle valve is controlled by the solenoid to provide the necessary fuel flow for smooth engine operation.

Feedback Carburetor (FBC) System

The feedback carburetor system provides the capability to perform closed loop fuel control. It also provides the capability to control the secondary air system, the deceleration spark control system and the throttle opener system. The basic functions of this system are depicted below. Imput signals from a variety of sensors are fed to a microprocessor based electronic control unit (ECU). The ECU then generates output signals for all of the controlled functions. The feedback carburetor is a 2-barrel, downdraft carburetor designed for closed loop system. When used with the closed loop system of mixture control, this carburetor includes special design features for optimum air/fuel mixtures during all ranges of engine operation. Fuel metering is accomplished through the use of three solenoid-operated on/off valves (jet mixture, enrichment and deceleration solenoids), adding or reducing fuel to the engine.The activation of the on/off valve is controlled by the length of the time current is supplied to the solenoid. The solenoid operates at a fixed frequency. By varying the amount of time the solenoid is energized during each cycle (defined as duty cycle), the air/fuel mixture delivered to the engine can be precisely controlled. The duty cycle to the solenoid is controlled by the ECU in response to the signals from the exhaust oxygen sensor, throttle position sensor. Incorporated in the feedback carburetor are eight basic systems of operation: fuel inlet, primary metering, secondary metering, accelerating pump, choke, jet mixture, enrichment and fuel cut-off.

ELECTRONIC CONTROL UNIT (ECU)

The electronic control unit is mounted in the passenger compartment and consists of a printed circuit board mounted in a protective metal box. It receives analog inputs from the sensors and converts them into

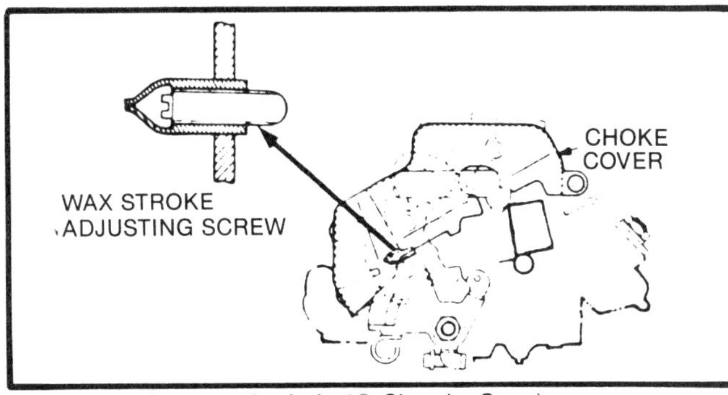

Tamper-proof automatic choke (© Chrysler Corp.)

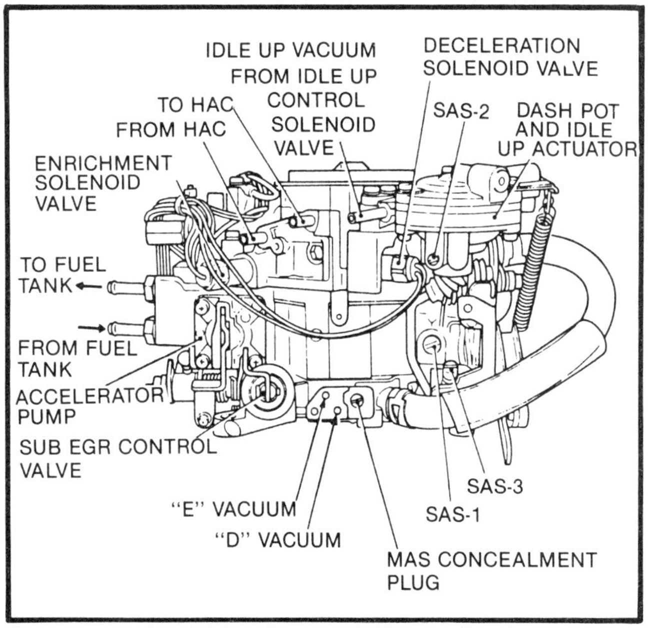

Side view (A) of the FBC carburetor (© Chrysler Corp.)

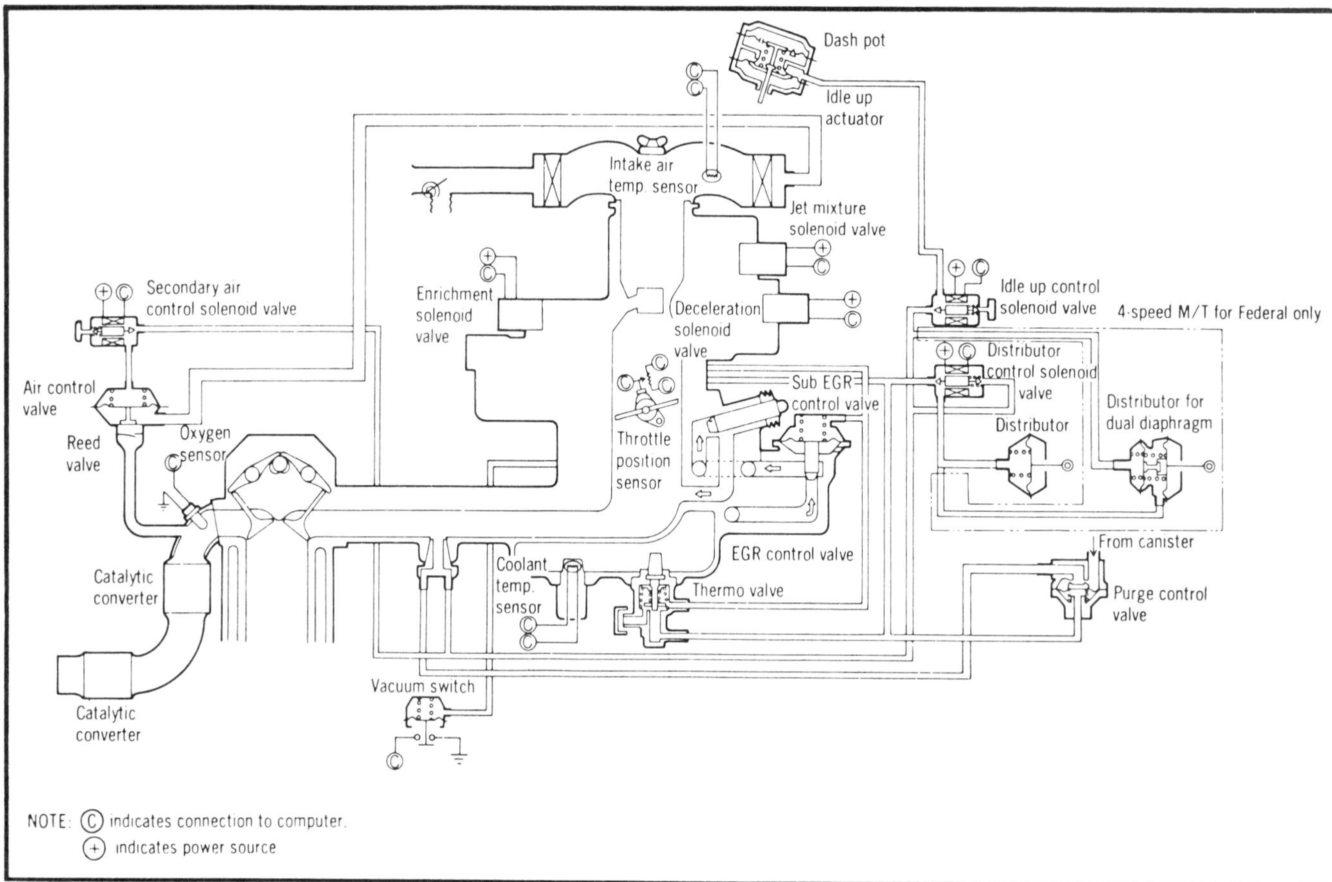

Schematic of Feedback Carburetor (FBC) system (© Chrysler Corp.)

digital signals. These signals and various discrete inputs are processed and used by the ECU in controlling the fuel delivery, secondary air, deceleration spark and throttle opener managements.

FEEDBACK CARBURETOR AIR/FUEL CONTROL

The feedback carburetor air/fuel ratio is controlled by the Electronic Control Unit (ECU). The ECU monitors the throttle position, engine

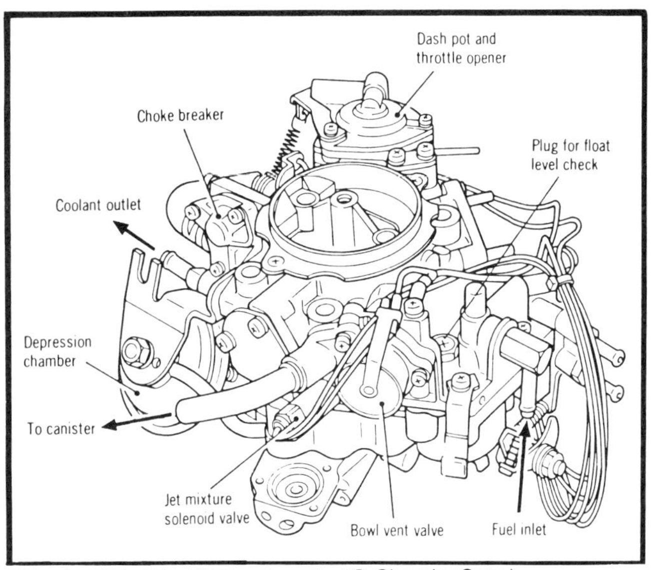

Side view (B) of the FBC carburetor (© Chrysler Corp.)

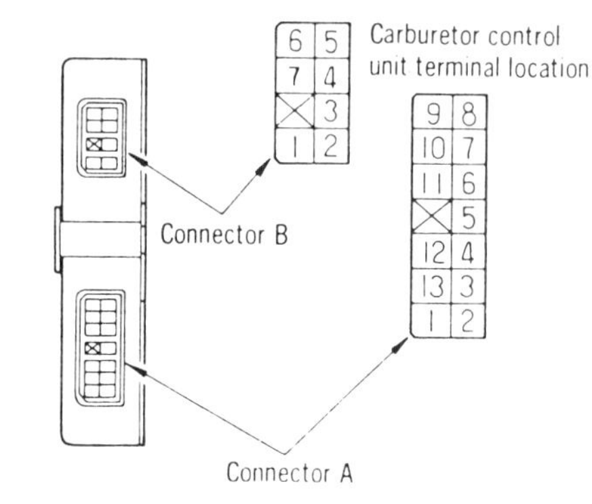

A1	Oxygen sensor	B1	Jet mixture solenoid valve
A2	Ground for sensor	B2	Deceleration solenoid valve
A3	Throttle position sensor (+)	B3	Idle up control solenoid valve
A4	Intake air temperature sensor	B4	Secondary air control solenoid
A5	Idle position switch	B5	Enrichment solenoid valve
A6	Ground	B6	Air conditioner cut relay
A7	Ignition switch	B7	Distributor advance vacuum exchange solenoid valve
A8	Ignition switch		
A9	Battery (+)		
A10	Ignition coil (−)		
A11	Ground		
A12	Coolant temperature sensor		
A13	Throttle position sensor (output)		

Terminal location for the FBC system (© Chrysler Corp.)

speed, coolant temperature, intake air temperature and exhaust oxygen concentration to calculate the fuel flow required to yield the desired air/fuel ratios for all operating conditions. Closed loop control is used to adjust the fuel flow to yield a near stoichiometeric air/fuel ratio when required. The fuel flow is modified to account for special operating conditions, such as hot starts, acceleration and deceleration.

ADAPTIVE MEMORY CONTROL

During the closed loop operation, the ECU controls the duty cycle of the jet mixture control solenoid valve, based on the output voltage signal from the exhaust sensor. The mean values of the duty cycle is stored in a Random Access Memory (RAM) and the last ones are stored, even if the ignition switch is turned off.

SECONDARY AIR CONTROL

A solenoid is used to control the air control valve signal vacuum. The solenoid is controlled by the ECU, based on the engine speed, idle position and coolant temperature. This valve sends air to the exhaust manifold.

DECELERATION SPARK CONTROL

In order to decrease the hydrocarbon (HC) emissions emitted during vehicle deceleration, ignition timing is advanced by the solenoid operated vacuum valve on the distributor, changing the vacuum supplied to the valve from carburetor ported vacuum to intake manifold vacuum. The solenoid valve is controlled by the electronic controlled unit, based on engine speed.

SENSORS

EXHAUST OXYGEN SENSOR

The oxygen sensor is mounted in the exhaust manifold. The output signal from thE sensor, which varies with the oxygen content of the exhaust gas stream, is provided to the ECU for use in controlling closed loop compensation of fuel delivery.

COOLANT TEMPERATURE SENSOR

The coolant temperature sensor is installed in the intake manifold. This sensor provides data to the ECU for use in controlling fuel delivery and secondary air management.

ENGINE SPEED SENSOR

The engine speed sensor signal comes from the ignition coil. Electric signals are sent to the ECU, where the time between these pulses is used to calculate engine speed, which is used in controlling fuel delivery, secondary air management, deceleration spark and throttle opener management.

THROTTLE POSITION SENSOR

This is a potentiometer mounted to the carburetor. The TPS provides throttle angle information to the ECU to be used in controlling the fuel delivery and secondary management.

VACUUM SWITCH

The switch is installed on the floor board or the fender and is turned on when the throttle valve is closed (idling) position. Information from this switch is provided to the ECU for use in controlling fuel delivery and secondary air management

INTAKE AIR TEMPERATURE SENSOR

This sensor is installed in the air cleaner. The function of this sensor is to measure the the temperature of the intake air into the air cleaner and provide this information to the ECU for use in controlling fuel delivery.

JET MIXTURE SYSTEM

The jet mixture system supplies fuel to the engine through jet mixture passages and jet valves for optimum air/fuel mixtures. This system is

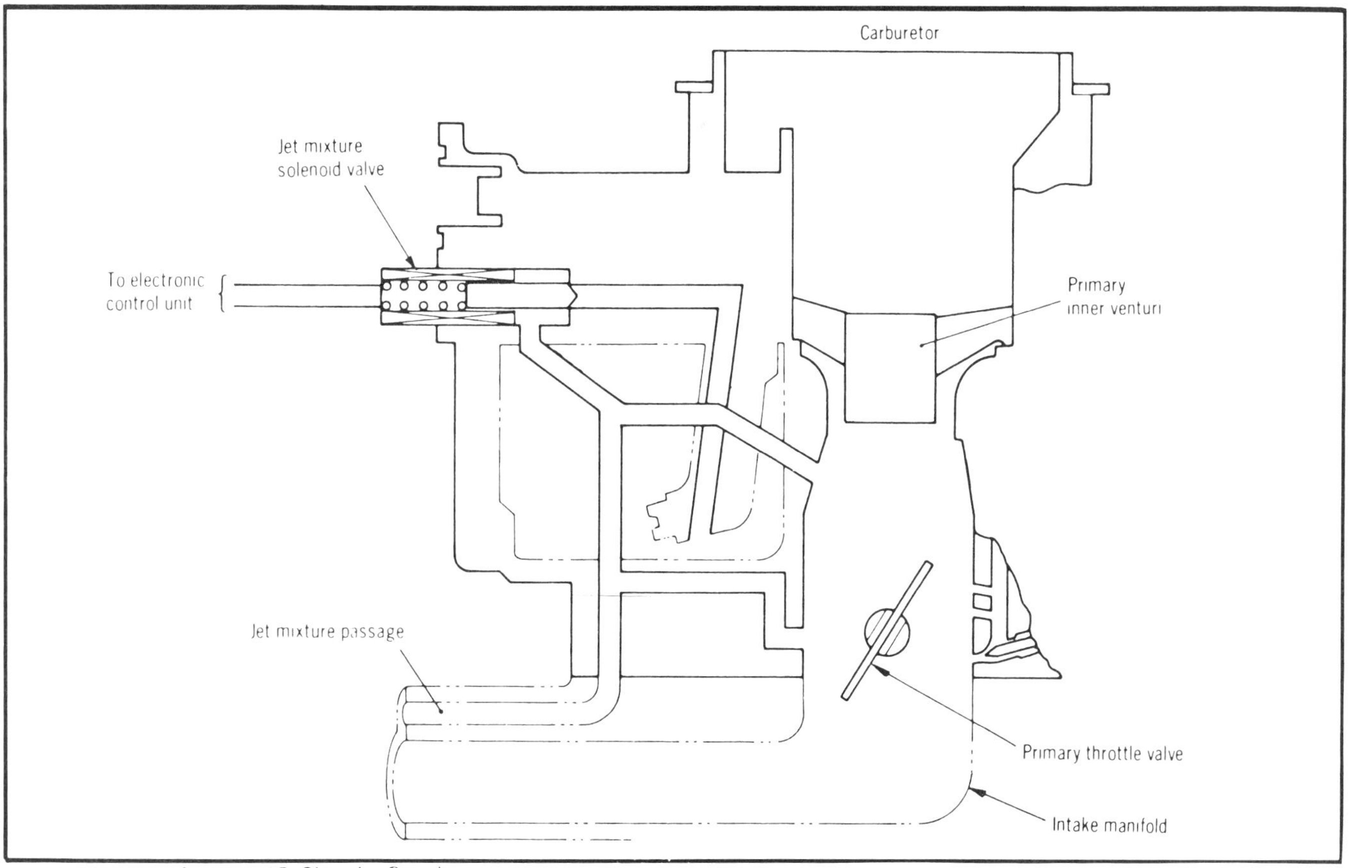

Jet mixture control system (© Chrysler Corp.)

calibrated by the jet mixture solenoid, which responds to an electrical impulse from the ECU. If the exhaust oxygen sensor detects a lean condition, the ECU energizes the solenoid at increasing duty cycles to enrich the mixture. If the exhaust sensor detects a rich condition, the solenoid receives a signal from the ECU to decrease the duty cycle to lean out the mixture. Thus, the solenoid is constanly responding to an electrical signal from the ECU to provide efficient control of air/fuel mixtures.

ENRICHMENT SYSTEM

Enrichment system consists of metering jet and an enrichment solenoid-operated on/off valve, which constantly provides additional fuel for the main metering system. The activation of the on/off valve is controlled by the length of time current is supplied to the solenoid. When additional fuel is required, such as heavy acceleration, heavy engine loads, cold start or warm -up operation, the ECU energizes the solenoid at the pre-set duty cycles.

Electronically Controlled Injection (ECI) System

The fuel control system is the Electronically Controlled Injection (ECI) system, which consists of an electronic control unit (ECU), two fuel injectors, an air flow sensor and other components.

The amount of fuel metered by the two fuel injectors is determined by an electric signal suppplied by the ECU. The ECU monitiors various engine and vehicle parameters, needed to calculate the fuel delivery time (the frequency and duration of injection), of the fuel injectors.

The fuel delivery time is modified by the ECU, according to such operating conditions as cranking, cold starting, altitude, acceleration, deceleration and so on. The fuel is drawn from the fuel tank and forced by the electric fuel pump to the pressure line through a fuel filter. At the end of the fuel line, the fuel pressure regulator controls the fuel pressure at the pre-set value so as for the pressure difference between fuel pressure and turbocharged intake air pressure above the throttle valve to be constant, and thus the amount of fuel injected is only dependent on the fuel delivery time controlled by the ECU.

The two fuel injectors installed in the fuel injection mixer assembly are alternately energized by the ECU, once every specified electric pulses from the air flow sensor almost all over the operation zone and at the other specified intervals in the remaining zones. Each injector features its swirl nozzle that atomizes fuel at higher efficiency, and facilitates the combustion of fuel.

The air flow sensor, installed in the air cleaner assembly, detects the air flow rate, with utilizing the Karman vortex phenomenon, in the form of electric pulses converted by the modulator from the counted Karman vortexes.

Enrichment

STARTING ENRICHMENT

To ensure the starting performance, the fuel delivery time is determined by the ECU which takes account of information on coolant temperature and other except electric pulses from the air flow sensor because the air flow rate is unstable under the starting conditions. The starting

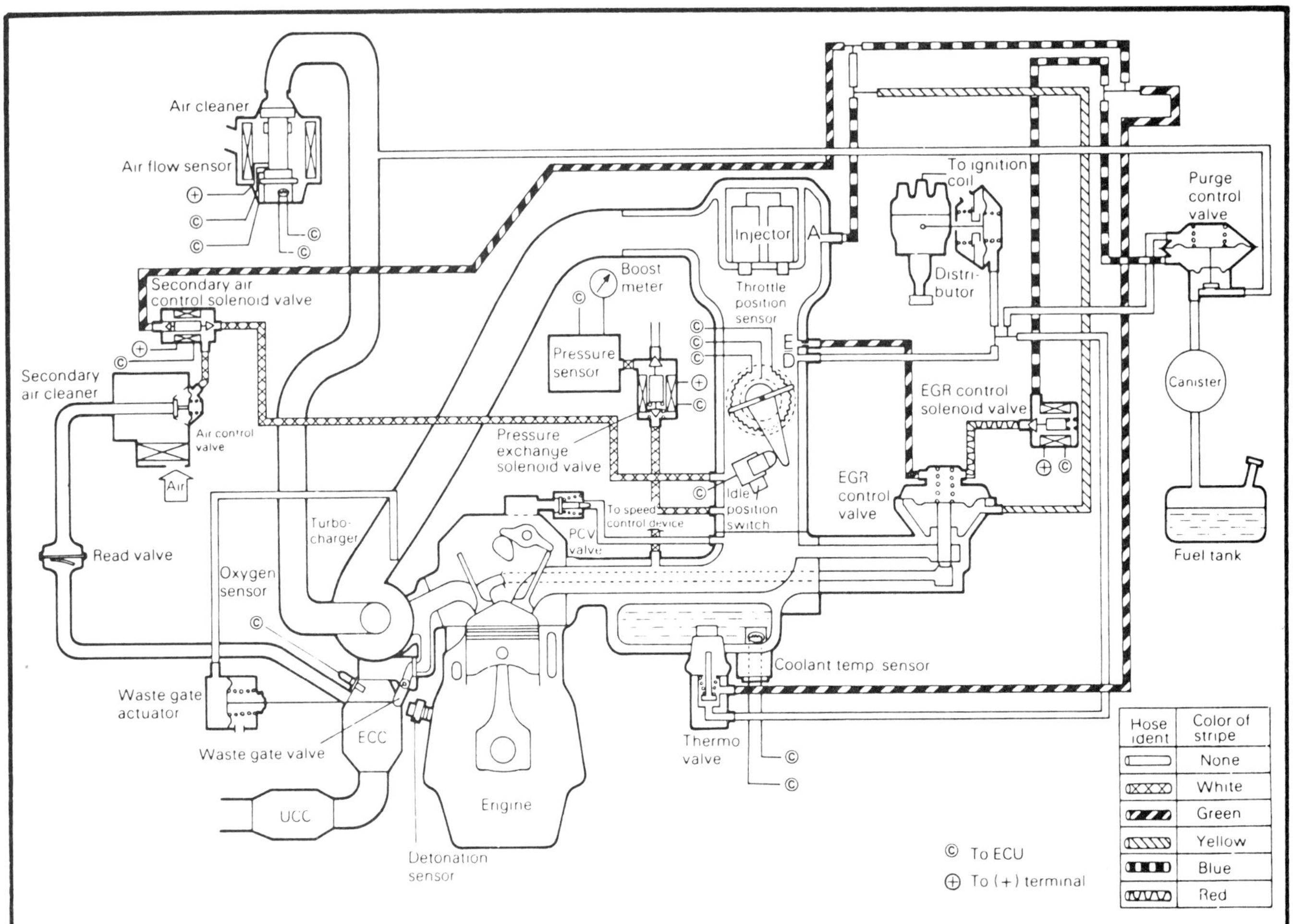

Schematic of Electronically Controlled Injection (ECI) system—1.6L turbocharged engine (© Chrysler Corp.)

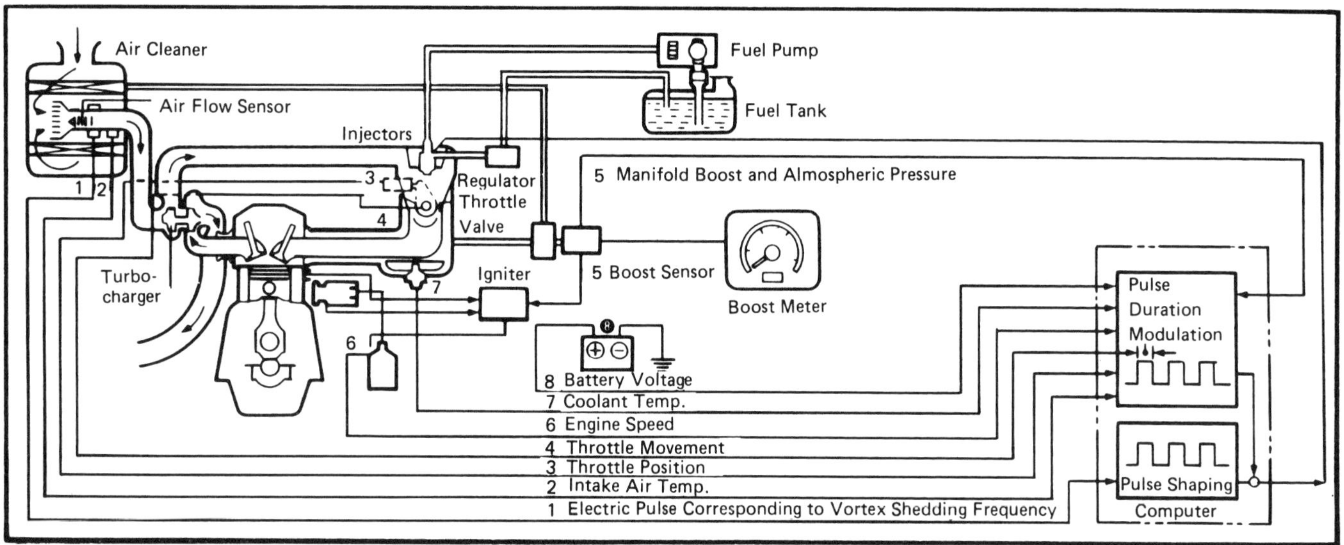

Electronically Controlled Injection Schematic for the 2.6L turbocharged engine (© Chrysler Corp.)

enrichment control is over when the ignition switch is released from the start position and the engine's speed gets over the specified one.

AFTER STARTING ENRICHMENT

For a very short period of time just after starting the engine, another enrichment is provided to obtain a stable combustion. To get proper enrichment for that short period of time, the ECU processes information on the coolant temperature as a parameter to determined the enrichment characteristic.

WARM-UP ENRICHMENT

The warm-up enrichment is provided for proper car operation by a signal from the ECU which processes information on the coolant temperature, until the coolant temperature rises at the pre-set level. This enrichment is provided during the engine warm up period, the open loop operation, and part of the closed loop operation.

ACCELERATION ENRICHMENT

The acceleration enrichment is provided for ensuring car driveability during acceleration both in the open and closed loop operation, by a signal from the ECU which processes information from the throttle position sensor.

FAST IDLE

The fast idle is controlled by one of the functions of the Idle Speed Control System (ISC).

DECELERATION FUEL CUT-OFF

To decrease the hydrocarbon emissions emitted during vehicle deceleration, the fuel delivery time is dramatically decreased by the electronic control unit, changing the injection intervals, which are determined by the specific pulses from the air flow sensor, with the specified duration of injection. During the vehicle deceleration, while the idle position switch is not yet in the ON position, the fuel delivery time is determined by the electronic control unit responding to the throttle valve closing speeds.

OVER PRESSURE FUEL CUT-OFF

In order to protect the turbocharged engine from over loading, the fuel injectors are energized by the ECU, only every ignition spark timing, when the pressure sensor detects higher intake manifold pressure than the pre-set value.

ECI SYSTEM AND PARAMETERS

The ECI system is a pulse time system that injects fuel into the fuel injection mixer above the throttle valve. Fuel is metered to the engine through the two electronically controlled fuel injection. The basic functions of this system are tabulated below. Input signals from a variety of sensors are fed to the ECU and then the ECU generates output signals for all of the controlled functions.

Parameters Sensed

1. Intake air flow
2. Intake air temperature
3. Intake manifold and barometric pressures
4. Coolant temperature
5. Engine speed
6. Exhaust oxygen concentration
7. Throttle position
8. Idle position
9. Battery voltage
10. Engine cranking
11. Park/Neutral mode
12. Vehicle speed
13. A/C system switch 'ON'
14. Motor position switch

Parameters Controlled

1. Fuel injection signal
2. Secondary air control valve signal
3. EGR control signal
4. Solenoid valve (for pressure control signal)
5. Idle speed controlsignal

Data Sensor and Parameters Sensed

AIR FLOW SENSOR

The air flow sensor installed in the air cleaner assembly consists of the device for generating Karman vortexs, the ultrasonic wave transmitter, the receiver and the modulator. Ultrasonic waves with a constant frequency are transmitted across the air flow containing Karman vortexes,

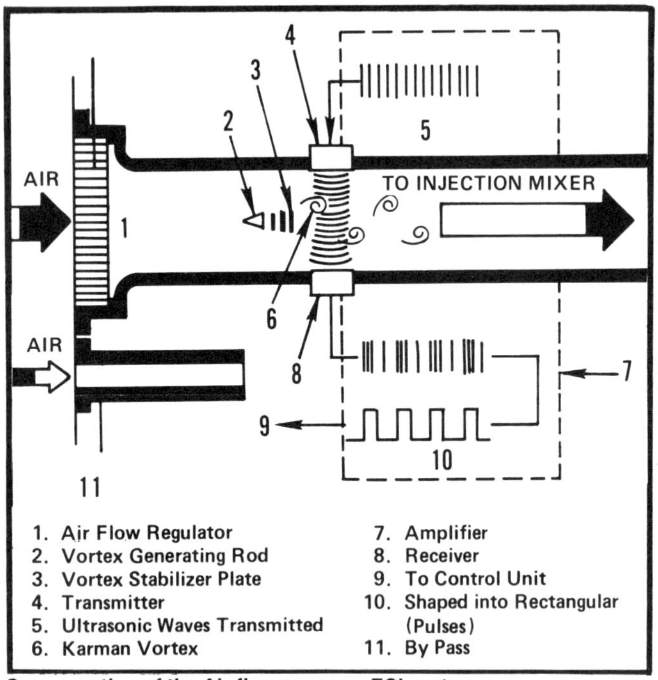

Cross-section of the Air-flow sensor—ECI system (© Chrysler Corp.)

which are generated proportionally to air flow rate, and then, the ultrasonic frequency is modulated by the vortexes. The receiver detects the modulated waves and the modulator converts them into electric pulses. The electric pulse information is transmitted to the ECU for use in controlling the fuel delivery time and secondary air management.

INTAKE AIR TEMPERATURE

This sensor is installed in the air cleaner. The function of this sensor is to measure the the temperature of the intake air in the air cleaner and provides this information to the ECU for use in controlling fuel delivery.

PRESSURE SENSOR

This sensor is located on the firewall and its function is to sense ambient barometric pressure and absolute pressure in the intake manifold.

Ambient barometeric pressure is sensed by the sensor energizing the solenoid valve by the ECU for the specified period immediately after engine starting and thereafter once every specified period. Information on ambient pressure changes due to weather and or altitude is provided to the ECU for controlling the fuel delivery time.

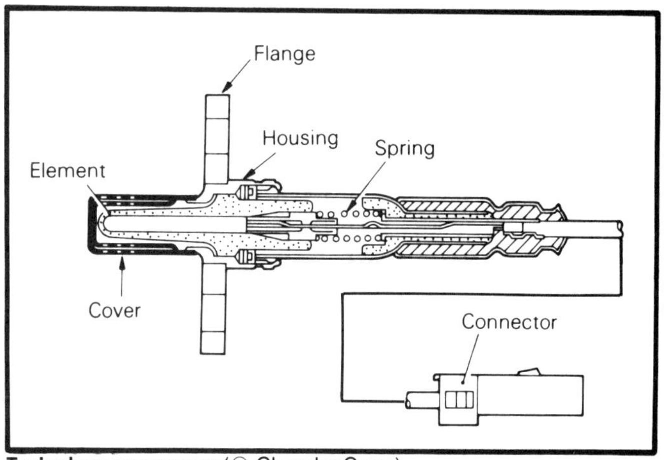

Typical oxygen sensor (© Chrysler Corp.)

During the remaining period, other than the above specified periods, the solenoid valve is de-energized and this sensor detects intake manifold pressure (absolute pressure), and the information is provided to the ECU for controlling the fuel delivery time, and idle speed.

OXYGEN SENSOR

The oxygen sensor is mounted in the exhaust manifold. The output signal from this sensor, which varies with the oxygen content of the exhaust gas stream, is provided to the ECU for use in controlling closed loop compensation of fuel delivery time.

COOLANT TEMPERATURE SENSOR

The coolant temperature sensor is installed in the intake manifold. This sensor provides data to the ECU for use in controlling fuel delivery time, EGR, secondary air management and idle speed.

ENGINE SPEED SENSOR

The engine speed sensor signal comes from the ignition coil. Electric signals are sent to the ECU, where the time between these pulses is used to calculate engine speed, which is used in controlling fuel delivery time, EGR, secondary air management, and idle speed.

THROTTLE POSITION SENSOR

This is rotary potentiometer mounted on the fuel injection mixer. The

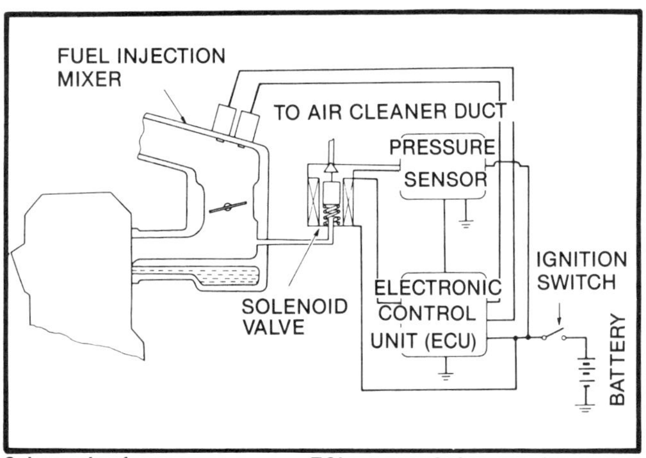

Schematic of pressure sensor—ECI system (© Chrysler Corp.)

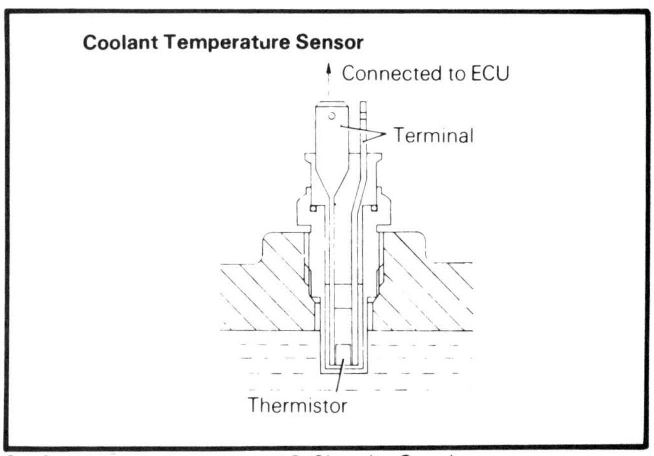

Coolant temperature sensor (© Chrysler Corp.)

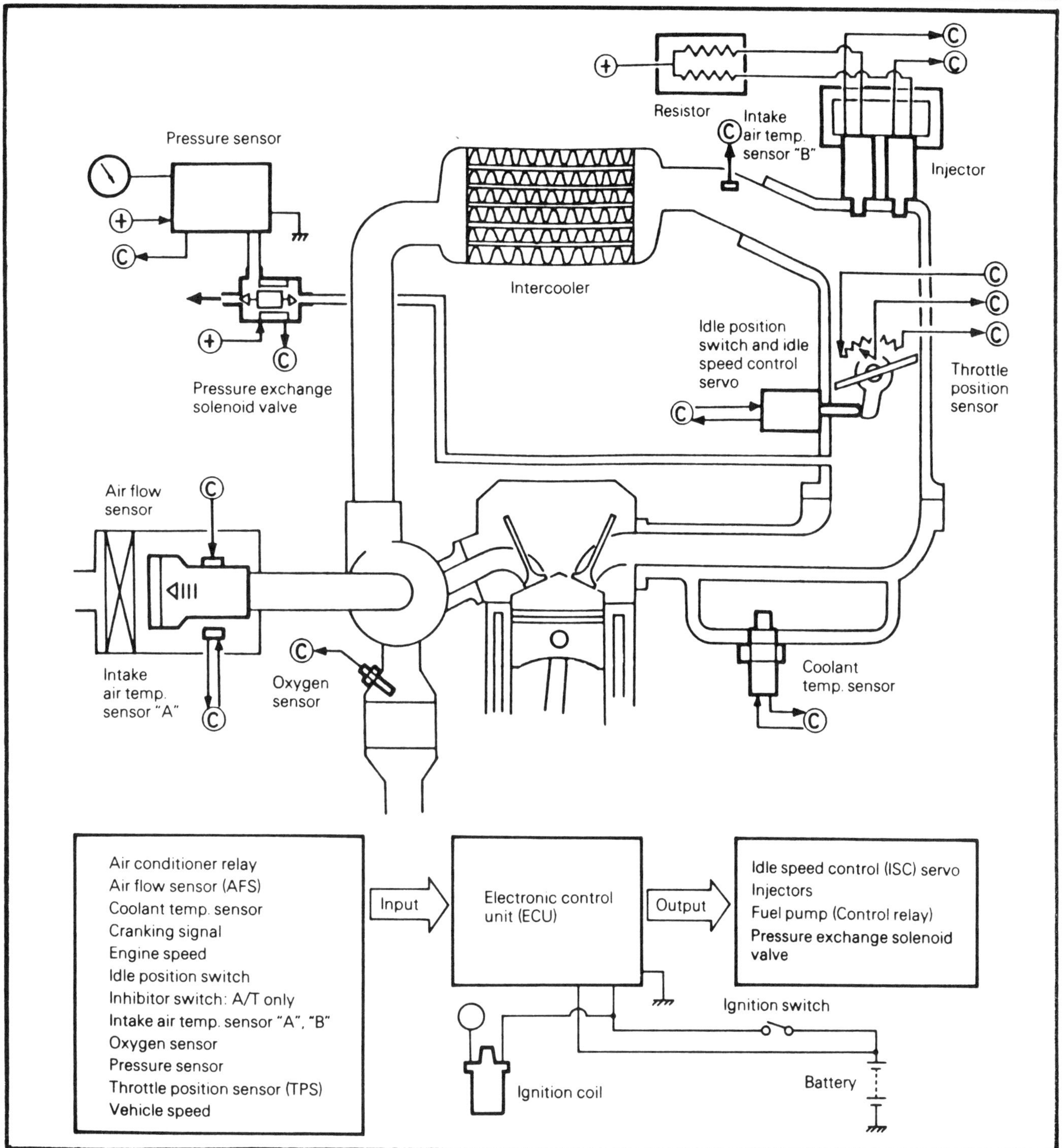

ECI Schematic—1986 Conquest (© Chrysler Corp.)

TPS provides throttle angle information to the ECU, to be used in controlling the fuel delivery and idle speed control.

IDLE SPEED POSITION SWITCH

This switch is installed on the fuel injection mixer assembly and turned on when the throttle valve is at closed (idling) position. Information from this switch is provided to the ECU for use in controlling fuel delivery time (during vehicle deceleration), idle speed and secondary air management. This switch is also used as an idle speed adjusting device.

BATTERY VOLTAGE

Battery voltage is detected by the ECU. The battery voltage signal provides information to the ECU to allow for voltage compensation of the controlled functions.

A/C SYSTEM SWITCH

The A/C system switch signal indicates when the A/C mode switch is in the A/C on position.

ENGINE CRANK SWITCH

The engine crank switch provides a signal to the ECU when the engine is cranked.

PARK/NEUTRAL SWITCH

The park neutral switch signal indicates when the automatic transmission gear selector is in the Park or Neutral position.

TIME

Time is generated by the ECU internal microprocessor clock.

VEHICLE SPEED SENSOR

The vehicle speed signal comes from the reed switch, which senses speedometer cable speed. Pulses from this switch are sent to the ECU where the time between these pulses is used to calculate vehicle speed, which is used in controlling idle speed.

ELECTRONIC CONTROL UNIT

The electronic control unit is mounted in the passenger compartment and consists of a printed circuit board mounted in a protective metal box. It receives analog inputs from the sensors and converts them into digital signals. These signals and various discrete inputs, are processed and used by the ECU in controlling the fuel delivery, secondary air, deceleration spark and throttle opener managements.

Parameters Controlled

FUEL INJECTION CONTROL

There are two ways to control fuel injection. One is the open loop control and the other is the closed loop control, which are switched from each other, by the ECU, based on the information from the coolant temperature sensor, oxygen sensor, air flow sensor and engine speed sensor. Under open loop control, the air fuel ratio is determined by the fuel delivery time (the frequency and duration of injection), which is controlled by the ECU. Under the closed loop control, the air fuel ratio is feedback controlled by the ECI system, including the exhaust oxygen sensor.

EGR CONTROL

The pressure applied to the EGR control valve is controlled by the solenoid valve motion which mixes the throttle ported pressure with the turbocharged pressure. The solenoid valve motion is controlled by the ECU, based on the information from the coolant temperature sensor and engine speed sensor.

SECONDARY AIR CONTROL

The pressure applied to the secondary air valve is switched by the solenoid valve from the intake manifold pressure to the turbocharger pressure or vice versa. The solenoid is controlled by the ECU, based on the engine speed, idle position, air flow and coolant temperature.

ADAPTIVE MEMORY CONTROL

During the closed loop operation, the ECU monitors the output voltage signal from the exhaust oxygen sensor and stores it in a Random Access Memory (RAM), the mean values of feedback gain (pro-portional and intergral gains). The last means are always stored, even if the ignition switch is turned off. During open loop operation, the fixed pre-programmed fuel delivery time is modified by the ECU with the mean values of feedback gain to improve emission performance.

NOTE: RAM is always powered from a standby power source.

IDLE SPEED CONTROL (ISC) SYSTEM

Engine idle speed, crank throttle angle and deceleration throttle angle are controlled by an electric motor driven actuator, which changes the throttle angle by acting as a movable idle stop. The ECU controls the ISC acuator by providing the appropriate outputs to yield the idle speed or throttle angle required for the particular operating condition. The electronic components for the ISC system are integral with the ECU.

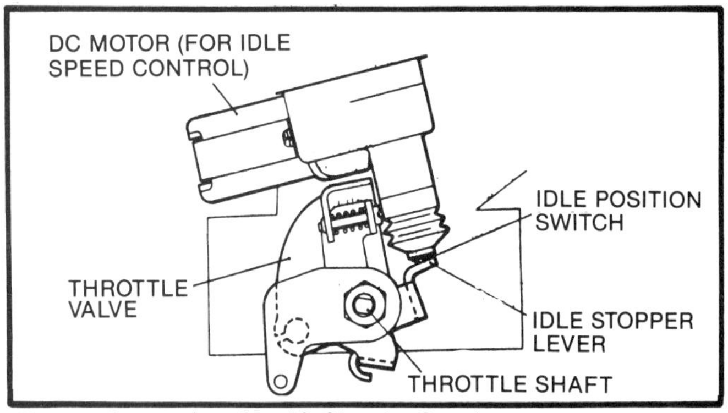

Idle speed control system (© Chrysler Corp.)

HIGH ALTITUDE COMPENSATION SYSTEM

The fuel injected vehicles are equipped with a pressure sensor and a solenoid valve which are utilized as a high altitude compensator. When the ignition switch is in the on position the solenoid valve is energized by the ECU for the specific time, which in turn allows the pressure sensor to detect the ambient atmospheric pressure. The fuel delivery time is then determined by the ECU in response to the ambient atmospheric pressure so that the air/fuel mixture at high altitudes can be maintained at the same ratio as at sea level.

Diagnosis Of The ECI System

Self-diagnosis is a system in which the input signal from each sensor is monitored by the ECU (computer) and, should any abnormality happen in the input signal, the abnormal item is memorized by the computer. The diagnosis items are eight items including that for normal condition and can be confirmed using a voltmeter.The abnormailty-diagnosis memory is kept by direct power supply from the battery. Therefore, the memory of diagnosis result is not erased by turning off the ignition switch. However, it is erased if the back-up power supply is turned off by disconnection of the battery cable or ECU connector.

NOTE: The memory is not erased if the power supply is turned on within ten seconds after turning the off the power supply to the ECU.

DIAGNOSIS ITEM

If there are two or more items found abnormal, they are indicated in the order of increasing code numbers.

ABNORMALITY DIAGNOSIS

Code Number	Diagnosis
1	Oxygen sensor and computer
2	Ignition pulse
3	Air flow sensor
4	Pressure sensor
5	Throttle positioner sensor
6	ISC Motor position switch
7	Coolant temperature sensor

INDICATION METHOD

The indication method is performed by deflection of the pointer of a voltmeter. Connect a voltmeter to the ECI diagnosis connector, which is located on the right side of the engine compartment, mounted near the shock tower and the following indication will be made.

1. When normal, there will be 12 volts constantly indicated on the voltmeter.
2. When abnormal, the needle on the voltmeter scale will fluctuate between 0 volts and 12 volts, every 0.4 seconds.
3. Engine speed at 12 volts is indicated. When there are two or more abnormal items, the low-code-numbered item is first indicated. Then, after indication of 0 volts for 2 seconds, subsequent indication is made in the same manner as previously stated above.

PRECAUTIONS FOR OPERATION

1. When the battery voltage is low, no detection of abnormality is made. Be sure to check the battery for conditions before starting the test.
2. Diagnosis item is erased if power supply from the battery or the ECU connector is disconnected. Do not disconnected the power supply from the battery before the diagnosis result is completely read.
3. Warm up the engine and drive the vehicle for a good distance before the oxygen sensor is diagnosed. Do not set the ignition switch in the off position after driving the vehicle. If the ignition switch is set to the off position, the result detected by the diagnosis is erased.
4. After the checks and correction are over, disconnect the negative battery cable for 15 seconds or more and connect it again to make sure the abnormal code is erased.

INSPECTION OF ECI SYSTEM

If the ECI components fail, interruption of the fuel supply or failure to supply the proper amount of fuel for the engine operating condition will be the result and these following conditions could be encountered.

a. Engine is hard to start or will not start at all.
b. Very unstable idle
c. Poor driveabilty

If any of these situations do occur, first perform the inspection by self-diagnosis and subsequent basic engine checks (incorrect engine adjustments, ignition system failure, etc.), and then inspect the ECI components by diagnosing the ECI system with a voltmeter.

Service Precautions

1. Before the battery terminals are disconnected, make sure that the ignition switch is set to the off position. If the battery cables are disconnected while the engine is running or when the ignition switch is in the on position, malfunction of the ECU or damage to the semiconductor could result.

Self diagnosis function indicating normal on volt meter scale—ECI (© Chrysler Corp.)

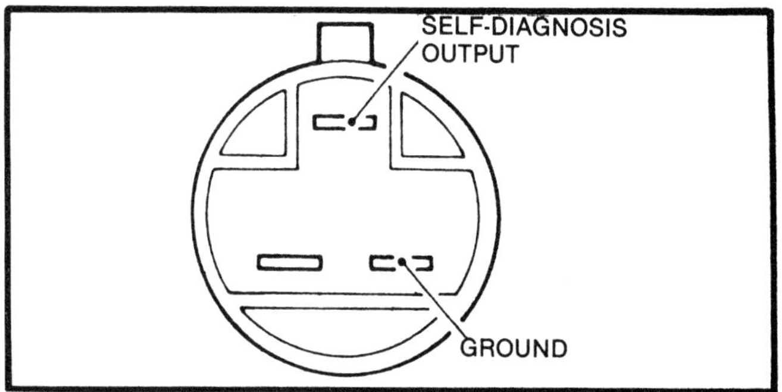

Connecting a voltmeter to the ECI diagnosis connector (© Chrysler Corp.)

2. Disconnect the battery cables before charging the battery and when reconnecting the cables be sure not to reverse the polarity.
3. Make sure that the harness connectors are securely connected. Use care not to allow entry of water or oil into the connectors.

Inspection Procedure by Self-Diagnosis

1. With the ignition switch in the off position connect a voltmeter between the self-diagnosis and the ground terminals of the ECI diagnosis connector, which is located on the right side of the engine compartment near the right shock tower.
2. Place the ignition switch in the on position and indications of the ECU memory contents will immediatley start. If the system is in normal condition, the needle on the voltmeter will constantly indicate 12 volts.
3. If any abnormality is in the memory, the needle of the voltmeter will fluctuate, indicating an abnormal item as previously outlined in the Indication Method in this section.
4. Abnormal item can be known from voltage waveform, that is the number of times the needle fluctuates.
5. After recording the abnormal item, check and repair each part, according to the check items in the diagnosis chart.
6. Turn the igniton switch off and if the defective parts have been repaired, disconnect the negative battery cable for 15 seconds or more and reconnect it again to be sure that the abnormal code has been erased.

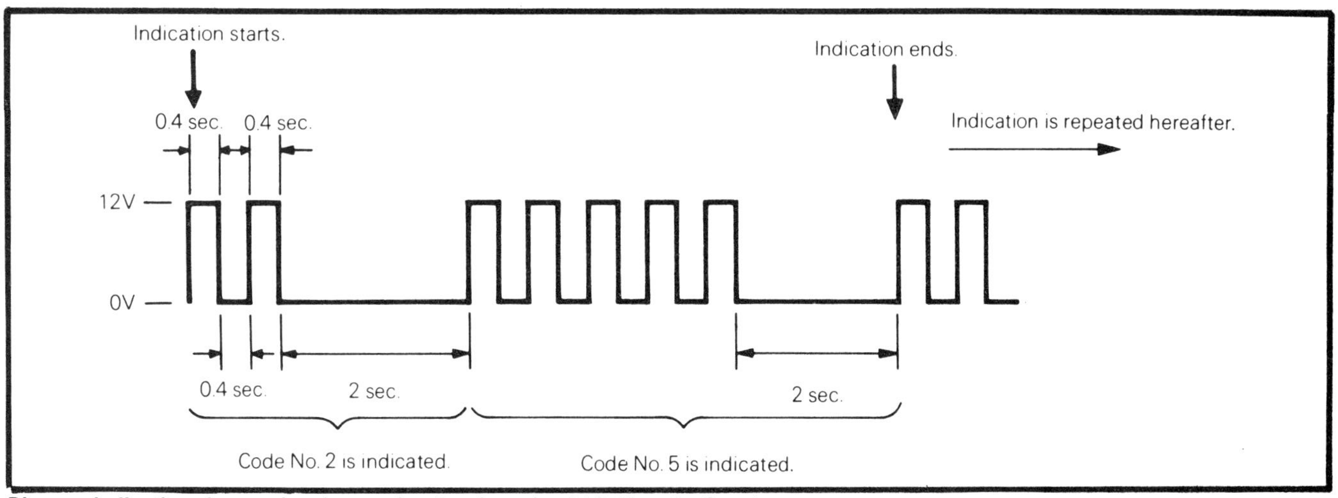

Diagram indicating abnormal codes number 2 and number 5 (© Chrysler Corp.)

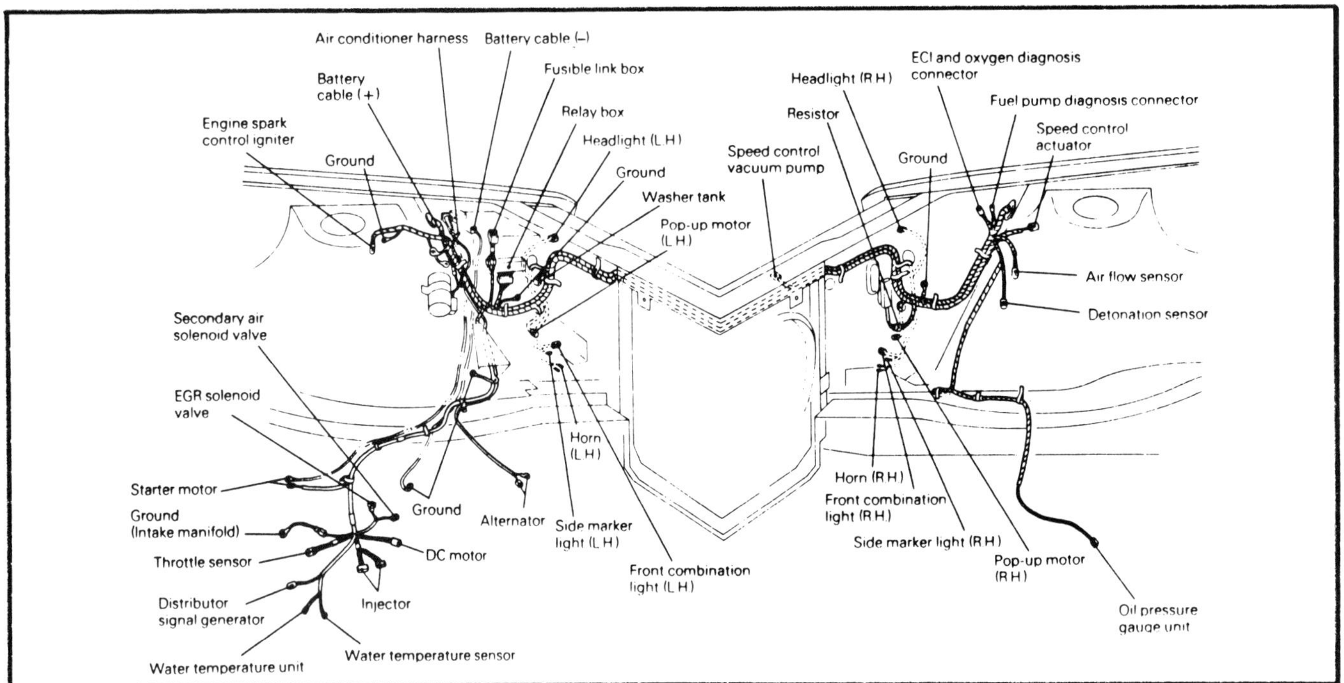

Location of the ECI diagnosis connector and other various connectors (© Chrysler Corp.)

Code No.	Diagnosis item	Voltage waveform (abnormal code)	Contents of diagnosis	Check item
1	Oxygen sensor	12V 0V	In city driving mode, oxygen sensor signal does not change for 20 seconds or more or engine stalls. When engine stalls, turn ignition switch from OFF position to ON position. If abnormal code is indicated, computer is normal. If no indication is made, computer is not in normal condition.	• Wire harness and connector • Oxygen sensor • Computer
2	Ignition signal	12V 0V	While cranking the engine, input of ignition signal is not applied to computer for 3 seconds or more.	• Wire harness and connector • Igniter • Computer
3	Air flow sensor	12V 0V	Air flow sensor output is 10 Hz or less while engine is idling, or it is 100 Hz or more when engine stalls.	• Wire harness and connector • Air flow sensor • Computer
4	Pressure sensor	12V 0V	Boost sensor output is 1,460 mmHg (4.5V) or more, or it is 65 mmHg (0.2V) or less.	• Wire harness and connector • Boost sensor • Computer
5	Throttle position sensor	12V 0V	Throttle sensor output is 0.2V or less, or it is 4V or more while engine is idling (idle switch is ON).	• Wire harness and connector • Throttle sensor • Computer
6	ISC motor position switch	12V 0V	Throttle sensor output is 0.89V with L switch OFF.	• Wire harness and connector • ISC servo • Computer
7	Water temperature sensor	12V 0V	Water temperature sensor output is 4.5V or more, or it is 0.1V or less.	• Wire harness and connector • Water temperature sensor • Computer
8	Car speed signal	12V 0V	Air flow sensor output is 500 Hz or more and car speed is 2.5 km/h (1.6 mph) or less.	• Wire harness and connector • Car speed sensor • Computer

Diagnosing the ECI system (© Chrysler Corp.)

Check item	Condition		Voltmeter reading when normal	Terminal location of computer
Power supply	Ignition switch OFF → ON		11 – 13V	B-1
Secondary air control solenoid valve	Ignition switch OFF → ST after warming up the engine		30 seconds 0.2 – 1V → 13 – 15V	A-10
Throttle position switch	Ignition switch OFF → ON	Accelerator closed	0.4 – 1.5V	A-1
		Accelerator wide opened	4.5 – 5.0V	
Coolant temperature sensor	Ignition switch OFF → ON	0°C (32°F)	3.5V	A-3
		20°C (68°F)	2.6V	
		40°C (104°F)	1.8V	
		80°C (176°F)	0.5V	
Intake air temperature sensor	Ignition switch OFF → ON	0°C (32°F)	3.5V	A-4
		20°C (68°F)	2.6V	
		40°C (104°F)	1.8V	
		80°C (176°F)	0.6V	
Idle position switch	Ignition switch OFF → ON	Accelerator closed	0 – 0.4V	A-5
		Accelerator wide opened	11 – 13V	
ISC motor position switch	Ignition switch OFF → ON		11 – 13V *1	A-14
EGR control solenoid valve	Ignition switch OFF → ON	Return to 0 volt after 1 sec.	0 – 0.5V after 1 sec.	B-4
ISC motor for extension	Ignition switch OFF → ON		Pointer C swings a moment after 15 sec.	B-6
*2 A/C relay	Ignition switch OFF → ON	A/C switch OFF → ON	11 – 13V	B-12
Reed switch for vehicle speed	Start engine, transmission in first and operate vehicle slowly		0.2 – 1V 4 – 5V	A-15
ISC motor for retraction	Ignition switch OFF → ON		Pointer swings a moment after 15 sec.	B-11
Cranking signal	Ignition switch OFF → ST		Over 8V	A-13
Control relay	Idling		0 – 1V	B-5
ISC motor position switch	Idling		11 – 13V	A-14

Check item	Condition		Voltmeter reading when normal	Terminal location of computer
Ignition pulse	Idling		12–15V	A-8
	3,000 rpm	When engine speed is increased, voltage drops slightly.	11–13V	
Air flow sensor	Idling		2.7–3.2V	A-7
	3,000 rpm			
Injector No. 1	Idling		13–15V	B-9
	3,000 rpm	When engine speed is increased, voltage drops slightly.	12–13V	
Injector No. 2	Idling		13–15V	B-10
	3,000 rpm	When engine speed is increased, voltage drops slightly.	12–13V	
Oxygen sensor	Keep 1,300 rpm after warming up the engine		Flashing 0–1V ↔ 2.7V	A-6
EGR control solenoid valve	Keep idling after warming up the engine Over 3,000 rpm, drops to zero volts.		13–15V	B-4
Pressure sensor	Ignition switch OFF → ON		1.5–2.6V	A-17 Switches back to 2.0V evey 2 min.
	Idling		0.2–1.2V	
ISC motor for extension	Idling	A/C switch OFF → ON	Over 5V *3	B-6
ISC motor for retraction	Idling	A/C switch ON → OFF	Over 6V *3	B-11

NOTE
*1: If ignition switch is turned on for 15 seconds or more, the reading drops to 1V or less momentarily and returns to 6 to 13V.
*2: A/C = Air conditioner
*3: It indicates 6V a moment (very short time). If it is difficult to read the indication, repeat to turn the air conditioner switch off and on or on and off several times. If voltmeter swings, the components are in good condition.

Output signals from the voltmeter tester (cont.) (© Chrysler Corp.)

Chrysler Truck Specifications

VEHICLE IDENTIFICATION NUMBER AND ENGINE IDENTIFICATION

Vehicle Identification Number (VIN)

The vehicle identification number (VIN) plate is mounted on the instrument panel, adjacent to the lower corner of the windshield on the left side of the vehicle and is visible through the windshield. The VIN number is composed of seventeen (17) digits. The tenth (10) digit of the VIN number represents the year of the vehicle, with 'F' representing 1985 and 'G' representing 1986.

Engine Identification

The eighth (8) digit of the VIN number represents the engine model. The engine identification number for the 2.0 liter engine is stamped on the engine block, adjacent to the alternator and engine oil dipstick. On the 2.6 liter engine, the number is stamped on the engine block, next to the idler pulley and above the bottom radiator hose. The 2.3 liter engine has the number stamped on the engine block, right next to the engine oil dipstick.

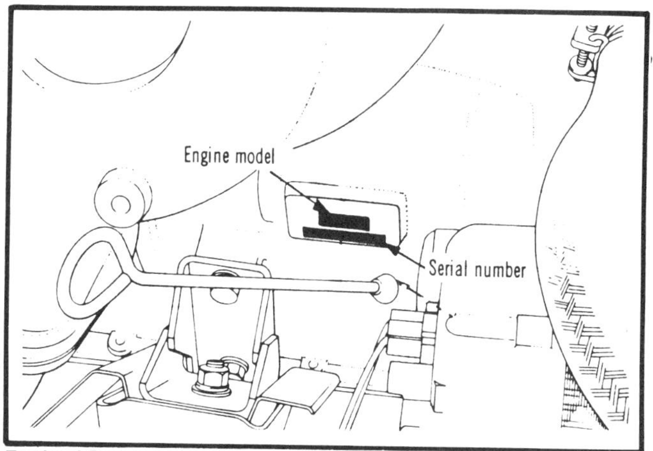

Engine I.D. number location 2.0 liter engine

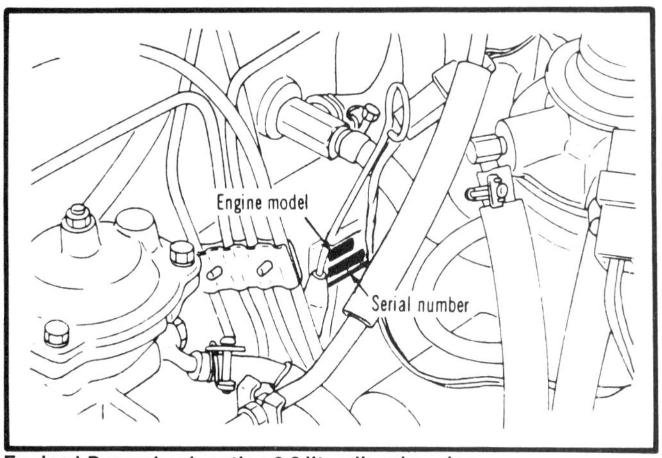

Engine I.D. number location 2.3 liter diesel engine

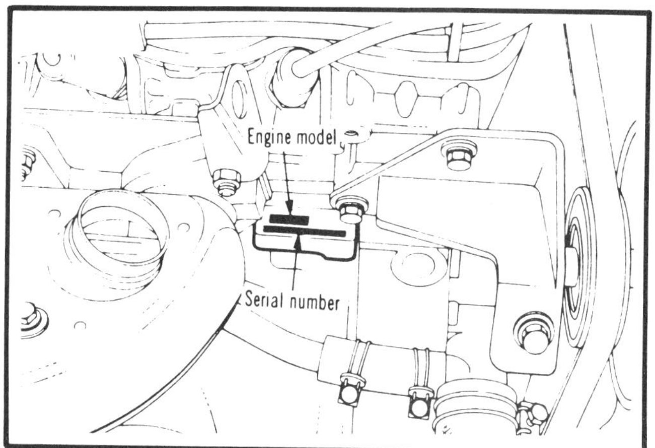

Engine I.D. number location 2.6 liter engine

TUNE-UP SPECIFICATIONS

(When analyzing compression test results, look for uniformity among cylinders, rather than specific pressures.)

Year	Engine Displace. cu. in. (cc)	Spark Plugs Type	Spark Plugs Gap (in.)	Distributor Point Dwell (deg)	Distributor Point Gap (in.)	Ignition Timing (deg) MT	Ignition Timing (deg) AT	Intake Valve Opens (deg) BTDC	Fuel Pump Pressure (psi)	Idle Speed (rpm)	Valve Clear (in) ● In.	Valve Clear (in) ● Ex.
'85–'86	121.9 (2000)	BPR6ES-11 ①	0.039 0.043	Electronic		5B	5B	19	N/A	750	0.010 ②	0.010
'85–'86	155.9 (2600)	BUR6EA-11 ①	0.039 0.043	Electronic		7B	7B	25	N/A	750 ③	0.010 ②	0.010

TUNE-UP SPECIFICATIONS

(When analyzing compression test results, look for uniformity among cylinders, rather than specific pressures.)

Year	Engine Displace. cu. in. (cc)	Spark Plugs Type	Spark Plugs Gap (in.)	Distributor Point Dwell (deg)	Distributor Point Gap (in.)	Ignition Timing (deg) MT	Ignition Timing (deg) AT	Intake Valve Opens (deg) BTDC	Fuel Pump Pressure (psi)	Idle Speed (rpm)	Valve Clear (in) ● In.	Valve Clear (in) ● Ex.
'85–'86 Ⓒ	155.9 (2600)	BPR5EA ①	0.039 0.043	Electronic		7B	7B	25	N/A	750 ③	0.010 ②	0.010
'85–'86	143.2 (2300)	Turbo Diesel	—	—		5 ATDC ④		20	306	750	0.010	0.010

NOTE: The underhood sticker often reflects tune-up specification changes made in production. Sticker figures must be used if they disagree with those in this chart.

● All adjustments done with engine hot
C Canada
B Before top dead center
N/A Not available at the time of publication
ATDC After top dead center

① See text
② Jet valve clearance: 0.010 in.
③ 800 rpm on automatic transaxle models
④ 5° ATDC at 0.0394 in. (1 mm) plunger stroke

IDLE SPEED SPECIFICATIONS

Engine	Transaxle	Curb Idle Speed (rpm)
2.0 Liter	M/T & A/T	750 [+ −] 100
2.6 Liter	M/T	750 [+ −] 100
2.6 Liter	A/T	800 [+ −] 100
2.6L Canada	A/T	800 [+ −] 100
2.3L Diesel	M/T	750 [+ −] 100

M/T Manual Transaxle
A/T Automatic Transaxle

IGNITION TIMING SPECIFICATIONS

Engine	Transaxle	Ignition Timing
2.0 Liter	M/T & A/T	5 BTDC [+ −] 2 at curb idle
2.6 Liter	M/T & A/T	7 BTDC [+ −] 2 at curb idle
2.3L Diesel	M/T	5 ATDC [+ −] 2 at curb idle ①

M/T Manual Transaxle
A/T Automatic Transaxle
① At 0.0394 in. (1mm) plunger stroke

Honda

INDEX

VEHICLE IDENTIFICATION NUMBER AND ENGINE IDENTIFICATION

Vehicle Identification Number

The vehicle indentification number (VIN) is mounted on the top edge of the instrument panel and is visible through the windshield from the outside of the vehicle. In addition, there is a Vehicle/Engine Identification plate under the hood on the cowl and on the Civic CRX-Si models on the firewall.

Engine Identification

The engine serial number is stamped on a metal tab which is riveted to the right front radiator support. On the Civic CRX-Si model, it is rivited to the left front side of the radiator. The engine serial number is a ten digit number, with the first three digits representing the engine model and the fourth digit representing the model year.

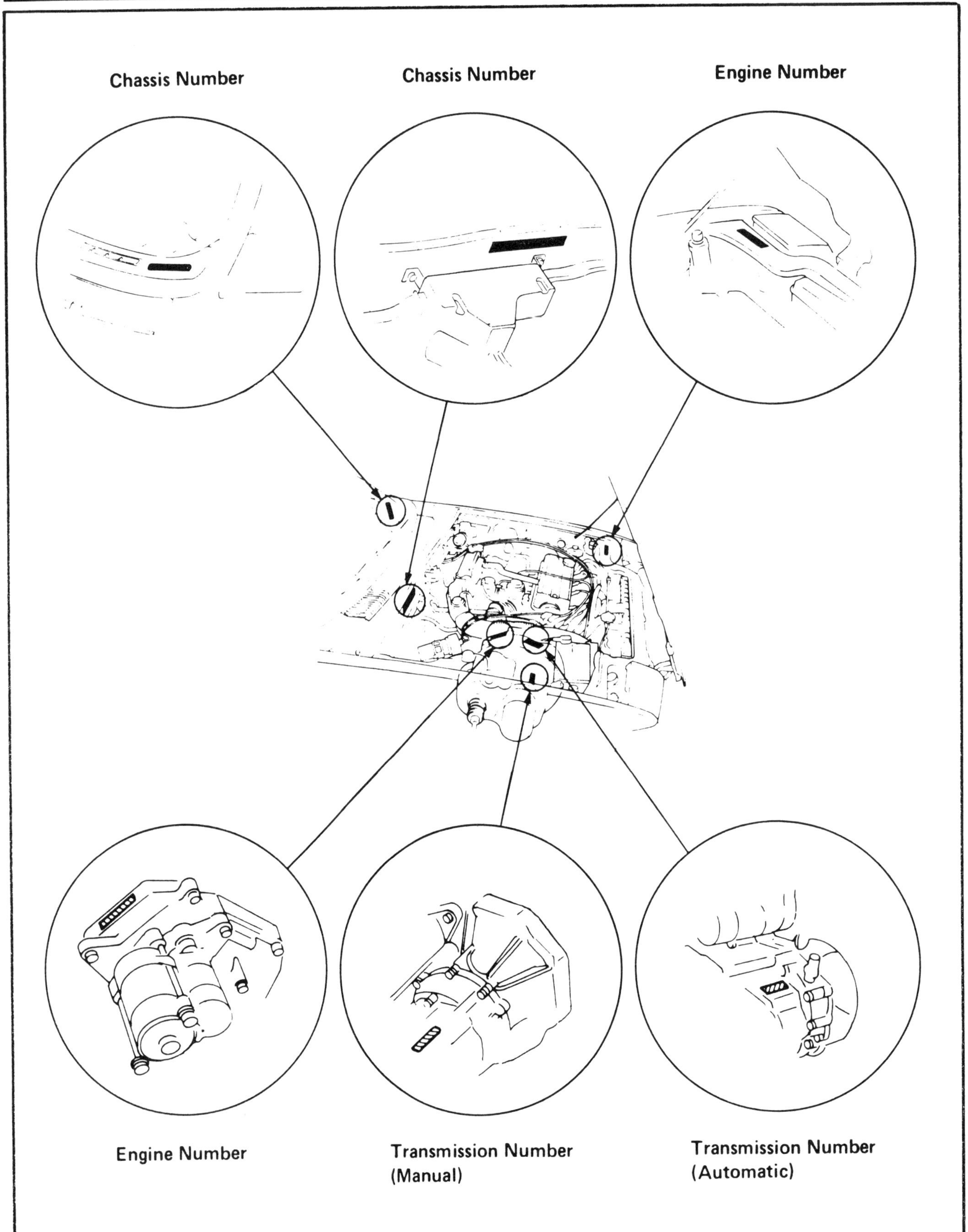

Identification number locations

TUNE-UP SPECIFICATIONS

(When analyzing compression test results, look for uniformity among cylinders, rather than specific pressures.)

Year	Model	Engine Displacement (cc)	Original Equipment Spark Plugs Type	Gap (in.)	Distributor Point Dwell (deg)	Distributor Point Gap (in.)	Basic Ignition Timing (deg) MT	Basic Ignition Timing (deg) AT	Intake Valve Fully Opens (deg)	Fuel Pump Pressure (psi)	Idle Speed (rpm) MT	Idle Speed (rpm) AT	Valve Clearance (in.) Intake (cold)	Valve Clearance (in.) Auxiliary (cold)	Valve Clearance (in.) Exhaust (cold)
'85	Civic	1488	BUR5EB-11 ⑲	0.042	Electronic		20B ⑳㉒	15B ㉑㉒	N.A.	3.0	650–750	650–750	0.007–0.009	0.007–0.009	0.009–0.011
	Civic CRX Si	1488	BPR6EY-11	0.042	Electronic		16B ㉒⑬	—	N.A.	3.5	550–650	—	0.007–0.009	0.007–0.009	0.009–0.011
	Civic	1342	BUR5EB-11	0.042	Electronic		21B ⑭㉒	—	N.A.	3.0	650–750	—	0.007–0.009	0.007–0.009	0.009–0.011
	Accord	1829	BUR5EB-11	0.042	Electronic		22B ⑤⑱	18B ⑤	N.A.	2.5	700–800	650–750	0.005–0.007	0.005–0.007	0.010–0.012
	Accord SE-i	1829	BPR6EY-11	0.042	Electronic		18B ⑤	18B ⑤	N.A.	3.5	700–800	700–800	0.005–0.007	0.005–0.007	0.010–0.012
	Prelude	1829	BPR6EY-11	0.042	Electronic		20B ⑤	12B ⑤	N.A.	2.5	750–850	750–850	0.005–0.007	0.005–0.007	0.010–0.012
'86	ALL	See Underhood Specifications Sticker													

NOTE: The underhood specifications sticker often reflects tune-up specification changes made in production. Sticker fiqures must be used if they disagree with those in this chart.

TDC—Top Dead Center
B—Before top dead center
A—After top dead center
—Not applicable
N.A. Not available

⑤ Aim timing light at red mark (yellow mark, '79 Accord M/T) on flywheel or torque converter drive plate distributor vacuum hose connected at specified idle speed
⑬ Calif.: 12B
⑭ Calif.: 16B
⑱ California models: 18B
⑲ CRX HF model: BUR4EB-11
⑳ CRX HF model: 21B
㉑ Models w/power steering: 17B
㉒ Aim timing light at red mark on crankshaft pulley

FIRING ORDERS

NOTE: To avoid confusion, always replace spark plug wires one at a time.

Firing order—1829cc Accord

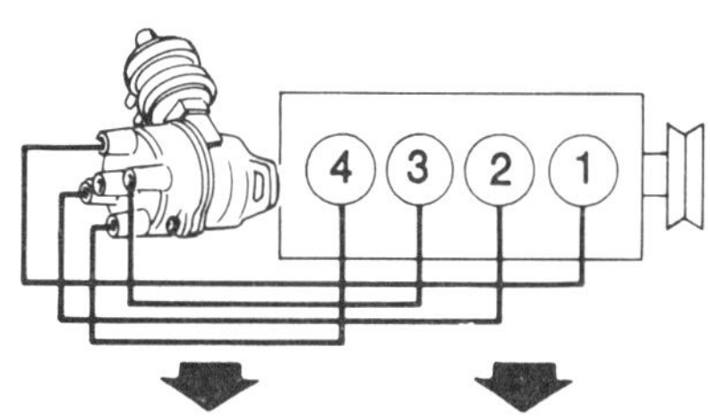

Firing order—1342 & 1488cc Civic

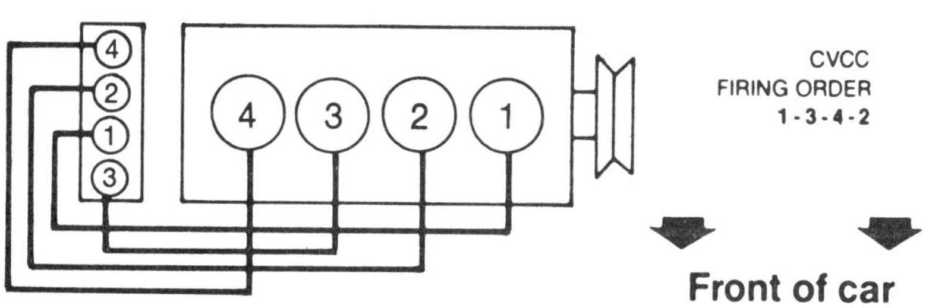

Firing order—1829cc Prelude

EMISSION EQUIPMENT USED

ALL MODELS
Crankcase Emission Control
Closed system
Positive crankcase vent valve
Evaporative Emission Control
Charcoal Canister—single
Air vent cut-off solenoid valve
Purge control valve
Fuel filler cap—with relief valve
Overfill-limiter—two way valve
Fuel check valve
Exhaust Emission Control
Main air jet control system
Catalytic converter—dual three way type
Secondary air supply system Exhaust gas recirculation (EGR) system
EGR valve
Throttle control system
Fuel control system—Electronic Controlled Carburetor System (ECC)
Electronic Controlled Injection System (Civic CRX-Si and Accord models only)
Feedback carburetor system
Thermostatic air cleaner
Oxygen sensor
Dash pot
Ignition timing controls
High altitude compensation
Anti-afterburning valve
Air jet controller
Throttle closer system (Civic 1.5L HF engine only)

REQUIRED EMISSION CONTROL MAINTENANCE
Emission Maintenance Intervals

Procedure	Mileage in Thousands
Air filter change	30,000
Idle speed adjustment	15,000
Oxygen sensor replacement	30,000
PCV system check	15,000
Evaporative canister replacement	15,000
Ignition timing check	45,000
Fuel filter replacement	15,000
Clean carburetor choke mechanism	30,000
Spark plug replacement	15,000
Valve lash adjustment	15,000
EGR valve cleaning	60,000
Anti-afterburning valve test	60,000

IDLE SPEED AND MIXTURE ADJUSTMENT

NOTE: The following specifications are published from the latest information available. If the published information differs from the information on the vehicle emission label, use the data on the emission label. The idle mixture adjustments are factory set and sealed and no adjustments are required. The mixture control adjustment screw opening is plugged to prevent any adjustment. Mixture control adjustments should be done when the mixture control unit is replaced or when the vehicle fails the emissions test.

IDLE SPEED AND MIXTURE
Adjustment

ACCORD AND CIVIC CARBURETED MODELS

NOTE: This procedure will require a propane enrichment kit. Do not smoke during this procedure and keep all open flame away from the work area.

1. Turn off all accessories and start the engine. Let the engine run until it reaches normal operating temperature (when the cooling fan comes on).
2. Remove the vacuum hose from the intake air control diaphragm and clamp the hose end.
3. Connect a tachometer to the engine. Check the idle speed and adjust the idle speed as necessary by turning the throttle stop screw.

NOTE: The idle boost solenoid valve will be operated if the steering is turned on those vehicles equipped with power steering. So it is important to adjust the idle speed with the steering in the straight ahead position. On all Civic models equipped with power steering, disconnect the power steering oil pressure switch wires and connect the wire terminals with a jumper wire to operate the idle controller. Also, if the idle speed is excessively high, check the dashpot system for correct operation.

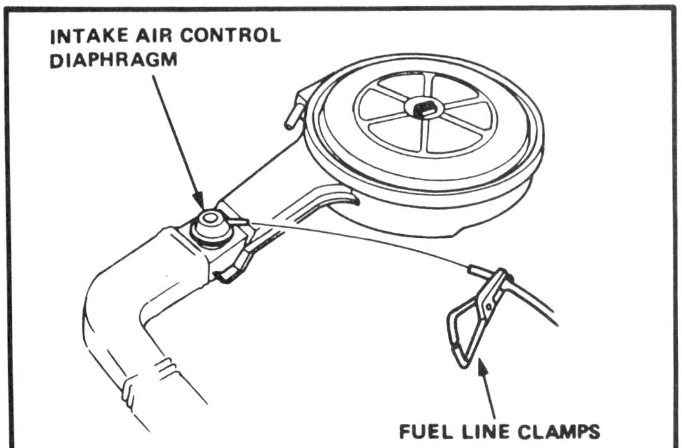

Disconnect and clamp—vacuum hose—from intake air control

4. Disconnect the air cleaner intake tube from the air duct on the radiator bulkhead. Insert the propane enrichment hose into the intake tube about four inches. Be sure that the propane enrichment kit has enough gas before going any further with the adjustment.
5. While the engine is idling, depress the button on top of the propane device. Then slowly open the propane control valve to obtain maximum engine speed. The engine speed should increase, as the percentage of the propane injected goes up.

NOTE: Open the propane control valve slowly, because a sudden surge of propane could stall the engine.

a. If the engine speed does not increase per specifications, the mixture is improperly adjusted, so go on to the next step.

b. If the engine speed increases per specification, go to Step 13.

6. Remove the air cleaner and disconnect the vacuum hose to the fast idle unloader. Pull the throttle cable out of its bracket.

7. Remove the carburetor nuts and the bolt securing the steel tubing vacuum manifold.

8. Lift the carburetor up off of the studs and tilt it back so the throttle controller bracket screws can be removed. Remove the idle controller bracket.

9. Remove the mixture adjusting screw hole cap and reinstall the throttle controller bracket.

10. Reinstall the carburetor. Reconnect the vacuum hose to the fast idle unloader and reinstall the air cleaner.

11. Start the engine and let it run until it reaches normal operating temperature. Remove and plug the vacuum hose from the air intake control diaphragm.

12. Reinstall the propane enrichment kit and recheck the maximum propane enrichment rpm.

a. If the propane enriched speed is to low, the mixture is to rich. Turn the mixture screw ¼ turn clockwise and recheck.

b. If the propane enriched speed is too high, the mixture is to lean. Turn the mixture screw ¼ turn counterclockwise and recheck.

13. Close the propane control valve and recheck the idle speed. Run the engine at 2500 rpm for ten seconds to stabilize the condition:

a. If the idle speed is within specifications, go on to Step 15.

b. If the idle speed is out of specifications, go to next step.

step.

14. Recheck the idle speed and if necessary, adjust by turning the throttle stop screw. Then repeat Steps 12 and 13.

15. Remove the propane enrichment kit and reconnect the air cleaner intake tube on the radiator bulk head.

16. Reinstall the mixture adjusting screw hole cap. Check the idle controller booster speed.

NOTE: There is no idle controller on automatic transmission vehicles without air conditioning and power steering.

a. On the 1500 automatic transmission vehicle with power steering, check the idle speed with the steering turned completely to the left or right.

b. On all other models except the 1500 automatic transmission with power steering, check the idle speed with the headlights on and the heater blower on high.

17. Adjust the idle speed as necessary, by turning the idle control screw.

PRELUDE

NOTE: This procedure will require a propane enrichment kit. Do not smoke during this procedure and keep all open flame away from the work area.

1. Turn off all accessories and start the engine. Let the engine run until it reaches normal operating temperature (when the cooling fan comes on).

2. Remove the vacuum hose from the intake air control diaphragm and clamp the hose end.

3. Connect a tachometer to the engine. Check the idle speed and adjust the idle speed as necessary by turning the throttle stop screw.

4. On vehicles equipped with automatic transmissions, remove the attaching bolt. Then remove the frequency solenoid valve A and air control valve A.

5. Disconnect the vacuum tubes and connect the lower hose to the air control valve A. Insert the propane enrichment hose into the intake tube about four inches. Be sure that the propane enrichment kit has enough gas before going any further with the adjustment.

NOTE: It is not necessary to disconnect the intake tube, because its opening is just behind the right headlight.

6. While the engine is idling, depress the button on top of the propane device. Then slowly open the propane control valve to obtain maximum engine speed. The engine speed should increase as the percentage of the propane injected goes up.

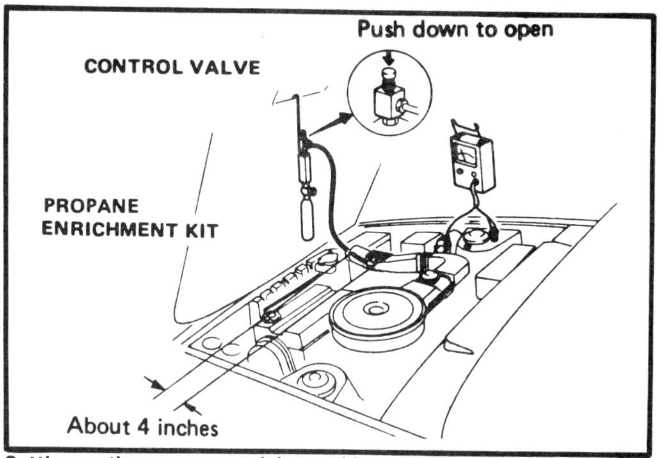

Setting up the propane enrichment kit

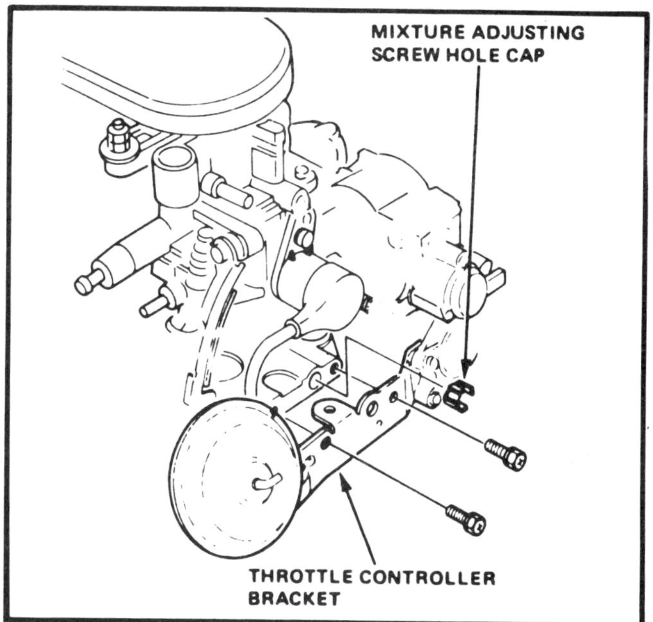

Removing the mixture adjusting screw hole cap

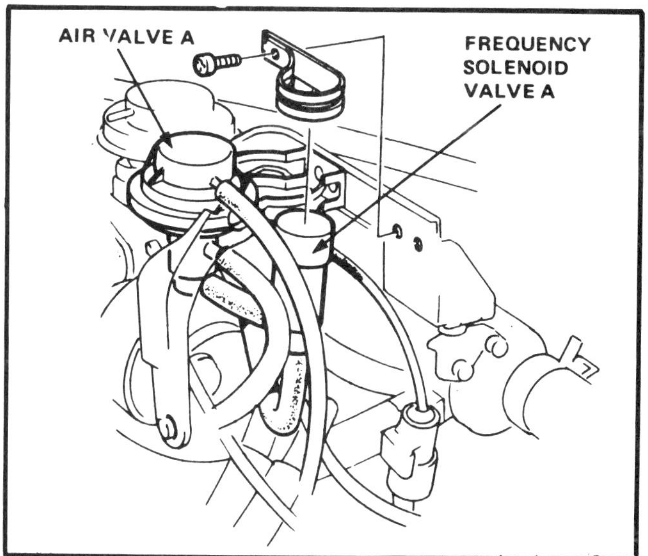

Removing air control valve A and frequency solenoid valve A

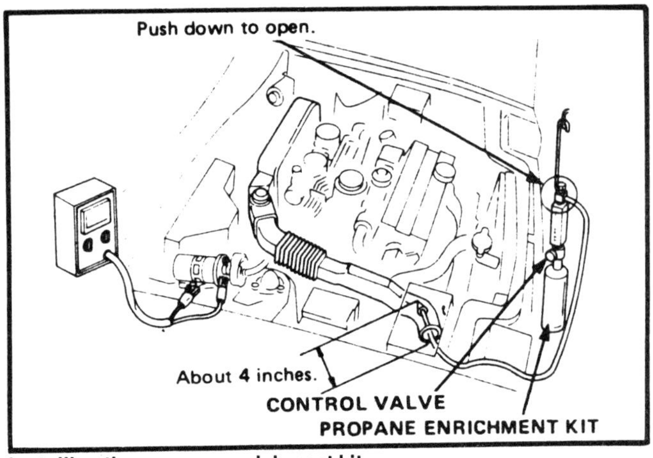

Installing the propane enrichment kit

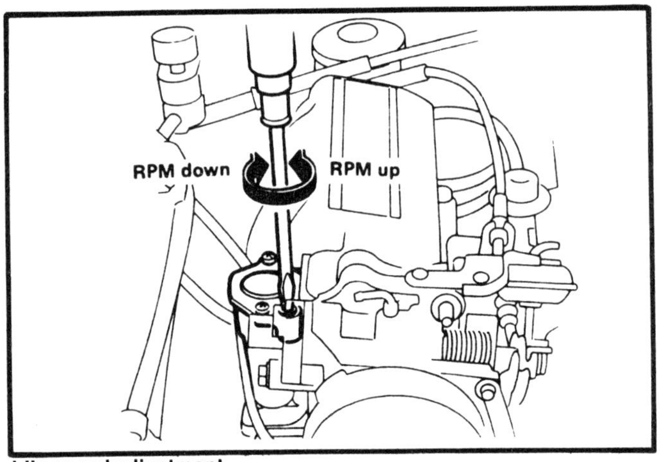

Idle speed adjustment

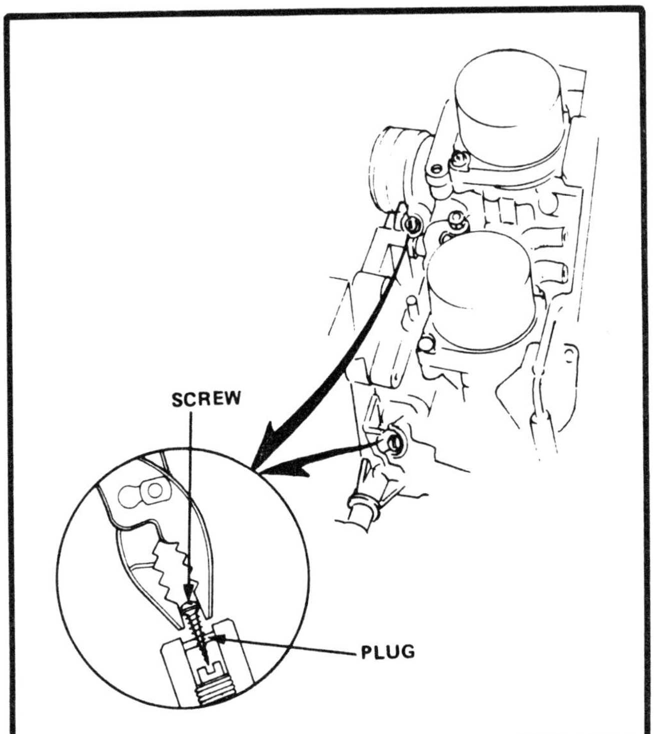

Removing the mixture adjusting screw plug

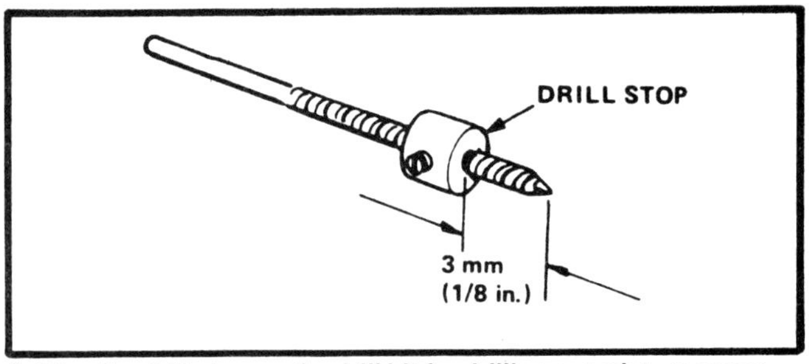

Setting the drill stop on the drill bit for drilling out mixture screw plug

NOTE: Open the propane control valve slowly, because a sudden surge of propane could stall the engine.

a. If the engine speed does not increase per specifications, the mixture is improperly adjusted. Go on to the next step.

b. If the engine speed increases per specification, the test is completed.

7. Remove the carburetors and place them on a suitable work bench.
8. Place a drill stop on a $^1/_8$ in. drill bit. Then drill through the center of the mixture screw hole plugs.

NOTE: Do not drill past the drill stop, because the drill bit could cause damage to mixture adjusting screw.

9. Remove the mixture adjusting screw hole caps, by inserting a 5mm sheet metal screw into the drill hole and pulling on the screw with a pair of pliers.
10. Reinstall the carburetors. Start the engine and let it run until it reaches normal operating temperature.
11. Reinstall the propane enrichment kit and recheck the maximum propane enrichment rpm.

a. If the propane enriched speed is to low, the mixture is to rich. Turn the mixture screw $^1/_4$ turn clockwise and recheck.

b. If the propane enriched speed is too high, the mixture is to lean. Turn the mixture screw $^1/_4$ turn counterclockwise and recheck.

13. Close the propane control valve and recheck the idle speed. Run the engine at 2500 rpm for ten seconds to stabilize the condition:

a. If the idle speed is within specifications, go on to Step 15.

b. If the idle speed is out of specifications, go on to the next step.

14. Recheck the idle speed and if necessary, adjust by turning the throttle stop screw. Repeat Steps 12 and 13.
15. Remove the propane enrichment kit and reconnect the intake air control diaphragm hose.
16. Reinstall the mixture adjusting screw hole caps. Adjust the idle speed as necessary, by turning the idle control screw.

ACCORD AND CIVIC CRX-Si WITH PROGRAMMED FUEL INJECTION (PGM-FI)

NOTE: The idle mixture on these models is pre-set and sealed at the factory. No adjustment should be necessary or attempted. The procedure below is for idle speed adjustment only.

1. With the engine running at normal operating temperature (or when the cooling fan comes on), connect a tachometer to the engine.
2. Check the idle speed and be sure to check it with all the electrical accessories off. The idle speed should be:

a. Manual—750 ± 50 rpm (in neutral)

b. Automatic—750 ± 50 rpm (in neutral or park)

NOTE: To prevent the idle control system from operating, disconnect and plug the number 27 vacuum hose on the Accord models and vacuum hose number 10 on the Civic CRX-Si models.

3. Adjust the idle speed if necessary, by turning the idle adjusting screw. If the idle can not be adjusted by use of the adjusting screw, check the operation of the fast idle valve.
4. Check the idle controller booster speed with the A/C on. The idle speed should be:

a. Manual—800 ± 50 rpm (in neutral)

b. Automatic—800 ± 50 rpm (in neutral or park)

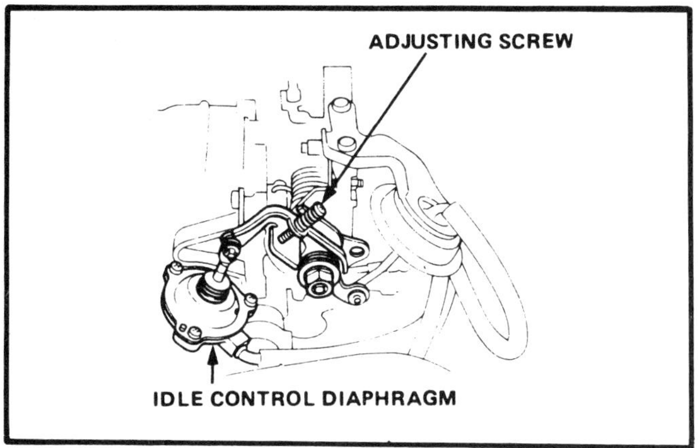

Adjusting screw on the idle control diaphragm

Adjust the idle speed if necessary, by turning the adjusting screw. Which is located on top of the idle control diaphragm.

FAST IDLE VALVE

Test

NOTE: The fast idle valve is adjusted at the factory and should not be disassembled. Be sure to check the PCV valve and circuit tubing for breakage, clogging or being disconnected. Check to see that the throttle valves are fully closed before inspecting the fast idle valve.

1. If the idle is to high after the engine has been warmed up, checked the following:
   a. Be sure that the engine has reached its normal operating temperature and check to see if the idling control is operating properly.
   b. Remove the cover of the fast idle valve and check to see if the valve is completely closed.
   c. If the valve is not fully closed, air will be sucked in from the valve seat area. This can be found by placing a finger in the valve seat area.
   d. If there is an air leak and sucking can be felt, replace the fast idle valve and adjust the idle speed as necessary.
2. If the idle speed is to low after the engine has warmed up, check the following:
   a. Remove the idle adjusting screw. Clean the screw and the air bypass channel with a suitable carburetor cleaner.
   b. After cleaning the adjusting screw, readjust the idle speed.

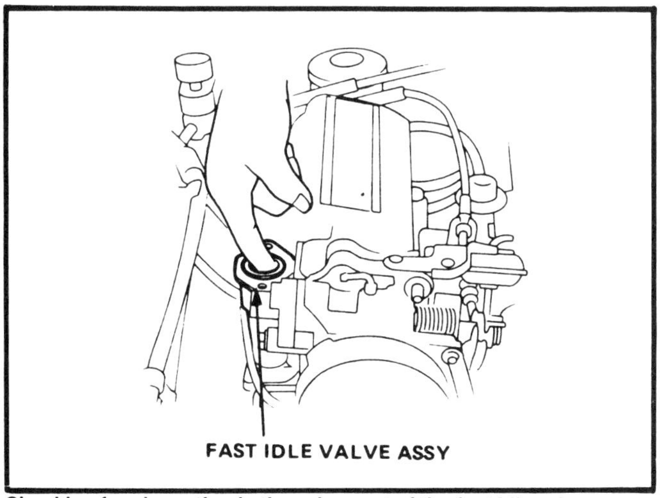

Checking for air suction in the valve seat of the fast idle valve

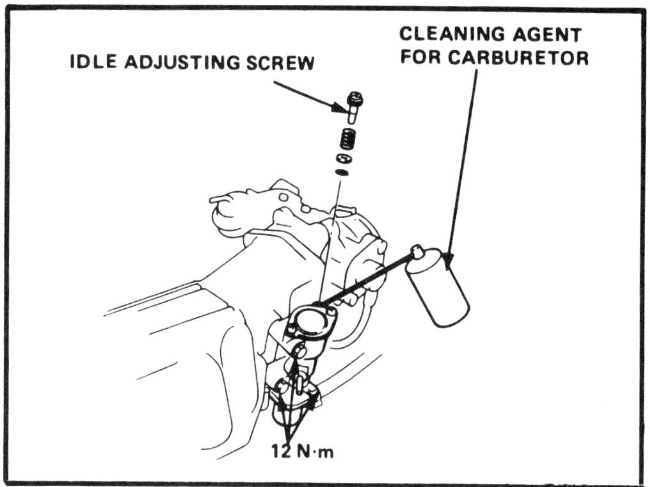

Cleaning the fast idle valve assembly

The fast idle speed should be:

1. 1000–1800 rpm for manual and automatic transmission models (Accord models).
2. 1250–2250 rpm for manual and automatic transmission models (Civic CRX-Si models).

NOTE: The fast idle speed is low when the engine is cold (below 122 degrees F. on the Accord models and 140 degrees F. on the Civic CRX-Si models).

3. To clean the fast idle valve assembly and further test it, do the following:
   a. Remove the fast idle valve assembly from the throttle body.
   b. Apply cold water to the valve and cool down the wax part of the fast idle valve to 41–86 degrees F.

Applying cold water to the fast idle valve assembly

   c. Blow through the lower part of the fast idle valve. Check to see if large amount of air flows without any resistance.
   d. If the air does not flow or the resistance is large, replace the valve and adjust the idle speed as necessary.

FAST IDLE CONTROL SOLENOID

Test

ACCORD

NOTE: When the valve is open, There should be 9 volts or more available between the black and yellow terminal and the blue and black terminal of the wire harness of the number one control box.

1. Disconnect the wire harness from the number one control box.

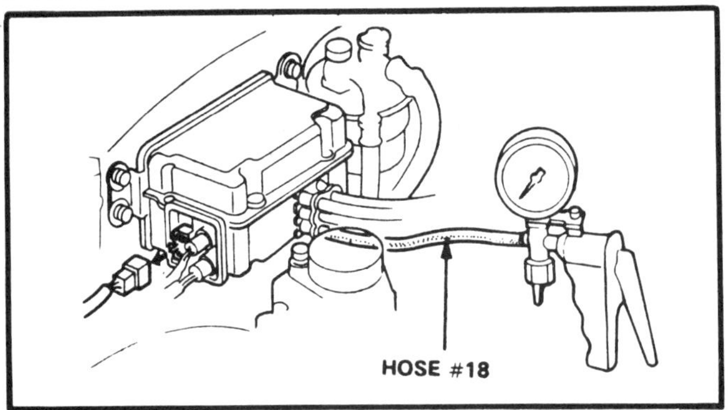

Applying vacuum to vacuum hose number eighteen

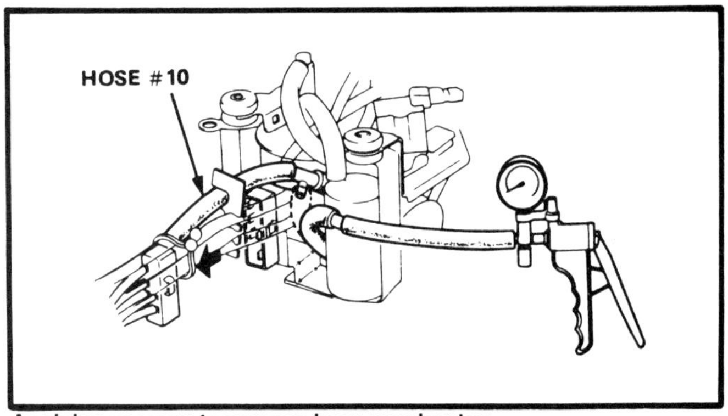

Applying vacuum to vacuum hose number ten

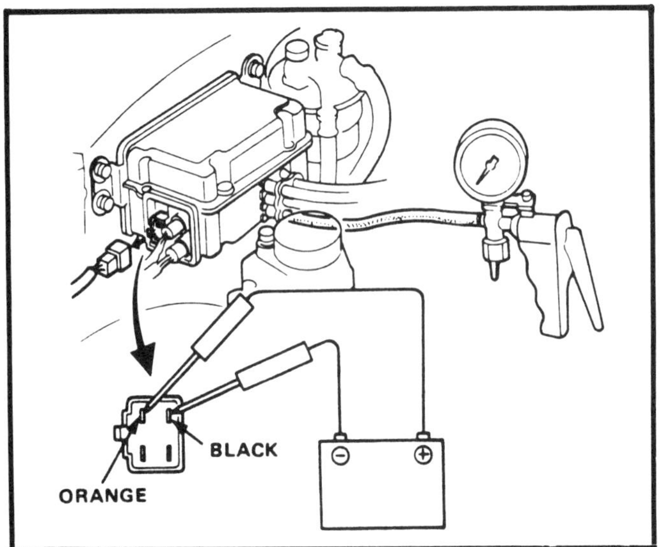

Connecting the hot lead and ground to the connector of the number one control box

2. Disconnect the vacuum hose number 18 from the air flow tube. Using a hand held vacuum pump, pump air into the number 18 vacuum hose. There should be no air flow.

3. Connect a hot lead wire from the positive battery terminal to the black terminal of the number one control box coupler. Connect the ground wire from the negative battery terminal to the orange terminal.

4. Disconnect the number 23 vacuum hose from the intake manifold. Pump air into the number 18 vacuum hose. There should be no air flow.

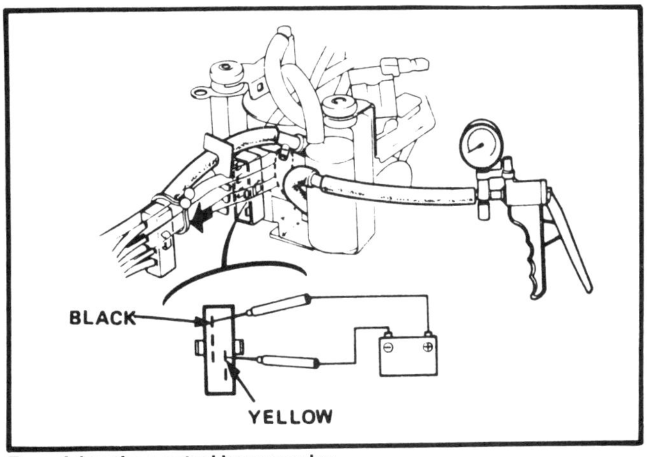

Energizing the control box coupler

5. If there is air flow, then there is a problem with the solenoid. The solenoid should be replaced.

Civic CRX-Si

NOTE: When the valve is open, there should be 9 volts or more available between the black and yellow terminal of the wire harness of the control box. The valve should open when the atmospheric pressure is 660 mmHg. or less. Vacuum is produced in the vacuum hose between the solenoid and the air cleaner.

1. Open the control box lid and disconnect the wire harness from the control box.
2. Disconnect the vacuum hose between the fast idle control solenoid valve and the number 10 vacuum hose.
3. Apply vacuum to the hose. It should not hold any vacuum.
4. Connect the battery positive terminal to the black terminal of the control box coupler. Connect the battery negative terminal to the yellow terminal.
5. Apply vacuum to the hose. It should not hold any vacuum. If it does hold vacuum, replace the valve.

IDLE SPEED SPECIFICATIONS

Engine	Transaxle	Curb Idle Speed (rpm)
1985-86 Accord and Accord SE-I		
1.8L	M/T	750 ± 50
1.8L	A/T	700 ± 50
1.8L — PGM-FI	M/T & A/T	750 ± 50
1985-86 Civic and Civic Wagon		
1.3L	M/T & A/T	650 ± 50
1.5L	M/T & A/T	650 ± 50
1.5L — HF	M/T & A/T	700 ± 50
1985-86 Civic CRX-Si		
1.5L — PGM-FI	M/T & A/T	750 ± 50
1985-86 Prelude		
1.8L — PGM-FI	M/T & A/T	850 ± 50

NOTE: These specifications are published from the latest information available. If the published information differs from the information on the vehicle emission label, use the data on the emission label.

PGM-FI—Programmed Fuel Injection
HF—High Fuel Efficiency
M/T—Manual Transaxle
A/T—Automatic Transaxle

HIGH ALTITUDE EMISSION REDUCTION

NOTE: This procedure is only necessary for the Civic models equipped with the 1.3L engine purchased outside of California. All other Civic's need not be adjusted for use at any altitude.

1. Remove the air cleaner and disconnect all the vacuum lines at the carburetor.
2. Remove the throttle cable from the bracket. Remove the carburetor bolts along with the vacuum manifold assembly. Remove the carburetor.
3. Place the carburetor in a position so that the throttle controller bracket screws are accessible. Remove the throttle controller screws.
4. Remove the throttle controller bracket and the mixture screw cap.
5. Turn the mixture screw ½ turn clockwise. The screw should be turned no less then one full turn from the seated position.
6. Reinstall the carburetor on the engine. Start the engine and allow it to reach normal operating temperature.
7. Check the idle speed with all accessories off and adjust as necessary.
8. Apply a new emission control information label under the hood.

IGNITION TIMING

ALL CIVIC MODELS

1. Connect a timing light and a tachometer to the engine. Start the engine and let it run until it reaches normal operating temperature.
2. Check the engine idle speed and adjust as necessary. While the engine is idling, aim the timing light at the pointer on the timing belt cover.
3. If the timing is out of specifications, loosen the distributor adjusting bolts. Turn the distributor counterclockwise to advance the timing and clockwise to retard the timing.
4. After setting the timing to specifications. Tighten the distributor adjusting bolts and recheck the timing and engine idle speed. Adjust as necessary.
5. After everything has been rechecked, remove the test equipment and install the cap on upper adjusting bolt.

NOTE: There are two bolts that can be loosened to adjust the timing. Do not loosen the smaller bolt under the distributor swivel mounting plate unless a satisfactory adjustment, using the larger bolt, cannot be obtained. The smaller bolt provides an extra range of adjustment in the event of the distributor being installed one tooth off.

ACCORD AND PRELUDE MODELS

1. Connect a timing light and a tachometer to the engine. Remove the rubber cap from the inspection window of the cylinder block. Start the engine and let it run until it reaches normal operating temperature.
2. Check the engine idle speed and adjust as necessary. While the engine is idling, aim the timing light at the pointer at the flywheel or drive plate.
3. If the timing is out of specifications, loosen the distributor adjusting bolts. Turn the distributor counterclockwise to advance the timing and clockwise to retard the timing.
4. After setting the timing to specifications, tighten the distributor adjusting bolts (or nut for fuel injected engine). Recheck the timing and engine idle speed. Adjust as necessary.

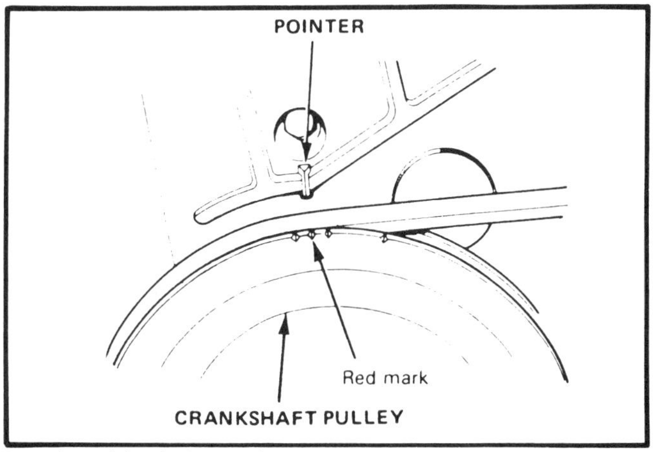

Location of the timing marks

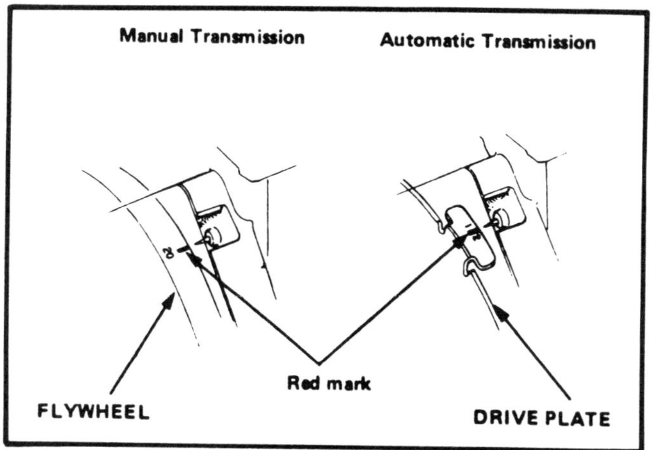

Timing mark location

5. After everything has been rechecked, remove the test equipment and install the cap on upper adjusting bolt.

IGNITION TIMING CONTROLS

A variety of ignition timing control systems are used on various models to reduce oxides of nitrogen and hydrocarbon emissions. These timing controls, combined with centrifugal advance, affects the time at which each spark plug ignites the air/fuel mixture. The ignition spark is controlled according to the engine coolant temperature, engine load and speed.

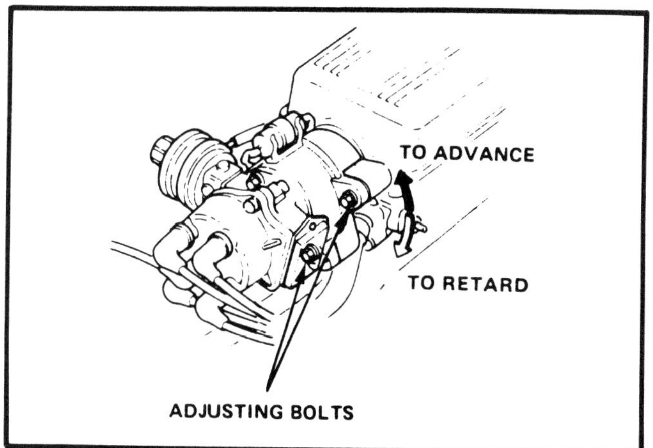

Distributor adjusting bolts

IGNITION TIMING SPECIFICATIONS

Engine	Transaxle	Ignition Timing
1985-86 Accord and Accord SE-i		
1.8L	M/T & A/T	18 BTDC ± 2 at curb idle
1.8L — PGM-FI	M/T	22 BTDC ± 2 at curb idle
1.8L — PGM-FI	A/T	18 BTDC ± 2 at curb idle
1985-86 Civic and Civic Wagon		
1.3L	M/T & A/T	21 BTDC ± 2 at curb idle
1.5L	M/T & A/T	21 BTDC ± 2 at curb idle
1.5 — HF	M/T	20 BTDC ± 2 at curb idle
1.5 — HF	A/T	15 BTDC ± 2 at curb idle
1985-86 Civic CRX-Si		
1.5L — PGM-FI	M/T & A/T	16 BTDC ± 2 at curb idle
1.5L — PGM-FI	M/T & A/T Cal.	12 BTDC ± 2 at curb idle
1985-86 Prelude		
1.8L	M/T	20 BTDC ± 2 at curb idle
1.8L	A/T	12 BTDC ± 2 at curb idle

PGM-FI—Programmed Fuel Injection
HF—High Fuel Efficiency
M/T—Manual Transaxle
A/T—Automatic Transaxle
Cal.—California

HOT START CONTROL

This system is designed to prevent an over rich mixture condition in the intake manifold, due to vaporization of residual fuel when starting a hot engine. This reduces carbon monoxide and hydrogen emissions.

ANTI-AFTERBURN VALVE

Various models are equipped with an anti-afterburn valve. The function of this valve is to allow fresh air into the intake manifold when there is a sudden increase in the manifold vacuum. The valve will respond only to a sudden increase in vacuum. The amount of time that this valve will stay open is determined by an internal diaphragm, which is acted on by the vacuum level.

AIR JET CONTROLLER

This is an atmospheric pressure sensing device used on California models and high altitude models only. As the altitude changes, a valve opens and closes to maintain the proper air flow to the carburetor.

CVCC ENGINE MODIFICATIONS

The most important part of the CVCC engine emission control system is the Compound Vortex Controlled Combustion (CVCC) cylinder head. Each cylinder has three valves, a conventional intake and exhaust valve and a smaller auxiliary intake valve. There are two combustion chambers per cylinder, a pre-combustion chamber and a main chamber. During the intake stroke, an extremely lean mixture is drawn into the main combustion chamber. At the same time, a very rich mixture is drawn into the smaller pre-combustion chamber through the auxiliary intake valve. The spark plug, which is located in the pre-combustion chamber, ignites the rich pre-mixture and then spreads into the main combustion chamber, igniting the lean mixture.

VACUUM ADVANCE

The distributor has two means of advancing the ignition timing. One is the centrifugal advance and is operated by the distributor weights. The other is the vacuum advance and is controlled by the vacuum advance diaphragm, located on the side of the distributor.

SPARK PLUGS

Model	Spark Plug Type	Gap (in.)
Accord	BUR5EB-11 (NGK)	(0.039–0.043)
Accord SE-I	BPR6EY-11 (NGK)	(0.039–0.043)
Civic and Civic Wagon	BUR5EB-11 (NGK)	(0.039–0.043)
Civic 1.5 HF Engine	BUR4EB-11 (NGK)	(0.039–0.043)
Civic CRX-Si	BPR6EY-11 (NGK)	(0.039–0.043)
Prelude	BPR6EY-11 (NGK)	(0.039–0.043)

NOTE: The underhood sticker often reflects tune-up specification changes made in production. Sticker figures must be used if they disagree with those in the chart.
HF—High Fuel Efficiency

EMISSION CONTROL SYSTEMS

CRANKCASE EMISSION CONTROL SYSTEM

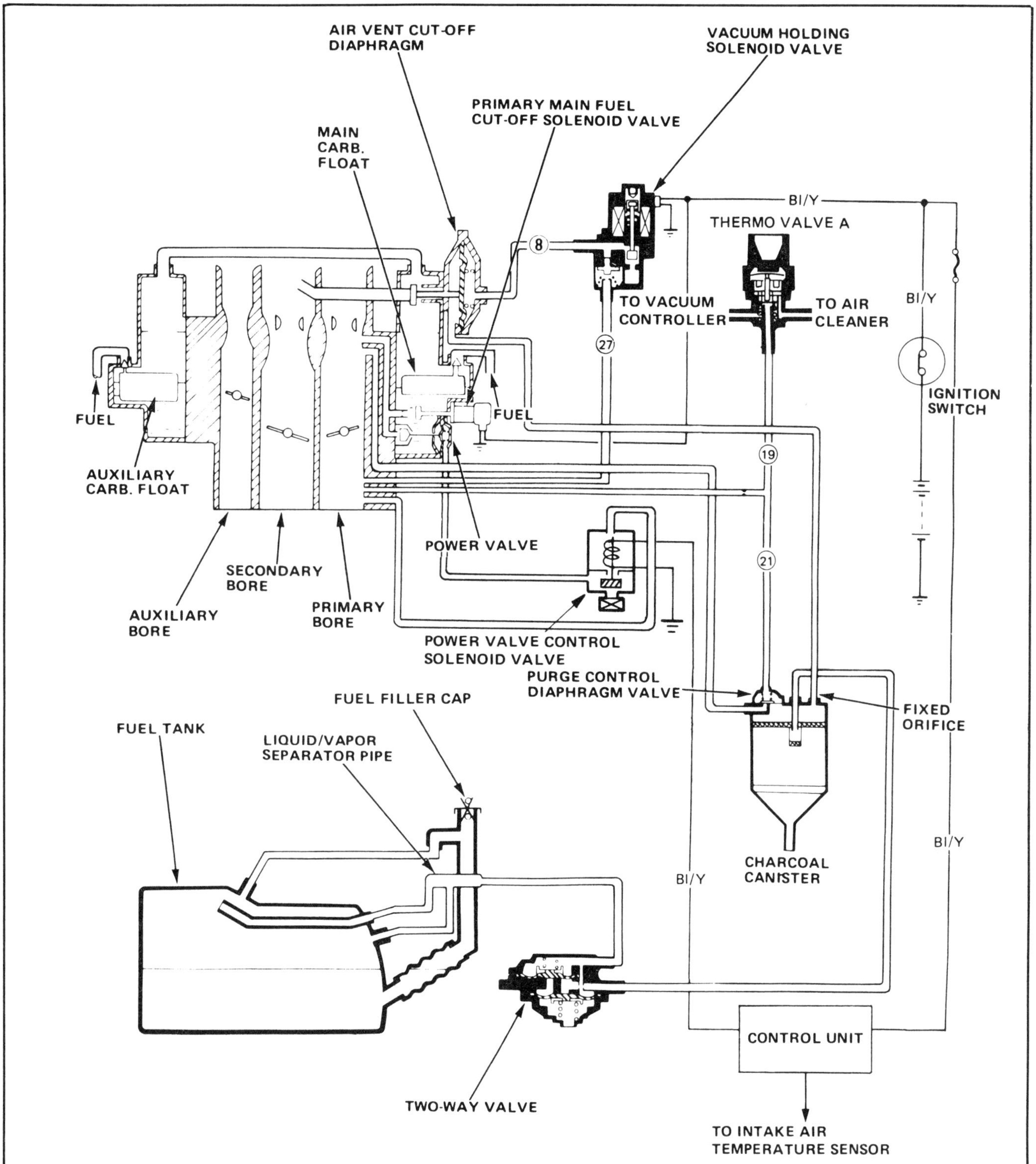

Fuel evaporative control system schematic

The PCV system is designed to prevent blow-by gas from escaping to the atmosphere. The PCV valve contains a spring loaded plunger. When the engine starts, the plunger in the PCV valve is lifted in proportion to intake manifold vacuum. Blow-by gas is drawn directly into the intake manifold. When the throttle valves are closed (idle) or partially open, blow-by vapor is returned directly to the intake manifold through the breather chamber and positive crankcase ventilation (PCV) valve with fresh air. When the throttle valves are wide open, the intake manifold vacuum decreases and vacuum in the air cleaner increases. In response to intake manifold vacuum, PCV valve increases the blow-by vapor flow. A small amount of vapor is returned through the valve cover breather hose and into the air cleaner.

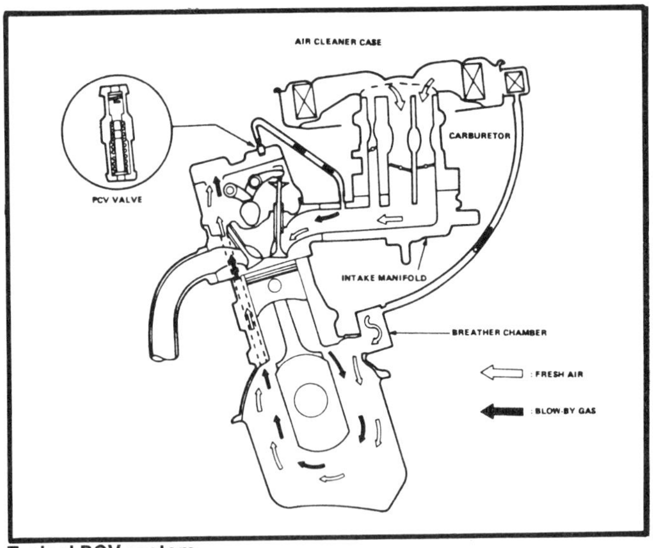

Typical PCV system

System Service

NOTE: The PCV valve should be checked every 15,000 miles and replaced as required. If the valve is found to be defective, do not attempt to clean it. Replace it.

1. With the engine running, remove the valve from its mounting. A hissing sound should be heard and vacuum should be felt from the inlet side of the valve.
2. Reinstall the valve. Remove the crankcase inlet air cleaner. Loosely hold a piece of stiff paper over the opening in the rocker cover. Allow one minute for the crankcase pressure to reduce itself. The paper should then suck itself against the rocker cover with noticeable force. Replace the inlet air cleaner in the rocker cover.
3. With engine stopped, remove the PCV valve. Shake the valve; a clicking sound should be heard indicating that the valve is not stuck.
4. If the valve fails any of the above tests, it should be replaced.

FUEL EVAPORATIVE EMISSION CONTROL SYSTEM

The function of the evaporation control system is to prevent the emission of gas vapors from the fuel tank and carburetor which could be expelled into the atmosphere. When fuel evaporates in the gas tank or the carburetor float chamber, the vapors pass through the vent hoses to the charcoal canister where they are stored until they can be drawn into the intake manifold when the engine is running. All vehicles are equipped with the charcoal canister for the storage of fuel vapors. The fuel bowls of all carburetors are vented internally and on some applications, do not require venting the canister. If this is the case, the bowl vent port on the canister will be capped. Most carburetors are also externally vented to the charcoal canister.

System Service

The charcoal canister itself is a non-serviceable component. The only service required for the system is to replace the filter pad on the bottom of the canister. The filter requires replacement every 12 months or 12,000 miles. If the vehicle is driven under severe conditions, the filter should be replaced more often.

OVERFILL LIMITER (TWO-WAY VALVE)

The overfill limiter consists of a pressure valve and a vacuum valve. The pressure valve is designed to open when the fuel tank internal pressure has increased over the normal pressure and the vacuum valve opens when a vacuum has been produced in the tank.

PURGE CONTROL VALVE

The purge control valve is kept closed during idling in order to prevent vaporized fuel from entering into the intake manifold for positive control of of high idle CO emission, which is a particular problem under high ambient temperatures. When the carburetor vacuum working on the diaphragm of the valve exceeds the pre-set value, the purge control valve is opened.

THERMO VALVE A

When the coolant temperature is higher than the pre-set value of thermo valve A, it closes and the purge control diaphragm valve in the canister is opened by intake manifold vacuum. Fuel vapor is purged from the charcoal canister through the purge control diaphragm valve by a venturi valve.

CARBURETOR FUEL CUT-OFF SOLENOID VALVE

When the engine is not running, the fuel passages for the main and slow primary fuel metering system are cut off by solenoid valves, so the fuel in the float chamber cannot enter the carburetor bore. When the ignition switch is off, the power valve control solenoid valve is de-energized. Vacuum is then held in the hose between the power valve and the check valve, closing the power valve and cutting off the fuel passage.

FUEL FILLER CAP

The fuel filler cap is equipped with relief valve to prevent the escape of fuel vapor into the atmosphere.

EXHAUST EMISSION CONTROL SYSTEM

DASHPOT SYSTEM

To improve combustion, a throttle controller holds the throttle slightly open to admit additional air during periods of gear shifting and deceleration. When the engine is running above idle, ported vacuum in the carburetor is applied to the throttle controller through a dashpot check valve. On deceleration, the vacuum bleeds off through the orifice in the dashpot check valve. It then gradually decreasing until the throttle closes completely. Throttle closing speed is determined by the size of the dashpot check valve orifice, tension of the throttle return spring and the amount of vacuum available at the carburetor port.

THROTTLE CONTROLLER

The throttle controller is provided in the system for easy starting of the engine. When cranking the engine to start. The cranking opener solenoid valve is activated to allow intake manifold vacuum into the diaphragm, so that appropriate throttle opening angle is obtained.

CATALYTIC CONVERTER

Catalytic converter require the use of unleaded fuel only. Leaded fuel will destroy the effectiveness of the catalyst as an emission control device. Under normal operating conditions, the converter will not require any maintenance. However, it is important to keep the engine properly tuned. If the engine is not kept properly tuned, engine misfiring may cause overheating of the cataylst. This may cause heat damage to the converter or vehicle components. This situation can also occur during diagnostic testing, if any spark plug cables are removed and the engine is allowed to idle for a long period of time.

EGR SYSTEM

The EGR system is designed to reduce oxides of nitrogen (NOx), by recirculating the exhaust gas through the EGR valve and the intake manifold into the combustion chambers. It is composed of the EGR valve, EGR control valves (A) and (B) and thermovalve (C). The EGR valve is operated by vacuum from the carburetor port and provides EGR volume proportional to engine loads (intake air volume) by the operation of the EGR control valves (A) and (B). The vacuum signal is ported above the idle throttle valve position to eliminate the EGR at idle. In cold engine operation, the thermovalve is open so ported vacuum is bled from the thermovalve, keeping the EGR valve closed for EGR cut-off. When the engine coolant temperature exceeds the set temperature of thermovalve (C), the thermovalve is closed and ported vacuum is applied to the EGR valve and the EGR control valve (A). This opens the EGR valve allowing exhaust gas into the intake manifold. The EGR control valve (B) is normally closed. When manifold vacuum from the carburetor insulator reaches a set level, the valve opens allowing venturi vacuum to enter the control valve (A) and the EGR valve.

EGR Valve and Flow Passages

1. Viewing through an opening in the EGR valve body, check if

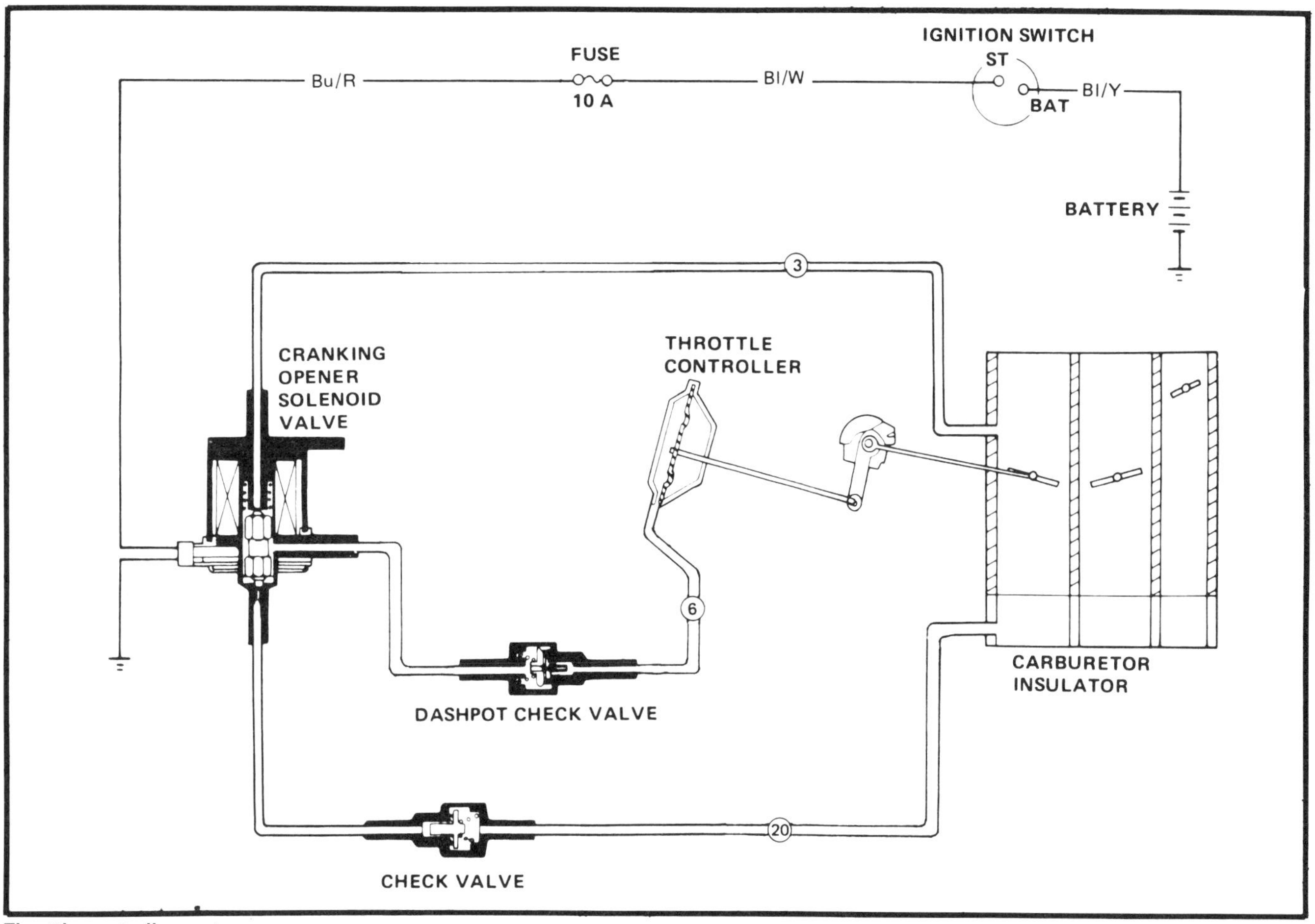

Throttle controller system

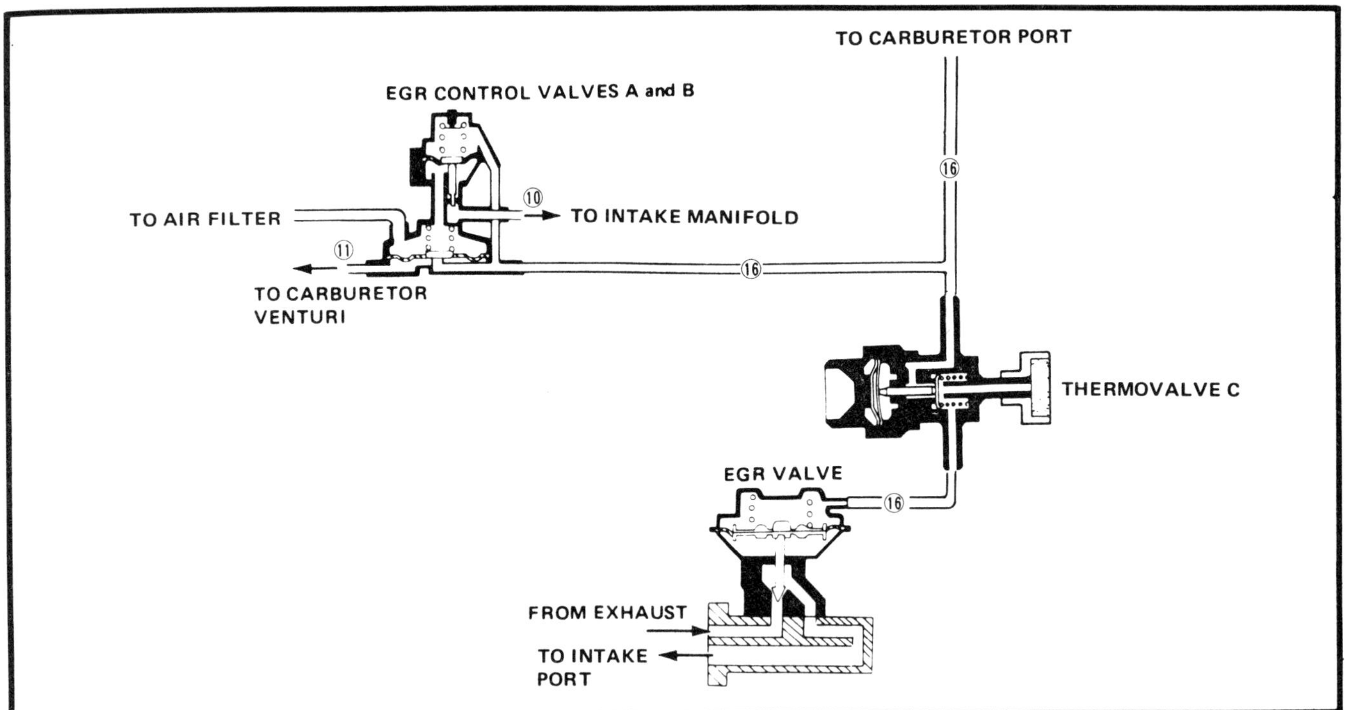

EGR system—typical

the valve shaft moves when the engine speed reaches 3000–3500 rpm under no-load conditions, after engine warming up.

2. Check if the EGR valve moves, when vacuum (6 in. Hg.) is applied to the EGR valve by a hand held vacuum pump or equivalent.

3. Either rough idling or engine stall should occur, when vacuum (6 in. Hg.) is applied to the EGR valve by a hand held vacuum pump or equivalent, after engine warming up.

4. If none of the above operate properly, remove the EGR valve. Clean the passages, both vacuum and exhaust and/or replace the EGR valve.

SECONDARY AIR SUPPLY SYSTEM

This system is designed to improve emission control performance, by supplying fresh air from the air cleaner into the exhaust manifold, through the air suction valve. The air suction control solenoid valve is open by signals from the control unit. When the intake manifold vacuum is above the set value of vacuum switch (C), manifold vacuum is applied to the air suction diaphragm valve to open the secondary air passage. When negative pressure induced by pulsation in the exhaust manifold makes the air suction reed valve open, fresh air enters into the exhaust manifold. An air bleed valve is used to prevent overheating in the exhaust system. Which could otherwise be caused by catalyst reaction to such extreme condition as rapid increase in air temperature inside the air cleaner case. When the intake air temperature rises excessively, the air bleed valve opens and manifold vacuum, which would normally be applied to the air suction diaphragm valve is bled off. This closes the secondary air passage.

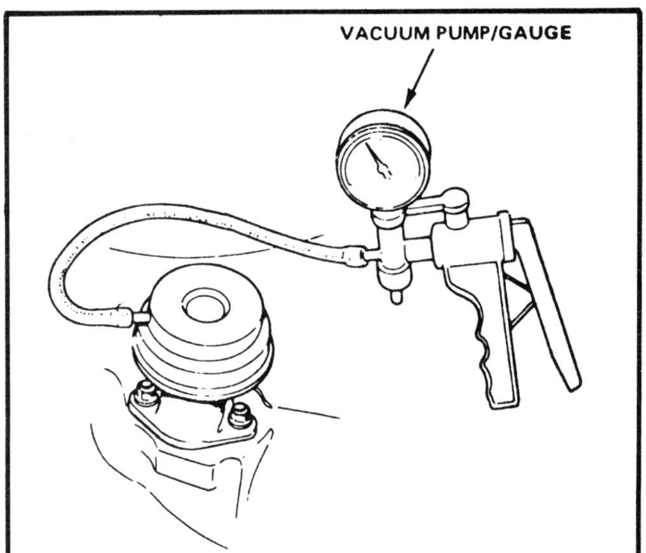

Applying vacuum to the EGR valve

ANTI-AFTERBURN VALVE

Various models are equipped with an anti-afterburn valve. The function of this valve, is to allow fresh air into the intake manifold when there is a sudden increase in the manifold vacuum. The valve will respond only to a sudden increase in vacuum. The amount of time that this valve will stay open, is determined by an internal diaphragm which is

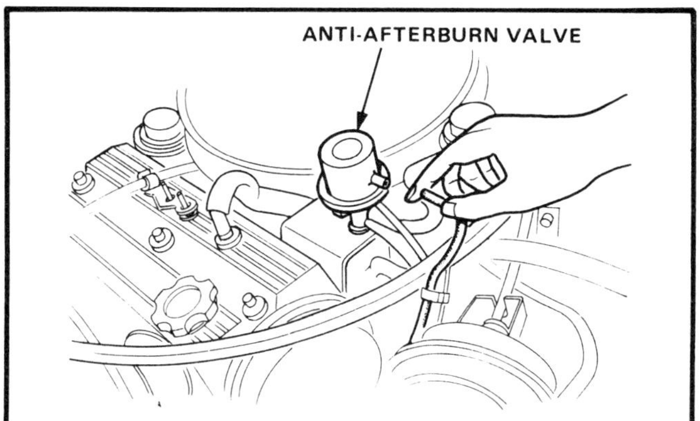

Typical—anti-afterburning valve

acted on by the vacuum level.The anti-afterburn control solenoid valve is provided to control the operation of the ant-afterburn valve by introducing manifold vacuum to the top of the anti-afterburn valve diaphragm. When the engine coolant temperature is below the set temperature of thermosenor (A) and the vehicle speed is below the preset value of the speed sensor, the anti-afterburn control solenoid valve is activated to open the vacuum passage. When the anti-afterburn control solenoid valve opens, the anti-afterburn valve does not open because there is no vacuum differential on the diaphragm.

Anti-Afterburn System Test

1. Disconnect the vacuum hose from the Anti-Afterburn Valve (AAV), then run the engine.
2. Check and see that air is being sucked into the vacuum hose. If there is no air, the vacuum line is defective.
3. Hold a piece of paper under the AAV and run the engine up to 3000 rpm, close the throttle valve quickly.
4. The paper should be sucked up at the moment the throttle valve is released. If it does not get sucked up, the AAV is defective and should be replaced.

THROTTLE CLOSER SYSTEM

CIVIC 1.5L HF ENGINE ONLY

The function of this system, is to close the throttle valve below its idle position for fuel economy, when decelerating the vehicle under certain conditions. When the throttle control solenoid valve is activated by the control unit. Manifold vacuum is routed through the throttle control solenoid valve, to the throttle closer diaphragm of the throttle controller which retracts the throttle stop screw.

MAIN AND SLOW AIR JET CONTROL SYSTEM

This system is designed to control air flow volume into the main and slow air jets of the carburetor according to the driving conditions. It also maintains optimum air/fuel ratio for achieving good fuel economy. When the solenoid valves are activated by the control unit, the primary slow and main air cut off solenoid valves supply additional air to each air passage of the carburetor, allowing the air/fuel mixture to lean.

FEEDBACK CONTROL SYSTEM

The feedback control system maintains the proper air/fuel mixture ratio by allowing air into the intake manifold, as is necessary, to adjust temporarily fuel-rich conditions. The system is made up of two subsystems, the X system and the M system. The X system consists of air control valve (B), frequency solenoid valve (B), feedback control solenoid valve , check valve (B) surge tanks (A) and (B) and control unit. When frequency solenoid valve (B) and feedback control solenoid valve are activated by the control unit, manifold vacuum is applied to air control (B). When air control valve (B) opens, it allows the correct amount of air to be fed into the intake manifold. Surge tank (A) acts as a vacuum reservoir while surge tank (B) dampens the vacuum pulses so that relatively steady vacuum is applied to the air control valve (B).

The M-system consists of an air control valve (A), frequency solenoid valve (A) and control unit. When frequency solenoid valve (A) is opened by the control unit, air control valve (A), which has been already opened by vacuum from the carburetor port, feeds the correct amount of air into the intake manifold. The amount of air is proportional to intake air volume.

OXYGEN SENSOR

The oxygen sensor is mounted in the exhaust manifold. It is used to sense oxygen concentration in the exhaust gas. If the fuel/ratio is leaner than the stoichiometric ratio in the mixture (i.e. excessive amount of air), the exhaust gas contains more oxygen. On the other hand, if the fuel/ratio is richer than the stoichiometric ratio, the exhaust gas hardly contains any oxygen.

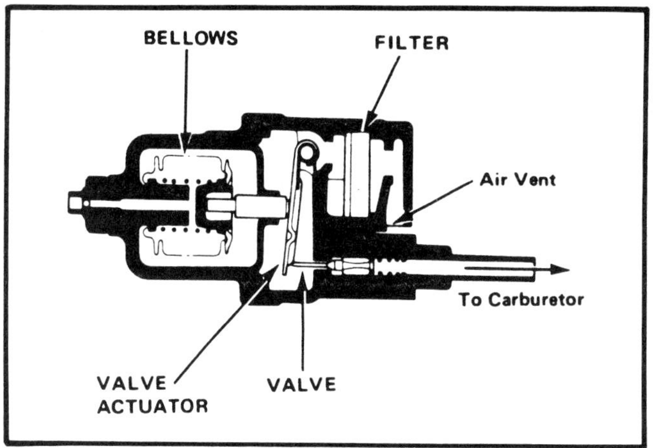

Air jet controller valve

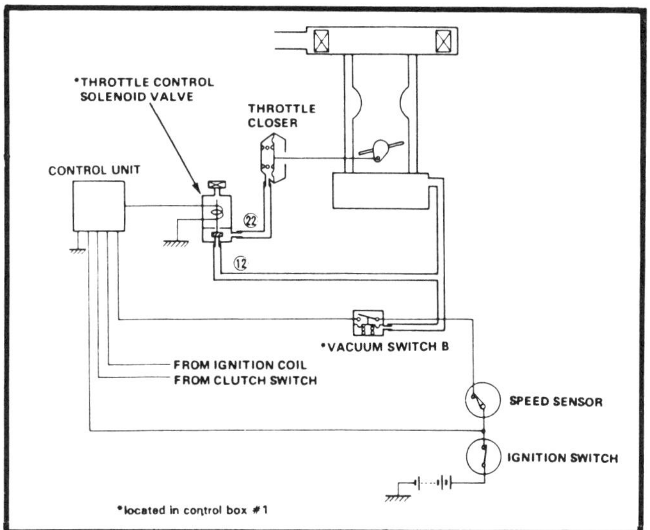

Exploded view of the throttle closer system

THERMOSTATIC AIR CLEANER

All carbureted models are equipped with a temperature regulated air cleaner which improves engine warm up characteristics and helps to minimize carburetor icing. This air cleaner has an air control door inside the snorkel to modulate the temperature of the air entering the carburetor. The air control door is controlled by an air control diaphragm, which reacts to intake manifold vacuum and the temperature inside the air horn.

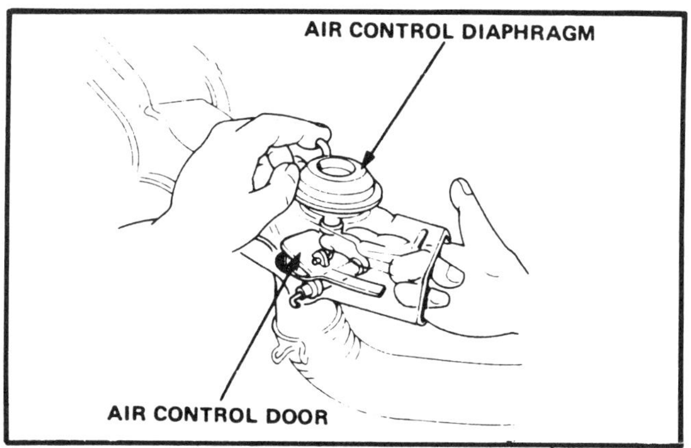

Typical thermostatic air cleaner

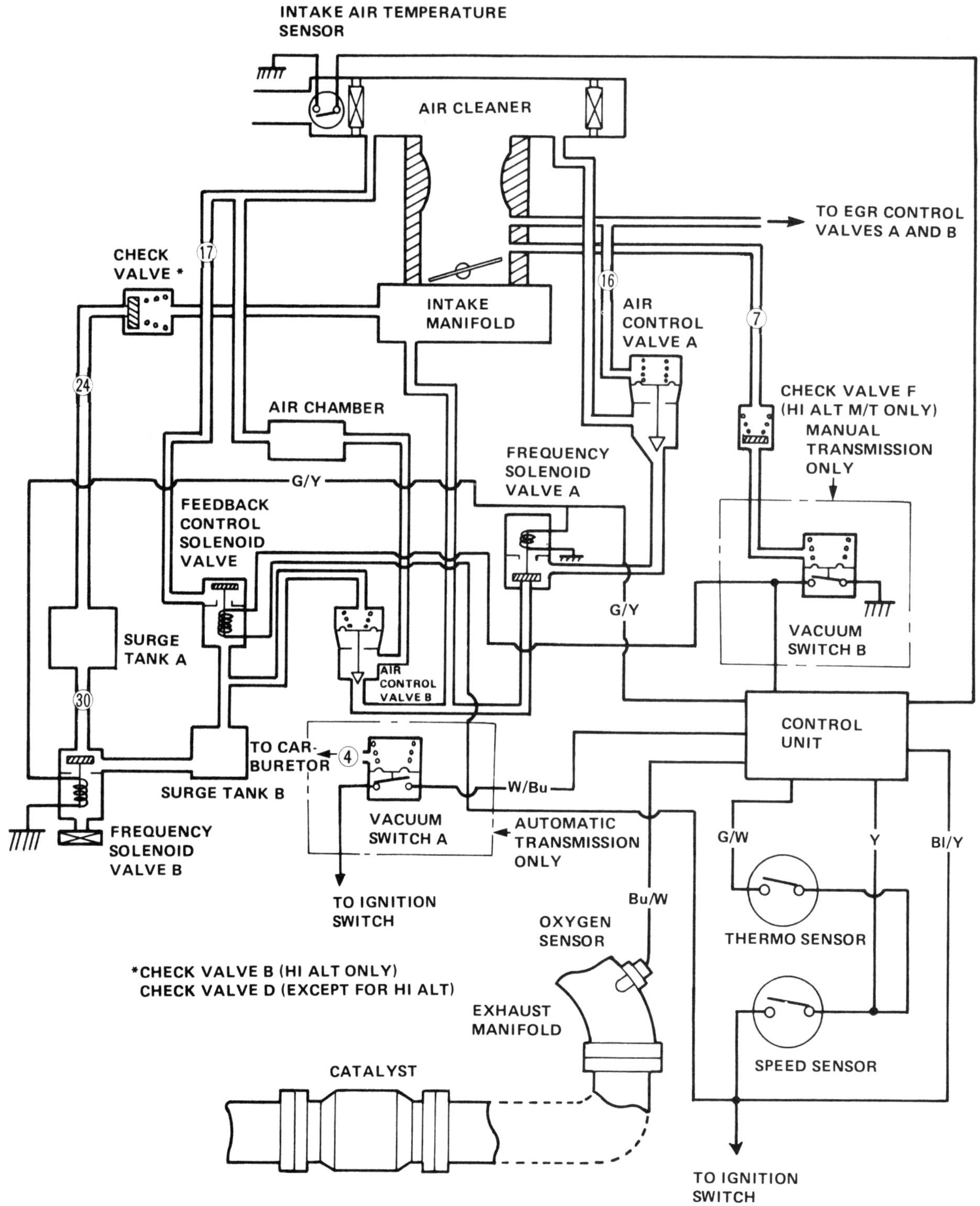

Feedback control system schematic

PROGRAMMED FUEL INJECTION (PGM-FI)

HONDA ACCORD SEI AND CIVIC CRX-Si

The new Honda programmed fuel injection system is made up of three sub-systems; Electronic Control, Air Intake and Fuel. The new PGM-FI system uses engine speed and absolute pressure in the manifold to provide the correct air-fuel ration and also to determine the amount of fuel to be injected. There is an eight-bit microprocessor and various sensors which provide the most accurate control of the air-fuel mixture under any operating conditions. The fuel injection on this system is a type of sequential injection. Which at the precise time a piston is on intake stroke, will inject fuel into the correct intake manifold runner.

The air intake system supplies the air for all the engine needs. The system is made up of the air cleaner, air intake pipe, throttle body, idle control system, fast idle mechanism and the intake manifold. The system incorporates a resonator in the air intake tub, to provide additional silencing as air is drawn into the system. The throttle body is a two barrel side draft type with a primary air horn at the top. The throttle valves and the air horn walls in the lower portion of the throttle body are heated by engine coolant (this will prevent icing). A dashpot is added to the throttle body, on the models equipped with a manual transmission, to slow the movement of the throttle valve as it comes close to the fully closed position.

HONDA MODIFICATIONS AND CORRECTIONS

1985 ACCORD

All Models

PROBLEM:

Some of these models are stalling at idle, when the engine is at normal operating temperature.

CAUSE:

The probable cause of this problem is the slow mixture cut-off solenoid is sticking.

CORRECTION:

1. Run the engine until it reaches normal operating temperature and let it idle.
2. Switch the ignition switch on and off ten times without allowing the engine to die until the tenth time.
3. Start the engine without touching the throttle. If the engine will not start without touching the throttle, the primary slow mixture cut-off solenoid is bad and should be removed and replaced. If the engine does start, go on to the next step.

NOTE: To be able to perform this next step a dummy solenoid is needed. Use a failed solenoid that has had its plunger cut off flush with its body.

4. Remove the primary slow mixture cut-off solenoid from the vehicle and put the dummy solenoid in its place.
5. Test drive the car. If the engine no longer stalls, the original solenoid is bad and should be replaced with the updated model.

NOTE: The diagnostic procedure used in steps 1 to 3 will identify most bad solenoids. Use Steps 4 and 5 only to verify the condition of an apparently 'good' solenoid.

1985 PRELUDE

PROBLEM:

Some Preludes have been experiencing hesitation on acceleration before the engine has reached normal operating temperature.

CORRECTION:

Install a cold driveability kit (available from the manufacturer), so as to hold full vacuum advance when the engine is cold.

1. Disconnect the two connectors from the side of emissions control box number one, which is located on the right side of the firewall.
2. Disconnect the number 4 hose from the T fitting near the intake duct, so it will not interfere with removing the control box.
3. Loosen the two bolts that hold the control box to the body. Then lift the box up and forward.
4. Remove the rubber insulator from the end of the lower bracket and set it aside, so it won't be lost if it falls off later.
5. Remove the four screws from the back of the control box and the one screw from the front, then remove the cover.
6. Detach the vacuum control solenoid valve from its mount, by removing the retaining screw.
7. Intall the new hose from the cold driveability kit in the control box. Be sure not to kink any of the hoses in the control box.
8. Re-attach the solenoid valve to its mount. Then reinstall the cover and the five screws. Put the rubber insulator back on the end of the lower bracket.
9. Reinstall the control box and tighten its two mounting bolts. Reconnect the number 4 hose and the two connectors.

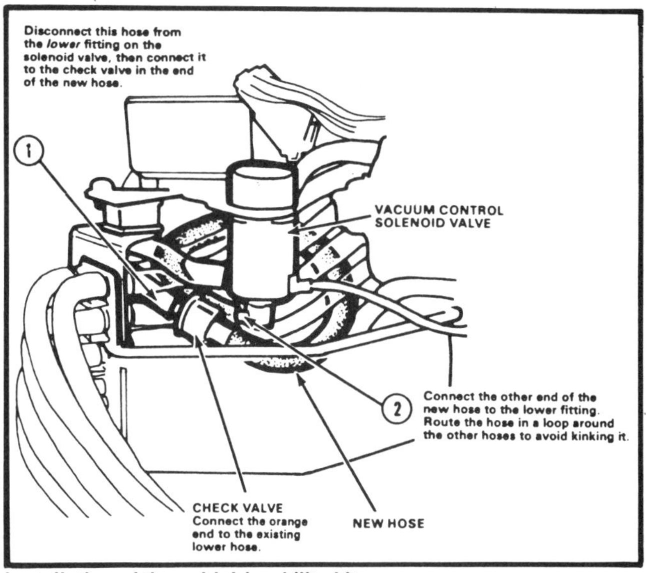

Installation of the cold driveability kit

Interconnect Diagram

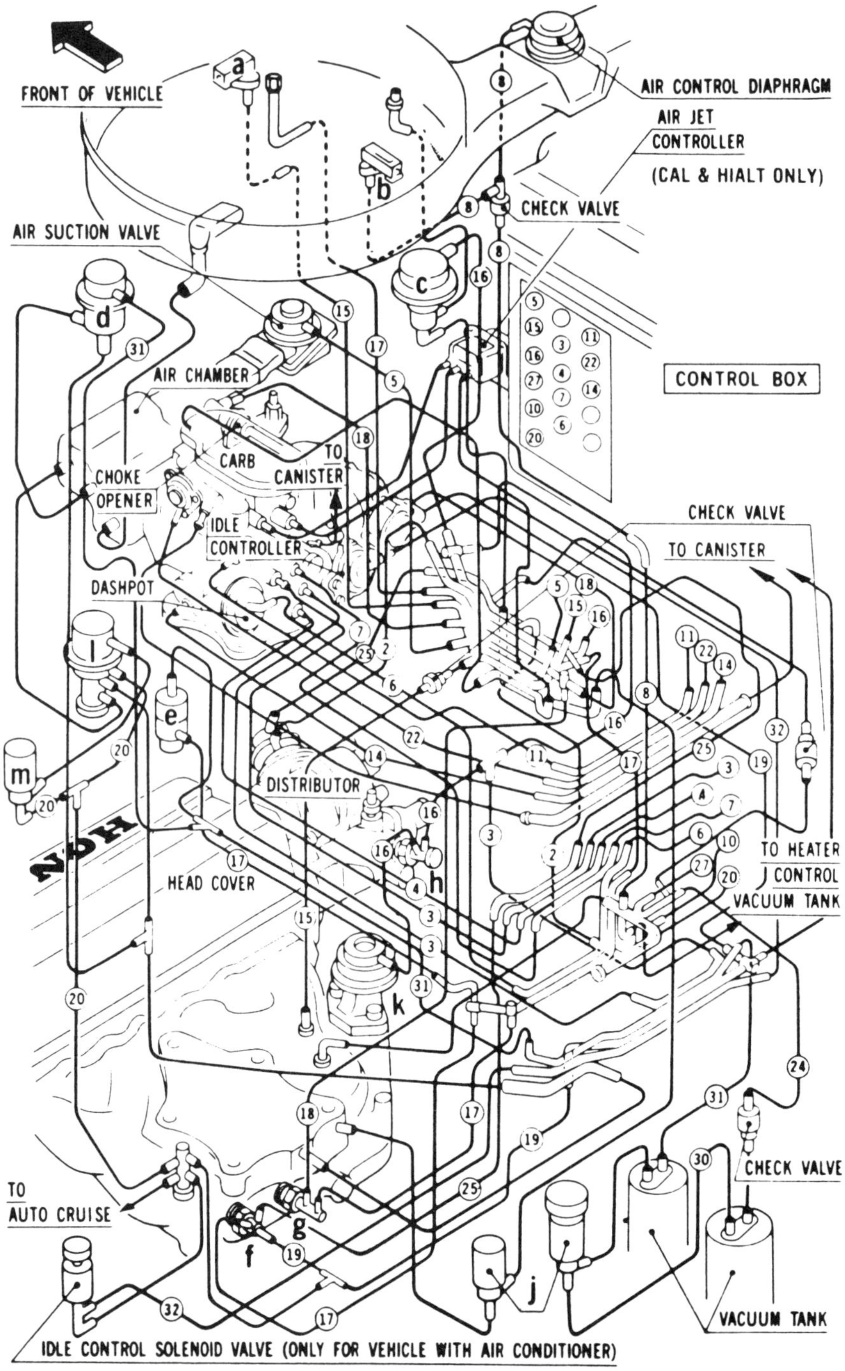

a. AIR BLEED VALVE A
b. AIR BLEED VALVE
c. AIR CONTROL VALVE A
d. AIR CONTROL VALVE B
e. PX LEAK SOLENOID VALVE
f. THERMO VALVE A
g. THERMO VALVE B
h. THERMO VALVE C
j. FREQUENCY SOLENOID VALVE
k. E.G.R VALVE
l. SHOT AIR VALVE
m. SHOT AIR VALVE CONTROL SOLENOID

85 Accord carbureted

Interconnect Diagram (cont'd)

Control Box

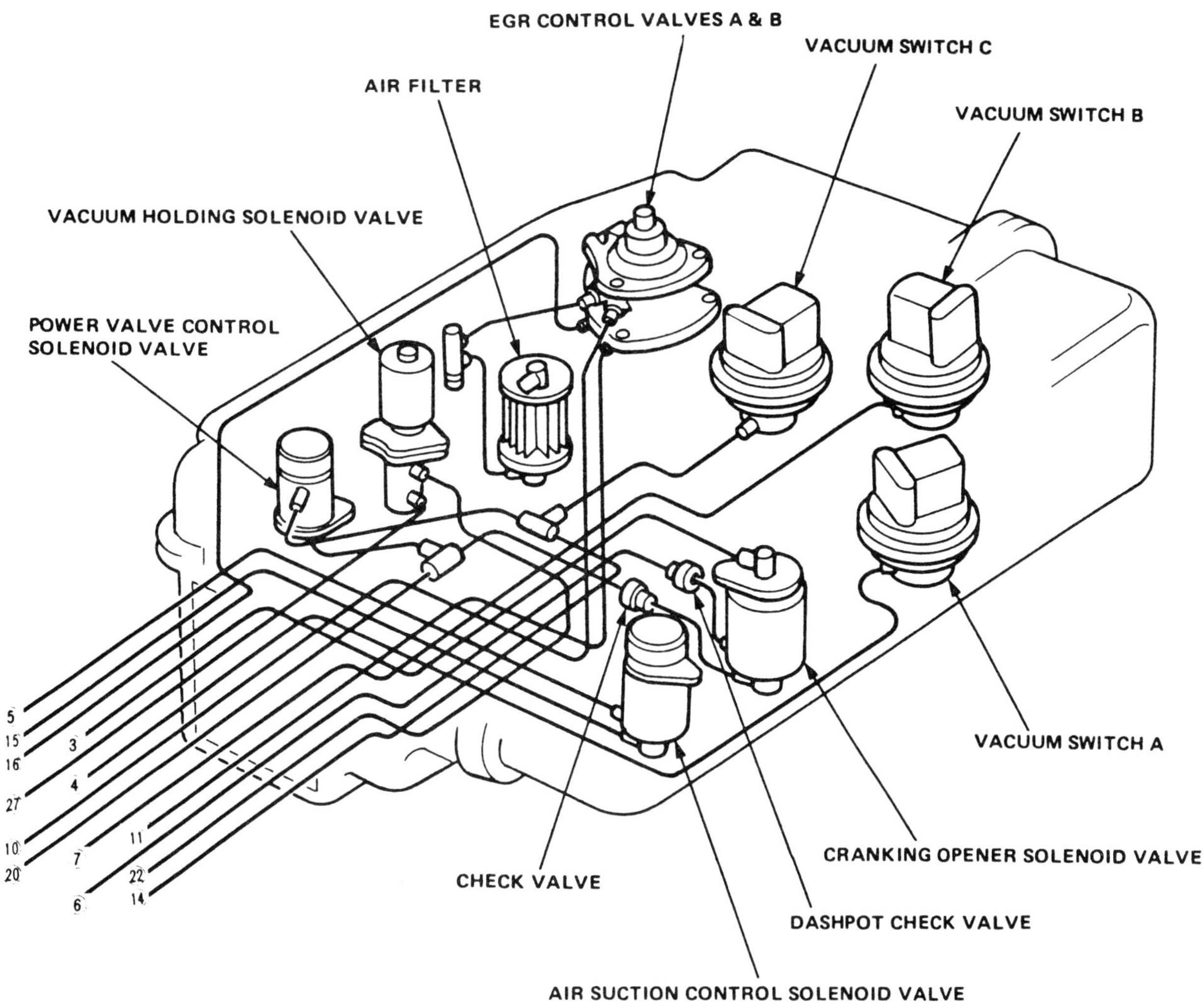

85 Accord

Vacuum and Electrical Connections

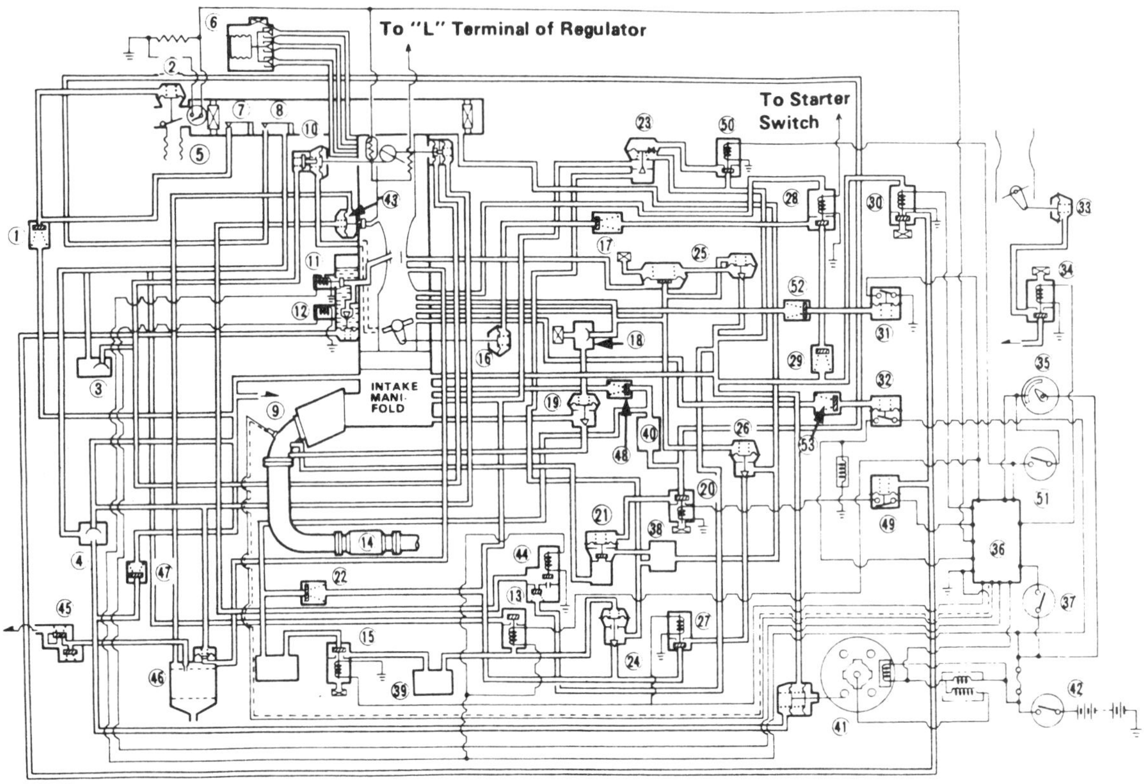

1. CHECK VALVE (INTAKE AIR TEMP.)
2. AIR CONTROL DIAPHRAGM
3. THERMOVALVE B
4. THERMOVALVE A
5. INTAKE AIR TEMPERATURE SENSOR
6. AIR JET CONTROLLER (CAL AND HI ALT ONLY)
7. AIR BLEED VALVE
8. AIR BLEED VALVE A
9. OXYGEN SENSOR
10. CHOKE OPENER
11. PRIMARY MAIN FUEL CUT-OFF SOLENOID VALVE
12. PRIMARY SLOW MIXTURE CUT-OFF SOLENOID VALVE
13. FEEDBACK CONTROL SOLENOID VALVE
14. CATALYTIC CONVERTER
15. FREQUENCY SOLENOID VALVE B
16. THROTTLE CONTROLLER
17. DASHPOT CHECK VALVE
18. THERMOVALVE C
19. EGR VALVE
20. AIR SUCTION CONTROL SOLENOID VALVE
21. AIR SUCTION VALVE
22. CHECK VALVE B (49 ST AND HI ALT ONLY)
23. ANTI-AFTERBURN VALVE
24. AIR CONTROL VALVE B
25. EGR CONTROL VALVES A AND B
26. AIR CONTROL VALVE A
27. FREQUENCY SOLENOID VALVE A
28. CRANKING OPENER SOLENOID VALVE
29. CHECK VALVE A
30. POWER VALVE CONTROL SOLENOID VALVE
31. VACUUM SWITCH A
32. VACUUM SWITCH B
33. IDLE CONTROLLER
34. IDLE CONTROL SOLENOID VALVE
35. SPEED SENSOR
36. CONTROL UNIT
37. CLUTCH SWITCH FOR MANUAL
NEUTRAL SWITCH FOR AUTOMATIC
38. AIR CHAMBER
39. SURGE TANK A
40. SURGE TANK B
41. DISTRIBUTOR
42. IGNITION SWITCH
43. AIR VENT CUT OFF DIAPHRAGM
44. VACUUM HOLDING SOLENOID VALVE
45. TWO-WAY VALVE
46. CHARCOAL CANISTER
47. CHECK VALVE C
48. CHECK VALVE D
49. VACUUM SWITCH C
50. ANTI-AFTERBURN CONTROL SOLENOID VALVE
51. THERMOSENSOR
52. CHECK VALVE E (HI ALT A/T ONLY)
53. CHECK VALVE F (HI ALT M/T ONLY)

85 Accord carbureted

Interconnect Diagram

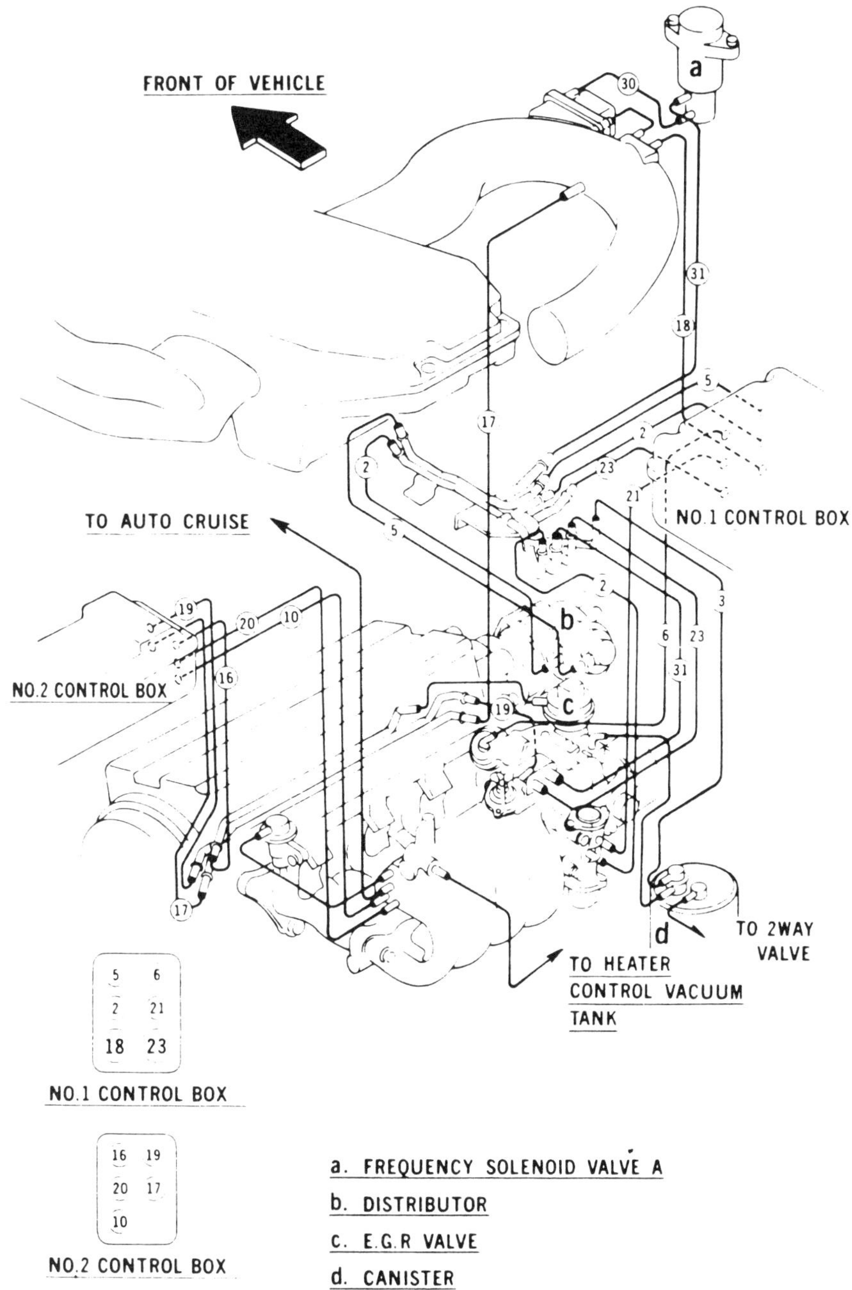

a. FREQUENCY SOLENOID VALVE A
b. DISTRIBUTOR
c. E.G.R VALVE
d. CANISTER

85 Accord fuel injection

Interconnect Diagram (cont'd)

Control Box #1

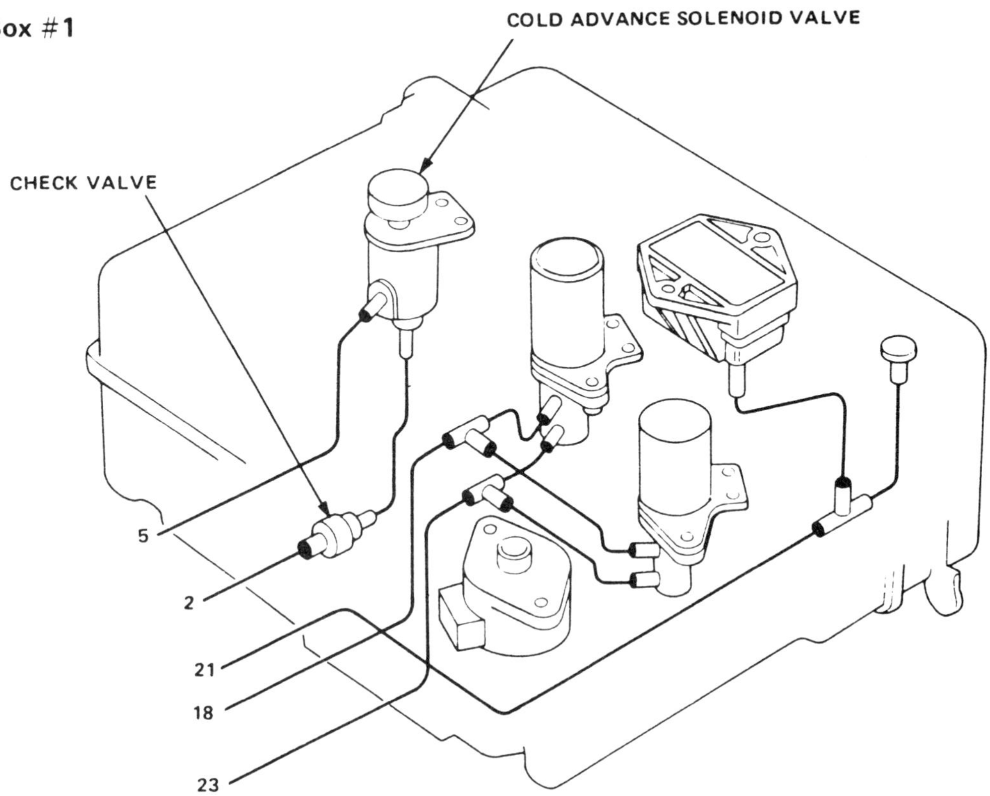

Control Box #2

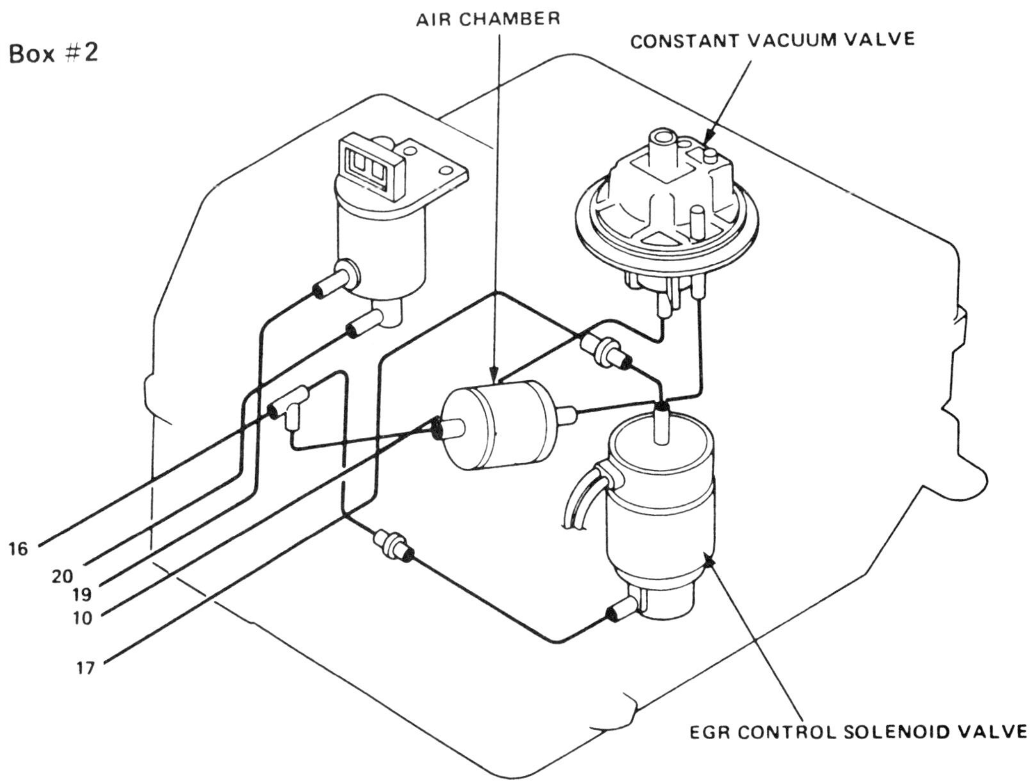

85 Accord fuel injection

Vacuum and Electrical Connections

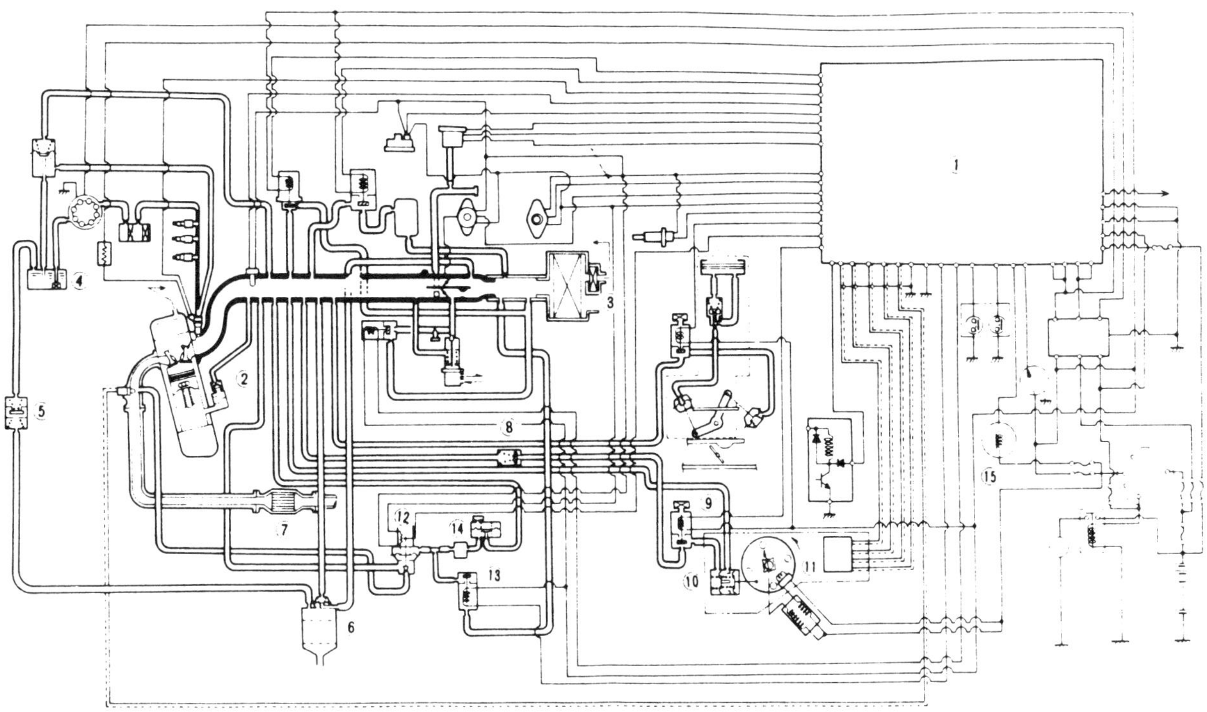

1 ELECTRONIC CONTROL UNIT
2 PCV VALVE
3 BLOW-BY FILTER
4 FUEL TANK
5 TWO-WAY VALVE
6 CHARCOAL CANISTER
7 CATALYTIC CONVERTER
8 CHECK VALVE
9 COLD ADVANCE SOLENOID VALVE
10 VACUUM CONTROLLER
11 DISTRIBUTOR
12 EGR VALVE LIFT SENSOR
13 EGR CONTROL SOLENOID VALVE
14 CONSTANT VACUUM CONTROL VALVE
15 PGM-FI WARNING LIGHT

85 Accord fuel injection

Interconnect Diagram

VACUUM HOSE ROUTING DIAGRAM — HONDA PRELUDE

ENGINE FAMILY IDENTIFICATION — EHN1.8VOFGF4, EHN1.8VOFHC3, DISPLACEMENT 112 in³
EVAPORATIVE FAMILY IDENTIFICATION — 85 FC
TRANSMISSION — 5 SPEED
— AUTOMATIC

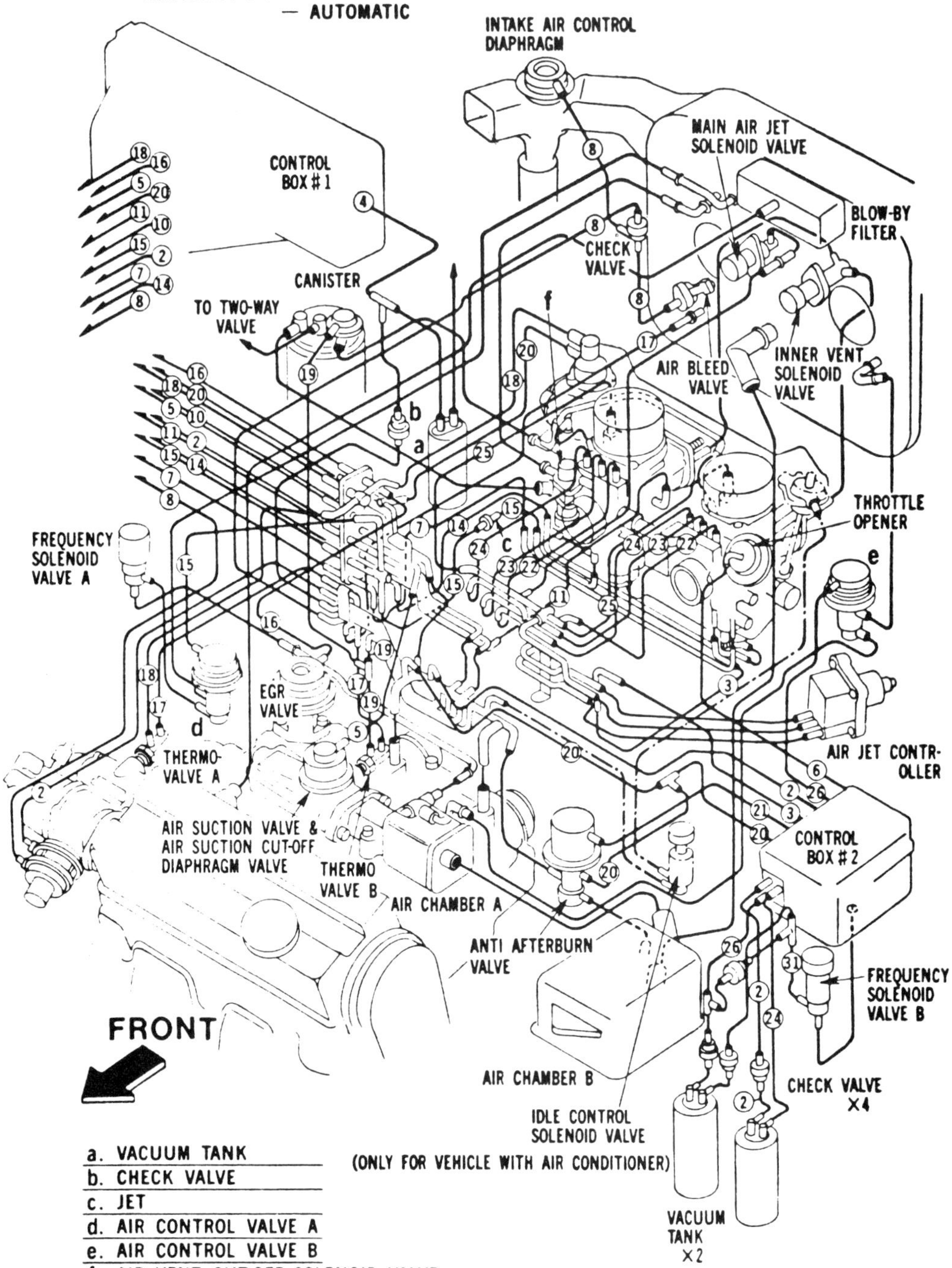

a. VACUUM TANK
b. CHECK VALVE
c. JET
d. AIR CONTROL VALVE A
e. AIR CONTROL VALVE B
f. AIR VENT CUT-OFF SOLENOID VALVE

•FORCE TUBES MORE THAN 1/2 in. (12 mm)

85 Prelude

Interconnect Diagram (cont'd)

Control Box ≠1

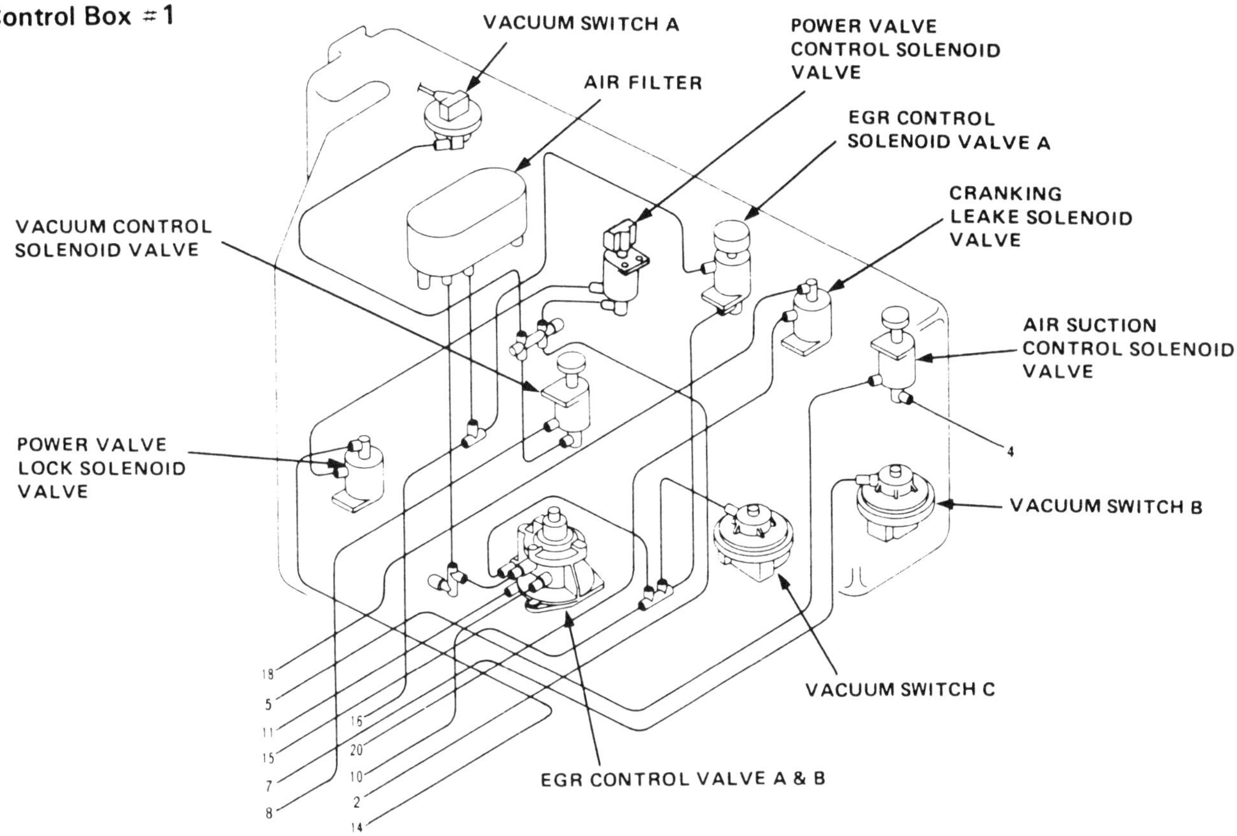

Control Box ≠2

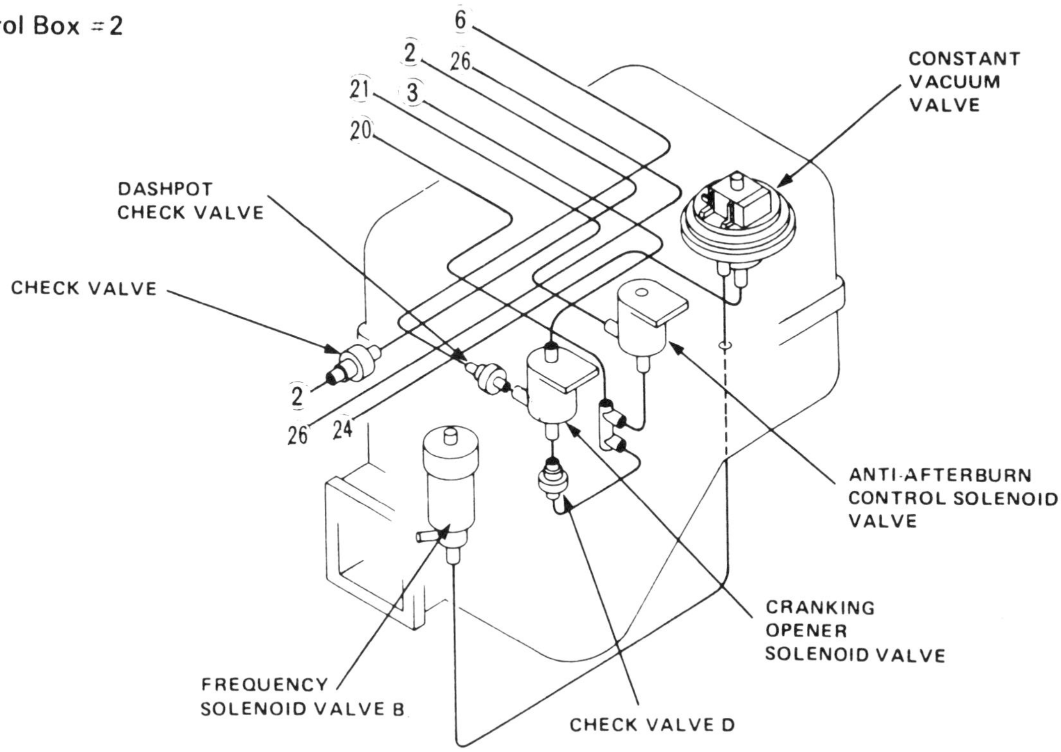

85 Prelude

— Vacuum and Electrical Connections

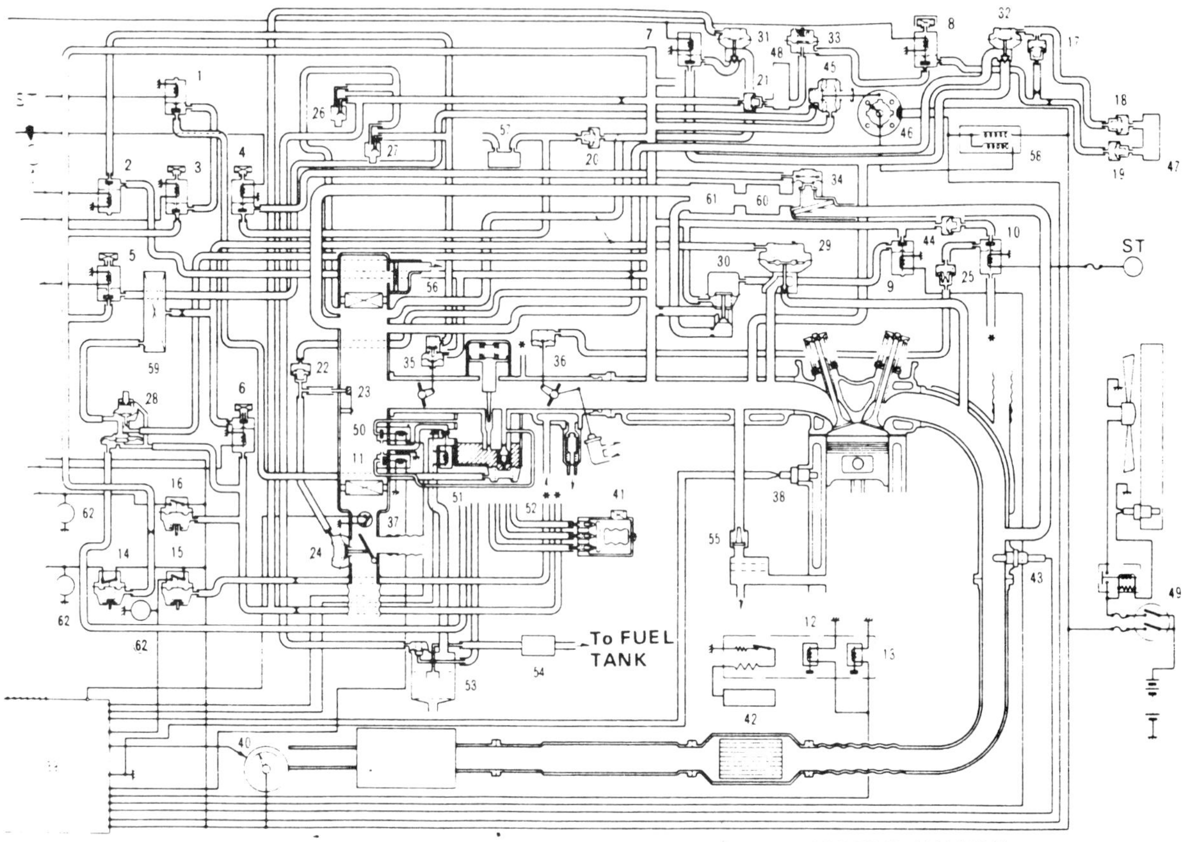

1 POWER VALVE LOCK SOLENOID VALVE
2 CRANKING LEAK SOLENOID VALVE
3 POWER VALVE CONTROL SOLENOID VALVE
4 AIR SUCTION CONTROL SOLENOID VALVE
5 VACUUM CONTROL SOLENOID VALVE
6 EGR CONTROL SOLENOID VALVE A
7 FREQUENCY SOLENOID VALVE A
8 FREQUENCY SOLENOID VALVE B
9 ANTI-AFTERBURN CONTROL SOLENOID VALVE
10 CRANKING OPENER SOLENOID VALVE
11 MAIN AIR JET CONTROL SOLENOID VALVE
12 RIGHT PRIMARY SLOW MIXTURE CUT-OFF SOLENOID VALVE
13 LEFT PRIMARY SLOW MIXTURE CUT-OFF SOLENOID VALVE
14 VACUUM SWITCH A
15 VACUUM SWITCH B
16 VACUUM SWITCH C
17 CHECK VALVE A
18 CHECK VALVE B
19 CHECK VALVE C
20 CHECK VALVE F
21 CHECK VALVE E
22 CHECK VALVE (INTAKE AIR TEMP.)
23 AIR BLEED VALVE
24 INTAKE AIR CONTROL DIAPHRAGM
25 DASHPOT CHECK VALVE
26 THERMOVALVE B
27 THERMOVALVE A
28 EGR CONTROL VALVES A & B
29 EGR VALVE
30 ANTI-AFTERBURN VALVE
31 AIR CONTROL VALVE A
32 AIR CONTROL VALVE B
33 CONSTANT VACUUM VALVE
34 AIR SUCTION VALVE
35 CHOKE OPENER
36 THROTTLE CONTROLLER
37 INTAKE AIR TEMPERATURE SWITCH
38 THERMOSENSOR
39 CONTROL UNIT
40 SPEED SENSOR
41 AIR JET CONTROLLER
42 CATALYTIC CONVERTER
43 OXYGEN SENSOR
44 CHECK VALVE D
45 DISTRIBUTOR VACUUM ADVANCE
46 DISTRIBUTOR
47 SURGE TANK A
48 SURGE TANK B
49 IGNITJON SWITCH
50 INNER VENT SOLENOID VALVE
51 AIR VENT CUT-OFF SOLENOID VALVE
52 POWER VALVE
53 CANISTER
54 TWO-WAY VALVE
55 PCV VALVE
56 BLOW-BY FILTER
57 VACUUM TANK
58 IGNITION COIL
59 AIR FILTER
60 AIR CHAMBER A
61 AIR CHAMBER B
62 AUXILIARY COIL

85 Prelude

Interconnect Diagram

Car / Trans	CAL		49 ST and HI ALT		
	1300/ 1500HF	1500	1300/ 1500HF	1300/ 1500HF	1500
Manual	■		■		
Automatic					

1300 (49ST and CAL)

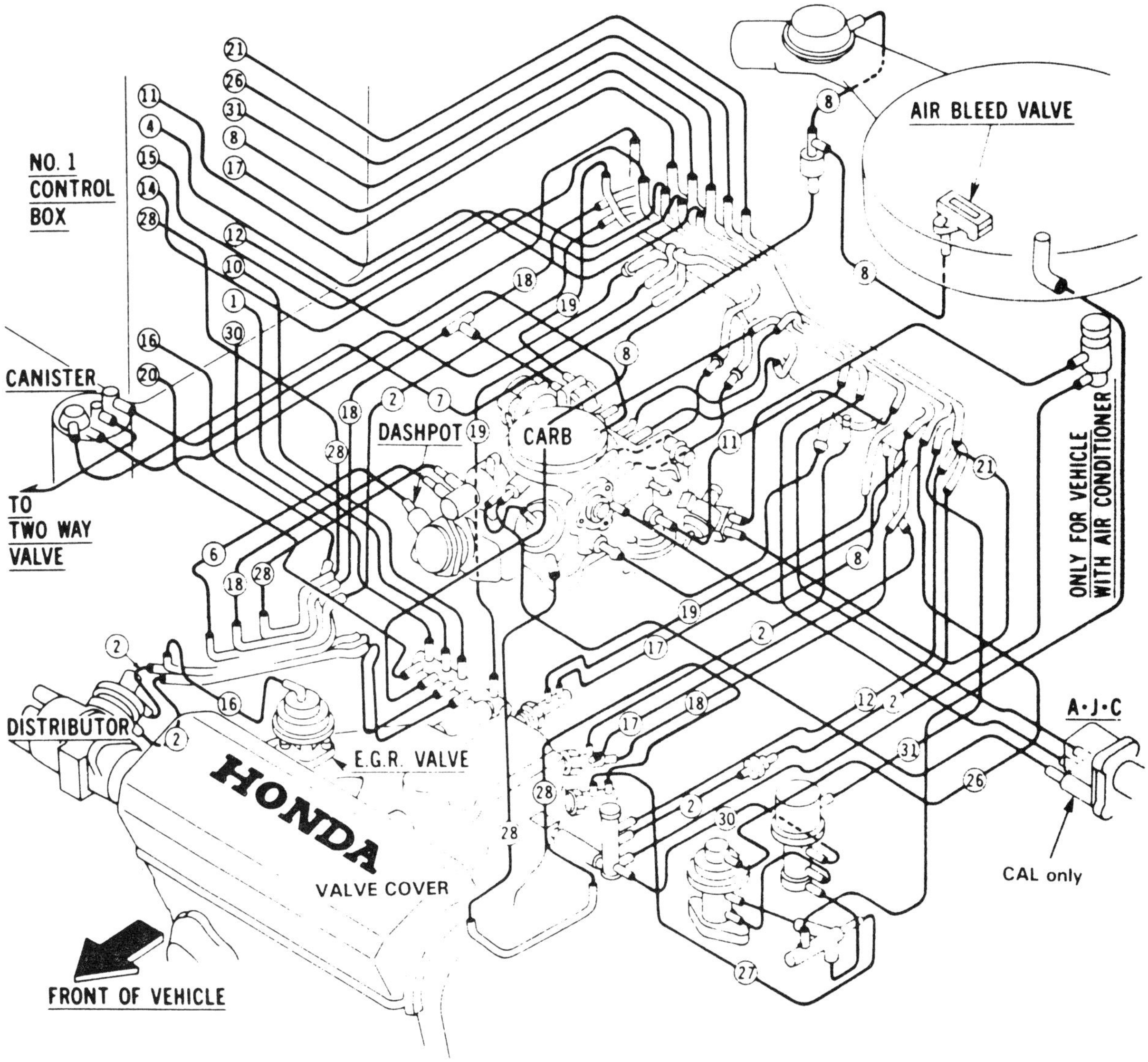

Interconnect Diagram (cont'd)

Car / Trans	CAL		49 ST and HI ALT		
	1300/ 1500HF	1500	1300/ 1500HF	1300/ 1500HF	1500
Manual				■	
Automatic					

1300 (HIGH ALTITUDE)

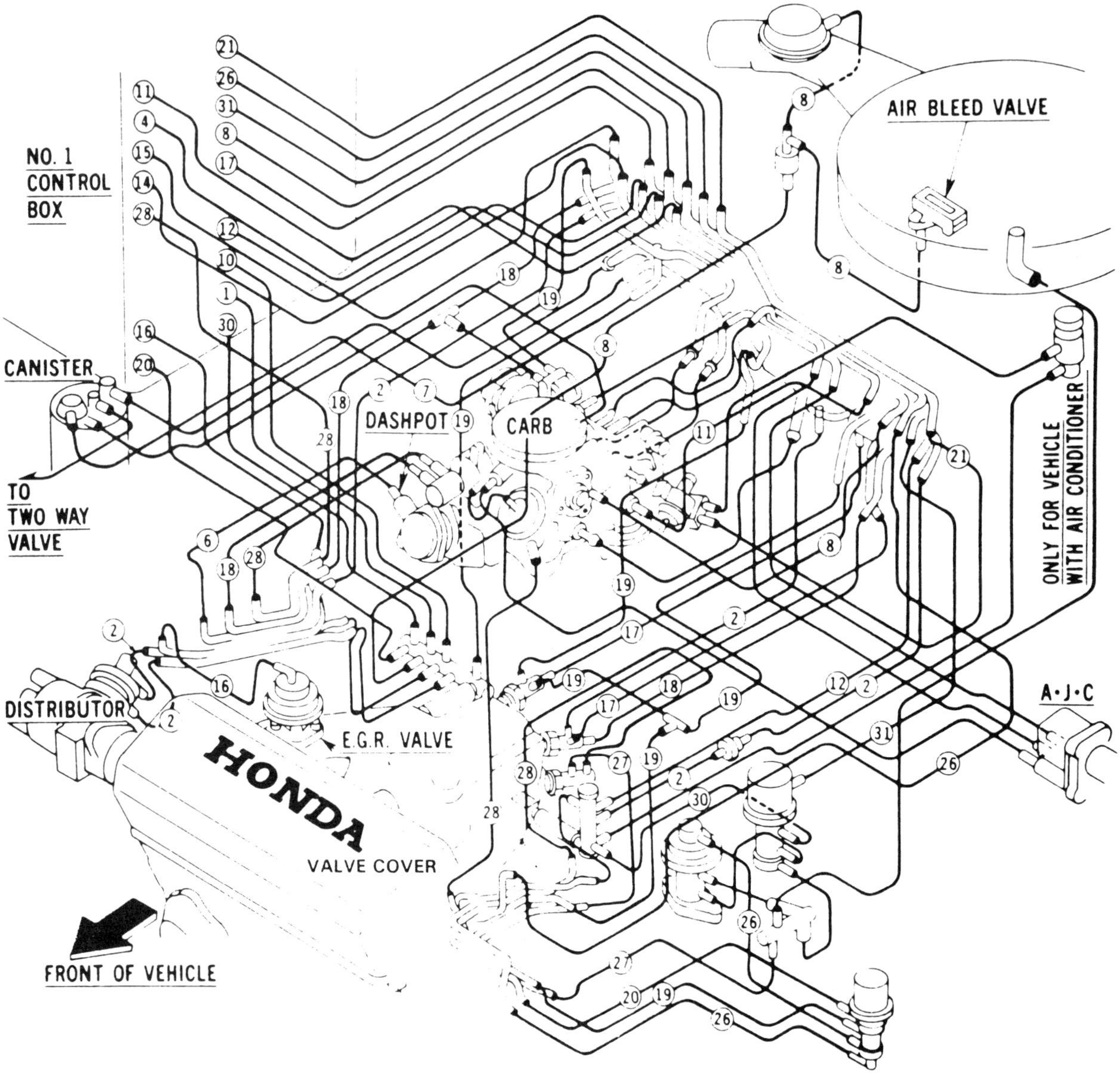

85 Civic

Interconnect Diagram

Car / Trans	CAL		49 ST and HI ALT		
	1300/ 1500HF	1500	1300/ 1500HF	1300/ 1500HF	1500
Manual			■		
Automatic					

1500HF (49ST)

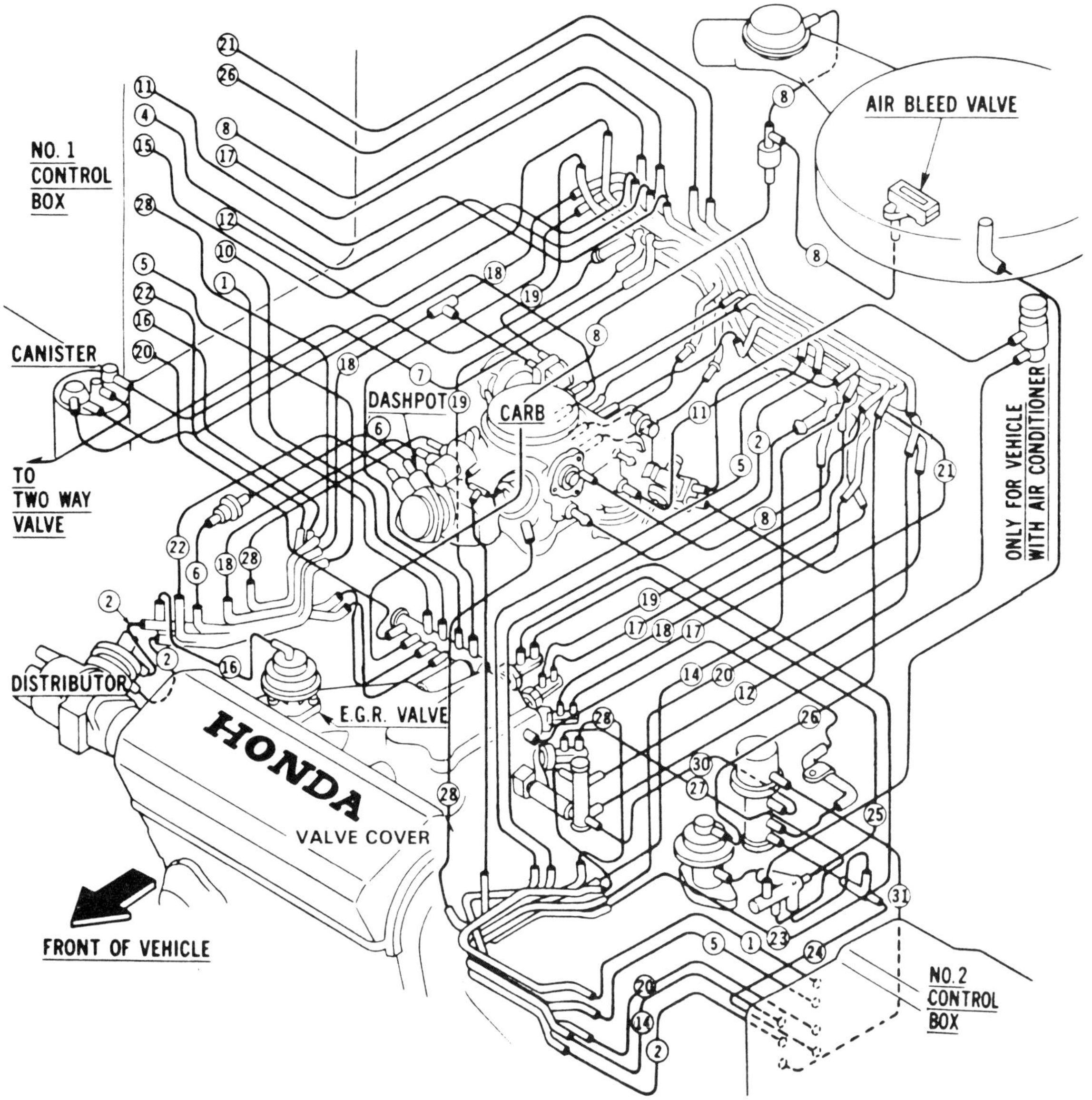

85 Civic

Interconnect Diagram (cont'd)

Trans \ Car	CAL		49 ST and HIALT		
	1300/ 1500HF	1500	1300/ 1500HF	1300/ 1500HF	1500
Manual	■				
Automatic					

1500HF (CAL)

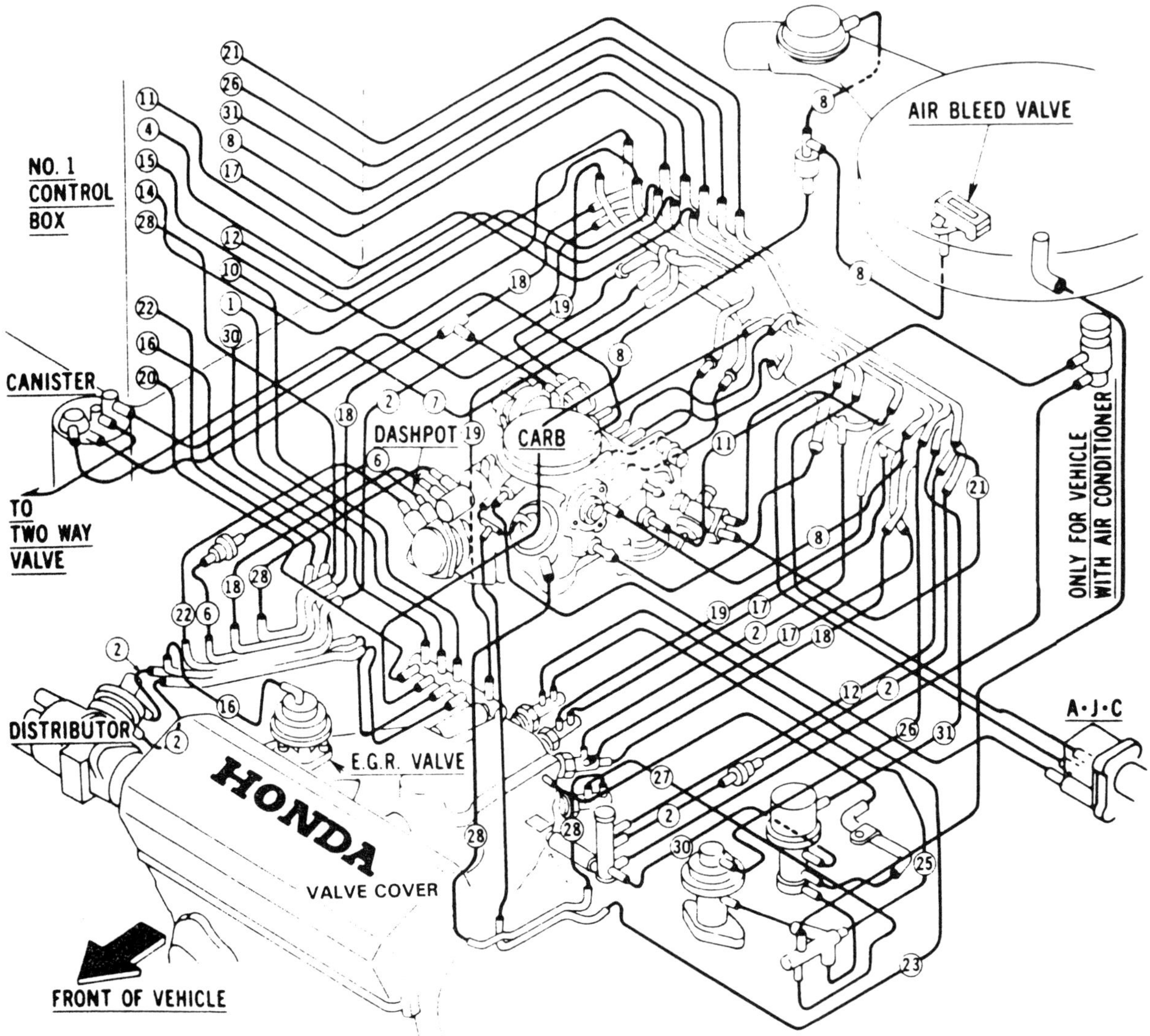

85 Civic

Interconnect Diagram

Trans \ Car	CAL		49 ST	and HI ALT	
	1300/ 1500HF	1500	1300/ 1500HF	1300/ 1500HF	1500
Manual				■	
Automatic					

1500HF (HIGH ALTITUDE)

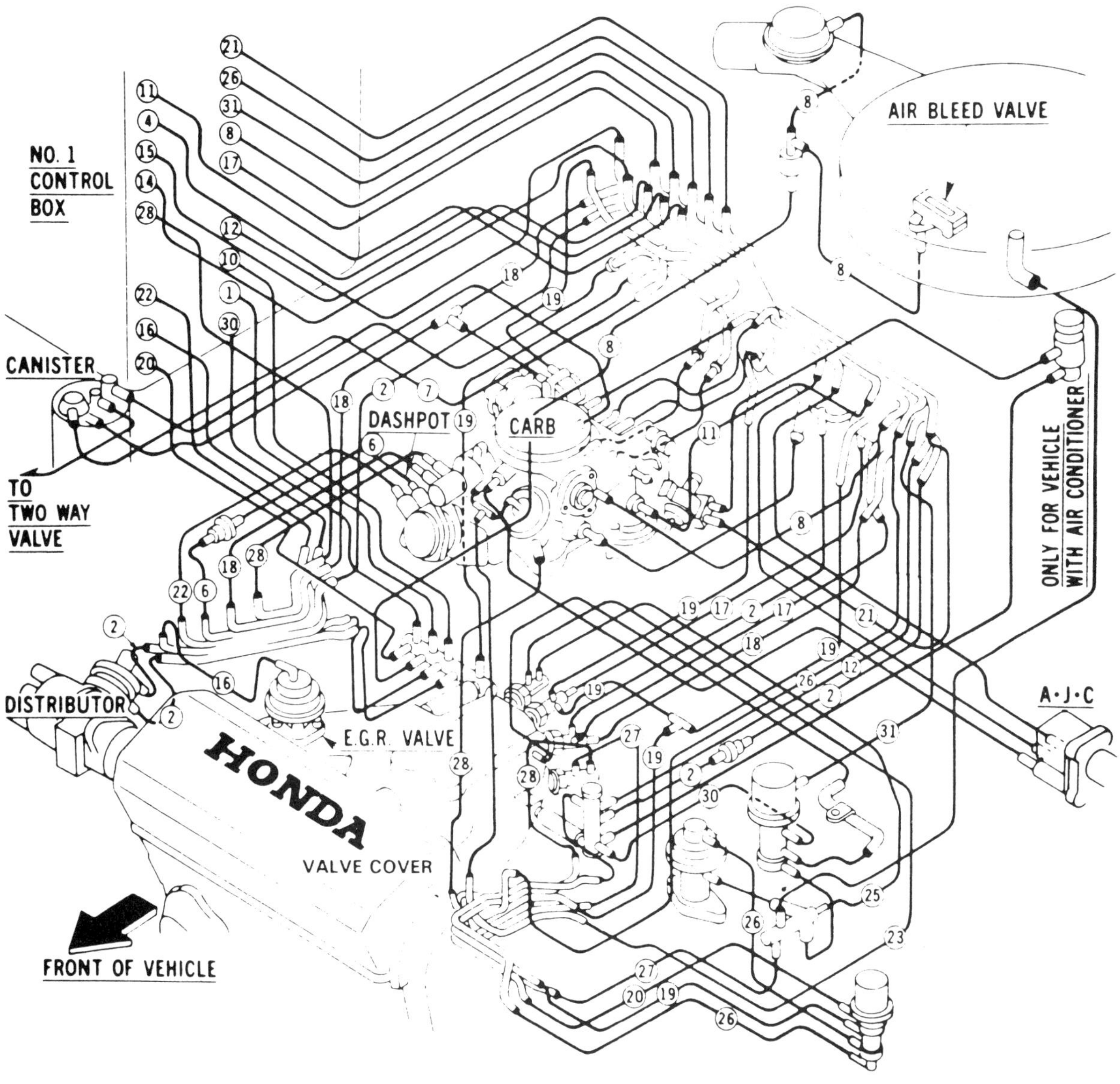

85 Civic

Interconnect Diagram (cont'd)

Car / Trans	CAL		49 ST and HI ALT		
	1300/ 1500HF	1500	1300/ 1500HF	1300/ 1500HF	1500
Manual					■
Automatic					

1500 M/T (49ST and HIGH ALTITUDE)

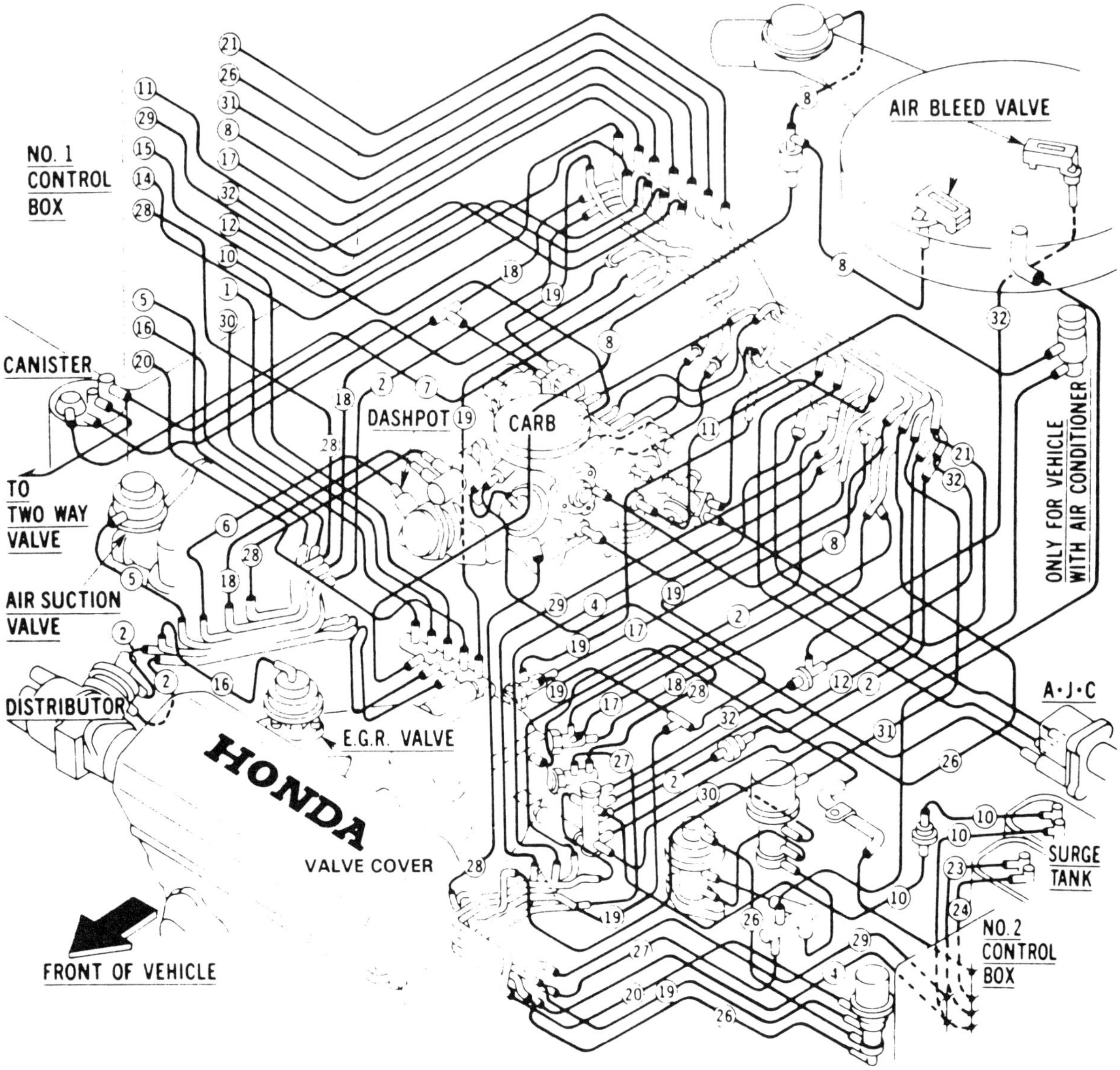

Interconnect Diagram

Car / Trans	CAL		49 ST	and HI ALT	
	1300/ 1500HF	1500	1300/ 1500HF	1300/ 1500HF	1500
Manual		■			
Automatic					

1500 M/T (CAL)

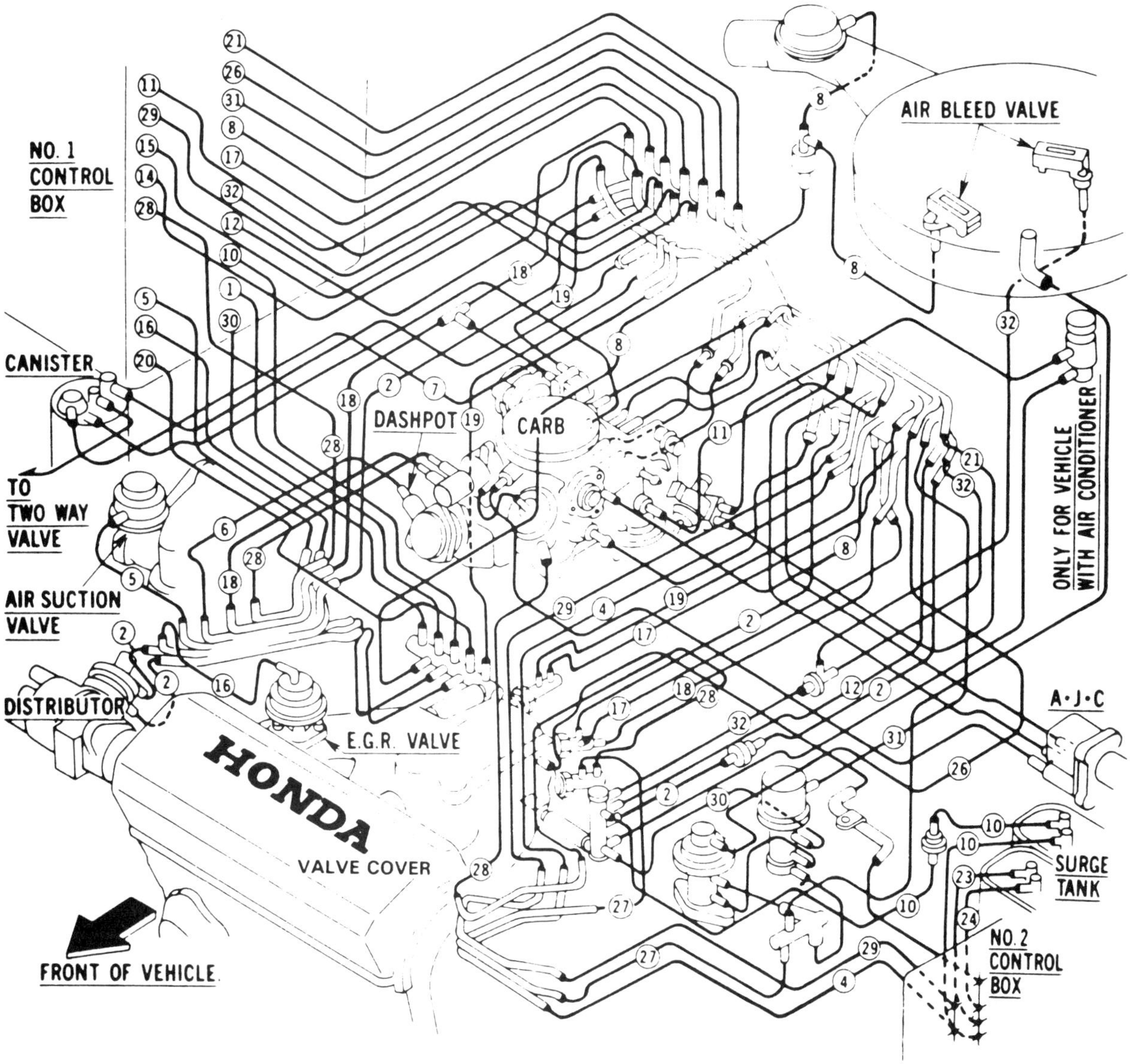

Interconnect Diagram (cont'd)

Trans \ Car	CAL 1300/1500HF	CAL 1500	49 ST and HI ALT 1300/1500HF	49 ST and HI ALT 1300/1500HF	49 ST and HI ALT 1500
Manual					
Automatic		*			*

*manual steering only

1500 A/T with manual steering

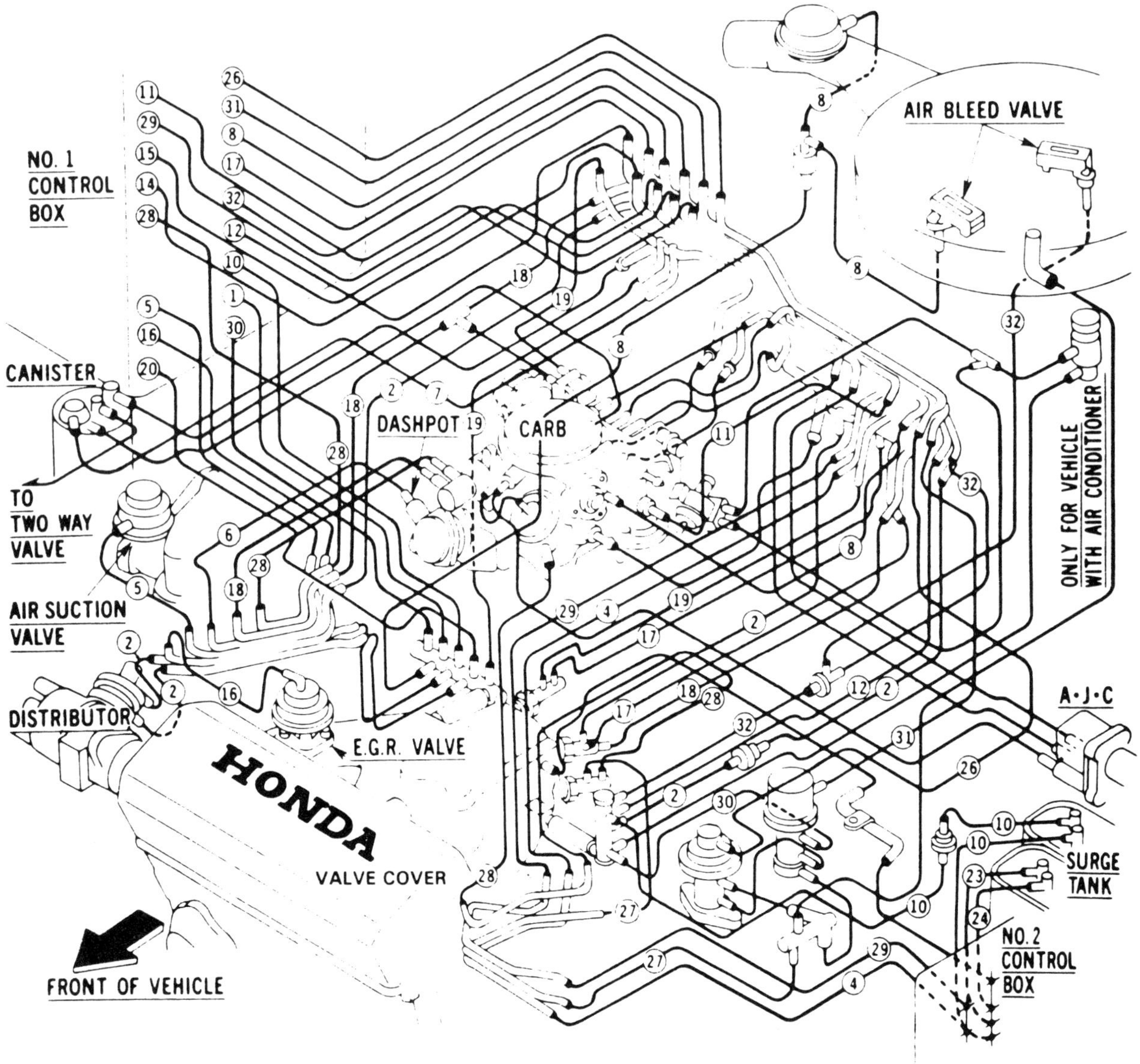

85 Civic

Interconnect Diagram

Car / Trans	CAL		49 ST	and HI ALT	
	1300/ 1500HF	1500	1300/ 1500HF	1300/ 1500HF	1500
Manual					
Automatic		*			*

*power steering only

1500 A/T with power steering

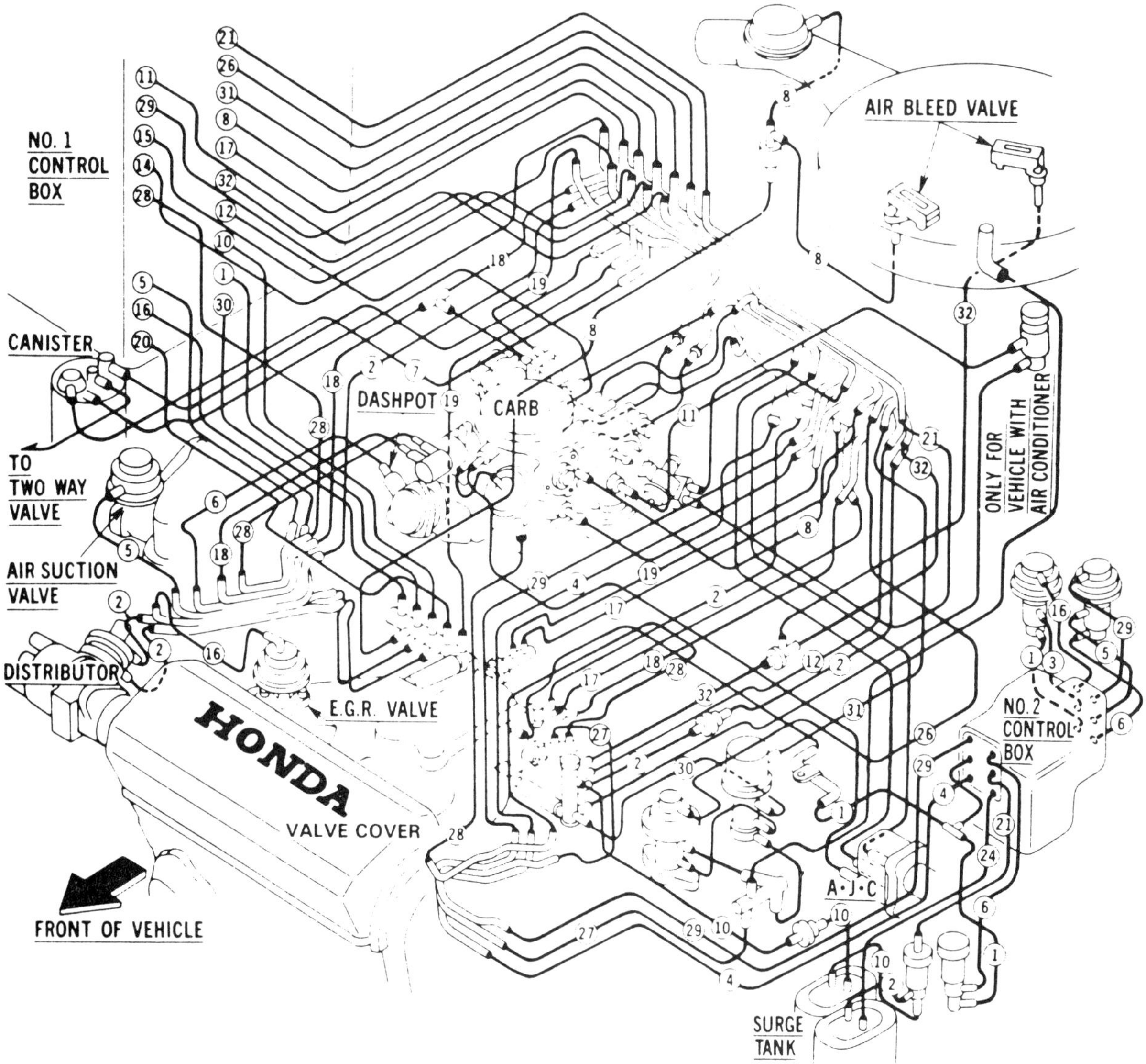

Interconnect Diagram (cont'd)

1300 and CAL/HI ALT 1500HF

Control Box #1

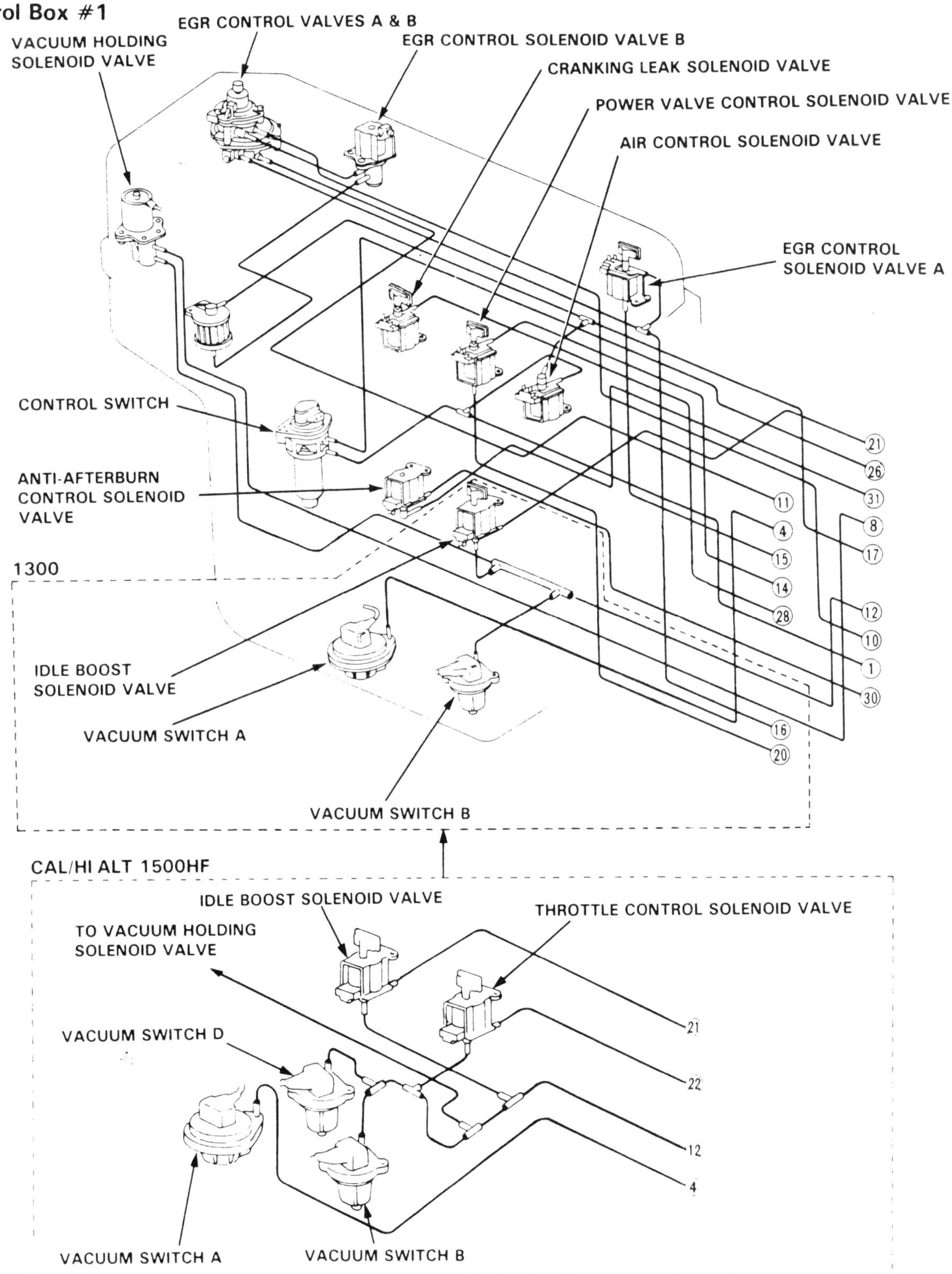

85 Civic

Interconnect Diagram

1500 except HF

Control Box #1

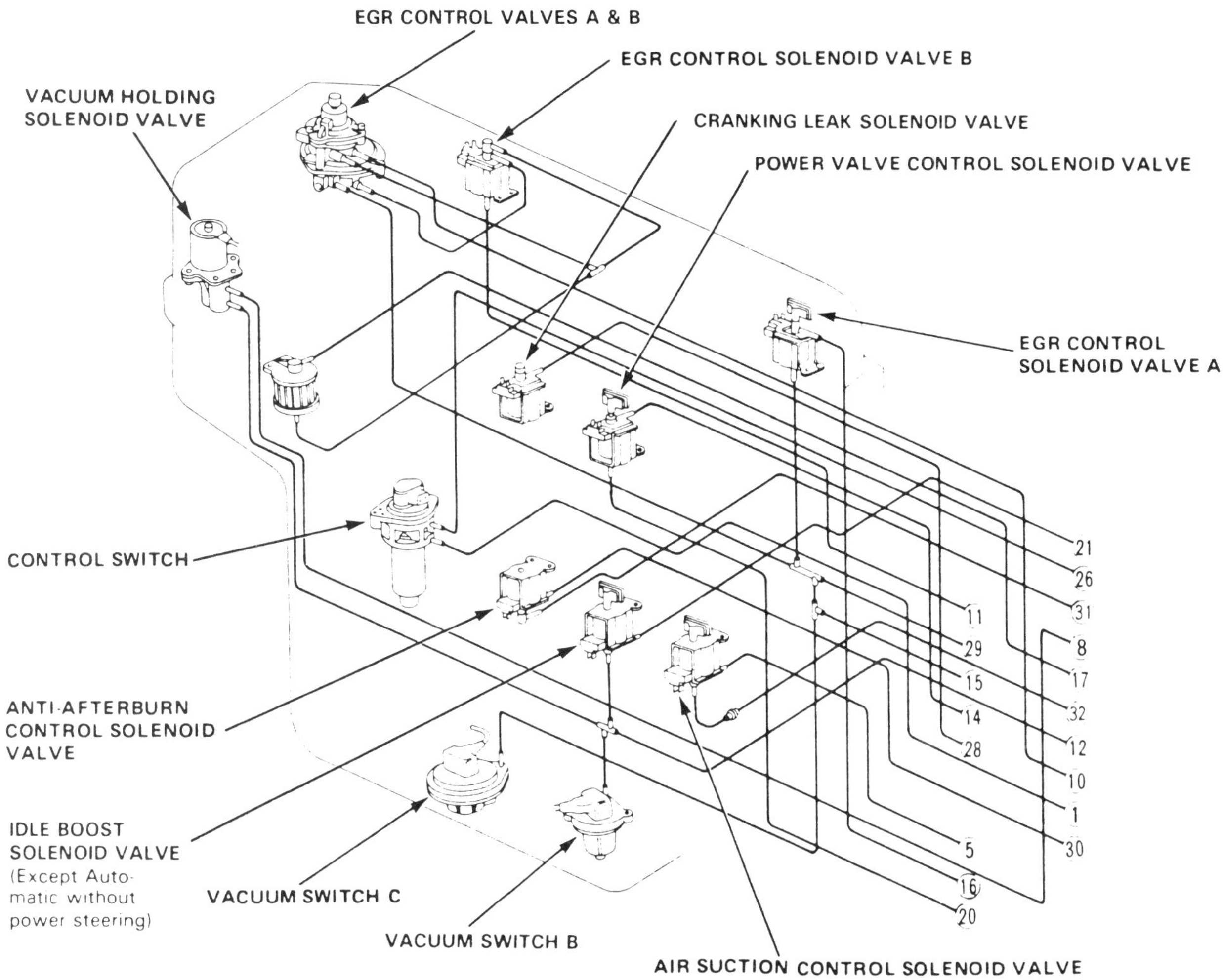

Interconnect Diagram (cont'd)

Control Box #2

Manual

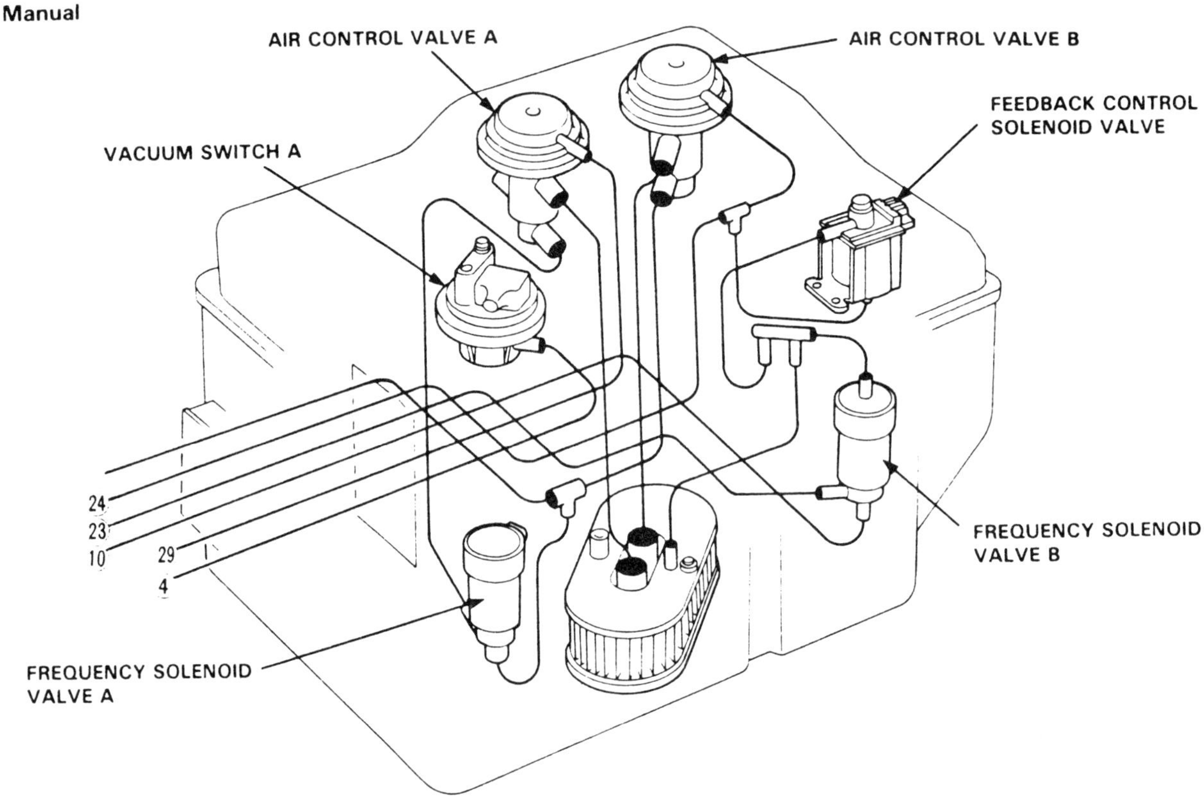

Automatic

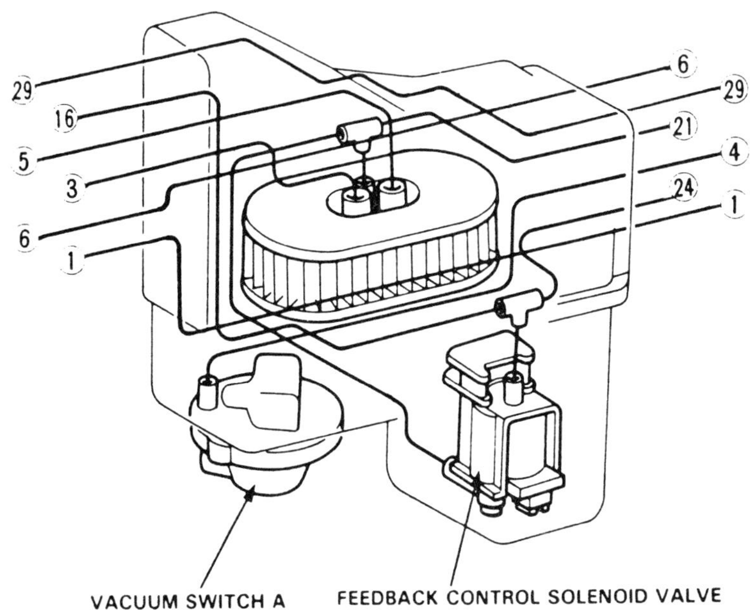

85 Civic

Interconnect Diagram

49 ST 1500HF

Control Box #1

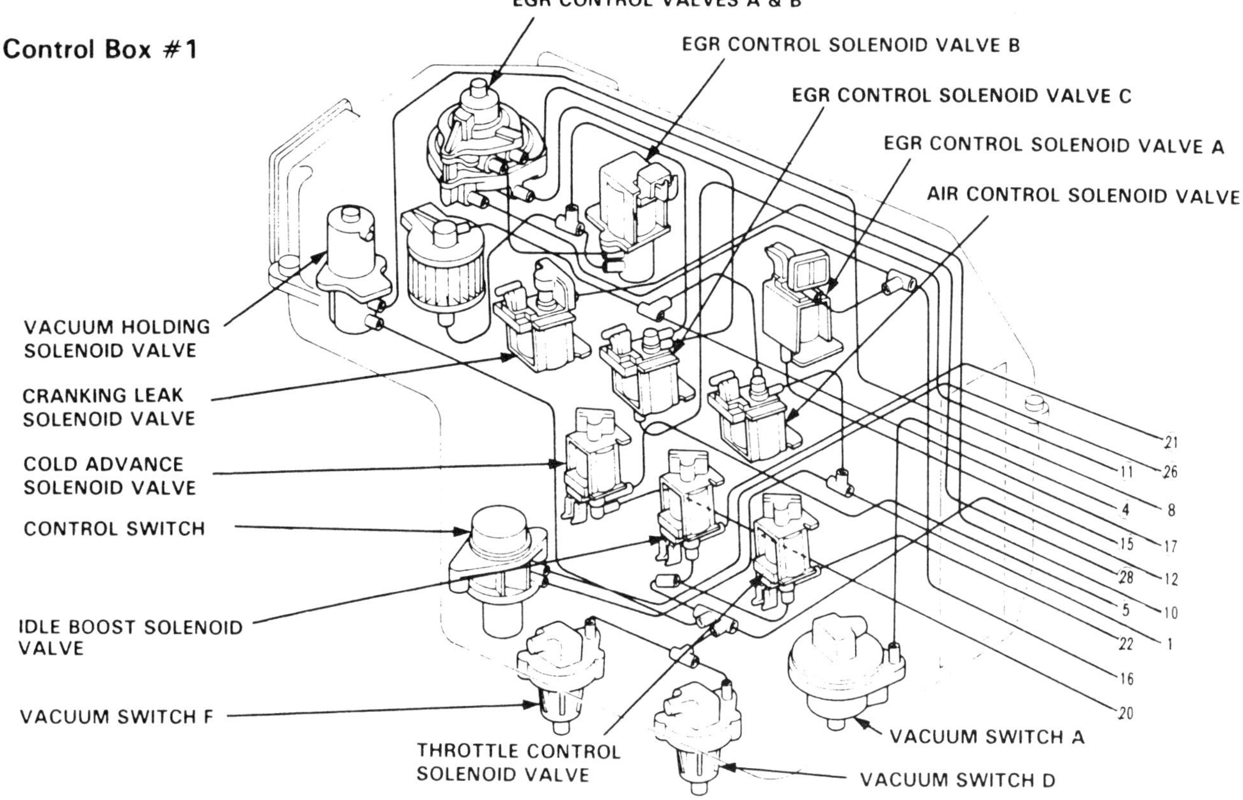

Control Box #2

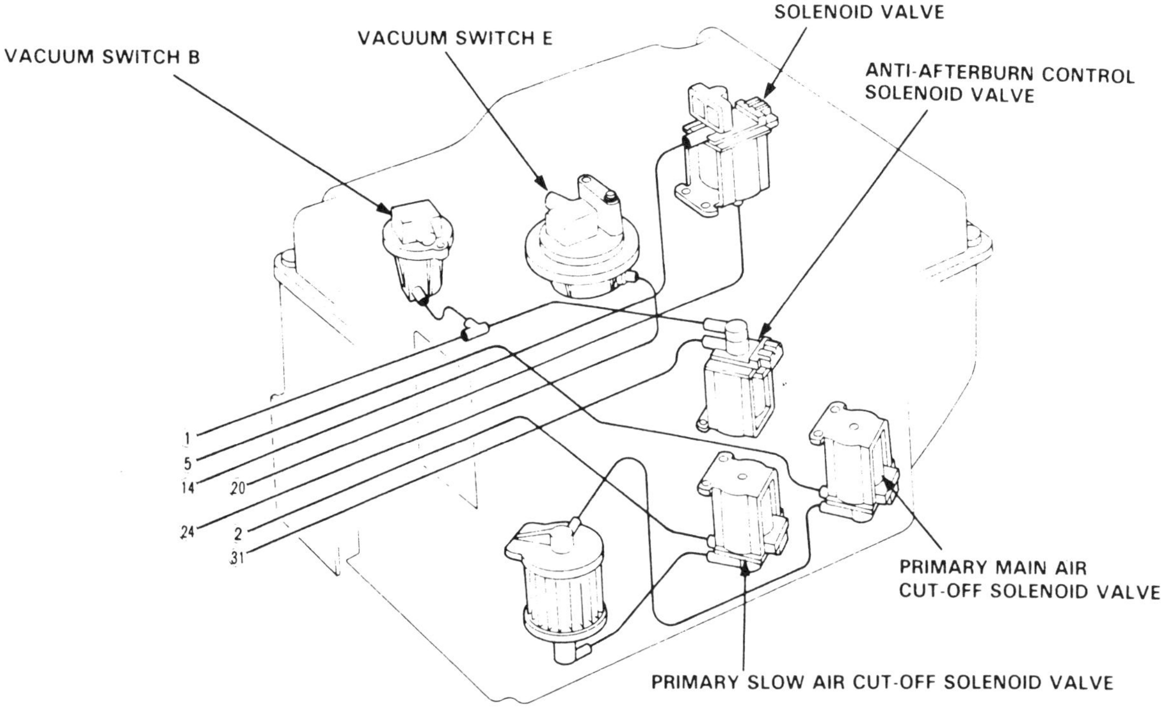

Vacuum and Electrical Connections

1300

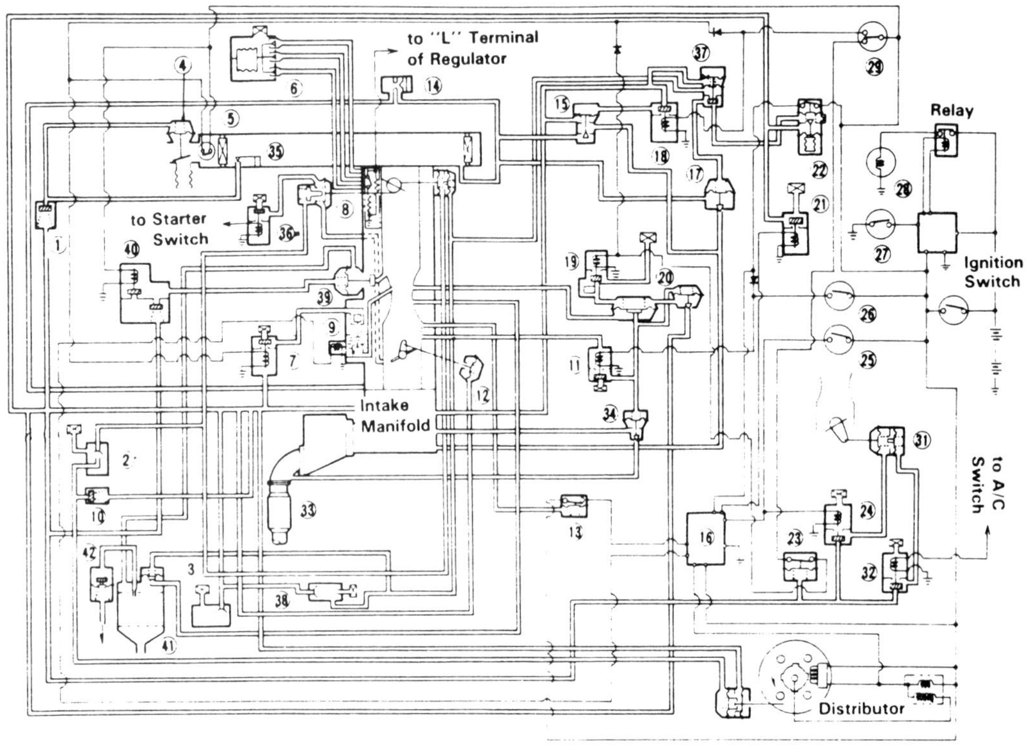

1500 except HF

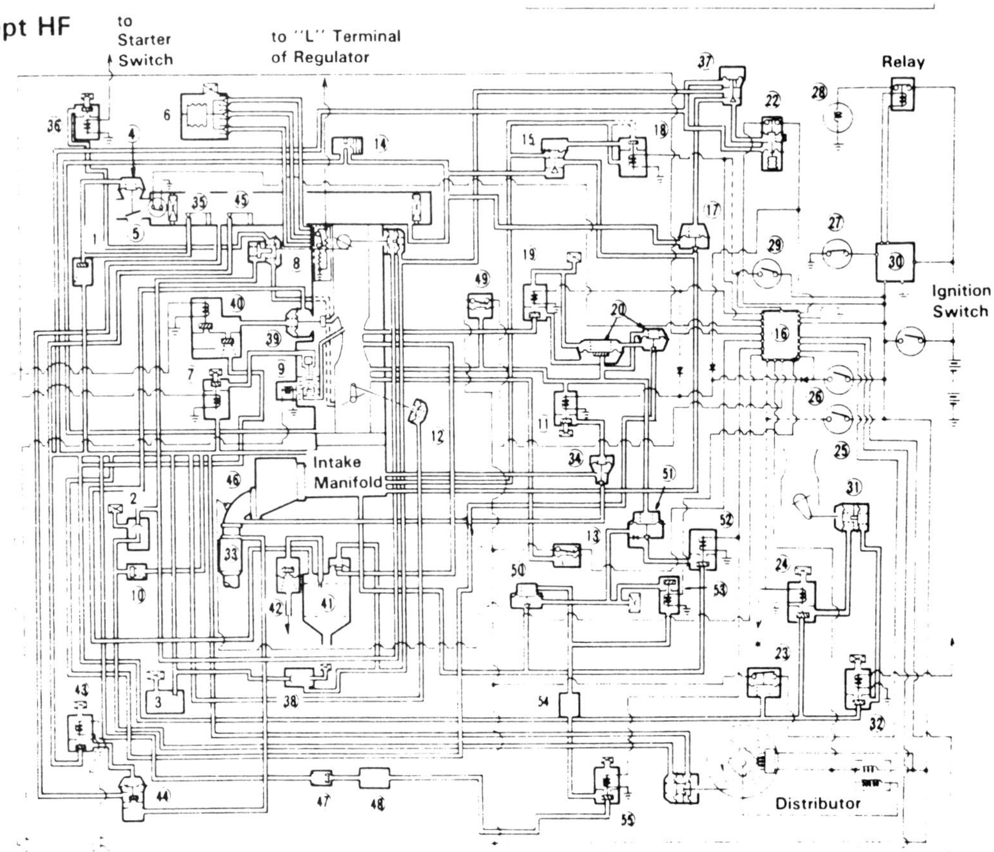

Manual only

*To power steering oil pressure switch

Vacuum and Electrical Connections

49 ST 1500HF

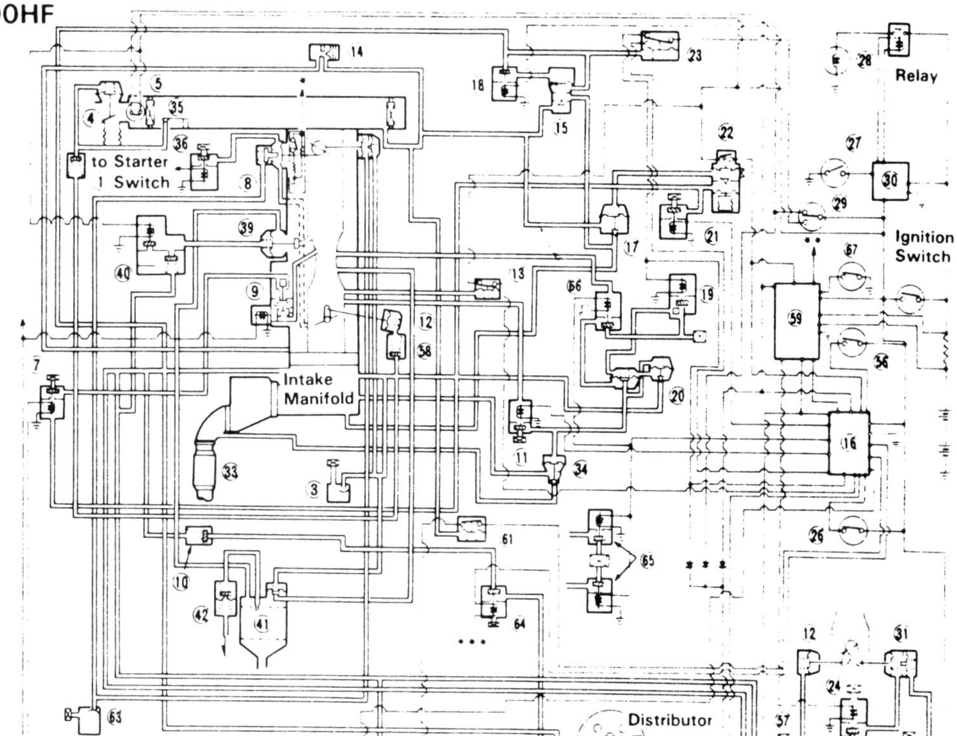

* to "L" Terminal of Regulator
** to Air Con. Switch
*** to Radiator Fan Control unit

CAL 1500HF

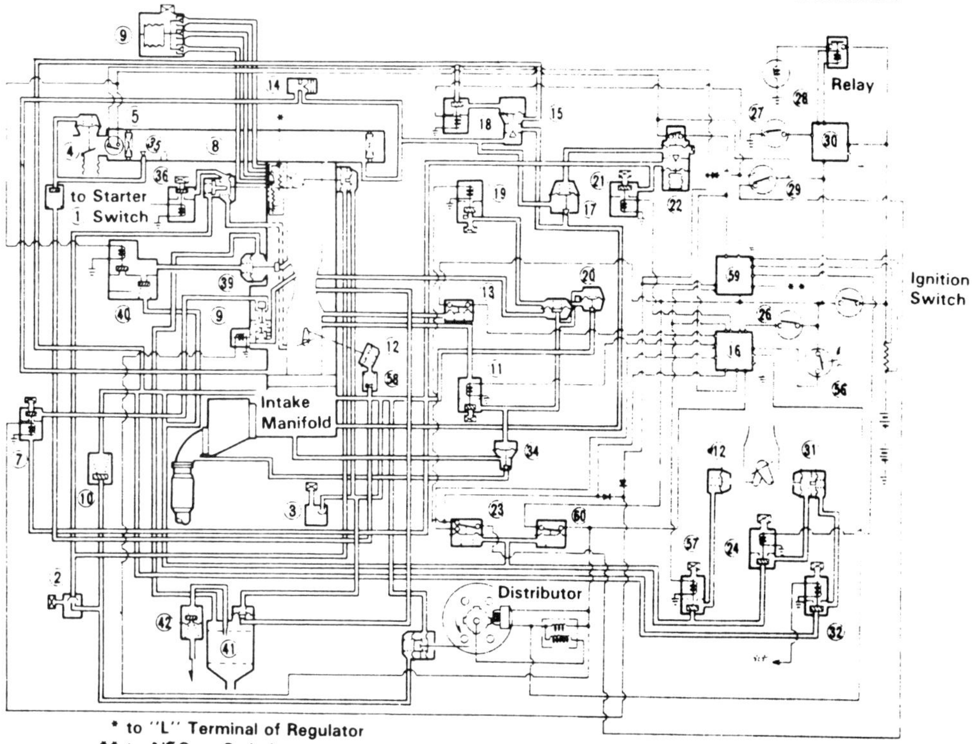

* to "L" Terminal of Regulator
** to Aif Con. Switch

HI ALT 1500HF

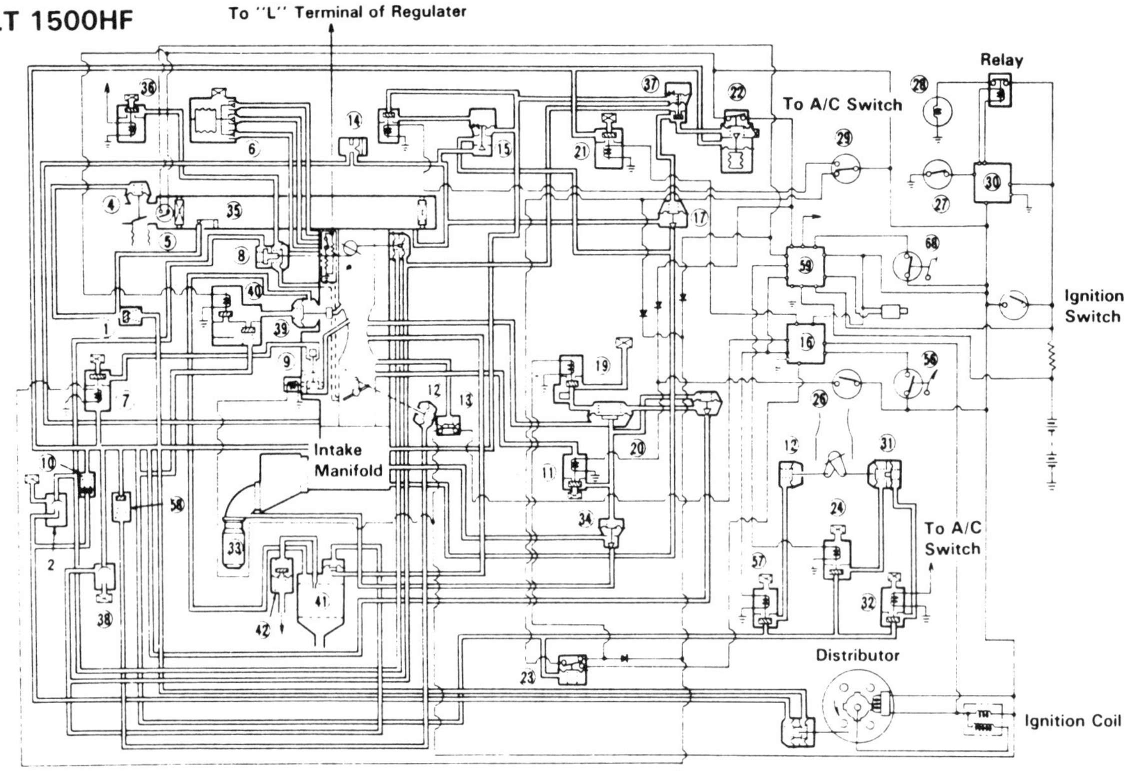

1 CHECK VALVE (INTAKE AIR TEMP. CONTROL)
2 THERMOVALVE B (Except 49 ST 1500HF)
3 THERMOVALVE A (Except HI ALT 1300/1500HF and 49 ST/HI ALT 1500 Manual except HF)
4 AIR CONTROL DIAPHRAGM
5 INTAKE AIR TEMP. SENSOR
6 AIR JET CONTROLLER (Except 49 ST 1300/1500HF)
7 POWER VALVE CONTROL SOLENOID VALVE
8 CHOKE OPENER
9 PRIMARY SLOW MIXTURE CUT-OFF SOLENOID VALVE
10 CHECK VALVE E
11 EGR CONTROL SOLENOID VALVE A
12 THROTTLE CONTROLLER
13 VACUUM SWITCH A
14 THERMOVALVE C
15 ANTI-AFTERBURN VALVE
16 CONTROL UNIT
17 AIR VALVE
18 ANTI-AFTERBURN CONTROL SOLENOID VALVE
19 EGR CONTROL SOLENOID VALVE B (Except 1500 Manual except HF)
20 EGR CONTROL VALVES A & B
21 AIR CONTROL SOLENOID VALVE (1300 and 1500HF)
22 CONTROL SWITCH
23 VACUUM SWITCH B
24 IDLE BOOST SOLENOID VALVE (Except 1500 Automatic without power steering)
25 THERMOSENSOR A (Except 1500HF and Automatic)
26 THERMOSENSOR B
27 AIR TEMP. SENSOR
28 RADIATOR FAN
29 SPEED SENSOR
30 RADIATOR FAN TIMER
31 IDLE CONTROLLER
32 IDLE CONTROL SOLENOID VALVE
33 CATALYTIC CONVERTER
34 EGR VALVE
35 AIR BLEED VALVE A
36 CRANKING LEAK SOLENOID VALVE
37 VACUUM CONTROL VALVE (HT ALT 1300/1500HF and 49 ST/HI ALT 1500 Manual except HF)
38 THERMOVALVE D (HT ALT 1300/1500HF and 49 ST/HI ALT 1500 Manual except HF)
39 AIR VENT CUT-OFF DIAPHRAGM
40 VACUUM HOLDING SOLENOID VALVE
41 CHARCOAL CANISTER
42 TWO-WAY VALVE
43 *AIR SUCTION CONTROL SOLENOID VALVE
44 *AIR CONTROL VALVE
45 *AIR BLEED VALVE B
46 *OXYGEN SENSOR
47 *CHECK VALVE B
48 *SURGE TANK A
49 *VACUUM SWITCH C
50 *AIR CONTROL VALVE B
51 *AIR CONTROL VALVE A
52 *FREQUENCY SOLENOID VALVE A
53 *FEEDBACK CONTROL SOLENOID VALVE
54 *SURGE TANK B
55 *FREQUENCY SOLENOID VALVE B
56 CLUTCH SWITCH (1500HF)
57 THROTTLE CONTROL SOLENOID VALVE (1500HF)
58 DASHPOT CHECK VALVE (1500HF)
59 ALTERNATOR CONTROL UNIT (1500HF)
60 VACUUM SWITCH D (49 ST/CAL 1500HF)
61 VACUUM SWITCH E (49 ST 1500HF)
62 VACUUM SWITCH F (49 ST 1500HF)
63 THERMOVALVE E (49 ST 1500HF)
64 COLD ADVANCE SOLENOID VALVE (49 ST 1500HF)
65 PRIMARY AIR CUT-OFF SOLENOID VALVE (49 ST 1500HF)
66 EGR CONTROL SOLENOID VALVE C (49 ST 1500HF)
67 STEERING SWITCH (49 ST 1500HF)
68 BRAKE SWITCH (HI ALT 1500HF)

*1500 except HF

Interconnect Diagram

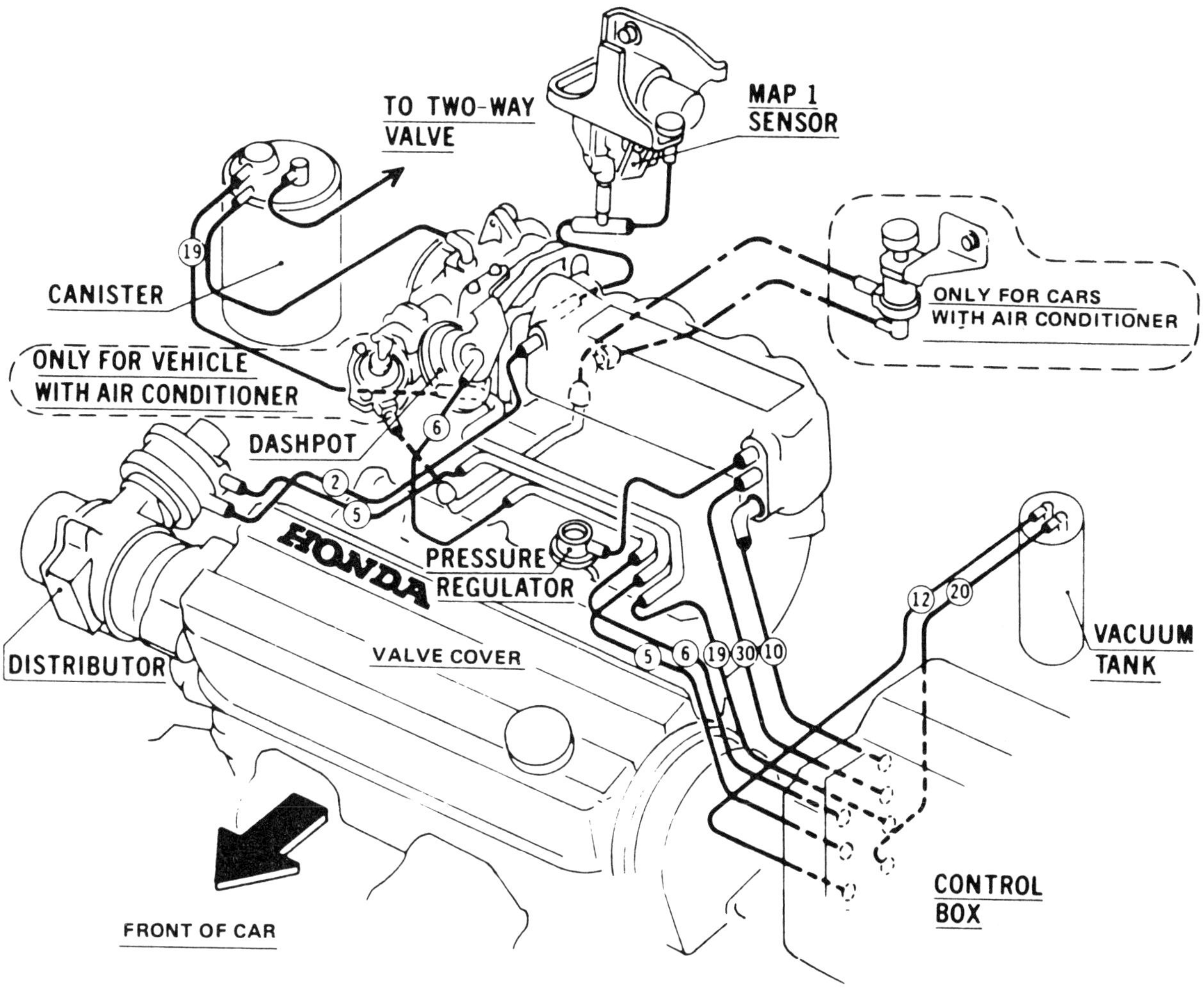

85 Civic CRX-Si

Interconnect Diagram (cont'd)

Control Box

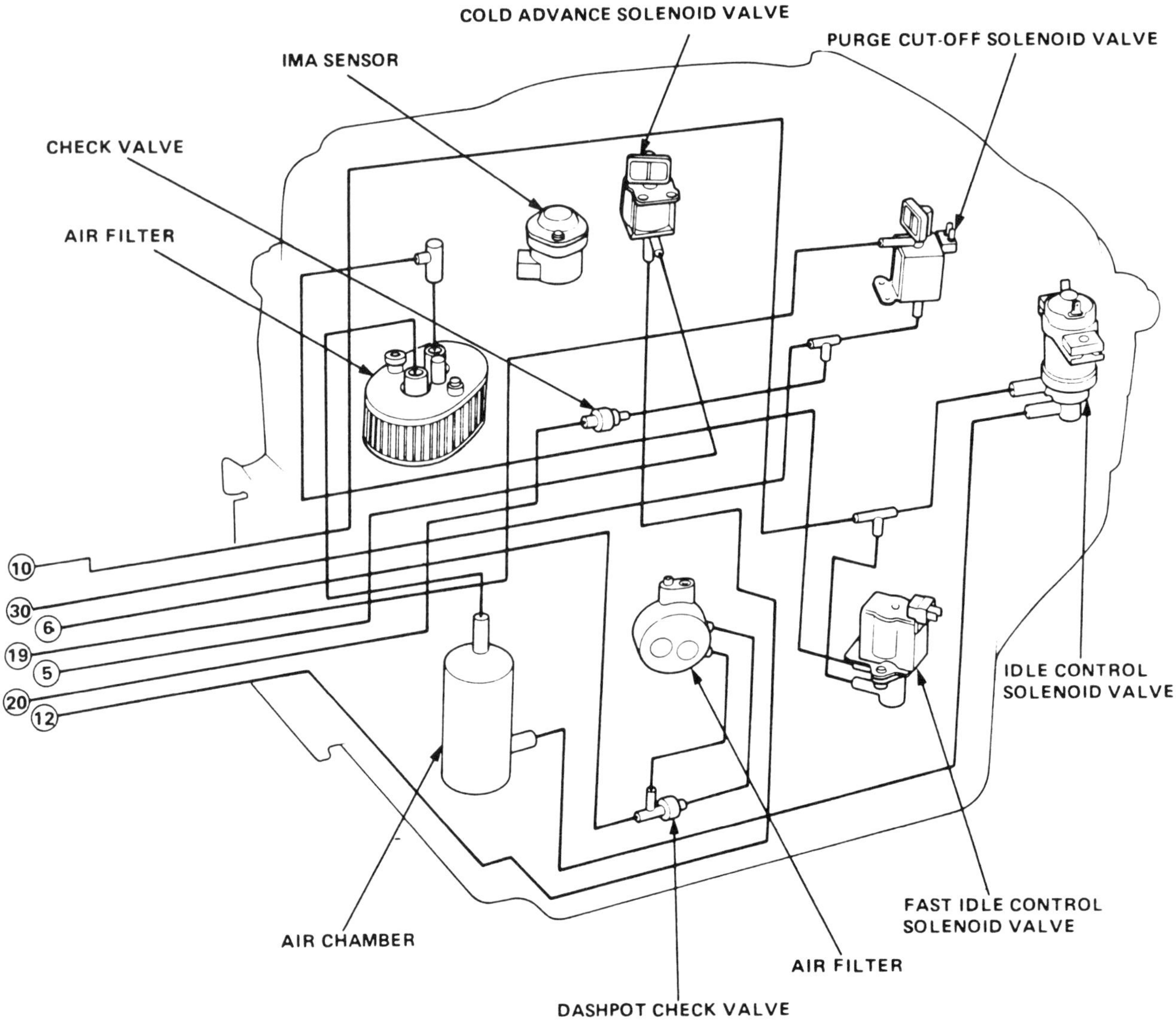

85 Civic CRX-Si

Vacuum and Electrical Connections

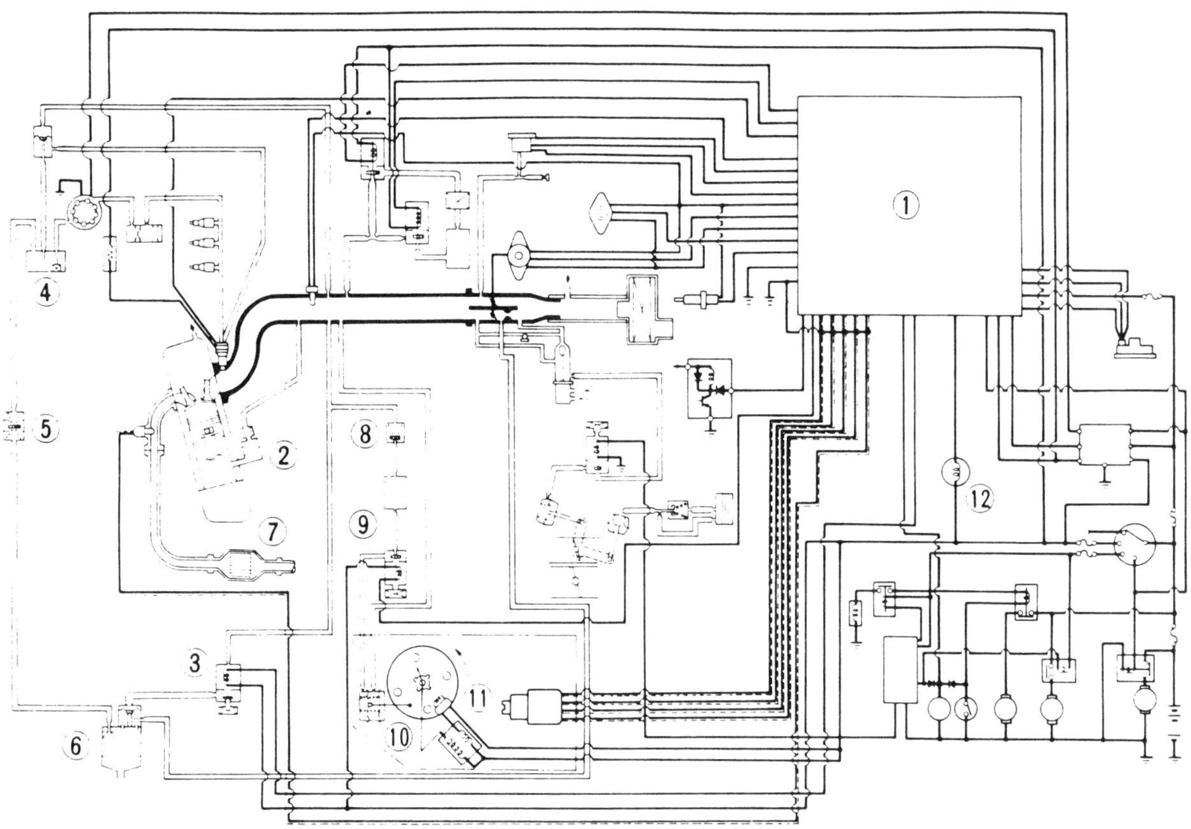

1 ELECTRONIC CONTROL UNIT
2 PCV VALVE
3 PURGE CUT-OFF SOLENOID VALVE
4 FUEL TANK
5 TWO-WAY VALVE
6 CHARCOAL CANISTER
7 CATALYTIC CONVERTER
8 CHECK VALVE
9 COLD ADVANCE SOLENOID VALVE
10 VACUUM CONTROLLER
11 DISTRIBUTOR
12 PGM-FI WARNING LIGHT

85 Civic CRX-Si

Interconnect Diagram

49ST/HI ALT

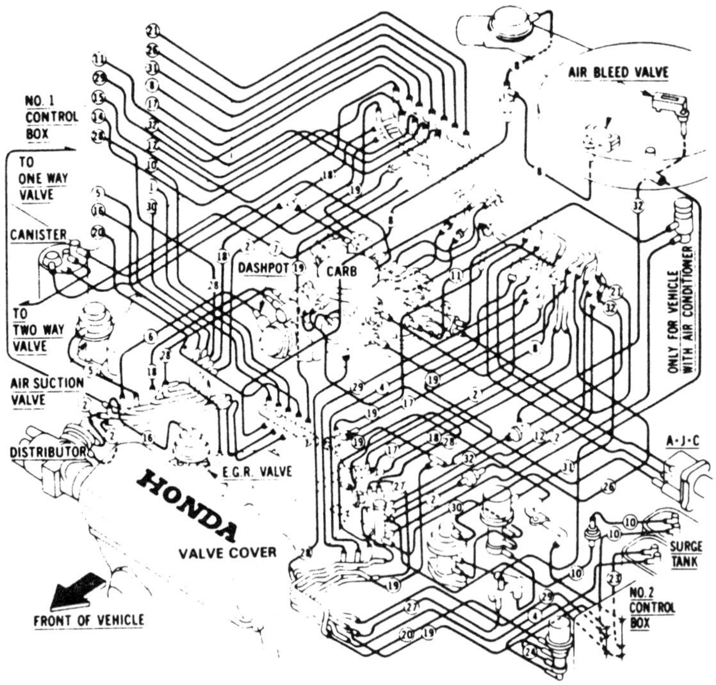

CAL

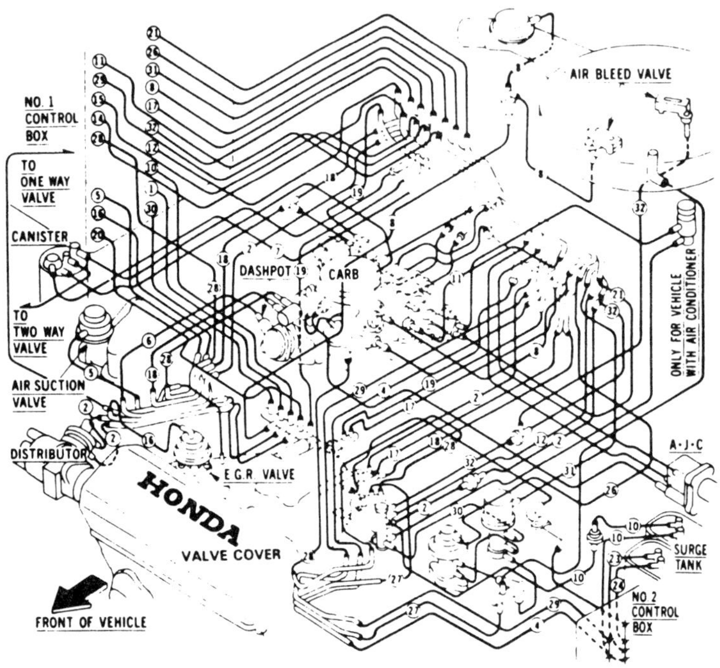

85 Civic 4-WD wagon

Interconnect Diagram (cont'd)

Control Box #1

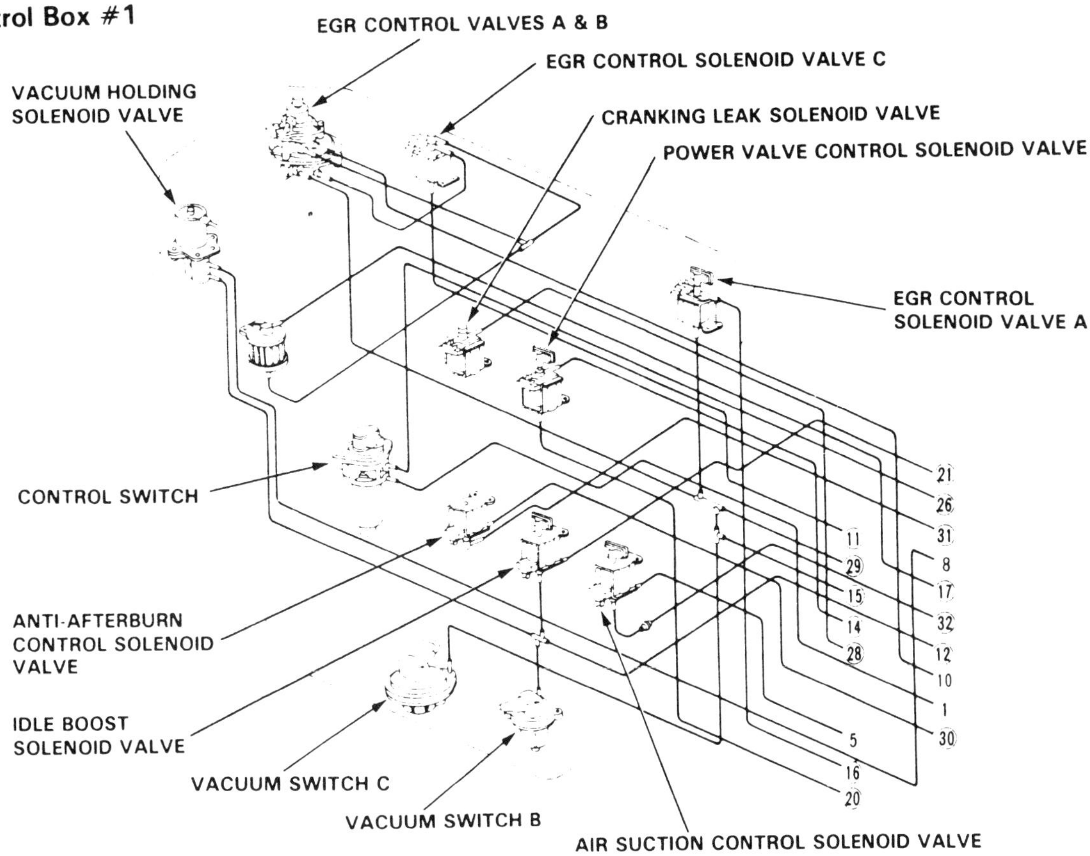

Control Box #2

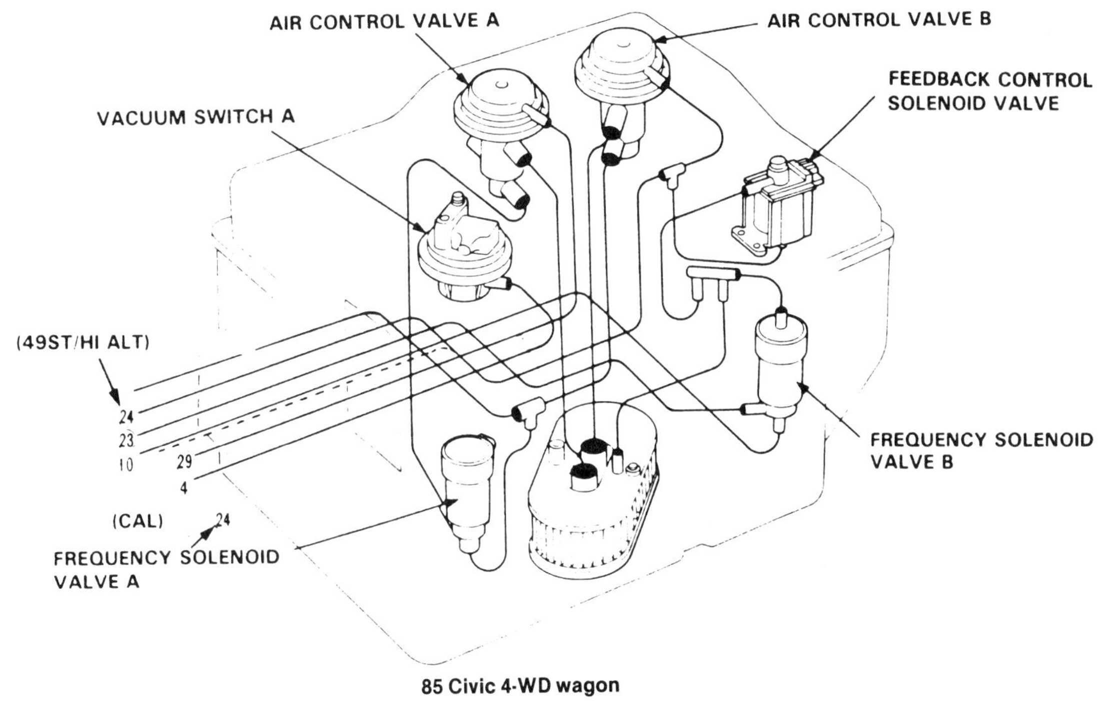

85 Civic 4-WD wagon

Vacuum and Electrical Connections

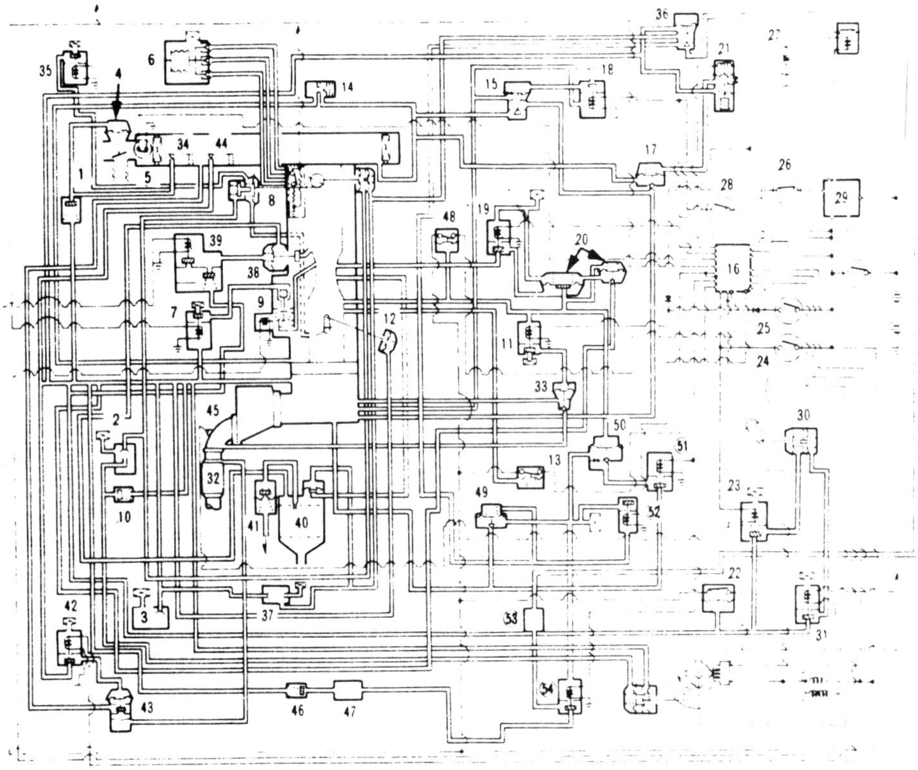

1 CHECK VALVE (INTAKE AIR TEMP. CONTROL)
2 THERMOVALVE B
3 THERMOVALVE A
4 AIR CONTROL DIAPHRAGM
5 INTAKE AIR TEMP. SENSOR
6 AIR JET CONTROLLER
7 POWER VALVE CONTROL SOLENOID VALVE
8 CHOKE OPENER
9 PRIMARY SLOW MIXTURE CUT-OFF SOLENOID VALVE
10 CHECK VALVE E
11 EGR CONTROL SOLENOID VALVE A
12 THROTTLE CONTROLLER
13 VACUUM SWITCH A
14 THERMOVALVE C
15 ANTI-AFTERBURN VALVE
16 CONTROL UNIT
17 AIR VALVE
18 ANTI-AFTERBURN CONTROL SOLENOID VALVE
19 EGR CONTROL SOLENOID VALVE C
20 EGR CONTROL VALVES A & B
21 CONTROL SWITCH
22 VACUUM SWITCH B
23 IDLE BOOST SOLENOID VALVE
24 THERMOSENSOR A
25 THERMOSENSOR B
26 AIR TEMP. SENSOR
27 RADIATOR FAN
28 SPEED SENSOR
29 RADIATOR FAN TIMER
30 IDLE CONTROLLER
31 IDLE CONTROL SOLENOID VALVE
32 CATALYTIC CONVERTER
33 EGR VALVE
34 AIR BLEED VALVE A
35 CRANKING LEAK SOLENOID VALVE
36 VACUUM CONTROL VALVE (49ST/HI ALT)
37 THERMOVALVE D (49ST/HI ALT)
38 AIR VENT CUT-OFF DIAPHRAGM
39 VACUUM HOLDING SOLENOID VALVE
40 CHARCOAL CANISTER
41 TWO-WAY VALVE
42 AIR SUCTION CONTROL SOLENOID VALVE
43 AIR CONTROL VALVE
44 AIR BLEED VALVE B
45 OXYGEN SENSOR
46 CHECK VALVE B
47 SURGE TANK A
48 VACUUM SWITCH C
49 AIR CONTROL VALVE B
50 AIR CONTROL VALVE A
51 FREQUENCY SOLENOID VALVE A
52 FEEDBACK CONTROL SOLENOID VALVE
53 SURGE TANK B
54 FREQUENCY SOLENOID VALVE B

85 Civic 4-WD wagon

Isuzu

INDEX

VEHICLE IDENTIFICATION NUMBER AND ENGINE IDENTIFICATION

Vehicle Identification Number (VIN)

The vehicle identification number (VIN) is embossed on a plate which is attached at the top left of the instrument panel. The number is visible through the windshield from outside the vehicle.

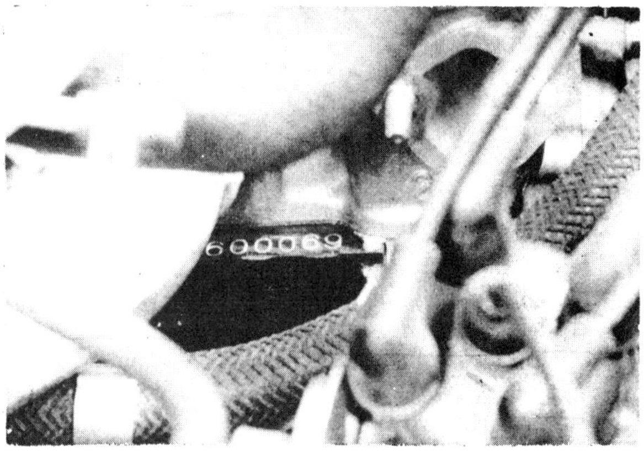

Diesel engine serial number location (© Isuzu Motor Corporation)

Engine Identification Number

The engine serial number is stamped on the top right front corner of the gasoline engine block. On diesel engines, the engine serial number is stamped on the left rear corner of the engine block.

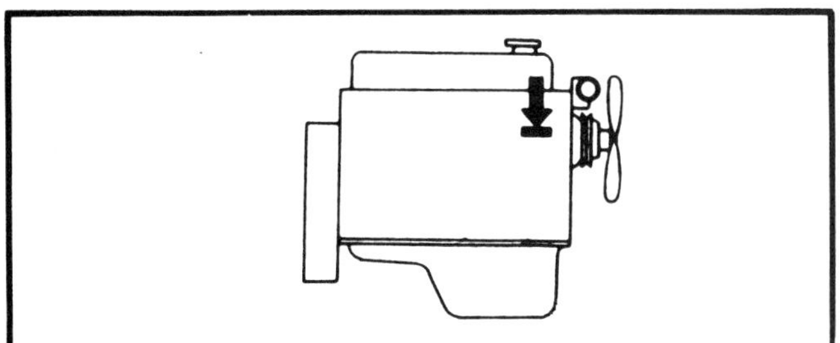

Gas engine serial number location (© Isuzu Motor Corporation)

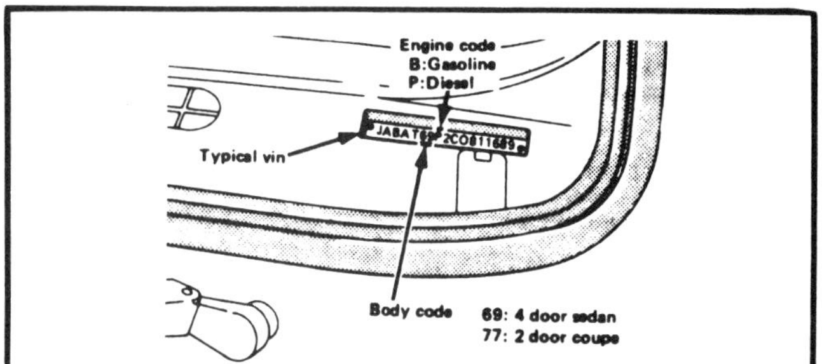

Vehicle identification plate location (© Isuzu Motor Corporation)

GENERAL ENGINE SPECIFICATIONS

Year	Engine Displacement (cu. in.)	Carburetor Type	Horsepower (@ rpm)	Torque @ rpm (ft. lbs.)	Bore × Stroke (in.)	Compression Ratio	Oil Pressure @ rpm (psi)
'85–'86	110.8	Hitachi DCH340	80 @ 4800	95 @ 3000	3.31 × 3.23	8.5:1	57 @ 1400
	111	Diesel	51 @ 5000	72 @ 3000	3.31 × 3.23	22.0:1	64 @ 1400
'85–'86	118.9	E.F.I.	90 @ 5000	108 @ 3000	3.43 × 3.29	9.3:1 ①	60 @ 1400

E.F.I.—Electronic Fuel Injection
① 9.2:1—1985 and later

GASOLINE ENGINE TUNE-UP SPECIFICATIONS

(When analyzing compression test results, look for uniformity among cylinders, rather than specific pressures)

Year	Engine Displace-ment (cu.in.)	Spark Plugs		Distributor		Ignition Timing (deg)		Intake Valve Opens (deg)	Fuel Pump Pressure (psi)	Idle Speed (rpm)	Valve Clear (in.) (cold)	
		Type	Gap (in.)	Point Dwell (deg)	Point Gap (in.)	MT	AT				In	Ex
'85–'86	110.8	BPR6ESII	0.040	Electronic		6B	6B	21	3.6	900	0.006	0.010
'85–'86	118.9	BPR6ESII	0.040	Electronic		12B	12B	28	2.0–2.5	900	0.006	0.010

NOTE: The underhood specifications sticker often reflects tune-up specification changes made in production. Sticker figures must be used if they disagree with this chart.
B Before top dead center

DIESEL ENGINE TUNE-UP SPECIFICATIONS

Year	Injector Opening Pressure (psi)	Low Idle (rpm)	Valve Clearance		Intake Valve Opens (deg.)	Injection Time (deg.)	Firing Order
			Intake	Exhaust			
'85–'86	2133 ③	625 ①	0.010	0.014	32B	18B ②	1-3-4-2

① A/T: 725 rpm
② 12B on 1982 and later
③ 1770—1984 and later

FIRING ORDERS

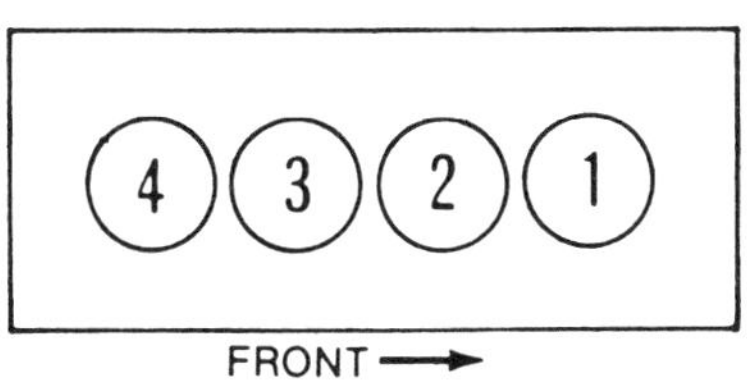

Diesel engine firing order 1-3-4-2
(© Isuzu Motor Corporation)

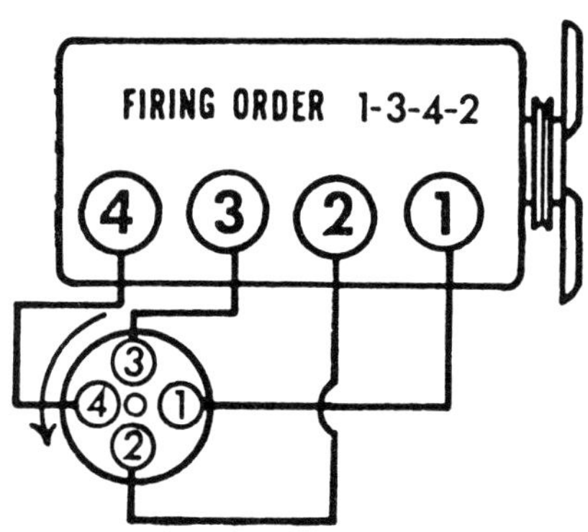

Gas engine firing order 1-3-4-2
(© Isuzu Motor Corporation)

EMISSION EQUIPMENT USED

1985-86

Air injection reactor system
Exhaust gas recirculation system
Thermostatic controlled air cleaner system
Catalytic converter
Positive crankcase ventilation system
Evaporative emission control system
Early fuel evaporation system
Closed loop emission control system
High altitude emission control system

REQUIRED EMISSION CONTROL MAINTENANCE

AIR INJECTION SYSTEM

This system should be checked every 30,000 miles for proper operation. Be sure to check all hoses, fittings and connections.

IDLE SPEED AND MIXTURE ADJUSTMENT

Check the choke mechanism for binding and all hoses and connections every 30,000 miles. Reset the engine idle speed after the first 5,000 miles and then every 15,000 miles thereafter, according to the vehicle emission control information label.

CATALYTIC CONVERTER

Check the converter for damage and tight connections, every 60,000 miles.

EXHAUST GAS RECIRCULATION SYSTEM

Remove the EGR valve and clean the carbon deposits from the valve, every 60,000 miles. Check the system for proper operation, every 15,000 miles.

FUEL EVAPORATION SYSTEM

Replace the fuel filter, every 30,000 miles. Check all hoses and connections and replace defective components as required. Replace the carbon canister if it is saturated with fuel.

PCV SYSTEM

Check and clean all hose connections and replace the PCV valve every 30,000 miles. If the valve is defective, do not attempt to clean it. Replace it.

SPARK PLUGS

Check and replace the engine spark plugs every 30,000 miles. Check and adjust the engine timing every 15,000 miles.

THERMOSTATIC AIR CLEANER SYSTEM

Replace the air filter element every 15,000 miles or at least once a year. Check the TAC system for proper operation every 30,000 miles. Be sure that all ducts and connections are functioning properly. Repair or replace defective components.

RUBBER EMISSION HOSES

Inspect hose surfaces for evidence of heat or mechanical damage. Hard or brittle rubber, cracking, checking, tears, cuts, abrasion and excessive swelling indicates deterioration of the rubber. Inspect all hose connections, couplings and clamps for leakage. Repair or replace all defective components as required.

IDLE SPEED AND MIXTURE ADJUSTMENT

NOTE: The following information is published from the latest data available. If this published information differs from the information found on the vehicle emission control information label, use the data on the label.

IDLE SPEED

Adjustment

CARBURATED ENGINES

1. Set the parking brake and block the drive wheels. Position the transmission selector lever in the neutral position.
2. Run the engine until normal operating temperature has been reached, be sure that the choke is fully open. Be sure that the air conditioner is off and the air cleaner installed.
3. Disconnect and plug the distributor vacuum advance line, the canister purge line, the EGR vacuum line and the idle compensator vacuum line.
4. If the vehicle is not equipped with air conditioning adjust the idle speed screw to 850–950 rpm.
5. If the vehicle is equipped with air conditioning turn the air conditioner on to maximum cold and the high blower on high.
6. Open the throttle about one third its full travel and then allow it to close. Adjust the idle speed controller adjusting screw to set the idle to 850–950 rpm.

IDLE MIXTURE

Adjustment

CARBURATED ENGINES

1. Set the parking brake and block the drive wheels. Position the transmission selector lever in the neutral position.
2. Remove the carburetor from the engine. Remove the idle mixture screw plug. Reinstall the carburetor.
3. Run the engine until normal operating temperature has been reached, be sure that the choke is fully open. Be sure that the air conditioner is off and the air cleaner installed.
4. Disconnect and plug the distributor vacuum advance line, the canister purge line, the EGR vacuum line and the idle compensator vacuum line.
5. Turn the idle mixture adjusting screw all the way in and then all the way out 1 ½ turns. Adjust the idle mixture to 850–950 rpm.
6. Adjust the dwell to about 36 degrees, using the dwell meter. Be sure that it is set on the four cylinder scale.

7. After the dwell adjustment has been made, recheck the throttle adjustment.

8. If the vehicle is equipped with air conditioning turn the air conditioner on to maximum cold and the high blower on high.

9. Open the throttle about one third its full travel and then allow it to close. Adjust the idle speed controller adjusting screw to set the idle to 850–950 rpm.

10. Remove the carburetor and reinstall the idle mixture screw plugs.

IDLE SPEED

Adjustment

DIESEL ENGINES

1. Set the parking brake and block the drive wheels. Position the transmission selector lever in the neutral position.

2. Run the engine until normal operating temperature has been reached and the coolant temperature is above 176 degrees F.

3. Install the proper tachometer and check the idle speed. Adjust to specification as required.

4. If the vehicle is equipped with automatic transmission the specification is 675–775 rpm. If the vehicle is equipped with manual transmission, the specification is 575–675 rpm.

FAST IDLE SPEED

Adjustment

DIESEL ENGINES

1. Set the parking brake and block the drive wheels. Position the transmission selector lever in the neutral position.

2. Run the engine until normal operating temperature has been reached and the coolant temperature is above 176 degrees F.

3. Install the engine tachometer according to the manufacturers instructions.

4. Disconnect the vacuum hoses from the vacuum switch. Join the two vacuum hoses together using a piece of pipe.

5. Loosen the adjusting nut on the injection pump and adjust the engine idle speed to specification by moving the nut. The proper specification is 900–950 rpm.

6. Tighten the adjusting nut. Remove the engine tachometer. Reconnect the vacuum hose to the vacuum switch.

IDLE SPEED

Adjustment

FUEL INJECTED ENGINES

1. Run the engine until normal operating temperature has been reached. Be sure that the throttle valve is completely closed and the idle contact is in the ON position.

2. Turn the air conditioner off. Position the transmission selector lever in the neutral or park position.

3. Disconnect the pressure regulator V.S.V. harness.

4. Adjust the engine idle speed to the specified speed, using the idle adjusting screw on the throttle valve body. The proper specification is 850–950 rpm.

5. Before adjusting the idle speed be sure that the idle port is clean, as restrictions in this port may cause idle speed fluctuations.

INITIAL TIMING SETTINGS

GASOLINE ENGINE

Adjustment

1. Run the engine until normal operating temperature is reached.

2. Connect the timing light according to the manufacturers instructions.

3. Loosen the distributor hold down bolt. Adjust the timing as required.

4. The timing specification for the I-Mark is 6 degrees B.T.D.C. at 900 rpm. The timing specification for the Impulse is 12 degrees B.T.D.C. at 900 rpm.

DIESEL ENGINE

Adjustment

1. Check that the notched line on the injection pump flange is in alignment with the notched line on the injection pump front bracket.

2. Bring the piston in number one cylinder to TDC on the compression stroke. The correct notch must be used for alignment, as the damper pulley has seven notches.

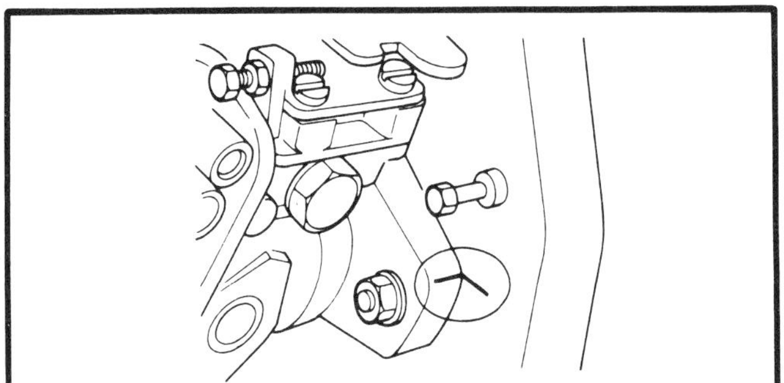

Diesel engine injection pump flange alignment (© Isuzu Motor Corporation)

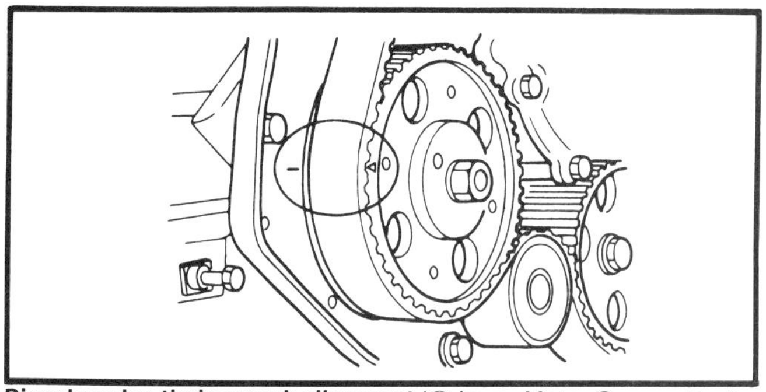

Diesel engine timing mark alignment (© Isuzu Motor Corporation)

3. With the timing pulley housing removed, check to be sure that the timing belt is properly tensioned and that the timing marks are aligned.

4. Remove the cam cover and the rear plug. Check that the fixing plate tool number J–29761 or equivalent fits smoothly into the slit at the rear of the camshaft. Remove the fixing plate.

5. Disconnect the injection pipe from the injection pump. Remove the distributor head screw and gasket and install the static timing gauge tool number J–28827 or equivalent. Set the lift to about 0.04 in.

6. Bring number one piston to a point about 45–60 degrees BTDC, by turning the crankshaft. Calibrate the dial indicator gauge to zero.

7. Turn the crankshaft pulley slightly in both directions and check that the gauge indication is stable.

NOTE: The damper pulley is provided with a total of seven notch lines. Four lines are at one end and three lines are at the other end of the pulley. The four lines are for static timing and the three lines are for dynamic timing.

8. Turn the crankshaft clockwise and take a reading on the dial indicator when the timing mark (12 degrees BTDC) on the crankshaft is in alignment with the pointer. The standard reading is 0.020 in.

9. If the reading on the dial indicator is not within specification, hold the crankshaft in position (12 degrees BTDC) and loosen the two nuts on the injection pump flange. Move the injection pump to a point where the dial indicator reads 0.020 in. Tighten the pump flange nuts.

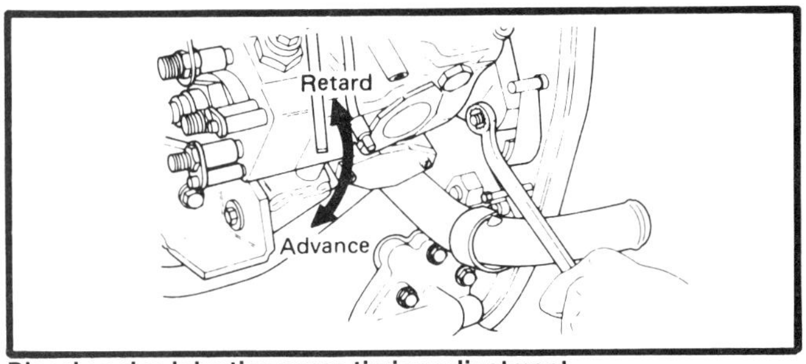

Diesel engine injection pump timing adjustment
(© Isuzu Motor Corporation)

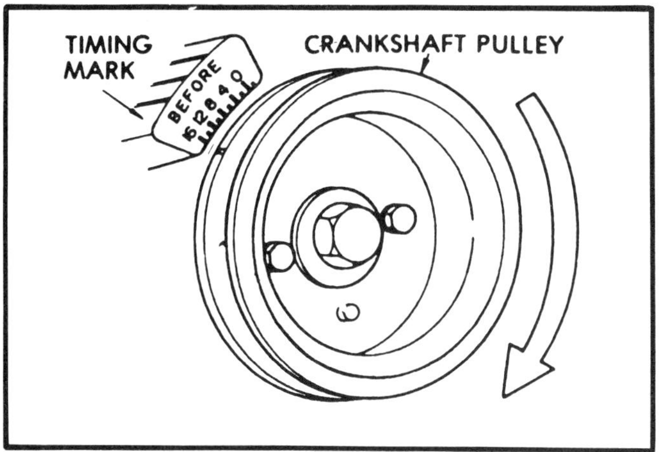

Diesel engine timing marks (© Isuzu Motor Corporation)

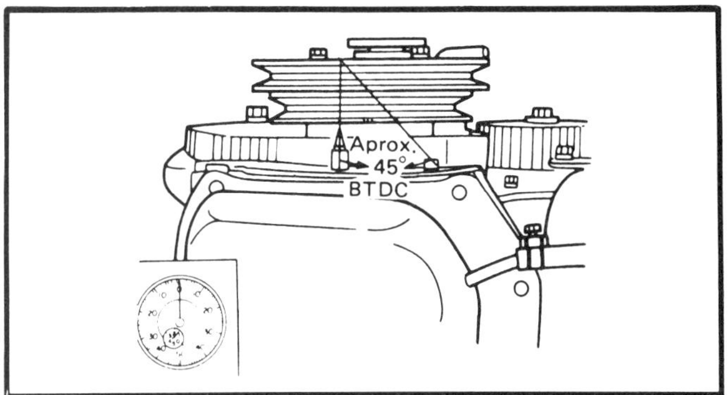

Turning the crankshaft pulley 45 degrees BTDC—diesel engine
(© Isuzu Motor Corporation)

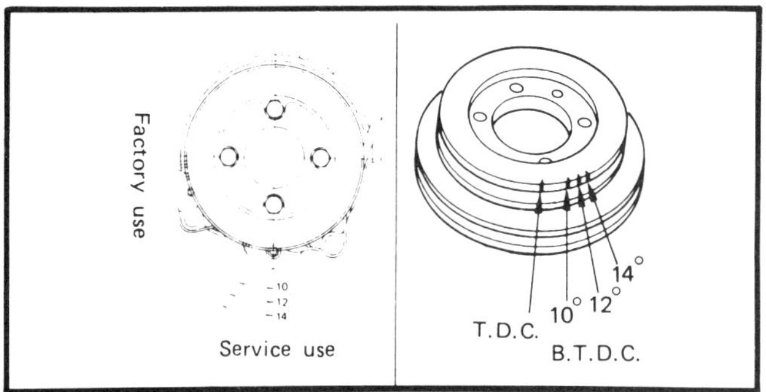

Multiple notches provided on the diesel engine damper pulley
(© Isuzu Motor Corporation)

EMISSION CONTROL SYSTEMS
CRANKCASE EMISSION CONTROL SYSTEM

PCV System

ALL EXCEPT CALIFORNIA DIESEL ENGINE

The PCV system is designed to force blow-by gases generated in the engine crankcase back into the intake manifold then deliver them together with the fuel mixture to the combustion chambers. This system is a closed type and consists of a baffle plate in the rocker arm cover for separating oil particles from blow-by gases, an orifice in the intake manifold for controlling suction of the blow-by gases and a hose from the air cleaner to allow fresh air into the system. Under normal operating conditions blow-by gases passing between the piston rings and fuel

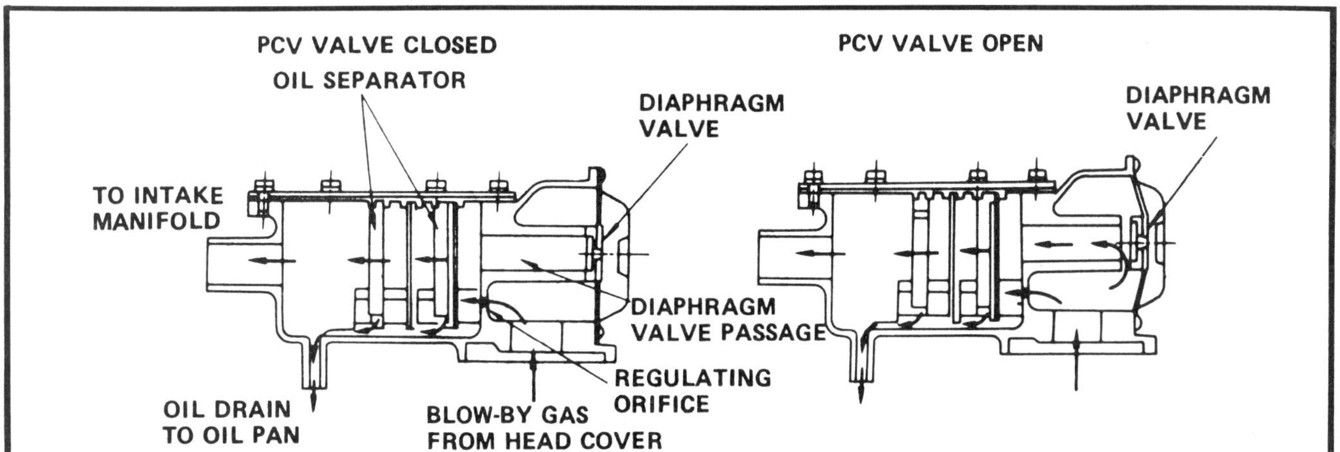

PCV valve operation—California diesel engines (© Isuzu Motor Corporation)

vapor from the fuel tank are mixed with the ambient temperature supplied from the air cleaner. This mixture is than drawn through the regulating orifice into the intake manifold for burning. When the engine is operating with the throttle wide open part of the blow-by generated is drawn directly into the air cleaner via the rear of the rocker arm cover.

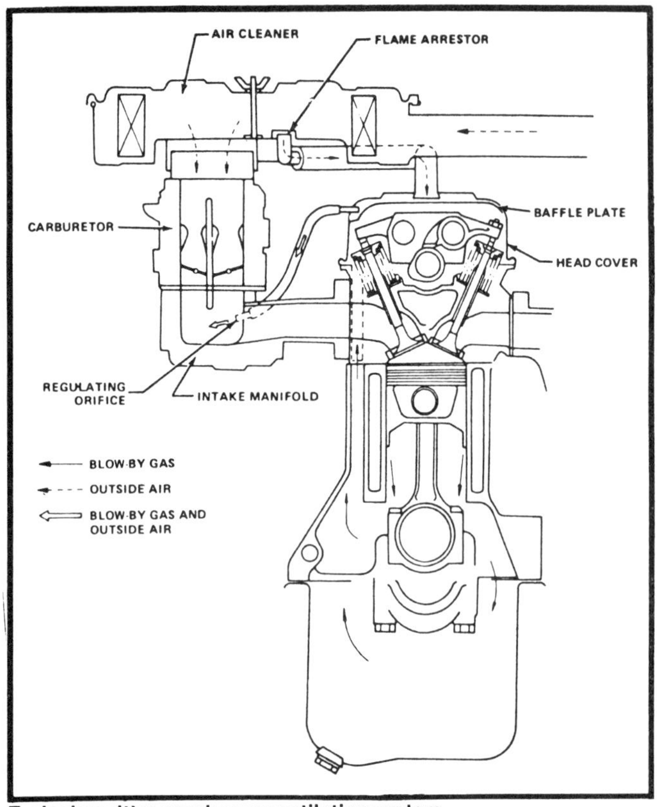

Typical positive crankcase ventilation system
(© Isuzu Motor Corporation)

CALIFORNIA DIESEL ENGINE

The PCV system used in California vehicles equipped with the diesel engine is the closed type system. It is designed to force blow-by gas, which is generated in the engine crankcase, back into the intake manifold and return together with fresh air back into the combustion chamber. In this system the blow-by gas is circulated back into the intake manifold for recombustion through an oil separator in the PCV valve assembly. The PCV valve assembly consists of the valve, which controls the crankcase pressure, and an oil separator which separates oil particles from blow-by gas and returns to the oil pan through a check valve. The check valve prevents the oil from flowing back from the oil pan. The function of the oil separator is to prevent the forming of carbon deposits produced by the mixture of EGR gas and oil mist contained in the blow-by gases. The PCV valve has a regulating orifice and a diaphragm valve which opens the diaphragm valve passage. When the engine is running at high speed, the high negative pressure from the intake manifold makes the diaphragm valve close. As a result, the blow by gas passes through the regulating orifice. When the engine is running at low speed, the negative pressure from the intake manifold is so small that the cylinder head cover pressure makes the diaphragm valve open. As a result the blow-by gas passes through the regulating orifice and the diaphragm passage which is opened by the diaphragm valve.

System Service

1. Clean the internal parts of the hoses and the regulating orifice with solvent and blow out any obstructions using compressed air.
2. Check the hoses for cracks, swelling and fatigue, replace defective hoses as required. Make sure that all hose connections are tight.
3. Never clean and reuse a PCV valve always replace the valve if service life is questionable.

PCV SYSTEM DIAGNOSIS

Symptom	Possible Cause	Repair
Rough idle, hard starting	Clogged PCV valve, vacuum leak in system	Replace PCV valve, repair vacuum leak
No PCV flow	Defective or clogged PCV valve, blocked line	Clean or replace line and PCV valve
Engine stalls at idle	Excessive PCV flow, dirty air filter element	Check PCV valve and air filter
Surging at road speed, hesitation	Improper PCV flow	Check PCV system, clean all lines.
Excessive oil consumption	PCV line blocked	Clear line

FUEL EVAPORATIVE EMISSION CONTROL SYSTEM

Evaporative Emission Control System

The evaporative control system is designed to lead the fuel vapor emitted from the fuel tank into the engine crankcase and also the fuel vapor emitted from the carburetor into the carbon canister. The fuel vapor that is drawn into the engine crankcase is mixed with blow-by gases and drawn into the intake manifold for combustion. The fuel vapor that is stored in the canister and mixed with ambient air is drawn directly to the intake manifold for combustion when the engine is running.This system consists of the vapor separator tank, check and relief valves, charcoal canister, vent switching valve, ventilation valve and tubes connecting these parts. The vapor separator tank separates the fuel from the vapor given off by the fuel in the fuel tank. It then returns the fuel to the fuel tank and carries the vapor into the engine crankcase.

The check and relief valve operates as follows; with the engine not running, the fuel vapor is drawn into the vapor separator and is fed into the check and relief valve. When the pressure of the fuel vapor becomes as high as 0.2–0.6 in. Hg., the check valve will open, allowing the vapor to flow into the engine crankcase. While the check valve is open, the valve at the air filter side remains closed, thus preventing vapor flow to the atmosphere. When the engine is running, a vacuum is formed in the fuel tank or the engine crankcase and the difference in pressure between the relief side of the valve and the fuel tank or the engine crankcase, becomes 0.2–0.6 in. Hg. The relief valve then opens, allowing ambient air from the air filter into the fuel tank or engine crankcase. This function brings the fuel tank or the engine crankcase back into an atmospheric pressure condition.

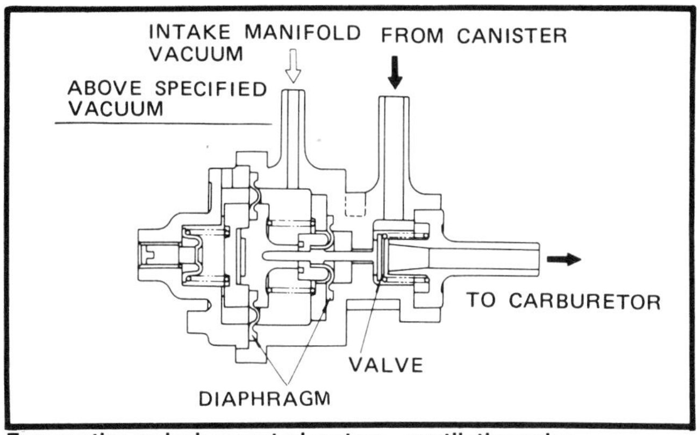

Evaporative emission control system—ventilation valve
(© Isuzu Motor Corporation)

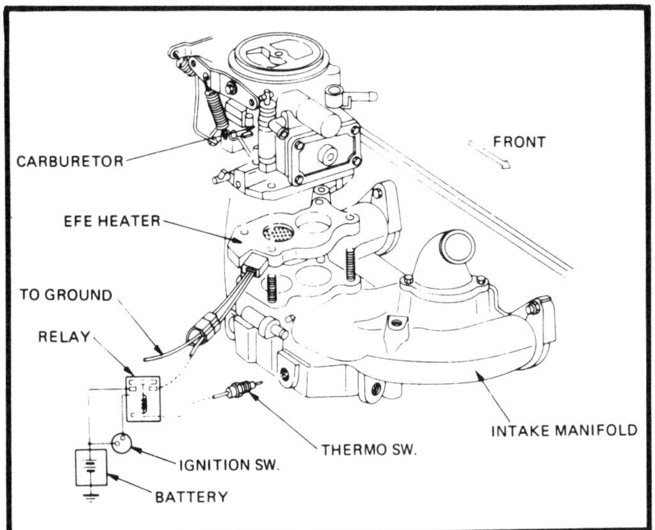

Early fuel evaporation system—exploded view
(© Isuzu Motor Corporation)

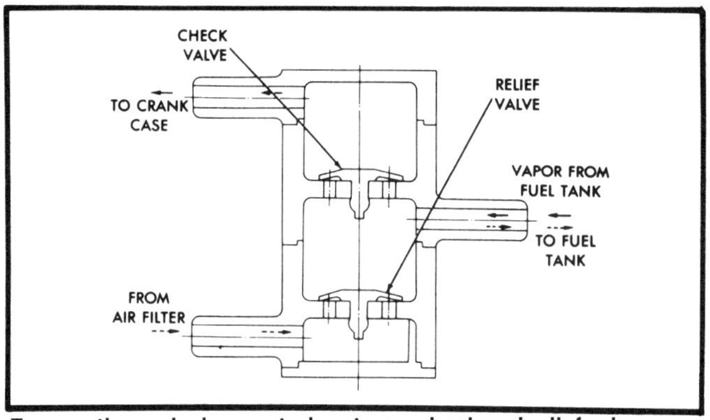

Evaporative emission control system—check and relief valve
(© Isuzu Motor Corporation)

The vent switching valve is an electrically operated solenoid valve built into the carburetor. When the engine is not running, the vent switching valve is opened to allow fuel vapor into the canister. When the engine is running, the vent switching valve is closed to prevent ambient air from flowing into the carburetor through the canister. The canister contains activated carbon to store up fuel vapor coming from the carburetor and the fuel tank. It also incorporates a purge control valve to purge fuel vapor from the carbon in the canister for delivery into the intake manifold. The diaphragm with the purge control valve is activated by a distributor vacuum advance signal. When the engine is off, the force of the diaphragm spring closes the purge vapor passage. When the engine is operating, the distributor vacuum advance signal is transmitted to the diaphragm. When the vacuum signal goes over the specified limits, the purge control valve opens the passage by moving the diaphragm and the fuel vapor in the canister is transposed into the intake manifold. The fuel vapor purge rate is controlled by the intake manifold vacuum pressure and the orifice in the purge control valve.

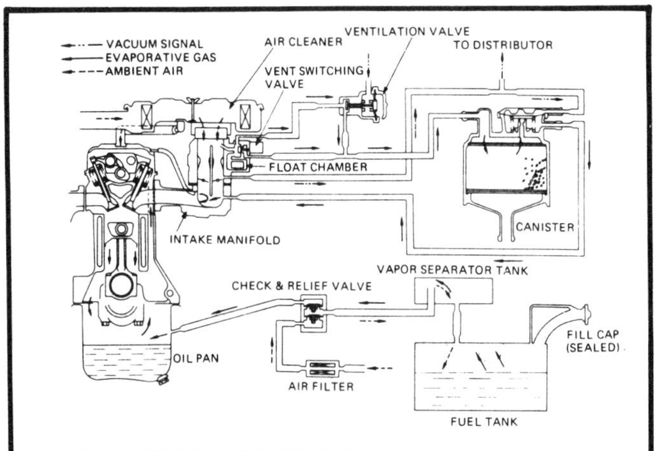

Fuel evaporation control system (© Isuzu Motor Corporation)

Early Fuel Evaporation System

The electric early fuel evaporation system uses a ceramic heater grid which is located under the primary bore of the carburetor and is considered part of the carburetor insulator gasket. It's function is to heat the in-coming fuel/air mixture for improved vaporization and drivability on cold driveaway. When the ignition switch is in the ON position, voltage is applied to the thermo switch. If the engine coolant temperature is below a specified value, the thermo switch is in the ON condition and a circuit is completed to the heater and current begins to flow. The heater, which incorporates a positive temperature coefficient semiconductor element, increases in temperature and then self regulates at a specified temperature, except at high engine speeds when the air/fuel flow will reduce the temperature below the regulated value. When coolant temperature reaches the specified value, the thermo switch turns off and current to the heater is cut off.

FUEL EVAPORATIVE SYSTEM DIAGNOSIS

Symptom	Possible Cause	Repair
Noticeable fuel odor or leaks.	Damaged or loose lines, saturated canister, broken or leaking two-way valve or liquid/vapor separator.	Repair lines, replace canister or two-way valve. Check liquid/vapor separator.
Fuel tank deformed.	Canister clogged, tank cap defective, two-way valve not relieving pressure or vacuum.	Replace defective canister or cap, clear vapor lines.
Insufficient fuel delivery or vapor lock.	Blocked or collapsed vapor lines, two-way valve or canister. Blocked fuel feed line.	Repair as necessary.
Tank won't take fuel.	Clogged or defective two-way valve.	Replace two-way valve.
Charcoal canister saturated with fuel.	Defective thermosensor, purge control/unloader solenoid or diaphragm valve. Blocked or disconnected vacuum or vapor lines.	Check operation of all FES components. Check all lines and connections. Replace charcoal canister.

EXHAUST EMISSION CONTROL SYSTEM

Thermostatic Air Cleaner System

The engines are equipped with a thermostatically controlled air cleaner system. This system consists of a thermo sensor, vacuum motor, hot air control valve and a hot idle compensator. These components are mounted on and around the air cleaner body and snorkel. When the engine is off, there is no vacuum present at the sensor unit or at the vacuum motor. The force of the vacuum motor spring closes off the heater air passage from the exhaust manifold heat stove. When the engine is first started, the thermo sensor is cool, thus allowing maximum vacuum to the vacuum motor. Maximum vacuum at the vacuum motor opens the hot air control valve which in turn closes off the ambient passage through the snorkel and opens the air passage from the manifold heat stove. Should the engine be heavily accelerated in this mode, the vacuum level in the system will drop to a low level so that the diaphragm spring will overcome the vacuum and open the snorkel passage.

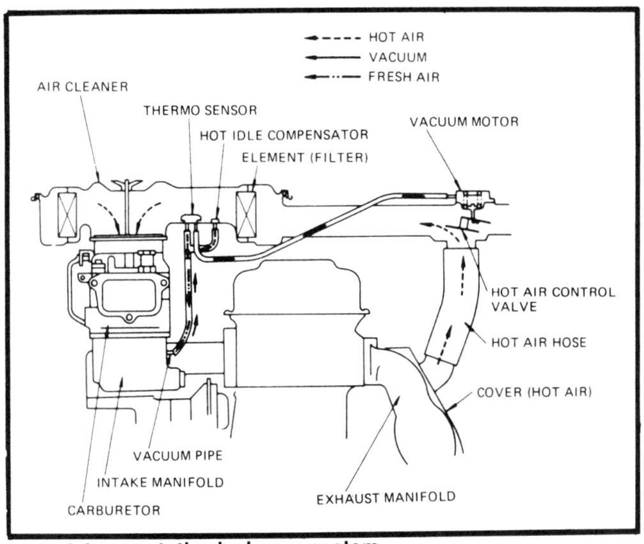

Typical thermostatic air cleaner system
(© Isuzu Motor Corporation)

As the engine heats up and the air past the thermo sensor reaches 100–111 degrees F., the thermo sensor comes into operation and begins to bleed off the supply of vacuum from the intake manifold. At about 111 degrees F., the thermo sensor completely bleeds off all vacuum to the vacuum motor so that the diaphragm spring closes the hot air control valve to the heat stove passage and opens the ambient air passage through the snorkel. With heat build up such as extended idle, climbing a slope or high speed driving, excessive fuel vapors enter the intake manifold causing an over rich fuel mixture that results in rough idle and increased carbon monoxide emission. To prevent this, the air cleaner assembly is equipped with a hot idle compensator. As a result, the engine heats up and as the air moving past the hot idle compensator reaches a specific temperature, the compensator opens to feed ambient air into the intake manifold, in order to lean out the temporarily rich mixture.

Air Injection System

The air injection system is designed so that ambient air is pressurized by an air pump, which in turn injects the air through the air injection nozzles, which are located near each exhaust valve, into the exhaust gas. The exhaust gases are high in temperature and self ignite when they come in contact with the oxygen of the ambient air that is being processed by the air pump. The main components of the system are the air pump, a check valve, the air switching valve, the vacuum switching valve, which is used on California vehicles, the mixture control valve and the air manifold with the air injection nozzles.

AIR PUMP

The air pump consists of a pump body, cover, rotor, vanes and a press fitted pressure relief valve. The pump is belt driven and the air pump pulley is mounted on the water pump shaft. Air is drawn in through the air cleaner suction hose into the pump suction chamber, where it is trapped between two vanes and the pump body. As the rotor turns, these vanes carry the air to the outlet chamber and then to the air manifold. The relief valve is press fitted into position on the outer chamber. It is held closed under normal operating conditions by the spring. However, when the pressure of the air at the outlet overcomes the tension of the spring, the valve is then pushed open and excess air is released so that it will not exceed the outlet air pressure of the air pump.

CHECK VALVE

The check valve is designed to allow air to pass through the system in only one direction. The valve is pushed open when the pressure of air supplied from the air pump overcomes the valve spring tension, but closes with the counterflow of exhaust gas from the manifold. This is done in order to prevent damage to the air pump and hoses, when the air supply from the air pump is stopped, due to a broken drive belt or exhaust backfiring.

AIR SWITCHING VALVE

This valve is designed to switch air flow from the air pump as the system requires it. The valve is operated by manifold vacuum and air pump pressure, which are switched by a three way vacuum switching valve. When manifold vacuum flows from pipe number one, the air check switching valve allows air from the air pump to the check valve. When air pump pressure flows to pipe number one , which is switched from the vacuum switching valve, the air valve shuts the air passage to the air manifold and at the same time opens the port to the air cleaner.

VACUUM SWITCHING VALVE (CALIFORNIA ONLY)

The vacuum switching valve has three ports, two of which are opened or closed, by electrically controlling the solenoid plunger. The solenoid plunger is energized when the water temperature is below 122 degrees F., or for ten seconds after the WOT switch is turned on when the water temperature is over 122 degrees F.

However, the solenoid plunger is de-energized at once when the WOT switch is turned off within ten seconds. When energized, the vacuum switching valve connects the diaphragm chamber "B" of the air switching valve to the intake manifold, which permits the manifold vacuum to be applied to chamber "B". When the solenoid plunger is de-energized, it plugs the ports so that chamber "B" of the air switching valve is connected to diaphragm chamber "A" of the valve.

MIXTURE CONTROL VALVE

The purpose of the mixture control valve is to prevent the backfiring of the exhaust system during deceleration. This component is designed to supply air into the intake manifold in order to prevent over enrichment of the air/fuel mixture, when the throttle valve in the carburetor is suddenly closed. The mixture control valve is held closed under normal operating conditions. When the vacuum in the intake manifold increases rapidly, the valve opens, allowing air into the intake manifold.

AIR MANIFOLD AND AIR INJECTION NOZZLES

Pressurized air from the air injection reactor pump is fed through the check valve and into the air manifold where it is distributed to each nozzle. These nozzles are positioned near the exhaust valves and are pointed toward the valve. Because of this positioning, air is constantly being injected into the exhaust manifold while the engine is running.

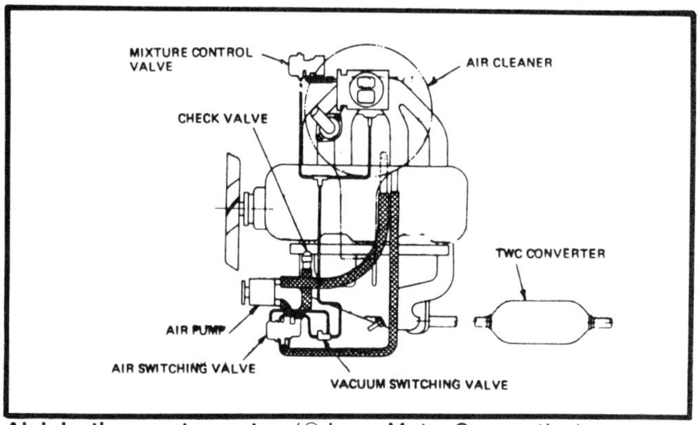

Air injection reactor system (© Isuzu Motor Corporation)

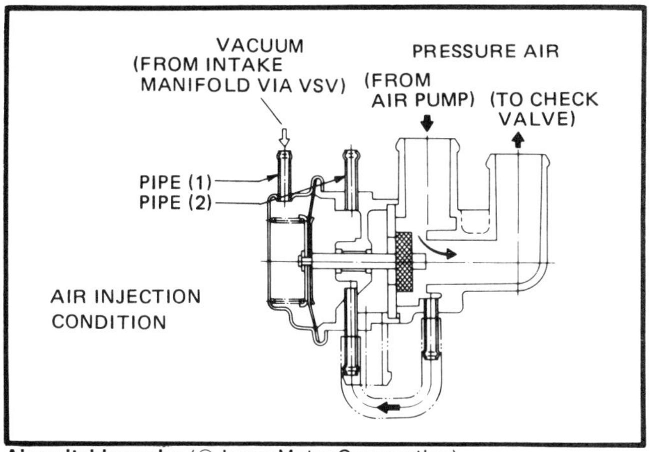

Air switching valve (© Isuzu Motor Corporation)

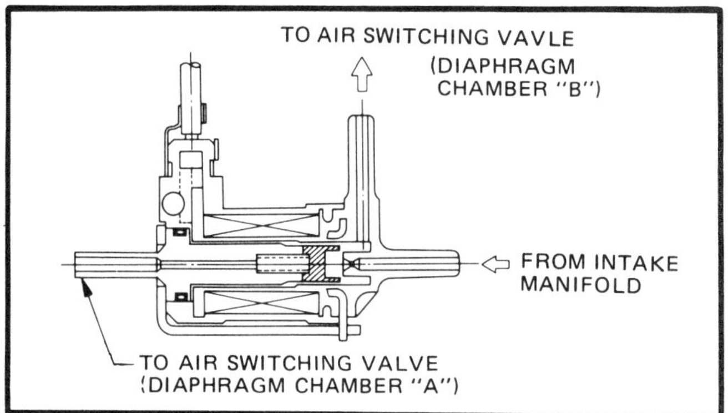

Vacuum switching valve—California vehicles (© Isuzu Motor Corporation)

Exhaust Gas Recirculation System

GAS ENGINE

The exhaust gas recirculation system is used to reduce combustion temperatures in the combustion chamber, thereby reducing oxides of nitrogen emissions. The exhaust gas is drawn into the intake manifold through the exhaust gas recirculation valve from the exhaust manifold. The vacuum diaphragm of the exhaust gas recirculation valve is connected to a signal port at the carburetor flange through a back pressure transducer, which is responsive to exhaust pressure. This is done in order to modulate the vacuum signal and the thermal vacuum valve, which operates the exhaust gas recirculation valve cold override.

As the carburetor throttle valve is opened , vacuum is applied to the diaphragm which in turn opens the exhaust gas recirculation valve and allows exhaust gas to be metered into the intake manifold. The exhaust gas recirculation system consists of an EGR valve, a back pressure transducer and a thermal vacuum valve.

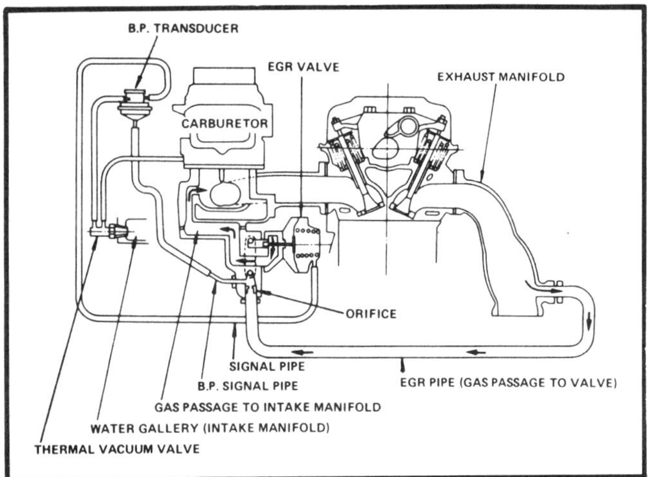

I-Mark gas engine EGR system (© Isuzu Motor Corporation)

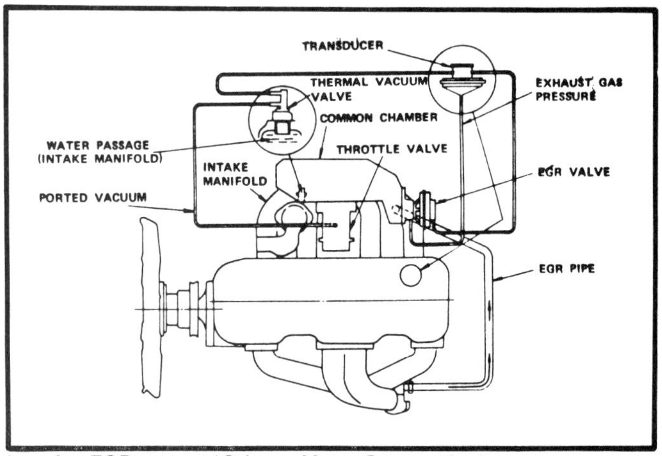

Impulse EGR system (© Isuzu Motor Corporation)

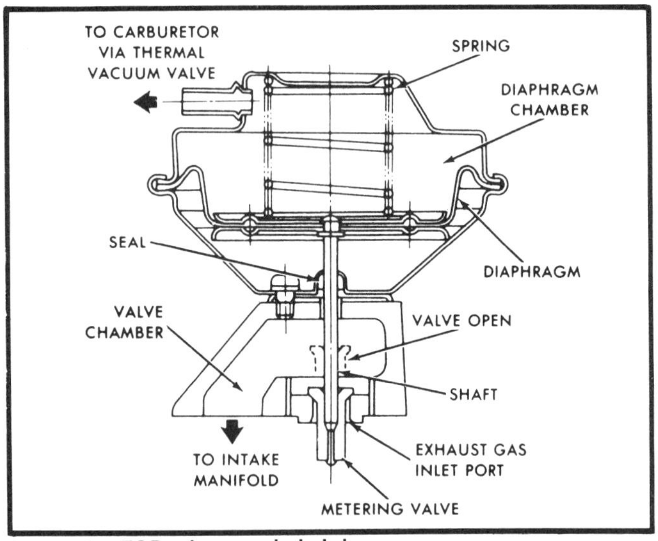

Gas engine EGR valve—exploded view
(© Isuzu Motor Corporation)

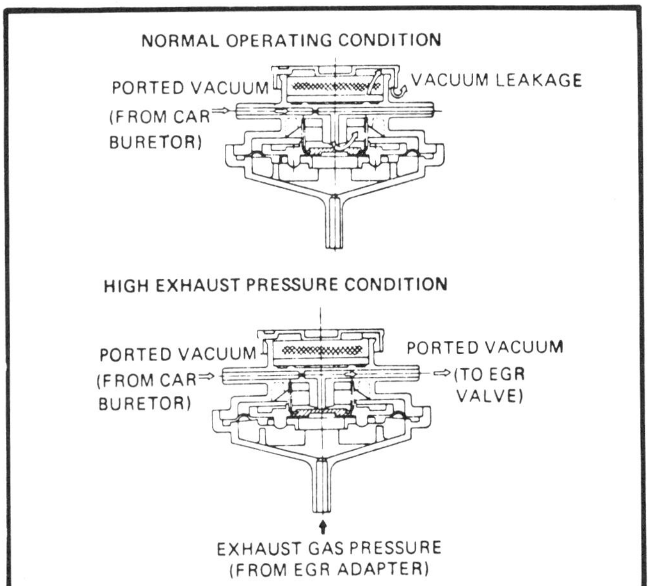

Gas engine back pressure transducer
(© Isuzu Motor Corporation)

EGR VALVE

The exhaust gas recirculation valve is mounted under the intake manifold . This component controls the exhaust gas flow that is drawn into the intake manifold from the exhaust manifold. The vacuum diaphragm chamber is connected to the vacuum port in the carburetor flange through the thermal vacuum valve and the back pressure transducer. As the throttle valve is opened, vacuum is applied to the vacuum diaphragm. When the vacuum reaches the specified value, the diaphragm moves against a spring force and is in the fully up position. As the diaphragm moves up, it opens the exhaust gas metering valve, which in turn allows exhaust gas to be pulled into the engine intake manifold.

BACK PRESSURE TRANSDUCER

The back pressure transducer valve is mounted in line with the thermal vacuum valve. This component is responsive to exhaust gas pressure. Under normal operating conditions, ported vacuum leaks into the atmosphere and is applied to the exhaust gas recirculation valve under high pressure conditions, thereby modulating the flow of recirculating exhaust gas.

THERMAL VACUUM VALVE

The thermal vacuum valve is mounted on the intake manifold and is connected in series between the vacuum port in the carburetor and the exhaust gas recirculation valve. The motive force of the valve operation is provided by the distortion of a bimetal tab. When coolant temperature is below approximately 122 degrees F., the valve is closed and the exhaust gas recirculation system is not in operation. When the coolant temperature is above approximately 122 degrees F., the valve is open and the exhaust gas recirculation system is operating.

Exhaust Gas Recirculation System

DIESEL ENGINE (CALIFORNIA ONLY)

The exhaust gas recirculation is used to reduce combustion temperature in the combustion camber, thereby reducing oxides of nitrogen emission. Recirculation gas is drawn into the intake manifold through a pipe from the exhaust gas recirculation valve. This system includes a thermo switch, a vacuum switching valve, an EGR controller, an engine speed sensor and a control lever position sensor.

EXHAUST GAS RECIRCULATION VALVE

The EGR valve controls the exhaust gas flow into the intake manifold from the exhaust manifold. As the vacuum switching valve is opened,

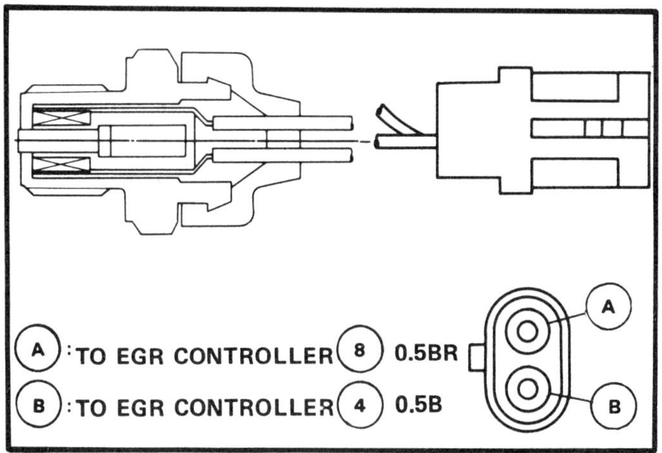

Diesel engine speed sensor (© Isuzu Motor Corporation)

vacuum is applied to the diaphragm. As the diaphragm moves up, it opens the exhaust gas valve, thus allowing exhaust gas to enter the intake manifold.

System Service

1. Check the EGR diaphragm by applying an outside vacuum source. The diaphragm should not leak down and should hold about 13.8 in. Hg. of vacuum.

VACUUM SWITCHING VALVE

The vacuum switching valve allows vacuum to reach the exhaust gas recirculation valve when it is switched on by the EGR controller, thus allowing the valve to open. This valve is mounted on the left side of the engine compartment and switches the EGR valve on by an electrical signal from the EGR controller.

System Service

Proper operation of the vacuum switching valve can be checked by carefully listening for the noise that is accompanied with the electrical

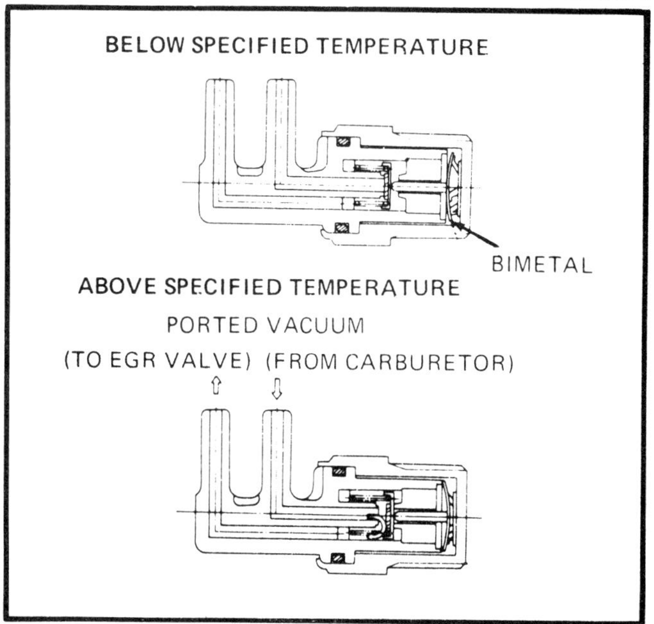

Gas engine thermal vacuum valve (© Isuzu Motor Corporation)

operation of the plunger. The plunger can be operated electrically by connecting the connector terminals directly to the battery, using the proper cables.

EGR CONTROLLER

The EGR controller sends an electrical "on-off" signal to the vacuum switching valve, according to coolant temperature , engine speed, and injection pump control· lever angle, which are sensed by the thermo switch, engine speed sensor and the control lever position sensor respectively. Injection pump control lever angle is transformed into voltage output by the control lever position sensor. The EGR controller is mounted together with the U.Q.O.S. controller.

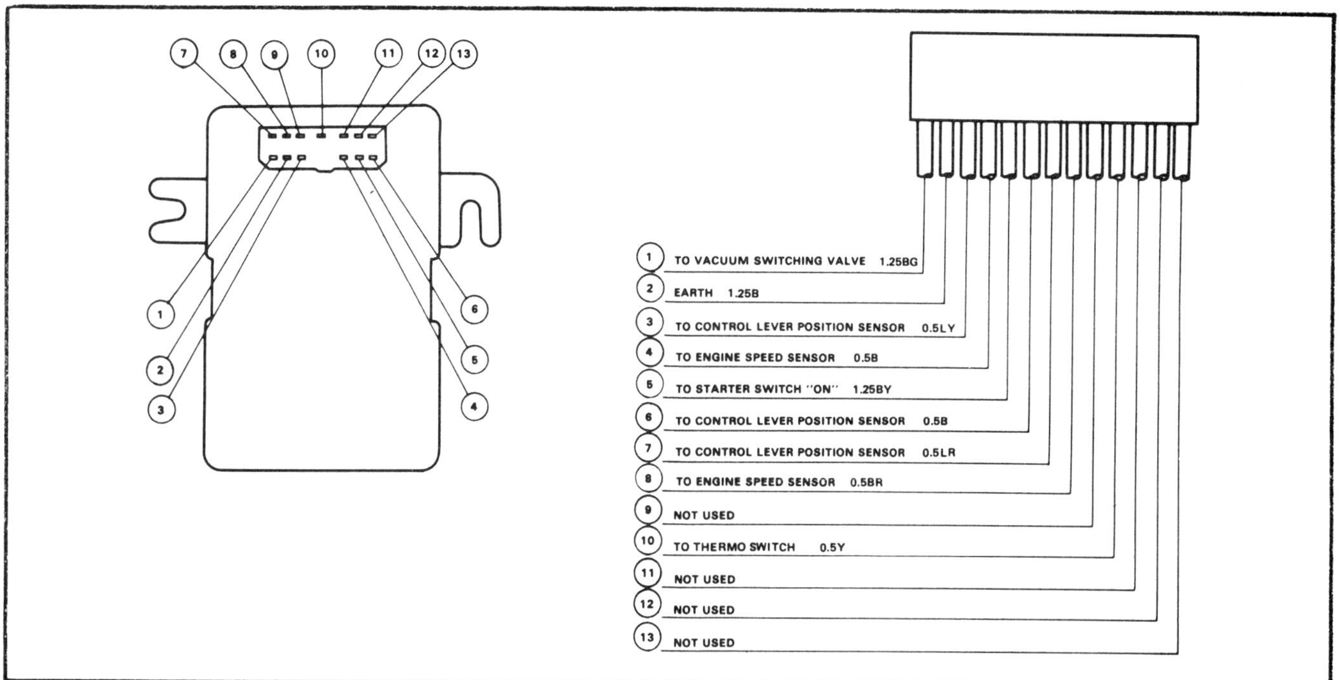

Diesel engine EGR controller (© Isuzu Motor Corporation)

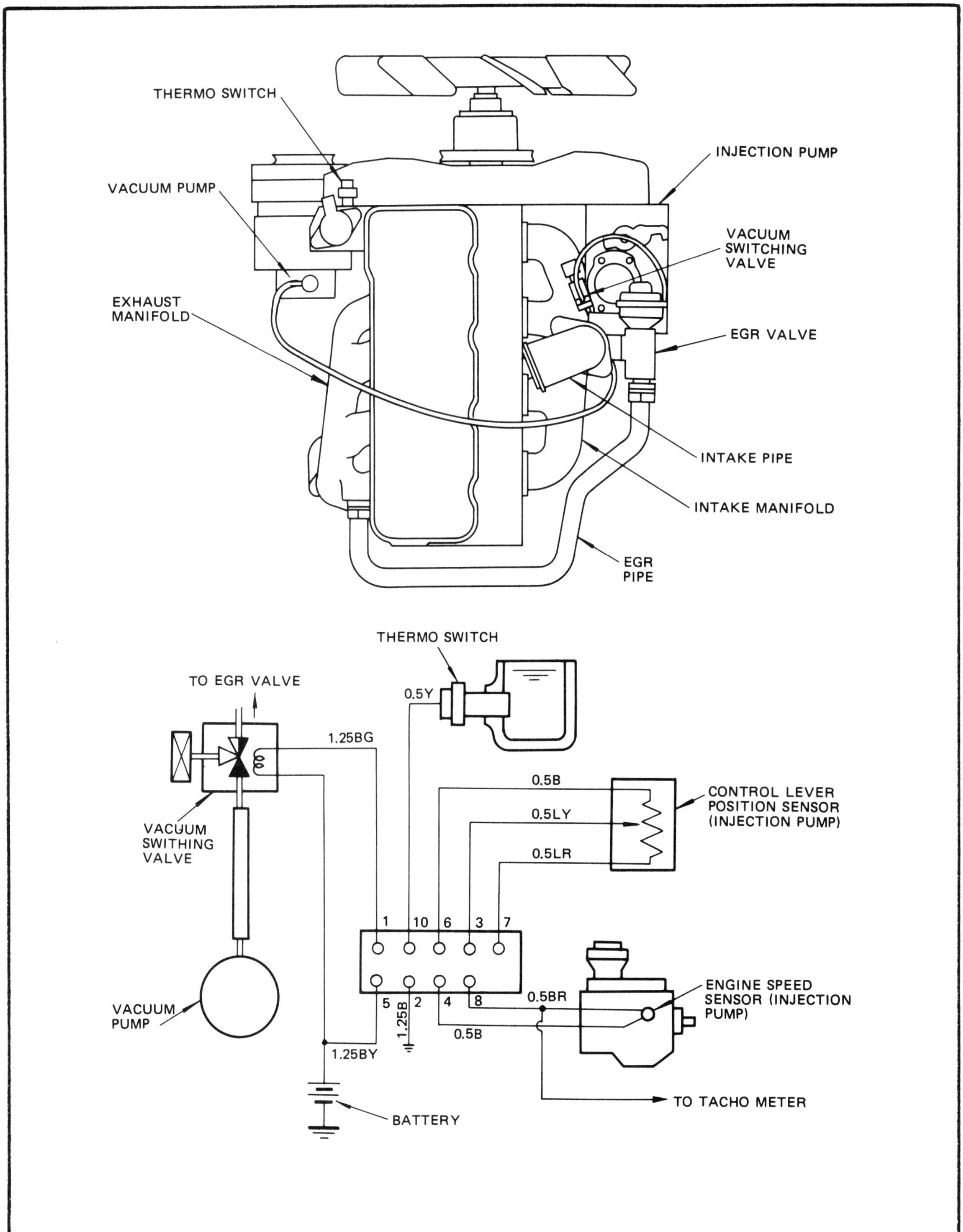

Diesel engine EGR system—California vehicles (© Isuzu Motor Corporation)

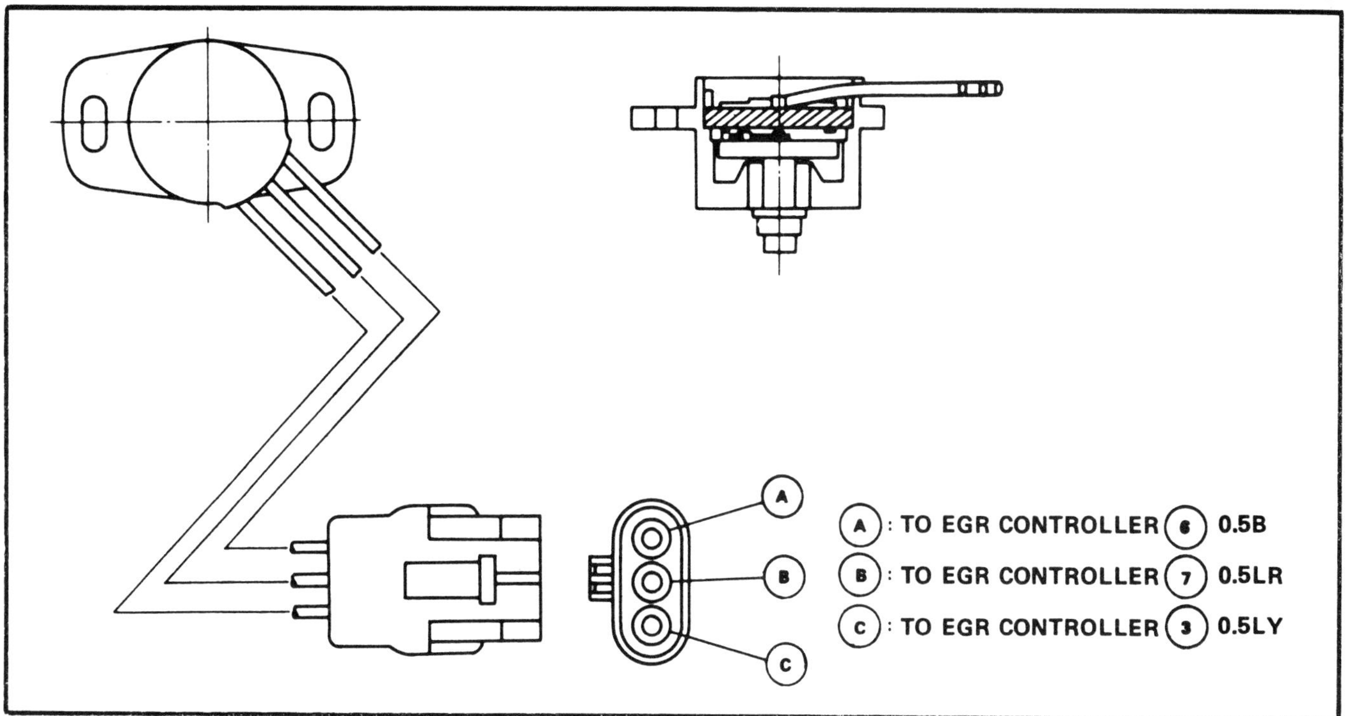

Diesel engine control lever position sensor (© Isuzu Motor Corporation)

System Service

1. Connect a voltmeter to the "black/green" and "yellow/black" color coded wire terminals at the vacuum switching valve wiring connector.
2. The "black/green" wire is positive and the other wire is negative.
3. The EGR controller is functioning properly if the voltage is about 12 volts when the engine speed is over 1200 rpm.

ENGINE SPEED SENSOR

The engine speed sensor is mounted on the right side of the fuel injection pump. It is an electromagnetic pickup and sends signals corresponding to the movement of the gear teeth to the EGR controller.

CONTROL LEVER POSITION SENSOR

The control lever position sensor is mounted on the front upper part of the fuel injection pump. It is a potentiometer and the resistance from terminal "B" to terminal 'A" and "C" varies, depending on the control lever position.

System Service

1. Connect a voltmeter to the "blue/yellow" and "blue/red" color coded terminals, without disconnecting the control lever position sensor connector.
2. The "blue/yellow" is positive and the other wire is negative.
3. Move the control lever so as to set the clearance between the control lever and the idle stopper bolt to 7mm.
4. The control lever position sensor is normal if the voltage is approximately 3.3–4.4 volts, after starting the engine.

THERMO SWITCH

The thermo switch detects coolant temperature and shuts the function of the EGR valve off during engine warm up. This function maintains acceptable cold driveability. The switch is mounted on the thermostat housing. While coolant temperature is below 131 degrees F., the circuit is open and the switch does not send electrical current to the EGR controller. When the temperature is above 131 degrees F., the switch sends electrical current to the EGR controller.

System Service

Submerge the end of the thermo switch in water and raise the temperature of the water gradually, while making a continunity test across the terminal and body, using a circuit testor.

Closed Loop Emission Control System

The closed loop emission control system precisely controls the air/fuel ratio near stoichiometry, which allows the use of a three way catalyst in order to reduce exhaust emissions. The major components of this system are, an exhaust gas oxygen sensor, an electronic controller, a vacuum controller which incorporates a duty solenoid, a controlled air/fuel ratio carburetor and a three way catalytic converter.

OXYGEN SENSOR

The oxygen sensor that is used in this system consists of a closed end zirconia sensor which is mounted in the exhaust gas stream of the exhaust manifold. The sensor generates a voltage which varies with the oxygen content in the exhaust gas stream. As the oxygen content rises, voltage falls and as the oxygen content falls, voltage rises.

ELECTRONIC CONTROL MODULE

The electronic control module generates a control signal to the vacuum controller solenoid which controls the carburetor air/fuel mixture through vacuum signals. The control signal is constantly cycling the solenoid between on and off as a function of the input voltages from the sensors. The signal that is generated by the electronic control module is selected from four operational modes. A description of these modes are as follows.

1. Inhibit Mode indicates no signal to the vacuum controller solenoid.
2. Enrichment Mode indicates a fixed duty cycle to the vacuum controller solenoid.

3. Open Loop Mode indicates a fixed duty cycle to the vacuum controller.

4. Closed Loop Mode indicates that a calculated duty cycle is generated based on the oxygen sensor and other sensor outputs.

During closed loop operation, the electronic control module monitors the voltage output of the oxygen sensor. The sensor voltages increases and passes through the electronic control module threshold set point and the porportional gain immediately changes the duty cycle on the output signal. Again the duty cycle is changed when, at a constant rate, sensor input voltage decreases and passes through the electronic control module threshold set point. The selection of integral and porportional gain rates by the electronic control module is based on engine operating conditions.

At idle condition, different gain rates are required for optimum air/fuel ratio control than those at partial load condition. The electronic control module also stores the current duty cycle being used for either idle or off idle engine operation. When the electronic control module senses a transition from idle to off idle, it steps to the duty cycle last recorded for stoichiometric operation. From then on, while the engine is at that operating condition, the system uses the basic proportional and integral gain controls to function. The electronic control module outputs the signal to control the slow cut solenoid valve, which is incorporated in the carburetor.

The electronic control module senses the coasting condition by means of the signals from the transmission gear position switch, clutch pedal position switch, and the idle position switch, which cuts of the flow of electrical current to the slow cut solenoid valve in such an instance when the engine running speed is beyond the limit of the specified speed, thus stopping the fuel flow to the carburetor. The flow of electric current to the slow cut solenoid valve is cut off only at the time when the vacuum signal of the vacuum switch is below the specified vacuum and when the engine speed exceeds the limit of the specified speed. It is noted that these two conditions must exist simultaneously for this to occur.

CATALYTIC CONVERTER

A three way catalytic converter is used which reduces oxides of nitrogen while oxidizing hydrocarbons and carbon monoxide. The converter is also used, in this system, to maintain the high conversion efficiency that is necessary to closely control the air/fuel mixture ratio near stoichiometry.

VACUUM CONTROLLER

The function of the vacuum controller is to convert the electrical signals that are transmitted from the electronic control module into vacuum signals by using the vacuum control solenoid. These vacuum signals make the fuel control actuators in the carburetor function, thus control the air/fuel ratios. The vacuum controller consists of a vacuum regulator and a vacuum control solenoid. Of these two, the first regulates the fluctuating vacuum from the intake manifold to the constant specified vacuum, while the latter regulates the opening and closing of the passage from the vacuum regulator by means of the signals coming from the electronic control module.

AIR SWITCHING VALVE

The electronic control module provides the signals to actuate an air switching valve via a vacuum switching solenoid valve. The air switching valve is open to allow the injection of the air into the exhaust ports of the engine when the temperature of the coolant is below a specified level and during a given period of time after the accelerator is depressed to the wide open throttle condition.

VACUUM SWITCHES

Electronic vacuum switches sensing vacuum are connected to the electronic control module. These switches enable the electronic control module to discriminate between closed throttle and open throttle.

TEMPERATURE SWITCH

The temperature switch is a thermo type switch which provides engine coolant temperature data to the electronic control module microprocessor. This temperature information is used as one of the criteria to determine when a system is ready to enter the closed loop mode of operation. The information from this switch is also used to determine the secondary air injection system operation.

SYSTEM MALFUNCTION LAMP

This system uses a dashboard mounted malfunction indicator lamp, which, for some failure modes, will inform the driver of the need for maintenance. In the event of a system failure, the "check engine" light will come on and remain on as long as the fault function occurs and the engine is running. The electronic control module incorporates a diagnostic program which will help in solving closed loop control system discrepancies. When activated, the diagnostic program will flash a code through the malfunction lamp, which in turn isolates the source of the system discrepancy.

TROUBLE DIAGNOSIS PROCEDURE

The location of a problem can be determined by the blinking of the "check engine" light. The light will flash, indicating a malfunction code when the diagnostic connector on the control unit is grounded with the engine running. The engine must be running in order to ground the terminal. The light will blink in patterns that will indicate the malfunction code. Malfunction codes 12 through 15 may be indicated even though nothing is wrong with the system. For example, trouble code 14 will be indicated without any trouble in the system if the wide open throttle condition is not brought about through the diagnostic procedure, but does not indicate a fault in the system. There are nine possible patterns indicating trouble within the system.

High Altitude Emission Control System

The high altitude compensation system consists of an altitude switch, solenoid valve and the electronic control module. This switch senses the altitude from atmospheric pressure and energizes the solenoid valve and the electronic control module at high altitude. The solenoid valve opens and leans the air/fuel mixture by increasing the air flow to the carburetor altitude compensation passage. The electronic control module changes the open duty cycle, also leaning the air/fuel mixture at high altitude and reducing the hydrocarbon and carbon monoxide emissions.

Incorporated within this system is a solenoid valve and an altitude switch. The solenoid valve is activated by current which is sent from the altitude switch. Under high altitude conditions the solenoid valve is opened to send air from the air cleaner to the three altitude compensation passages which are located in the carburetor. The altitude switch consists of a micro switch and a bellows. The bellow is expanded or contracted due to changes in the atmospheric pressure, or at a pressure lower than the calibrated atmospheric pressure.

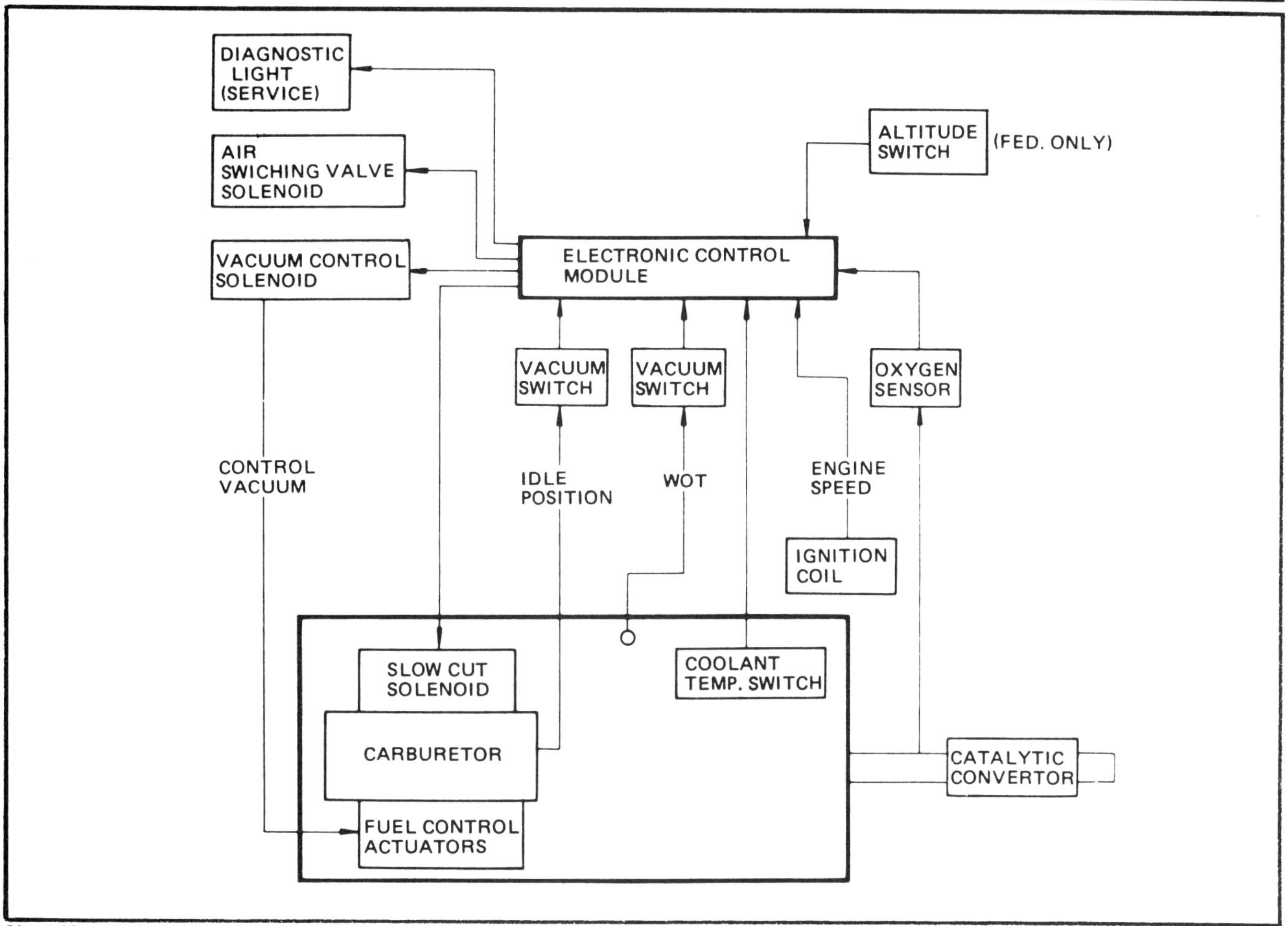

Closed loop control system schematic (© Isuzu Motor Corporation)

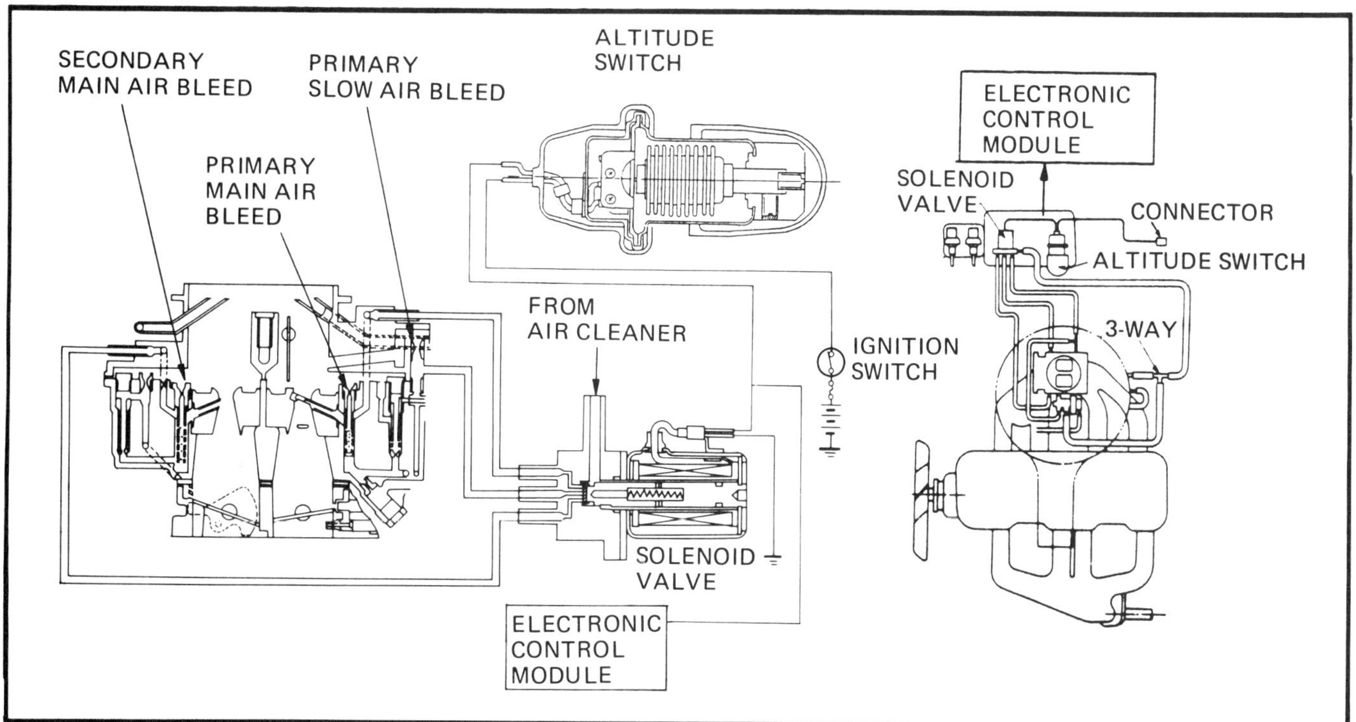

High altitude emission control system components (© Isuzu Motor Corporation)

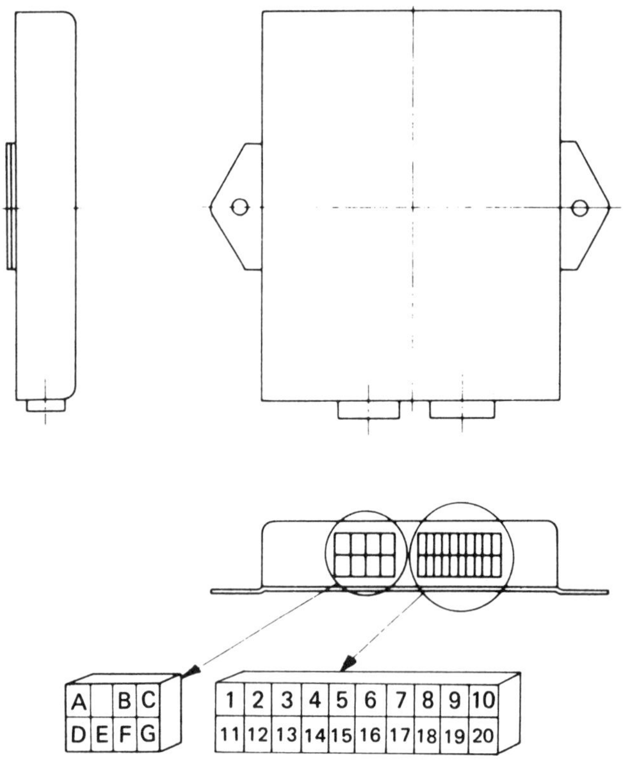

ELECTRONIC CONTROL MODULE CONNECTIONS IDENTIFICATION

Pin	Function	Pin	Function
A	Power Ground	7	WOT Switch Input
B	System Malfunction Lamp	8	Coolant Temperature Switch Input
C	Ignition Coil Tach Input	9	Not Used
D	AIR Solenoid Output	10	Not Used
E	Vacuum Control Solenoid	11	Idle Position Switch Ground
F	Carburetor Fuel Cut Solenoid	12	WOT Switch Ground
G	+12V to Ignition	13	Oxygen Sensor Ground
1	Control Ground	14	Not Used
2	Oxygen Sensor Input	15	Not Used
3	Not Used	16	Not Used
4	+12V to Ignition	17	Not Used
5	Not Used	18	Diagnose Decode
6	Idle Position Switch Input	19	Diagnose Decode Ground
		20	Not Used

Location of pin connectors on the electronic control unit (© Isuzu Motor Corporation)

DIAGNOSIS CIRCUIT CHECK

Rubber hose and wire should be checked before this check

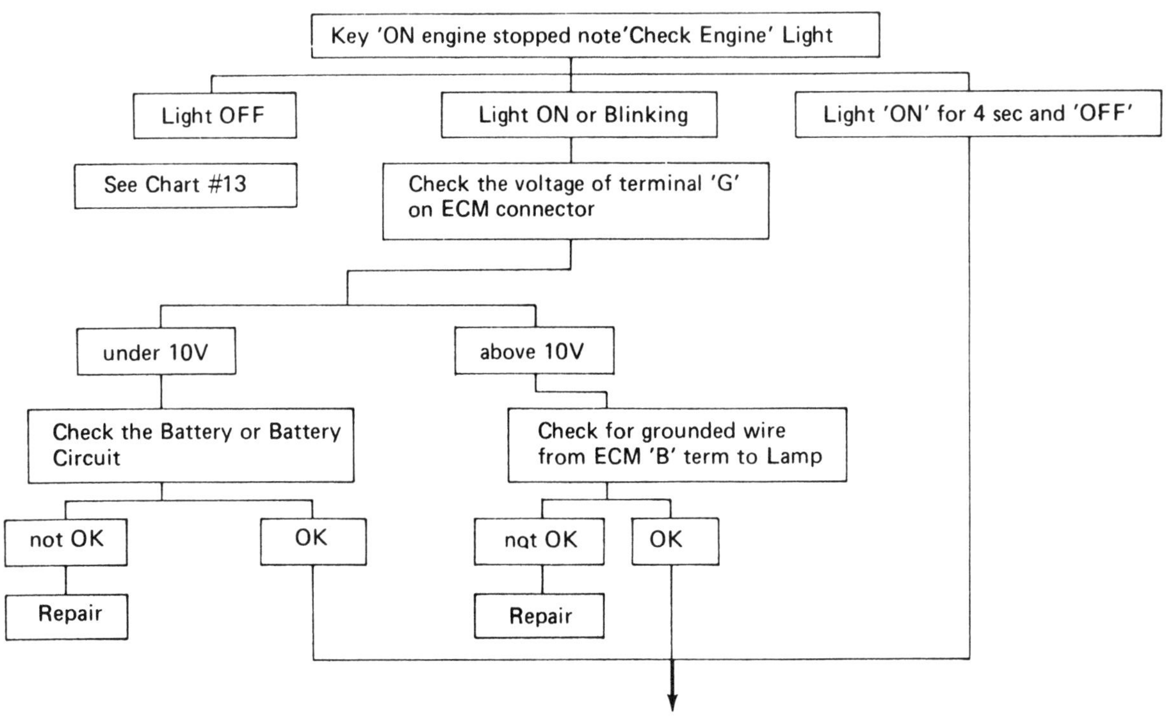

DIAGNOSIS CIRCUIT CHECK

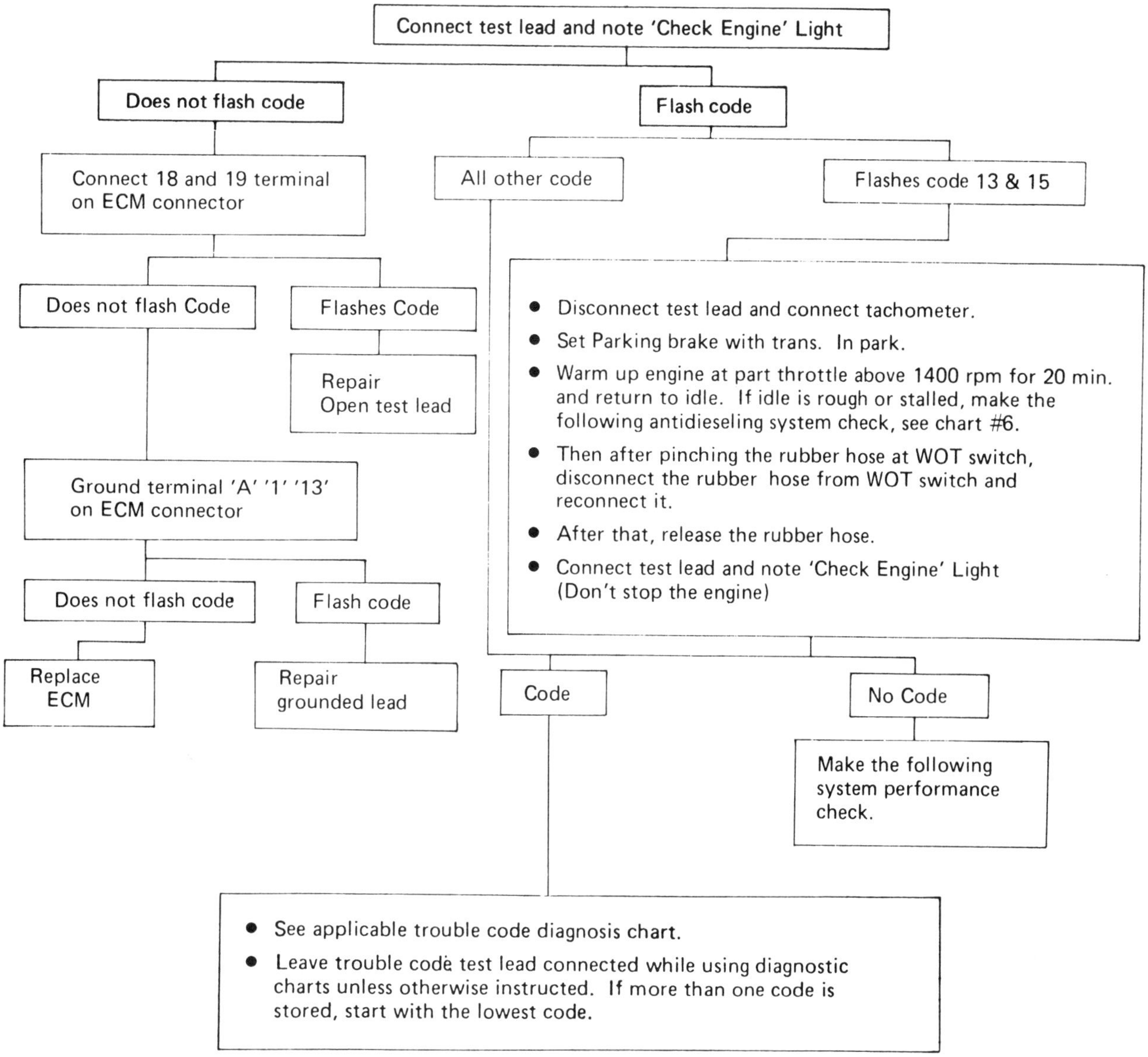

SYSTEMS PERFORMANCE CHECK

Driver complaint or emission failure engine performance problem, (ODOR, SURGE, FUEL ECONOMY)

Inspect for harness and hose disconnects at coolant sensor duty solenoid, etc. repair as necessary.

- Cold operation complaint and Full throttle performance complaint see chart #5.
- All other complaints - follow chart below on warm engine (Upper radiator hose hot)

NOTICE: The system performance check should be performed after any repairs to this system have been made.

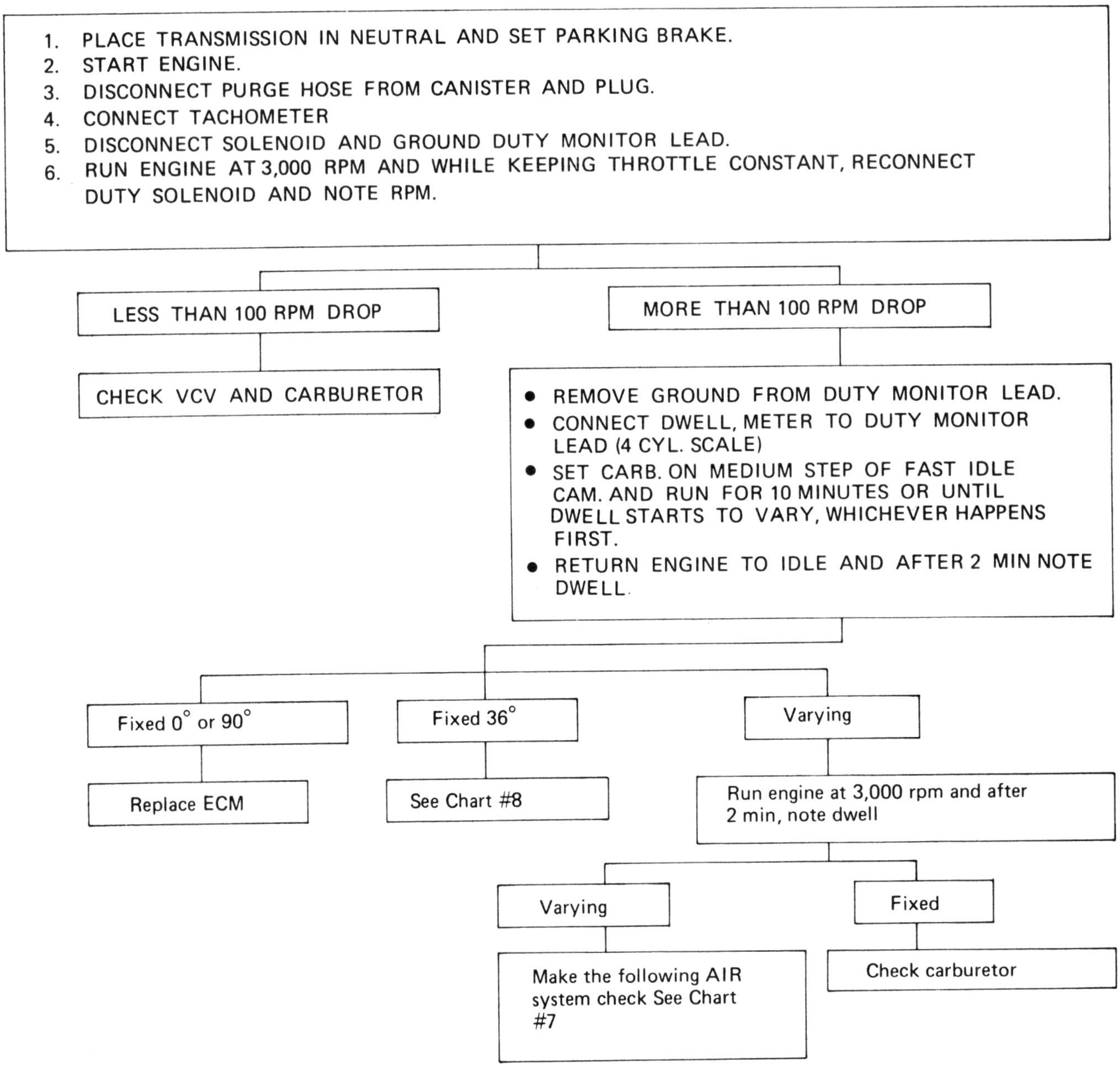

CHART NO. 1

Check Procedure for VCV

Before this check, confirm all vacuum hoses for splits, kinks, clogged and proper connection.

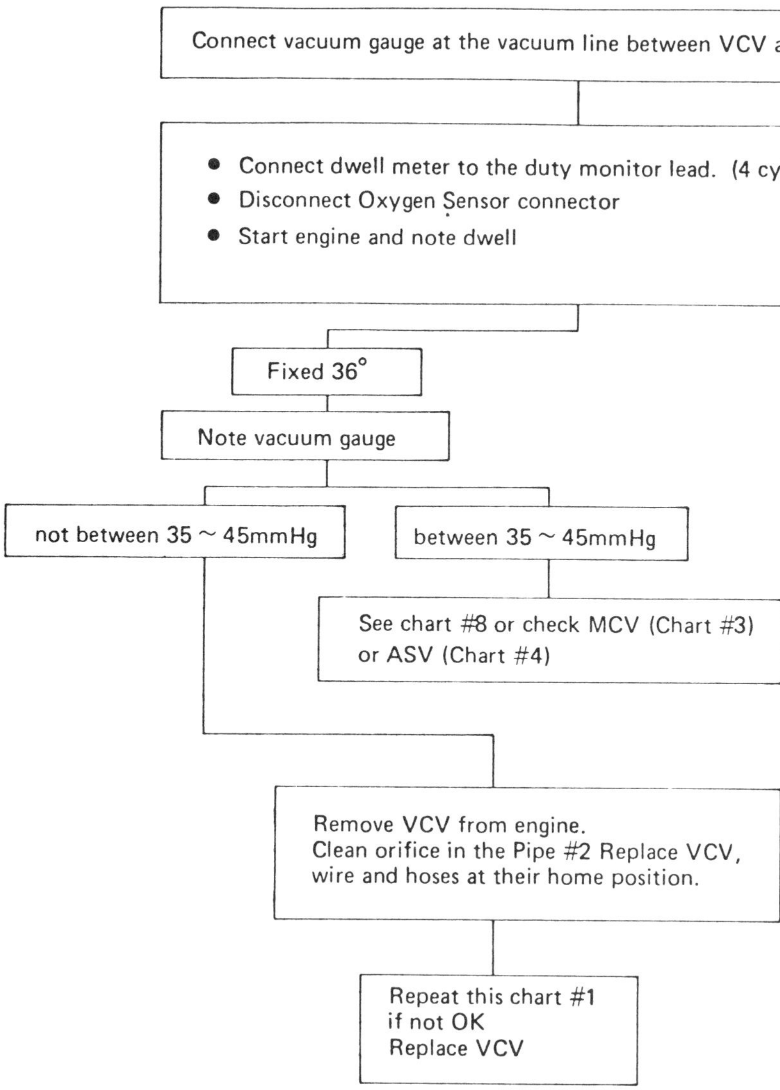

CHART NO. 2

Before this check confirm all connections of rubber hoses and wires.

1. Place transmission in Neutral and set parking brake and block drive wheels.
2. Disconnect purge hose at Canister and plug it.
3. Connect dwell meter to duty monitor lead (4 cyl. scale).
4. Set carb on high step of fast idle cam and run engine until dwell starts to vary.
5. Return engine to idle and note dwell.
6. Confirm dwell varying.
7. Disconnect oxygen sensor connector and measure time required for the dwell meter to read a fixed 36°.

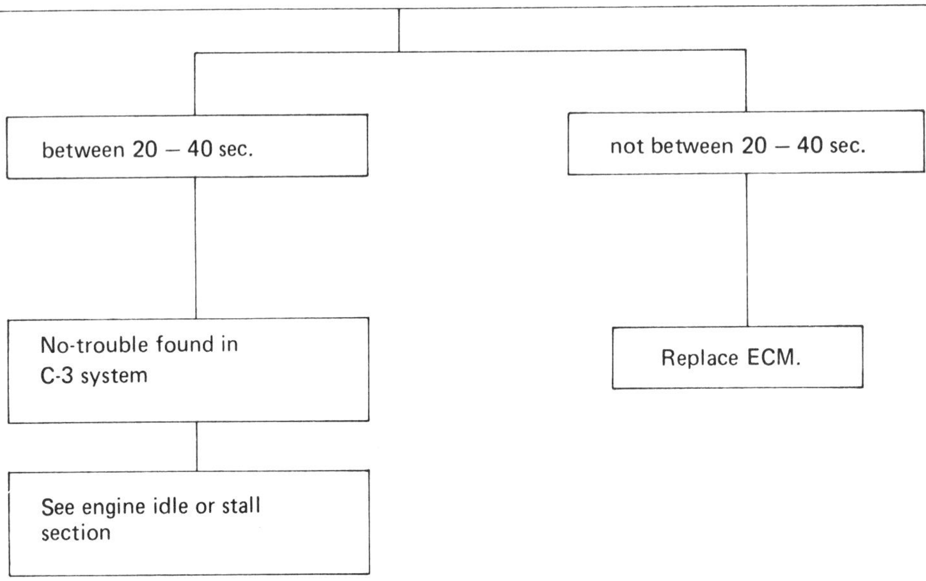

CHART NO. 3

Before this check, confirm the connection of rubber hose and tube.

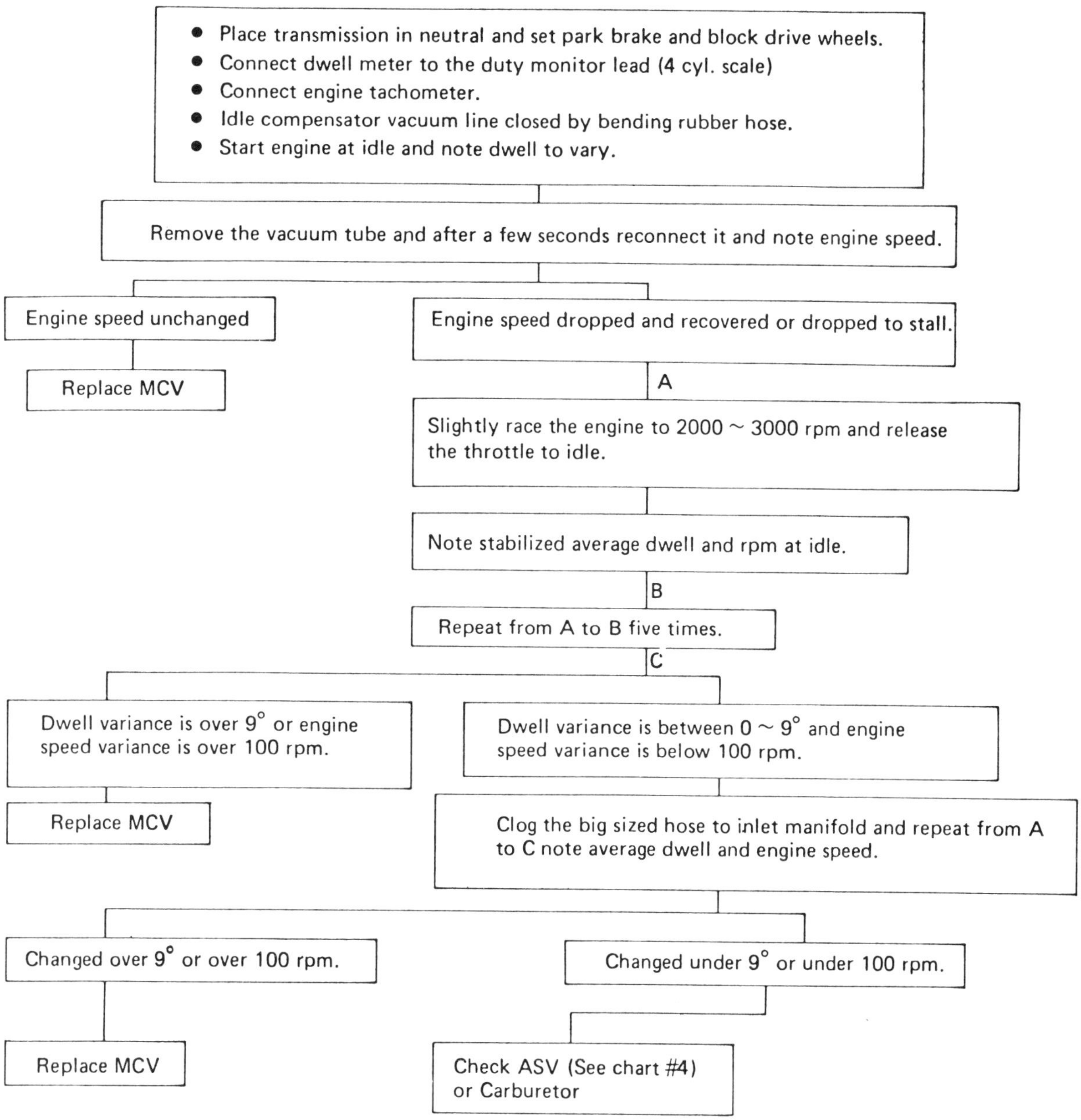

CHART NO. 4

Before this check, confirm the rubber hose for splits, kinks and proper connections.

- Place transmission in Neutral and set park brake and block drive wheels.
- Disconnect the rubber hose from nipple A and B.
- Connect vacuum pump to nipple A.
- Supply vacuum over 100mmHg to diaphragm chamber.
- Note vacuum pump.

Hold | Not hold → Replace ASV

(Hold:)

- Start engine.
- Disconnect the rubber hose from ASV to C/V and from ASV to A/C.
- Check air switching function.

(Diaphragm chamber vacuum)		(Air outlet)
Supplied over 100mHg	→	from Pine B'.
Atmosphere	→	from Pipe A'.

OK | not OK → Replace ASV

(OK:)

- Disconnect vacuum pump from nipple A' and connect nipple A and B by rubber hose.
- Connect dwell meter to duty moniter lead (4 cyl. scale)
- Plug C/V and warm up engine until dwell starts to vary.
- With engine idling, note average dwell.
- Connect the rubber hose from ASV to C/V, note average dwell.
- Compare with those dwells.

Difference over 9° → Replace ASV

Difference under 9° → Check MCV or Carburetor (See chart #3)

ASV : Air Switching Valve
C/V : Check Valve
A/C : Air Cleaner

CHART NO. 5

Open Loop Mode and WOT Enrichment Mode Circuit Check.
Before this check, confirm the operation of coolant temp switch itself and the proper connection of rubber hose and wire.

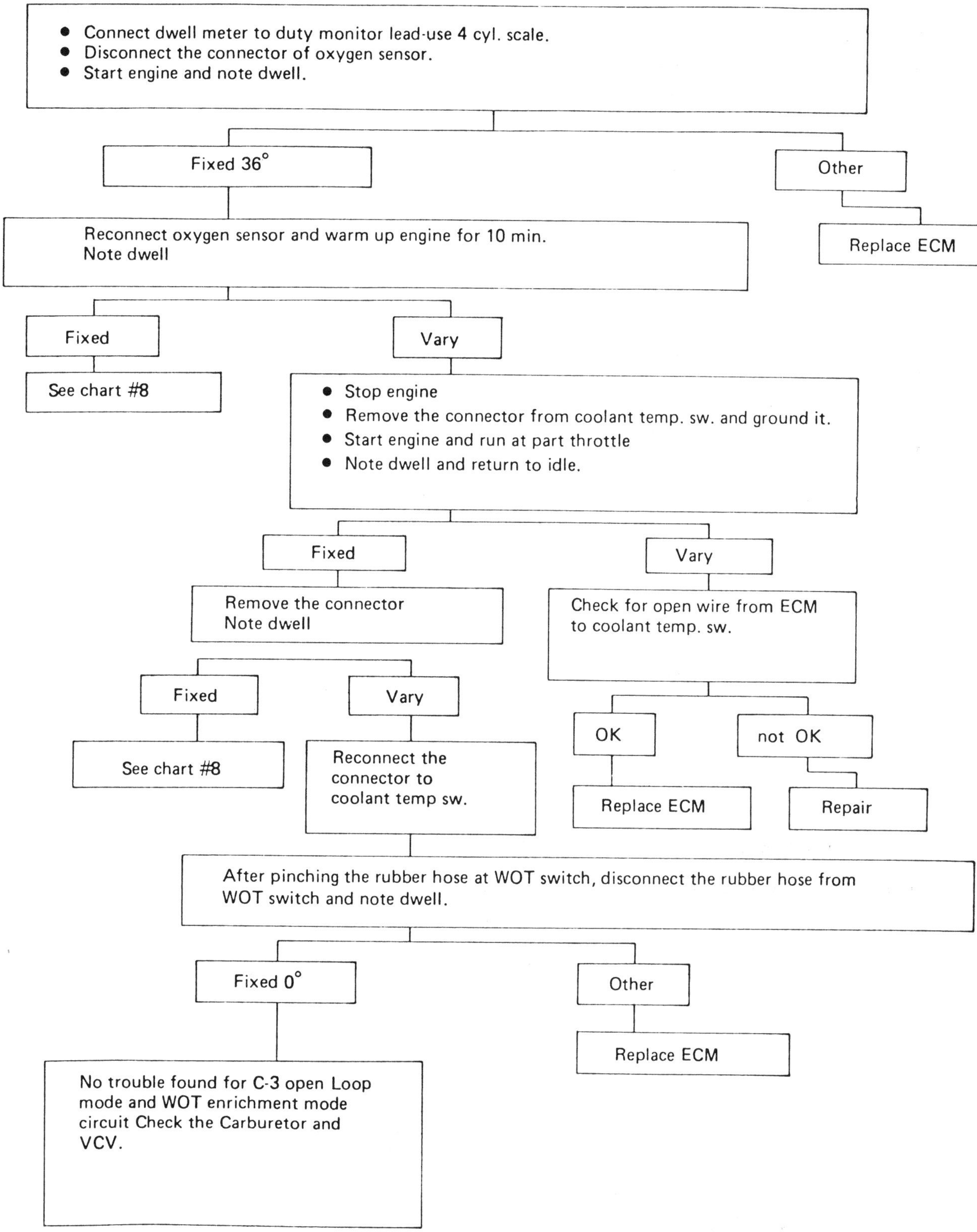

CHART NO. 6

Antidieseling system check

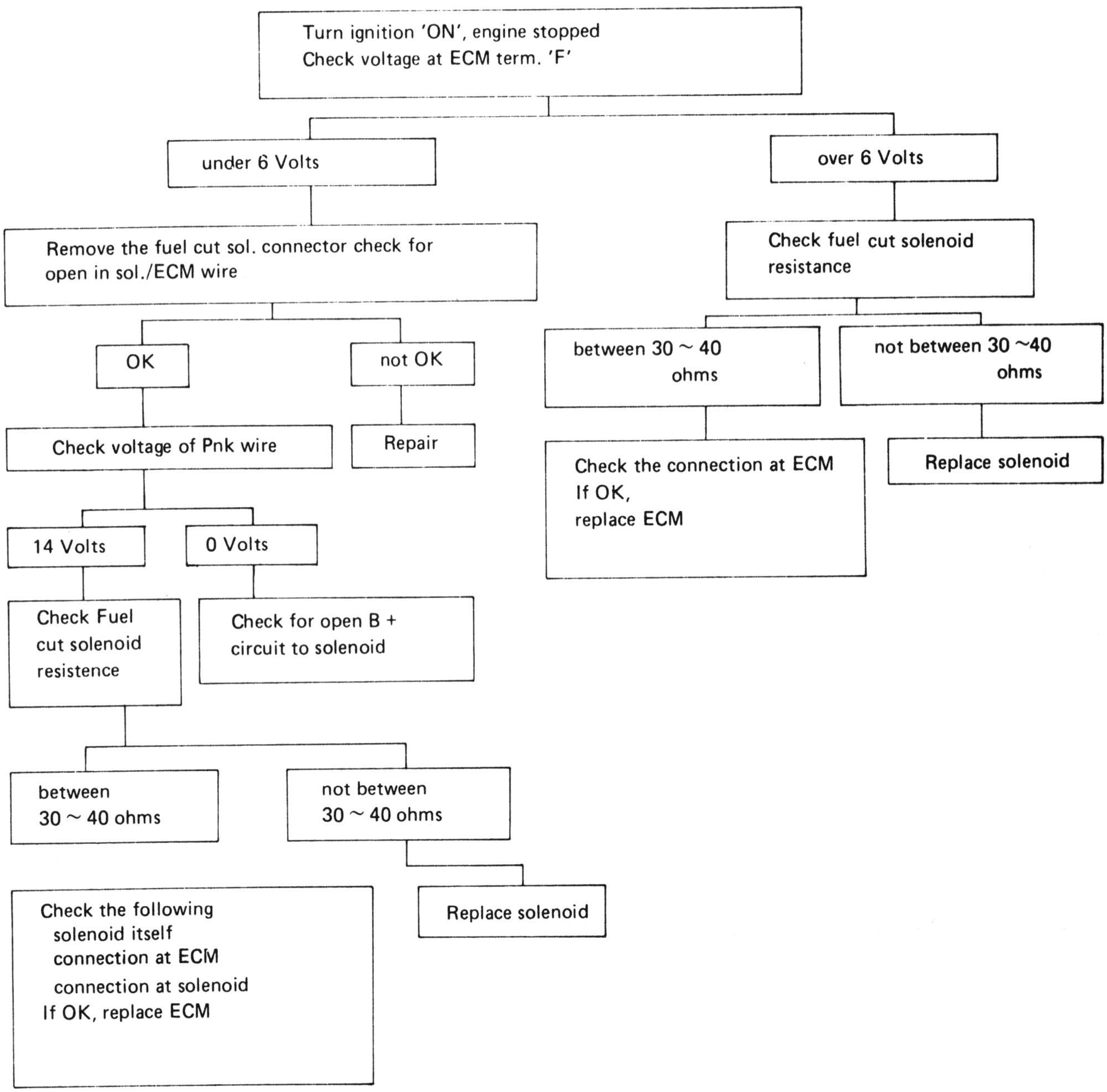

CHART NO. 7

AIR System Check

Make visual inspection of hoses and connectors for leaks and proper connections.
Repair as necessary.

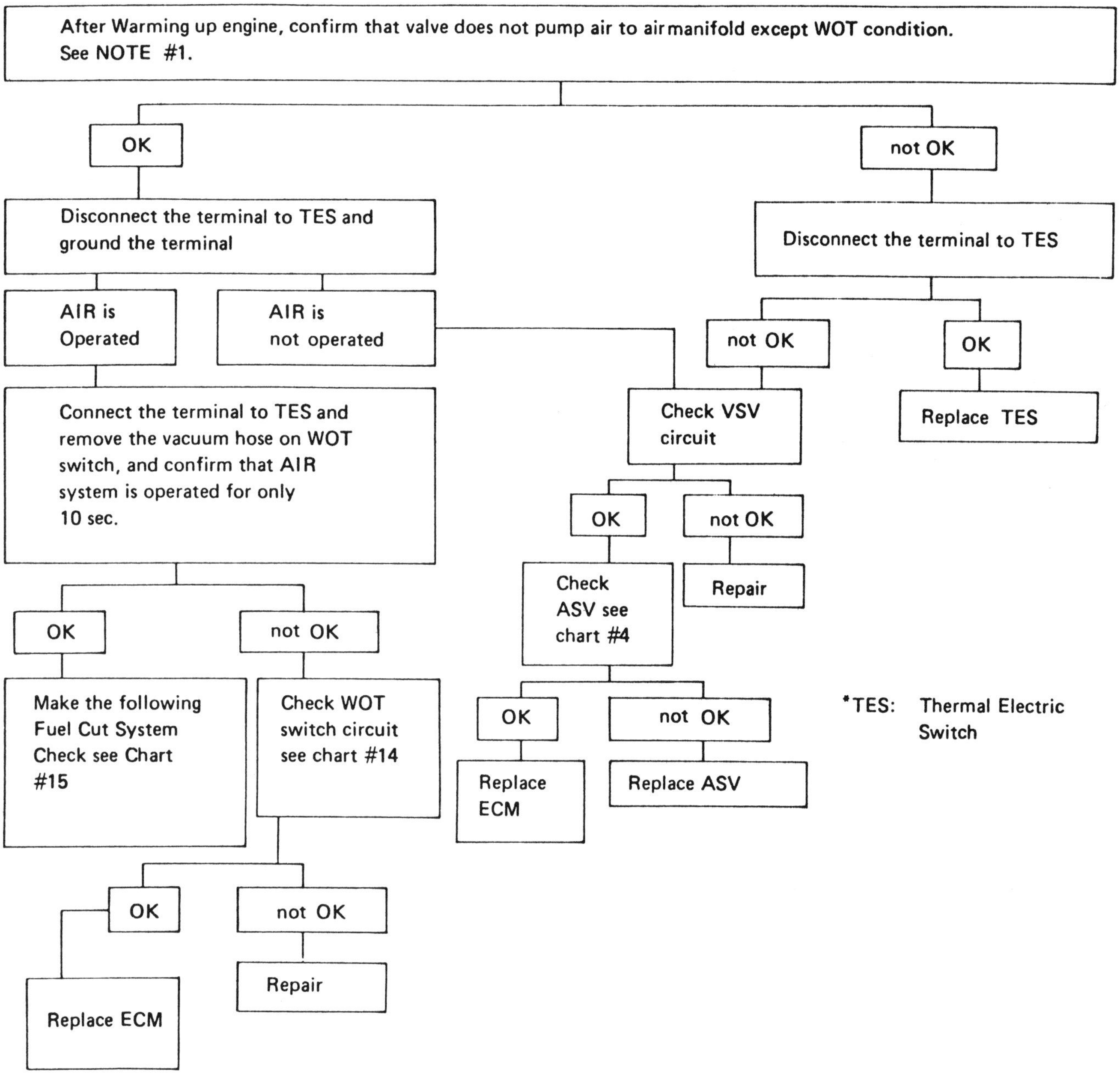

*TES: Thermal Electric Switch

The check for the operation of AIR system should be done by disconnecting the rubber hose from ASV to Air Manifold.

CHART NO. 8

Fixed 36°

Rubber hose and wire should be checked before this test

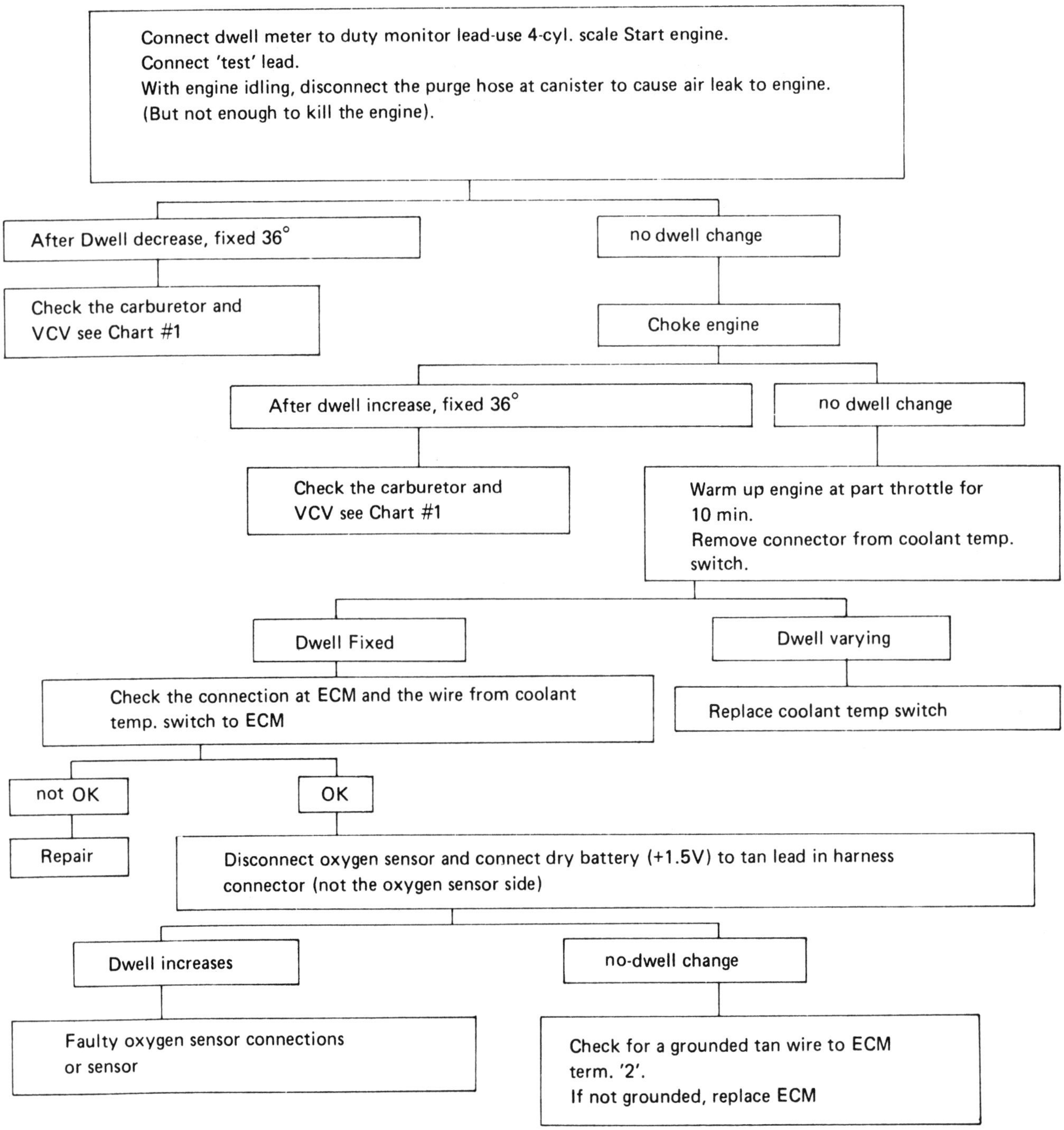

CHART NO. 9

Malfunction Code 21, 22

Check connections at duty solenoid. If O.K.:

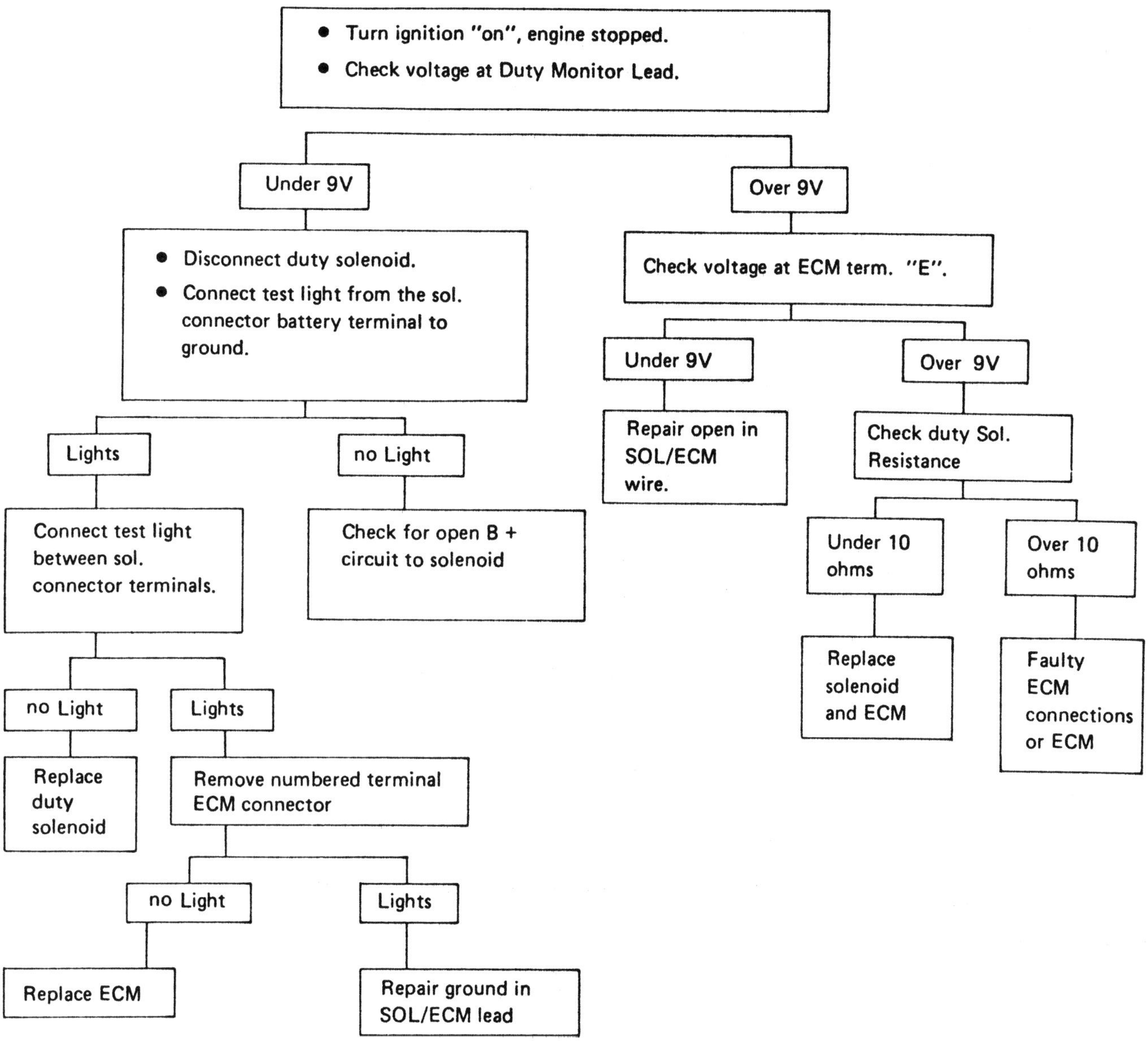

CHART NO. 10

Malfunction Code 23

When this code is indicated, don't stop the engine before this check.
If the engine was stopped, before this check, warm up engine for 20 min.

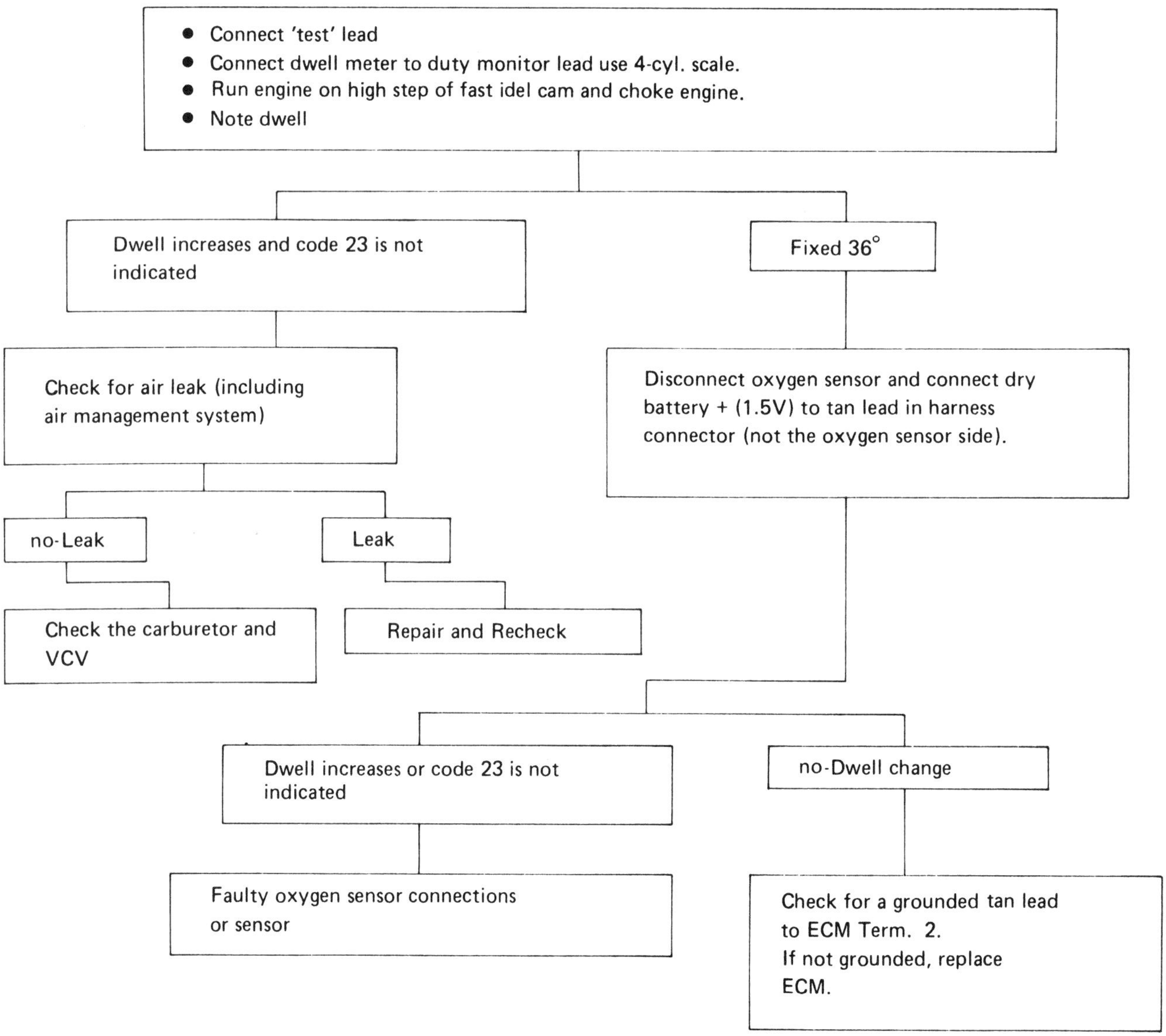

CHART NO. 11

Malfunction Code 24

Coolant Temperature switch (Thermal Electric switch) has failed.

When this code is indicated, don't stop the engine before this check.
If the engine was stopped, before this check, warm up engine above 1400rpm for 20min. and confirm that code 24 is indicated.

- With engine idling
- Connect 'test' lead
- Connect dwell meter to duty monitor lead
- Run the warm engine at part throttle for 5min.
- With engine idling, remove connector from coolant temp. sw.

- Dwell varied and code 24 is not indicated
  - Replace coolant temp. sw.
- Dwell fixed
  - Check the connections at ECM and the wire from coolant temp. sw. to ECM.
    - not OK
      - Repair
    - OK
      - Replace ECM

CHART NO. 12

Malfunction Code 25

Check that all pins are fully inserted in the socket and that 'A' '1' '13' is grounded completely.
If O.K., replace ECM.

CHART NO. 13

"CHECK ENGINE" LIGHT INOPERATIVE

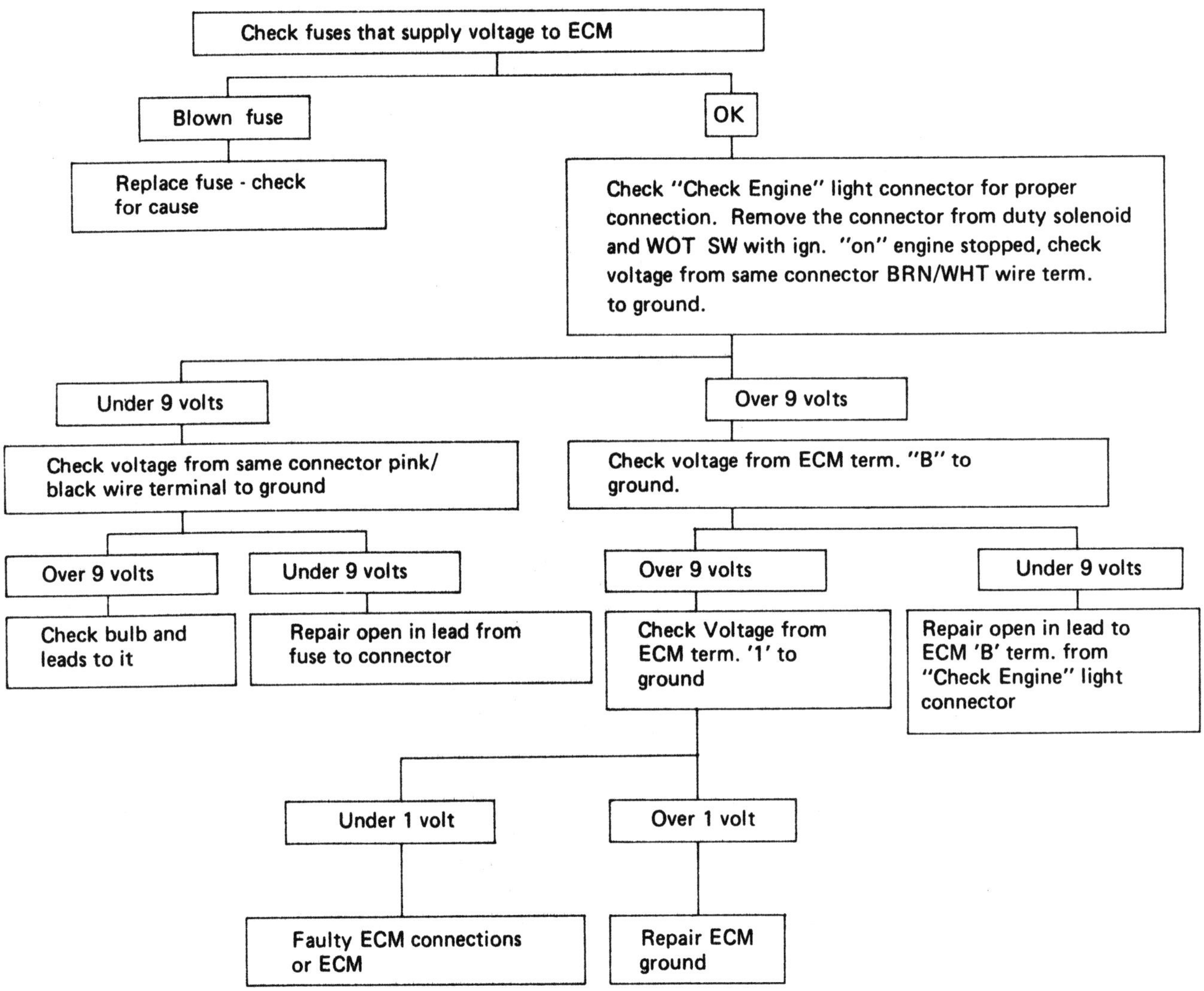

CHART NO. 14

Malfunction Code 12 ~ 15 — Idle or WOT switch is failed

Before this check, confirm the proper connection of rubber hose and wire

- Stop engine
- Check the operation of Idle (Code 12, 13) or WOT (code 14, 15) switch by using ohmmeter and vacuum pump

OK →

Check the following

- Connection of the wire at ECM
- Connection of the wire at the switch.
- Rubber hose itself.
- Wire itself.

  - **OK** → Replace ECM
  - **not OK** → Repair

not OK → Replace switch

CHART NO. 15

Fuel Cut system check

- Connect tachometer and warm up engine for 10 min.
- Run the vehicle in the first gear.
- While decelerating (Don't push clutch and accel pedal), confirm the following operation by checking engine speed and the voltage of ECM term. 'F'.

Engine speed	Voltage of ECM term 'F'
above 2000 ~ 2200 rpm	above 6V
under 2000 ~ 2200 rpm	under 6V

OK → Make the following ECM Check see Chart #2

not OK → Check the clutch switch or the Neutral switch and wire or the wire from ECM term 'C' to Ig Coil.

- **OK** → Replace ECM
- **not OK** → Repair

CHART NO. 16

Check for Canister

Before this check, confirm all vacuum hoses for splits kinks, clogged and proper connections.

Disconnect the purge signal line at the canister Connect vacuum pump to purge signal pipe and supply vacuum 5 ~ 8 in. Hg

- **Hold**
  - Disconnect vacuum pump and reconnect the rubber hose
- **not hold**
  - Replace canister

Place transmission in Neutral and set park brake and block drive wheels.
Connect dwell meter to the duty monitor lead (4 cyl. scale)
Connect vacuum gauge to the rubber hose from canister to TVV (near the canister)
Connect tachometer
Idle compensator vacuum line closed by bending rubber hose.
Warm up the engine until dwell starts to vary.
Return to idle and confirm vacuum gauge is under 1 in. Hg

A

Race the engine and confirm the gauge moves over 5 in. Hg.
Return to idle and after 1 min, note dwell and engine speed.

B

Repeat from A to B five times

- **Dwell vary**
  - Dwell variation is over 9° or engine speed variation is over 100rpm
    - Replace canister
  - Dwell variation is under 9° or engine speed variation is under 100rpm
    - Pinching the rubber house from canister to Inlet Manifold, note dwell and compare with average dwell before pinching
      - Difference over 9°
        - Replace canister
      - Difference under 9°
        - Canister is OK
- **Dwell Fixed**
  - Engine speed variation is over 100rpm
    - Replace canister
  - Engine speed variation is under 100rpm
    - Pinching the rubber hose from canister to Inlet Manifold, slightly racing the engine to 2000 ~ 3000rpm and release the throttle to idle and after 1 min, note engine speed and compare before pinching

CHART NO. 17

Check for carburetor slow-cut solenoid C-3 System

Place transmission in Neutral and set park brake and block drive wheels.

- Before proceeding to the steps shown below, check to see if harness connector connecting ECM to carburetor is connected securely. If found loose, re-connect securely.
- Warm up the engine sufficiently enough and proceed to the steps shown in the chart below.

Stop engine

Disconnect harness connector at ECM and carburetor sides, both. By using ohmmetor, check to see if the lead wire connecting carburetor slow-cut solenoid to ECM F-terminal is opened.

- If found not opened → Connect harness connector at ECM side as it was originally.
- Open → Replace or repair the harness

A

By using a jumping cord connected to the terminal for carburetor slow-cut solenoid, conduit the 12V voltage from battery to carburetor lead wire, and by using jumping cord, the grounding lead wire is grounded to the engine block.

With the grounding lead wire alternately grounded or opened for several times, listen to the evidence of operating sound of slow-cut solenoid and feeling for vibration.

- No operating sound heard → Replace slow-cut solenoid
- Operating sound heard → Start the engine, continuing operation of the idle speed.
  - Engine idle stall → Replace slow-cut solenoid
  - OK

B

Disconnect the grounding lead wire connected to the terminal for carburetor slow-cut.

- Engine stop → Repeat check from A to B 3 times → No trouble found
- Idling continued → Replace slow-cut solenoid

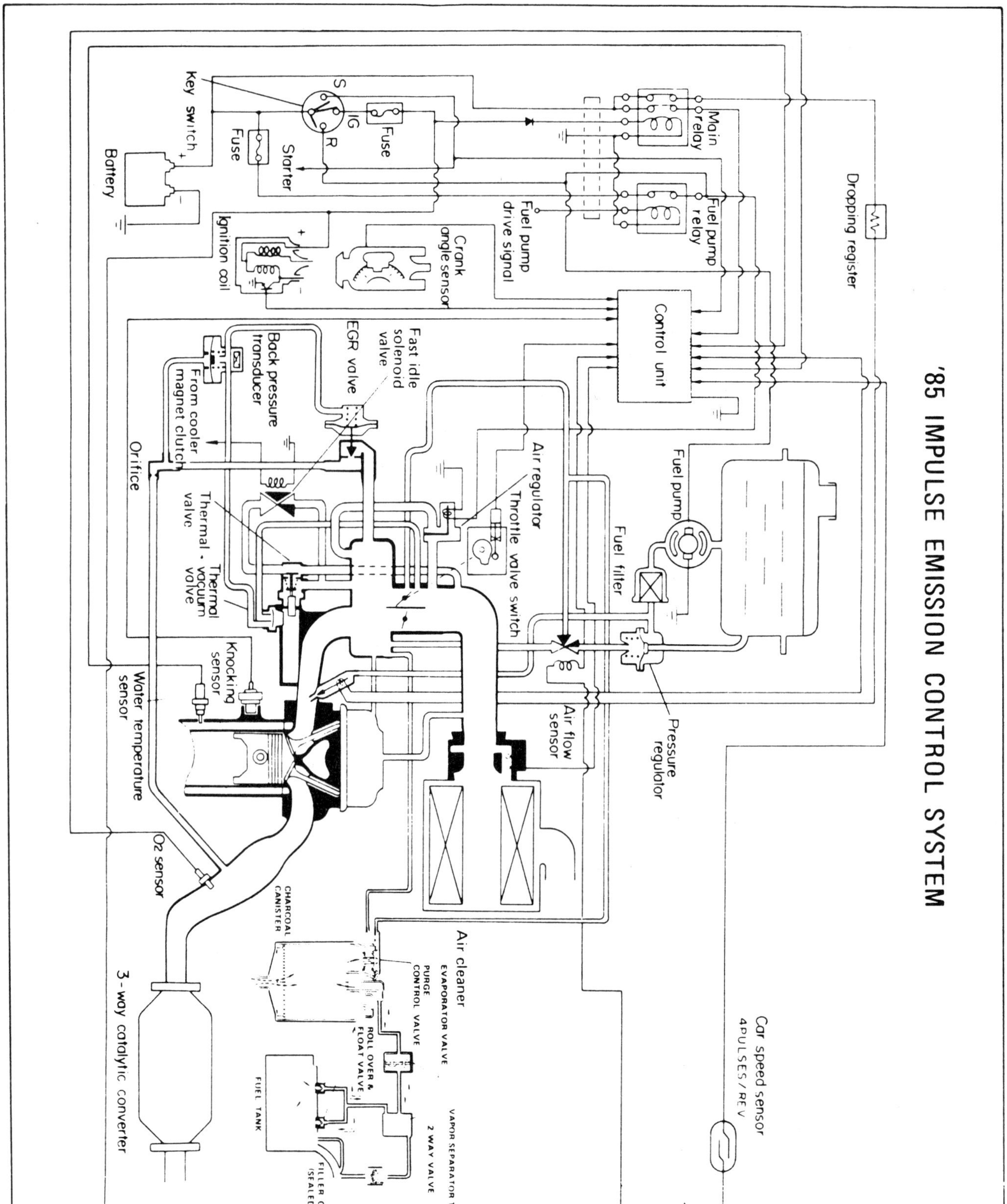

ISUZU IMPULSE (JR) EMISSION CONTROL SYSTEM (© Isuzu Motor Corporation)

EXHAUST GAS RECIRCULATION SYSTEM

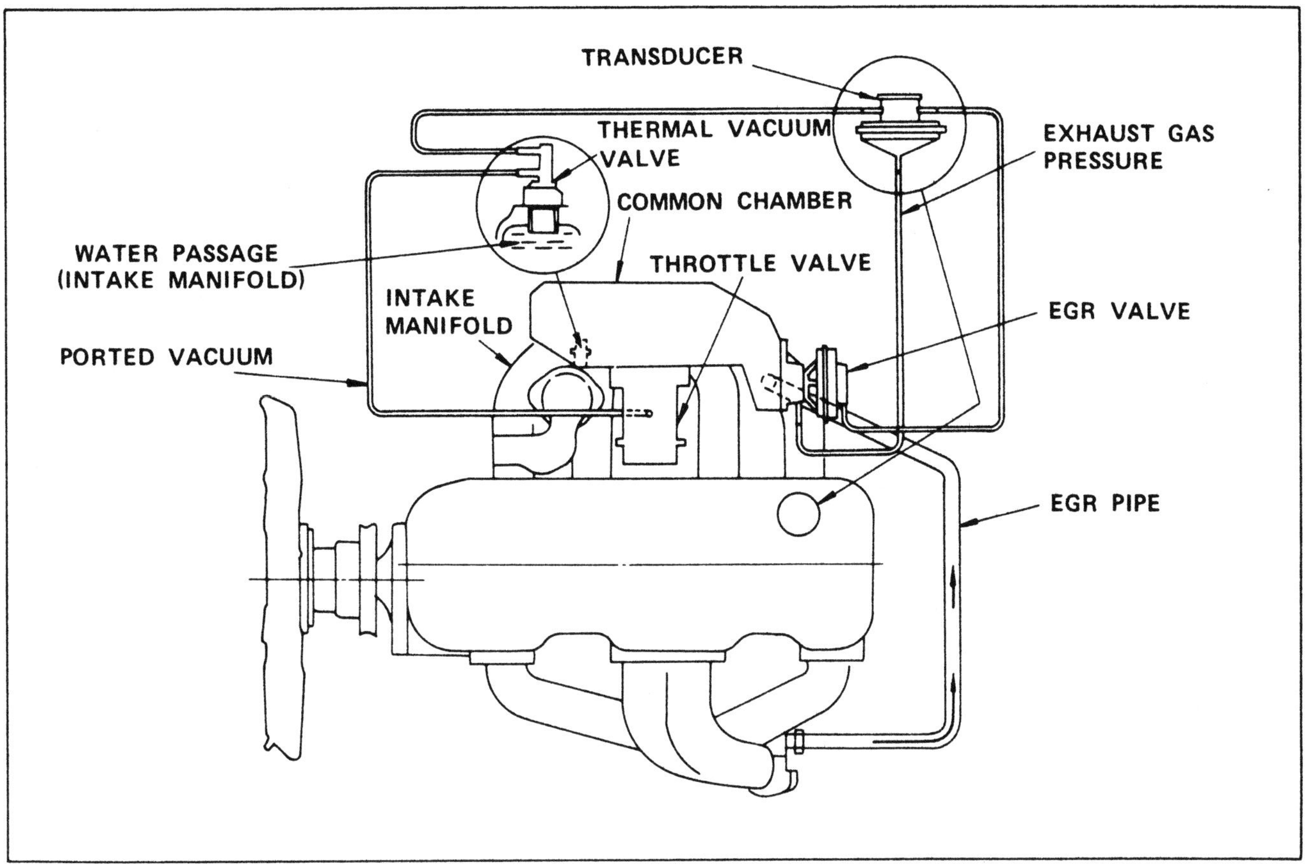

MAIN DATA AND SPECIFICATIONS

E.G.R. valve operating negative pressure	
Starting	51 mmHg (2.0 in.Hg)
Wide open	124 mmHg (4.9 in.Hg)
Thermal vacuum valve opening temperature	Over 36° to 44°C (97° to 111°F)
Back pressure transducer coasting pressure	50 mmAq (2 in.Aq)

ISUZU IMPULSE EXHAUST GAS RECIRCULATION SYSTEM (© Isuzu Motor Corporation)

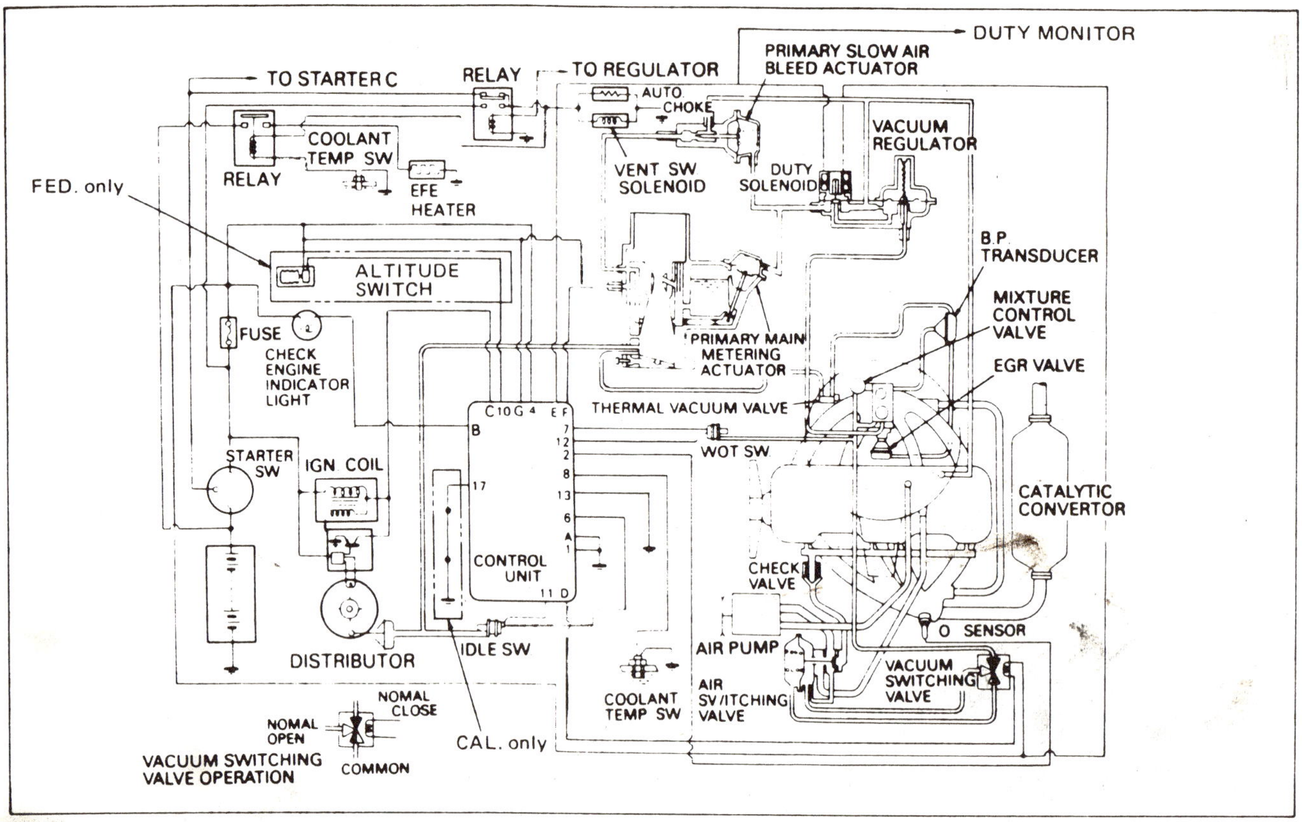

ISUZU I-MARK EMISSION CONTROL SYSTEM (© Isuzu Motor Corporation)

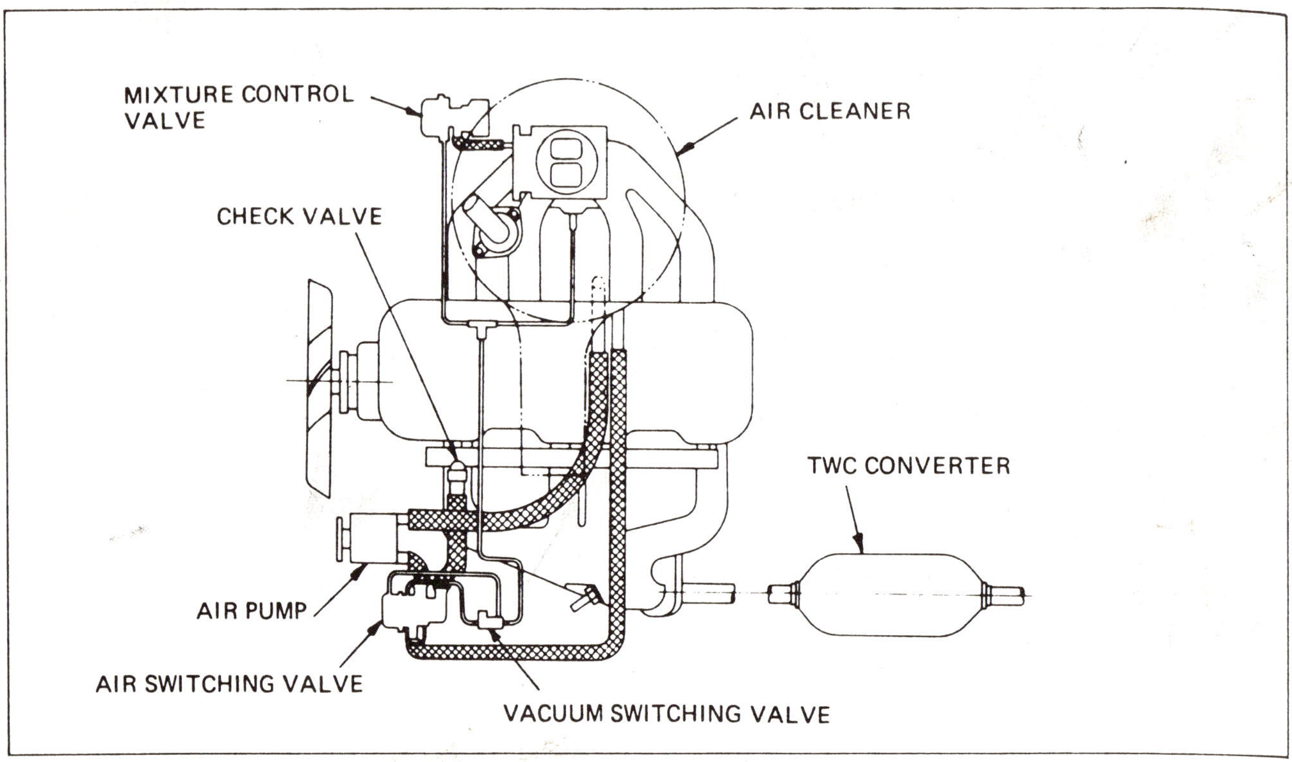

ISUZU I-MARK AIR INJECTION REACTOR SYSTEM (© Isuzu Motor Corporation)

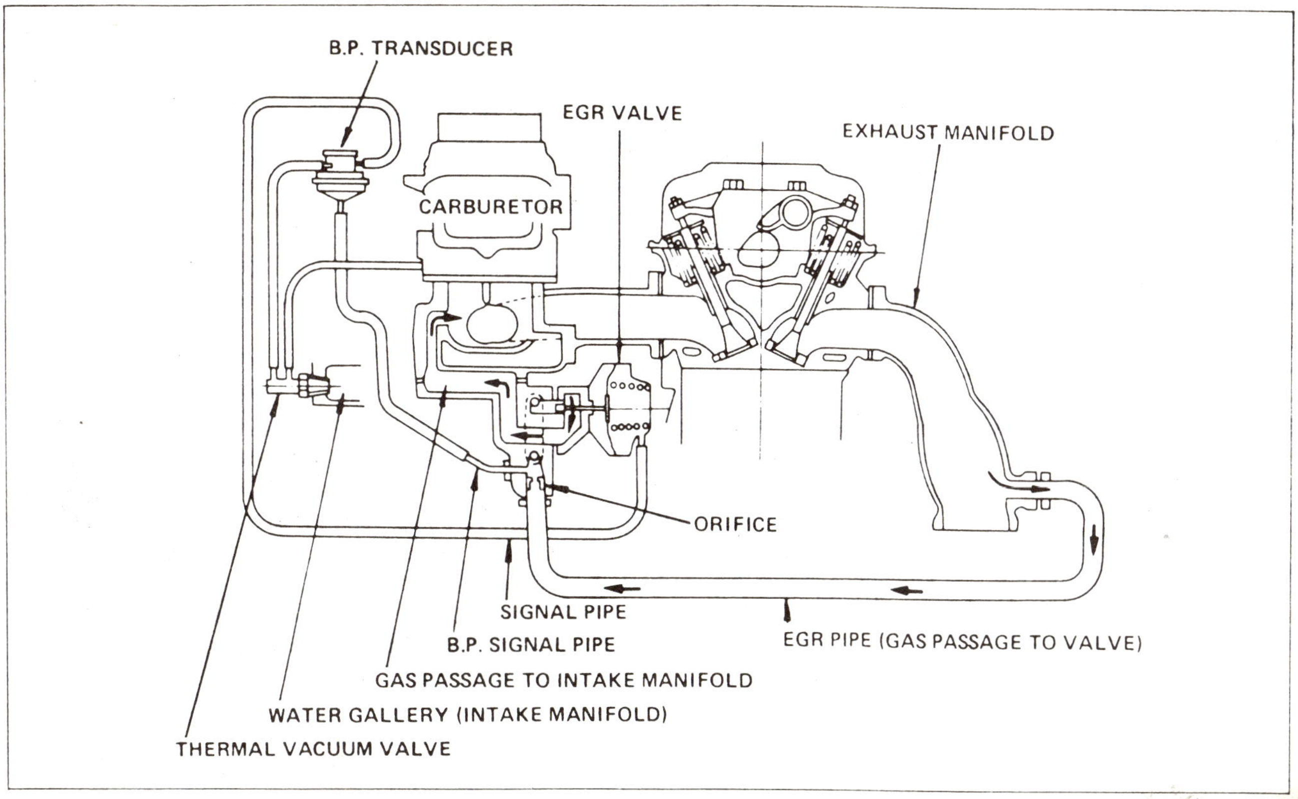

ISUZU I-MARK EXHAUST GAS RECIRCULATION SYSTEM (© Isuzu Motor Corporation)

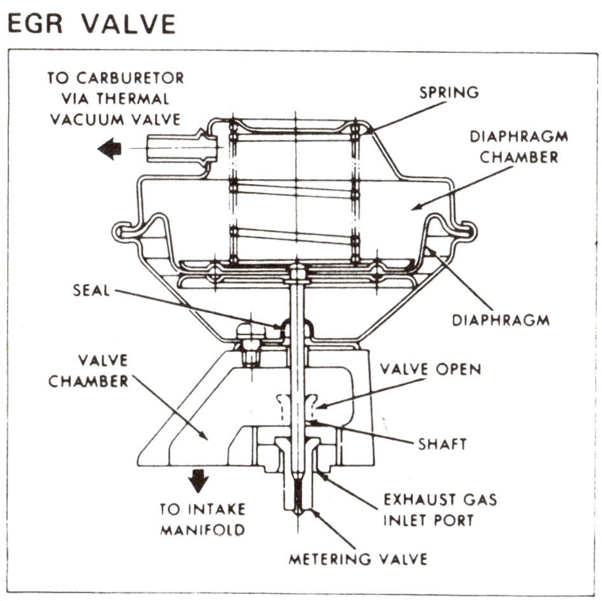

ISUZU I-MARK EGR VALVE (© Isuzu Motor Corporation)

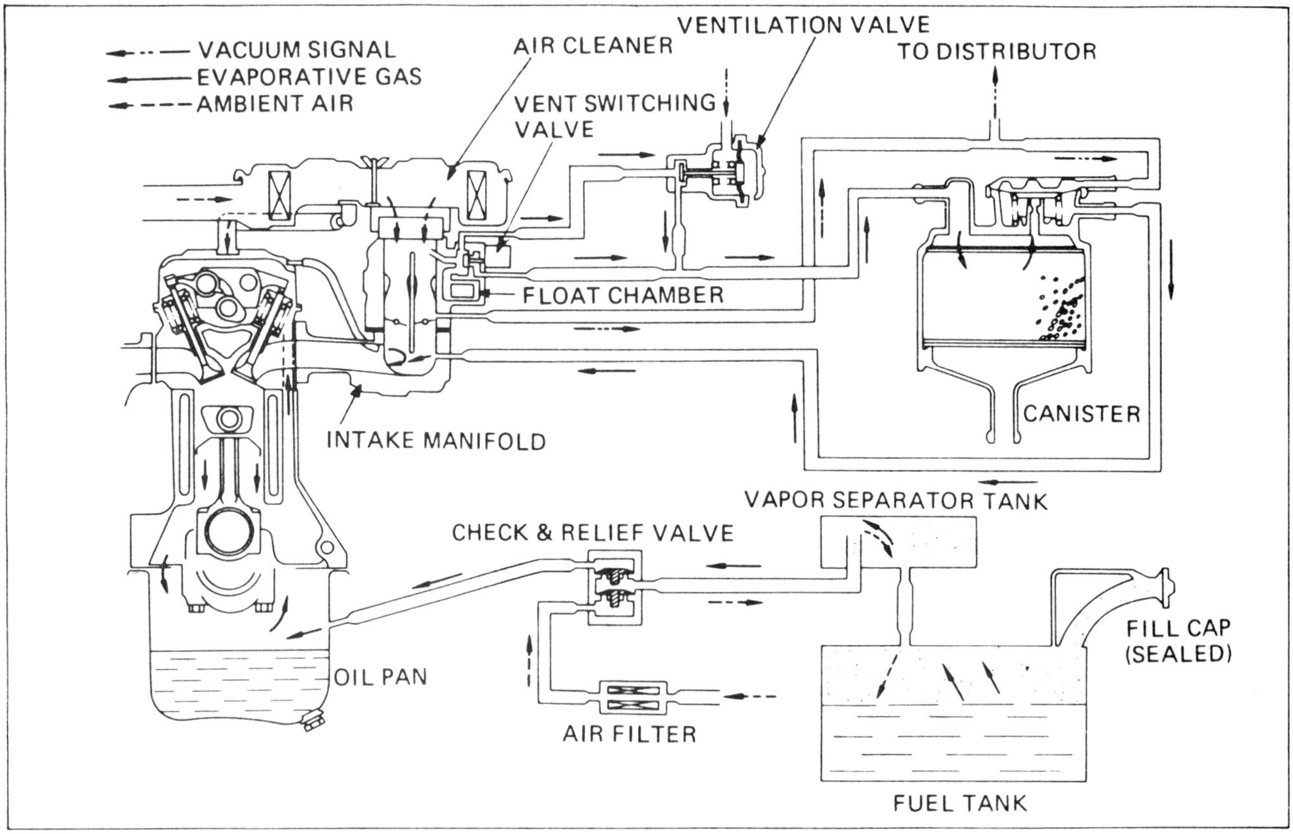

ISUZU I-MARK EVAPORATIVE EMISSION CONTROL SYSTEM (© Isuzu Motor Corporation)

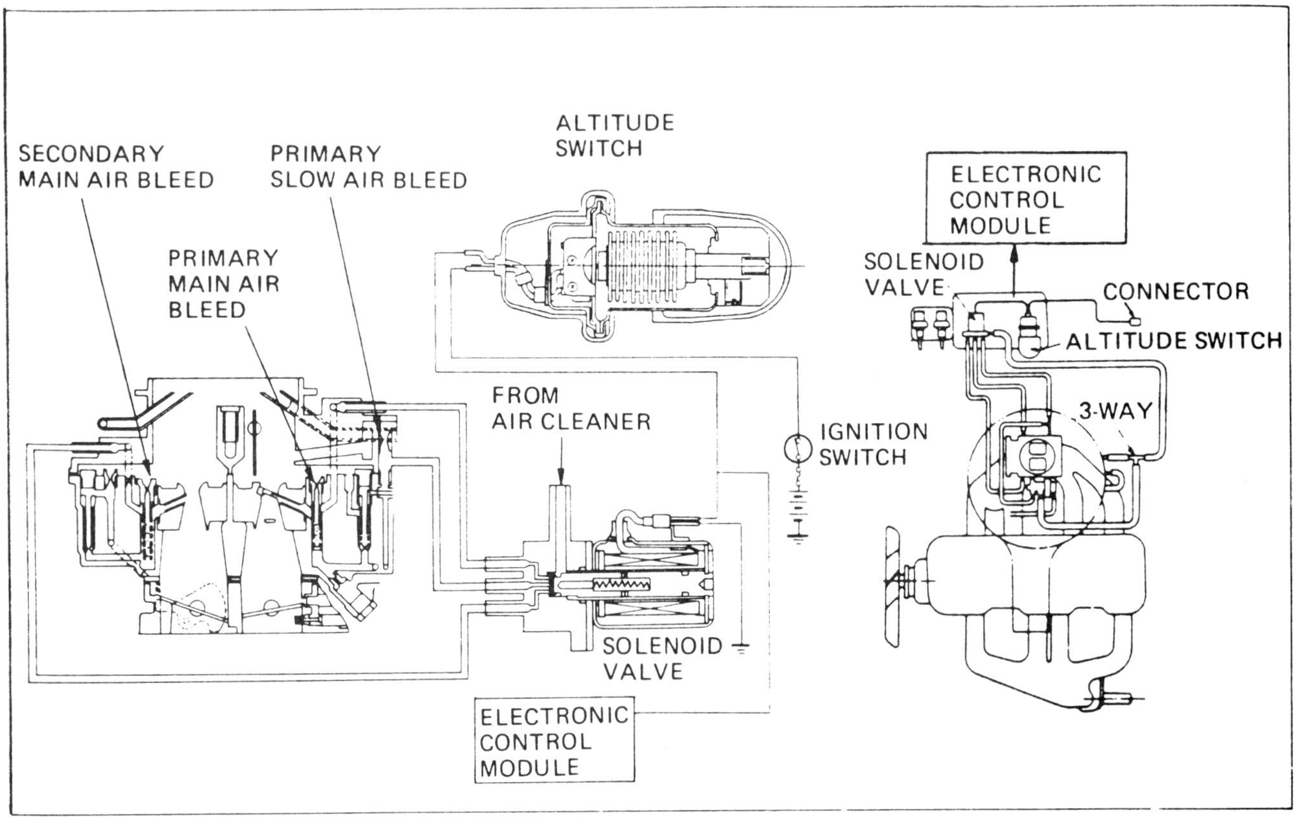

ISUZU I-MARK HIGH ALTITUDE EMISSION CONTROL DEVICE (© Isuzu Motor Corporation)

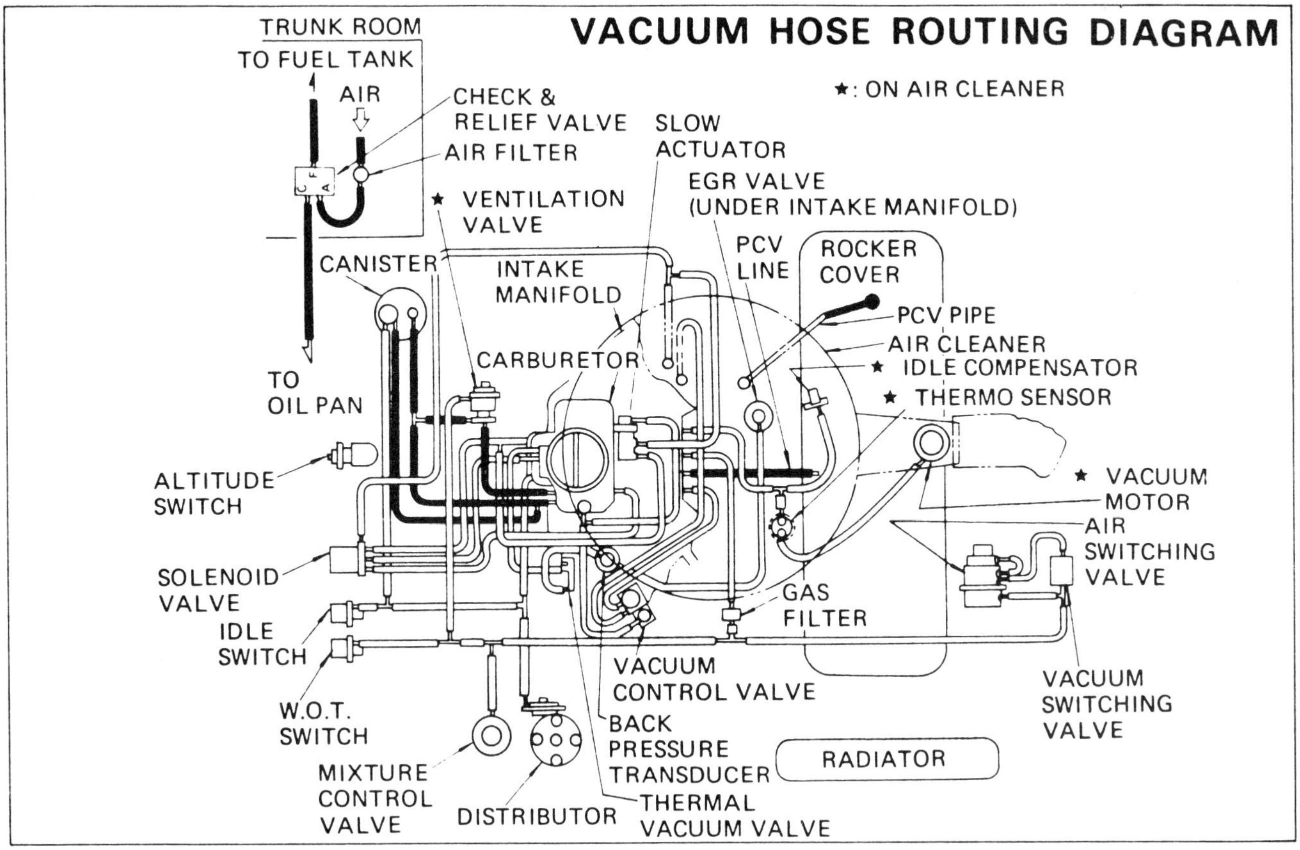

ISUZU I-MARK VACUUM HOSE ROUTING DIAGRAM (FEDERAL) (© Isuzu Motor Corporation)

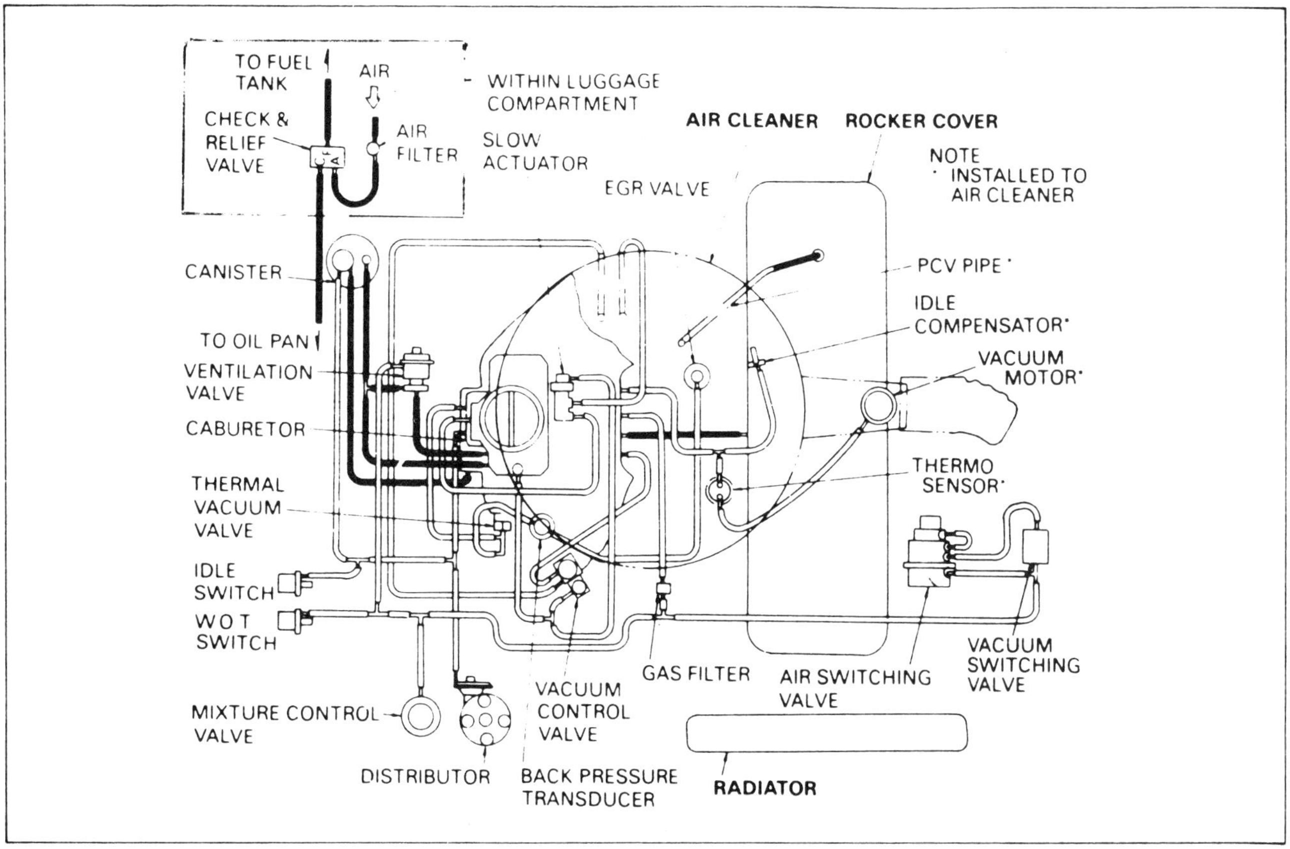

ISUZU I-MARK VACUUM HOSE ROUTING DIAGRAM (CALIFORNIA) (© Isuzu Motor Corporation)

EMISSION CONTROL SYSTEM

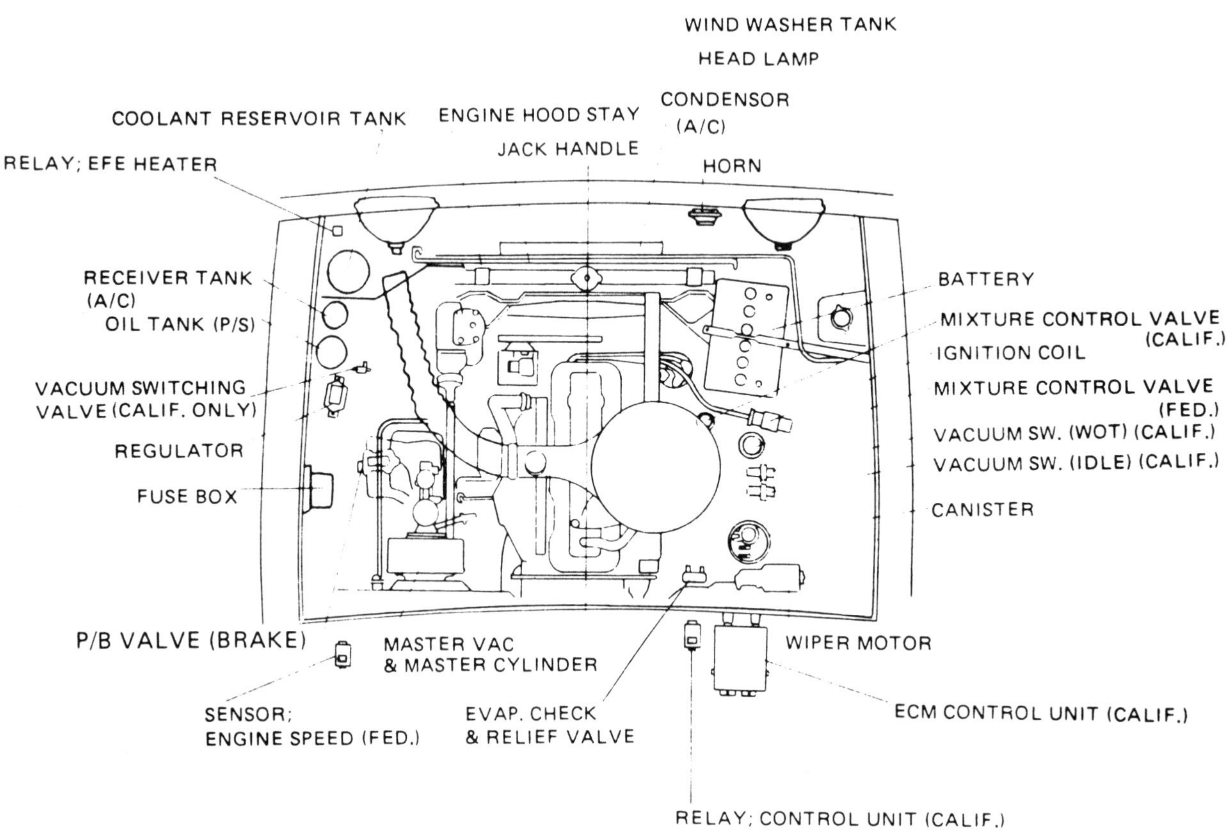

ISUZU TROOPER II (KB82) EMISSION CONTROL SYSTEM COMPONENTS (© Isuzu Motor Corporation)

SYSTEM DIAGRAM

FEDERAL

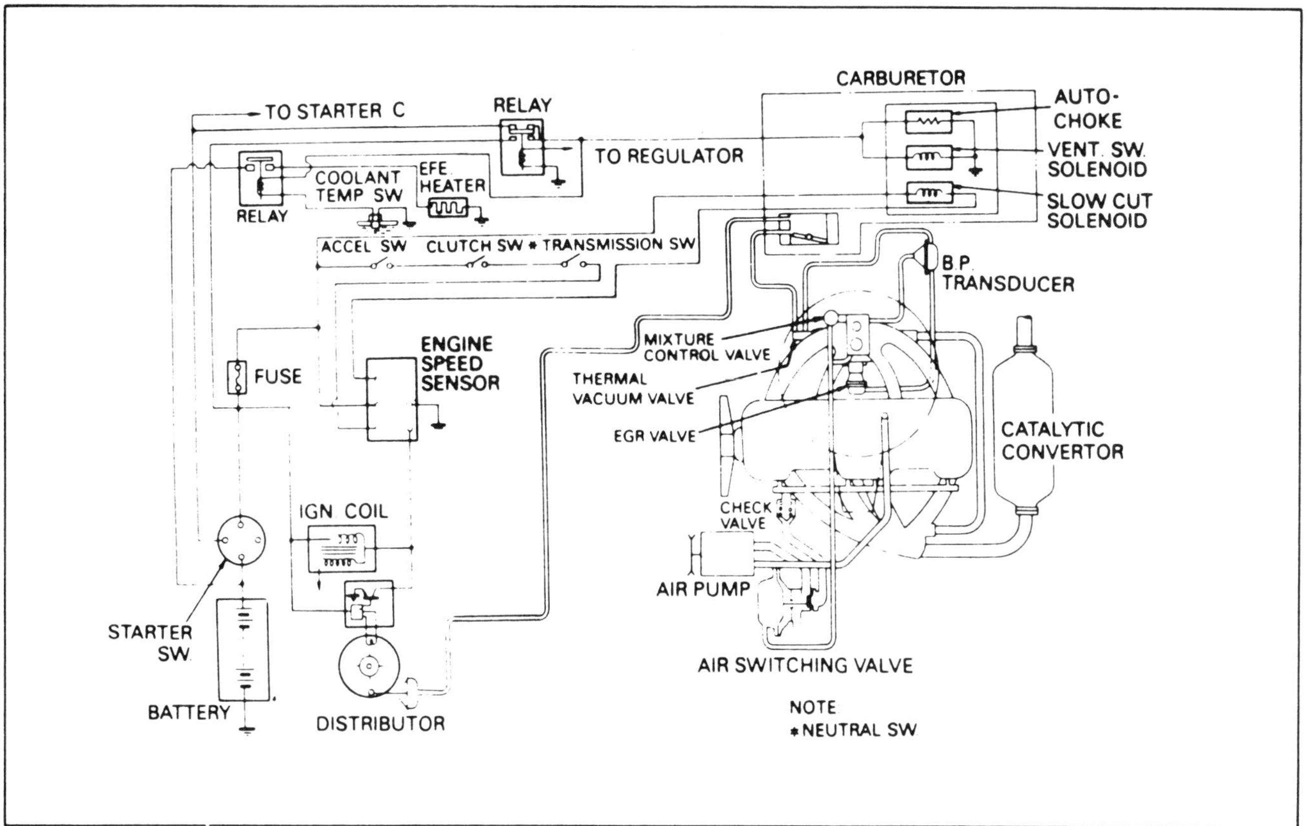

ISUZU TROOPER II EMISSION CONTROL SYSTEM DIAGRAM (FEDERAL) (© Isuzu Motor Corporation)

VACUUM HOSE ROUTING

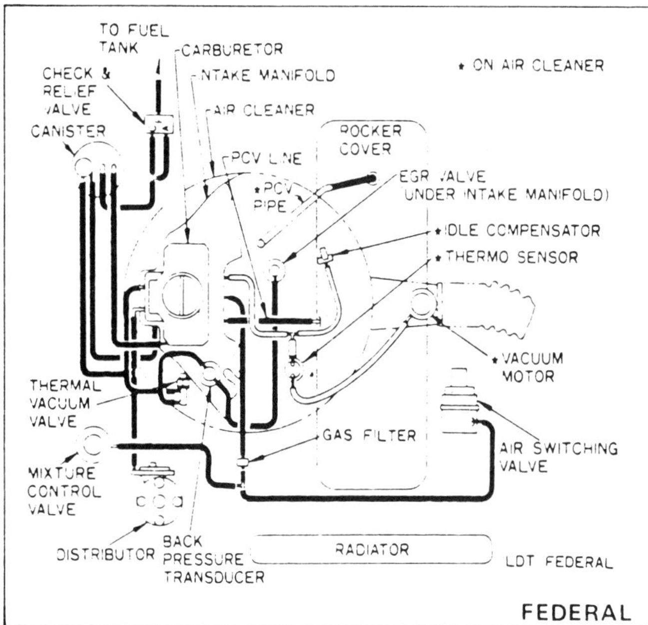

FEDERAL

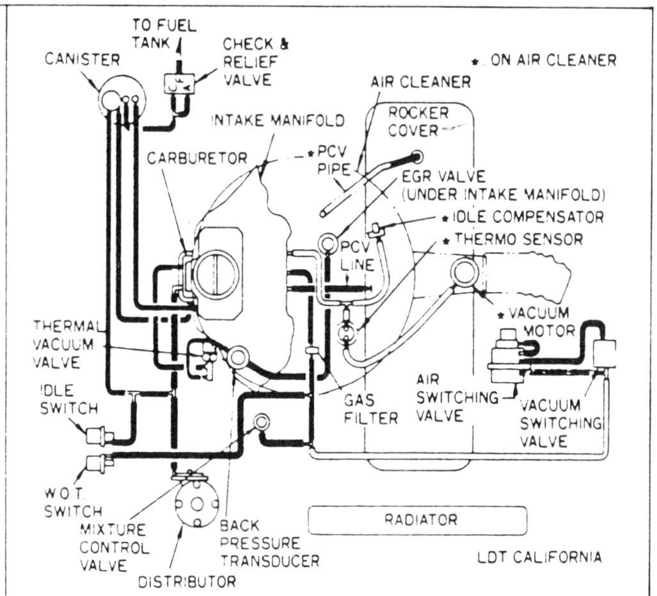

CALIFORNIA

ISUZU TROOPER II VACUUM HOSE ROUTING DIAGRAM (© Isuzu Motor Corporation)

CALIFORNIA

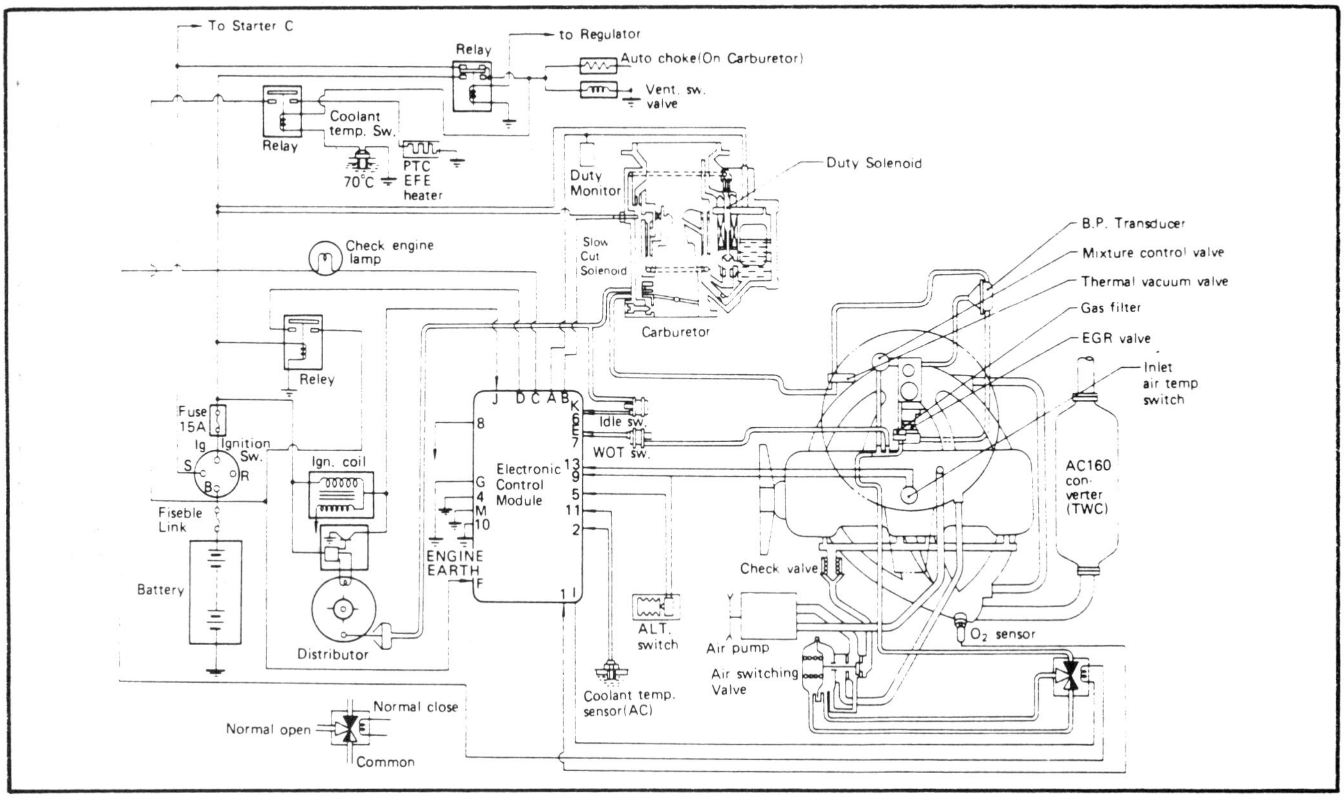

The California emission control system consists of an Air Injection Reactor (AIR) System, (ECM controlled), an Exhaust Gas Recirculation (EGR) System, a Thermostatically Controlled Air (TCA) Induction System, Positive Crankcase Ventilation (PCV) System, an Evaporative Emission Control System (EEC) and a Closed Loop emission control system.

The Closed Loop emission control is a system that precisely controls the air-fuel ratio near stoichiometry, allowing the use of a three-way catalyst to reduce oxides of nitrogen and oxidize hydrocarbons and carbon monoxide.

The essential components are an exhaust gas oxygen sensor, an electronic control module (ECM), a controlled air fuel ratio carburetor and a three-way catalytic converter.

ISUZU TROOPER II EMISSION CONTROL SYSTEM DIAGRAM (CALIFORNIA) (© Isuzu Motor Corporation)

EVAPORATION CONTROL SYSTEM

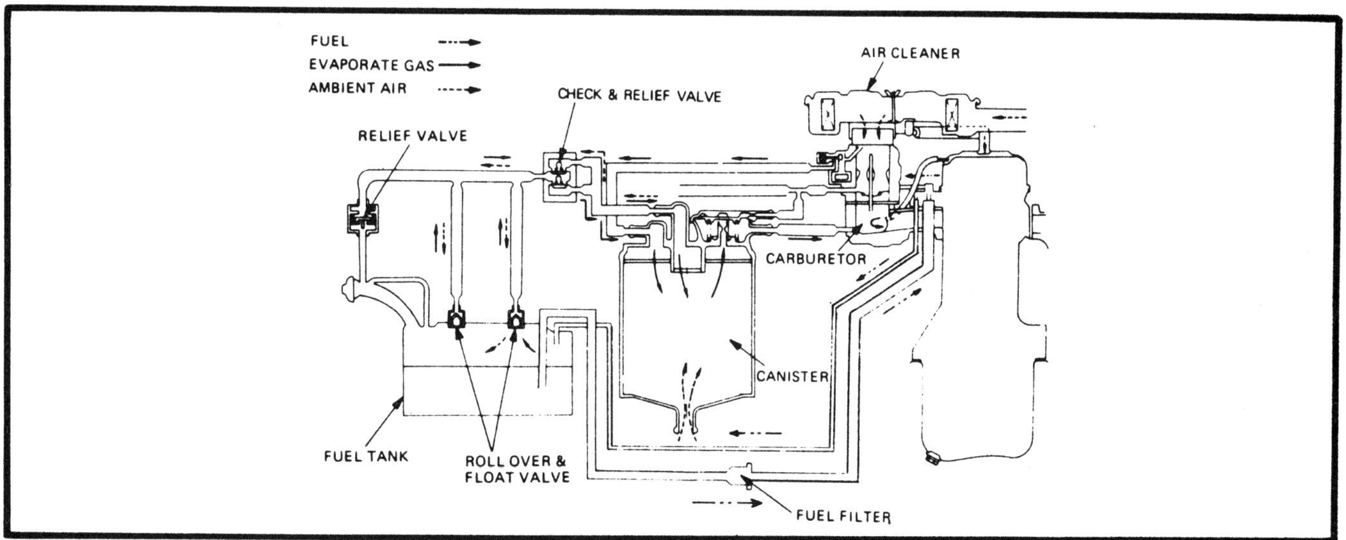

The evaporation control system is designed to lead the fuel vapor emitted from the fuel tank and the carburetor into the carbon canister. This system consists of the roll over & float valves, relief valve, check and relief valve, charcoal canister, vent switching valve and tubes connecting these parts.

The fuel vapor, stored in the canister and mixed with ambient air, is led directly to the intake manifold for combustion when the engine is running.

ISUZU TROOPER II EVAPORATION CONTROL SYSTEM (© Isuzu Motor Corporation)

EXHAUST GAS RECIRCULATION SYSTEM (EGR)

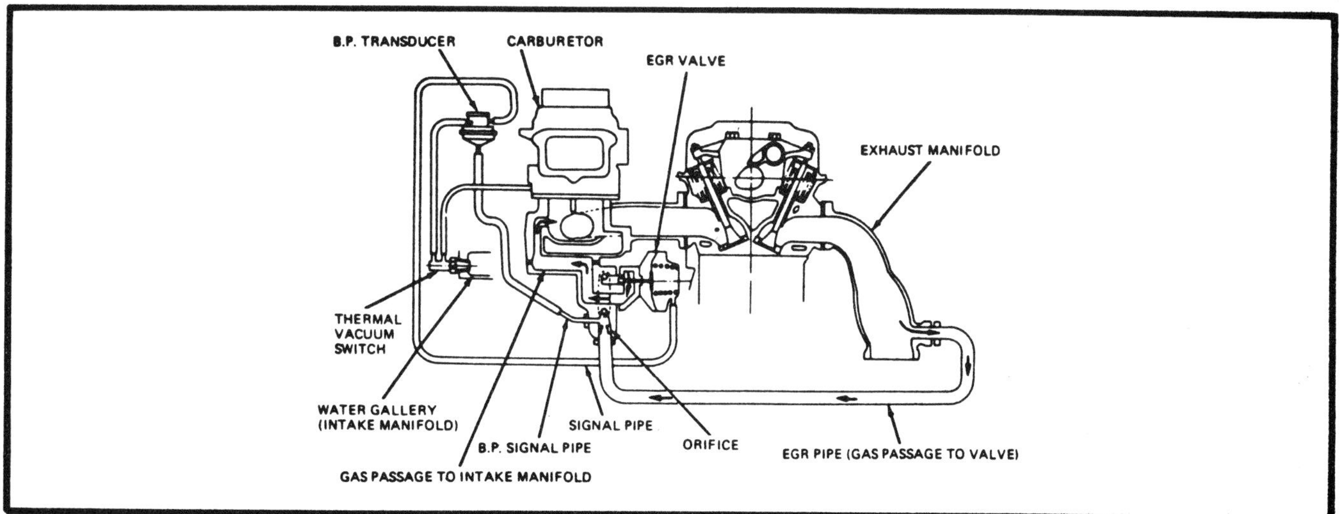

The Exhaust Gas Recirculation (EGR) system is used to reduce combustion temperatures in the combustion chamber, thereby reducing oxides of nitrogen emissions.

The exhaust gas is drawn into the intake manifold through a steel pipe and EGR valve from the exhaust manifold. The vacuum diaphragm of the EGR valve is connected to a signal port at the carburetor flange through a Back Pressure Transducer responsive to exhaust pressure to modulate the vacuum signal and a thermal vacuum valve which operates for EGR cold override.

As the carburetor throttle valve is opened, a vacuum is applied to the diaphragm, which opens the EGR valve and allows exhaust gas to be metered into the intake manifold.

ISUZU TROOPER II EXHAUST GAS RECIRCULATION SYSTEM (EGR) (© Isuzu Motor Corporation)

HIGH ALTITUDE EMISSION CONTROL SYSTEM

If the place of a vehicles's principal use is changed from low altitude to high altitude or vice versa, altitude adjustment will be necessary for improvement of engine performance and exhaust emission control performance.

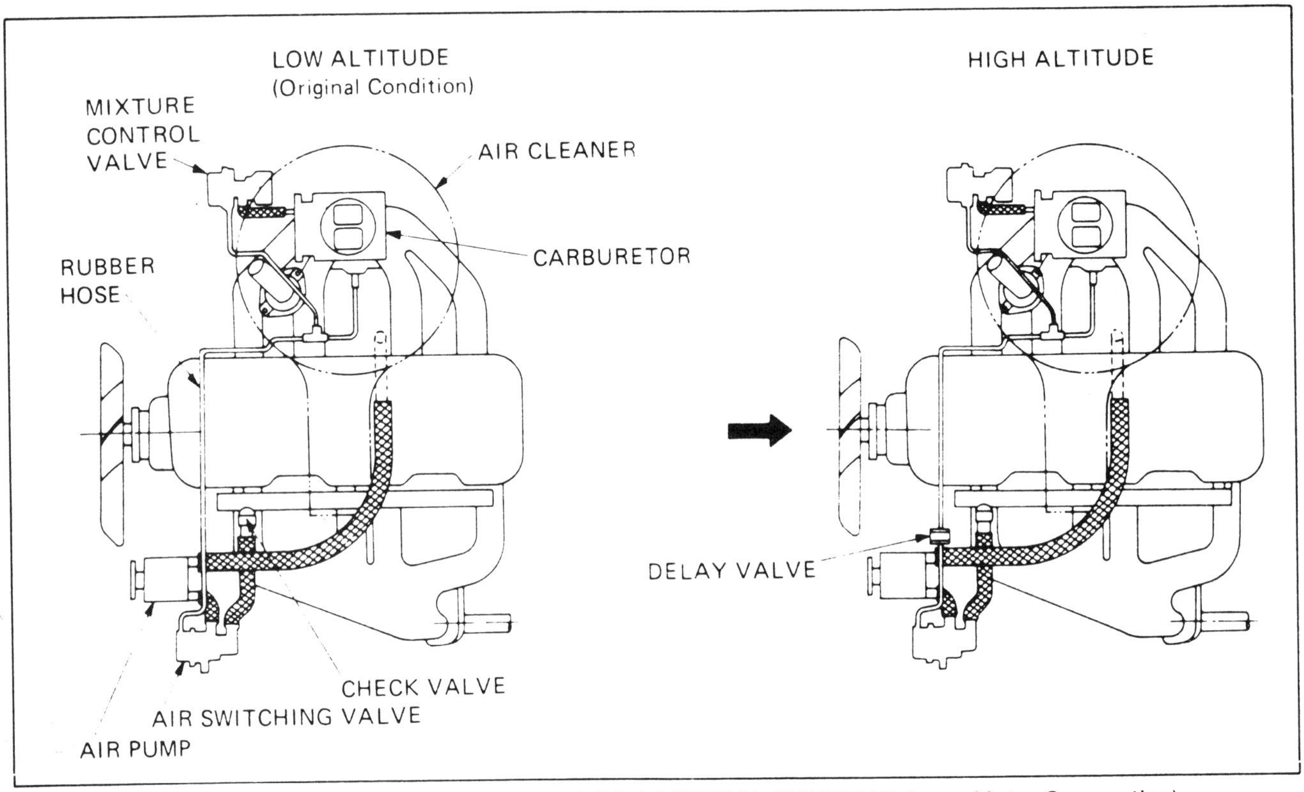

ISUZU TROOPER II HIGH ALTITUDE EMISSION CONTROL SYSTEM (© Isuzu Motor Corporation)

ALTITUDE MODIFICATION AND/OR ADJUSTMENT PROCEDURE

New vehicle

1. Low altitude to high altitude procedure
   1) Stop the engine and apply the parking brake.
   2) Disconnect the rubber hose from air switching valve located on the front left side of the engine.
   3) Connect the delay valve to the above disconnected hose and valve. Ensure that the rubber hose on the mark "A" side of the delay valve is connected to the 3-way, and the rubber hose on the mark "B" to the air switching valve.
   4) Adjust the engine speed to the specification.
   5) Check the ignition timing.
   6) Attach the label.

   High altitude kit

Device & Parts	● Delay valve with hose ● Label

2. High altitude to low altitude procedure
   (Original destination is high altitude specification)
   1) Disconnect the rubber hose from the air switching valve located on the front left side of the engine.
   2) Remove the delay valve.
   3) Connect the rubber hose to the air switching valve.
   4) Adjust the engine speed to the specification.
   5) Check the ignition timing.
   6) Remove the label.

Isuzu Truck Specifications

VEHICLE IDENTIFICATION NUMBER AND ENGINE IDENTIFICATION

Vehicle Identification Number (VIN)

The vehicle identification number (VIN) is embossed on a plate which is attached to the top left instrument panel. The number is visible through the windshield from outside the vehicle.

Engine Identification Number

The engine serial number is stamped on the top right front corner of the engine block on gas engines. On diesel engines, the engine serial number is stamped on the left rear corner of the engine block.

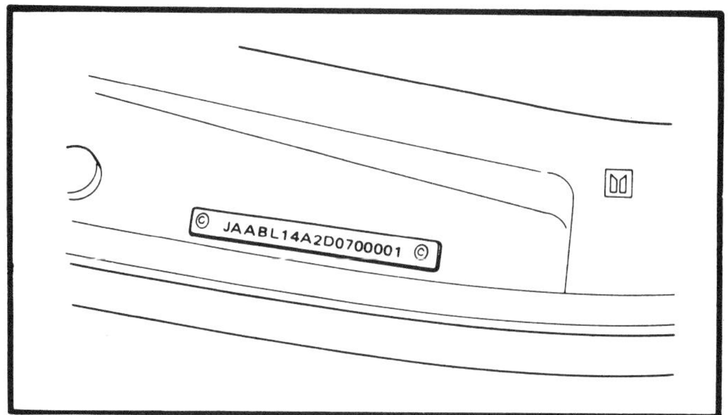

Vehicle identification plate location (© Isuzu Motor Corporation)

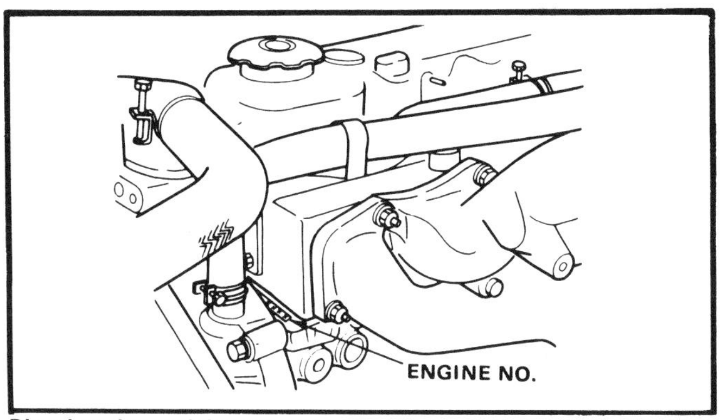

Diesel engine serial number location (© Isuzu Motor Corporation)

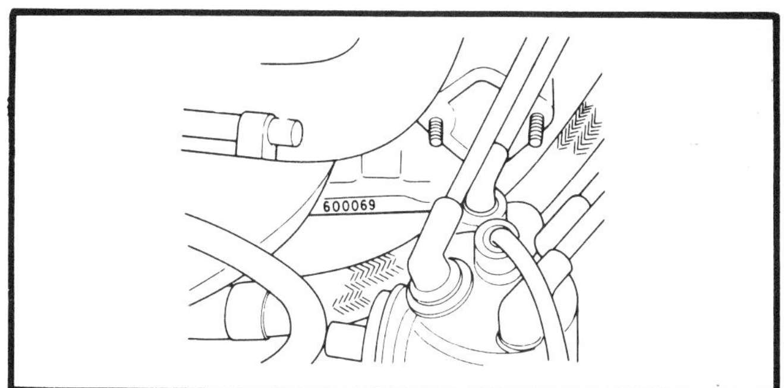

Gas engine serial number location (© Isuzu Motor Corporation)

GASOLINE ENGINE TUNE-UP SPECIFICATIONS

(When analyzing compression test results, look for uniformity among cylinders, rather than specific pressures)

Year	Engine Displacement (cu.in.)	Spark Plugs Type	Spark Plugs Gap (in.)	Distributor Point Dwell (deg)	Distributor Point Gap (in.)	Ignition Timing (deg) MT	Ignition Timing (deg) AT	Intake Valve Opens (deg)	Fuel Pump Pressure (psi)	Idle Speed (rpm)	Valve Clear (in.) (cold) In	Valve Clear (in.) (cold) Ex
'85	110.8	BPR6ESII	0.040	Electronic		6B	6B	21	3.6	900	0.006	0.010
'85	118.9	BPR6ESII	0.040	Electronic		12B	12B	28	2.0–2.5	900	0.006	0.010
'85	See Underhood Specifications Sticker											

NOTE: The underhood specifications sticker often reflects tune-up specification changes made in production. Sticker figures must be used if they disagree with this chart.
B Before top dead center

DIESEL ENGINE TUNE-UP SPECIFICATIONS

Injector Opening Pressure (psi)	Low Idle (rpm)	Valve Clearance Intake	Valve Clearance Exhaust	Intake Valve Opens (deg.)	Injection Time (deg.)
1770	625 ①	0.010	0.014	32B	12B

① A/T: 725 rpm

FIRING ORDERS

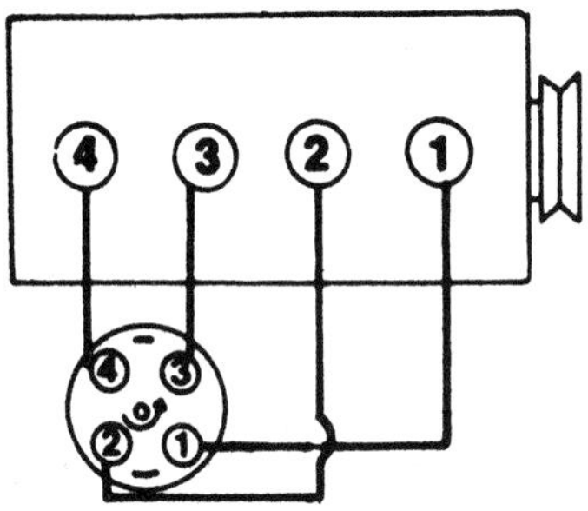

Firing order—gas engine 1-3-4-2 (© Isuzu Motor Corporation)

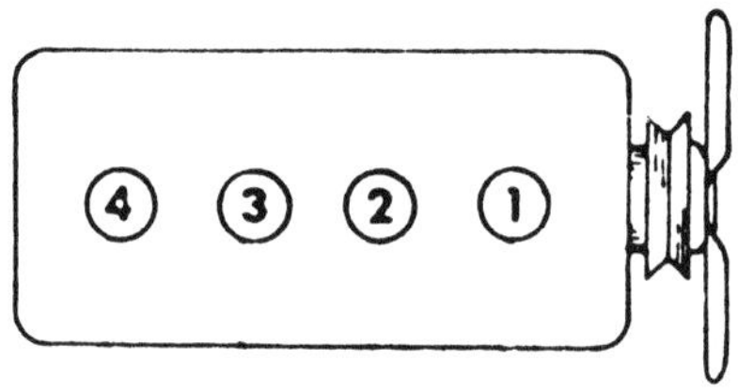

Firing order—diesel engine 1-3-4-2 (© Isuzu Motor Corporation)

Mazda

INDEX

MAZDA PICK-UP TRUCKS

VEHICLE IDENTIFICATION NUMBER AND ENGINE IDENTIFICATION

Vehicle Identification Number (VIN)

This number is on a plate located on the drivers side windshield pillar and is visible through the glass. A chassis number is stamped on a piece of metal which is riveted to the firewall on the passenger side of the vehicle.

Engine Model and Serial Number

The Mazda 626 and RX-7 engine serial number is located to the rear of the alternator bracket, stamped on the engine block. The 626 diesel engine serial number is located between the first and second fuel injectors, stamped on the engine block. The GLC engine serial number is stamped on the left hand side of the block, just below the head.

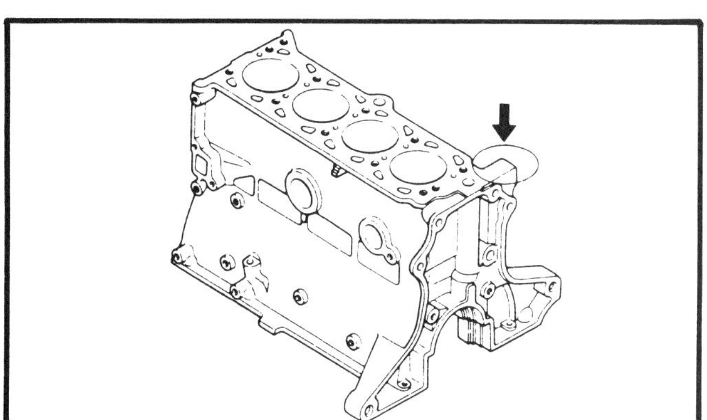

Engine serial number location for the Mazda GLC

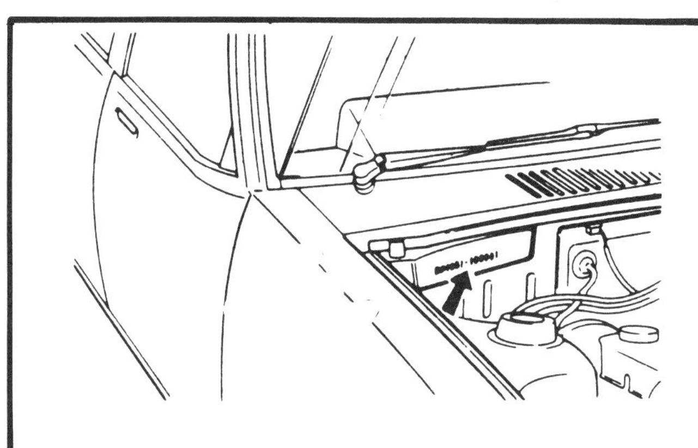

Typical chassis number location

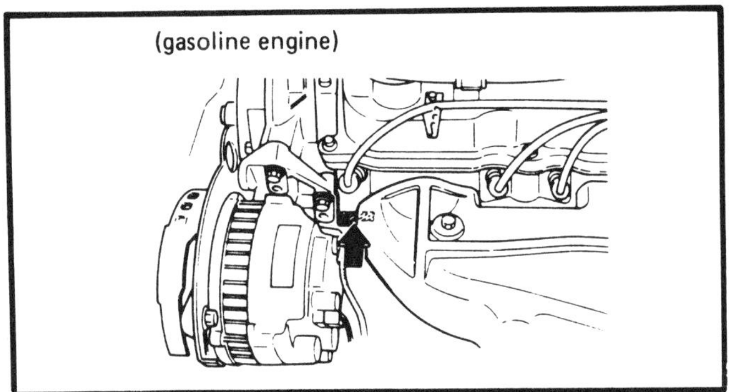

Engine serial number location for the Mazda 626 gas engine

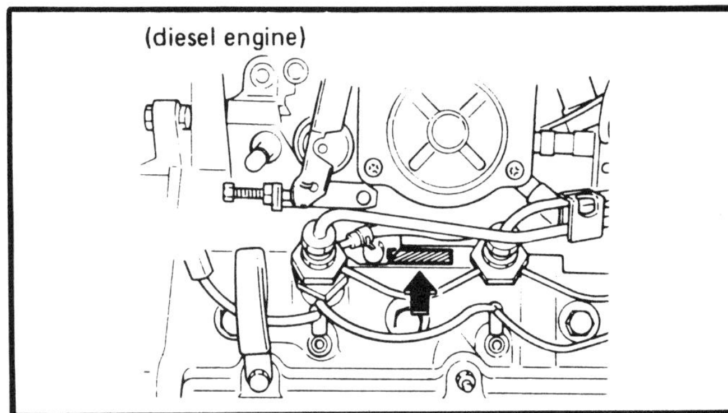

Engine serial number location for the Mazda 626 diesel engine

TUNE-UP SPECIFICATIONS—PISTON ENGINE

(When analyzing compression test results, look for uniformity among cylinders, rather than specific pressures)

Year	Engine Displacement (cu. in.)	Spark Plugs Type	Spark Plugs Gap (in.)	Distributor Point Dwell (deg)	Distributor Point Gap (in.)	Ignition Timing (deg) MT	Ignition Timing (deg) AT	Intake Valve Opens (deg)	Fuel Pump Pressure (psi)	Idle Speed (rpm)	Valve Clearance (in.) In	Valve Clearance (in.) Ex
'85	90.9	⑧	.031	Electronic		8B④	8B④	15	2.8–⑥ 3.8	850③	.010	.012
'85 (Gas)	121.9	BPR5ES, BPR6ES	.031	Electronic		6B	6B	17	2.8–⑦ 3.5	750⑤	.012	.012
'85–'86 (Diesel)	121.9	—	—	—	—	—	—	13	1920	800–850	.010	.014
'86	See Underhood Specifications Sticker											

NOTE: The underhood specifications sticker often reflects tune-up specification changes made in production. Sticker figures must be used if they disagree with those in this chart.

③ Automatic 750 in Drive
④ '83 & later—6° BTDC
⑤ Automatic: 700 in Drive
⑥ '84 & later—4.27–5.97
⑦ '84 & later–2.8–4.27
⑧ BPR5ES, BPR6ES ('81–'85)
AGR-22, AGR-32 ('83–'85)
RN11YC ('85)

TUNE-UP SPECIFICATIONS—ROTARY ENGINE

(When analyzing compression test results, look for uniformity among cylinders, rather than specific pressures)

Year	Engine Displacement (cu. in.)	Spark Plugs Type	Spark Plugs Gap (in.)	Distributors Point Dwell (deg)	Distributors Point Gap (in.)	Ignition Timing (deg) Leading Normal	Ignition Timing (deg) Leading Retarded	Ignition Timing (deg) Trailing Normal	Idle Speed (rpm) MT	Idle Speed (rpm) AT
'84–'85	70④	③	0.053–0.057⑤	Electronic		0⑥	—	20A	750⑦	750①
'86	See Underhood Specifications Sticker									

NOTE: The underhood specifications sticker often reflects tune-up specification changes made in production. Sticker figures must be used if they disagree with those in this chart.

TDC—Top dead center
A—After top dead center
B—Before top dead center
MT—Manual transmission
AT—Automatic transmission
deg—degrees
① Transmission in Drive

③ 1980: BR7ET, W22EBR
BR8ET, W25EBR
BR9ET, W27EBR
1981–85: BR7EQ14, W22EDR14
BR8EQ14, W25EDR14
BR9EQ14, W27EDR14

④ 13B
eng: 80
⑤ 13B
eng: .055
⑥ 13B
eng: Leading-5A
⑦ 13B
eng: 800

FIRING ORDER

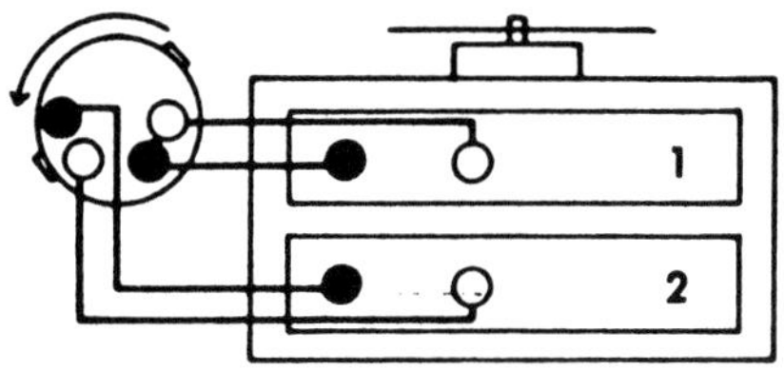

Rotary engine

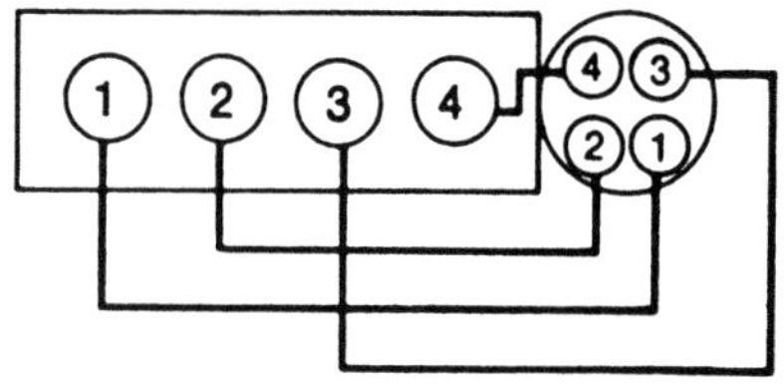

1490 cc engine—GLC 1998 cc engine—626

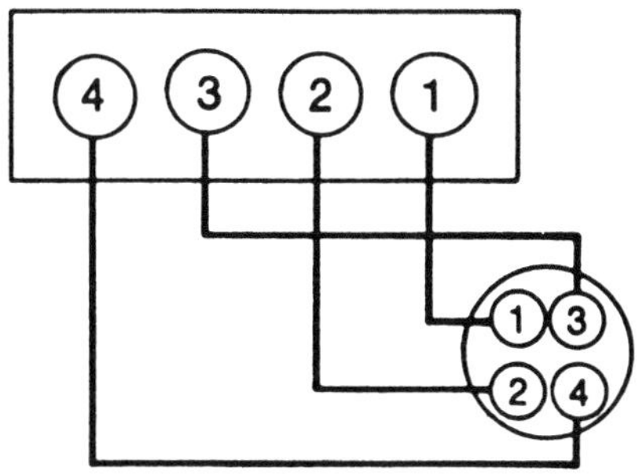

1415 cc engine—GLC engine—626

EMISSION EQUIPMENT USED

MAZDA 626

Control unit
Air cleaner
Reed valve
Idle compensator valve
Air vent solenoid valve
Idle switch
Dashpot
Coasting richer solenoid
Slow fuel cut-off solenoid valve
Clutch switch and neutral switch
EGR valve
EGR position sensor
Duty solenoid valve
Vacuum sensor
Water thermo sensor
Oxygen sensor
Air/fuel solenoid valve
Three-way solenoid valve
Altitude compensator
Vacuum switch charcoal canister
Purge control valve number one
Water thermo valve
Purge control valve number two
Check valve
Catalytic converters
Servo diaphragm
Power steering switch
Anti-afterburn valve

MAZDA GLC

Reed valve A and B
Air control valve
Water temperature sensor
EGR modulator valve
Check valve
Choke solenoid valve
PCV Valve
High altitude compensator
Oxygen sensor
Water thermo switch
Catalytic converters
Neutral switch
EGR thermo valve
Charcoal canister
Control unit
Fuel pump
Clutch switch
Servo diaphragm
ACV solenoid valve
Throttle positioner solenoid valve
EGR solenoid valve
Number one vacuum switch
Number two vacuum switch
Number three vacuum switch
Idle switch
Air Vent solenoid valve
Slow fuel cut-off valve
Air vent solenoid valve
Idle compensator
Number two choke diaphragm
Number one choke diaphragm
EGR control valve
Vacuum delay valve (automatic transmission only)
Anti-afterburn valve (automatic transmission only)

MAZDA RX-7

- Anti-afterburn valve number one
- Anti-afterburn valve number two
- A/C solenoid valve
- Air cleaner
- Air control valve
- Air pump
- Air vent solenoid valve
- ALC valve
- Charcoal canister
- Choke diaphragm number one
- Choke diaphragm number two
- Catalyst thermo sensor
- Check and cut valve
- Choke switch
- Clutch switch
- Neutral switch
- Coasting valve
- Control unit
- Dash pot
- Delay valve
- Heat hazard sensor
- Idle compensator
- Main air bleed control solenoid valve
- Catalytic converters
- Port air solenoid valve
- Port air switching valve
- Purge valve
- Relief solenoid valve
- Richer solenoid valve (manual transmission only)
- Shutter solenoid valve
- Shutter valve
- Split air solenoid valve
- Switching solenoid valve
- Throttle opener
- Throttle sensor
- Vacuum control solenoid valve
- Vacuum switch (manual transmission only)
- Water temperature switch number one
- Water temperature switch number two

REQUIRED EMISSION CONTROL MAINTENANCE

Emission Maintenance Intervals

Procedure	Mileage in Thousands
Air filter change	30,000
Idle speed adjustment	15,000
Oxygen sensor replacement	30,000
PCV system check	15,000
Evaporative canister replacement	15,000
Ignition timing check	45,000
Fuel filter replacement	15,000
Clean carburetor choke mechanism	30,000
Spark plug replacement	15,000
Valve lash adjustment	15,000
EGR valve cleaning	60,000
Anti-afterburning valve test	60,000

IDLE SPEED AND MIXTURE ADJUSTMENT

NOTE :The following specifications are published from the latest information available. if the published information differs from the information on the vehicle emission label, use the data on the emission label. The idle mixture adjustments are factory set and sealed and no adjustments are required. The mixture control adjustment screw opening is plugged to prevent any adjustment. Mixture control adjustments should be done when the mixture control unit is replaced or when the vehicle fails the emissions test.

Idle Speed Specifications

IDLE SPEED

Adjustment

MAZDA RX-7—ALL EXCEPT MODELS W/13B ENGINE

1. Start the engine and let it run until it reaches normal operating temperature.
2. Connect a tachometer to the engine and adjust the idle speed as necessary, by turning the throttle adjusting screw.
3. Inspect and adjust the throttle sensor by disconnecting the brown connector and connect a dual test light to the green connector and the battery.
4. Quickly decelerate the engine speed from 3000 rpm to idling speed and be sure that both test lights illuminate at the same time.
5. If the test lights do not light up at the same time, remove the cap from the throttle sensor adjusting screw and turn the screw left or right until both test light illuminate at the same time.
6. Disconnect the test light and reconnect the brown connector.

MAZDA RX-7—13B FUEL INJECTED ENGINE

1 .Start the engine and let it run until it reaches normal operating temperature, connect a tachometer to the engine.

2. .Check and adjust the throttle sensor as necessary (as outlined above), disconnect the vent and vacuum solenoid connector.

3. Adjust the idling speed to 800 rpm by turning the air adjust screw (AAS).

MAZDA 626 AND GLC

1. Set the parking brake, block the drive wheels and turn off all accessories.

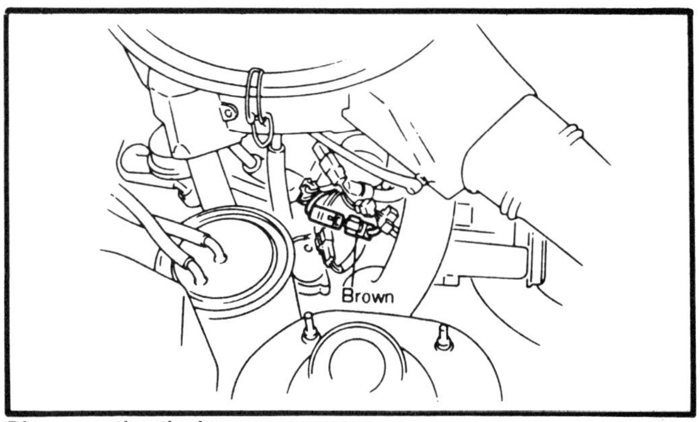

Disconnecting the brown connector

2. Connect a tachometer to the engine. Run the engine until normal operating temperature has been reached.

3. If so equipped, remove the plug from the electric cooling fan motor. Position the selector lever in the drive position.

4. Adjust the idle speed to specifications, by turning the throttle adjusting screw.

NOTE: If the idle speed specification differ from the specification on the vehicle emission label, use the data on the label.

MAZDA 626 DIESEL ENGINE

1. Check the throttle pedal free-play, it should be 0.04–0.12 inch. If the free-play is not within specifications, adjust the cable by loosening the lock-nut on the cable bracket and turning the adjusting nut. Tighten the lock-nut.

2. Connect a tachometer to the engine. Run the engine until the normal operating temperature has been reached.

3. Check the idle speed. Adjust the idle speed to 800–850 rpm, by loosening the lock-nut on the idle adjusting bolt. Turn the bolt clockwise to increase the idle speed and counterclockwise to decrease the idle speed. Tighten the lock-nut.

4. Accelerate engine and recheck the idle speed. Be sure that the accelerator cable returns to idle position.

IDLE MIXTURE

Adjustment

MAZDA RX-7—ALL EXCEPT MODELS W/13B ENGINE

1. Remove the carburetor from the vehicle. Separated the the main body from the throttle body.

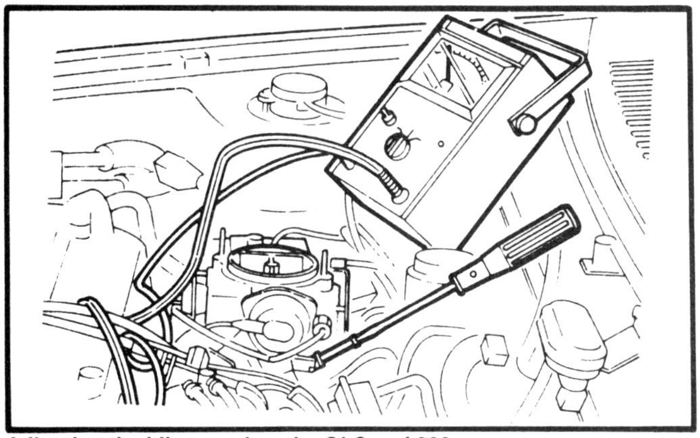

Adjusting the idle speed on the GLC and 626

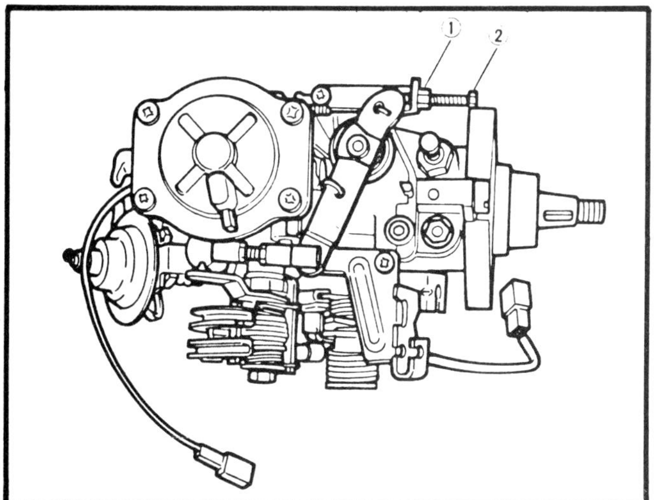

Adjusting the idle speed on the 626 diesel engine

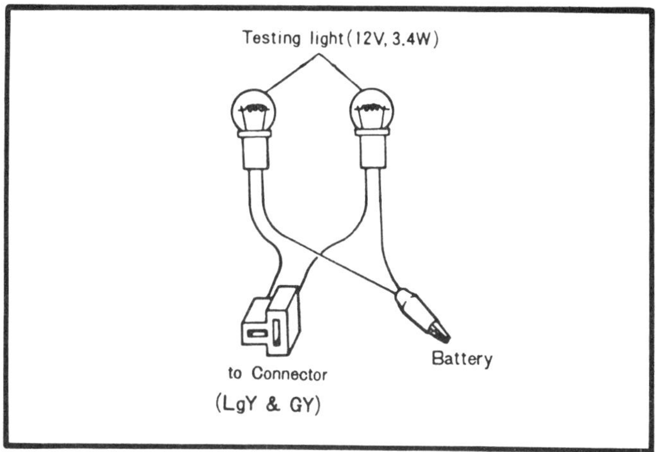

Connecting the test lights to the green connector

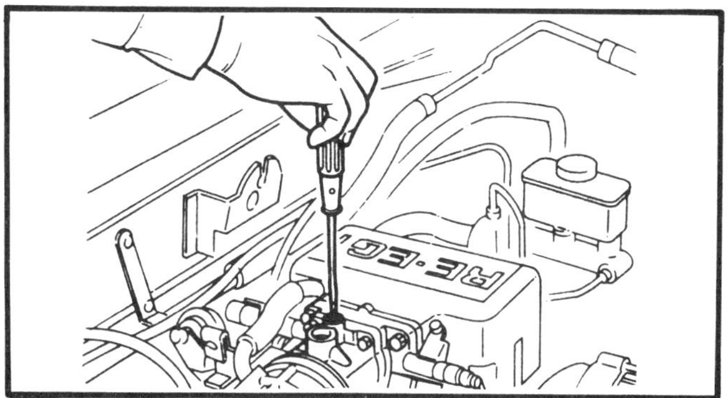

Idle speed adjustment screw location

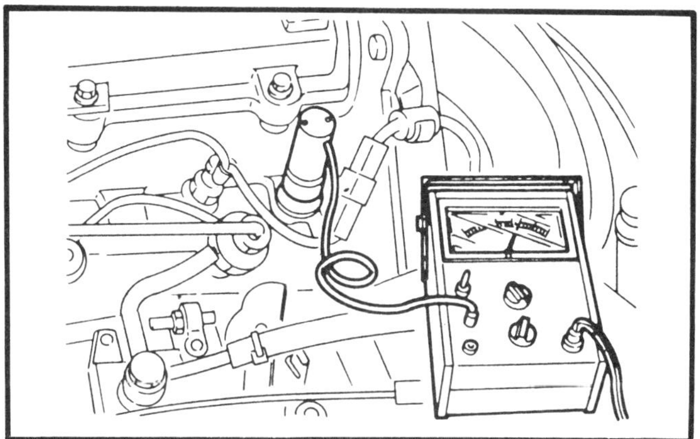

Connecting the diesel tachometer

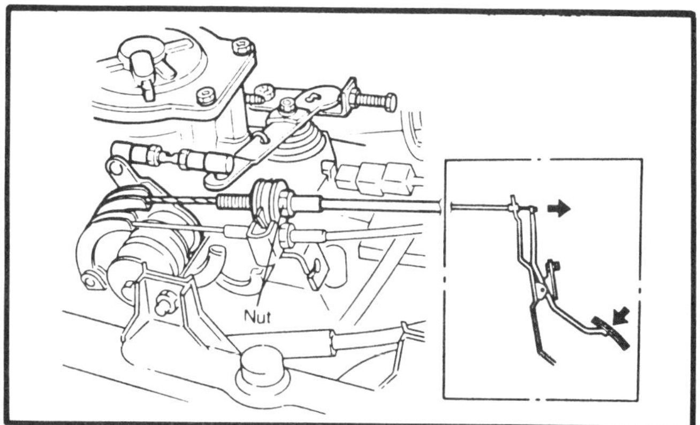

Adjusting the throttle pedal free-play

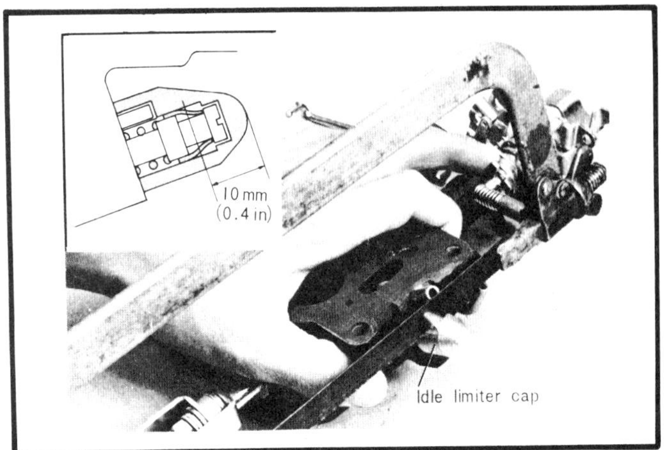

Cutting through the mixture limiter cap

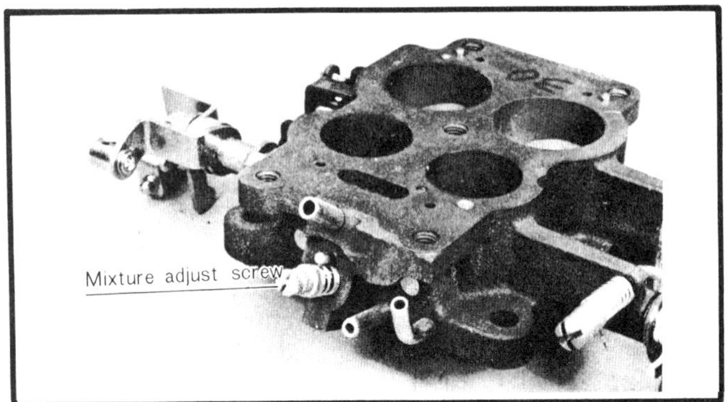

New mixture screw installation

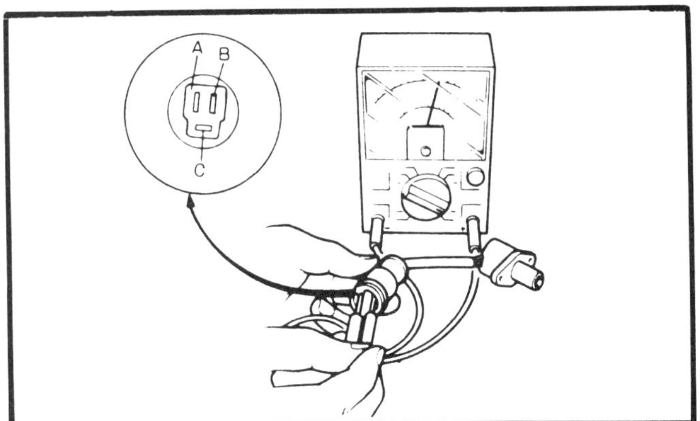

Connecting the ohmmeter to the variable resistor

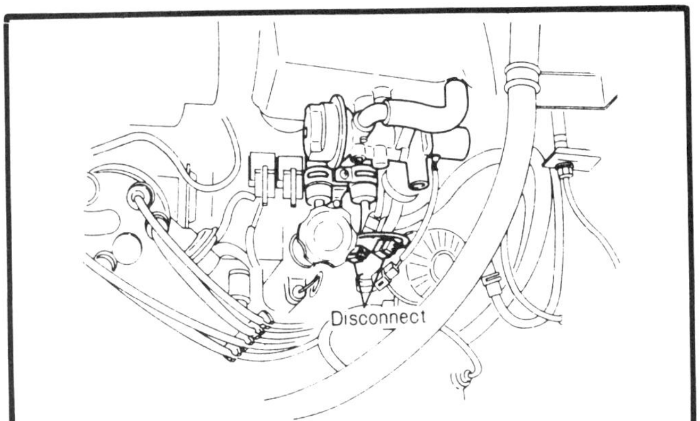

Disconnecting the vent/vacuum solenoid valve connector

2. Use a hack saw and cut through the limiter cap and mixture screw about 0.4 in. (10mm) from the end of the cap.

3. Remove the mixture screw. Install a new mixture screw and tighten it until it it is fully seated. Back the screw out three turns.

4. Install the carburetor back on the vehicle and start the engine, letting it run until normal operating temperature has been reached. Check the idle speed and adjust as necessary by turning the throttle adjust screw.

5. Set the idle speed at the highest rpm by backing out the mixture adjusting screw.

6. Reset the idle speed by turning the throttle adjust screw, 770 rpm in neutral for manual transmission 870 rpm in neutral for automatic transmission.

7. Screw in the idle mixture screw and adjust the idle speed to 750 rpm for manual transmission and 840 rpm for automatic transmission.

8. On vehicles equipped with automatic transmission, shift the selector to the drive position and adjust the idle speed to 750 rpm by turning the throttle adjust screw.

9. After the idle mixture adjustment is completed, fit an idle limiter cap onto the mixture adjust screw securely.

10. After adjusting the idle speed, the throttle sensor on the carburetor should be adjusted as previously outlined in this section.

MAZDA RX-7—13B FUEL INJECTED ENGINE

NOTE: The idle mixture adjustment usually is not necessary. The idle mixture adjustment should be adjusted when the variable resistor is replaced. Disconnect the variable resistor connector and connect an ohmmeter to the variable resistor. If the continuity (resistance) does not exist, replace the variable resistor and adjust the idle mixture. Resistance from terminal A to C should be 0.5–4.5 ohms and from terminal B to C, 0.5–4.5 ohms.

1. Start the engine and let it run until it reaches normal operating temperature, connect a tachometer to the engine.

2. Check and adjust the throttle sensor as necessary (as outlined above). Disconnect the vent and vacuum solenoid connector.

3. Adjust the idling speed to 800 rpm by turning the air adjust screw (AAS).

4. Set the idle speed at the highest by turning the variable resistor and then readjust the idle speed to 800 rpm by turning the air adjust screw (AAS).

5. Turn the variable resistor counter-clockwise until the engine speed reaches 780 rpm then turn it clockwise until the engine speed reaches 800 rpm.

6. Connect the vent and vacuum solenoid connector, fill up the head of the adjust screw with the adhesive agent part # N304–23–795 or equivalent.

MAZDA 626

NOTE: Before adjusting the idle mixture, check the the idle mixture with a tachometer as described below.

1. Start the engine and let it run until it reaches normal operating temperature.

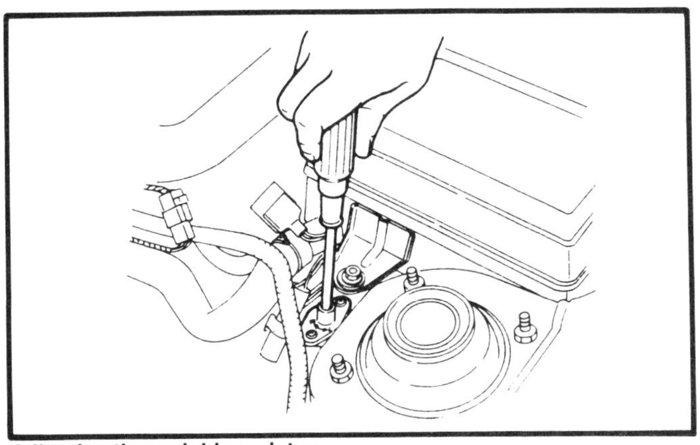
Adjusting the variable resistor

2. Connect a dwell meter (90 degrees, 4 cylinder) to the BrY wire in the check connector.

3. Check the idle mixture. With the idle speed set at specifications, the idle mixture should be 20–70 degrees on the dwell meter. If the idle mixture is not within specifications, adjust the idle mixture as follows.

4. Remove the carburetor from the vehicle and remove the spring pin from the base of the carburetor.

5. Install the carburetor on the vehicle and start the engine, letting it run until it reaches normal operating temperature.

6. Be sure that the air cleaner is installed and the idle compensator is closed. Connect a tachometer to the engine.

7. Connect a dwell meter (90 degrees, 4 cylinder) to the BrY wire in the check connector of the A/F solenoid valve and read the meter.

8. The reading should be 32–40 degrees at idle. If the reading is not within specification, adjust the idle mixture by turning the idle mixture adjustment screw.

NOTE: If the adjustment cannot be made, it is possible the oxygen sensor is not operating properly or either a broken wire or short in the the wiring between the oxygen sensor and the control unit.

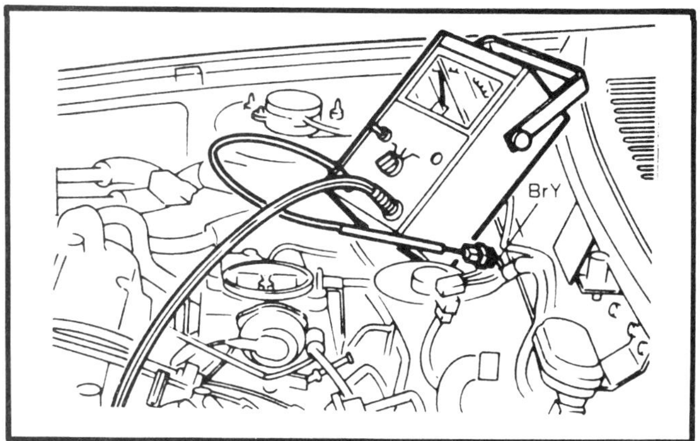

Connecting a dwellmeter to the BrY terminal

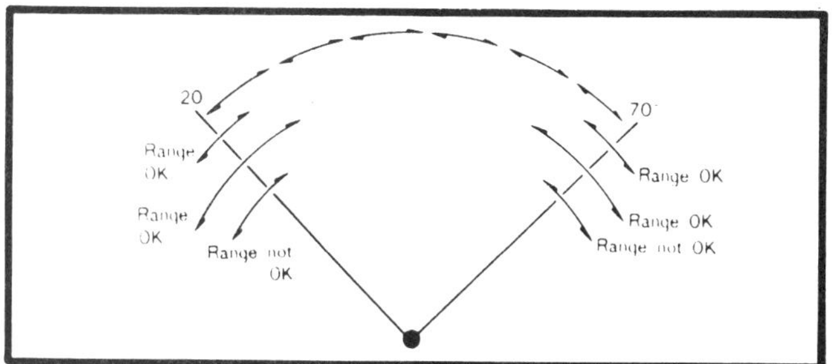

Idle mixture range on the dwellmeter

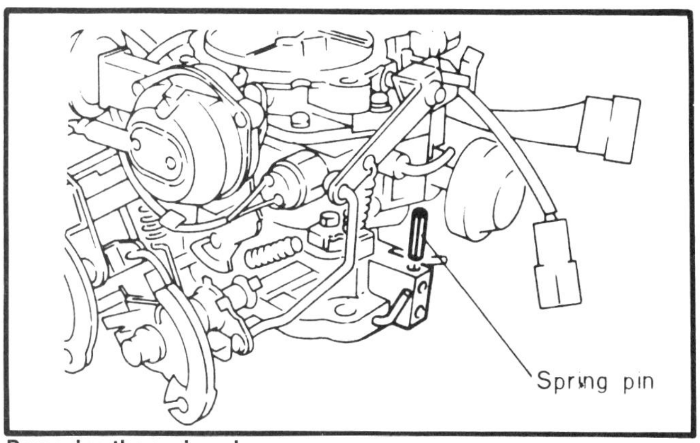

Removing the spring pin

9. Drive the spring pin back into to position in the carburetor.

MAZDA GLC

NOTE: Do not fix the spring pin to lock the mixture adjust screw until the adjustment is performed.

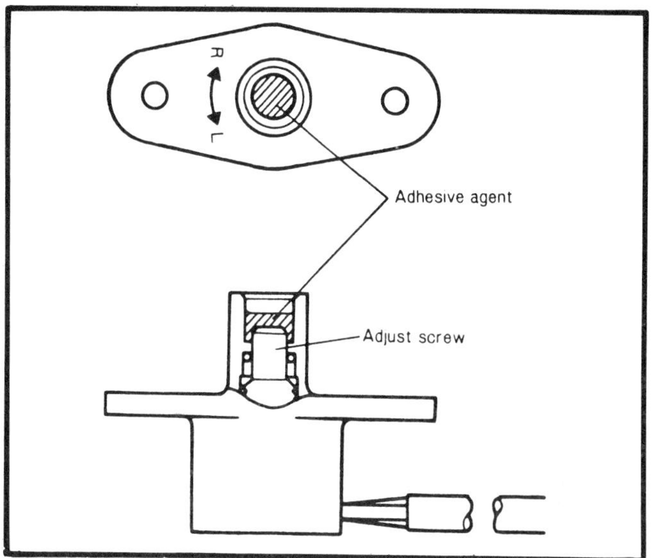

Sealing the mixture adjustment screw

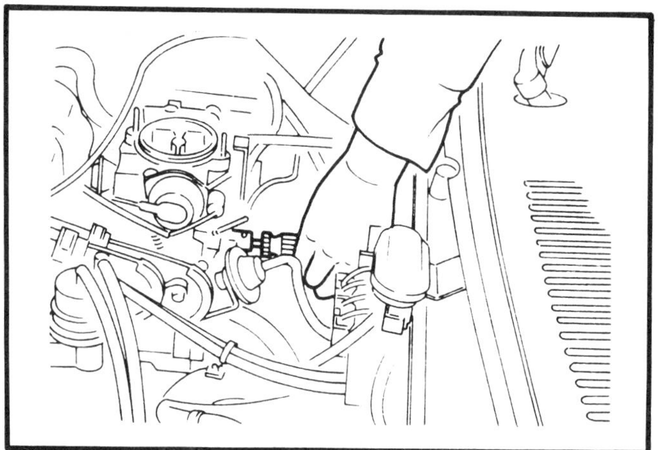
Adjusting the idle mixture

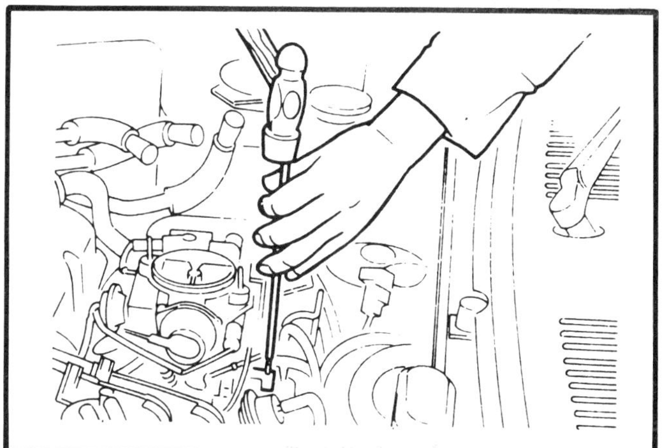
Installing the drive pin

1. Start the engine, letting it run until it reaches normal operating temperature. Connect a tachometer to the engine.

2. Connect a dwell meter (90 degrees, 4 cylinder) to the Y wire in the check connector of the A/F solenoid valve and read the meter.

3. The reading should be 32–40 degrees at idle. If the reading is not within specification, adjust the idle mixture by turning the idle mixture adjustment screw.

4. Be sure that the idle speed is set at specification. If not, adjust as necessary by using the throttle adjust screw. Fix the spring pin to lock the mixture adjust screw.

IDLE SPEED SPECIFICATIONS

Engine	Transaxle	Curb Idle Speed (rpm)
1985–86 RX-7		
12A Rotary	M/T & A/T	700 ± 100
13B Rotary	M/T & A/T	800 ± 100
1985–86 626		
2.0L	M/T	750 ± 100
2.0L	A/T	700 ± 100
2.0L Diesel	M/T & A/T	800 ± 100
1985–86 GLC		
1.5L	M/T	850 ± 100
1.5L	A/T	750 ± 100

M/T—Manual Transaxle
A/T—Automatic Transaxle

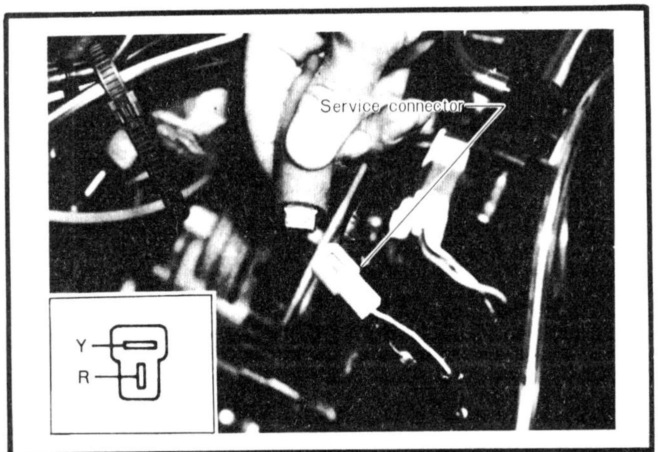

Dwellmeter connection on the Y terminal of the check connector

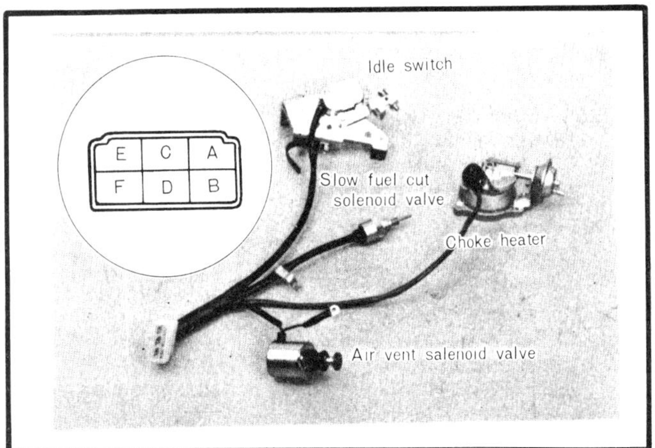

Idle switch terminal identification

IDLE SWITCH

Inspection and Adjustment

MAZDA GLC and 626

1. Disconnect the idle switch and connect an ohmmeter to the terminals C and D on the idle switch.

2. The tester should show continuity when the accelerator pedal is depressed.

3. Start the engine and let it run until it reaches normal operating temperature.

4. With the engine running at idle, connect a tachometer and connect a voltmeter to the idle switch terminals.

5. Increase the engine speed to above 2000 rpm and then decelerate gradually. The voltmeter reading at idle should be 12 volts and with the rpm above 1200–1300, the voltmeter should read 1.5 volts.

6. If the reading is out of specifications, adjust as necessary by turning the adjusting screw.

Glow Plug Inspection

MAZDA 626 DIESEL ENGINE

1. Disconnect the wiring from the glow plug.

2. Using an ohmmeter, check the continuity between each terminal of the glow plug and the cylinder head.

3. If continuity does not exist, remove and replace the glow plug.

Dash Pot Inspection and Adjustment

MAZDA 626

1. Quickly move the throttle lever and make sure that the dashpot rod comes out quickly to its full stroke, accompanying the movement of the throttle lever.

2. Release the throttle lever and make sure that it returns slowly to the idle position, after it has contacted the dashpot rod.

3. Start the engine and let it run until it reaches normal operating temperature.

4. With the engine running at idle, connect a tachometer to the engine and slowly increase the engine speed. Make sure the dashpot rod separates from the lever at approximately 2000 rpm.

5. If the function is not correct, loosen the locknut and adjust it by turning the dashpot diaphragm.

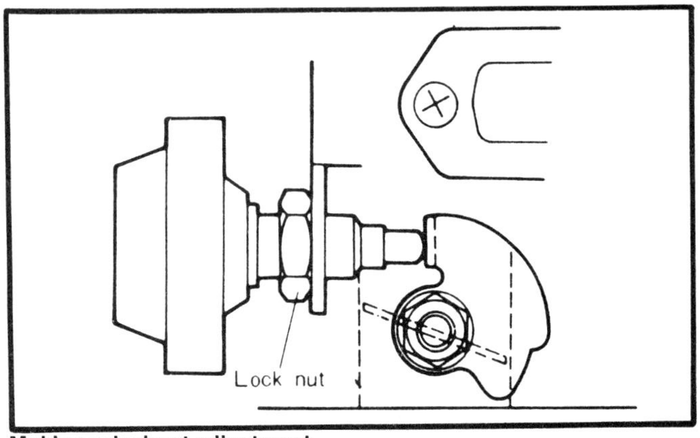

Making a dashpot adjustment

MAZDA RX-7

1. Quickly move the throttle lever and make sure that the dashpot rod comes out to its full stroke, accompanying the movement of the throttle lever.

2. Release the throttle lever and make sure that it returns slowly to the idle position, after it has contacted the dashpot rod.

3. Start the engine and let it run until it reaches normal operating temperature.

4. With the engine running at idle, connect a tachometer to the

engine and slowly increase the engine speed until the dash pot rod separates from the throttle lever.

5. Slowly decrease the engine speed and check the engine speed at which the throttle lever just touches the dash pot rod.

6. The engine speed should be 2350–2650 rpm. If the engine speed is not within specifications, loosen the locknut and adjust the engine speed by turning the dash pot diaphragm.

IDLE UP SYSTEM

Testing

GLC AND 626

1. Warm the engine up to normal operating temperature and let it idle. Turn the air conditioner on and check to see that the servo diaphragm stem is raised.

2. The servo diaphragm is activated when the A/C is on and the engine speed is 1300 rpm for 626 models and 1400–1900 rpm on GLC models.

3. Turn the A/C off. On the 626 models, turn the steering wheel all the way to the right or left to ensure the servo diaphragm stem is raised.

4. If the system does not react as described in this test, check the 3-way solenoid valves, servo diaphragm and control unit signal.

Ignition Timing

MAZDA RX-7

1. Start the engine and let it run until it reaches normal operating temperature. Connect a tachometer to the engine.

2. Check the idle speed and adjust as necessary. Connect a timing light to the high-tension wire of the leading spark plug (bottom plug) on the front rotor.

3. Aim the timing light at the timing indicator (yellow pointer) located to the rear of the crankshaft pulley.

4. If the leading timing is not correct, loosen the distributor locknut and rotate the distributor housing until the correct leading timing is obtained (12A engine—0 degrees ATDC, 13B engine—5 degrees ATDC).

5. Tighten the distributor locknut, recheck the leading timing and check the idle speed adjust as necessary.

6. To check the trailing timing, connect the timing light to high tension wire of the trailing spark plug (top plug) on the front rotor.

7. Aim the timing light at the timing indicator (red pointer) located to the rear of the crankshaft pulley and ahead of the leading timing indicator.

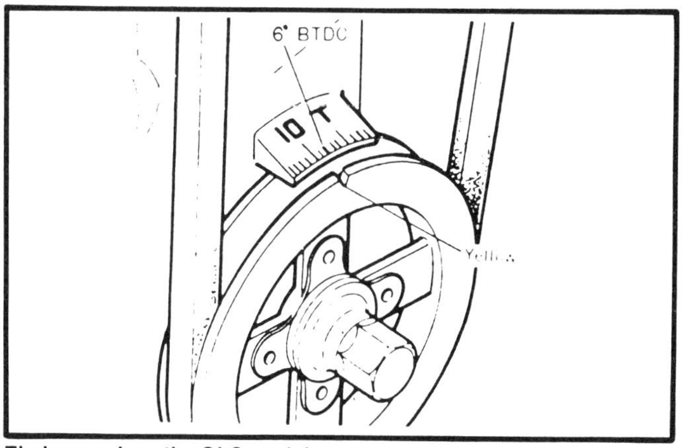

Timing mark on the GLC models

8. If the trailing timing is not correct, loosen the trailing vacuum advance unit attaching screws and move the vacuum unit until the correct timing is obtained.

9. Tighten the vacuum unit attaching screws, recheck the trailing timing and the idle speed and adjust as necessary.

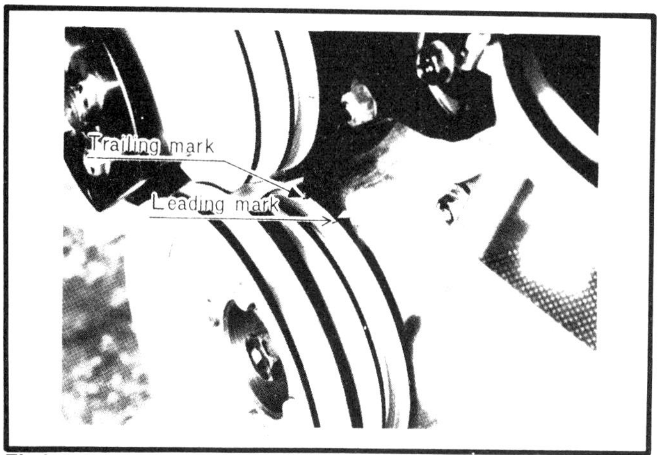

Timing mark locations

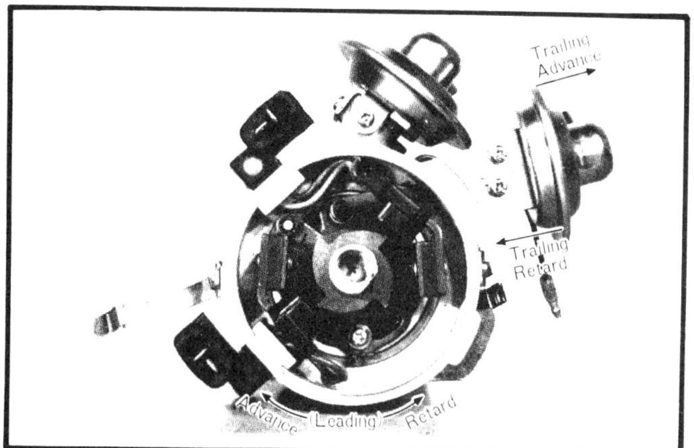

Moving the distributor for timing adjustment

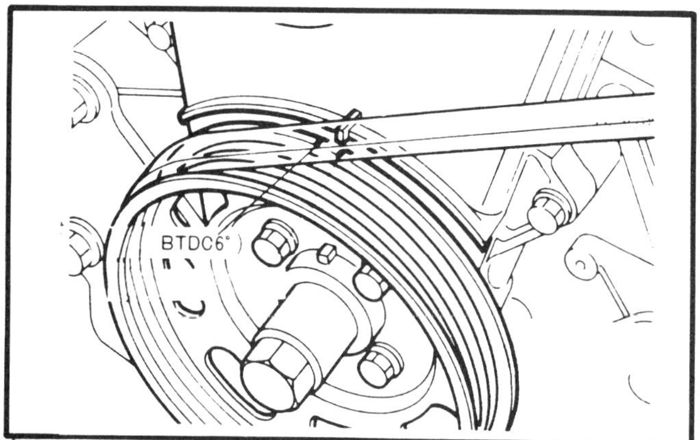

Timing mark location on the 626 (gas) models

MAZDA 626 AND GLC

1. Turn off all accessories and start the engine, let it run until it reaches normal operating temperature.

2. Disconnect the wiring coupler to the engine cooling fan on the 626 models only.

3. Connect a tachometer and connect timing light, check the idle speed and adjust as necessary.

NOTE: For models with an automatic transaxle, block the wheel and set the parking brake and shift the selector to the Neutral position.

4. Aim the timing light at the timing yellow timing mark on the crankshaft pulley and the timing mark (6 degrees BTDC) on the timing belt cover.

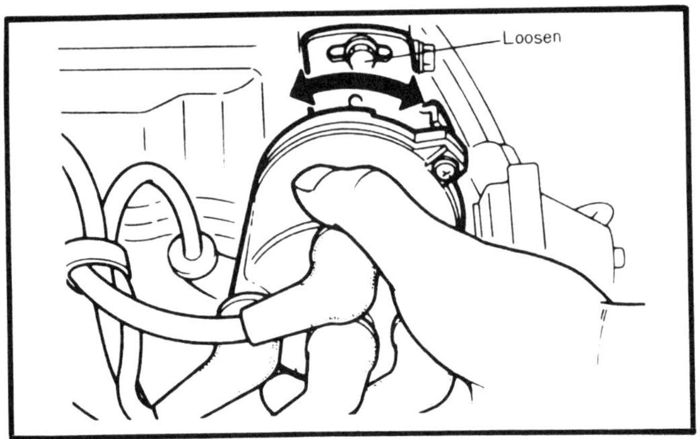

Moving the distributor to adjust the timing

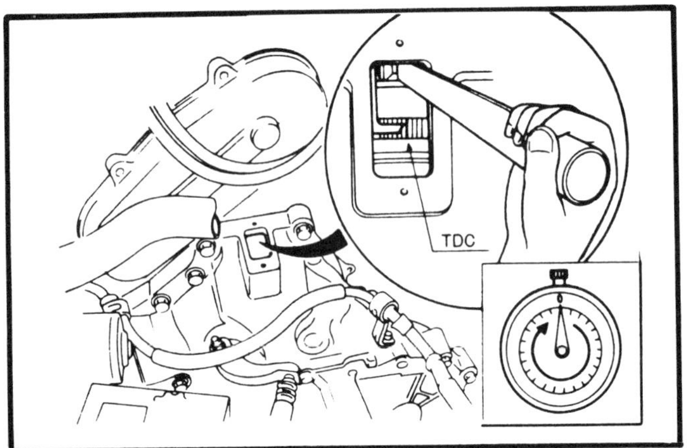

Aligning the timing mark on the flywheel

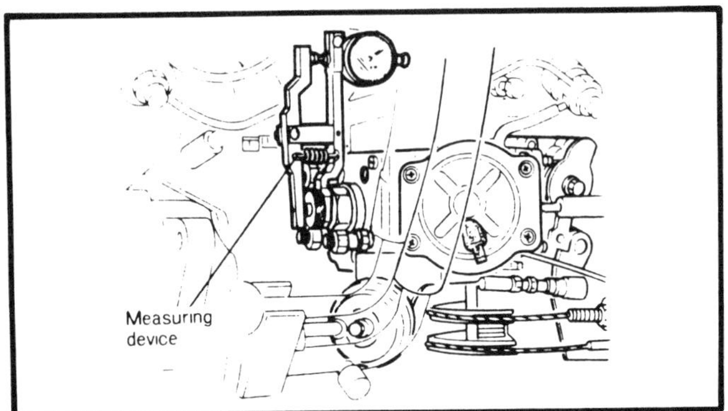

Installing the measuring device (dial indicator)

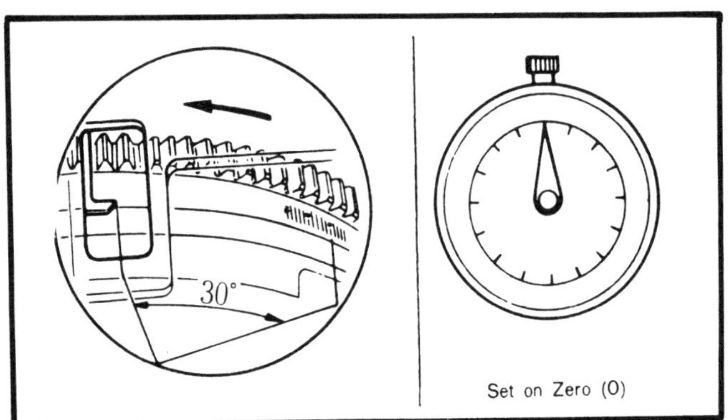

Positioning the flywheel 30 degrees to the counterclockwise side (set the dial indicator at 0)

5. If the timing marks do not align, loosen the distributor body installation nut, and turn the distributor body until the timing marks align on the 6 degree mark.

6. Tighten the distributor mounting bolt, recheck the timing and the idle speed and adjust as necessary. Remove all testing equipment and reconnect the wiring coupler to the engine cooling fan.

Checking and Adjusting Injection Timing

MAZDA 626 DIESEL ENGINE

1. Remove the fuel injection pipes and remove the service hole cover on the clutch housing.

2. Align the timing mark (TDC) on the flywheel with the indicator pin by turning the crankshaft.

3. Remove the hydraulic head plug on the injection pump, mount the measuring device # 49–9140–074 (dial indicator) or equivalent into the plug hole on the hydraulic head, so the tip of the dial gauge touches the plunger end of the pump and the dial gauge indicates approximately 0.008 in. (2.0mm).

NOTE: The delivery valve should be removed as an assembly shown in order to mount the measuring device. When installing the delivery valve, the gasket should be replaced and the valve should be torqued 3–4 ft. lbs.

4. Turn the flywheel slowly counterclockwise (in the reverse direction of the engine rotation), until the timing mark on the crankshaft pulley moves from the original position (TDC) to the counterclockwise side by 30–50 degrees and make sure the dial indicator pointer stops.

5. Set the dial indicator to zero. When setting the dial indicator to zero, confirm that the pointer does not deviate from the scale mark of zero, by slightly turning the crankshaft to the right and left.

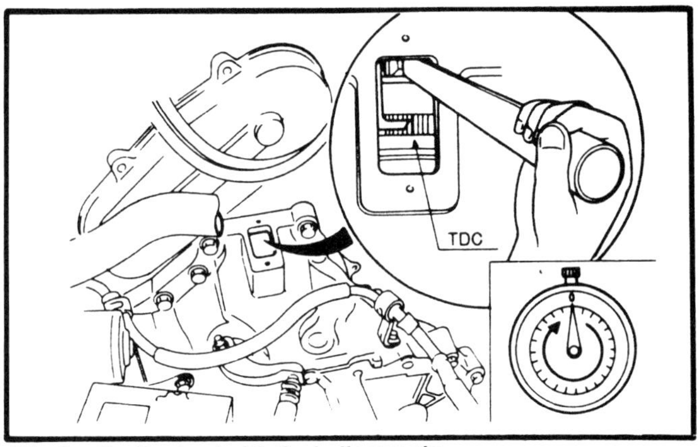

Aligning the flywheel with the indicator pin

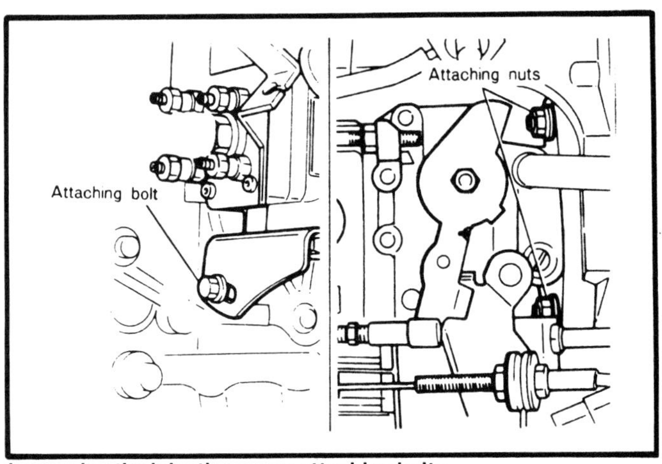

Loosening the injection pump attaching bolts

IGNITION TIMING SPECIFICATIONS

Engine	Transaxle	Ignition Timing
1985-86 RX7		
12A Rotary	M/T & A/T	12 ATDC Trailing
12A Rotary	M/T & A/T	0 ATDC Leading
13B Rotary	M/T & A/T	20 ATDC Trailing
13B Rotary	M/T & A/T	5 ATDC Leading
1985-86 626		
2.0L	M/T & A/T	6 BTDC at curb idle
1985-86 GLC		
1.5L	M/T & A/T	6 BTDC at curb idle

M/T—Manual Transaxle
A/T—Automatic Transaxle

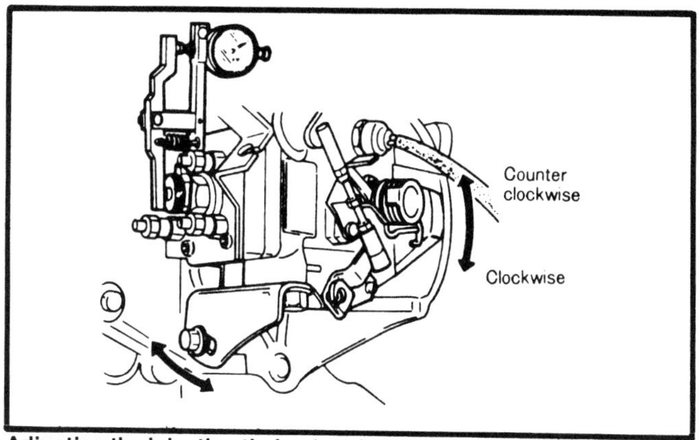

Adjusting the injection timing (moving the pump)

6. Turn the flywheel clockwise (in direction of the engine rotation), to align the timing mark (TDC) on the flywheel with the indicator pin.
7. If the dial indicator pointer indicates 0.004 ± 0.0008 in. (1.0 ± 0.02mm) when the timing mark (TDC) is aligned with the indicator pin, the injection timing is correctly adjusted. If it does not, adjust the pump timing as outlined in Steps 8 and 9.
8. Start the engine and run it until it reaches normal operating temperature.
9. Stop the engine and loosen the injection pump attaching nuts and bolts. Adjust the injection timing by moving the pump until the cam lift becomes 0.004 ± 0.0008 in. (1.0 ± 0.02mm) on the dial indicator.
10. Tighten the injection pump nuts and bolts, recheck the cam lift and remove all test equipment.

SPARK PLUGS

ORIGINAL EQUIPMENT

1985–86 RX-7

Nippondenso
W22EDR–14
W25EDR–15
W27EDR–14NGK
BPR7EQ–14
BPR8EQ–14
BPR9EQ–14GAP
0.053–0.055 in.

ORIGINAL EQUIPMENT

1985–86 MAZDA 626

Motorcraft
AGR–22
AGR–32NGK
BP5ES
BPR6ESGAP
0.031–0.033 in.

ORIGINAL EQUIPMENT

1985–86 MAZDA GLC

CHAMPION
RN–11YCNGK
BP5ES
BPR6ESMOTORCRAFT
AGR–22
AGR–32GAP
0.031–0.033 in.

EMISSION CONTROL SYSTEMS

CRANKCASE EMISSION CONTROLS

Positive Crankcase Ventilation (PCV)

The PCV system is designed to prevent blow-by gas from escaping to the atmosphere. The PCV valve contains a spring loaded plunger. When the engine starts, the plunger in the PCV valve is lifted in proportion to intake manifold vacuum and blow-by gas is drawn directly into the intake manifold. When the throttle valves are closed (idle) or partially open, blow-by vapor is returned directly to the intake manifold through the breather chamber and positive crankcase ventilation (PCV) valve with fresh air. When the throttle valves are wide open, the intake manifold vacuum decreases and vacuum in the air cleaner increases. In response to intake manifold vacuum, PCV valve increases the blow-by vapor flow. A small amount of vapor is returned through the valve cover breather hose and into the air cleaner.

System Service

NOTE: The PCV valve should be checked every 15,000 miles and replaced as required. If the valve is found to be defective do not attempt to clean it. Replace it.

1. With the engine running, remove the valve from its mounting. A hissing sound should be heard and vacuum should be felt from the inlet side of the valve.
2. With the engine stopped, remove the PCV valve. Shake the valve, a clicking sound should be heard indicating that the valve is not stuck.
3. If the valve fails any of the above tests, it should be replaced.

FUEL EVAPORATIVE EMISSION CONTROL SYSTEM

Evaporation Control System

The function of the evaporation control system is to prevent the emission of gas vapors from the fuel tank and carburetor to be expelled into the atmosphere. When fuel evaporates in the gas tank or the carburetor float chamber the vapors pass through the vent hoses to the charcoal canister where they are stored until they can be drawn into the intake manifold when the engine is running. All vehicles are equipped with the charcoal canister for the storage of fuel vapors. The fuel bowls of all carburetors are vented internally and on some applications do not require venting the canister. If this is the case the bowl vent port on the canister will be capped. Most carburetors are also externally vented to the charcoal canister.

System Service

The charcoal canister itself is a non-serviceable component. The only service required for the system is to replace the filter pad on the bottom of the canister. The filter requires replacement every 12 months or 12,000 miles. If the vehicle is driven under severe conditions the filter should be replaced more often.

OVERFILL LIMITER (TWO-WAY VALVE)

The overfill limiter consists of a pressure valve and a vacuum valve. The pressure valve is designed to open when the fuel tank internal pressure has increased over the normal pressure and the vacuum valve opens when a vacuum has been produced in the tank.

PURGE CONTROL VALVE

The purge control valve is kept closed during idling in order to prevent vaporized fuel from entering into the intake manifold for positive control of of high idle CO emission, which is a particular problem under high ambient temperatures. When the carburetor vacuum working on the diaphragm of the valve exceeds the pre-set value, the purge control valve is opened.

Number One Purge Control Valve Test

1. Remove the air cleaner and plug the hoses of the idle compensator, thermo-sensor and reed valves.

Testing the check and cut valve

2. Disconnect the air vent hose from the carburetor and plug the hose.
3. Warm up the engine to the normal operating temperature and run at idling speed.
4. Cover the bottom of the charcoal canister and increase the engine speed slowly, making sure that air is sucked into the canister.
5. If air is sucked into the number one purge control valve and the water thermo-valve, both are in proper operating condition.

PURGE CONTROL VALVE

Testing

RX-7

1. Disconnect the vacuum hose from the purge valve and start the engine.
2. With the engine running at idle speed, place a finger on the port opening and check that the air is not drawn into the port.
3. Increase the engine speed to 2000 rpm, air should be drawn into the port.

CHECK AND CUT VALVE

Testing

RX-7

1. Remove the check and cut valve, which is located just before the fuel tank.
2. Connect a pressure gauge to the passage of the fuel tank and blow through the valve.
3. The valve should open with pressure 0.78–1.0 inch lbs.,remove the pressure gauge and connect it to the passage open to the atmosphere.
4. Blow through the valve and if the valve opens at a pressure of 0.14–0.71 inch lbs., the valve is in proper operating condition. If the valve fails any of these test, remove and replace the valve.

AIR CONTROL VALVE

Testing

GLC AND RX-7

1. The air control valve is located on the air cleaner next to the reed valve. Disconnect the vacuum tube from the air control valve. Run the engine at idle.
2. Apply intake manifold vacuum to the air control valve. Air should be sucked in through the reed valve inlet port of the air cleaner. The amount of air should decrease when manifold vacuum is not applied.

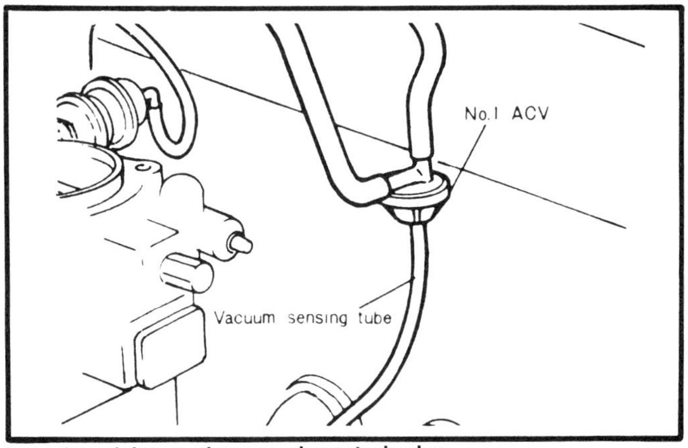

Location of the number one air control valve

626

1. Remove the air cleaner cover and element. Set engine speed to 1500 rpm. Place a piece of paper over the air control valve inlet port and check that air is sucked in.

2. Disconnect the vacuum sensing tube from the air control valve and plug the tube. Air flow through the reed valve should stop.

NOTE: Some vehicles are equipped with two control valves. The following procedure is used to check valve number two.

3. Warm up the engine to normal operating temperature. Remove the air cleaner and element. Set the engine speed at 1500 rpm. Place a piece of paper over the number two inlet port and check to see that no air is sucked in.

4. Ground the green/blue terminal of the three way solenoid valve with a jumper wire. Check that air is sucked into the valve. Do not ground the black/white terminal.

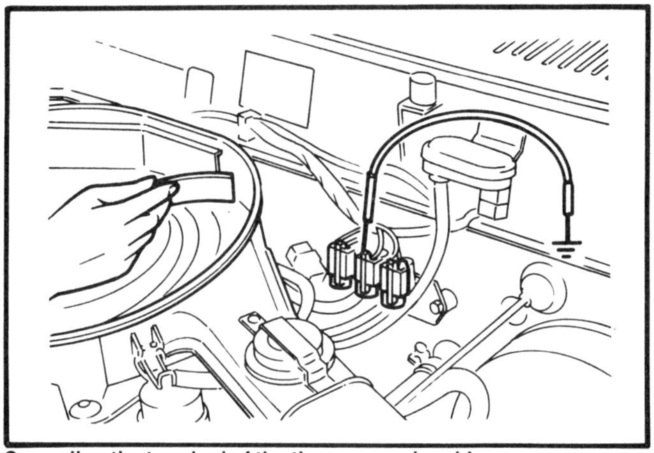

Grounding the terminal of the three-way solenoid

AIR CONTROL VALVE SOLENOID

Testing

GLC

1. The air control valve solenoid is located on the firewall behind the carburetor. The solenoid valve controls the vacuum signal to the air control valve. The solenoid valve is energized by the control unit at idle.

2. The air control valve is open to vacuum flow when not energized. Unplug the connector and apply 12 volts to the terminal with the yellow/black wire. Ground the terminal with the light green wire.

3. Blow air through the solenoid valve and check that the air is expelled out of the valve air filter.

AIR VENT SOLENOID VALVE

Testing

ALL MODELS

1. Disconnect the air vent hose from the canister and check it for cracking or other damage.

2. Blow through the hose and make sure that the air passes through the air vent solenoid valve.

3. Turn the ignition switch on and blow through the air vent hose. Make sure the air does not pass through the air vent solenoid valve. If it does, replace the valve.

THREE-WAY SOLENOID VALVES

Testing

GLC AND 626

1. On the 626 models, unplug the connector from the air conditioning solenoid valve. Ground the black/red terminal and apply 12 volts to the black/white terminal. With the power applied, the solenoid valve should be open.

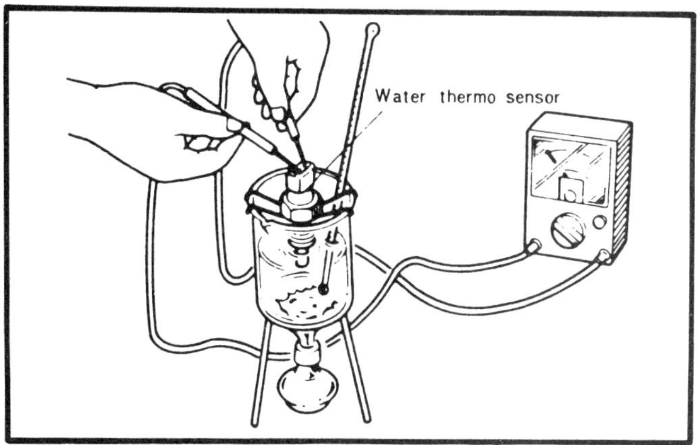

Testing the water thermo sensor

2. Unplug the connector from the power steering solenoid valve, ground the green/red terminal apply 12 volts to the black/white terminal. The solenoid valve should be open.

3. On the GLC models, unplug the connector from the throttle positioner solenoid valve. The valve should be located on the firewall behind the carburetor.

4. Apply 12 volts to the yellow/black terminal and ground the light green/red terminal.

5. The throttle positioner sensor should be open to manifold vacuum signal. If the switch does not react to the test as described, replace the valve.

WATER TEMPERATURE SENSOR SWITCH

Testing

GLC AND 626

1. The water temperature sensor switch is located on the intake manifold.

2. Remove the temperature sensor and check the calibration. If the calibration is not within specifications, replace the sensor.

TEMPERATURE SENSOR CALIBRATION CHART

Coolant Temperature (Degrees F.)	Resistance (Ohms)
-4	14.6–17.8
68	2.21–2.69
176	0–.64

WATER THERMO VALVE

Testing

GLC AND 626

1. Remove the water thermo valve and immerse the valve into a container of water.

2. Heat the water gradually and observe the temperature, blow through the valve and if air passes through the valve at 129 degrees F. or higher, the valve is operating properly.

WATER THERMO SWITCHES

Testing

GLC AND 626

1. Remove the switch from the lower left hand corner of the radiator.

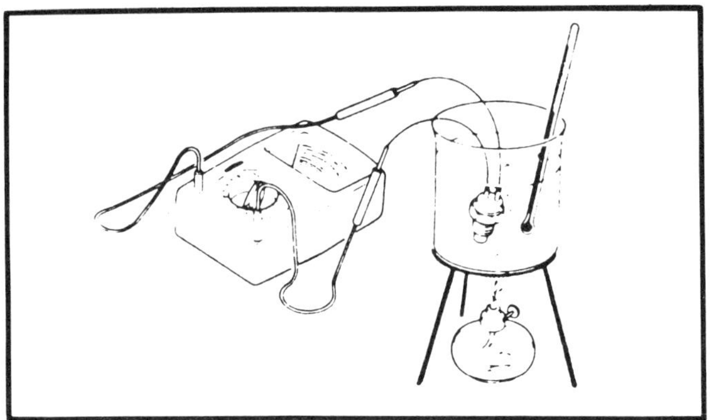

Testing the water thermo switch

NOTE: On the GLC models, the switch will have green/red and black wires attached to it. On the 626 models, green/black and black wires are attached to the switch.

2. Place the switch in water with a thermometer and heat the water gradually.
3. Using an ohmmeter, check for continuity at the switch terminals. Also take note of the temperature when continuity is found at the switch terminals.
4. The switch should show continuity at 66 degrees F. If the switch does not meet these specifications, replace the switch.

EXHAUST EMISSION CONTROL SYSTEM

Air Injection System

DESCRIPTION AND OPERATION

The air injection system is made up of the basic components, a belt driven air pump with integral relief valve, air control valve with relief valve, air by-pass valve, water thermo valves, vacuum delay valve, check valve and various hoses and piping. A heat hazard sensor system is used to prevent exhaust system overheating. The air injection system is designed to reduce hydrocarbon and carbon monoxide content of exhaust gases by injecting a controlled amount of compressed air into the exhaust manifold.

Air is drawn from the clean side of the air cleaner by the air pump and directed under pressure to the air control valve. The number one relief valve of the air control valve diverts part of the air pump air back into the air pump when the engine is operating under heavy load conditions. The air injection system also contains thermo valves, delay valves, check valves and a vehicle speed sensor to control and protect the air injection system.

REED VALVE

Testing

1. Warm up the engine to normal operating temperature and then shut off the engine.
2. Remove the air cleaner cover and element, start the engine and let it idle.
3. Place a thin piece of paper over the reed valve inlet port and make sure that air is sucked in.

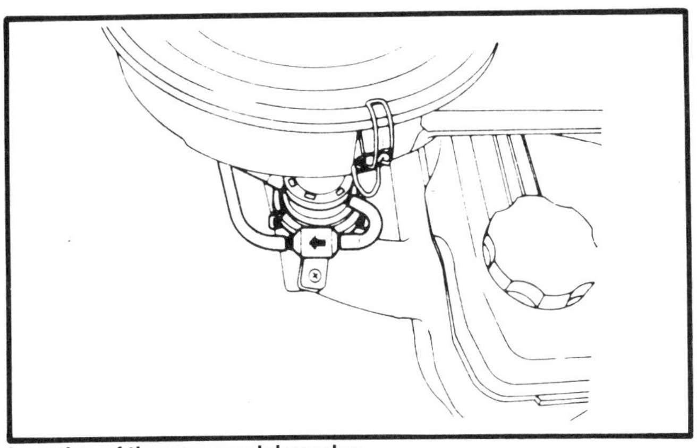

Location of the vacuum delay valve

4. Increase the engine speed to 1500 rpm and check the exhaust leakage at the air inlet fitting by placing a thin piece of paper over the port.
5. Replace the reed valve as necessary.

CHECK VALVES

Testing

1. Run the engine until normal operating temperature has been reached. Stop the engine and disconnect the air hose from the check valve connector.
2. Slowly increase the engine idle speed to 1500 rpm while holding your hand over the open connection.
3. Check for exhaust leak from the check valve. If it leaks, replace the valve.

VACUUM DELAY VALVE

Testing

GLC AND 626

1. Disconnect the vacuum delay valve and connect a vacuum tube to a hand held vacuum pump.
2. Connect the vacuum delay valve to the vacuum tube from the vacuum pump and shut off the other side of the vacuum delay valve with your thumb.
3. Apply a vacuum of over 19.7 in. Hg. with the vacuum pump, release your thumb from the vacuum delay valve and check the time required for the vacuum reading to decrease to 3.9 in. Hg. from 15.7 in. Hg. (specified time: 2.3 ± 1 second).

Exhaust Gas Recirculation System

EGR VALVE AND EGR THERMO VALVE

Testing

GLC

1. Warm the engine up to normal operating temperature and let the engine idle.
2. Disconnect the vacuum sensing tube from the EGR valve and using a hand held vacuum pump or equivalent, apply vacuum directly to the EGR valve.
3. When the vacuum is apply to the EGR valve, the engine should run rough or stall. If it does not, remove and replace the EGR valve or the thermo valve.

EGR VALVE AND EGR POSITION SENSOR

Testing

626

1. Warm up the engine to the normal operating temperature, stop the engine and remove the air cleaner.
2. Plug the hoses of the idle compensator, thermo sensor and reed valves. Disconnect the vacuum sensing tube from the EGR valve and plug the tube.
3. Connect a vacuum pump to the EGR valve. Start the engine and let it idle.
4. Apply vacuum to the EGR valve with the vacuum pump. If the engine starts to run rough or stall when the vacuum exceeds approximately 1.97 in. Hg., the EGR valve is operating properly.
5. Disconnect the position sensor connector. Connect a multi-tester to the top single terminal and the bottom left hand terminal in the connector.
6. If the multi-tester shows some resistance, the position sensor is not defective.

DUTY SOLENOID VALVE

Testing

626

1. Remove the air cleaner and disconnect the vacuum sensing tube from the EGR valve.
2. Blow through the vent tube and make sure air passes. Disconnect the duty solenoid valve.
3. Apply battery voltage to the bottom left terminal of the solenoid valve connector.
4. Blow through the vent tube and make sure air is not passed.

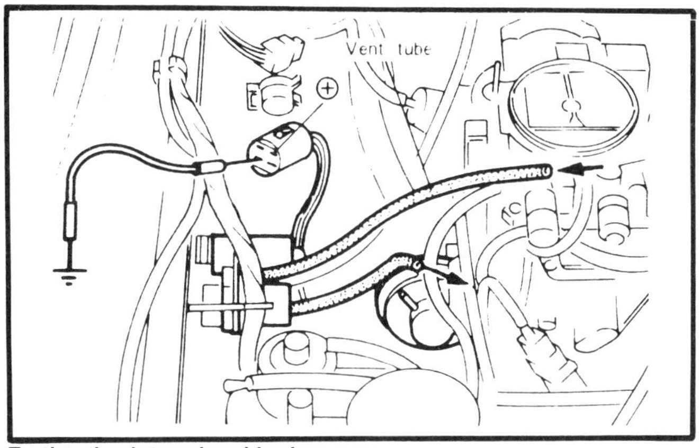

Testing the duty solenoid valve

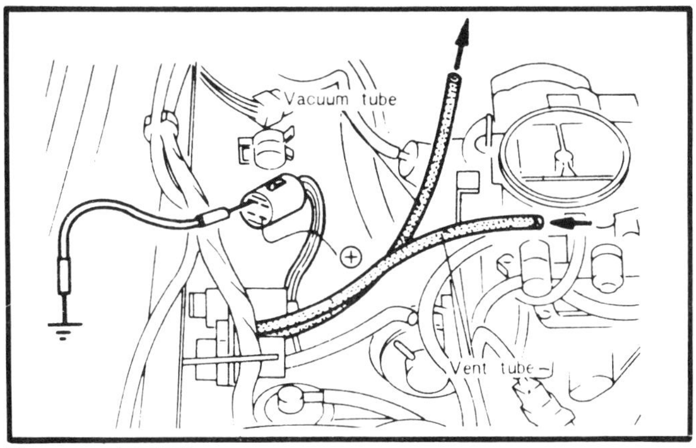

Vacuum valve test

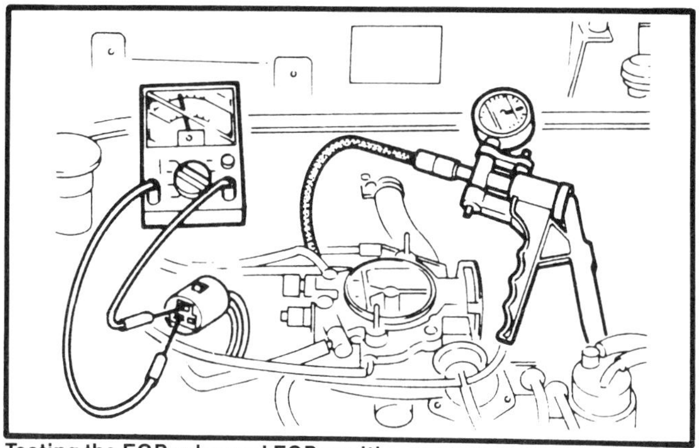
Testing the EGR valve and EGR position sensor

VACUUM VALVE

Testing

626

1. Disconnect the vacuum sensing tube between the duty solenoid valve and the intake manifold at the intake manifold.
2. Blow through the vacuum sensing tube and make sure air is not passed. Apply battery voltage to the top left terminal of the solenoid valve connector.
3. Blow through the vacuum sensing tube and make sure that air passes through it. Replace the duty solenoid valve, if necessary.

VACUUM SENSOR

Testing

626

1. Warm the engine up to normal operating temperature and then let the engine idle.
2. Connect a voltmeter to the vacuum sensor L terminal (the terminal to the far right looking at the sensor) and take a reading.
3. The reading should be one volt in the above conditions. With the vacuum tube disconnect from the vacuum sensor, the voltage reading should be four volts.
4. If the voltage readings on the vacuum sensor are out of specifications, replace the sensor.

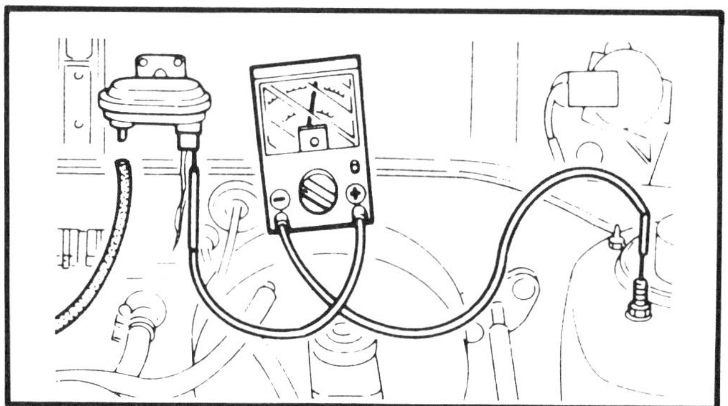
Connecting the voltmeter to the L terminal

EGR SOLENOID VALVE

Testing

GLC

1. Warm up the engine to normal operating temperature and let the engine idle.
2. Remove the EGR solenoid valve plug and blow air through from the top port (A). Be sure that air comes through the front port (B).

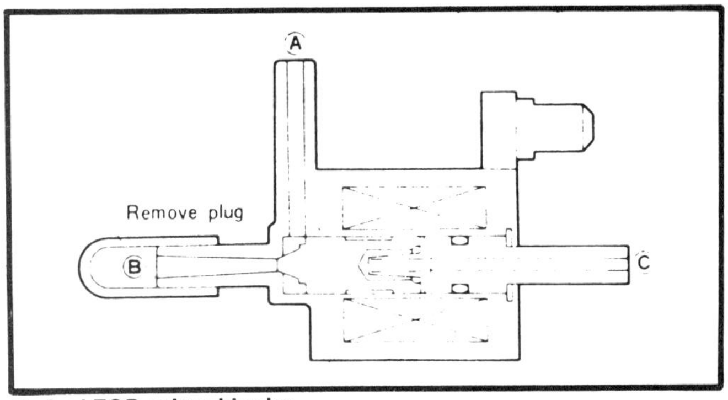

Typical EGR solenoid valve

3. Disconnect the coupler from the solenoid valve. Blow air through the solenoid valve from the top port (A) and make sure hat air comes out of the back port (C).

EGR Modulator Valve

Testing

GLC

1. Warm up the engine to normal operating temperature and let the engine idle.
2. Disconnect the vacuum from the single port of the EGR modulator valve.
3. Increase the engine speed to 2000 rpm and be sure air is sucked in at the inlet fitting. Replace the valve as necessary.

Slow Fuel Cut System

Testing

GLC AND 626

1. Warm up the engine to normal operating temperature and let the engine idle.
2. Disconnect the neutral safety switch connector. Connect a voltmeter to the check-connector Lg terminal (terminal on the right side of the connector). The voltmeter reading should be 1.5 volts.
3. On the manual transmission models, connect an ohmmeter to the terminals of the neutral switch. The switch should show continuity when the gearshift lever is in the neutral position and no continuity when the lever is in any other gear position.
4. On the automatic transmission models connect a tachometer to the engine and increase the engine speed to 3000 rpm and then decelerate rapidly.
5. Take a reading on the voltmeter during the deceleration period. The reading should be 12 volts at 2000 rpm and 1.5 volts below 2000 rpm.

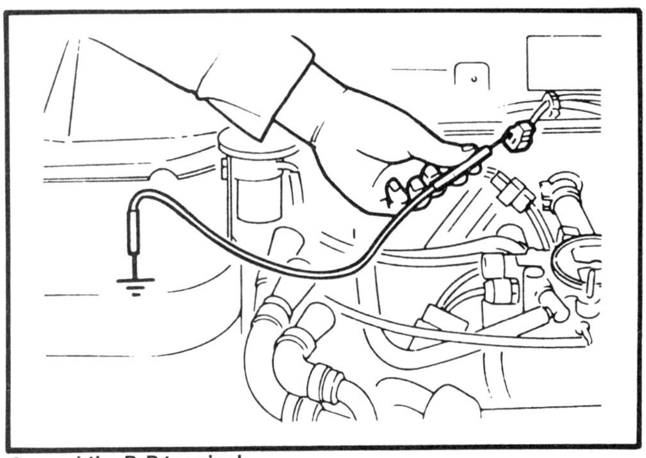

Ground the BrB terminal

SLOW FUEL CUT SOLENOID SYSTEM

Testing

GLC AND 626

1. Run the engine at idle speed and disconnect the slow fuel cut solenoid valve connector.
2. If the engine stalls, the solenoid is operating properly. If the engine does not stall, replace the solenoid.

CLUTCH SWITCH

Testing

ALL MANUAL TRANSAXLES ONLY

1. Disconnect the clutch switch connectors and connect an ohmmeter to the terminals of the switch.
2. The ohmmeter should show continuity when the clutch pedal is fully depressed. If it does not, replace the clutch switch.

POWER STEERING SWITCH

Testing

626

1. Start the engine and let it run at idle. Unplug the power steering switch connector from the steering gear housing.
2. Using a ohmmeter, check to see if there is continuity between the terminals of the power steering switch connector when the steering wheel is turned all the way to the right or left.
3. When the oil pressure is above 355 psi, the switch should be on. At pressures below 355 psi, the switch should be off.

COASTING RICHER SYSTEM

Testing

626

1. Warm the engine up to normal operating temperature and let the engine idle. Disconnect the neutral switch connector.
2. Connect a voltmeter to the check-connector BrB terminal. The reading should be 12 volts.
3. Connect a tachometer and increase the engine speed while pushing the idle switch button.
4. Take the reading of the voltmeter during deceleration. The reading should be 12 volts at 2000 rpm, 1.5 volts below 2000 rpm and 12 volts below 1400 rpm.

COASTING RICHER SOLENOID VALVE

Testing

626

1. Run the engine at idle speed and ground the connector BrB terminal with a lead wire.
2. The engine speed should increase. If the engine speed fails to increase, the solenoid valve should be replaced.

NEUTRAL SWITCH

Testing

626 WITH MANUAL TRANSAXLE

1. Disconnect the neutral switch connectors and connect an ohmmeter to the terminals of the switch.
2. The ohmmeter should show continuity when the vehicle is in neutral and show no continuity in any of the other position.

626 WITH AUTOMATIC TRANSAXLE

1. Position the vehicle on a rolling road tester (dynomometer), If equipped. Secure the vehicle so it does not move forward or to the side.

2. Start the engine and increase the engine speed to 9 mph, if on the dynomometer. Disconnect the neutral switch connector.

3. Connect a ohmmeter to the neutral switch terminal and ground and check the continuity between the switch and the ground.

4. There should be continuity when the vehicle is stopped and no continuity when the engine speed is increased. If the switch fails this test, remove and replace the switch.

SERVO DIAPHRAGM

Adjustment

GLC

1. Connect a tachometer to the engine, warm up the engine and let it idle.

2. Stop the engine and remove the air cleaner. Disconnect the vacuum sensing tube from the servo diaphragm.

3. Connect a vacuum line from a source of intake manifold vacuum to the servo diaphragm.

4. Start the engine and increase the engine speed to 2000 rpm with the throttle lever and release the throttle lever.

5. The engine speed should be 1200 + 100 rpm, if the engine speed does not fall into specifications, adjust the diaphragm until the engine speed is within specifications.

626

1. Remove the air cleaner, plug the hoses of the idle compensator, thermo sensor and reed valves. With all accessories off, warm up the engine to normal operating temperature.

2. On the models equipped with A/C, disconnect the A/C solenoid valve to servo vacuum sensing tube from the servo diaphragm.

3. With the engine idling, apply manifold vacuum to the servo diaphragm. The engine speed should increase to 1200–1400 rpm. If the engine speed is not up to specifications, increase it or decrease it by turning the adjustment screw on the back of the servo diaphragm.

4. On the models equipped with A/C and power steering, repeat step one and disconnect the power steering solenoid valve to servo vacuum sensing tube.

5. Apply manifold vacuum to the servo. The engine should increase 800–1000 rpm on the manual transaxle models and 10501250 rpm on the automatic transaxle models.

6. If the engine speed has to be adjusted, turn the adjusting screw on the servo diaphragm head and recheck the engine speed.

THROTTLE POSITIONER SOLENOID VALVE

Testing

GLC

1. Disconnect the vacuum sensing tube (A) from the servo diaphragm and disconnect the vacuum sensing tube (B) from the solenoid valve.

2. Disconnect the connector of the solenoid valve. Blow air through the solenoid valve from the vacuum sensing tube (A) and make sure air comes out of port A (lower front on the valve).

3. Apply battery voltage to the solenoid valve with a suitable jumper wire. Blow air through the solenoid valve from vacuum tube (A) and make sure the air comes out of the valve's air filter.

4. If the solenoid valve does not operate properly, replace it with a new one.

5. To further test the solenoid valve, start the engine and let it run at idle.

6. Connect a tachometer to the engine and increase the engine speed gradually, checking to see that the servo diaphragm operates at a the specified engine speed. The specified engine speed is 1400–1900 rpm. If the valve fails to meet the specifications, replace it with a new one.

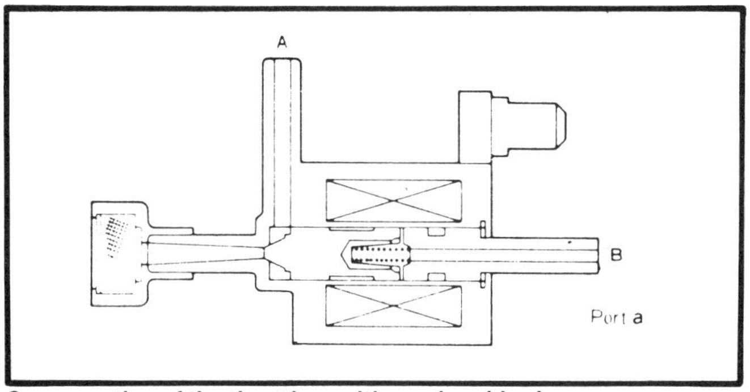

Cross section of the throttle position solenoid valve

Catalytic Converter System

ALL MODELS

Catalytic converters require the use of unleaded fuel only. Leaded fuel will destroy the effectiveness of the catalysts as an emission control device. Under normal operating conditions, the converter will not require any maintenance. However, it is important to keep the engine properly tuned. If the engine is not kept properly tuned, engine misfiring may cause overheating of the cataylsts. This may cause heat damage to the cônverters or vehicle components. This situation can also occur during diagnostic testing if any spark plug cables are removed and the engine is allowed to idle for a long period of time.

Idle Compensation System

Testing

ALL MODELS

1. Check to see that the valve is in a closed position when the bimetal temperature is lower than operating temperature (153 degrees F.).

2. Apply vacuum to the vacuum hose. If excessive leakage is found, replace the idle compensator valve assembly.

3. When the bimetal temperature is more than 153 degrees F., check to see if the valve is in the open position. If the valve is not in the open position, remove and replace the valve.

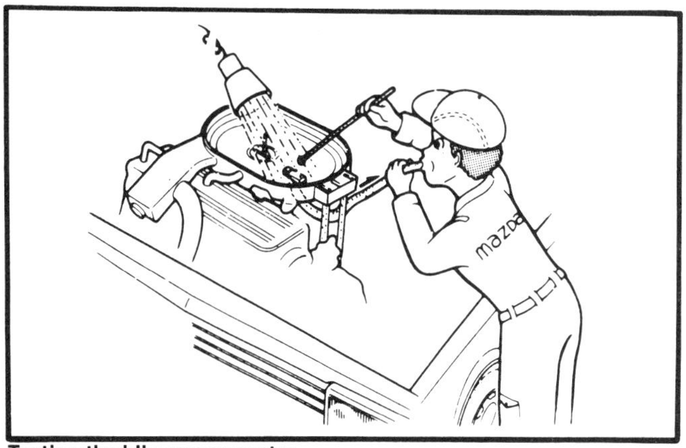

Testing the idle compensator

ANTI-AFTERBURN VALVE

Testing

GLC AND 626

1. Operate the engine at idling speed and close the air intake port of the anti-afterburn valve and make sure there is no air being drawn in.

2. Increase the engine speed and quickly decelerate. Air should be drawn into the intake valve for 1–2 seconds after the accelerator is released.

3. If there is no vacuum after the accelerator has been released, remove and replace the valve.

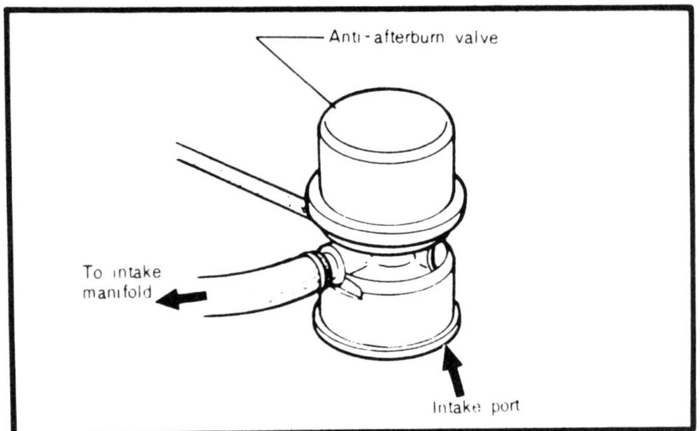

Typical anti-afterburn valve

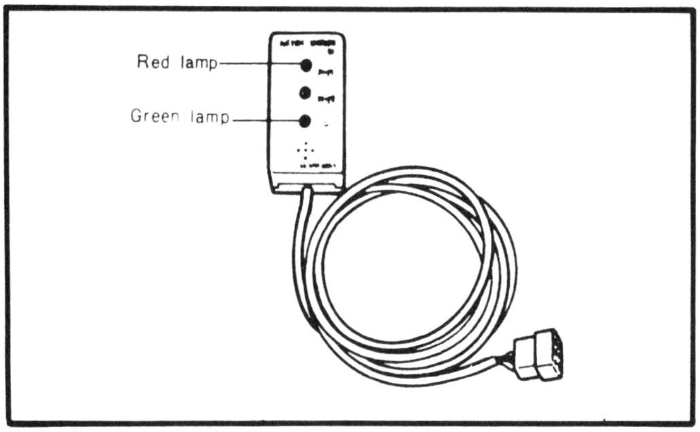

Typical system checker 83

OXYGEN SENSOR

ALL MODELS

The oxygen sensor is mounted in the exhaust manifold. The output signal from this sensor, which varies with the oxygen content of the exhaust gas stream, is provided to the electronic control unit for use in controlling closed loop compensation of fuel delivery.

HIGH ALTITUDE COMPENSATOR

Testing

ALL MODELS

1. Apply vacuum through each port on the high altitude compensator valve.
2. Air should pass through at a high altitude of 3,608 feet and air should be blocked at a low altitude of less than 3,608 feet.

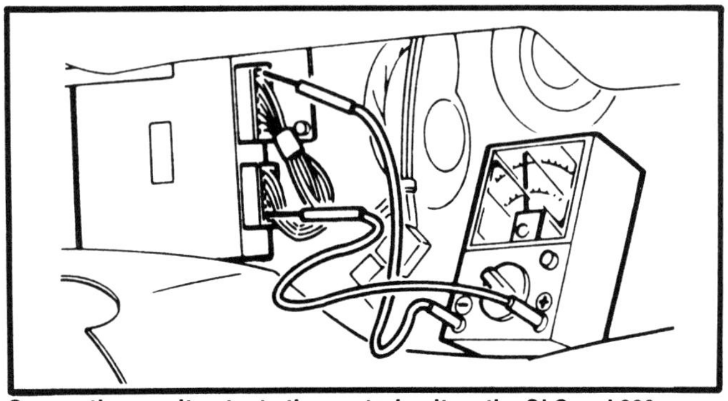

Connecting a voltmeter to the control unit on the GLC and 626 models

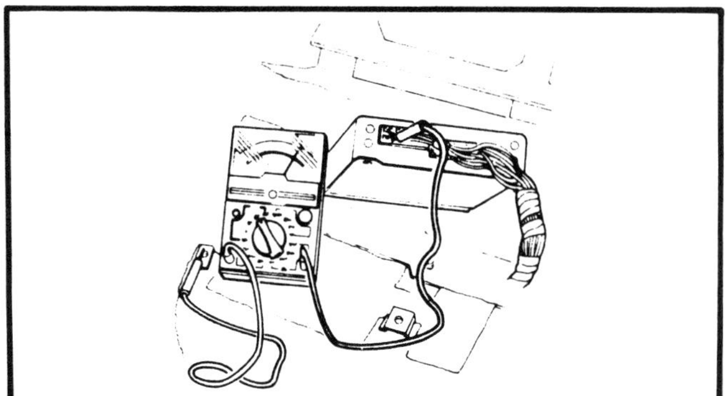

Connecting a voltmeter to the control unit on the Mazda RX-7 models

ALTITUDE COMPENSATOR

Testing

ALL MODELS

1. Remove the air cleaner and start the engine, making sure the engine is idling smoothly.
2. Block the slow port on the carburetor air horn by placing a finger over the port. The idle speed should drop at altitudes of between 1,640–4,920 feet. If it does not, go on to the next step.
3. Remove the altitude compensator valve and blow air through the valve's two ports. Check that air passes through the valve when the altitude is between 1,640–4,920 feet. If air fails to pass through the valve at this altitude, remove and replace the valve.

Control Device

ALL MODELS

The control unit detects the engine speed, the intake manifold vacuum, the coolant temperature , the oxygen sensor concentration in the exhaust gas, the position of the EGR control system and the functions of other various sensors. It also controls the fuel control system, the deceleration

	Code No.	Checking item	Indicated pattern (Lamp ON: –, OFF: · · · ·)
STEP 1 (RED)	1	Engine speed (rpm signal)	ON: 0.4 sec. OFF: 2.0 sec. 1 CYCLE
	2	Water thermo sensor	0.4 0.4 0.4 / 0.4 0.4 0.4 2.0 / 1 CYCLE
	*3	Feed back system	2.0 2.0 1 CYCLE

*3 If the troubles have been found, do not turn off the ignition switch.
If it is turned off, feed back system can not be checked with the system checker.

System checker 83 trouble code cycle for the Mazda GLC

control system, the secondary air control system and the EGR control system.

Control Unit Inspection

1. Connect a voltmeter to the control unit (which in most cases is located under the dash panel on the right side kick panel) as shown in the illustration provided.

2. Turn the ignition switch to the ON position (do not start the engine) and measure the voltage at each terminal with the aid of the chart provided. If the proper voltage is not indicated on the voltmeter, check all wiring, connections and finally check the involved component.

Control Unit Inspection With System Checker 83

NOTE: The system checker 83 (part # 49–G030–920) can be used to detect and indicate any problems of each sensor, damaged wiring, poor contact or a short circuit between each of the sensor-control units. Trouble is indicated by a red lamp and a buzzer.

1. Warm up the engine to the normal operating temperature and run it at idle.

2. Connect the system checker 83 to the control unit check connector.

Terminal		Condition			
		IG switch: ON	Idle	1700 rpm	2500 rpm
A (YG)	IG. signal	12V	–	–	–
B (YB)	Battery power (+)	12V	–	–	–
C (BW)	Air cond. signal	Air cond: ON = 12V, Air cond: OFF = 0V			
D (BrW)	Slow fuel cut solenoid	–	0V	–	12V (With terminals m, o disconnected)
E		It is not necessary to check.			
F (Y)	A/F solenoid	–	0 ~ 12V	–	–
H (LgR)	Throttle positioner solenoid	–	0V	12V	–
I		It is not necessary to check.			
J	Blank	–	–	–	–
K (YG)	EGR solenoid	–	0V	–	12V During rapid acceleration
L	Blank	–	–	–	–
M (Lg)	ACV solenoid	–	12V	–	–
N (B)	Earth	0V	–	–	–
a (GY)	O2 sensor	–	0 ~ 1V	–	–
b (GY)	O2 sensor sealed	0V	–	–	–
c (LB)	Water temp. sensor	–	about 1.2V	–	–
d (RW)	Earth (Water temp. sensor)	0V	–	–	–
e	Blank	–	–	–	–
f	Blank	–	–	–	–
g	Blank	–	–	–	–
h	Blank	–	–	–	–
j (LG)	*P/S signal	P/S: ON = 0V, P/S: OFF= 12V			
k (LW)	Vacuum switch	about 10V	0V	–	–
l (LR)	Vacuum switch	about 10V	0V	–	–
m (GB)	Neutral switch, clutch switch	–	0V	–	–
n		It is not necessary to check.			
o (BrR)	Idle switch	–	about 10V	0V	–
p (GY)	Earth	0V	–	–	–
q (GR)	Water thermo switch	–	0V	–	–
r (B)	Earth	0V	–	–	–

* P/S = Power steering

a	c	e	g		k	m	o	q
b	d	f	h	j	l	n	p	r

A	C	E		I	K	M
B	D	F	H	J	L	N

GLC control unit terminal chart

Code no.	Location of problem	Indication
1	Engine speed	
2	Water thermo-sensor	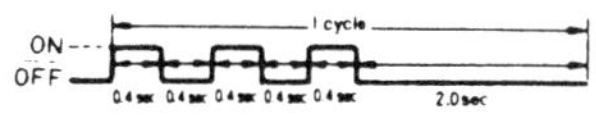
3	Feed back system	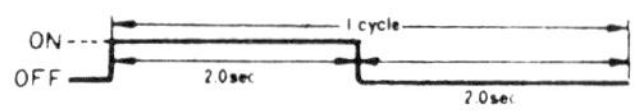
4	Vacuum sensor	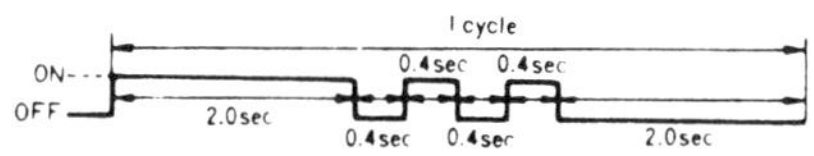
5	EGR position sensor	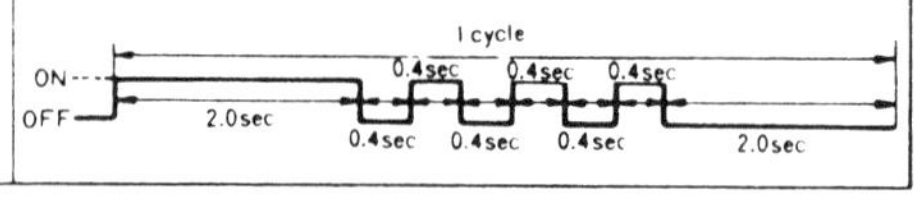

System checker 83 trouble code cycle for the Mazda 626

3. Check the trouble-indication lamp. If the lamp lights up, check for the cause of the problem.

SWITCHING SOLENOID VALVE

Test and Adjustment Procedures

RX-7

1. Warm up the engine and run it at idle speed. Connect a tachometer to the engine.
2. Disconnect the connector from the throttle sensor and connect a jumper wire to the A and C terminals.
3. Disconnect the vacuum sensing tube at the vacuum pipe and place a finger over the vacuum tube opening to make sure that there is vacuum present in the tube.
4. Gradually increase the engine speed and make sure that there is no vacuum present in the tube when the engine speed is 1000–1200 rpm or higher.
5. Disconnect the connector from the number one water temperature switch and connect a jumper wire to both terminals of the connector. Pull the choke knob out about 0.6 in.
6. Increase the engine speed and make sure that air is sucked into the tube at any engine speed. Disconnect the jumper wire connected in the previous step and reconnect the connector to the water temperature switch.
7. Disconnect the jumper wire that was connected in step number two and reconnect the connector to the throttle sensor.

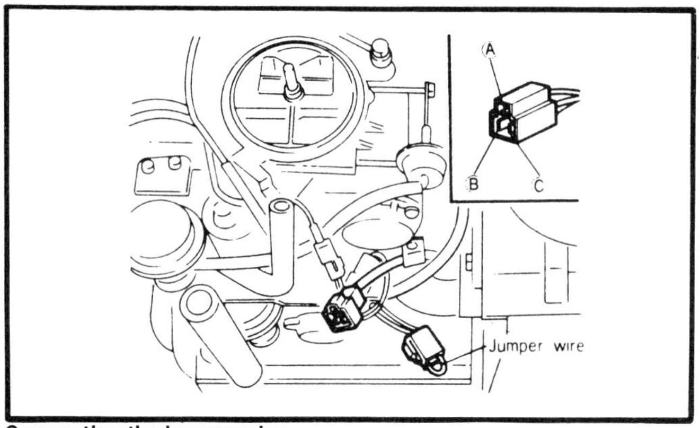

Connecting the jumper wire

	Indicated pattern	Remarks
GREEN LAMP	 ON	Air-fuel mixture is rich.
	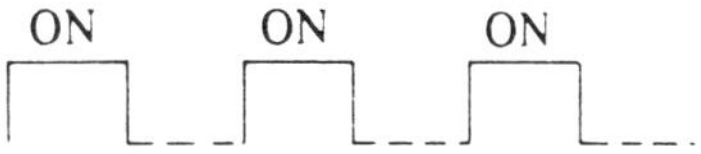 ON ON ON OFF OFF OFF	Air-fuel mixture is correct. If the lamp is turned on and off more than 8 times for 10 seconds at the speed of 2,500 rpm, the O_2 sensor is normal.
	 OFF	Air-fuel mixture is lean.

System checker 83 green lamp trouble code for the Mazda GLC

Terminal	Connection to	Voltage with ignition ON (when functioning properly)
A (input)	Ignition coil ⊖ terminal	approx. 12 V
B (input)	Ignition power supply	approx. 12 V
C (input)	Air-conditioner relay	below 1.5 V (approx. 12 V with blower switch ON)
E, I (output)	System check terminal	E: approx. 12 V; after 0.4 sec: below 1.5 V; after 2 sec: I = below 1.5 V
F (output)	A/F solenoid valve	approx. 12 V
H (output)	Coasting richer solenoid valve	approx. 12 V
D (output)	Slow-fuel-cut solenoid valve	below 1.5 V
J (output)	Solenoid for the air conditioner	approx. 12 V
K, L (output)	Duty solenoid valve	approx. 12 V
M (output)	Solenoid valve for ACV	approx. 12 V
N, R (ground)	Engine ground	0 V
a (input)	O_2 sensor	below 1.5 V
b (ground)	O_2 sensor	below 1.5 V
c (output)	Water thermo-sensor	approx. 3.4 V at 20°C, approx. 1.1 V at 80°C
d (ground)	Water thermo-sensor ground	below 1.5 V
e (output)	Vacuum sensor	approx. 4 V
f (ground)	Vacuum sensor ground	below 1.5 V
g (power supply)	Vacuum sensor	approx. 5 V
h (ground)	EGR position sensor	below 1.5 V
j (output)	EGR position sensor	below 1.5 V
l (power supply)	EGR position sensor	approx. 5 V
k (input)	Vacuum switch	approx. 12V
m (input)	Clutch switch, neutral switch	below 1.5 V
n (input)	Ignition switch "START" terminal	below 1.5 V (10 V at "START")
o (input)	Idle switch	approx. 12 V
p (ground)	Idle switch and vacuum switch ground	below 1.5 V
q (input)	Water thermo-switch	below 1.5 V (at 17°C or higher)

626 control unit terminal chart

8. Gradually increase the engine speed and make sure that vacuum is present in the vacuum tube at any engine speed. There should be no vacuum when the engine is accelerated quickly. Reconnect the vacuum sensing tube to the vacuum pipe.

RELIEF SOLENOID VALVE

Testing

RX-7

1. Disconnect the vacuum sensing tube from the solenoid valve and vacuum pipe, blow through the solenoid valve from the vacuum sensing tube B and make sure air passes through the valve and comes out from port C.
2. Disconnect the connector from the relief solenoid valve and connect 12 volts to the terminals on the valve.
3. Blow through the valve from the vacuum sensing tube B. Make sure air passes through the valve and comes from port A on the valve.

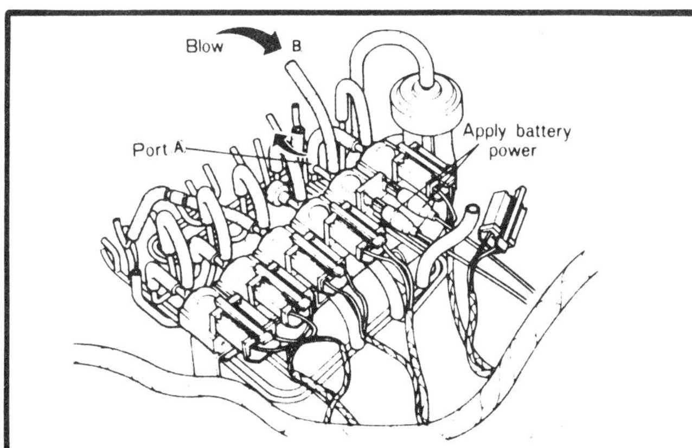

Testing the relief solenoid valve

Terminal	Connection to	Voltage with ignition ON (When functioning properly)
A	Ignition coil ⊖ terminal	approx. 12V
B	Ignition switch	approx. 12V
C	Choke switch	below 1.5V
D	Ignition coil ⊖ terminal	approx. 12V
E	Throttle sensor	approx. 8V
F	Switching solenoid valve	approx. 12V
G	Throttle sensor	1 ~ 2V
H	Ignition switch "START" terminal	below 1.5V (above 8V at "START")
I	Throttle sensor	0V
J	No. 2 water temperature switch	below 1.5V
L	Air-con. solenoid valve	approx. 12V
M	Main air bleed solenoid valve	approx. 12V
N	Relief solenoid valve	below 2V
O	Heat hazard sensor	below 2V
P	Ground	0V
Q	Shutter solenoid valve	below 1.5V
R	Idle richer solenoid valve	0V
S	Vacuum control solenoid valve (L)	approx. 12V (in neutral)
T	Vacuum control solenoid valve (T)	approx. 12V (in neutral)
U	Port air solenoid valve	below 1.5V
V	Ignition switch "START" terminal	below 1.5V (above 8V at "START")
a	Catalyst thermo sensor	below 1.5V
b		
c	Catalyst thermo sensor	below 1.5V
d		
e	Port air switching valve	approx. 12V
f		
h	Vehicle speed sensor	0V: Reed switch (speedometer) . . . closed 6 ~ 12V: Reed switch (speedometer) . . . open
i	Air-con. cut relay	below 1.5V . . . air con. switch OFF
j		
k	Neutral switch	0V . . . in gear approx. 12V . . . in neutral
l	Clutch switch	0V . . . pedal released approx. 12V . . . pedal depressed
m		
n	Fuel pump cut relay	below 1.5V

Control unit connector

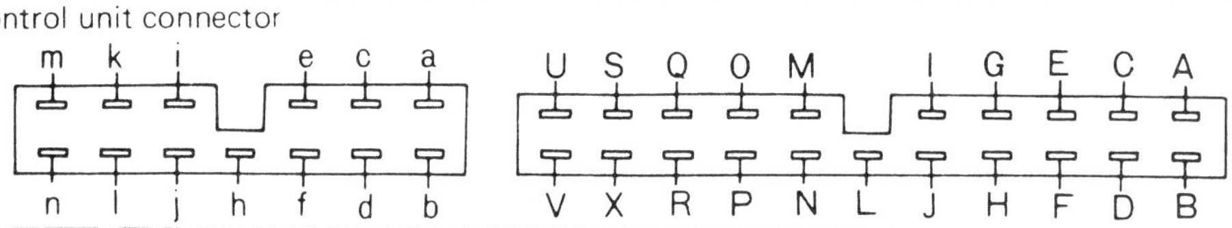

RX-7 12A engine control unit terminal chart

PORT AIR SOLENOID VALVE

Testing

RX-7

1. Warm up the engine to normal operating temperature and run it at idle. Connect a tachometer to the engine.

2. Disconnect the connector from the throttle sensor and connect a jumper wire to the A and C terminals of the connector.

3. Connect a voltmeter to the port air solenoid valve (GB) terminal and ground. Increase the engine speed and observe the voltmeter reading. The voltage should be as follows; at idle to 3000 rpm, voltage should be below 2 volts. At 3000–4000 rpm, the voltage should be 12 volts. More than 4000 rpm, the voltage should be below 2 volts.

4. Disconnect the jumper wire that was connected to the throttle sensor and reconnect the connector to the throttle sensor.

5. Position the vehicle on a rolling road tester (dynomometer). Be sure to use a wire rope to secure the vehicle so it does not move forward. Increase the vehicle speed and observe the voltmeter reading.

6. The voltage should be below 2 volts at 50 mph and 12 volts above 50 mph, if the port air solenoid valve fails this test, replace it.

Terminal	Connection to	Voltage with ignition ON (when functioning properly)
A	Main relay	approx. 12V
B	Ground	0V
C	Water thermo sensor	1 ~ 2V (warm engine)
D	Ground	0V
E	Air flow meter	4 ~ 6V . . . at 20°C 1.5 ~ 3.5V . . . at 50°C
F	Injector (#20)	approx. 12V
G	Throttle sensor & Atmospheric pressure sensor	4.5 ~ 5.5V
H	Injector (#10)	approx. 12V
I	Throttle sensor	approx. 1V
J	Vacuum switch	approx. 12V
L	Variable resistor (V/R)	0 ~ 12V (Varies according to the V/R adjustment)
M	Ignition switch "START" terminal	below 1.5V
N	O_2 sensor	0V
O	Air flow meter	approx. 12V
P	Atmospheric pressure sensor	approx. 4V
Q	Air flow meter	approx. 2V
R	Air flow meter	approx. 7.5V
S	Ground	0V
T	Ground	0V
U	Ignition coil (T) - terminal	approx. 12V
V	Main relay	approx. 12V
a	Switching solenoid valve	approx. 12V
b	Relief solenoid valve control unit	approx. 12V
c	Checking connector	0V
d	Vacuum control solenoid valve (T/L)	approx. 12V
e	Pressure regulator control valve	below 1.5V
f	Checking connector	0V
h	Vent solenoid valve	below 1.5V (throttle sensor is adjusted properly)
i	Clutch switch	below 1.5V . . . pedal released approx. 12V . . . pedal depressed
j	Neutral switch	below 1.5V . . . in neutral approx. 12V . . . in gear
k	Water temperature switch	below 1.5V . . . above 15°C
l	Intake air temperature sensor	8.5 ~ 10.5V . . . at 20°C 5 ~ 7V . . . at 50°C
m	Air-con. switch	below 1.5V . . . air-con. switch OFF
n	Vacuum control valve	approx. 12V (throttle sensor is adjusted properly)

Control unit connector

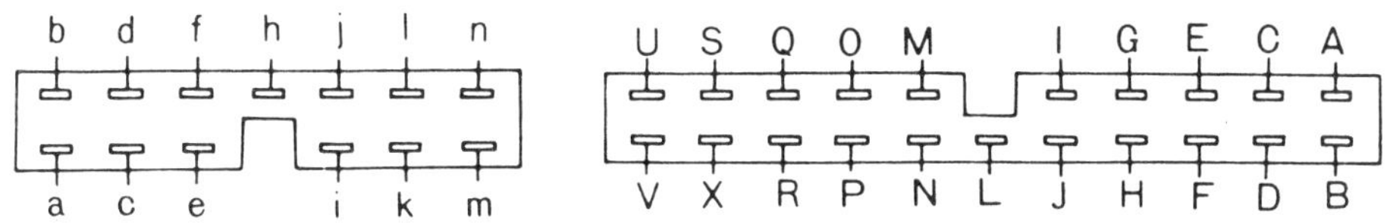

RX-7 13B engine control unit terminal chart

PORT AIR SWITCHING VALVE

Testing

RX-7

1. Warm the engine to normal operating temperature and run it at idle. Connect a tachometer to the engine.
2. Disconnect the connector from the throttle sensor and connect a jumper wire to the A and C terminals of the connector.
3. Connect a voltmeter to the port air switching valve (R) terminal and ground. Increase the engine speed and observe the voltmeter reading. At idle, the voltage should be below 2 volts at 3000 rpm. At 3000–4000 rpm, the voltage should be 12 volts and at more than 4000 rpm, the voltage should be below 2 volts.
4. Disconnect the jumper wire that was connected to the throttle sensor and reconnect the connector to the throttle sensor.

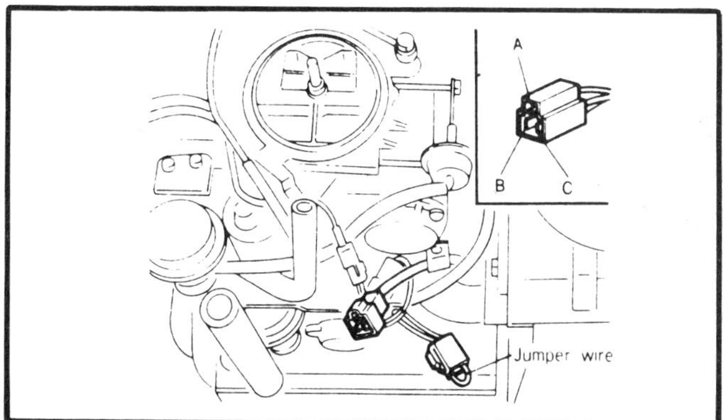

Bypassing the throttle sensor

Code No.	Location of problem	Indication	Checking procedure
1	Engine speed	1cycle; ON / OFF; 0.4 2.0 sec	Disconnect the trailing coil – terminal crank engine at least 1.5 seconds, with IG "ON" code should be heard.
2	Air flow meter	ON / OFF; 0.4 0.4 0.4 2.0 sec	Disconnect air flow meter connector, turn IG "ON" code should be heard.
3	Water thermo sensor	ON / OFF	Disconnect the water thermo sensor connector, turn IG "ON" code should be heard.
4	Oxygen (O_2) sensor	ON / OFF; 2.0 2.0 sec	
5	Throttle sensor	ON / OFF; 2.0 0.4 0.4 2.0 sec	Disconnect throttle sensor connector, turn IG "ON" code should be heard.
6	Atmospheric pressure sensor	ON / OFF; 2.0 0.4 0.4 0.4 0.4 sec	Disconnect the atmospheric pressure sensor, turn IG "ON" code should be heard.

Note
1) If there is trouble in 2 or more places, the indication will be for the smaller code number first.
2) Even if the problem is corrected during indication, 1 cycle will be indicated.

System checker 83 trouble code cycle for the Mazda RX-7

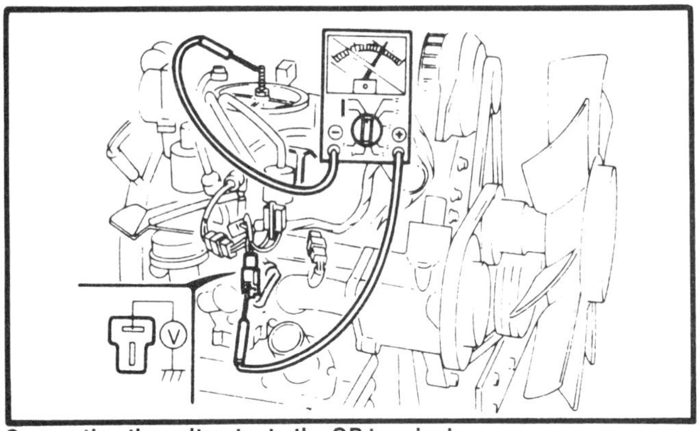

Connecting the voltmeter to the GB terminal

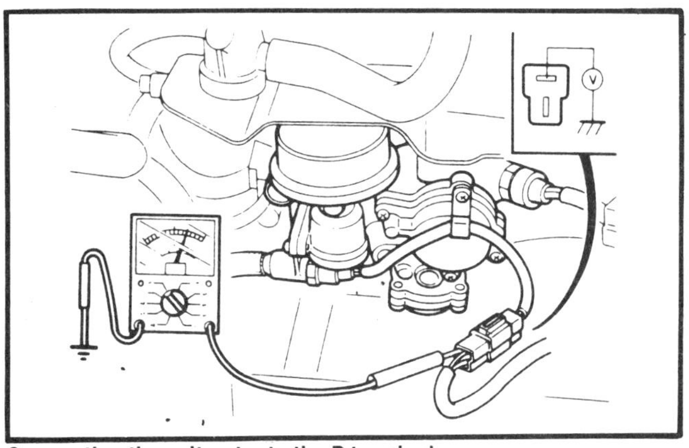

Connecting the voltmeter to the R terminal

5. Position the vehicle on a rolling road tester (dynomometer). Be sure to use a wire rope to secure the vehicle so it does not move forward. Increase the vehicle speed and observe the voltmeter reading.

6. The voltage should be below 2 volts at 50 mph and 12 volts above 50 mph. If the port air solenoid valve fails this test, replace it.

CATALYST THERMO SENSOR

Testing

RX-7

1. Disconnect the connector for the catalyst thermo sensor. Connect an ohmmeter to the catalyst thermo sensor terminals and check for continuity between the terminals.

2. If there is no continuity between the terminals, remove and replace the catalyst thermo sensor.

HEAT HAZARD WARNING SYSTEM

Testing

RX-7

1. The heat hazard warning light should come on when the ignition switch is turned ON. Start the engine and the warning light should go off.

2. Disconnect the heat hazard sensor connector and connect a jumper wire to both terminals. The heat hazard warning light should come on. If it does not, check the wiring to the light and perform the heat hazard sensor test.

HEAT HAZARD SENSOR

Testing

RX-7

1. Remove the sensor (located in the floor board under the mat).

Wrap the sensor and a thermometer with aluminum foil to prevent oil penetration and place the assembly in a container of oil.

2. Connect a test light and battery to the sensor terminals in the connector.

3. Gradually heat the oil. The test light should light up when the temperature of the aluminum foil reaches 266 ± 18 degrees F.

4. If the sensor does not operate within the specification, replace the sensor.

NUMBER ONE AND NUMBER TWO WATER TEMPERATURE SWITCHES

Testing

RX-7

1. Remove the number one water temperature switch from the water pump and remove number two water temperature switch from the lower radiator tank.

2. Place the number one water temperature switch in a container of water with a thermometer and gradually heat up the water.

3. Using an ohmmeter and observing the thermometer, check the temperature at which continuity does not exist between the terminals of the switch.

4. The number one water temperature switch should show no continuity above the temperature of 158 ± 11.7 degrees F. The number two water temperature switch should show no continuity above the temperature of 59 ± 7 degrees F.

5. If the switches do not fall into the specifications of the test, replace the switches.

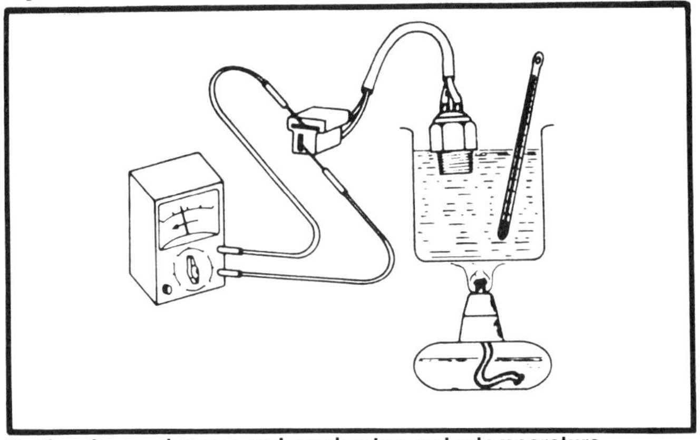

Testing the number one and number two water temperature switches

CHOKE SWITCH AND CHOKE MAGNET

Testing

RX-7

1. Disconnect the connector from the choke switch and check the continuity between the numbered terminals in the connector, using an ohmmeter.

2. With the choke knob pulled out 0.4 ± 0.08 inches, there should be continuity between terminals three and seven and with the choke knob in any position, there should be continuity between terminals six and eight.

CHOKE RELAY

Testing

RX-7 WITH 12A OR 13B ROTARY ENGINES

1. Disconnect the choke relay connector and check the continuity between the numbered terminals, using an ohmmeter.

2. Without any power applied to the relay, there should be continuity between terminals one and two and none between terminals three and four. With a power lead from the battery to terminal six and a

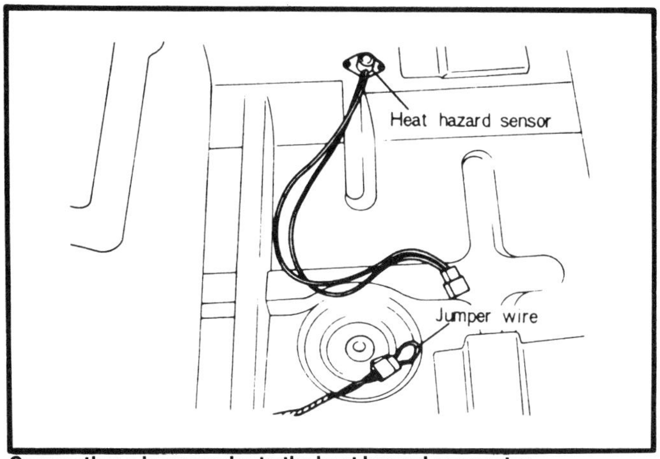

Connecting a jumper wire to the heat hazard connector

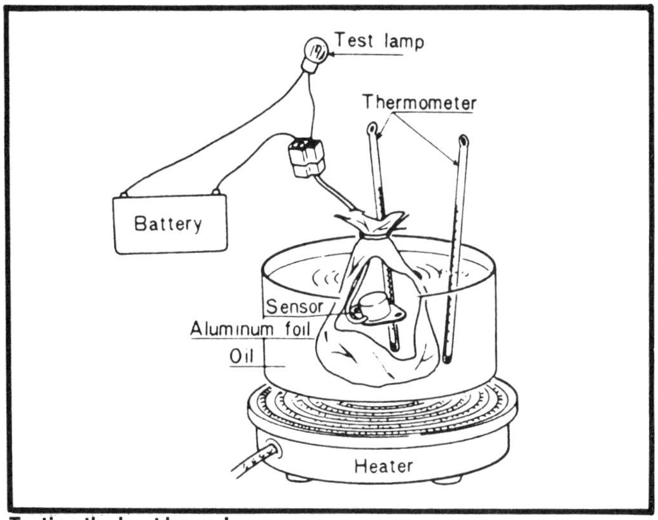

Testing the heat hazard sensor

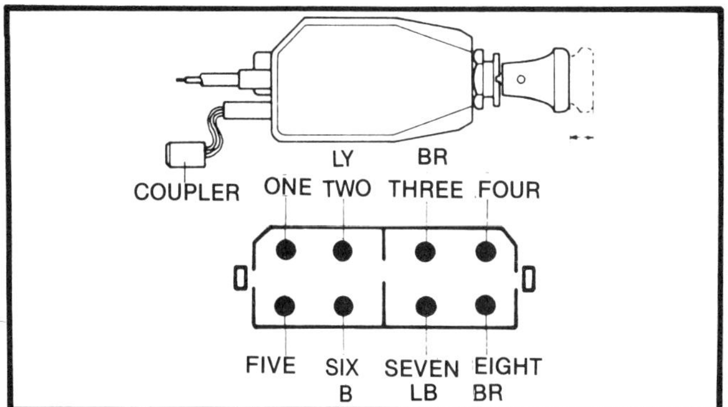

Exploded view of the choke switch and magnet

ground from the battery to terminal five, there should be continuity between terminals three and four and none between one and two.

POWER STEERING SWITCH

Testing

RX-7

1. Start the engine and let it run at idle. Unplug the power steering switch connector from the steering gear housing.

2. Using an ohmmeter, check to see if there is continuity between the terminals of the power steering switch connector when the steering wheel is turned all the way to the right and left.

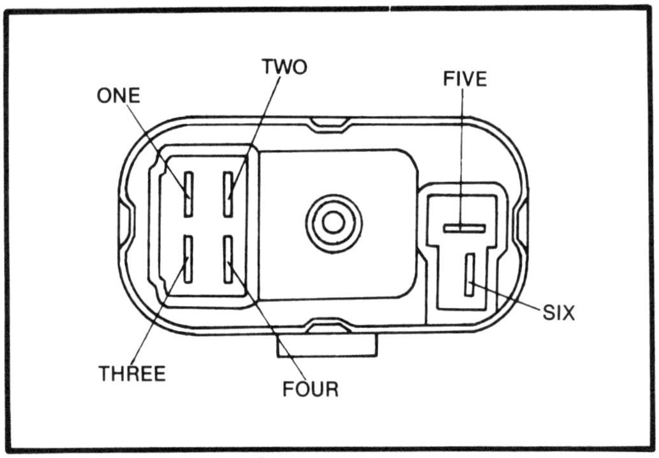

Choke relay terminal location

3. When the oil pressure is above 427 psi, the switch should be on. At pressures below 427 psi, the switch should be off.

NEUTRAL SWITCH

Testing

RX-7

1. Disconnect the neutral switch connectors and connect an ohmmeter to the terminals of the switch.
2. The ohmmeter should show continuity when the vehicle is in neutral and show no continuity in any of the other position.

VACUUM DELAY VALVE

Testing

RX-7

1. Disconnect the vacuum delay valve and connect a vacuum tube to a hand held vacuum pump.
2. Connect the vacuum delay valve to the vacuum tube from the vacuum pump and shut off the other side of the vacuum delay valve with your thumb.
3. Apply a vacuum of over 19.7 in. Hg. with the vacuum pump. Release your thumb from the vacuum delay valve and check the time required for the vacuum reading to decrease to 3.9 in. Hg. from 15.7 in. Hg. (specified time: 12 ± 1 second).

ANTI-AFTERBURN VALVE ONE AND TWO

Testing

RX-7

1. Operate the engine at idling speed and close the air intake port of the anti-afterburn valve. Make sure there is no air being drawn in.
2. Increase the engine speed more than 3000 rpm and quickly decelerate. Air should be drawn into the intake valve for 1–2 seconds after the accelerator is released.
3. If there is no vacuum after the accelerator has been released, remove and replace the valve. Repeat this test for the number two anti-afterburn valve.

THROTTLE SENSOR

Testing

RX-7

1. Disconnect the connector for the throttle sensor and connect an ohmmeter to the throttle sensor.
2. Open the throttle valve and observe the ohmmeter reading. At idle it should read one kilo-ohm and fully open it should read five kilo-ohms.

VACUUM CONTROL SOLENOID VALVE

Testing

RX-7

1. Disconnect the vacuum sensing tube from the solenoid and vacuum pipe.
2. Blow through the solenoid valve from the port B and make sure the air passes through the valve and comes out from the air filter C.
3. Disconnect the connector from the vacuum control solenoid valve and connect 12 volts to the terminals on the valve.
4. Blow through the valve from port B and make sure that the air passes through the valve and comes out from the port A of the valves.

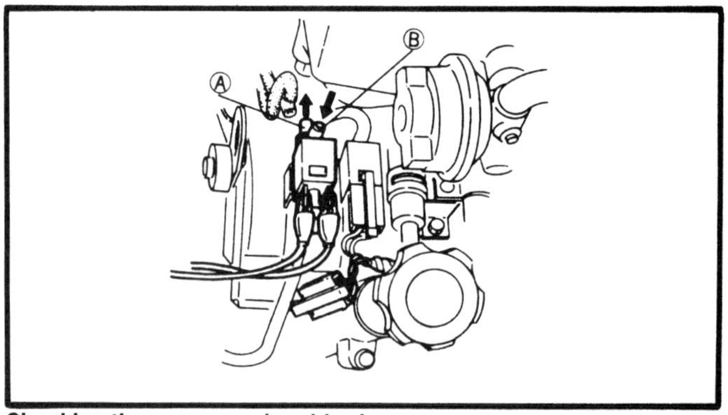

Checking the vacuum solenoid valve

COASTING SHUTTER VALVE

Testing

RX-7

1. Warm the engine up to normal operating temperature. Stop the engine and disconnect the air cleaner to the coasting valve hose at the air cleaner. Start the engine and run it at idle. Place your finger over the disconnected hose. Air should not be drawn into the hose.
2. Disconnect the electrical connector from the shutter solenoid valve (yellow dot). Air should be drawn into the disconnected hose and the idle should fluctuate.
3. At the same time, the coasting valve rod should be pulled into the coasting valve about 0.4 in., opening the shutter valve. Replace the coasting/shutter valve if it does not respond properly.

SHUTTER SOLENOID VALVE

Testing

RX-7

1. Disconnect the vacuum sensing tubes from the shutter solenoid valve, which is identified with a yellow dot.
2. Blow air through the solenoid valve hose. Air should go through the valve and escape at the front port.
3. Disconnect the electrical connector and apply battery voltage to the terminals on the solenoid valve.
4. Blow through the hose again. Air should escape through the air filter at the rear of the solenoid valve.
5. If the shutter solenoid valve does not perform to specification, replace the valve.

AIR CONDITIONER SOLENOID VALVE

Testing

RX-7

1. Disconnect the vacuum sensing tubes from the solenoid valve and vacuum port at the rear of the valve. Blow through the vacuum hose and ensure air passes through the valve and escapes from the air filter.

2. Disconnect the valve electrical connector and apply battery voltage to the terminals on the switch. Blow through the hose again. Air should pass through the valve and escape from the rear port. If the valve does not respond, perform signal check.

AIR CONDITIONER SOLENOID VALVE SIGNAL CHECK

Testing

RX-7

1. Warm up the engine to normal operating temperature. Stop the engine and connect a tachometer. Connect a voltmeter to the air conditioner solenoid valve (white color dot), negative terminal. Start the engine and turn the A/C on.
2. On the automatic transaxle models, current should not flow to the solenoid at any engine speed. On manual transaxle models, increase the engine speed to 2000 rpm with the throttle.
3. Slowly decrease the engine speed and watch the voltmeter. The voltmeter should read near 0 when the engine speed is 1000–1200 rpm. If the valve does not respond, replace the air conditioner solenoid valve.

SPLIT AIR SOLENOID VALVE

Testing

RX-7

1. Connect a voltmeter to the split air solenoid (LR) terminal and ground. Turn the ignition switch to the ON position.
2. Shift into 5th gear and observe the voltmeter. It should read 0 volts in 5th gear and 12 volts in all other gears. If it does not meet these specifications, replace the valve.

MAIN AIR BLEED CONTROL SOLENOID VALVE

Testing

RX-7

1. Warm up the engine to normal operating temperature and connect a tachometer to the engine.
2. Disconnect the connector from the throttle sensor and connect a jumper wire to the A and C terminals of the connector.

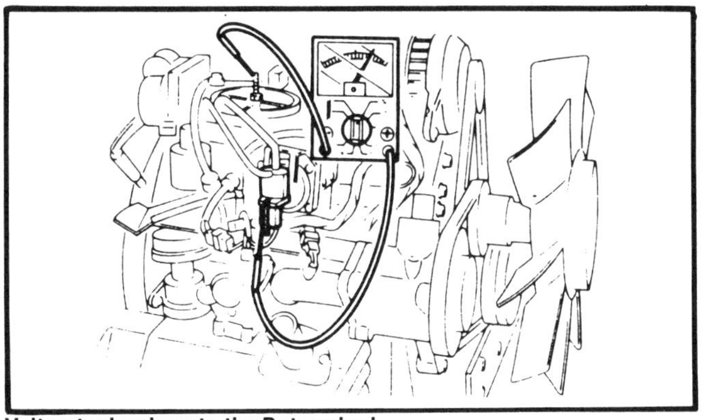

Voltmeter hook-up to the Br terminal

3. Connect a voltmeter to the main air bleed control solenoid (Br) terminal and ground. Increase the engine speed and observe the voltmeter reading.
4. The voltmeter should read 12 volts at idle to 3000 rpm, below 2 volts at 3000 to 4000 rpm and 12 volts at more than 4000 rpm.
5. Disconnect the jumper wire connected to the throttle sensor connector and reconnect the connector to the throttle sensor.

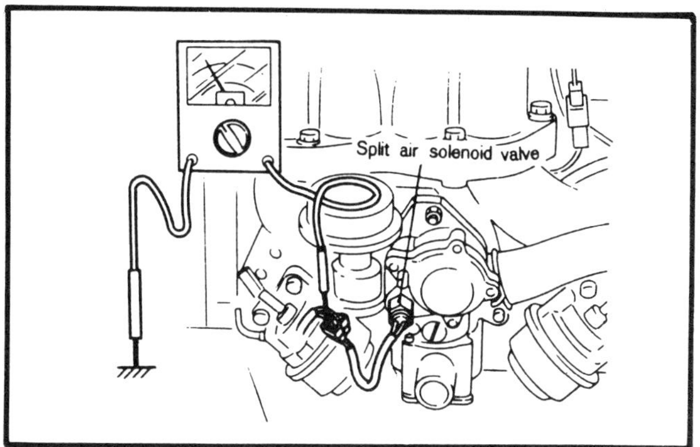

Connecting a voltmeter to the split air solenoid valve

RICHER SOLENOID VALVE (MANUAL TRANSAXLE ONLY)

Testing

RX-7

1. Start the engine and let it idle. Connect a voltmeter to the richer solenoid (WR) terminal and ground. Take a voltmeter reading. The reading should be 12 volts.
2. Increase the engine speed more than 1500 rpm and decrease it, while observing the voltmeter reading. The reading should be below 2 volts for thirty seconds when the engine rpm drops to 1100 rpm or less.
3. Disconnect the vacuum sensing tube of the vacuum switch at the idle compensator and take a voltmeter reading. The reading should be 0 volts. Reconnect the vacuum sensing tube to the idle compensator.

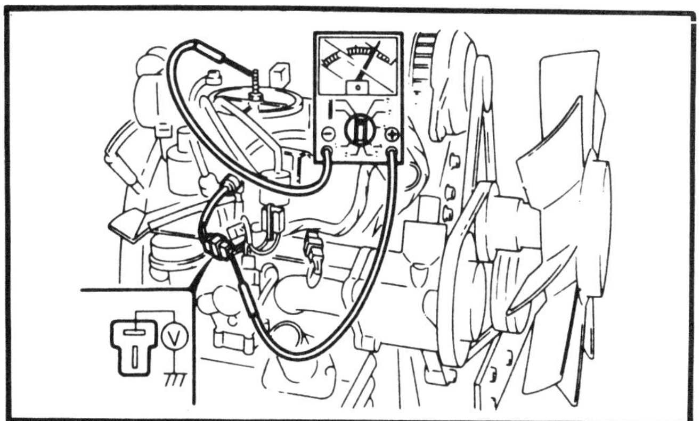

Connecting a voltmeter to the WR terminal

4. Disconnect the clutch switch connector. Depress the clutch pedal and shift into 1st through 5th gear.
5. Increase the engine speed to more than 1500 rpm and then decrease it. The voltmeter reading should be 12 volts. If the richer solenoid control valve fails any of these tests, replace it.

THROTTLE OPENER

Testing

RX-7

1. Switch off all accessories, remove the fuel filler cap, disconnect and plug the idle compensator tube at the air cleaner. Connect a tachometer to the engine and warm up the engine to normal operating temperature.
2. Disconnect the electrical connector from the switching solenoid valve, disconnect and plug the vacuum sensing tubes from the vacuum

control units on the distributor, except California models with manual transaxles.

3. Turn the air conditioner off, disconnect the electrical connector from the air conditioner solenoid valve and connect battery voltage to the solenoid terminals.

4. When the battery voltage is applied to the solenoid terminals, throttle opener should increase engine speed from idle to 1150–1250 rpm, in neutral. If the engine speed is not within specifications, turn the adjusting nut on the throttle opener arm until the engine speed is within specification.

CHOKE DIAPHRAGM NUMBER ONE AND TWO

Testing

RX-7

1. Remove the air cleaner assembly, start the engine and let it run at idle speed.
2. Disconnect the vacuum sensing tubes from the number one and two choke diaphragms.
3. Each diaphragm shaft should come out from the choke diaphragm. If either diaphragm fails this test, replace it.

HOT START SYSTEM

Testing

RX-7

1. Inspect the hot start assist cable and linkage for proper installation, no binding or sticking and full return. Warm the engine up to normal operating temperature and then shut the engine off.

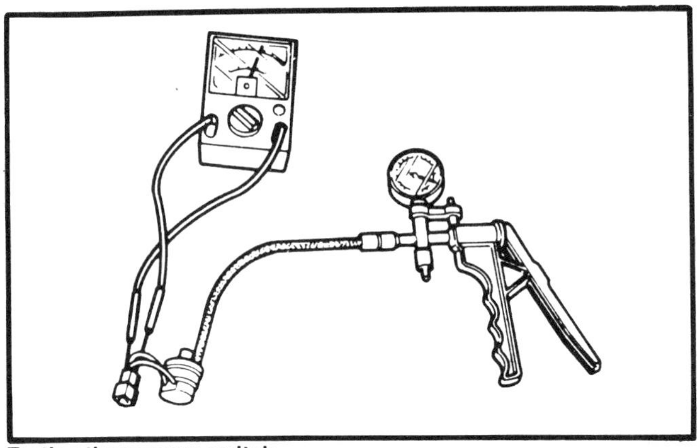

Testing the vacuum switch

2. Disconnect the leading and trailing primary wires from the ignition coils. Crank the engine. The hot start lever should open the throttle valve. If the hot start system does not respond as outlined, check the hot start assist relay.
3. Before replacing the hot start motor relay, check the number one water temperature switch first.

HOT START RELAY ASSIST

Testing

RX-7

1. Disconnect the electrical connector from the hot start relay. Using an ohmmeter, check the continuity between the terminals.
2. Continuity should exist between terminals number one and five with no power applied. Continuity should not exist between terminals number one and three with no power applied.
3. Connect battery voltage to the relay by connecting the positive terminal of the battery to the number two terminal of the relay and the negative terminal of the battery to the number four terminal of the relay.
4. With the battery voltage applied, continuity should exist between the number one and three terminals. Continuity should not exist between terminals number one and five with the power applied.

HOT START ASSIST CABLE

Adjustment

RX-7

1. Remove hot start assist cable lock spring from the cable bracket. Slowly pull the outer cable until the hot start lever just touches the stopper lever. Check the clearance between the cable bracket and cable locknut.
2. If the clearance is not 0.02–0.08 inch, adjust clearance by turning the locknut. Recheck the clearance and install the lock spring.

VACUUM SWITCH

Testing

RX-7

1. Remove the vacuum switch and connect a hand held vacuum pump or equivalent to the vacuum switch.
2. Connect an ohmmeter to the vacuum switch and check the continuity between the switch terminals.
3. When the vacuum switch is holding 4.7 in. Hg. of vacuum, there should be no continuity at the terminals. When the switch is holding more than 4.7 in. Hg. of vacuum the terminals should show continuity. If the switch fails this test, replace it.

VACUUM DIAGRAMS

EMISSION COMPONENTS LOCATION

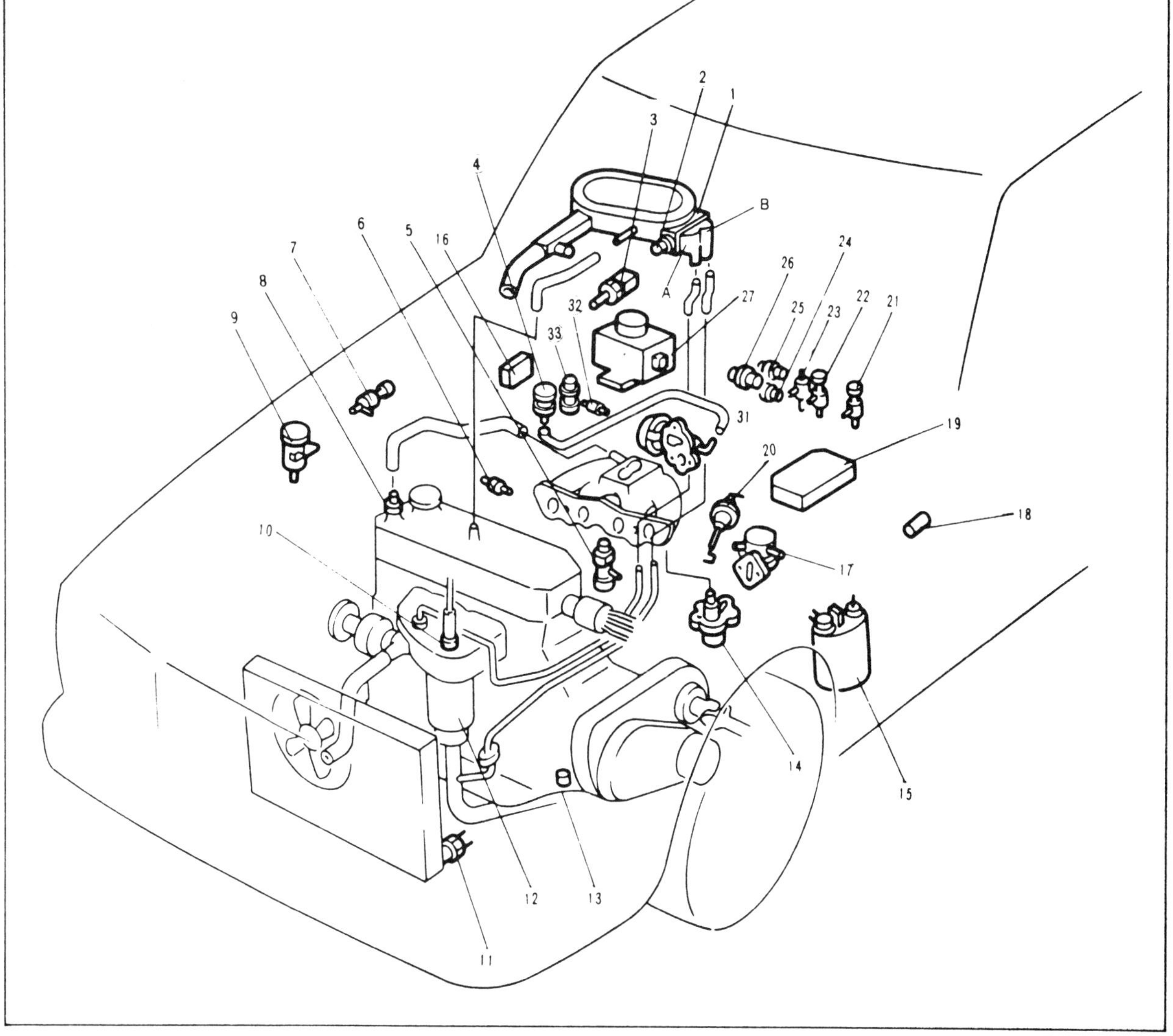

1. Reed valves (A & B)
2. Air control valve (Federal)
3. Water temperature sensor
4. EGR modulator valve
5. Water thermo valve
6. Check valve
7. Choke solenoid valve
8. PCV valve
9. High altitude compensator (Federal)
10. O_2 sensor
11. Water thermo switch
12. Front catalyst (Federal & California)
13. Neutral switch
14. EGR thermo valve
15. Canister
16. Control unit
17. Fuel pump
18. Clutch switch
19. Rear catalyst
20. Servo diaphragm
21. ACV solenoid valve (Federal)
22. Throttle positioner solenoid valve
23. EGR solenoid valve
24. No. 1 vacuum switch
25. No. 2 vacuum switch
26. No. 3 vacuum switch
27. Carburetor
    - Idle switch
    - A/F solenoid valve
    - Slow fuel cut solenoid valve
    - Air vent solenoid valve
28. Idle compensator
29. No. 2 choke diaphragm
30. No. 1 choke diaphragm
31. EGR control valve
32. Vacuum delay valve (only ATX)
33. Anti-Afterburn valve (only ATX)

MAZDA—GLC MODELS—COMPONENTS LOCATIONS

SYSTEM DIAGRAM (CALIFORNIA)

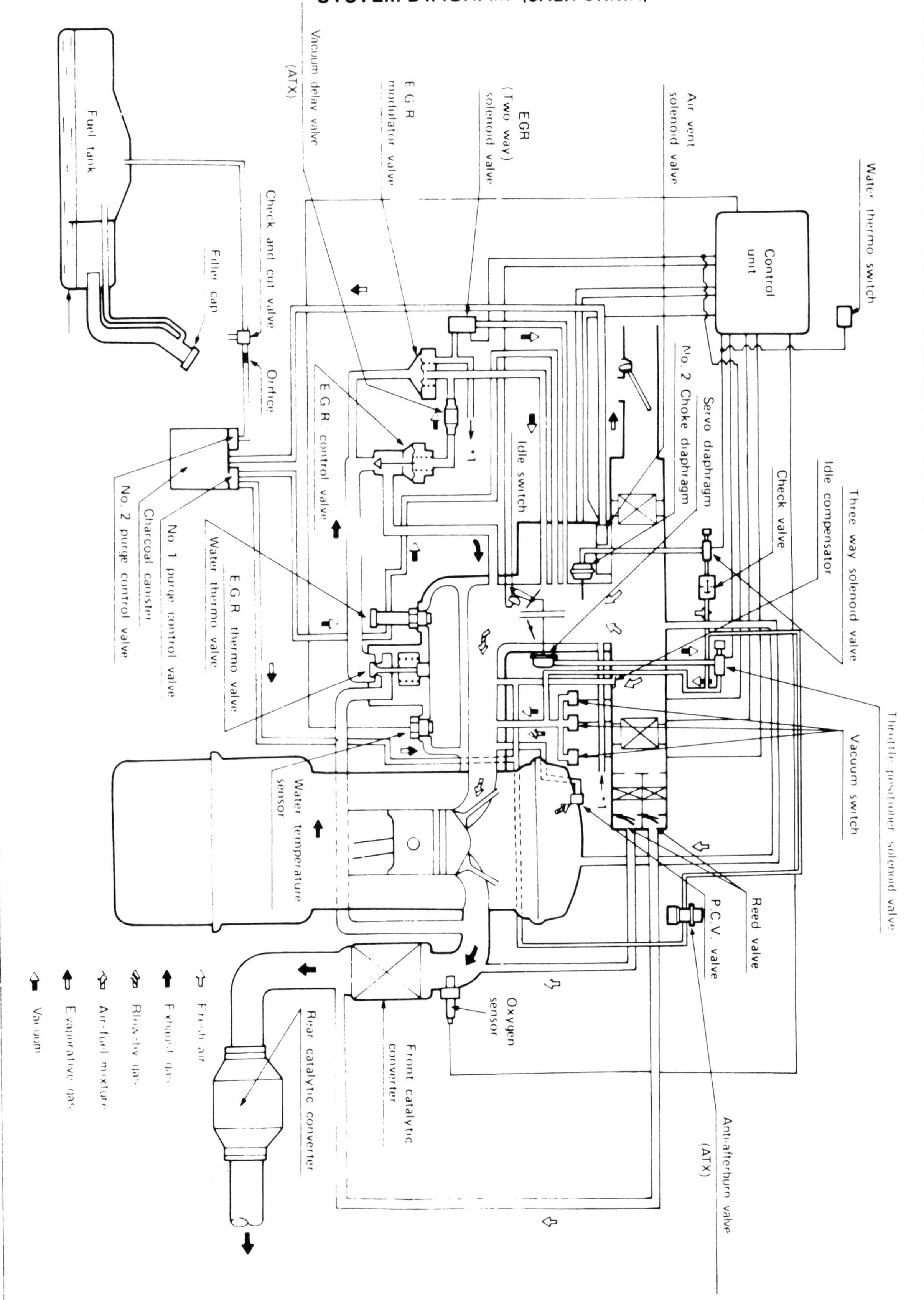

MAZDA—GLC MODELS—CALIFORNIA—SYSTEM DIAGRAM

SYSTEM DIAGRAM (FEDERAL)

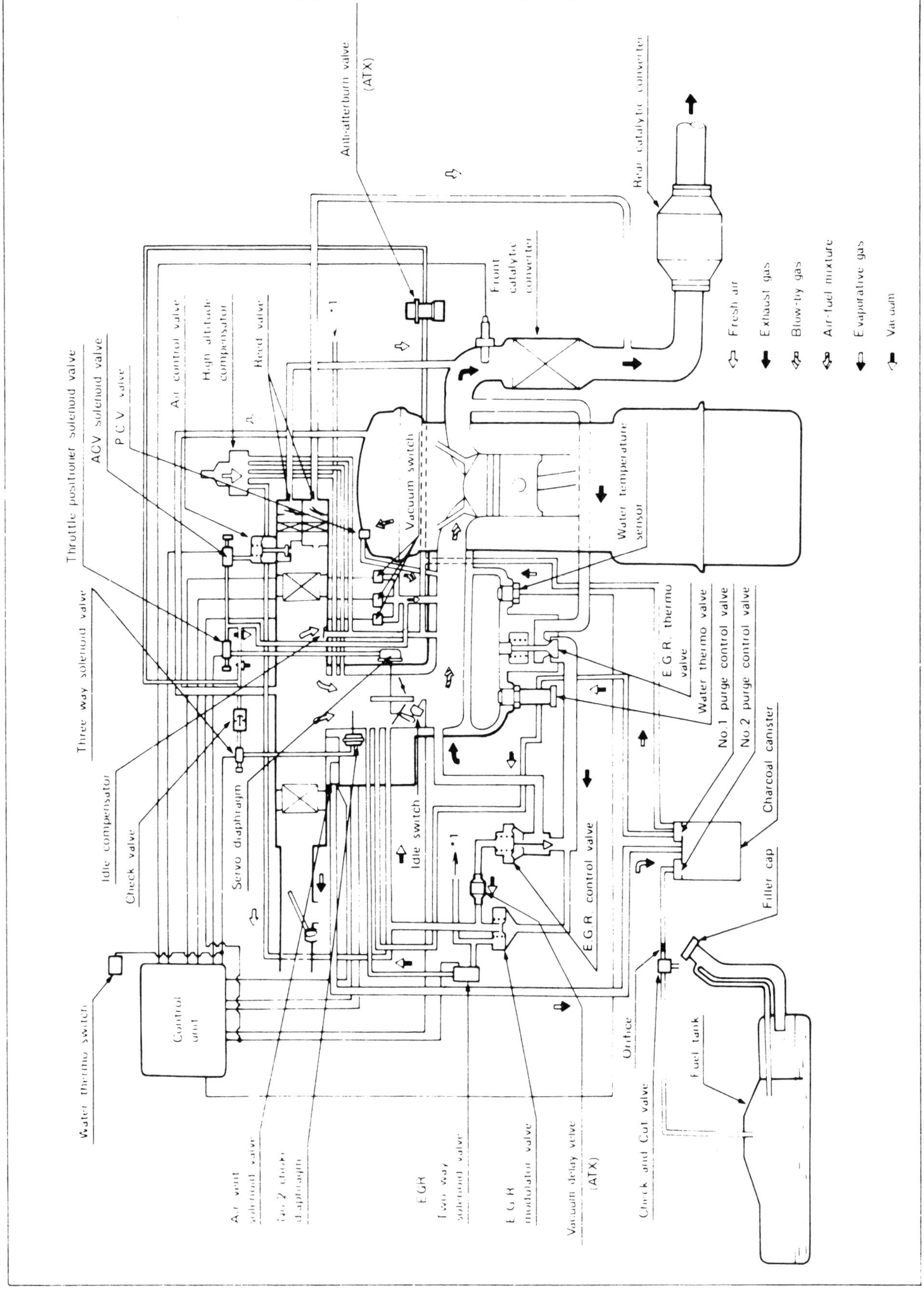

MAZDA—GLC MODELS—FEDERAL—SYSTEM DIAGRAM

SYSTEM DIAGRAM (CANADA)

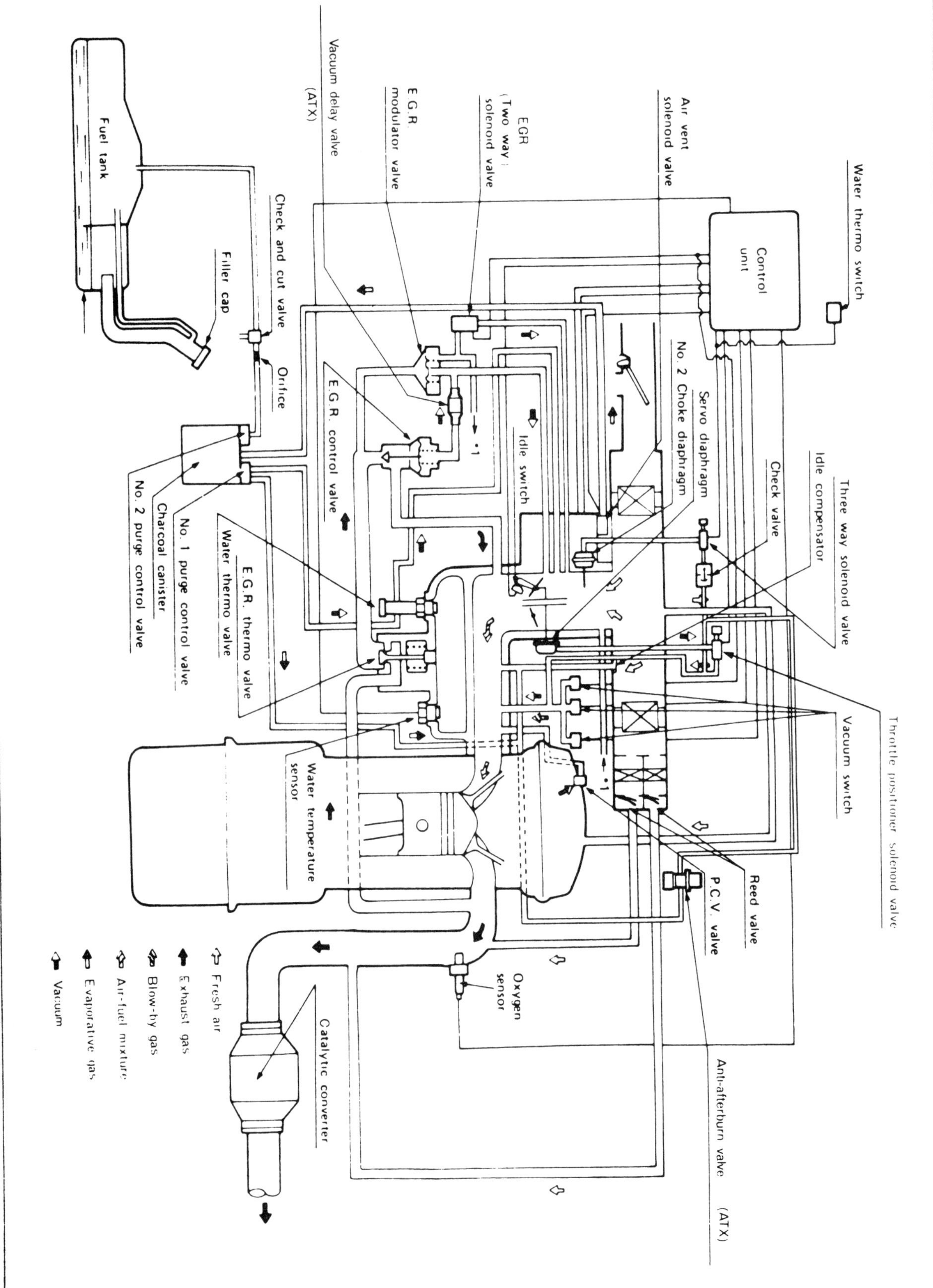

MAZDA—GLC MODELS—CANADA—SYSTEM DIAGRAM

PIPING (FEDERAL)

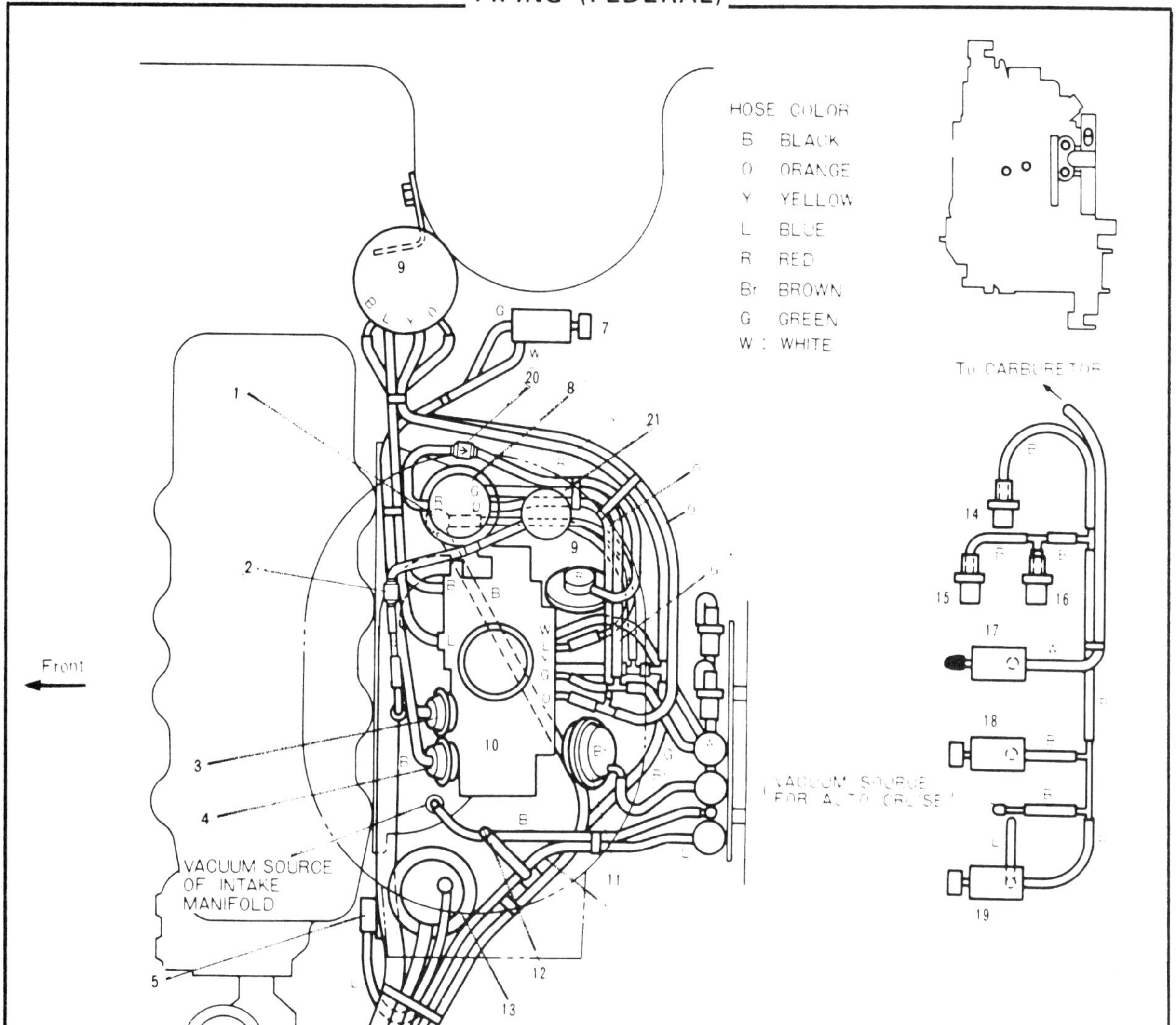

1. Water thermo valve
2. Check valve
3. No. 1 choke diaphragm
4. No. 2 choke diaphragm
5. Air control valve
6. Canister
7. Choke solenoid valve
8. EGR modulator valve
9. EGR valve
10. Carburetor
11. Servo diaphragm
12. Idle compensator
13. Fuel pump
14. No. 3 vacuum switch
15. No. 2 vacuum switch
16. No. 1 vacuum switch
17. EGR solenoid valve
18. Throttle positioner solenoid valve
19. ACV solenoid valve
20. Vacuum delay valve (ATX)
21. Anti-afterburn valve (ATX)

MAZDA—GLC MODELS—FEDERAL

PIPING (CALIFORNIA AND CANADA)

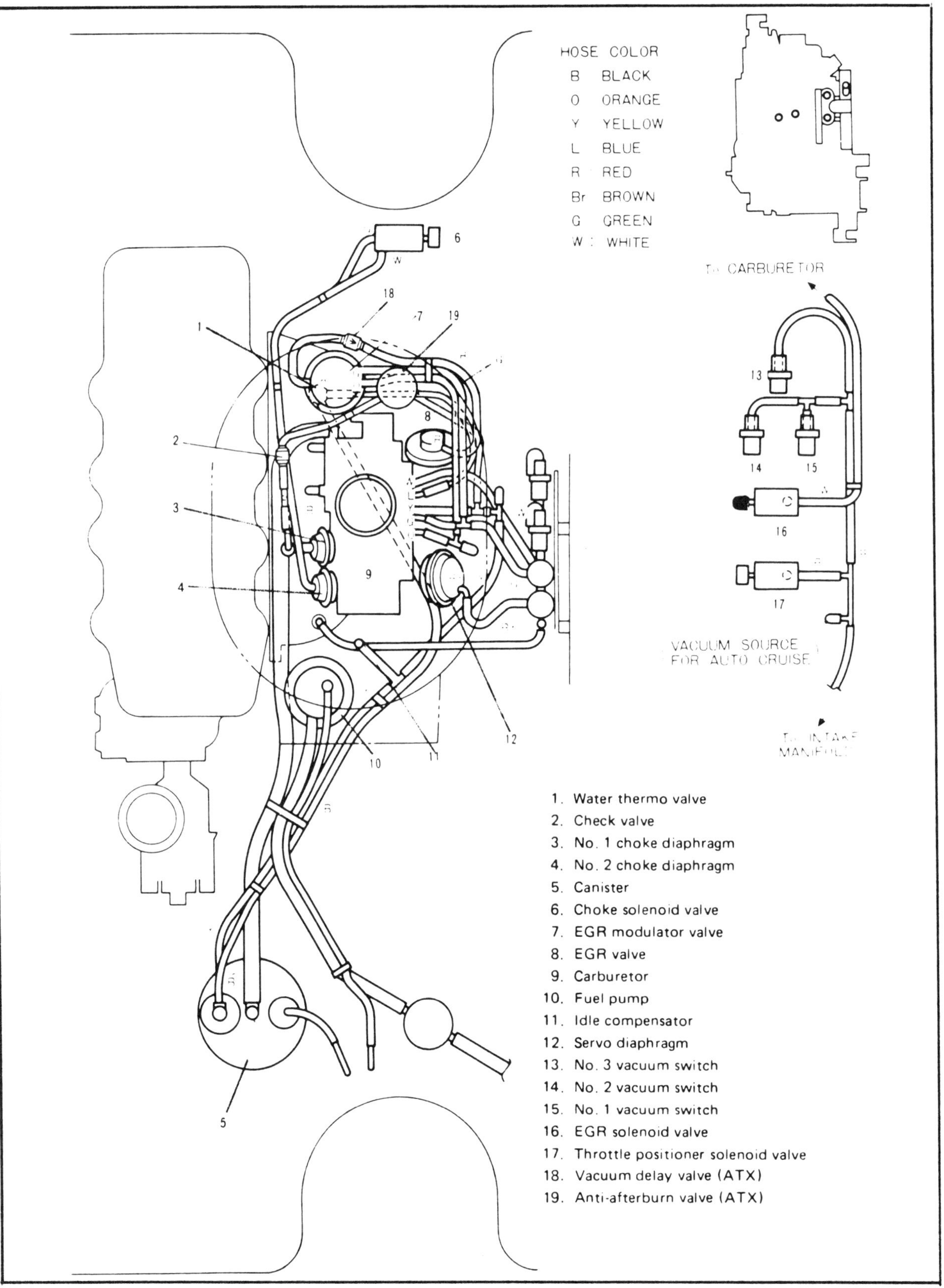

MAZDA—GLC MODELS—CALIFORNIA AND CANADA

EMISSION COMPONENTS LOCATION

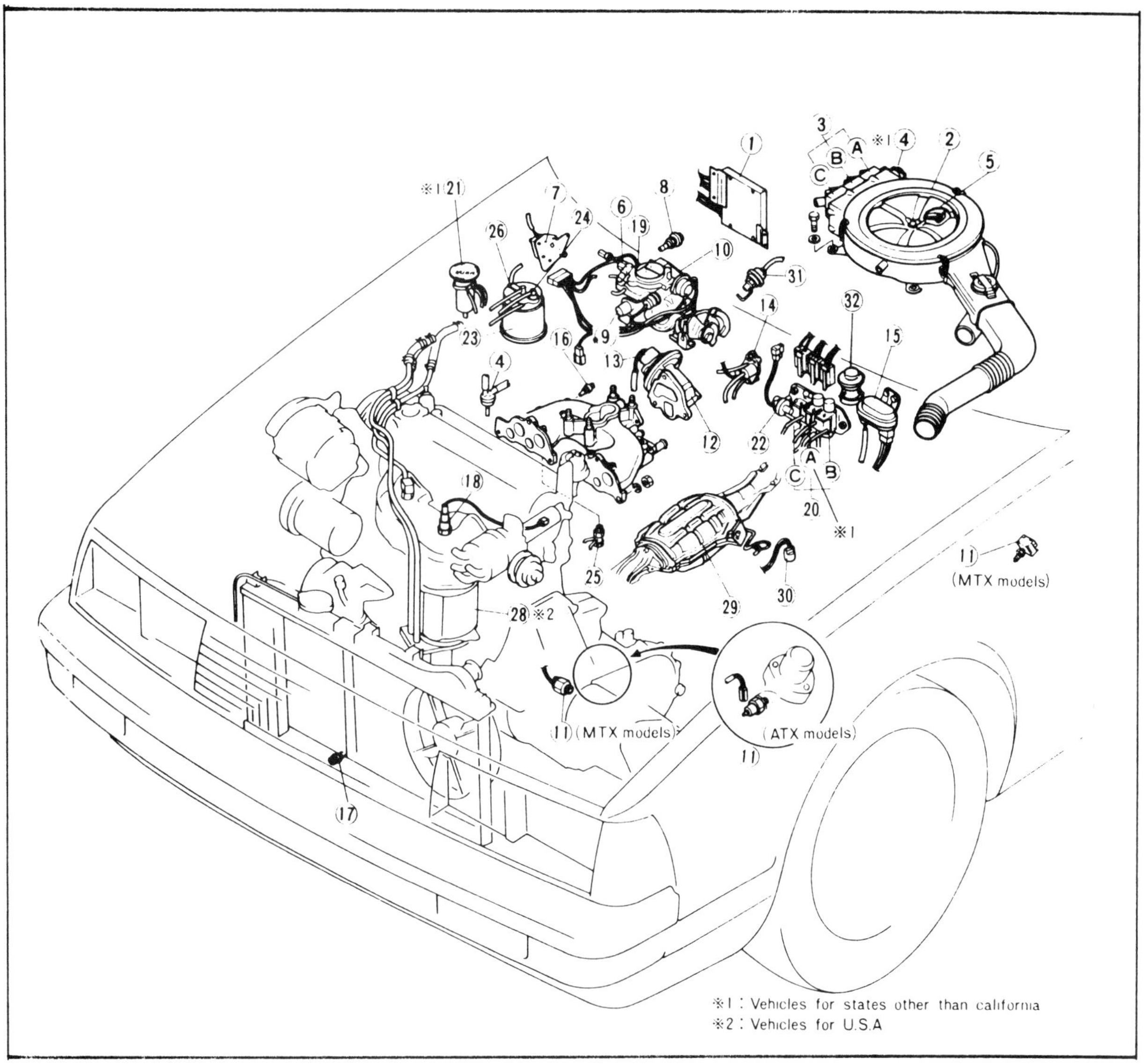

1. Control unit
2. Air cleaner
3. Reed valve
4. No.1 ACV
   No.2 ACV *1
5. Idle compensator valve
6. Air vent solenoid valve
7. Idle switch
8. Dashpot
9. Coasting richer solenoid
10. Slow fuel cut solenoid valve
11. Clutch switch and neutral switch
12. E.G.R. valve
13. E.G.R. position sensor
14. Duty solenoid valve
15. Vacuum sensor
16. Water thermo sensor
17. Water thermo switch
18. O_2 sensor
19. A/F solenoid valve
20. Three-way solenoid valve
21. Altitude compensator
22. Vacuum switch
23. Canister
24. Purge control valve No.1
25. Water thermo valve
26. Purge control valve No.2
27. Check valve
28. Front catalyst
29. Rear catalyst
30. Power steering switch
31. Servo diaphragm
32. Anti-afterburn valve (ATX)

MAZDA—626 MODELS—COMPONENTS LOCATIONS

SYSTEM DIAGRAM

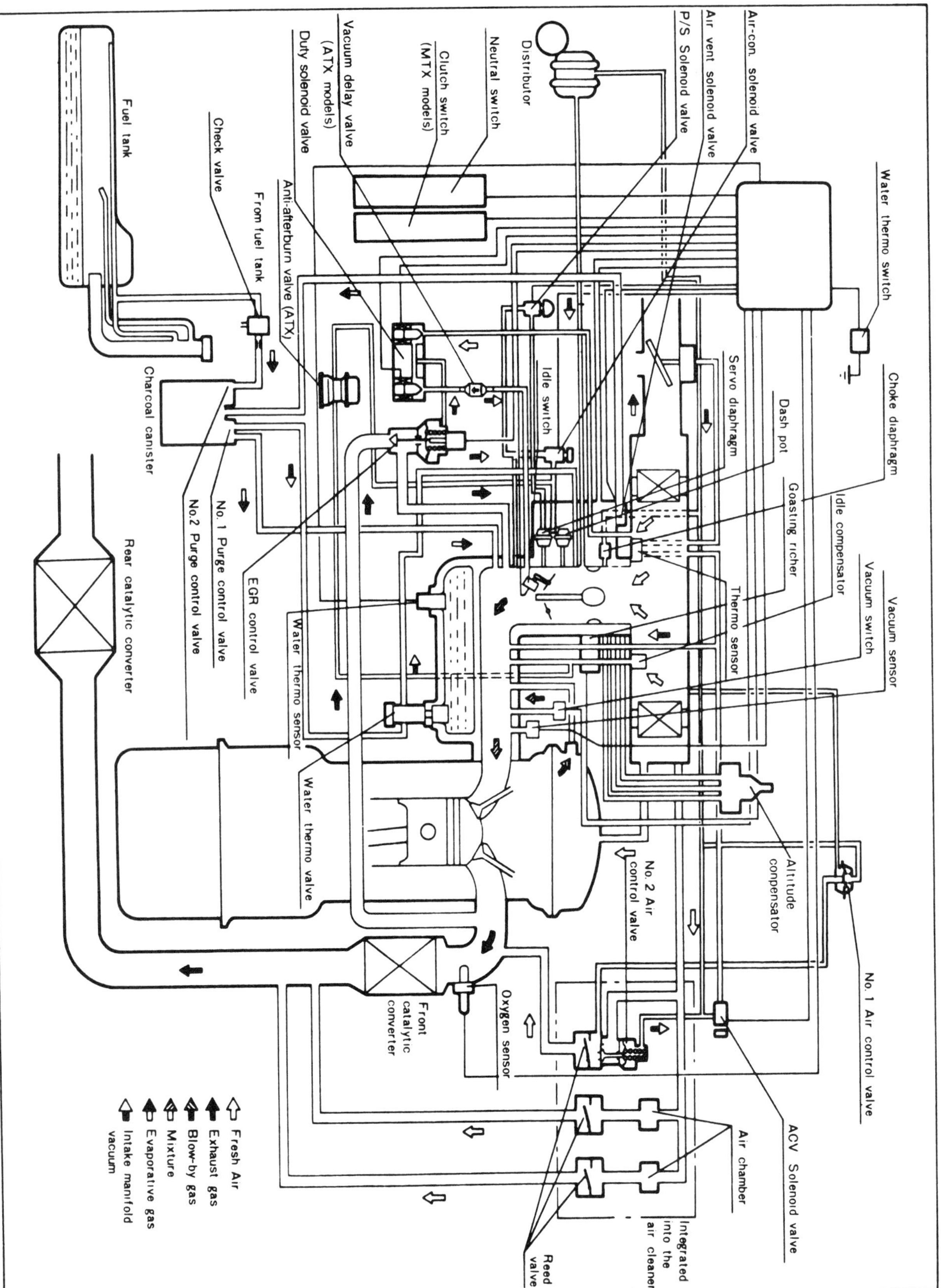

MAZDA—626 MODELS—EXC. CALIFORNIA—SYSTEM DIAGRAM

SYSTEM DIAGRAM

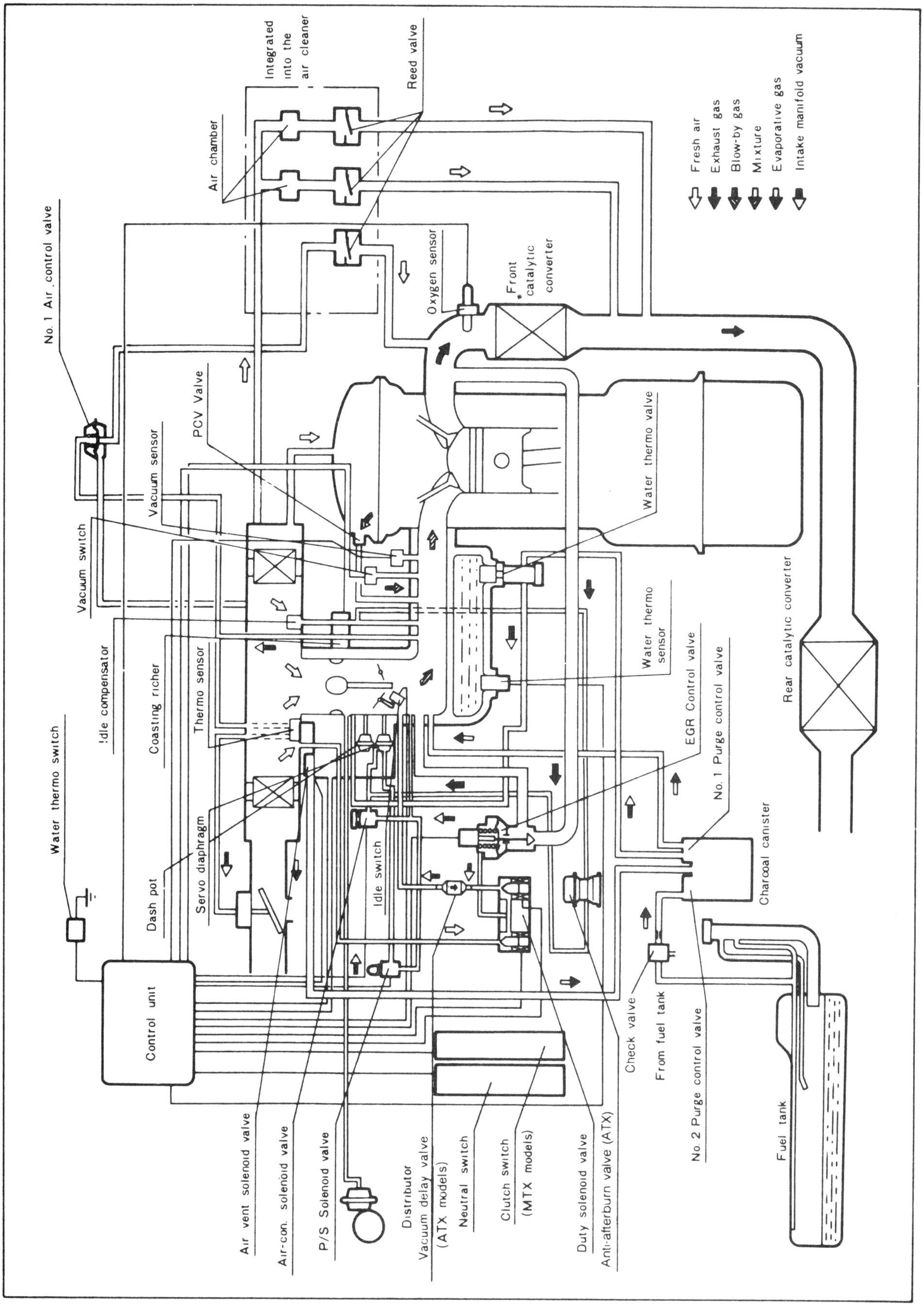

MAZDA—626 MODELS—CALIFORNIA—SYSTEM DIAGRAM

SYSTEM DIAGRAM

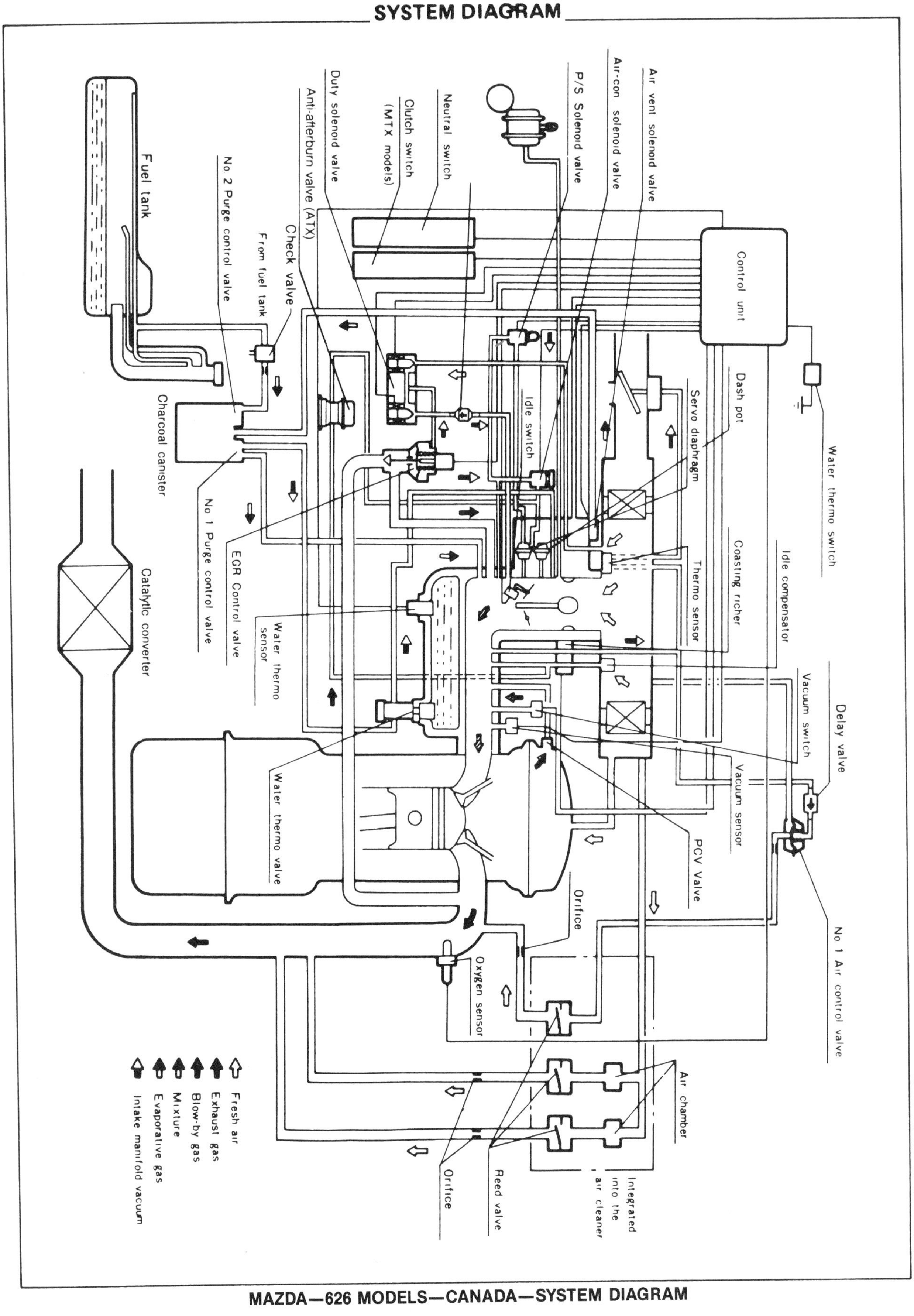

MAZDA—626 MODELS—CANADA—SYSTEM DIAGRAM

VACUUM DIAGRAMS

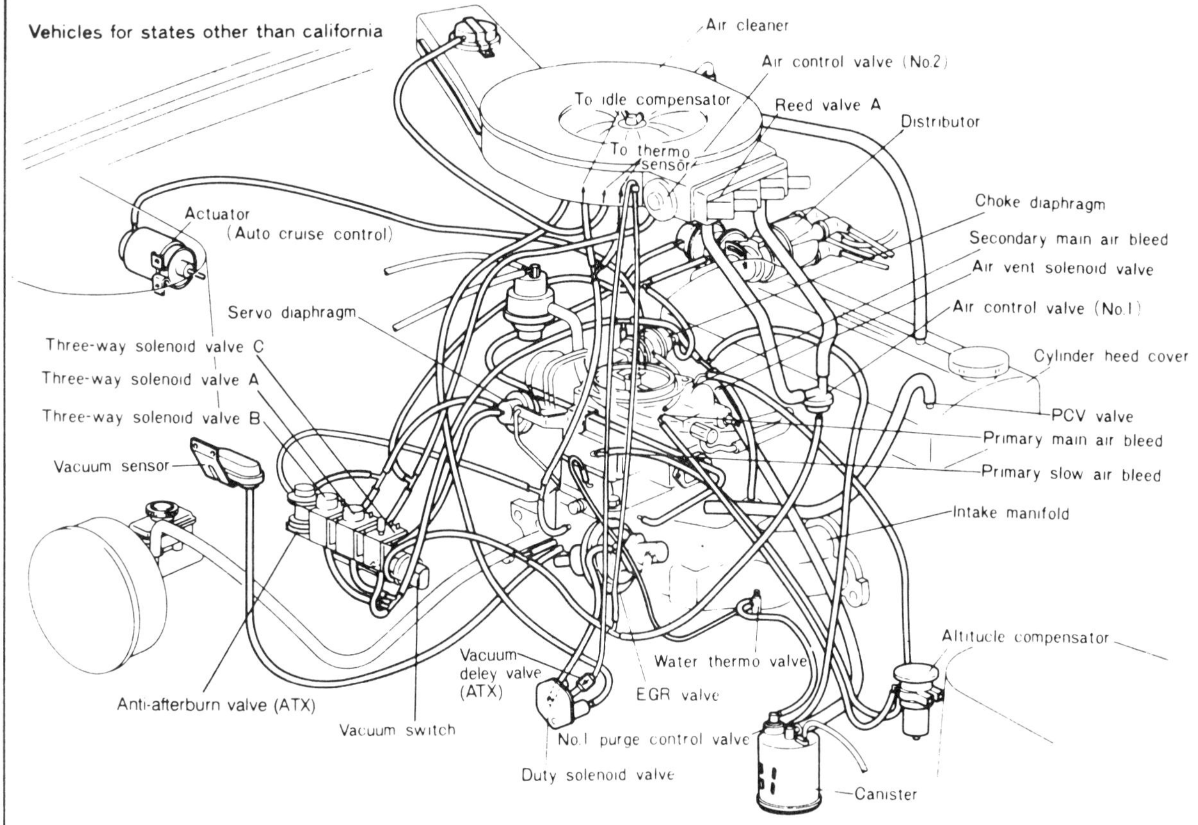

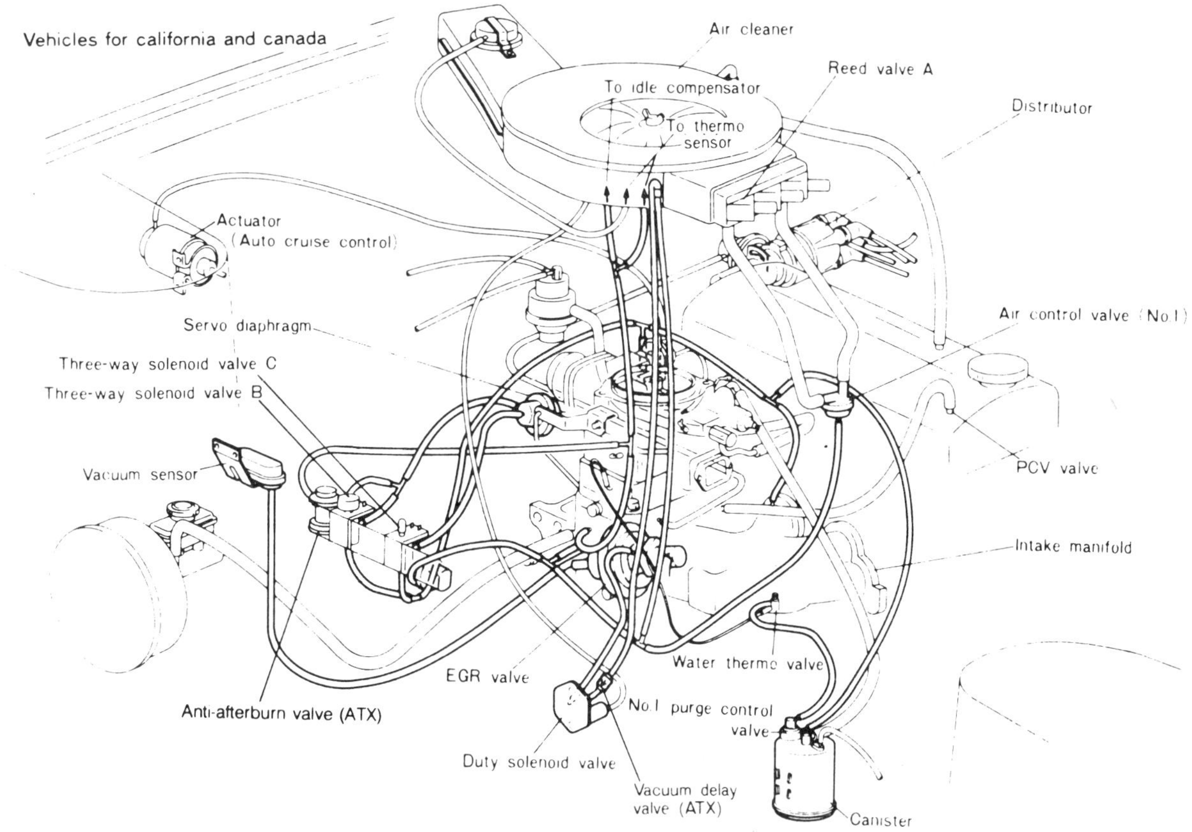

MAZDA—626 MODELS—VACUUM HOSE ROUTING

HOSE COLOR CODES

Component	Color	Connected to:
Air cleaner	• Green • Green • White • Black	Intake manifold (~ idle compensator) Intake manifold (~ thermo sensor) Duty solenoid valve Cylinder head cover
Air control valve No.1	• Black • Black • Black	Air cleaner Reed valve A Intake manifold
Air control valve No. 2 (Vehilces for states other than California)	• Orange	Three-way solenoid valve A
PCV valve	• Black	Intake manifold
High altitude compensator (Vehicles for states other than California)	• Blue • Yellow • Brown	Primary main air bleed Primary slow air bleed Secondary main air bleed
Canister	• Black • Black • Black • Blue	Evaporative pipe Air vent solenoid valve (carburetor) Intake manifold Water thermo valve
Duty solenoid value	• White • Red • Blue	Air cleaner (vent) EGR valve Intake manifold (vacuum)
Vacuum switch	• Black	Intake manifold
Three-way solenoid valve A (Vehicles for states other than California)	• Black • Orange	Intake manifold Air control valve No.2, distributor and carburetor (choke diaphragm)
Three-way solenoid valve B	• Black • Yellow	Intake manifold Servo diaphragm (air-con)
Three-way solenoid valve C	• Black • Green	Intake manifold Servo diaphragm (power-steering)
Vacuum sensor	• Black	Intake manifold
Actuator (auto cruise control – if so equipped)	• Black	Intake manifold
Water thermo valve	• Blue • Nylon Tube (white)	No.1 purge control valve Carburetor (throttle vacuum)

MAZDA—626 MODELS—HOSE COLOR CODES

SYSTEM DIAGRAM

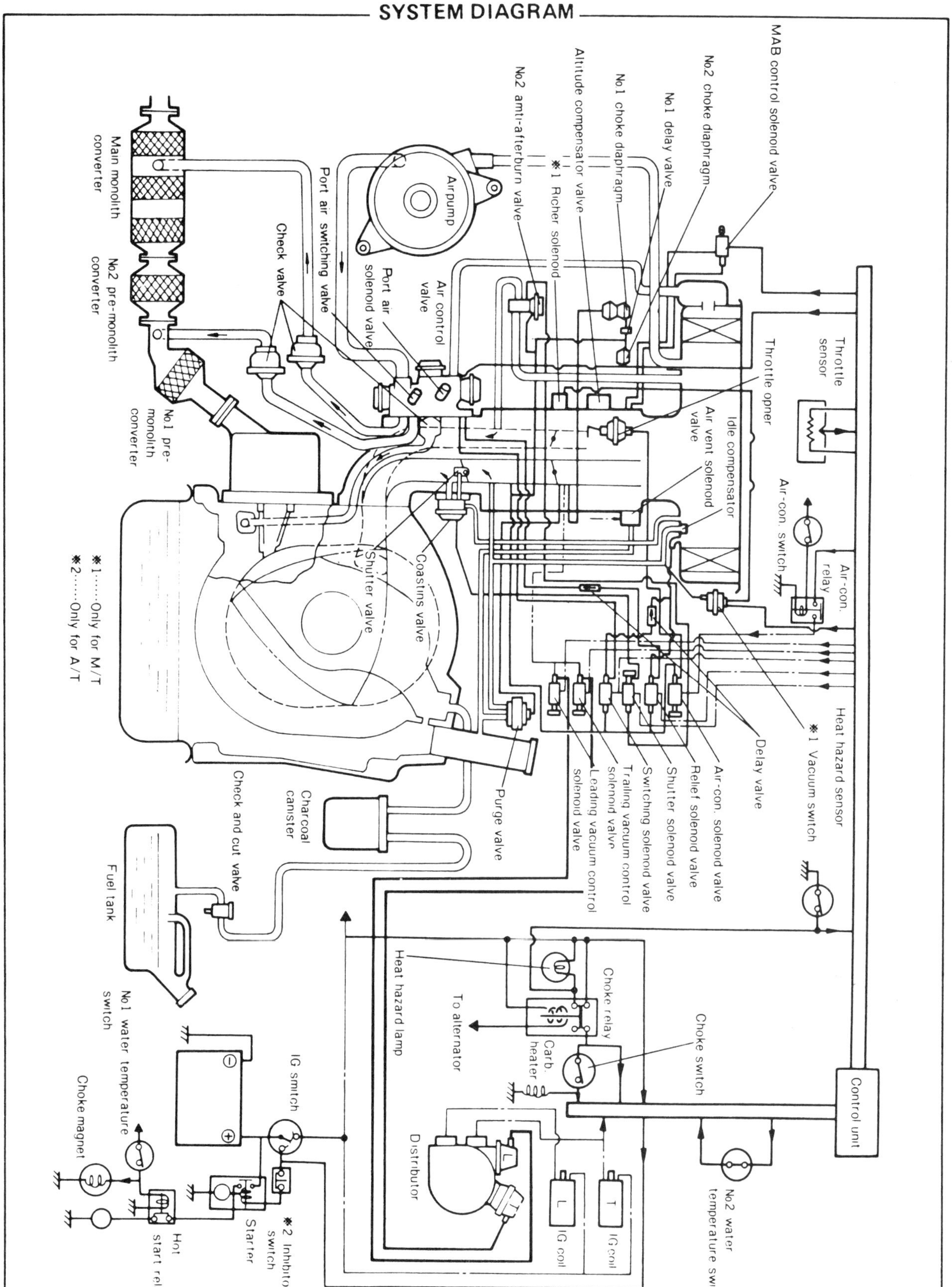

MAZDA—RX7 MODELS—EMISSION SYSTEM DIAGRAM

EMISSION CONTROL SCHEMATIC DIAGRAM

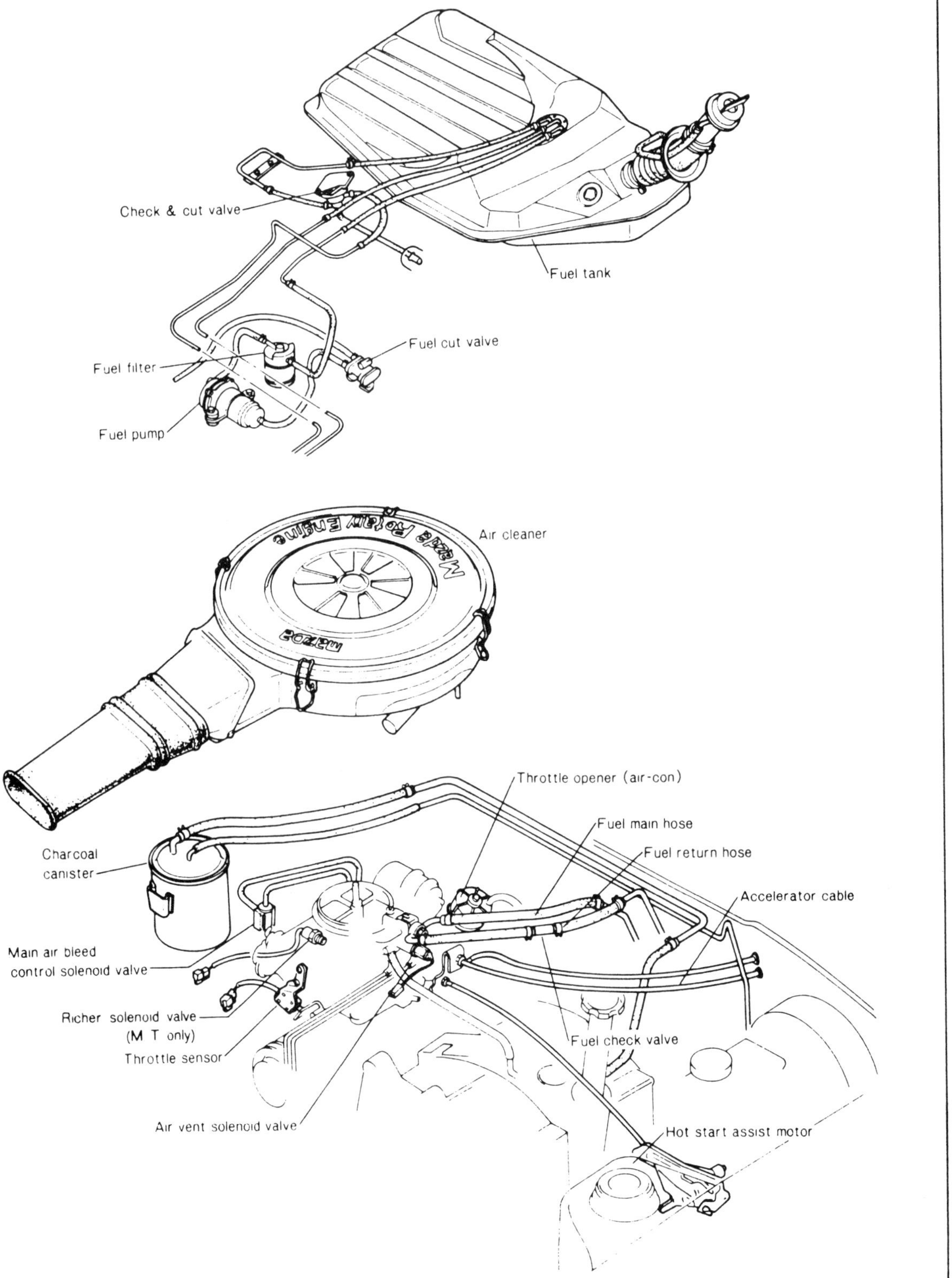

MAZDA—RX7 MODELS—EMISSION SYSTEM SCHEMATIC

EMISSION CONTROL SCHEMATIC DIAGRAM

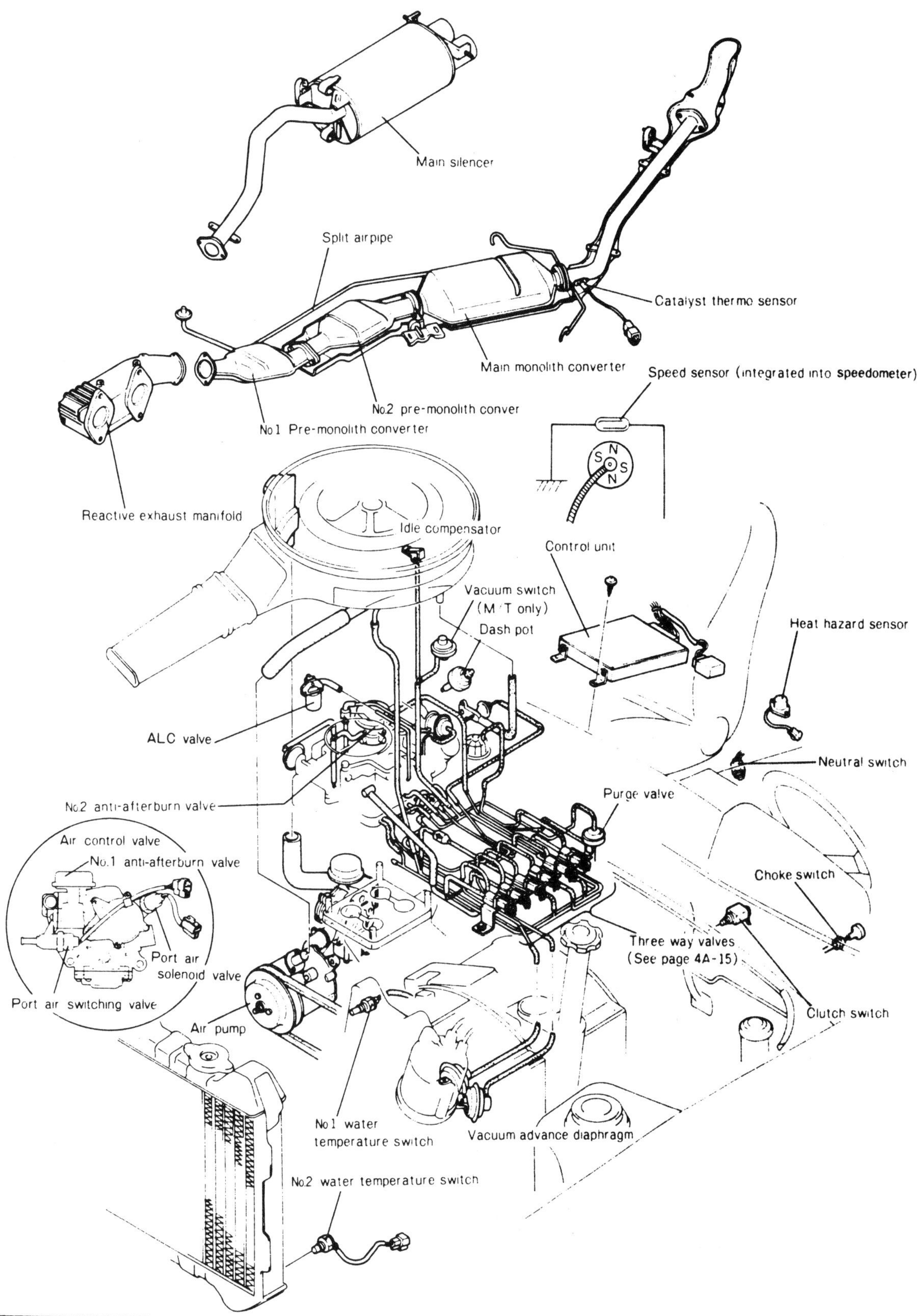

MAZDA—RX7 MODELS—EMISSION SYSTEM SCHEMATIC

VACUUM HOSE ROUTING DIAGRAM

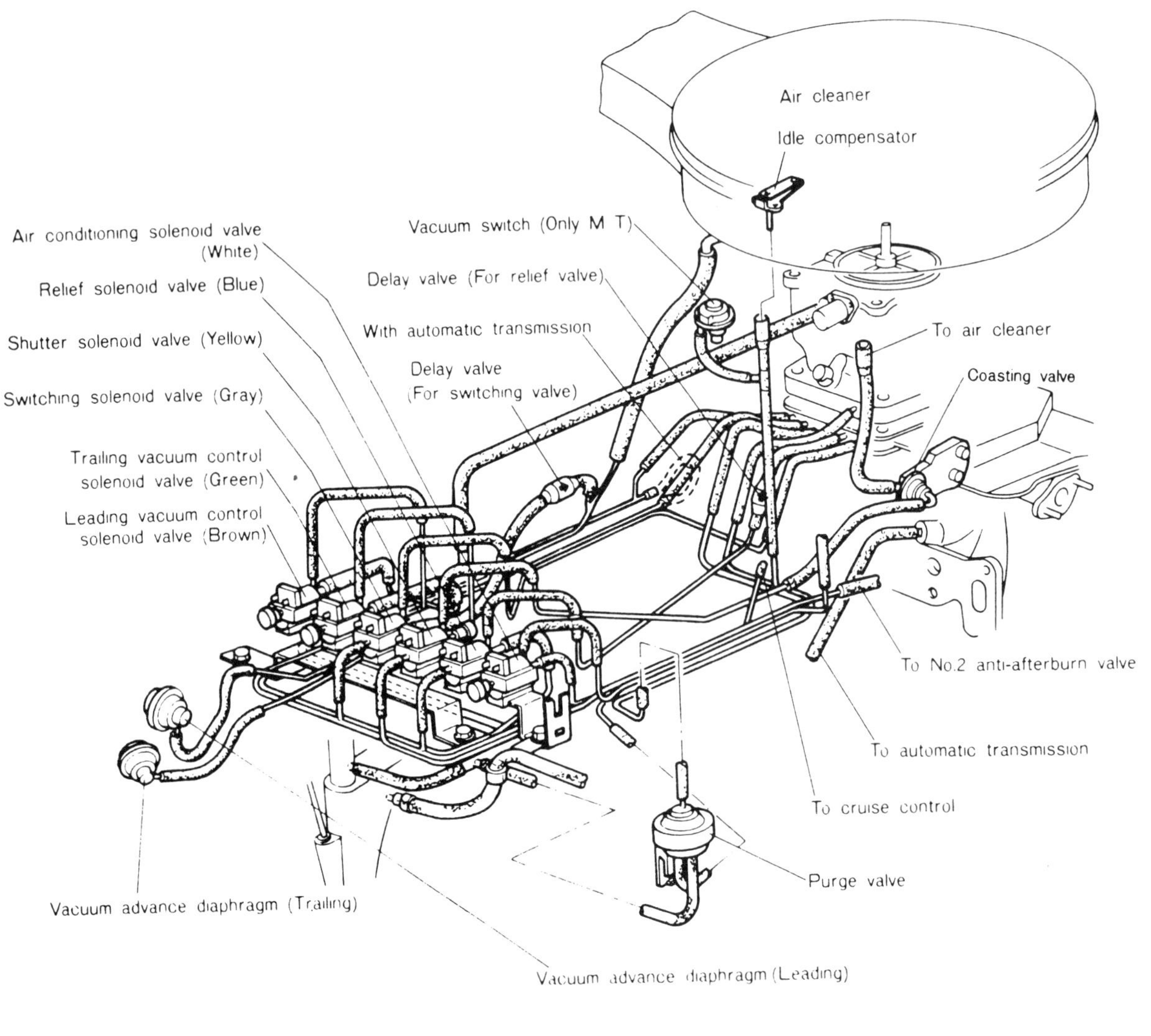

MAZDA—RX7 MODELS—VACUUM HOSE ROUTING

EMISSION CONTROL SCHEMATIC DIAGRAM

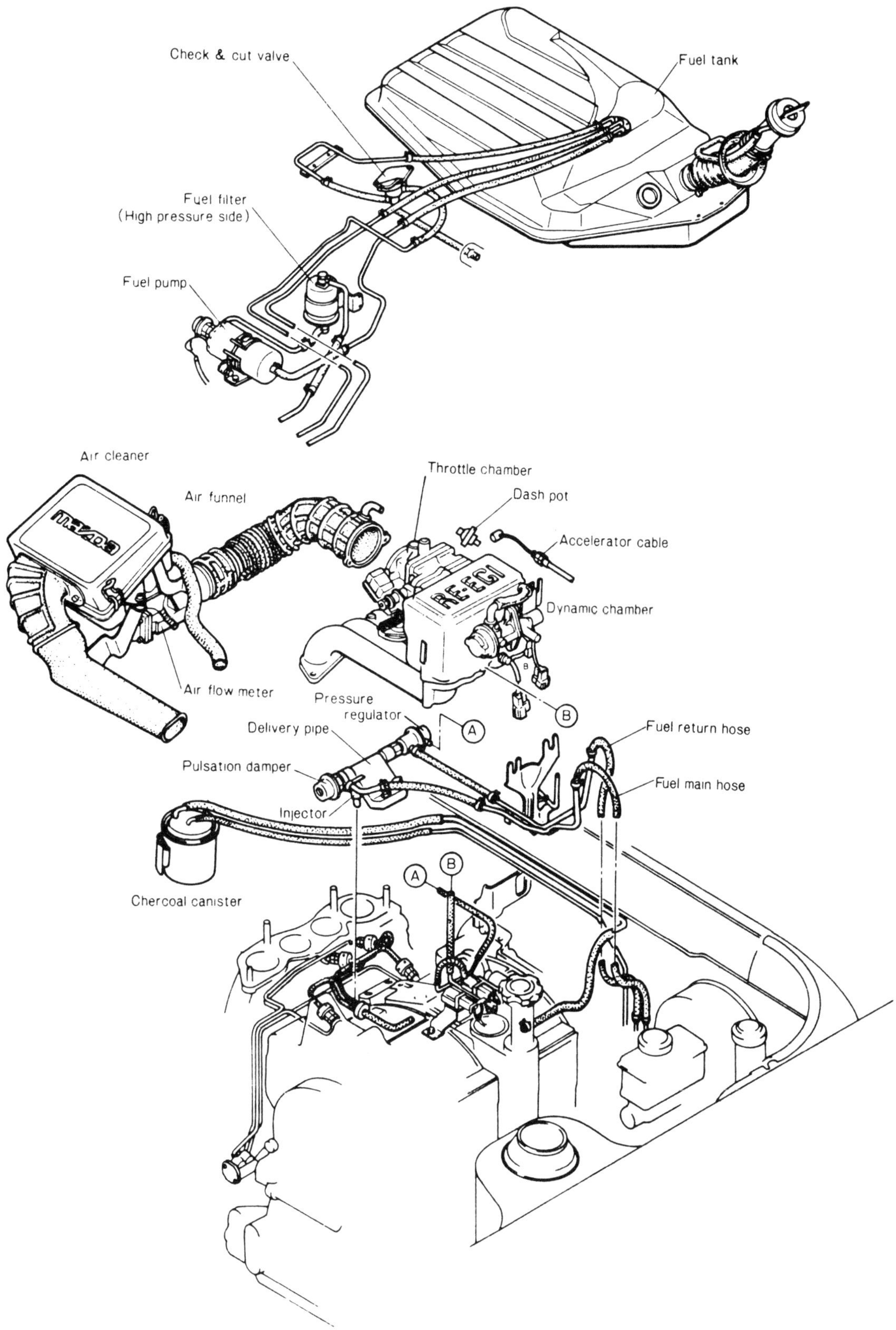

MAZDA—RX7 MODELS—EMISSION SYSTEM SCHEMATIC

SYSTEM DIAGRAM

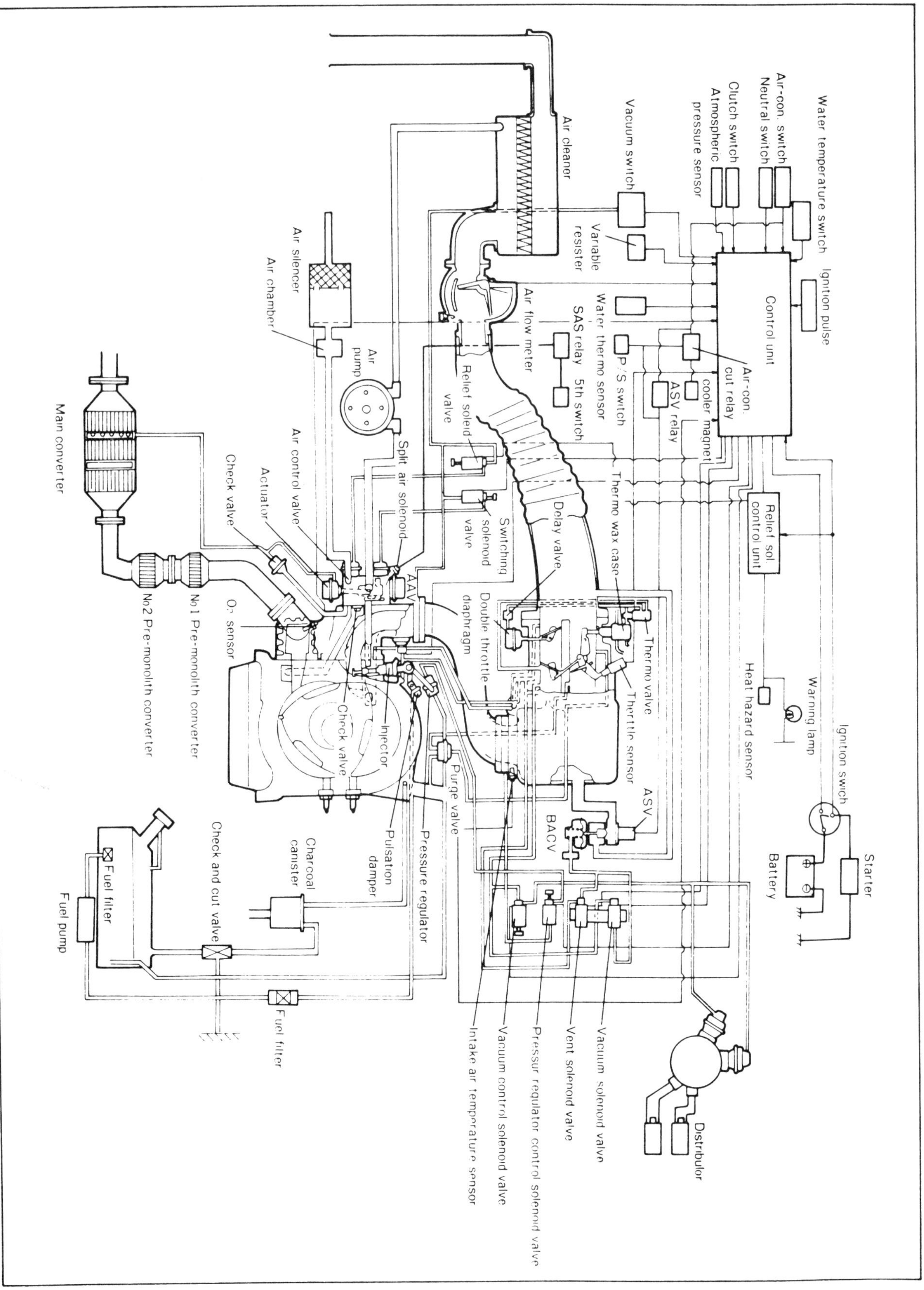

MAZDA—RX7 MODELS—EMISSION SYSTEM DIAGRAM

EMISSION CONTROL SCHEMATIC DIAGRAM

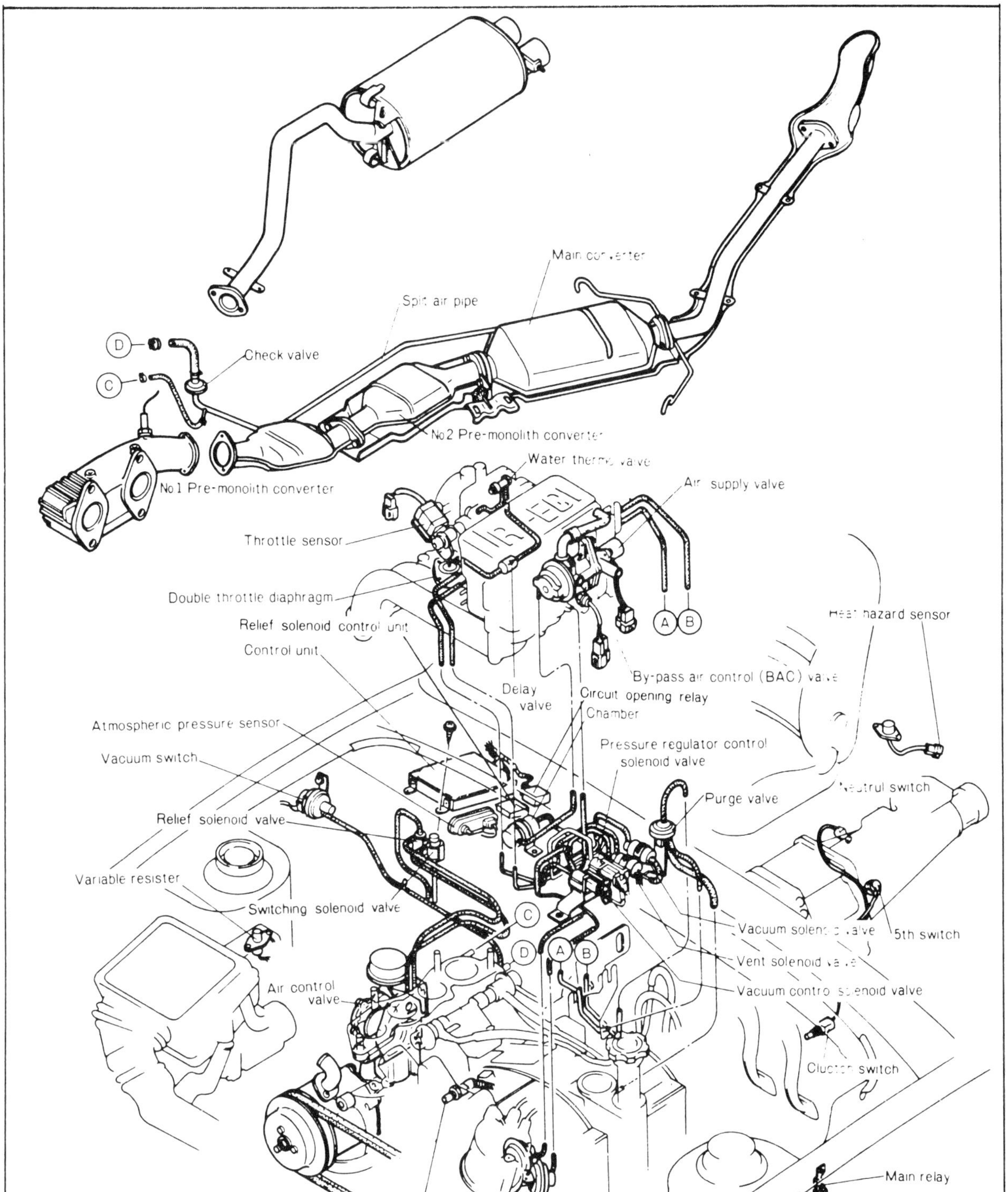

MAZDA—RX7 MODELS—EMISSION SYSTEM SCHEMATIC

VACUUM HOSE ROUTING DIAGRAM

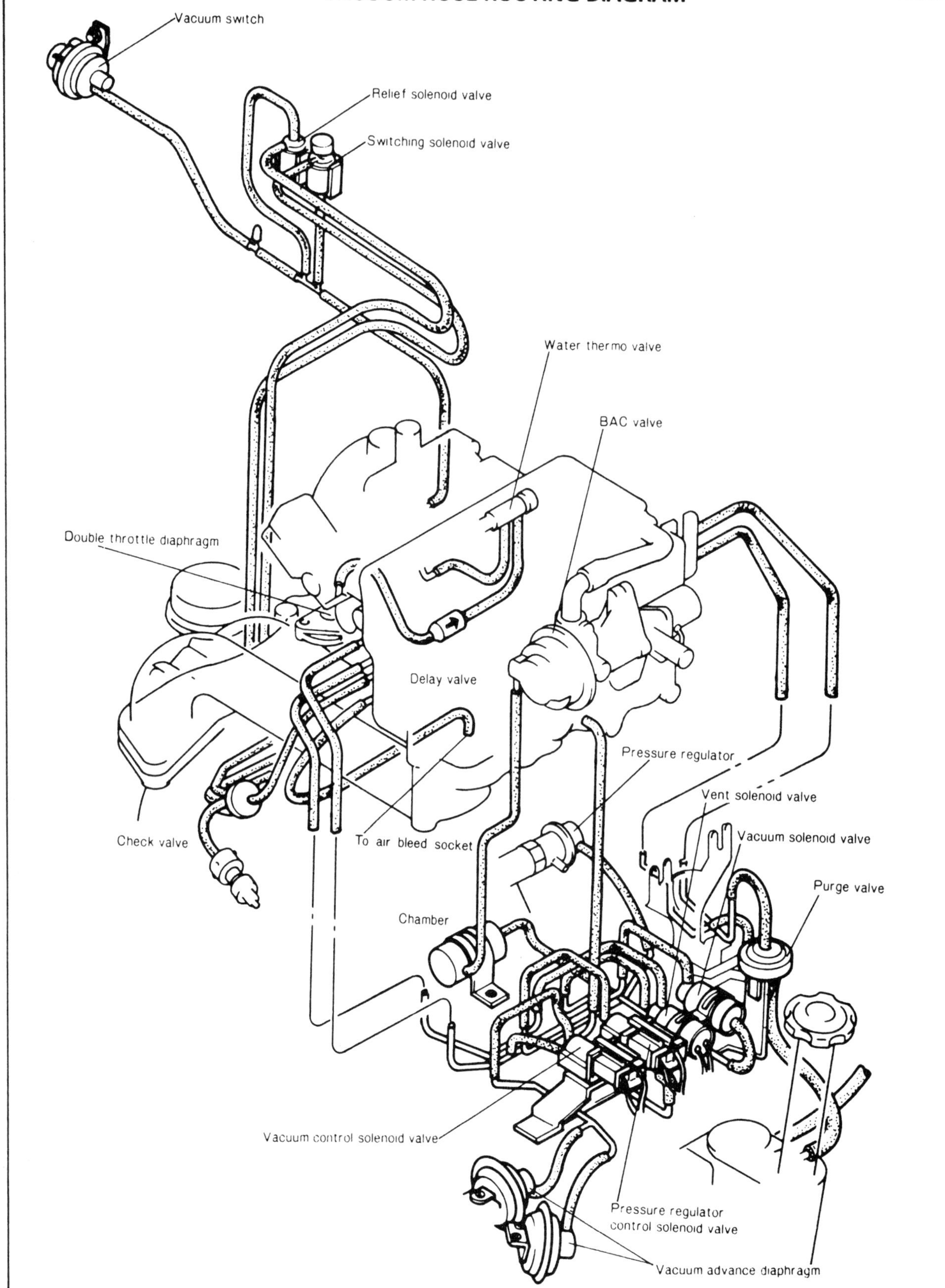

MAZDA—RX7 MODELS—VACUUM HOSE ROUTING

Mazda Truck Specifications

VEHICLE IDENTIFICATION NUMBER AND ENGINE IDENTIFICATION

Vehicle Identification Number (VIN)

The Vehicle Identification Number (VIN) is stamped on a plate located on the drivers side windshield pillar and is visible through the glass. A chassis number is stamped on a plate, which is riveted to the frame in the right front wheel-well area of the vehicle.

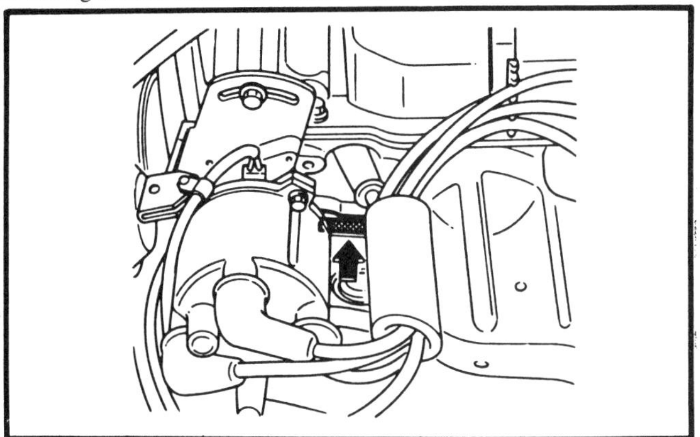

Location of the engine identification number

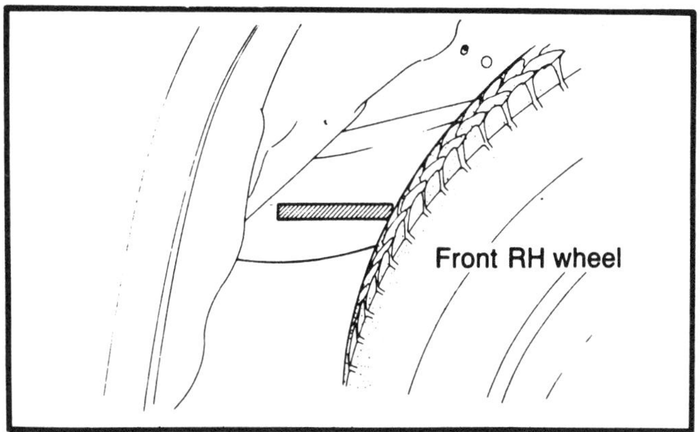

Chassis number location

Engine Identification

The engine identification number is stamped on the engine block, just below the distributor housing.

TUNE-UP SPECIFICATIONS—PISTON ENGINE

(When analyzing compression test results, look for uniformity among cylinders, rather than specific pressures)

Year	Engine Displacement (cu. in.)	Spark Plugs Type	Spark Plugs Gap (in.)	Distributor Point Dwell (deg)	Distributor Point Gap (in.)	Ignition Timing (deg) MT	Ignition Timing (deg) AT	Intake Valve Opens (deg)	Fuel Pump Pressure (psi)	Idle Speed (rpm)	Valve Clearance (in.) In	Valve Clearance (in.) Ex
'85	90.9	⑧	.031	Electronic		8B④	8B④	15	2.8–⑥ 3.8	850③	.010	.012
'85 (Gas)	121.9	BPR5ES, BPR6ES	.031	Electronic		6B	6B	17	2.8–⑦ 3.5	750⑤	.012	.012
'85–'86 (Diesel)	121.9	—	—	—	—	—	—	13	1920	800–850	.010	.014
'86	See Underhood Specifications Sticker											

NOTE: The underhood specifications sticker often reflects tune-up specification changes made in production. Sticker figures must be used if they disagree with those in this chart.

③ Automatic 750 in Drive
④ '83 & later—6° BTDC
⑤ Automatic: 700 in Drive
⑥ '84 & later—4.27–5.97
⑦ '84 & later-2.8–4.27
⑧ BPR5ES, BPR6ES ('81–'85) AGR-22, AGR-32 ('83–'85) RN11YC ('85)

TUNE-UP SPECIFICATIONS—ROTARY ENGINE

(When analyzing compression test results, look for uniformity among cylinders, rather than specific pressures)

Year	Engine Displacement (cu. in.)	Spark Plugs		Distributors		Ignition Timing (deg)			Idle Speed (rpm)	
		Type	Gap (in.)	Point Dwell (deg)	Point Gap (in.)	Leading Normal	Leading Retarded	Trailing Normal	MT	AT
'84–'85	70④	③	0.053–0.057⑤	Electronic		0⑥	—	20A	750⑦	750①
'86	See Underhood Specifications Sticker									

NOTE: The underhood specifications sticker often reflects tune-up specification changes made in production. Sticker figures must be used if they disagree with those in this chart.

TDC—Top dead center
A—After top dead center
B—Before top dead center
MT—Manual transmission
AT—Automatic transmission
deg—degrees
① Transmission in Drive

③ 1980: BR7ET, W22EBR
BR8ET, W25EBR
BR9ET, W27EBR
1981–85: BR7EQ14, W22EDR14
BR8EQ14, W25EDR14
BR9EQ14, W27EDR14

④ 13B
eng: 80

⑤ 13B
eng: .055

⑥ 13B
eng: Leading-5A

⑦ 13B
eng: 800

FIRING ORDER

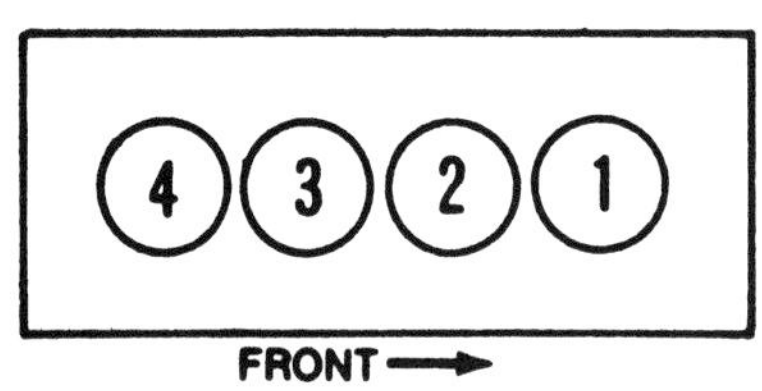

Firing order 1-3-4-2

Mercedes-Benz

INDEX

VEHICLE IDENTIFICATION NUMBER AND ENGINE IDENTIFICATION

Vehicle Identification Number (VIN)

The vehicle identification number (VIN) is stamped on a metal tab that is located on the top left side of the instrument panel. The metal tab is mounted so that the vehicle identification number can be seen through the windshield from outside the vehicle. The first digit represents the model year. "F" represents 1985 and "G" represents 1986.

ENGINE IDENTIFICATION

The third digit of the VIN number represents the piston displacement in cubic centimeters.

Location of important information on the 190D and 190E

Example: FMB 2.3 V 6 F A 1 X

F MB 2.3 V 6 F A 1 X

Model year
F = 1985 G = 1986 etc.

Manufacturer's code
Mercedes-Benz

Piston displacement
2300 cc

Vehicle class
D = passenger car with diesel engine
V = passenger car with gasoline engine

Type of fuel feed
6 = mechanical injection
9 = mechanical injection with turbocharger

Type of catalyst
F = 3-way catalyst in combination with lambda control
J = no catalyst (diesel)
K = trap oxidizer

For free use of manufacturer
A = Federal and California, F = Federal, C = California

For free use of manufacturer for purpose of certification
Internal control digit

Check digit

The key to 1985 and later engine identification is a 10 digit number

1 Certification Tag (left door pillar)
2 Identification Tag (left window post)
3 Chassis No.
4 Engine No.
5 Body No. and Paintwork No.
6 Information Tag California version Vacuum line routing for emission control system
7 Emission Control Tag
8 Emission Control Tag Catalyst Information

Location of important information on the 300TD

1 Certification Tag (left door pillar)
2 Identification Tag (left window post)
3 Chassis No.
4 Engine No.
5 Body No. and Paintwork No.
6 Emission Control Tag
7 Information Tag California version Vacuum line routing for emission control system
8 Emission Control Tag Catalyst Information

Location of important information on the 380SE, 500SEC and 500SEL

1. Certification tag (left door pillar)
2. Identification tag (left window post)
3. Chassis no.
4. Body no. and paintwork no.
5. Engine no.
6. Emission control tag
7. Emission control tag catalyst information

Location of important information on the 300SD

1. Catalyst and certification tag (left door pillar)
2. Identification tag (left window post)
3. Chassis no.
4. Body no. and paintwork no.
5. Engine no. on rear engine block
6. Emission control information

Location of important information on the 380SL

DIESEL ENGINE TUNE-UP SPECIFICATIONS

Model	Valve Clearance (cold) ① Intake (in.)	Valve Clearance (cold) ① Exhaust (in.)	Intake Valve Opens (deg)	Injection Pump Setting (deg)	Injection Nozzle Pressure (psi) New	Injection Nozzle Pressure (psi) Used	Idle Speed (rpm) ②	Cranking Compression Pressure (psi)
190D	Hyd.	Hyd.	⑤	15A	1564–1706	1422–1706	700–800	284–327
240D	0.004	0.016	13.5B	24B	1564–1706	1422–1706	750–800	284–327
300D, 300CD, 300TD (5-cylinder, non-turbo)	0.004	0.012	13.5B	24B ④	1635–1750 ③	1422	700–800	284–327

DIESEL ENGINE TUNE-UP SPECIFICATIONS

Model	Valve Clearance (cold) ① Intake (in.)	Valve Clearance (cold) ① Exhaust (in.)	Intake Valve Opens (deg)	Injection Pump Setting (deg)	Injection Nozzle Pressure (psi) New	Injection Nozzle Pressure (psi) Used	Idle Speed (rpm) ②	Cranking Compression Pressure (psi)
300D, 300CD, 300SD, 300TD (5-cylinder, turbo)	0.004	0.014	13.5B	24B ④⑥	1958–2074	1740	650–850 ⑦	284–327

B Before Top Dead Center
① In cold weather (below 5°F.), increase valve clearance 0.002 in.
② Manual transmission in Neutral; Automatic in Drive.
③ Difference in opening pressure on injection nozzles should not exceed 71 psi.
④ The injection pump is in start of delivery position when the mark on the pump camshaft is aligned with the mark on the injection pump flange.
⑤ New timing chain: 11A
Used timing chain (12,000 miles): 12A
⑥ 1984–86: 15A
⑦ 1984–86: 700–800

GASOLINE ENGINE TUNE-UP SPECIFICATIONS

(When analyzing compression test results, look for uniformity among cylinders, rather than specific pressures)

Year	Model	Spark Plugs Type	Spark Plugs Gap (in.)	Distributor Point Dwell (deg)	Ignition Timing (deg)	Intake Valve Opens (deg)	Fuel Pump Pressure (psi) Idle ●	▲ Idle Speed (rpm)	Valve Clearance* (in.) In (cold)	Valve Clearance* (in.) Ex (cold)
'85	190E	S12YC	0.032	Elec.	5B	③	77–80	700–600	Hyd.	Hyd.
	380SL	N10Y ④	0.032	Elec.	TDC w/o vacuum	24A	①	500–600	Hyd.	Hyd.
	380SE 380SEC, 380SEL	N10Y ⑤	0.032	Elec.	TDC w/o vacuum	24A	①	500–600	Hyd.	Hyd.
	500SEC, 500SEL	N10Y ⑤	0.032	Elec.	TDC w/o vacuum	②	①	600–700	Hyd.	Hyd.
'86	All	See Underhood Specification Sticker								

CAUTION: If the specifications listed above differ from those on the tune-up decal in the engine compartment, use those listed on the tune-up decal.

NOTES: 1. On transistor ignitions, only a transistorized dwell meter can be used. Transistor ignitions are recognizable by the "Blue" ignition coil, 2 series resistors and the transistor switchgear.
2. To counteract wear of the fiber contact block, adjust the dwell to the lower end of the range.

A After Top Dead Center
B Before Top Dead Center
w/vacuum—vacuum advance connected
w/o vacuum—vacuum advance disconnected
*Below 0°F; increase valve clearance by 0.002 in.
—Not Available
▲In Drive
● timing for test measurements @ 2mm valve lift

① Approximately 1 quart in 30 seconds
② Left side—22° ATDC
Right side—20° ATDC
③ 1985-86: New timing chain—11 A
Used timing chain (12,500 miles)—12A
④ 1985: N12Y
⑤ 1985: N11Y

FIRING ORDERS

NOTE: The position of the number one tower on the distributor cap may vary. To avoid confusion when replacing wires, always replace wires one at a time. The notch cut into the rim of the distributor body always indicates number one cylinder.

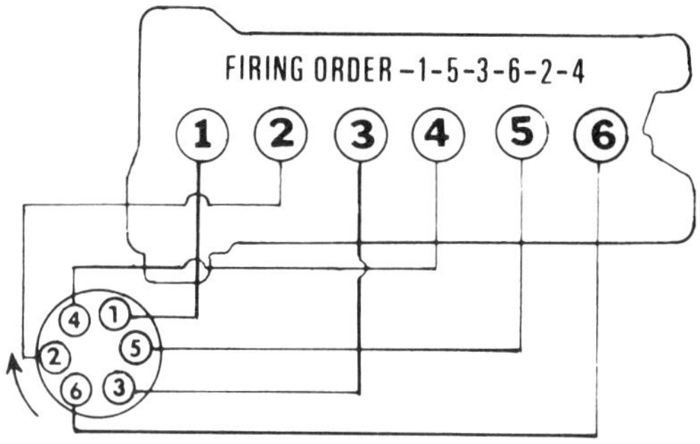

6-cylinder gas engine

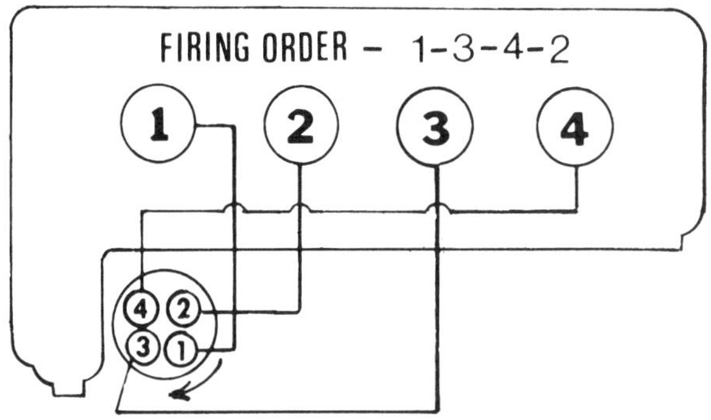

4-cylinder gas engine

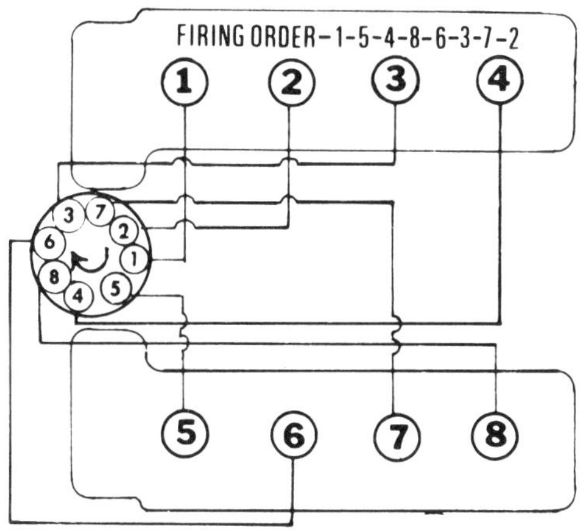

V8 engine

EMISSION EQUIPMENT USED

GASOLINE ENGINES

1985–1986

Air Injection System (AIS)
Catalytic Converter (Cat)
Constant Injection System (CIS-E)
Fuel Evaporation System (FES)
Idle Speed Control (ISC)
KE-Jetronic Fuel Injection
Lambda Control Unit (LCU)
Oxygen Sensor (OS)
Positive Crankcase Ventilation (PCV)

DIESEL ENGINES

1985–1986

Exhaust Gas Recirculation (EGR)-California vehicles only
Positive Crankcase Ventilation (PCV)

REQUIRED EMISSION CONTROL MAINTENANCE

NOTE: The following information is being published from the latest information available at the time of publication. If this information from any data located on the underhood emissions label, or in the owners manual, use the information in the owners manual or on the emissions label, since this information reflects the latest changes or modifications to the vehicle. All maintenance intervals are at 15,000 miles unless otherwise noted.

PCV SYSTEM

Clean and check all hoses and connection and replace the valve every 30,000 miles.

FUEL EVAPORATION SYSTEM

Replace the fuel filter every 15,000 miles. Clean and inspect all hoses and connections. Replace the canister if saturated with fuel.

AIR INJECTION SYSTEM

Check and inspect all hoses and connections. Check the system for proper operation.

CATALYTIC CONVERTER

Check for damage and tight connection every 30,000 miles.

EGR SYSTEM

Check and inspect all hoses and connections. Clean any carbon desposits from all the lines and the throttle valve.

IGNITION TIMING

Check and adjust ignition timing every 15,000 miles.

IDLE SPEED AND MIXTURE

Check and adjust if necessary idle speed mixture every 15,000 miles.

SPARK PLUGS

Inspect spark plugs every 15,000 miles. Replace if necessary.

OXYGEN SENSOR

Replace every 30,000 miles. Reset instrument panel reminder light.

AIR CLEANER

Replace the air cleaner every 15,000 miles.

DRIVE BELTS

Inspect the power steering, oil pump, air conditioning compressor and alternator drive belts for cracks, fraying or other signs of deterioration. Replace or adjust as necessary.

NOTE: Replace all filters more frequently if the vehicle is used for trailer towing or driven in heavy traffic or where ambient temperatures are extremely high or low.

IDLE SPEED AND MIXTURE ADJUSTMENTS

NOTE: The following information is being published from the latest information available at the time of publication. If the information contained herein differs from that which is listed on the vehicles emission label, use the specifications given on the label.

IDLE SPEED

Adjustment

ALL DIESEL ENGINES EXCEPT 1985 AND LATER 190D

1. Run the engine to normal operating temperature.
2. On normally aspirated engines, turn the idle speed adjuster on the instrument panel completely to the right.
3. Disconnect the pushrod at the angle lever.
4. Check the idle speed. Adjust by loosening the locknut and adjusting the idle speed screw. Tighten the locknut.
5. On all models except turbo diesels, adjust the pushrod so that a clearance of 0.2 in. exists between the cam on the lever and the actuator on the switchover valve. The lever on the fuel injection pump must rest against the idle stop.
6. Depress the stop lever as far as possible. The cruise control bowden cable should be free of tension against the angle lever. Use the adjusting screw to adjust the tension. Release the stop lever. The bowden cable should have a slight amount of play.
7. On turbo diesels, adjust the pushrod so the roller in the guide lever rests free of tension against the stop.
8. Put the automatic transmission in DRIVE and turn the steering wheel to full lock in either direction. The engine should run smoothly. If not, adjust the idle speed. Disconnect the cruise control connecting rod and push the lever clockwise to the idle stop. Attach the connecting rod, making sure the lever is approximately 0.4 in. from the idle speed stop.

1985 AND LATER 190D

NOTE: Testing the idle speed on the 190D will require two special tools. A digital tester (Sun-1019, 2110 or All-Test 3610-MB) and a TDC impulse transmitter. Without these special tools, idle speed adjustment is impossible and should not be attempted.

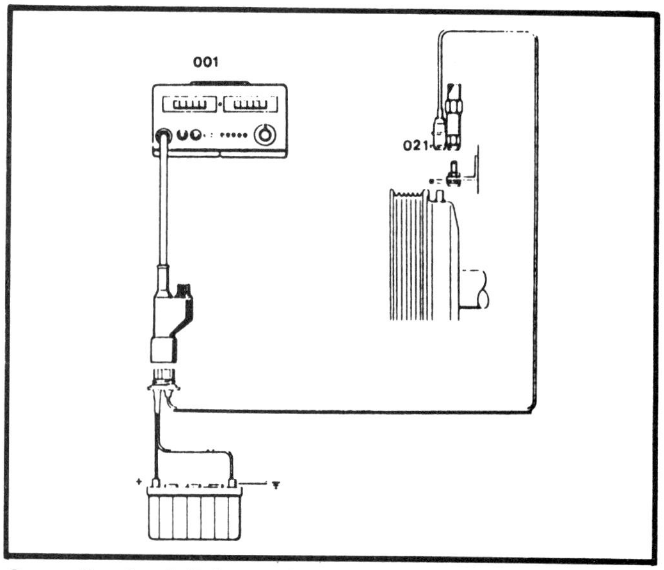

Connecting the digital testor (001) and the TDC impulse transmitter (021) on the 190D

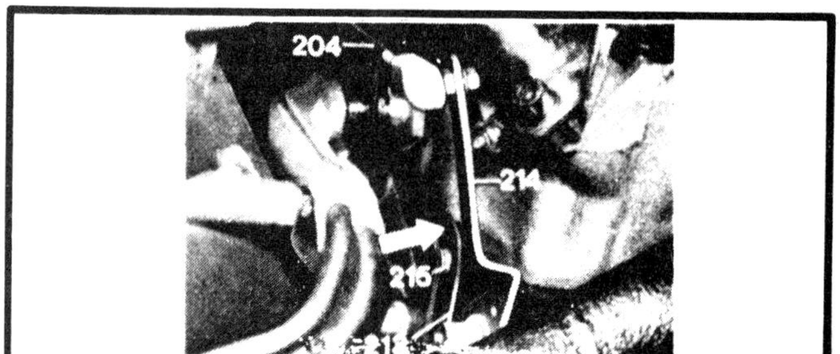

Throttle linkage on the 190D

1. Start the engine and allow it to reach normal operating temperature.
2. Connect the digital testor and the TDC impulse transmitter.
3. Inspect all linkages for ease of operation.
4. Disconnect the pushrod (204) from the adjuster lever (214).
5. Start the engine and check the idle speed. If required, adjust by loosening the locknut on the vacuum control unit and then turn the unit itself in or out.
6. After the idle speed is correct, tighten the vacuum control unit locknut and reconnect the pushrod so that it is tension free when the lever is against the idle speed stop.
7. Switch on all the power accessories and check that the engine continues to rum smoothly. Readjust the idle speed if necessary.
8. Disconnect the special tools and turn off the engine.

GASOLINE ENGINES

1985 AND LATER (4 CYLINDER AND V8)

These engine have electronically controlled idle speed, using a solenoid connected to the control unit of the vehicle.

IDLE MIXTURE

Adjustment

GASOLINE ENGINES

NOTE: A tamper proof plug is used over the mixture control unit adjustment screw. Adjustment is not a normal maintenance procedure and should only be performed in the event of control unit replacement or the vehicle failing an emissions test.

1. Start the engine and allow it to reach normal operating temperature. Check that the idle speed is within normal range and then adjust the ignition timing.
2. Remove the cover from the diagnostic plug on the fender apron located near the hood hinge.
3. Connect the negative lead of the voltmeter to pin number 3 of the plug. Connect the positive lead to the battery.
4. Disconnect the oxygen sensor connector. Note the needle position of the voltmeter (the needle should remain steady with no visible oscillation).

NOTE: In order to gain access to the oxygen sensor plug, unscrew the shroud bolt and push the plug out of the holder.

5. Reconnect the oxygen sensor plug. Upon reconnection, the voltmeter needle should begin to vibrate evenly around the mark noted in step 4. If not, mixture adjustment will be required.
6. To adjust, remove the tamper proof plug from the adjustment screw and insert a hex wrench. Adjust the screw with the hex wrench until the needle vibration on the voltmeter is centered around the original position.
7. Remove the test equipment and replug the adjustment screw hole.

IDLE SPEED
Gasoline Engines

Model	Idle Speed
190E	600–700 rpm in Drive or Neutral
380SE	500–600 rpm in Drive
380SEC	500–600 rpm in Drive
380SEL	500–600 rpm in Drive
380SL	500–600 rpm in Drive
500SEC	600–700 rpm in Drive
500SEL	600–700 rpm in Drive

NOTE: The above figures are for testing purposes only. All gasoline engines utilize electronically controlled idle speed, using a solenoid connected between terminals 1 and 5 of the control unit. No adjustment is either necesary or possible.

Diesel Engines

Model	Idle Speed
190D	700–800 rpm in Drive or Neutral
300D	700–800 rpm in Drive
300CD	700–800 rpm in Drive
300SD	700–800 rpm in Drive
300TD	700–800 rpm in Drive

INITIAL TIMING SETTINGS

GASOLINE ENGINES

Adjustment

1. Connect a timing light and tachometer according to the manufacturer's instructions to the engine.
2. Run the engine at the above specified speed and read the firing point on the balancing plate or vibration damper while shining the timing light at it.

NOTE: The balancer on some engines have two timing scales. If in doubt as to which scale to use, rotate the crankshaft in the direction of rotation only until the distributor rotor is aligned with the notch on the distributor housing (No. 1 cylinder). In this position, the timing pointer should be at TDC on the proper timing scale.

Typical vibration damper is marked with a pin

3. Adjust the ignition timing by loosening the distributor clamp bolt and rotating the distributor. To advance the timing, rotate the distributor in the opposite direction of normal rotation. To retard the timing, rotate the distributor in the direction of normal rotation.
4. Once the timing has been adjusted, recheck the timing once again to be sure it was not disturbed when tightening the distributor clamp bolt.
5. Disconnect the timing light and tachometer.

INJECTION TIMING

Adjustment

DEISEL ENGINES

1. Turn the crankshaft in the direction of normal direction until the No. 1 cylinder in on the compression stroke and the timing mark is aligned with the pointer.
2. Clean all pump connections, tag and disconnect all vacuum hoses from the pump and then disconnect the No. 1 injection line.
3. Unscrew the pipe connection and remove the compression spring and the pressure valve. Leave the pressure valve carrier and the copper gasket in position.

CAUTION

Never unscrew the element connection below the pipe connection or else the pump will require recalibration.

4. Install the pipe connection and the overflow pipe.

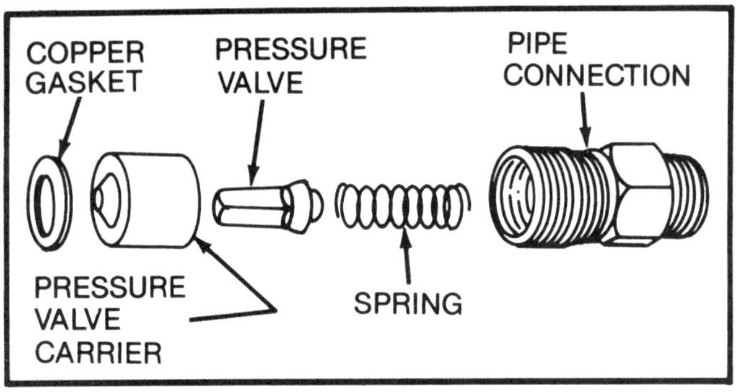

Injection pump pipe connection fittings

5. Rotate the crankshaft to 15° ATDC on the compression stroke of No. 1 cylinder. Open the vent or hollow screw on the filter. Use the hand pump to pump the fuel until it comes out of the overflow pipe.

NOTE: While adjusting the timing or checking the overflow, always make sure that the lever is at the full throttle position.

6. Rotate the crankshaft until the fuel stops dripping from the overflow pipe. One drop should fall approximately 3 seconds later. Note the position on the crankshaft pulley and then adjust the pump position so that the injection stops at top dead center.
7. Turn the crankshaft at least two more revolutions to ensure that the pump and crankshaft marks align properly.
8. Tighten the pump mounting bolts. Remove the overflow pipe and then install the pressure valve, spring and pipe connection.
9. Install the injection line and bleed the fuel system. Run the engine and check for leaks. If the fitting leaks, replace the connection and its gasket under the pressure valve carrier.

NOTE: The above information is being published from the latest information available at the time of publication. If this information differs from the data located on the underhood emissions sticker, use the data on the underhood emissions sticker.

SPARK PLUGS

Year	Model	Spark Plug Type	Gap in. (mm)
	Champion		
1985-86	190E	RS12YC	0.032 (0.08)
	380SE, 380SEC 380SEL, 380SL	N11YC	0.032 (0.08)
	500SEC, 500SEL	N9YC	0.032 (0.08)
	Bosch		
1985-86	190E	H8D, H8DC	0.032 (0.08)
	380SE, 380SEC 380SEL, 380SL	W8D	0.032 (0.08)
	500SEC, 500SEL	W7D0	0.032 (0.08)
	NGK		
1985-86	190E	①	0.032 (0.08)
	380SE, 380SEC 380SEL, 380SL	BP5EC	0.032 (0.08)
	500SEC, 500SEL	BP5EY	0.032 (0.08)

① Unavailable at time of publication.

EMISSION CONTROL SYSTEMS

CRANKCASE EMISSION CONTROL SYSTEM

Positive Crankcase Ventilation (PCV)

PCV VALVE

The PCV valve functions to prevent crankcase vapors from entering the atmosphere. Filtered air is routed into the crankcase, the vapors drawn out and routed into the intake manifold in order to be burned along with the air/fuel mixture. Should the crankcase vapors exceed the flow capacity of the PCV valve, the air flow in the system reverses and the vapors are drawn through the air cleaner element and into the carburetor to be burned along with the air/fuel mixture.

Trouble Diagnosis

When rough engine idling is encountered, check for a clogged PCV valve or a pinched or plugged hose. Inspect the PCV valve assembly for any unusual conditions or damage. Repair or replace as required. The following procedure can be followed.

1. Remove the PCV valve from the rocker arm cover.
2. Start the engine and allow it to reach normal operating temperature. Allow the engine to run at idle speed.
3. Check for vacuum at the end of the valve. If there is no vacuum at the end of the valve, inspect the system for clogged hoses or a clogged manifold port.
4. Turn off the engine and remove the PCV valve. Replace the valve using a new valve.
5. Check the systems hoses and clamps and replace any that show signs of deterioration.

EVAPORATIVE EMISSION CONTROL SYSTEM

The Evaporative Emission Control System is designed to trap fuel vapors emitted from the fuel tank during normal engine operation. This will prevent gasoline vapor discharge into the atmosphere. Gasoline vapors are absorbed through the use of a fuel vapor charcoal canister. The charcoal canister absorbs the gasoline vapors and stores them until they can be removed and burned in the engine.

Charcoal Canister

This fuel vapor canister is used to absorb and store fuel vapors emitted from the fuel tank. While the engine is idling, a small amount of fuel vapor is drawn from the canister and routed to the engine for combustion. This will allow purge of the canister through the PCV hose while the engine is idling. Certain models may use a replaceable filter located on the canister which prevents contaminants from entering the canister. Normal servicing requires that the filter element be replaced periodically.

Trouble Diagnosis

Incorrect operation of the above system will result in poor engine idling, poor driveability or engine stalling. Inspect the system for damaged charcoal canister, split or dry rotted hoses or hoses not connected. Repair or replace as necessary.

EXHAUST EMISSION CONTROL SYSTEM

The Exhaust Emission Control System for gasoline engine Mercedes-Benz vehicles includes air injection, oxygen sensor, catalytic converter, idle speed control and exhaust gas recirculation systems.

Diesel engine models use only exhaust gas recirculation (EGR) and positive crankcase ventilation (PCV) systems. The PCV system is covered under the Crankcase Emission Control section.

Air Injection System

All gasoline engine models utilize this system. The air injection system works in conjunction with the feedback system to reduce exhaust emissions and also to improve all around drivability. The system consists of an air pump, relief valve, diverter valve, air filter, check valve, delay valve, thermal vacuum valves and all connecting lines and hoses. The air is injected through ports in the cylinder head, directly into the engine.

Operation

The air injection system is used to improve the catalyst operation while the engine is in its warm up phase. Air injected only when the engine coolant temperature is below 108°F the oil temperature is above 61°F and the oxygen sensor is not in operation. When all these conditions exist at the same time, the air injection relay will provide power to the air pump clutch and the switchover valve. When the switchover valve has been activated, it will provide vacuum to open the air injection shut-off valve, thus allowing air from the pump to pass through the check valve and into the exhaust manifold.

Testing

1. Run the engine until it reaches normal operating temperature. The air pump clutch should be disengaged and the air pump should not be operating.
2. Disconnect the plug at the temperature coolant switch and then insert a jumper wire into the connector terminals. Turn the ignition switch off and on without actually starting the engine; the clutch and the switchover valve should operate.
3. Connect the negative lead of a voltmeter to terminal number 3 of the diagnostic plug located on the inner fenderwell, connect the positive lead to the battery. With the engine running, at normal operating temperature, voltage should be present. Take note of the voltmeter needle position and then turn the engine off.
4. Disconnect the connector from the magnetic clutch on the air pump and then connect the terminals to the ground and the battery. Start the engine and observe the voltmeter. The reading should be in the approximate position that it was in Step 3.
5. Disconnect the vacuum lines from the switchover valve and reconnect them to each other. The voltmeter reading should increase noticeably, if not, refer to Component Testing.

Component Testing

1. If the air pump clutch operated when the engine was running at normal operating temperature, check the coolant temperature switch using an ohmmeter. If zero resistance is shown, replace the relay.
2. With the coolant temperature at approximately 212°F and the engine at idle, unplug the oxygen sensor plug (near the exhaust pipe bracket under the engine) and the coolant temperature switch. Bridge the terminals in the temperature switch using a jumper wire. The clutch should operate, if not, check for voltage at the temperature switch connector and repair any wire as necessary.
3. Reconnect the temperature switch. With the oxygen sensor still disconnected, remove the vacuum line from the air shut off valve and connect a vacuum gauge. There should be no vacuum present. If there is vacuum unplug the temperature switch. If the vacuum drops, replace the temperature switch. If vacuum does not drop, repair or replace the wiring , relay or switch over valve as necessary.
4. Bridge the temperature switch connector terminals with a jumper wire; vacuum should be present. If not, check and repair the vacuum lines as necessary. Inspect the shut off valve for leaks.
5. Disconnect the hose leading to the air pump from the air cleaner, at the air cleaner. If there is strong suction, system testing is complete. If there is no suction, check the hoses and/or replace the air pump. If there is very light suction, disconnect the purple/blue vacuum line at the air switchover valve. If vacuum is present at the valve, replace it.

Exhaust Gas Recirculation (EGR) System

240 AND 300 SERIES ONLY

All diesel engined models are equipped with an Exhaust Gas Recirculation (EGR) system. The system reduces oxides of nitrogen (NOx) by allowing a small quantity of exhaust gas to enter the intake manifold. This in turn lowers the combustion chamber temperatures while at the same time reducing NOx formation.

The 5 cylinder system consists of an EGR valve, thermal vacuum valve, switch over valves, vacuum control valve, several orifices and connecting hoses.

The 4 cylinder system consists of a control unit, electric switch over valve, throttle valve housing, corrugated tube to return gases, several orifices and a TDC transmitter.

OPERATION

5 CYLINDER ENGINES

When the engine coolant temperature is below 104°F, no recirculation of the exhaust gas takes place. When engine coolant is above this temperature, EGR is permitted, except when the engine is at idle or full throttle. The vacuum control valve and an orifice regulate vacuum flow to the EGR valve at different engine speeds. This ensures that the EGR flow is correct.

4 CYLINDER ENGINES

With the coolant temperature below 104°F, no EGR takes place. The thermal vacuum valve opens at temperatures above this level. Vacuum is then available for the electric switchover valve. At idle speed, there is no EGR and the switchover is not energized. The EGR valve is vented externally.

Starting at speeds of 1300 rpm, the impulses from the TDC transmitter are converted into electrical signals. The switchover valve is energized and then opens. Vacuum arrives at the EGR valve and then opens the valve. The EGR valve opening is dependent upon engine load. As rpm increases, the throttle valve will open and pressure between the intake and exhaust manifolds is reduced; as a result, less exhaust gas is drawn into the intake manifold.

The vacuum control valve determines the amount of vacuum at the EGR valve depending on engine load. This amount of vacuum serves as a controlling factor for the EGR position and also for shifting of the automatic transmission.

Shortly before full throttle, a lever with a cam switches a mechanical switchover valve to external venting, thus venting EGR vacuum and eliminating all EGR operation.

At speeds above 45 mph, the impulses transmitted from the speedometer impulse transmitter are converted to a control signal. Current to the electric switchover valve is interrupted, exhaust gases are vented, and no further EGR takes place.

EGR Valve

ALL MODELS

The EGR valve is operated by vacuum signals. The valve controls the amount of recirculation, depending on the strength of the vacuum signal. A tube from the exhaust manifold provides exhaust gases to the EGR valve which is located on the intake manifold.

Thermal Vacuum Valve

ALL MODELS

The thermal vacuum valve is installed in the thermostat housing. On certain vehicles it is color coded blue. The valve is closed below 104°F and open above this temperature. When the valve is open , EGR can take place.

SWITCHOVER VALVE

ALL MODELS

The switchover valve controls vacuum flow to the EGR valve. Two valves are used in the normal EGR system (a third valve controls full throttle downshifts on non-turbocharged models). One valve vents vacuum when the engine is at idle, while the other vents vacuum when the engine is at full throttle operation. All valves are located under a cover on the cylinder head, They are actuated by a throttle linkage cam.

VACUUM CONTROL VALVE

ALL MODELS

The vacuum control valve is located on the fuel injection pump. It vents vacuum to the atmosphere as the throttle opens. With increasing engine load, vent size increases, thus decreasing the vacuum signal to the EGR valve. This reduces EGR as the engine load increases.

ORIFICE

ALL MODELS

Several vacuum restrictors are used to control vacuum levels throughout the system. These restrictors should not be changed unless adjustment is required.

THROTTLE VALVE HOUSING

4 CYLINDER ENGINES

The throttle valve housing is required to increase the amount of EGR in the partial load range. The outside diameter of the throttle valve housing is designed so that in its closed position, a small section will always remain open. This allows the engine to idle smoothly at high altitudes, with the air conditioning on and the power steering being used.

The lower half of the housing is provided with guide vanes. these vanes ensure a complete mixture of the intake air and the recirculated exhaust gas, so that each cylinder is provided with exactly the same quantity of exhaust gas.

ELECTRIC SWITCHOVER VALVE

4 CYLINDER ENGINES

The switchover valve controls vacuum to the EGR valve. It opens opens or closes by way of an electronic control unit depending on engine rpm and vehicle speed.

ELECTRIC CONTROL UNIT

4 CYLINDER ENGINES

The electronic control unit processes input data such as engine rpm from the TDC transmitter and driving speed from the impulse transmitter of the speedometer.

TDC TRANSMITTER

4 CYLINDER ENGINES

A TDC transmitter is attached to the front of the cylinder block and transmits engine rpm information to the control unit for all EGR operation.

TESTING

NOTE: The vacuum diagrams in this section are designed to be a base guide. Due to manufacturer's changes made during production, actual vacuum hose routings may differ from the illustrations. Refer to the underhood emissions sticker and use that data if it differs from the illustrations here.

EGR VALVE

4 CYLINDER ENGINES

1. With yellow orifice installed, connect a vacuum gauge between the EGR valve and the electric switchover valve. Run the engine at idle and note the reading on the vacuum gauge. If vacuum is present, check all vacuum lines for blockage and proper connection. Connect a volt/ohmmeter to the plug at the switchover valve and measure the voltage. If no voltage is present, replace the valve. If voltage is present, check all electrical wirings and components.
2. If no vacuum is present in Step 1, increase engine speed to 1300 rpm and note vacuum. If the gauge reads 9.4–10.7 in. Hg, stop the engine and connect a hand vacuum pump to the EGR valve and apply 10.7 in. Hg. Pull off the vacuum line and the EGR valve should be hard to close. If not, replace the valve. If it closes, proceed to vacuum control check.
3. If in Step 2, no vacuum, or low vacuum, was noted, check vacuum lines for proper connections or leaks. Disconnect the white and brown line from the angled connection of the thermal vacuum valve. Pull the white/purple/brown line from the EGR valve and blow through it; if no air escapes, replace the thermal vacuum valve.
4. Place your hand on the switchover valve and run the engine to 1300 rpm. The valve should switch. If not, inspect the electrical wiring and components. If it does switch, pull off the vacuum line and check for vacuum. If there is no vacuum, replace the switchover valve.

VACUUM CONTROL

EXCEPT 5 CYLINDER ENGINES

1. Connect a vacuum testor between the EGR valve and the switchover valve. Disconnect the connecting rod at the ball head. Start the engine and run to 1300 rpm. Place an adjusting roller (916 589 00 21 00) on the vacuum control valve and move the lever against its stop. Check the vacuum reading.
2. If the gauge reads 5.3–6.4 in.Hg, proceed to Step 3. If the reading is above or below this value, inspect the orifice for obstruction. If the vacuum reading was too low, install a smaller orifice. If this does not increase vacuum, or if the reading was too high and a yellow orifice was installed, replace the vacuum control valve.
3. Pull off the central plug. Connect a test light between the tapping point on the black orifice and the valve plate connection. Connect a vacuum testor to probe 3 on the central plug. Start the engine and read the vacuum gauge. If vacuum does not drop, replace the switchover valve. If vacuum is 18.8–21.5 in.Hg. at idle, and zero when the regulating linkage is moved to its full throttle stop, stop the engine.
4. Vacuum should remain constant for about 2 minutes. If not, replace the switchover valve. If vacuum holds remove the connections and proceed to step five.
5. Connect a vacuum testor between the EGR valve and the switchover valve. Operate the vehicle on a dynamometer or road, but not at full throttle. Read vacuum. If there is no vacuum above 40 mph, and vacuum is present below 40 mph the system is okay and the test is finished. If vacuum does not drop at speeds above 40 mph, proceed to Step 6.

6. Connect a voltmeter to the plug at the switchover valve and drive the vehicle. If 12 volts are present at speeds above 40 mph, check electrical wirings and components and replace any defective components. If there is no voltage, replace the switchover valve.

5 CYLINDER ENGINES

1. Connect a vacuum gauge between the thermal vacuum valve and the EGR valve using a "T" connection. With the engine idling and the throttle against its stop, no vacuum should be present. Open the throttle until all slack in the free travel rod is taken up. The vacuum level should be 5–7 in. Hg. If the system operates correctly, test the EGR valve. If not, proceed with the system check with the engine at normal operating temperature.
2. Check all vacuum lines for tight connections. Blow through the lines to ensure that none are plugged. Clean the vacuum source fitting on the pump.
3. Disconnect the vacuum line at the EGR valve, then pull off the white/purple/brown hose at the thermal vacuum valve. Blow through the hoses; if no air passes, replace the thermal vacuum valve.
4. Disconnect the vacuum hose plug at the switchover valve plate. Connect a jumper hose between the vacuum source point and point 1 at the switchover valve plug. Plug point 2 and connect a vacuum gauge to point 3.
5. With the engine idling, no vacuum should be present. Open the throttle until all slack is taken up in the free travel rod. About 10–12 in. Hg. should be present. Return the engine to idle speed, then watch the vacuum gauge. Vacuum should remain constant for at least 2 minutes.
6. Disconnect the plug from point 2. Vacuum should drop to zero. If the valve does not operate properly, replace it. Connect a vacuum source jumper hose to point 2 and plug point 1. Leave the vacuum gauge connected to point 3.
7. With the engine idling, about 10–12 in. Hg. vacuum should be present. Clamp a vacuum source jumper hose tightly and stop the engine. Vacuum should remain constant for 2 minutes. Move the throttle against the full throttle stop, pull off the hose at point 1 and observe the vacuum gauge.
8. Vacuum should remain constant. Return the throttle to idle and pull off the jumper hose. Vacuum should drop to zero. If the switch does not operate properly, replace it. If the switch is okay, check the EGR valve.

EGR VALVE

Start the engine. With the engine at idle, eliminate all slack in the free travel rod. Disconnect and connect the vacuum line at the EGR valve and listen for valve operation. If the valve does not operate, replace it.

VACUUM CONTROL VALVE

1. Connect a vacuum gauge between the thermal vacuum valve and the EGR valve. Increase engine speed to 900 rpm (do not pull on STOP lever).
2. Vacuum should measure 5.7–6.7 in. Hg. on non-turbo models, and 4.9–5.7 in. Hg. on turbo models. If not, check the orifice to ensure that it is clean. If the level is too high, install a larger orifice. If the level is too low, install a smaller orifice. If the vacuum level cannot be correctly adjusted, replace the vacuum control valve.

AIR FLOW SENSOR

Adjustment

NOTE: If the air flow sensor is replaced on gasoline engine models, the following procedure should be used to adjust the on-off ratio.

1. Run the engine until it reaches normal operating temperature. Remove the purge line at the purge valve and plug it. Run the engine at 2500 rpm and determine the average on-off ratio of the frequency valve.
2. Run the engine at 500 rpm and set the frequency valve on-off ratio to the value found in the above step, if different. Repeat these steps until the average difference in the frequency valve on-off ratio is less than 10. Reconnect the purge line.

OXYGEN SENSOR

Replacement of the Oxygen Sensor

The oxygen sensor must be replaced at 30,000 mile intervals. When the reminder light on the dashboard is lit, disconnect the oxygen sensor plug (slide the plug out of its holding clip if necessary). Loosen the cable clamps and remove the sensor. Apply anti-seize compound to the threads of the new sensor and install. Tighten to 37–44 ft. lbs.

REMINDER LIGHT

Reset

380 AND 500 SERIES

1. The instrument cluster must be partially removed on certain models. Using a steel wire with a small hook on the end, slip the wire between the right side of the cluster and the dashboard. Turn the hook to engage the cluster and the dashboard. Turn the hook to engage the cluster and gently pull the edge of the cluster out of the retaining clips.
2. Remove the oxygen sensor bulb at the extreme lower corner of the cluster. Press the cluster back into position. No reset switch is provided.

VACUUM CIRCUITS

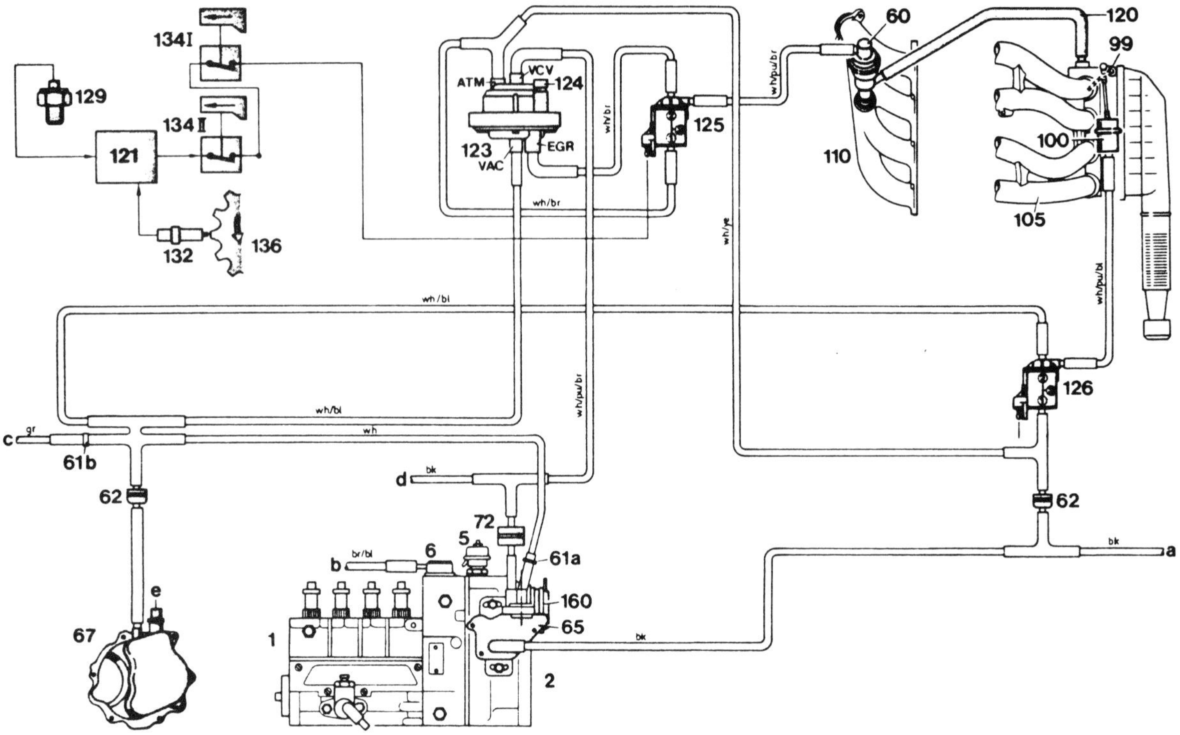

1. Injection pump
2. Governor
5. Altitude compensation capsule
6. Vacuum shut-off unit
60. EGR valve
61a. Orifice
61b. Orifice 0.5 mm
62. Vent filter
65. Vacuum control valve
67. Vacuum pump
72. Vacuum damper (autom. transmission only)
99. Vacuum control flap
100. Actuator for (99)
105. Intake manifold
110. Exhaust manifold
120. Exhaust gas return line
121. Control unit
123. Vacuum amplifier
124. Adjusting screw for (123)
125. Switchover valve, electric, EGR valve
126. Switchover valve, electric, vacuum control flap
129. Coolant temperature switch (100°C)
131. Speed sensor from speedometer (manual transmission only)
132. Rpm sensor, engine
134/I. Microswitch
134/II. Microswitch
136. Ring gear
140. Check valve—brake booster
160. Magnetic actuator, injection pump

a = Vent line to passenger compartment
b = Key, engine stop
c = Remaining consumers
d = Automatic transmission

VAC = Vacuum
VCV = Vacuum control valve
EGR = Exhaust gas recirculation
ATM = Atmosphere

bk = black
bl = blue
br = brown
gr = green
pu = purple
re = red
ye = yellow

Function diagram for automatic and manual transmissions

VACUUM CIRCUITS

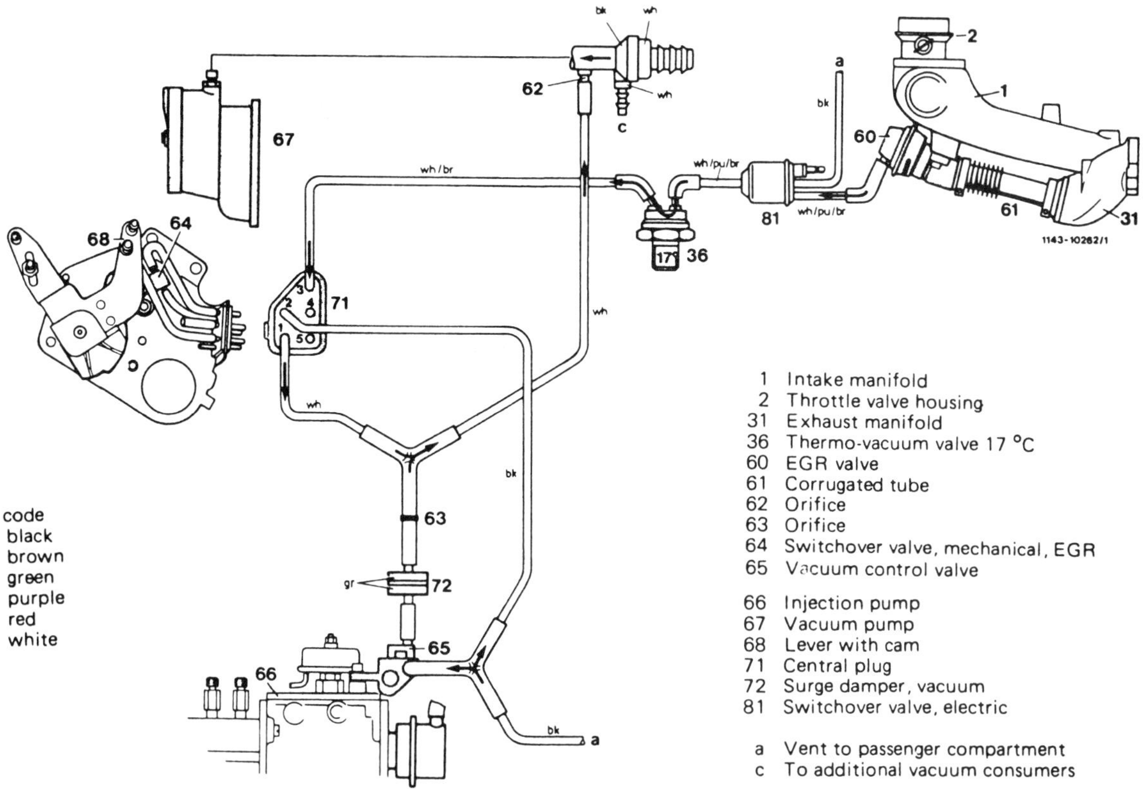

1 Intake manifold
2 Throttle valve housing
31 Exhaust manifold
36 Thermo-vacuum valve 17 °C
60 EGR valve
61 Corrugated tube
62 Orifice
63 Orifice
64 Switchover valve, mechanical, EGR
65 Vacuum control valve
66 Injection pump
67 Vacuum pump
68 Lever with cam
71 Central plug
72 Surge damper, vacuum
81 Switchover valve, electric

a Vent to passenger compartment
c To additional vacuum consumers

Color code
bk = black
br = brown
gr = green
pu = purple
re = red
wh = white

Mercedes-Benz 240D—diesel (man. trans.)

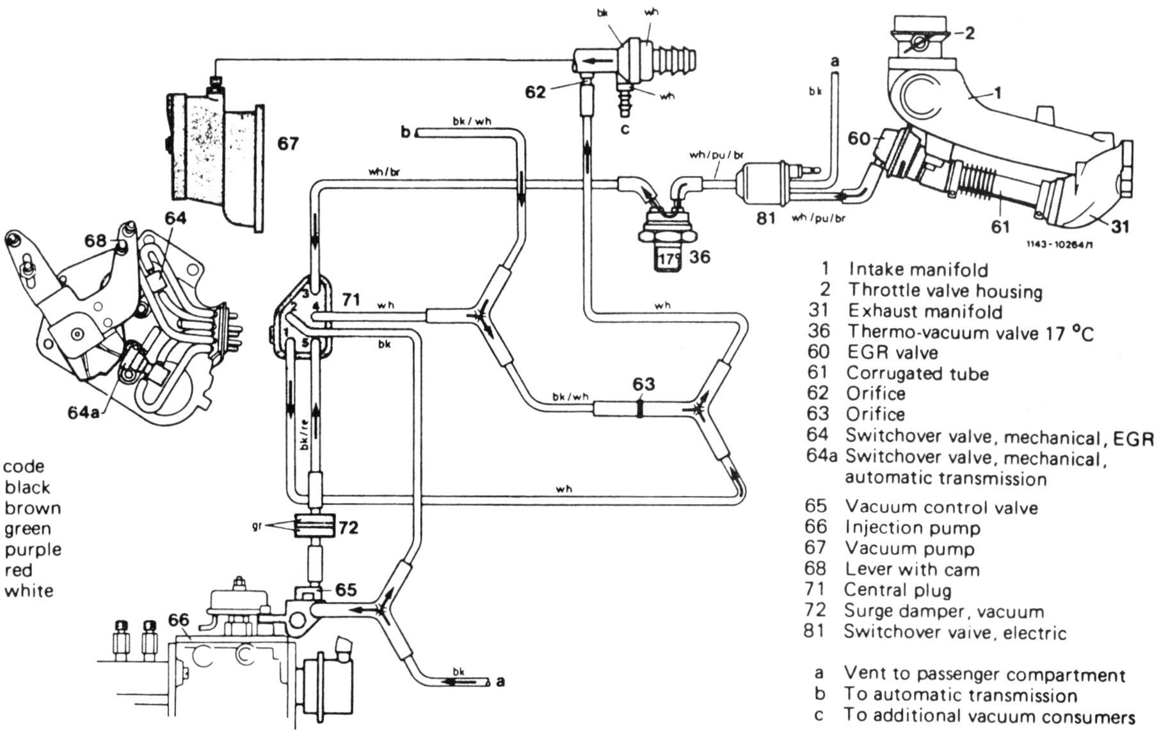

1 Intake manifold
2 Throttle valve housing
31 Exhaust manifold
36 Thermo-vacuum valve 17 °C
60 EGR valve
61 Corrugated tube
62 Orifice
63 Orifice
64 Switchover valve, mechanical, EGR
64a Switchover valve, mechanical, automatic transmission
65 Vacuum control valve
66 Injection pump
67 Vacuum pump
68 Lever with cam
71 Central plug
72 Surge damper, vacuum
81 Switchover valve, electric

a Vent to passenger compartment
b To automatic transmission
c To additional vacuum consumers

Color code
bk = black
br = brown
gr = green
pu = purple
re = red
wh = white

Mercedes-Benz 240D—diesel (auto. trans.)

VACUUM CIRCUITS

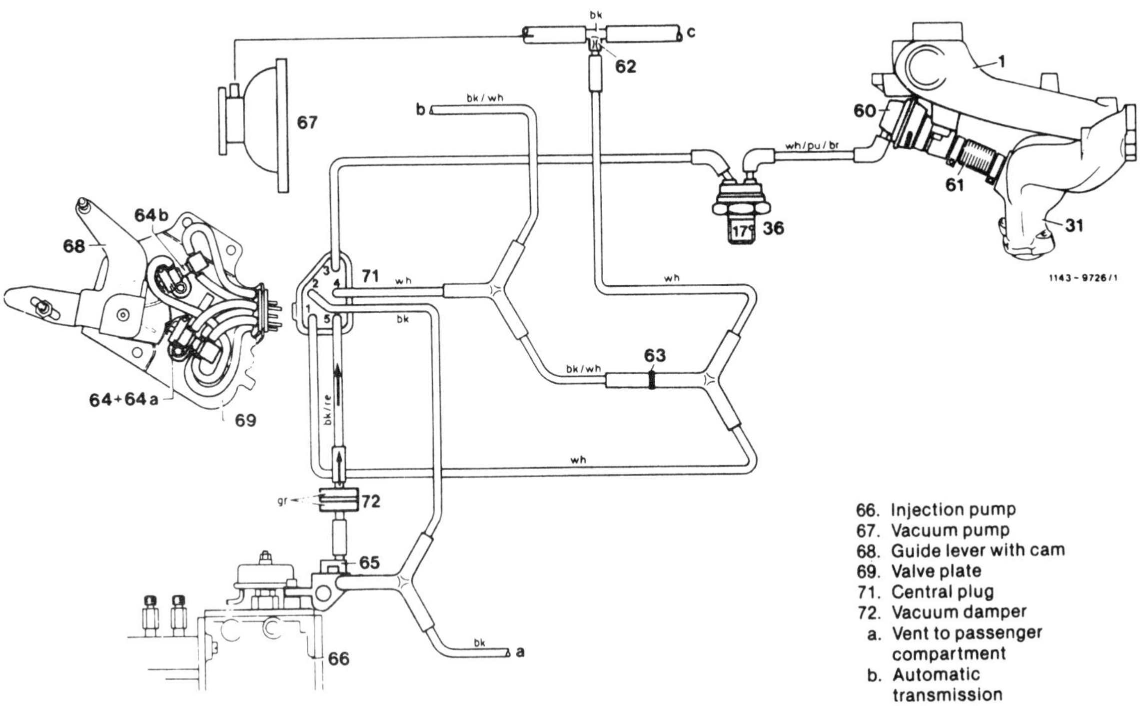

1. Intake manifold
31. Exhaust manifold
36. Thermo valve 17°C/ 63°F
60. EGR valve
61. Corrugated tubing
62. Orifice
63. Orifice
64. Switchover valve, automatic transmission
64a. Switchover valve, idle speed shutoff—EGR
64b. Switchover valve, full throttle shutoff—EGR
65. Vacuum control valve
66. Injection pump
67. Vacuum pump
68. Guide lever with cam
69. Valve plate
71. Central plug
72. Vacuum damper
a. Vent to passenger compartment
b. Automatic transmission
c. Brake unit

Color code:
bk = black
br = brown
gr = green
pu = purple
re = red
wh = white

Mercedes-Benz 300—diesel (wo/turbo)

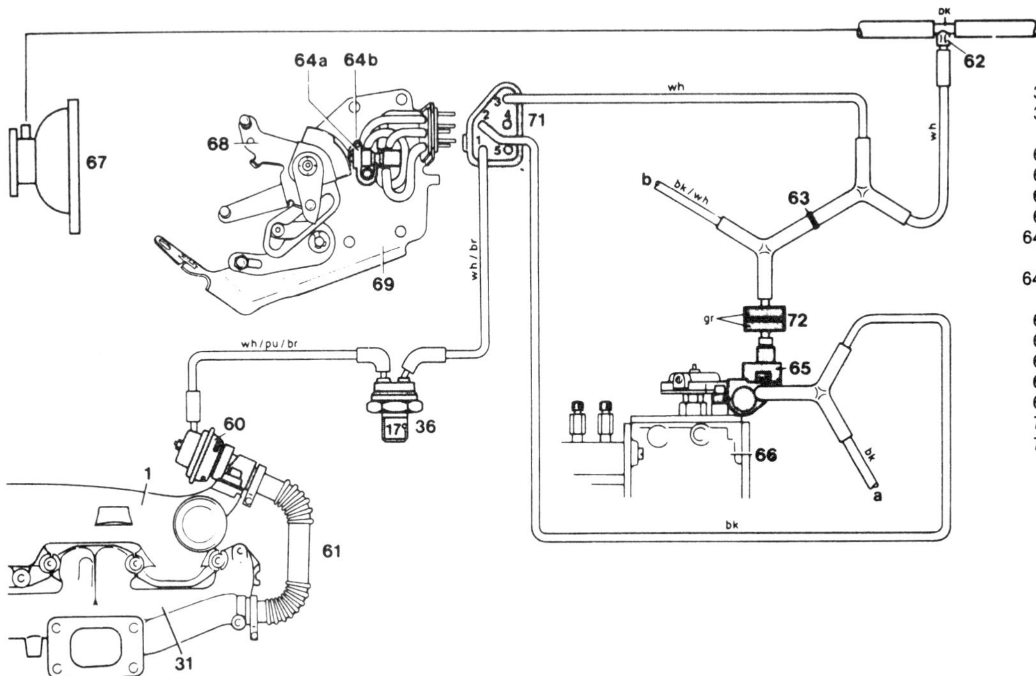

1. Intake manifold
31. Exhaust manifold
36. Thermo valve 17°C/ 63°F
60. EGR valve
61. Corrugated tubing
62. Orifice
63. Orifice
64a. Switchover valve, idle speed shutoff—EGR
64b. Switchover valve, full throttle shutoff—EGR
65. Vacuum control valve
66. Injection pump
67. Vacuum pump
68. Guide lever with cam
69. Valve plate
71. Central plug
72. Vacuum damper
a. Vent to passenger compartment
b. Automatic transmission
c. Brake unit

Color code:
bk = black
br = brown
gr = green
pu = purple
re = red
wh = white

Mercedes-Benz 300 diesel w/turbo

VACUUM CIRCUITS

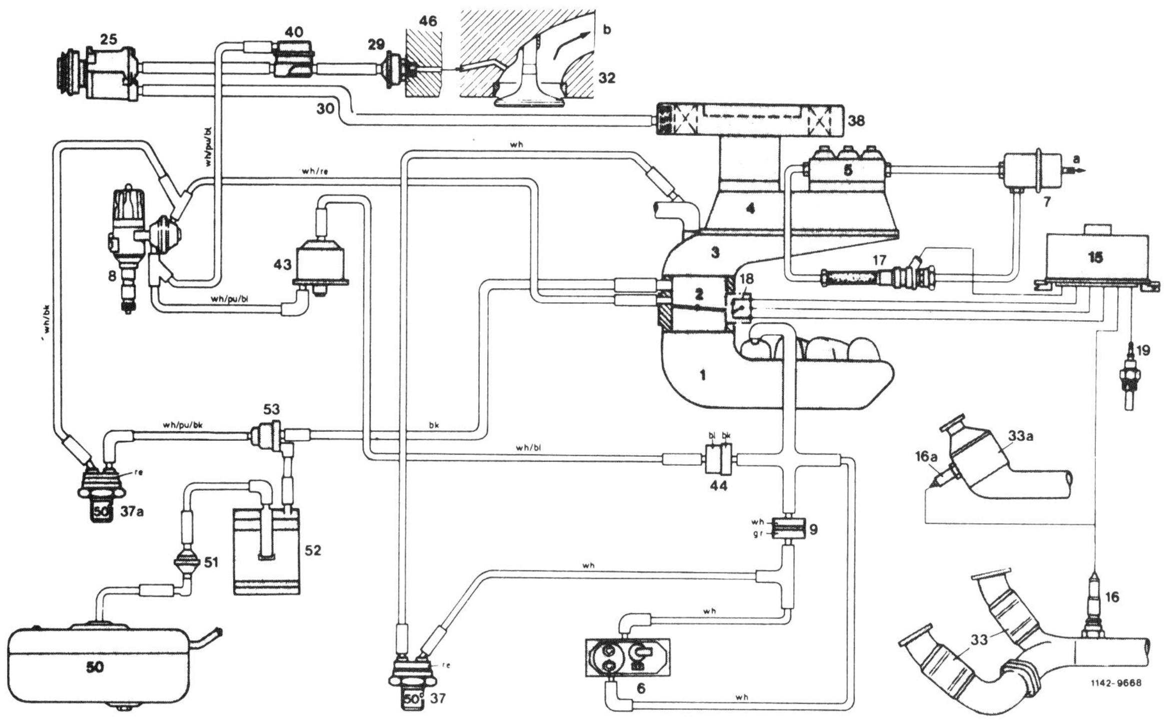

1. Intake manifold
2. Throttle valve housing
3. Aire duct housing
4. Air flow sensor
5. Fuel distributor
6. Warm-up compensator
7. Silencer (damper)
8. Ignition distributor
9. Orifice
15. Control unit
16. Oxygen sensor (model 107)
16a. Oxygen sensor (model 126)
17. Frequency valve
18. Throttle valve switch
19. Temperature switch, oil 16°C/60°F
25. Air pump
29. Check valve (injected air)
30. Air injection line
32. Cylinder head
33. Primary catalyst (model 107)
33a. Primary catalyst (model 126)
37. Thermo valve 50°C/122°F
37a. Thermo valve 50°C/122°F
38. Air cleaner
40. Air injection shutoff valve
43. Switchover valve
44. Check valve (vacuum)
46. Timing housing cover
50. Fuel tank
51. Vent valve
52. Charcoal canister
53. Purge valve
a. Leak-off connection
b. To exhaust manifold

Color code
- bk = black
- bl = blue
- gr = green
- pu = purple
- re = red
- wh = white

Mercedes-Benz 380 series

VACUUM CIRCUITS

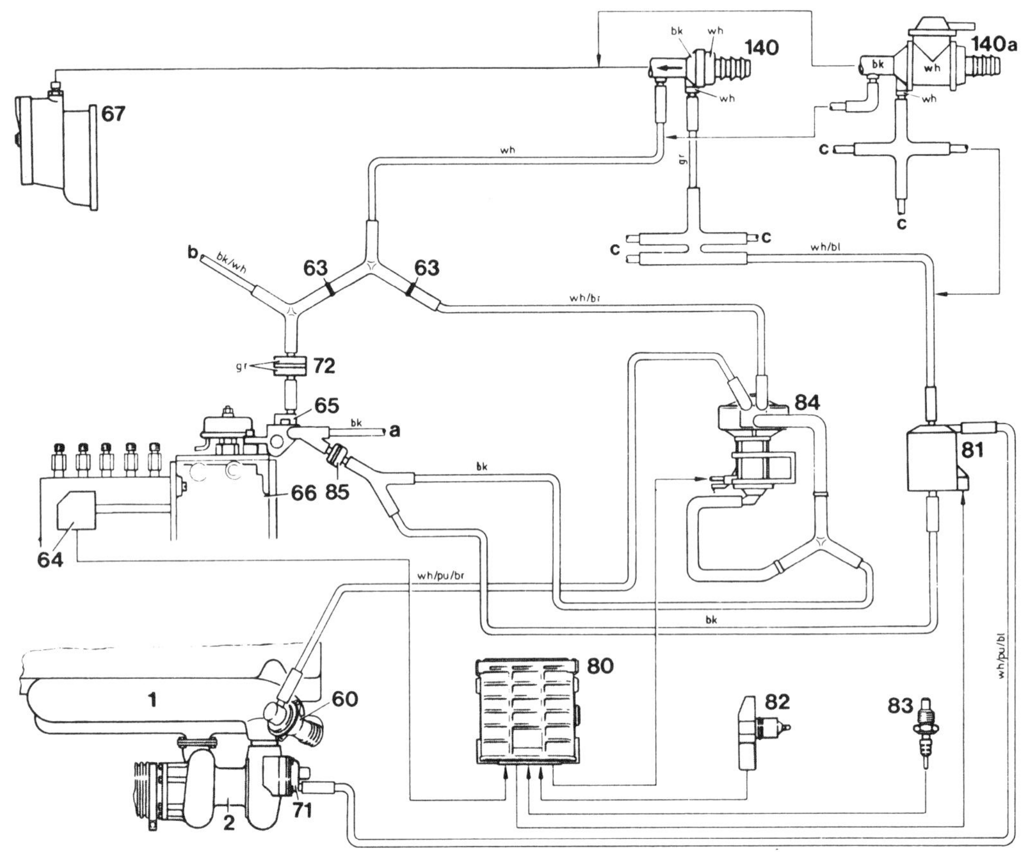

1	Intake manifold	71	Recirculating air safety valve	140	Check valve, model 123	bk = black
2	Exhaust gas turbocharger	72	Vacuum damper	140a	Check valve, model 126	bl = blue
60	EGR valve	80	Control unit			br = brown
63	Throttle (orifice) 0.5 mm	81	Switchover valve	a	Vent line to passenger compartment	gr = green
64	Fuel rack position sensor	82	Rpm sensor	b	To automatic transmission	pu = purple
65	Vacuum control valve	83	Temperature sensor, coolant	c	Remaining consumers	re = red
66	Injection pump	84	Vacuum converter			wh = white
67	Vacuum pump	85	Positive vent filter			

Emission control system—5 cylinder diesel engines

Merkur

INDEX

CAR SERIAL NUMBER AND ENGINE IDENTIFICATION

Vehicle Identification Number (VIN)

The vehicle identification number (VIN) is stamped on a metal tag, for registration purposes, and is fastened to the instrument panel close to the windshield on the driver's side.

A vehicle certification label is affixed on the left left front door post. The upper half of the label contains the name of the manufacturer, month and year of manufacture, gross vehicle weight rating (GVWR), gross axle weight rating (GAWR) and the certification statement.

The seventeen character VIN number is used for warranty purposes and indicates manufacturer, type of restraint system, line, series, body type, engine, model year and consecutive unit number. The remaining portion of the vehicle certification label is restricted to coded information as to the color and body type, the interior trim, air conditioning, radio type, sun roof type as well as axle, transmission, springs, district and special order codes.

NOTE: Some of the listed codes are not used if the vehicle is not equipped with specific accessories.

W F I	B	P 8 0	W	1	F	M	6 0 6 7 3 3
Positions 1-3	Position 4	Positions 6-7	Position 8	Position 9	Position 10	Position 11	Positions 12-17
World Manufacturer Identifier	Restraint System Type	Line Series, Body Type	Engine Type	Check Digit	Model Year	Assembly Plant	Production Sequence Number
	Passenger Car Only	Constant-P (Passenger)			1985	Karmann	
	Active Belt-B Passive Belt-P						

Example of Merkur XR4Ti vehicle identification number decoding (© Ford Motor Co.)

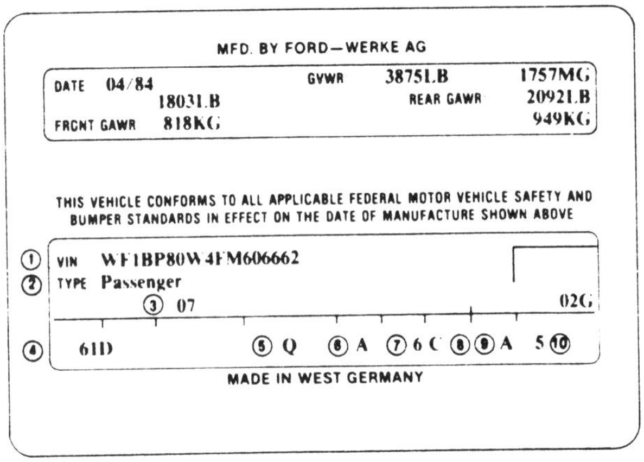

1. Vehicle Identification Number
2. Vehicle Type
3. Paint
4. Body Type Code
5. Interior Trim
6. Air Conditioning
7. Radio
8. Sunroof
9. Axle Ratio
10. Transmission

Example of Merkur XR4Ti vehicle certification label (© Ford Motor Co.)

Engine Identification Number

The engine code is included in the vehicle identification number and is identified by the letter 'W' representing the 2.3 Liter, OHV, fuel injected engine.

TUNE-UP SPECIFICATIONS

1985–86 MERKUR XR4Ti

Spark Plugs—BSFC-32, AWSF-32 (14mm taper seat)
Gap—0.034 in. (0.086mm)
Ignition timing—13 BTDC (Manual Transmission), 10 BTDC (Automatic Transmission)

Coil Primary Resistance—0.31–1.0 ohms
Coil Secondary Resistance—8,000–11,500 ohms
Ignition Coil reserve—24,000 volts
Secondary wire resistance—5,000 ohms/inch maximum
Engine idle speed—900 ± 75 rpm (Electronically controlled)

TUNE-UP SPECIFICATIONS

Year	Model	Engine Displacement cu. in. (cc)	Spark Plugs Gap (in.)	Distributor Point Dwell (deg)	Distributor Point Gap (in.)	Ignition Timing MT (deg)	Ignition Timing AT (deg)	Idle Speed MT (rpm)	Idle Speed AT (rpm)	Valve Clearance Intake (in.)	Valve Clearance Exhaust (in.)
'85	XR4Ti	140 (2.3)	.034	Electronic	Electronic	13①	10①	900①	900①	Hyd.	Hyd.
'86	See Underhood Specifications Sticker										

NOTE: Refer to Emissions/Tuneup information provided in folder under the hood. Use the "specs" provided if they differ from the chart. Production changes and calibrations are reflected in the "specs" provided in the folder.
① Controlled by EEC-IV system. See text.

FIRING ORDER

1–3–4–2 ——— Front to rear.

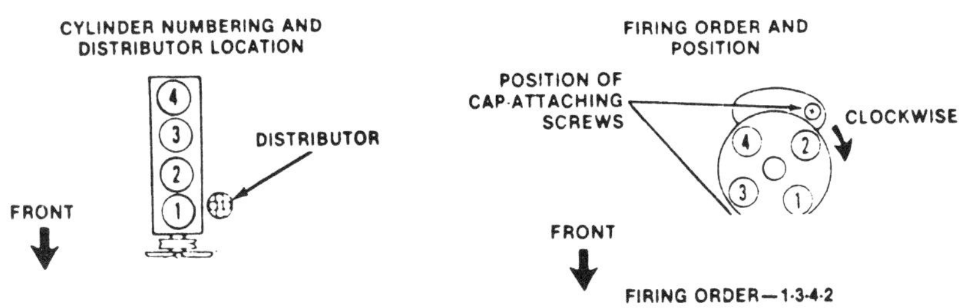

Firing order, cylinder numbering sequence and distributor location (© Ford Motor Co.)

FIRING ORDERS

NOTE: To avoid confusion when replacing spark plug wires, always remove them one at a time.

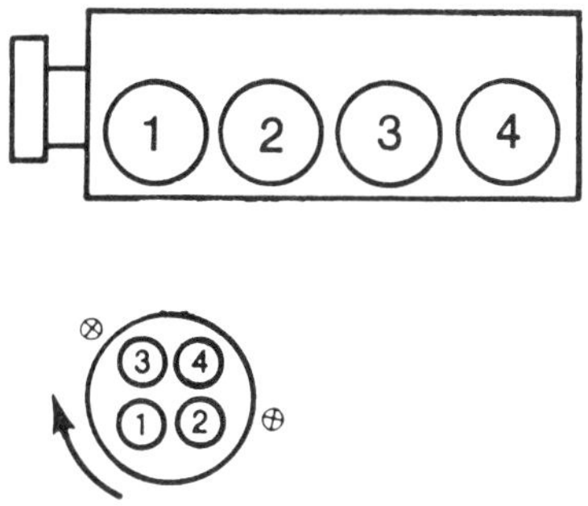

FORD MOTOR CO. 2300 cc 4-cyl.
Engine firing order: 1-3-4-2
Distributor rotation: clockwise

EMISSION EQUIPMENT USED

Positive Crankcase Ventilation System
Oil separators mounted to the valve cover and engine block
Fuel Evaporation Control system
Pressure/vacuum relief fuel cap
Fuel tank vapor venting
Carbon canister with hoses routed to air cleaner assembly
Exhaust Control System
Exhaust Gas Recirculation System
Catalytic Converter (TWC)
Electronic Ignition System (ECC-IV)
Electronic Fuel Injection System
Microprocessor Control Unit (MCU)
Universal Distributor with changeable octane rod (TFI–IV)
Turbo Charger
Electronic Timing Advance/retard knock sensor
Exhaust gas oxygen sensor

NOTE: Emission information appears on the Vehicle Emission Control Information decal, located in the service information folder, mounted on the left-hand side of the hood's inner panel. This decal identifies the engine displacement and provides certain tune-up specifications.

REQUIRED EMISSION CONTROL MAINTENANCE

MILES (000)	5	10	15	20	25	30	35	40	45	50
KM (000)	8	16	24	32	40	48	56	64	72	80
EMISSION CONTROL SERVICE										
Change Engine Oil (every 12 months) OR	X	X	X	X	X	X	X	X	X	X
Change Oil Filter (every 12 months) OR	X	X	X	X	X	X	X	X	X	X
Replace Spark Plugs			(X)			X			(X)	
Check Accessory Drive Belt(s)	(X)					X				
Replace Air Cleaner Filter						X				
Change Engine Coolant (every 36 months) OR						X				
Check Engine Coolant Protection	ANNUALLY									

MERKUR (FORD)
REQUIRED EMISSION MAINTENANCE

	MILES (000)	5	10	15	20	25	30	35	40	45	50
	KM (000)	8	16	24	32	40	48	56	64	72	80
GENERAL MAINTENANCE											
Rotate Tires		X		X			X			X	
Check Exhaust Heat Shields							X				
Lubricate Steering and Front Suspension							X				

(X) All items with either a "X" or a "(X)" code are required to be performed in all states except California. For cars sold in California, only "X" items are REQUIRED to be performed. However, Ford recommends that you also perform maintenance on items designated by a "(X)" in order to achieve best vehicle operation.

If driving habits include use of the vehicle on a daily basis for several miles and none of the Unique Driving Conditions, perform the maintenance as prescribed on the chart.

MILES (000)	3	6	9	12	15	18	21	24	27	30	33	36	39	42	45	48
KM (000)	4.8	9.6	14.4	19.2	24	28.8	33.6	38.4	43.2	48	52.8	57.6	62.4	67.2	72	76.8
EMISSION CONTROL SERVICE																
Change Engine Oil (every 3 months) OR	X	X	X	X	X	X	X	X	X	X	X	X	X	X	X	X
Change Engine Oil Filter (every 3 months) OR	X	X	X	X	X	X	X	X	X	X	X	X	X	X	X	X
Spark Plugs: Check/Regap		X		X		X		X				X		X		X
Spark Plugs: Replace					(X)					X					(X)	
Check Accessory Drive Belt(s)	(X)									X						
Replace Air Cleaner Filter (1)										X						
Replace Engine Coolant, Check Hoses & Clamps EVERY 36 Months OR										X						
Check Engine Coolant Protection	A N N U A L L Y															
GENERAL MAINTENANCE																
Rotate Tires	X				X					X					X	
Inspect Exhaust Heat Shields										X						
Change Automatic Transmission Fluid (2)										X						
Lubricate Steering and Front Suspension Linkage										X						

(1) If operating in severe dust, suggest proper replacement intervals.
(2) If your vehicle accumulates 5,000 miles (8000 km) or more per month or is used in CONTINUOUS stop and go service, change every 30,000 miles (48000 km)—not necessary for severe dust, short trips or extensive idling.
(X) All items with either a "X" or a "(X)" code are required to be performed in all states except California. For cars sold in California, only "X" items are REQUIRED to be performed. However, Ford recommends that you also perform maintenance on items designated by a "(X)" in order to achieve best vehicle operation.

If driving habits frequently include the following, perform the maintenance as prescribed on the chart.

1. Operating when outside temperatures remain **below freezing** and most trips are less than 5 miles (8 km)
2. Operating during HOT WEATHER (above 90°F, 32°C) and;
   a. Driving continuously in excess of normal highway speeds
   b. Driving in stop-and-go "rush hour" traffic
3. Using a car-top carrier or otherwise carrying heavy loads
4. Operating in severe dust conditions
5. Extensive idling, such as police, taxi or door-to-door delivery use

IDLE SPEED AND MIXTURE ADJUSTMENT

IDLE SPEED

Adjustment

2.3L EFI TURBO

NOTE: This procedure is to be performed only if the curb idle is not within the 900 ± 75 rpm specifications. The curb idle speed (rpm) is controlled by the EEC–IV processor and the idle speed control air bypass valve assembly. If the engine cub idle rpm is not within specifications after performing this procedure, further diagnosis must be performed on the Electronic System.

1. Position the transmission in Neutral and turn the A/C–HEAT selector to the OFF position.
2. Have the engine at normal operating temperature and shut down.
3. Disconnect the idle speed control air bypass valve power lead.
4. Start the engine and operate at 2000 rpm for 120 seconds.
5. If the electric cooling fan begins to operate during the idle speed set procedure, disconnect the power lead and proceed.
6. Allow the engine to idle and check the base idle speed rpm of 750 ± 50.
7. Adjust the engine rpm to specifications by turning the throttle valve plate stop screw, located on the air intake charge throttle body assembly.
8. Turn off the engine and reconnect the power lead to the idle speed control air bypass valve and to the electric cooling fan, if removed.
9. Verify that the throttle plate is not stuck in the throttle body bore by moving the throttle plate control.

NOTE: If excessive engine idle speeds are experienced when driving the vehicle on the idle system, turn the ignition switch to the OFF position and restart.

AIR MIXTURE

Adjustment

No adjustment of the air/fuel mixture is provided. The injected fuel mixture is controlled by the microprocessor control unit, part of the Electronic Engine Control system, through input sensors reaction to specific engine requirements.

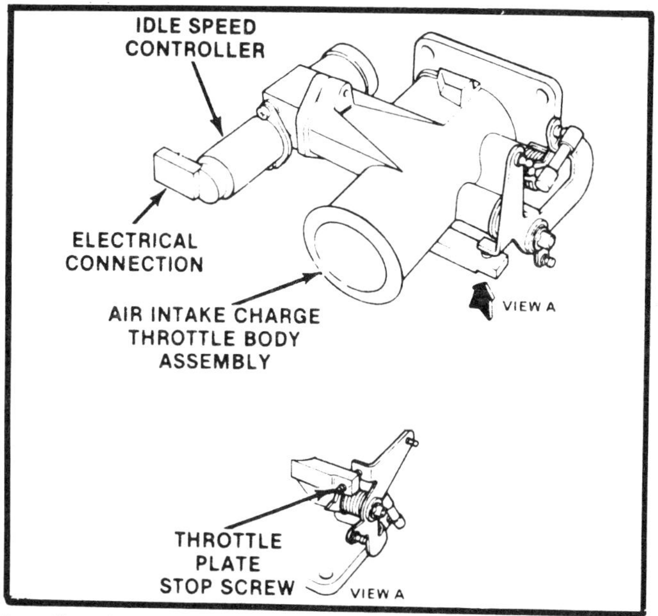

Location of idling speed adjusting screw (© Ford Motor Co.)

INITIAL TIMING

Adjustment

NOTE: Make all adjustments with the engine at normal operating temperature, the transmission in NEUTRAL (manual) or PARK (automatic), parking brake applied, wheels blocked and all accessories turned off.

1. With the engine off, clean and highlight the timing mark on the crankshaft pulley and on the belt cover. Connect a timing light and tachometer to the engine.
2. Disconnect the single wire connector near the distributor and start the engine.
3. The engine is equipped with an electronic idle speed control and the engine should be running at 900 ± 75 rpm.

NOTE: If idle adjustment is necessary, refer to the Idle Speed Adjustment outline for correct procedures.

4. With the aid of the timing light, verify that the ignition timing is 10 degrees (Manual transmission) or 13 degrees (automatic transmission) BTDC.

NOTE: If the underhood Vehicle Emission Control Information decal specifications differ from this procedure, follow the decal instructions, as it will reflect manufacture product changes.

5. If necessary to adjust the ignition timing, loosen the distributor hold-down bolt and rotate the distributor housing either clockwise or counterclockwise until the correct timing is achieved. Tighten the hold-down bolt (5.9–8.5 ft. lbs.) and recheck the ignition timing.

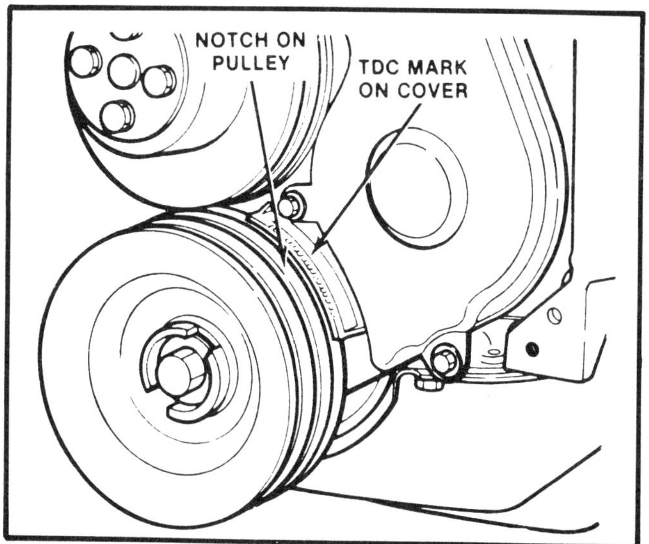

Engine timing scale (© Ford Motor Co.)

6. After the timing has been verified, stop the engine and remove all test equipment. Reconnect the single wire connector at the distributor.

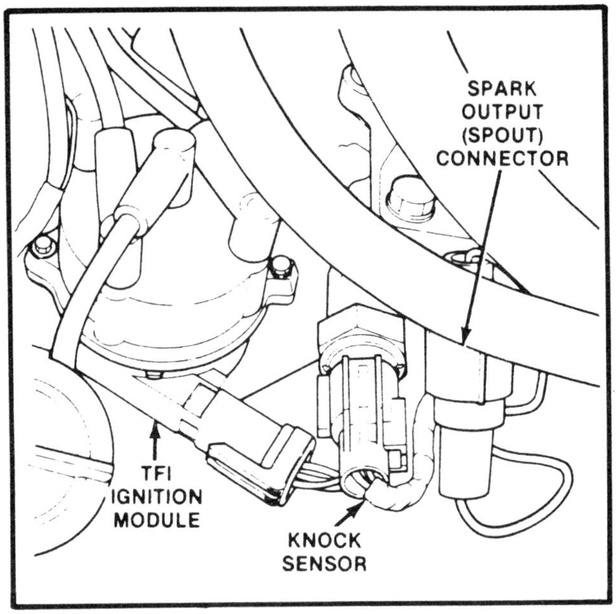

Spark output (spout) connector (© Ford Motor Co.)

EMISSION CONTROL SYSTEMS

SPARK PLUGS

NOTE: The following spark plug type and gap specifications are obtained from the latest information available at time of publication. If this information differs from the information listed on the Vehicle Emission Control Information Decal, use the data from the Emission decal.

Factory recommended Spark Plugs—BSFC-32, AWSF-32-C

Factory Recommended Gap—0.034 in. (0.086mm)

CRANKCASE EMISSION CONTROL SYSTEM

Positive Crankcase Ventilation (PCV) System

The PCV system is considered a closed induction system, with the air inlet located on the turbo housing and the air outlet located on the upper intake manifold. Two oil separators are used, one located on the valve cover and the second located on the side of the engine block. An internal oil baffle plate is used to prevent excessive oil splash on the block oil separator and is bolted between the 3rd and 4th main bearing webs. The system operates on the pressure/vacuum principle, with no PCV flow valve used.

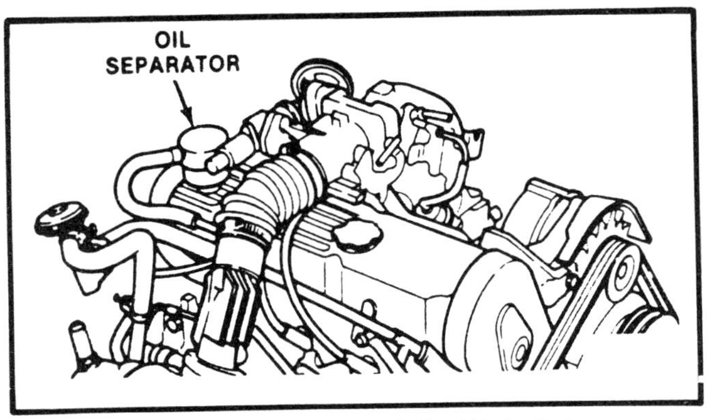

Oil separator located on valve cover (© Ford Motor Co.)

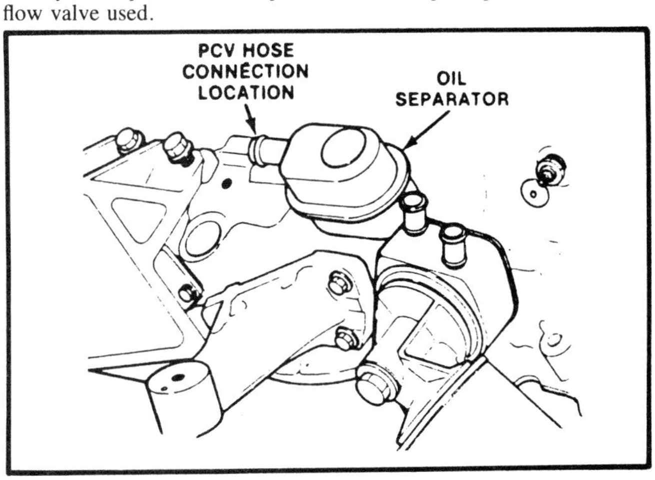

Oil separator located on engine block. Note its location near the oil filter assembly (© Ford Motor Co.)

EVAPORATIVE EMISSION CONTROL SYSTEM

CARBON CANISTER

A carbon canister is used to collect the fuel vapors from the fuel tank and to store them until the engine has been started. After the engine is operating, the canister is purged of stored vapors via connecting hoses to the induction air cleaner system of the engine.

PRESSURE/VACUUM RELIEF FUEL CAP

The fuel cap contains an integral pressure and vacuum relief valve. The vacuum valve acts to allow air into the fuel tank to replace the fuel as sit is used.The vacuum valve will open after a vacuum of 0.5 psi is attained. The pressure valve acts as a backup pressure relief valve in the event the normal venting system is overcome by excessive generation of internal pressure or restriction of the system. The pressure relief valve operating range is 1.6-2.1 psi.

FUEL TANK VAPOR VENTING

Fuel vapors in the fuel tank are vented to the carbon canister through the vapor valve assembly. The valve is located in a central location in the upper surface of the fuel tank. A vapor space between the fuel level and the tank upper surface is combined with a small orifice and front float-off valve in the vapor valve assembly to prevent liquid fuel from passing to the carbon canister. The vapor space also allows for thermal expansion of the fuel.

FUEL FILLER NECK RESTRICTOR

A fuel filler neck restrictor is used to prevent the use of all other fuels except unleaded gasoline.

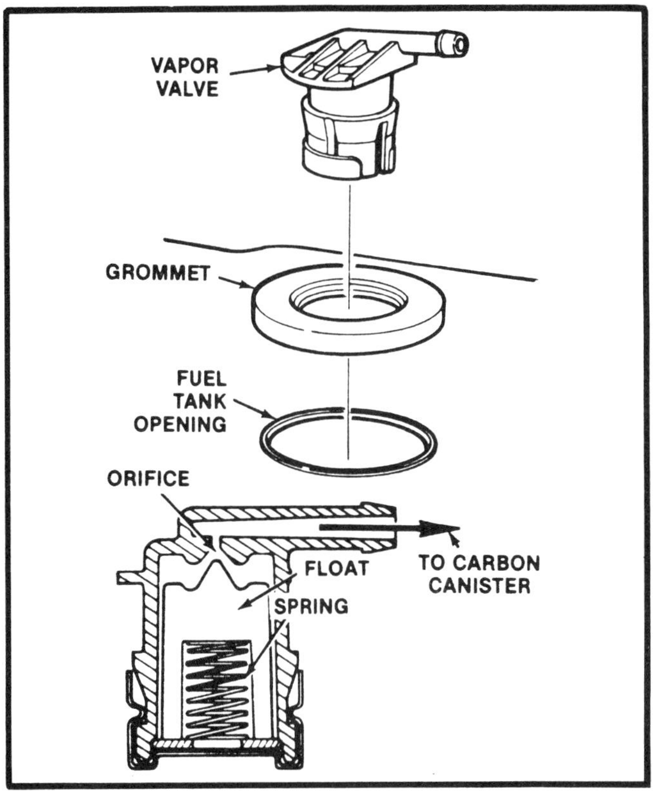

Cross section of vapor vent valve assembly (© Ford Motor Co.)

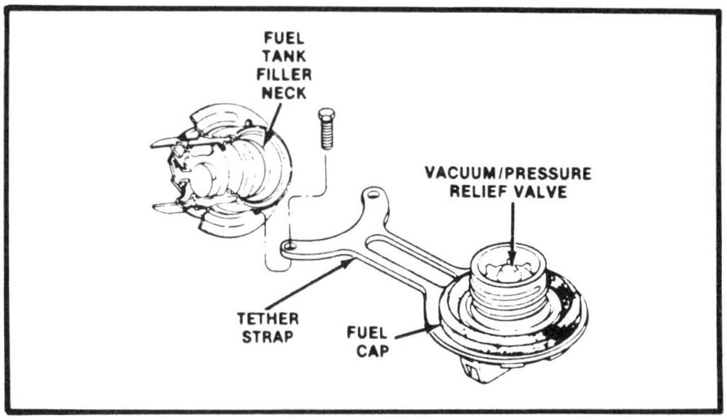

Vacuum/pressure relief fuel cap assembly (© Ford Motor Co.)

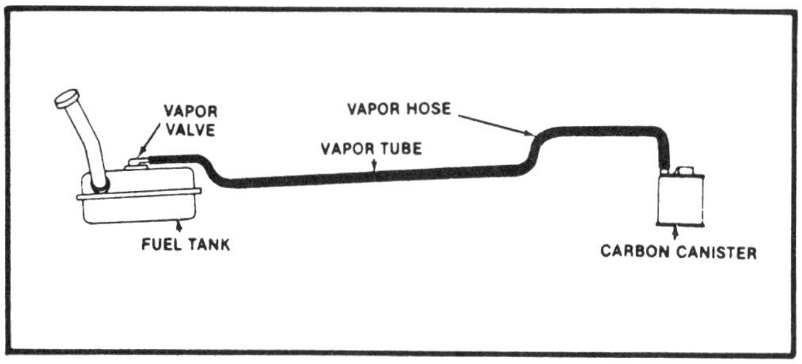

Evaporative Emission Control system hose routing and connections from fuel tank to carbon canister (© Ford Motor Co.)

EXHAUST EMISSION CONTROL SYSTEM

Exhaust Gas Recirculation (EGR) System

A vacuum operated EGR valve is used to allow exhaust gases to recirculate into the induction system along with the air/fuel mixture, thereby lowering the combustion temperatures. An EGR control solenoid is used to direct ported vacuum to the EGR valve. The solenoid operates a specific time after the engine starts. With a higher coolant temperature at start, the time delay is shorter. The solenoid closes the vacuum port to the EGR valve at times of high temperature, high load (boost) and high engine speed.

Catalytic Converter

The catalytic converter is an emission control device added to the exhaust system to chemically alter elements contained in the exhaust gases. The catalytic converter contains ceramic honeycomb (monolithic) that is coated with a rhodium/platinum catalyst. These chemicals will react with oxides of nitrogen (NOx), unburned hydrocarbons (HC), carbon monoxide (CO), to form nitrogen (N2), water vapor (H2O) and carbon dioxide (CO2). Because the catalytic converter is able to chemically alter these three elements, it is referred to a a three-way catalyst (TWC). During the chemical process, additional heat is generated. Heat shields are installed between the converter and the vehicle body for this reason and must not be removed.

Electronic Engine Control (EEC) System

The engine ignition system has three distinct operating areas. Two of these areas have remained basically the same through out the evolution of electronic engine control systems, but have been refined to fit into the world of electronics. These two systems are the primary and secondary ignition systems. The third area is the electronic replacement of the breaker points, along with added sensors to monitor engine operating information, road speed and other pertinent operating information of the vehicle, imputing this information into a micro-processor called the Electronic Control Assembly or ECA. This information is then processed and commands are sent out to varied components of the engine, directing them to perform in a prescribed manner for greater engine and vehicle operating efficiency.

Since the introduction of the Electronic Engine Control system, much refinement has been done, allowing for more ignition and fuel control to be included into the Electronic Control Assembly, requiring the use of added sensor equipment on the vehicle. In the EEC–IV Electronic Engine Control System used on the Merkur, the following components are required.

1. Exhaust gas oxygen sensor (EGO).
2. Idle speed controller.
3. Throttle positioner sensor.
4. Exhaust gas recirculation valve (EGR).

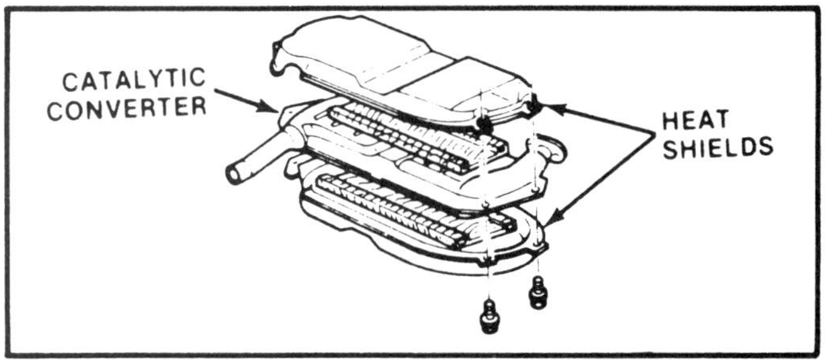

Typical catalytic converter (© Ford Motor Co.)

5. Exhaust gas recirculation control solenoid.
6. Knock sensor.
7. Thick film ignition (TFI) module.
8. Engine coolant temperature sensor (ECT).
9. Carbon canister.
10. Vane air meter/air charge temperature sensor.
11. Turbocharger boost control solenoid.
12. Wide open throttle A/C cutout relay.
13. Electronic Control Assembly (ECA) self-test connector.
14. Barometeric absolute pressure sensor (BAP).
15. Electronic control assembly (ECA).

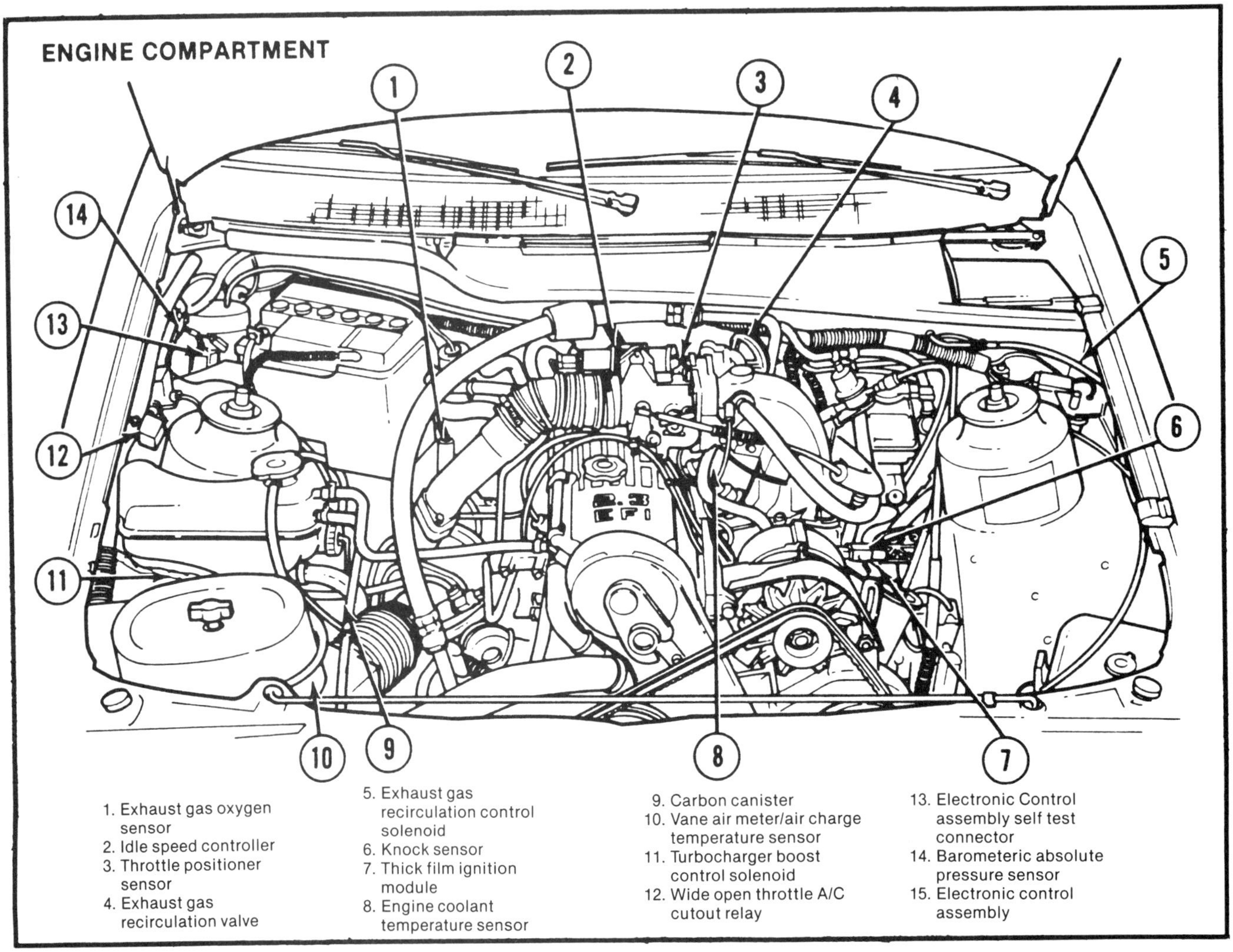

Location of the varied components of the EEC-IV Electronic Control system (© Ford Motor Co.)

Ignition System (EEC–IV)

In the EEC–IV ignition system, the spark output and the timing are controlled by an electronic ignition module, mounted on a universal type distributor and connected to the overall engine control system, which monitors and directs the module to perform its functions during the varied vehicle operation modes.

The universal type distributor incorporates an integrally mounted Thick Film ignition module (TFI–IV), a Hall Effect vane switch-stator assembly and an octane rod. No distributor calibration is required and initial timing is not a normal adjustment. The octane rod can be replaced, if the need is required and proper authorization has been received, due to federal and/or state emission requirements. A standard zero degree rod is installed in the distributors initially, with three degree and six degree rods available through Ford Motor Company service outlets.

CAUTION

Notification must be placed on an emission information decal regarding the change of the octane rod, authorization number and the decal placed either on the engine or in the engine compartment where it can readily be seen during future service.

The primary function of the universal distributor, used with the EEC–IV system, is to direct the high secondary voltage to the spark plugs and to supply crankshaft position and frequency information to the Electronic Control Assembly (ECA). This is accomplished with the use of a Hall effect device, commonly called a Profile Ignition Pick-up switch. The switch consists of a Hall effect device on one side and a magnet on the other. The rotary cup, which has windows and vanes, rotates and passes through the space between the Hall effect device and the magnet.

When a window is between the Hall effect device and the magnet, the magnetic flux path is not completed from the magnet, through the Hall effect and back to the magnet. In this case, the switch is off and the Hall effect device sends no signal.

As the vane passes between the device and the magnet, the magnetic lines of flux increase. The flux lines are shunted through the vane, back to the magnet, increasing the effect of the magnet on the Hall effect device and the device is turned on. When the next window passes between the device and the magnet, the device is turned off and the signal is stopped.

This voltage pulse is used by the EEC–IV system for sensing crankshaft position and computing the desired spark advance based on engine demands and calibration. The EEC–IV engine control system automatically senses and compensates for changes in altitude, such as from sea level to mountains.

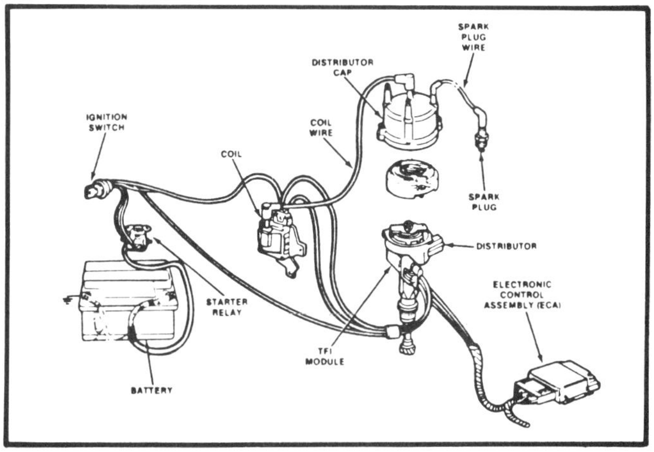

Electronic ignition system used with the Merkur XR4Ti models
(© Ford Motor Co.)

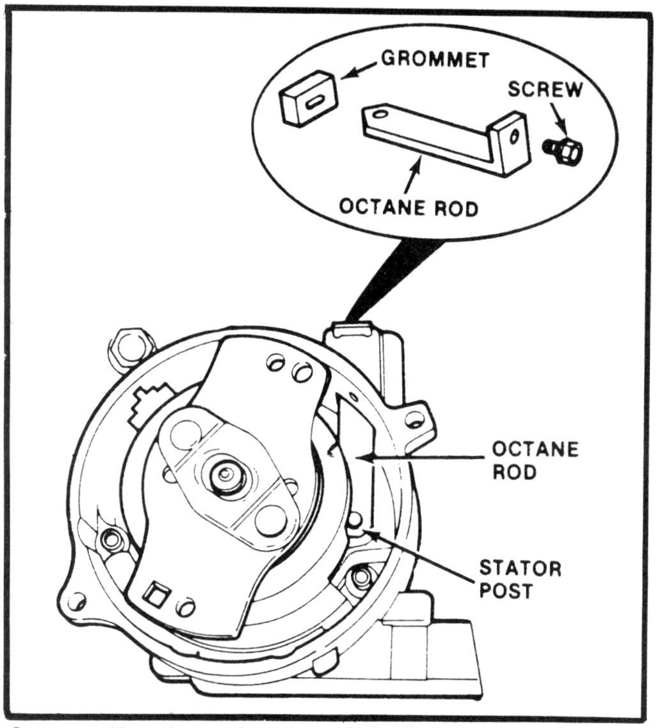

Octane rod location on the universal distributor
(© Ford Motor Co.)

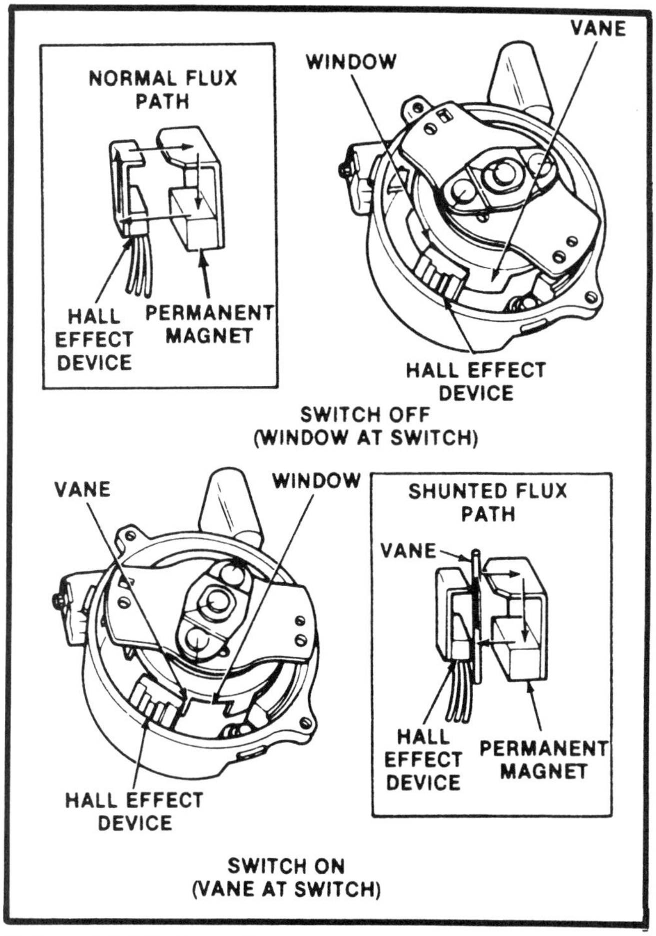

Universal distributor equipped with Hall effect mechanism
(© Ford Motor Co.)

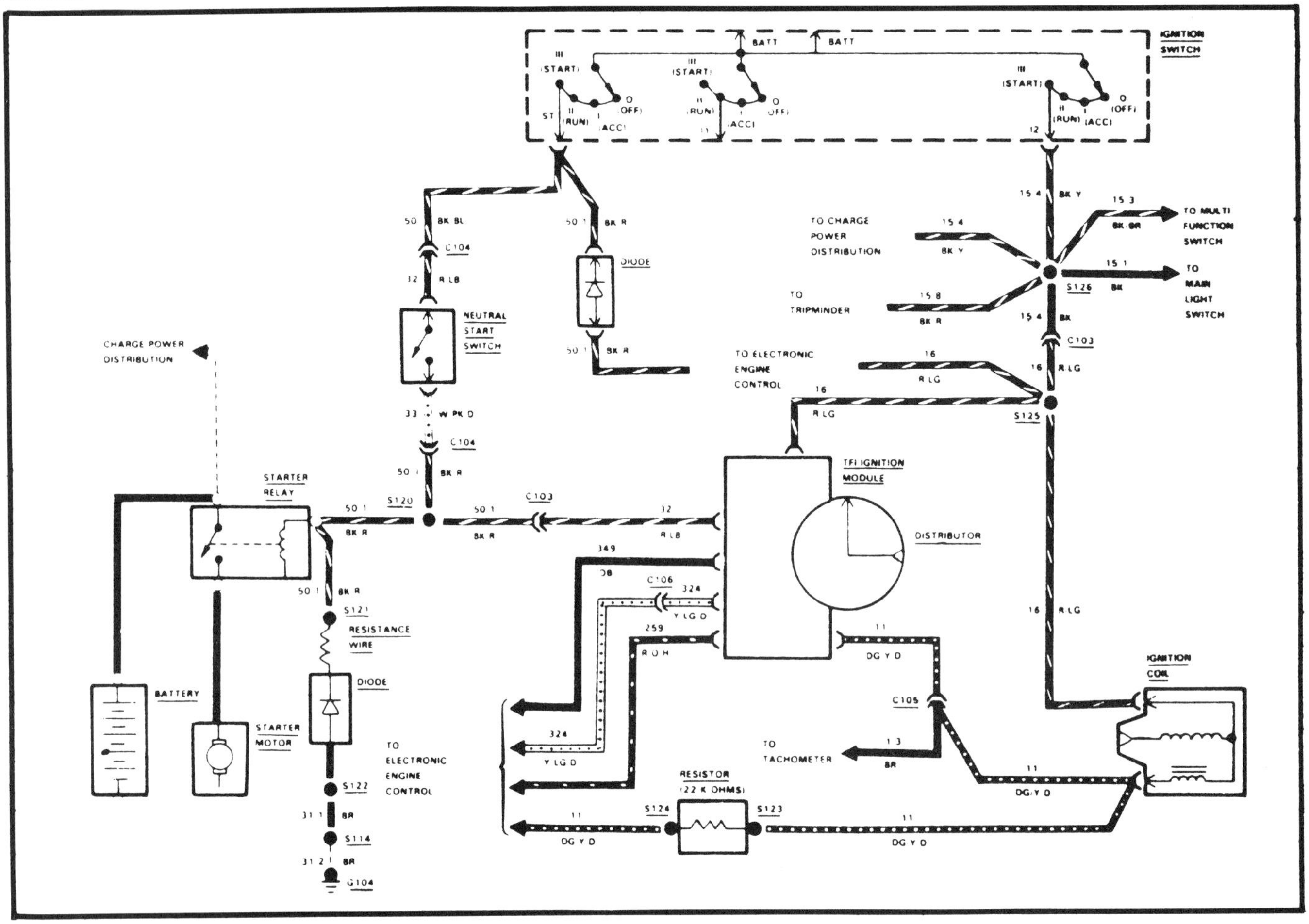

Schematic of electronic ignition system (© Ford Motor Co.)

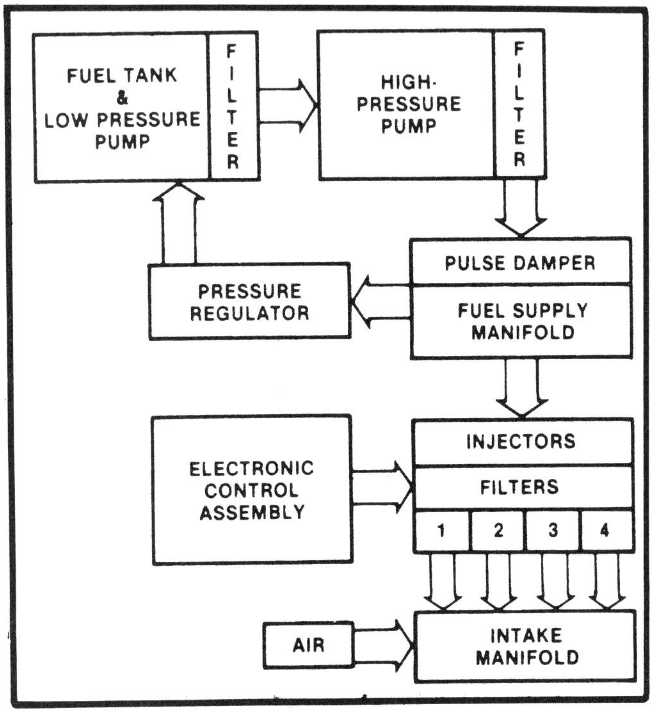

Basic description of fuel injection system operation (© Ford Motor Co.)

The electronic engine control system will operate the engine with a 10 degree BTDC constant spark timing should a problem arise in the calibration assembly or the Electronic Control Assembly (ECA). The constant 10 degree advance is a fail-safe mode which permits the vehicle to be driven in for service when the electronics are not functioning properly. Should this occur, it is necessary to go into the full electronics diagnosis routine.

NOTE: The distributor TFI–IV module has a 'push-start' mode, allowing for a 'push-start' on manual transmission equipped vehicles, should it be necessary. Earlier TFI modules did not contain this feature. The normal safety precautions should be observed during any push-start attempts.

Electronic Fuel Injection System

The Electronic Fuel Injection System (EFI) is a multi-point, pulse timed, mass air flow, fuel injection system. Fuel is metered into the intake air stream in accordance with engine demands through four fuel injectors, mounted on a tuned intake manifold. An on-board vehicle electronic engine control (EEC–IV) computer accepts inputs from the numerous engine sensors to compute the required fuel flow rate necessary to maintain a prescribed air/fuel ration through out the entire engine operational range. The computer then outputs commands to the fuel injectors to meter the required quantity of fuel.

FUEL DELIVERY SYSTEM

The fuel delivery system consists of an in-tank low pressure fuel pump. A high pressure chassis mounted electric fuel pump and the necessary interconnecting tubing. The fuel drawn from the tank is passed through a filter before it enters the low pressure pump. The fuel is filtered again

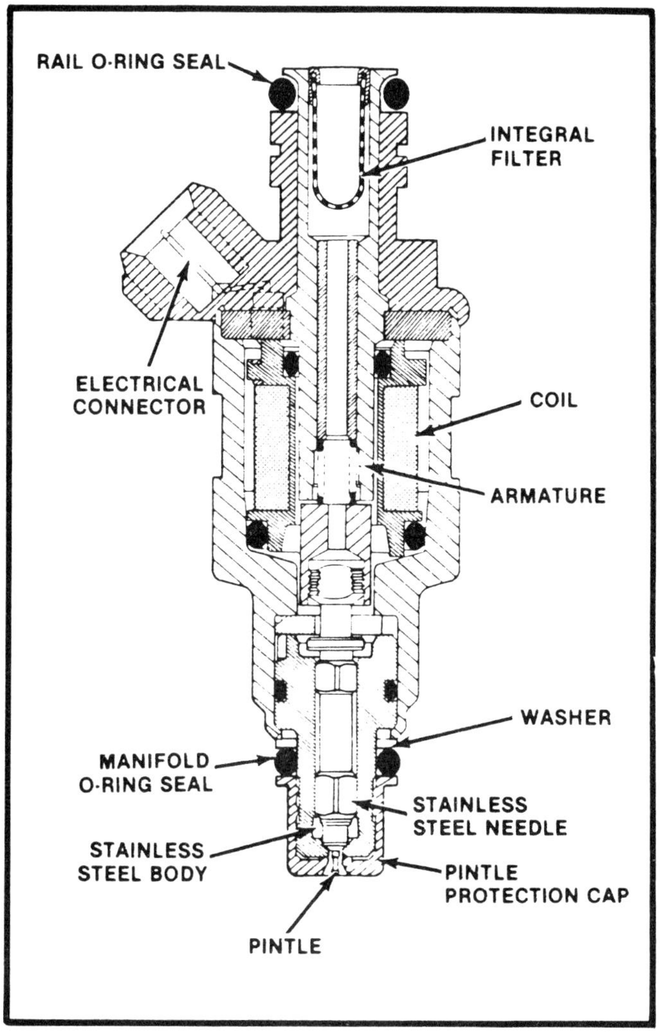

Cross section of fuel injector assembly (© Ford Motor Co.)

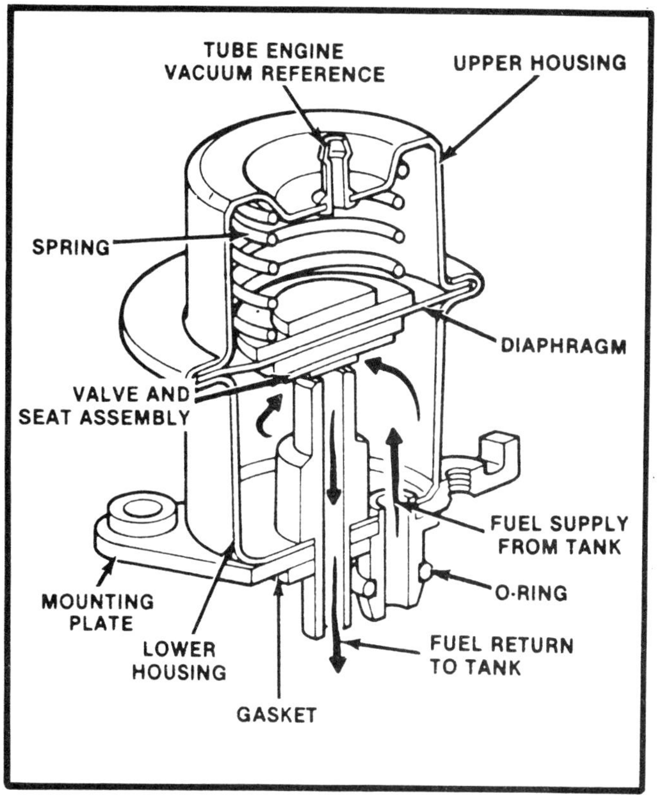

Cross section of fuel pressure regulator (© Ford Motor Co.)

as it leaves the high pressure pump, on its way to the pulse damper and fuel supply manifold. The pulse damper smooths the pressure variations generated by the opening and closing of the fuel injectors. The fuel supply manifold incorporates the electrically controlled fuel injectors directly above the engine's four intake ports. The injectors, when energized, sprays a metered quantity of fuel into the the intake air stream.

A constant fuel pressure drop is maintained across the injector nozzles by a pressure regulator. The regulator is connected in series with the fuel injectors and is positioned downstream from them. Excess fuel supplied by the pump, but not required by the engine, passes through the regulator and returns to the fuel tank through a fuel return line.

Two injectors are energized simultaneously, once every crankshaft revolution. The period of time that the injectors are energized (injector on-time or pulse width) is controlled by the Electronic Engine Control (EEC) computer (ECA). Air entering the engine is measured by a vane air flow meter located between the air cleaner and the turbocharger. The air flow is then compressed by the turbocharger before introduction into the intake manifold. This air flow information, along with the input from various other engine sensors, is used to compute the required fuel flow rate.

Turbocharger

The turbocharger is basically an air compressor that is connected to the air induction system to increase the air flow into the engine. The turbocharger used with the Merkur's 2.3 liter engine is a blow-thru system,

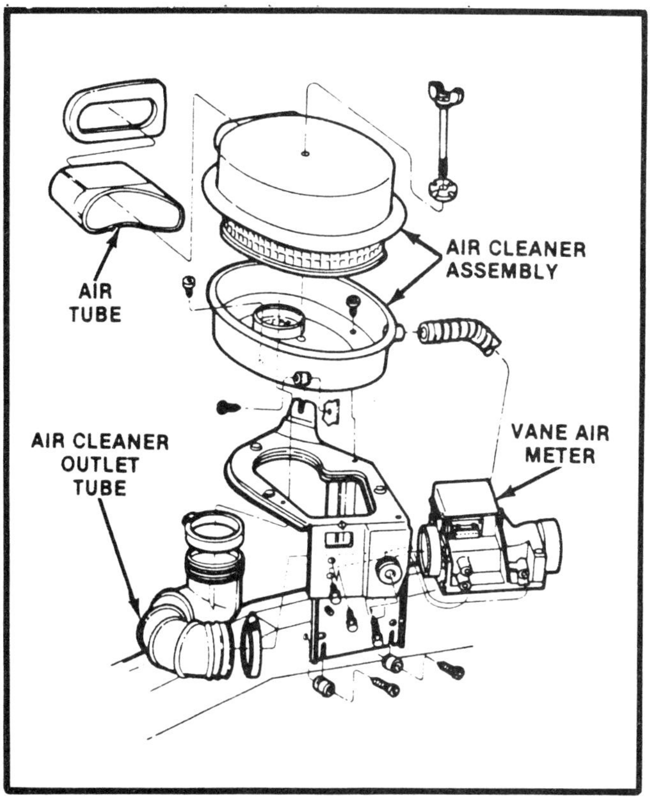

Vane air meter and air cleaner assembly (© Ford Motor Co.)

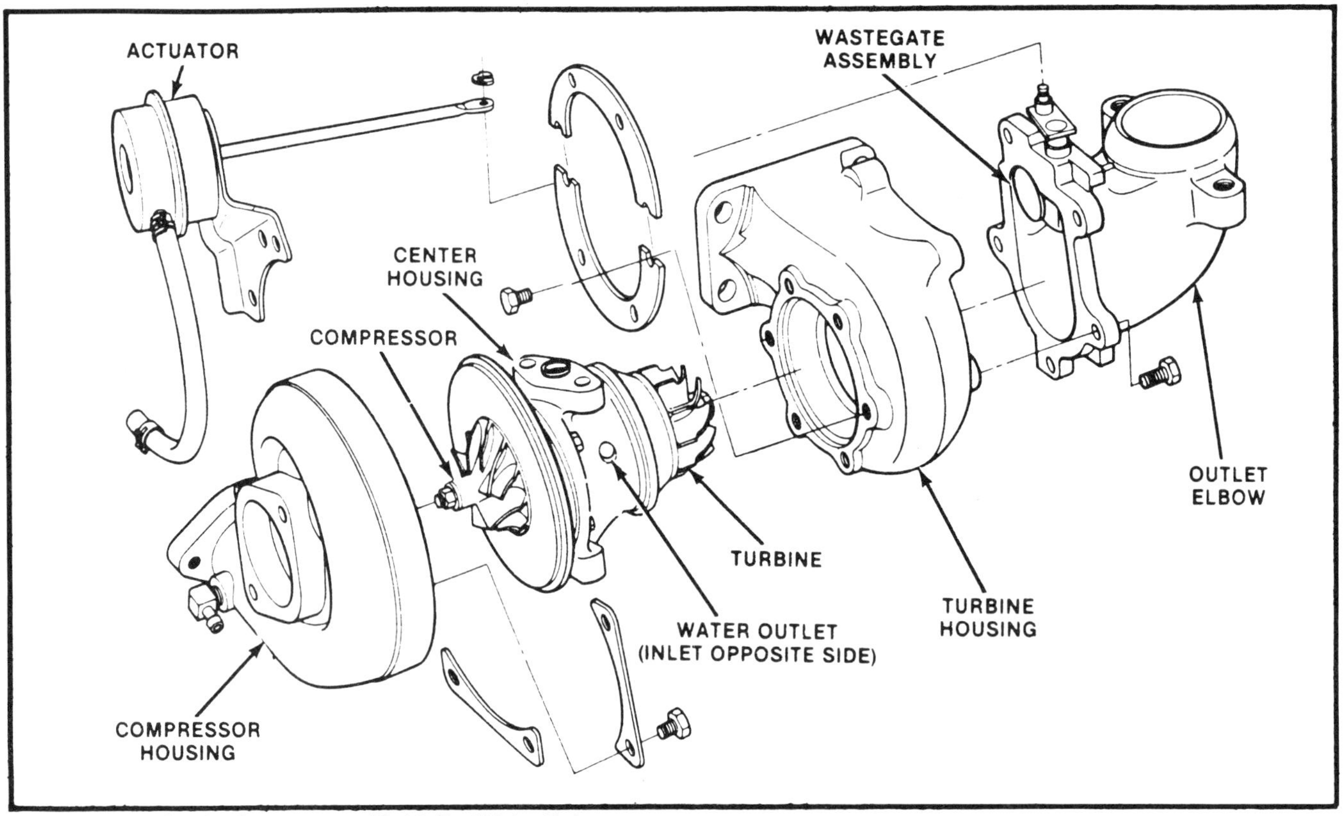

Exploded view of turbocharger assembly (© Ford Motor Co.)

which means the fuel injectors are mounted downstream from the turbocharger, rather than upstream as they would be in a draw-thru system. The energy required to compress the air is taken from the engine's exhaust gases. By using heat and pressure usually expelled through the exhaust system, the turbocharged engine can increase heavy throttle and wide-open throttle power levels, while maintaining, at part throttle, fuel economies of smaller displacement engines.

The turbine converts this normally wasted energy into rotating mechanical force. The rotational force of the turbine is transferred to the compressor side of the turbocharger through the interconnecting shaft. At operating speed, the spinning compressor wheel creates its own suction or vacuum at the air inlet elbow. This vacuum draws more air into the engine than normal vacuum created by piston movement. The additional air is pumped into the intake manifold where it is mixed with fuel supplied by the electronically controlled fuel injectors. As the turbocharger pressure forces the air/fuel mixture into the cylinders, the mixture becomes tightly packed. This denser and heavier mixture burns with greater force that increases horsepower and torque in comparison with non-turbocharged engines of the same displacement. This power increase is referred to as 'boost'.

TURBOCHARGER COMPONENTS

1. Wastegate actuator is a spring loaded diaphragm device which senses and controls the pressure from the compressor outlet.
2. The compressor is a centrifugal, radial outflow type unit.
3. The turbine is a centripetal, radial inflow type unit, which drive the compressor through an interconnecting shaft.
4. The outlet elbow contains the wastegate assembly which allows a portion of the exhaust gases to bypass the turbine wheel which limits the compressor speed.
5. The center housing supports the bearings, the compressor, turbine wheel and the oil seals. The center housing is water cooled to aid in controlling the high amount of heat generated through the turbocharger assembly.

NOTE: For greater in-depth coverage of Electronic Engine Control Systems and the Fuel Injection Systems, refer to Chilton's Electronic Engine Control Service Manual.

EMISSION CONTROL SYSTEMS

Electronic Engine Control Component Location

Barometric pressure sensor—RH fender Apron
EEC power relay—Attached to lower RH cowl near ECA
EGR control solenoid—LH fender apron
EGR valve position sensor—Top of RH front of engine
EGR vent solenoid—LH front of engine
Electronic control assembly (ECA)—Lower RH cowl
Engine coolant temperature sensor—Top front of engine
Exhaust gas oxygen sensor—LH rear of engine
Fuel injectors—Upper LH side of engine
Fuse link A—At starter relay
Idle speed actuator—LH side of engine
Throttle positioner sensor—LH side of engine
Turbo boost solenoid—RH side of engine at turbocharger
Vane air flow meter—RH front of engine
Knock sensor—Bottom, LH rear of engine
Manifold charge temperature sensor—RH side of engine on manifold
TFI ignition module—Connected to RH side of distributor

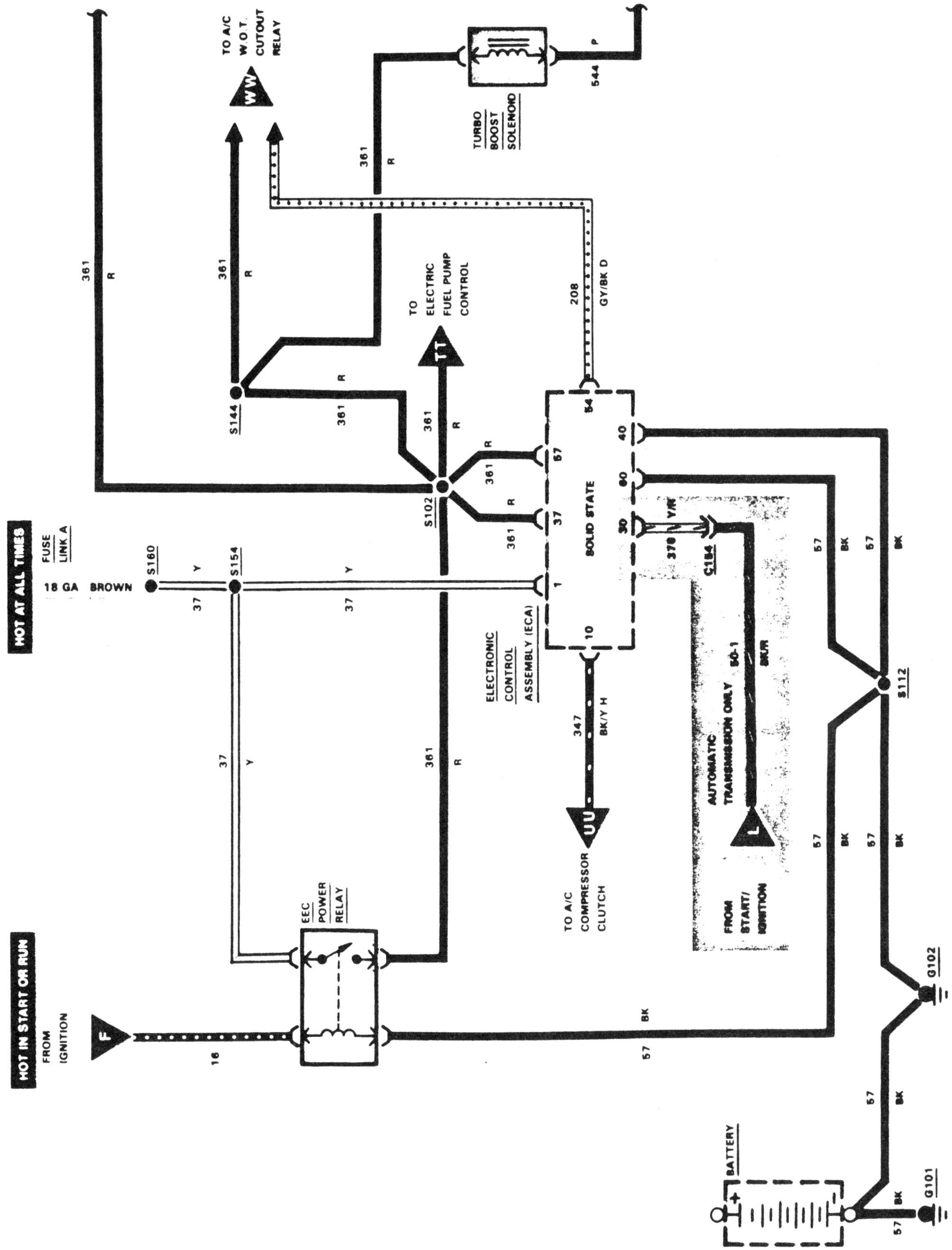

Electronic Engine Control electrical schematic, part one (© Ford Motor Co.)

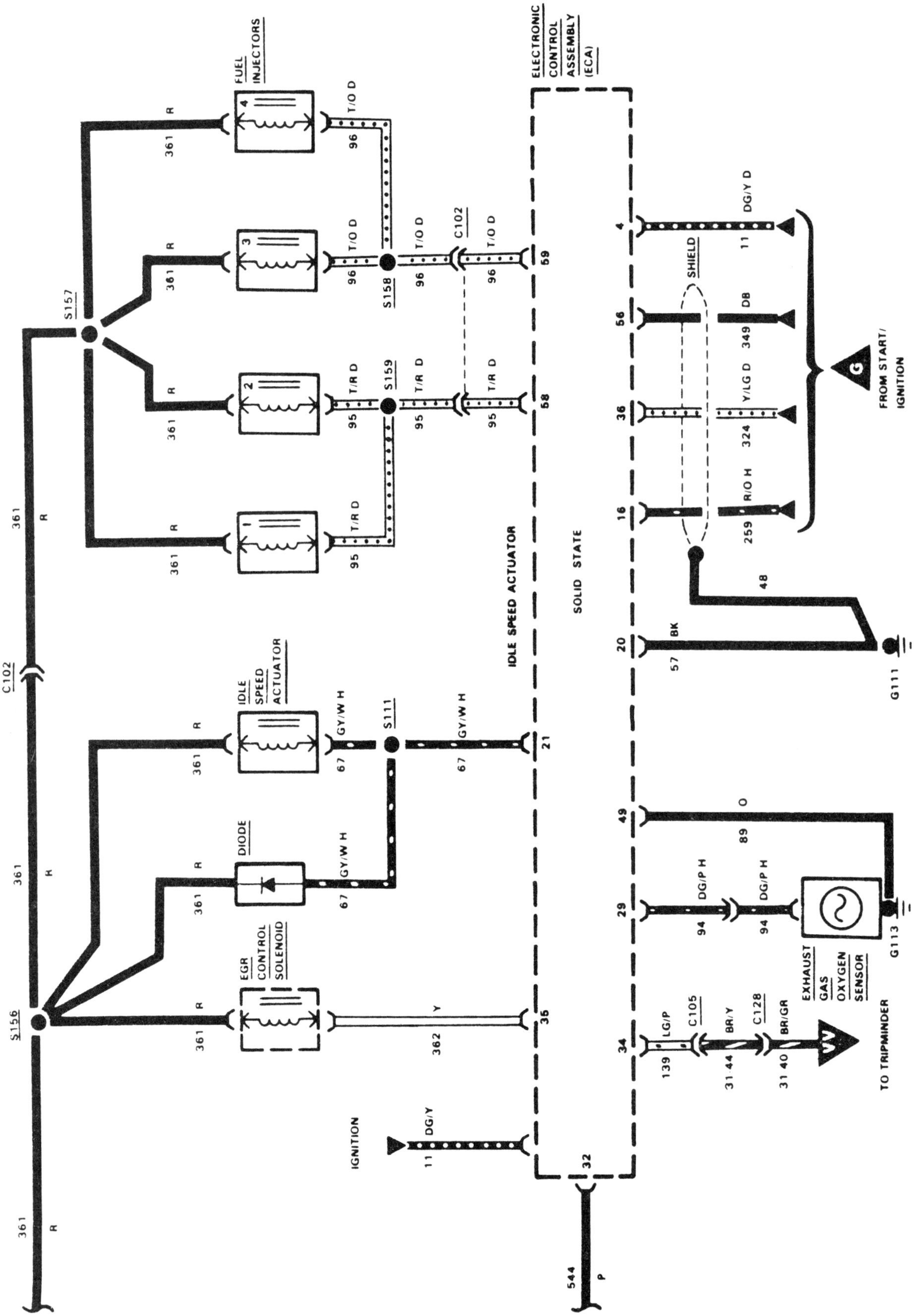

Electronic Engine Control electrical schematic, part two (© Ford Motor Co.)

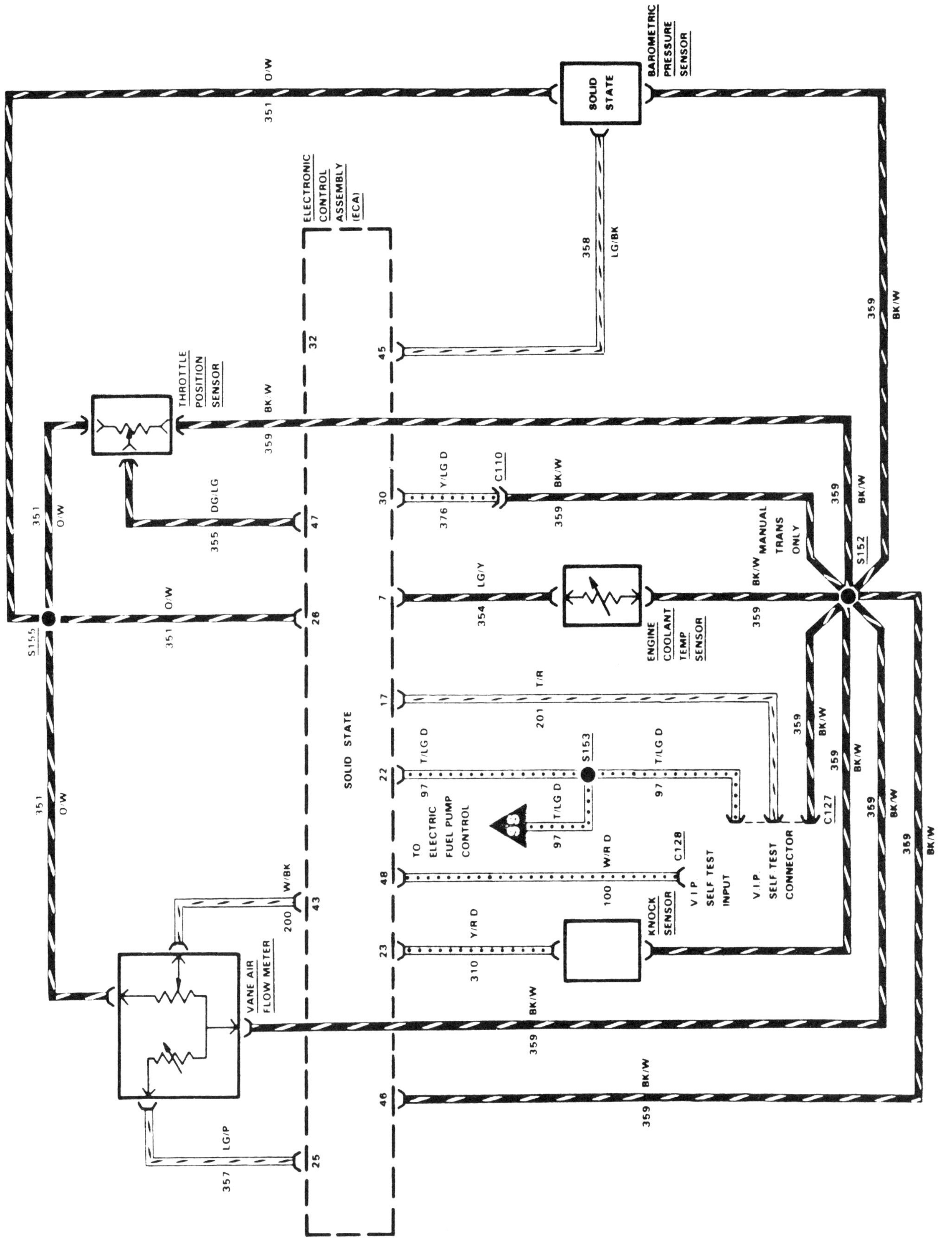

Electronic Engine Control electrical schematic, part three (© Ford Motor Co.)

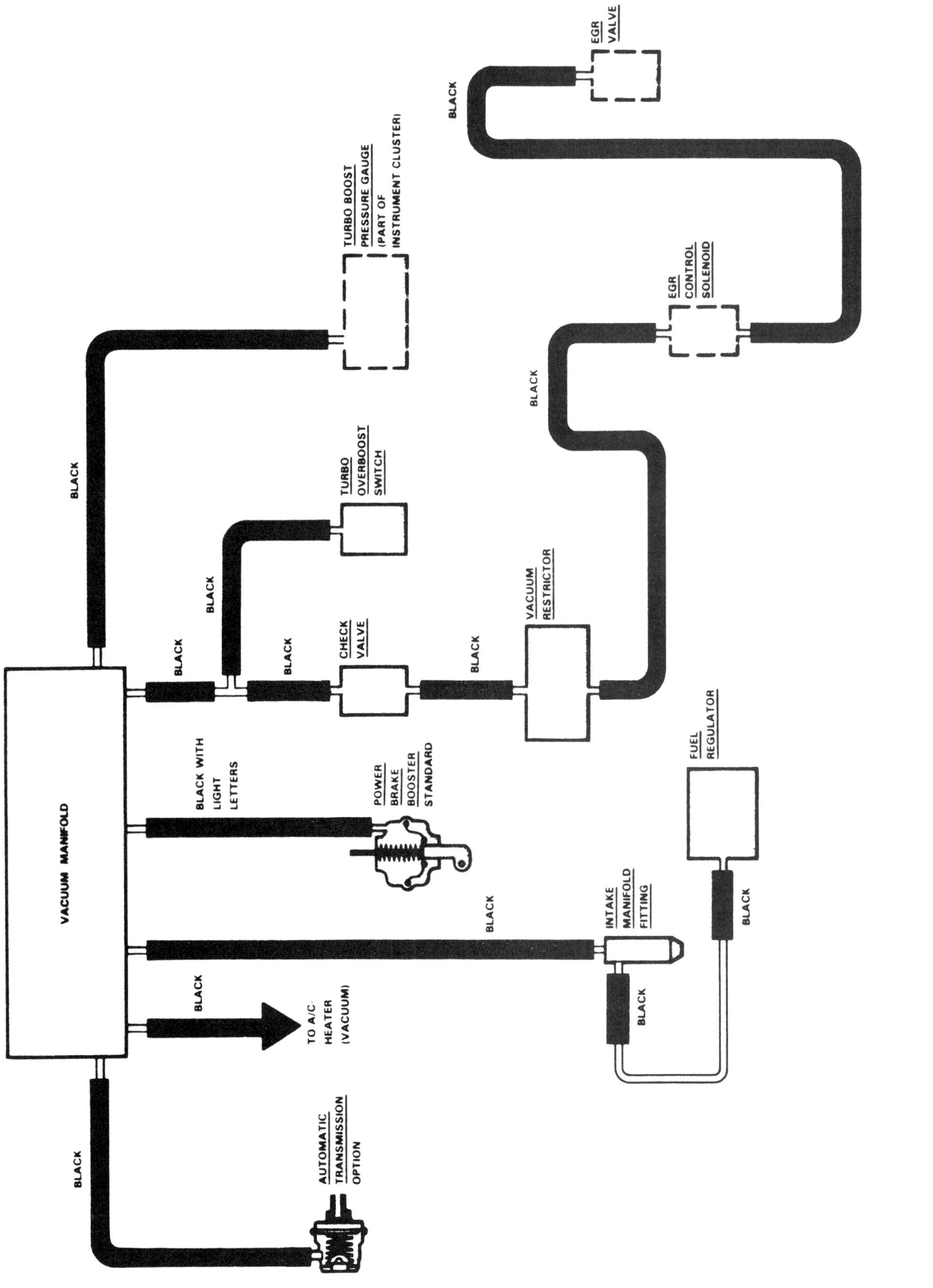

Engine vacuum schematic (© Ford Motor Co.)

Mitsubishi

INDEX

VEHICLE IDENTIFICATION NUMBER AND ENGINE IDENTIFICATION

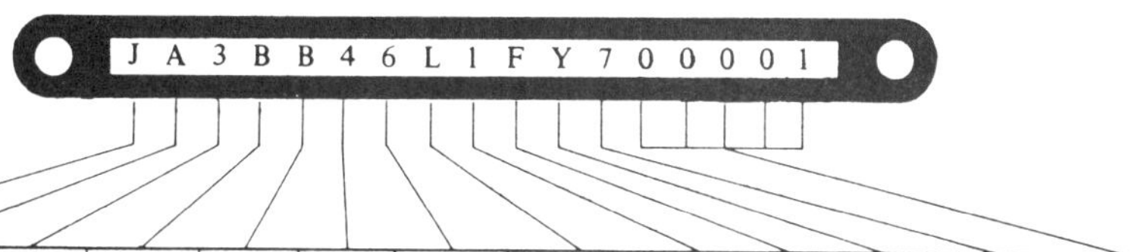

1st Digit	2nd Digit	3rd Digit	4th Digit	5th Digit	6th Digit	7th Digit	8th Digit	9th Digit	10th Digit	11th Digit	12th Digit	13th to 17th Digits
Country	Make	Vehicle type	Others	Line	Price class	Body	Engine	•Check digits	Model year	Plant	Transaxle	Serial number
J- Japan	A- Mitsubishi	3- Passenger Car	B- Manual Seat Belt	B- Galant	4- High	6- 4 door Sedan	L- 2.4 liter (143.4CID)	1 2 3 . . . 9 X	F- 1985 Year	Y- Nagoya Plant	7- 4 speed A/T Federal 8- 4 speed A/T California	00001 to 99999

NOTE • "Check digit" means a single number of letter X used to verify the accuracy of transcription of vehicle identification number.

Typical 17 digit vehicle identification number (© Mitsubishi Motor Sales of America, Inc.)

Vehicle Identification Number (VIN)

The vehicle identification number (VIN) consists of seventeen digits. This number is embossed on a plate, located on the upper left corner of the instrument panel near the windshield. The eight digit of the number represents the engine code and the tenth digit of the number represents the year of the vehicle, 'F' equals 1985 and 'G' equals 1986.

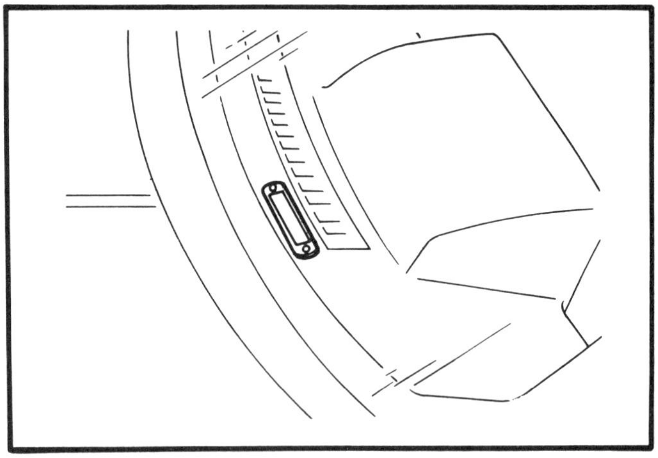

VIN location (© Mitsubishi Motor Sales of America, Inc.)

Engine serial number	Number cycling
64AA0201 to 64YY9999	AA0201 → AA9999
	AB0001 → AY9999
	BA0001 → YY9999

Typical engine serial number
(© Mitsubishi Motor Sales of America, Inc.)

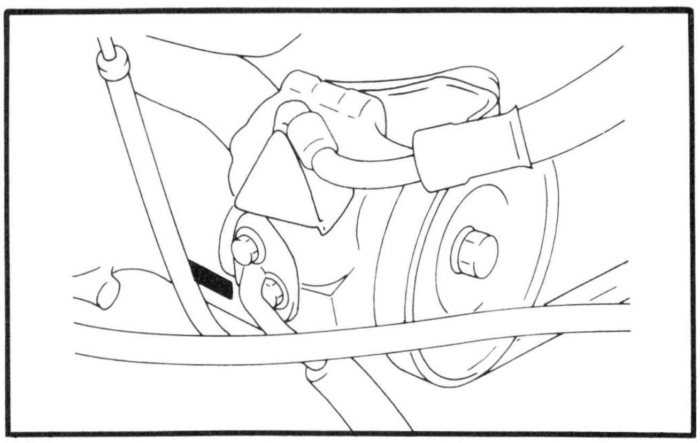

Galant 2.4L engine model number (© Mitsubishi Motor Sales of America, Inc.)

Engine Identification

The Engine Identification number for all models is stamped at the right front side on the top edge of the cylinder block. The engine serial number is stamped near the engine model number.

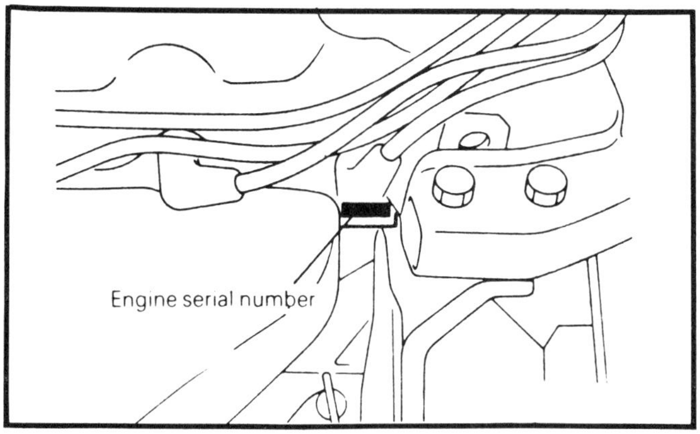

Mirage 1.5L and 1.6L engine serial number
(© Mitsubishi Motor Sales of America, Inc.)

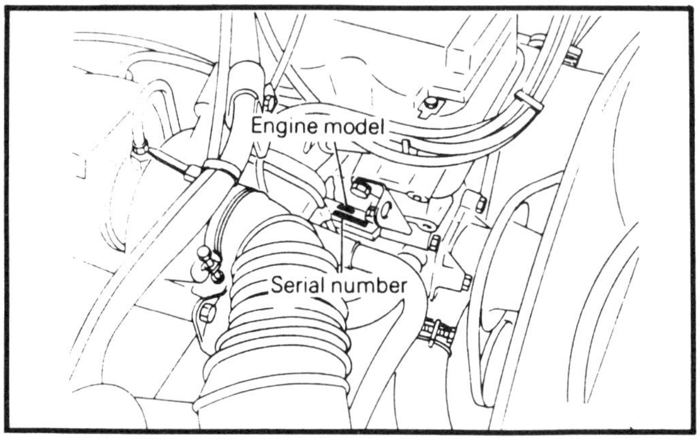

Starion 2.6L engine model and serial number
(© Mitsubishi Motor Sales of America, Inc.)

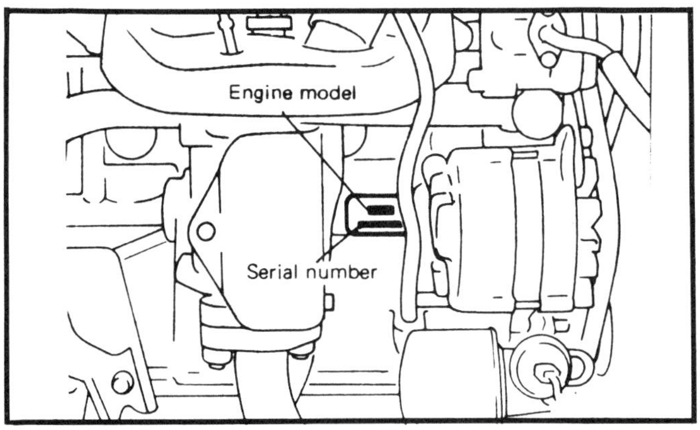

Cordia/Tredia 1.8L and 2.0L engine model and serial number
(© Mitsubishi Motor Sales of America, Inc.)

TUNE-UP SPECIFICATIONS

(When analyzing compression test results, look for uniformity among cylinders, rather than specific pressures.)

Year	Engine Displacement (cc)	Spark Plugs Type	Spark Plugs Gap (in.)	Distributor	Ignition Timing (deg) MT	Ignition Timing (deg) AT	Intake Valve Opens (deg) BTDC	Fuel Pump Pressure (psi)	Idle Speed (rpm)	Valve Clear (in.) ● In.	Valve Clear (in.) ● Ex.
'85–'86	1500	W20EP-U10 ①	0.039 0.043	Electronic	3B	3B ②	18	2.7–3.7	③	0.006	0.010 ④
	1600 T/C	BUR7EA-11 ①	0.035 0.039	Electronic	8B	8B ②	20	2.4–3.4	⑤	0.006	0.010 ④
	1800 T/C	W20EP-U10 ⑦	0.039 0.043	Electronic	10B	10B ②	57	35–47	⑧	0.006	0.010 ④
	2000	W20EP-U10 ⑦	0.039 0.043	Electronic	5B	5B ②	19	2.4–3.4	⑥	0.006	0.010 ④
	2400	BP6ES-11 ①	0.039 0.043	Electronic	5B	5B ②	25	35–47	⑨	0.006	0.010 ④
	2600	BP6ES-11 ①	0.039 0.043	Electronic	10B	10B ②	25	35–47	⑩	0.006	0.010 ④

NOTE: The underhood sticker often reflects tune-up specification changes made in production. The sticker figures must be used if they disagree with those in this chart.

B Before top dead center
T/C Turbocharger
● All clearances set hot
① See text
② ± 2° at curb idle
③ Man. Trans: 700 ± 100 rpm
Auto. Trans: 750 ± 100 rpm
④ Jet valve Clearance: 0.010
⑤ 700 ± 100 rpm with Idle Speed Control
⑥ 700 ± 100 rpm with Idle Speed Control
⑦ Vehicles without turbocharger—W-22EPR-S11 0.035 to 0.039 in. gap
⑧ 700 ± 100 rpm with Idle Speed Control
⑨ 750 ± 100 rpm
⑩ 850 ± 100 rpm

FIRING ORDERS

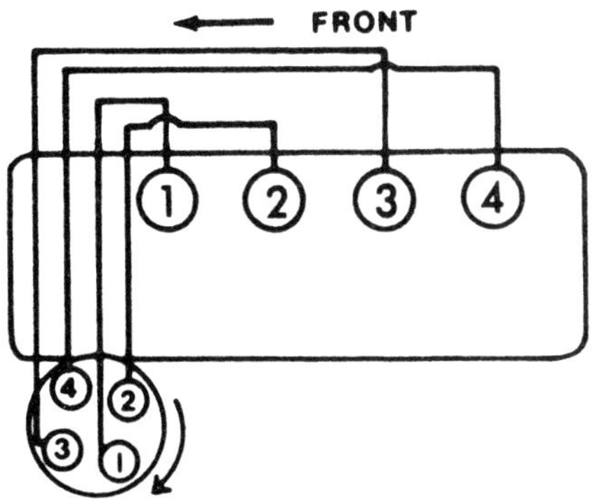

1.5L and 1.8L engine firing order
(© Mitsubishi Motor Sales of America, Inc.)

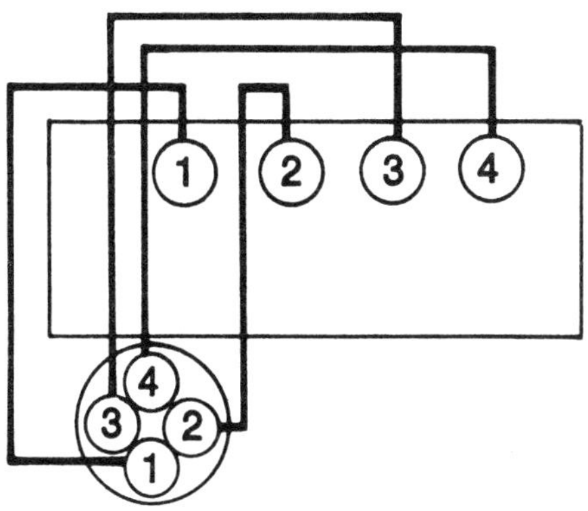

1.6L engine firing order
(© Mitsubishi Motor Sales of America, Inc.)

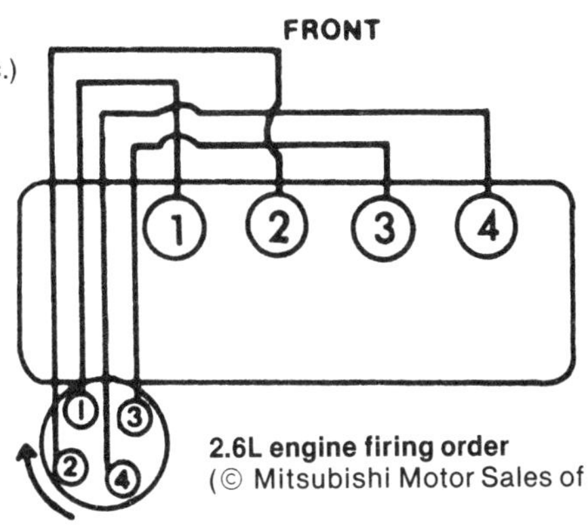

2.6L engine firing order
(© Mitsubishi Motor Sales of America, Inc.)

EMISSION EQUIPMENT USED

1985–86 STARION

Crankcase Emission Control
- Closed system
- Positive crankcase vent valve

Evaporative Emission Control
- Canister—Single
- Bowl vent valve
- Purge control valve
- Fuel filler cap—with relief valve
- Overfill-limiter—two way valve
- Fuel check valve

Exhaust Emission Control
- Jet valve
- Catalytic converter—dual three way type
- Secondary air supply system—reed valve
- Exhaust gas recirculation (EGR) system
- EGR valve—single
- Thermo valve
- Fuel control system—Electronic Controlled Injection System (ECI)
- High altitude compensation
- Idle speed control system (for A/C unit only)

1985–86 CORDIA/TREDIA AND MIRAGE

Crankcase Emission Control
- Closed system
- Positive crankcase vent valve

Evaporative Emission Control
- Canister—Single
- Bowl vent valve
- Purge control valve
- Fuel filler cap—with relief valve
- Overfill-limiter—two way valve
- Fuel check valve

Exhaust Emission Control
- Jet valve
- Catalytic converter—dual three way type
- Secondary air supply system—reed valve
- Exhaust gas recirculation (EGR) system
- EGR valve—single + sub
- Thermo valve—Single wax pellet type
- Fuel control system—Electronic Controlled Injection System (ECI)
- Feedback carburetor system
- Heated air intake system—vacuum type (ECI system not available)
- Deceleration device
- Dash pot—Incorporated in the ECI system
- Deceleration spark advance
- High altitude compensation—Incorporated into the ECI system
- Throttle opener
- Tamper-proof (mixture, choke-ECI system not available)

1985–86 GALANT

Crankcase Emission Control
- Closed system
- Positive crankcase vent valve

Evaporative Emission Control
- Canister—Single
- Bowl vent valve
- Purge control valve
- Fuel filler cap—with relief valve
- Overfill-limiter—two way valve
- Fuel check valve

Exhaust Emission Control
- Jet valve
- Catalytic converter—dual three way type
- Secondary air supply system—reed valve
- Exhaust gas recirculation (EGR) system
- EGR valve—single
- Thermo valve
- Fuel control system—Electronic Controlled Injection System (ECI)
- High altitude compensation—controlled by computer

REQUIRED EMISSION CONTROL MAINTENANCE

Emission Control System Maintenance	Service Intervals	Mileage in Thousands	7.5	15	22.5	30	37.5	45	50
		Kilometers in Thousands	12	24	36	48	60	72	80
Change engine oil every 12 months	OR		X	X	X	X	X	X	
Replace engine oil filter every 12 months	OR			X		X		X	
Change engine oil every 6 months*	OR		Every 3,000 miles (4,800 km)						
Replace engine oil filter every 12 months*	OR		Every 6,000 miles (9,600 km)						
Check drive belt (for water pump and alternator) for condition; adjust tension as required	AT			X				X	
Replace drive belt (for water pump and alternator)	AT					X			
Check valve clearance; adjust as required	AT			X		X		X	
Check ignition timing; adjust as required every 5 years	OR								X
Check engine idle speed; adjust as required	AT			X		X		X	
Clean carburetor choke mechanism and linkage	AT					X			

REQUIRED EMISSION CONTROL MAINTENANCE

Emission Control System Maintenance	Service Intervals	Mileage in Thousands	7.5	15	22.5	30	37.5	45	50
		Kilometers in Thousands	12	24	36	48	60	72	80
Check throttle position system; adjust as required	Initial	AT		X					
	After initial check at 50,000 miles (80,000 km), thereafter every 50,000 miles (80,000 km)								
Replace fuel filter (except filter in fuel tank) every 5 years		OR							X
Check fuel system (tank, line and connections) for leaks every 5 years		OR							X
Replace vacuum hoses, secondary air hoses and crankcase ventilation hoses every 5 years		OR							X
Replace fuel hoses, water hoses and fuel vapor hoses every 5 years		OR							X
Replace turbocharger air intake hoses every 5 years*		OR							X
Replace turbocharger oil hoses every 5 years*		OR							X
Replace air cleaner filter		AT				X			
Clean crankcase emission control system (PCV valve) every 5 years		OR							X
Check evaporative emission control system (except canister) for leaks and clogging every 5 years		OR							X
Replace canister		AT							X
Replace spark plugs		AT				X			
Replace ignition cables every 5 years		OR							X
Replace oxygen sensor		AT							X

NOTE: *Cars with a turbocharger

IDLE SPEED AND MIXTURE ADJUSTMENTS

NOTE: The following specifications are published from the latest information available. If the published information differs from the information on the vehicle emission label, use the data on the emission label.

Idle Speed Specifications

IDLE SPEED

Adjustment

1985–86 CORDIA/TREDIA, MIRAGE AND GALANT

NOTE: The improper setting (throttle valve opening) will increase the exhaust gas temperature and deceleration, which in turn will reduce the life of the catalyst greatly and deteriorate the exhaust gas cleaning performance. It will also effect the fuel consumption and the engine braking.

1. With the vehicle in park, the drive wheels blocked and all the accessories off, Run the engine until it reaches normal operating temperature.
2. Bring the engine rpm up to 2000–3000 rpm for about ten seconds, then let the engine idle for a least two minutes.

IDLE SPEED SPECIFICATIONS

Engine	Transaxle	Curb Idle Speed (rpm)
1985-86 Mirage		
1.5L	M/T	700 ± 100
1.5L	A/T	750 ± 100
1.6L-T/C	M/T & A/T	700 ± 100 ①
1985-86 Cordia/Tredia		
1.8L	M/T & A/T	700 ± 100 ①
2.0L	M/T & A/T	700 ± 100 ①
1985-86 Galant		
2.4L	M/T & A/T	750 ± 100
1985-86 Starion		
2.6L EFI	M/T & A/T	850 ± 100

M/T—Manual Transaxle
A/T—Automatic Transaxle
T/C—Turbocharger
EFI—Electronic Fuel Injection
① With Idle Speed Control (ISC)

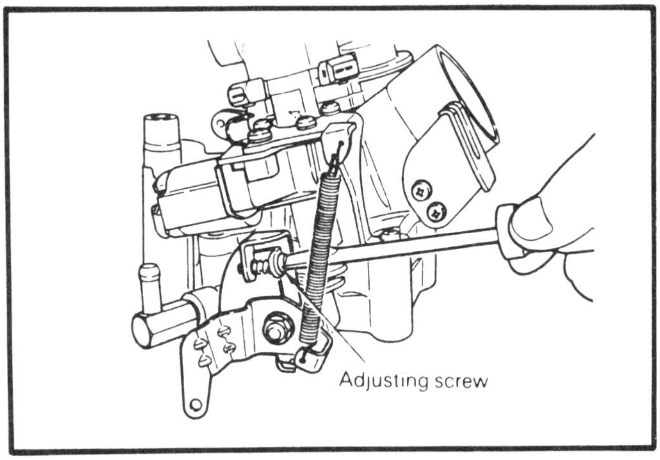

Idle adjusting screw (© Mitsubishi Motor Sales of America, Inc.)

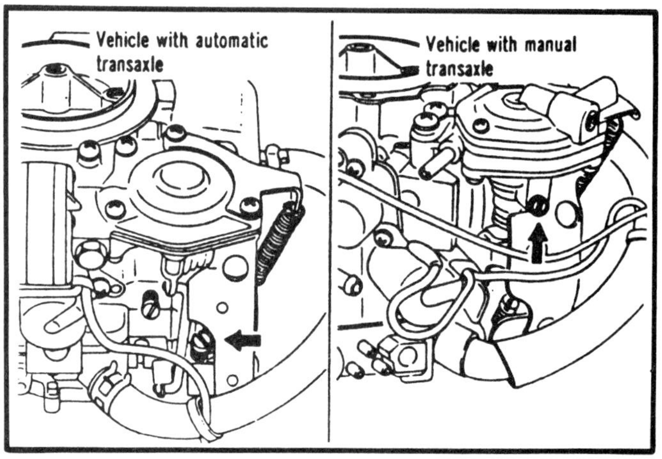

Location of speed adjusting screw—typical with A/C (© Mitsubishi Motor Sales of America, Inc.)

3. Connect a tachometer to the engine and check the idling speed. If it does not meet specifications, readjust the idle speed to the nominal specification, using the idle speed adjusting screw, which is located closest to the primary throttle valve shaft.

1985–86 STARION

Idle mixture adjustment cannot be accomplished with Electronically Controlled Injection system. The fuel delivery is controlled by the Electronic Control Unit (ECU), reacting to varied operating conditions.

IDLE-UP SPEED

Adjustment

1985–86 CORDIA/TREDIA AND MIRAGE—CARBURATOR EQUIPPED ENGINES WITHOUT A/C

NOTE: Adjustment condition: lights, electric cooling fan and all accessories are off and transaxle is in neutral.

1. Make sure the curb idle speed is within specifications and adjust if necessary.
2. By using the auxiliary lead wire, activate the idle up solenoid valve. Apply the intake manifold vacuum to the throttle opener and activate the throttle opener.
3. Open the throttle slightly (to engine speed of about 2000 rpm) and then slowly close it.
4. Adjust the engine speed to the specifications with the idle-up adjusting screw.
5. After repeating Step 3, check the engine speed.
6. Remove the auxiliary lead wire used in Step 2 and reconnect the idle-up solenoid valve wiring.

IDLE-UP SPEED

Adjustment

1985–86 CORDIA/TREDIA AND MIRAGE—CARBURATOR EQUIPPED ENGINES

1. With the vehicle in park, the drive wheels blocked and all the accessories off. Run the engine until it reaches normal operating temperature.
2. Disconnect the electric cooling fan connector. On vehicles equipped with power steering, set the tires in the straight ahead position to prevent the pump from being loaded. Set the steering wheel in the stationary position.
3. Be sure that the curb idle speed is within the specifications and adjust if necessary.
4. With the air conditioner on, adjust the engine speed to the specified speed with the throttle opener setting screw (idle-up adjusting screw).
5. Reconnect the electric cooling fan connector and turn the A/C on and off several times to check the operation of the throttle opener.

IDLE SPEED CONTROL (ISC) SERVO AND THROTTLE POSITION SENSOR

Adjustment

1985–86 STARION, GALANT AND MIRAGE WITH 1.6L TURBOCHARGED ENGINE

NOTE: This adjustment is very important, since the vehicle driveability depends upon it. If the ISC servo, throttle position sensor, mixing body, or thottle body has been replaced or removed for any reason, use the following procedure to make the adjustments.

1. With the vehicle in park, the drive wheels blocked and all the accessories off, run the engine until it reaches normal operating temperature.
2. Stop the engine and disconnect the accelerator cable from the throttle lever of the injection mixer. Loosen the two throttle position sensor mounting screws, turn the throttle position sensor clockwise as far as it will go and then temporarily tighten the screws.
3. Turn the ignition switch to the ON position for 15 seconds and then turn it off. This will set the ISC servo to the specified position.
4. Disconnect the ISC servo harness connector. Start the engine, check the engine speed and adjust to specifications.

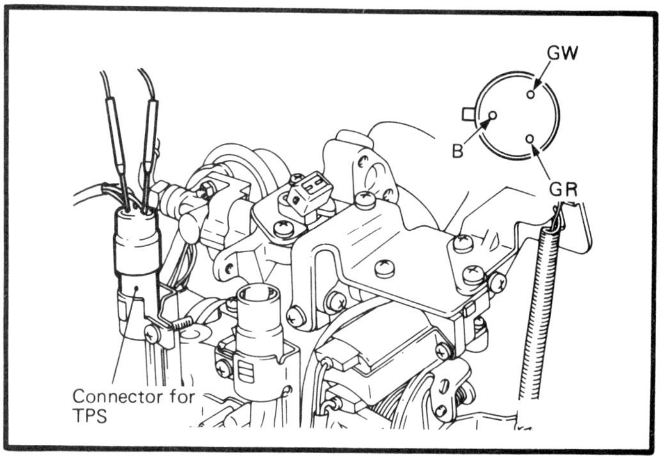

Location of the terminals for the voltmeter hook-up (© Mitsubishi Motor Sales of America, Inc.)

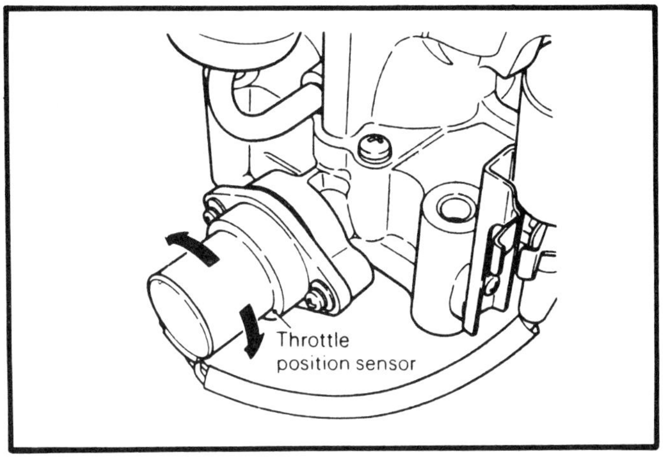

Adjusting the throttle position sensor
(© Mitsubishi Motor Sales of America, Inc.)

5. Stop the engine and disconnect the throttle position sensor harness connector.

6. Connect an adapter and a digital voltmeter between the throttle position sensor connector. Insert the probes of the digital voltmeter into the GW lead (TPS outlet) and the B lead (ground) of the body side harness. Place the ignition switch to the ON position, but do not start the engine.

7. Read the throttle position sensor output voltage.

8. If the measurement of the output voltage does not agree with the 0.48 ± 0.03 volts, loosen the throttle position sensor screws and turn the sensor left or right to bring the sensor into specifications. After applying sealant to the sensor, tighten the mounting screws.

9. Fully open the throttle valve one and confirm that the output voltage is correct when it is returned.

10. Remove the adapter and the digital voltmeter, reconnect the ISC servo harness connector and confirm that the curb idle is correct.

IDLE SPEED CONTROL (ISC)

Adjustment

1985–86 CORDIA/TREDIA 2.0L ENGINE W/AUTOMATIC TRANSAXLE

NOTE: When replacing the ISC servo, the engine speed should be adjusted.

1. With the vehicle in park, the drive wheels blocked and all the accessories off, Run the engine until it reaches normal operating temperature.

2. Remove the carburetor from the engine. Remove the concealment plug from the carburetor as described in the Idle Mixture Adjustment in this section.

3. Reinstall the carburetor onto the engine and relax the tension on the accelerator cable.

4. Place the ignition switch to the on position and wait for a least 18 seconds. Turn the ignition switch off and disconnect the ISC actuator connector and the oxygen sensor connector.

5. Start the engine check the ignition timing and adjust if necessary. Increase the engine speed between 2000–3000 rpm, two or three times and then let the engine idle for 30 seconds.

6. Adjust the mixture adjusting screw (MAS) for a CO concentration of 0.1–0.3 %. Adjust the engine rpm to the specified speed by using the ISC adjustment screw.

7. Turn the idle mixture screw (secondary air supply screw) until the engine reaches its highest rpm. Turn the screw ⅔ of turn in the reverse direction from that point.

8. Race the engine two or three times. Check to be sure that the CO and engine rpm are still adjusted to specifications. If they are not, re-adjust as necessary.

10. Adjust the tension of the accelerator cable. The cable should have enough play so as not to interfere with idle switch.

11. Reconnect the ISC actuator connector and the oxygen sensor connector.

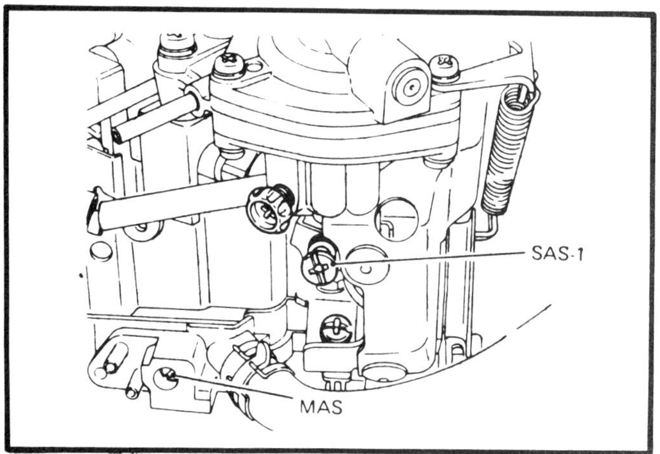

Location of the idle mixture screw #1 and the mixture adjusting screw (MAS) (© Mitsubishi Motor Sales of America, Inc.)

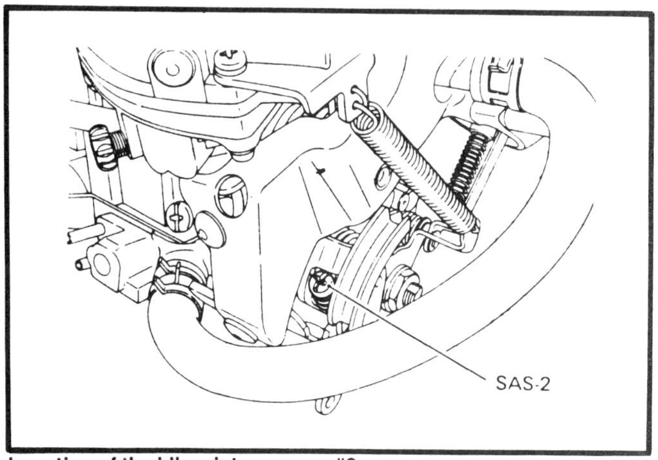

Location of the idle mixture screw #2
(© Mitsubishi Motor Sales of America, Inc.)

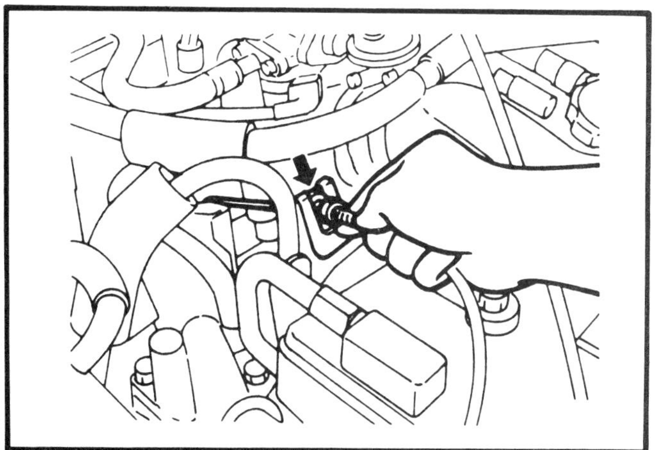

Adjusting the accelerator cable tension
(© Mitsubishi Motor Sales of America, Inc.)

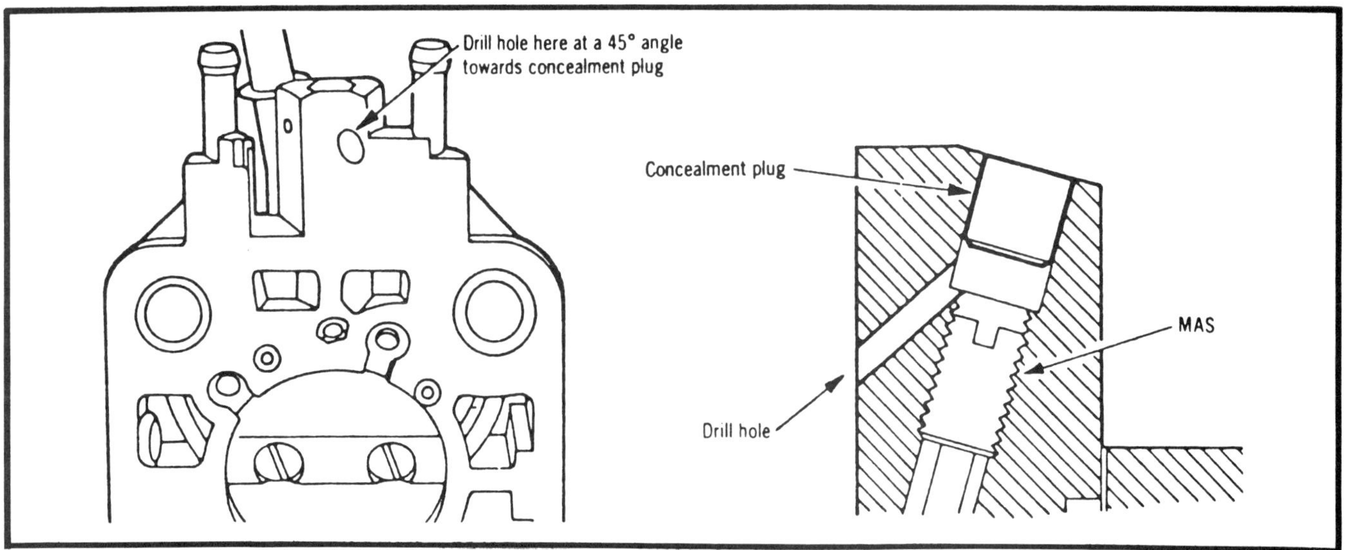

Removing the concealment plug (© Mitsubishi Motor Sales of America, Inc.)

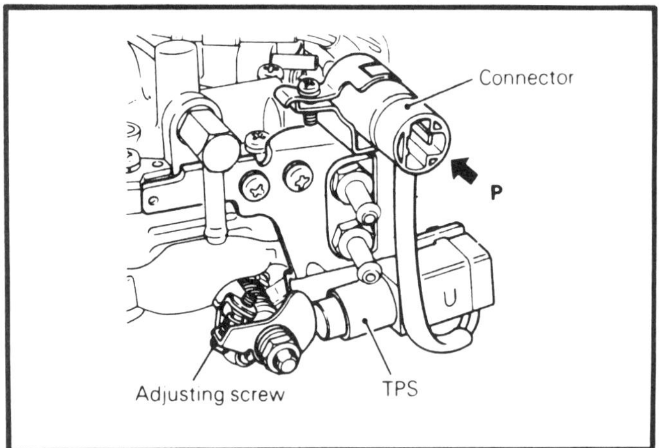

TPS outlet voltage adjusting screw
(© Mitsubishi Motor Sales of America, Inc.)

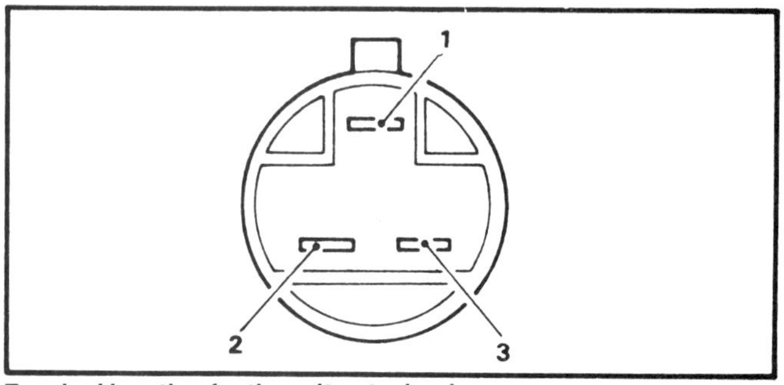

Terminal location for the voltmeter hook-up
(© Mitsubishi Motor Sales of America, Inc.)

THROTTLE POSITION SENSOR (TPS)

Adjustment

1985–86 CORDIA/TREDIA WITH 2.0L ENGINE AND MANUAL TRANSAXLE

MIRAGE WITH 1.5L ENGINE

1. Start the engine and let it run until the engine reaches normal operating temperature. Be sure that the fast idle cam is released.
2. Stop the engine. Back off the SAS (secondary air supply) screw #1 and SAS screw #2 sufficiently and close the throttle valve fully. Make sure to count the number of turns taken on the SAS screws.
3. Connect a digital voltmeter the bottom two slots of the TPS connector. Turn the ignition switch on and adjust the TPS outlet voltage to 250 millivolts (mV) with the adjusting screw.
4. Tighten the SAS screws by giving the same number of turns to the screws as recorded in Step 2.

NOTE: The Cordia/Tredia 2.0L engine with automatic transaxle TPS adjustment is the same, except for the following:

a. Make the Idle speed control adjustment first.
b. The TPS outlet voltage should be 445 millivolts (mV).

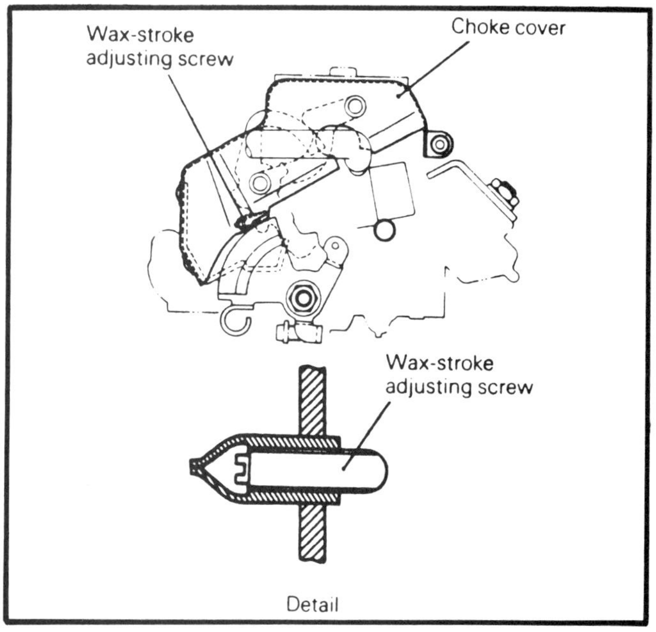

Typical—Tamper-proof automatic choke
(© Mitsubishi Motor Sales of America, Inc.)

IDLE MIXTURE

Adjustment

1985–86 CORDIA/TREDIA AND MIRAGE WITH FEEDBACK CARBURETORS

1. Remove the carburetor from the engine and place the carburetor in a suitable fixture in order to remove the concealment plug.
2. Drill a $^{5}/_{64}$ in. (2mm) pilot hole in the casting surrounding the idle mixture adjusting screw. Then re-drill the hole to $^{1}/_{8}$ in. (3mm) and insert a punch into the hole to drive out the concealment plug.
3. Reinstall the carburetor on the engine without the concealment plug.
4. With the vehicle in park, the drive wheels blocked and all the accessories off, run the engine until it reaches normal operating temperature.
5. Turn off the engine and disconnect the negative battery cable for about three seconds, and reconnect the cable.
6. Disconnect the connector of the exhaust oxygen sensor. Run the vehicle for five minutes at a speed of 30 mph or run the engine for more than five minutes at the engine speed of 2,000–3,000 rpm.
7. Run the engine at idle for two minutes and set the idle CO and the engine speed to specifications (the idle CO: 0.1–0.3 % at nominal curb idle speed).
8. Reconnect the oxygen sensor connector. Readjust the engine speed, if necessary and install the concealment plug into the hole to seal the idle mixture adjusting screw.

1985–86 STARION AND GALANT

Idle mixture adjustment cannot be accomplished with Electronically Controlled Injection system. The fuel delivery is controlled by the Electronic Control Unit (ECU), reacting to varied operating conditions.

TAMPER-PROOF AUTOMATIC CHOKE

All carburetors have a tamper-proof choke. The choke related parts are factory adjusted. Neither removal of the choke cover or tampering with the W.A.S. wax-stroke adjustment screw is required in service except when major carburetor overhaul or adjustment of choke-calibration-related parts is required by state or local inspections.

INITIAL TIMING PROCEDURE

ALL MODELS

1. Connect a suitable timing light to the number one cylinder.
2. Connect the positive lead of the test tachometer to the negative primary terminal of the ignition coil. Connect the negative lead of the test tachometer to a good ground.
3. Start the engine and run until operating temperature has been reached.
4. With the engine at operating temperature and fast idle off, momentarily open the throttle and release it to make sure that there is no bind in the linkage and that the idle speed screw is against its stop.
5. Disconnect and plug the vacuum line at the distributor.
6. Adjust the curb idle speed to specification.
7. Aim the timing light and check the engine timing. Advance or retard the ignition timing as required.
8. Once the timing has been set, recheck the curb idle specification and adjust as required. If the curb idle speed needs adjusting, be sure to recheck the ignition timing again.

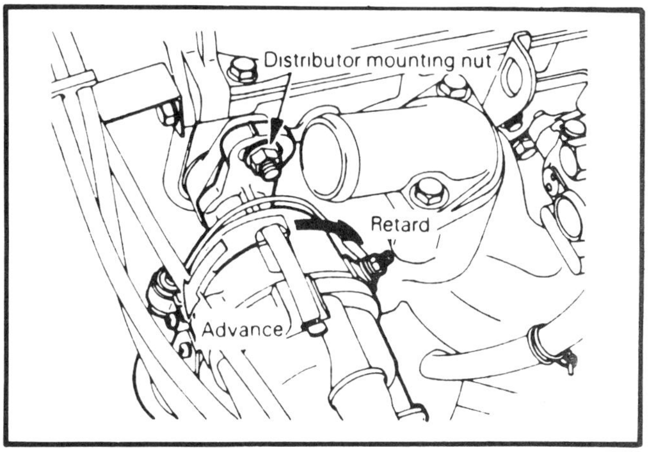

Adjusting the ignition timing
(© Mitsubishi Motor Sales of America, Inc.)

VACUUM ADVANCE

1985–86 MIRAGE WITH 1.6L TURBOCHARGER ENGINE AND MANUAL TRANSAXLE (FEDERAL ONLY)

The distributor is equipped with a dual diaphragm type vacuum advance mechanism, each being independent of each other. The main diaphragm utilizes carburetor ported vacuum to advance the ignition timing, and the sub diaphragm is actuated by manifold vacuum below the throttle valve to provide additional.

DISTRIBUTOR SPECIFICATIONS
1985–86 Mirage

	Federal and California		California
	1.5L engine with 4-speed M/T and 5-speed M/T	1.5L engine with A/T	1.6L turbo engine
Model No.	T3T62673/MD088686	T3T62671/MD088687, MD088688 (for Calif.) with A/T	T3T63175/MD074647 T3T62672
Type	Contact pointless type with electronic control unit (Electronic ignition system)		

DISTRIBUTOR SPECIFICATIONS
1985-86 Mirage

	Federal and California		California	
	1.5L engine with 4-speed M/T and 5-speed M/T	1.5L engine with A/T	1.6L turbo engine	
Advance mechanism	Centrifugal type and vacuum type	Centrifugal type and vacuum type	Centrifugal type and vacuum type	
Firing order	1—3—4—2			
Centrifugal advance—Crankshaft degrees at Crankshaft speed rpm				
Initial	0°/2,000	0°/2,000	0°/1,000	
Middle	9°/3,500	9°/3,500	12°/2,500	
Final	17°/6,000	17°/6,000	26°/5,300	
Vacuum advance—Crankshaft degrees at mmHg (in. Hg)				
Initial	0°/-60 (-2.36)	0°/-80 (-3.15)	0°/-80 (-3.15)	0°/-80 (-3.15)
Final	25°/-210 (-8.27)	23°/-280 (-11.02)	23°/-280 (-11.02)	23°/-280 (-11.02)
Idle advance—Crank degrees at mmHg (in. Hg) (High-altitude)	2.5°	2.5°	2.5°	2.5°
Ignition coil model No.	E-064	E-064	LB-119	E-064

DISTRIBUTOR SPECIFICATIONS
1985-86 Cordia/Tredia

Engine	2.0L for Fed.	2.0L for Calif.	1.8L
Type	Breaker pointless	Breaker pointless	Breaker pointless
Model No./Part No.	T3T61877/MD073529	T3T61876/MD073527	T3T63171/MD073445
Rotation	Clockwise	Clockwise	Clockwise
Firing order	1-3-4-2	1-3-4-2	1-3-4-2
Advance-Centrifugal (Distributor degrees at distributor rpm)	0° at 500 rpm 6° at 1,400 rpm 10° at 3,000 rpm	0° at 400 rpm 6° at 1,400 rpm 10° at 3,000 rpm	0° at 500 rpm 6° at 1,250 rpm 11° at 2,250 rpm
Advance-Vacuum [Distributor degrees as mm (in.) mercury]	0° at -80 mm (-3.15 in.) 6° at -150 mm (-5.9 in.) 11.5° at -280 mm (-11.02 in.)	0° at -80 mm (-3.15 in.) 4° at -150 mm (-5.9 in.) 10° at -360 mm (-14.17 in.)	0° at -60 mm (-2.36 in.) 8.5° at -150 mm (-5.9 in.) 12.5° at -210 mm (-8.27 in.)
Idle Advance [Distributor degrees at mm (in.) mercury]	— —	— —	0° at +80 mm (+3.15 in.) -7.5° at +360 mm (+14.17 in.)
Signal generator coil resistance Ω	—	—	920-1,120 at 20°C (68°F)
Signal rotor gap mm (in.)	0.8 (.031)	0.8 (.031)	0.2 (.008) or more

DISTRIBUTOR SPECIFICATIONS
1985-86 Galant

Type	Breaker pointless
Model No./Part No.	MD084893 (NIPPON DENSO)
Firing order	1—3—4—2
Advance-Centrifugal (Distributor degrees at distributor rpm)	0° at 1,000 rpm 14° at 2,800 rpm 23° at 5,000 rpm
Advance-Vacuum [Distributor degrees at mm (in.) mercury]	0° at -60 mmHg (-2.36 in.) 25° at -210 mmHg (-8.27 in.)
Signal rotor gap mm (in.)	.8 (.031)

DISTRIBUTOR SPECIFICATIONS
1985-86 Starion

Identification model No./part No.	T4T63371/MD061593
Type	Breaker pointless type
Firing order	1—3—4—2
Basic timing	10° ± 2° BTDC
Curb idle speed rpm	850 ± 100
Advance—Centrifugal (Distributor degrees at distributor rpm)	0° at 1,200 rpm 13° at 2,300 rpm 25° at 5,000 rpm
Advance—Vacuum [Distributor degrees at mm (in.) of mercury]	0°/-80 (3.1) 6°/-150 (5.9) 23°/-280 (11.0)
Retard—Pressure [Distributor degrees at mm (in.) of mercury]	0°/100 (3.9) 7°/450 (17.7)
Signal generator coil resistance Ω	920-1,120 at 20°C (68°F)
Signal rotor gap mm (in.)	0.2 (.008) or more

1985-86 CORDIA/TREDIA AND MIRAGE

The vacuum advance is operated by ported vacuum during periods of acceleration and cruising. During deceleration, a switching solenoid and speed sensor are used to change the vacuum source from ported to manifold vacuum at predetermined speeds, preventing excessive HC emissions. However, when engine speeds drop to or below below a specified level, the vacuum source is changed from manifold vacuum to ported vacuum, in order to maintain a smooth engine operation.

1985-86 STARION AND GALANT

The distributor is equipped with a vacuum advance unit, which is controlled by the Electronic Spark Controller unit, to allow advance and retard at pre-determined speeds and engine operating conditions.

IGNITION TIMING SPECIFICATIONS

Engine	Transaxle	Ignition Timing
1985-86 Mirage		
1.5L	M/T & A/T	3 BTDC ± 2 at curb idle
1.6L-T/C	M/T & A/T	8 BTDC ± 2 at curb idle ①
1985-86 Cordia/Tredia		
1.8L	M/T & A/T	10 BTDC ± 2 at curb idle ①
2.0L	M/T & A/T	5 BTDC ± 2 at curb idle ①

IGNITION TIMING SPECIFICATIONS

Engine	Transaxle	Ignition Timing
1985-86 Galant		
2.4L	M/T & A/T	5 BTDC ± 2 at curb idle
1985-86 Starion		
2.6L EFI	M/T & A/T	10 BTDC ± 2 at curb idle

M/T—Manual Transaxle
A/T—Automatic Transaxle
T/C—Turbocharger
EFI—Electronic Fuel Injection
① With Idle Speed Control (ISC)

SPARK PLUGS

ORIGINAL EQUIPMENT

1985–86 MIRAGE, GALANT AND STARION

NIPPONDENSO
W20EPR–U10
W22EPR–S11 (1.6L only)

ALTERNATE PLUG

NGK
BP6ES–11
BPR6ES–11
BUR7EA–11 (1.6L only)CHAMPION
RN–9Y
N–9Y

GAP

Champion and Nippondenso
0.035–0.039 in. (0.9–1.0mm)NGK
0.039–0.043 in. (1.0–1.1mm)

ORIGINAL EQUIPMENT

1985–86 CORDIA/TREDIA

NIPPODENSO
W20EP–U10
W22EPR–S11—turbocharged models

ALTERNATE

NGK
BP6ES–11
BPR6ES–11
BUR7EA–11—turbocharged models

GAP

NGK
0.035–0.039 in. (0.9–1.0mm)Nippondenso
0.039–0.043 in. (1.0–1.1mm)

CARBURETOR SPECIFICATIONS

1985–86 MIRAGE WITH 1.5L ENGINE

Type—Down-draft, 2 barrel, feedback typeModel No.:
28–32DIDTF–407 M/T 1.5L Federal
28–32DIDTF–408 A/T 1.5L Federal
28–32DIDTF–405 M/T 1.5L California
28–32DIDTF–406 A/T 1.5L California
Throttle bore:
Primary—1.102 in.(28mm)
Secondary—1.260 in.(32mm)
Fast idle opening at 73 F.(23 C.):
M/T—0.019 in.(0.47mm)
A/T and Canada—0.021 in.(0.54mm)
Additional Information:
Deceleration solenoid valve (DSV)
Jet mixture solenoid valve (JSV)
Enrichment solenoid valve (ESV)
Throttle position sensor (TPS)
Type—Variable resistor
Resistance—5 kilo-ohms
Bowl vent valve (BVV)
Dash pot
Choke type—Full-automatic type

1985–86 CORDIA/TREDIA WITH 2.0L ENGINE

Type—Down-draft, 2 barrel, feedback.Model No.:
32–35DIDTF–200 M/T 2.0L Federal
32–35DIDTF–201 A/T 2.0L Federal
32–35DIDTF–195 M/T 2.0L California
32–35DIDTF–196 A/T 2.0L California
Throttle bore:
Primary—1.260 in. (32mm)
Secondary—1.378 in. (35mm)
Fast idle opening at 73 F.(23 C.):
M/T—0.025 in. (0.63mm)
A/T—0.028 in. (0.71mm)
Additional Information:
Deceleration solenoid valve (DSV)
Jet mixture solenoid valve (JSV)
Enrichment solenoid valve (ESV)
Throttle position sensor (TPS)
Type—Variable resistor
Resistance—5 kilo-ohms
Bowl vent valve (BVV)
Dash pot
Choke type—Full-automatic

ELECTRONICALLY CONTROLLED INJECTION SYSTEM

1985–86 MIRAGE 1.6L ENGINE

Injection mixer
Model No.:
46EID–602 (M/T)
46EID–652 (A/T)
Injector:
Type—Electromagnetic
No.of injectors—2
Coil resistance—2 ohms
Injector identification mark—B
Throttle bore—1.811 in. (46mm)
Fuel pressure regulator
Regulated pressure—35.6 psi (245 kPa)
Throttle position sensor (TPS):
Type—Variable resistor
Resistance—5 kilo-ohms
Idle position switch Type—Contact point
Idle speed control (ISC)
Electronic control unit (ECU):
Model No.—E2T13271 (M/T)
Air Flow Sensor Type—Karman vortex type with intake air
Intake air temperature sensor Type—Thermistor
Resistance—2.2 kilo-ohms at 77 F. (25 C.)
Resistor:
Model No.—E8T00271
Resistance—6 ohms
Oxygen sensor Type—Zirconia sensor
Coolant temperature sensor Type—Thermistor
Resistance:
16,200 ohms at -4 F. (-20 C.)
2,450 ohms at 68 F. (20 C.)
296 ohms at 176 F. (80 C.)
Pressure sensor:
Model No.—E1T15271
Control relay:
Model No.—E8T00171
Accelerator control method—Cable type

1985–86 STARION

Injection mixer:
Model No.—46EID–603

Injector Type—Electromagnetic type
No. of injectors—2
Coil resistance—2 ohms
Injector identification mark—H
Throttle bore—1.811 in. (46mm)
Fuel pressure regulator:
Regulated pressure—35.6 psi (245 kPa)
Throttle position sensor (TPS) Type—Variable resistor
Resistance—5 kilo-ohms
Idle position switch Type—Contact point
Idle speed control (ISC)
Electronic control unit (ECU):
Model No.—E2T13481 (M/T)
Air Flow Sensor Type—Karman vortex with intake air temperature sensor
Intake air temperature sensor Type—Thermistor
Resistance—2.2 kilo-ohms at 77 F. (25 C.)
Resistor Model No.—E8T00271:
Resistance—6 ohms
Oxygen sensor Type—Zirconia sensor
Coolant temperature sensor Type—Thermistor
Resistance:
16,200 ohms at -4 F. (-20 C.)
2,450 ohms at 68 F. (20 C.)
296 ohms at 176 F. (80 C)
Pressure sensor:
Model No.—E1T15271
Control relay:
Model No.—E8T00571
Accelerator control method—Cable type

1985–86 CORDIA/TREDIA 1.8L ENGINE

Injection mixer:
Model No.—46EID-601 (46EID-604 with auto-cruise lever)
Injector Type—Electromagnetic type
No. of injectors—2
Coil resistance—2 ohms
Injector identification mark—L
Throttle bore—1.811 in. (46mm)
Fuel pressure regulator:
Regulated pressure—35.6 psi (245 kPa)
Throttle position sensor (TPS)Type—Variable resistor
Resistance—5 kilo-ohms
Idle position switch Type—Contact point
Idle speed control (ISC)
Electronic control unit (ECU):
Model No.—E2T13381
Air Flow Sensor Type—Karman vortex type with intake air temperature sensor
Intake air temperature sensor Type—Thermistor
Resistance—2.2 kilo-ohms at 77 F. (25 C.)
Resistor:
Model No.—E8T00271
Resistance—6 ohms
Oxygen sensor Type—Zirconia sensor
Coolant temperature sensor Type—Thermistor
Resistance:
16,200 ohms at -4 F. (-20 C.)
2,450 ohms at 68 F. (20 C.)
296 ohms at 176 F. (80 C.)
Pressure sensor
Model No.—E1T15271
Control relay Model No.—E8T00171
Accelerator control method—Cable type

1985–86 GALANT

Injection mixer:
Model No.—50EIS-100
Injector Type—Electromagnetic
No. of injectors—2
Coil resistance—2 ohms
Throttle bore—1.811 in. (46mm)
Fuel pressure regulator:
Regulated pressure—35.6 psi (245 kPa)
Throttle position sensor (TPS) Type—Variable resistor
Resistance—5 kilo-ohms
Idle position switch Type—Contact point
Idle speed control (ISC)
Electronic control unit (ECU):
Model No.—E2T13572
Air Flow Sensor Type—Karman vortex type with intake air temperature sensor
Model No.—E5T00174
Intake air temperature sensor Type—Thermistor
Resistance—2.2 kilo-ohms at 77 F. (25 C.)
Oxygen sensor Type—Zirconia sensor
Coolant temperature sensor Type—Thermistor
Resistance:
16,200 ohms at -4 F. (-20 C.)
2,450 ohms at 68 F. (20 C.)
296 ohms at 176 F. (80 C.)
Atmospheric Pressure sensor—Located in the ECU
Control relay Model No.—E8T00171
Accelerator control method—Cable type

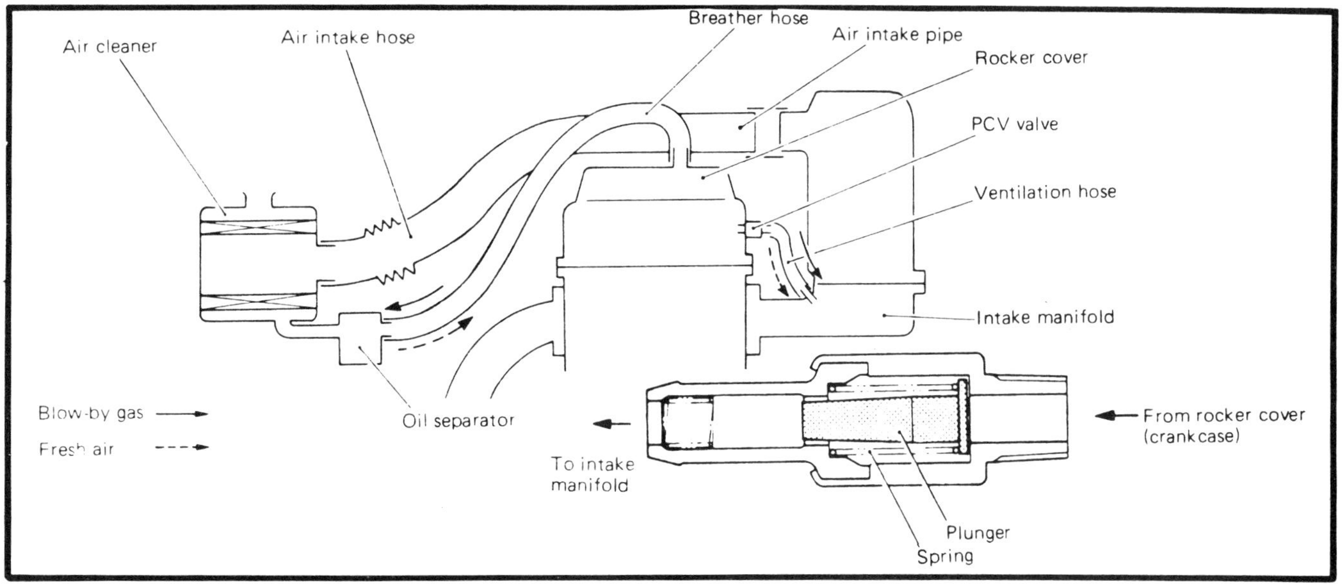

Typical—crankcase emission control system (© Mitsubishi Motor Sales of America, Inc.)

EMISSION CONTROL SYSTEMS

CRANKCASE EMISSION CONTROL SYSTEM

PCV System

All vehicles are equipped with a closed crankcase ventilation system. The system consists of a PCV valve mounted on the cylinder head cover, with a hose extending from the valve to the base of the carburetor. A closed engine oil inlet air cleaner with a hose connecting it to the carburetor air cleaner housing, provides a source of air to the system. The ventilation system operates by manifold vacuum, which is air drawn from the carburetor air cleaner through the crankcase air cleaner and into the engine crankcase. It is then circulated through the engine and drawn out through the PVC valve hose and into a passage at the base of the carburetor where it becomes part of the air/fuel mixture.It is then drawn into the combustion chamber where it is burned and expelled with the exhaust gases.

This system is essentially the same as the system used on vehicles equipped with electronic fuel injection. If the vehicle is equipped with electronic fuel injection, the engine, when operating, generates crankcase pressure that is used to purge crankcase vapors through the PCV valve to a port in the throttle body. When the engine is shut down the cooling gases in the crankcase causes a partial vacuum which is relieved by a breather hose that draws filtered air from the air cleaner assembly to the right valve cover. However, all air entering the intake manifold must be measured by the air flow meter to insure proper air/fuel ratio.

It is very important that the PCV system on a vehicle equipped with electronic fuel injection remain a closed system and be free from any disconnects or breaks in the hoses which may allow unmetered air to enter the intake system. Such conditions would cause poor driveability, fuel economy and performance.

System Service

NOTE: The PCV valve should be checked every 15,000 miles and, replaced as required. If the valve is found to be defective, do not attempt to clean it, replace it.

1. With the engine running remove the valve from its mounting. A hissing sound should be heard and vacuum should be felt from the inlet side of the valve.
2. Reinstall the valve. Remove the crankcase inlet air cleaner. Loosely hold a piece of stiff paper over the opening in the rocker cover. Allow one minute for the crankcase pressure to reduce itself. The paper should then suck itself against the rocker cover with noticeable force. Replace the inlet air cleaner in the rocker cover.
3. With the engine stopped, remove the PCV valve. Shake the valve. A clicking sound should be heard, indicating that the valve is not stuck.
4. If the valve fails any of the above tests, it should be replaced.

FUEL EVAPORATIVE EMISSION

Evaporation Control System

The function of the evaporation control system is to prevent the emission of gas vapors from the fuel tank and carburetor to be expelled into the atmosphere. When fuel evaporates in the gas tank or the carburetor float chamber, the vapors pass through the vent hoses to the charcoal canister, where they are stored until they can be drawn into the intake manifold when the engine is running. All vehicles are equipped with the charcoal canister for the storage of fuel vapors. The fuel bowls of all carburetors are vented internally and on some applications, do not

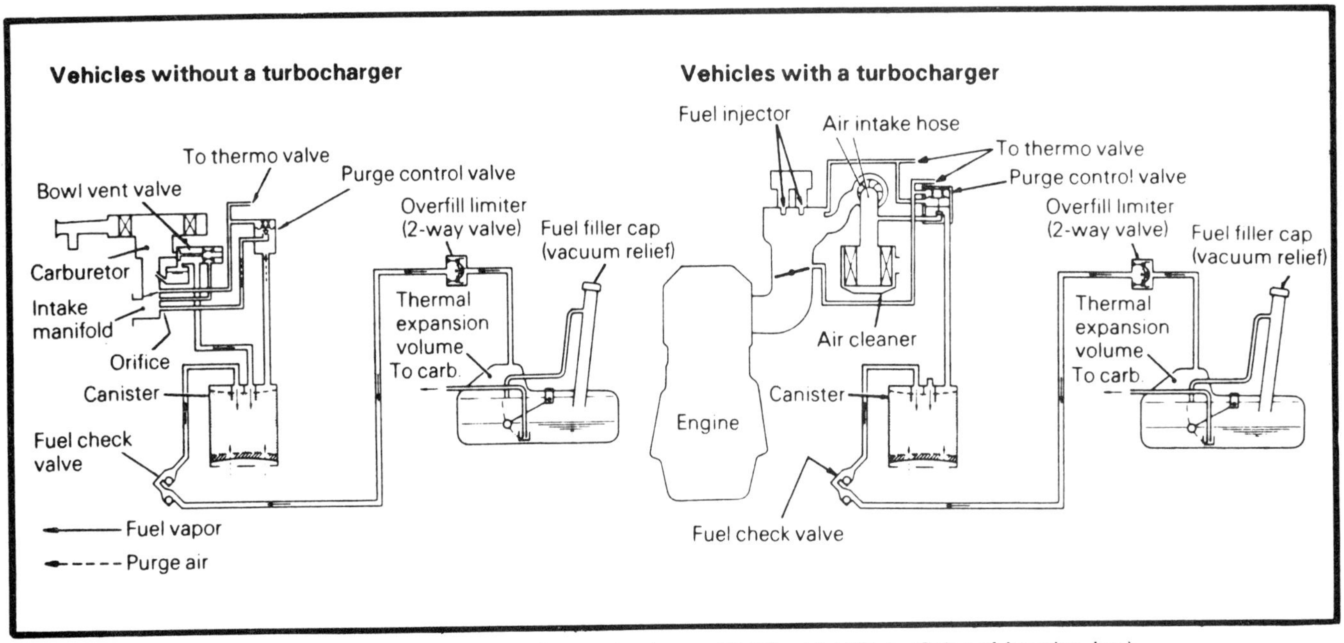

Fuel evaporative emission control system—with and without turbocharger (© Mitsubishi Motor Sales of America, Inc.)

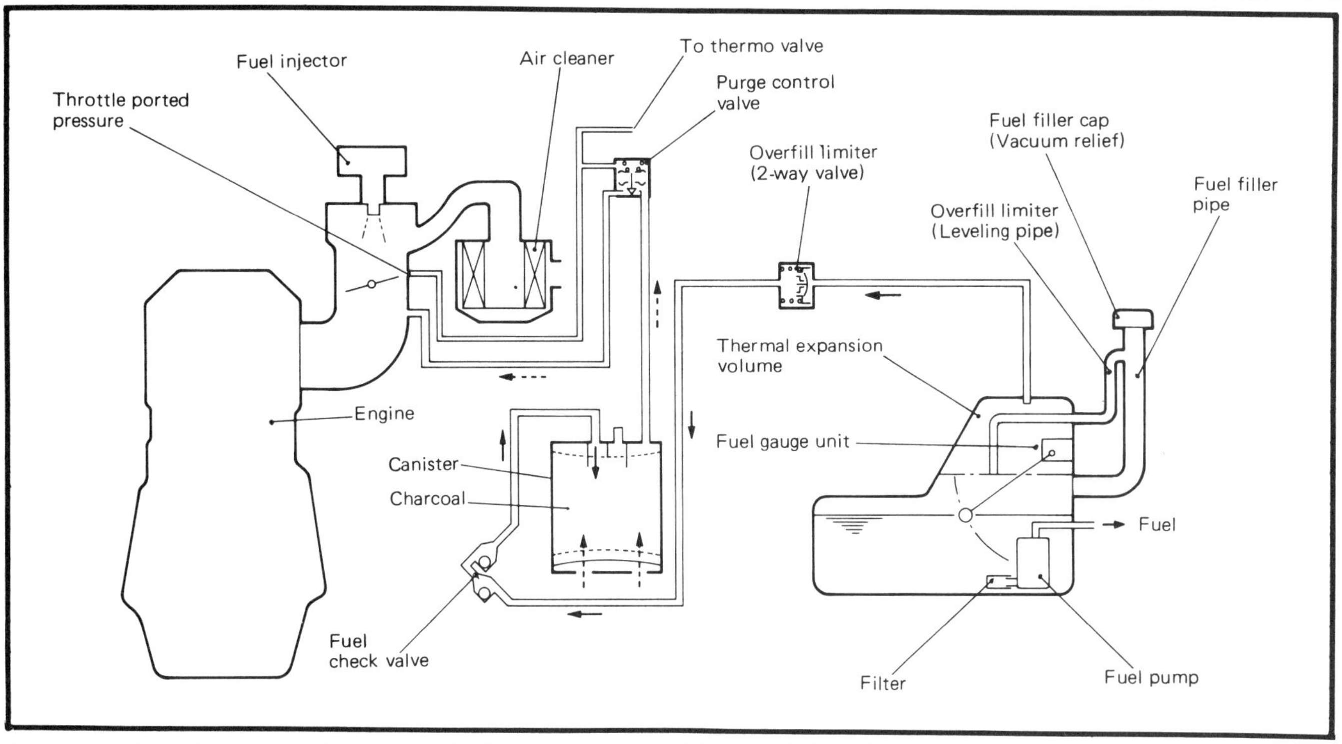

Fuel evaporative emission control system with fuel injection (© Mitsubishi Motor Sales of America, Inc.)

require venting the canister. If this is the case, the bowl vent port on the canister is plugged. However, most carburetors are externally vented to the charcoal canister.

System Service

The charcoal canister itself is a non serviceable component. The only service required for the system is to replace the filter pad on the bottom of the canister. The filter requires replacement every 12 months or 12,000 miles.If the vehicle is driven under severe conditions, the filter should be replaced more often.

BOWL VENT VALVE

The bowl vent valve is connected to the carburetor fuel bowl, the charcoal canister and the air pump discharge. When the engine is not running, there is no air pump pressure applied. There is a direct connection between the fuel bowl and the canister. When the engine is shut down, the valve air pressure bleeds down and the fuel bowl is allowed to vent into the canister.

OVERFILL LIMITER (TWO-WAY VALVE)

The overfill limiter consists of a pressure valve and a vacuum valve. The pressure valve is designed to open when the fuel tank internal pressure has increased over the normal pressure and the vacuum valve opens when a vacuum has been produced in the tank.

PURGE CONTROL VALVE

Carbureted Models

The purge control valve is kept closed during idling in order to prevent vaporized fuel from entering into the intake manifold, for positive control of of high idle CO emission, which is a particular problem under high ambient temperatures. When the carburetor vacuum, working on the diaphragm of the valve, exceeds the pre-set value, the purge control valve is opened.

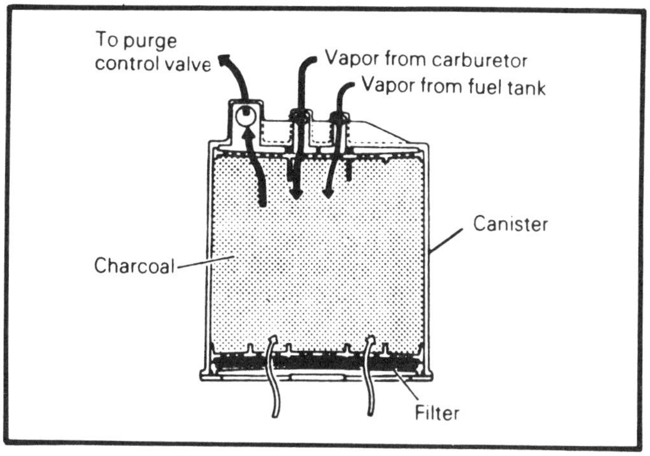

Typical fuel vapor canister
(© Mitsubishi Motor Sales of America, Inc.)

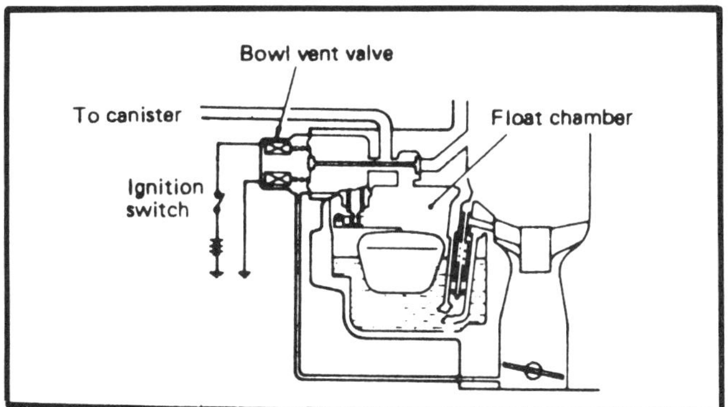

Bowl vent valve system
(© Mitsubishi Motor Sales of America, Inc.)

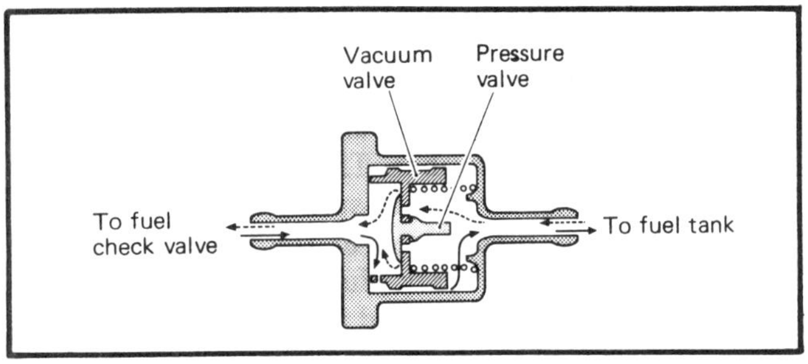

Overfill limiter valve—typical
(© Mitsubishi Motor Sales of America, Inc.)

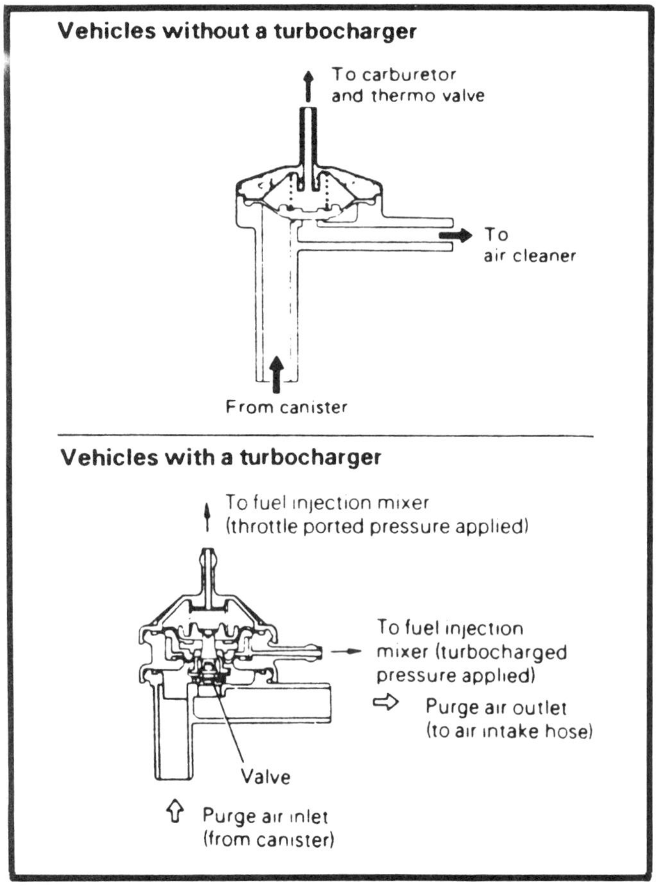

Purge control valve—with and without turbocharger
(© Mitsubishi Motor Sales of America, Inc.)

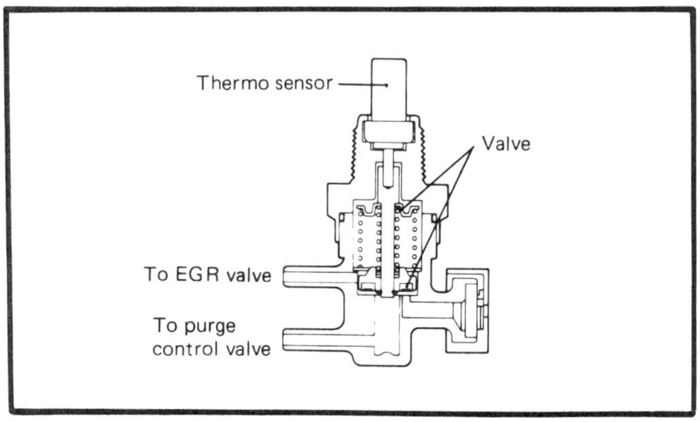

Thermo valve (© Mitsubishi Motor Sales of America, Inc.)

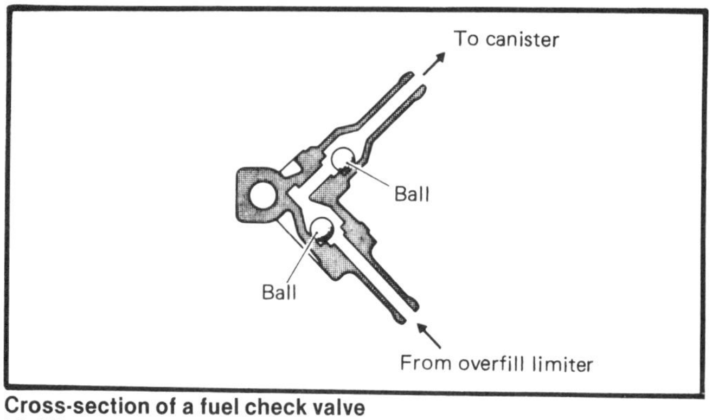

Cross-section of a fuel check valve
(© Mitsubishi Motor Sales of America, Inc.)

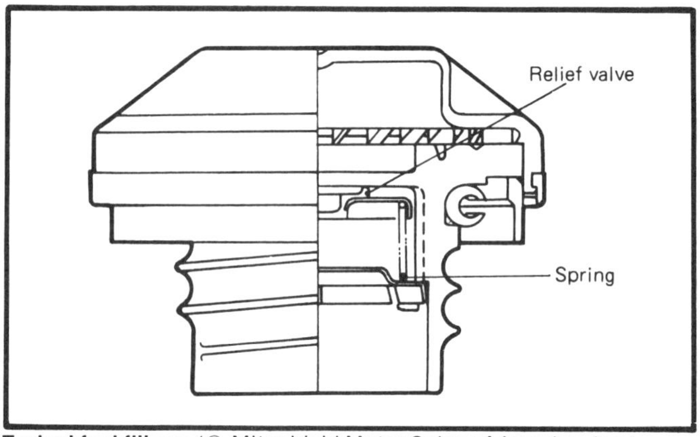

Typical fuel fill cap (© Mitsubishi Motor Sales of America, Inc.)

ECI AND TURBOCHARGED ENGINES

The purge control valve is closed at idling to prevent vaporized fuel from entering into the air intake hose, for positive control of high idle CO emissions which is a particular problem under high ambient temperature. Once the pressure difference between the turbocharger and the throttle ported pressures exceeds the pre-set value, the purge control valve is opened.

THERMO VALVES

A thermo valve, incorporated in the EGR system for sensing the coolant temperature at the intake manifold, closes the purge control valve when the coolant temperature is lower than a pre-set value, in order to reduce CO and HC emissions under warm-up conditions and opens the purge control valve when the coolant temperature becomes above the preset temperature.

FUEL CHECK VALVE

The fuel check valve is used to prevent fuel leaks, should the vehicle suddenly roll over. The valve is connected in the fuel vapor line (between canister and overfill limiter) and is installed on the firewall. The fuel check valve contains two balls and under normal conditions, the gasoline vapor passage in the valve is opened, but if a vehicle roll-over occurs, one of the balls closes the fuel passage, thus preventing fuel leaks.

FUEL FILLER CAP

The fuel filler cap is equipped with relief valve to prevent the escape of fuel vapor into the atmosphere.

EXHAUST EMISSION CONTROL SYSTEM

CARBURETED ENGINES

Exhaust emission (carbon monoxide, Hydrocarbons and oxides of nitrogen) are controlled by a combination of engine modifications and the addition of special control components. Modifications to the combustion chamber, intake manifold, camshaft, carburetor and ignition system form a basic control system. Additional control devices include an exhaust gas recirculation system (EGR valves and thermo valves), a jet air control valve, a deceleration spark-advance system, a coasting air valve, an air switching valve, a dashpot, dual catalytic converters, a secondary air supply system (pulse air feeder) and a heated air intake system. These systems have been integrated into a highly effective system which controls exhaust emissions while maintaining good vehicle performance.

ECI TURBOCHARGED ENGINES

Exhaust emission (carbon monoxide, Hydrocarbons and oxides of nitrogen) are controlled by a combination of engine modifications and the addition of special control components. Modifications are to the combustion chamber, intake manifold and camshaft. Additional control devices include an exhaust gas recirculation system (EGR), dual catalytic converters, a secondary air supply system and a fuel control system. These systems have been integrated into a highly effective system which controls exhaust emissions while maintaining good vehicle performance.

JET AIR SYSTEM

In addition to the intake valve and the exhaust valve, a jet valve has been added to draw jet air (super lean mixture or air) into the combustion chamber. The jet valve assembly consists of a jet valve, jet body and spring, which is all screwed into the jet piece and press fitted into the cylinder head with the jet opening toward the spark plug. The jet valve draws its air from an air passage in the carburetor, intake manifold and cylinder head. As the air flows through the two intake openings, the air goes through the carburetor passage into the intake manifold and cylinder head and flows through the jet valve and opening into the combustion chamber. The jet valve is operated by the same cam as the intake valve and are joined by a common rocker arm so that the jet valve and intake valve open and close at the same time.

On the intake stroke, the air-fuel mixture flows through the intake valve port into the combustion chamber. At the same time, the jet air is forced into the combustion chamber and the jet air running out of the jet opening, scavanges the residual gases around the spark plug and creates a good ignition condition. It also produces a strong swirl in the combustion chamber, which continues through out the compression stroke and improves the flame spread after ignition, assuring a high combustion efficiency. The jet air swirl dwindles with the increase of the throttle valve opening, but the intensified inflow of normal intake air-mixture can satisfactorily promote combustion.

JET VALVE

Clearance Adjustment

ALL MODELS

NOTE: An incorrect jet valve clearance would affect the emission levels and could also cause engine troubles, so the jet valve clearance must be correctly adjusted. Adjust the jet valve clearance before adjusting the intake valve clearance. Furthermore, the cylinder head bolts should be retightened before making this adjustment. The jet valve clearance should be adjusted with the adjusting screw on the intake valve side fully loosened.

1. Start the engine and let it run at idle until it reaches normal operating temperature.
2. With the piston in the cylinder positioned at top dead center on the compression stroke, loosen the adjusting screw for the intake valve two or more turns.
3. Loosen the locknut on the adjusting screw for the jet valve. Turn the adjusting screw counterclockwise and insert a 0.010 in. (0.25mm) feeler gauge between the jet valve stem and the adjusting screw.
4. Tighten the adjusting screw until it touches the feeler gauge. Turn the locknut to secure it, while holding the rocker arm adjusting screw with a suitable tool to keep it from turning.
5. Be sure that the 0.010 in. (0.25mm) feeler gauge can be easily inserted and then adjust the intake valve clearance to 0.006 in. (0.15mm).

NOTE: The exhaust valve clearance is 0.010 in. (0.25mm) and all the valve clearances are to be set with the engine hot.

6. Check the idle speed and Co. Adjust if necessary.

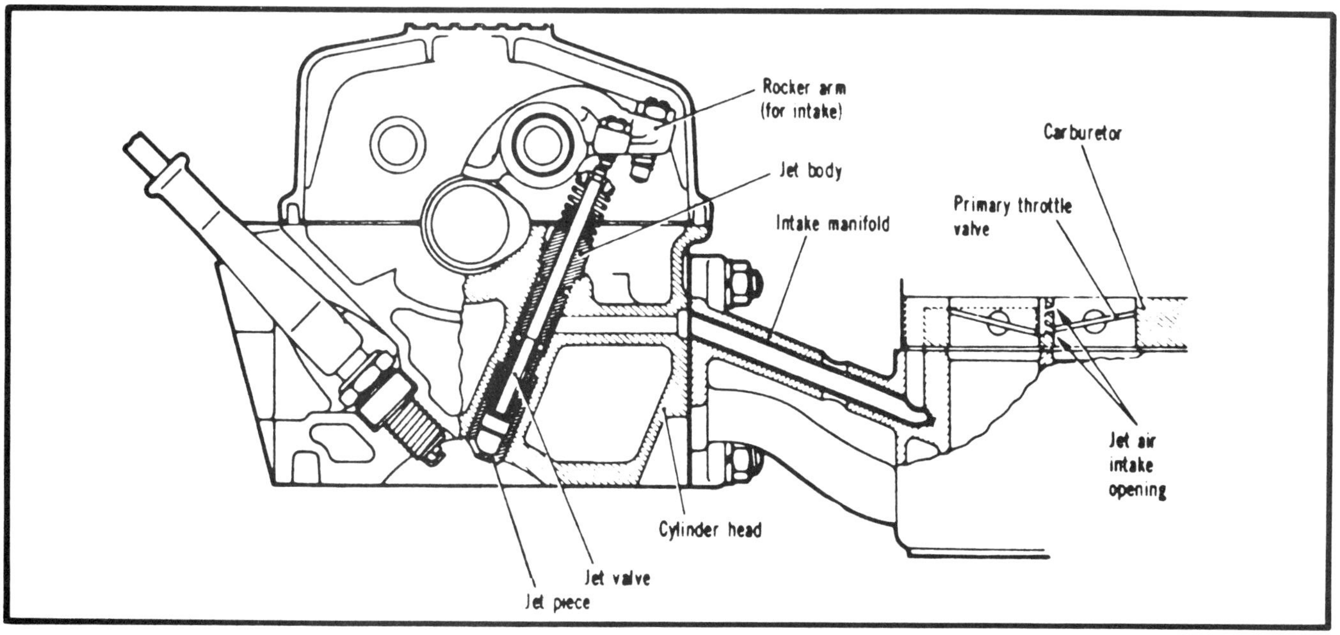

Exploded view of the jet valve system (© Mitsubishi Motor Sales of America, Inc.)

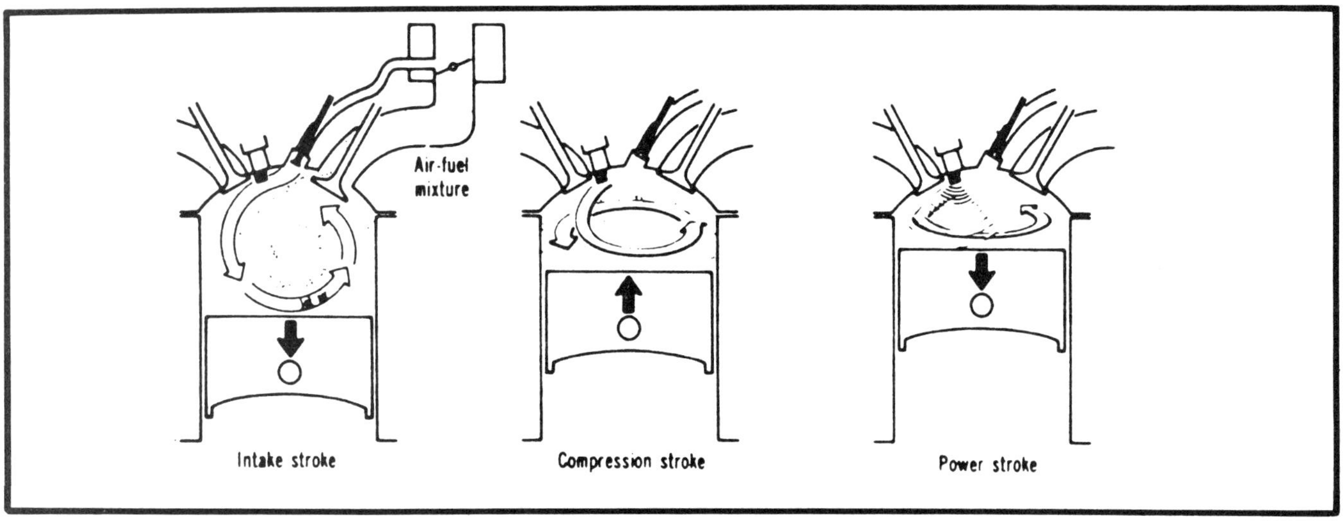

Operation of the jet valve system (© Mitsubishi Motor Sales of America, Inc.)

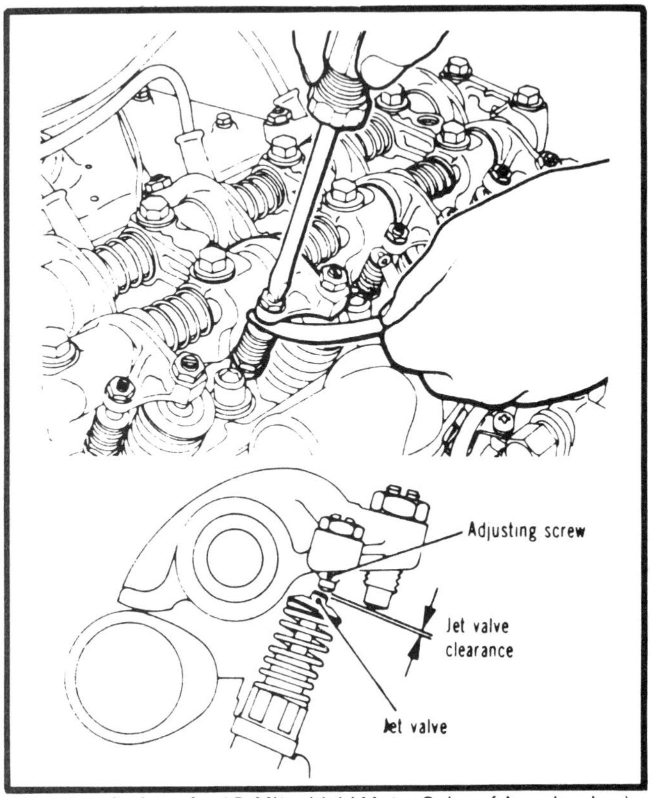

Adjusting the jet valve (© Mitsubishi Motor Sales of America, Inc.)

Removal

ALL MODELS

1. Disconnect the negative battery cable and remove the rocker arm cover, rocker arm and the rocker shaft. Using special tool, jet valve socket wrench MD998310 or its equivalent, remove the jet valve.

NOTE: When the jet valve socket wrench is used, be sure the wrench is not tilted with respect to the center of the jet valve. If the tool is tilted, the valve stem might be bent by the force exerted on the valve spring retainer, resulting in defective jet valve operation.

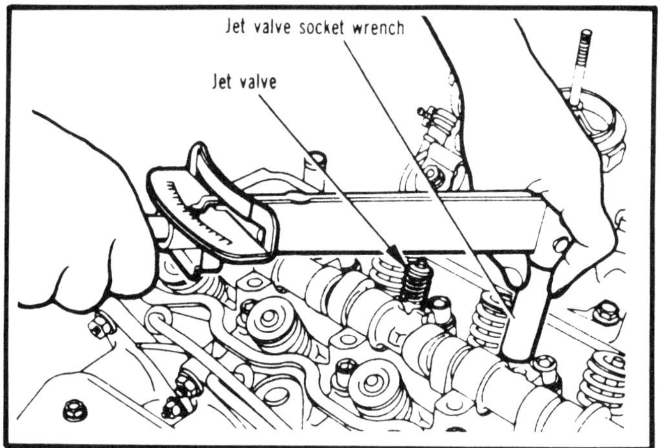

Removing the jet valve
(© Mitsubishi Motor Sales of America, Inc.)

2. Using a hand held spring compressor, remove the jet valve spring retainer lock, valve spring retainer and the valve spring.
3. Check the jet valve assembly for deterioration, cracks or damage, and replace if defective.

Installation

1. Using a seal installer, install a new jet valve stem seal. Do not reuse the old jet valve stem seal.
2. Apply engine oil to the stem of the jet valve and insert the jet valve into the jet body. Be sure that the valve slides smoothly.
3. Using the spring compressor, compress the jet valve spring and install the jet valve spring retainer. Be careful not to damage the valve stem with the spring compressor.
4. Install a new O-ring in the groove around the jet body and apply engine oil to the O-ring.
5. Screw the jet valve assembly into the cylinder head, using special tool MD 998310 jet valve socket wrench or equivalent. Torque the jet valve assembly to 13–15 ft.lbs.(18–21 Nm).
6. Install the rocker shaft, rocker arm, rocker arm cover and install the negative battery cable.

Catalytic Converter

Catalytic converters require the use of unleaded fuel only. Leaded fuel will destroy the effectiveness of the catalysts as an emission control

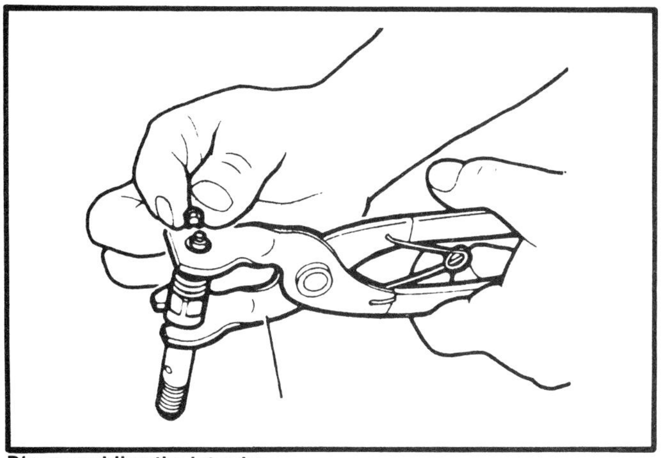

Disassembling the jet valve
(© Mitsubishi Motor Sales of America, Inc.)

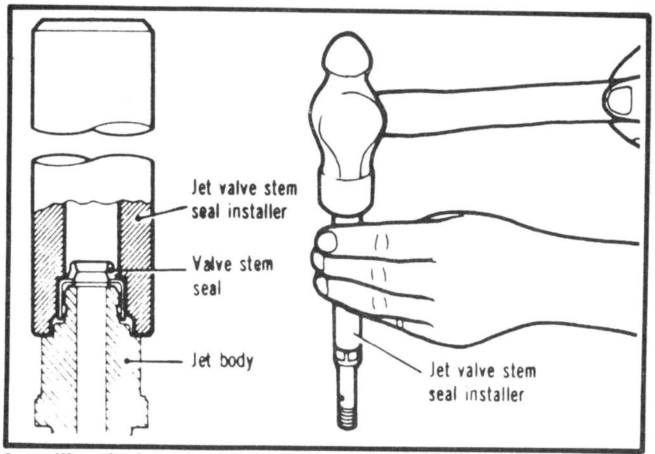

Installing the jet valve (© Mitsubishi Motor Sales of America, Inc.)

device. Under normal operating conditions, the converter will not require any maintenance. However, it is important to keep the engine properly tuned. If the engine is not kept properly tuned, engine misfiring may cause overheating of the cataylst. This may cause heat damage to the converters or vehicle components. This situation can also occur during diagnostic testing, if any spark plug cables are removed and the engine is allowed to idle for a long period of time.

Secondary Air Supply System

REED VALVE

EXCEPT VEHICLES EQUIPPED WITH ECI SYSTEM

The air injection system consists of a reed valve with a secondary air control valve, a solenoid valve, a check valve ECU and some sensors. The reed valve supplies secondary air into the front catalytic converter for the purpose of promoting oxidation of exhaust emissions during the engine warm-up operation, the vehicle deceleration and heavy engine load operation zones. The reed valve is actuated by exhaust vacuum being generated from pulsation in the exhaust manifold and extra air is supplied into the exhaust manifold through the secondary air control valve. The secondary control valve is opened by the intake manifold pressure when the solenoid valve is energized by the ECU, based on the information on coolant temperature, engine speed, time, throttle position and idle position.

VEHICLES EQUIPPED WITH ECI SYSTEM

The air injection system consists of a reed valve with a secondary air control valve, and a solenoid valve. The reed valve supplies secondary air into the front catalytic converter for the purpose of promoting oxidation of exhaust emissions during the engine warm-up operation, the

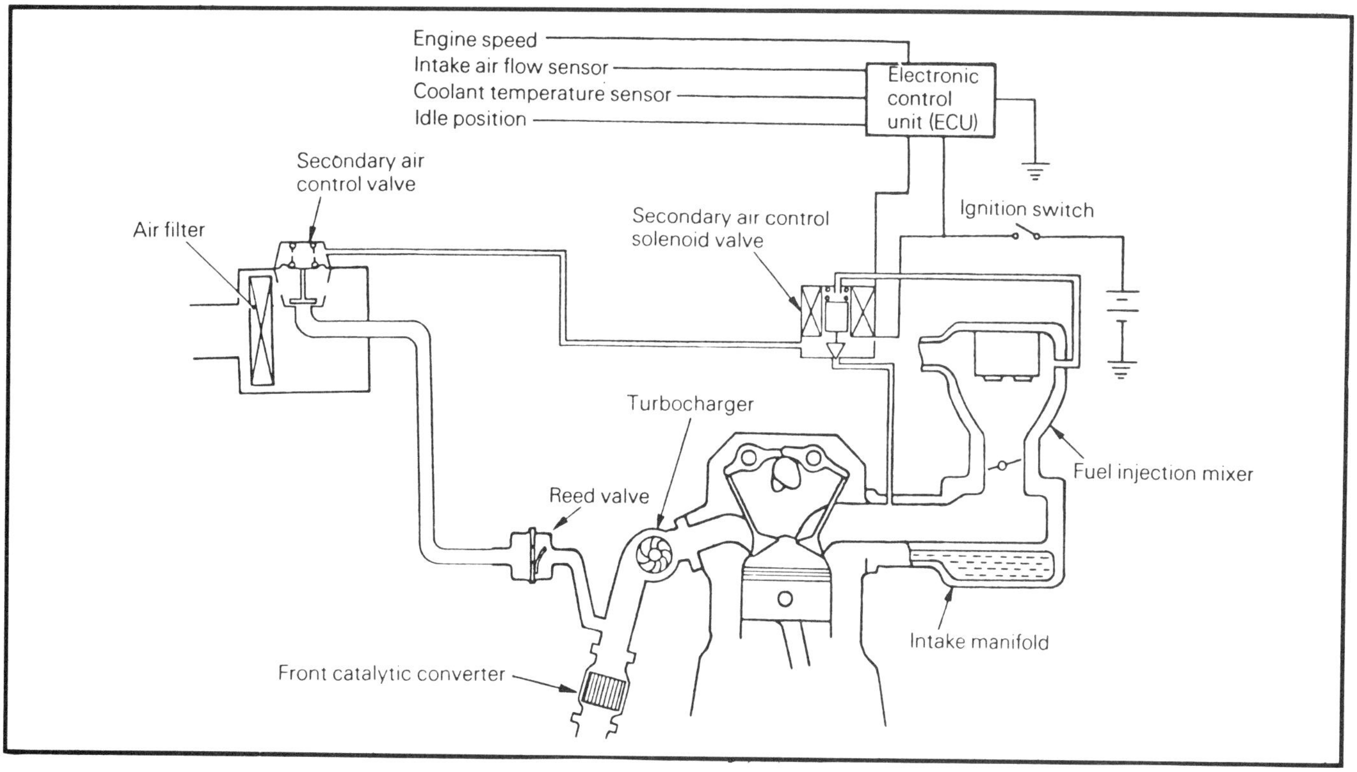

Secondary air supply system—typical (© Mitsubishi Motor Sales of America, Inc.)

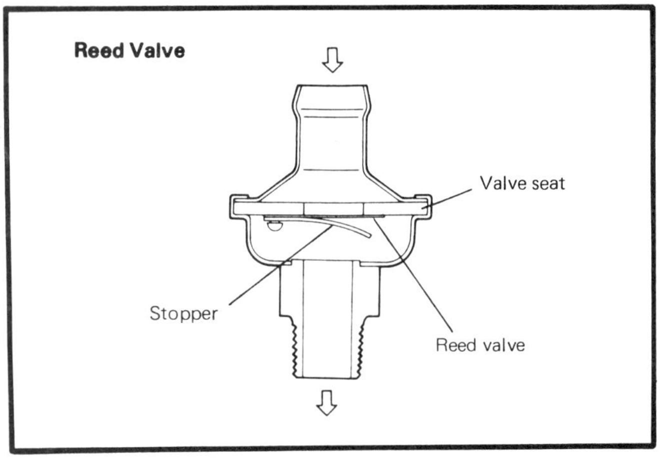

Exploded view of the reed valve
(© Mitsubishi Motor Sales of America, Inc.)

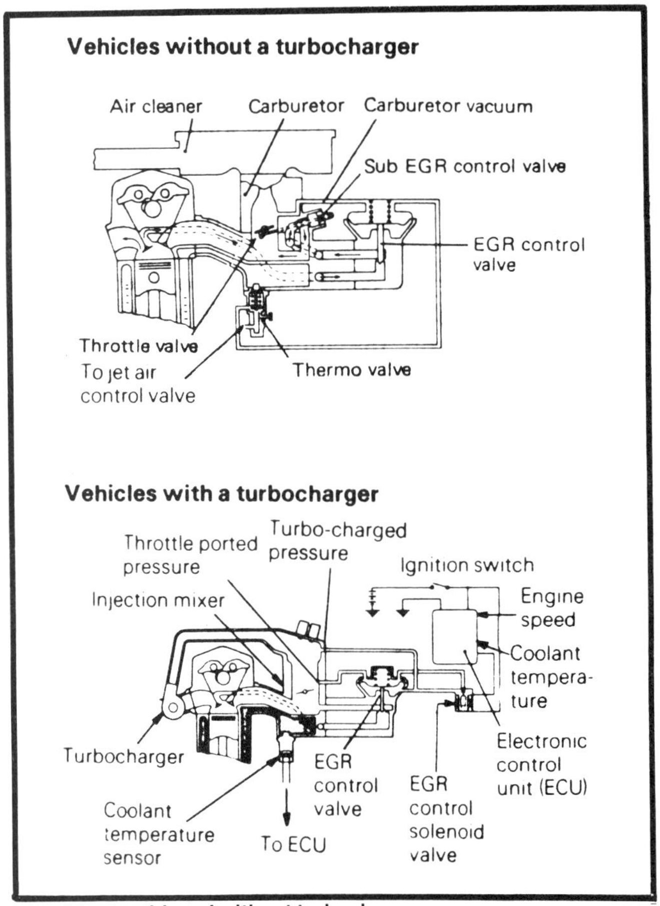

EGR system with and without turbocharger
(© Mitsubishi Motor Sales of America, Inc.)

engine hot-start operation and the vehicle deceleration. The reed valve is actuated by exhaust vacuum being generated from pulsation in the exhaust manifold and extra air is supplied into the exhaust manifold through the secondary air control valve. The secondary control valve is opened by the intake manifold pressure when the solenoid valve is energized by the ECU, based on the information of coolant temperature, intake air flow, engine speed, ignition switch start and idle positon.

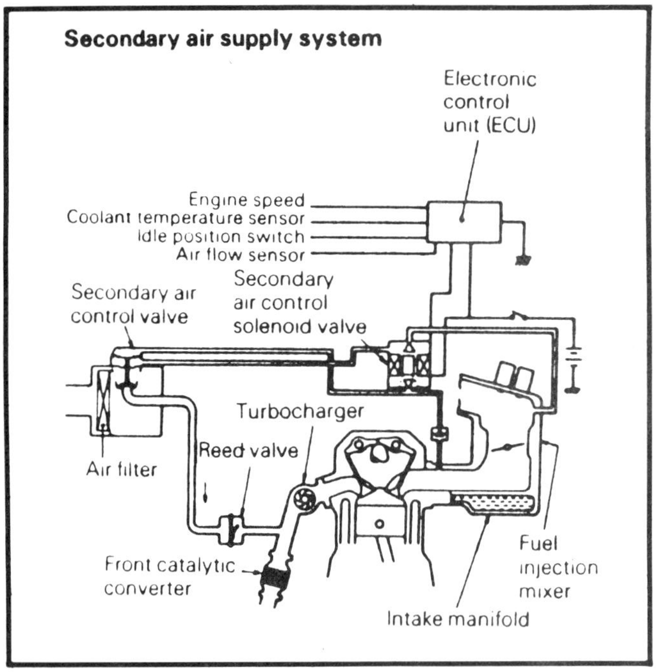

Reed valve for ECI equipped vehicles
(© Mitsubishi Motor Sales of America, Inc.)

Exhaust Gas Recirculation (EGR) System

CORDIA/TREDIA AND MIRAGE

In this exhaust gas recirculation system, the exhaust gas is partially recirculated from an exhaust port of the cylinder head, into a port located at the intake manifold below the carburetor, while the exhaust gas recirculation flow is controlled the the EGR control valve, a sub EGR valve and the thermo valve. Stringent oxides of nitrogen standard requires increased exhaust gas recirculation flow, which in turn affects driveability. To stop this problem, the EGR flow is increased to attain effective oxides of nitrogen reduction.

STARION AND GALANT

In this exhaust gas recirculation system, the exhaust gas is partially recirculated from an exhaust port of the cylinder head, into a port located at the intake manifold below the fuel injection mixer, while the exhaust gas recirculation flow is controlled the the EGR control valve and a solenoid valve. The pressure applied to the EGR control valve is controlled by the solenoid valve motion, which mixes the throttle ported pressure with the turbo charged pressure. The solenoid valve motion is controlled by the ECU, based on the information from the coolant temperature and engine speed.

System Service

The valve should be inspected every 15,000 miles for deposits. Particular attention should be given to the poppet and seat area of the valve. If deposits are present, the valve should be cleaned. Manifold heat control valve solvent or equivalent should be used when cleaning the valve, this compound will allow the deposits on the poppet and seat area to soften. The deposits may be removed by opening the valve, using an external vacuum source and scrapping the deposits from the poppet and seat with a sharp suitable tool. If wear to the stem or other moving components is noticed, the EGR valve should be replaced. Care should be used when cleaning the valve, so as not to damage the diaphragm. Once the diaphragm has be damaged, the valve will have to be replaced.

System Testing

ALL MODELS

1. Inspect all hose connections between the carburetor, intake man-

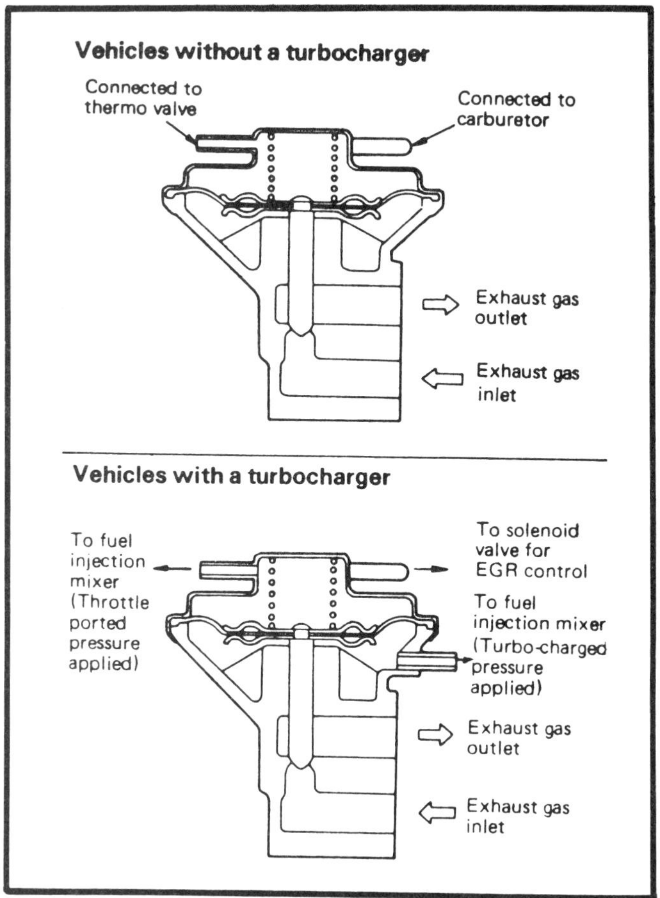

Typical EGR valves with and without turbocharger
(© Mitsubishi Motor Sales of America, Inc.)

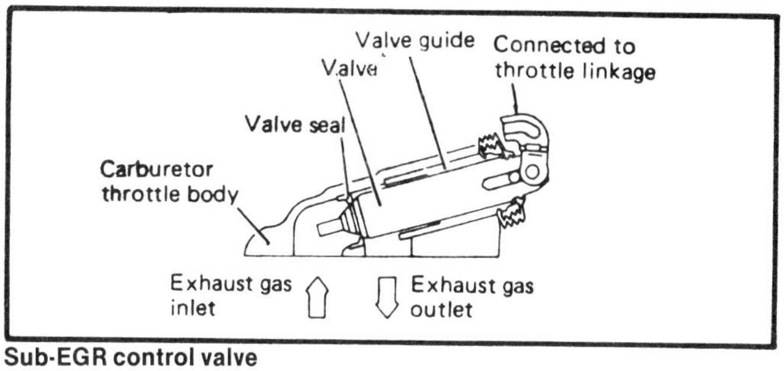

Sub-EGR control valve
(© Mitsubishi Motor Sales of America, Inc.)

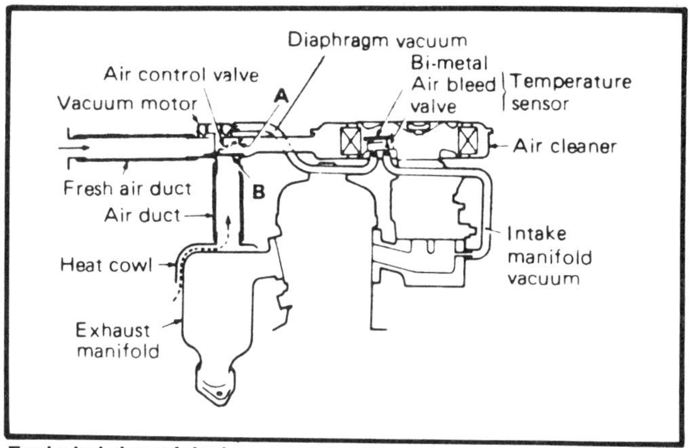

Exploded view of the heated air intake system
(© Mitsubishi Motor Sales of America, Inc.)

ifold and the EGR control valve. Replace any and all hoses that are found to be defective.

2. With the engine warmed up and running, position the selector lever in the neutral position and accelerate the engine to about 2000 rpm, but not over 3,000 rpm.

3. Visible movement in the EGR valve stem should be seen. During this operation, a change in the relative position of the groove on the EGR valve stem will occur.

4. This operation can be repeated several times to confirm EGR operation. Movement of the stem indicates the the valve is functioning properly.

5. If the stem does not move accordingly, the EGR valve is not functioning properly and the valve should be checked.

SUB-EGR CONTROL VALVE

CORDIA/TREDIA AND MIRAGE

The Sub EGR control valve is directly opened and closed with the motion of the throttle valve, through a linkage set up, in order to closely modulate the EGR flow, controlled by the EGR control valve, in response to the throttle valve opening.

THERMO VALVE

ALL MODELS

The Thermo valve in the EGR vacuum supply line is a switch sensitive to engine temperature. The thermo valve may eliminate the vacuum signal to the EGR valve during warm-up when less oxides of nitrogen NOx is generated and less EGR is needed to maintain acceptable driveability.

SOLENOID VALVE

STARION

When the solenoid valve is incorporated in the EGR system, it is energized by the electronic control unit (ECU). The EGR control valve is closed due to the throttle ported pressure mixed with the turbocharged pressure and is opened with the aid of the throttle ported vacuum applied to its diaphragm, when the sòlenoid valve is de-energized by the ECU.

Solenoid Valve Test

1. Start the engine and let it idle until it reaches normal operating temperature.

2. Remove the green stripe hose from the injector mixer and connect a hand held vacuum pump or equivalent, to the end of the hose.

3. Disconnect the connector of the solenoid valve and apply a vacuum of 9.8 in. Hg. If a unstable idling occurs or engine stalls at this time, the EGR system is good.

4. Reconnect the connector of the solenoid valve and perform the test in step four again. This time, the engine should remain unaffected as negative pressure chamber of the EGR valve is opened to the atmosphere.

5. If an unstable idle occurs, it means that the solenoid valve is not operating properly and it is staying in the close position. Therefore the valve should be removed and replaced.

Heated Air Intake System

CORDIA/TREDIA AND MIRAGE, EXCEPT WITH TURBOCHARGER

These models are equipped with a temperature regulated air cleaner which improves engine warm up characteristics and helps to minimize carburetor icing. This air cleaner has a door inside the snorkel to modulate the temperature of the air entering the carburetor. The door is controlled by a vacuum motor and a bi-metal temperature sensor combination system, which reacts to intake manifold vacuum and the temperature inside the air horn. When the bi-metal senses the inside air horn temperature at or below 86 degrees F., the air bleed valve portion of the temperature sensor remains closed. Intake manifold vacuum is applied to the vacuum moto,r which opens the air control door to let the pre-heated air flow through the heat cowl and air duct, into the air cleaner.

When the bi-metal sensor senses the inside the air horn temperature at or above 113 degrees F., the air bleed valve portion of the temperature sensor is fully open, this allows the intake air to go directly to the carburetor through the outside air duct, closing the heated air duct regardless of the intake manifold vacuum.

At intermediate temperatures, the air entering the carburetor is a blend of outside air and pre-heated air, as regulated by the bi-metal sensor controlling the degree of opening on the air control door.

System Service

1. Make sure that all vacuum hoses and the stove to the air cleaner flexible connector are properly installed and in good condition.
2. With a cold engine and ambient temperature less than 84 degrees F., the heat control door or valve should be in the up or heat on position.
3. With the engine warm and running, check the air temperature entering the snorkel or at the sensor. When the air temperature entering the outer end of the snorkel is 113 degrees F. or higher, the door should be in the down or heat off position.
4. Remove the air cleaner from the engine and allow it to cool down to 84 degrees F. With 15 inches of vacuum applied to the sensor, the door or valve should be in the up or heat on position. Should the door not rise to the heat on position, check the diaphragm or motor for proper operation.
5. To test the diaphragm or motor, apply 10 in. of vacuum with vacuum pump tool number C-4207 or equivalent.
6. The valve should be in the full up position. Should this not be the case, replace the air cleaner and body assembly.

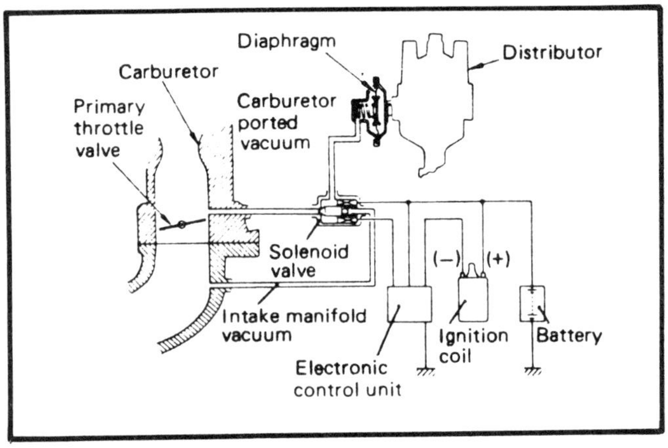

Deceleration spark advance system
(© Mitsubishi Motor Sales of America, Inc.)

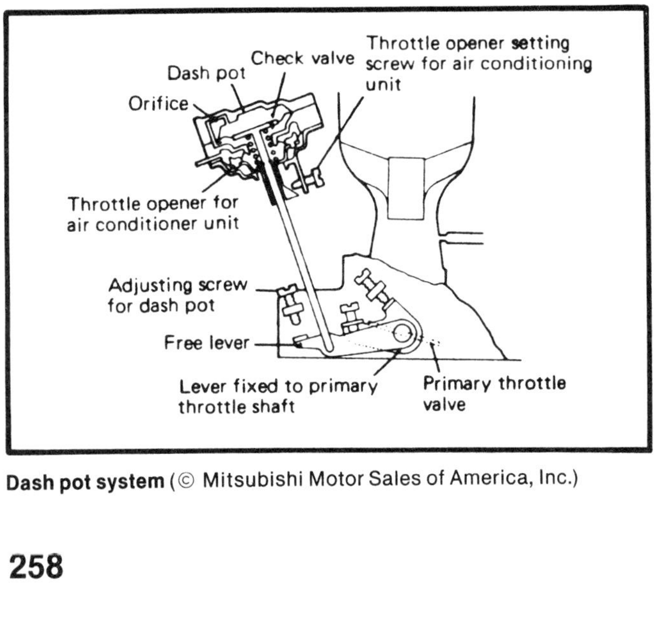

Dash pot system (© Mitsubishi Motor Sales of America, Inc.)

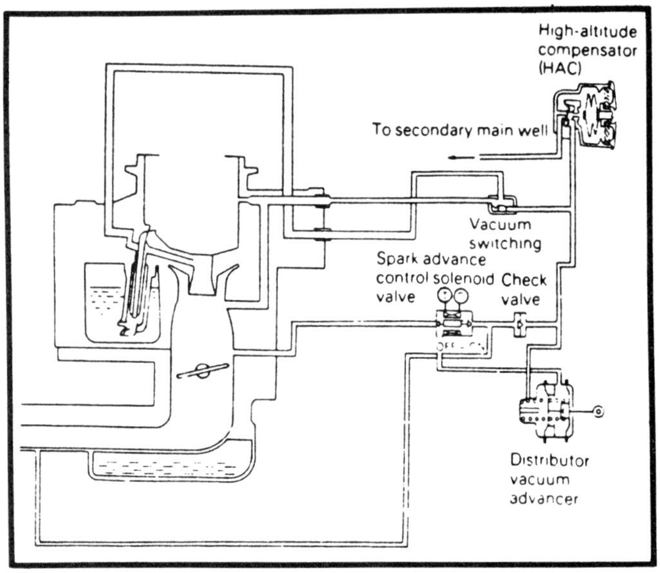

High altitude compensation system
(© Mitsubishi Motor Sales of America, Inc.)

Deceleration Devices

CORDIA/TREDIA AND MIRAGE

The deceleration device is used to decrease hydro-carbon (HC) emissions during vehicle deceleration.

DECELERATION SPARK ADVANCE SYSTEM (EXCEPT VEHICLES WITH A TURBOCHARGER)

CORDIA/TREDIA AND MIRAGE

In order to decrease the hydro-carbon (HC) emissions emitted during vehicle deceleration, ignition timing is advanced by the solenoid operated vacuum valve on the distributor, changing the vacuum supplied to the valve from carburetor ported vacuum to intake manifold vacuum. The solenoid valve is controlled by the electronic controlled unit, based on engine speed.

DECELERATION FUEL CUT-OFF

STARION AND GALANT

To decrease the hydro carbon emissions emitted during vehicle deceleration, the fuel delivery time is dramatically decreased by the electronic control unit, changing the injection interval, which is determined by the specific pulses from the air flow sensor with the specified duration of injection. During the vehicle deceleration, while the idle position switch is not yet in the ON position, the fuel delivery time is determined by the electronic control unit responding to the throttle valve closing speeds.

DASH POT

CORDIA/TREDIA AND MIRAGE

The carburetor is equipped with a dash pot, which delays the throttle valve closure onto its normal idling position, thereby reducing the amount of hydro carbon (HC) emissions emitted.

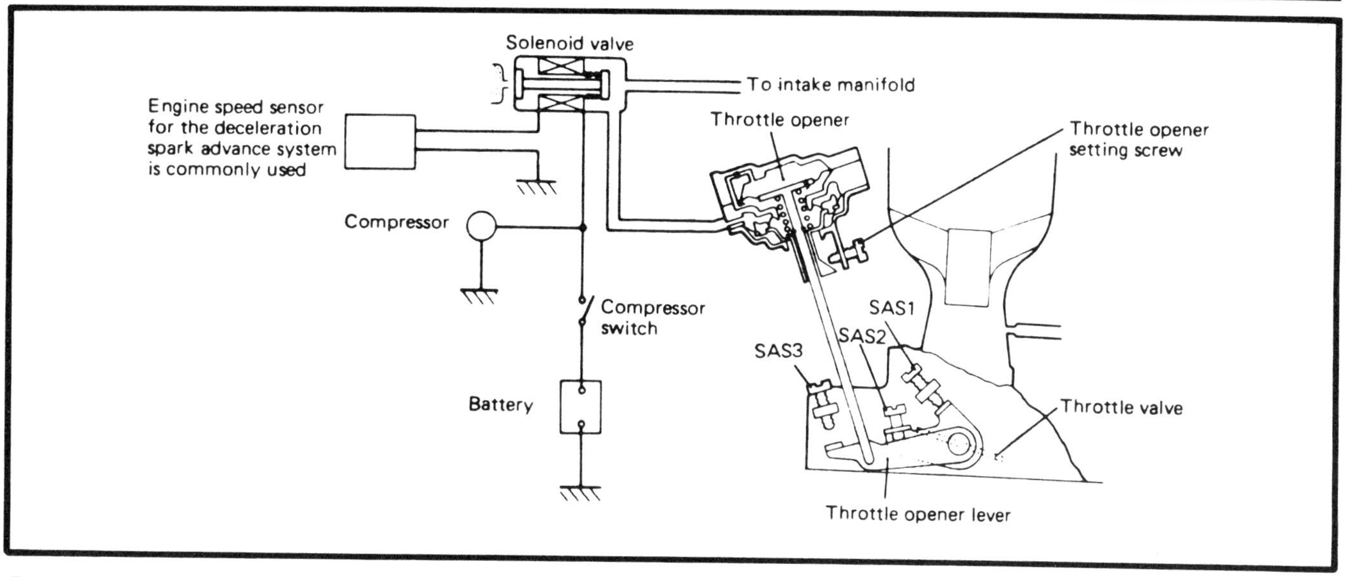

Typical throttle opener system (© Mitsubishi Motor Sales of America, Inc.)

HIGH ALTITUDE COMPENSATION SYSTEM

FEDERAL CARBURETED MODELS

So that all vehicles will meet the federal requirements at all altitudes, all the carbureted vehicles are equipped with high altitude compensation system, in addition to the feedback carburetor system. The high altitude compensation system is made up of the following components; a high altitude compensator (HAC), a vacuum switching valve and a distributor equipped with a high altitude advance system. Air/fuel ratio at high altitude is controlled by the HAC and to be approximately the same value as the ratio at sea level, by supplying additional bleed air into the primary and secondary main wells, through HAC and the vacuum switching valve, controlled by the high altitude compensator.

In order to reduce hydro carbon and carbon monoxide emissions and to get better driveability at high altitude, the ignition timing is advance by specified degrees at high altitude. The spark advance signal is sent to the sub-diaphragm chamber of the distributor through the high altitude compensator. On the California models, instead of using the vacuum switching valve, a check valve is used in its place.

Electronically Controlled Injection Models

The fuel injected vehicles are equipped with a pressure sensor and a solenoid valve, which are utilized as a high altitude compensator. When the ignition switch is in the ON position, the solenoid valve is energized by the ECU for a specific time, which in turn allows the pressure sensor to detect the ambient atmospheric pressure. The fuel delivery time is then determined by the ECU in response to the ambient atmospheric pressure, so the air/fuel mixture at high altitudes can be maintained at the same ratio as at sea level.

THROTTLE OPENER SYSTEM

CORDIA/TREDIA AND MIRAGE (WITH A/C)

This system contains the following components; a solenoid valve, an engine speed sensor and the compressor switch of the A/C unit. When the A/C compressor is turned on, the engine speed will fall below specified value. When this occurs, the solenoid valve will open and in turn transfer the intake manifold vacuum to the throttle opener. The result is the engine running at a higher speed, due to the setting of the new throttle valve opening against the compressor's load. When the A/C compressor is shut off, the throttle opener system stops working and the engine runs at the original engine speed.

TAMPER RESISTANT IDLE MIXTURE

All carburetors will have a tamper resistance idle mixture adjustment. The CO setting has been done as a factory adjustment. Neither removal of the plug nor adjustment of the mixture screw is required in service, unless a major carburetor overhaul, throttle body replacement, or high CO is present and adjustments are needed by state or local inspections.

TAMPER-PROOF AUTOMATIC CHOKE

All carburetors also have tamper-proof choke. The choke-related parts are factory adjusted. Neither removal of the choke cover, nor tampering with wax-stroke adjusting screw (WAS) is required in service, except when major carburetor overhaul or adjustment of choke-calibration-related-parts is required by state or local inspection.

FUEL CUT-OFF SYSTEM

When the ignition key is turned off, the deceleration solenoid valve cuts off the fuel flow to prevent engine run-on. During certain deceleration, the deceleration solenoid valve reduces the fuel flow, in order to decrease hydrocarbon emissions and improve fuel economy. Under normal engine operation, the needle valve is drawn open by the solenoid, to provide the necessary fuel flow for smooth engine operation.

Feedback Carburetor (FBC) System

The feedback carburetor system provides the capability to perform closed loop fuel control. It also provides the capability to control the secondary air system, the deceleration spark control system and the throttle opener system. The basic functions of this system are depicted below. Imput signals from a variety of sensors are fed to a microprocessor based electronic control unit (ECU). The ECU then generates output signals for all of the controlled functions. The feedback carburetor is a 2-barrel, downdraft carburetor designed for the closed loop system. When used with the closed loop system of mixture control, this carburetor includes special design features for optimum air/fuel mixtures during all ranges of engine operation. Fuel metering is accomplished through the use of three solenoid-operated on/off valves (jet mixture, enrichment and deceleration solenoids) adding or reducing fuel to the engine.The activation of the on/off valve is controlled by the length of the time current is supplied to the solenoid. The solenoid operates at a fixed frequency. By varying the amount of time the so-

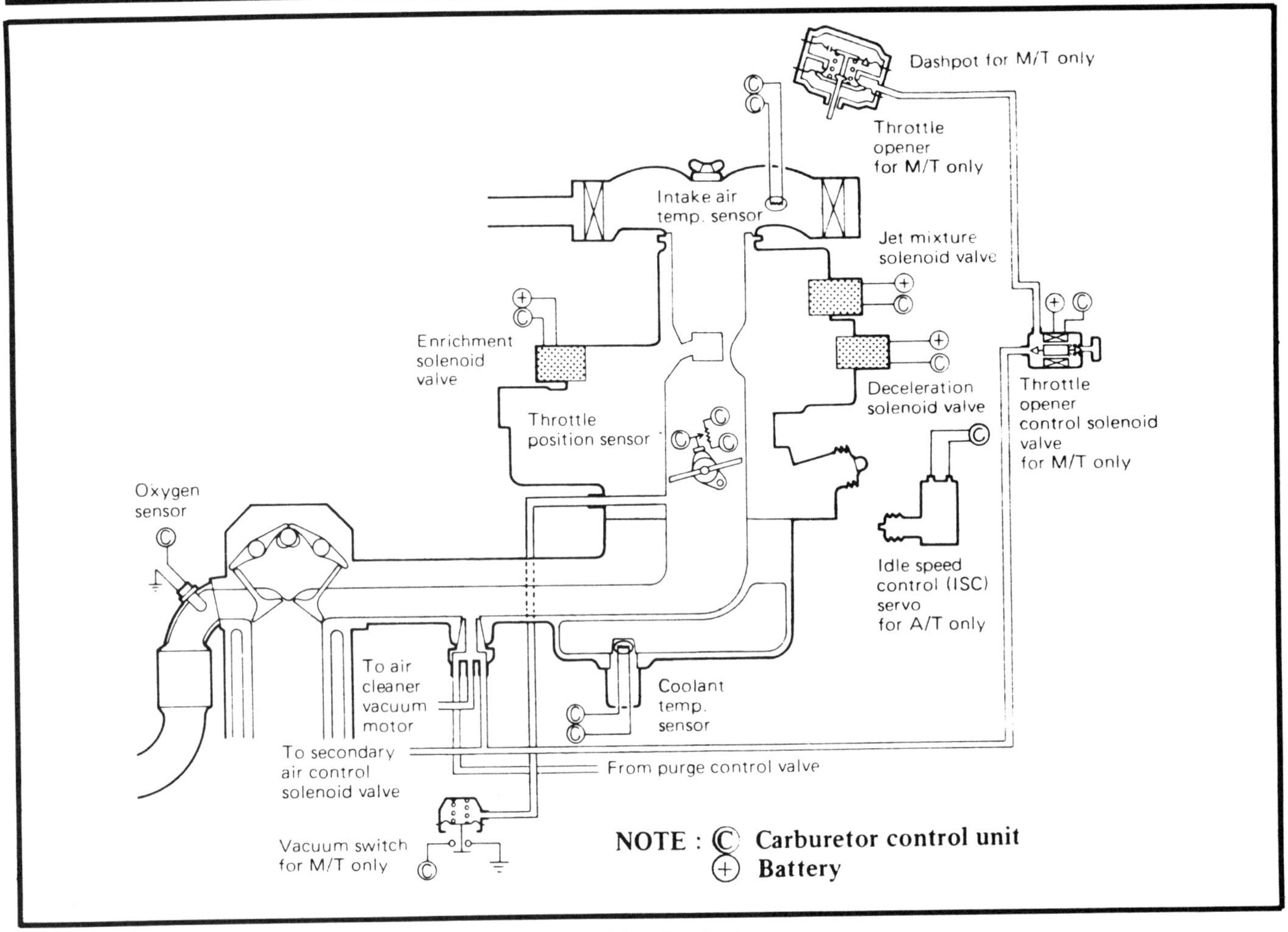

Typical Feedback carburetor system (© Mitsubishi Motor Sales of America, Inc.)

lenoid is energized during each cycle (defined as duty cycle), the air/fuel mixture delivered to the engine can be precisely controlled. The duty cycle to the solenoid is controlled by the ECU in response to the signals from the exhaust oxygen sensor and throttle position sensor. Incorporated in the feedback carburetor are eight basic systems of operation: fuel inlet, primary metering, secondary metering, accelerating pump, choke, jet mixture, enrichment and fuel cut-off.

ELECTRONIC CONTROL UNIT

The electronic control unit is mounted in the passenger compartment and consists of a printed circuit board mounted in a protective metal box. It receives analog inputs from the sensors and converts them into digital signals. These signals and various discrete inputs, are processed and used by the ECU in controlling the fuel delivery, secondary air, deceleration spark and throttle opener managements.

FEEDBACK CARBURETOR AIR/FUEL CONTROL

The feedback carburetor air/fuel ratio is controlled by the ECU. The ECU monitors the throttle position, engine speed, coolant temperature, intake air temperature and exhaust oxygen concentration to calculate the fuel flow required to yield the desired air/fuel ratios for all operating conditions. Closed loop control is used to adjust the fuel flow to yield a near stoichiometeric air/fuel ratio when required. The fuel flow is modified to account for special operating conditions, such as hot starts, acceleration and deceleration.

ADAPTIVE MEMORY CONTROL

During the closed loop operation, the ECU controls the duty cycle of the jet mixture control solenoid valve, based on the output voltage signal from the exhaust sensor. The mean values of the duty cycle is stored in a Random Access Memory (RAM) and the last ones are stored, even if the ignition switch is turned off.

SECONDARY AIR CONTROL

A solenoid is used to control the air control valve signal vacuum. The solenoid is controlled by the ECU, based on the engine speed, idle position and coolant temperature. This valve sends air to the exhaust manifold.

DECELERATION SPARK CONTROL

In order to decrease the hydro-carbon (HC) emissions emitted during vehicle deceleration, ignition timing is advanced by the solenoid operated vacuum valve on the distributor, changing the vacuum supplied to the valve from carburetor ported vacuum to intake manifold vacuum. The solenoid valve is controlled by the electronic controlled unit, based on engine speed.

SENSORS

EXHAUST OXYGEN SENSOR

The oxygen sensor is mounted in the exhaust manifold. The output

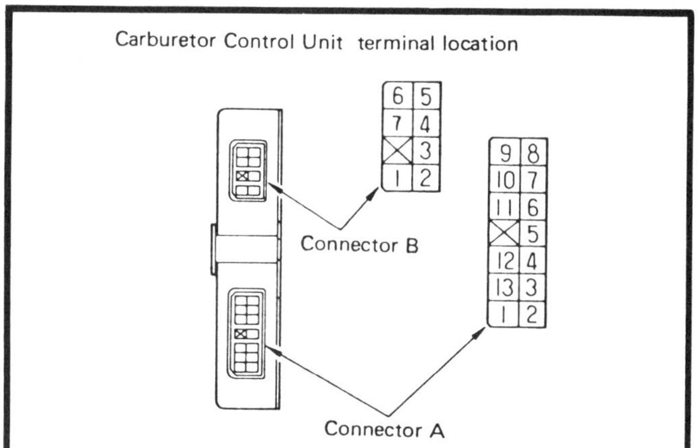

A-1. Oxygen sensor
A-2. Ground for sensor
A-3. Throttle position sensor +
A-4. Intake air temperature sensor
A-5. Idle position switch
A-6. Ground
A-7. Ignition switch
A-8. Ignition switch
A-9. Battery ⊕
A-10. Ignition coil ⊖
A-11. Ground
A-12. Coolant temperature sensor
A-13. Throttle position sensor (output)
B-1. Jet mixture solenoid valve
B-2. Deceleration solenoid valve
B-3. Idle up control solenoid valve
B-4. Secondary air control solenoid
B-5. Enrichment solenoid valve
B-6. Air conditioner cut relay
B-7. Distributer advance vacuum exchange solenoid valve

Terminal location for the FBC system
(© Mitsubishi Motor Sales of America, Inc.)

signal from this sensor, which varies with the oxygen content of the exhaust gas stream, is provided to the ECU for use in controlling closed loop compensation of fuel delivery.

COOLANT TEMPERATURE SENSOR

The coolant temperature sensor is installed in the intake manifold. This sensor provides data to the ECU for use in controlling fuel delivery and secondary air management.

ENGINE SPEED SENSOR

The engine speed sensor signal comes from the ignition coil. Electric signals are sent to the ECU, where the time between these pulses is

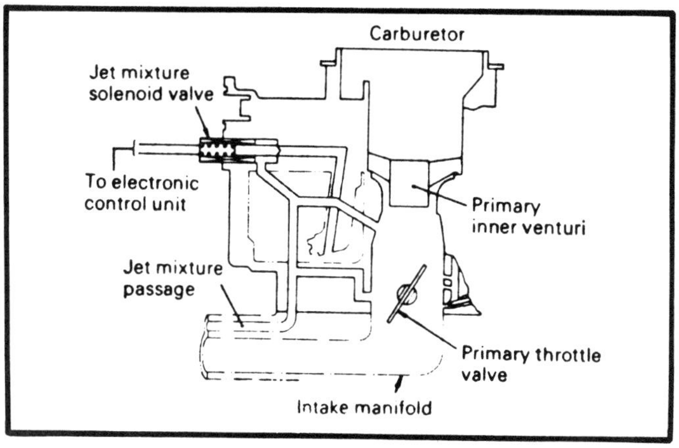

Exploded view of the jet mixture system
(© Mitsubishi Motor Sales of America, Inc.)

used to calculate engine speed, which is used in controlling fuel delivery, secondary air management, deceleration spark and throttle opener management.

THROTTLE POSITION SENSOR

This is a potentiometer mounted to the carburetor. The TPS provides throttle angle information to the ECU, to be used in controlling the fuel delivery and secondary management.

VACUUM SWITCH

The switch is installed on the floor board or the fender and is turned on when the throttle valve is closed (idling) position. Information from this switch is provided to the ECU, for use in controlling fuel delivery and secondary air management.

INTAKE AIR TEMPERATURE

This sensor is installed in the air cleaner. The function of this sensor is to measure the the temperature of the intake air in the air cleaner and provides this information to the ECU, for use in controlling fuel delivery.

JET MIXTURE SYSTEM

The jet mixture system supplies fuel to the engine through jet mixture passages and jet valves for optimum air/fuel mixtures. This system is calibrated by the jet mixture solenoid, which is responding to an electrical impulse from the ECU. If the exhaust oxygen sensor detects a lean condition, the ECU energizes the solenoid at increasing duty cycles to enrich the mixture. If the exhaust sensor detects a rich condition,

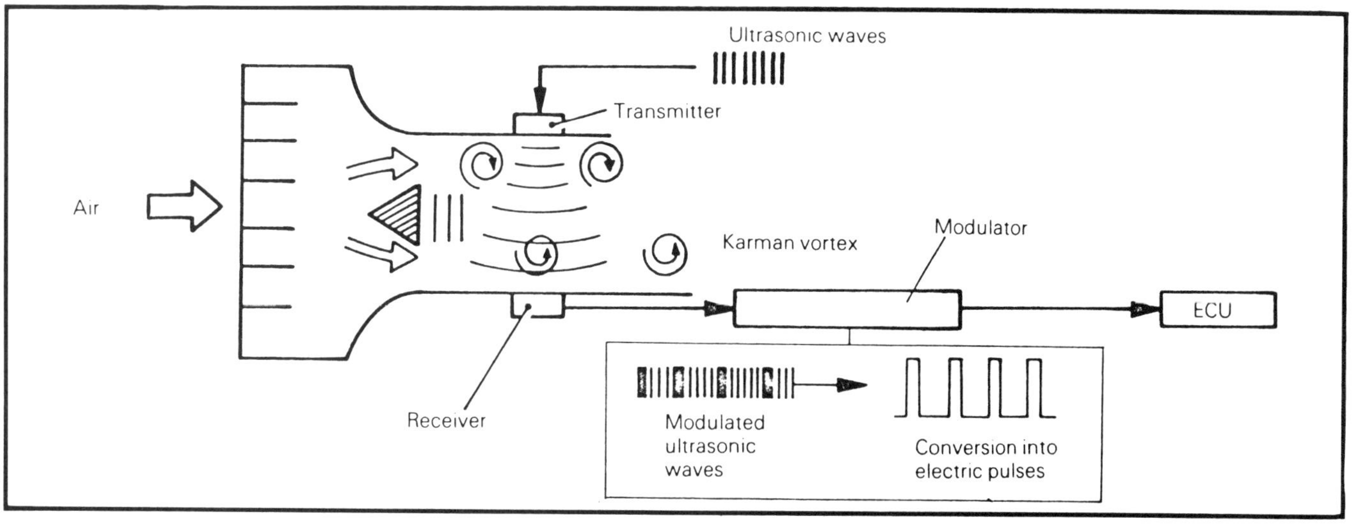

Exploded view of the ECI system (© Mitsubishi Motor Sales of America, Inc.)

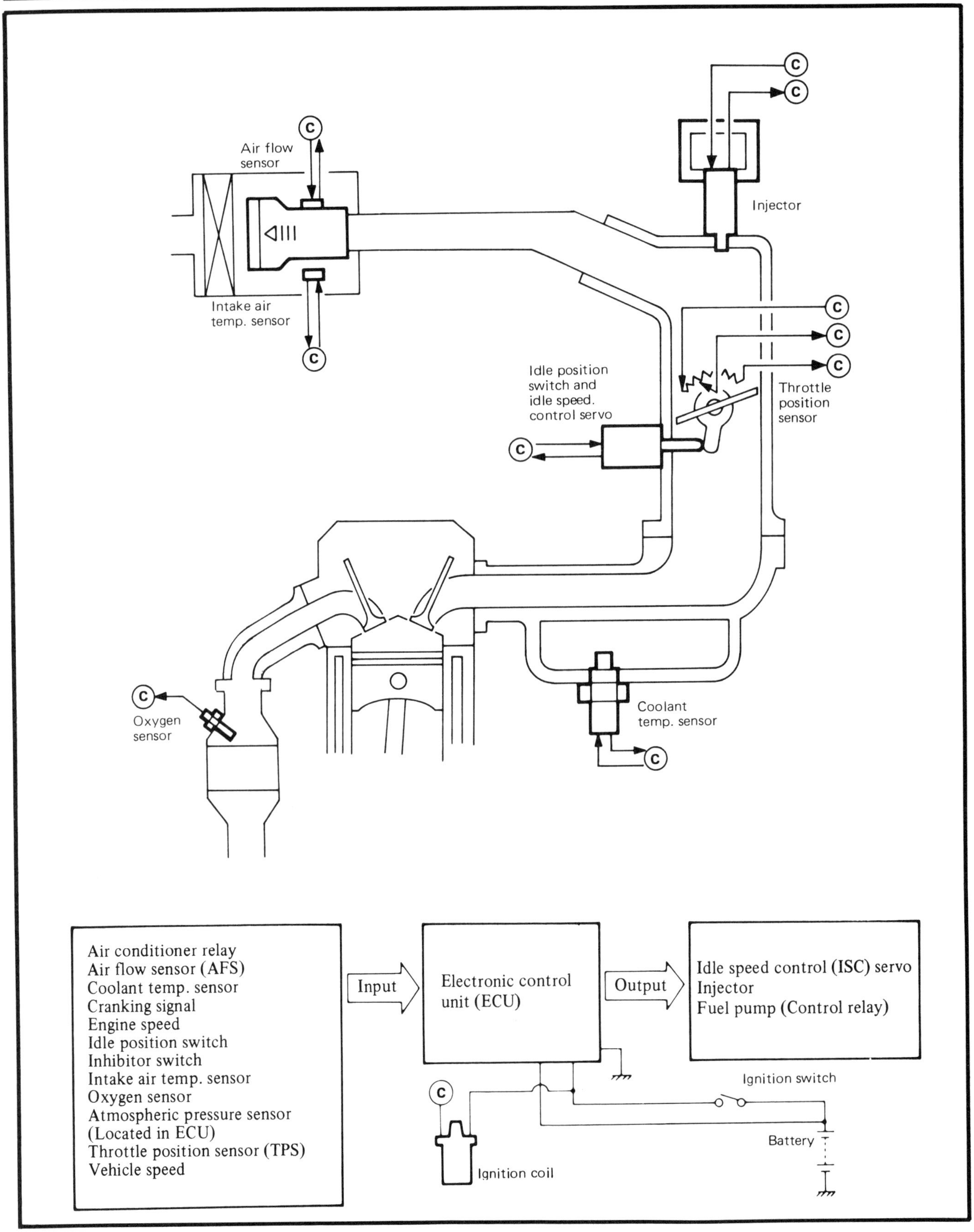

Typical oxygen sensor (© Mitsubishi Motor Sales of America, Inc.)

the solenoid receives a signal from the ECU at decreasing duty cycle to lean out the mixture. Thus, the solenoid is constantly responding to an electrical signal from the ECU, to provide efficient control of air/fuel mixtures.

ENRICHMENT SYSTEM

The enrichment system consists of metering jet and an enrichment solenoid operated on/off valve, which constantly provides additional fuel for the main metering system. The activation of the on/off valve is controlled by the length of time current is supplied to the solenoid. When additional fuel is required, such as during heavy acceleration, heavy engine loads, cold start or warm -up operation, the ECU energizes the solenoid at the pre-set duty cycles.

Electronically Controlled Injection (ECI) System

The fuel control system is the Electronically Controlled Injection (ECI) system, which consists of an electronic control unit (ECU), two fuel injectors, an air flow sensor and other components. The amount of fuel metered by the two fuel injectors is determined by an electric signal supplied by the ECU. The ECU monitiors various engine and vehicle parameters needed to calculate the fuel delivery time (the frequency and duration of injection) of the fuel injectors. The fuel delivery time is modified by the ECU, according to such operating conditions as cranking, cold starting, altitude, acceleration, deceleration and so on. The fuel is drawn from the fuel tank and forced by the electric fuel pump to the pressure line through a fuel filter. At the end of the fuel line, the fuel pressure regulator controls the fuel pressure at the preset value, so for the pressure difference between fuel pressure and turbocharged intake air pressure above the throttle valve to be constant, the amount of fuel injected is only dependent on the fuel delivery time controlled by the ECU. The two fuel injectors installed in the fuel injection mixer assembly are alternately energized by the ECU, once every third electric pulses from the air flow sensor, almost all over the operation zone and at the other specified intervals in the remaining zones. Each injector features its swirl nozzle that atomizes fuel at higher efficiency and facilitates the combustion. The air flow sensor installed in the air cleaner assembly, detects the air flow rate, utilizing the Karman vortex phenomenon, in the form of electric pulses converted by the modulator from the counted Karman vortexes.

ECI SYSTEM

The ECI system is a pulse time system that injects fuel into the fuel injection mixer above the throttle valve. Fuel is metered to the engine through the two electronically controlled fuel injection. The basic functions of this system are tabulated below. Input signals from a variety of sensors are fed to the ECU and then the ECU generates output signals for all of the controlled functions.

PARAMETERS SENSED

1. Intake air flow.

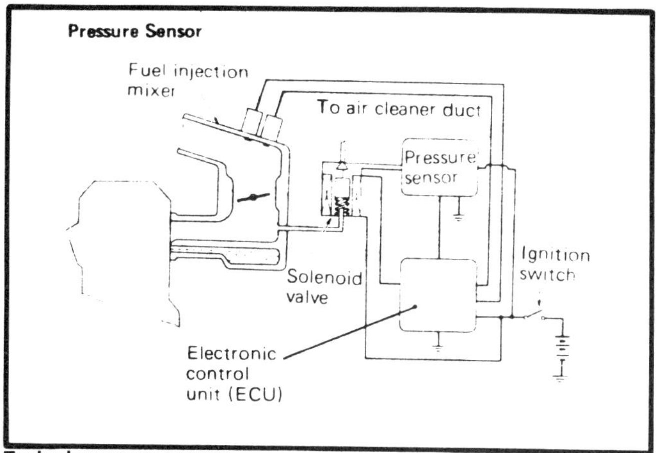

Typical pressure sensor
(© Mitsubishi Motor Sales of America, Inc.)

2. Intake air temperature.
3. Intake manifold pressure. (and barometeric pressure).
4. Coolant temperature.
5. Engine speed
6. Exhaust oxygen concentration
7. Throttle position
8. Idle position
9. Battery voltage
10. Engine cranking
11. Park/Neutral mode
12. Vehicle speed
13. A/C system switch 'ON'
14. Motor position switch

PARAMETERS CONTROLLED

1. Fuel injection signal
2. Secondary air control valve sisgnal
3. EGR control signal
4. Solenoid valve for pressure sensor control signal
5. A/C control signal

Data Sensor and Parameters Sensed

AIR FLOW SENSOR

The air flow sensor installed in the air cleaner assembly consists of the device for generating Karman vortexs, the ultrasonic wave transmitter, the receiver and the modulator. Ultrasonic waves with a constant frequency are transmitted across the air flow containing Karman vortexes,

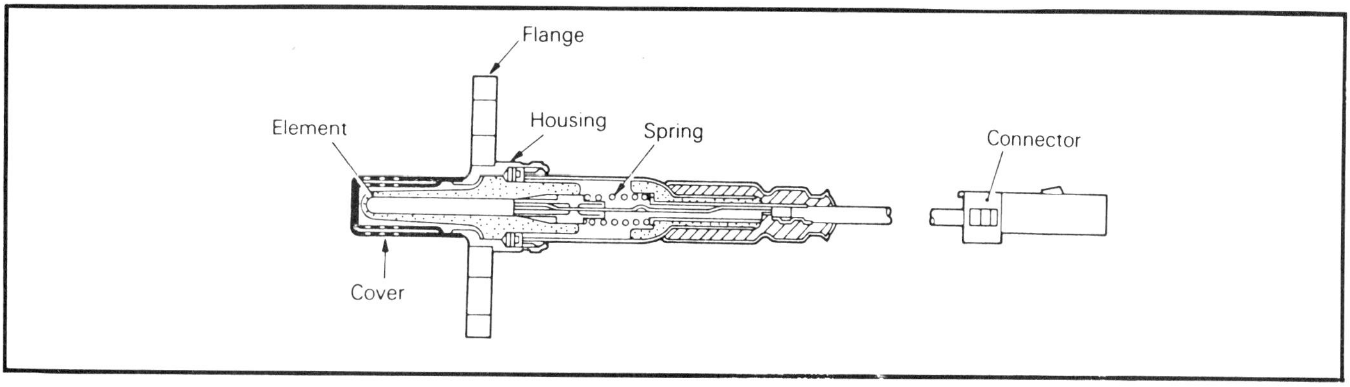

Cross-section of the air flow sensor (© Mitsubishi Motor Sales of America, Inc.)

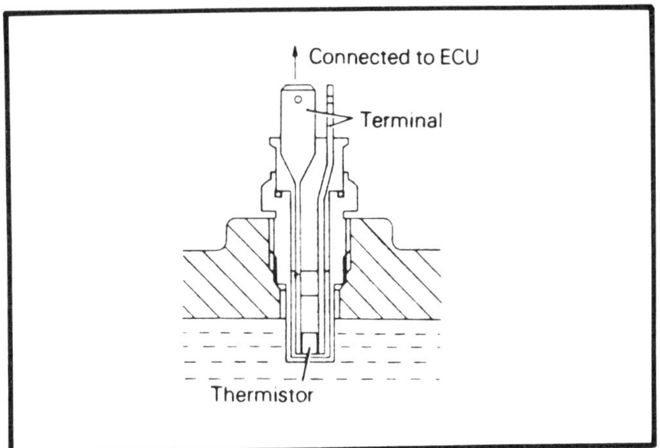

Coolant temperature sensor
(© Mitsubishi Motor Sales of America, Inc.)

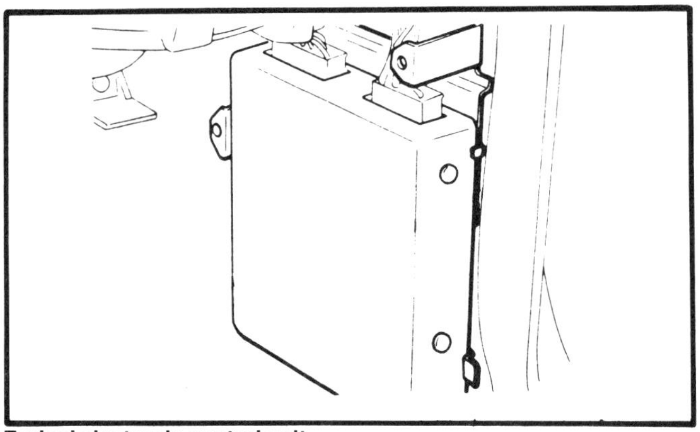

Typical electronic control unit
(© Mitsubishi Motor Sales of America, Inc.)

which are generated proportionally to air flow rate and then, the ultrasonic frequency is modulated by the vortexes. The receiver detects the modulated waves and the modulator, converts them into electric pulses. The electric pulse information is transmitted to the ECU for use in controlling the fuel delivery time and secondary air management.

INTAKE AIR TEMPERATURE

This sensor is installed in the air cleaner. The function of this sensor is to measure the the temperature of the intake air in the air cleaner and provides this information to the ECU for use in controlling fuel delivery.

PRESSURE SENSOR

This sensor is located on the firewall and its function is to sense ambient barometric pressure and absolute pressure in the intake manifold. Ambient barometeric pressure is sensed by the sensor, energizing the solenoid valve by the ECU for the specified period immediately after engine starting and thereafter, once every specified period. Information on ambient pressure changes due to weather and or altitude, is provided to the ECU for controlling the fuel delivery time. During the remaining period, other than the above specified periods, the solenoid valve is de-energized and this sensor detects intake manifold pressure (absolute pressure) and the information is provided to the ECU for controlling the fuel delivery time and idle speed.

OXYGEN SENSOR

The oxygen sensor is mounted in the exhaust manifold. The output signal from this sensor, which varies with the oxygen content of the exhaust gas stream, is provided to the ECU for use in controlling closed loop compensation of fuel delivery time.

COOLANT TEMPERATURE SENSOR

The coolant temperature sensor is installed in the intake manifold. This sensor provides data to the ECU for use in controlling fuel delivery time, EGR, secondary air management and idle speed.

ENGINE SPEED SENSOR

The engine speed sensor signal comes from the ignition coil. Electric signals are sent to the ECU, where the time between these pulses is used to calculate engine speed, which is used in controlling fuel delivery time, EGR, secondary air management and idle speed.

THROTTLE POSITION SENSOR

This is rotary potentiometer mounted on the fuel injection mixer. The TPS provides throttle angle information to the ECU, to be used in controlling the fuel delivery and idle speed control.

IDLE POSITION SWITCH

This is installed on the fuel injection mixer assembly and turned on when the throttle valve is at closed (idling) position. Information from this switch is provided to the ECU for use in controlling fuel delivery time (during vehicle deceleration), idle speed and secondary air management. This switch is also used as an idle speed adjusting device.

BATTERY VOLTAGE

Battery voltage is detected by the ECU. The battery voltage signal provides information to the ECU to allow for voltage compensation of the controlled functions.

A/C SYSTEM SWITCH

The A/C system switch signal indicates when the A/C mode switch is in the A/C on position.

ENGINE CRANK SWITCH

The engine crank switch provides a signal to the ECU when the engine is cranked.

PARK/NEUTRAL SWITCH

The park neutral switch signal indicates when the automatic transmission gear selector is in the Park or Neutral position.

VEHICLE SPEED SENSOR

The vehicle speed signal comes from the reed switch, which senses speedometer cable speed. Pulses from this switch, are sent to the ECU, where the time between these pulses is used to calculate vehicle speed, which is used in controlling idle speed.

ELECTRONIC CONTROL UNIT

The electronic control unit is mounted in the passenger compartment and consists of a printed circuit board mounted in a protective metal box. It receives analog inputs from the sensors and converts them into digital signals. These signals and various discrete inputs are processed and used by the ECU in controlling the fuel delivery, secondary air, deceleration spark and throttle opener managements.

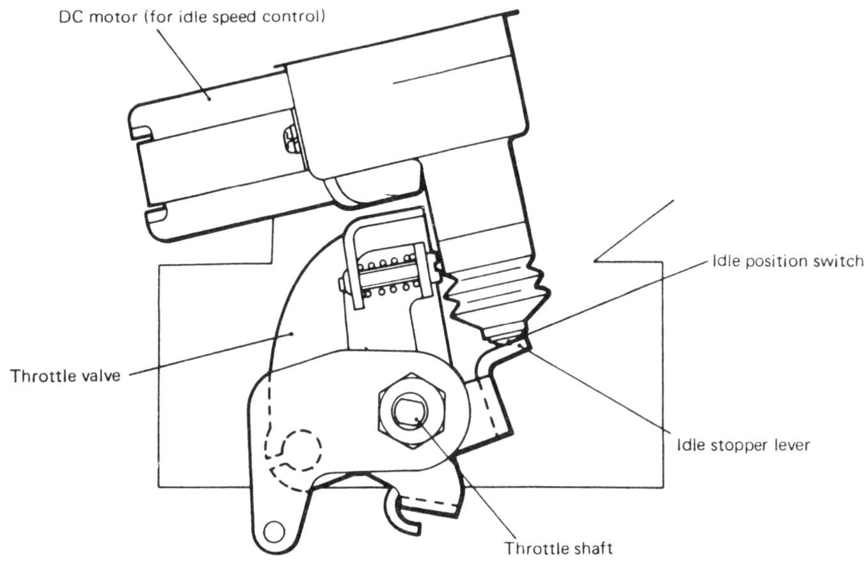

Exploded view of the idle speed control system (© Mitsubishi Motor Sales of America, Inc.)

Parameters Controlled

FUEL INJECTION CONTROL

There are two ways to control fuel injection. One is the open loop control and the other is the closed loop control, which are switched by the ECU, based on the information from the coolant temperature sensor, oxygen sensor, air flow sensor and engine speed sensor.

Under open loop control, the air fuel ratio is determined by the fuel delivery time (the frequency and duration of injection) which is controlled by the ECU.

Under the closed loop control, the air fuel ratio is feedback controlled by the ECI system including the exhaust oxygen sensor.

EGR CONTROL

The pressure applied to the EGR control valve is controlled by the solenoid valve motion, which mixes the throttle ported pressure with the turbocharged pressure. The solenoid valve motion is controlled by the ECU, based on the information from the coolant temperature sensor and engine speed sensor.

SECONDARY AIR CONTROL

The pressure applied to the secondary air valve is switched by the solenoid valve from the intake manifold pressure to the turbocharger pressure or vice versa. The solenoid is controlled by the ECU, based on the engine speed, idle position, air flow and coolant temperature.

ADAPTIVE MEMORY CONTROL

During the closed loop operation, the ECU monitors the output voltage signal from the exhaust oxygen sensor and stores it in a Random Access Memory (RAM). The last means are always stored, even if the ignition switch is turned off. During open loop operation, the fixed pre-programmed fuel delivery time is modified by the ECU with the mean values of feedback gain to improve emission performance.

NOTE: RAM is always powered from a standby power source.

IDLE SPEED CONTROL (ISC) SYSTEM

Engine idle speed, crank throttle angle and deceleration throttle angle are controlled by an electric motor driven actuator, which changes the throttle angle by acting as a movable idle stop. The ECU controls the ISC acuator by providing the appropriate outputs to yield the idle speed or throttle angle required for the particular operating condition. The electronic components for the ISC system are integral with the ECU.

HIGH-ALTITUDE COMPENSATION SYSTEM

The fuel injected vehicles are equipped with a pressure sensor and a solenoid valve which are utilized as a high altitude compensator. When the ignition switch is in the on position the solenoid valve is energized by the ECU for the specific time, which in turn allows the pressure sensor to detect the ambient atmospheric pressure. The fuel delivery time is then determined by the ECU, in response to the ambient atmospheric pressure, so the air/fuel mixture at high altitudes, can be maintained at the same ratio as at sea level.

DECELERATION FUEL CUT-OFF

To decrease the hydrocarbon emissions emitted during vehicle deceleration, the fuel delivery time is dramatically decreased by the electronic control unit changing the injection interval, which are determined by the specific pulses from the air flow sensor with the specified duration of injection. During the vehicle deceleration, while the idle position switch is not yet in the ON position, the fuel delivery time is determined by the electronic control unit responding to the throttle valve closing speeds.

OVER PRESSURE FUEL CUT-OFF

In order to protect the turbocharger engine engine from over loading, the fuel injectors are energized by the ECU only every ignition spark timing when the pressure sensor detects higher intake manifold pressure than the pre-set value.

Enrichment

STARTING ENRICHMENT

To ensure the starting performance, the fuel delivery time is determined by the ECU which takes account of information on coolant temperature and others, except electric pulses from the air flow sensor, because the air flow rate is unstable under the starting conditions. The starting enrichment control is over when the ignition switch is released from the start position and the engine's speed gets over the speed rpm specified.

AFTER STARTING ENRICHMENT

For a very short period of time, just after starting the engine, another enrichment is provided to obtain a stable combustion. To get proper enrichment for that short period of time, the ECU processes information on the coolant temperature as a parameter to determined the enrichment characteristic.

WARM-UP ENRICHMENT

The warm-up enrichment is provided for proper car operation by a signal from the ECU, which processes information on the coolant temperature until the coolant temperature rises at the pre-set level. This enrichment is provided during the engine warm up period, the open loop operation and part of the closed loop operation.

ACCELERATION ENRICHMENT

The acceleration enrichment is provided for ensuring car driveability during acceleration, both in the open and closed loop operation, by a signal from the ECU which processes information from the throttle position sensor.

FAST IDLE

The fast idle is controlled by one of the functions of the Idle Speed Control System (ISC).

Diagnosis Of The ECI System

ALL MODELS

Self-diagnosis is a system in which the input signal from each sensor is monitored by the ECU (computer) and, should any abnormality happen in the input signal, the abnormal item is memorized by the computer. The diagnosis items are eight items including that for normal condition and can be confirmed using a voltmeter or the special ECI tester tool# MD-998406. The abnormailty-diagnosis memory is kept by direct power supply from the battery. Therefore, the memory of diagnosis result is not erased by turning off the ignition switch. However, it is erased if the back-up power supply is turned off by disconnection of the battery cable or ECU connector.

NOTE: The memory is not erased if the power supply is turned on within ten seconds after turning off the power supply to the ECU. Also, when using the special ECI tester tool# MD–998406, a special ECI harness tool# MD–998407 and a special ECI harness connector 'B' tool# MD–998429 have to be used with the special tester.

DIAGNOSIS ITEM

If there are two or more abnormality-diagnosis items found abnormal, they are indicated in the order of increasing code numbers.

INDICATION METHOD

The indication method is performed by deflection of the pointer of a voltmeter or the ECI tester. Connect a voltmeter or the ECI tester (if using the ECI tester set to position six) to the ECI diagnosis connector, which is located on the right side of the engine compartment, mounted near the shock tower (or in the glove compartment on the Galant models) and the following indication will be made.

1. When normal, there will be 12 volts constantly indicated on the voltmeter or the ECI tester.

ABNORMALITY DIAGNOSIS ITEMS

Code Number	Diagnosis Item
1	Oxygen sensor and Computer
2	Ignition pulse
3	Air Flow sensor
4	Pressure sensor
5	Throttle position sensor
6	ISC motor position switch
7	Coolant temperature sensor
8	Car speed

2. When abnormal, the needle on the voltmeter scale or the ECI tester will fluctuate between 0 volts and 12 volts every 0.4 seconds.
3. Engine speed at 12 volts is indicated, when there are two or more abnormal items, the low-code-numbered item is first indicated.
4. After indication of 0 volts for 2 seconds, subsequent indication is made in the same manner as previously stated above.

PRECAUTION FOR OPERATION

1. When the battery voltage is low, no detection of abnormality is made. Be sure to check the battery for conditions before starting the test.
2. Diagnosis item is erased if power supply from the battery or the ECU connector is disconnected. Do not disconnected the power supply from the battery before the diagnosis result is completely read.
3. Warm up the engine and drive the vehicle for a good distance before the oxygen sensor is diagnosed. Do not set the ignition switch in the off position after driving the vehicle. If the ignition switch is set to the off position, the result detected by the diagnosis is erased.
4. After the checks and correction are over, disconnect the negative battery cable for 15 seconds or more and connect it again to make sure the abnormal code is erased.

INSPECTION OF ECI SYSTEM

If the ECI components fail, interruption of the fuel supply or failure to supply the proper amount of fuel for the engine operating condition will be the result and these following conditions could be encountered.

1. Engine is hard to start or will not start at all.
2. Very unstable idle
3. Poor driveabilty

If any of these situations do occur, first perform the inspection by self-diagnosis and subsequent basic engine checks, for incorrect engine adjustments, ignition system failure, etc, and then inspect the ECI components by diagnosing the ECI system with a voltmeter or the ECI tester.

Service Precautions

1. Before the battery terminals are disconnected, make sure that the ignition switch is set to the off position. If the battery cables are disconnected while the engine is running or when the ignition switch is in the ON position, malfunction of the ECU or damage to the semiconductor could result.
2. Disconnect the battery cables before charging the battery and when reconnecting the cables to be sure not to reverse the polarity.
3. Make sure that the harness connectors are securely connected. Use care not to allow entry of water or oil into the connectors.

Inspection Procedure by Self-Diagnosis

ALL MODELS

1. With the ignition switch in the OFF position, connect a voltmeter or the ECI tester (set on number six) between the self-diagnosis and

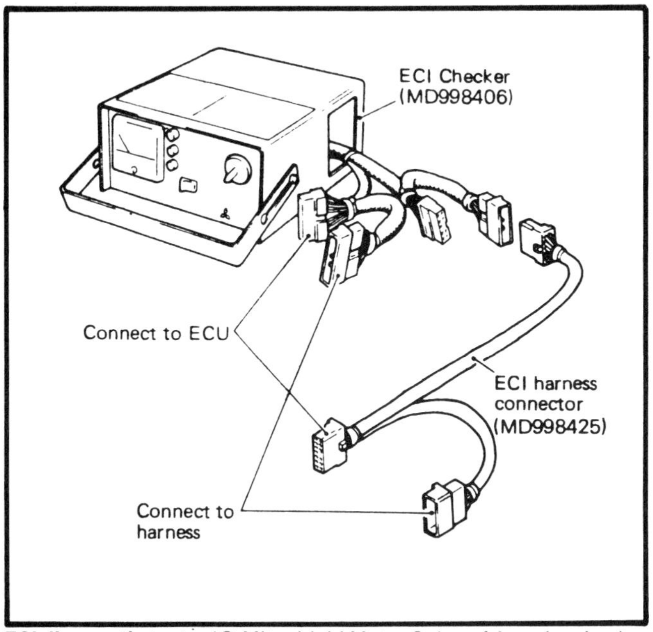

ECI diagnostic tester (© Mitsubishi Motor Sales of America, Inc.)

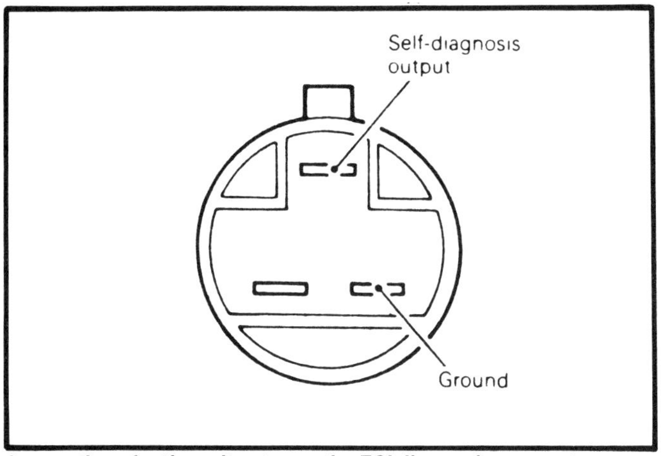

Connections for the voltmeter on the ECI diagnosis connector (© Mitsubishi Motor Sales of America, Inc.)

the ground terminals of the ECI diagnosis connector, which is located on the right side of the engine compartment, near the right shock tower (the ECI diagnosis connector is located in the glove compartment on the Galant models).

2. Place the ignition switch in the ON position and indications of the ECU memory contents will immediately start. If the system is in normal condition, the needle on the voltmeter or the ECI tester will constantly indicate 12 volts.

3. If any abnormality is in the memory, the needle of the voltmeter or the ECI tester will fluctuate, indicating an abnormal item as previously outlined in the Indication Method in this section.

4. Abnormal item can be known from voltage waveform, that is the number of times the needle fluctuates, which is shown in the diagnosis chart provided in this section.

5. After recording the abnormal item, check and repair each part, according to the check items in the diagnosis chart.

6. Turn the ignition switch off and if the defective parts have been repaired, disconnect the negative battery cable for 15 seconds or more and reconnect it again to be sure that the abnormal code has been erased.

Inspection Procedure for the ECI System

NOTE: This test should only be performed after completion of all steps in the Inspection by Self-Diagnosis, that was previously outlined in this section. Also use only special ECI tester tool# MD–998406 when performing this test.

1. With the ignition switch off, remove the ECU cover and the body side harness connector A (17 poles) and connector B (13 poles) from the ECU.

2. Set the ECI tester to OFF and set the select switch of the tester to A.

3. Connect the connectors of the ECI tester by using the ECI harness connector to ECU and body harness connectors and place the ECI tester on the front passengers seat.

4. Check output monitor operating condition (on or off) by switching the ignition switch from OFF to ON. Monitor operating conditions should be as outlined below:

Air flow sensor—ON Injector pulse—OFF
Oxygesn sensor—OFF

NOTE: Perform the ECI checks in relationship to the checks listed in the check procedure chart for the ECI-system.

5. If the ECI tester shows anything out of specifications, check the corresponding sensor and related wiring. Repair or replace the malfunction.

6. After any or all malfunctions have been repaired or replaced, recheck them with the ECI tester to confirm that the repair or replaced part is operating properly.

NOTE: Air flow sensor and injector pulse may be inspected by output monitors as well, as shown in the chart below with the ECI tester select switch set to B:

7. Set the check switch of the ECI tester to OFF and set the ignition switch to OFF.

8. Disconnect the connectors of the ECI tester from the ECU and body side harness connectors. Connect the body side harness connectors to the ECU and install the ECU cover.

ECI OUTPUT MONITOR CHART

LED Output Monitor	Operation When Normal
Air flow sensor	Always flashing with check switch at 1 through 12 ①
Injector pulse	Always flashing with check switch at 1 through 12 ①
Oxygen sensor	Flashing at closed loop zone

① Flashing are so short that light appears to be continuously on.

Code No.	Diagnosis items	Voltage waveform (abnormal code)	Contents of diagnosis	Check item
1	Oxygen sensor	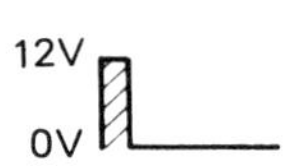	No change (for 20 seconds or more) of oxygen-sensor signal within feedback range of oxygen sensor.	• Wire harness and connector • Oxygen sensor • ECU
2	Ignition signal		While cranking the engine input of ignition signal is not applied to ECU for 3 seconds or more.	• Wire harness and connector • Igniter • ECU
3	Air flow sensor		Air flow sensor output is 10 Hz or less while engine is idling, or it is 100 Hz or more when engine stalls.	• Wire harness and connector • Air flow sensor • ECU
4	Atmospheric pressure sensor	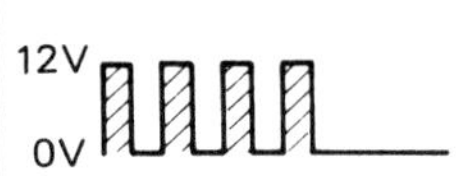	Atmospheric pressure sensor output is 855 mmHg (4.5V) or more, or it is 38 mmHg (0.2V) or less.	• Wire harness and connector • ECU
5	Throttle position sensor		Throttle sensor output is 0.2V or less, or it is 4V or more while engine is idling (idle switch is ON).	• Wire harness and connector • Throttle sensor • ECU
6	ISC motor position switch		Throttle sensor output is 0.4V with L switch OFF.	• Wire harness and connector • ISC servo • ECU
7	Coolant temperature sensor		Water temperature sensor output is 4.5V or more, or it is 0.1V or less.	• Wire harness and connector • Water temperature sensor • ECU

Diagnosing the ECI system (© Mitsubishi Motor Sales of America, Inc.)

Select switch	Check switch	Check item	Condition		Check meter reading when normal	Terminal location of computer
A	1	Power supply	Ignition switch OFF → ON		11-13V	B-1
	2	Secondary ari control solenoid valve	Ignition swtich OFF → ST after warming up the engine		After 15 seconds 0-0.5V ↓ 13-15V	A-10
	3	Throttle position switch	Ignition switch OFF → ON	Accelerator closed	0.4-1.5V	A-1
				Accelerator wide opened	4.5-5.0V	
	4	Coolant temperature sensor	Ignition switch OFF → ON	0°C (32°F)	3.5V	A-3
				20°C (68°F)	2.6V	
				40°C (104°F)	1.8V	
				80°C (176°F)	0.6V	

Select switch	Check switch	Check item	Condition		Check meter reading when normal	Terminal location of computer
A	5	Intake air temperature sensor	Ignition swtich OFF → ON	0°C (32°F)	3.5V	A-4
				20°C (68°F)	2.6V	
				40°C (104°F)	1.8V	
				80°C (176°F)	0.6V	
	6	Idle position swtich	Ignition switch OFF → ON	Accelerator closed	0-0.4V	A-5
				Accelerator wide opened	11-13V	
	7	ISC motor position switch	Ignition swtich OFF → ON		11-13V*[1]	A-14
	8	–	–		–	–
	9	–	–		–	–
	10	A/C (Air conditioner) relay	Ignition switch OFF → ON	A/C switch OFF	0-0.5V	B-12
				A/C switch ON	11-13V	
	11	Lead switch for vehicle speed	Start engine, transaxle in first and operate vehicle slowly		Flasing 0-0.5V ↕ Over 2V	A-15
	12	–	–		–	–

Check procedure chart for the ECI system (© Mitsubishi Motor Sales of America, Inc.)

Select switch	Check switch	Check item	Condition	Check meter reading when normal	Terminal location of computer
B	1	Cranking signal	Ignition switch OFF → ST	Over 8V	A-13
	2	Control relay	Idling	0-0.5V	B-5
	3	–	–	–	–
	4	Ignition pulse	Idling	12-14V	A-8
			3000 rpm	11-13V	
	5	Air flow sensor	Idling	2.7-3.2V	A-7
			3000 rpm		
	6	Injector (+)	Idling	6-8V	B-9
			3000 rpm	7-9V	
	7	Injector (–)	Idling	6-8V	B-10
			3000 rpm	7-9V	

Select switch	Check switch	Check item	Condition		Check meter reading when normal	Terminal location of computer
B	8	Oxygen sensor	Keep 1300 rpm after warming up the engine		Flashing 0.4-1V ↕ 2.7V	A-6
	9	EGR control solenoid valve	Keep idling after warming up the engine		13-15V	B-4
			Raise the engine 3500 rpm		0-0.5V	
	10	–	–		–	
	11	ISC motor for extension	Idling		0.2V	B-6
				A/C switch: OFF → ON	Momentarliy Over 6V	
	12	ISC motor for retraction	Idling		0.2V	B-11
				A/C switch: ON → OFF	Momentarily Over 6V	

NOTE: *[1]: If ignition switch is turned to ON for 15 seconds or more, the reading drops below 5V momentarily.

Check procedure chart for the ECI system (cont.) (© Mitsubishi Motor Sales of America, Inc.)

Mitsubishi Truck Specifications

VEHICLE IDENTIFICATION NUMBER AND ENGINE IDENTIFICATION

Vehicle Identification Number (VIN)

The vehicle identification number (VIN) plate is mounted on the instrument panel, adjacent to the lower corner of the windshield on the left side of the vehicle, and is visible through the windshield. The VIN number is composed of seventeen (17) digits. The tenth (10) digit of the VIN number represents the year of the vehicle, with 'F' representing 1985 and 'G' representing 1986.

Engine Identification

The eighth (8) digit of the VIN number represents the engine model. The engine identification number for the 2.0 liter engine is stamped on the engine block, adjacent to the alternator and engine oil dipstick. On the 2.6 liter engine, the number is stamped on the engine block, next to the idler pulley and above the bottom radiator hose. The 2.3 liter engine has the number stamped on the engine block, right next to the engine oil dipstick.

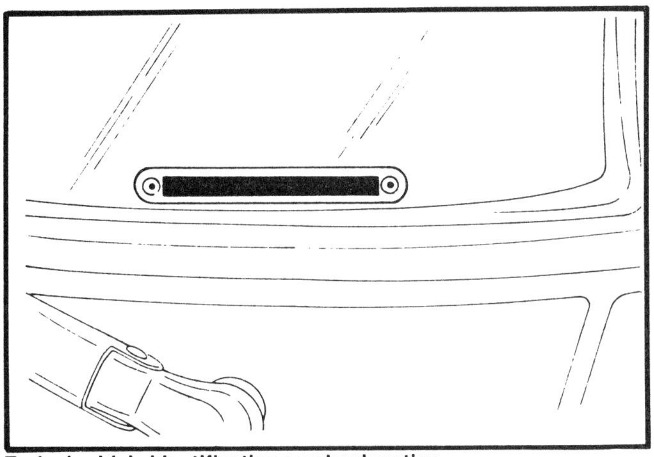

Typical vehicle identification number location

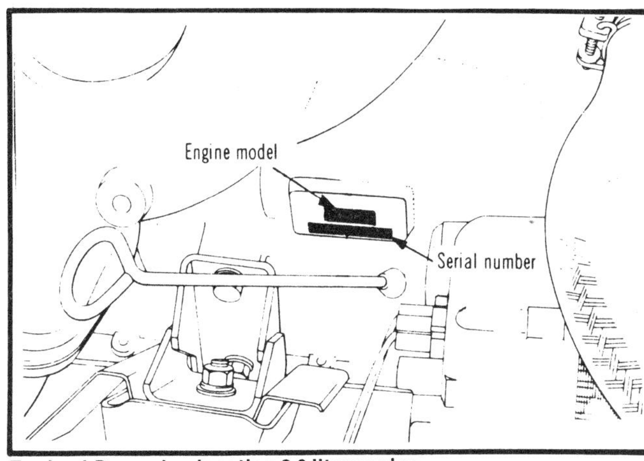

Engine I.D. number location 2.0 liter engine

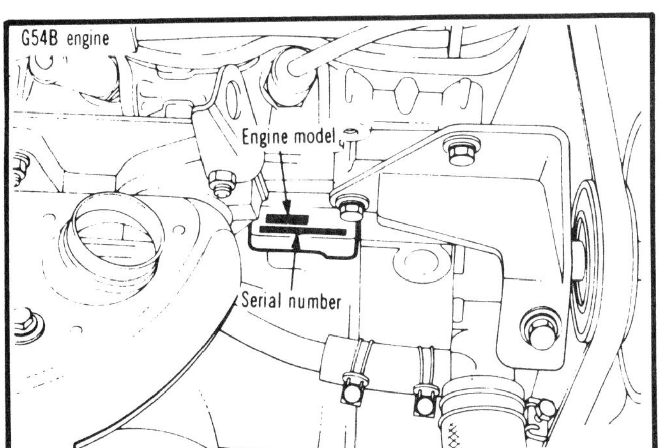

Engine I.D. number location 2.6 liter engine

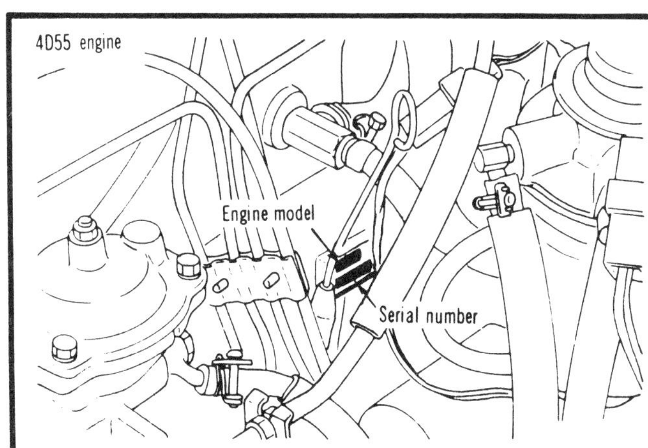

Engine I.D. number location 2.3 liter diesel engine

TUNE-UP SPECIFICATIONS

(When analyzing compression test results, look for uniformity among cylinders, rather than specific pressures.)

Year	Engine Displace. cu. in. (cc)	Spark Plugs Type	Spark Plugs Gap (in.)	Distributor Point Dwell (deg)	Distributor Point Gap (in.)	Ignition Timing (deg) MT	Ignition Timing (deg) AT	Intake Valve Opens (deg) BTDC	Fuel Pump Pressure (psi)	Idle Speed (rpm)	Valve Clear (in) ● In.	Valve Clear (in) ● Ex.
'85–'86	121.9 (2000)	BPR6ES-11 ①	0.039 0.043	Electronic		5B	5B	19	N/A	750	0.010 ②	0.010
'85–'86	155.9 (2600)	BUR6EA-11 ①	0.039 0.043	Electronic		7B	7B	25	N/A	750 ③	0.010 ②	0.010
'85–'86	143.2 (2300)	Turbo Diesel	—	—		5 ATDC ④		20	306	750	0.010	0.010

NOTE: The underhood sticker often reflects tune-up specification changes made in production. Sticker figures must be used if they disagree with those in this chart.

● All adjustments done with engine hot
B Before top dead center
N/A Not available at the time of publication
ATDC After top dead center

① See text
② Jet valve clearance: 0.010 inches
③ 800 rpm on automatic transaxle models
④ 5° ATDC at 0.0394 in. (1 mm) plunger stroke

FIRING ORDERS

NOTE: To avoid confusion, always replace spark plug wires one at a time.

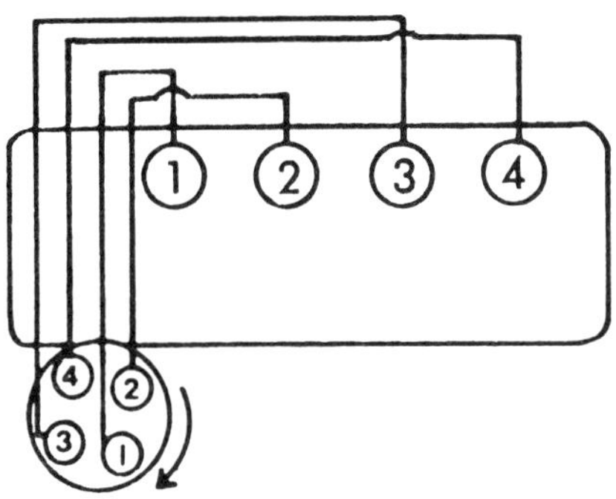

1468, 1597, 1795, and 2350cc engines

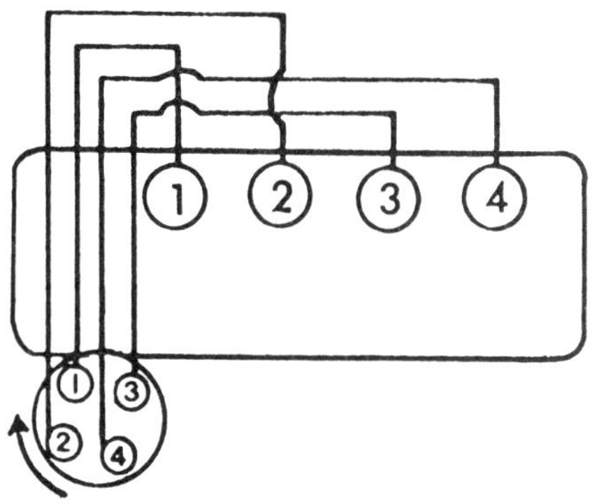

2555cc engine

EMISSION EQUIPMENT USED

2.0L AND 2.6L ENGINES

Crankcase Emission Control
Closed system
Positive crankcase vent valve

Evaporative Emission Control
Canister-Single
Bowl vent valve
Purge control valve
Fuel filler cap-with relief valve
Vapor separator tank
Overfill-limiter-two way valve
Fuel check valve

Exhaust Emission Control
Jet valve
Catalytic converter—dual three way type (single Canada)
Secondary air supply systemreed valve
Exhaust gas recirculation—(EGR) system
EGR valve—single + sub (single Canada)
Thermo valve—Single wax pellet type
Fuel control system—Electronic Controlled Injection System (ECI)
Feedback carburetor system
Heated air intake system—vacuum type
Deceleration device
Coasting air valve
Dash pot
Deceleration spark advance
High altitude compensation
Throttle opener
Tamper-proof (mixture, choke-ECI system not available)

2.3 LITER DIESEL ENGINE

Crankcase Emission Control
Closed system with a breather hose

Evaporative Emission Control
Idle-up solenoid for vehicle with air conditioning

Exhaust Emission Control
Exhaust gas recirculation EGR valve
EGR control valve

REQUIRED EMISSION MAINTENANCE

EMISSION CONTROL SYSTEM MAINTENANCE		SERVICE INTERVALS	7.5	15	22.5	30	37.5	45	50	52.5	60	67.5	75	82.5	90	97.5	100	105	112.5	120
		MILEAGE IN THOUSANDS / KILOMETERS IN THOUSANDS	12	24	36	48	60	72	80	84	96	108	120	132	144	156	160	168	180	192
ENGINE OIL CHANGE EVERY 12 MONTHS		OR	x	x	x	x	x	x		x	x	x	x	x	x	x		x	x	x
ENGINE OIL FILTER CHANGE EVERY 12 MONTHS		OR		x		x		x			x		x		x			x		x
DRIVE BELT (FOR WATER PUMP, ALTERNATOR)	CHECK CONDITION AND ADJUST TENSION AS REQUIRED	AT		x		x		x			x		x		x			x		x
	REPLACE	AT				x					x				x					x
CHECK VALVE CLEARANCE (JET VALVE): ADJUST AS REQUIRED		AT		x		x		x			x		x		x			x		x
CHECK IGNITION TIMING: ADJUST AS REQUIRED EVERY 5 YEARS		OR							x								x			
CHECK ENGINE IDLE SPEED: ADJUST AS REQUIRED		AT		x		x		x			x		x		x			x		x
CLEAN CARBURETOR CHOKE MECHANISM AND LINKAGE		AT				x					x				x					x
CHECK THROTTLE POSITION SYSTEM; ADJUST AS REQUIRED	INITIAL	AT		x																
	AFTER INITIAL CHECK, AT 50,000 MILES (80,000 KM), THEREAFTER EVERY 50,000 MILES (80,000 KM)																			
REPLACE FUEL FILTER EVERY 5 YEARS		OR							x								x			
CHECK FUEL SYSTEM (TANK, LINE AND CONNECTION) FOR LEAKS EVERY 5 YEARS		OR							x								x			
REPLACE VACUUM HOSES, SECONDARY AIR HOSES AND CRANKCASE VENTILATION HOSES EVERY 5 YEARS		OR							x								x			
REPLACE FUEL HOSES, WATER HOSES, VAPOR HOSES AND FUEL FILLER CAP EVERY 5 YEARS		OR							x								x			
REPLACE AIR CLEANER FILTER		AT				x					x				x					x
CHECK CRANKCASE EMISSION CONTROL SYSTEM (PCV VALVE); CLEAN EVERY 5 YEARS		OR							x								x			
CHECK EVAPORATIVE EMISSION CONTROL SYSTEM (EXCEPT CANISTER) FOR LEAKS AND CLOGGING EVERY 5 YEARS		OR							x								x			
REPLACE CANISTER		AT							x								x			
REPLACE SPARK PLUGS		AT				x					x				x					x
REPLACE IGNITION CABLES EVERY 5 YEARS		OR							x								x			
REPLACE EGR VALVE*		AT							x								x			
CLEAN SUB EGR VALVE*		AT							x								x			
REPLACE OXYGEN SENSOR		AT							x								x			
CHECK DISTRIBUTOR* EVERY 5 YEARS		OR							x								x			
CHECK INTAKE TEMPERATURE CONTROL SYSTEM* EVERY 5 YERAS		OR							x								x			
CHECK SECONDARY AIR SYSTEM* EVERY 5 YEARS		OR							x								x			
REPLACE SOLENOID VALVE AIR FILTER OF VACUUM CONTROL SYSTEM EVERY 5 YEARS		OR							x								x			

Note: *Except for California.

EMISSION CONTROL SYSTEM MAINTENANCE	SERVICE INTERVALS	MILEAGE IN THOUSANDS	7.5	16	22.5	30	37.5	45	50	52.5	60	67.5	75	82.5	90	97.5	100	105	112.5	120
		KILOMETERS IN THOUSANDS	12	24	36	48	60	72	80	84	96	108	120	132	144	156	160	168	180	192
ENGINE OIL CHANGE EVERY 6 MONTHS	OR	FOR FEDERAL	EVERY 5,000 MILES (8,000 KM)																	
		FOR CALIFORNIA	EVERY 3,000 MILES (4,800 KM)																	
ENGINE OIL FILTER CHANGE EVERY 12 MONTHS	OR	FOR FEDERAL	EVERY 5,000 MILES (8,000 KM)																	
		FOR CALIFORNIA	EVERY 6,000 MILES (9,600 KM)																	
CHECK VALVE CLEARANCE; ADJUST AS REQUIRED		AT		x		x		x			x		x		x			x		x
CHECK ENGINE IDLE SPEED; ADJUST AS REQUIRED		AT				x					x				x					x
REPLACE FUEL FILTER EVERY 2 YEARS		OR				x					x				x					x
CHECK FUEL SYSTEM FOR LEAKS EVERY 5 YEARS		OR							x								x			
REPLACE AIR CLEANER FILTER		AT				x					x				x					x
REPLACE FUEL HOSES, WATER HOSES, BOOST HOSE AND CRANKCASE VENTILATION HOSE EVERY 5 YEARS		OR							x								x			
REPLACE TURBOCHARGER AIR INTAKE HOSES EVERY 2 YEARS		OR				x					x				x					x
REPLACE TURBOCHARGER OIL HOSE EVERY 5 YEARS		OR							x								x			
REPLACE VACUUM PUMP OIL HOSE EVERY 2 YEARS		OR				x					x				x					x

INITIAL TIMING PROCEDURE

2.0L AND 2.6L ENGINES

1. Connect a suitable timing light to the number one cylinder.
2. Connect the positive lead of the test tachometer to the negative primary terminal of the ignition coil. Connect the negative lead of the test tachometer to a good ground.
3. Start the engine and run until operating temperature has been reached.
4. Have the engine at operating temperature and fast idle off. Momentarily open the throttle and release it to make sure that there is no bind in the linkage and that the idle speed screw is against its stop.
5. Disconnect and plug the vacuum line at the distributor.
6. Adjust the curb idle speed to specification.
7. Aim the timing light at the timing marks and check the engine timing. Loosen the distributor hold down bolt and rotate the distributor left or right to advance or retard the timing as necessary.
8. Once the timing has been set, recheck the curb idle specification and adjust as required. If the curb idle speed needs adjusting, be sure to recheck the ignition timing again.

INJECTION PUMP TIMING

2.3L Diesel Engine

1. Turn the crankshaft so as to position the piston in the number one cylinder, at top dead center on compression stroke.
2. Loosen but do not remove the four injection pipe union nuts, on the injection side of the pump. While loosening the union nuts, hold the delivery valve holder with a spanner wrench or equivalent, to prevent it from rotating with the union nut.
3. Loosen but do not remove the two nuts and two bolts securing the injection pump.
4. The next step requires special tool # MD998384, which is a Prestroke Measuring Adapter. Before installing this adapter, make sure that the push rod on the adapter is protruding by 0.4 in. (10mm). The protrusion of the push rod can be adjusted with the inner nut on the adapter.
5. Remove the timing check plug from the head of the injection pump. Install the special prestroke measuring adapter, along with a dial indicator.
6. Turn the crankshaft to such a position that the notch on the pulley is at approximately 30 degrees before top dead center on compression stroke of the piston in the number one cylinder.
7. Set the dial indicator to zero. Slightly turn the crankshaft clockwise and counterclockwise to be sure that the dial indicator pointer does not deviate from the zero position.
8. If the dial indicator pointer moves from the zero position, readjust the pulley position so that the notch on the pulley is at 30 degrees before top dead center.
9. Turn the crankshaft in the normal direction to bring the notch to 5 degrees ATDC. Check the dial indicator, it should read 0.0394 ± 0.0011 in. (1 ± 0.03mm).
10. If the dial indicator does not fall within specifications, tilt the injection pump body right or left until the dial indicator comes within the specified range. Tighten the injection pump mounting nuts and bolts to 22–25 ft. lbs.
11. Repeat Steps 6–9 to check and make sure that the adjustment has been made correctly. Readjust if necessary.
12. Remove the special prestroke measuring adapter and install the new copper gasket on the timing check plug. Install the plug into the head of the injection pump and torque the plug to 6–7 ft. lbs.
13. While holding the the delivery valve holder with a spanner wrench or equivalent, tighten the four injection pipe union nuts to 22–25 ft. lbs.

Nissan (Datsun)

INDEX

VEHICLE IDENTIFICATION NUMBER AND ENGINE IDENTIFICATION

Vehicle Identification Number (VIN)

The vehicle identification number (VIN) plate is attached to the left top of the dash, behind the windshield. There is an additional vehicle identification plate located on the firewall on all models except the 300ZX, where it is located on the drivers side strut tower. This plate contains vehicle type, chassis number, model number, color codes, engine model and displacement, transmission/transaxle model, and axle model.

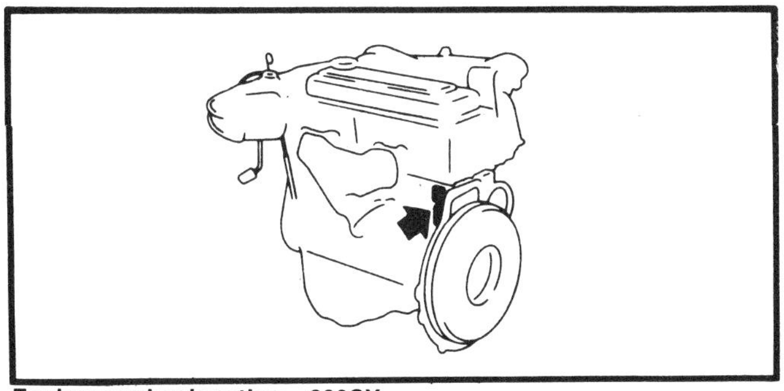

Engine number location—200SX

Engine Identification

The engine number is stamped on the right side top edge of the cylinder block, where it bolts to the bellhousing, on all models except the 200SX, Maxima, and the 300ZX. The 200SX with the CA20E/CA18ET engines, has the number stamped on the left rear edge of the block, next to the bellhousing. The engine serial number is preceded by the engine model code. The Maxima and 300ZX V6 engines (VG30E/VG30ET), has the number stamped on the right rear edge of the right cylinder bank, facing upward.

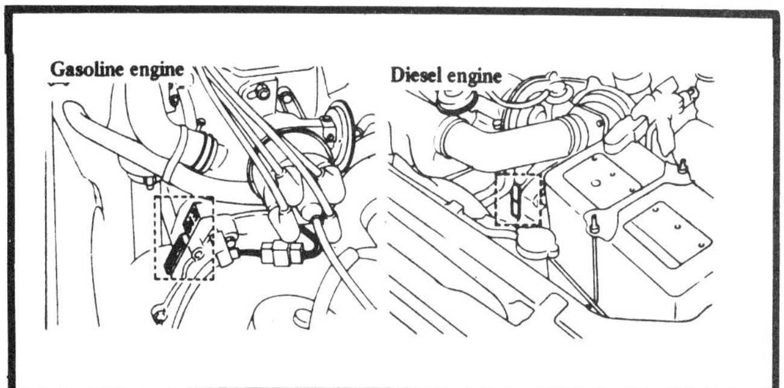

Engine number location—all except 200SX, 300ZX and Maxima

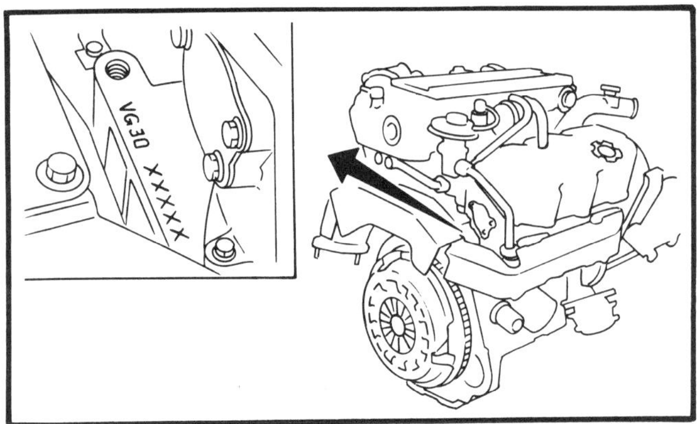

Engine number location—300SX and Maxima

Chassis Number

The chassis number plate is on the firewall under the hood. The chassis serial number is preceded by the model designation. The Emission Control information label is attached to the under side of the hood.

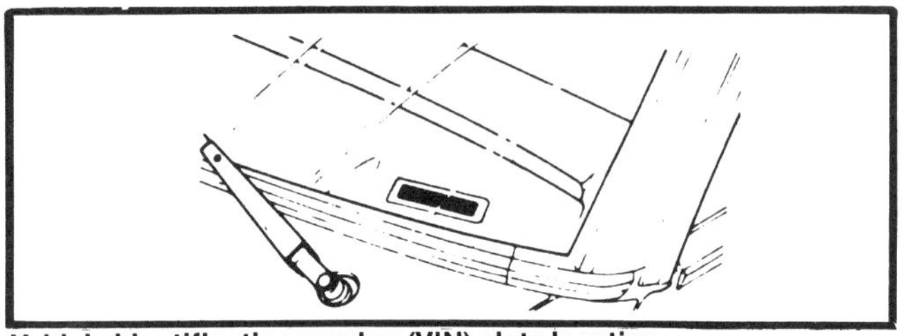

Vehicle identification number (VIN) plate location

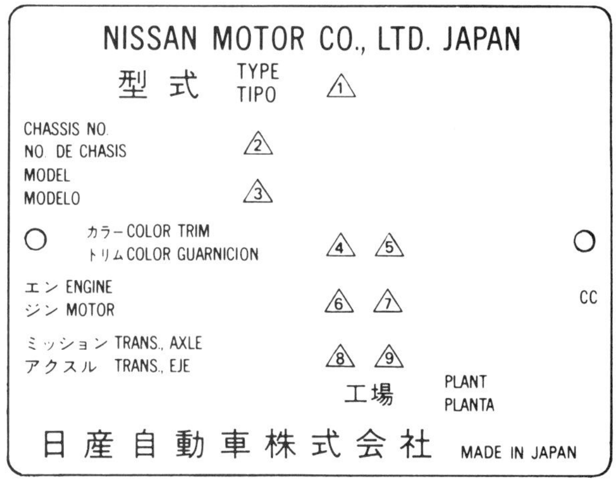

1 Type
2 Vehicle identification number (Chassis number)
3 Model
4 Body color code
5 Trim color code
6 Engine model
7 Engine displacement
8 Transmission model
9 Axle model

Vehicle identification plate

ENGINE I.D. TABLE

Model	Number of Cylinders	Displacement cu. in. (cc)	Type	Engine Model Code
Pulsar, Sentra	4	97.6 (1597)	OHC	E16
Sentra (Diesel)	4	103.7 (1680)	OHC	CD17
200SX, Stanza	4	120.4 (1974)	OHC	CA20E, CA20S
200SX (Turbo)	4	110.3 (1809)	OHC	CA18ET
300ZS, Maxima	V6	180.6 (2960)	OHC	VG30E, VG30ET

GASOLINE ENGINE TUNE-UP SPECIFICATIONS

(When analyzing compression test results, look for uniformity among cylinders, rather than specific pressures)

Year	Model	Spark Plugs Type	Spark Plugs Gap (in.)	Distributor Dwell (deg)	Distributor Air Gap (in.)	Ignition Timing (deg) MT	Ignition Timing (deg) AT	Fuel Pump Pressure (psi)	Idle Speed (rpm) MT	Idle Speed (rpm) AT▲	Valve Clearance (in.)● In	Valve Clearance (in.)● EX
'85	200SX	BCPR6ES-11 ③	.039–.043	Electronic	.012–.020	4B	0B	37	750	700	.012	.012
'85	200SX Turbo	BCPR6ES-11 ③	.039–.043	Electronic	.012–.020	15B	15B	37	750 ④	—	.012	.012
'85	300ZX	BCPR6ES-11	.039–.043	Electronic	NA	20B	20B	37	700 ⑥	700	⑤	⑤

GASOLINE ENGINE TUNE-UP SPECIFICATIONS

(When analyzing compression test results, look for uniformity among cylinders, rather than specific pressures)

Year	Model	Spark Plugs		Distributor		Ignition Timing (deg)		Fuel Pump Pressure (psi)	Idle Speed (rpm)		Valve Clearance (in.) ●	
		Type	Gap (in.)	Dwell (deg)	Air Gap (in.)	MT	AT		MT	AT▲	In	EX
'85	300ZX Turbo	BCPR6E-11	.039–.043	Electronic	NA	20B	20B	37	700	650	⑤	⑤
'85	Sentra, Pulsar (E16)	BPR5ES-11 ①	.039–.043	Electronic	.012–.020	15B ②	8B ②	3.8	800 ⑦	650	.011	.011
'85	Stanza (CA20E)	BCPR6ES-11	.039–.043	Electronic	NA	4B	0B	3.8	750	700	.012	.012
'85	Stanza (CA20S)	BCPR6ES-11	.039–.043	Electronic	NA	0B	0B	3.8	650	650	.012	.012
'85	Maxima (VG30E)	BCPR6ES-11	.039–.043	Electronic	NA	20B	20B	37	700 ⑥	700	⑤	⑤
'86	All	See Underhood Specification Sticker										

NOTE: The underhood specifications sticker often reflects tune-up specification changes made in production. Sticker figures must be used if they disagree with those in this chart.

● Set hot
▲ In Drive
MT: Manual trans.
AT: Automatic trans.
① Canada: BPR5ES; gap: .031–.035
② 5ATDC: Calif., Canada
③ Intake side; Exhaust side BCPR5ES-11
④ 680 rpm high altitudes
⑤ Hydraulic valve lifters—no adjustment necessary

DIESEL ENGINE TUNE-UP SPECIFICATIONS

Year Model	Engine Displacement cu. in. (cc)	Warm Valve Clearance (in.)		Injection Pump Setting (deg)	Injection Nozzle Pressure (psi)		Idle Speed (rpm)	Compression Pressure (psi)
		In	Ex		New	Used		
'85	103.7 (1680)	0.012	0.020	See Text	1,920–2,033	1,778–1,920	750	455
'86	All	See Underhood Specification Sticker						

NOTE: The underhood specification sticker often reflects tune-up specification changes made in production. Sticker figures must be used if they disagree with those in this chart.

FIRING ORDERS

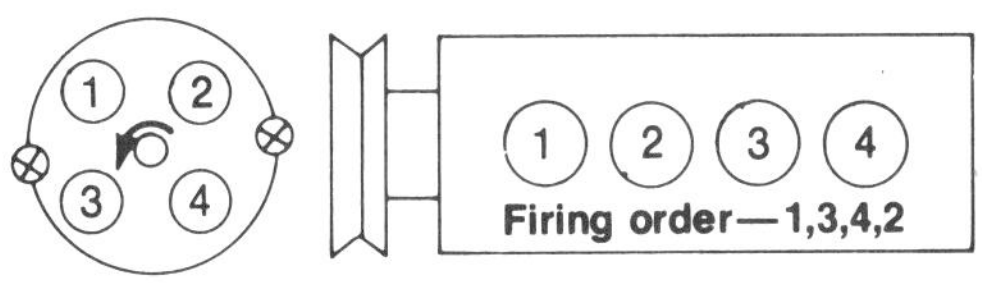

E16 engine

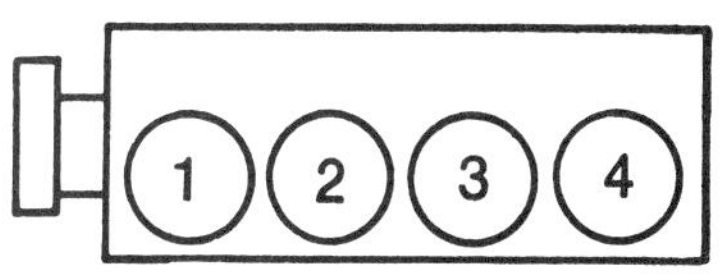

Firing order 1-3-4-2—CD 17

FIRING ORDERS

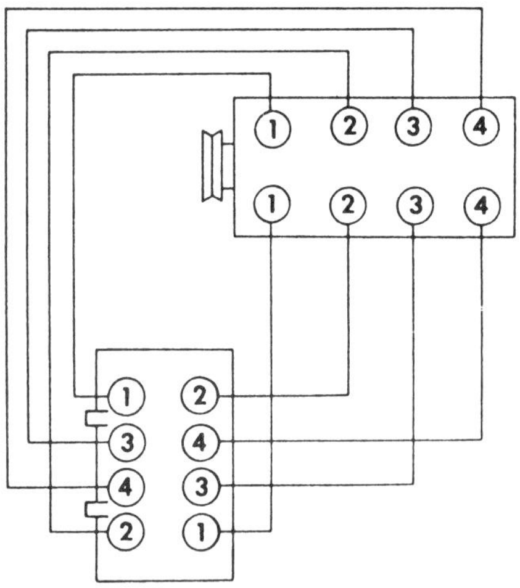

CA20 engine

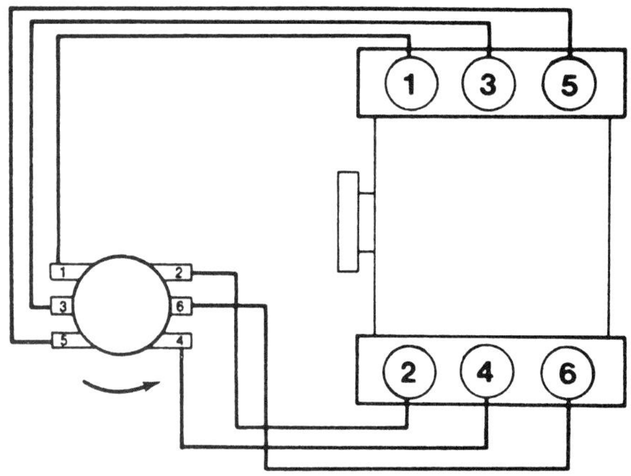

V6 engine firing order: 1-3-5-4-6-2

EMISSION EQUIPMENT USED

1985–86 200SX AND 200SX TURBO

Air flow controlled fuel injection (AFC)
Catalytic converter (CAT)
Electronic concentrated control system (ECCS)
Electronic control unit (ECU)
Exhaust gas recirculation (EGR)
Exhaust gas sensor (EGS)
Fuel evaporation system (FES)
Fuel shut-off system (FSS)
Intake manifold vacuum control (IMVC)
Spark timing control system (STCS)
Transmission control system (TCS)

1985–86 300ZX and 300ZX TURBO

Air flow controlled fuel injection (AFC)
Catalytic converter (CAT)
Electronic concentrated control system (ECCS)
Electronic control unit (ECU)
Exhaust gas recirculation (EGR)
Exhaust gas sensor (EGS)
Fuel evaporation system (FES)
Fuel shut-off system (FSS)
Intake manifold vacuum control (IMVC)
Positive crankcase ventilation (PCV)
Spark timing control system (STCS)

1985–86 MAXIMA

Air flow controlled fuel injection (AFC)
Air injection valve (AIV)
Boost controlled decel valve (BCDD)
Catalytic converter (CAT)
Electronic concentrated control system (ECCS)
Electronic control unit (ECU)
Exhaust gas recirculation (EGR)
Exhaust gas sensor (EGS)
Fuel evaporation system (FES)
Positive crankcase ventilation (PCV)
Spark timing control system (STCS)

1985–86 PULSAR NX

Air injection valve (AIV)
Catalytic converter (CAT)
Electronic controlled carburetor (ECC)
Exhaust gas recirculation (EGR)
Exhaust gas sensor (EGS)
Fuel evaporation system (FES)
Feedback Carburetor (FC)
Fuel shut-off system (FSS)
Mixture Heating System (MHS)
Positive crankcase ventilation (PCV)
Spark timing control system (STCS)
Thermostatic air cleaner (TAC)
Throttle opener control system (TOCS)

1985–86 SENTRA

GASOLINE ENGINES

Air injection valve (AIV)
Catalytic converter (CAT)
Electronic controlled carburetor (ECC)
Exhaust gas recirculation (EGR)
Exhaust gas sensor (EGS)
Fuel evaporation system (FES)
Feedback Carburetor (FC)
Fuel shut-off system (FSS)
Mixture Heating System (MHS)
Positive crankcase ventilation (PCV)
Spark timing control system (STCS)
Thermostatic air cleaner (TAC)
Throttle opener control system (TOCS)

DIESEL ENGINES

Positive carankcase ventilation (PCV)
High altitude compensation (HAC)
Exhaust gas recirculation (EGR)

1985–86 STANZA

Air flow controlled fuel injection (AFC)
Air injection valve (AIV)
Catalytic converter (CAT)
Electronic concentrated control system (ECCS)
Exhaust gas recirculation (EGR)
Exhaust gas sensor (EGS)
Fuel evaporation system (FES)
Intake manifold vacuum control (IMVC)
Positive crankcase ventilation (PCV)
Spark timing control system (STCS)

ENGINE AND EMISSION CONTROL SYSTEM DIAGRAM

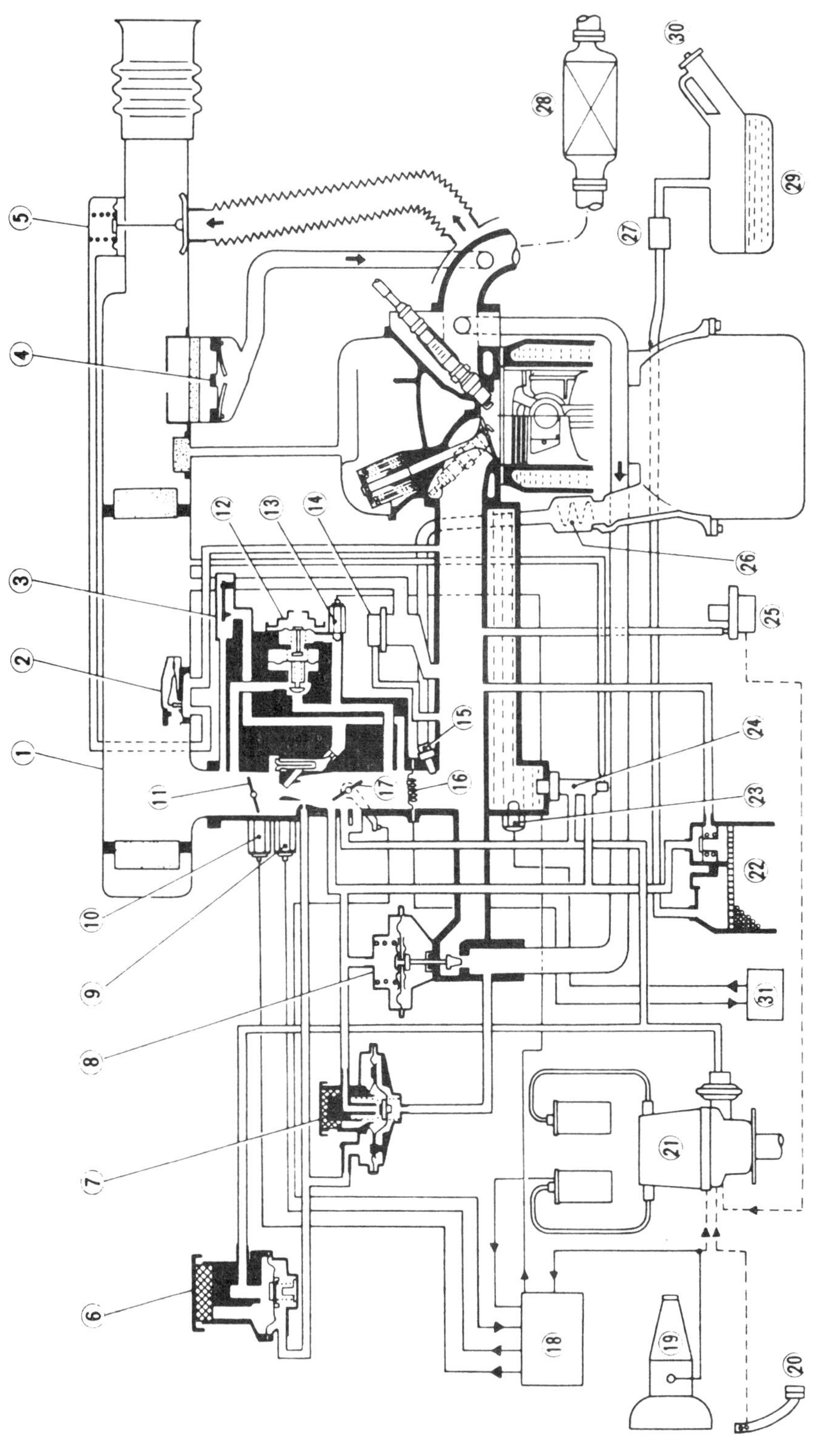

1 A.T.C. air cleaner
2 Air temperature sensor
3 Hot idle compensator
4 A.I.V. valve
5 Vacuum motor
6 Vacuum control valve
7 V.V.T. valve
8 E.G.R. valve
9 Anti-dieseling solenoid
10 Idle speed control solenoid
11 P.T.C. auto-choke
12 B.C.D.D.
13 B.C.D.D. control solenoid (M/T)
14 A.B. valve
15 P.C.V. valve
16 Mixture heater
17 Throttle valve switch
18 Engine revolution unit
19 Neutral switch (M/T)/Inhibitor switch (A/T)
20 Clutch switch (M/T)
21 Distributor (IC ignition unit)
22 Carbon canister
23 Water temperature switch
24 T.V.V.
25 Vacuum switch
26 Oil separator
27 Fuel check valve
28 Oxidation catalytic converter
29 Fuel tank
30 Fuel filler cap (Vacuum relief valve)
31 Mixture heater relay

COMPONENT PARTS LOCATION

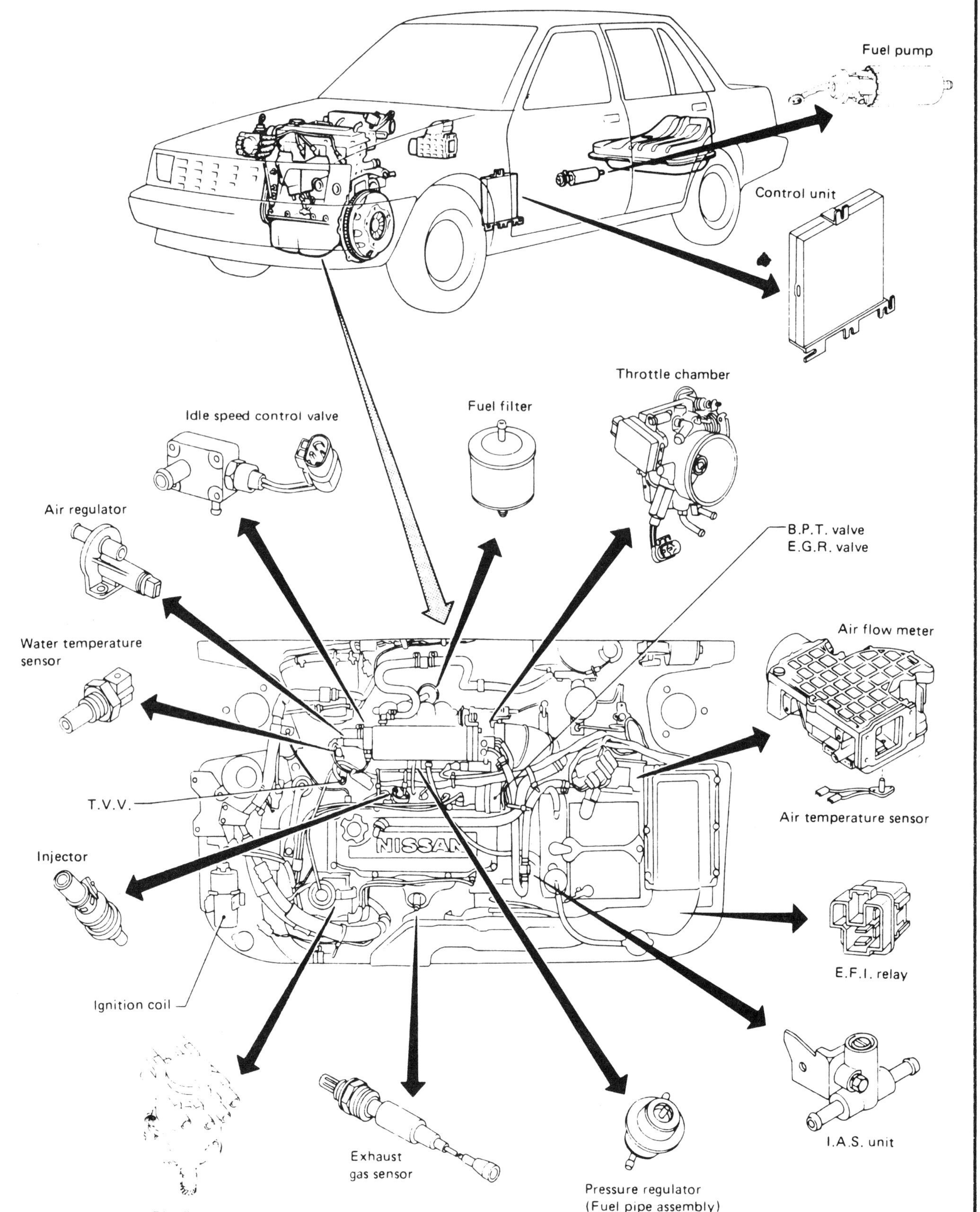

COMPONENT PARTS LOCATION—CA20E Engine

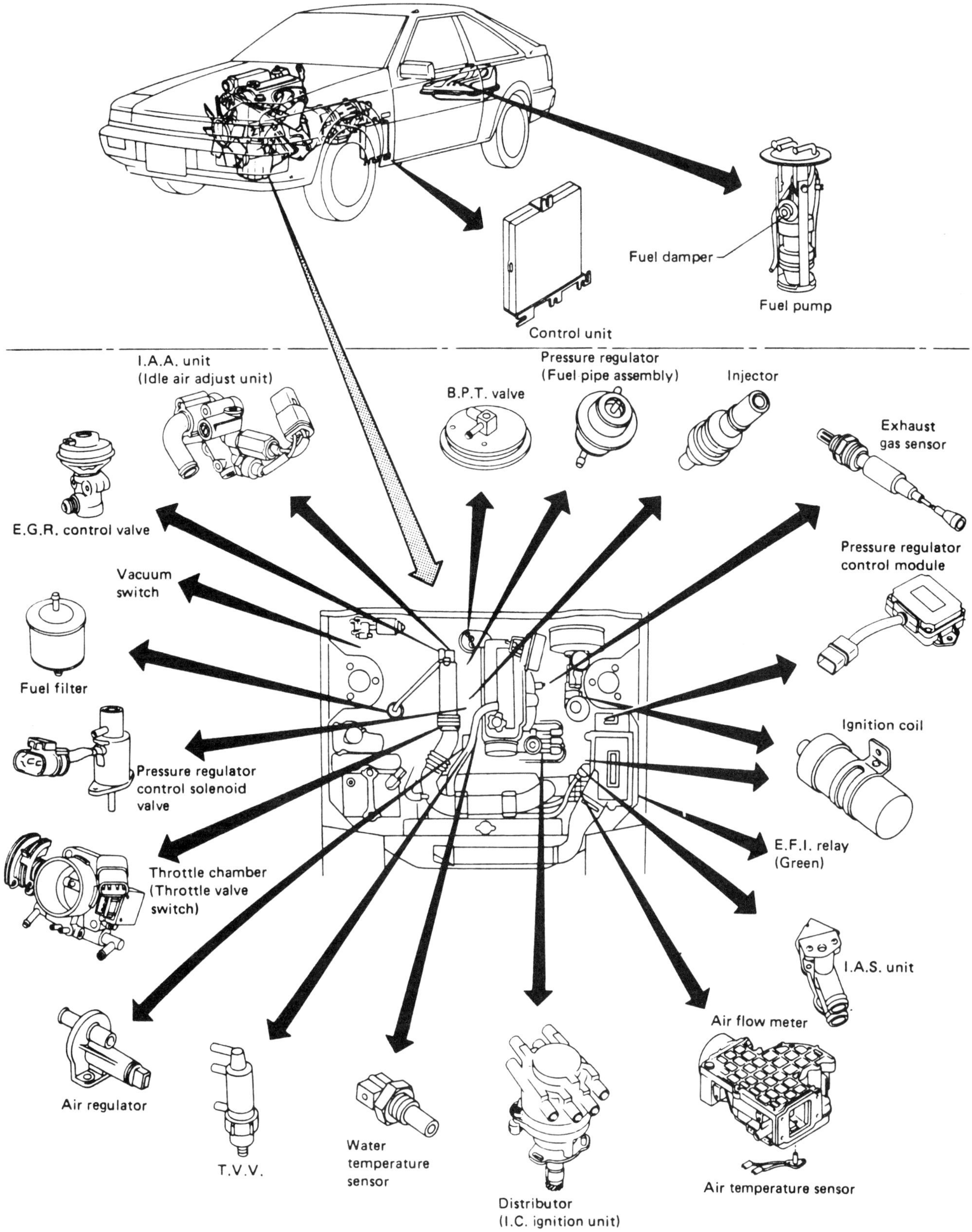

COMPONENT PARTS LOCATION—CA18ET Engine

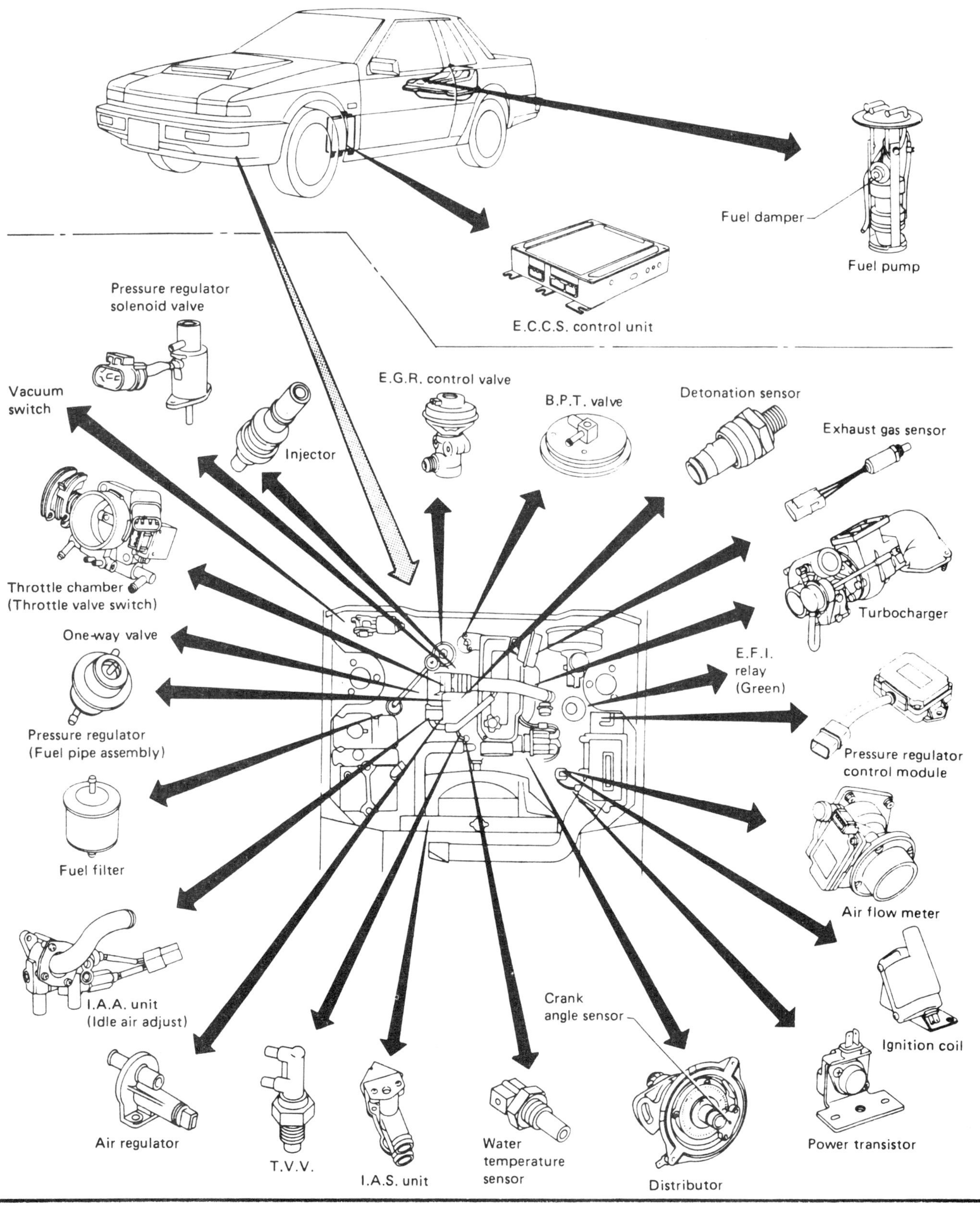

COMPONENT PARTS LOCATION

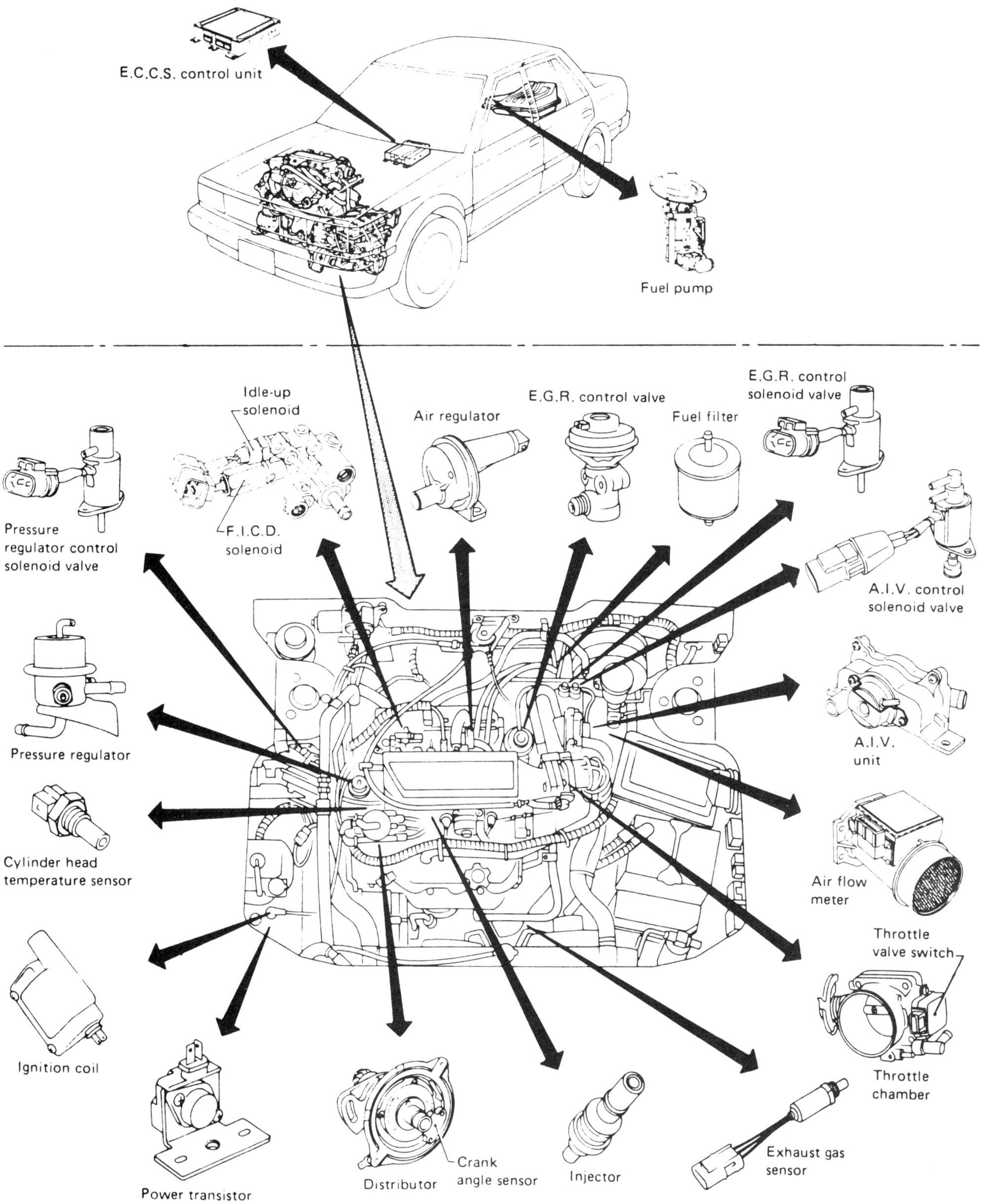

COMPONENT PARTS LOCATION

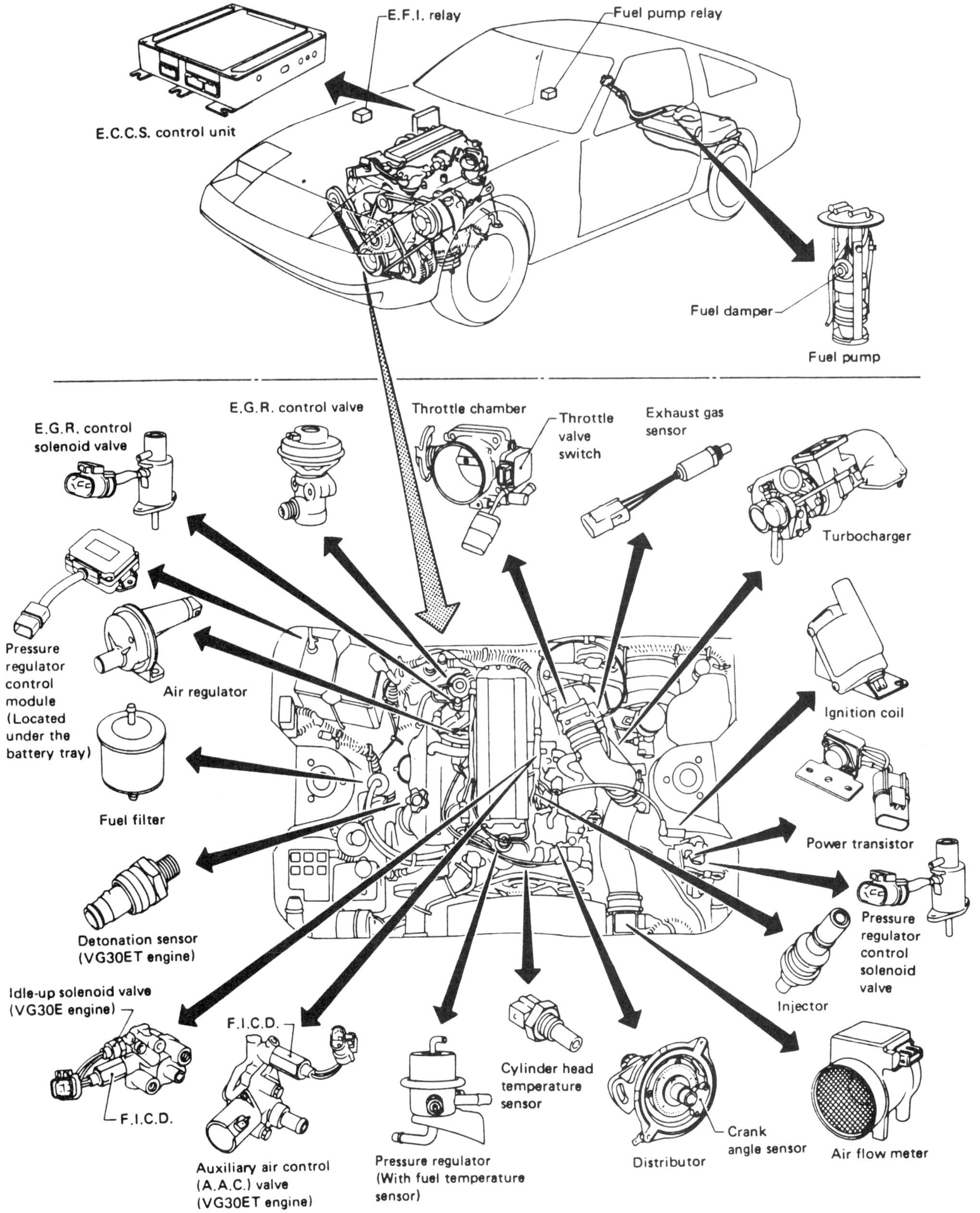

REQUIRED EMISSION CONTROL MAINTENANCE

Crankcase Ventilation System: Inspect and replace the filter every 15,000 miles.

Fuel Evaporation System: Check all hoses and connections every 30,000 miles. Replace filter.

Thermostatic Air Cleaner: Check operation and change the filter every 15,000 miles.

Air Induction Filter: Replace every 30,000 miles.

Idle RPM and Mixture Ratio System: Check the exhaust CO% with an analyzer every 15,000 miles.

Valve Adjustment: Every 15,000 miles

Ignition Timing Control System: Inspect for proper operation every 15,000 miles.

Choke Mechanism: Inspect for proper operation every 15,000 miles.

Exhaust Gas Sensor: Replace every 30,000 miles

EGR System: Check for proper operation every 30,000 miles. Clean all carbon deposits from EGR valve and lines.

NOTE: Although not required, the manufacturer recommends that all emission systems be periodically inspected for proper operation. Change the filters more frequently if the vehicle is operated under extremely dusty conditions, or ambient temperatures are extremely low or high.

IDLE SPEED AND MIXTURE ADJUSTMENTS

1985–86

NOTE: The following information is being published from the latest information available at the time of publication. If the information differs from the information on the vehicle emissions label, use the information on the emissions label.

Idle Speed and Mixture Adjustment

CARBURETED MODELS

NOTE: The idle mixture adjusting screw which was preset at the factory should be adjusted only in the event of major carburetor overhaul, throttle body replacement or to lower the exhaust emissions. The idle mixture adjusting screw should not be removed during routine maintenance.

Preparation

Before adjusting the idle speed, be certain the following items are operating correctly:

- Battery.
- Ignition System
- Engine oil & coolant levels
- Fuses
- E.C.C. components
- Main harness connector
- Hoses
- Oil filler cap and oil level gauge
- Valve clearance
- Connect the engine timing light and tachometer in their proper locations.
- Apply the parking brake and block the wheels
- Check that all accessories and switches are in the OFF position.
- Check that the radiator fan has stopped and if it is operating, wait until it stops.

IDLE SPEED SPECIFICATION

Model / Engine	Idle Speed
200SX / CA20E engine	
Curb idle A/T	700 rpm in drive
Curb idle M/T	750 rpm in neutral
200SX Turbo / CA18ET	
Curb idle M/T (Sea Level)	750 rpm in neutral
Curb idle M/T (High Alt.)	680 rpm in neutral
300ZX / VG30E engine	
Curb idle A/T	700 rpm in drive
Curb idle M/T (Sea Level)	700 rpm in neutral
Curb idle M/T (High Alt.)	650 rpm in neutral
300ZX Turbo / VG30ET engine turbocharged	
Curb idle A/T	650 rpm in drive
Curb idle M/T	700 rpm in neutral
Maxima / VG30E engine	
Curb idle A/T	700 rpm in drive
Curb idle M/T (Sea Level)	700 rpm in neutral
Curb idle M/T (High Alt.)	650 rpm in neutral
Pulsar/NX / E16 engine	
Curb idle A/T (exc. Calif. & Canada)	650 rpm in drive
Curb idle M/T (exc. Calif. & Canada)	800 rpm in neutral
Curb idle A/T (Calif. & Canada)	650 rpm in drive
Curb idle M/T (Calif. & Canada)	750 rpm in neutral

IDLE SPEED SPECIFICATION

Model / Engine	Idle Speed
Sentra / E16S engine	
Curb idle A/T (exc. Calif. & Canada)	650 rpm in drive
Curb idle M/T (exc. Calif. & Canada)	800 rpm in neutral
Curb idle A/T (Calif. & Canada)	650 rpm in drive
Curb idle M/T (Calif. & Canada)	750 rpm in neutral
Sentra Diesel / CD17 engine	
Curb idle All	750 rpm in neutral
Stanza / CA20E engine	
Curb idle A/T	700 rpm in drive
Curb idle M/T	750 rpm in neutral
Stanza / CA20S engine	
Curb idle A/T	650 rpm in drive
Curb idle M/T	650 rpm in neutral

• When checking "CO" percentage, insert the probe at least 15.7 inches into the tail pipe.

For the adjustment of the idle mixture, the following steps must be performed to remove the idle mixture seal plug:

a. Remove the carburetor from the engine.

b. Carefully drill the idle adjusting screw seal plug and remove it from the plug hole with a suitable tool.

c. After adjusting the carburetor idle rpm and mixture, turn off the engine and install a new seal plug on the carburetor.

NOTE: Inspections and adjustments should be carried out while the shift lever is in "D" position on the automatic models or in "Neutral" on the manual transaxle equipped models. When increasing the engine speed, be certain the automatic transaxle models are in "N" or "P", with the brakes applied.

1. Warm-up the engine until the water temperature gauge registers normal operating temperature (middle of gauge) and the engine speed is below 1,000 rpm.
2. Run the engine at idle speed for 2 minutes.
3. Increase the engine speed between 2,000–3,000 rpm, two or three times under no load, then idle the engine for one minute.

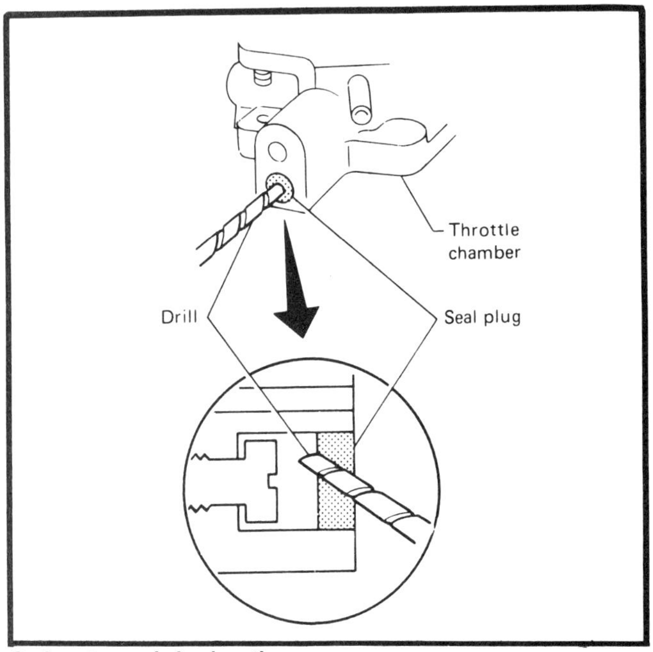

Carburetor seal plug location

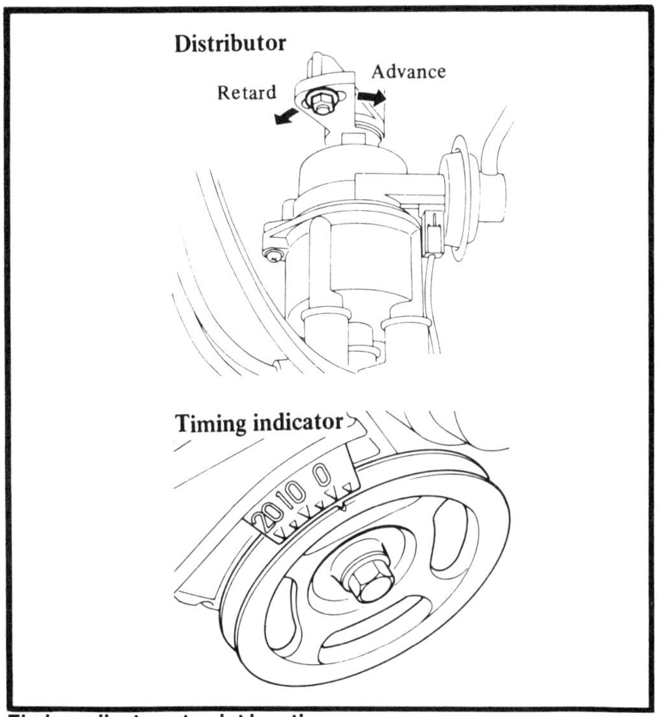

Timing adjustment point location

4. Check the ignition timing. If necessary, adjust it to specifications by loosening the distributor hold-down bolt and turning the distributor.
5. Increase the engine speed between 2,000–3,000 rpm, two or three times under no load, then idle the engine for one minute.
6. Check the idle speed. If necessary, adjust it to specifications by turning the throttle adjusting screw. If the idle speed will not adjust, replace the control unit, carburetor, or the vacuum control modulator.
7. Turn off the engine and disconnect the air-fuel ratio harness connector.
8. Restart and increase the engine speed between 2,000–3,000 rpm, two or three times under no load, then idle the engine for one minute.
9. Properly calibrate the CO meter and insert the probe at least 15.7 in. into the tail pipe.
10. Check the CO percentage at idle. It should be 3.0% ± 1.0%. If necessary, repeat Step 8 and adjust the idle CO percentage by turning the idle adjusting screw.
11. Turn off engine and connect the air-fuel solenoid harness connector.

12. Disconnect the vacuum control modulator harness connector.
13. Restart and increase the engine speed between 2,000–3,000 rpm, two or three times under no load, then idle the engine for one minute.
14. Check the idle speed. If necessary, adjust it to specifications by turning the throttle adjusting screw.
15. Turn off the engine, and connect the vacuum control modulator harness connector.
16. Restart and increase the engine speed between 2,000–3,000 rpm, two or three times under no load, then idle the engine for one minute.
17. Check the idle speed. If necessary, adjust it to specifications by turning the throttle adjusting screw. If the idle speed will not adjust replace the control unit, carburetor, or the vacuum control modulator.

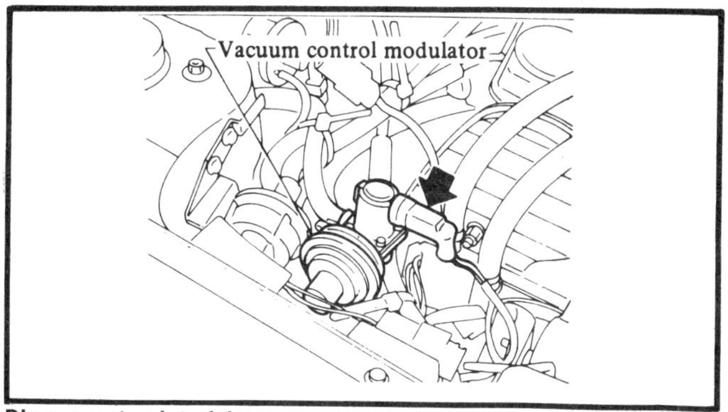

Disconnect point of the vacuum control modulator

Idle Speed Adjustment

FUEL INJECTED MODELS

Before adjusting the idle speed, be certain the following items are operating correctly:
- E.G.R valve operation
- Air cleaner clogging
- Ignition system
- Engine oil and coolant levels
- Valve clearance
- E.F.I./E.C.C.S. component parts
- E.F.I./E.C.C.S. harness connectors
- Vacuum hoses
- Air intake system (Oil filler cap, oil level gauge,etc.)

1. The air conditioner and all electrical accessories should be turned "OFF".
2. Apply the parking brake and block the wheels.
3. Make adjustments only while the cooling fan has stopped.

NOTE: Inspections and adjustments should be carried out while the shift lever is in "D" position on the automatic models or in "Neutral" on the manual transaxle equipped models.

4. Warm-up the engine until the water temperature gauge registers normal operating temperature (middle of the temperature gauge).
5. Stop the engine and attach an adapter harness, VG30E, VG30ET, and CA18ET engines only, between the ignition coil primary terminals and the harness connector. Then connect the tachometer to the adapter harness.
6. Start the engine and run at about 2,000 rpm for about 2 minutes under no-load.

Idle-up solenoid—300ZX shown, Maxima similar

7. Disconnect the idle up solenoid harness connector on VG30E and VG30ET engines.
8. Increase the engine speed two or three times under no-load, then run the engine for one minute at idle speed.
9. Check the idle speed. If necessary, adjust it to specifications by turning the idle speed adjusting screw.
Figs. NI-15, 16, 17
10. Stop the engine, reconnect the idle up solenoid harness connector on VG30E and VG30ET engines.
11. Disconnect the adapter harness on VG30E, VG30ET, and CA18ET engines. Connect the ignition coil harness connector to the ignition coil.

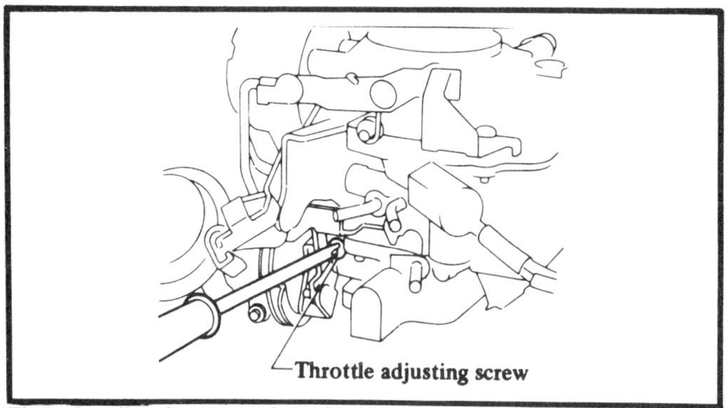

Throttle adjusting screw location

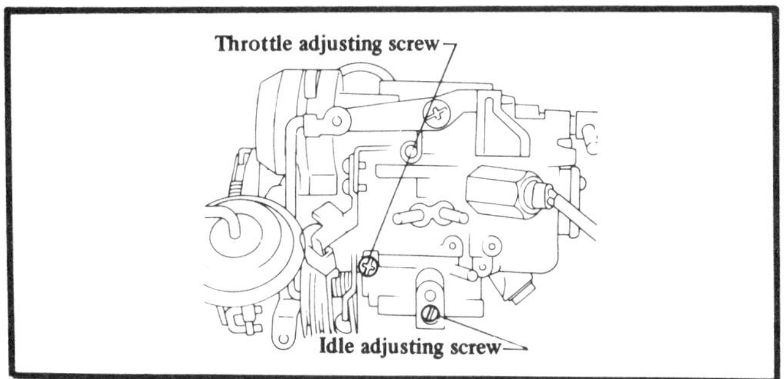

Idle speed and mixture adjusting screw locations

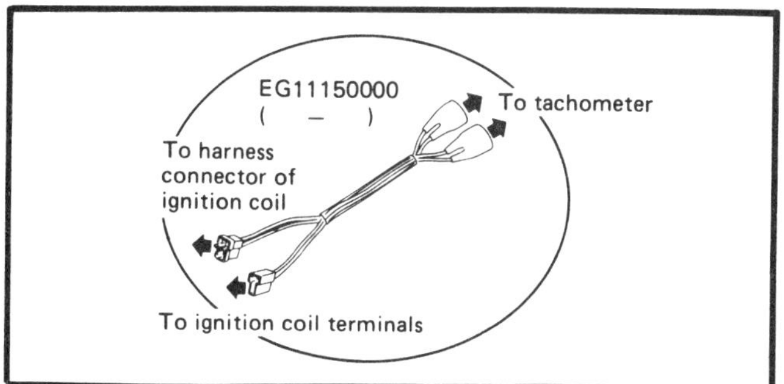

Tachometer adapter harness—VG30E, VG30ET and CA18EF engines

Mixture Ratio Feedback System

Inspection

CA20E ENGINE

Before adjusting be certain the following items are operating correctly:
- Battery

- E.G.R valve operation
- Air cleaner clogging
- Ignition system
- Engine oil and coolant levels
- Valve clearance
- E.F.I./E.C.C.S. component parts
- E.F.I./E.C.C.S. harness connectors
- Vacuum hoses
- Air intake system (Oil filler cap, oil level gauge,etc.)

1. The air conditioner, and all electrical accessories should be turned "OFF".
2. Apply the parking brake and block the wheels.
3. Make adjustments only while the cooling fan has stopped.

NOTE: Inspections and adjustments should be carried out while the shift lever is in "D" position on the automatic models or in "Neutral" on the manual transaxle equipped models.

4. Warm-up the engine until the water temperature gauge registers normal operating temperature (middle of the temperature gauge).
5. Start the engine and run at about 2,000 rpm for about 2 minutes under no-load.
6. Perform the E.F.I. self-diagnosis. Check, correct or replace any malfunctioning parts.
7. Does the engine idle smoothly? If not, check and clean the injectors and replace the injectors, if necessary.
8. Disconnect the distributor vacuum hose from the distributor vacuum controller and plug the hose.
9. Start the engine and run at about 2,000 rpm for about 2 minutes under no-load.
10. Run the engine at idle speed and check that the engine speed is below 1,000 rpm.
11. Increase the engine speed between 2,000–3,000 rpm, two or three times under no load, then idle the engine for one minute.
12. Check the ignition timing. If necessary, repeat Step 11 and adjust

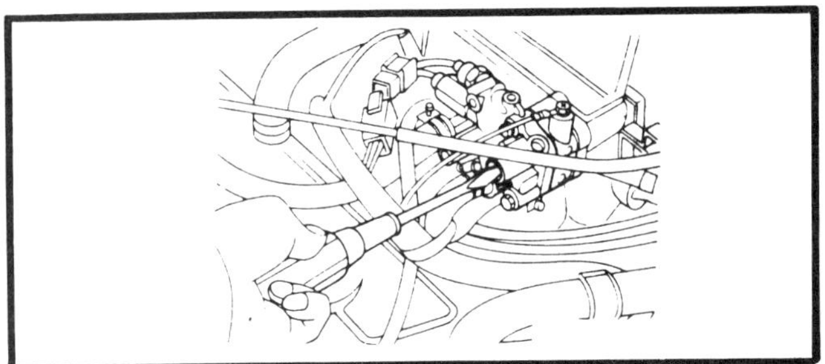

Idle speed adjusting screw—Maxima, 300ZX similar

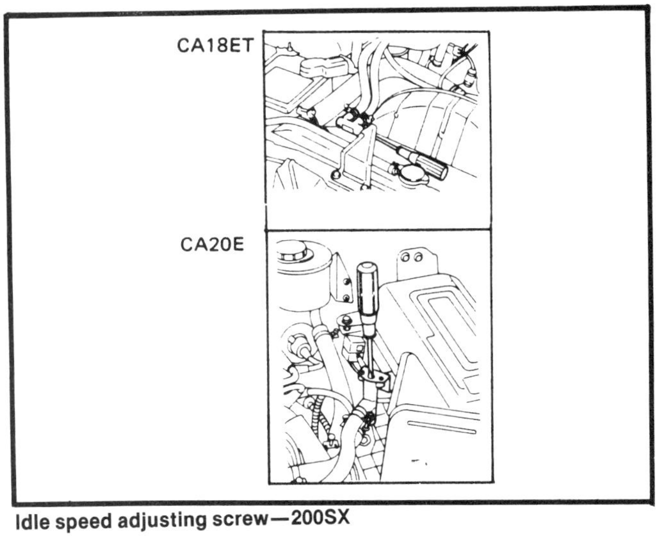

Idle speed adjusting screw—200SX

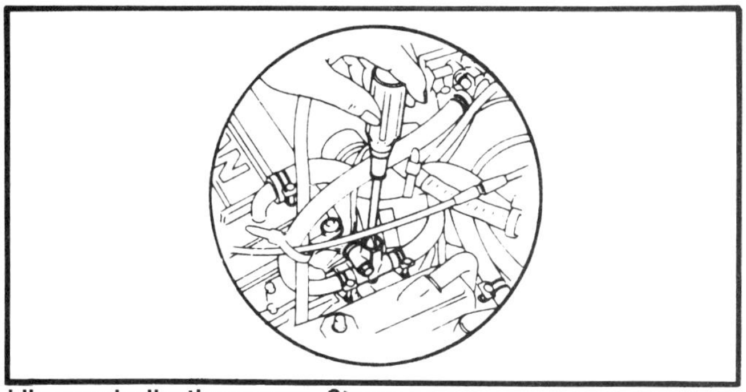

Idle speed adjusting screw—Stanza

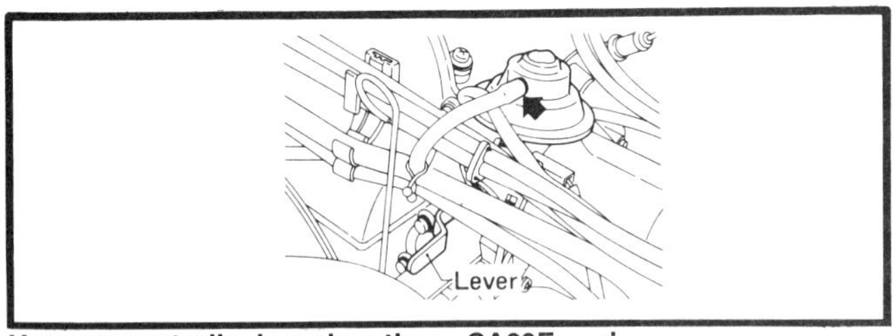

Vacuum controller hose location—CA20E engine

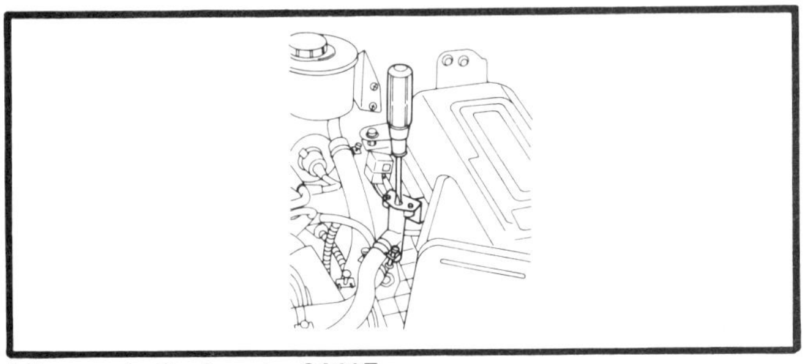

Idle speed adjustment—CA20E

it to specifications by loosening the distributor hold-down bolt and turning the distributor.

13. Connect the distributor vacuum hose to the distributor vacuum controller.
14. Run at about 2,000 rpm for about 2 minutes under no-load.
15. Check to see that even if the inspection lamp (RED) on the control unit goes on, the green lamp goes on and off more than 9 times during 10 seconds at 2,000 rpm. If O.K., go to the following step. If NO, go to Step 20.
16. Increase the engine speed between 2,000–3,000 rpm, two or three times under no load, then idle the engine for one minute.
17. Check that the inspection lamps (RED and GREEN) on the control unit blinks. They should blink simultaneously. If they DO NOT, go to Step 27.
18. Increase the engine speed between 2,000–3,000 rpm, two or three times under no load, then idle the engine for one minute.
19. Check the idle speed. If necessary, repeat Step 18. and adjust it to specifications by turning the idle speed adjusting screw. If the idle speed meets specifications, this is the END of this procedure.
20. Check the exhaust gas sensor harness using the following steps:
    a. Turn off the engine and disconnect the negative battery cable.
    b. Disconnect the 20-pin connector from the Control Unit.
    c. Disconnect the exhaust gas sensor harness connector and connect the terminal for the exhaust gas sensor harness connector to ground with a jumper wire.
    d. Check the continuity between terminal No.4 of the 20-pin connector and ground of the vehicle. If continuity exists proceed to Step 21. If continuity does not exist, repair or replace the E.F.I. harness and connect the battery ground. Return to Step 4.
21. Connect E.F.I. 20-pin connector to the control unit. Connect the negative battery cable and disconnect the jumper wire.

22. Disconnect the water temperature sensor harness connector. Connect a resistor (2.5 k) between the terminals of the water temperature sensor harness connector.

23. Start the engine and warm-up the engine until the water temperature gauge registers normal operating temperature (middle of the temperature gauge).

24. Increase the engine speed between 2,000–3,000 rpm, two or three times under no load, then idle the engine for one minute.

25. Check the CO percentage at idle. It should be less than 5.0%. After checking the CO percentage:

a. Disconnect the resistor from the terminals of the water temperature sensor connector.

b. Connect the water temperature sensor harness connector to the water temperature sensor. If the CO percentage meets specifications, turn off the engine, replace the exhaust gas sensor, and proceed to Step 33. If the CO percentage does NOT meet specifications, proceed to the next step.

26. Connect the exhaust gas sensor harness to the gas sensor.

27. Turn off the engine and remove the air flow meter from the vehicle.

28. Drill a hole in the seal plug which seals the air by-pass screw of the air flow meter and remove the seal plug with a suitable tool.

29. Install the air flow meter on the vehicle. Start the engine and warm-up the engine until the water temperature gauge registers normal operating temperature (middle of the temperature gauge).

30. Adjust the idle mixture ratio by turning the air by-pass screw of the air flow meter so that the inspection lamps (green and red) on the E.F.I. control unit blink simultaneously. If the idle mixture is adjustable, proceed to the next step. If the idle mixture is NOT adjustable, replace the air flow meter and return to Step 4.

31. Turn off the engine and remove the air flow meter from the vehicle. Insert a new seal plug into the air by-pass screw hole. Tap the seal plug with a suitable tool, thereby installing the seal plug into the air flow meter.

32. Install the air flow meter on the vehicle and return to Step 4.

33. Start the engine and warm-up the engine until the water temperature gauge registers normal operating temperature (middle of the temperature gauge).

34. Run the engine at about 2,000 rpm for about 2 minutes under no-load.

35. Make certain that even if the (RED) lamp goes on, the inspection lamp (GREEN) goes on and off more than 9 times during 10 seconds at 2,000 rpm. If this checks OK, return to Step 4. If it checks no good, replace the E.F.I. control unit and then return to Step 4.

CA18ET ENGINE

Before adjusting be certain the following items are operating correctly:

- Battery
- E.G.R valve operation
- Air cleaner clogging
- Ignition system
- Engine oil and coolant levels
- Valve clearance
- E.F.I./E.C.C.S. component parts
- E.F.I./E.C.C.S. harness connectors
- Vacuum hoses
- Air intake system (Oil filler cap, oil level gauge,etc.)

1. The air conditioner, and all electrical accessories should be turned "OFF".

2. Apply the parking brake and block the wheels.

3. Make adjustments only while the cooling fan has stopped.

NOTE: Inspections and adjustments should be carried out while the shift lever is in "D" position on the automatic models or in "Neutral" on the manual transaxle equipped models.

4. Warm-up the engine until the water temperature gauge registers normal operating temperature (middle of the temperature gauge).

5. Start the engine and run at about 2,000 rpm for about 2 minutes under no-load.

6. Perform the E.C.C.S. self-diagnosis. Check, correct or replace any malfunctioning parts.

7. Does the engine idle smoothly? If not, check and clean the injectors, and replace the injectors if necessary.

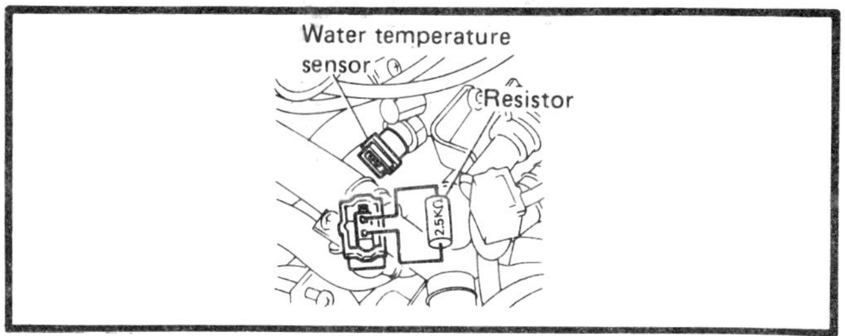

Connecting point of the 2.5k Ω resistor—CA20E

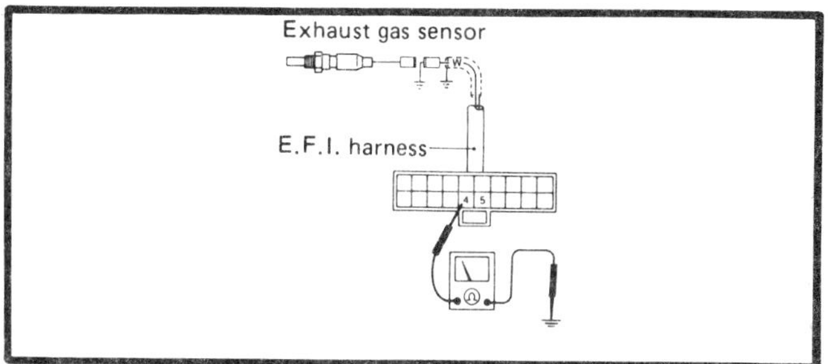

Continuity test of the exhaust gas sensor—CA20E

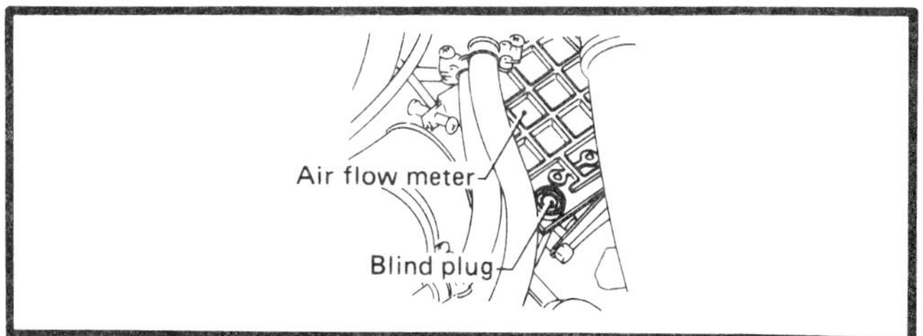

Air flow meter plug location—CA20E

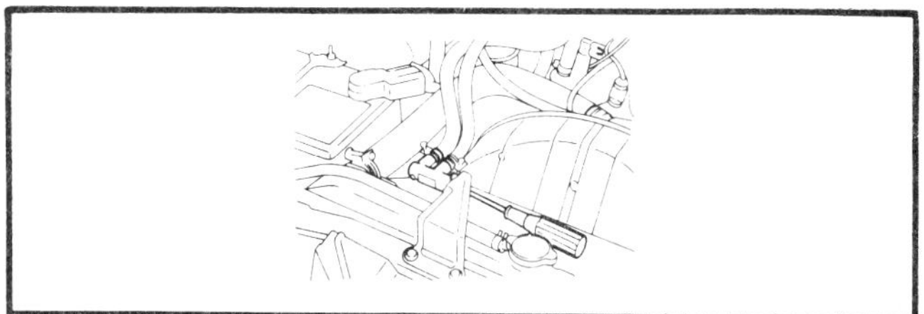

Idle speed adjustment—CA18ET

8. Start the engine and run at about 2,000 rpm for about 2 minutes under no-load.

9. Increase the engine speed between 2,000–3,000 rpm, two or three times under no load, then idle the engine for one minute.

10. Run the engine at idle speed and check that the engine speed is below 1,000 rpm.

11. Check the idle speed. If necessary, repeat Step 9 and 10 and adjust it to specifications by turning the idle speed adjusting screw.

12. Check the ignition timing. If necessary, repeat Step 9 and 10 and adjust it to specifications by loosening the distributor hold-down bolt and turning the distributor.

13. Run at about 2,000 rpm for about 2 minutes under no-load.

14. Check to see that even if the inspection lamp (RED) on the control unit goes on, the green lamp goes on and off more than 9 times during 10 seconds at 2,000 rpm. If O.K., go to the following step. If NO, go to Step 19.

15. Increase the engine speed between 2,000–3,000 rpm, two or three times under no load, then idle the engine for one minute.

16. Check that the inspection lamps (RED and GREEN) on the control unit blinks. They should blink simultaneously. If they DO NOT go to Step 26.

17. Increase the engine speed between 2,000–3,000 rpm, two or three times under no load, then idle the engine for one minute.

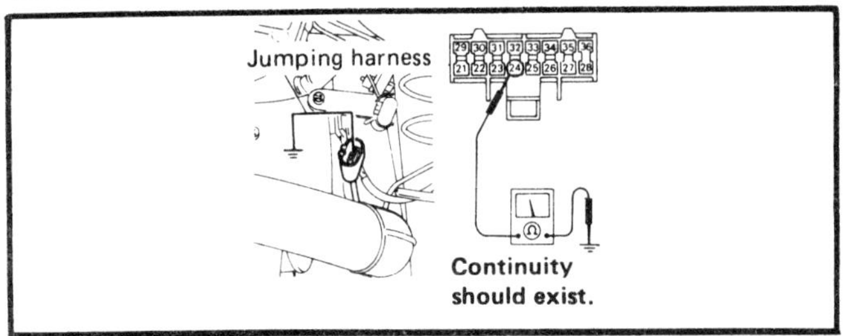

Testing the exhaust gas sensor harness—CA18ET

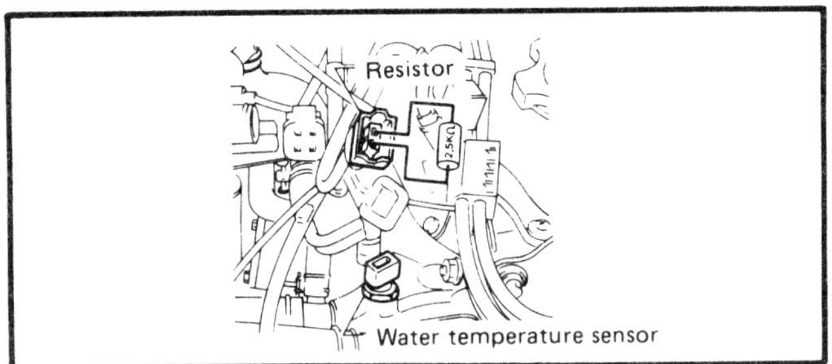

Connecting point of the 2.5k Ω resistor—CA18ET

18. Check the idle speed. If necessary, repeat Step 17 and adjust it to specifications by turning the idle speed adjusting screw. If the idle speed meets specifications, this is the END of this procedure.

19. Check the exhaust gas sensor harness using the following steps:
   a. Turn off the engine and disconnect the negative battery cable.
   b. Disconnect the 16-pin connector from the Control Unit.
   c. Disconnect the exhaust gas sensor harness connector and connect the terminal for the exhaust gas sensor harness connector to ground with a jumper wire.
   d. Check the continuity between terminal No.24 of the 16-pin connector and ground of the vehicle. If continuity exists, proceed to Step 20. If continuity does not exist, repair or replace the E.C.C.S. harness. Return to Step 8.

20. Connect 16-pin connector and the negative battery cable. Disconnect the jumper wire.

21. Disconnect the water temperature sensor harness connector. Connect a resistor (2.5 k) between the terminals of the water temperature sensor harness connector.

22. Start the engine and warm-up the engine until the water temperature gauge registers normal operating temperature (middle of the temperature gauge).

23. Increase the engine speed between 2,000–3,000 rpm, two or three times under no load, then idle the engine for one minute.

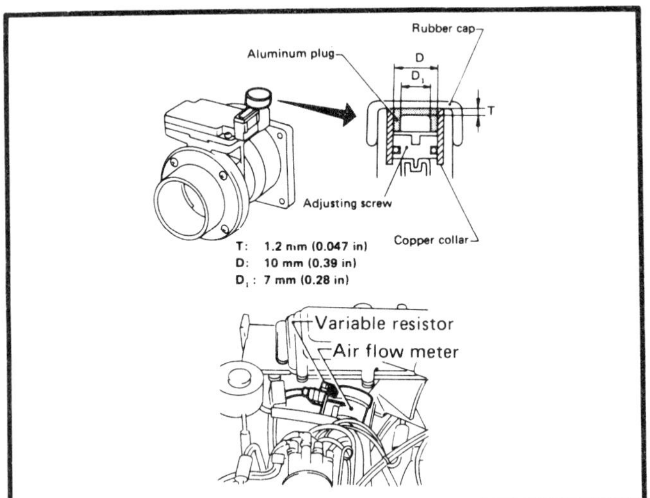

Airflow meter plug location—CA18ET

24. Check the CO percentage at idle. It should be less than 8.0%. After checking the CO percentage:
   a. Disconnect the resistor from the terminals of the water temperature sensor connector.
   b. Connect the water temperature sensor harness connector to the water temperature sensor. If the CO percentage meets specifications, turn off the engine, replace the exhaust gas sensor, and proceed to Step 32. If the CO percentage does NOT meet specifications, proceed to the next step.

25. Connect the exhaust gas sensor harness to the gas sensor.

26. Turn off the engine and remove the air flow meter from the vehicle.

27. Drill a hole in the seal plug which seals the air by-pass screw of the air flow meter and remove the seal plug with a suitable tool.

28. Install the air flow meter on the vehicle. Start the engine and warm-up the engine until the water temperature gauge registers normal operating temperature (middle of the temperature gauge).

29. Adjust the idle mixture ratio by turning the variable resistor of the air flow meter so that the inspection lamps (green and red) blink simultaneously. If the idle mixture is adjustable, proceed to the next step. If the idle mixture is NOT adjustable, replace the air flow meter and return to Step 8.

30. Turn off the engine and remove the air flow meter from the vehicle. Insert a new seal plug into the air by-pass screw hole. Tap the seal plug with a suitable tool, thereby installing the seal plug into the air flow meter.

31. Install the air flow meter on the vehicle and return to Step 8.

32. Start the engine and warm-up the engine until the water temperature gauge registers normal operating temperature (middle of the temperature gauge).

33. Run the engine at about 2,000 rpm for about 2 minutes under no-load.

34. Replace the exhaust gas sensor and make sure that even if the (RED) lamp goes on, the inspection lamp (GREEN) goes on and off more than 9 times during 10 seconds at 2,000 rpm. If this checks OK, return to Step 8. If it checks no good, replace the C.E.C.U. and then return to Step 8.

VG30E ENGINE

Before adjusting be certain the following items are operating correctly:

- Battery
- E.G.R valve operation
- Air cleaner clogging
- Ignition system
- Engine oil and coolant levels
- E.C.C.S. harness connectors
- Vacuum hoses
- Air intake system (Oil filler cap, oil level gauge,etc.)
- Fuses
- Engine compression
- Throttle valve and throttle valve switch operation

1. The air conditioner and all electrical accessories should be turned "OFF".
2. Apply the parking brake and block the wheels.
3. Make adjustments only while the cooling fan has stopped.

NOTE: Inspections and adjustments should be carried out while the shift lever is in "D" position on the automatic models or in "Neutral" on the manual transmission equipped models.

4. Check that the following items are normal:
   a. Engine operating temperature
   b. Idle speed
   c. Ignition timing
5. Verify that the diagnosis mode selector is turned fully counterclockwise.
6. Start the engine and run at about 2,000 rpm for about 2 minutes under no-load.
7. Check to see that even if the inspection lamp (GREEN) on the control unit goes on, the green lamp goes on and off more than 5 times during 10 seconds at 2,000 rpm. If O.K., go to the following step, if NO, go to Step 12.
8. Disconnect the throttle valve switch harness connector.

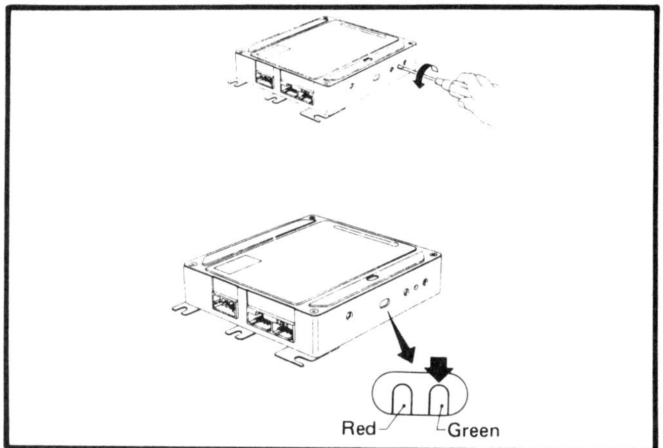

Turning the diagnosis mode selector counterclockwise, and the inspection lamp locations

9. Increase the engine speed between 2,000–3,000 rpm, two or three times under no load, then idle the engine for one minute.

10. Check to see that even if the inspection lamp on the control unit blinks. If O.K., go to the following step If NO, go to Step 18.

11. Connect the throttle valve switch harness connector. This ends this procedure.

12. Check the exhaust gas sensor harness using the following steps:

a. Turn off the engine and disconnect the negative battery cable.

b. Disconnect the 16-pin connector from the Control Unit.

c. Disconnect the exhaust gas sensor harness connector and connect the terminal for the exhaust gas sensor harness connector to ground with a jumper wire.

d. Check the continuity between terminal No.24 of the 16-pin connector and ground of the vehicle. If continuity exists proceed to Step 13. If continuity does not exist, repair or replace the E.C.C.S. harness. Return to Step 4.

13. Disconnect the cylinder head temperature sensor harness connector. Connect a resistor (2.5 k) between the terminals of the water temperature sensor harness connector.

14. Start the engine and warm-up the engine until the water temperature gauge registers normal operating temperature (middle of the temperature gauge).

15. Increase the engine speed between 2,000–3,000 rpm, two or three times under no load, then idle the engine for one minute.

16. Check the CO percentage at idle. It should be 0.2–4.0%. After checking the CO percentage:

a. Disconnect the resistor from the terminals of the water temperature sensor connector.

b. Connect the cylinder head temperature sensor harness connector to the water temperature sensor. If the CO percentage meets specifications, turn off the engine and proceed to Step 24. If the CO percentage does NOT meet specifications, proceed to the next step.

17. Connect the exhaust gas sensor harness and disconnect the throttle valve switch harness connector.

18. Turn off the engine and remove the air flow meter from the vehicle.

19. Drill a hole in the seal plug which seals the variable resistor of the air flow meter and remove the seal plug with a suitable tool.

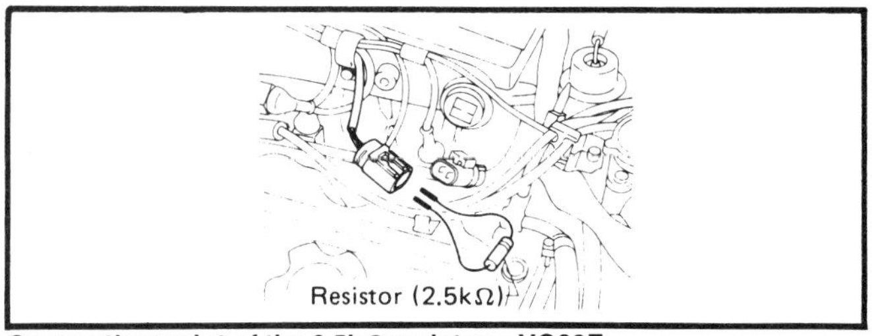

Connecting point of the 2.5k Ω resistor—VG30E

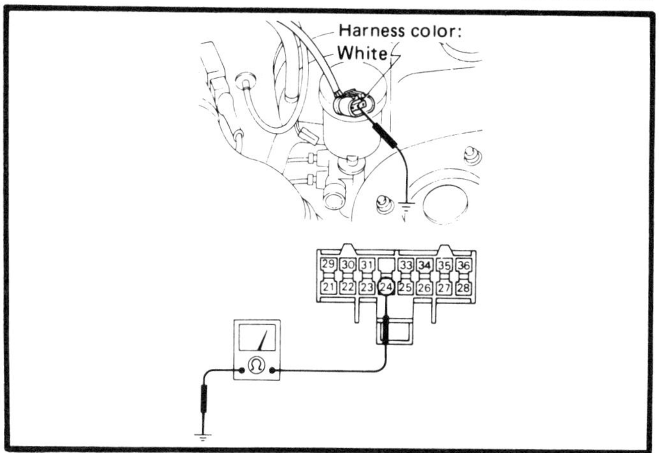

Testing the exhaust gas sensor harness—VG30E

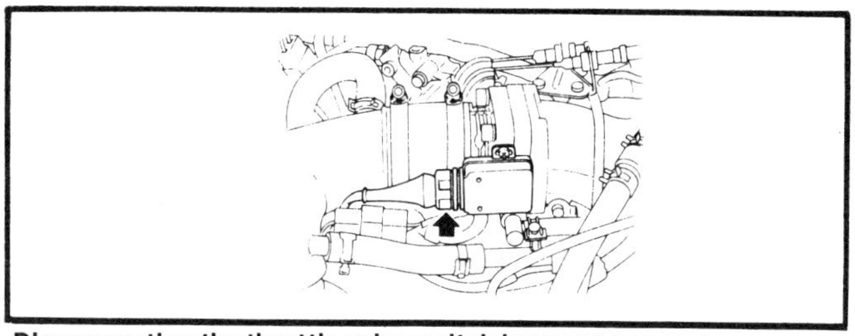

Disconnecting the throttle valve switch harness connector

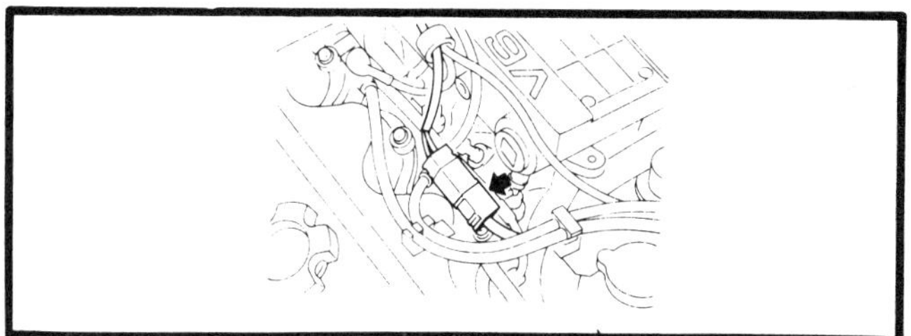

Removing the cylinder head temperature harness

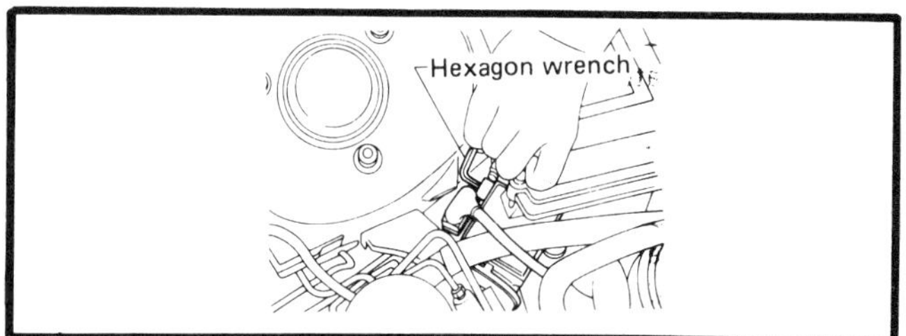

Idle mixture adjustment—VG30E

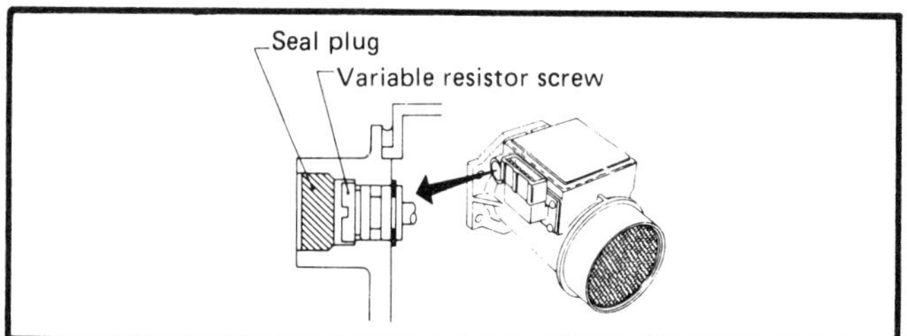

Airflow meter seal plug location—VG30E

20. Install the air flow meter on the vehicle. Start the engine and warm-up the engine until the water temperature gauge registers normal operating temperature (middle of the temperature gauge).

21. Adjust the idle mixture ratio by turning the variable resistor of the air flow meter so that the inspection lamps blink simultaneously. If the idle mixture is adjustable, proceed to the next step. If the idle mixture is NOT adjustable, replace the air flow meter and return to Step 4.

22. Turn off the engine and remove the air flow meter from the vehicle. Insert a new seal plug into the air by-pass screw hole. Tap the seal plug with a suitable tool, thereby installing the seal plug into the air flow meter.

23. Install the air flow meter on the vehicle. Connect the throttle valve switch connector and return to Step 4.

24. Replace the exhaust gas sensor and make certain that the inspection lamp on the control unit goes on and off more than 5 times during 10 seconds, at 2,000 rpm. If this checks OK, return to Step 4. If it checks no good, replace the E.C.C.S. control unit and then return to Step 4.

IDLE SPEED

Adjustment

DIESEL ENGINE

1. Turn off lights, heater fan and all electrical accessories.
2. Apply the parking brake and block the wheels.
3. Set the transmission shift lever in "Neutral". Start the engine and warm it up until the water temperature indicator points to the middle of the gauge.
4. Attach a diesel tacho tester's pick-up to the No.1 injection tube.

NOTE: In order to obtain a more accurate reading of engine speed, remove the clamps on the NO.1 injection tube.

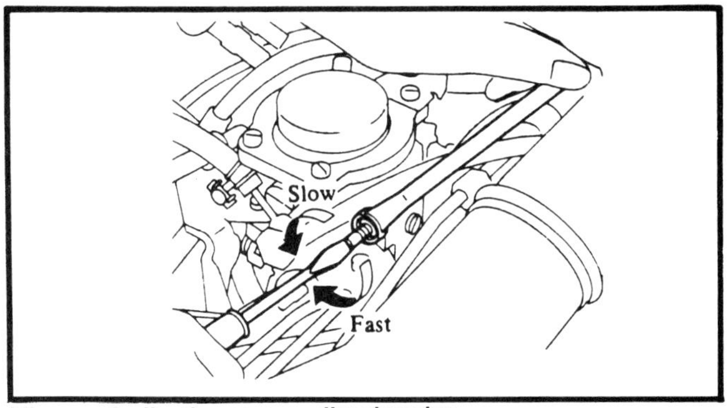

Idle speed adjusting screw—diesel engine

5. Run the engine at about 2,000 rpm for two minutes under load. Increase the engine two or three times. Make sure that it returns to idle speed. If not, check the accelerator linkage for binding.
6. Run the engine for one minute at idle speed.
7. Check the idle speed. If necessary, adjust to specifications using the following procedures.
8. Turn off the engine.
9. Disconnect the accelerator wire from the injection pump control lever.
10. Move the control lever to WOT wide open throttle and loosen the idle screw lock nut while holding the control lever.
11. Turn the idle screw to obtain the specified rpm.
12. Turn off the engine.
13. Tighten the idle adjusting screw locknut while holding the control lever at wide open throttle.
14. Connect the accelerator wire.

INITIAL TIMING SETTING

1. Run the engine until it reaches normal operating temperature and turn all off all the accessories, including the air conditioning.
2. With the transmission in drive on the automatic transmission models, or neutral on the manual models and the parking brake set, adjust the idle speed to specifications.
3. Disconnect and plug the distributor vacuum hose and on the Maxima models, disconnect the grey wire harness connector at the distributor.
4. Loosen the distributor hold down clamp, connect a timing light and adjust the timing to specification.

DIESEL INJECTION TIMING

Kiki/Bosch VE-Type Injection

1. Remove the under cover.
2. Drain the coolant lower than the cold start device and remove the cold start device water hoses.

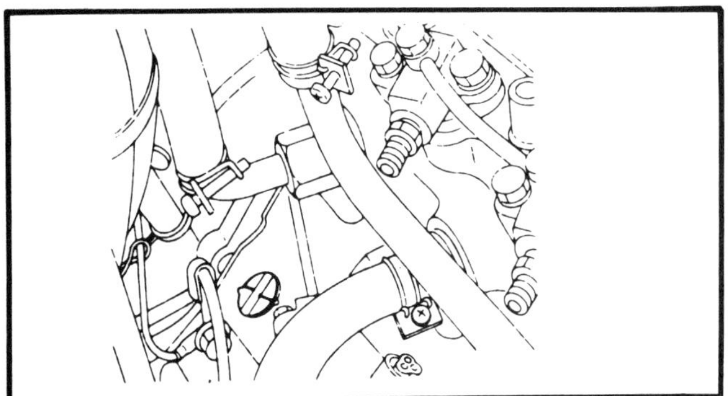
Aligning grooves in rear plate and flywheel

3. Remove the power steering pump.
4. Set the number 1 cylinder at Top Dead Center on its compression stroke.

NOTE: Make sure the grooves in the rear plate are aligned.

5. Remove the fuel injection tubes.
6. Loosen the fork screw on the cold start device. Turn the fork 90° and set the cold start device in the free position.

CAUTION

Never remove the screw on the cold start device wire. If it should be removed, the pump assembly should be readjusted at a service shop specified by the pump manufacturer.

7. Remove the plug bolt from the rear side of the injection pump, and in its place, attach special tool KV11229352 or equivalent.
8. Check the plunger lift adjustment using the following procedure:
   a. Loosen the injection pump fixing nuts and bracket bolt.
   b. Turn the crankshaft counterclockwise 15–20 degrees from the No. 1 cylinder at Top Dead Center.

c. Find the dial needle rest point, then set the gauge to zero.

d. Turn the crankshaft clockwise two complete rotations in order to remove the play in the cam mechanism. Loosen the tensioner and retighten.

NOTE: The belt tension is automatically set by the tension spring.

e. Turn crankshaft clockwise until the No. 1 cylinder is set at Top Dead Center on the compression stroke.

f. Read the dial gauge indication.

g. If the dial indication is not within the range shown in the chart, turn the pump body until it does. If the indication is smaller than the specification, turn the pump body counterclockwise. If the indication is larger, turn the pump body clockwise.

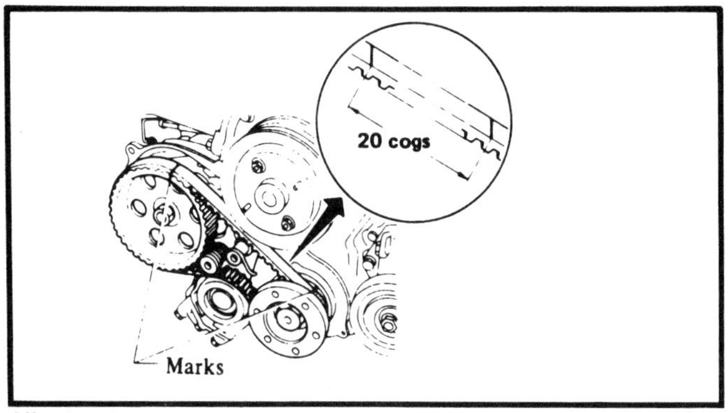

Aligning timing marks

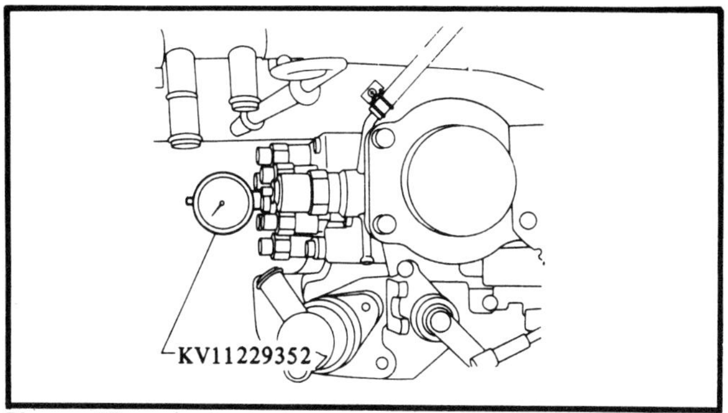

Plunger lift measuring tool

h. Tighten the injection pump fixing nut and bracket bolt. Torque the fixing nut to 7–15 ft.lbs. Torque the bracket bolt to 22–26 ft.lbs.

9. Tighten the injection securely.
10. Disconnect the Tool and reinstall the plug bolt with a new washer. Torque the bolt to 10–14 ft.lbs.
11. Set the fork in the original position by pulling the cold start device wire, then tighten the fork screw.
12. Connect the fuel injection tube.
13. Install the power steering pump.
14. Connect the cold start device water hoses.
15. Refill cooling system.
16. Install the under cover.

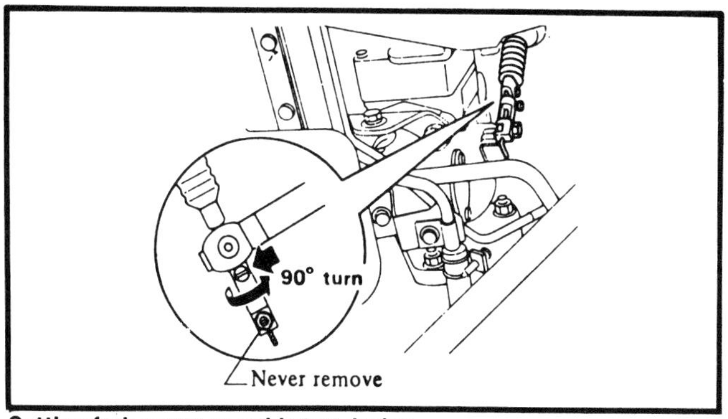

Setting fork screw on cold start device

PLUNGER LIFT

mm (in)

For Low Altitudes

M/T	0.85 ± 0.03 (0.0335 ± 0.0012)
A/T	0.81 ± 0.03 (0.0319 ± 0.0012)

For High Altitudes (Non-California Model Only)

M/T	0.90 ± 0.03 (0.0354 ± 0.0012)
A/T	0.85 ± 0.03 (0.0335 ± 0.0012)

IGNITION TIMING SPECIFICATION

Model / Engine	Ignition Timing
200SX / CA20E engine	
A/T	0° ± 2° BTDC
M/T	4° ± 2° BTDC
200SX Turbo / CA18ET	
All	15° ± 2° BTDC
300ZX / VG30E engine	
All	20° ± 2° BTDC
300ZX Turbo / VG30ET engine turbocharged	
All	20° ± 2° BTDC
Maxima / VG30E engine	
All	20° ± 2° BTDC

IGNITION TIMING SPECIFICATION

Model / Engine	Ignition Timing
Pulsar/NX / E16 engine	
A/T (exc. Calif. & Canada)	8° ± 2° BTDC
M/T (exc. Calif. & Canada)	15° ± 2° BTDC
All (Calif. & Canada)	5° ± 2° ATDC
Sentra / E16S engine	
A/T (exc. Calif. & Canada)	8° ± 2° BTDC
M/T (exc. Calif. & Canada)	15° ± 2° BTDC
All (Calif. & Canada)	5° ± 2° ATDC
Stanza / CA20E engine	
A/T	0° ± 2° BTDC
M/T	4° ± 2° BTDC
Stanza / CA20S engine	
All	0° ± 2° BTDC

SPARK PLUGS
1985-86 All Models

Model	Engine	Spark Plug	Gap
200SX	CA20E & CA18ET	BCPR6ES-11 ① BCPR5ES-11 ②	.039-.043
300ZX	VG30E	BCPR6ES-11	.039-.043
300ZX (Turbo)	VG30E	BCPR6ES-11	.039-.043
Maxima	VG30E	BCPR6ES-11	.039-.043
Pulsar	E16	③	.039-.043
Sentra	E16S	③	.039-.043
Stanza	CA20E & CA20S	BCPR6ES-11 ① BCPR6ES-11 ②	.039-.043 .039-.043

① Intake side
② Exhaust side
③ Non-California M/T and all Calif models: BPR5ES-11
Non-California A/T and Canada models: BPR5ES

CARBURETOR SPECIFICATIONS
1985-86 Models

Model	Float level mm (in.)	Fast idle speed (rpm)	I.S.C.A. (rpm)	Dash pot mm (in.)	Throttle opener mm (in.)
DCR342-35	12 (0.47)	2,400-2,700	—	—	—
DCR342-36	12 (0.47)	2,800-3,100	—	—	—
DCF328-1F	12 (0.47)	2,600-3,400	—	—	—
DFC328-2F	12 (0.47)	2,900-3,700	—	0.56 (0.0220)	—
DFE2832-5	12 (0.47)	—	2,800-3,200 ①	—	—

CARBURETOR SPECIFICATION
1985-86b Models

Model	Float level mm (in.)	Fast idle speed (rpm)	I.S.C.A. (rpm)	Dash pot mm (in.)	Throttle opener mm (in.)
DFE2832-6	12 (0.47)	—	2,800–3,200 ①	—	—
DCZ328-11G	12 (0.47)	1,900–2,700	—	0.72 (0.0283)	0.52 (0.0205)
DCZ328-12G	12 (0.47)	2,400–3,200	—	0.56 (0.0220)	0.52 (0.0205)

I.S.C.A.—Idle speed control actuator
① At full stroke, at standard set point—1,500–1,700 rpm

EMISSION CONTROL SYSTEMS

CRANKCASE EMISSION CONTROL SYSTEM

Positive Crankcase Ventiliation System

The positive crankcase ventilation (PCV) system is designed to return the blow-by gases to the intake manifold and to charge fresh air into the crankcase. The P.C.V. valve is provided to accomplish this.

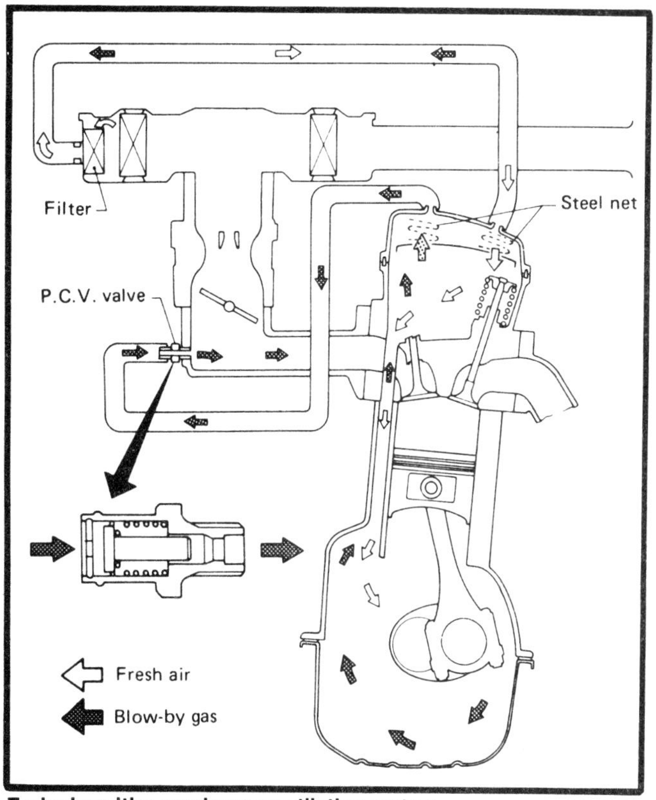

Typical positive crankcase ventilation system

PCV VALVE

System Service

With the engine running at idle, remove the P.C.V. hose from the valve. If the valve is working correctly, a hissing or sucking noise should be heard as air passes through it and a strong vacuum should be felt when a finger is placed over the valve.

Ventilation Hose

Check all hoses for vacuum leaks. Disconnect all hoses and blow out with compressed air. If any hose cannot be cleared, replace it.

PCV SYSTEM DIAGNOSIS

Symptom	Possible Cause	Repair
Rough Idle	Clogged PCV valve, leaking vacuum hose	Replace valve, clean or replace hose
No PCV Flow	Defective PCV valve, clogged lines	Clean or replace lines and PCV valve

PCV SYSTEM DIAGNOSIS

Symptom	Possible Cause	Repair
Engine Stalls At Idle	Excessive PCV flow, clogged air cleaner	Check PCV valve and air cleaner element
Surging At Road Speed	Improper PCV flow	Check PCV valve or reduction piece. Clean all lines and tighten connections.
Excessive exhaust smoke (diesel engine)	Malfunction in control valve	Check or replace control valve

FUEL EVAPORATIVE EMISSION CONTROLS

Evaporative Emission Control System

The evaporative emission control system is used to reduce hydrocarbons emitted to the atmosphere from the fuel system. This is accomplished by activated charcoals in the carbon canister.

When the engine is not running, the fuel vapor from the sealed fuel tank is led into the canister which contains activated carbon and is stored there until the engine is started.

The canister retains the fuel vapor until it is purged by the air drawn through the purge line to the intake manifold, when the engine is running. When running at idle, the purge control valve is closed.

A small amount of purge air flows into the intake manifold through the constant purge orifice. As the engine speed is increased and the throttle vacuum rises higher, the purge control valve opens and the vapor is sucked into the intake manifold through both the constant purge orifice and the fixed orifice.

VAPOR VENT LINE

System Service

CARBURETED ENGINES

1. Inspect the vapor lines for loose connections, proper routing, cracks, damage, chafing and deterioration.
2. Check the fuel tank vacuum relief valve to see if it's clogged or stuck.

FUEL INJECTED ENGINES

1. Check the hoses and fuel tank filler cap.

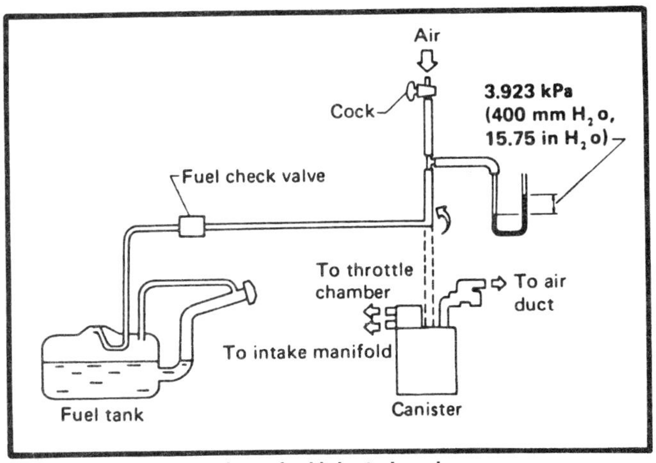

Vapor vent line inspection—fuel injected engines

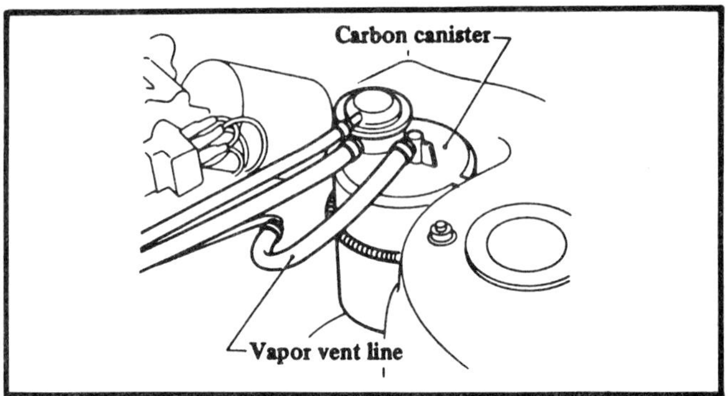

Fuel vapor lines

2. Disconnect the vapor vent line connecting the carbon canister to the fuel tank.
3. Connect a three way connector, a manometer and a cock (or an equivalent 3-way cock) to the end of the vent line.
4. Supply fresh air into the vapor vent line through the cock, a little at a time until the pressure becomes 3.923 kPa (15.75 in. H_2O).
5. Shut the cock completely and leave it unattended.
6. After 2 ½ minutes, check the height of the liquid in the manometer.
7. Variation in height should remain at 0.245 kPa (0.98 in. H_2O).
8. When the filter cap does not close completely, the height should drop to zero in a short time. If it does not drop to zero, the cause is a clogged hose.

NOTE: If the vent line is clogged, the fuel tank is not able to breath which causes fuel starvation or vapor lock. The line must be repaired or replaced.

CARBON CANISTER

Purge Control Valve Test

CARBURETED ENGINES

1. Disconnect the hose, in line, between the T-connector and the carbon canister at the T-connector.
2. Using a hand vacuum pump or equivalent, draw air from the hose running to the vacuum hole in the carbon canister and ensure that there is no leak. If there is a leak, replace the canister.

Vent Switching Valve Test

1. Disconnect the vent hose from the carburetor float chamber.
2. With the engine running, use a hand vacuum pump or equivalent, draw air from the vent line of the carbon canister and check that there are no leaks.
3. Check again with the engine off.

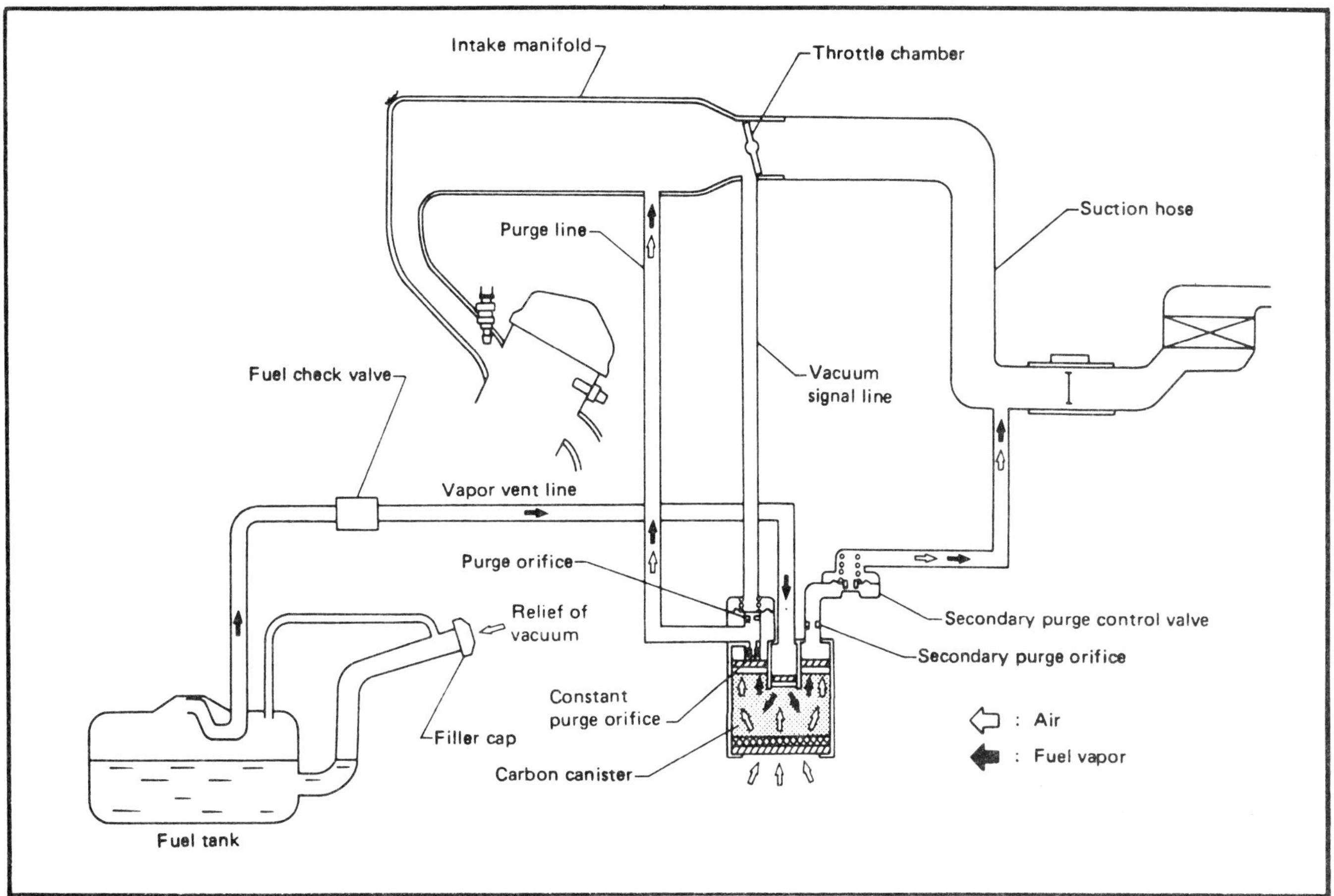

Evaporative emission control system—VG30E shown, others similar

FUEL INJECTED ENGINES

Test the carbon canister using the following procedure:

1. Using a hand held vacuum pump or equivalent, draw air through port(s) "A" and ensure that there is NOT a vacuum leak.
2. Using a hand held vacuum pump or equivalent, draw air through port(s) "B" and ensure that there IS a vacuum leak. If any valves are malfunctioning, replace the faulty parts.

Fuel Tank Vacuum Relief Valve Test

Remove the filler cap and perform the following tests.

1. Wipe clean the filler cap and place it in a vacuum test receptacle.

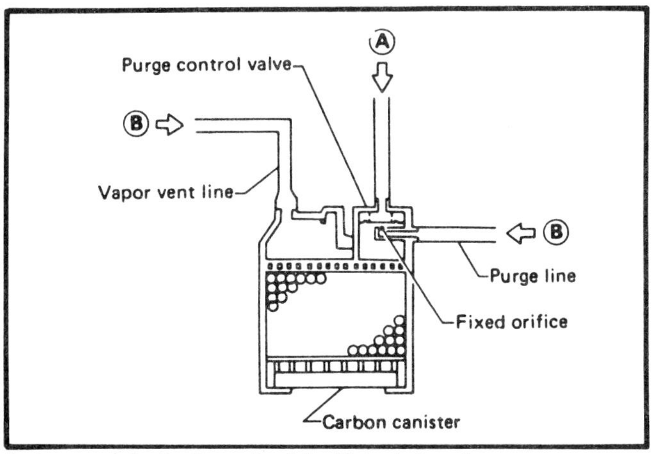

Carbon canister testing—4 cyl. fuel injected engines

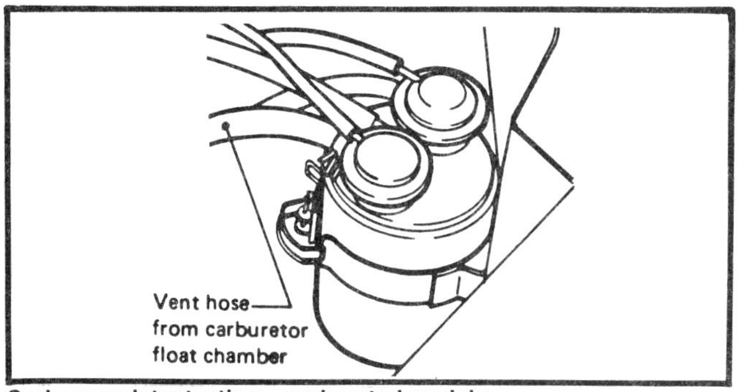

Carbon canister testing—carbureted models

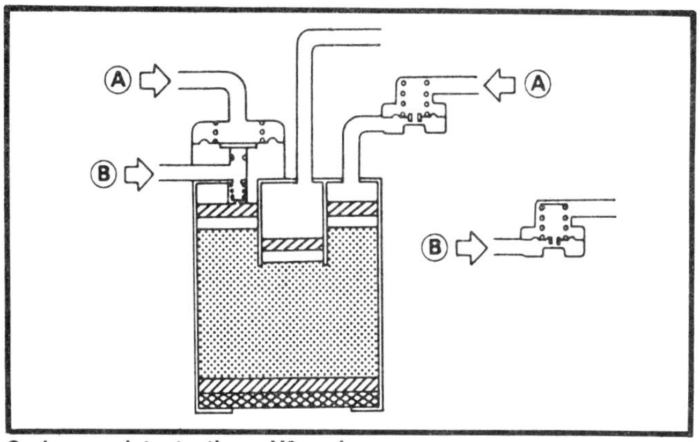

Carbon canister testing—V6 engines

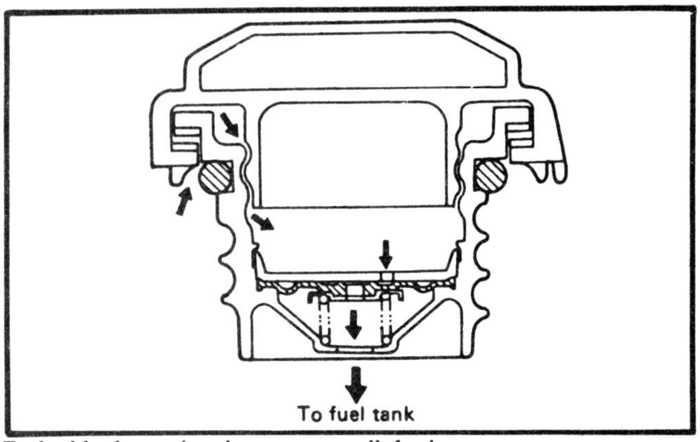

Typical fuel cap showing vacuum relief valve

2. Withdraw the air. A slight resistance accompanied by the valve indicates that the valve is operating properly. Also adding by further vacuum to the valve, the resistance should disappear when the valve clicks.

3. If the valve is clogged, or if no resistance is felt, replace the filler cap.

Fuel Check Valve Test

1. Blow air through the connector on the fuel tank side of the valve. A substantial resistance should be felt and a portion of the air flow should be directed towards the engine.

2. Blow air through the connector on the fuel tank side of the valve. The air flow should be smoothly directed toward the fuel tank.

3. If the check valve does not perform to the standards above, replace the valve.

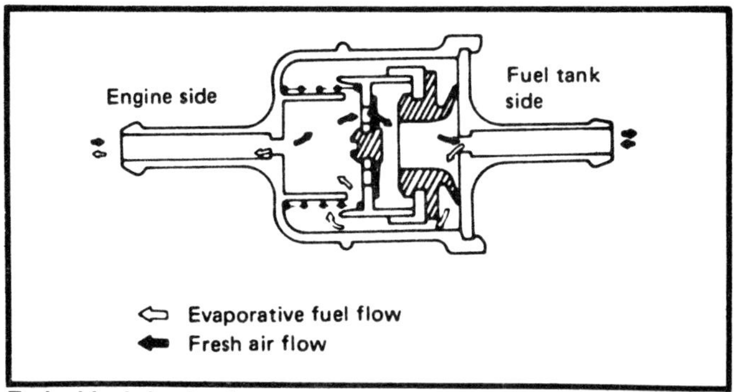

Typical fuel check valve

EXHAUST EMISSION CONTROLS

Thermostatic Air Cleaner (TAC)

The rate of fuel atomization varies with the temperature of the air that the fuel is being mixed with. The air/fuel ratio cannot be held constant for efficient fuel combustion with a wide range of air temperatures. Cold air being drawn into the engine causes a denser and richer air/fuel mixture, inefficient fuel atomization and thus, more hydrocarbons in the exhaust gas. Hot air being drawn into the engine causes a leaner air/fuel mixture and more efficient atomization and combustion for less hydrocarbons in the exhaust gases.

The automatic temperature controlled air cleaner is designed so that the temperature of the ambient air being drawn into the engine is automatically controlled to hold the temperature of the air and consequently, the air/fuel ratio at a constant rate for efficient fuel combustion.

A temperature sensing vacuum switch controls vacuum applied to a vacuum motor operating a valve in the intake snorkle of the air cleaner. When the engine is cold or the air being drawn into the engine is cold, the vacuum motor opens the valve, allowing air heated by the exhaust manifold to be drawn into the engine. As the engine warms up, the temperature sensing unit shuts off the vacuum applied to the vacuum motor which allows the valve to close, shutting off the heated air and allowing cooler, outside (under hood) air to be drawn into the engine.

Maintenance

No maintenance is necessary, except for periodic inspection of all air cleaner components for proper function. If hesitation, excessive fuel consumption or lack of power are encountered, check the hot air control system before checking the carburetor. Check all vacuum hoses and lines for connection and deterioration.

Testing

When the air around the temperature sensor of the unit mounted inside the air cleaner housing reaches 100° F., the sensor should block the flow of vacuum to the air control valve vacuum motor. When the temperature around the sensor is below 100° F., the sensor should allow vacuum to pass onto the air valve vacuum motor, thus blocking off the air cleaner snorkle to underhood (unheated) air.

When the temperature around the sensor is above 118° F., the air control valve should be completely open to under hood air.

If the air cleaner fails to operate properly, check for loose or broken vacuum hoses. If the hoses are not the cause, replace the vacuum motor in the air cleaner.

Air Injection Valve System (AIV)

The air injection valve (AIV) system is designed to send secondary air to the exhaust manifold, utilizing a vacuum caused by exhaust pulsation in the exhaust manifold.

The exhaust pressure in the exhaust manifold usually pulsates in response to the opening and closing of the exhaust valve and it decreases below the atmospheric pressure periodically.

If a secondary air intake is opened to the atmosphere under vacuum conditions, the secondary air can be drawn into the exhaust manifold in proportion to the vacuum.

On Federal models, it is not necessary to operate this system under normal engine conditions because a 3-way catalytic converter is used. Therefore, this system is designed to send secondary air to the exhaust manifold when the engine is cold.

Preliminary Inspection

Check the hose for damage, looseness, flatting or faulty conditions, and check each part for proper installation. Replace as necessary.

AIR INJECTION VALVE

Testing

1. Disconnect the air injection hose at the air injection pipe side. Suck or blow on the hose to make sure that air flow does not exist.

2. Connect a hand held vacuum pump to the air injection control valve.

3. Suck or blow on the hose to make sure that air flows only on the injection pipe side while the control valve is operated.

NOTE: On Canadian models, the air flow should continuously flow only on the air injection pipe side. There is no control valve.

4. Check the air injection valve and the air injection control valve for damage or binding. Replace as necessary.

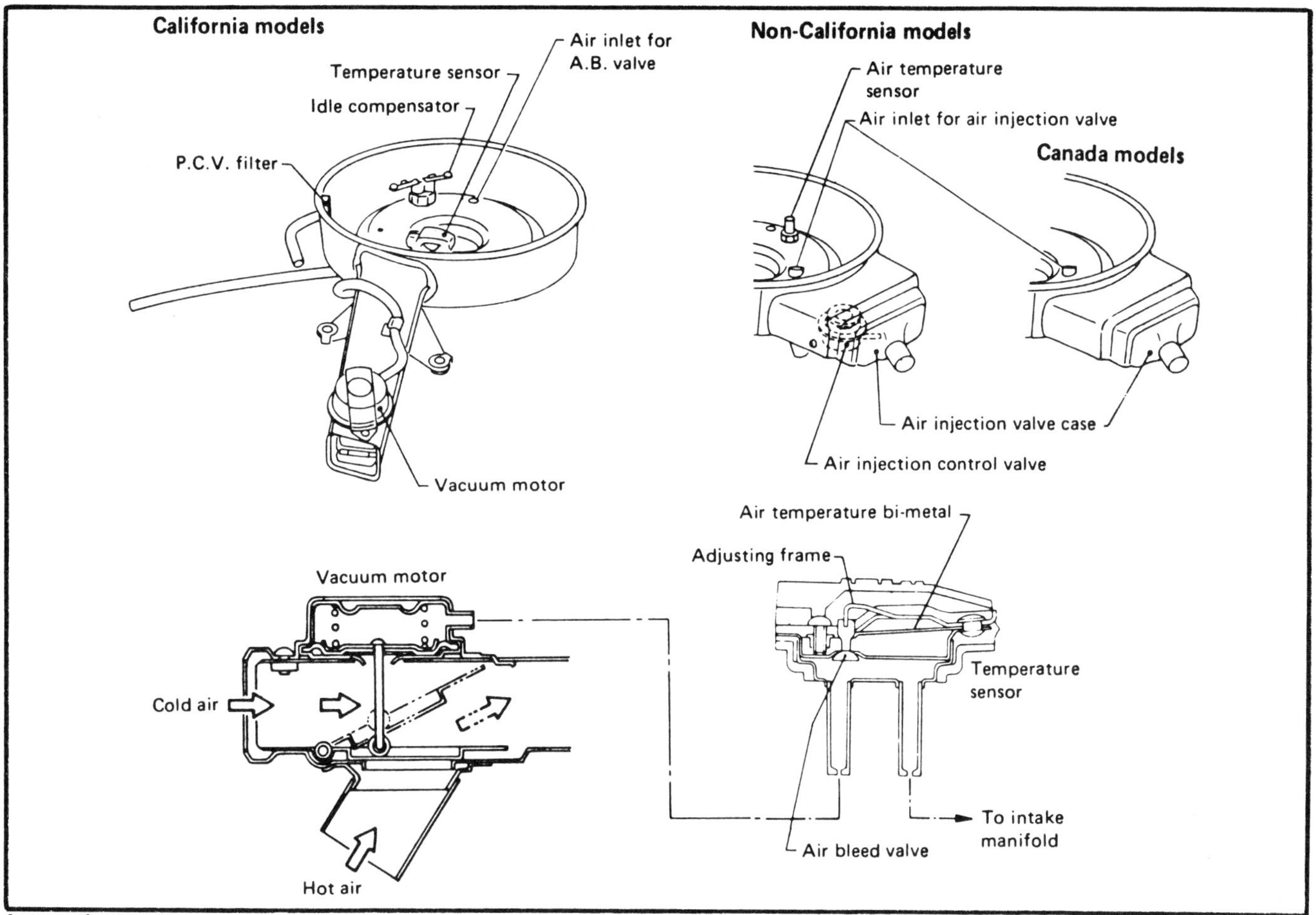

Automatic temperature control (ATC) air cleaner

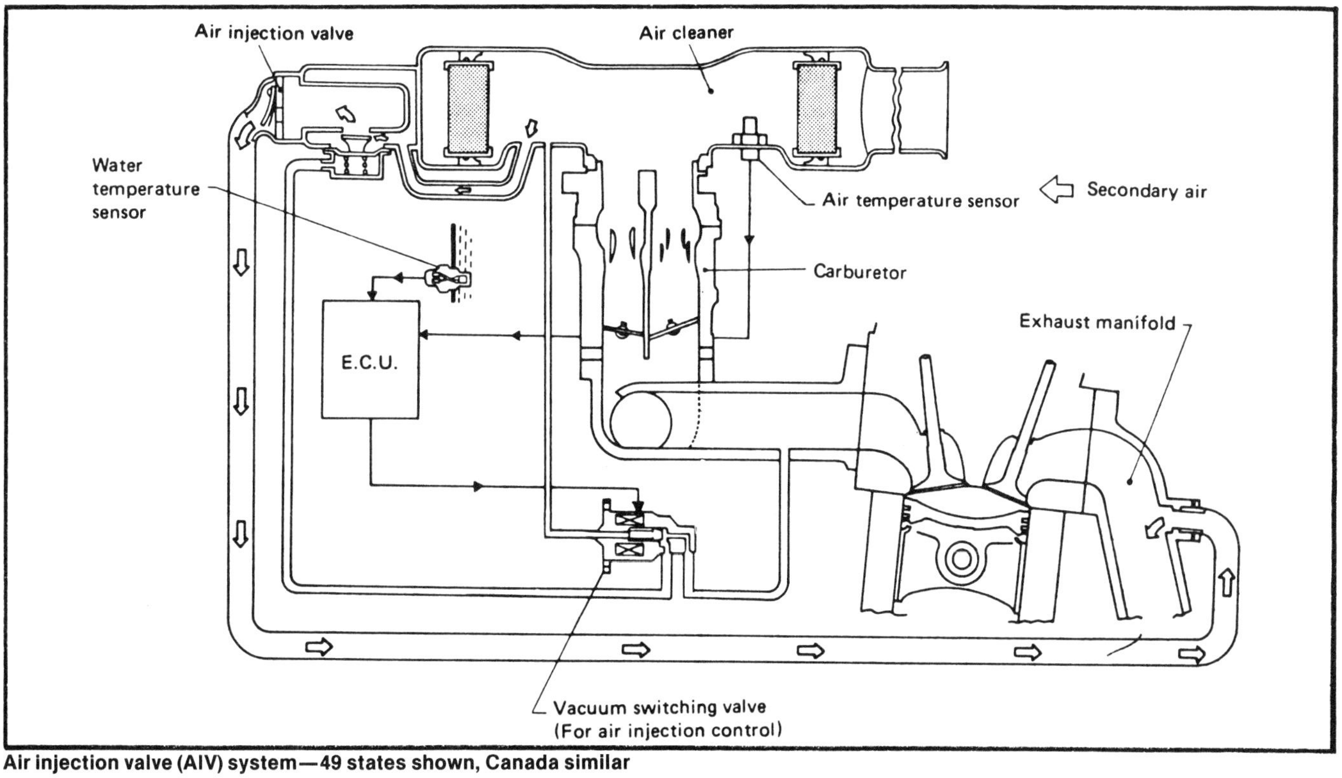

Air injection valve (AIV) system—49 states shown, Canada similar

A.B. VALVE

Testing

1. Warm up the engine to operating temperature.
2. Disconnect the hose from the air cleaner and place a finger near the outlet.
3. Run the engine to about 3,000 rpm in neutral, then quickly return to idle. If a pull of suction force on your finger is felt, the A.B. valve is working properly. If no suction is felt, replace the valve.

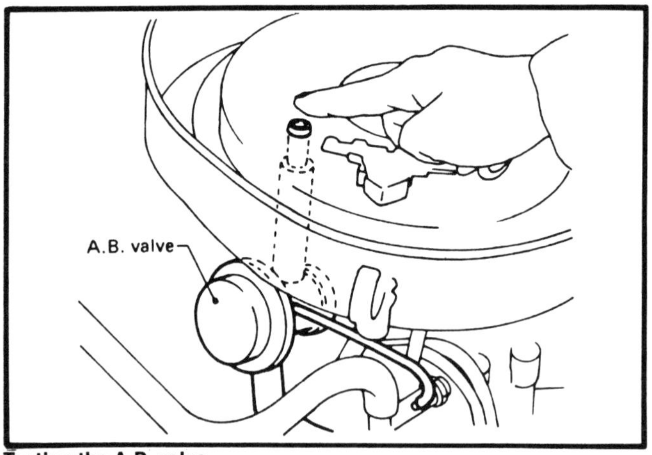

Testing the A.B. valve

Exhaust Gas Recirculation (EGR) System

The function of the exhaust gas recirculation system is to return part of the exhaust gas to the combustion chamber to reduce the spark flame temperature during combustion. The result is a lower nitrogen oxide (NOx) content in the exhaust gas.

E.G.R. CONTROL VALVE

The EGR control valve controls the quantity of the exhaust gas to be led into the intake manifold through the vertical movement of the valve, connected to the diaphragm to which vacuum is applied in the response to the opening of the carburetor throttle valve.

B.P.T. VALVE

The B.T.P. valve monitors exhaust pressure to activate the diaphragm controlling the throttle chamber vacuum, applied to the E.G.R. control valve. Which is to say, recirculated exhaust gas is controlled in response to the positioning of the E.G.R. control valve or to the engine operation.

THERMAL VACUUM VALVE (2–PORT BIMETAL TYPE)

This thermal vacuum valve is mounted on the front side of the intake manifold. It detects engine coolant temperature by the means of a built–in bimetal and opens and closes the vacuum passage in the thermal vacuum valve.

When the vacuum passage is open, the carburetor vacuum signal is applied to the diaphragm of the E.G.R. control valve to actuate the taper valve connected to the diaphragm.

THERMAL VACUUM VALVE (3–PORT WAX TYPE)

This type of valve detects engine coolant temperature by means of wax expansion and opens or closes the air passage from the air cleaner. When the vacuum passage is open, the carburetor vacuum signal is applied to the diaphragm of the E.G.R. control valve to actuate the taper valve connected to the diaphragm. This valve is also used as a component for the Catalyst Warm-up System and Evaporative Emission Control System.

TESTING

Total System Check

1. Make a total visual check of the E.G.R. control system. If necessary clean any components to facilitate inspection. If any hoses are found cracked or broken, replace as necessary.
2. With the engine stopped, check the E.G.R. control valve for any indication of binding or sticking by moving the diaphragm of the control valve upwards with your finger.
3. With the engine running, check the E.G.R. control system referring to the Operation chart. When checking the E.G.R. control valve operation, place a finger on the diaphragm of the valve.

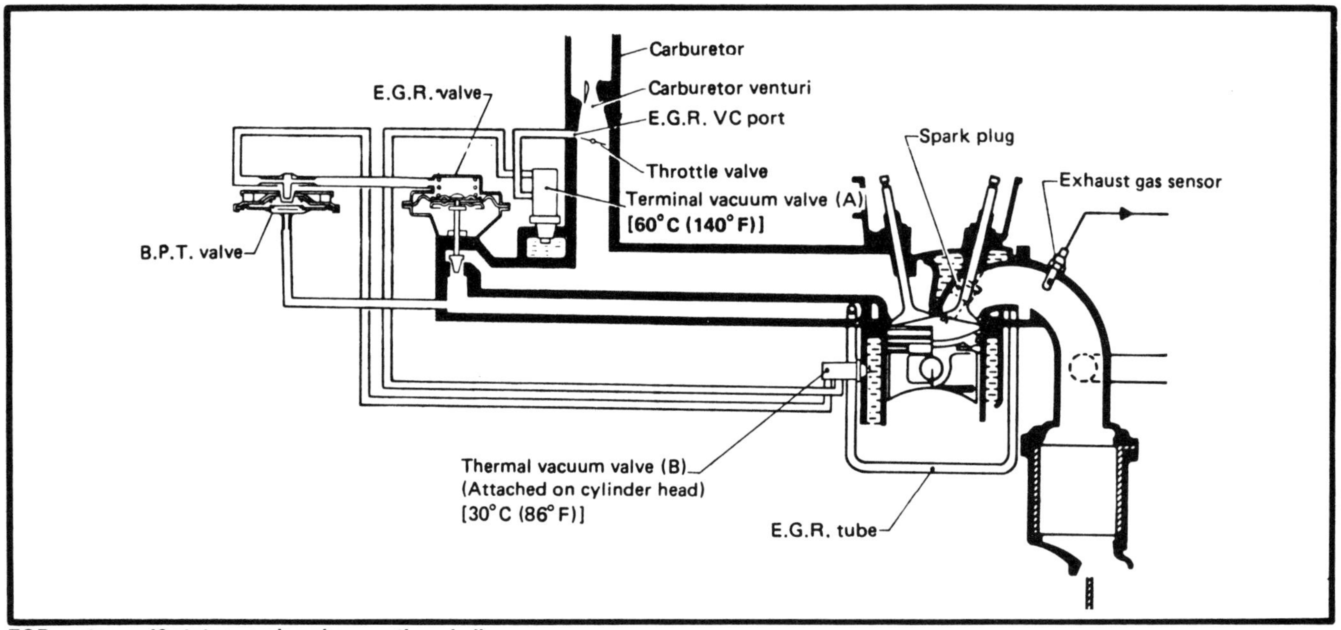

EGR system—49 states version shown, other similar

OPERATION

Thermal vacuum valve (A) Water temperature °C (°F)	Thermal vacuum valve (A) Operation	Thermal vacuum valve (B) Cylinder head wall temperature °C (°F)	Thermal vacuum valve (B) Operation	E.G.R. valve vacuum line	B.P.T. valve Exhaust gas pressure	B.P.T. valve Operation	E.G.R. system
Below 30 (86)	Close	Below 30 (86)	Close	Close	Any condition		Not actuated
30–60 (86–140)	Close	30–60 (86–140)	Open	Close	Any condition		Not actuated
Above 60 (140)	Open	Above 60 (140)	Open	Open	Low	Open*	Not actuated
					High	Close*	Actuated

*To atmospheric pressure

E.G.R. CONTROL VALVE

Testing

1. Apply vacuum the the E.G.R. control valve diaphram. If the valve moves to the full position, it is operating properly. The E.G.R. control valve will remain open for more than 30 seconds after the vacuum has been shut off.
2. Visually check the E.G.R. control valve for damage, warpage or deformation.

B.P.T. VALVE

Testing

1. Disconnect the B.P.T. valve from the engine.
2. Apply a pressure of more than 0.490 kPa (1.97 in. H_2O) to the valve.
3. Orally suck the port back as shown, to check for leakage. If a leak is detected, replace the valve.

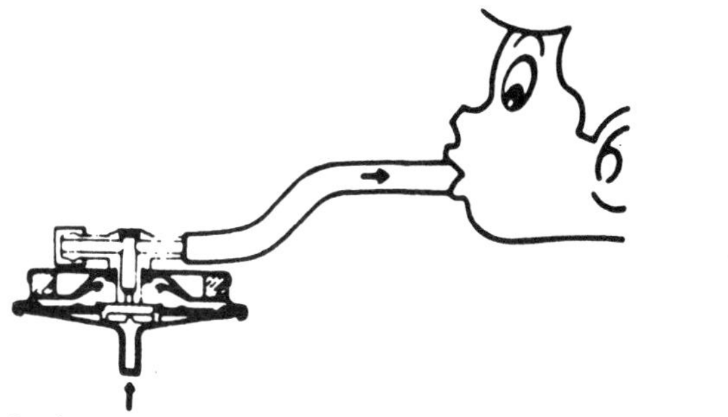

B.T.P. valve testing

THERMAL VACUUM VALVE

Testing

NOTE: Before removing, drain about 1 ⅛ qts. of engine coolant.

1. Remove the thermal vacuum valve from the engine.
2. Inhale air from the port of the E.G.R. system and check to be sure that the thermal vacuum valve opens or closes in response to its temperature.

THERMAL VACUUM VALVE SPECIFICATIONS— 49 STATE MODEL

Thermal Vacuum Valve (A)

Water temperature °C (°F)	Valve
Above 60 (140)	Open
Below 60 (140)	Closed

Thermal Vacuum Valve (B) (at cylinder head)

Water temperature °C (°F)	Valve
Above 30 (86)	Open
Below 30 (86)	Closed

THERMAL VACUUM VALVE SPECIFICATIONS— CALIF. AND CANADA

Water temperature °C (°F)	Valve
Below 50 (122)	Open
Above 50 (122)	Closed

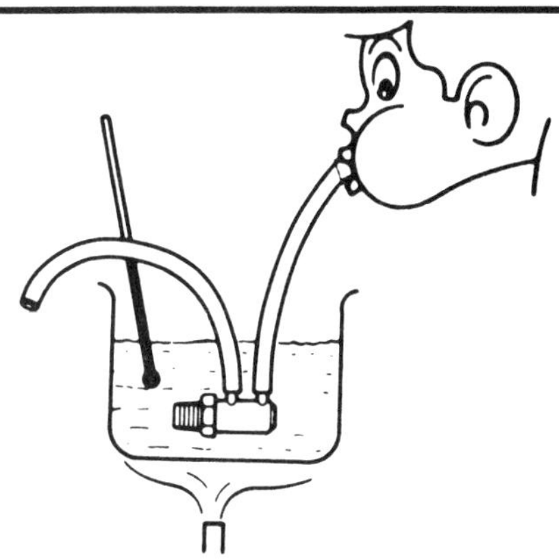

Testing the thermal vacuum valve

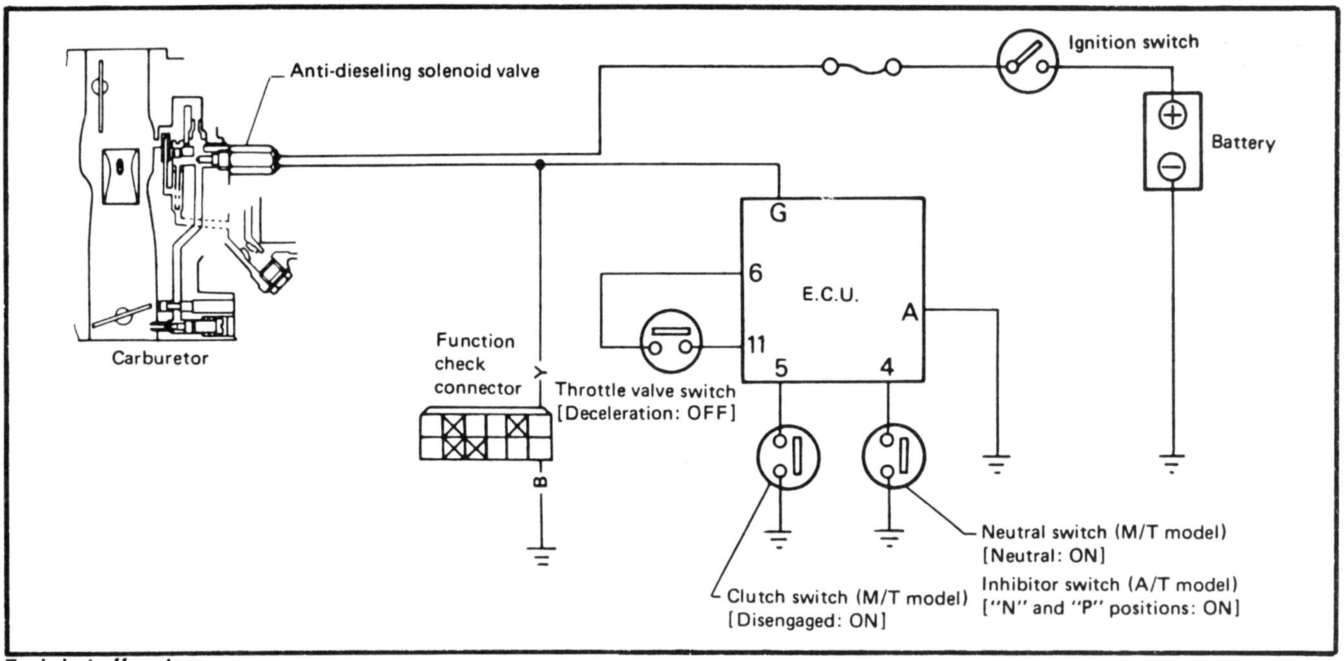

Fuel shut-off system

Fuel Shut-off System

The main job of the fuel shut-off system is to reduce fuel consumption and hydro-carbon (HC) emissions during deceleration. The system consists of switches; a anti-dieseling solenoid valve, vacuum switch, neutral switch, clutch switch on the manual transmission, inhibitor or neutral safety switch on the automatic transmission and a throttle valve switch.

Testing

NOTE: Refer to the "Check E.C.C. system" in the Electronic Ignition Section of the 1986 Chilton Import Service Manual for diagnosis.

Mixture Heating System

In order to reduce emissions during engine warm-up, the positive tem-

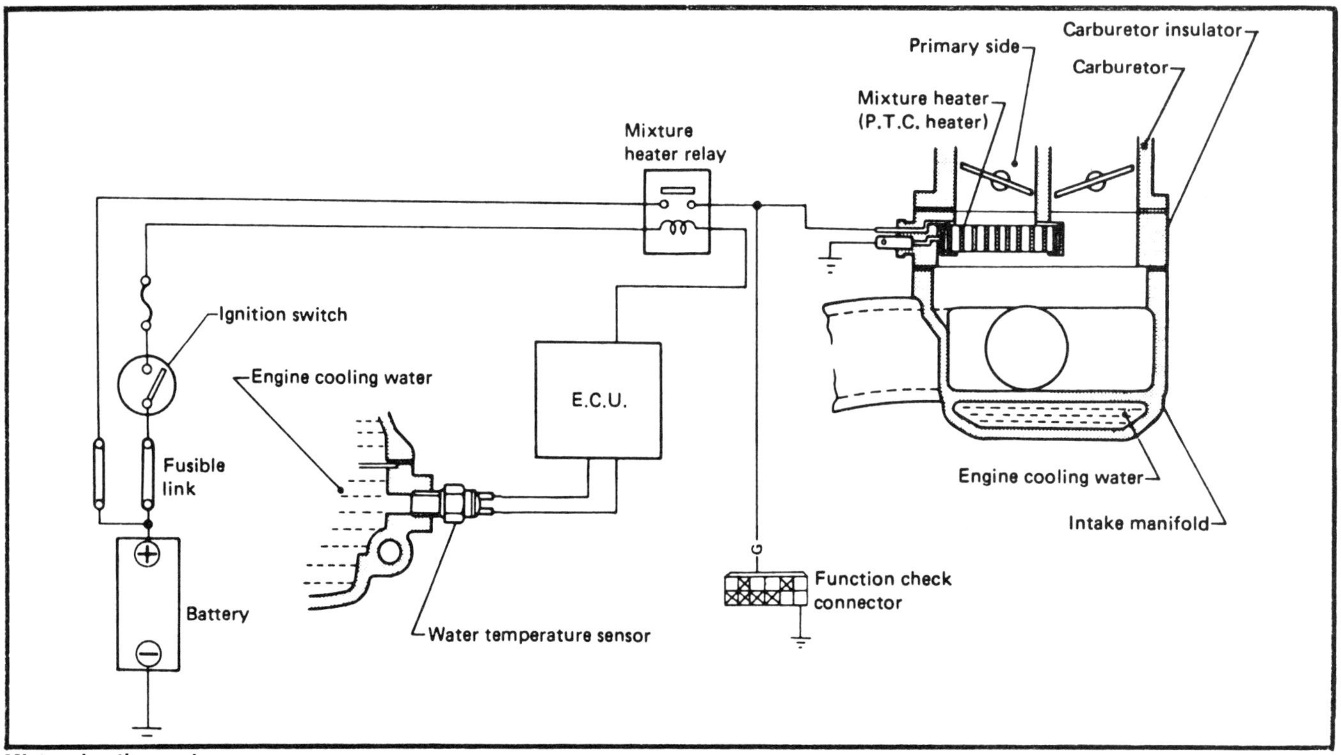

Mixture heating system

perature coefficient (P.T.C.) heater is installed between the carburetor and the intake manifold.

During an engine cold start, the P.T.C. heater is electrically energized and the air-fuel mixture which passes through this heater is heated.

After the engine reaches operating temperature, the heater current is cut off by the E.C.C. control unit.

MIXTURE HEATER

The mixture heater uses a honeycomb P.T.C. heater design and is situated in the primary side of the carburetor insulator. With this design, resistance to current flow increases as the temperature increases, and vice versa. As a result, the current flow is maintained constant.

TESTING

1. Check the continuity between the terminals of the mixture heater. If the resistance is too high, replace the carburetor insulator.
2. Total system check:
   a. When the ignition switch is turned "ON" while the engine is "OFF", confirm the presence of battery voltage at the mixture heater harness connector. If no battery voltage is present, check the water temperature sensor and mixture heater relay.
   b. Check the continuity between the terminals of the water temperature sensor at the connector.
   c. Check the mixture heater relay. If necessary, check the wiring harness and automatic choke relay.

Spark Timing Control System (STCS)

The ignition timing is controlled by the central electronic control unit which adjusts to the engine operating conditions; that is, as the best ignition timing in each driving condition has been memorized in the unit, the ignition timing is determined by the electric signal calculated in the unit.The signals used for the determination of the ignition timing are water temperature, engine rpm, engine load, etc.

The signal from the central electronic control unit is transmitted to the power transistor of the ignition coil and controls the ignition timing.

TESTING

NOTE: Refer to the "Check E.C.C. system" in the Electronic Ignition Section of the 1986 Chilton Import Service Manual for diagnosis.

VEHICLE SPEED SENSOR

Testing

1. Remove the speedometer cable at the transaxle.
2. Disconnect the E.E.C. 20-pin harness.
3. Turn the ignition switch to "ON", rotate the speedometer pinion

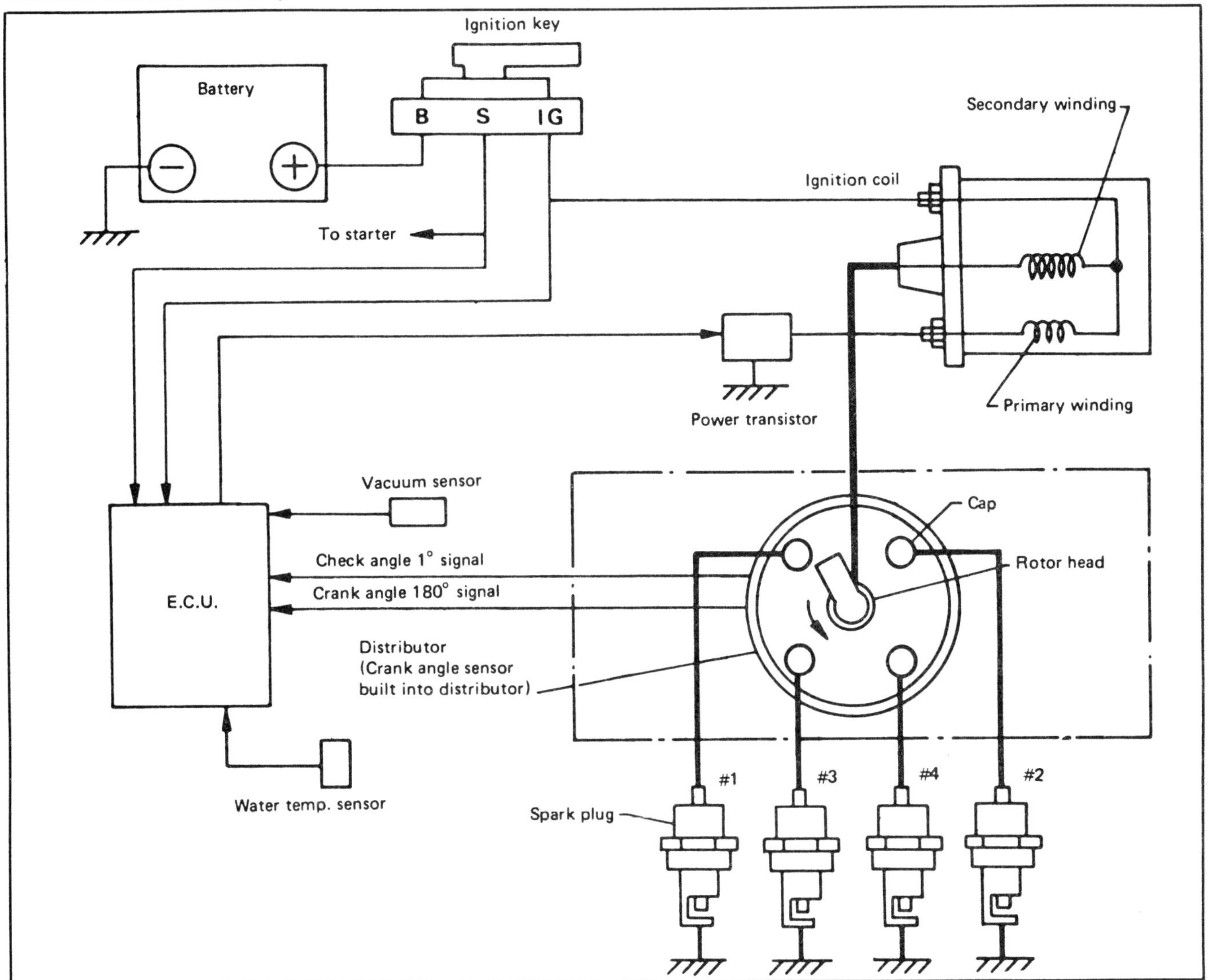

Spark timing control system—49 state models

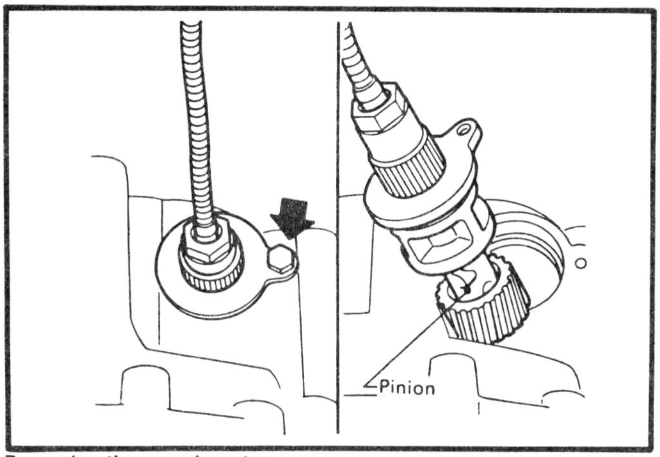

Removing the speedometer

and check that voltage exists between terminal No. 10 and the body ground.

POWER TRANSISTOR

Check the resistance between the power transistor terminal (E.E.C. unit side), body ground and the ignition coil negative (-) terminal. Replace as necessary.

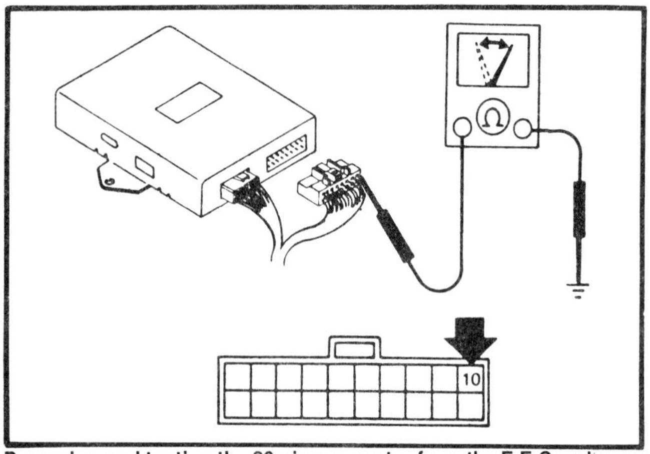

Removing and testing the 20-pin connector from the E.E.C. unit

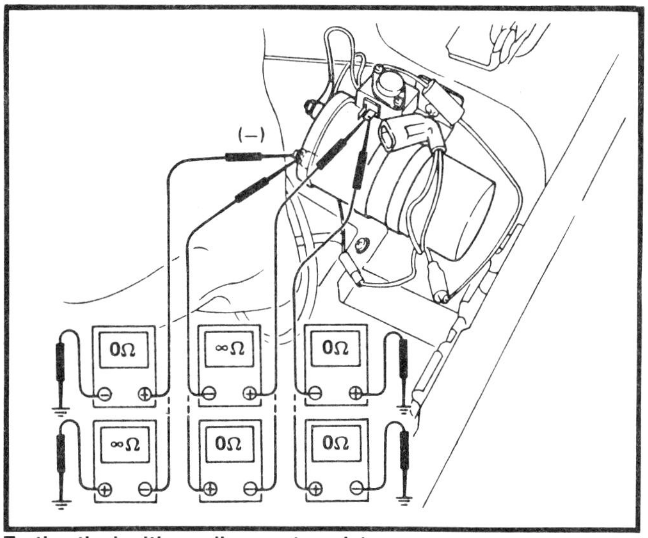

Testing the ignition coil power transistor

Throttle Opener Control System (TOCS)

The function of the throttle opener is to open the throttle valve of the carburetor slightly while the car is decelerating. During deceleration, the manifold vacuum rises and the quantity of the mixture in the engine is not sufficient for normal combustion to continue, consequently, a large amount of unburned HC is emitted.

Carburetors equipped with the throttle opener supply the engine with an adequate charge of combustible mixture to maintain proper combustion during deceleration, resulting in a drastic reduction in HC emission.

TESTING

Total System Test

NOTE: When the idling speed is too high and does not drop to idling speed, the throttle opener control system should be checked.

1. Check the continuity between the "G/R" and "B" terminals specified in the function check connector with the ignition switch OFF. If continuity does not exist, the solenoid may be faulty. Replace the throttle opener control valve assembly.
2. Turn on the ignition switch and check the voltage across the "G/R" and "B" terminals.

MANUAL TRANSMISSION MODELS

Remove the speedometer cable from the combination meter. Then spin the speedometer in the combination meter with your fingers and confirm that the speedometer pointer indicates more than 10 mph temporarily. Voltage between the "G/R" and the "B" terminals should be as follows:

- Above 10 MPH.....0 Volts
- Below 10 MPH.....12 VoltsIf not, amplifier or speed detecting switch may be faulty. Replace as necessary.

AUTOMATIC TRANSMISSION MODELS

Voltage between the two terminals should be changed as follows:

- "N" or "P" position.....12 Volts
- Other position.....0 Volts If not, replace the inhibitor switch.

Catalytic Converter System (U.S.A. models)

The 3-way catalytic converter utilizes a catalyst to accelerate the re-combustion of HC and CO and reduce NOx in the exhaust gas, changing

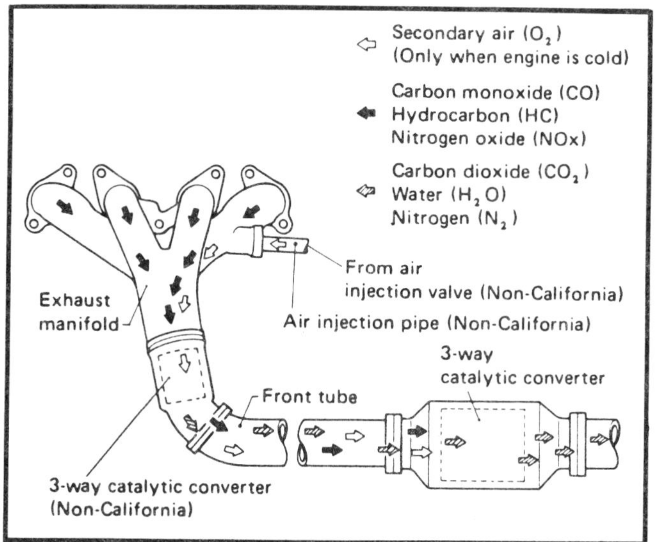

Catalytic converter system

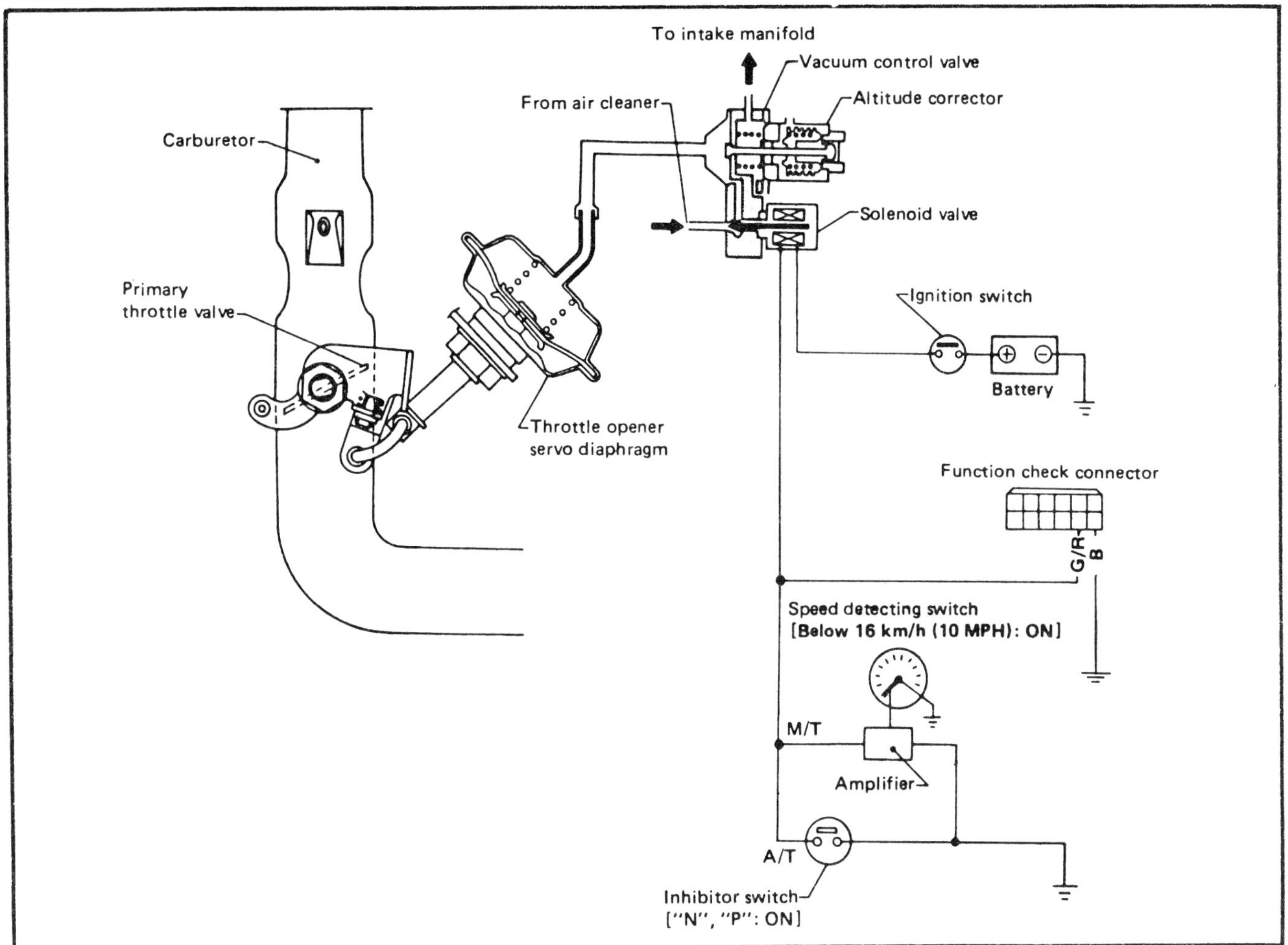

Throttle opener control system—Canadian models only

them into harmless CO_2, H_2O and N_2.To accomplish the oxidation and reduction of such harmful contents, the exhaust gas sensor monitors the O_2 level, feeds it back to the E.C.C. control unit and maintains the mixture ratio to the stoichiometric points at all times.

OPERATION

The exhaust gas from the engine contains unburned, harmful components. The mixture ratio feedback system reduces such harmful components in the exhaust gas. In this system, an exhaust gas sensor monitors the contents of O_2 density to determine the combustion condition and maintains the mixture to the stoichometric point. While the mixture ratio is so maintained, the 3-way catalytic converter activates to change the harmful components (HC, CO, and NOx) into harmless CO_2, H_2O and N_2. In this way, the catalytic converter cleans the exhaust gas and discharges CO_2, H_2O and N_2 into the atmosphere.

TESTING

Preliminary Inspection

Visually inspect the condition of all components including hoses, tubes, and wires, replace if necessary.

Converter Inspection

Check to see if the converter in operating normally or not by checking the variation in CO percentage using the following procedure:

1. Set the parking break. Shift the transmission into "Neutral" (manual trans.) or "Park" (automatic trans.).
2. Inspect the convertor for damage or cracks.
3. Adjust the engine idle speed.
4. Increase the engine speed between 2,000–3,000 rpm, two or three time under no-load.
5. If the idle speed increases, readjust it to specifications using the throttle adjusting screw.
6. Warm up the engine at 2,000 rpm for about four minutes under no-load.
7. Measure the CO percentage at idle speed.

NOTE: After Step 6 has been completed, wait for one minute before checking the CO percentage.

8. If the CO percentage measured is less than 0.3%, the converter is operating properly.
9. If the CO percentage measured is over 0.3%, check the mixture ratio feedback system to see if its operating properly. Then, recheck the CO percentage.
10. If the CO percentage is still above 0.3%, the converter is malfunctioning. replace the converter.

Boost Controlled Deceleration Device (B.C.D.D.)

The B.C.D.D. serves to reduce HC emission during coasting. The high manifold vacuum during coasting prevents the complete combustion of the mixture gas due to the reduced amount of intake air.

As a result, an excess amount of HC is emitted into the atmosphere. When the manifold vacuum exceeds the set value, the B.C.D.D. operates to supply additional air, bypassing the throttle valve. Complete combustion of the fuel is assisted by this additional air and HC emission are thereby reduced.

Testing

1. Connect a vacuum gauge to the intake manifold.
2. Start the engine and observe the vacuum gauge with the engine racing.
3. If the B.C.D.D. is in good condition, the vacuum gauge will indicate correct vacuum readings on the gauge.

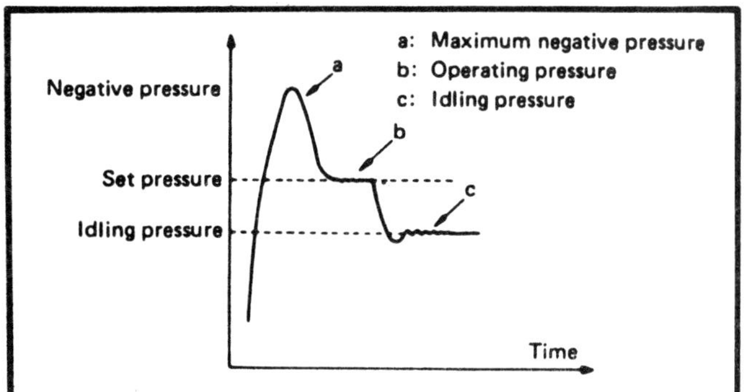

Characteristic curve of B.C.D.D.

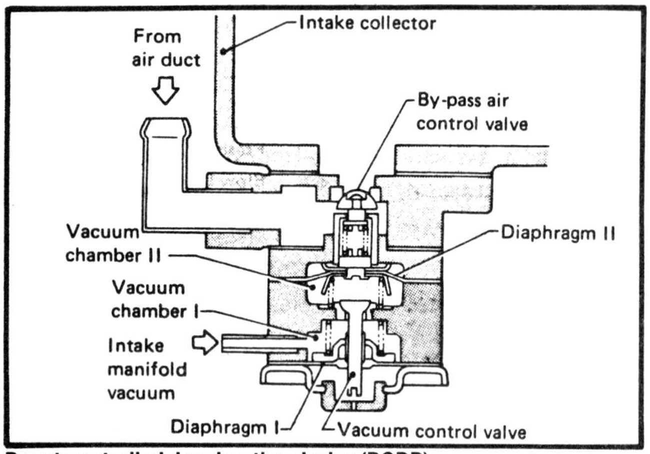

Boost controlled deceleration device (BCDD)

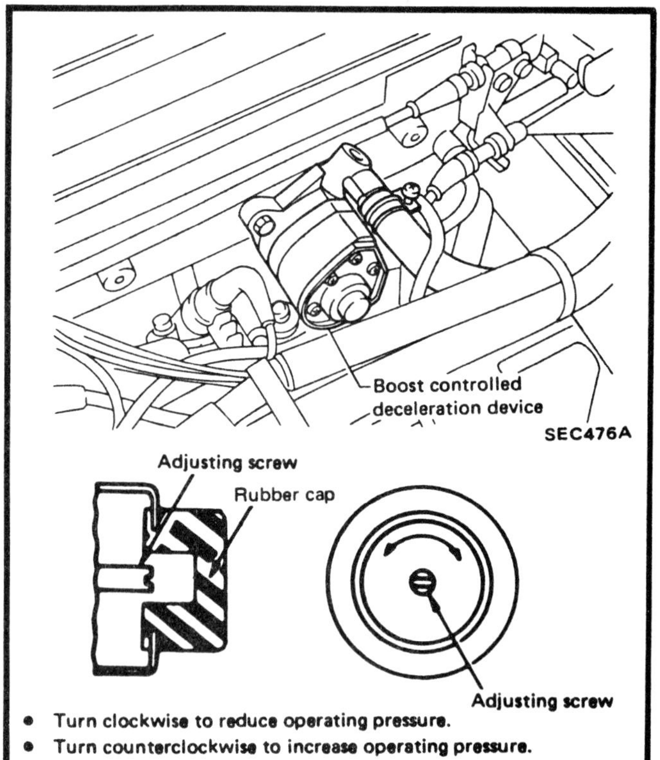

Adjusting the B.C.D.D. pressure

4. If it does not perform as described above, adjust the operating pressure using the following procedure:
   a. Remove the protective cap on the B.C.D.D.
   b. While increasing the engine speed, turn the adjusting screw until the specified set pressure is obtained. B.C.D.D. set pressure: 22.44 ± 0.79 in. Hg.

NOTE: Turning the adjusting screw ¼ turn will cause a change in operating pressure of about 2.7 kPa (0.79 in. Hg.). Do not fit the tip of a screwdriver tightly into the screw slot.

VACUUM CIRCUITS

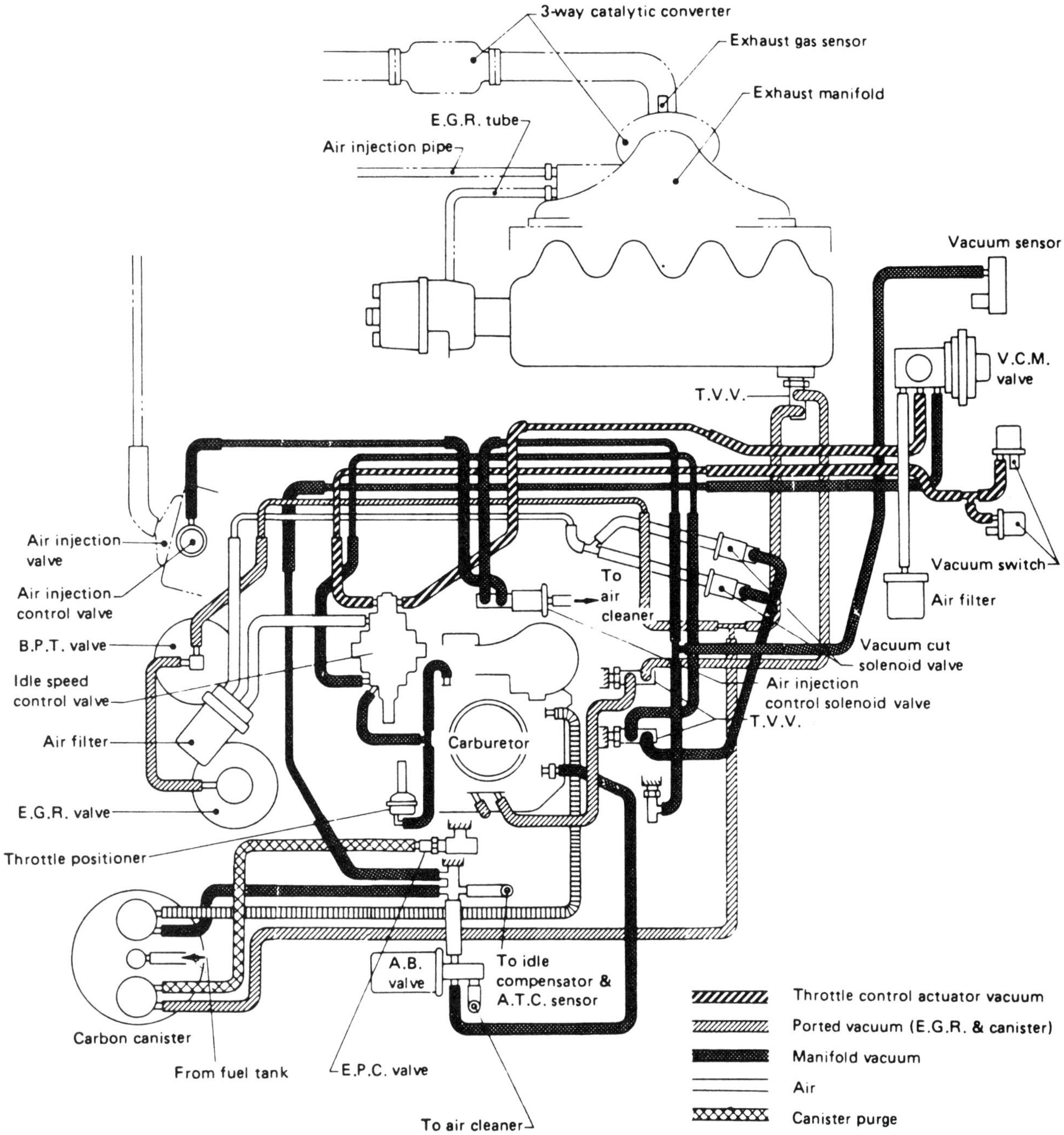

NISSAN SENTRA—FEDERAL

VACUUM CIRCUITS

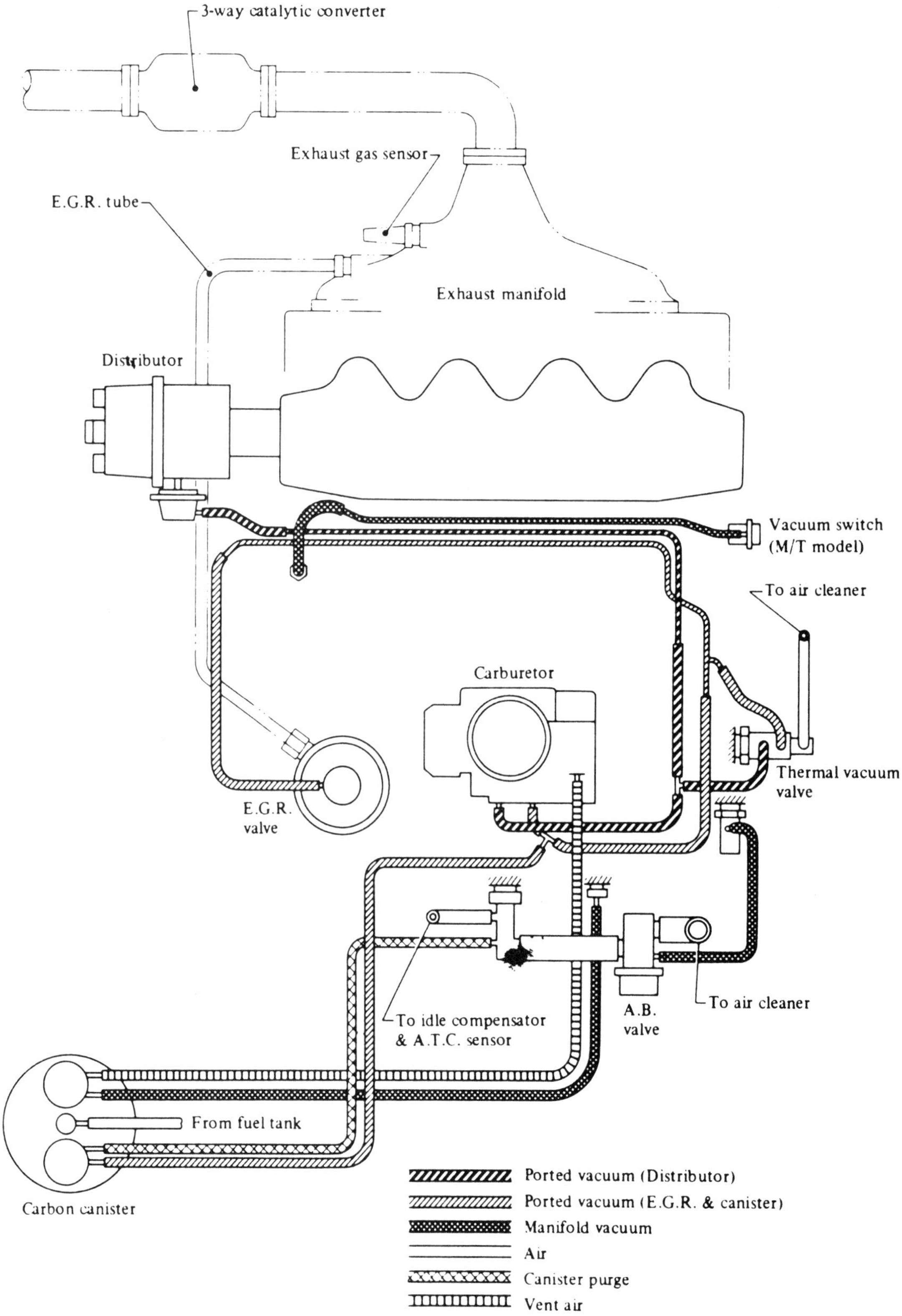

NISSAN SENTRA— CALIFORNIA

VACUUM CIRCUITS

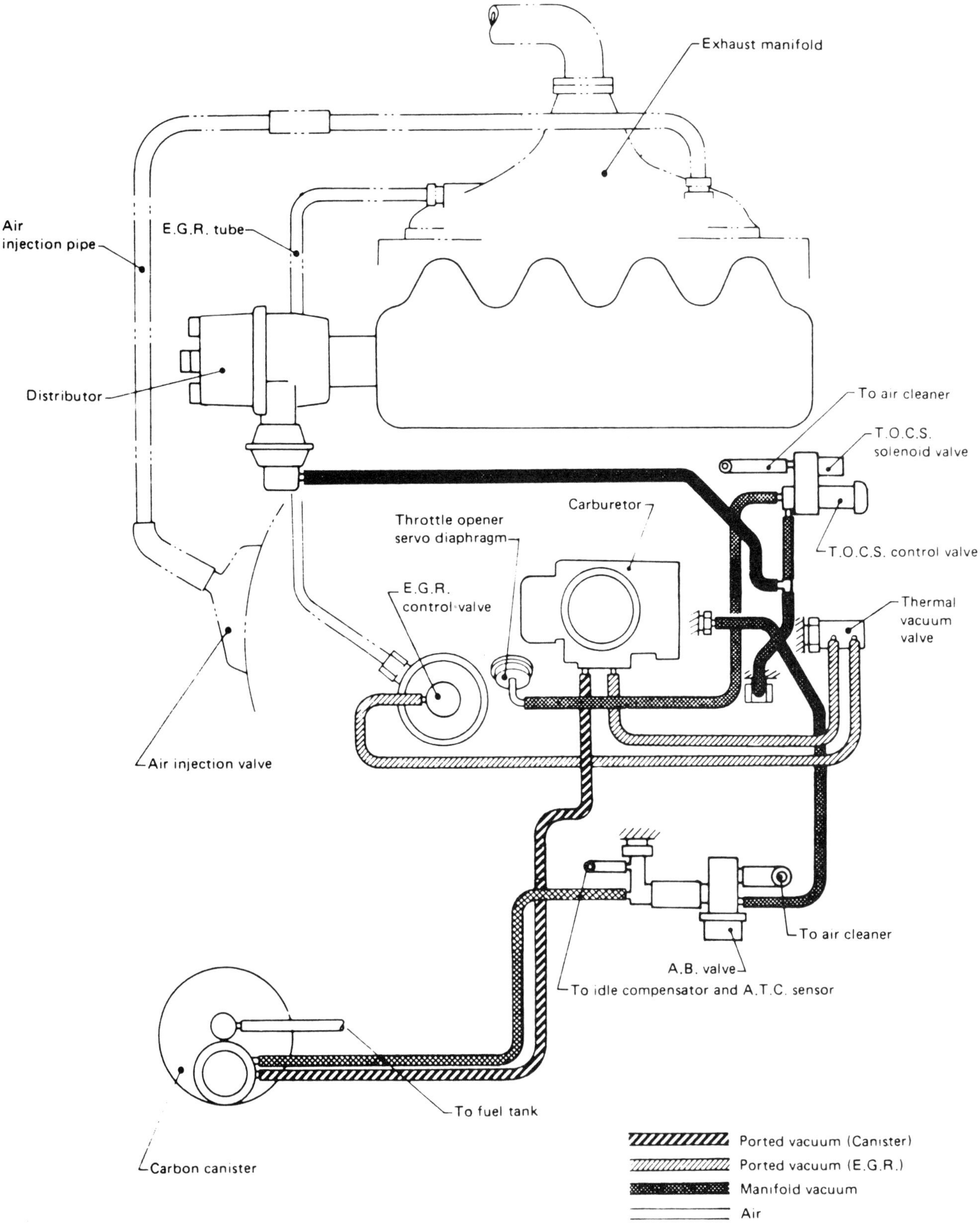

NISSAN SENTRA—CANADA

VACUUM CIRCUITS

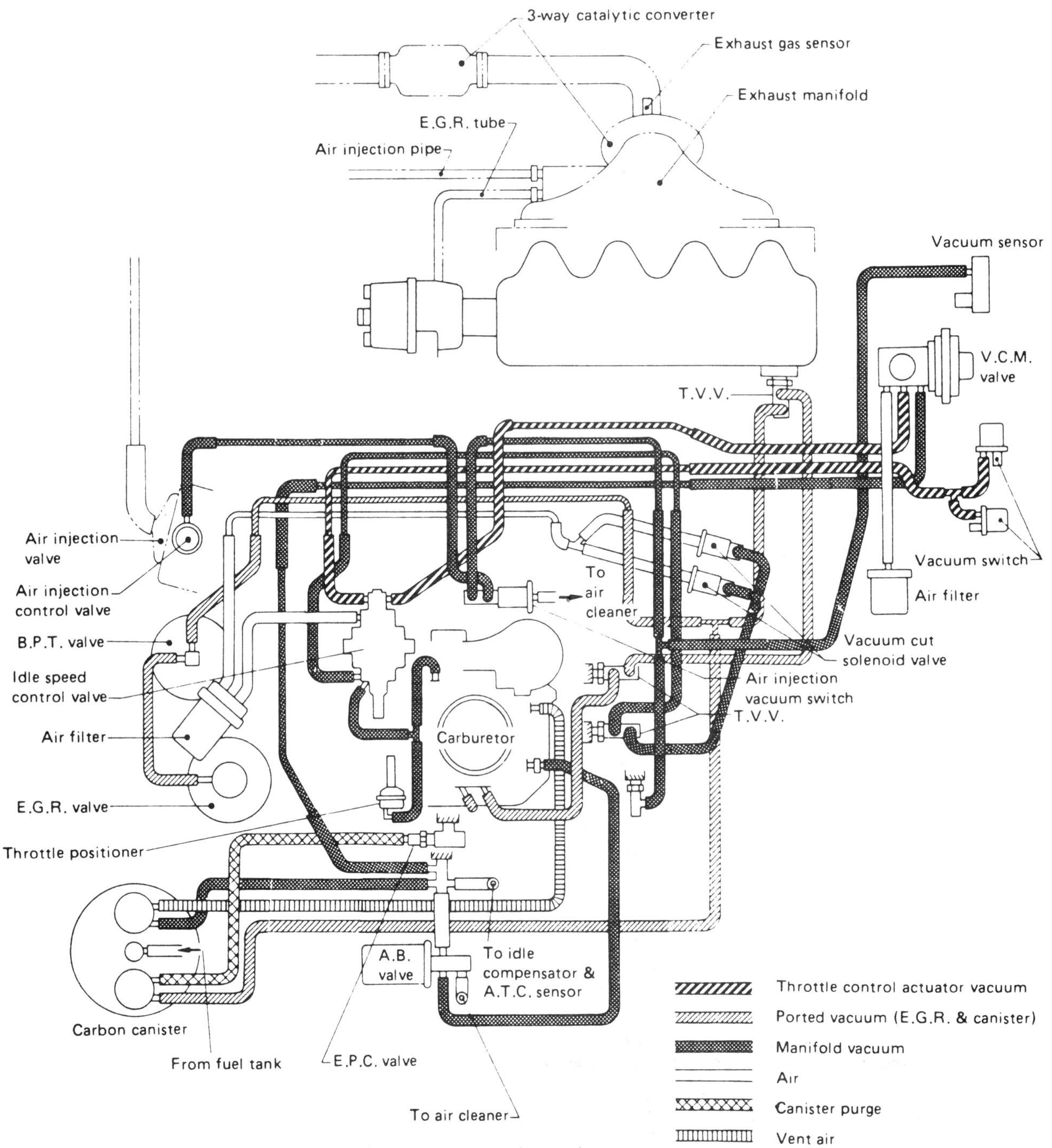

NISSAN PULSAR—EXC CALIFORNIA AND CANADA

VACUUM CIRCUITS

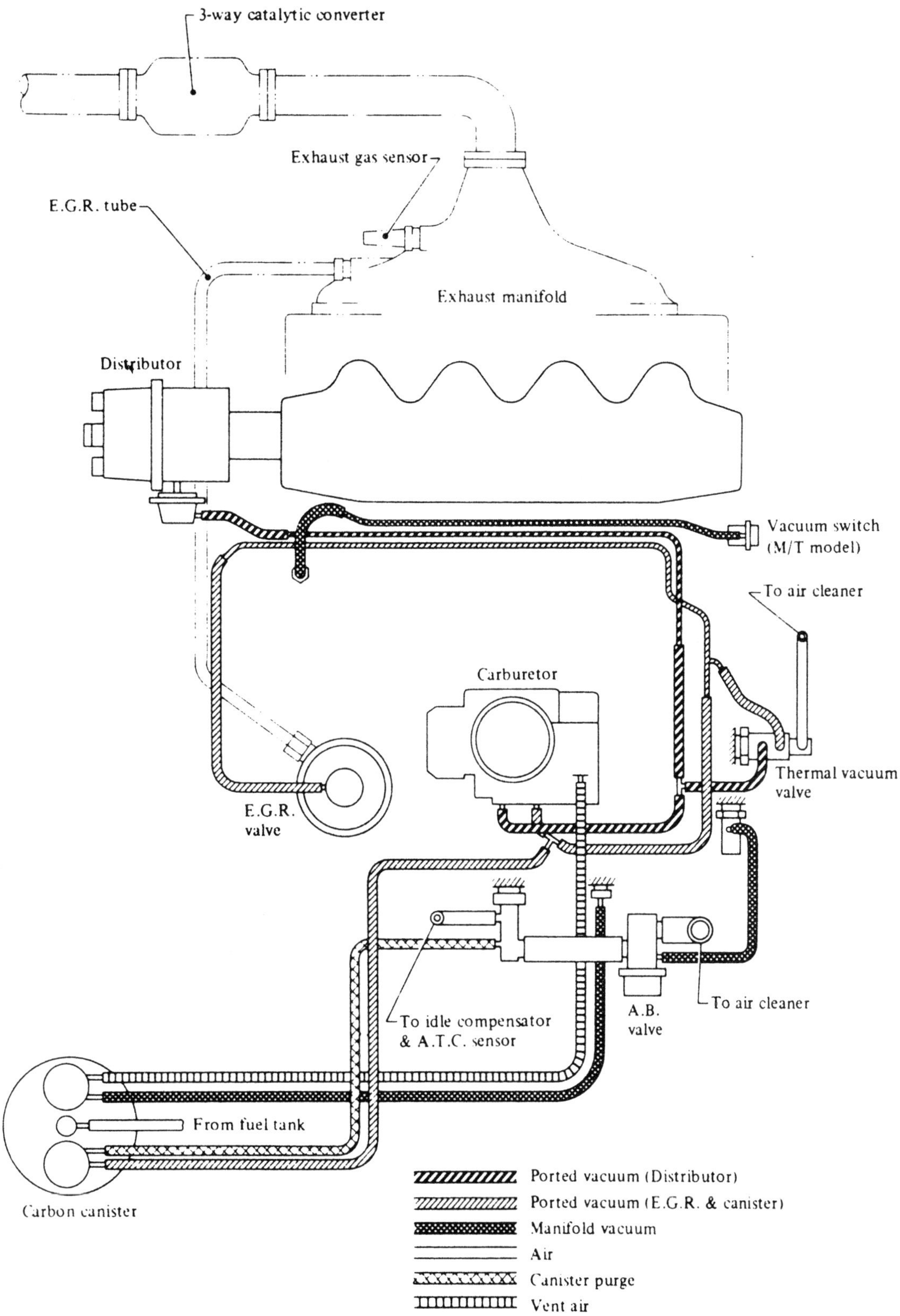

NISSAN PULSAR—CALIFORNIA

VACUUM CIRCUITS

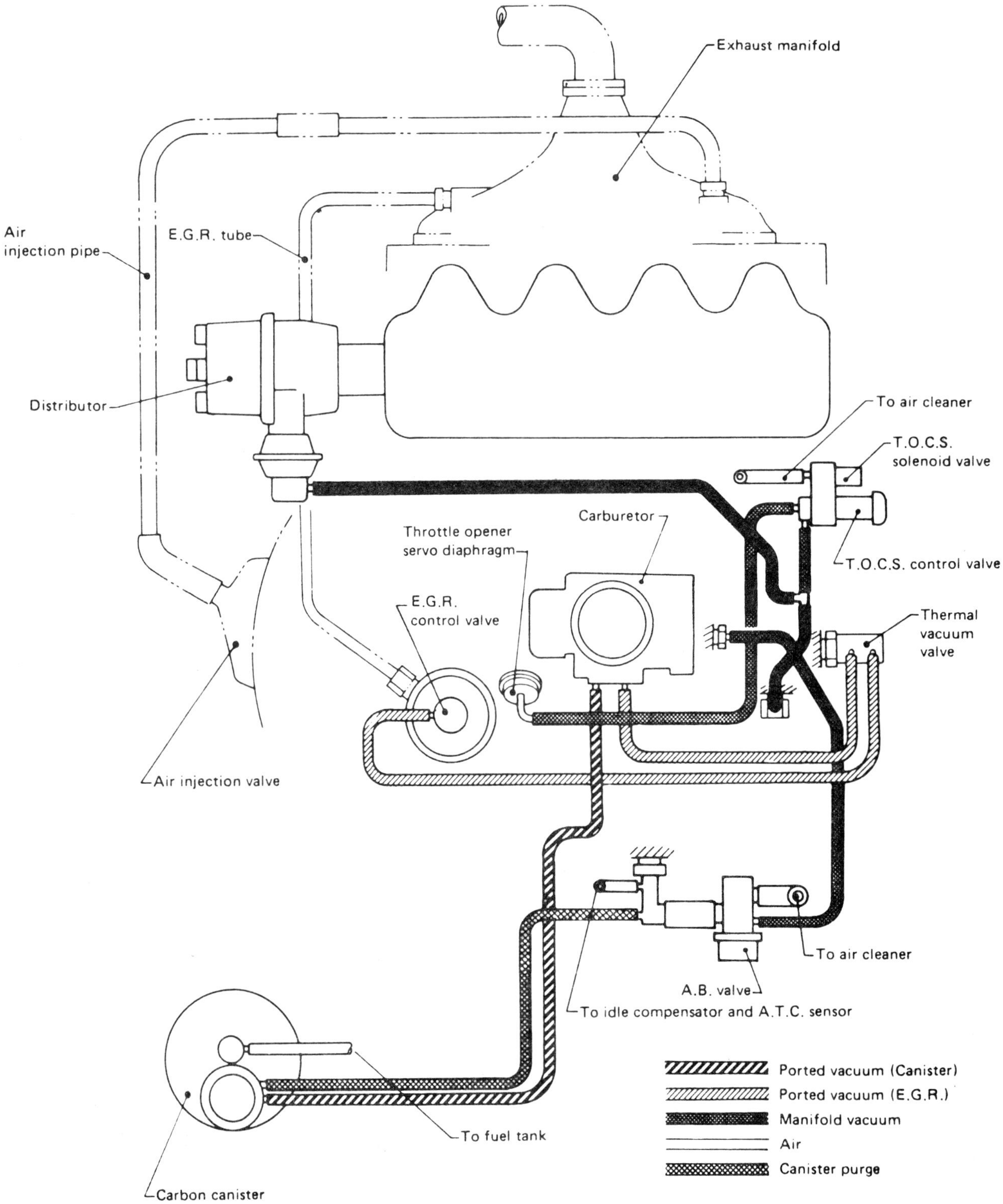

NISSAN PULSAR—CANADA

VACUUM CIRCUITS

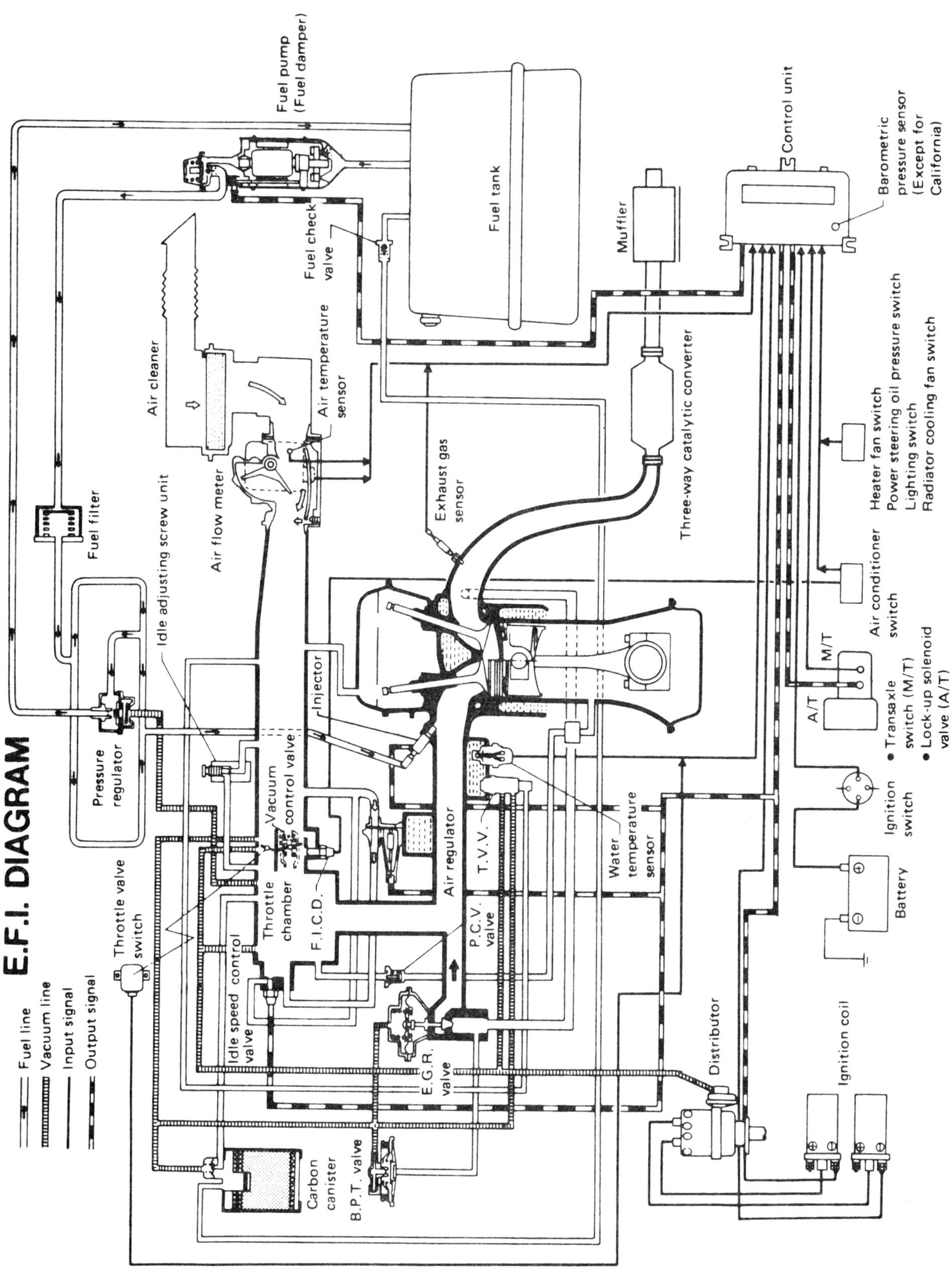

NISSAN STANZA

VACUUM CIRCUITS

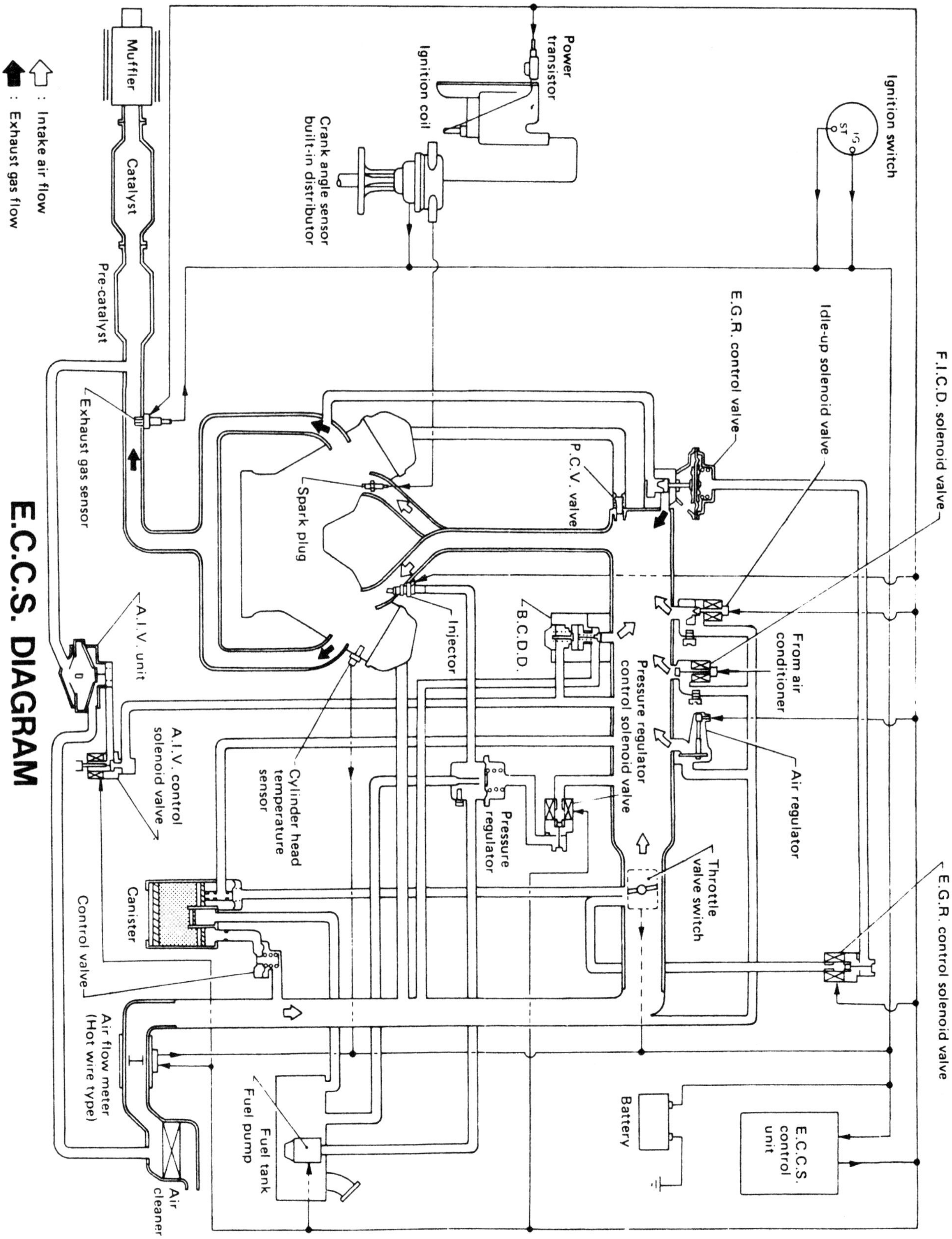

NISSAN MAXIMA

VACUUM CIRCUITS

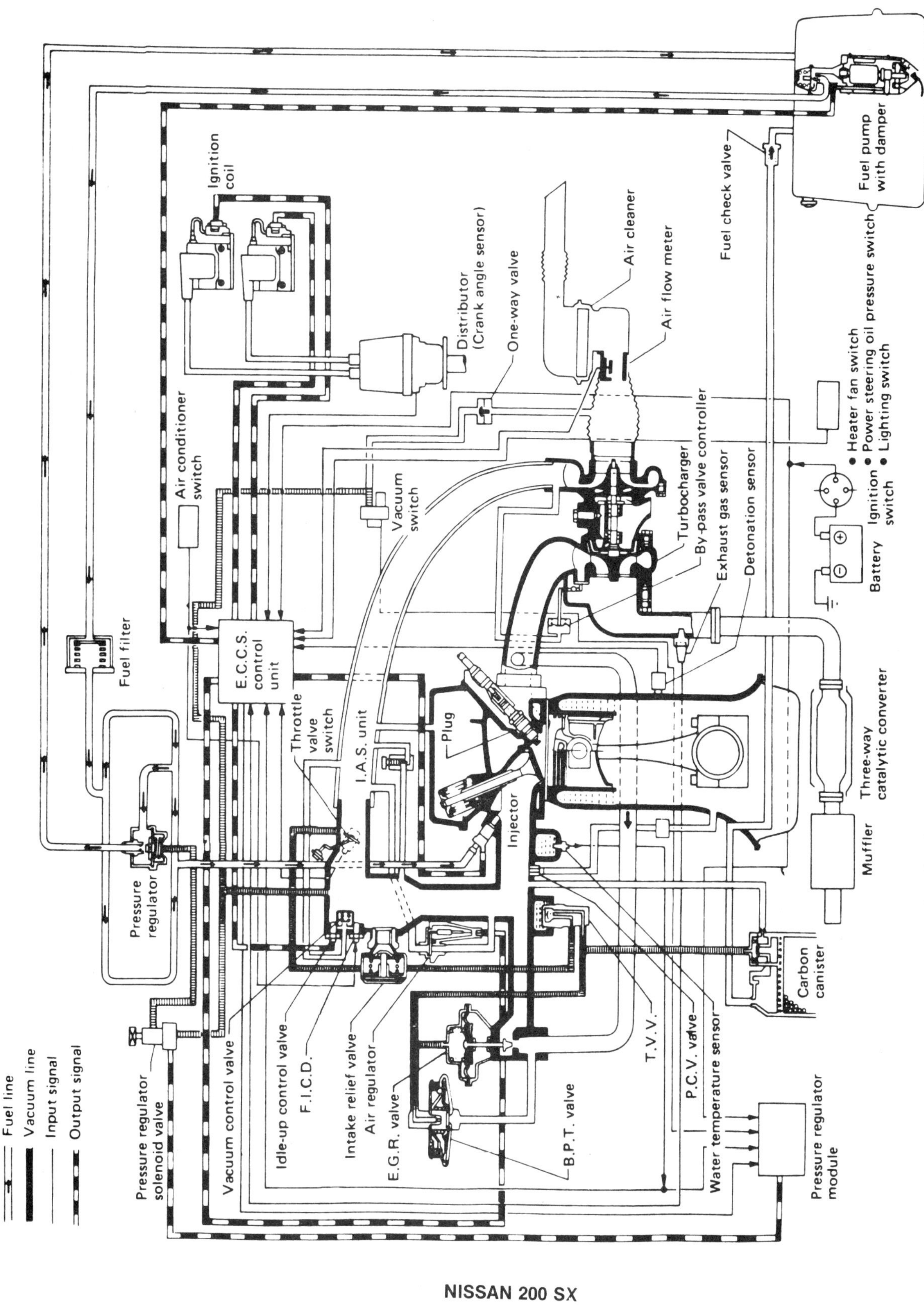

E.C.C.S. DIAGRAM—CA18ET Engine

VACUUM CIRCUITS

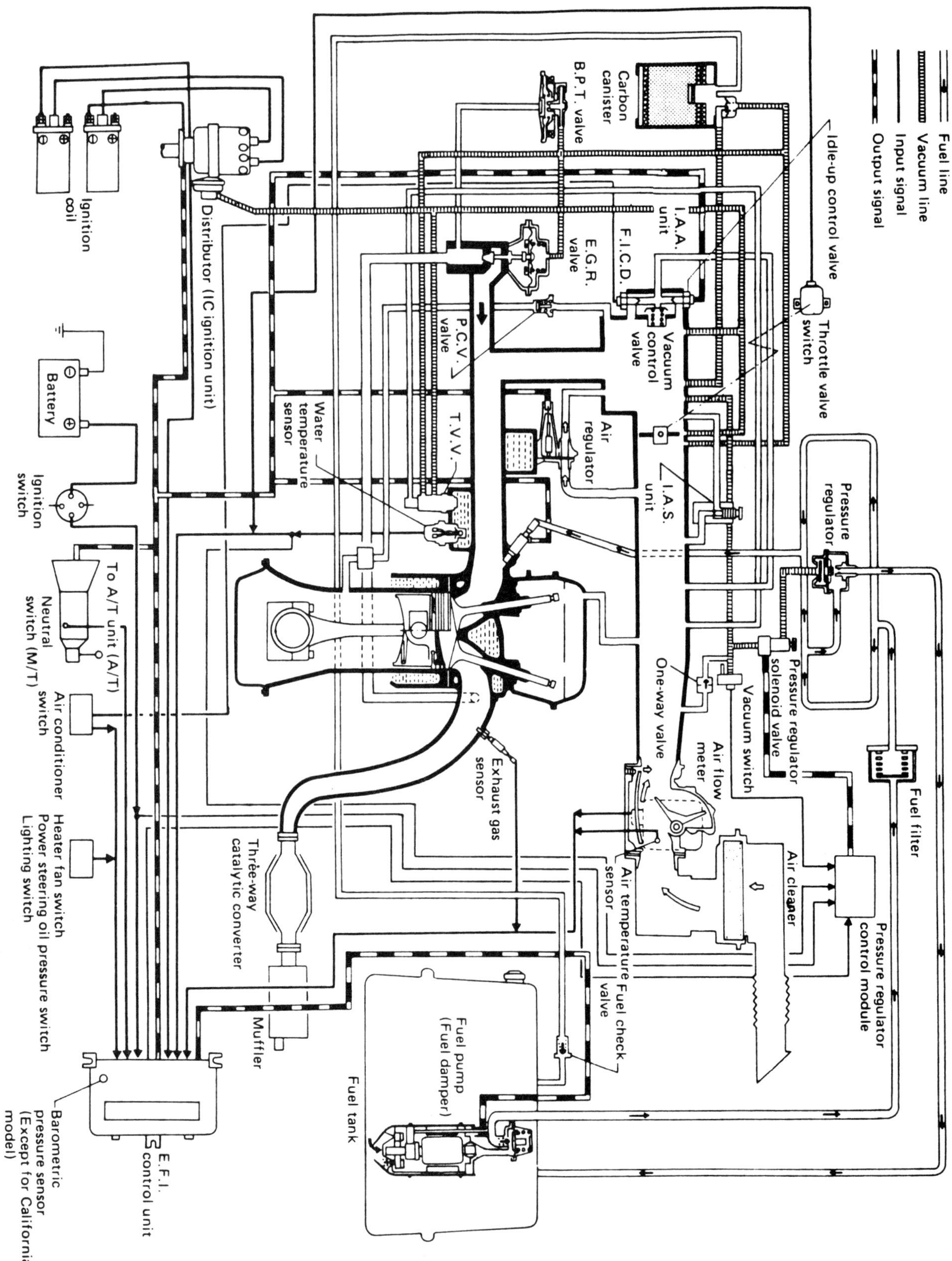

E.F.I. DIAGRAM—CA20E Engine

VACUUM CIRCUITS

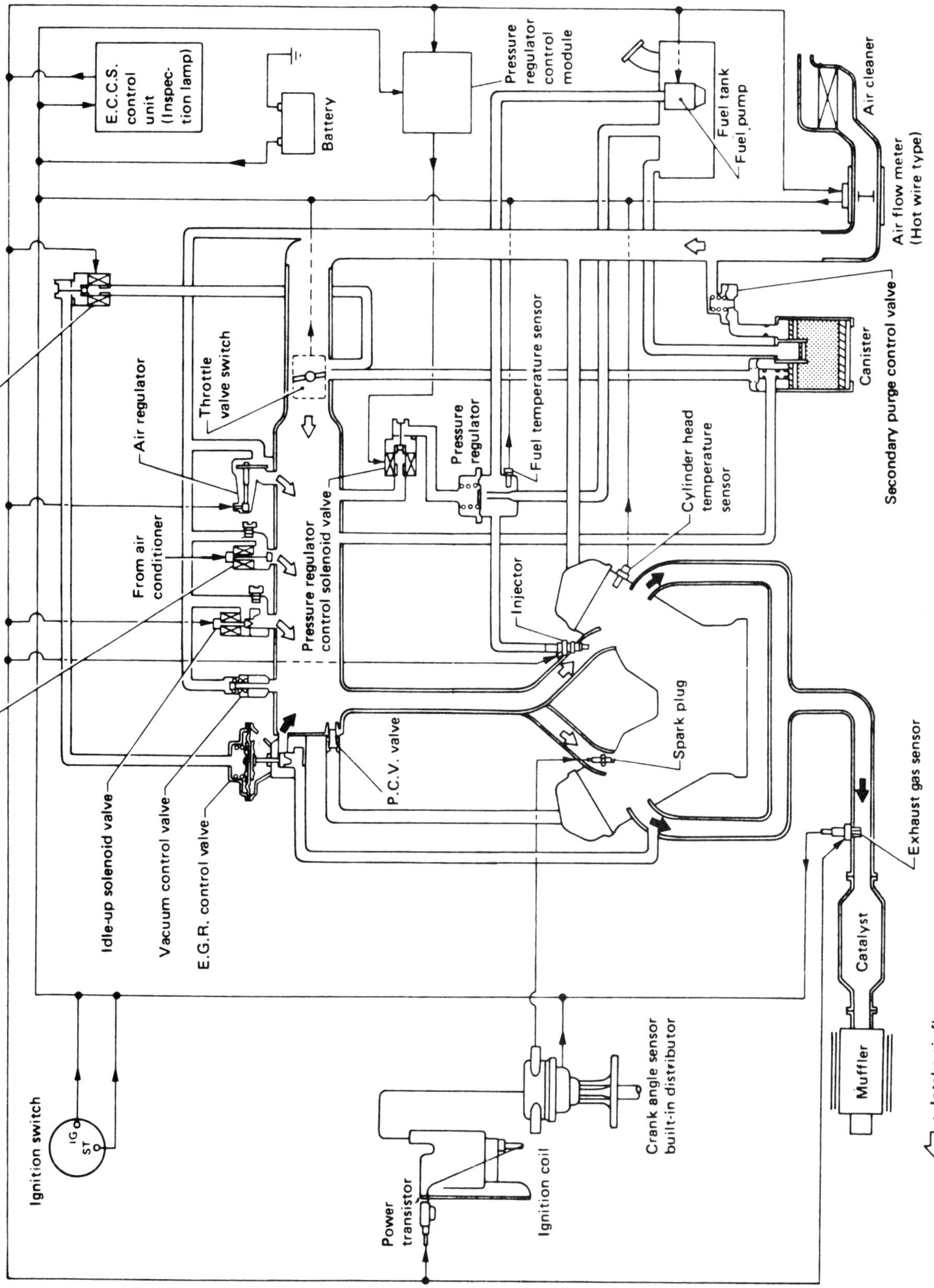

E.C.C.S. DIAGRAM
VG30E Engine (Without turbocharger)

VACUUM CIRCUITS

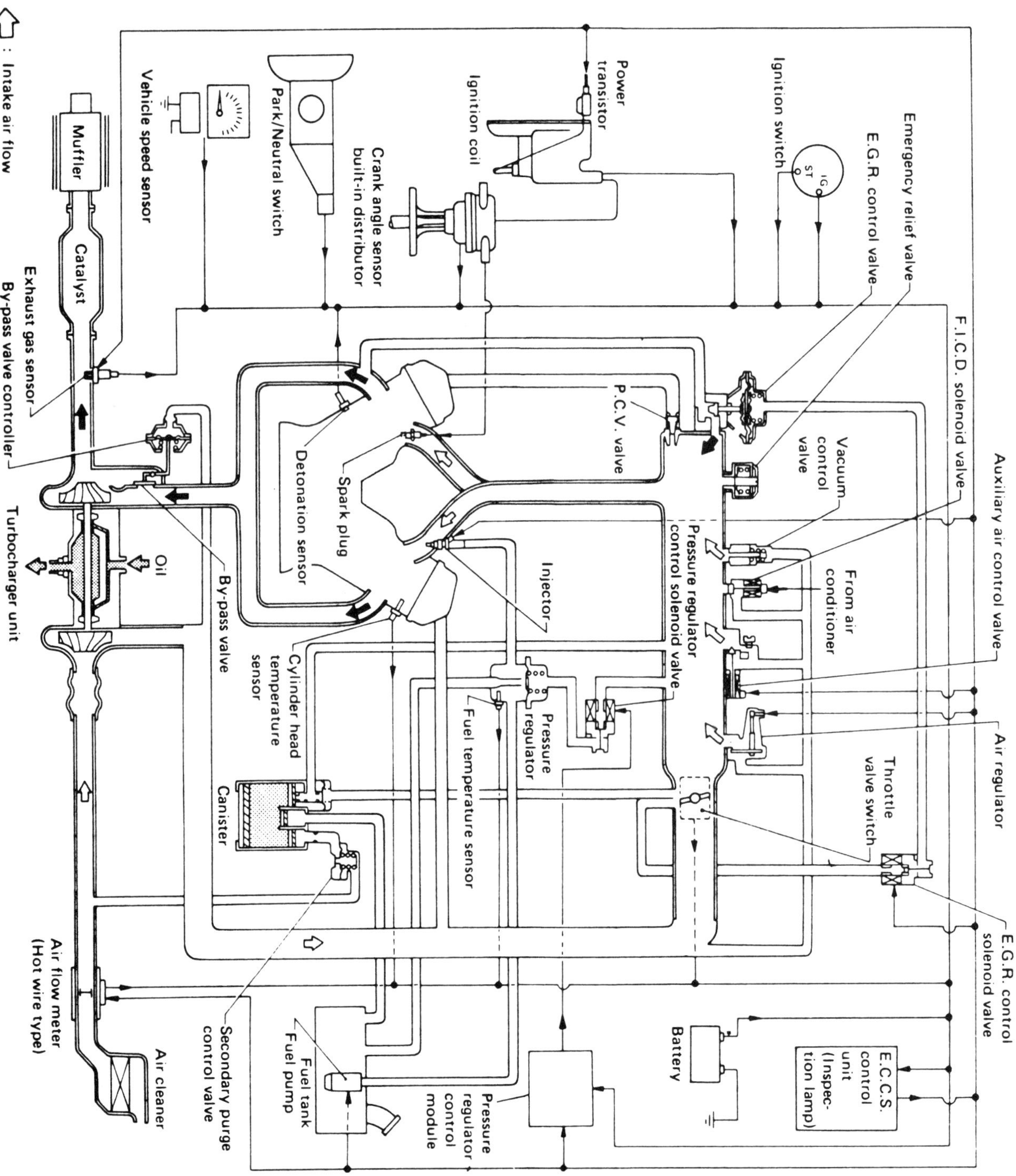

E.C.C.S. DIAGRAM
VG30ET Engine (With turbocharger)

VACUUM CIRCUITS

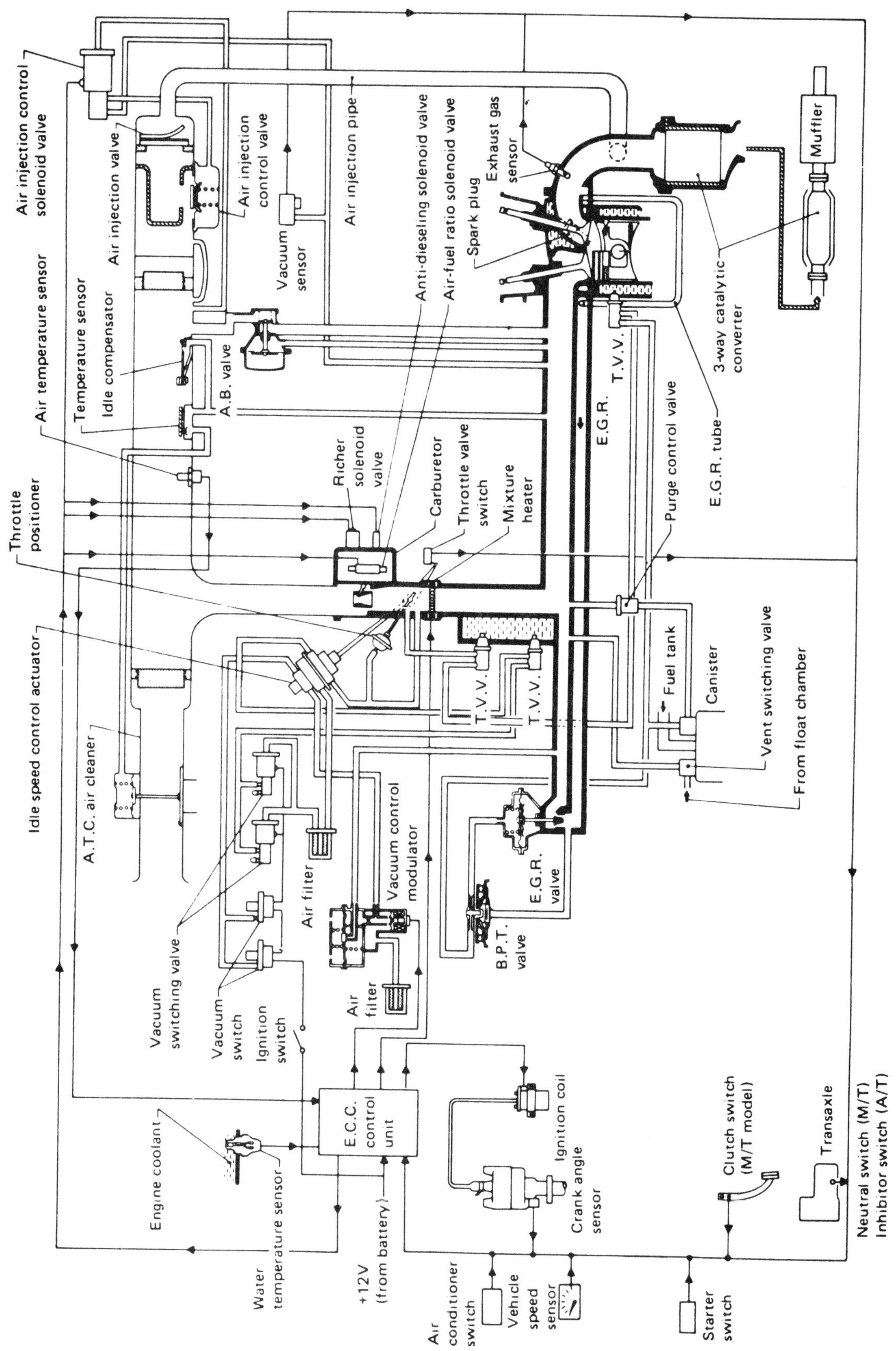

NISSAN SENTRA—EXC. CALIFORNIA AND CANADA

VACUUM CIRCUITS

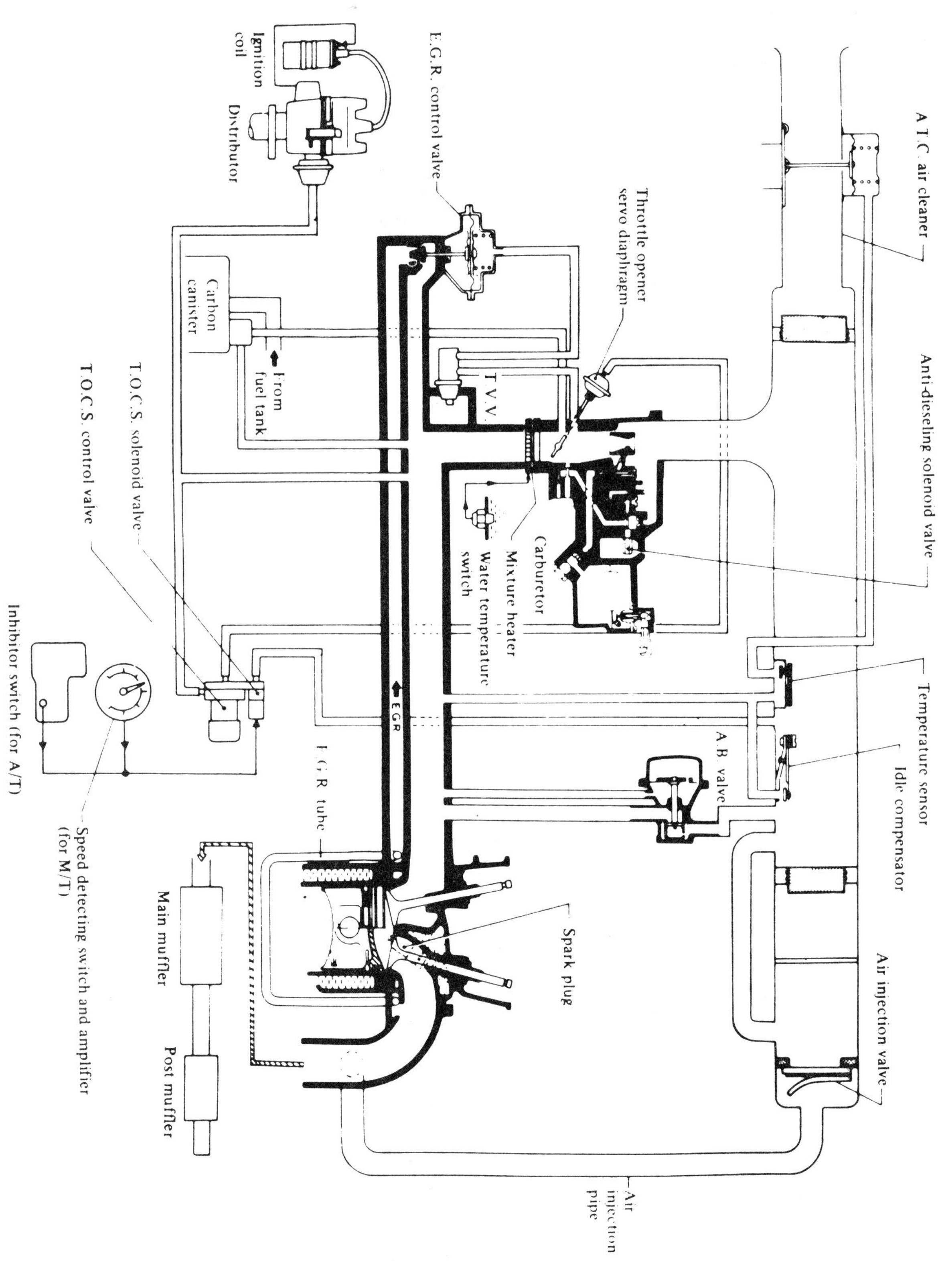

VACUUM CIRCUITS

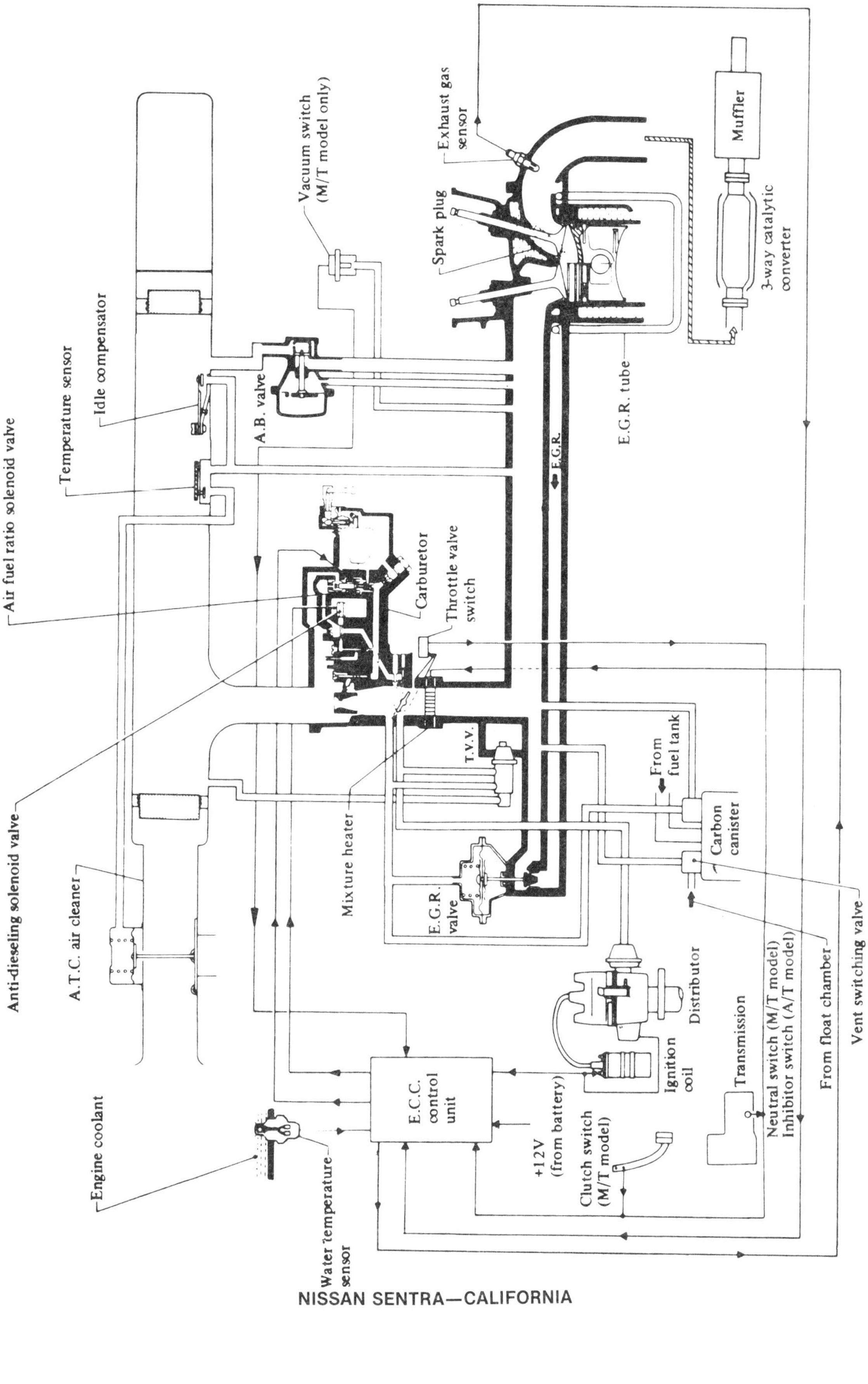

NISSAN SENTRA—CALIFORNIA

VACUUM CIRCUITS

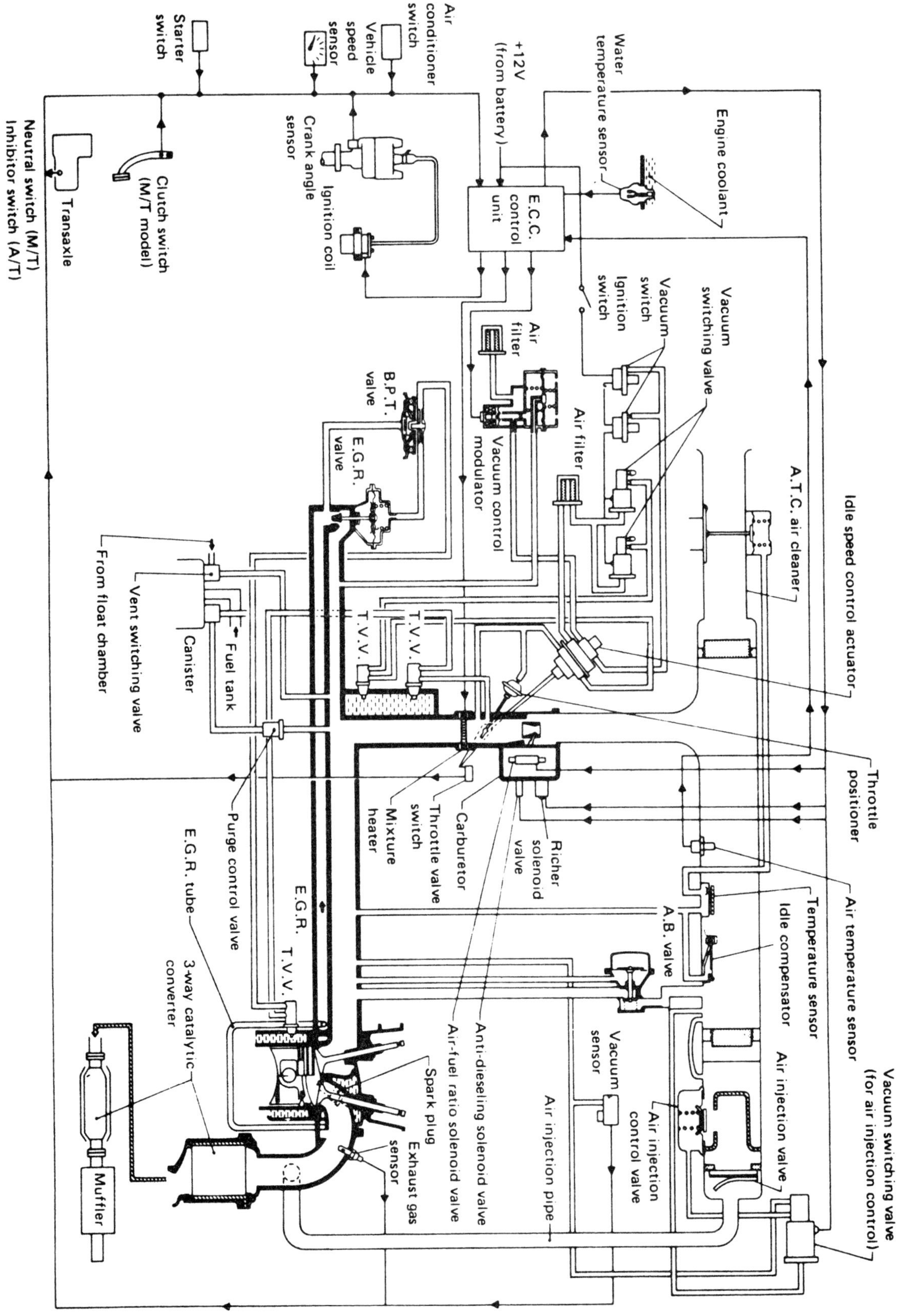

NISSAN PULSAR—EXC. CALIFORNIA AND CANADA

VACUUM CIRCUITS

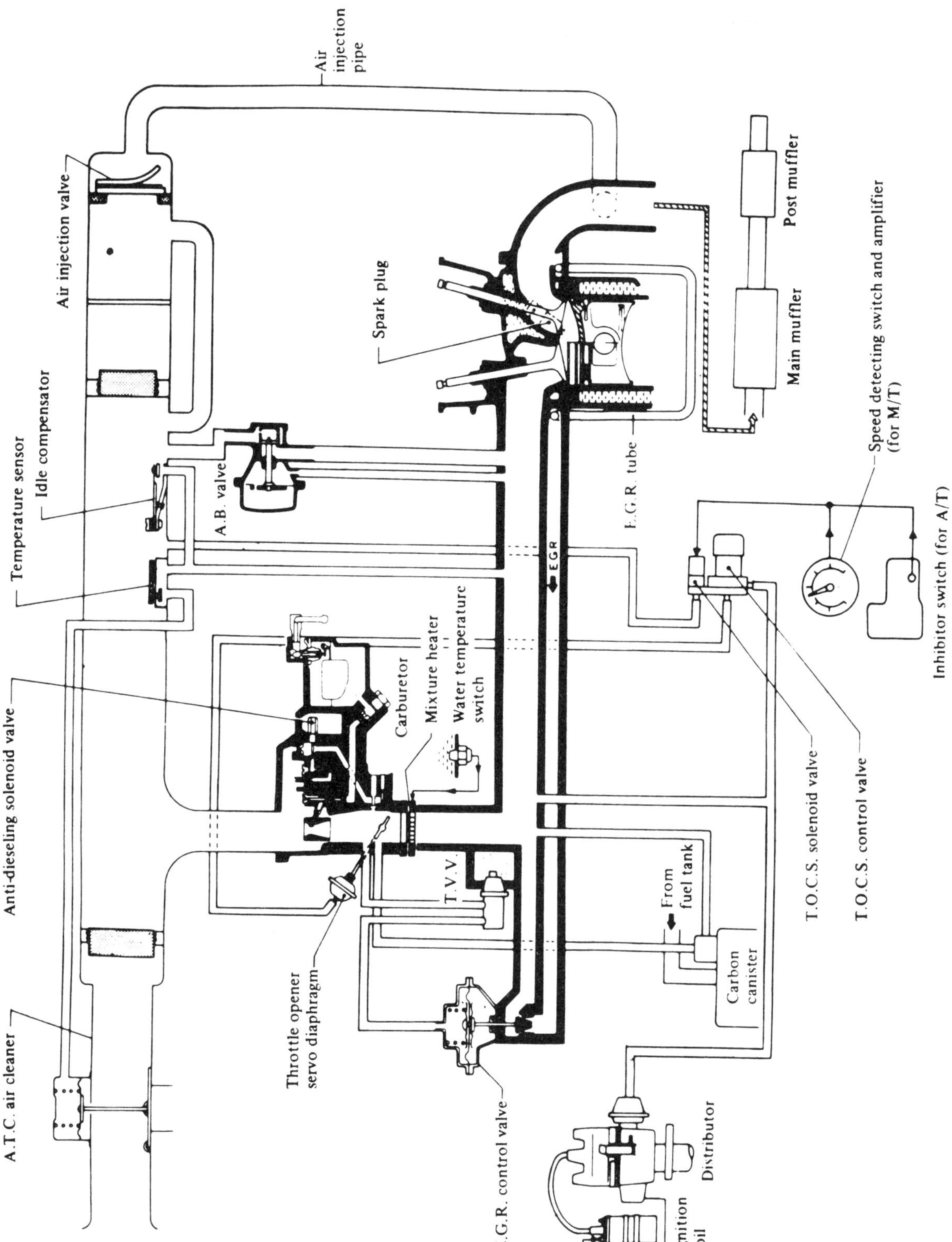

NISSAN PULSAR—CANADA

VACUUM CIRCUITS

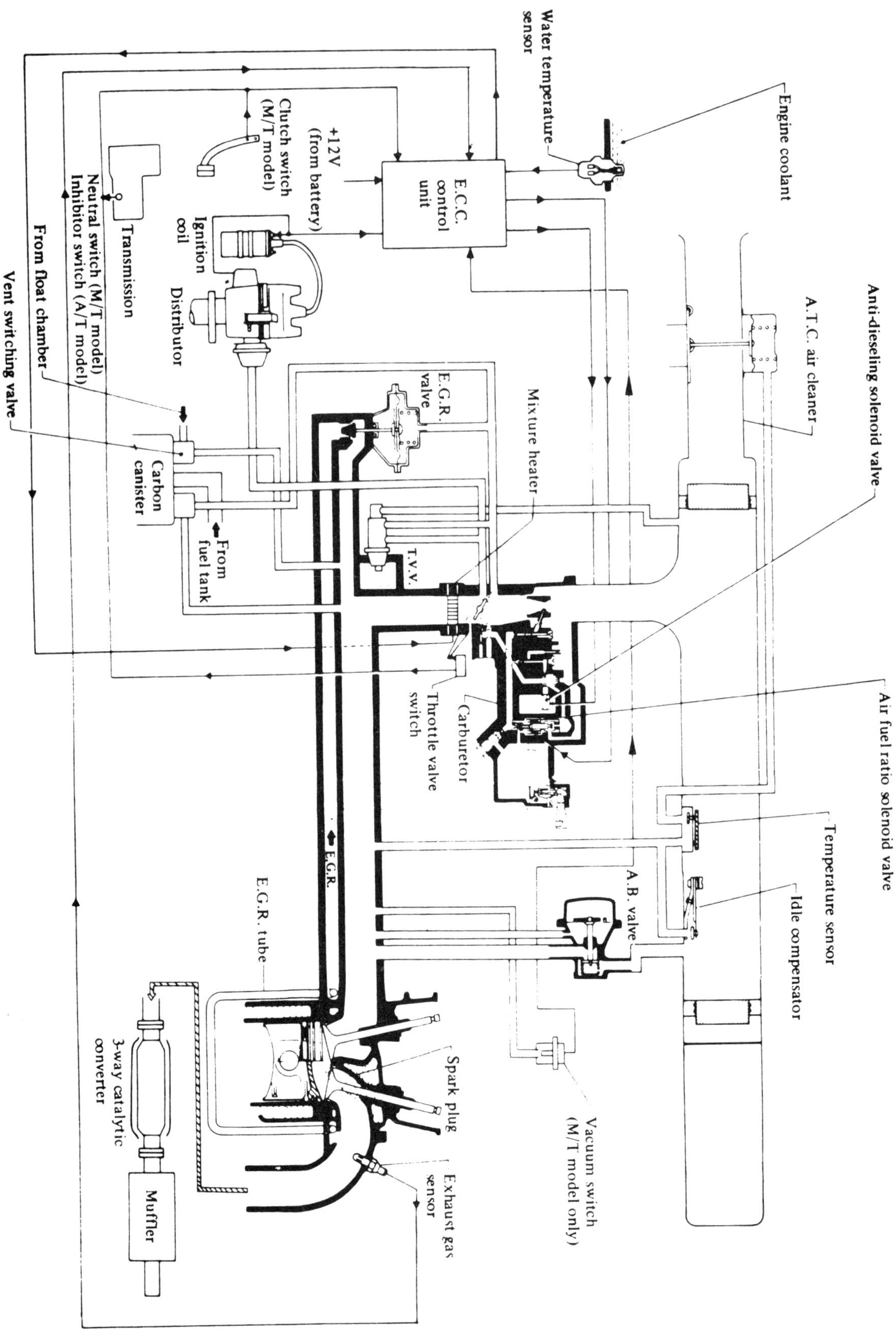

NISSAN PULSAR—CALIFORNIA

Datsun/Nissan Truck Specifications

NOTE: During the model year 1986, Nissan will introduce a completely new truck, available in 4WD and with an optional 3.0L V6 engine with single port fuel injection. The V6 engine that is reported to be used in this new truck, is the same V6 available in the Nissan 300ZX models. The specifications for this new truck are not available at the time of this publication, but the specifications for the V6 engine in the 300ZX are located in the Nissan car section of this manual.

VEHICLE IDENTIFICATION NUMBER AND ENGINE IDENTIFICATION

Vehicle Identification Number (VIN)

The vehicle identification number (VIN) plate is attached to the left top of the dash, behind the windshield. There is an additional vehicle identification plate located on the firewall. This plate contains vehicle type, chassis number, model number, color codes, engine model and displacement, transmission/transaxle model, and axle model.

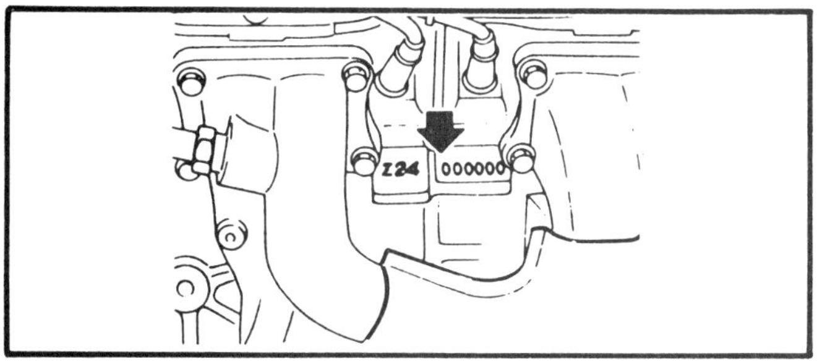

Engine I.D. number location for the gasoline engine

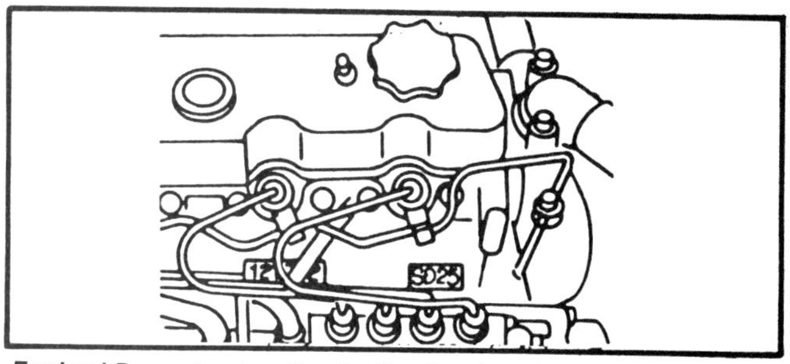

Engine I.D. number location for the diesel engine

Engine Identification Number

The engine identification number for the Z20 and the Z24 engines, is stamped on the left side of the engine block, below the exhaust manifold. On the SD25 engine, the number is stamped on the right side of the engine block, under the fuel injection lines.

GASOLINE ENGINE TUNE-UP SPECIFICATIONS

(When analyzing compression test results, look for uniformity among cylinders, rather than specific pressures)

Year	Engine Displacement cu in. (cc)	Spark Plugs Type	Spark Plugs Gap (in.)	Distributor Air Gap (in.)	Ignition Timing (deg) MT	Ignition Timing (deg) AT	Fuel Pump Pressure (psi)	Idle Speed (rpm) MT	Idle Speed (rpm) AT	Valve Clearance (in.) In	Valve Clearance (in.) Ex
'85–'86	4-119 (1952)	①	0.031–0.035	0.012–0.020	5B	5B	2.7–3.4	700	700	0.012	0.012
'85–'86	4-146 (2389)	①	0.031–0.035	0.012–0.020	3B	3B	2.7–3.4	650 ②	650 ②	0.012	0.012

NOTE: All idle speed adjustments are ±100 rpm. The ignition timing should be checked out with the distributor vacuum advance hose disconnected.

B Before top dead center
① BPR6ES on the intake side
BPR5ES on the exhaust side
② 4WD—800 ± 100 rpm

DIESEL ENGINE TUNE-UP SPECIFICATIONS

Year Model	Engine Displacement cu. in. (cc)	Warm Valve Clearance (in.)		Intake Valve Opens (deg)	Injection Pump Setting (deg)	Injection Nozzle Pressure (psi)		Idle Speed (rpm)	Compression Pressure (psi)
		In	Ex			New	Used		
1985	4-151 (2488)	0.014	0.014	16	See Text	1493–1607	1422–1493	700 ± 100	427 ①
1986	4-151 (2488)	0.014	0.014	16	See Text	1493–1607	1422–1493	700 ± 100	427 ①

① Minimum compression pressure is 356 psi.

FIRING ORDERS

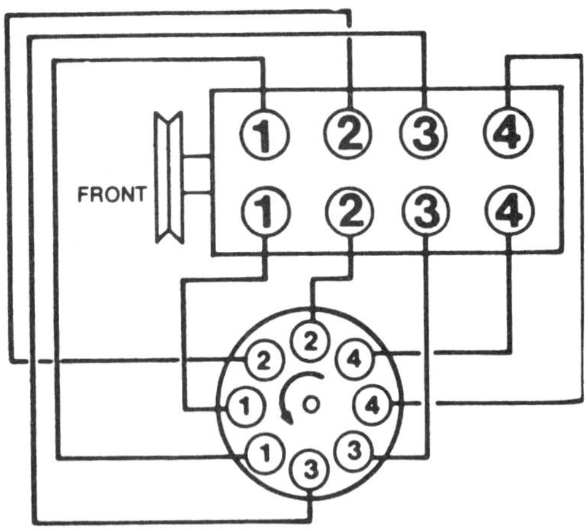

FIRING ORDER: 1-3-4-2
DISTRIBUTOR ROTATION: COUNTERCLOCKWISE

Firing order for the Z20 and Z24 engines

Firing order for the SD25 diesel engine—1, 3, 4, 2

EMISSION EQUIPMENT USED

Z20 AND Z24 ENGINES

Crankcase Emission Control
Closed system
Positive crankcase vent valve

Evaporative Emission Control
Canister-Single
Thermal vacuum valve (3 port wax type)

Exhaust Emission Control
Air injection system
Catalytic converter
Secondary air supply system-reed valve
Exhaust gas recirculation (EGR) system
EGR control system
Fuel shut-off system
Intake manifold vacuum control system
Spark timing control system
Spark plug switching control system
Detonation control system

SD25 DIESEL ENGINE

Crankcase Emission Control
Closed system with a breather hose

Exhaust Emission Control
Exhaust gas recirculation EGR valve
EGR control valve
Venturi vacuum transducer (VVT) valve
Vacuum amplifier
Check valves

SPARK PLUG APPLICATION CHART

Engine	Intake Side	Exhaust Side	Gap (in)
Z20 (Hot)	BPR6ES	BPR5ES	0.031–0.035
Z24 (Hot)	BPR6ES	BPR5ES	0.031–0.035
Z20 (Cold)	BPR7ES	BPR6ES–BPR7ES	0.031–0.035
Z24 (Cold)	BPR7ES	BPR6ES–BPR7ES	0.031–0.035
Z20 (Standard)	BPR6ES	BPR5ES	0.031–0.035
Z24 (Standard)	BPR6ES	BPR5ES	0.031–0.035

REQUIRED EMISSION CONTROL MAINTENANCE

CRANKCASE VENTILATION SYSTEM

Inspect and replace the filter every 15,000 miles.

FUEL EVAPORATION SYSTEM

Check all hoses and connections every 30,000 miles. Replace filter.

THERMOSTATIC AIR CLEANER

Check operation and change the filter every 15,000 miles.

AIR INDUCTION FILTER

Replace every 30,000 miles.

IDLE RPM AND MIXTURE RATIO SYSTEM

Check the exhaust CO% with an analyzer every 15,000 miles.

VALVE ADJUSTMENT

Every 15,000 miles.

IGNITION TIMING CONTROL SYSTEM

Inspect for proper operation every 15,000 miles.

CHOKE MECHANISM

Inspect for proper operation every 15,000 miles.

EXHAUST GAS SENSOR

Replace every 30,000 miles.

EGR SYSTEM

Check for proper operation every 30,000 miles. Clean all carbon deposits from EGR valve and lines.

NOTE: Although not required, the manufacturer recommends that all emission systems be periodically inspected for proper operation. Change the filters more frequently if the vehicle is operated under extremely dusty conditions, or ambient temperatures are extremely low or high.

IDLE SPEED AND MIXTURE ADJUSTMENTS

1985–6

NOTE: The following information is being published from the latest information available at the time of publication. If the information differs from the information on the vehicle emissions label, use the information on the emissions label.

Idle Speed Adjustment

CARBURETED ENGINES

NOTE: The idle mixture adjusting screw (preset at the factory), should be adjusted only in the event of major carburetor overhaul, throttle body replacement or to lower the exhaust emissions. The idle mixture adjusting screw should not be removed during routine maintenance.

IDLE SPEED SPECIFICATIONS

Engine	Transaxle	Curb Idle Speed (RPM)
Z20	M/T & A/T	700 ± 100
Z24 (2WD)	M/T	700 ± 100
Z24 (2WD)	A/T	650 ± 100
Z24 (4WD)	M/T	800 ± 100
Z24 (2WD)	A/T	650 ± 100 Canada
Z24 (4WD)	M/T	800 ± 100 Canada
SD-25 (Diesel)	M/T & A/T	700 ± 100

M/T Manual Transmission
A/T Automatic Transmission
2WD Two wheel drive
4WD Four wheel drive

Preparation

Before adjusting the idle speed, be certain the following items are operating correctly:

- Battery
- Ignition System
- Engine oil & coolant levels
- Fuses
- Oil filler cap and oil level gauge
- Valve clearance
- Connect the engine timing light and tachometer in their proper locations.
- Apply the parking brake and block the wheels
- Check that all accessories and switches are in the OFF position.
- Check that the radiator fan has stopped and if it is operating, wait until it stops.

NOTE: Inspections and adjustments should be carried out while the shift lever is in "D" position on the automatic models or in "Neutral" on the manual transaxle equipped models. When increasing the engine speed, be certain the automatic transaxle models are in "N" or "P", with the brakes applied.

Adjustment

1. Warm-up the engine until the water temperature gauge registers normal operating temperature (middle of gauge) and the engine speed is below 1,000 rpm.
2. Disconnect and plug the distributor vacuum hose from the distributor vacuum controller. Run the engine at idle speed for 2 minutes.
3. Increase the engine speed between 2,000–3,000 rpm, two or three times under no load, then idle the engine for one minute.
4. Check the ignition timing. If necessary, adjust it to specifications by loosening the distributor hold-down bolt and turning the distributor. Reconnect the vacuum hose to the distributor vacuum controller.5.

Increase the engine speed between 2,000–3,000 rpm, two or three times under no load, then idle the engine for one minute. 6.

Make sure that the inspection lamp on the ECC control unit goes on and off more than four times during a ten second interval. 7.

Check the idle speed. If necessary, adjust it to specifications by turning the throttle adjusting screw. If the idle speed will not adjust, replace the control unit, carburetor, or the vacuum control modulator.

FUEL INJECTED ENGINES

Before adjusting the idle speed, be certain the following items are operating correctly:

- E.G.R valve operation
- Air cleaner clogging
- Ignition system
- Engine oil and coolant levels
- Valve clearance
- E.F.I./E.C.C.S. component parts
- E.F.I./E.C.C.S. harness connectors
- Vacuum hoses
- Air intake system (Oil filler cap, oil level gauge,etc.)

Adjustment

1. The air conditioner and all electrical accessories should be turned "OFF".
2. Apply the parking brake and block the wheels.
3. Make adjustments only while the cooling fan has stopped.

NOTE: Inspections and adjustments should be carried out while the shift lever is in "D" position on the automatic models or in "Neutral" on the manual transaxle equipped models.

4. Warm-up the engine until the water temperature gauge registers normal operating temperature (middle of the temperature gauge).
5. Start the engine and run at about 2,000 rpm for about 2 minutes under no-load.
6. Increase the engine speed two or three times under no-load, then run the engine for one minute at idle speed.
7. Check the idle speed. If necessary, adjust it to specifications by turning the idle speed adjusting screw.

DIESEL ENGINE

Adjustment

1. Turn off lights, heater fan and all electrical accessories.
2. Apply the parking brake and block the wheels.
3. Set the transmission shift lever in "Neutral". Start the engine and warm it up until the water temperature indicator points to the middle of the gauge.
4. Attach a diesel tacho tester's pick-up to the No.1 injection tube.

NOTE: In order to obtain a more accurate reading of engine speed, remove the clamps on the No.1 injection tube.

5. Run the engine at about 2,000 rpm for two minutes under load. Increase the engine two or three times. Make sure that it returns to idle speed. If not, check the accelerator linkage for binding.
6. Run the engine for one minute at idle speed.
7. Check the idle speed. If necessary, adjust to specifications using the following procedures (700 ± 100 rpm).
8. Turn off the engine.
9. Disconnect the accelerator wire from the injection pump control lever.
10. Move the control lever to wide open throttle (WOT) and loosen the idle screw locknut while holding the control lever.11.

Turn the idle screw to obtain the specified rpm.

12. Turn off the engine.
13. Tighten the idle adjusting screw locknut while holding the control lever at wide open throttle.
14. Connect the accelerator wire.

Idle Mixture Ratio Feedback System

Inspection

CARBURETED ENGINESWhen checking "CO" percentage, insert the probe at least 15.7 inches into the tail pipe.

1. Disconnect and plug the vacuum hose from the vacuum switch. Race the engine two or three times under a no-load conditions for one minute at idle speed.2.

Check the idle CO mixture with a suitable CO meter. If the CO reading is within specifications, the inspection is complete. But if the CO reading is less than four percent, check the following and go on to the following procedure.

   a. Check the vacuum hoses for cracks and for proper connection.
   b. Check for air leaks at the carburetor and the intake manifold.

3. Turn off the engine and remove the carburetor from the engine.
4. Carefully drill the idle mixture adjusting screw seal plug and remove it from the plug hole with a suitable tool.
5. After drilling out the plug, be sure to remove the shaving with compressed air. Reinstall the carburetor on the engine
6. Start the engine and let it run until it reaches normal operating temperature. Then race the engine two or three times at 2000 rpm, under a no-load condition.
7. With the engine idling, adjust the idle CO to specifications by turning the idle mixture adjusting screw (1.0 ± 0.7 %).
8. If the proper idle CO percentage can not be obtained by adjustment. Remove and overhaul the carburetor or replace it.
9. After adjusting the idle CO to specifications, install a new concealment plug over the idle mixture screw.

Peugeot

INDEX

VEHICLE IDENTIFICATION NUMBER AND ENGINE IDENTIFICATION

Vehicle Identification Number (VIN)

The vehicle identification number (VIN) is located on a stamped riveted plate, mounted in front of the radiator on early models and later models have the VIN number stamped on the right front inner fender. The tenth digit represents the model year, "F" for 1985 and "G" for 1986.

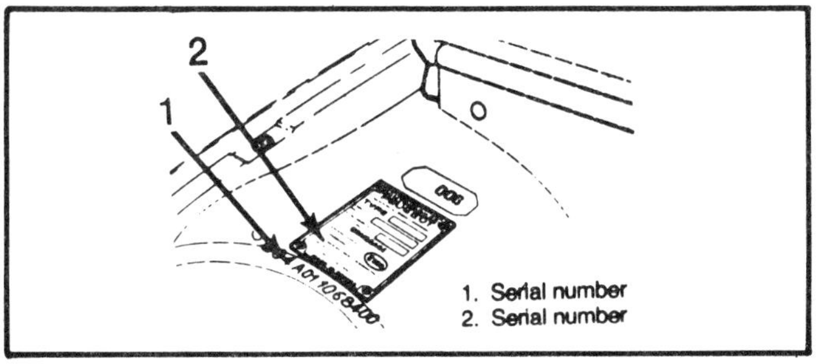

The Vehicle Identification Number (VIN) is located on the right inner fender panel

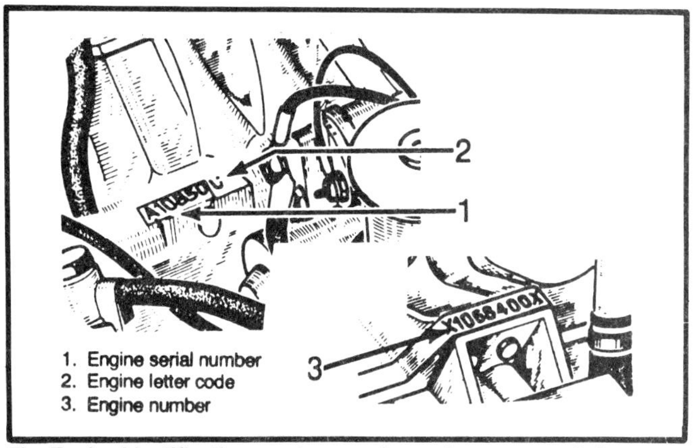

The Engine Identification numbers are located on two metal plates, attached to the engine block

Engine Identification

The engine identification number and engine type are found on two metal plates, riveted on the lower left side of the engine block.

ENGINE IDENTIFICATION

Year	Car Model	Engine Code	Engine Displacement cu. in. (cc)	Fuel Delivery
'85–'86	505 ①	XN6	120.3 (1971)	CIS fuel injection
'85–'86	504D, 505D	XD2, XD2C	140.6 (2304)	Diesel fuel injection
'85–'86	505TD ②	XD2S	140.6 (2304)	Diesel fuel injection

CIS Constant Injection System (K-Jetronic)
① GI, S and STI gasoline engine models
② Turbodiesel

GASOLINE ENGINE TUNE-UP SPECIFICATIONS

Year	Model	Engine Displacement cu. in. (cc)	Spark Plugs Type	Spark Plugs Gap (in.)	Distributor Point Dwell (deg)	Distributor Point Gap (in.)	Ignition Timing ② MT ● (deg)	Ignition Timing ② AT (deg)	Idle Speed MT (rpm)	Idle Speed AT (rpm)	Valve Clearance ① Intake (in.)	Valve Clearance ① Exhaust (in.)
'85–'86	505	120.3 (1971)	WR 7 DS	.024	Electronic		8 ③	8 ③	900–950 ④	900–950 ④	.004	.010
	505	131.5 (2155)	WR 7 DS	.036	Electronic		10	10	910	910	.008	.012

NOTE: The underhood specifications sticker often reflects tune-up specification changes made in production. Sticker figures must be used if they disagree with those in this chart.

BTDC Before top dead center
① Valves must be adjusted in proper sequence
② All timing degrees BTDC
③ 8° ± 2 BTDC @ 800 RPM
④ 800–850 in Neutral

DIESEL ENGINE TUNE-UP SPECIFICATIONS

Year	Model	Valve Clearance (cold) Intake (in.)	Valve Clearance (cold) Exhaust (in.)	Intake Valve Opens (deg)	Injection Pump Setting (deg)	Injection Nozzle Pressure (psi) New	Injection Nozzle Pressure (psi) Used	Idle Speed (rpm)	Compression Pressure (psi)
'85–'86	504D, 505D	.010	.010	12B	13B	1740–1813	1668–1813	780–830 ①	261 ④
'85–'86	505TD, 604TD	.006	.010	NA	.016 ②③	1900	NA	830–860	261 ④

NA Not available at time of publication
① With A/C: 830–860
② Figure is for in. BTDC
③ Canada: .031
④ Maximum variation between cylinders: 72 psi

FIRING ORDERS

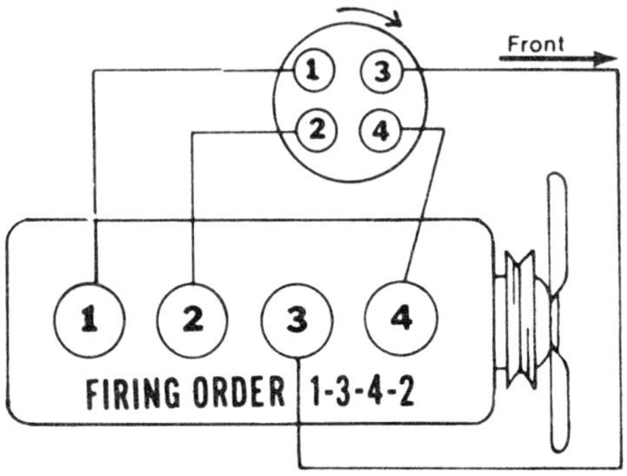

Four cylinder firing order, including diesel engines

EMISSION EQUIPMENT USED

505 GASOLINE MODELS

Positive Crankcase Ventilation System (PVC).
Fuel Evaporation System (FES).
K-Jetronic (CIS) Fuel Injection with Oxygen Sensor.
Air Injection System (AIS) with Pulsair valves.
Deceleration Control System.
Catalytic Converter (CAT).

505 DIESEL (NON-TURBO) MODELS

Exhaust Gas Recirculation System.
Electro-valve–EGR (Green).
Electro-valve–EGR (Blue).
Electro-valve–EGR (Red).
Upper load sensor.
Lower load sensor.

Engine Speed (RPM) Sensor.
Vacuum Controlled Throttle Flap.
Two-Stage EGR Valve.
Electronic Control Box (ECB).

505 TURBO DIESEL MODELS

Exhaust Gas Recirculation System.
Engine Speed (RPM) Sensor.
Coolant Temperature Thermocontact.
Vacuum Converter.
Electronic Control Box.
EGR Valve.
Vacuum Damper.
Prefilter.
Electro-valve–EGR.
Non-return Valve.

7. 18 C thermocontact (64.5 F)
9. Leaning out delay tachymetric relay
12. 40 C thermocontact (104 F)
13. Pulsair electrovalve
14. Fuel injection cut-out electrovalve
15. Throttle flap micro-switch
16. Deceleration tachymetric relay
17. Vacuum advance relay
20. Fuel vapor recovery electrovalve
21. 50% OCR relay
22. Ignition relay
23. Electronic control box relay
24. Twin-speed relay (vacuum advance relay)
25. Cold cut-out relay
26. Leaning out delay relay
27. Fuel pump tachymetric relay
28. Coil and module
29. OCR test wire connector

Model 505 gasoline engine emission component locations

REQUIRED EMISSION CONTROL MAINTENANCE

(ES) VEHICLE EMISSION CONTROL INFORMATION
AutomobilesPEUGEOT

1985 STANDARDS

	HC	CO	NOx	Evap.
CALIFORNIA	0.41g/mile	7.0g/mile	0.4g/mile	2.0g/test
FEDERAL	0.41g/mile	3.4g/mile	1.0g/mile	2.0g/test

- ENGINE FAMILY : FPE2.1V5FAB8 **CATALYST**
- EVAPORATIVE FAMILY : 2.0A
- ENGINE CODES : 2.1AB.M5 - 2.1AB.M5F
- DISPLACEMENT : 2.155 liters – (131.48 ci)
- EMISSION CONTROL SYSTEM : EFI - TWC - EGS
- ENGINE TUNE UP SPECIFICATIONS :
- SPARK PLUGS : WR7DS
- GAP : $0.9^{+0.1}_{+0}$
- INITIAL TIMING ADVANCE : 10° ± 2° BTDC at 910 ± 40 RPM
- IDLE SPEED : 910 ± 40 RPM

- IDLE CO 1.3 ± 0.3% : See workshop manual
- ADVERTISED HORSE POWER : 142 HP at 5600 RPM (DIN)

IDLE SPEED ADJUSTMENT PROCEDURE :
- TRANSMISSION : NEUTRAL
- ENGINE HOT
- ALL ELECTRICAL ACCESSORIES OFF

1) ENGINE WARM (SPEED STABILIZED AT 3000 RPM WAIT FOR 2 OPERATION CYCLES OF THE SELF DISENGAGING FAN)
2) FAN MUST BE DISENGAGED
3) ACT ON THE PILOT SCREW TO GET AN ENGINE SPEED OF 910 ± 40 RPM

- VALVE ADJUSTMENT
ENGINE COLD (6 HOURS REST MINIMUM)

INTAKE $0.20^{+0.05}_{+0}$ mm EXHAUST $0.30^{+0.05}_{+0}$ mm

ADJUSTMENT PROCEDURES SEE WORKSHOP MANUAL
"THIS VEHICLE CONFORMS TO US E.P.A. AND CALIFORNIA ARB REGULATIONS APPLICABLE TO 1985 MODEL YEAR NEW MOTOR VEHICLES".

01.84

Typical Vehicle Emission Control Information label

A Common manifold
B Distributor
5. Vacuum switch
6. Delay valve
11. Air injection shut-off valve
13. Pulsair electrovalve
14. Fuel injection cut-out electrovalve
17. Vacuum advance electrovalve

Model 505 gasoline engine emission component locations

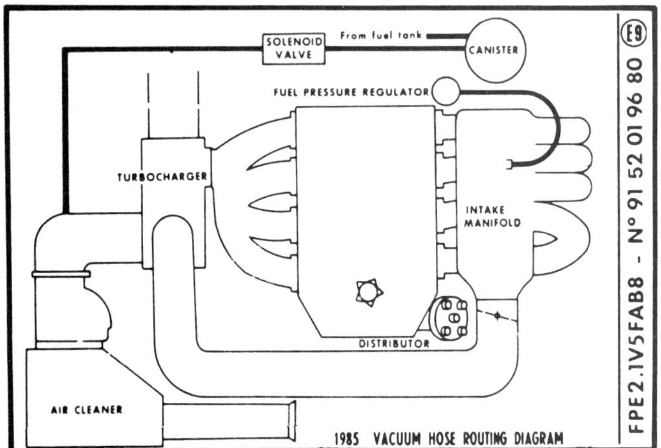

Typical 1985 turbocharged gasoline engine vacuum hose routing

The Peugeot emission control components are not designed to be maintained at regular intervals; they are to be replaced when diagnosed as faulty. However, the oxygen (Lambda) sensor must be replaced every 30,000 miles and the idle checked with a CO meter. Carefully check the condition of all vacuum hoses on both gasoline and diesel engines regularly and replace if necessary.

IDLE SPEED AND MIXTURE ADJUSTMENTS

NOTE: Manual transmission in Neutral, automatic in Drive, both with parking brake ON and wheels chocked.

505 GASOLINE MODELS SPECIFICATIONS

Man./Auto. Trans., 50 States
- Idle—800–850 rpm.
- Fast idle—1500 rpm.
- Idle CO—0.8%.

Fuel supply pressure.
- Test—4.5–5.2 bars.
- Adjustment—4.7–4.0 bars.

NOTE: The following special tools or their equivalents are necessary for the completion of this procedure: Dwell/Tachometer; HC/CO analyzer; richness adjusting tool compatible with K-Jetronic CIS injection systems; CO sampling tube (which connects the CO analyzer to the catalytic converter); and an idle speed adjusting screwdriver (Peugeot special tool No. 8.0140 BA).

1. Run the engine up to "hot" operating temperature, to the point where the electric fan comes on. Make sure the throttle flap initial opening is correctly adjusted. Turn off the air conditioning and all electrical accessories.

2. Connect the dwell/tachometer according to the tach manufacturer's instructions to test wire No. 153 on the driver's side of the engine compartment. Using the idle speed adjusting screwdriver, adjust the air bleed screw on the throttle body so that the idle is stable at 800–850 rpm.

3. Read the scale of the dwell/tachometer. The value should be 58.5 degrees at idle. If the reading is incorrect, adjust the CO.

4. Adjust the CO by first removing the tamper-proof plug which covers the mixture adjustment screw in the fuel distributor. Connect the CO sampling tube to the front tap of the catalytic converter, and connect the HC/CO analyzer.

5. Check the idle speed, making sure it is set at 800–850 rpm. Adjust the mixture using the richness adjusting tool and turning the mixture screw in the fuel distributor. The CO % should be 0.8%.

NOTE: Before each reading, remove the adjusting tool.

6. Recheck the idle speed and fine-tune if necessary. Install a new tamper-proof plug in the fuel distributor. Remove the CO sampling tube from the converter and reinstall the plug. Disconnect all test equipment.

505 DIESEL (NON-TURBO) MODELS

SPECIFICATIONS

Man./Auto. Trans., 50 States.
- Idle without A/C—780–830 rpm.
- Idle with A/C—830–860 rpm.

505 TURBODIESEL MODELS

SPECIFICATIONS

Man./Auto. Trans., 50 States.
- Idle—830–860 rpm.

Adjustment

The idle on the diesel and turbodiesels is adjusted while the engine is hot, after the electric cooling fan has come on.

1. Check that the fast idle stop on the front of the injection pump is not in contact with the throttle lever and that the accelerator cable is released.

2. Connect a tachometer which is compatible with diesel engines. Adjust the set screw on the injection pump to obtain the correct idle speed.

INITIAL TIMING

505 GASOLINE MODELS

NOTE: The timing plate should first be checked before the timing is adjusted.

1. Set the piston of No. 1 cylinder to TDC on its compression stroke. Adjust the timing plate so that the 0 degree reference notch is aligned with the notch in the crankshaft pulley. The engine can now be accurately timed.

2. Disconnect the distributor vacuum advance hose and plug it.

3. Connect a timing light according to the light manufacturer's instructions. Check the ignition timing with the engine idling at 800–850 rpm. The 8 degree reference notch on the timing plate should be aligned with the notch in the pulley. Reconnect the vacuum hose.

DIESEL INJECTION TIMING

NOTE: An injection pump dial indicator designed for use with the Bosch injection pump, a dial indicator support (Peugeot special tool No. 8.0117F or equivalent), and a feeler extension for the dial indicator are required for this procedure.

1. Remove the plug located in the center of the four fuel line outlets on the injection pump. Install the dial indicator support, attach the proper feeler extension to the dial indicator, and install the dial indicator onto the injection pump. Zero the needle.

2. Loosen the injection pump mounting bolts and turn the injection pump as necessary (by means of the slotted bolt holes) to obtain the following lift reading on the dial indicator.

NOTE: On models equipped with a gear driven injection pump, if there is not enough spacing in the pump mounting bolt hole slots to attain the proper pump lift reading, the injection pump drive gear teeth are incorrectly meshed by at least one tooth. If this situation occurs, the pump must be removed, retimed, and reinstalled (in that order).

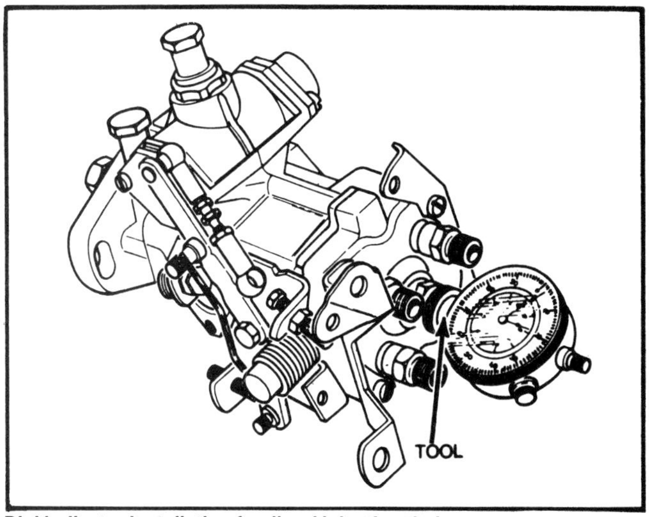

Dial indicator installation for diesel injection timing

3. The setting should be rechecked at this point. Make sure that the No. 1 cylinder piston is still at TDC on the compression stroke.

4. Rotate the crankshaft counterclockwise 7 dial indicator revolutions. As the indicator needle nears the end of the 7th revolution, check that the pump-mounted indicator has stabilized in the BDC range. Check the indicator reading. Slowly rotate the crankshaft clockwise and check that the piston position and the pump lift correspond to the setting for the particular model listed in the chart above.

5. Tighten the injection pump mounting bolts. Remove the dial indicator and dial indicator support. Install the plug in the injection pump.

SPARK PLUGS

Year	Model	Engine	AC	Autolite	Bosch	Champion
'85–'86	505	XN6, N9T	R43XLS	64	WR7DS	RN9GY

EMISSION CONTROL SYSTEMS

CRANKCASE EMISSION CONTROL SYSTEM

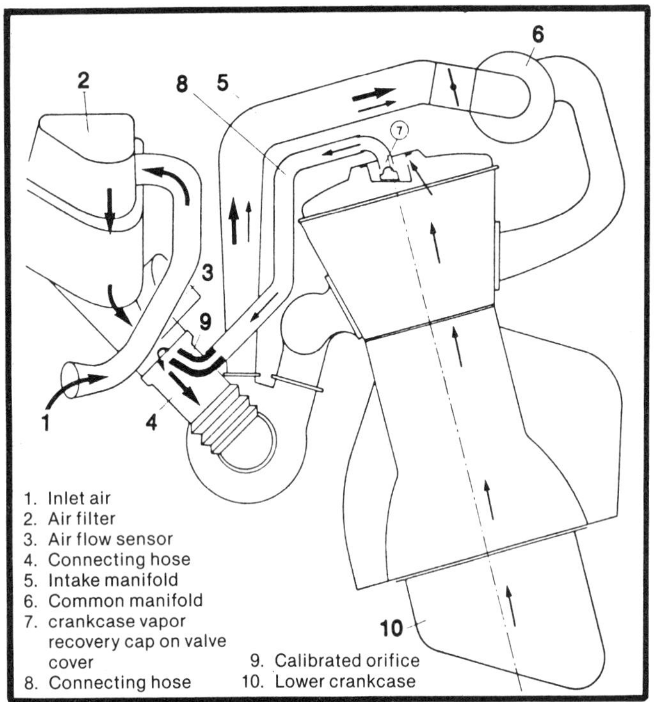

Crankcase emission control system

Closed Positive Crankcase Ventilation

The crankcase ventilation system prevents crankcase gases from being vented into the atmosphere. The system also limits pressure in the crankcase. Excess pressure could cause leaks at the bearings and seals.

Two circuits are used to control the emissions. A primary circuit (connecting the PCV line to the intake manifold) contains a metal screen which separates the oil mist from the blowby gases and a "sonic jet" which senses the pressure differential and controls the volume of crankcase gases, thus avoiding an unbalanced idle mixture.

The secondary circuit (PCV line to air cleaner) prevents entry of air into the crankcase and oil from entering the carburetor air intake circuit at higher engine speeds. A calibrated tube limits crankcase vacuum when throttles are wide open.

When the primary circuit (closed throttle), and the secondary circuit (open throttle) are in the midrange, both circuits operate at the same time.

PCV SYSTEM TROUBLE DIAGNOSIS CHART

Condition	Possible Cause	Correction
Slow, unstable idle, frequent stalling.	1. Valve completely plugged or stuck.	1. Replace valve
	2. Restricted filter.	1. Replace filter, clean system.
Oil in air cleaner	1. PCV system plugged.	1. Replace valve.
	2. Leak in closed ventilation system.	1. Clean system as required.
		2. Inspect for leaks to atmosphere and correct as necessary.

FUEL EVAPORATIVE EMISSION CONTROLS

In order to control the evaporation of fuel into the atmosphere from the fuel tank and the carburetor when the engine is shut off, both are vented into a charcoal canister. A calibrated valve in the fuel tank line meters the flow of vapors into the charcoal canister to an amount that the canister can handle. The station wagons have an additional vapor-liquid separator located near the fuel tank. When the engine is started, the system pulls the vapors out of the charcoal canister and into the intake manifold, thereby purging the canister of fuel vapors.

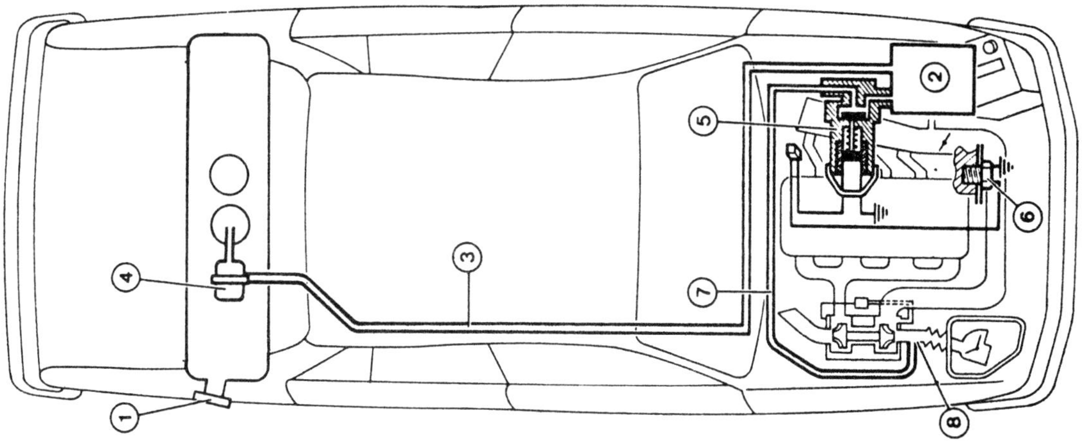

1. Non-vented threaded fuel cap
2. Canister
3. hose circuit
4. One-way valve
5. Electrovalve
6. Thermocontact (60° C.)
7. Hose
8. Metered jet

Fuel evaporative emission control system

EVAPORATIVE EMISSIONS TROUBLE DIAGNOSIS CHART

Possible Cause	Correction
EVAPORATIVE SYSTEM	
1. Liquid fuel leaking from fuel lines, fuel pump, or carburetor.	1. Tighten connections or repair or replace as necessary.
2. Canister cracked or damaged.	1. Replace canister
3. Inoperative bowl vent valve (Refer to Bowl Vent Valve Test Procedure)	1. Repair or replace hoses. 2. Replace canister.
4. Inoperative purge valve (Refer to Canister Purge Valve Test Procedure)	1. Repair or replace hoses. 2. Replace canister.
5. Disconnected, misrouted, kinked, deteriorated, or damaged vapor lines.	1. Check for proper connections, routings, and condition. Repair or replace as necessary.
6. Bowl vent hose misrouted.	1. Route hose properly, with no low spots.
7. Air cleaner or air cleaner gasket improperly installed.	1. Install air cleaner properly, replacing gasket if necessary.
POOR IDLE OR POOR DRIVEABILITY	
1. Inoperative purge valve (Refer to Canister Purge Valve Test Procedure).	1. Repair or replace hoses. 2. Replace canister.
2. Inoperative bowl vent valve (Refer to Bowl Vent Valve Test Procedure).	1. Repair or replace hoses. 2. Replace canister.
COLLAPSED OR EXPANDED FUEL TANK	
1. Plugged or pinched vapor pipe, hoses, or defective fuel cap.	1. Inspect all lines from tank to canister. Repair or replace as necessary. 2. Replace fuel cap.
2. Canister filter plugged.	1. Replace filter.

NOTE: Under hood area—Perform Evaporative System Pressure Test to determine possible causes.

EXHAUST EMISSION CONTROL SYSTEM

Air Injection System

The air injection system adds oxygen into the exhaust manifold, allowing unburned hydrocarbons (HC) in the hot exhaust gases to be burned more completely. The system consists of a belt driven rotary vane air pump, a diverter valve, an electro-valve, an intake manifold injection valve and non-return valves.

AIR PUMP

Some precautions should be taken when working on the air pump.

1. Do not operate the engine with the pump removed or drive belt disconnected.
2. Do not lubricate the pump.
3. If the filter is contaminated, replace it. Do not attempt to clean it.
4. When adjusting the drive belt, never pry on the pump body.
5. Never clamp the pump body in a vise.
6. The internal parts of the air pump are not serviceable. Replace the pump as an assembly.
7. The following external parts may be replaced on the pump: drive pulley, filter and pressure relief valve.
8. Do not hand rotate the pulley on a new pump and never rotate the pulley in the direction opposite to the normal direction.

EGR Valve

In order to reduce the nitrogen oxide (NOx) in the exhaust, a controlled amount of exhaust gas is recirculated into the intake manifold. The main component of the system is the valve, which opens between the exhaust and intake manifolds. The valve opening is controlled by vacuum, according to engine conditions, by a vacuum control electro-valve and a vacuum amplifier. California engines have an additional vacuum switch.

ELECTRO-VALVE

During engine start up and warm up, the electro-valve stops the flow of vacuum to the EGR valve, to prevent the EGR valve from opening when the choke is operating.

VACUUM AMPLIFIER

This valve controls the amount of vacuum to the EGR valve according to engine conditions.

VACUUM SWITCH

This switch reacts to close the EGR valve during high intake manifold vacuum situations. The valve is only used on California vehicles.

Air Injection System

505 GASOLINE

The air injection system fitted to all 1985–86 Peugeot gasoline engines is basically similar to the earlier system. It injects oxygen into the exhaust system, creating additional combustion and allowing hydrocarbons (HC) to be burned more completely. The major addition to the air injection system is Pulsair, a series of four one-way valves which deliver fresh air from the air filter to each of the four exhaust ports, but do not allow exhaust gases to return to the intake. Because the Pulsair system operates from the cyclical engine vacuum at the exhaust ports, the air pump is eliminated from the system.

TROUBLESHOOTING

The operation of the Pulsair system can be checked in the following manner. If a Pulsair valve is diagnosed as faulty, the valve must be replaced; it cannot be repaired.

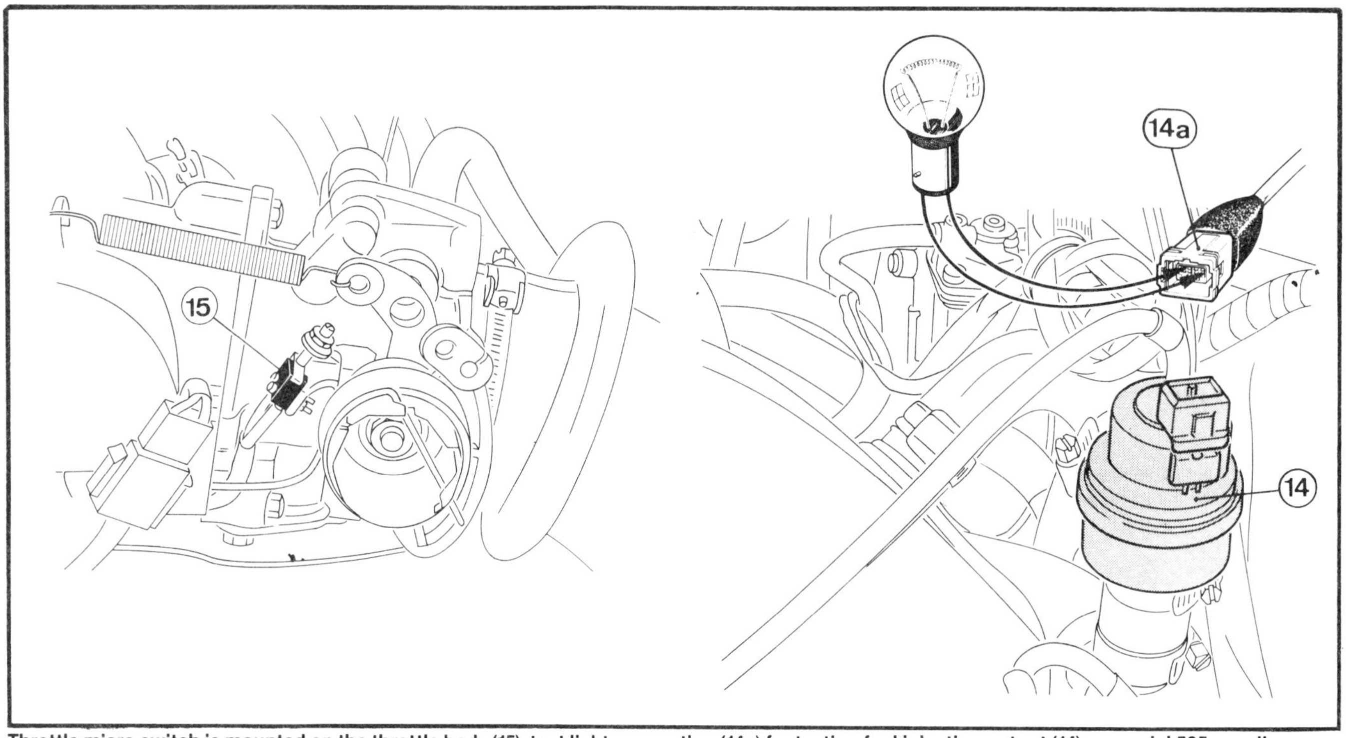

Throttle micro-switch is mounted on the throttle body (15), test light connection (14a) for testing fuel injection cutout (14) on model 505 gasoline engines

1. Start the engine. Disconnect wire No. 157 from the 40°C thermocontact ("12" in the accompanying illustration) near the passenger side firewall. Put this wire in contact with wire No. M157, which is the other wire on the thermocontact.

2. The operation of the Pulsair valves is now detectable, the valves emitting a slight "hum". Movement of the air injection shut-off valve diaphragm can also be seen through the upper section of the valve. Reconnect wire No. 157 if operation is satisfactory.

3. If operation is not satisfactory, check the operation of the electrovalve ("13" in the accompanying illustration) and its electrical current supply. Also check the vacuum hose connections to the Pulsair valves. If both of these checks are positive, replace the Pulsair valve.

NOTE: The arrow which is on the body of the air injection shutoff valve should point to the Pulsair valves. The arrow should not be visible when facing the engine.

Fuel Injection Cut-Out

505 Gasoline

A fuel injection cut-out is employed on 1985–86 models to control emissions during engine deceleration. The injection is cut-out by equalizing the pressures on both sides of the air flow sensor. This pressure equalization is accomplished with a Bosch electrovalve ("14" in the accompanying illustration; located next to the driver's side MacPherson strut tower), which acts on an auxiliary circuit to the mixture regulator. When the pressures on both sides of the air flow sensor plate are equal, the fuel metering distributor piston is no longer activated by the air flow sensor plate lever, thus interrupting fuel output to the injectors.

TROUBLESHOOTING

1. Disconnect one of the wires on the 40°C thermocontact if the engine is cold (coolant temperature less than 40°C).

2. Start the engine and run it up to 1,800 rpm. Release the throttle. You should hear the electrovalve operate.

3. If no sound is heard, check the current supply to the electrovalve as follows.

4. Disconnect the connector which plugs into the top of the valve. Connect a test light to the terminals of the connector. Run the engine up to 1,800 rpm and release the throttle. If the test light comes on then goes off, check the vacuum hose routing and condition. If the connections and hoses are good, replace the electrovalve.

5. If the test light does not come on, check the electrical connections to the electrovalve connector. Check the condition and adjustment of the throttle micro-switch located on the side of the throttle body. Check the operation of the twin-speed relay ("24" in the accompanying illustration) on the relay panel. If these are OK, replace the tachymetric relay (the big relay, "16" in the accompanying illustration, on the upper part of the relay panel).

Catalytic Converter

The catalytic converter reduces hydrocarbons (HC) and carbon monoxide (C) by converting them to carbon dioxide and water.

EXHAUST OVERHEATING WARNING SYSTEM

An exhaust overheat warning light is located on the dash to warn the operator that an abnormal overheat problem exists under the vehicle, at the catalytic converter. When the warning light comes on, allow the engine to cool down, then resume journey at reduced speed and have the vehicle inspected as soon as possible.

LAMBDA® SENSOR

The 505 fuel injected models use the Lambda® oxygen sensor system to check the exhaust gases for unburned hydrocarbons. The sensor sends signals from the exhaust pipe to the control unit. Under certain con-

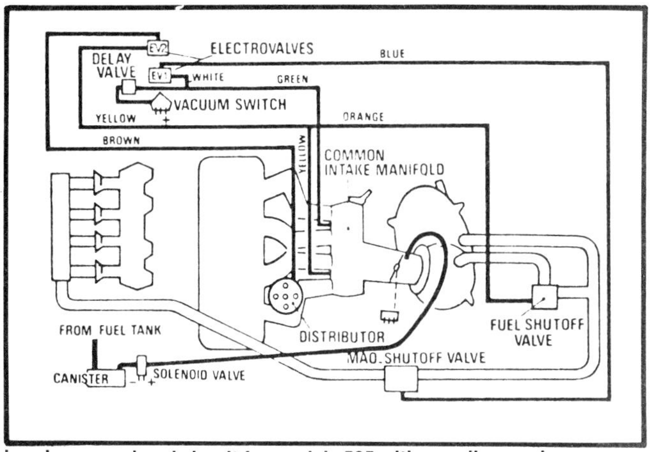

Load sensor signal circuit for models 505 with gasoline engines

ditions, the signals will cause the control unit to alter the fuel ratio to bring it into specifications. If the engine has not reached operating temperature, the signals will be overridden and the enrichment system will operate.

Diesel EGR

This system is used on the XD2C and XD2S engines to recirculate a small percentage of exhaust gas into the engine's intake to reduce peak combustion temperatures and therefore nitrogen oxide emissions. Because of the relative absence of intake manifold vacuum in the normally unthrottled diesel, the system is more complex than that which is used on most gasoline engines.

XD2C ENGINES

The EGR valve itself contains two vacuum diaphragms and, therefore, has two stages of operation. The system incorporates a vacuum operated air throttle which operates in the first stage of operation to ensure enough intake vacuum exists so that exhaust will be drawn effectively into the intake manifold. Electrically operated vacuum valves coordinate the staged operation of the EGR valve and the medium-throttle operation of the air throttle with the travel of the injection pump throttle lever and the engine rpm, as measured by a flywheel mounted speed sensor.

XD2S ENGINES

This system is generally similar to the one on the XD2C. Instead of a staged EGR valve and associated solenoid operated vacuum valves, it uses an injection pump mounted vacuum converter, a regulator that uses vacuum produced by the engine's accessory vacuum pump. The converter converts the injection pump throttle rod's position to a vacuum signal which then operates the EGR valve according to engine load. An electric valve also cuts EGR off when the engine is cold, is below 1,500 rpm, or is above 3,100 rpm.

Diesel engines are not as sensitive as gasoline engines to a small excess of EGR, that is, they usually don't misfire if the EGR system is providing a little too much exhaust. But, if the system is providing full EGR under the wrong conditions the engine will show black smoke in the exhaust and will lack full power.

Checking Out the XD2C EGR

1. A quick check for incorrect control of the EGR valve can be made by disconnecting the connector (C3) for the speed sensor. If the engine now runs without black smoke and with full power, the EGR system is at fault. You should make the next test to at least ensure that the EGR valve stem moves before suspecting a problem with the engine's injection system or basic mechanical parts.

2. Next, with the engine warm, watch the EGR valve stem and the throttle flap stem. Accelerate the engine using the link-rod going to the

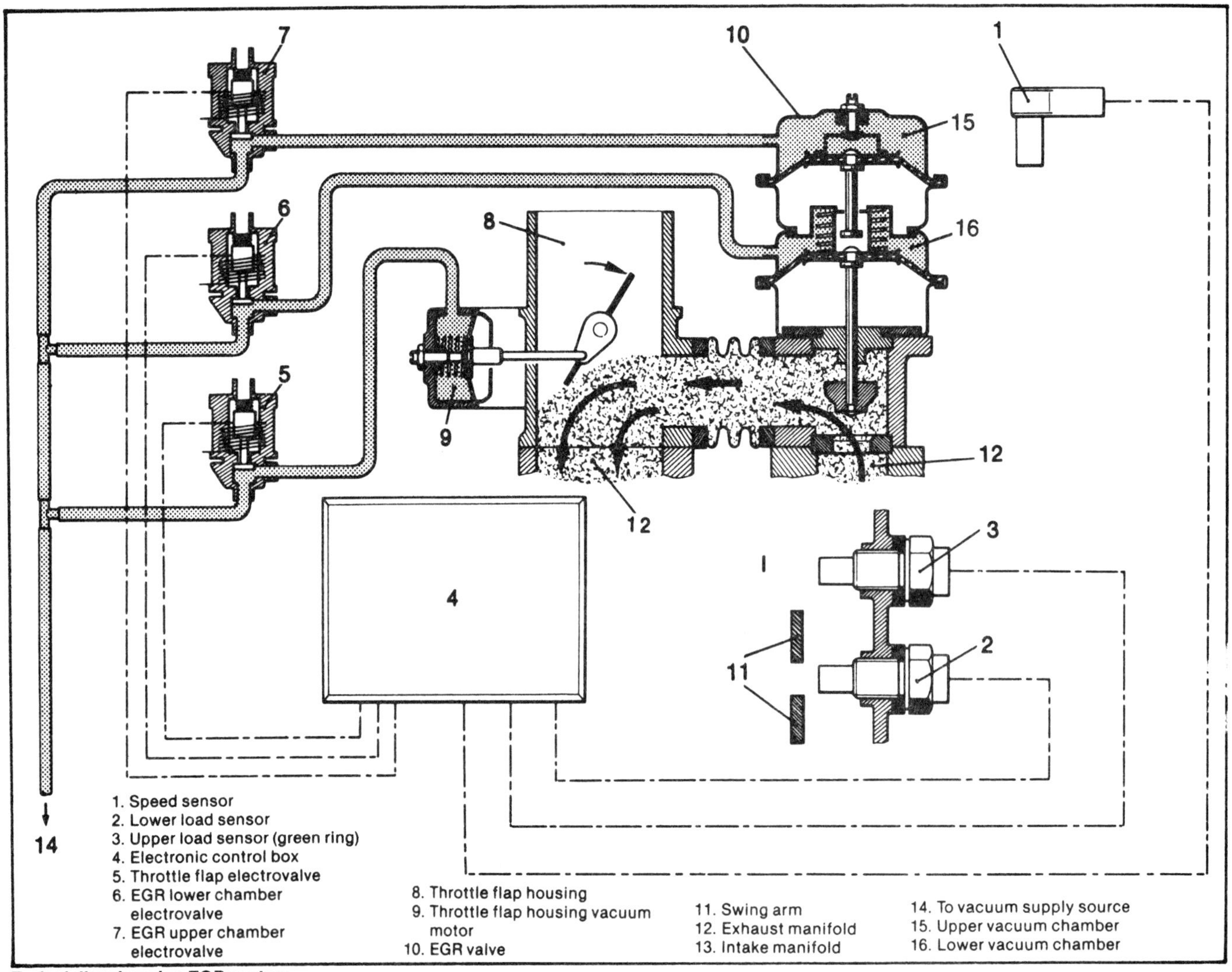

Typical diesel engine EGR system

injection pump. At about 1,300-1,500 rpm, the throttle flap stem should open and the EGR valve should open all the way. The vacuum tubes to both EGR chambers should have vacuum at this point; so should the vacuum line to the throttle flap diaphragm. As you accelerate further, the throttle flap will first close, then the EGR valve lower chamber will lose its vacuum and, finally at near maximum engine speeds, the EGR valve will lose all actuating vacuum and will close.

3. If there is any doubt about the response of the EGR valve, you should remove it, labeling vacuum hoses first. Clean it with a safe solvent, and reinstall it with a new gasket. Make sure the stem moves freely. If the valve still does not respond and you're getting proper vacuum supply, it probably has one or two perforated diaphragms and should be replaced.

4. If you're not getting the vacuum supply in the proper sequence, check the speed sensor. It's located on the top of the flywheel housing. Check to see if the sensor is tightly mounted, and if it is, proceed to the next test. If not, adjust it by first pushing it against the flywheel, then pulling it back about 0.020 in. Tighten the mounting clamp. Check the sensor's resistance with an ohmmeter. Pull the connector shown in the illustration and measure the resistance as shown. It must be 40–60 ohms, or the sensor requires replacement.

5. You can also check the resistance of any or all of the solenoid operated vacuum valves, located under the EGR computer, near the air cleaner. Simply disconnect both wiring connectors (noting how they connect) and measure resistance with an ohmmeter. It should be 40 ohms. Otherwise, replace the valve.

Checking the XD2S EGR

1. Start with the engine cold (well under 120 degrees F.). Unplug the vacuum line to the EGR valve. Start the engine and accelerate it up to about 2,000 rpm with the injection pump rod while checking for vacuum in the vacuum line. You can plug a vacuum gauge into the line to test for vacuum, if you have one. There should be no vacuum while the engine is cold.

2. Run the engine until it is warm. As the engine passes 120 degrees F. (well below normal operating temperature), you should begin to get vacuum at the EGR vacuum line when the engine is accelerated. With the line plugged in, the EGR valve should open as the engine passes 1,500 rpm with manual transmission, and 1,300 rpm with automatics. If the EGR valve does not open with the line connected, disconnect the line and measure for vacuum with a gauge or check with your finger. If there is vacuum even when the engine is cold, check the wiring to the thermosensor. If the wiring is OK, replace the thermosensor. If you have found that vacuum is getting to the EGR valve (under the proper conditions), and the valve stem still did not move, remove the valve and clean it with a safe solvent. If you do not find that the valve responds to cleaning, test for adequate vacuum to the valve by measuring the vacuum as the engine is accelerated warm. The valve must respond to a little over 4 in.Hg. of vacuum, or it must be replaced.

3. Disconnect the ground wire from the vacuum solenoid valve (wire No. 146). Connect a jumper wire from the ground wire connector

on the valve to a good ground. Then, disconnect the link rod from the converter lever, removing just a little of the shrink tubing, if necessary (leave as much of the tubing on as possible). With the engine idling, move the lever on the vacuum converter steadily in the direction of acceleration and carefully observe the movement of the EGR valve stem. If the EGR valve moves smoothly, the vacuum converter is ok. If it moves in spurts, and you're sure it is clean and the stem is free to move, the vacuum converter is probably faulty. It can only be tested with specially designed equipment.

4. Reconnect the link rod and wiring. Accelerate the engine and watch the response of the EGR valve. The valve should begin opening at a moderate rpm (1,500 with manual transmission, 1,300 with automatic) and then close again past 3,100 rpm. If the valve does not respond this way there is a problem in the speed sensor, electronic control box, or the wiring. If the valve never responds, and it worked in doing the above test, check the electrical connection between the solenoid valve and the control box, and between the control box and the thermosensor. You can test the thermosensor for failing off by grounding the wire leading to it and repeating the test in this step. If the valve now works, replace the thermosensor. Check the speed sensor as described in Step 6.

If that is OK and the wiring is OK but there is still no vacuum to the EGR valve (the solenoid valve does not open), the electronic control box or its wiring is defective.

If the solenoid valve comes on as the engine is accelerated, but never goes off above 3,100 rpm, or if its response is inconsistent (coming on and going off at varying rpm) first check associated wiring and then replace the electronic control box.

5. If the solenoid valve does not respond, you can check its resistance with an ohmmeter. Disconnect both wires (noting the two connectors. Resistance is normally about 40 ohms. If it's much higher than this or infinite, replace the valve.

6. Check the speed sensor as in Step 4 of the EGR troubleshooting procedure for the XD2C engine above. They are identical.

Electronic Ignition

ECU EZ 200 K IGNITION SYSTEM

Principle of the Autodiagnosis

1. The EX 200 K ignition E.C.U. (171) incorporates an "Autodiagnostic" feature which enables it to identify some of the eventual faults which may occur in the system during operation. By the same token, it is also able to alert the driver of the existence of this fault. This is done through the indicator (173) in the lower left corner of the instrument cluster (orange LED).

2. The number of blinks of this indicator will determine which component within the system presents the fault.

Detection range of the ignition E.C.U. EZ 200 K

It controls and checks the following:

1. Continued detonation.
2. Battery voltage.
3. Electronic circuit which corrects detonation (internal to the E.C.U.
4. The detonation sensor.
5. The potentiometer and its electrical connections (194).
6. The engine load input signal emanating from the injection E.C.U.

Operation of LED System

1. Operation of the LED is as follows:
   a. + Ignition key "ON" — LED ON. b.
   Engine started — LED OFF — normal operation.

2. If the LED remains ON or OFF: check its power supply and its ground.
   a. Occasional sporadic "flashes" of the LED — NORMAL OPERATION.
   b. Constant blinking of the LED — FAULT IN THE IGNITION SYSTEM.

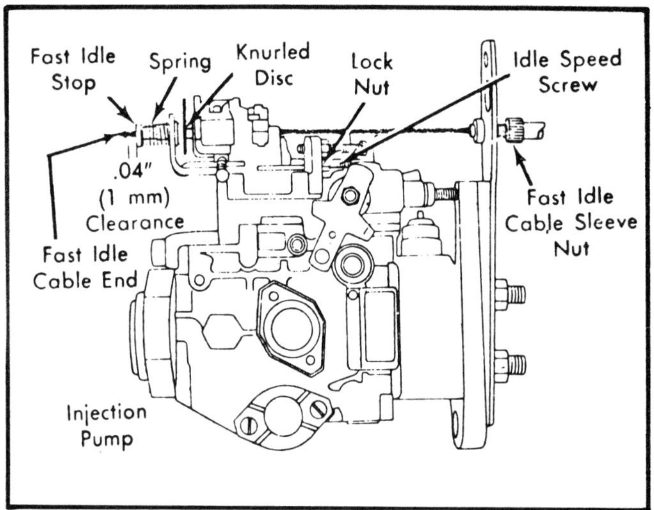

Bosch Diesel engine injection pump adjustment locations

3. As a safety measure, as soon as a fault in the system is detected, the system will automatically post a retard to the spark advance (-20° to -22°).

4. This retard will be reflected by a lack of power and performance of the engine.

5. The repair technician will be able to locate the failed component, based on the LED's number of blinks per cycle and the corresponding table.

NOTE: The speed of the display of the blink cycles will relate directly to engine rpm. It is therefore IMPERATIVE to make the determination of the diagnostic cycle at idle by counting the number of blinks.

CHECKING THE DETONATION SENSOR

Installation Precautions

1. Mounting to the block without any washers.
2. Use normal holding LOCTITE.
3. Tightening torque for the detonation sensor (172A): 1.3 mdaN ± 0.2 (9.43 ft.lbs. ± 1.45).
4. Positioning: connector facing downward.

Electrical checking

1. Detonation sensor connected.
2. Hook up an ohmmeter between pin (12) and pin (13) of the ignition E.C.U. (171) connector.
3. If this value is not reached after checking the continuity and insulation of wires 8 and M8, replace the detonation.

DETONATION AND IGNITION SYSTEM FAULT LED INDICATOR

Installation Precautions

1. When reinstalling the LED holder, make sure the LED is through its passage in the front part of the cluster before locking the holder (the connecting leads of the diode could get broken if it is forced while butting on inside of front face).

2. The position of the LED support must be respected for appropriate polarity. Red tag on the printed circuit is to line up with the red tag (2) on the LED holder.

NOTE: Should the diode be removed from its support, when reinstalling it, the long electrode should face the reference tag.

<table>
<tr><th colspan="2">Fault*</th><th colspan="2">Spark advance correction</th><th>Causes</th><th>Checks</th></tr>
<tr><td colspan="2">1</td><td colspan="2">−22°</td><td>Continued detonation
Maximum correction reached</td><td>Cooling system
Distributor timing (23)
Boost pressure</td></tr>
<tr><td colspan="2">2</td><td colspan="2">−20°</td><td>Battery voltage is under 10.5 volts.</td><td>Battery (12)
Charging circuit</td></tr>
<tr><td colspan="2">3</td><td colspan="2" rowspan="4">−20° non-revers-ible until ignition switch is turned off</td><td>Detonation correction circuit in the ignition E.C.U. defective.</td><td>Try with a new ignition E.C.U. (171).</td></tr>
<tr><td colspan="2">4</td><td>Erroneous signal received from detonation sensor (Check for engine speed above 3 200 rpm).</td><td>Detonation sensor (172A) (see corresponding chapter).
Electrical circuit continuity.</td></tr>
<tr><td colspan="2">5</td><td>Signal from potentiometer is greater than 4.3 volts.</td><td>Try with a new potentiometer (194)</td></tr>
<tr><td rowspan="2">6</td><td>A</td><td>Connections and wiring between ignition E.C.U. and potentiometer defective.</td><td>Continuity of the potentiometer's (194) electrical circuit.</td></tr>
<tr><td>B</td><td>Mirco-switch
Open
Closed</td><td>Advance
Full Load
!dle</td><td>NO LOAD SIGNAL EMANATING FROM THE INJECTION E.C.U.</td><td>Electrical circuit continuity between pin 8 of the ignition E.C.U. (171) and pin 6 of the injection E.C.U. (181) (wire 6).
Check with a new ignition E.C.U. (171).
Check with a new injection E.C.U. (181).</td></tr>
</table>

*The number of the FAULT represents the number of blinks per cycle of the LED (173) on the instrument cluster.

NOTE - *For the "irreversible" faults (3, 4, 5, 6A), it is necessary to turn ignition key "OFF" to "erase" the fault entered in memory.*

Diagnostic code list

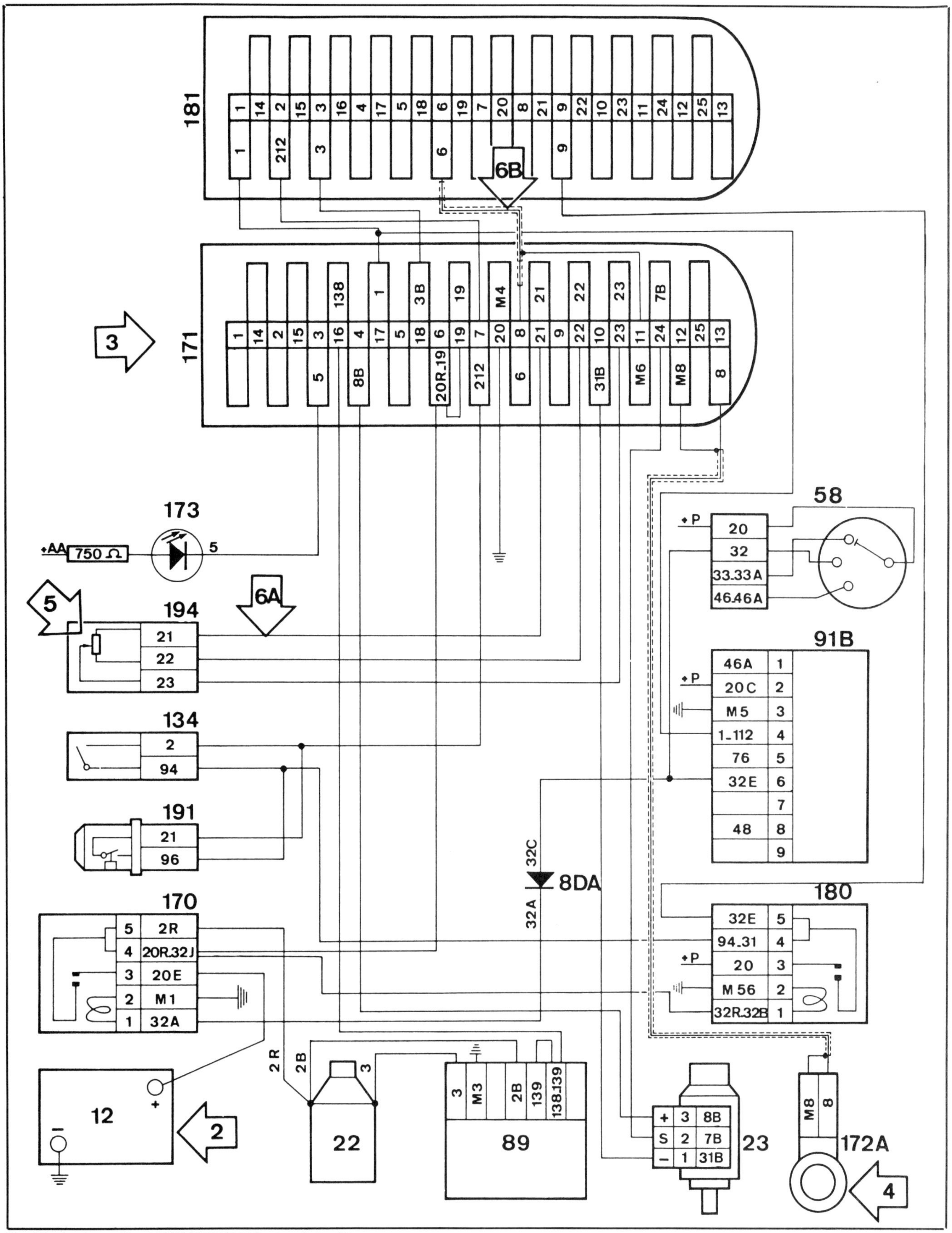

Electrical schematic of ECU EZ 200 K Electronic ignition

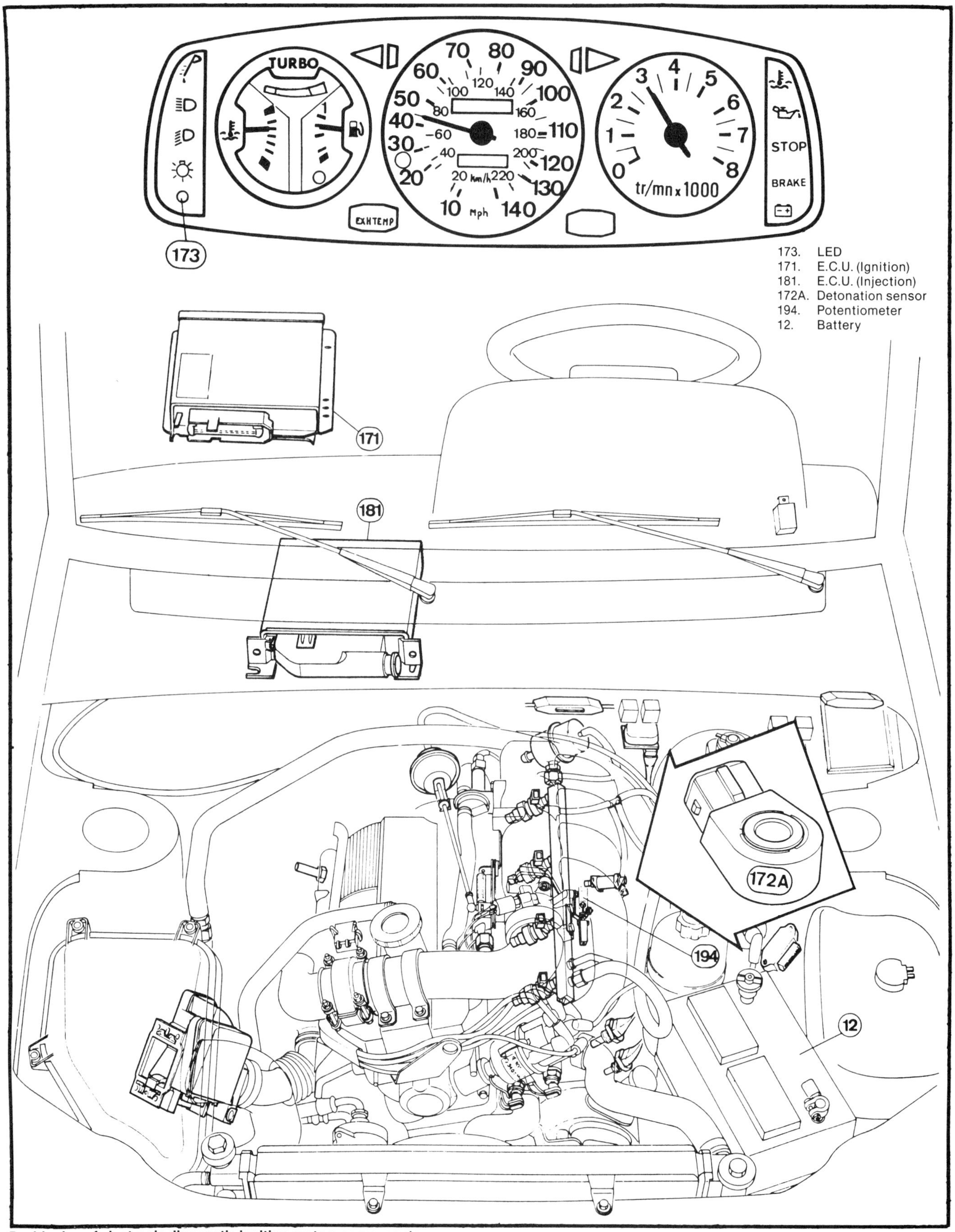

Positioning of electronic diagnostic ignition system components

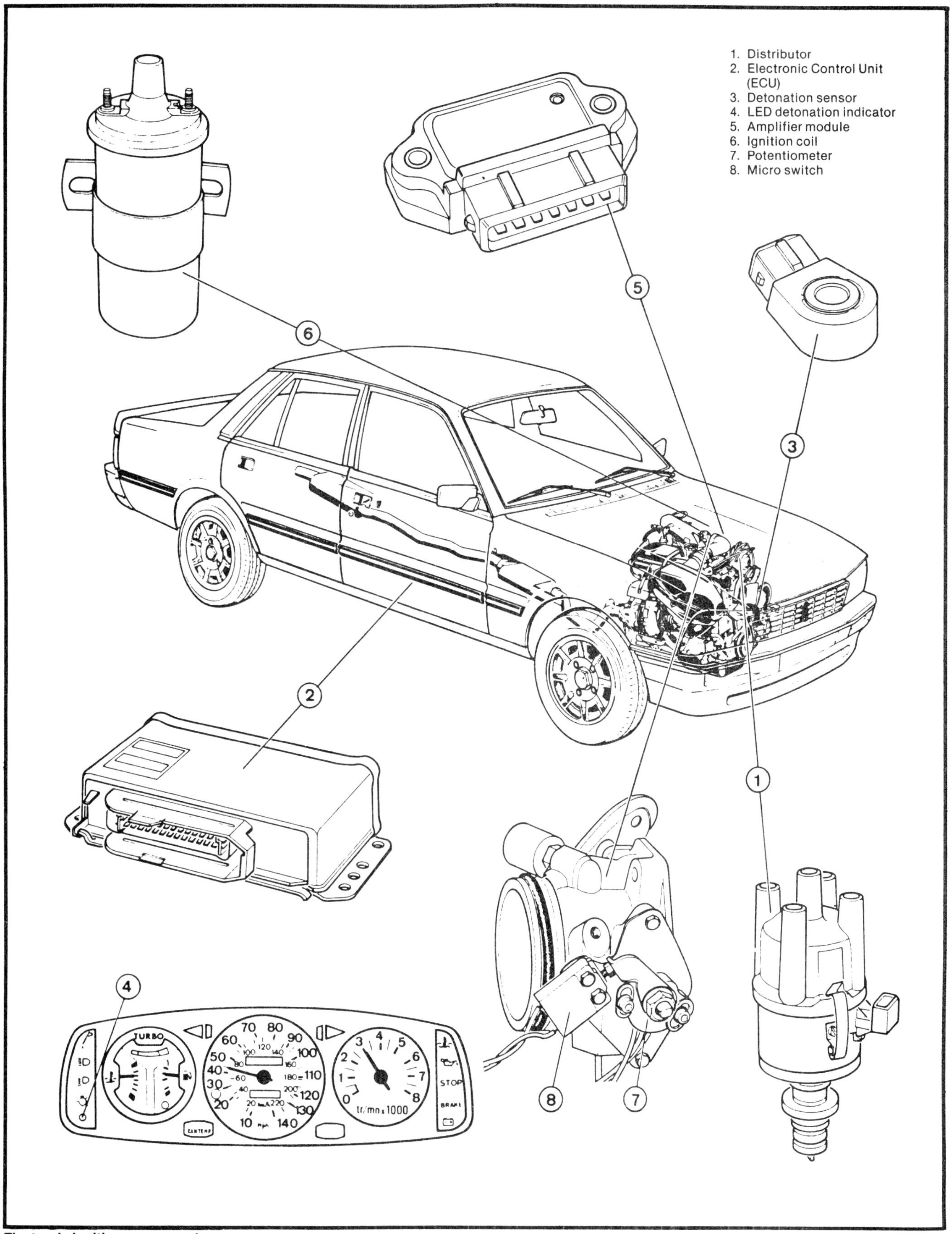

Electronic ignition components

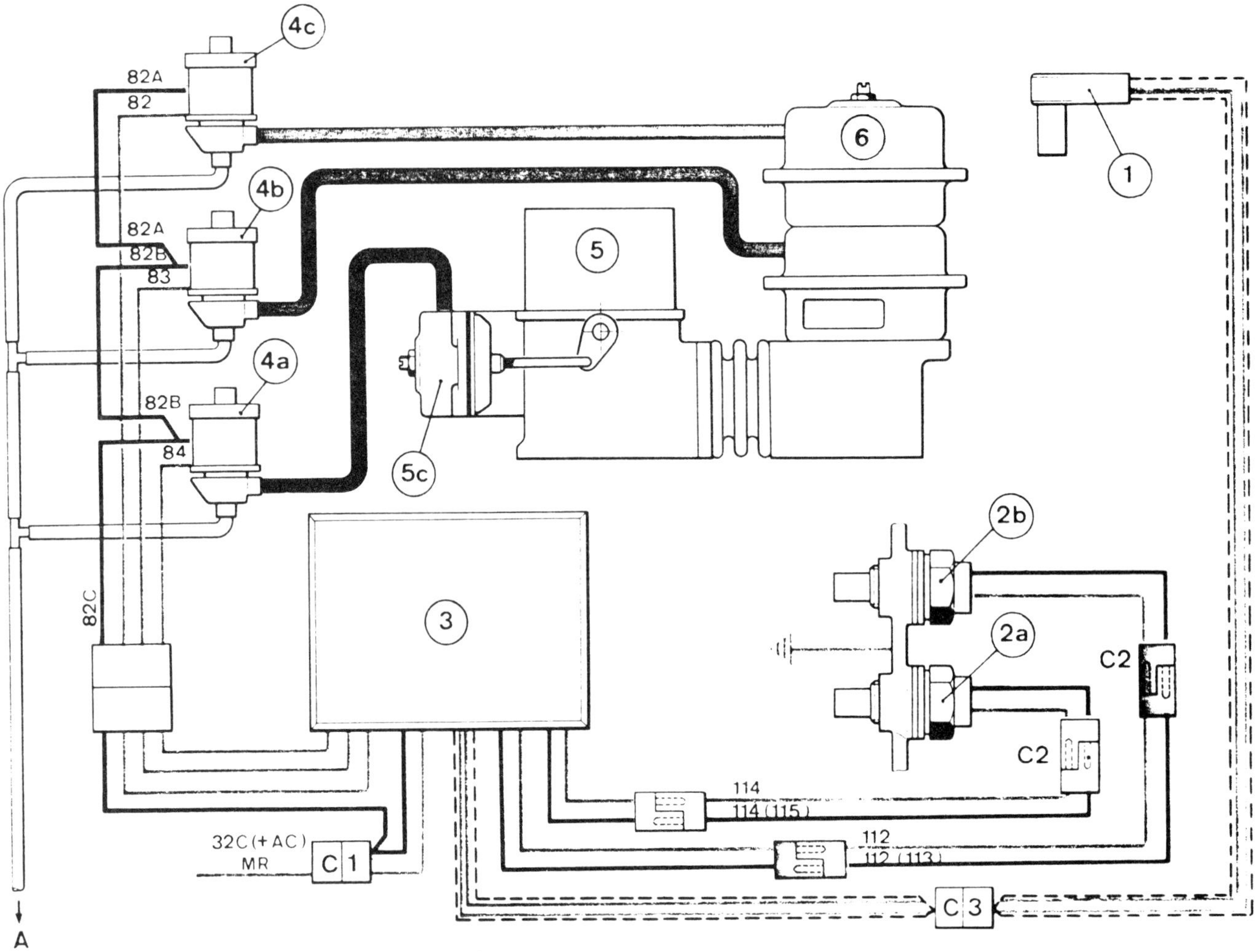

XD2-C ENGINE
EGR SYSTEM

Component legend

- **1** : speed sensor
- **2a - 2b** : load sensors
- **3** : electronic control box
- **4a - 4b - 4c** : electrovalves
- **5** : throttle flap housing
- **5c** : vacuum motor
- **6** : EGR valve
- **A** : to vacuum source

Electrical circuit legend

- Current supply (+)
- Ground circuit (–)
- Load sensor signal circuit

VACUUM AND ELECTRICAL DIAGRAM

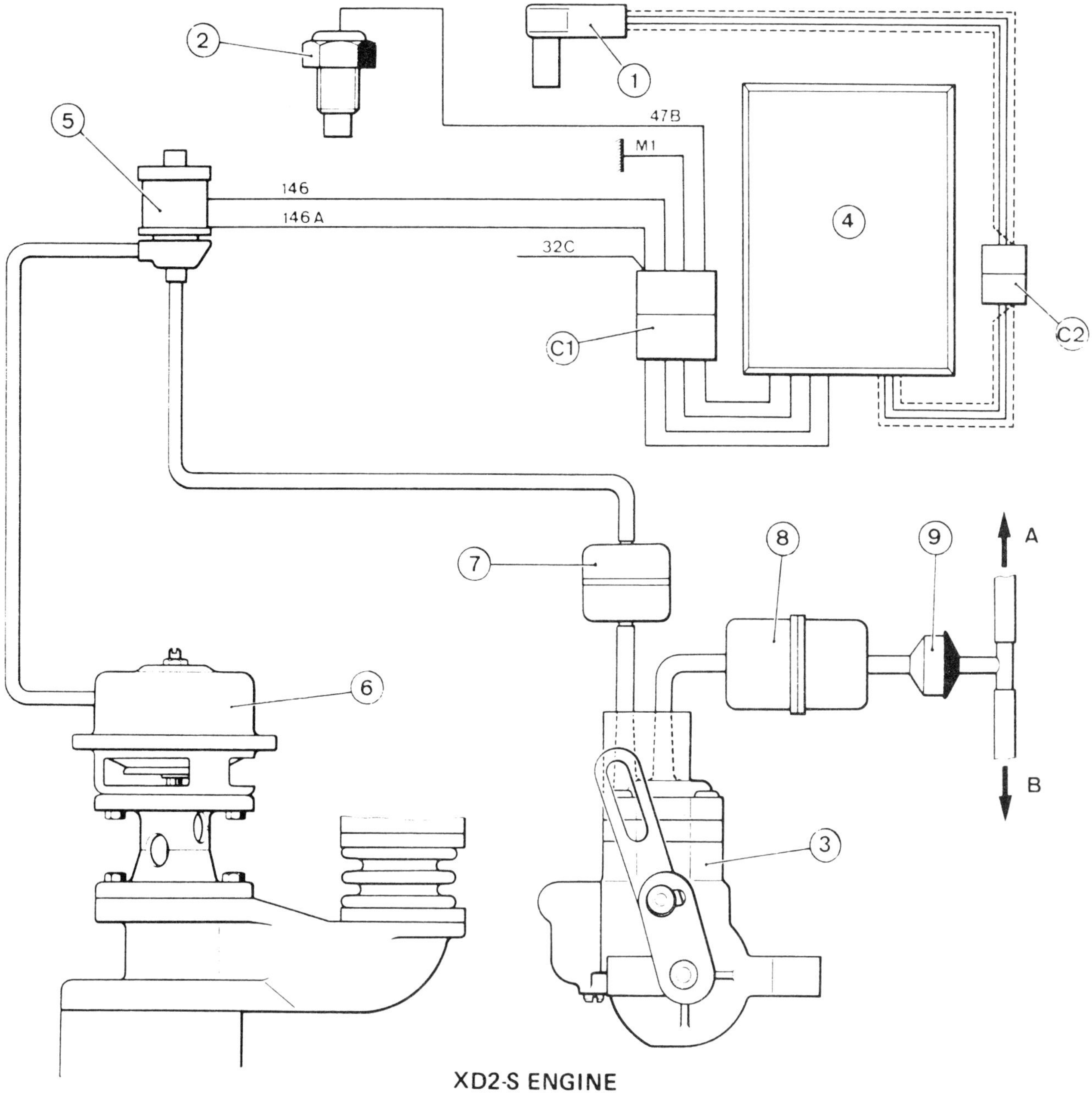

XD2-S ENGINE

EXHAUST GAS RECIRCULATION

1. Speed sensor
2. Thermocontact
3. Converter
4. Electronic control box
5. Electrovalve
6. EGR valve
7. Vacuum damper
8. Prefilter
9. Non return valve

A. To vacuum pump
B. Toward brake booster and vacuum reserve
C1 Main electrical control box connector
C2 Injection plug from electrical control box to speed sensor

Engine lacks power	Black exhaust smoke	Excessive engine oil consumption	Blue exhaust smoke	Turbocharger noisy	Cyclic sound from turbocharger	Oil leak from compressor seal	Oil leak from turbine seal	CAUSE	REMEDY
▲	▲	▲	▲			▲		Clogged air filter element	Replace element according to engine manufacturers recommendations
	▲	▲	▲	▲	▲	▲		Obstructed air intake duct to turbo compressor	Remove obstruction or replace damaged parts as required
▲	▲			▲				Obstructed air outlet duct from compressor to intake manifold	Remove obstruction or replace damaged parts as required
▲	▲			▲				Obstructed intake manifold	Refer to engine manufacturers manual & remove obstruction
				▲				Air leak in duct from air cleaner to compressor	Correct leak by replacing seals or tightening fasteners as required
▲	▲	▲	▲	▲				Air leak in duct from compressor to intake manifold	Correct leak by replacing seals or tightening fasteners as required
▲	▲	▲	▲	▲				Air leak at intake manifold to engine joint	Refer to engine manufacturers manual & replace gaskets or tighten fasteners as required
▲	▲	▲	▲	▲		▲		Obstruction in exhaust manifold	Refer to engine manufacturers manual & remove obstruction
▲	▲					▲		Obstruction in muffler or exhaust stack	Remove obstruction or replace faulty components as required
▲	▲			▲		▲		Gas leak in exhaust manifold to engine joint	Refer to engine manufacturers manual & replace gaskets or tighten fasteners as required
▲	▲			▲		▲		Gas leak in turbine inlet to exhaust manifold joint	Replace gasket or tighten fasteners as required
				▲				Gas leak in ducting after the turbine outlet	Refer to engine manufacturers manual & repair leak
		▲	▲			▲	▲	Obstructed turbocharger oil drain line	Remove obstruction or replace line as required
		▲	▲			▲	▲	Obstructed engine crankcase vent	Refer to engine manufacturers manual, clear obstruction
		▲	▲			▲	▲	Turbocharger center housing sludged or coked	Change engine oil & oil filter, overhaul or replace turbo as required
▲	▲							Fuel injection pump or fuel injectors incorrectly adjusted	Refer to engine manufacturers manual – replace or adjust faulty component(s) as required
▲	▲							Engine camshaft timing incorrect	Refer to engine manufacturers manual & replace worn parts
▲	▲	▲	▲			▲	▲	Worn engine piston rings or liners (blowby)	Refer to engine manufacturers manual & repair engine as required
▲	▲	▲	▲			▲	▲	Internal engine problem (valves, pistons)	Refer to engine manufacturers manual & repair engine as required
▲	▲	▲	▲	▲	▲	▲	▲	Dirt caked on compressor wheel and/or diffuser vanes	Clean using a Non-Caustic cleaner & Soft Brush. Find & correct source of unfiltered air & change engine oil & oil filter
▲	▲	▲	▲	▲		▲	▲	Damaged turbocharger	Analyze failed turbocharger, find & correct cause of failure, overhaul or replace turbocharger as required

Turbo Charger Troubleshooting and Diagnosis

Porsche

INDEX

VEHICLE IDENTIFICATION NUMBER AND ENGINE IDENTIFICATION

Vehicle Identification Number (VIN)

911 MODELS

The chassis number for all 911 models is located on the drivers side windshield post and is visible from outside the vehicle. The chassis number is also found in the luggage compartment under the rug and and on the identification plate near the front hood lock catch. The chassis identification number breaks down in the following manner: Serial Number Example—911-5-2-1-0001. The first number is the series

Chassis serial number location

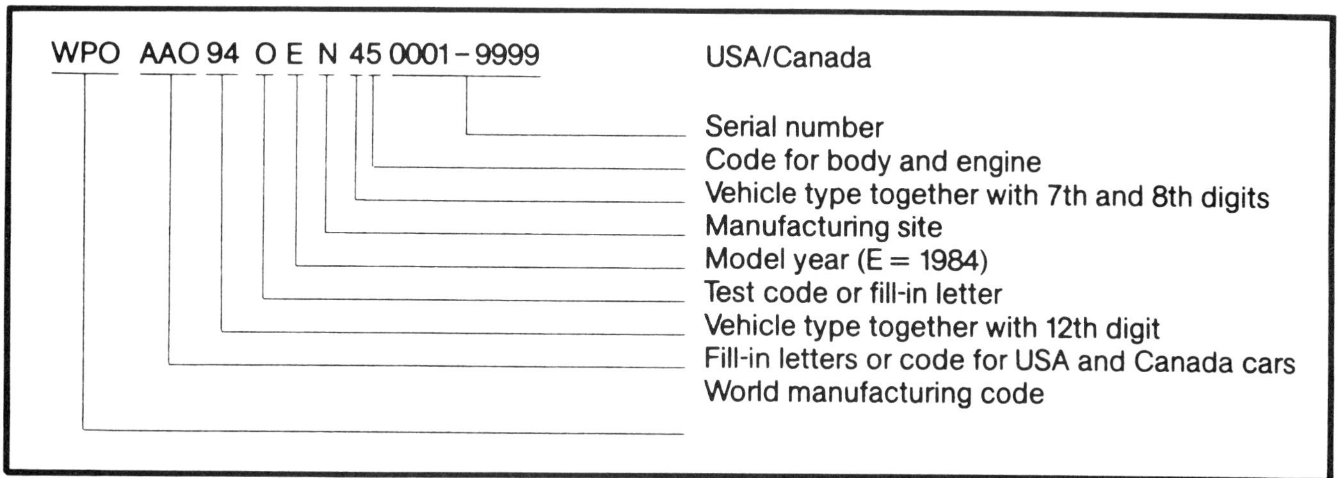

944 serial number

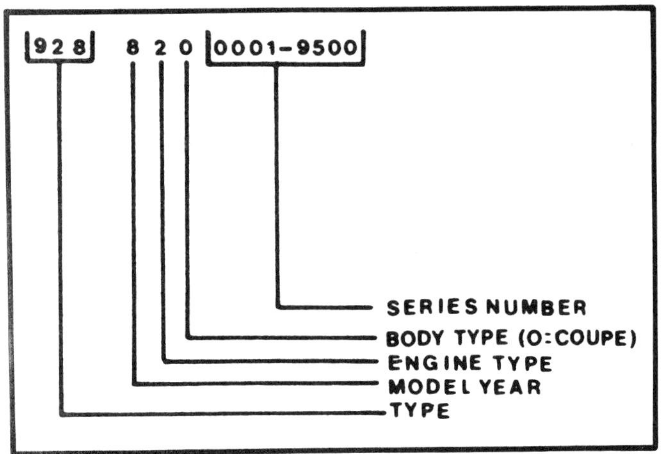

928 serial number

type, the second number is the model year (5 = 1985,6 = 1986 etc.), the third number is the engine type (1 = T, 2 = E or 2.7S, 3 = S, 4 = Carrera, 8 = Turbo), the fourth digit represents the body type (0 = Coupe, 1 = Targa) and the fifth digit being the sequential serial number.

928 AND 944 MODELS

The chassis serial number is located on the left windshield post and can be viewed from the outside of the vehicle. The vehicle identification plate is located in the engine compartment near the battery.

Engine Identification

911

The engine number is located on the right side of the crankcase adjacent to the blower. Engine numbers are divided as follows:

ENGINE NUMBER EXAMPLE

6–3–6–0001

ENGINE TYPE 6

6 = the number of cylinders.

ENGINE MODEL

1 = T-Japan-3.01 liter
2 = E-USA-3.01 liter
3 = S-Rest of world-3.01 liter
4 = Carrera
5 = Calif.-3.01 liter
8 = Turbo-3.31 liter

MODEL YEAR

F = 1985
G = 1986

928 AND 944 MODELS

The engine serial number is stamped on the left of the crankcase near the clutch housing.

TUNE UP SPECIFICATIONS
911

Year	Model	Engine Displacement cc (cu. in.)	Spark Plugs Type	Spark Plugs Gap (in.)	Distributor Point Dwell (deg)	Distributor Point Gap (in.)	Ignition Timing (deg) @ idle rpm	Intake Valve Opens (deg)	Compres. Press (psi)	Idle Speed (rpm)	Valve Clearance (in.) In	Valve Clearance (in.) Ex
85	911 Carrera	3164 (193)	WR7DC	0.028	Electronic		3A @ 800	9 ②	①	800	0.004	0.004
'86	SEE UNDERHOOD SPECIFICATIONS STICKER											

NOTE: The underhood specifications sticker often reflects tune-up specification changes made in production. Sticker figures must be used if they disagree from those in this chart.

A After Top Dead Center
① All cylinders should be within 22 psi of the highest reading. Compression test to be performed with the engine above 140°F
② With valve clearance of 0.1 mm

TUNE UP SPECIFICATIONS
944 and 928

Year	Model	Engine Displacement cc	Engine Displacement (cc. in)	Spark Plugs Type	Spark Plugs Gap (in.)	Ignition Timing ① (rpm)	Intake Valve Opens (deg)	Idle Speed (rpm)	Valve Clearance (cold) (in.) Intake	Valve Clearance (cold) (in.) Exhaust
'85	928S (4 Cam)	4957	(302.5)	WR7DS	.028–.032	10 ②	11°ATDC	680 ③	Hyd.	Hyd.
'85	944	2479	(151)	WR8DS	.028–.032	3°–7°@900 ⑤	1°ATDC	850–950	Hyd.	Hyd
'85	944 Turbo	2479	(151)	WR6DC	.028–.032	5° ②	—	840 ③	Hyd.	Hyd.
'86	All	See Underhood Specifications Sticker								

A—After Top Dead Center
B—Before Top Dead Center
① With vacuum hose disconnected, if so equipped.
② Checking specification only; timing is self-adjusting.
③ Not adjustable—electronically controlled.

FIRING ORDERS

NOTE: The position of the number one tower on the distributor cap may vary. To avoid confusion when replacing wires, always replace wires one at a time. The notch cut into the rim of the distributor body indicates number 1 cylinder.

FIRING ORDER: 1-6-2-4-3-5

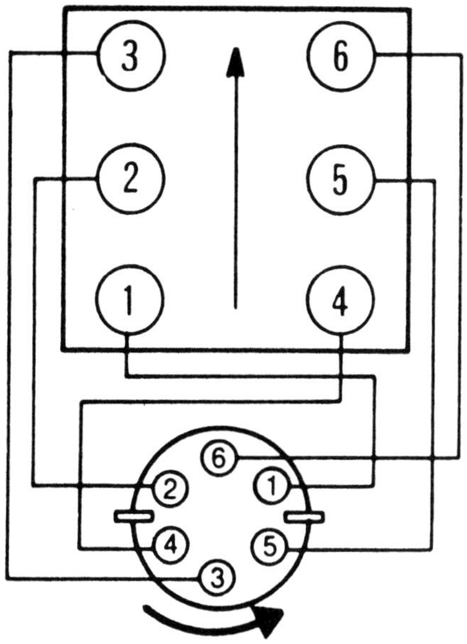

911 models

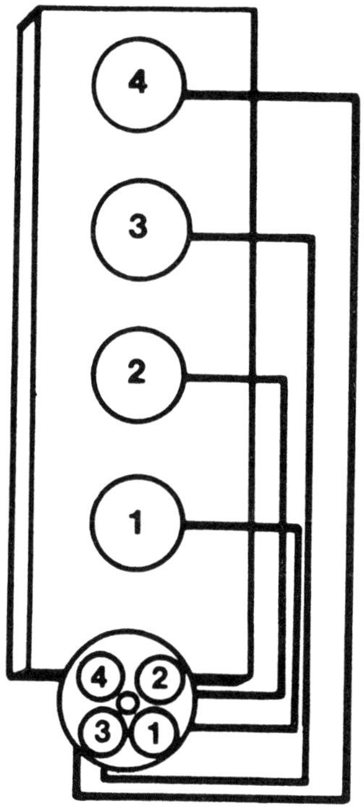

Firing order—944 engine
Distributor rotation—clockwise

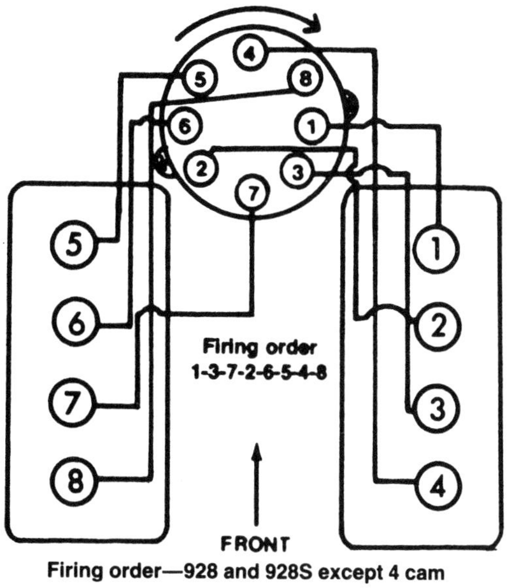

Firing order—928 and 928S except 4 cam

EMISSION EQUIPMENT USED

1985–86 911 MODELS
Catalytic Converter (CAT)
Continuous Injection System (CIS)
Fuel Evaporation System (FES)
Oxygen Sensor (OX)
Positive Crankcase Ventilation (PCV)

1985–86 944 AND 928 MODELS
Air Flow Controlled Fuel Injection (AFC)
Catalytic Converter (CAT)
Continuous Injection System (CIS)
Fuel Evaporation System (FES)
Oxygen Sensor (OX)
Positive Crankcase Ventilation (PCV)

REQUIRED EMISSION CONTROL MAINTENANCE

NOTE: The following information is being published from the latest information available at the time of publication. If this information differs from any data located on the underhood emissions sticker, or in the owners manual, use the information found on the emissions sticker or in the owners manual, since this information reflects the latest changes or modifications made to the vehicle during production.

SPARK PLUGS

On these vehicles, the spark plugs should last for 30,000 miles (48,000 km). If there is any indication of engine misfire, the plugs should be replaced immediately.

UNDERHOOD RUBBER AND PLASTIC COMPONENTS (EMISSION HOSES)

Inspect the hoses for evidence of heat or mechanical damage. Hard or brittle rubber, cracking, tears, cuts, abrasion and excessive swelling indicate deterioration of the rubber. Inspect all hose connections, couplings and clamps for leakage. Repair or replace as required.

POSITIVE CRANKCASE VENTILATION VALVE (PCV)

Every 15,000 miles (24,000 km), the valve should be checked. If the valve is found to be clogged, replace it with a new valve. Never attempt to clean an old PCV valve.

AIR FILTER

The filter in the carburetor air cleaner should be replaced every 30,000 miles (48,000 km).

FUEL EVAPORATION SYSTEM

Replace the fuel filter every 15,000 miles (24,000 km). Clean and inspect all hoses and connections.

CONTINUOUS INJECTION SYSTEM

Check and inspect all hoses and connections. Check the system for proper operation.

CATALYTIC CONVERTER

Certain models are equipped with a catalytic converter in the exhaust system. Check for damage and tight connection every 30,000 miles (48,000 km).

DRIVE BELTS

Inspect the power steering, air conditioning and alternator drive belts for signs of cracks, fraying or other signs of deterioration. Replace or adjust as necessary.

NOTE: Replace all filters more frequently if the vehicle is used for trailer towing, driven in heavy traffic, dusty operating conditions or where ambient temperatures are extremely high or low.

IDLE SPEED AND MIXTURE

Adjustments

NOTE: The following information is being published from the latest information available at the time of publication. If the information contained herein differs from that which is listed on the vehicles emission label, use the specifications given on the label. Before any adjustments are made be sure that the timing specification is correct and the engine is at normal operating temperature.

911 MODELS

1. Install the exhaust pickup line on the test connection of the catalytic converter.
2. Connect the CO tester according to the manufacturers instructions.
3. Disconnect the oxygen sensor plug, which is located on the left side of the engine compartment.

NOTE: Be sure that the oil tank cap and seal are installed properly before checking or adjusting the idle speed. Leaks can cause false readings when performing the adjusting procedure.

4. Turn the control screw on the throttle housing until the proper idle rpm is obtained.
5. Check the CO level. Correct the CO level as required.
6. If adjustment is necessary, remove the plug from the mixture control unit, which is located between the fuel distributor unit and the venturi, insert the adjusting tool.
7. Do not force the tool down while making the adjustments or the engine will stall. Turn the adjusting screw very slowly, as the slightest turn will change the CO reading.
8. Remove the adjusting tool. Accelerate the engine. Allow the engine to return to idle and recheck the idle speed and the CO level.

9. Disconnect the test equipment and plug all connections as required.

928 AND 928S MODELS

NOTE:

The following information is being published from the latest information available at the time of publication. If the information contained herein differs from that which is listed on the vehicles emission label, use the specifications given on the label. Before any adjustments are made be sure that the timing specification is correct and the engine is at normal operating temperature.

1. Disconnect the plug for the oxygen sensor. The sensor is located under the foot support on the passenger side of the vehicle.
2. Connect the test line at the test point which is located on the catalytic converter. Connect the CO testor and a tachometer.
3. If the CO mixture requires adjustment, insert the CO adjuster tool through the opening of the air flow sensor.
4. Turn the screw clockwise for a richer mixture and counterclockwise for a leaner mixture. Allow the engine to stabilize at idle.
5. Always make adjustments from lean to rich in very small increments.

944 MODELS

NOTE: The following information is being published from the latest information available at the time of publication. If the information contained herein differs from that which is listed on the vehicles emission label, use the specifications given on the label. Before any adjustments are made be sure that the timing specification is correct and the engine is at normal operating temperature.

1. Install the exhaust gas test line to the test point located on the catalytic converter.
2. Connect a CO testor and tachometer to the engine.
3. With the engine at normal operating temperature, turn the regulating screw on the throttle until the proper idle speed is reached.
4. Remove the oxygen sensor cap. Detach the plug. If the CO level is not correct, adjust the setting on the air flow sensor.
5. Connect the plug for the oxygen sensor.

IDLE SPEED AND MIXTURE ADJUSTMENT CHART

Vehicle	Year	Idle rpm	CO percentage
911 Carrera	'85–'86	780–820	.6–1.0
928S	'85–'86	700–800	.4–.8
944	'85–'86	850–950	.4–.8

INITIAL TIMING SETTINGS

NOTE: Refer to the vehicle emission control label located inside the engine compartment before making any adjustments. If the information on the label differs with the information given, use the data on the label. Some vehicles are equipped with the DME system and ignition timing is not possible.

911SC, 911 TURBO AND 911 CARRERA MODELS

To check ignition timing on these models, a Porsche special tool (VAG 1367 testor) is required. Terminals"B" and "C" of the test jack must be bridged with a jumper in order to stop operation of the idle regulator. The test jack is located on the drivers side of the vehicle next to the coil. Timing is then checked in the normal manner. At 800 rpm, ignition timing should be 3° ATDC.

Direct Porsche tool (VAG 1367) or stroboscopic timing light at the 25° timing mark on the crankshaft pulley. Make sure that the engine is at normal operating temperature and that all electrical accessories are OFF. Also, be certain that the distributor rotor is correctly installed in relation to the mark on the distributor housing.

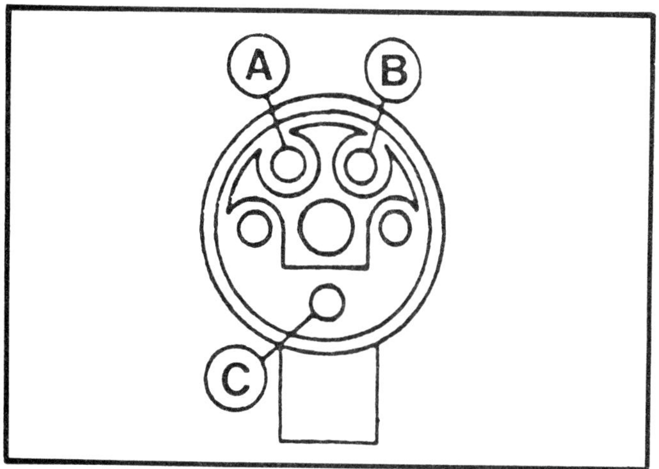

Ignition timing test connector

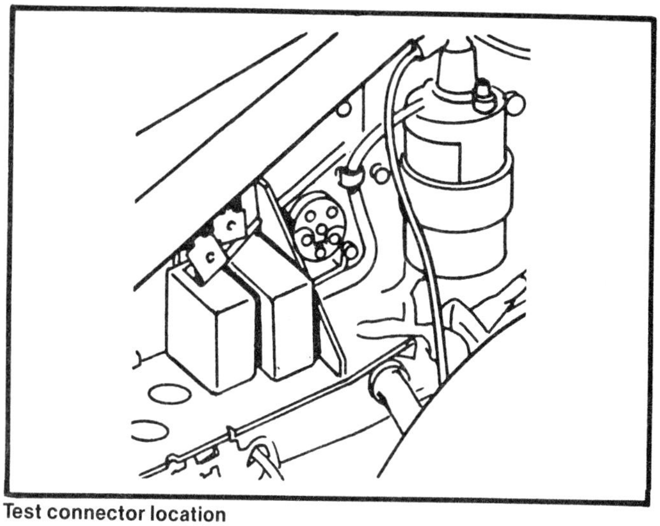
Test connector location

Bridge terminals "B" and "C" of the test jack; this simulates full throttle on the control unit and stops operation of the idle regulator. Full throttle timing should be 25° BTDC at 3800 rpm.

944 AND LATER MODELS

The 944 Porsche incorporates a Digital Motor Electronic (DME) ignition system. With this system, ignition timing is controlled electronically and no adjustment is necessary.

928 MODELS

NOTE: When connecting test equipment to the ignition system, the ignition key must be off or the battery disconnected due to the current that can be present in the primary and secondary circuits.

1. Start the engine and allow it to reach normal operating temperature.

IGNITION TIMING SPECIFICATION CHART

Vehicle	Timing (rpm)	Timing Specification (degrees)
911SC	900–1000	5 BTDC
911 Turbo	900–1000	0-4 BTDC
911 Carrera	780–820	3 ATDC
928	3000	23 BTDC
928S	3000	20 BTDC
944	850–950	3-7 BTDC

SPARK PLUGS

Year	Model	Type	Gap (in.)
'85	928S (4 Cam)	WR7DS	.029–.032
'85	944	WR8DS	.028–.032
'85	944 Turbo	WR8DS	.028–.032
'85	911 Carrera	W7DC	0.028
'86	All	See Underhood Specifications Sticker	

NOTE: The underhood specification sticker often reflects tune-up specification changes made in production. Sticker figures should always be used if they differ from those in this chart.

2. Connect a timing light to the engine according to the manufacturers instructions. A positive terminal for connecting the timing light is located in the engine compartment. Connect a tachometer.

3. With the timing light connected to the number 1 ignition cable, detach both distributor hoses at the distributor.

4. The timing marks are located on the crankshaft pulley and are colored for identification. With the engine at 3000 rpm, focus the timing light on the timing marks. Timing should be as specified. See the specification chart.

5. To adjust, loosen the distributor hold down bolt and turn the distributor as necessary. Tighten the hold down bolt and recheck the timing. Readjust if needed.

OXYGEN SENSOR LIGHT

Reset

911 MODELS

1. Disconnect the battery. Remove the speedometer assembly.
2. Working from the speedometer mounting hole in the instrument panel of the vehicle, press the reset button of the counter. It may be necessary to use a thin piece of metal rod to reset the counter.
3. The counter will return to zero and the lamp will go out. It is normal for the sensor light to come on when the ignition is turned on, and to go out when the engine is started.

928 MODELS

1. The counter is located on the right side of the passenger seat on the floor.
2. To reset the counter, loosen the knurled head screw and remove the cover of the counter switch.
3. Press the reset button of the switch against the stop in order to restore the counter indicator to normal operation.
4. It is normal for the sensor light to come on when the ignition is turned on, and go out when the engine is started.

OXYGEN SENSOR

Removal and Installation

NOTE: The oxygen sensor must be replaced every 30,000 miles on all vehicles except the 944 models. On the 944 models, the oxygen sensor must be replaced at 60,000 miles.

1. Raise the vehicle and support safely.
2. Remove the plug from the oxygen sensor terminal.
3. Pull the safety plug from the sensor and remove the oxygen sensor from its mounting.
4. Before installing the new sensor, coat the component with anti-seize paste. Be sure that the paste does not get into the slot of the sensor or it could malfunction. Install the new sensor and torque to 35 ft.lbs.

EMISSION CONTROL SYSTEMS

CRANKCASE EMISSION CONTROL SYSTEM

Positive Crankcase Ventilation (PCV)

PCV Valve

All models are equipped with a positive crankcase ventilation system (PCV). The purpose of the PCV system is to prevent harmful vapors from escaping into the atmosphere and the build up of crankcase pressures which could cause oil leaks.

The 924 model system carries vapors from the crankcase to the oil tank and then to the air cleaner. The crankcase vapors are then burned along with the air/fuel mixture.

The 928 and 944 model crankcase ventilation systems incorporates an oil separator, which is used as an oil filler tube. The crankcase fumes are routed through the separator pipe, where the liquid oil can settle and flow back into the crankcase. These fumes continue through a hose to the lower section of the air cleaner, where a flame arrestor is located. A coolant preheat line is placed along the portion of the vent hose to warm the crankcase fumes before entering the air cleaner.

Maintenance

The only maintenance required on the crankcase ventilation system is a periodic check. The PCV valve should be replaced at the recommended intervals. Check hoses for plugging and deterioration. Replace or repair as necessary.

PCV SYSTEM DIAGNOSIS

Symptom	Possible Cause	Repair
Rough idle, hard starting	Clogged PCV valve, vacuum leak in system	Replace PCV valve, repair vacuum leak
No PCV flow	Defective or clogged PCV valve, blocked line	Clean or replace line and PCV valve
Engine stalls at idle	Excessive PCV flow, dirty air filter element	Check PCV valve and air filter
Surging at road speed, hesitation	Improper PCV flow, oil soaked flame trap	Check PCV system, clean all lines
Excessive oil consumption	PCV valve blocked	Clear line

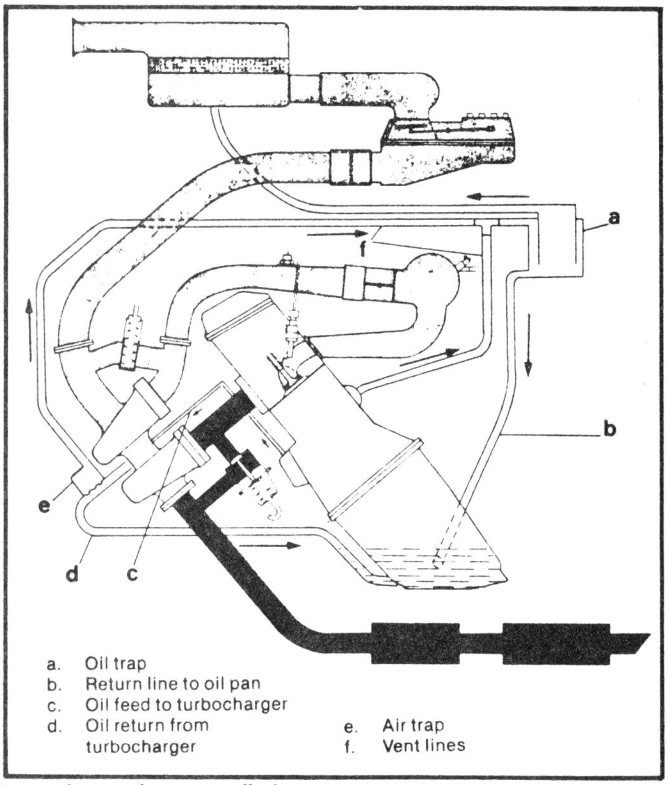

924 turbo crankcase ventilation

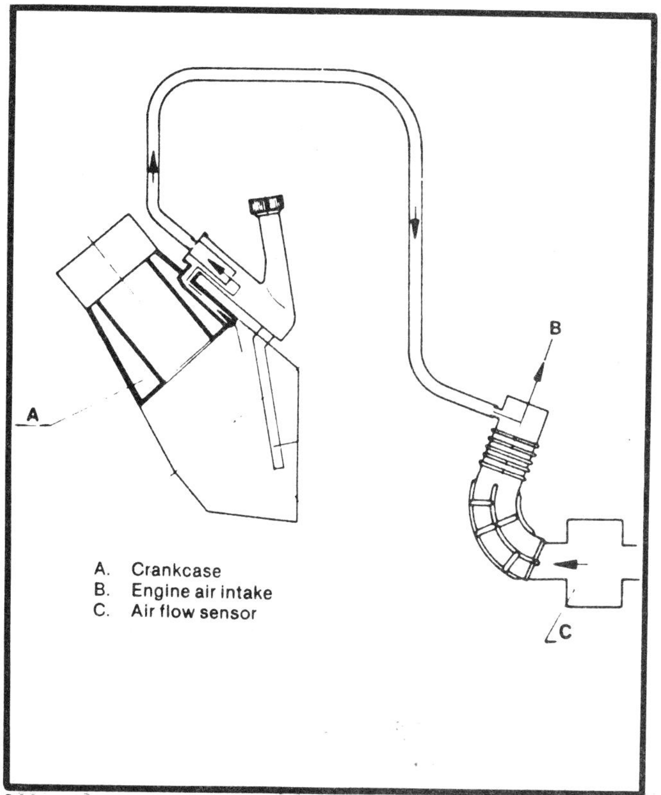

944 crankcase ventilation

FUEL EVAPORATIVE EMISSION CONTROL SYSTEM

The Fuel Evaporative Emission Control System prevents the escape of any raw fuel vapors into the atmosphere. The system consists of a charcoal canister and an expansion chamber. Vapors from the fuel tank are trapped in the canister. When the engine is running, fresh air is drawn in through the charcoal filter. The fresh air cleans the canister and routes the unburned hydrocarbons through the air cleaner to be burned during normal combustion. The fuel tank is vented to an expansion chamber which prevents fuel vapors from entering the atmosphere.

Trouble Diagnosis

Incorrect operation of the above system will result in poor engine idling, poor driveability or engine stalling. Inspect the system for a damaged charcoal canister, split or dry rotted hoses or hoses not connected. Repair or replace as necessary.

Maintenance

Maintenance of this system consists of checking the condition of the various connecting lines and hoses. The charcoal filter should also be inspected at 15,000 mile intervals and replaced at 50,000 miles.

FUEL EVAPORATIVE SYSTEM DIAGNOSIS

Symptom	Possible Cause	Repair
Noticeable fuel odor or leaks	Damaged or loose lines, saturated canister, broken or leaking two-way valve or liquid vapor separator	Repair lines, replace canister or two-way valve. Check liquid vapor separator
Fuel tank deformed	Canister clogged, tank cap defective, two-way valve not relieving pressure or vacuum	Replace defective canister or cap, clear vapor line
Insufficient fuel delivery or vapor lock	Blocked or collapsed vapor lines, two-way valve or canister. Blocked fuel feed line	Repair as necessary
Tank won't take fuel	Clogged or defective two-way valve	Replace two-way valve
Charcoal canister saturated with fuel	Defective thermosensor, purge control unloader solenoid or diaphragm valve. Blocked or disconnected vacuum or vapor lines	Check operation of all FES components. Check all lines and connections. Replace charcoal canister

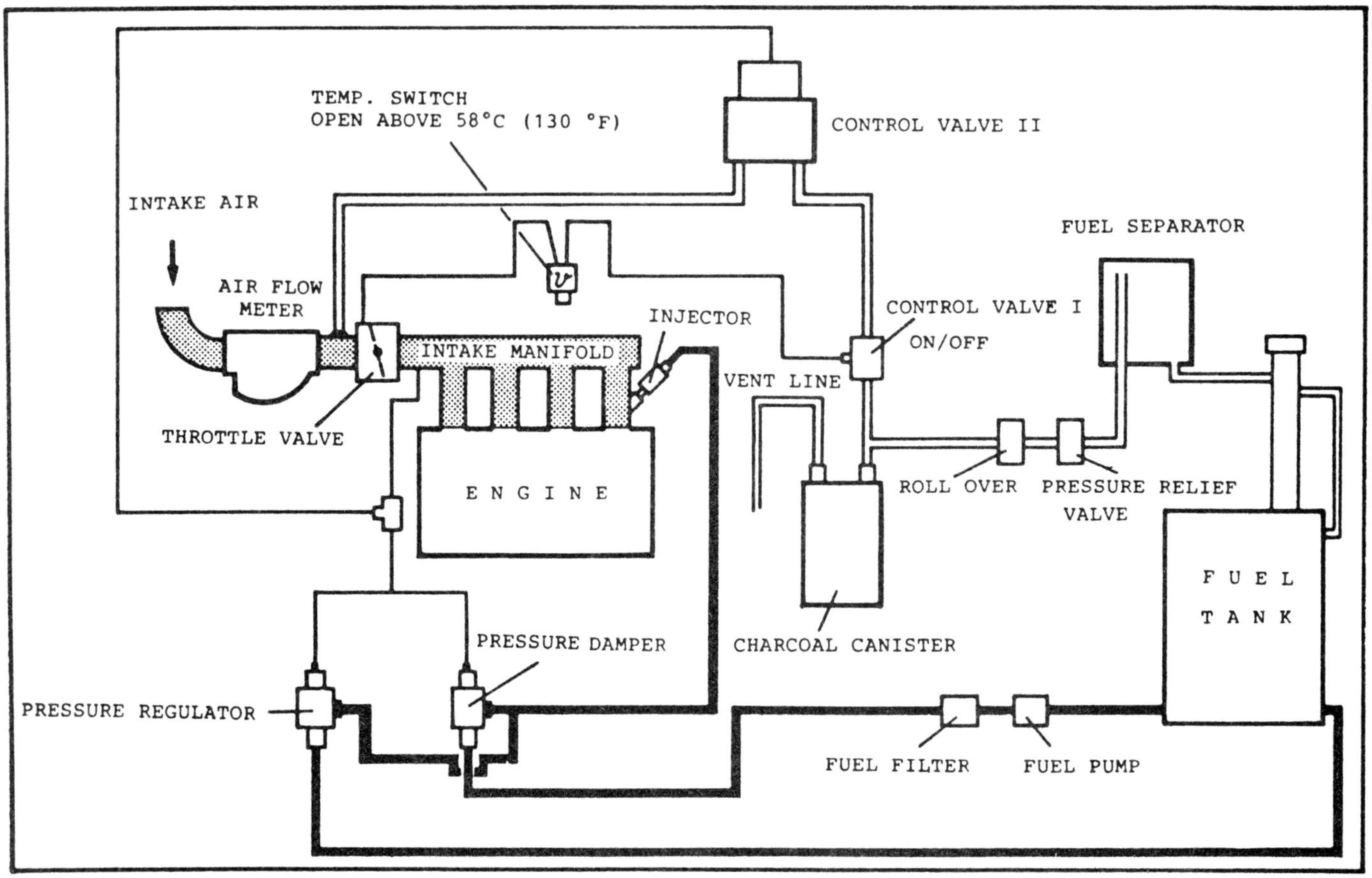

944 evaporative emission system—others similar

EXHAUST EMISSION CONTROL SYSTEM

All models are equipped with an Exhaust Gas Emission Control System which lowers NOx emissions. Its function is to allow metered amounts of cooling exhaust gases into the air/fuel mixture. The recirculated gas lowers the peak flame temperature during combustion in order to cut the output of oxides of nitrogen. Exhaust gas from the exhaust pipe pass through a filter element where it is cleaned. The vacuum operated Exhaust Gas Recirculation (EGR) Valve controls the amount of exhaust gas which enters the intake manifold.

Certain 911 models incorporate the use of a thermal reactor which also reduces HC and CO emissions by supplying an improved location for exhaust combustion. These vehicles are also equipped with an additional heater blower.

Catalytic coverters, located in the exhaust system, are standard on all models. This device contains metals which act as catalysts to cause a reaction to convert hydrocarbons and carbon monoxide into harmless water and carbon dioxide. Service on the converter consists of replacing it when it malfunctions. The catalytic converter should be inspected every 30,000 miles.

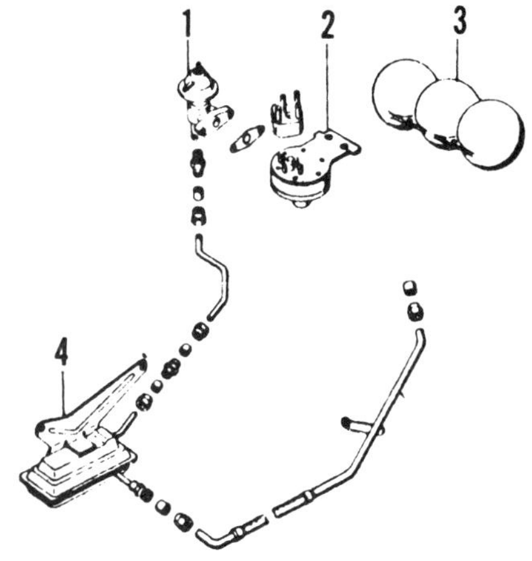

EGR system—typical

EGR VALVE

Testing

1. Disconnect the vacuum line from the EGR valve.
2. Disconnect the vacuum hose from the distributor vacuum unit and extend the hose.

EGR SYSTEM DIAGNOSIS

Symptom	Possible Cause	Repair
EGR valve does not move during test	Leaking or clogged vacuum line or hose connection. Defective EGR valve	Repair vacuum leak, replace EGR valve
Engine won't idle or stalls on return to idle	EGR control system blocked or inoperative. Defective speed sensor	Check all vacuum connections and control valves. Replace defective control valves or speed sensor
EGR valve leaks in closed position. Poor wide-open throttle performance	EGR valve defective or blocked with carbon. Defective EGR control valve	Clean or replace EGR valve, replace control valve
No vacuum to EGR valve	Thermosensor defective, leaking vacuum line	Replace thermosensor or repair vacuum hose connection
Excessive HC or CO levels in exhaust	Air injection system inoperative	Check all components for proper operation. Check all hoses for unrestricted flow

CATALYTIC CONVERTER DIAGNOSIS

Symptom	Possible Cause	Repair
Leaking exhaust gases	Leaks at pipe joints, damaged gaskets or rusted exhaust pipes	Tighten clamps, repair exhaust system as necessary
Loss of engine power, internal rattles in exhaust system	Dislodged baffles in muffler, broken ceramic insert in converter	Replace converter or muffler
Excessive HC or CO	Contaminated catalyst	Replace converter
Excessive catalyst temperature, warning light on constantly	Mixture set too rich, choke malfunction, malfunction in secondary air supply system	Reset air/fuel mixture, check choke operation, check air injection system

OXYGEN SENSOR SYSTEM DIAGNOSIS

Symptom	Possible Cause	Repair
Dash light on (OXS)	Normal service due	Replace oxygen sensor. Reset service reminder
Excessive CO level when warm	Malfunction in electrical circuit, defective control unit	Check circuit, replace control unit
CO cannot be adjusted	Malfunction in fuel system, contaminated oxygen sensor	Repair fuel system, replace oxygen sensor
Idle does not increase under load	Malfunction in idle stabilizer microswitch	Replace idle stabilizer microswitch

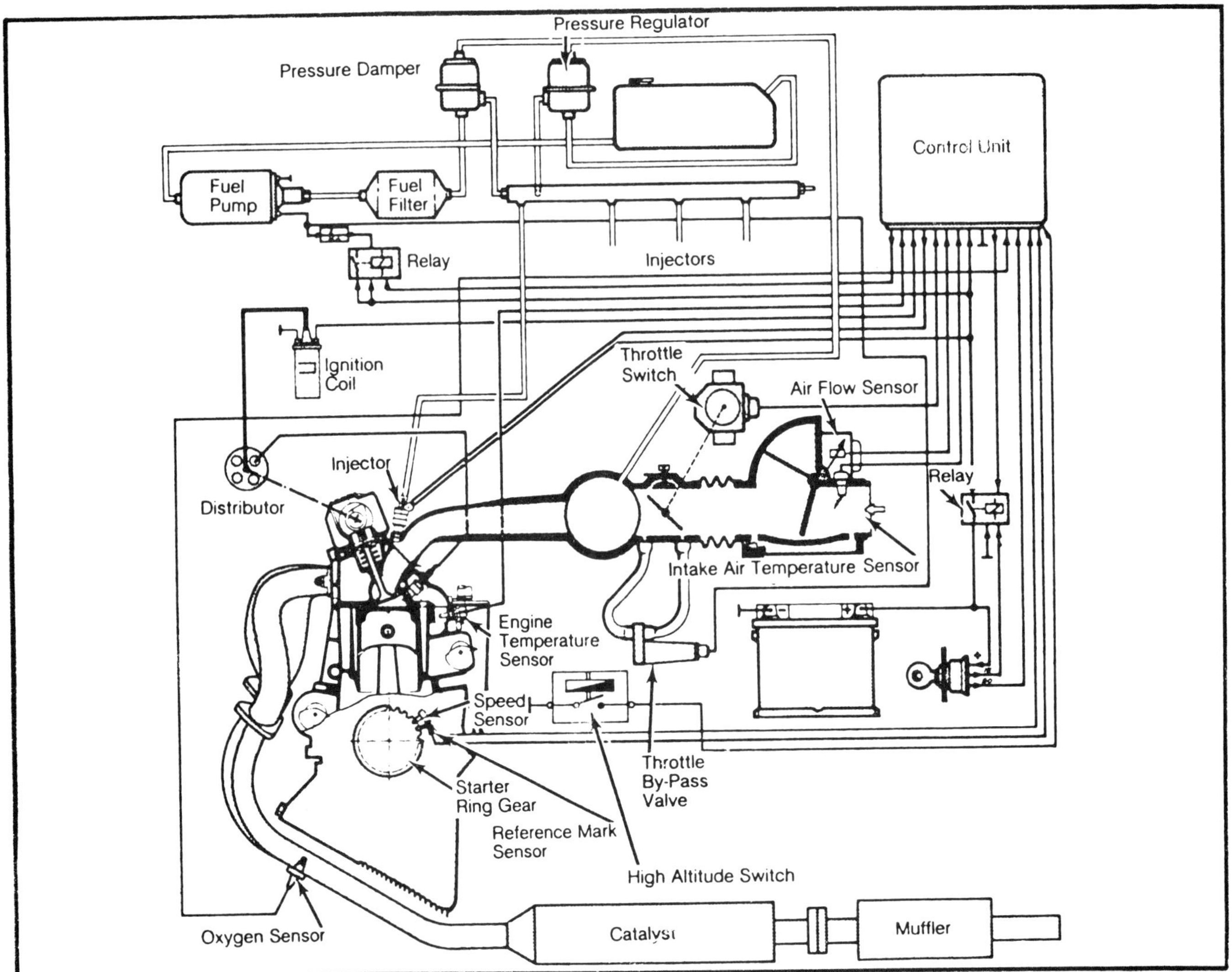

Schematic of AFC fuel injection—944 shown

3. Start the engine and allow it to idle.
4. Connect the distributor vacuum hose to the EGR valve. The engine should stumble or stall.
5. If the idle speed stays even, the EGR line is clogged or the EGR valve is defective. Repair or replace as required.

Service

The only required maintenance is that the EGR filter be replaced at the recommended intervals.

1. Disconnect the line fittings at each end of the filter.
2. Unbolt the bracket retaining screws. Remove the filter.
3. Install the bracket replacement filter. The rest of the installation is the reverse of the removal procedure.

C.I.S. FUEL INJECTION DIAGNOSIS

Engine does not start or starts poorly when **cold**	Engine does not start or starts poorly when **warm**	Irregular idle (engine shakes) during warm-up	Irregular idle (engine shakes) with engine **warm**	Engine does not draw fuel smoothly (backfires)	Engine misfires under full load	Insufficient power	Engine runs on (diesels)	Excessive fuel consumption	Flat spot during acceleration	Idle CO value too **high**	Idle CO value too **low**	Idle speed cannot be adjusted (too high)	Engine stalls immediately after starting	Cause(s)
►	►	►	►		►				►		►			Vacuum system leaking
►	►		►			►	►		►	►	►			Air flow sensor plate and/or control plunger not moving smoothly
	►						►							Air flow sensor plate stop incorrectly set
►		►												Auxiliary air valve does not open
												►		Auxiliary air valve does not close
►	►				►								►	Electric fuel pump not operating
►														Defective cold start system
		►	►				►	►		►				Leaking cold start valve
►		►												Incorrect cold control pressure
	►		►	►	►	►			►				►	Warm control pressure too high
			►	►		►		►	►	►			►	Warm control pressure too low
					►	►			►				►	Incorrect system pressure
	►													Fuel system pressure leakage
	►	►	►		►		►							Injection valve(s) leaking, opening pressure too low
		►	►			►			►					Unequal fuel delivery between cylinders
	►	►	►	►			►	►	►	►	►			Basic idle and/or CO adjustment incorrect
						►								Throttle plate does not open completely

VACUUM CIRCUITS

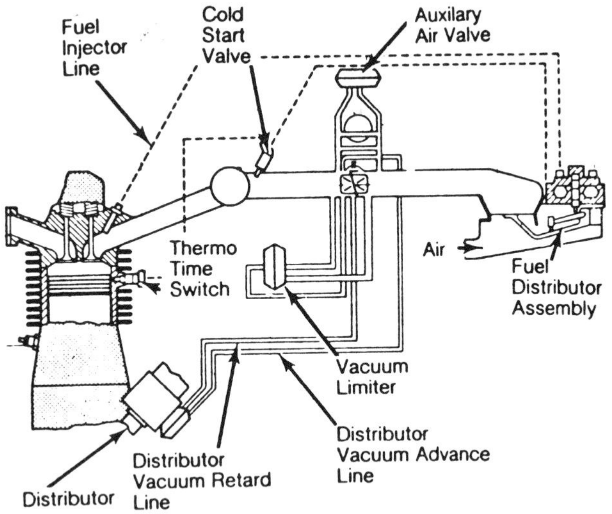

Porsche 911 vacuum circuit

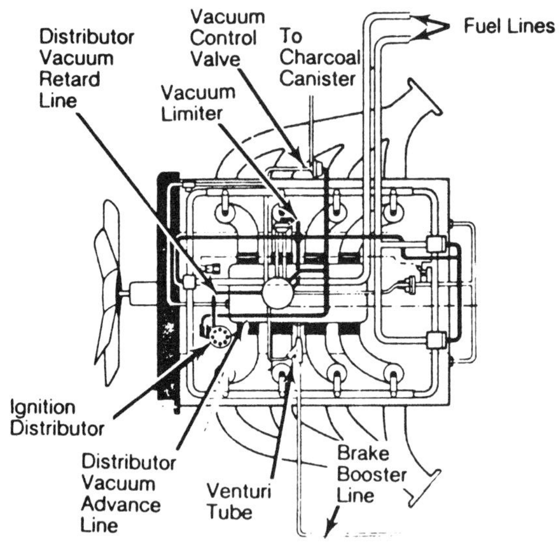

Porsche 928 vacuum circuit

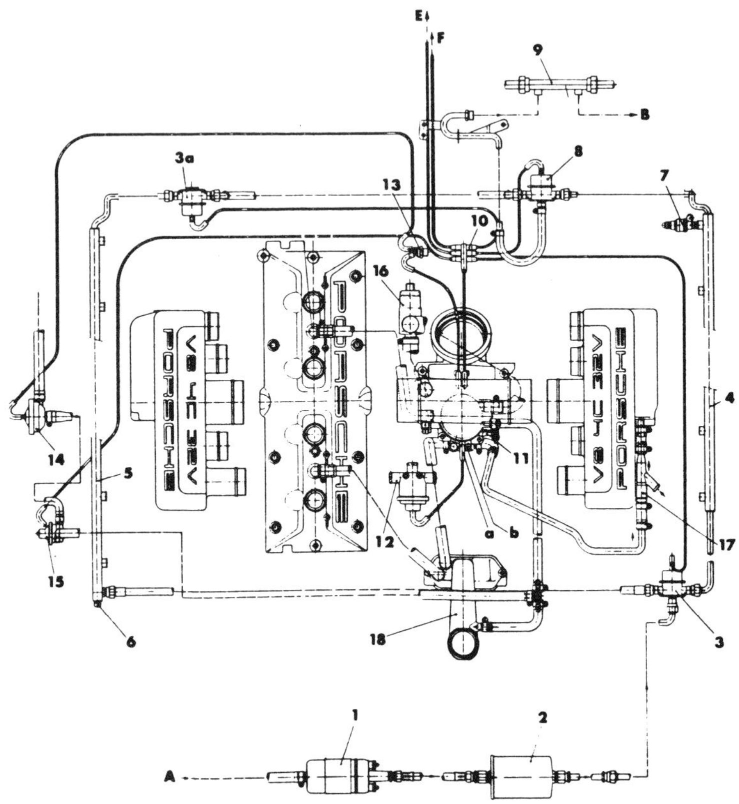

1 – Fuel pump
2 – Fuel filter
3 – Pressure damper
3a – Pressure damper
4 – Fuel line (cyl. 5 . . . 8)
5 – Fuel line (cyl. 1 . . . 4)
6 – Fuel pressure test connection
7 – Fuel injectors
8 – Pressure regulator
9 – Fuel cooler
10 – Vacuum distributor
11 – Throttle housing
12 – Diverter valve
13 – Thermo switch
14 – Shift valve
15 – Control valve
16 – Air regulating valve
17 – Vacuum booster venturi
18 – Oil filler neck

a – Connection 49 states
b – Connection California

A – From fuel tank
B – Return to fuel tank
E – To automatic transmission
F – To EZF control unit

PORSCHE 928 MODELS

Renault

INDEX

VEHICLE IDENTIFICATION NUMBER AND ENGINE IDENTIFICATION

Vehicle Identification Number (VIN)

Renault vehicles are identified by two plates in the engine compartment. The rectangular plate shows the model number, serial number, maximum gross vehicle weight (GVW), maximum gross axle weight rating, date of manufacturer and vehicle class. The second plate, oval in shape, indicates the model number, the transmission type, the basic equipment, the optional equipment code and the manufacturer's number.

Engine Identification

The engine identification plate is attached to the engine block on the left side at the rear, just below the cylinder head. The plate identifies the (A) engine type, (B) the French Ministry of Mines Homologation number, (C) the engine equipment, (D) the manufacturer's identification number, (E) the engine indexing, number, and (F), the manufacturing sequence number.

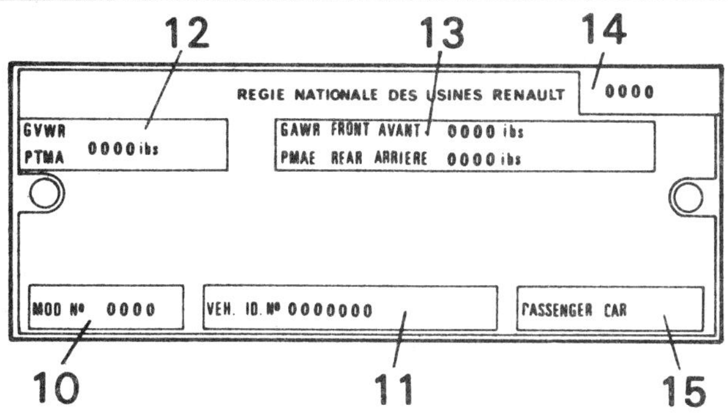

10. The model number
11. The serial number
12. The maximum allowable load for the vehicle
13. The maximum allowable load for the front and rear axles
14. The date of manufacture (month-year)
15. The vehicle class

Rectangular identification plate

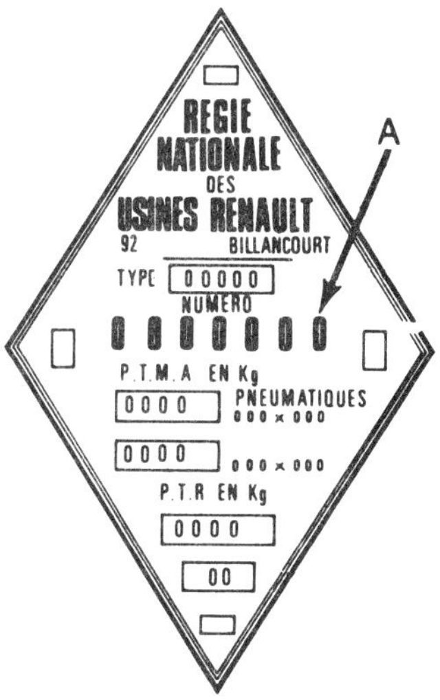

Diamond shaped identification plate

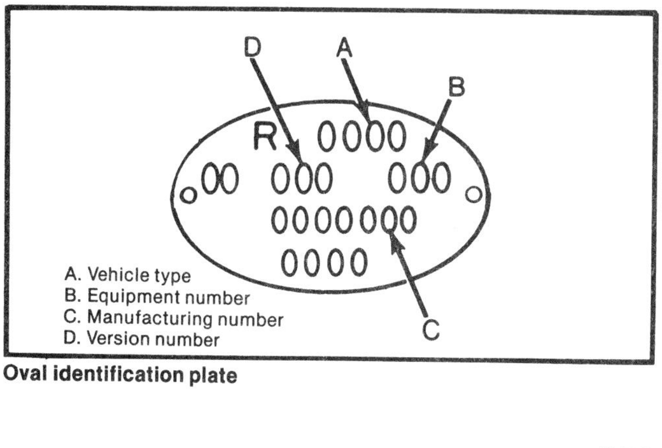

A. Vehicle type
B. Equipment number
C. Manufacturing number
D. Version number

Oval identification plate

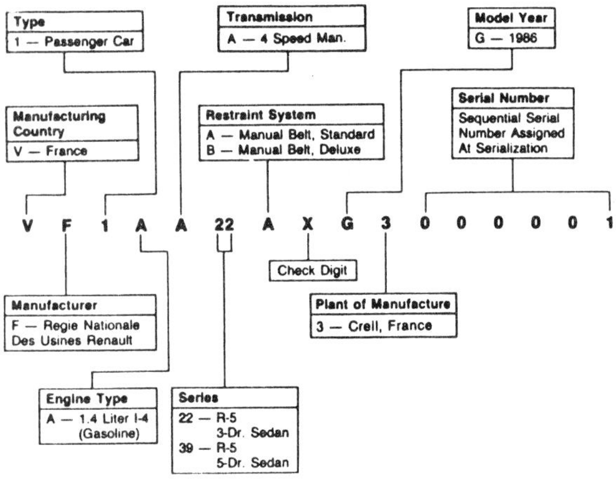

VIN number decoding

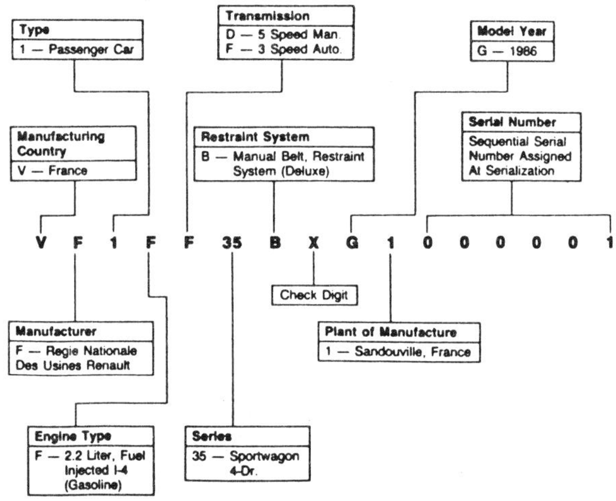

Renault Sport Wagon VIN number decoding

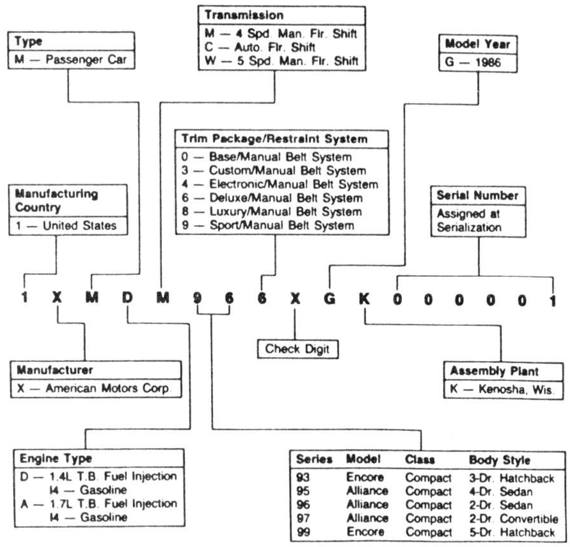

Renault Alliance/Encore VIN number decoding

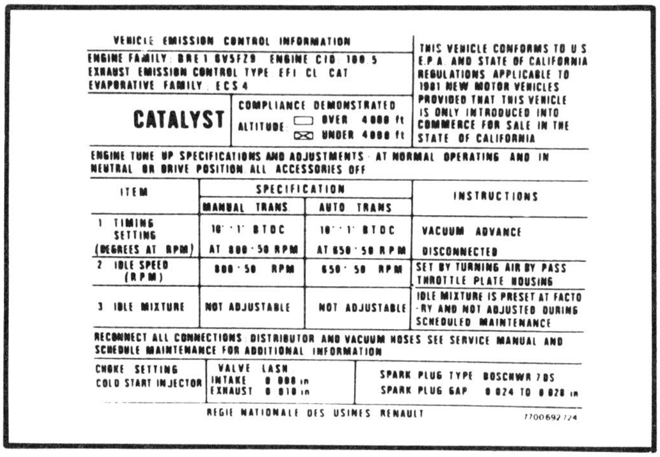

Typical Emission Control Information label

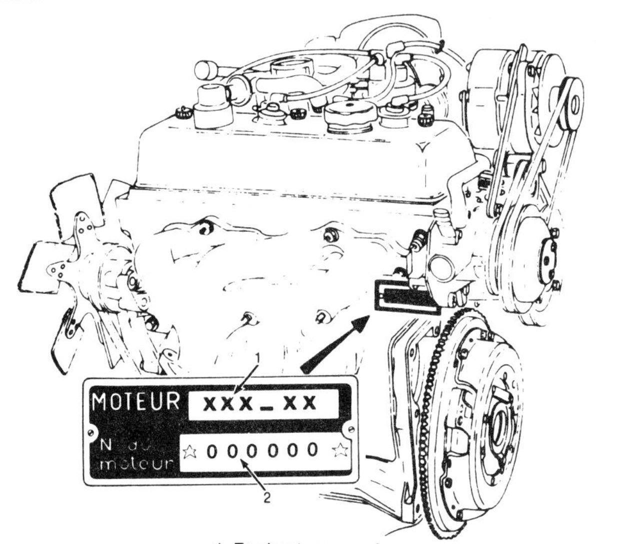

1. Engine type number
2. Engine fabrication number

Engine identification plate

TUNE-UP SPECIFICATIONS

Model	Year	Engine Displacement cu in. (cc)	Spark Plug Type	Spark Plug Gap (in.)	Distributor Point Dwell (deg)	Distributor Point Gap (in.)	Ignition Timing (deg) MT	Ignition Timing (deg) AT	Intake Valve Opens (deg)	Fuel Pump Pressure (psi) ③	Idle Speed (rpm) MT	Idle Speed (rpm) AT	Valve Clearance (in.) In	Valve Clearance (in.) Ex
LeCar, Alliance, Encore	'85–'86	85.2 (1397)	RN-12Y	.032	Electronic		8B	8B	12B	28–36	700	700	.006	.008
18i, Fuego, Sport Wagon	'85–'86	100.5 (1647) ②	WR7DS	.024–.028	Electronic		10B	10B	22B	28–36	800	650	.008	.010
Fuego	'85	95.5 (1565)	①	.022–.026	Electronic		10B	10B	22B	28–36	800	650	.008	.010

NOTE: The underhood sticker often reflects tune-up specification changes made in production. Sticker figures must be used if they disagree with those in this chart.
B: Before Top Dead Center
① CR42 LTS or RS 9YC
② Turbo
③ '86 operating pressure
1.4L–14.5 psi
1.7L–17.4 psi

FIRING ORDERS

NOTE: To avoid confusion, always replace spark plug wires one at a time.

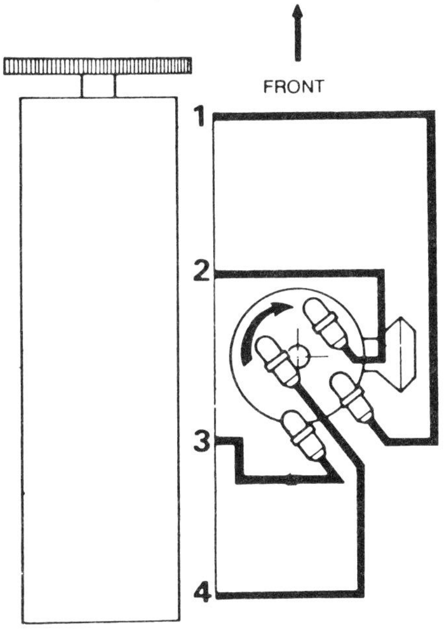

LeCar, Alliance/Encore spark plug wiring diagram

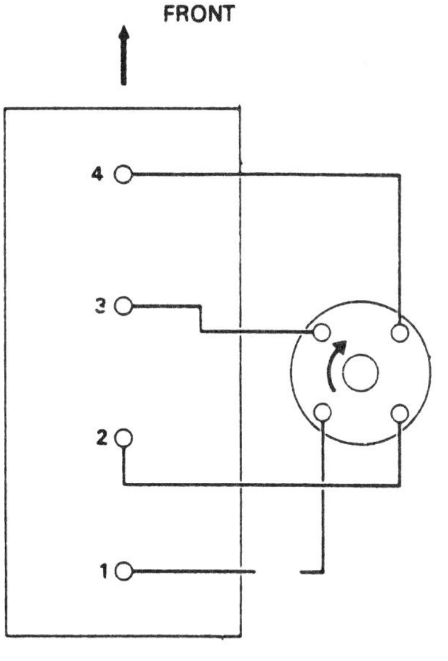

Sport Wagon and Fuego spark plug wiring diagram

EMISSION EQUIPMENT USED

1985–86 ALL MODELS

Air injection
Evaporative Emission Control
Ignition Timing Advance Control System
Positive Crankcase Ventilation System
Catalytic Converter
Exhaust Gas Recirculation System
Oxygen Sensor
Coolant temperature sensor
Manifold Air Temperature Sensor (TBI only)
Manifold Absolute Pressure Sensor (TBI only)
Wide Open Throttle Switch (TBI only)

REQUIRED EMISSION CONTROL MAINTENANCE

AIR INJECTION SYSTEM

Visually inspect the system every 12 months or 12,000 miles.

EVAPORATIVE EMISSION CONTROL

Visually inspect the system every 10,000 miles. Replace the filter every 50,000 miles.

POSITIVE CRANKCASE VENTILATION

Inspect the PCV valve every 12,000 miles or 12 months. Replace the PCV valve every 24,000 miles or 24 months.

CATALYTIC CONVERTER

Inspect converter element every 24,000 miles.

EXHAUST GASD RECIRCULATION (EGR)

Inspect hoses and diaphragm every 10,000 miles. Clean valve every 24,000 miles.

OXYGEN SENSOR

Remove and inspect every 25,000 miles.

IDLE SPEED AND MIXTURE ADJUSTMENT

Fuel Injection

IDLE SPEED ADJUSTMENT

FUEGO SPORT WAGON AND CALIFORNIA

ALLIANCE AND ENCORE

1. Start the engine and allow it to come to operating temperature.
2. Connect a tachometer to leads D1-1 and D1-3 of the diagnostic connector.
3. Turn all accessories off. Wait for the electric fan to shut off.
4. Turn the throttle plate bypass screw to obtain the correct idle speed.

ALLIANCE AND ENCORE, EXCEPT CALIFORNIA

These cars use Throttle Body Injection. Idle speed adjustment is necessary only if the Idle Speed Control motor (ISC) has been replaced.

1. Remove the air cleaner.
2. Start the engine and allow it to reach normal operating temperature. Make sure that the A/C control is off.
3. Connect a tachometer to terminals D1-1 and D1-3 of the diagnostic connector.
4. Turn the engine off. The ISC plunger should move to the fully extended position.
5. Disconnect the ISC motor wire connector and start the engine.
6. Engine speed should be 3300–3700 rpm. If not, turn the hex head bolt on the end of the plunger to obtain the 3500 rpm.
7. Fully retract the ISC motor plunger by holding the closed throttle open. The closed throttle switch plunger should not be touching the throttle lever when the throttle is returned to the closed position. If contact is noted, check the throttle linkage and/or cable for binding.
8. Connect the ISC wire.
9. Turn the engine off for 10 seconds. The ISC motor plunger should move to the fully extended position.
10. Start the engine. The engine should speed up to 3500 rpm, stay there for a brief time, then drop off to normal idle speed as shown on the underhood sticker and in the Tune-Up chart.
11. Shut off the engine, disconnect the tachometer and apply a penetrating thread sealant to the adjustment screw threads.

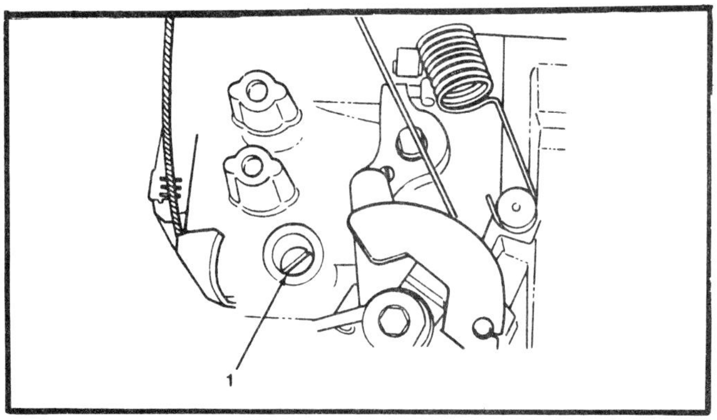

Idle speed adjustment on Fuego and California Alliance/Encore. Number one is the idle speed adjustment screw

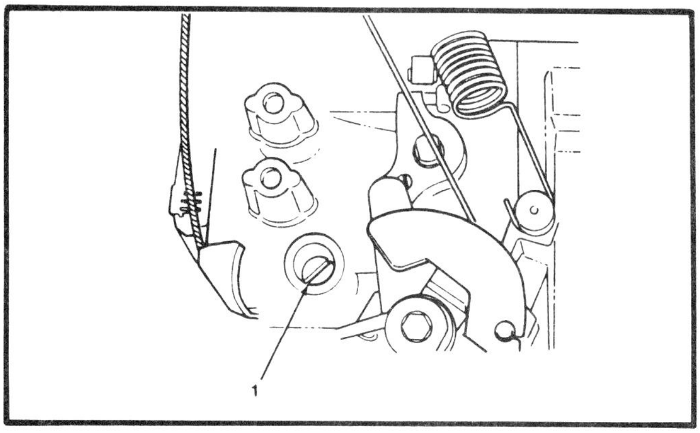

Idle speed adjustment points on Fuego and non-California Alliance/Encore models. Point is idle speed adjusting screw

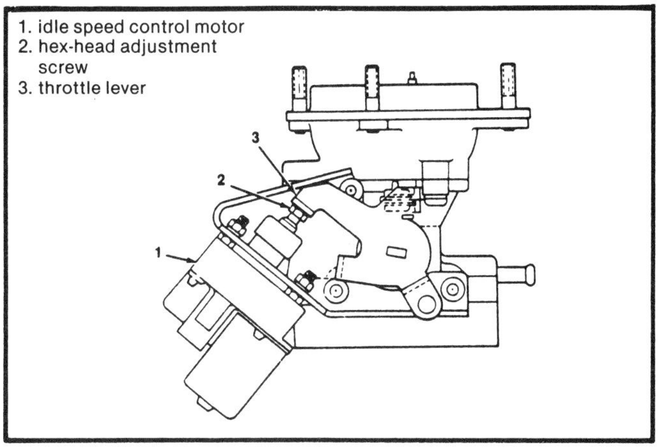

Idle speed adjustment point on 49 State Alliance/Encore models

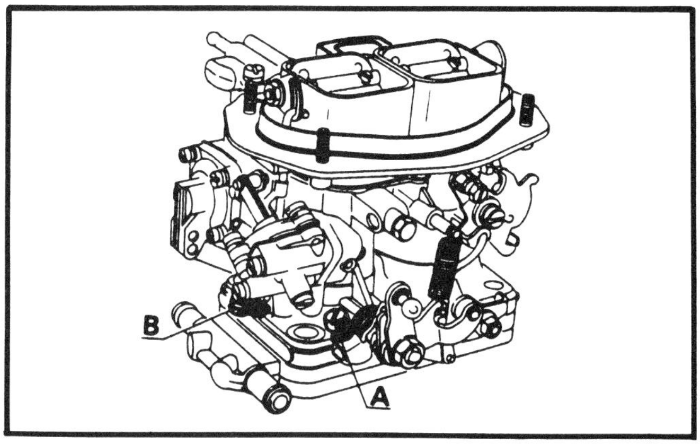

LeCar idle speed and mixture adjustment point

Idle Speed and Mixture Adjustment

CARBURETED ENGINES

NOTE: 1985–86 models must be adjusted with the use of a CO meter to obtain the correct emissions reading and idle speed. If a CO meter is not available and the idle speed and mixture must be adjusted, use the following procedure.

WITHOUT A CO METER (EXCEPT CALIFORNIA)

1. Clamp off the air injection hose between the diverter valve and the engine, using the appropriate special tool.
2. Turn the throttle plate screw so that the engine speed is 775 rpm.
3. Turn the idle mixture screw to obtain the highest possible idle speed.
4. Lower the engine speed 20–25 rpm by turning the idle mixture screw clockwise. Remove the air injection hose block-off clamp and check the idle speed which should be 850 ± 50 rpm. If the idle speed is incorrect, turn the throttle plate screw to adjust.

WITH A CO METER (EXCEPT CALIFORNIA)

1. Bring the engine to normal operating temperature.
2. Clamp or disconnect and plug, the air pump hose to injection manifold.
3. Turn the idle speed screw to obtain 675–725 rpm.
4. With a CO meter connected, adjust the fuel metering screw to obtain 0.5–2.0 CO value. Readjust as required.
5. Unclamp or connect the air injection hose. The idle should be 700–800 rpm. Adjust as necessary to obtain the idle speed.

CALIFORNIA

1. Bring the engine to normal operating temperature.
2. Connect a vacuum gauge to the line between the carburetor and the vacuum solenoid regulator.

ELECTRONIC IGNITION

Adjustments

Normally, distributor adjustments should not be necessary at each tune-up, but the distributor has some provisions for adjusting trigger wheel air gap and for ignition timing.

TRIGGER WHEEL AIR GAP

1. Loosen the impulse sender unit retaining screws 1, 2, 3 and 4 slightly.
2. Using a non-magnetic feeler gauge measure the air gap between an impulse sender coil stud and one of the arms of the trigger wheel. The gap must be 0.012–0.24 in. (0.3–0.6mm).
3. Move the impulse sender unit as needed to adjust the air gap and tighten the retaining screws. Check the gaps for all four arms of the trigger wheel and adjust as necessary.

NOTE: If the air gap of any arm of the trigger wheel cannot be correctly adjusted, replace the distributor.

IGNITION TIMING WITH SECONDARY IMPULSE SENDER

The Secondary impulse sender coil (9) is offset from the primary (7) by 3° in some models, and by 5° in others.

NOTE: This timing adjustment must be made after every adjustment of the trigger wheel air gap.

1. Loosen screws 2 and 11 as shown in the illustration.
2. Align an arm of the trigger wheel with the stud (8) of the primary impulse sender, so the centerline of the arm points to the middle of the stud.

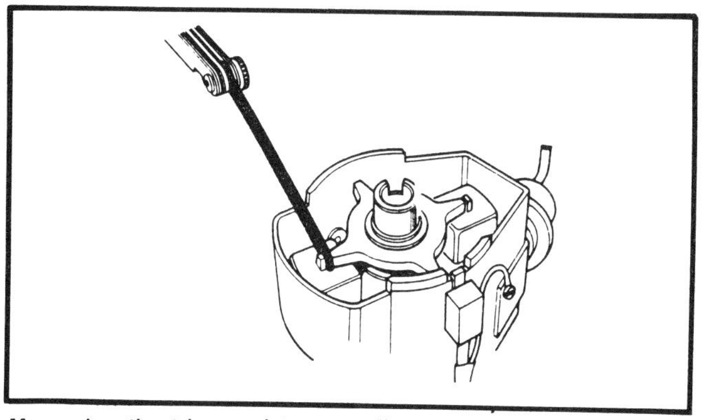

Measuring the trigger plate gap adjustment with a non-magnetic feeler gauge (plastic or brass)

3. To adjust the timing with a 3° offset secondary impulse sender, carefully move the secondary impulse sender so the centerline of its stud (10) aligns with the left edge of the nearest arm of the trigger wheel.
4. To adjust the timing with a 5° offset secondary impulse sender, carefully move the secondary impulse sender so the right edge of its stud aligns with the left edge of the nearest arm of the trigger wheel.
5. Tighten the screws 2 and 11.

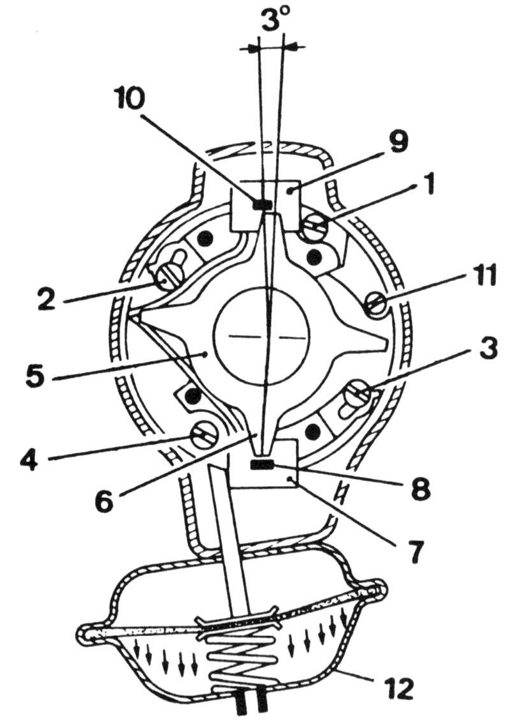

1. & 2. Adjustment screws for the secondary impulse sender-to-trigger wheel clearance
3. & 4. Adjustment screws for the primary impulse sender-to-trigger wheel clearance
5. Trigger wheel
6. Trigger wheel arm
7. Primary impulse sender coil
8. Primary impulse sender stud
9. Secondary impulse sender coil
10. Secondary impulse sender stud
11. Coil offset adjustment screw
12. Vacuum advance capsule

Electronic distributor parts are identified, showing the alignment of the impulse sender coils offset 3 degrees

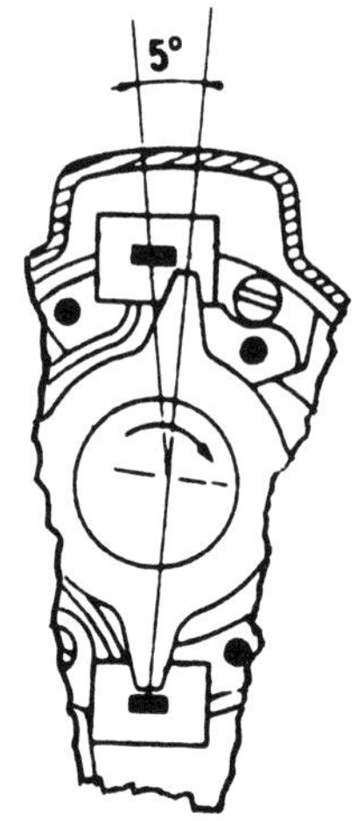

Alignment of impulse sender coils offset 5 degrees

IGNITION TIMING

Adjustment

1. Locate the timing marks on the flywheel and the flywheel housing.

2. Clean the timing marks so that you can see them.

3. Mark the timing marks with a piece of chalk or with paint. Color the mark on the scale that will indicate the correct timing when it is aligned with the mark on the flywheel. It is also helpful to mark the notch in the flywheel with a small dab of color.

4. Attach a tachometer to the engine.

5. Attach a timing light according to the manufacturer's instructions. If the timing light has three wires, one is attached to the No. 1 spark plug with an adapter. The other wires are connected to the battery. The red wire goes to the positive side of the battery and the black wire is connected to the negative terminal of the battery.

6. Disconnect the vacuum line at the distributor and plug it.

7. Make sure that all of the wires clear the fan and then start the engine.

8. Adjust the idle to the correct setting.

9. Aim the timing light at the timing marks. If the marks that you put on the flywheel and the flywheel housing are aligned when the light flashes, the timing is correct. Turn off the engine and remove the tachometer and the timing light. If the marks are not in alignment, proceed with the following steps.

10. Turn off the engine.11.

Loosen the distributor lockbolt just enough so that the distributor can be turned with a little effort.12. Start the engine. Keep the wires of the timing light clear of the fan.

13. With the timing light aimed at the flywheel and the marks on the engine, turn the distributor in the direction of rotor rotation to retard the spark, and in the opposite direction of rotor rotation to advance the spark. Align the marks on the flywheel and the engine with the flashes of the timing light.

14. When the marks are aligned, tighten the distributor lockbolt and recheck the timing with the timing light to make sure that the distributor did not move when you tightened the lockbolt.

15. Turn off the engine and remove the timing light.

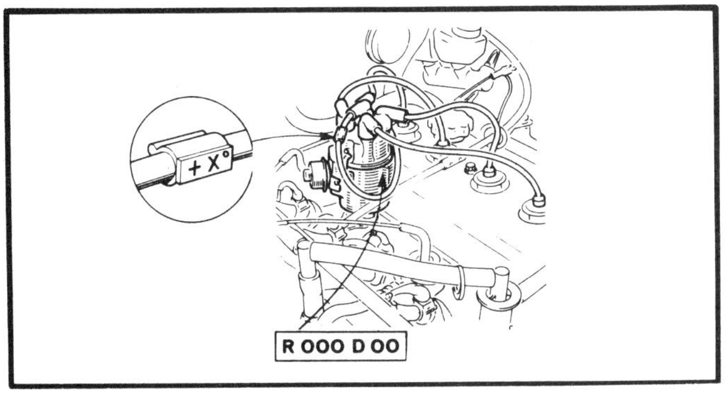

Identification of secondary wiring for initial timing—typical

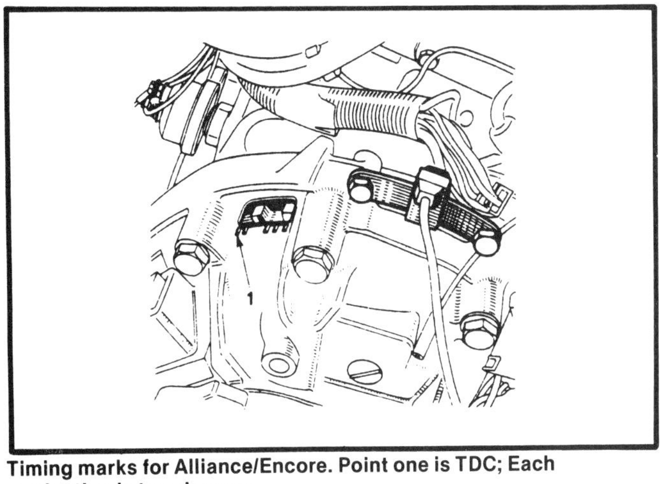

Timing marks for Alliance/Encore. Point one is TDC; Each graduation is two degrees

SPARK PLUG APPLICATION CHART

Year	Model	Spark Plug
'85–'86	LeCar, Encore, Alliance	RN-12YC
	Sport Wagon	CR42LTS or RS9YC

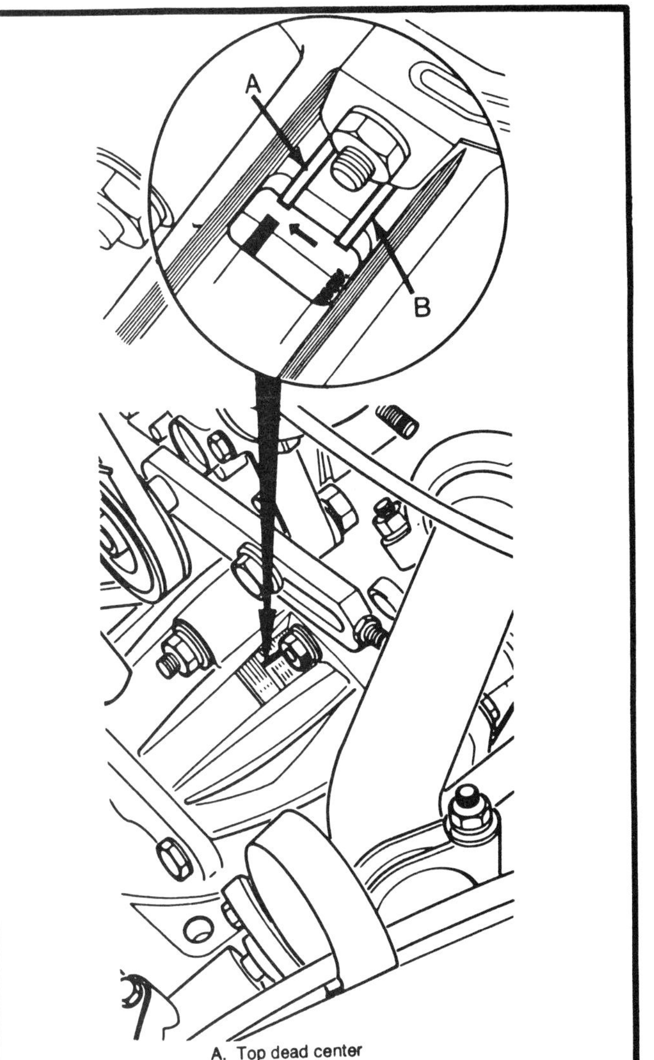

Ignition timing, LeCar

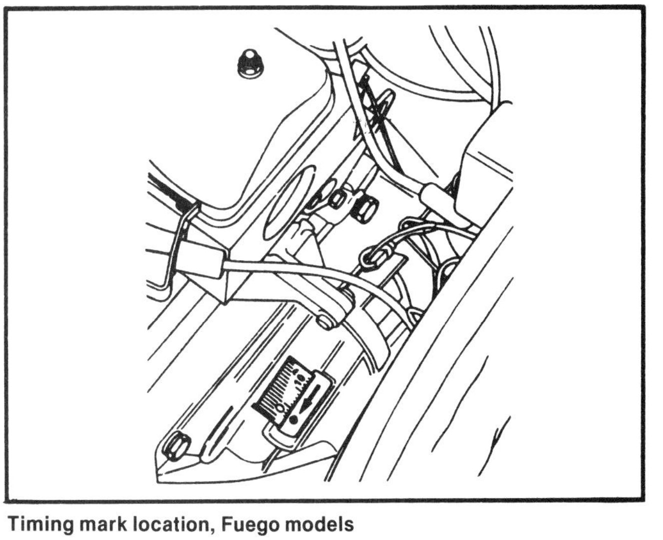

Timing mark location, Fuego models

EMISSION CONTROL SYSTEM

POSITIVE CRANKCASE EMISSION SYSTEM

Positive Crankcase Ventilation System

Crankcase vapors are routed to the induction system to be burned in the combustion chambers instead of merely being vented to the atmosphere.

EVAPORATIVE EMISSION CONTROL SYSTEM

The main component is a canister containing charcoal and a filter. When the engine is off, gasoline vapor settles in the charcoal instead of being vented to the atmosphere. When the engine is started, intake manifold vacuum pulls fresh air in from beneath the charcoal (filtered) and carries the unburned gases to the intake manifold to be burned in the combustion chambers. The vacuum will continue to draw a small amount of vapor from the tank (through a calibrated orifice) even while the engine is running. The process will cycle in this manner, beginning at the time the engine is turned off.

EXHAUST EMISSION CONTROL SYSTEM

Carburetor Air Intake Pre-Heating System

A thermostatically controlled air cleaner mixes heated and ambient incoming air to the carburetor, thereby improving fuel vaporization which assists in improving emissions and driveability during cold engine operation.

Accelerated Idle System

This system is intended to reduce the amount of hydrocarbon emissions during deceleration. The primary throttle plate is opened slightly during deceleration, between speeds of 15–20 miles per hour. Some models use a coolant temperature sensing switch to activate the solenoid.

Air Injection

A belt-driven pump delivers fresh, compressed air to the exhaust ports of the engine. The presence of oxygen will lengthen the combustion process, which reduces the amount of unburned gases in the exhaust. A relief valve and a diverter valve control the amount of air delivered to the exhaust ports. Depending upon whether the engine is accelerating/cruising or decelerating, the diverter valve either delivers the air to the exhaust ports, or closes off the air passage to the exhaust ports, respectively. A check valve is provided in the system which will prevent damage to the air injection pump should the engine backfire.

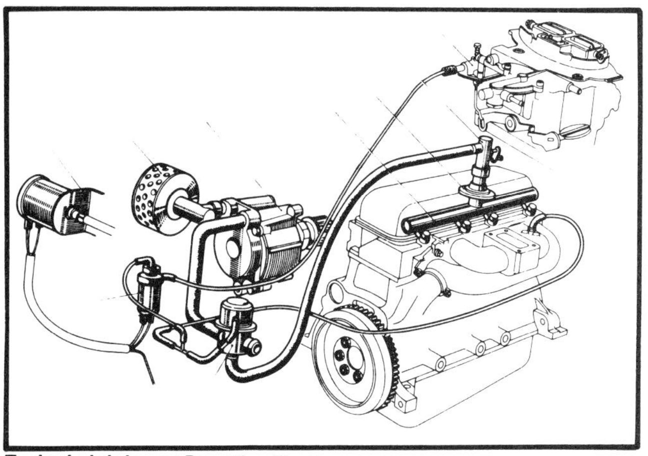

Typical air injector Reaction System

Catalytic Converter

The catalytic converter chemically alters the exhaust gases before the gases reach the atmosphere. A "two-way" catalyst uses either pellets or screens coated with platinum and palladium. The chemical content of the precious metals oxidizes (neutralizes) controlled amounts of carbon monoxide and hydrocarbons. A "three-way" catalyst uses platinum, palladium, and rhodium, which acts on oxides of nitrogen emissions. Some "three-way" converters also are ported to accept fresh air (from an injection pump) which further reduces oxides of nitrogen emissions.

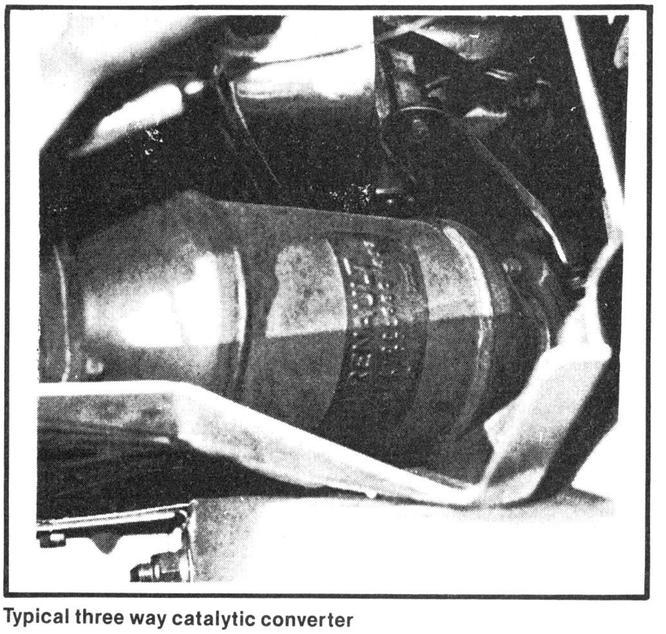

Typical three way catalytic converter

E.G.R. (Exhaust Gas Recirculation) System

The E.G.R. valve admits varying amounts of exhaust gases into the combustion chambers, thereby diluting the incoming air/fuel mixture to reduce oxides of nitrogen.

Ignition Timing Advance Control System

A contact switch which works in conjunction with the choke cable (on LeCar models) closes and activates a solenoid which admits full vacuum to the distributor vacuum advance unit for improved emissions and driveability when the choke valve is closed.

Oxygen Sensor

The ceramic-clad sensor is housed in the exhaust manifold. The ceramic detector is enclosed in a threaded base. The outer part of the detector is in contact with exhaust gases, while the inner part is in contact with ambient air. When exhaust gas temperature passes 300 degrees Centigrade, the sensor compares the oxygen content of the gases with the air. If the disparity exceeds specifications, the sensor is activated.

Service to the sensor is limited to replacement necessitated by exhaust gas erosion. Inspection of the sensor is by visually checking it for signs of wear. To remove the sensor, use a 22mm crow-foot wrench. Hand start the new sensor.

CAUTION

Use antiseize compound on the threads ONLY. Any foreign substance on the ceramic will impair its sensing abilities.

Tighten the sensor to 28–34 ft.lb.

Wide Open Throttle Switch

The WOTS is mounted on the side of the side of the throttle body and regulates mixture through the Electronic Control Unit. To replace the WOTS:

1. Remove the air cleaner.

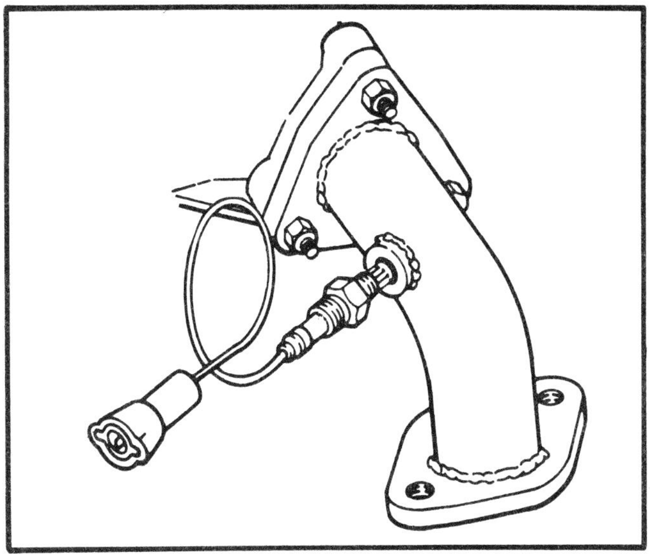

Oxygen sensor installation location

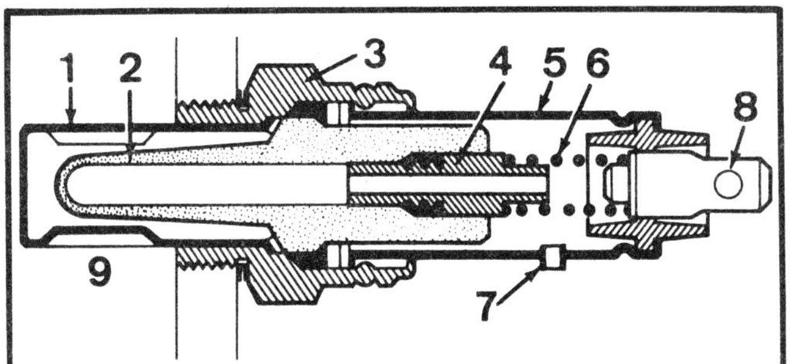

1. Protective sleeve
2. Ceramic sensor
3. Base
4. Contact cover
5. Protective cover
6. Contact spring
7. Ambient air inlet
8. Electrical connector
9. Exhaust gas

Oxygen sensor cross section

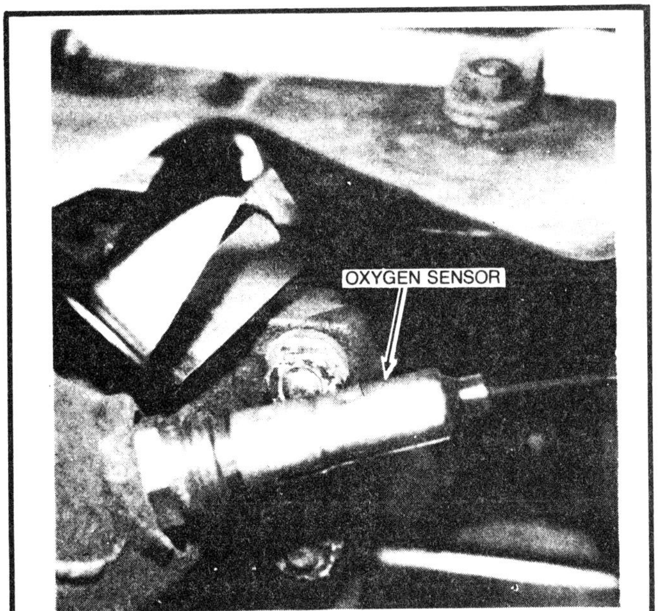

Typical oxygen sensor

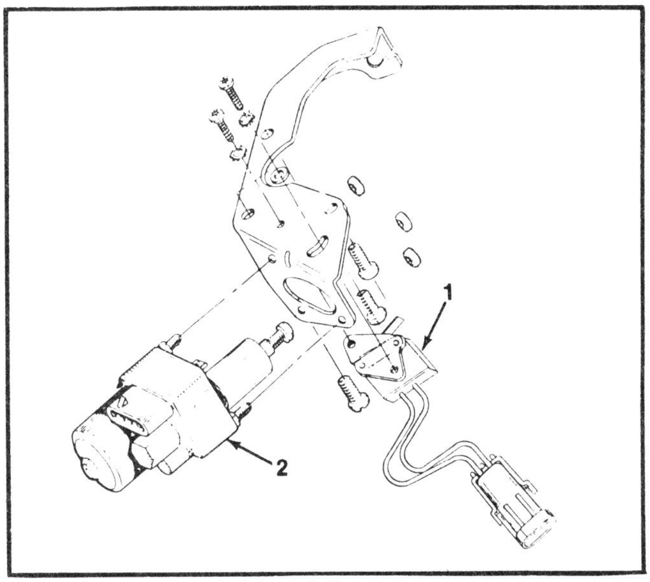

Wide open throttle switch installation

2. Disconnect the throttle return spring.
3. Disconnect the throttle cable.
4. Remove the ISC motor bracket-to-throttle body screws.
5. Disconnect the WOTS and ISC connectors.
6. Remove the bracket, ISC motor and WOTS from the throttle body.
7. Remove the ISC motor.
8. Unscrew the WOTS from the bracket.
9. Installation is the reverse of removal.

Coolant Temperature Sensor

This unit, located in the cylinder head, provides input to the injection control unit to provide the correct air/fuel ratio under changing ambient conditions.

Troubleshooting is by continuity testing with a volt/ohmmeter. Service is limited to replacement.

Manifold Air Temperature Sensor

The MAT sensor is installed in the intake manifold in front of the intake port. The sensor reacts to the temperature of the air/fuel mixture in the manifold and inputs the Electronic Control Unit.

Testing consists of a continuity check with a volt/ohmmeter. Service is limited to replacement. The sensor is fastened by two Torz-head screws.

Coolant Temperature Sensor

The CTS is located in the engine water jacket and inputs the Electronic Control Unit, regulating rich mixture during cold engine operation.

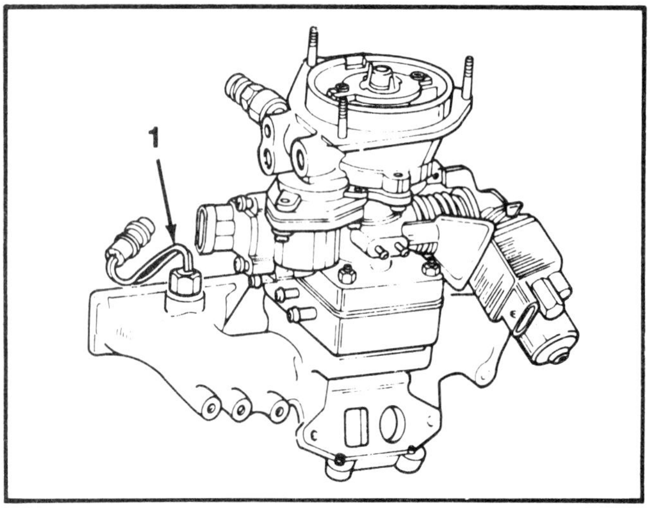

Manifold air temperature switch

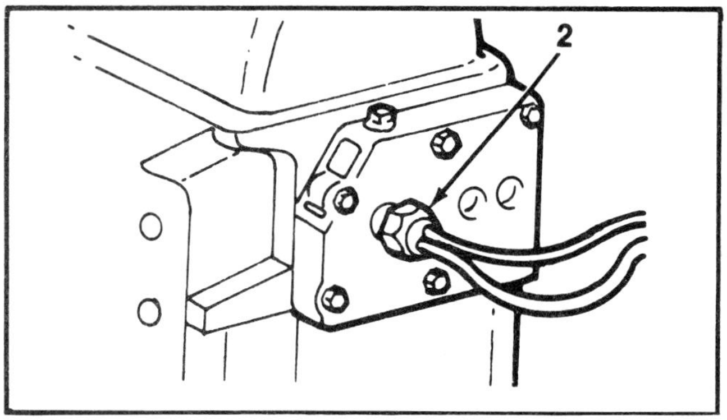

Coolant temperature switch

Testing is by continuity check with a volt/ohmmeter. Service is limited to replacement. Always replace the unit with the engine cold to avoid hot coolant burns.

Manifold Absolute Pressure Sensor

The MAPS reacts to pressure changes in the intake manifold and inputs the Electronic Control Unit, varying mixture according to barometric density. Testing is limited to continuity checks with a volt/ohmmeter.Service is limited to replacement. The MAP is located on the plenum chamber.

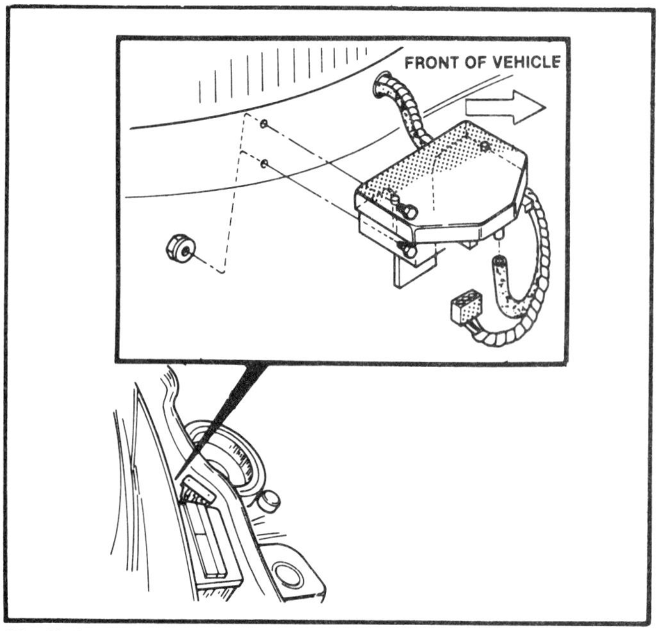

Manifold Absolute pressure switch

Throttle Body Fuel Injection System

DIAGNOSTIC CONNECTORS—1986 TBFI SYSTEM

On 1986 Alliance/Encore models, the diagnostic connectors (D1 and D2) are located on the heater plenum in the engine compartment.

The 1986 diagnostic connector pin terminal circuits are outlined in the following chart.

FUEL PUMP OPERATING PRESSURE '86

Engine	Fuel Pump Operating Pressure
1.4L (85.2)	1.0 bar (14.5 psi)
1.7L (100.5)	1.2 bar (17.4 psi)

ECU CONNECTOR

1986 TBFI SYSTEM

The ECU connector used for the 1986 TBFI system is similar to the connector used in previous systems. Terminal identification and specific engine application is detailed.

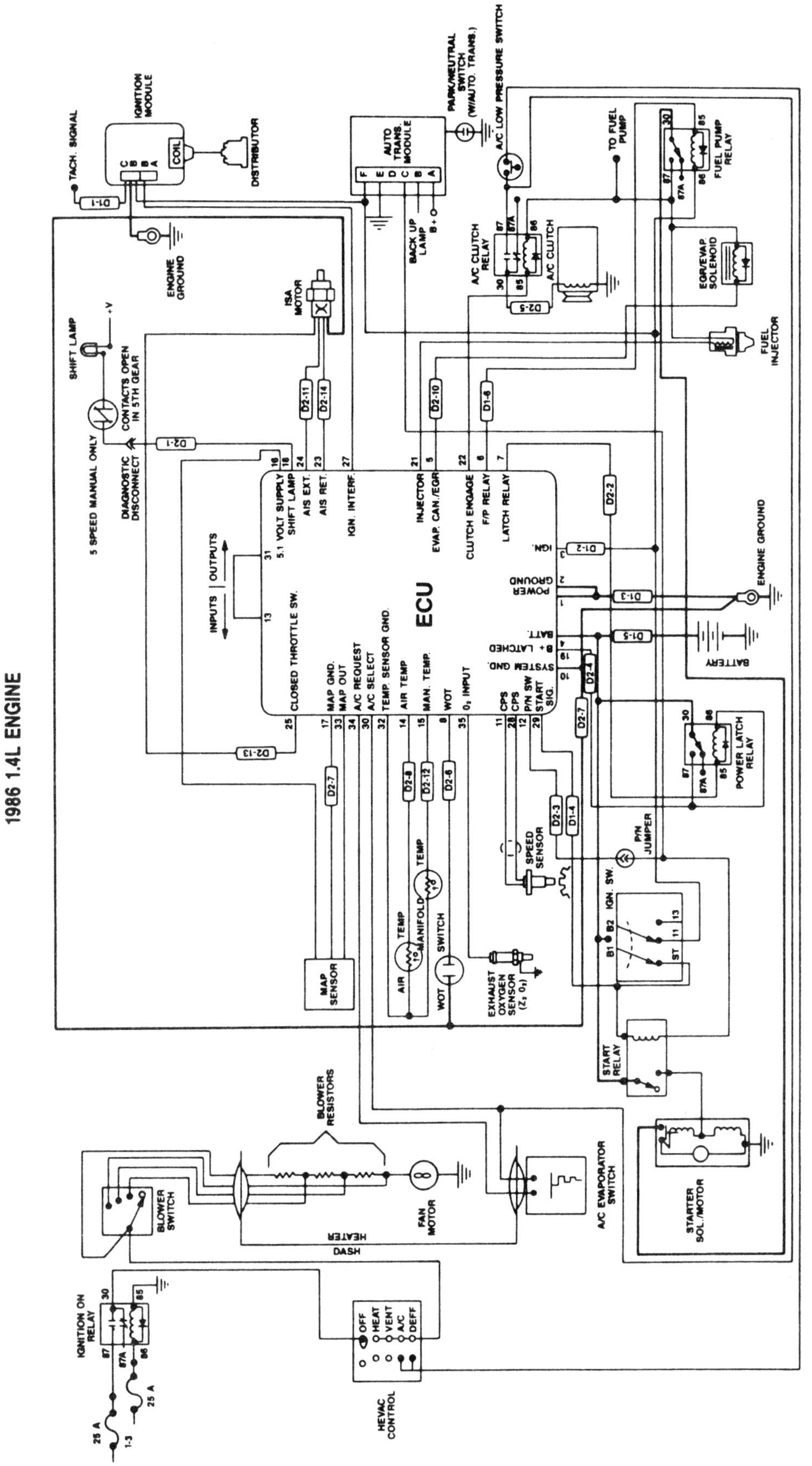

Electrical schematic for 1986 1.4L engine

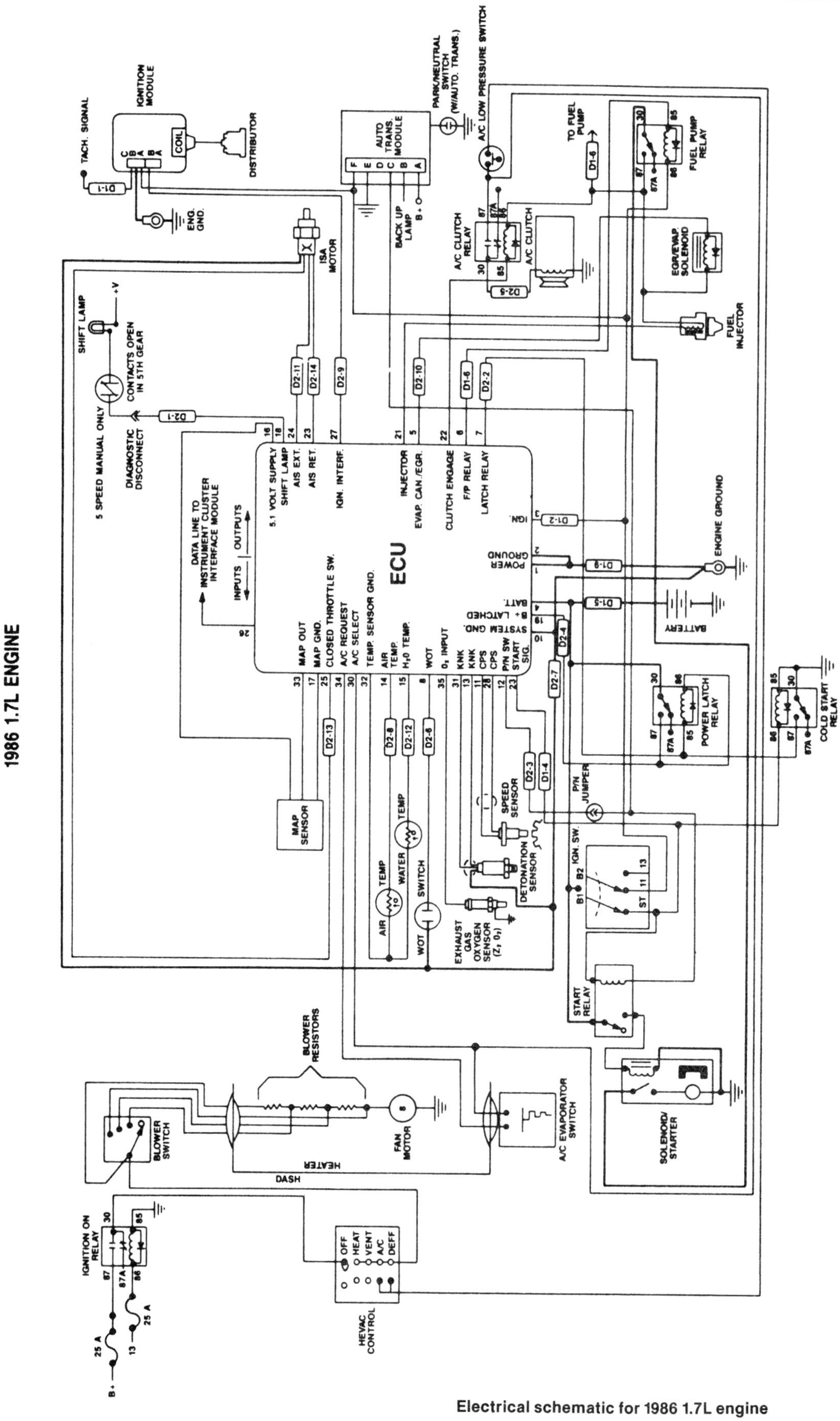

Electrical schematic for 1986 1.7L engine

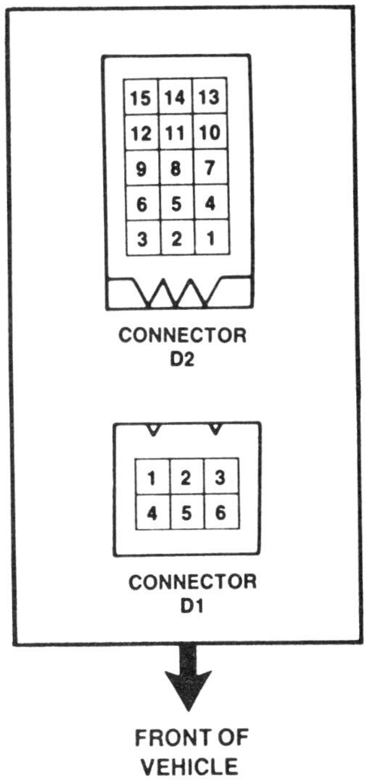

Connector D1

1. Tach Input
2. Ignition
3. Ground
4. Start Solenoid
5. Battery
6. Fuel Pump

Electronic diagnostic connector, 1986 TBFI system for Alliance/ Encore, mounted on the heater plenum

AIR/FUEL TEMPERATURE (MAT) SENSOR—ALL

A new air/fuel temperature sensor is used with 1986 Alliance/Encore fuel injection systems. The new sensor has an integral wire harness, a smaller diameter boss, and different calibration for use with the 1986 system.

BALLAST RESISTOR ELIMINATION

The fuel system ballast resistor used on previous models, is not required with 1986 Renix fuel injection systems. Although the resistor has been eliminated from the 1986 system, it is still required on prior year models.

COLD START RELAY—1.7L ENGINE

For 1986, a cold start relay will only be used with the Alliance/Encore 1.7L engine. Refer to the wiring diagrams in Tests and Diagnosis for circuitry details.

INTAKE MANIFOLD TEMPERATURE (MT) SENSOR–1.4L ENGINE

A new design intake manifold temperature sensor is used with the 1986 Alliance/Encore 1.4L engine fuel injection system. The new sensor (1) does not come into contact with engine coolant. Instead, it measures temperature of the intake manifold outer surface. The new sensor is located on the intake manifold.

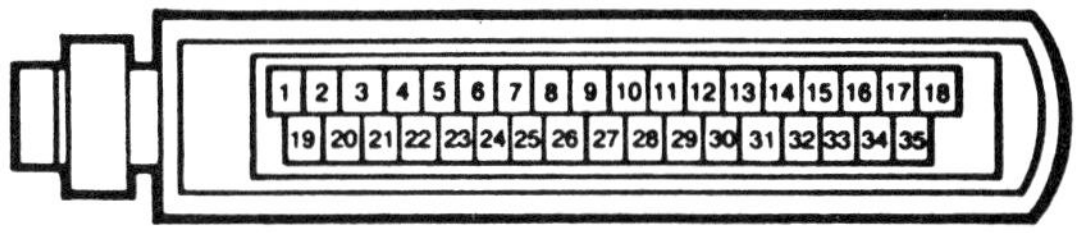

1. Ground
2. Ground
3. Ignition Switch
4. Battery
5. EGR Valve/Canister Purge
6. Fuel Pump Relay
7. System Power Relay (latch relay)
8. WOT Switch
9. Not Used
10. System Ground
11. Speed Sensor
12. Park/Neutral Switch (auto trans. only)
13. Knock Sensor (1.7L)
14. Manifold Air/Fuel Temperature Sensor
15. Coolant Temperature Sensor (1.7L) Manifold Temperature Sensor (1.4L)
16. Manifold Absolute Pressure—Supply Voltage
17. Manifold Absolute Pressure—Ground
18. Shift Lamp (manual trans. only)
19. System Power (B +)
20. Not Used
21. Injector
22. A/C Compressor Clutch
23. ISA Motor Retract (reverse)
24. ISA Motor Extend (forward)
25. Closed Throttle (idle) Switch
26. Not Used
27. Ignition Interference
28. Speed Sensor
29. Start Signal
30. A/C Select
31. Knock Sensor (1.7L)
32. Temperature Sensor Ground
33. MAP Sensor Output
34. A/C Request
35. O_2 Sensor Input

ECU connector used with the 1986 TBFI system

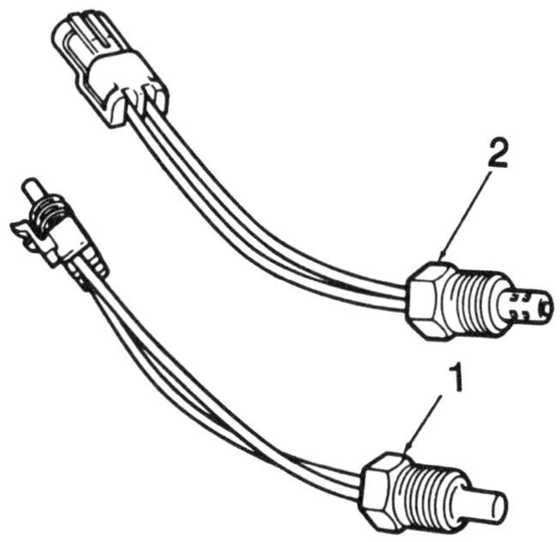

Comparison between new (1) and old style (2) intake manifold temperature sensor, used on the 1.4L engine

VACUUM CIRCUITS

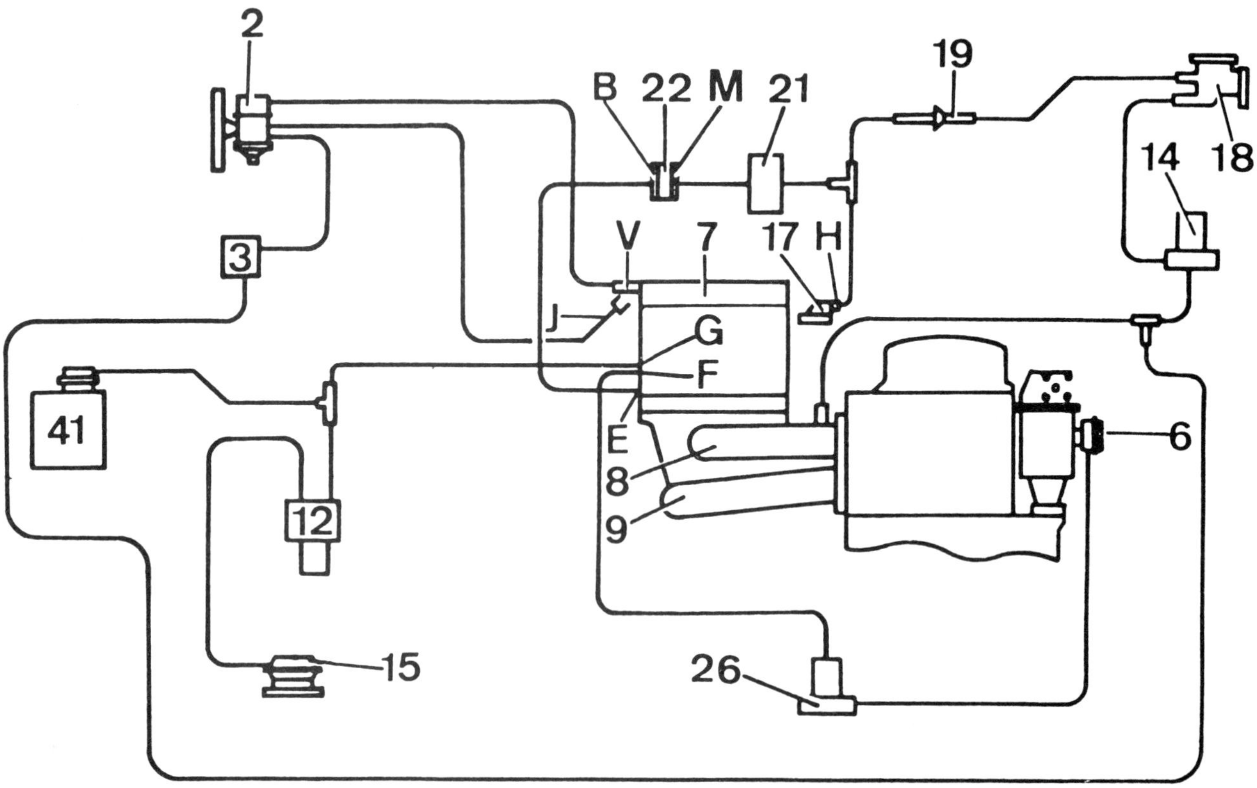

Item	DESCRIPTION
2	Vacuum solenoid regulator
3	Vacuum tank (fuel control system)
6	Distributor
7	Carburetor
8	Intake manifold
9	Exhaust manifold
12	Coolant thermovalve
14	Fast idle solenoid valve
15	E.G.R. valve
17	Throttle plate opener
18	Vacuum regulator
19	Check valve
21	Vacuum reservoir (fast idle system)
22	Delay valve
26	Vacuum advance solenoid valve

Item	DESCRIPTION
41	Charcoal cannister
Carburetor references	
J	Yellow ring
V	Green ring
E	Red ring
F	Black ring
G	Brown ring
H	Blue ring
M	Brown side
B	White side

RENAULT LeCAR—WO/AIR CONDITIONING—CALIFORNIA

VACUUM CIRCUITS

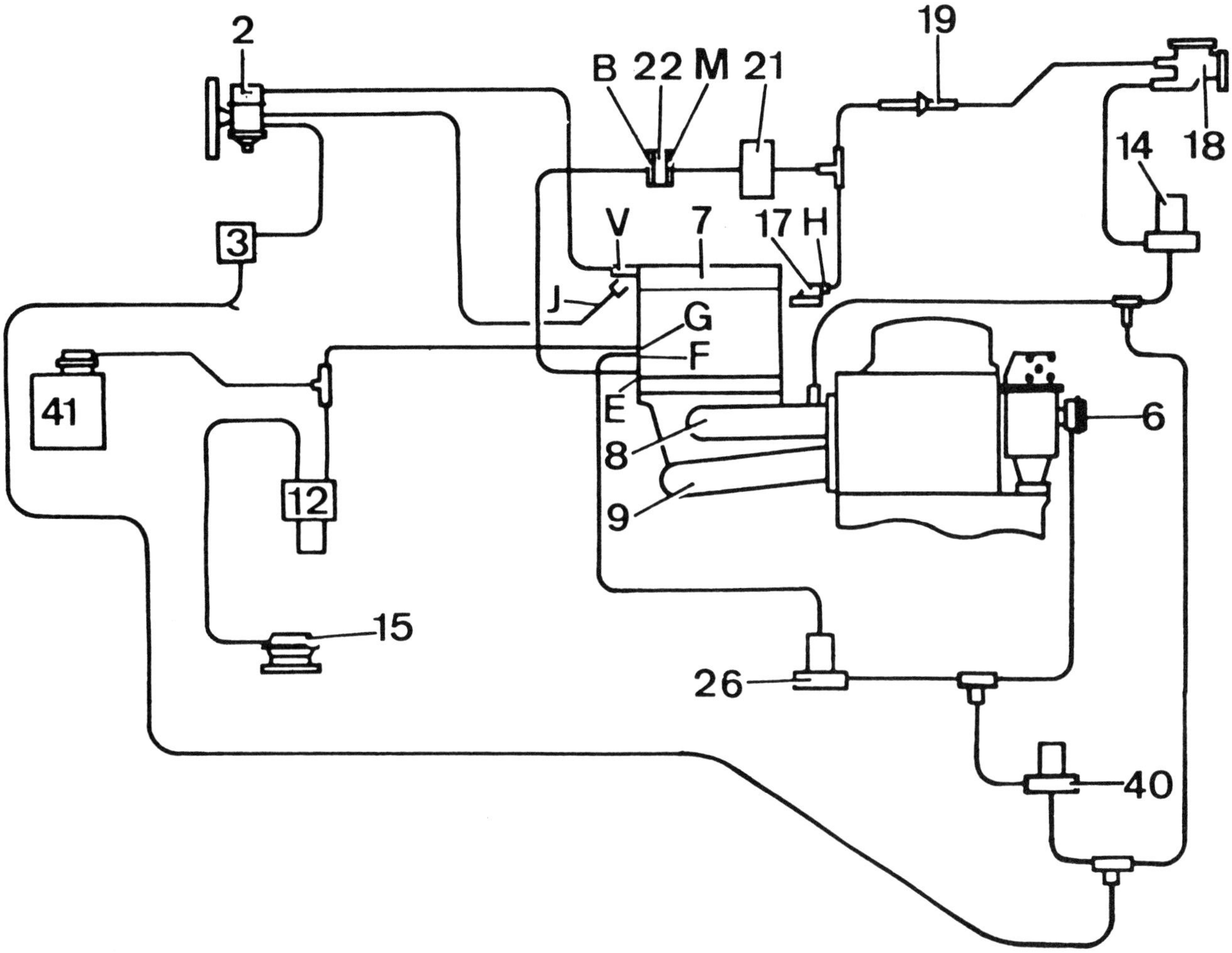

Item	DESCRIPTION
2	Vacuum solenoid regulator
3	Vacuum tank (fuel control system)
6	Distributor
7	Carburetor
8	Intake manifold
9	Exhaust manifold
12	Coolant thermovalve
14	Fast idle solenoid valve
15	E.G.R. valve
17	Throttle plate opener
18	Vacuum regulator
19	Check valve
21	Vacuum reservoir (fast idle system)
22	Delay valve
26	Vacuum advance solenoid valve

Item	DESCRIPTION
41	Charcoal cannister
Carburetor references	
J	Yellow ring
V	Green ring
E	Red ring
F	Black ring
G	Brown ring
H	Blue ring
M	Brown side
B	White side

RENAULT LeCAR—W/AIR CONDITIONING—CALIFORNIA

VACUUM CIRCUITS

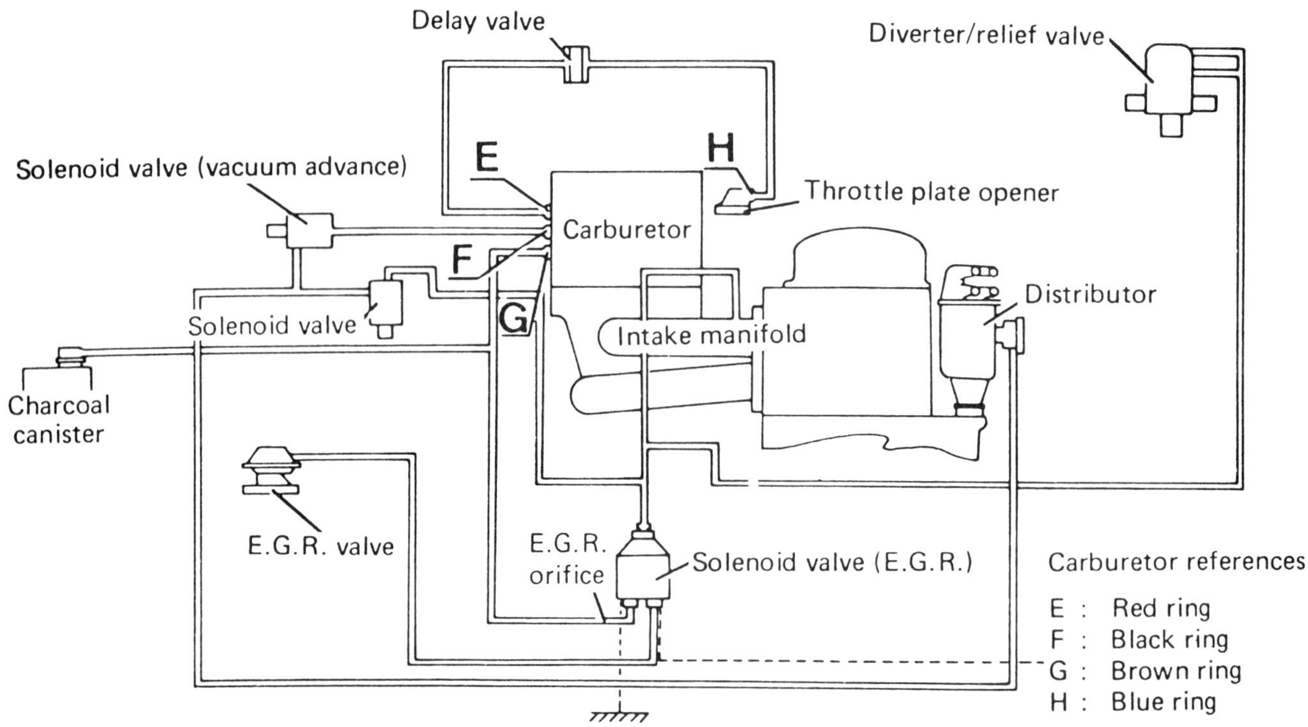

RENAULT LeCAR—W/AIR CONDITIONING—49 STATES

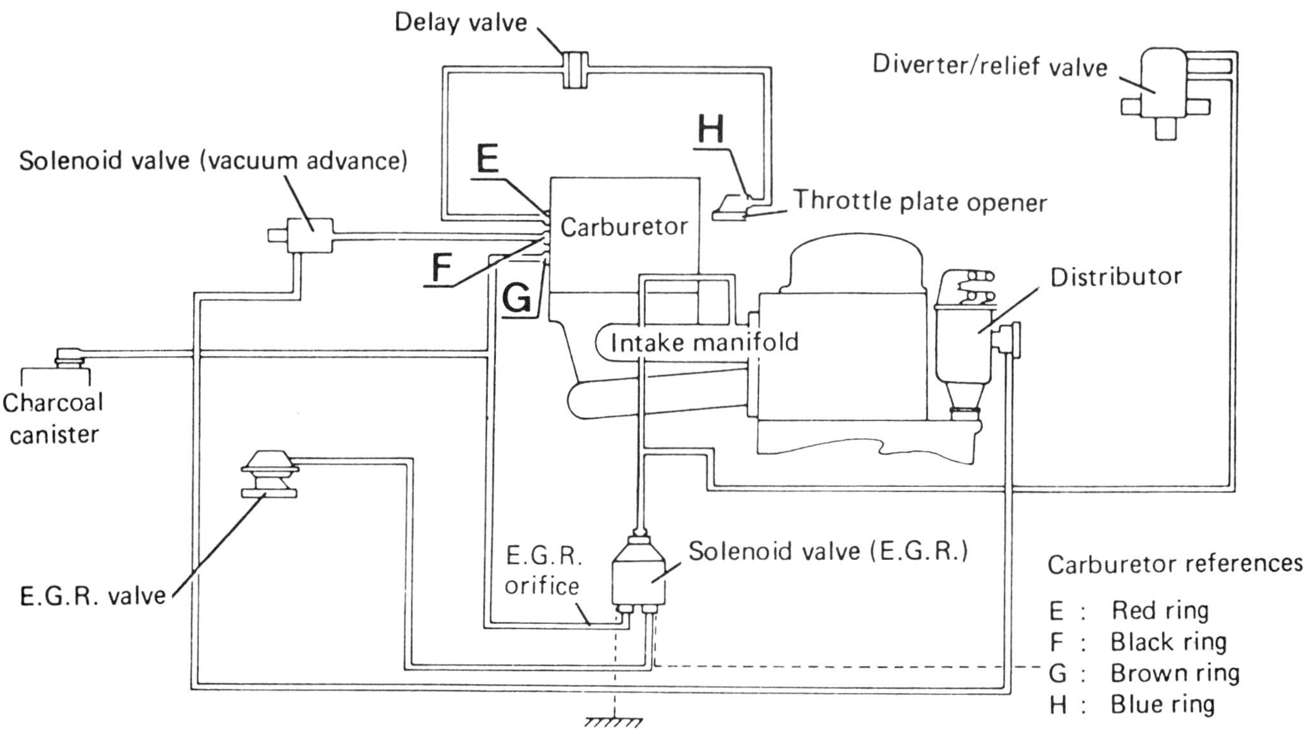

RENAULT LeCAR—WO/AIR CONDITIONING—49 STATES

SAAB

INDEX

VEHICLE IDENTIFICATION NUMBER AND ENGINE IDENTIFICATION

Vehicle Identification Number (VIN)

The Vehicle Identification Number (VIN) is located on the right inner fender panel and on the left top side of the dash panel, directly behind the windshield. A seventeen digit identification number is used to identify the manufacturer and included components. The chassis number is also punched in the rear crossbeam in the luggage compartment.

Engine Identification

The engine identification number is stamped on a pad, located on the left side of the engine block, forward of the fuel injection unit.

YS3 A M 3 5 J X E 1 000001

Positions 1-3 — Manufacturer
YS3 = Saab-Scania AB

Position 4 — Model
A = 900

Position 5 — Series
M = 900
H = 900 S
T = Turbo

Position 6 — Body type
2 = 2-door Sedan
3 = 3-door Combi Coupé
4 = 4-door Sedan
5 = 5-door Combi Coupé
6 = 4-door Sedan, extended length

Position 7 — Transmission type
4 = 4-speed manual
5 = 5-speed manual
6 = 3-speed automatic

Position 8 — Engine type
J = Fuel injection
S = Turbo
L = Turbo with charging air cooler

Position 9 — Check digit

Position 10 — Model year
E = 1984
F = 1985

Position 11 — Assembly plant
1 and 2 = Trollhättan
3 = Arlöv
6 and 7 = Nystad (Finland)

Positions 12-17 — Serial number

Decoding of Vehicle Identification Number (VIN)

TUNE-UP SPECIFICATIONS

Year	Engine Displacement (cc)	Spark Plugs Type	Spark Plugs Gap	Distributor Point Dwell (deg)	Distributor Reluctor Gap (in.)	Basic Ignition Timing (deg)	Intake Valve Opens (deg)	Fuel Pump Pressure (psi)	Idle Speed (rpm)	Valve Clearance (in.) Intake	Valve Clearance (in.) Exhaust
'85–'86	1985	③	0.024–0.028	Electronic ⑦	0.010	②⑥⑧	10 BTDC ⑤	①	875	0.008–0.010	0.016–0.018 ④

NOTE: If these specifications differ from those on the engine compartment stickers, use the sticker specifications.

① Fuel injected engines (all models): Fuel line pressure before the control pressure regulator is 66.9-69.7 (setting valve), and 48.5-54.0 PSI (warm engine) after the control pressure regulator (located in fuel distributor).

② '85–'86—20° @ 2000 rpm (non-Turbo)
'85–'86 Canada, manual transmission—20° BTCD @ 2000 rpm (non-Turbo)
'85–'86 Canada, automatic transmission—23° BTCD @ 2000 rpm (non-Turbo)

③ Turbo—NGK-BP7ES, Champion N7Y or N7YC; non-Turbo NGK BP6ES, Bosch W70 or Champion N9Y, N9YC

④ Turbo—Hydraulic Tappets—No adjustment

⑤ Turbo—12° BTDC

⑥ With vacuum hose disconnected

⑦ Canadian models with points ignition, 47-53° dwell

⑧ Tubo 16 valve—16° @ 850 rpm (USA and Canada)

FIRING ORDERS

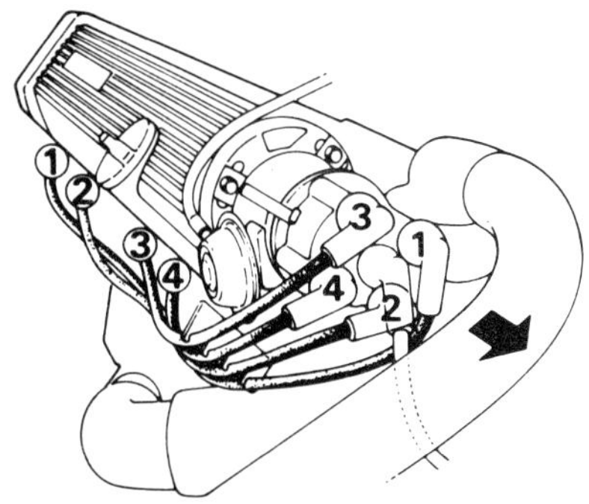

Distributor location and firing order, eight valve engines

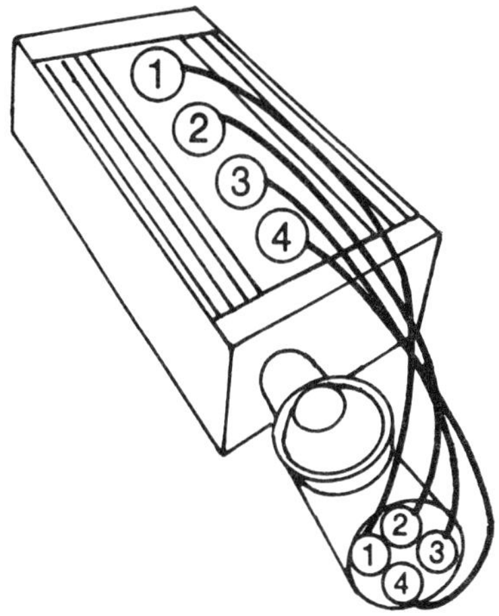

Firing order and ignition cable positioning—1985 and later 16 valve

EMISSION EQUIPMENT USED

SAAB 900 AND 900 TURBOCHARGED (1985–86)

Catalytic Converter (CAT)
Thermostatic Air Cleaner (THERMAC)
Oxygen Sensor(OS)
Dashpot(DP)
Fuel Evaporation System (FES)
Positive Crankcase Ventilation(PVC)
Deceleration Control System (DCS)
Exhaust Gas Recirculation (EGR)
Charcoal Canister (CC)
Delay Valve(DV)
Electronic Control Unit (ECU)
Fuel Shut-Off System (FSS)

Note: Model B201 = four cylinder, four stroke engine with overhead camshaft. Model B202 = four cylinder, four stroke engine with twin overhead camshaft.

REQUIRED EMISSION CONTROL MAINTENANCE

CATALYTIC CONVERTER

Check for body deterioration or leaks. Replace any defective parts or hanger brackets as required. Reset the indicator light on the dash panel, if equipped.

THERMOSTATIC AIR CLEANER

Replace the air cleaner very 15,000 miles, ore frequently if the vehicle is driven daily in heavy traffic or unusually dusty conditions. Check all system components for proper operation.

OXYGEN SENSOR

Replace the oxygen sensor every 30,000 miles and check the operation of the enrichment micro—switch. Replace if defective. Reset the service reminder lamp. Check the system for proper operation.

DASHPOT

The dashpot is located on the throttle housing and is set at the factory. It cannot be serviced.

FUEL EVAPORATION SYSTEM

Check the fuel filter cap, vent lines, canisters and connections for wear, deterioration and/or damage which could cause leakage. Tighten any loose connections and replace any damaged components. The system should be checked 60,000 miles or 48 months and every 12 months thereafter.

CONTINUOUS INJECTION SYSTEM

Inspect the injection hoses and the check valve contained in the injection system Replace defective items as required.

POSITIVE CRANKCASE VENTILATION

Every 15,000 miles, check all hoses and connection for leaks or deterioration. Remove and flush all lines then blow dry using compressed air to clear any blockage. Check the valve and if found to be defective or clogged replace with a new one. Do not attempt to clean an old PVC valve.

DECELERATION CONTROL SYSTEM

This system is used on manual transmission vehicles only. The valve operation should be checked every 15,000 miles. Inspect the by-pass hose, filter and electrical components for any damage or deterioration. Replace any defective parts as required.

EXHAUST GAS RECIRCULATION

Inspect the EGR valve very 15,000 miles. Clean the EGR valve assembly with a cleaning solvent and blow dry using compressed air, Check the rubber components and replace any hoses that are defective or damaged. Remove any carbon deposits.

DELAY VALVE (CANADIAN VEHICLES ONLY)

The delay valve is mounted in the vacuum passage between the throttle valve housing and the vacuum control unit of the distributor. To check the delay valve, a timing light, tachometer and stop watch are needed.

1. Connect the timing light and the tachometer to the engine.
2. Start the engine and run until it reaches normal operating temperature.
3. Run the engine at 3000 rpm and take the time from the time the throttle is closed to open.
4. Check the firing point using the timing light. The vacuum regulator should cut in after a few seconds, and the ignition advance should increase, if not replace the delay valve.

FUEL SHUT-OFF SYSTEM

The fuel shut-off system stops the injection of fuel during deceleration which in turn reduces the consumption of fuel and the hydrocarbon emissions. Check the system components for proper operation.

ELECTRONIC CONTROL UNIT

The electronic control unit is part of the ignition system which amplifies the pulse received from the distributor. The control unit is mounted in front of the left hand wheel housing. There is no periodic maintenance and if a fault is suspected with the unit the entire unit must be replaced.

CHARCOAL CANISTER

Replace the charcoal canister every 60,000 miles. Check all hoses leading to the canister making sure that they are not brittle or kinked preventing proper operation of the unit.

IDLE SPEED AND MIXTURE

NOTE: The following information is being published from the latest information available at the time of publication. If the information on the underhood emissions control label, use the data on the emissions label.

Adjustment

1. Run the engine until it reaches operating temperature.
2. Adjust the idle speed at 875 rpm.
3. If the vehicle is not equipped with a catalytic converter, remove the pulse air hose and plug the air intake to the non-return valve.

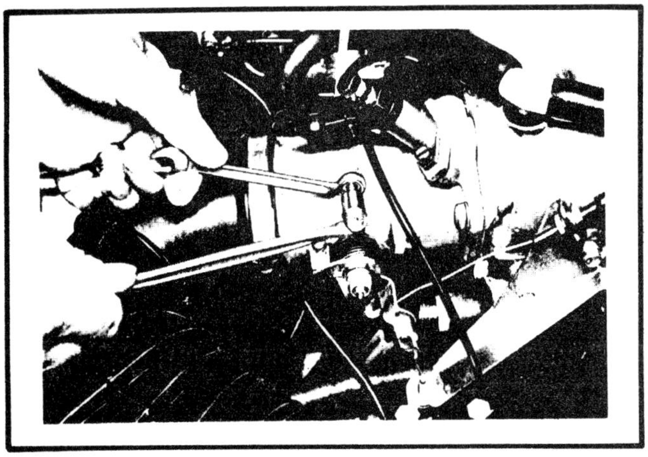

Adjusting the idle speed

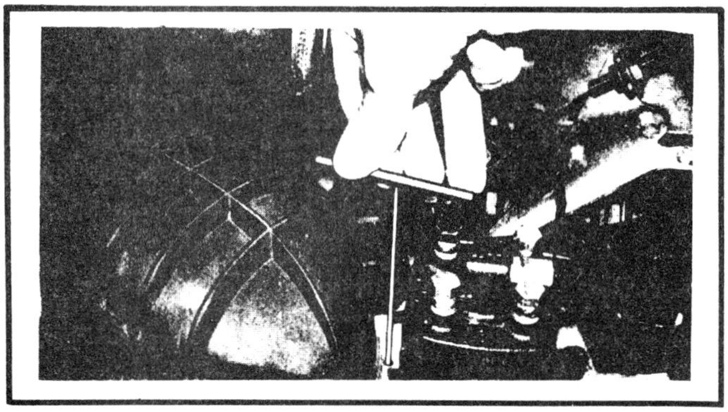

Adjusting the CO value

4. Connect the CO meter sensor to the exhaust pipe. Remove the oxygen sensor wire.
5. Read and adjust the idle speed and CO valve a required. Before each reading, accelerate the engine and allow the engine to return to curb idle. Wait 30 seconds before taking the next CO reading.
6. Adjust the idle speed by turning the idle adjusting screw on the throttle valve housing.
7. Adjust the CO level by turning the adjusting screw located on the fuel distributor. Turning the screw clockwise will make the mixture richer while turning it counterclockwise will make it easier.

NOTE: Be sure to check the emissions label provided on the vehicle for any changes before making the necessary adjustments.

PERMISSIBLE CO CONTENT OF EXHAUST GASES

Catalyst equipped	③④	**1.0 ± 0.25%**
Non-catalyst equipped	①②	0.75 ± 0.25%
Canada, except Turbo model ②		1.5 ± 0.5%

① Pulse air disconnected and plugged
② Charcoal canister disconnected
③ Check valve with sensor connected, after catalyst. Maximum 0.3%
④ 1985 and later; probe in front of catalyst—0.5 to 1.2% CO. probe behind catalyst—less than 0.3% CO.

INITIAL TIMING SETTINGS

IGNITION TIMING

Ignition timing is set in the conventional manner, using the makes that are located on the flywheel. The engine is also equipped for checking the timing using standard ignition service instruments.

NOTE: The SAAB ignition service instruments consist of a tachometer, cam angle motor, timing light and a switch for operating the starter. These are connected to the vehicle in the standard manner.

The equipment in the vehicle comprises a pin in the flywheel and a service socket in the clutch cover. The service instrument is connected to the clutch cover by means of a special connector and to the plug lead of the number 1 cylinder by means of a terminal. The ignition service instrument is also connected to ignition service socket at the fuse box and by means of a impulse transmitter at the plug lead for the number 1 cylinder.

NOTE: The following information is being published from the latest available information. If these specifications differ from those found on the emissions label, follow the data on the emissions label.

IGNITION TIMING

Engine	Degrees BTDC @ rpm
B201 (8 valve)	20 @ 2000
B202 (16 valve)	16 @ 850

NOTE: Verify timing setting with Vehicle Emission Information label. If specifications differ, use information from Emission label.

Vacuum Advance

PROPORTIONAL EGR

The EGR valve is controlled by means of a vacuum regulator. When the valve opens, a small amount of exhaust gases is recirculated through the EGR crosspipe the EGR valve to the inlet manifold. When the induction air passes through the venturi, a venturi signal which is proportional to the total air flow is obtained. The signal is transmitted to the EGR vacuum amplifier which amplifies the signal 14 times by means of the manifold vacuum reservoir.

The manifold vacuum reservoir is connected to the amplifer. The amplified signal then goes via the vacuum signal switch cuts out the EGR signal at engine speeds below 2500 + 500 rpm. This occurs by means of a hole drilled through the throttle valve housing (during the

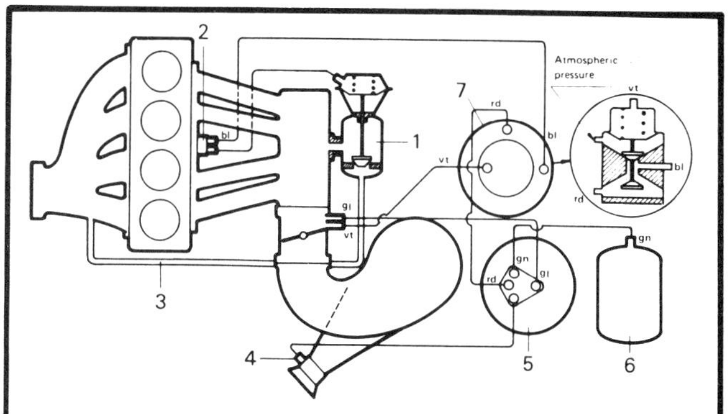

1. EGR valve
2. PVS valve
3. EGR crosspipe
4. Venturi with vacuum signal connection
5. Amplifier
6. Vacuum reservoir
7. Vacuum signal switch

EGR—Proportional system

running-in of the engine, the cut-off speed can be somewhat lower). The ported vacuum switch valve senses the temperature of the coolant and cuts off the EGR signal a temperatures lower than approximately 100°F., which results in improved driveability immediately after cold engine starting. At wide open throttle, the vacuum in the manifold reservoir disappears after a few seconds and the EGR valve closes.

TWO-PORT EGR

In the two-port EGR system, the opening of the EGR valve is regulated by two adjacent vacuum port in the throttle valve housing, a holding valve, a release valve and a ported vacuum switch valve. When the EGR valve opens, a small proportion of the exhaust gases are recirculated to the inlet manifold.

As the throttle valve is opened slightly and the valve passes the two vacuum ports, a gradual increase in the vacuum is obtained along with a gradual opening of the EGR valve.

When the throttle valve is opened wide and the vacuum in the ports diminishes, the earlier vacuum at the EGR valve is maintained for about six seconds by means of the holding valve.

When the throttle valve is closed, the EGR valve must also be closed to prevent rough idling of the engine. A release valve is fitted for this purpose and the valve is regulated by a additional port located inside the throttle valve housing. Thus, the vacuum that is maintained by the holding valve is released.

When the temperature of the engine is below approximately 104°F.,

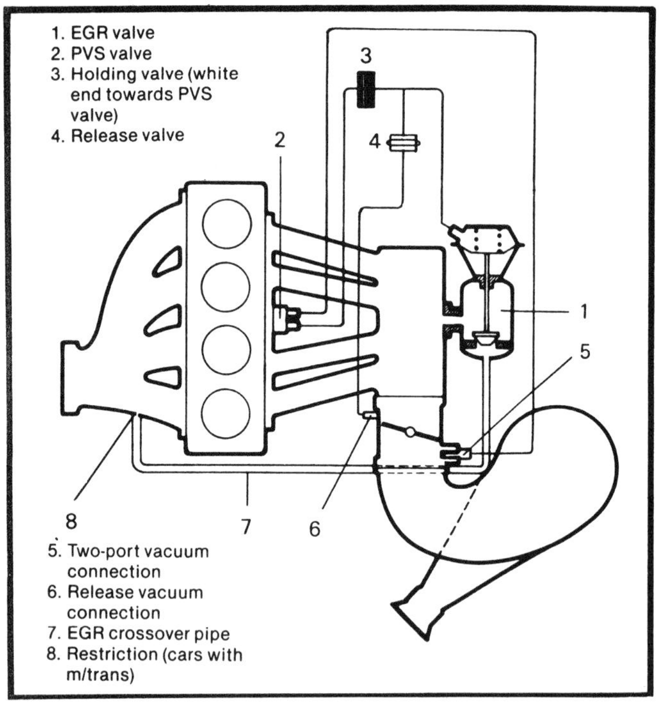

EGR—Two port system

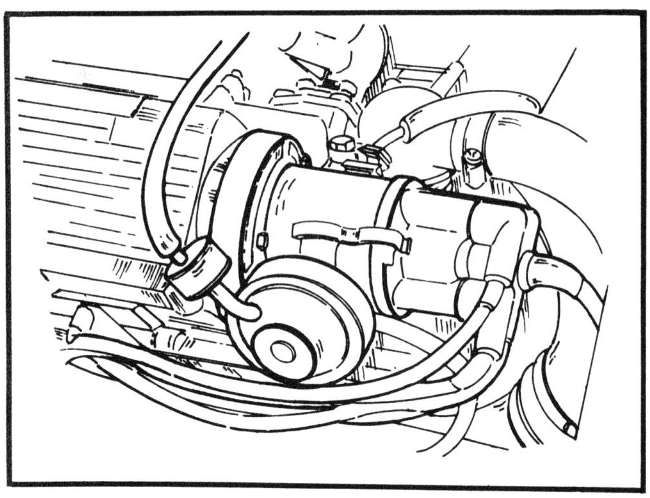

Use of delay valve for vacuum advance

the ported vacuum switch valve shuts off the vacuum between the throttle valve and EGR valve to improve the running of the engine during the warm-up period.

ON-OFF EGR

When this type of EGR valve opens, a small quantity of exhaust gases flow through the metering office from the exhaust manifold. The EGR valve is controlled by means of a vacuum from the throttle valve housing. The vacuum hole is located relative to the throttle valve so that

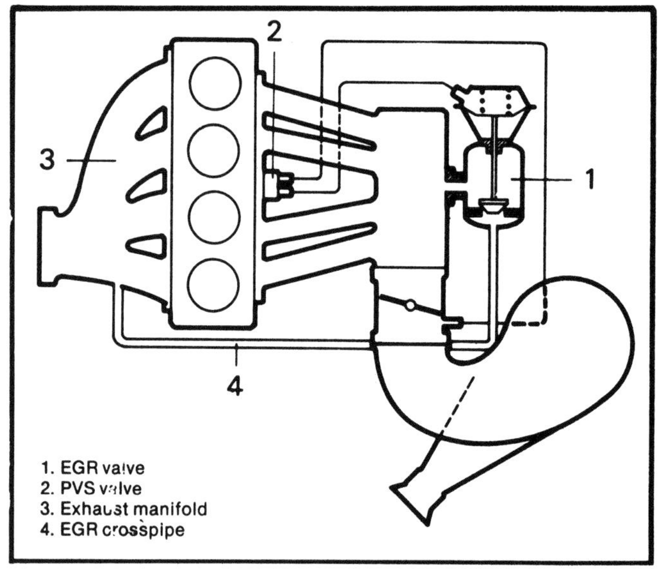

EGR—On-off system

DELAY VALVE OPERATION

Type	Engine speed (fast idling) at which the valve should open
EGR on-off	Approx. 1 900 rev/min
EGR two port, manual transmission	2 600-3 200 rev/min
EGR two port, automatic transmission	2 300-2 900 rev/min

vacuum signal is obtained when the engine speed is about 1,900 rpm. Even during conditions of small loads, a strong vacuum is obtained to open the valve completely. At full throttle and slightly below, the vacuum is so weak that the valve does not open.

The ported vacuum switch valve senses the temperature of the coolant and cuts out the vacuum at temperature lower than approximately 100°F. which means that improved driveability is obtained after starting a cold engine.

DELAY VALVE

A delay valve is mounted in the vacuum passage between the throttle valve housing and the vacuum control unit of the distributor. The valve delays the formation of vacuum by around six seconds. The ignition advance is therefore also delayed during acceleration and the emission of nitric oxide (NOx) is reduced.

The white end of the delay valve should be towards the vacuum control unit of the distributor. It is also important that the valve is fitted with the shorter hose running between the valve and the vacuum control unit of the distributor.

When the suction line is to be disconnected (e.g. in conjuction with checking the ignition timing), always disconnect the hose at the throttle housing. Otherwise there will be a risk of dirt or debris entering and clogging the delay valve.

SPARK PLUGS

Engine	Plug Type: Normal Driving	Plug Type: Hard Driving	Gap (in.)
Turbo 8-Valve	NGK BP 6 ES Champion N9YC Bosch W7DC	NGK BP 7 ES	.024–.028
Turbo 16-Valve	Champion C9GY NGK BCP 6 ES Champion C9YC Bosch F7DC	Champion C7GY	.024–.028
Non-Turbo	NGK BP 6 ES Champion N9YC Bosch W7DC		.024–.028

EMISSION CONTROL SYSTEMS

CRANKCASE EMISSION CONTROL SYSTEM

Positive Crankcase Emission System

The crankcase ventilation system is completely enclosed. The system is comprised of a three way nipple in the valve cover, from which a small hose is routed to the inlet manifold, and thicker hose routed to the air cleaner assembly. The sizes of the hoses are designed to regulate the removal of the crankcase gases and to route them into the engine to be mixed with the air/fuel mixture for combustion. The normal routing of the gases are to the inlet manifold, but at times of full acceleration, the gases are routed directly to the air cleaner assembly and mixed with the air as it is drawn into the inlet manifold.

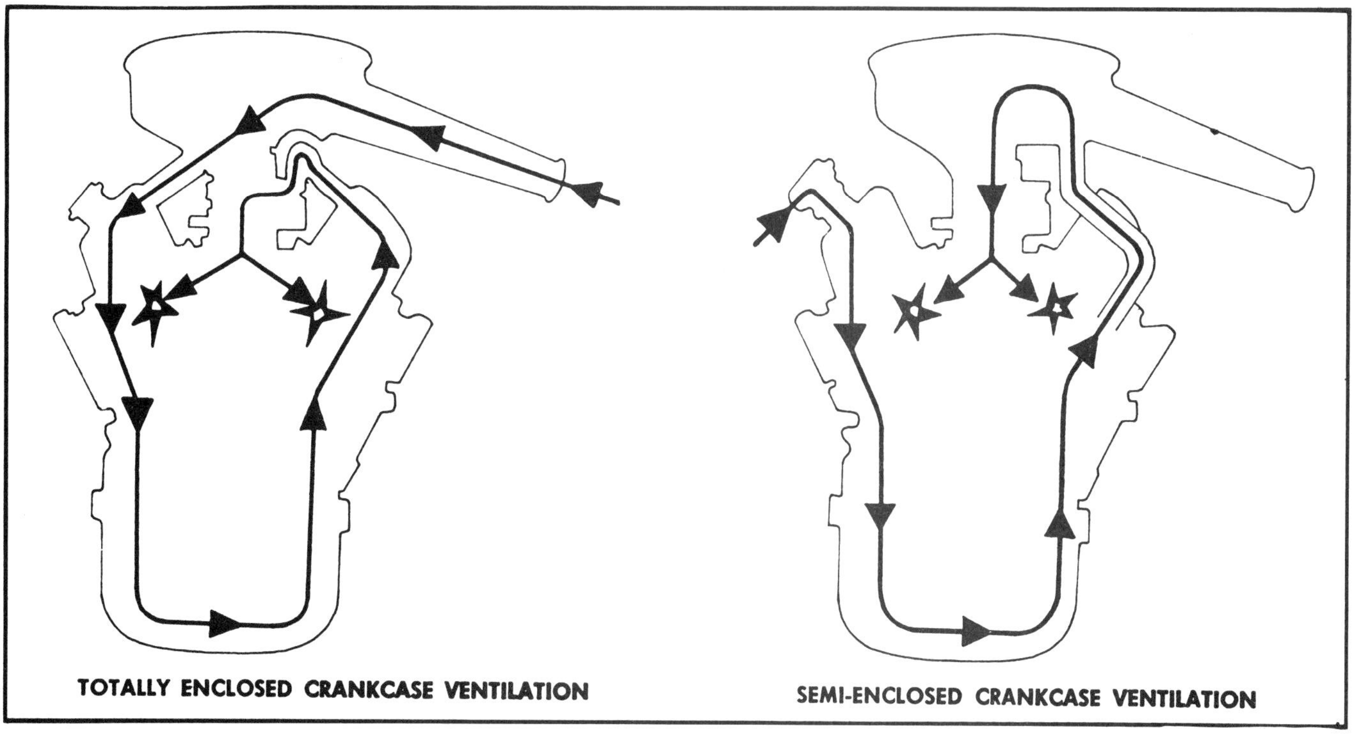

Comparison of semi-enclosed and totally enclosed crankcase ventilation system

POSITIVE CRANKCASE VENTILATION SYSTEM TROUBLE DIAGNOSIS

Symptom	Possible Cause	Repair
Rough engine idle	Defective canister, clogged or leaking vacuum hose.	Replace canister, clean or replace hose.
No PCV flow	Clogged lines or flame trap.	Clean or replace lines and flame trap.
Engine stalls at idle	Saturated carbon canister, clogged breather filter.	Replace canister or filter.
Surging at road speed	Improper PCV flow.	Check PCV system for proper function. Clean all lines and component.

EVAPORATIVE EMISSION CONTROL SYSTEM

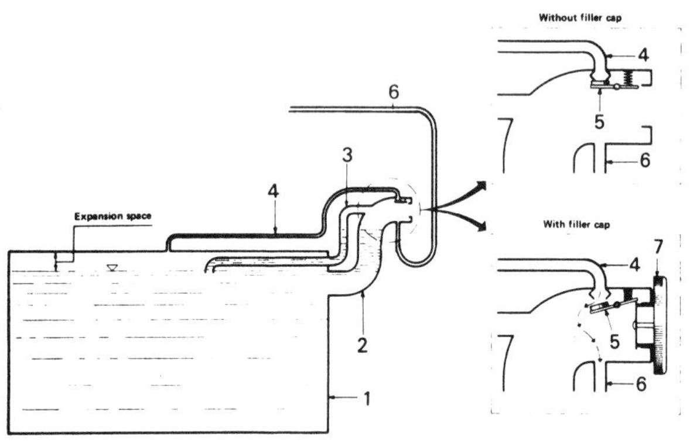

1. Fuel tank
2. Filler pipe
3. Ventilation tube
4. Ventilation tube
5. Spring-loaded valve
6. Ventilation hose
7. Filler cap

Fuel tank ventilation system

Vehicles for the USA-market are quipped with a charcoal filter, which absorbs the vapors from the fuel tank. The charcoal canister is placed in the engine compartment. It is connected to the vent hose of the fuel tank and with a hose to the air cleaner. When the engine is running, fresh air is sucked through the charcoal filter and further to the engine inlet system. The filter is then cleaned of gasoline.

The vehicles are equipped with a roll over valve. The valve is connected to the ventilation hose between the filler tube and the charcoal canister. In the event of the car rolling over or ending up on its side, a pendulum will actuate the valve, which will shut off the ventilation hose thereby preventing the escape of fuel form the fuel tank. A valve is located in the ventilation hose.

As the fuel in the tank expands, the valve causes an increases in pressure which reduces the amount of vapor (hydrocabons) evacuated from the tank.

Fuel Tank Ventilation

The fuel tank is designed so as to allow internal expansion of the fuel. The expansion space is opened by a valve which is actuated by the filler cap.

When fuel is poured in, the tank will not be completely filled, and instead the level rises only slightly above the lower opening on the venting tube. The reason for this is that an air cushion is formed above this level and prevents further filling of the tank.

When the tank cap is screwed on, a lever is actuated which opens the valve, thus providing a communication from the upper part of the tank to the surrounding air via the ventilation hose. The hose runs inside the roof channel, through the left wheel housing. The fuel, which increases in volume when the temperature rises, is now able to expand inside the tank instead of being pressed up through the filler pipe. As the fuel level becomes lower when driving, air is drawn into the tank via the ventilation hose.

EXHAUST EMISSION CONTROLS

Air Cleaner

The air cleaner is mounted at the front of the left hand wheel housing. In addition to cleaning the air, it also absorbs the suction noise. The air cleaner element consists of a paper of a cartridge which cannot be washed or moistened. The only service it requires is to blow the cartridge clean with compressed air or to replace it.

A thermostatically controlled valve which governs the air preheating according to the ambient air temperature is located in the intake of the air cleaner.

The valve housing has two air intakes, one for cold air and one for heated air which is drawn in through a hose running from a cover on the exhaust manifold. The valve is actuated by a thermostat in the cold air intake which senses the temperature of the ambient air. At temperatures below 43°F. (8°C.) turbo 13°F. (-5°C.) only heated air is inducted into the engine. At temperatures above 64°F. (18°C.) turbo 45°F. (5°C.), the valve admits cold air only into the engine.

Thermostatically Controlled Air Preheating

A thermostatically controlled valve which governs the air preheating

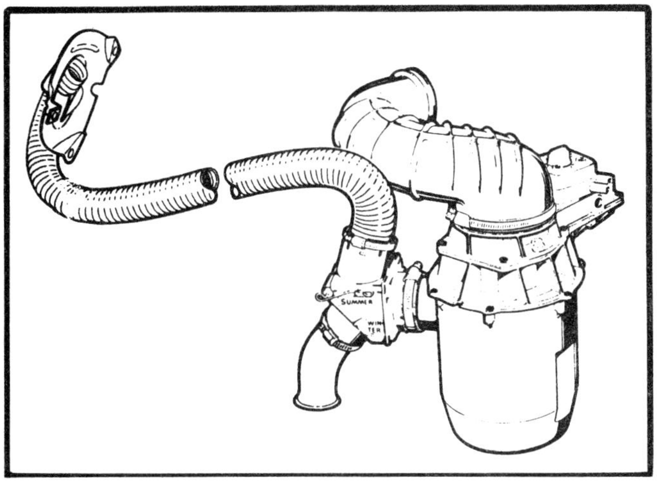

Typical air preheater used on Canadian vehicles

according to the ambient air temperature is located in the air cleaner intake.

The valve housing has two air intakes, one for cold air and one for heated air which is drawn in through a hose running from a cover from the exhaust manifold.

The valve is actuated by a thermostat in the cold air intake which senses the temperature of the ambient air.

Checking

A rough check of the operation of the valve is possible by comparison with ambient air temperature:

1. At temperature below + 43°F. (8°C.) Turbo 23°F. (- 5°C.), heated air only is inducted into the engine.
2. At temperatures between 43°F. (+8°C.) and 64°F. (+18°C.) Turbo 23°F. (-5°C.) and 40°F. (+5°C.), the valve admits varying proportions.
3. At temperatures above 64°F. (+ 18°C.) Turbo 40°F., (+5°C.), cold air only is inducted.

A more accurate check involves submerging the thermostat body in warm water (+43°F./+8°C. and +64°F./18°C., respectively), and checking the position of the valve.

Air Injection

In order to meet the regulations concerning emission of hydrocarbon and carbon monoxide, the engines have been equipped with an air injection system. The function of the air injection is to create afterburning in the exhaust pipes and exhaust manifold. The system includes the following components:

1. Air pump
2. Air inlet hose
3. Check valve
4. Air distribution pipe with injection tubes

The air pump is driven by a V-belt from the crankshaft pulley. /The air is sucked into the pump via a labyrinth seal at the pulley and is pumped out to the air hose and the distribution pipe. A relief valve opens if the pressure in the distribution pipe. A relief valve opens if the pressure in the distribution pipe becomes too great.

The function of the check valve is to prevent exhaust gases from entering the air pump if the V-belt should break. The air distribution pipe connects the check valve with the four injection tubes. The ends of injection tubes are located at the hottest part of the exhaust passage in order to achieve a maximum afterbuning effect.

The pump noise is partly absorbed by the labyrinth seal and partly by a small silencer located above the relief valve. In addition insulation is glued to the dash panel behind the pump.

NOTE: When CO adjustment is to be carried out, the air inlet hose must be disconnected and the inlet in the exhaust manifold bust be plugged. When the air system is disconnected, the engine idling speed drops 50–100 rpm. and this must be compensated for by means of the idling speed adjustment screw prior to the CO adjustment, reconnect air pump hose and readjust idle rpm according to specifications.

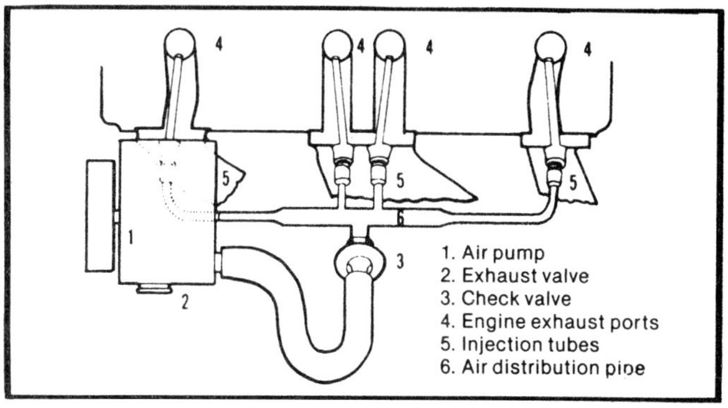

Typical air injection system

Inspection Service

1. Remove the air pump drive belt.
2. Remove the three pulley retaining screws and remove the air pump pulley.
3. Clean the pulley paying particular attention to the recess for the centrifugal cleaner.
4. Fit the pulley and the V-belt.
5. remove the air inlet hose and inspect the check valve at the air distribution pipe.
6. Check and fit the air inlet hose.

Removal

1. Disconnect the air inlet hose and remove the check valve.
2. Remove the exhaust manifold.
3. Loosen the cap nuts and remove the branch pipe.
4. Remove the fitting and remove the injection tubes. Remove the injection cones.
5. Installation is the reverse of the removal procedure.

Pulse Air System

The purpose of the pulse air system is to supply air to the exhaust gases from the engine to brig about continued oxidation of the hydrocabons and carbon monoxide in the exhaust system.

The system is composed of two check valves which are connected to the exhaust manifold by means of dual inlet pipes. The pipes open into the exhaust valves where the exhaust gases are hot, which is important to achieve oxidation of the exhaust system.

The check valves are grouped so that one goes to No.1 and No.4 cylinders and the other to No. 2 and No. 3 cylinders. Air is supplied to the valves through a hose from the air cleaner. The function of the pulse air system is based on the vacuum occurring in the exhaust system during the pulses. For a brief moment at the start of the suction stroke, a vacuum is produced in the exhaust manifold which will cause the check valve to open and allow a small amount of air into the exhaust system.

Inspection

Remove the hose between the air cleaner and check valve and check that it is free from dents or cracks. Run the engine at idling sped and make sure that air is being sucked through the check valve.

Ignition System

Two types of ignition systems are used: a conventional version and a breakerless version.

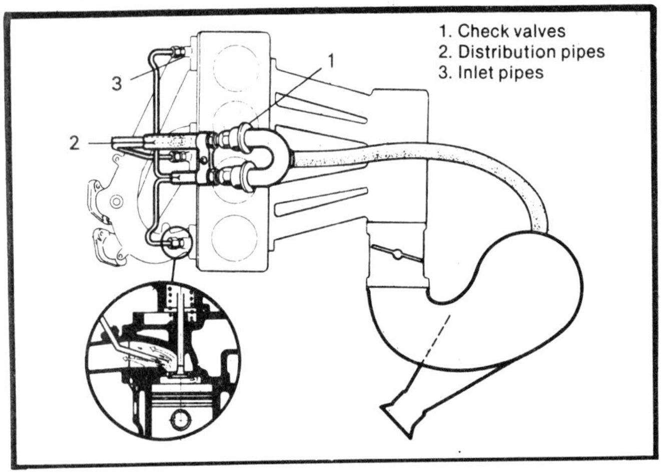

Pulse Air System

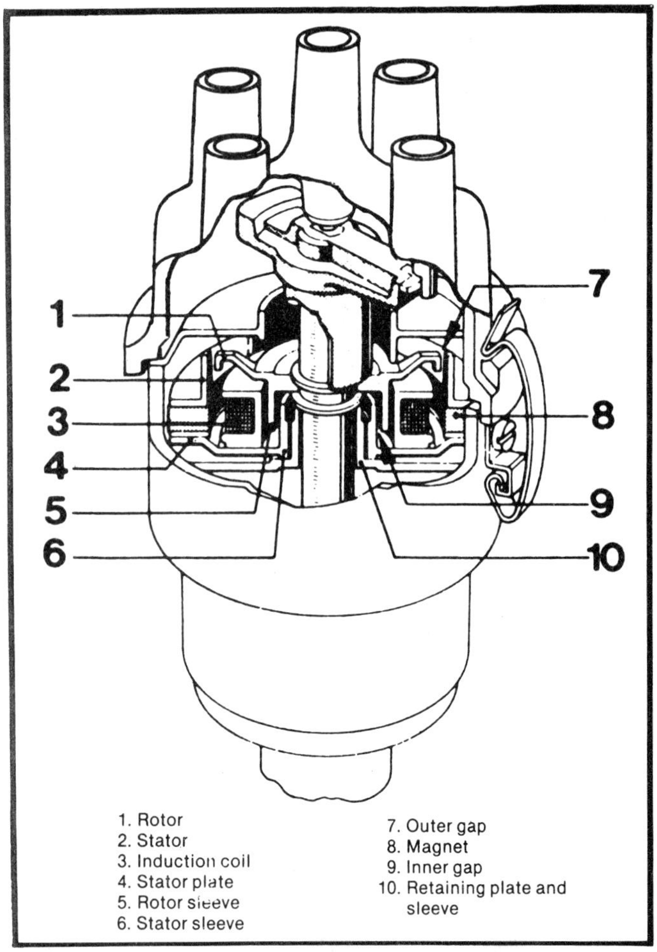

Breakerless distributor less Hall Effect

A conventional system with mechanical breaker contacts (points) in the distributor is installed in vehicles for the European markets (excluding Turbo vehicles), and early vehicles for the USA markets.

Breakerless ignition systems with electromagnetic impulse transmitter or a semi-conductor impulse transmitter (Hall transmitter) in the distributor with a control unit incorporating an electronic breaker function

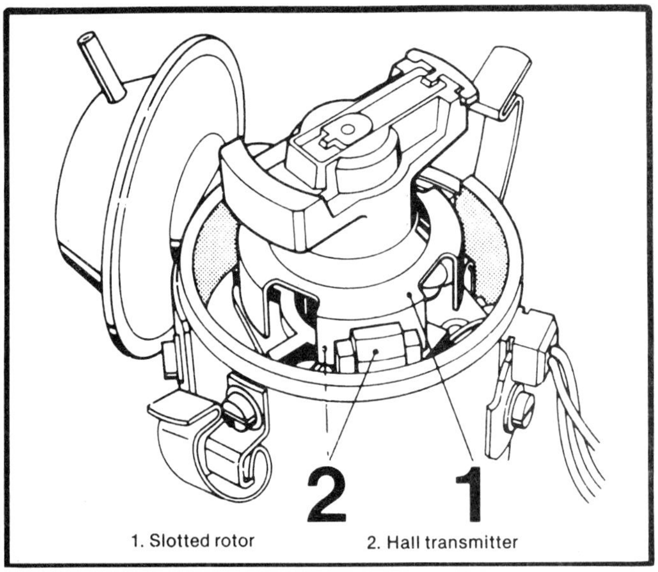

Hall Effect distributor

BREAKERLESS IGNITION SYSTEM

This system differs from conventional ignition system as follows.
The breaker points in the distributor have been replaced by a transmission unit comprising an impulse transmitter, an induction coil and a rotor disc. The impulse transmitter is connected to an electronic control unit in which the signal from the distributor is converted and amplified. The electronic control unit is connected to the ignition coil which is a high-voltage coil that has been specially adapted for the system.
The breaker function of the electronic control unit is therefore of the same character as that performed by the mechanical points in a conventional ignition system. Thus, the advantage of the breakerless system is that, thanks to the pulse generator, there is not mechanical wear affecting the setting of the timing.

Operating Principle

A sinusoidal control voltage is present in the induction coil, which alternates rapidly between positive and negative polarity. This alternation in polarity is utilized to transmit pulses. The pulse transmitted is governed by the engine speed and varies between 0.3 V and 100 V. The signal is then converted and amplified in the electronic control unit. When the sinusoidal voltage passes through zero, the ignition voltage is induced in the secondary circuit of the coil (when the rotor poles are in line with the stator poles).

This corresponds to the breaking of the points in a conventional ignition system. In the input stage, which is designed in the form of a Schmitt trigger, the sinusoidal signal from the distributor is amplified and converted to a rectangular pulse. The regulation of the dwell angle adapts the time of the current flow through the output transistor and the ignition coil to the engine speed. In contrast to the constant dwell angel in a conventional ignition system, the electronic system increases the dwell angle with increasing engine speed, thereby providing a high ignition voltage even a high engine speed. At the driver, the signal is amplified once again and proceeds to the Darlington output stage. Current now flows through the primary circuit of the ignition coil. At the moment of ignition, the ignition voltage is induced in the secondary circuit of the coil.

BREAKERLESS IGNITIONS SYSTEM WITH HALL TRANSMITTER

With the exception of the electric transmitter unit the distributor with the Hall transmitter is similar to the previously used distributor. The transmitter consists of a semi-conductor and a magnet which activates

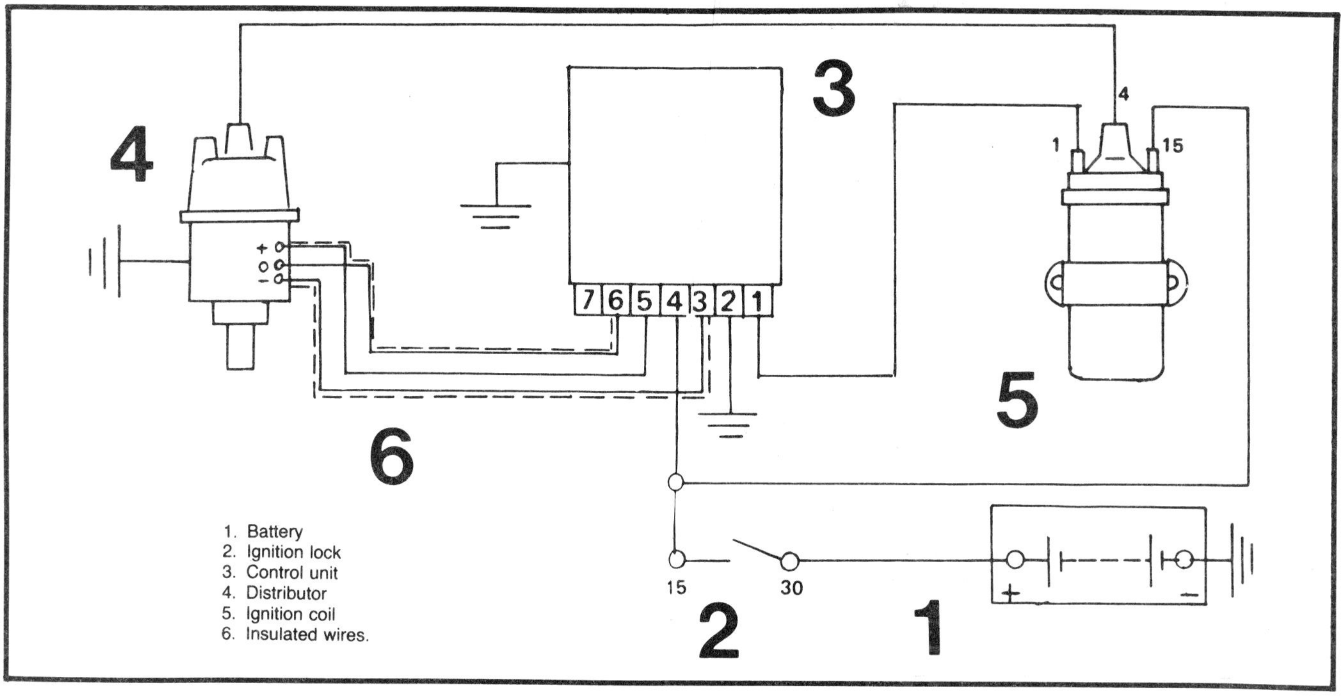

Electrical schematic of electronic ignition system

the semi-conductor. A slotted rotor makes and breaks the magnetic field in that way determines the ignition timing. The slotted rotor is not exposed to any wear which results in ignition timing remaining constant.

NOTE: The Hall transmitter and the rotor cannot be removed and consequently the entire distributor must be replaced if the transmitter unit is suspected of being faulty.

CONTROL UNIT

The control unit has the same function as that in the inductive system but is smaller because it is constructed using hybrid technology. The components in the control unit cannot be replaced. If a fault is suspected then the entire unit must be replaced. The control unit is located on a panel in front of the left wheel arch.

DISTRIBUTOR

Operation

The system with the Hall transmitter operates in principle in the same way as the inductive system, the difference being that the transmitter unit in the distributor consists of a Hall transmitter and a rotor. The control unit in this system is smaller, due mainly to it being constructed of integrated circuits.

CONTROL UNIT

Operation

The hall transmitter consists of a slotted rotor that rotates with the distributor shaft and a Hall transmitter which is mounted on a plate inside the distributor body.

The Hall transmitter consists of a semi-conductor layer on a magnetically conductive material plus a magnet mounted so that there is a gap between the magnet and the semi-conductor.

The Hall transmitter functions in accordance with the Hall effect. If a constant current is caused to flow through a semi-conductor and the semi-conductor is exposed to a magnetic field, then a low voltage will be generated between the edges of the semi-conductor.

The slotted rotor which as four slots, rotates in the gap between the magnet and the semi-conductor. This cuts the magnetic field and breaks up the voltage. When a slot in the rotor is positioned over the transmitter a voltage is generated which provides an impulse to the control unit. When the rotor cuts the magnetic field no Hall-effect is generated. The signal current to the control units is high, and the control unit is conducting. The primary winding is now charging the ignition coil.

When the slot of the rotor is in front of the Hall transmitter, Hall-effect is generated. The signal current to the control unit is low. The control unit is activated and opens the primary circuit. Ignition voltage is now induced in the coil's secondary circuit.

The connection resistance has been removed from this system and the final stage in the control unit is now responsible for current limitation. An ignition coil with a low resistance primary winding can herefore be employed. The dwell angle is also adjusted, which means hat the current limitation is only required for a very short time because the swell angle is adapted to suit the varying battery voltage, engine rpm and temperature. The control unit also contains a standing voltage shut-off. This means that the final stage is shut-off after 1 second approximately, if the engine is not running but the ignition is on. The final stages are reconnected when the engine starts.

Poor Engine Operation

Poor running of the engine is unlikely to be caused by faulty electronics. Check the following items first:

1. Good connections throughout ignition system.
2. Ignition setting and centrifugal and vacuum advance.
3. General condition of spark plugs.
4. Rotor: check operation, insulation (burns, dirt) and contacts.
5. Distributor cap: insulation (cracks, flashover burns, dirt) and contacts.
6. Ignition leads.
7. Ignition coil: spark length at starting revs 12mm minimum; insulation (burns, dirt).
8. Connect a dwell meter and check that the dwell angle varies with the speed as shown on the chart. At less than 2,000 rpm, the dwell angle is largely regulated by the control unit. At speeds greater than 2,000 rpm, the dwell angle is largely regulated by the shape of the impulse transmitter. Thus wide deviations in the dwell angle can be traced either to the control unit or the impulse control unit and may be faulty as often as in conventional ignition system.

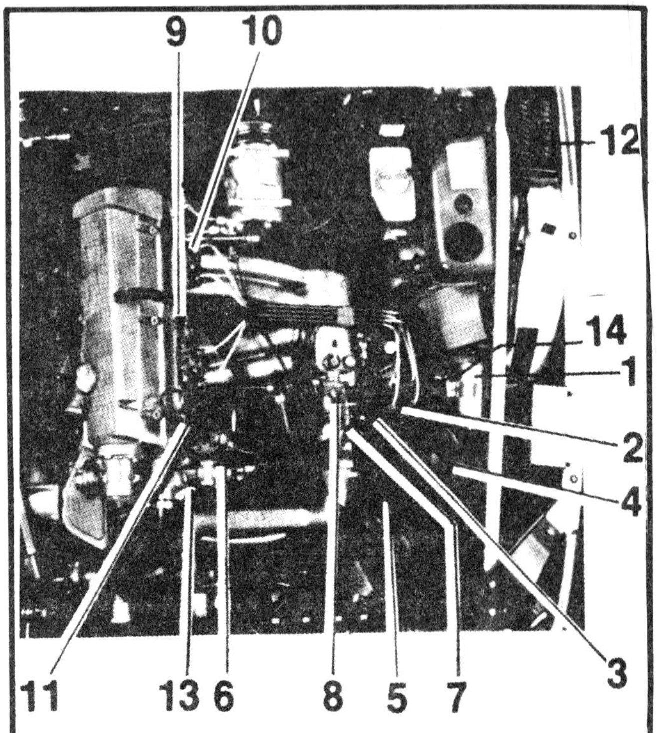

1. Fuel filter
2. Fuel distributor
3. Air flow sensor
4. Air cleaner
5. Rubber bellows
6. Warm-up regulator
7. Throttle valve housing
8. Cold start valve
9. Thermo-time switch
10. Injection valve
11. Auxiliary air valve
12. Fuses and relay box
13. Thermo switch
14. Pressure impulse contact

Fuel injection (K-Jetronic) component locations

Removal and Installation

1. Remove the distributor cap after marking the location of the number one spark plug wire on the distributor housing.

2. Disconnect the primary wire connector from the distributor and the hose from the vacuum advance unit.

3. Crank the engine until the flywheel marking is at TDC (0°) and the distributor rotor is pointing to the indicating or reference mark on the distributor housing for the number one cylinder.

4. Match mark the distributor housing ton the valve cover housing. Remove the distributor housing to the valve cover housing. Remove the distributor retaining bolts and pull the distributor forward from the end of the valve cover housing. Note the position of the distributor drive lugs.

CAUTION

Do not rotate the engine when the distributor is removed from the engine.

ELECTRONIC IGNITIONS CAUTIONS

1. When the engine is running, dangerously high voltage which may prove fatal may be present in the primary circuit of the coil and in all cables connected to terminal 1. This is because the spark energy in the system is considerably higher than in conventional ignition systems.

2. When the ignition switch is turned on, full primary current will flow through the primary circuit of the coil. Consequently, always disconnect terminal 15 on the ignition coil before commencing work with the ignition on.

3. In conjuction with work on the screened impulse cable between the impulse transmitter in the distributor and the control unit, strict attention must be given to polarity. Should the polarity be reversed, a stable basic ignition setting will not be possible and timing and dwell will also be affected.

Diagnosis of the Electronic Ignition System

The recommended steps must be performed one at a time and in the stated order. If on inspection components are found to be defective, these must be replaced before any further troubleshooting is carried out. If a fault is known to occur, for example, under certain temperature conditions, always attempt as far as possible to trace the faults as far as possible under the same conditions. Such procedure is necessary because defective electronic components may operate perfectly under normal temperature conditions before breaking down completely. Bad connections can also be affected by varying temperatures.

CATALYTIC CONVERTER

The catalytic converter is located in the exhaust system between the engine an muffler. The unit is composed of a ceramic material insert of honeycomb design. Prior to the use of the Lambda oxygen sensor in conjuction with the special continuous injection system, the cars were equipped with a dual-type catalytic converter, but with the use of the oxygen sensor an CIS system, a new three-way converter is now employed. This new three-way converter is capable of reducing the content of the exhaust gases of hydrocarbons, carbon monoxide and oxides of nitrogen, down to the prescribed Federal and State emission levels, on the conditions that the accurate regulation of the air/fuel ratio is maintained under all driving conditions.

Platinum and palladium are used in the dual type converter, while platinum, palladium and rhodium are used as the catalyst in the three-way converters.

Fuel Injection

All Saab vehicles sold in the U.S. are equipped with a Bosch CIS (continuous injection; also known as K-Jetronic) injection system. All models except Turbo also use this system.

An electric fuel pump which is mounted inside the gas tank provides fuel at a constant pressure to the mixture control unit. The latter consists of an air flow sensor which measures the flow of air to the engine and which acts mechanically on the fuel distributor. The fuel distributor provides the injection valves with the correct amount of fuel. The fuel is injected continuously into the intake manifold immediately upstream of the inlet valve. Vehicles are equipped with a special CIS system which is composed of a electronic control unit that is regulated by a oxygen sensor located in the exhaust manifold.

OXYGEN SENSOR REGULATED CI-SYSTEM

All 1980 and later vehicles are equipped with a special continuous injection system combined with an electronic control system which is regulated by an oxygen sensor that is located in the exhaust manifold. These vehicles are also equipped with a three-way catalytic converter which is located between the exhaust manifold and the muffler in the exhaust system. The sensor-regulated injection system ensures that the air/fuel mixture is continually kept within the required limits for vehicle operation and emission controls.

Spectrum (Chevrolet)

INDEX

VEHICLE IDENTIFICATION NUMBER AND ENGINE IDENTIFICATION

Vehicle Identification Number (VIN)

The vehicle identification number (VIN) is stamped on a metal tab that is mounted on the top left side of the instrument panel. The metal tab is mounted so that the VIN number can be seen through the windshield from outside the vehicle. The eighth digit of the VIN number is the engine identification code and the tenth digit represents the model year. "F" represents 1985 and "G" represents 1986.

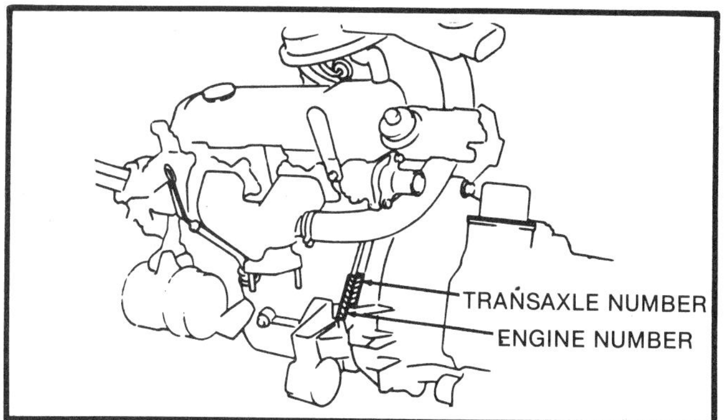

Plate number location (© General Motors Corp.)

ENGINE IDENTIFICATION

The engine identification number is stamped on the flange of the engine near the transaxle mounting. It is located toward the front of the engine on the left side.

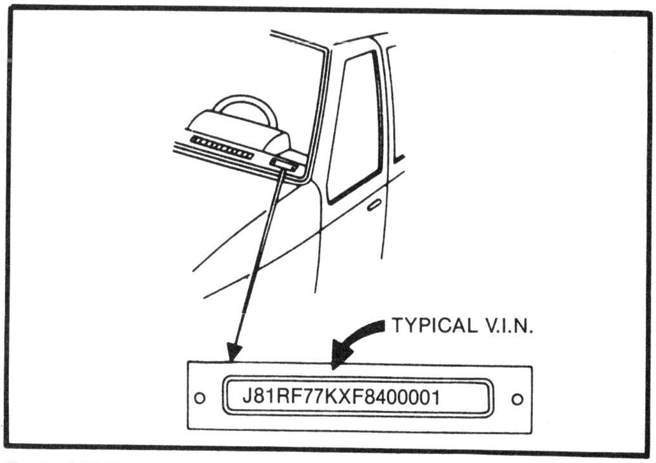

Typical V.I.N. (© General Motors Corp.)

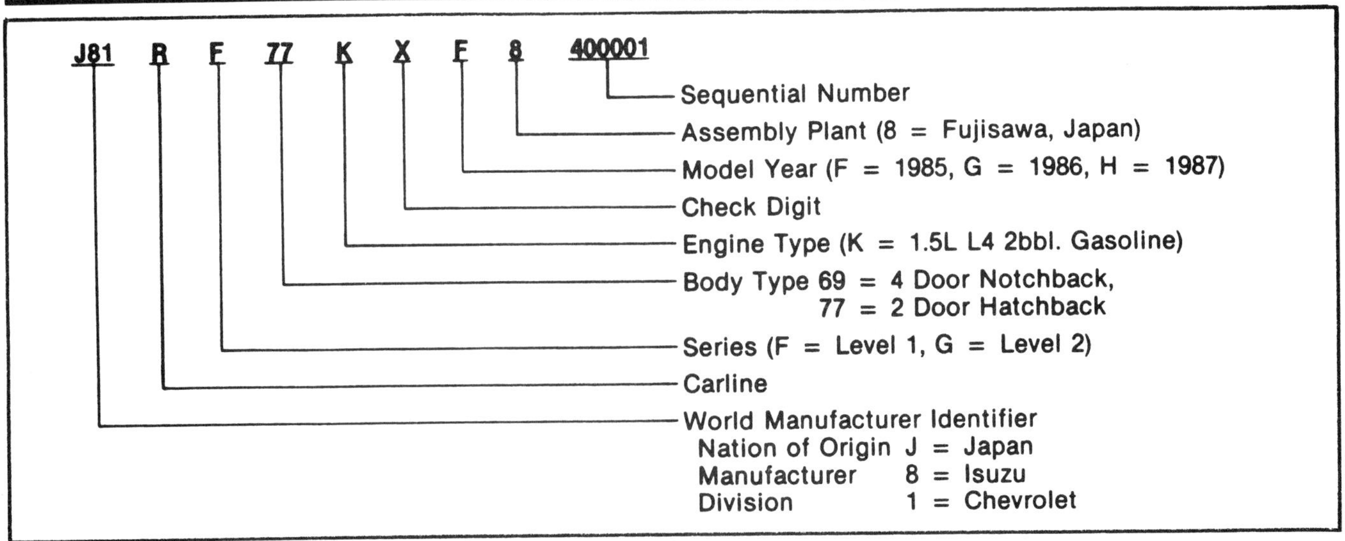

V.I.N. number breakdown (© General Motors Corp.)

TUNE-UP SPECIFICATIONS

Year	Eng. V.I.N. Code	Engine No. Cyl. Displacement (cu. in.)	Eng. Mfg.	hp.	Spark Plugs Orig. Type	Spark Plugs Gap (in.)	Ignition Timing (deg BTDC) Man. Trans.	Ignition Timing (deg BTDC) Auto. Trans.	Valves Intake Opens (deg)	Fuel Pump Pressure (psi)	Idle Speed (rpm) Man. Trans.	Idle Speed (rpm) Auto. Trans.
'85	K	4-94	Isuzu	70	BPR6ES-11	.040	15	10	17B	6–8	700	750
'86	G	4–94	Isuzu	SEE UNDERHOOD SPECIFICATIONS STICKER								

NOTE: The underhood specifications sticker often reflects tune-up specification changes made in production. Sticker figures must be used if they disagree with those in this chart.

FIRING ORDERS

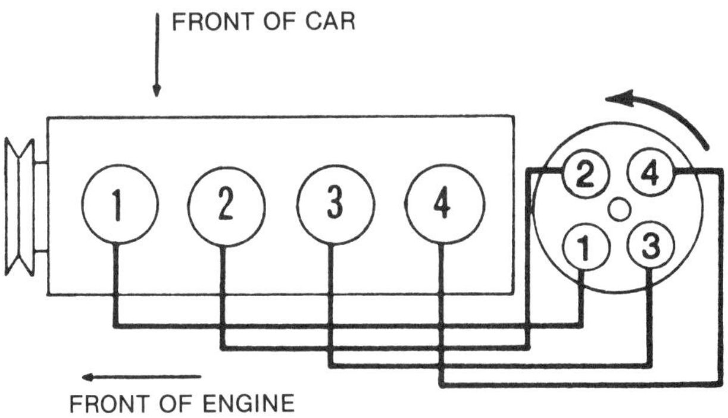

Engine firing order (© General Motors Corp.)

EMISSION EQUIPMENT USED

1985–86

- Oxygen Sensor
- Electronic Control Module
- Catalytic Converter
- Coolant Temperature Sensor
- Inlet Air Temperature Switch
- Throttle Position Sensor
- Positive Crankcase Ventilation
- Early Fuel Evaporation System
- Early Fuel Evaporation Heater
- Exhaust Gas Recirculation System
- Thermal Vacuum Valve
- Thermostatically Controlled Air Cleaner
- Mixture Control Valve
- Fuel Cut System
- Shift Indicator Light
- Transaxle Switch
- Clutch Switch
- Evaporative Emission Control System
- Tank Pressure Control Valve
- Vent Switching Valve
- Charcoal Canister

REQUIRED EMISSION CONTROL MAINTENANCE

MAINTENANCE SCHEDULE
Severe Driving Conditions

TO BE SERVICED	WHEN TO PERFORM Miles (Kilometers) or Months, Whichever Occurs First	The services shown in this schedule up to 48,000 miles (80 000 km) are to be performed after 48,000 miles at the same intervals															
	MILES (000)	3	6	9	12	15	18	21	24	27	30	33	36	39	42	45	48
	KILOMETERS (000)	5	10	15	20	25	30	35	40	45	50	55	60	65	70	75	80
Engine Oil Replacement*	Every 3,000 mi. (5 000 km) or 3 mos.	•	•	•	•	•	•	•	•	•	•	•	•	•	•	•	•
Oil Filter Replacement*	At first and then every other oil change	•		•		•		•		•		•		•		•	
Chassis & Body Lubrication	Every other oil change		•		•		•		•		•		•		•		•
Carburetor Choke Inspection*	Every 30,000 miles (50 000 km)										•**						
Valve Clearance Adjustment*	Every 15,000 miles (25 000 km)					•**					•**					•**	
Engine Idle Speed Adjustment*	At 3,000 miles (5 000 km) only	•**															
Drive Belt Inspection*	Every 30,000 miles (50 000 km)										•**						
Cooling System Inspection*	Every 15,000 miles (25 000 km)					•**					•**					•**	
Cooling System Refill*	Every 30,000 miles (50 000 km)										•						
Rear Wheel Bearing Repack	See explanation for service interval																
Manual Transaxle Fluid Replacement	At 6,000 miles (10 000 km) & every 30,000 miles (50 000 km)		•								•						
Automatic Transaxle Fluid Replacement	Every 15,000 miles (25 000 km)					•					•					•	
Spark Plug Replacement*	Every 30,000 miles (50 000 km)										•**						
Power Steering Fluid Replacement	Every 21,000 miles (35 000 km) or 24 months							•							•		
Power Steering Rubber Hose Replacement	Every 45,000 miles (75 000 km)															•	
Air Cleaner Element Replacement*	See explanation for service interval**										•**						
Fuel Cap. Lines & Tank Inspection*	Every 15,000 miles (25 000 km)					•					•					•	

*An Emission Control Service.

**In California, these are the minimum Emission Control Maintenance Services an owner must perform according to the California Air Resources Board.

MAINTENANCE SCHEDULE
Normal Driving Conditions

TO BE SERVICED	WHEN TO PERFORM Miles (Kilometers) or Months, Whichever Occurs First	The services shown in this schedule up to 45,000 miles (75 000 km) are to be performed after 45,000 miles at the same intervals						
	MILES (000)	5	7.5	15	22.5	30	37.5	45
	KILOMETERS (000)	8	12.5	25	37.5	50	62.5	75
Engine Oil Replacement*	Every 7,500 mi. (12 500 km) or 12 mos.		•	•	•	•	•	•
Oil Filter Replacement*	At first and then every other oil change		•		•		•	
Chassis & Body Lubrication	Every 7,500 miles (12 500 km) or 12 months		•	•	•	•	•	•
Carburetor Choke Inspection*	Every 30,000 miles (50 000 km)					•**		
Valve Clearance Adjustment*	Every 15,000 miles (25 000 km)			•**		•**		•**
Engine Idle Speed Adjustment*	At 5,000 miles (8 000 km) only	•**						
Drive Belt Inspection*	Every 30,000 miles (50 000 km)					•**		
Cooling System Inspection*	Every 15,000 miles (25 000 km)			•		•		•
Cooling System Refill*	Every 30,000 miles (50 000 km)					•		
Rear Wheel Bearing Repack	Every 30,000 miles (50 000 km)					•		
Manual Transaxle Fluid Replacement	At 7,500 miles (12 500 km) & every 30,000 miles		•			•		
Automatic Transaxle Fluid Replacement	Every 30,000 miles (50 000 km)					•		
Spark Plug Replacement*	Every 30,000 miles (50 000 km)					•**		
Power Steering Fluid Replacement	Every 22,500 miles (37 500 km) or 24 months				•			•
Power Steering Rubber Hose Replacement	Every 45,000 miles (75 000 km)							•
Air Cleaner Element Replacement*	Every 30,000 miles (50 000 km)					•**		
Fuel Cap. Lines & Tank Inspection*	Every 15,000 miles (25 000 km)			•		•		•

*An Emission Control Service.

**In California, these are the minimum Emission Control Maintenance Services an owner must perform according to the California Air Resources Board.

NOTE: The following information is being published from the latest information available at the time of publication. If this information differs from any data on the underhood emission label or in the owners manual, use the information in the owners manual or on the emissions label, since this information reflects the latest changes or modifications to the vehicle.

CARBURETOR CHOKE

Check the choke valve operation every 30,000 miles for proper operation. Any binding from gas build-up should be cleaned.

ENGINE IDLE SPEED ADJUSTMENT

Adjust the engine idle speed as necessary. Any adjustments must be made using equipment known to be accurate. Follow the specification located on the underhood emissions sticker.

DRIVE BELTS

Inspect the power steering, oil pump, air conditioning compressor and generator drive belts for cracks, fraying or other signs of deterioration. Replace or adjust as necessary.

SPARK PLUGS

Replace the spark plugs every 30,000 miles with the type of spark plug specified in the Tune-Up Specification chart.

AIR CLEANER ELEMENT

Replace the engine air cleaner element every 30,000 miles. Vehicle operation in dusty climates will mandate more frequent changes.

FUEL LINES, FUEL TANK AND FUEL CAP

Inspect the fuel lines, fuel tank and fuel cap for any damage which could result in leakage. Inspect the fuel cap and gasket for correct sealing. Replace or repair defective parts as required.

EXHAUST GAS RECIRCULATION

Every 30,000 miles (48,000 km) the EGR flow passages should be checked for deposits. If the EGR passages in the inlet manifold exhibit signs of excessive build-up, the passages should be cleaned. All loose particles should be removed in order to prevent the valve from clogging. All mounting surfaces of the intake manifold and valve assembly should be cleaned using a wire wheel to remove the exhaust deposits. Use a new gasket when installing a new EGR valve. Do not use any kind of solvent or cleaning fluids. They will damage the valve diaphragm.

UNDERHOOD RUBBER AND PLASTIC COMPONENTS (EMISSION HOSES)

Inspect the hose surfaces for evidence of heat and/or mechanical damage. Hard or brittle rubber, cracking, checking, tears, cuts, abrasions, and excessive swelling indicate deterioration of the rubber. Inspect all hose connections, couplings and clamps for leakage. Replace as required.

CRANKCASE INLET AIR CLEANER

The crankcase inlet air filter must be kept clean and lubricated. The inlet air cleaner should be washed in a cleaning solution or similar solvent. Lubricate the filter by inverting it and filling it with SAE 30 engine oil. Position the air cleaner to allow excess oil to drain through the vent nipple located in the top of the air cleaner. More frequent service may be required on vehicles that are operated extensively on short run, stop and go or extended idle service.

CHARCOAL CANISTER

Replace the filter element in the base of the canister with a new element every 30,000 miles (48,000 km). The filter should be replaced more often if the vehicle is driven under dusty or sandy conditions.

POSITIVE CRANKCASE VENTILATION (PCV) VALVE

Every 30,000 miles (48,000 km), the valve should be checked. If the valve is found to be clogged, replace it with a new valve. Never attempt to clean an old PCV valve.

IDLE SPEED AND MIXTURE ADJUSTMENT

Carburetor

The carburetor used is a two barrel, downdraft air and fuel metering device which has two stages of operation. The primary side of the carburetor has a small bore, a bridge nozzle and double venturis. Primary fuel metering is done with a duty solenoid also called the mixture control solenoid.

The second side of the carburetor has one large bore and a secondary main metering system which supplies fuel to the engine under certain operating conditions.

Idle Speed

Adjustment

1. Start the engine and allow it to reach normal operating temperature. Make sure that the choke valve is open. If the vehicle is equipped with air conditioning, make sure that it is off along with the cooling fan.
2. Inspect the float level to see that it is within the prescribed level.
3. Tag and disconnect the carburetor to distributor vacuum line, canister purge line and the EGR vacuum line. Plug all the lines.
4. Connect a tachometer according to manufacturers instructions.
5. Adjust the idle speed to 750 rpm for manual transmissions and 1000 rpm for vehicles equipped with automatic transmissions, by turning the idle speed adjusting screw.

FAST IDLE SPEED

Adjustment

1. Remove the air cleaner assembly from the vehicle.

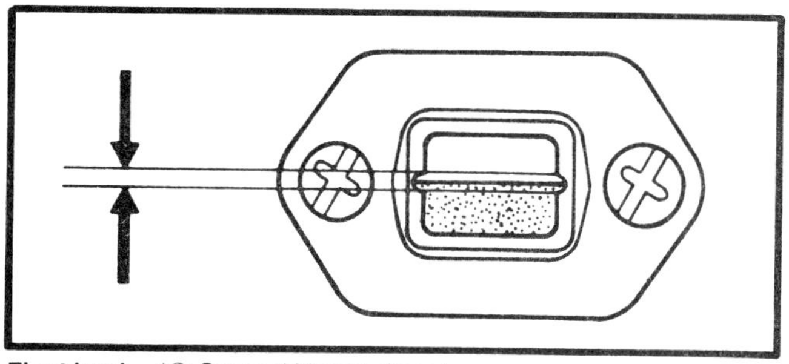

Float level (© General Motors Corp.)

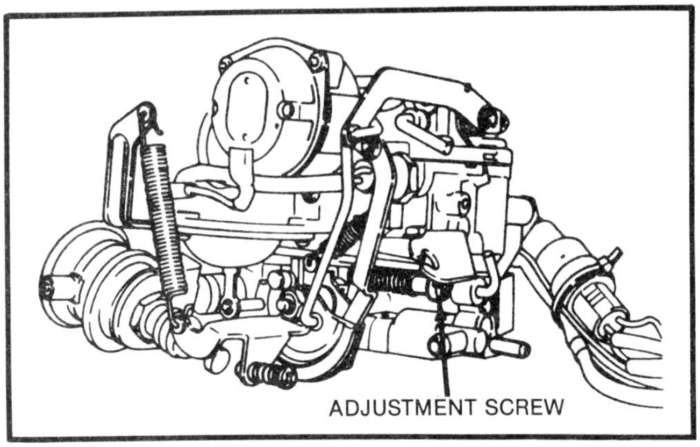

Idle speed adjustment screw (© General Motors Corp.)

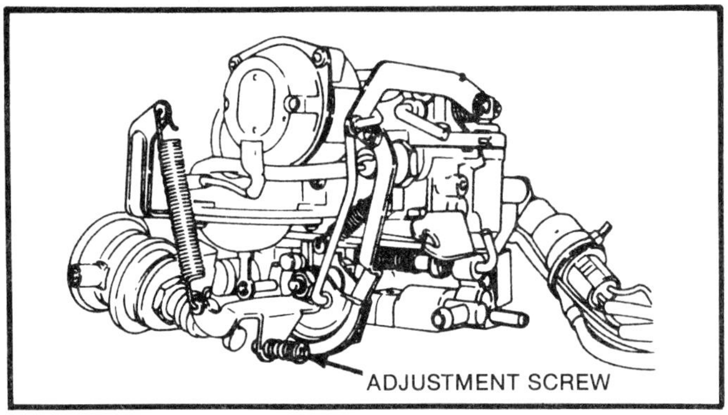

Dash pot adjustment screw (© General Motors Corp.)

2. While holding the throttle valve slightly, close the choke valve and release the throttle. Set to the first stage of the fast idle cam.
3. Start the engine without using the accelerator pedal.
4. Connect a tachometer according to the instructions supplied with it.
5. Fast idle speed should be 850 rpm for manual transmissions and 980 rpm for automatics. Adjust accordingly.
6. Stop the engine and reinstall the air cleaner assembly. Disconnect the tachometer.

Idle Mixture

Adjustment

NOTE: The idle mixture adjusting screw has been sealed, according to federal regulations, after adjustment to the optimum conditions at the factory. If adjustment becomes necessary, removal of the plug is possible, but it must be plugged after adjustment is completed.

1. Disconnect the negative battery cable. Drain the cooling system. Remove the air cleaner assembly.
2. Tag and disconnect the hoses and electrical connections from the carburetor.
3. Remove the necessary engine control cables in order to gain access to the carburetor bolts.
4. Remove the carburetor attaching bolts located on the bottom side of the inlet manifold. Remove the carburetor from the vehicle.
5. Mark the center of the mixture adjusting screw plug. Drill a small hole into the plug and thread a small screw into the hole. Pull on the screw in order to remove the plug.
6. Remove the idle mixture adjusting screw. Inspect the screw for any damage which may have resulted from screw removal. Replace the screw if necessary.
7. Reinstall the idle mixture adjusting screw. Lightly seat the screw. Back the screw out the following number of turns:
   a. Three turns from fully closed position for vehicles equipped with a manual transmission.
   b. Two turns from fully closed position for vehicles equipped with an automatic transmission.
8. Reinstall the carburetor and the air cleaner assembly on the vehicle. Set the parking brake and block the drive wheels securely. Position the gear selector in the neutral position.
9. Turn all accessories off. Make sure that the front wheels are facing in the straight forward position. Start the engine and allow it to reach normal operating temperature. Check the ignition timing.
10. Check the idle speed adjustment and correct to proper specifications if necessary. Refer to IDLE SPEED ADJUSTMENT procedures as outlined above.
11. Using a dwell meter, connect the positive side (+) to the duty monitor lead and ground the negative side (-).
12. Set the dwell accordingly, and adjust the idle mixture adjusting screw so that the dwell becomes 36 degrees.
13. Readjust the throttle adjusting screw to 750 rpm for (M/T) and 1000 rpm (A/T). Stop the engine and remove the tachometer.
14. Apply Locktite Number 262 or its equivalent to the new plug. Install the new plug flush to the carburetor surface. Check the fast idle speed and idle speed. Make adjustments as needed.

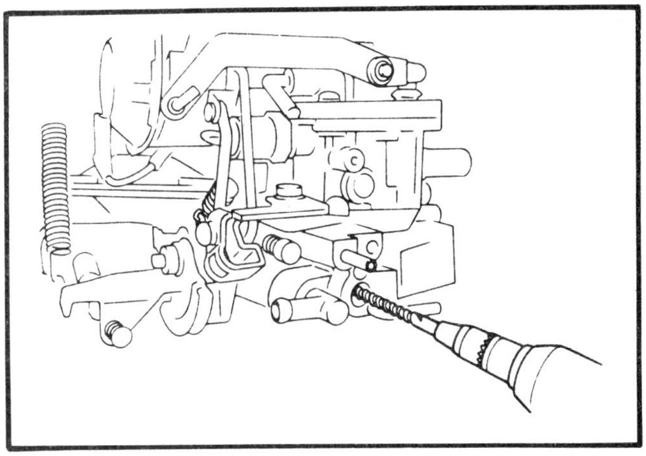

Drilling the mixture adjusting screw plug
(© General Motors Corp.)

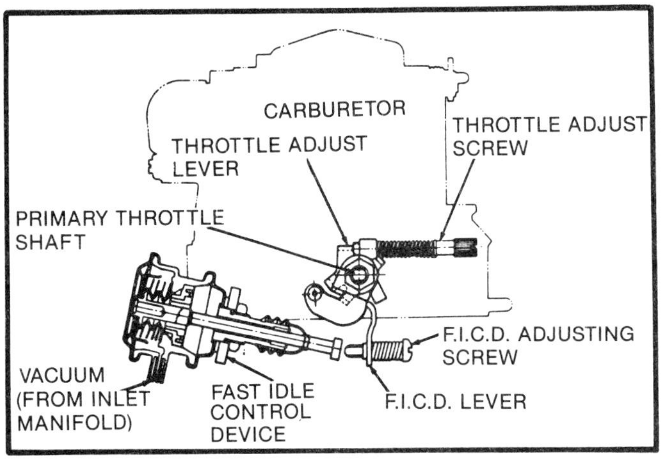

Carburetor adjusting locations (© General Motors Corp.)

IGNITION TIMING AND IGNITION SYSTEM

IGNITION SYSTEM

The Chevrolet Spectrum uses a solid state ignition system. It is comprised of the spark plugs, ignition coil, and distributor. The distributor has a rotor, a module, a pole piece, a vacuum advance and a centrifugal advance.

The signal generator is used to generate the ignition signal and consists of a pole piece, a magnet and a pick-up coil. The pole piece is attached to the distributor shaft, and the magnet and pick-up coil are attached to the pick-up coil base plate.

When the distributor shaft rotates, the magnetic flux passing through the pick up coil varies due to the change in the air gap between the pick-up coil and the pole piece. Because of this, the alternating current voltage is induced in the pick-up coil. The induced voltage turns the module on and off which switches off the ignition coil primary current. The high voltage is induced in the secondary winding of the ignition coil and ignition sparks are generated at the spark plugs.

CENTRIFUGAL ADVANCE

Testing

1. Disconnect the negative battery cable.
2. Remove the distributor cap.
3. Turn the rotor counterclockwise and release it.
4. The rotor should return smoothly by spring force. If not, replace the defective components as required.

VACUUM ADVANCE

Testing

1. Disconnect the negative battery cable. Remove the distributor cap.
2. Disconnect the vacuum hose from the vacuum advance unit.
3. Connect a vacuum tester to the advance unit.
4. Apply approximately 16 in. Hg. of vacuum and release it. Check to see that the pick-up base plate moves smoothly.
5. If the base plate does not move smoothly, inspect the plate and vacuum advance unit. Repair or replace as required.

IGNITION TIMING

Adjustment

1. Perform Steps 1–5 of the Idle Speed Adjustment procedure as outlined in this section
2. Connect a timing light to the number one spark plug lead.
3. Loosen the distributor flange bolt.
4. Align the notched line on the crankshaft pulley with the mark on the timing cover using the timing light in the conventional manner.
5. Adjust the ignition timing by turning the distributor clockwise or counterclockwise as required. Ignition timing on vehicles equipped with manual transaxles should be 15° BTDC
750 rpm. Automatic transaxle equipped vehicles should be 10° BTDC
1000 rpm.
6. Tighten the distributor flange bolt. Reconnect the vacuum lines and set the idle speed.
7. Check the ignition timing and if it deviates from the above specifications, repeat the ignition timing procedure.

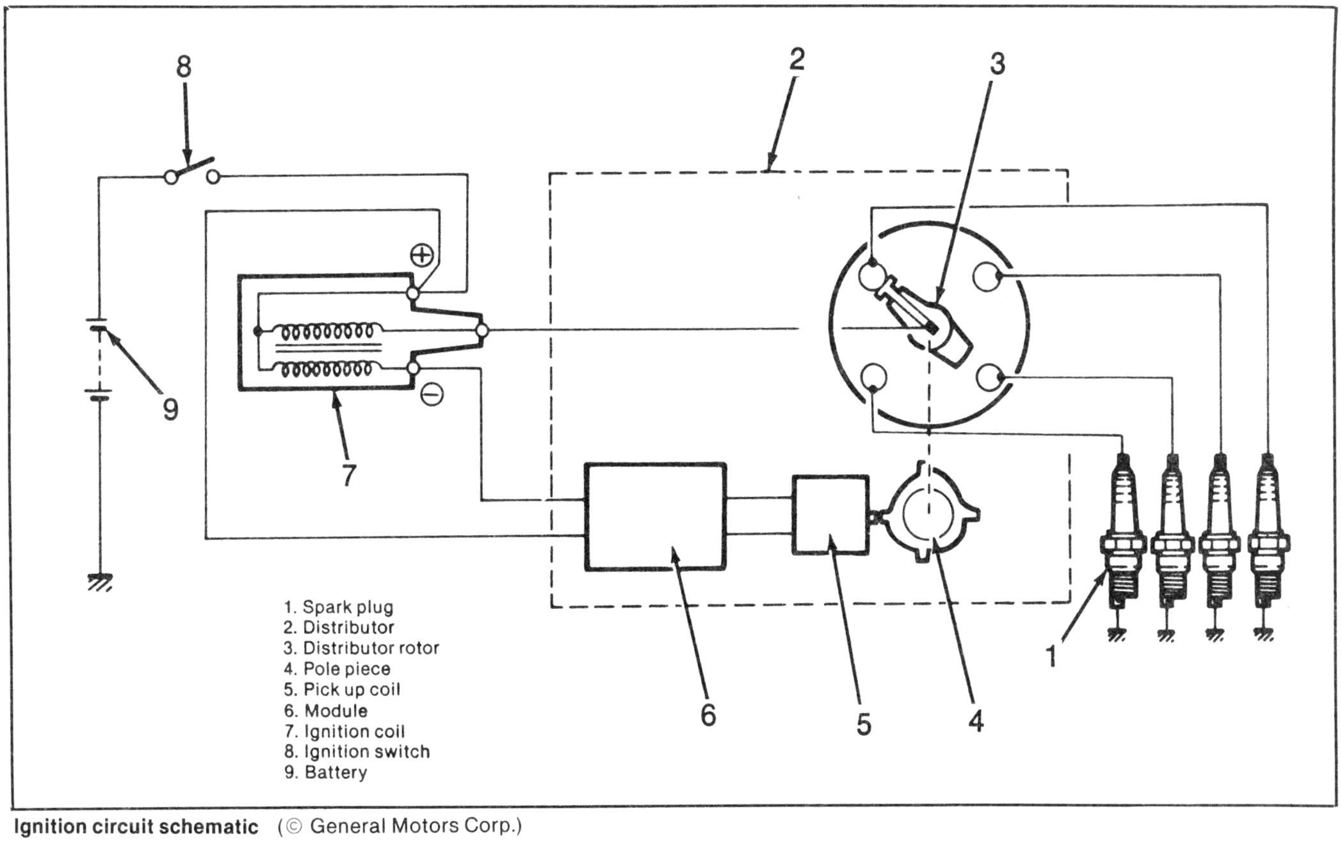

Ignition circuit schematic (© General Motors Corp.)

EMISSION CONTROL SYSTEMS

CRANKCASE EMISSION CONTROL SYSTEM

Positive Crankcase Ventilation (PCV) System

The PCV system forces blow-by gas, which originates in the engine crankcase, back to the intake manifold. It is then routed to the combustion chamber along with the air/fuel mixture.

The system is of the closed type variety. It is comprised of a head cover baffle plate, hose connections, the necessary hoses and a regulating orifice. The regulating orifice controls the blow-by gas suction. The hose and hose connections route fresh air from the air cleaner into the control system.

Under normal driving conditions, blow-by gas, which passes between the cylinder wall and pistons, is mixed in the engine along with fresh air from the air cleaner. The baffle plate separates the oil particles from the blow-by gas.

The blow-by gas is then drawn through the regulating orifice into the intake manifold where it is reburned in the combustion chamber. When the engine is running at maximum speed, intake manifold negative pressure will not be enough to handle all of the blow-by gas. Some of the gas will be drawn into the air cleaner, through the back end of the head cover.

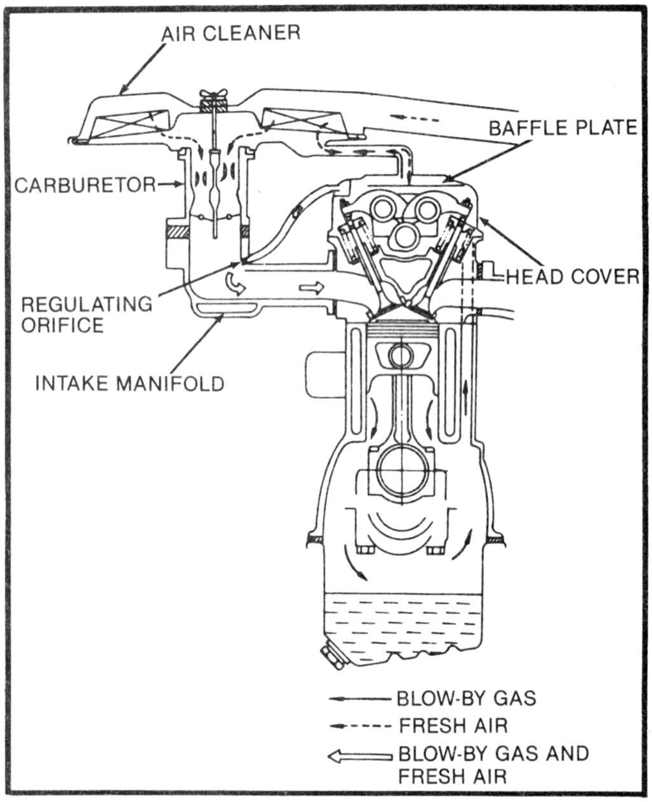

Positive crankcase ventilation air circulation
(© General Motors Corp.)

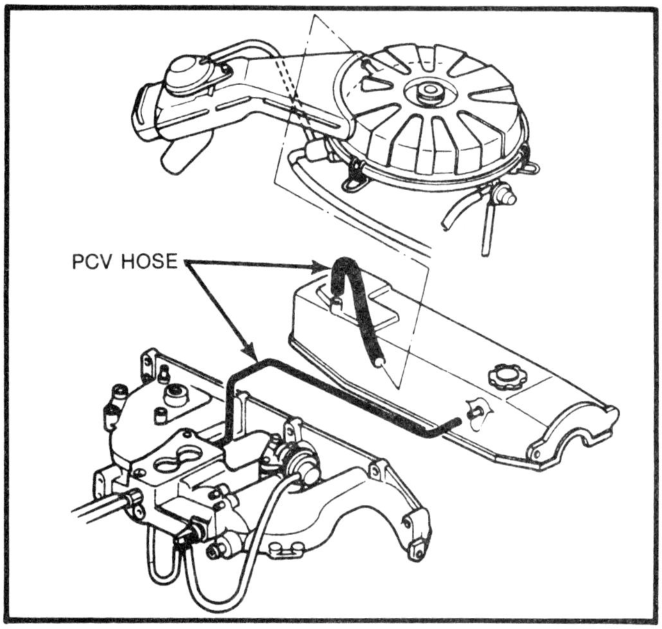

Positive crankcase ventilation hose routing
(© General Motors Corp.)

PCV SYSTEM

Inspection

Clean the inside of the hoses and regulating orifice using detergent oil. Using compressed air blow away any foreign matter.

Trouble Diagnosis

When rough engine idling is encountered, check for a clogged PCV valve or a pinched or plugged hose. Replace or repaired as required. The following procedure is recommended.

1. Remove the PCV valve from the rocker arm cover.
2. Start the engine and allow it to reach normal operating temperature. Allow the engine to run at idle speed.
3. Check for vacuum at the end of the valve. If there is no vacuum at the end of the valve, check for defective hoses or a clogged manifold port.
4. Turn off the engine and remove the PCV valve. Shake the valve and listen for the rattle of the check needle inside the valve. If no rattle is present, replace the valve.
5. Check the systems hoses and clamps and replace any that show signs of deterioration.

EVAPORATIVE EMISSION CONTROL SYSTEM

The evaporative emission control system routes fuel vapor to the charcoal canister. The fuel vapor is emitted from the fuel tank and the carburetor whenever the engine is not running. The system is composed of a pressure control valve, a charcoal canister, a vent switching valve, which is built into the carburetor, a thermal vacuum valve and the necessary connecting tubing. While the engine is off, the fuel vapor is stored in the charcoal canister. When the engine is started, the stored vapor is mixed with outside air and fed directly into the intake manifold to be burned during normal engine operation.

TANK PRESSURE CONTROL VALVE

When the engine is not running, fuel vapor from the fuel tank seeps toward the tank pressure control valve. The valve is in the closed position by means of a diaphragm spring. The fuel vapor then flows through the by-pass orifice to be stored in the charcoal canister. In the event that a negative pressure develops in the fuel tank, outside air will flow through the charcoal canister and the by-pass orifice into the fuel tank which will relieve the pressure.

When the engine is running, the manifold vacuum will force open the tank pressure control valve. Fuel vapor is then stored in the charcoal canister.

VENT SWITCHING VALVE

The vent switching valve is an electrically operated solenoid valve. It is built into the carburetor.

When the engine is running, the valve closes. The closed valve blocks the flow of outside air between the carburetor and the charcoal canister.

When the engine is off, the valve opens. This will allow fuel vapor to enter the charcoal canister.

CHARCOAL CANSISTER

The charcoal canister, which is filled with activated carbon, stores fuel vapors which are emitted from the fuel tank and carburetor. The canister is also equipped with a purge control valve. The valve serves to purge fuel vapor from the activated carbon. From here, it is delivered to the intake manifold for combustion.

The control valve diaphragm is activated by carburetor timed vacuum. When the engine is not running, the diaphragm spring holds the purge control valve closed. At this point, carbureted ported vacuum is applied to the diaphragm. If the vacuum exceeds the specified value, the diaphragm opens the purge control valve. The fuel vapor from the activated carbon is then delivered to the engine intake manifold.

The fuel vapor purge rate is controlled by the engine intake manifold vacuum and the purge control valve orifice.

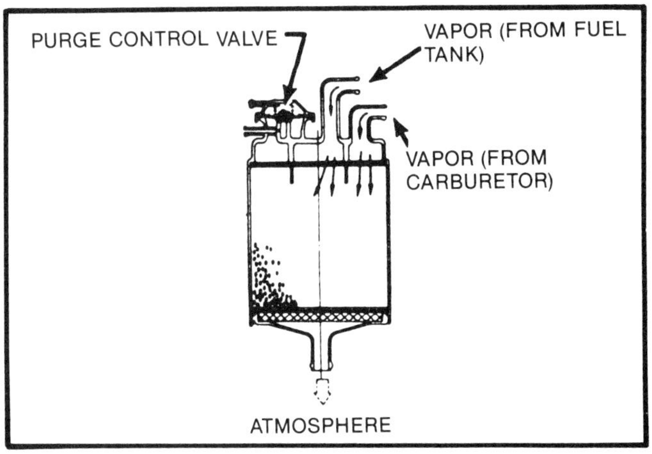

Charcoal canister air circulation (© General Motors Corp.)

THERMAL VACUUM VALVE

The thermal vacuum valve is connected in series to the purge signal line. This prevents vapor purge during the initial engine warm-up period. This will assist with the air/fuel mixing ratio.

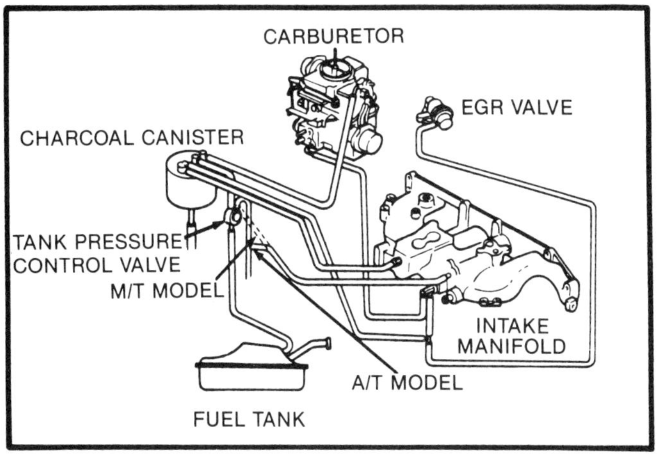

Vacuum hose routing (© General Motors Corp.)

FILLER CAP, FUEL TANK AND VAPOR LINES

Inspection

1. Inspect all lines and connection for proper sealing, loose connections or cracked components.
2. Inspect the fuel tank and fuel lines leading to the engine. Look for any deteriorated parts or leakage around connections.
3. Replace any components found to be defective.

TANK PRESSURE CONTROL VALVE

Inspection

1. Remove the rubber hoses from the pressure control valve.
2. Apply 7.5–8.5 kPa of vacuum to pipe A and blow air into inlet pipe "B" while outlet pipe "C" is plugged.
3. Air should not pass through the valve.
4. If air passes through, replace the valve.

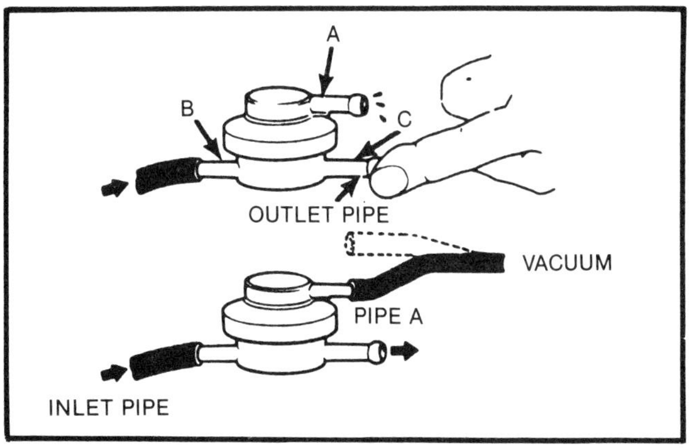

Tank pressure control valve (© General Motors Corp.)

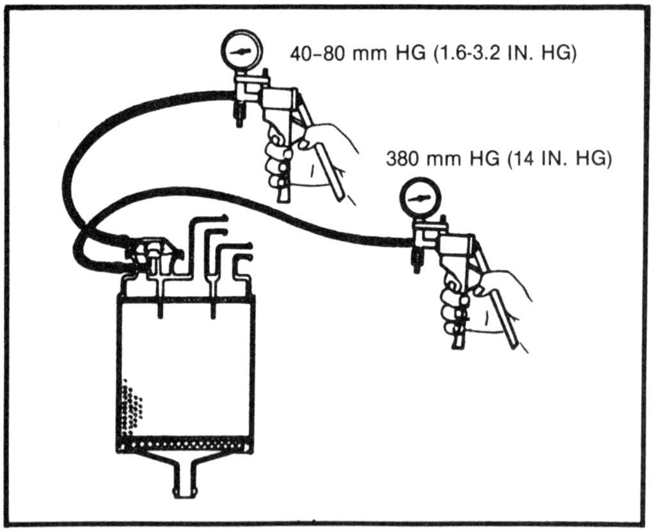

Checking the purge control valve (© General Motors Corp.)

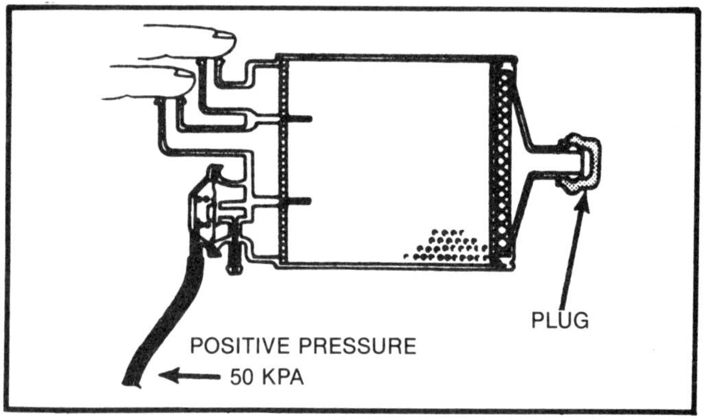

Purge control valve (© General Motors Corp.)

CHARCOAL CANISTER

Inspection

1. Remove the charcoal canister from the vehicle.

2. Visually inspect the charcoal canister for cracks or other signs of damage.

3. Apply 50 kPa of positive pressure to the port marked V.C. There should be no air leaking from the diaphragm.

4. Apply 14.96 in. Hg. vacuum to the port marked "PURGE" and maintain.

5. Apply vacuum gradually to the port marked "V.C.". If the purge control valve begins to open (the reading on the vacuum gauge on the "PURGE" port goes down), at pressure between 1.6 and 3.2 inches, the purge control valve is functioning normally.

6. If a problem is found, replace the charcoal canister.

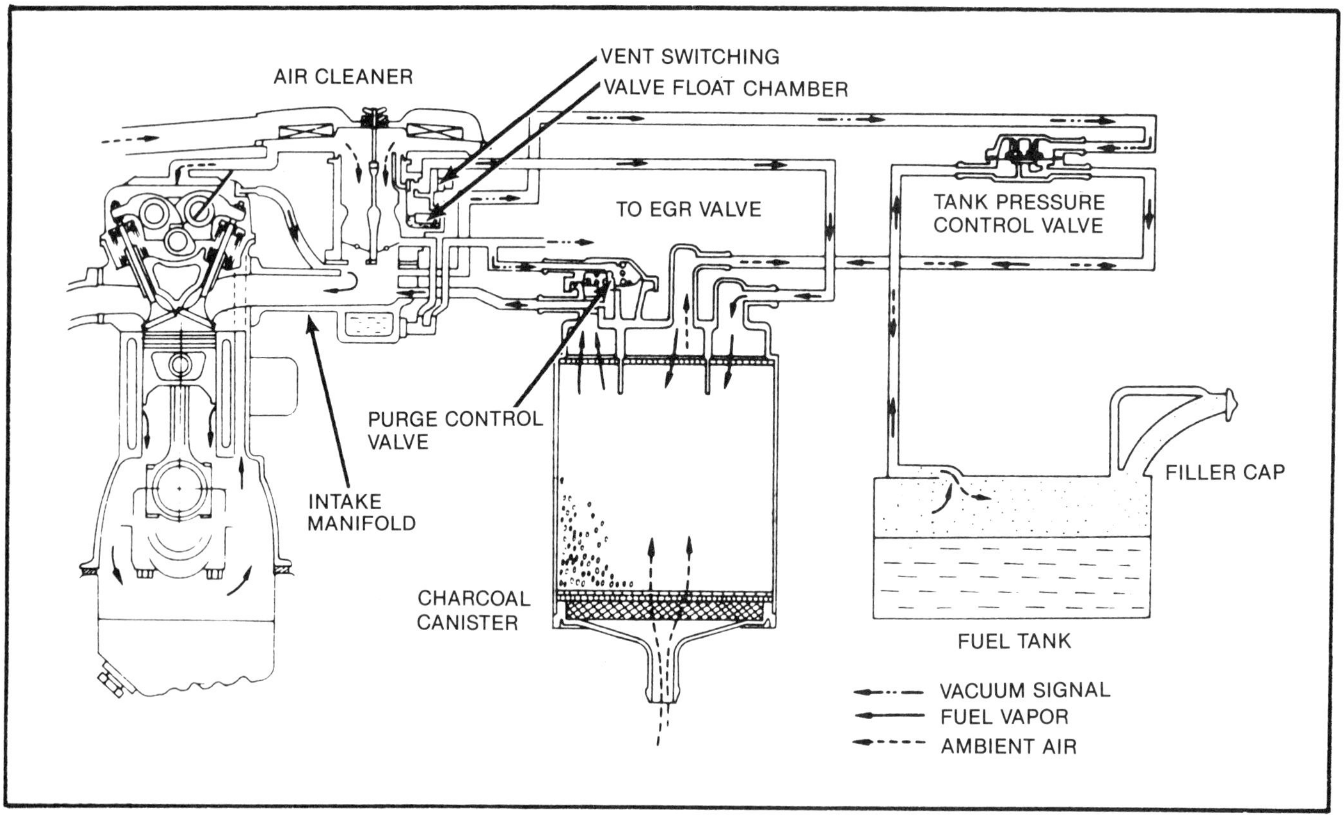

Evaporative emission control system (© General Motors Corp.)

EXHAUST EMISSION CONTROL SYSTEM

Exhaust Gas Recirculation (EGR) System

The EGR system lowers combustion temperatures in the combustion chamber. This action reduces the amount of nitrogen oxide emissions. Exhaust gas is drawn from the cylinder head exhaust port into the intake manifold through the passages in the cylinder head, EGR valve and intake manifold. The EGR valve vacuum diaphragm is connected to the carburetor by means of a thermal vacuum valve. This controls the EGR cold override.

Opening the carburetor throttle will apply a vacuum to the diaphragm. This in turn opens the EGR valve which will meter exhaust gas into the intake manifold.

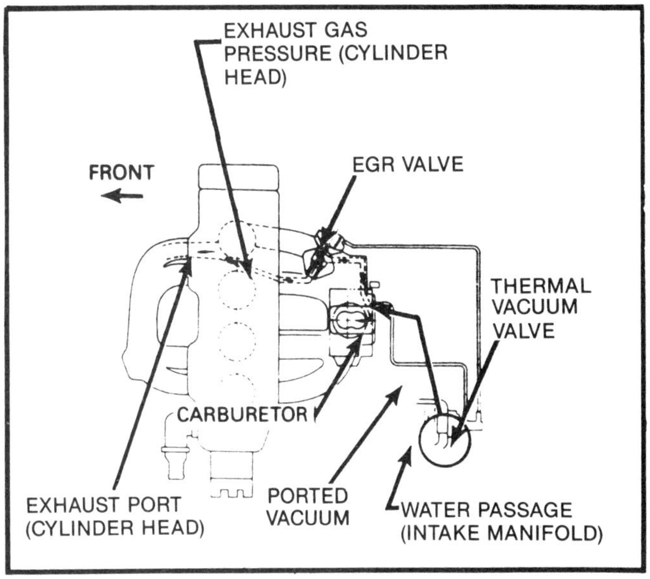

Exhaust gas recirculation system (© General Motors Corp.)

Exhaust Gas Recirculation (EGR)

The EGR valve is mounted to the intake manifold. It controls the amount of exhaust gas flowing into the intake manifold. Vacuum diaphragm vacuum is controlled by the vacuum diaphragm chamber. The vacuum diaphragm is connected to the carburetor flange vacuum port.

When the vacuum force reaches a specified force, the diaphragm overcomes the holding force of the spring and moves it to its fully raised position. The upward motion of the diaphragm opens the valve. This will allow exhaust gas to be routed to the engine intake manifold.

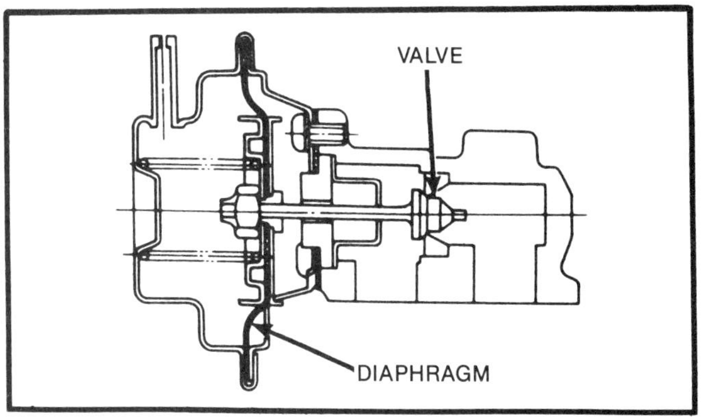

Exhaust gas recirculation valve (© General Motors Corp.)

THERMAL VACUUM VALVE (TVV)

The TVV is mounted on the intake manifold. It is connected to the carburetor vacuum port and the EGR valve. Valve operation is actuated through a bi-metal force. As long as the coolant temperature remains below a certain level, the valve remains closed and the EGR system will not operate. When the coolant temperature rises above the specified level, the EGR valve will open and the system operates.

Exhaust Gas Recirculation System

Inspection

1. Connect a tachometer to the engine according to the manufacturers instructions.
2. Connect a vacuum gauge between the three way connector, the EGR valve and the thermal vacuum valve.
3. With the engine cold, start the engine and check that there is no vacuum at any engine speed.
4. Allow the engine to reach normal operating temperature. Check that the vacuum gauge indicates 0 kPa at 3500 rpm. At 3500 rpm, disconnect and plug the hose to the EGR valve. Engine speed should increase.
5. Stop the engine and push the diaphragm plate by hand and release it, make sure that the diaphragm plate return to the original position smoothly.
6. If the above conditions do not exist, replace the defective components as required.

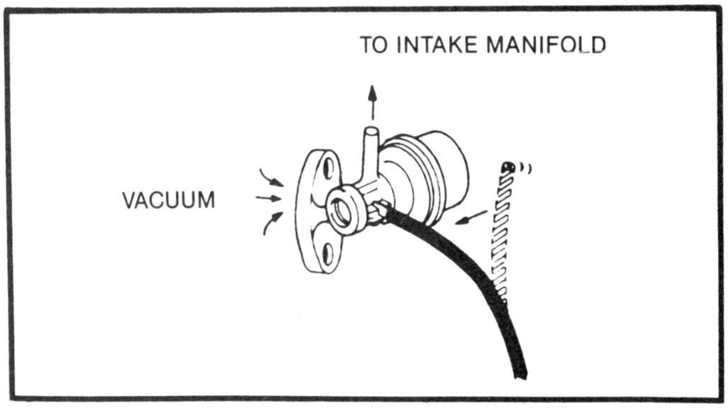

Checking the mixture control valve (© General Motors Corp.)

THERMAL VACUUM VALVE

Inspection

1. Heat the sensing portion of the thermal vacuum valve to above 130 degrees, using hot water.
2. Blow air into one of the pipes. The thermal vacuum valve should be open.
3. Cool the sensing portion of the thermal vacuum valve to below 95 degrees, using cool water.
4. Blow air into one of the pipes. Check that the thermal vacuum valve is closed.
5. If a problem is found, replace the thermal vacuum valve.

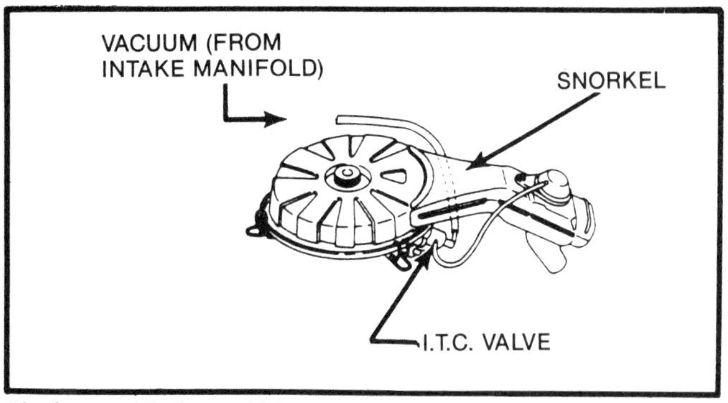

Air cleaner (© General Motors Corp.)

MIXTURE CONTROL VALVE

Inspection

1. Remove the air cleaner assembly from the vehicle.
2. Start the engine and allow it to reach normal operating temperature.
3. Disconnect the vacuum hose from the mixture control valve. Check to see that vacuum is not felt at the air inlet.
4. Reconnect the vacuum hoses and check to see that vacuum is felt.

NOTE: At this time, the engine should stall out or idle rough. This is a normal occurence.

5. If vacuum is or is not felt at the above times, replace the mixture control valve.

Thermostatically Controlled Air Cleaner (TCA)

The thermostatically controlled air cleaner (TCA) system helps to improve fuel vaporization by controlling the temperature of the intake air. The intake air is almost always at a constant level regardless of outside temperature and driving conditions. This helps to distribute the mixture evenly to each cylinder and to stabilize the air/fuel mixture ratio.

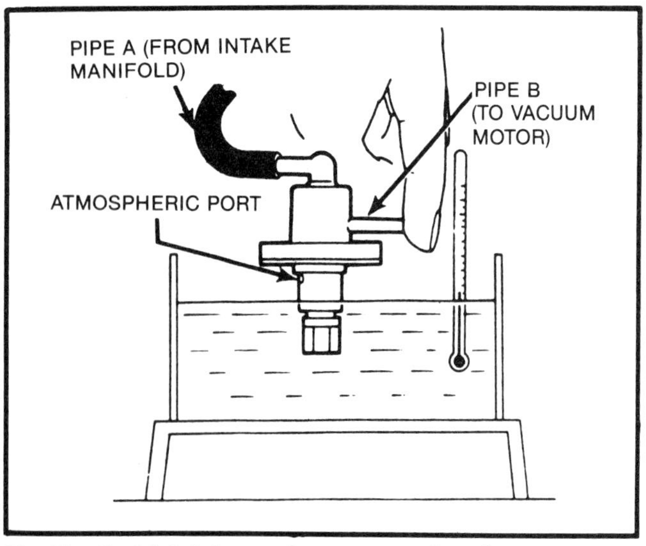

Checking the inlet air temperature compensator valve
(© General Motors Corp.)

MIXTURE CONTROL VALVE

The mixture control valve is mounted under the air cleaner cover. It routes fresh air to the intake manifold. This will prevent over enrichment of the air/fuel mixture. This action prevents engine back fire during periods of deceleration.

Inspection

1. Remove the air cleaner assembly from the vehicle.
2. Cool the inlet air compensator valve using compressed air.
3. Check that the air control valve closes the cool air passage at idle.
4. Reinstall the air cleaner assembly. Start the engine and allow it to reach normal operating temperature.
5. Check that the air control valve opens the cool air passage at idle.

Testing

1. Cool the sensing portion of the inlet air temperature compensator below 95 degrees. Blow air into pipe "A"(from intake manifold).

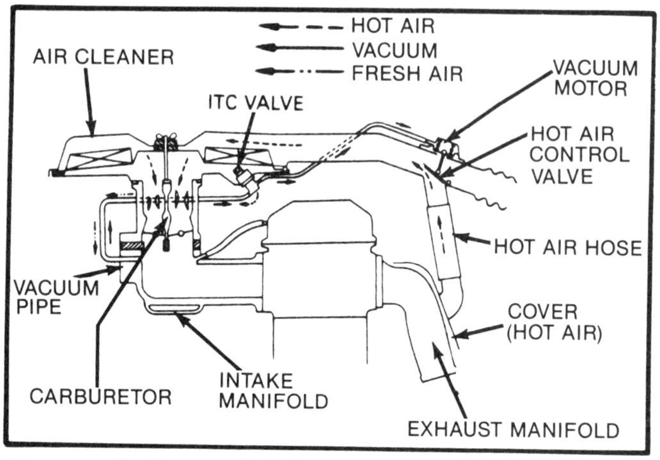

Thermostatically controlled air cleaner system
(© General Motors Corp.)

2. Check that a small amount of air flows from pipe "A" to pipe "B", to vacuum motor. Check that air doesn't flow through the atmospheric port, while closing pipe "B".
3. Using hot water, heat the inlet air temperature compensator valve to approximately 113 degrees F. Blow air into pipe "A" while pipe "B" is closed. Check that air flows through the remaining port.

NOTE: During the above procedure, be careful not to allow water to reach the inlet air temperature compensator.

4. If the test results are not as described above, replace the inlet air temperature compensator valve.

CATALYTIC CONVERTER

Inspection

1. Raise the vehicle and support safely.
2. Inspect all exhaust pipe rubber mountings for deterioration, cracks or other signs of damage.
3. Inspect all components for leaks, loose connections or dents.
4. Inspect nearby body areas for any structural damage such as open seams which could allow exhaust fumes into the passenger compartment.
5. Repair or replace damaged components as required.

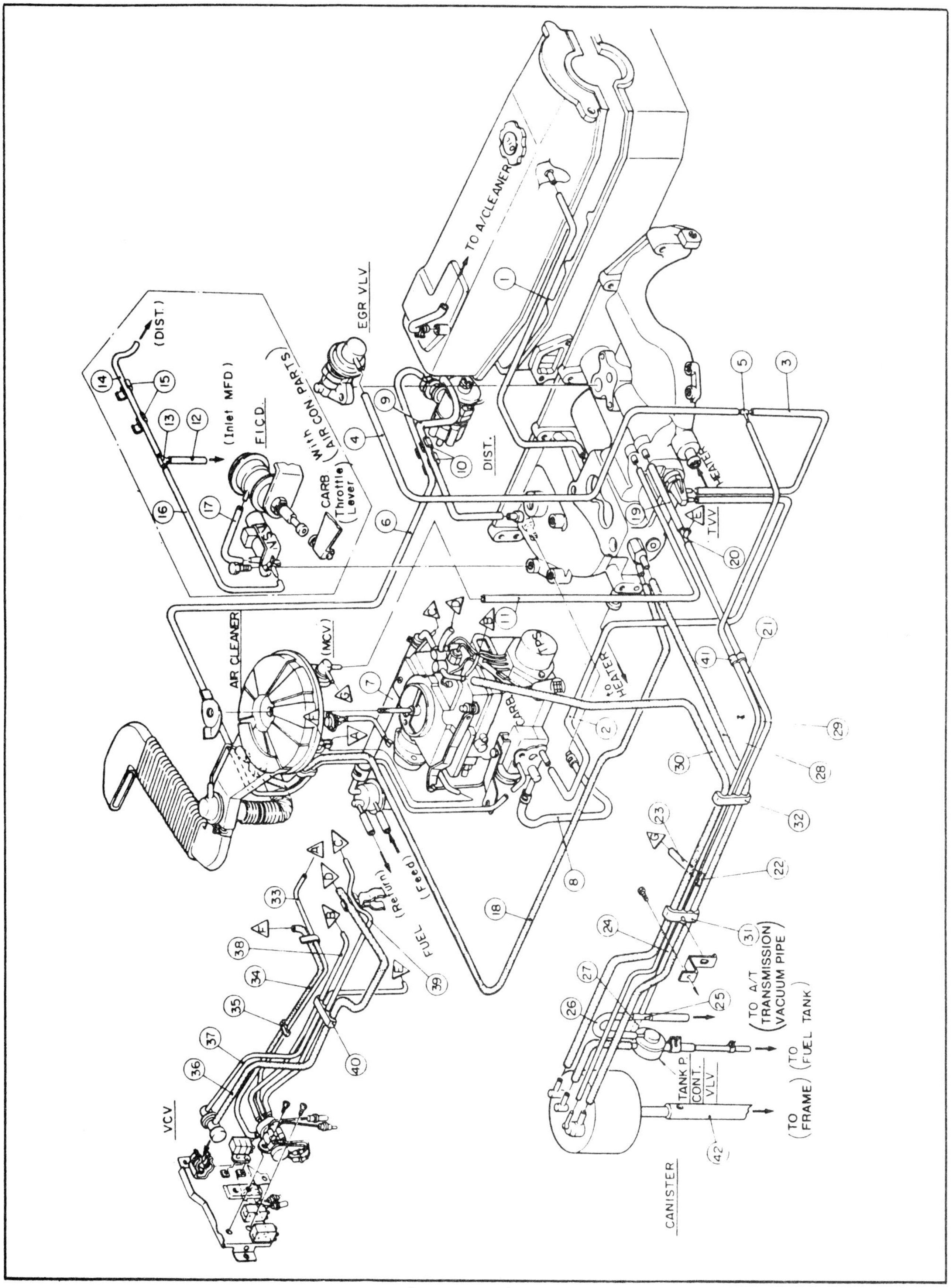

Component Layout

HOSE CONNECTIONS IDENTIFICATION			
NO.	CONNECTION	NO.	CONNECTION
1	Head Cover — Inlet	22	3 Way; MCV — VCV
2	Carb — TVV	23	3 Way — MCV
3	TVV — 3 Way	24	3 Way — 3 Way
4	3 Way — EGR VLV	25	3 Way; A/T Mission A/T Only
5	3 Way: EGR	26	3 Way — TPC VLV
6	Carb — Dist	27	3 Way — TPC VLV M/T Only
7	Carb — F/Pump	28	Canister — Inlet
8	Inlet — Carb (Hot Water)	29	3 Way — Canister
9	Inlet — Dist	30	Carb — Canister
10	Clip; 2 Way, Dist	31	Clip; 4 Way
11	MCV — Inlet	32	Clip; 4 Way
12	Inlet — 3 Way (With A/C)	33	Solenoid — Carb; Slow
13	3 Way; Dist — VSV (With A/C)	34	Solenoid — A/Cleaner
14	3 Way — Dist (With A/C)	35	Clip; 2 Way
15	Clip; 2 Way, Dist (With A/C)	36	VCV — Carb
16	3 Way — VSV (With A/C)	37	VCV — 3 Way
17	VSV — FICD (With A/C)	38	Solenoid — Carb; Main
18	ITC VLV — Inlet	39	Solenoid — Carb; SECO
19	Inlet — 3 Way	40	Clip; 4 Way
20	3 Way; Inlet — VCV	41	Clip; R/Hose
21	3 Way — 3 Way	42	Canister Drain

Hose Identification

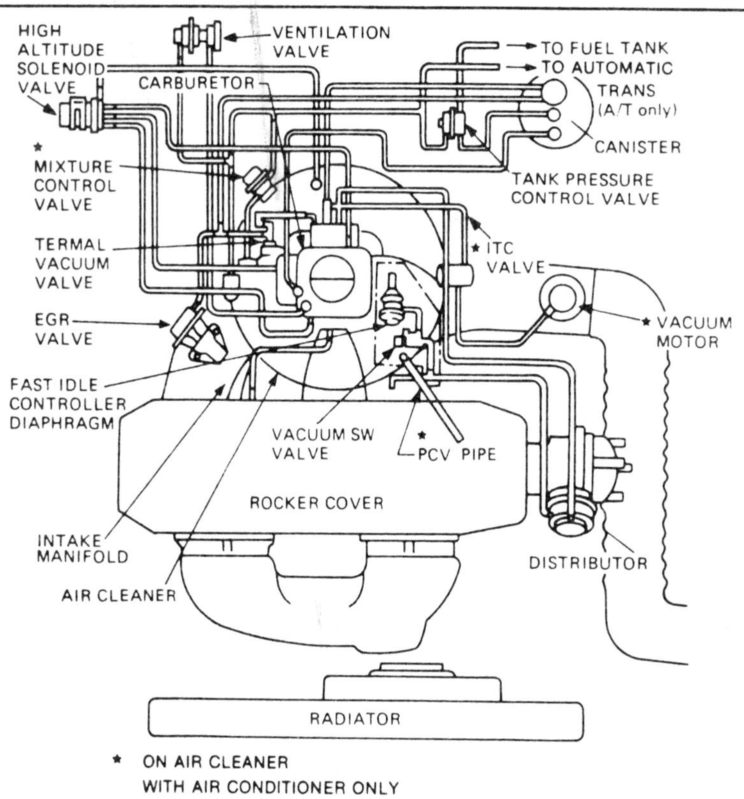

Vacuum Hose Routing

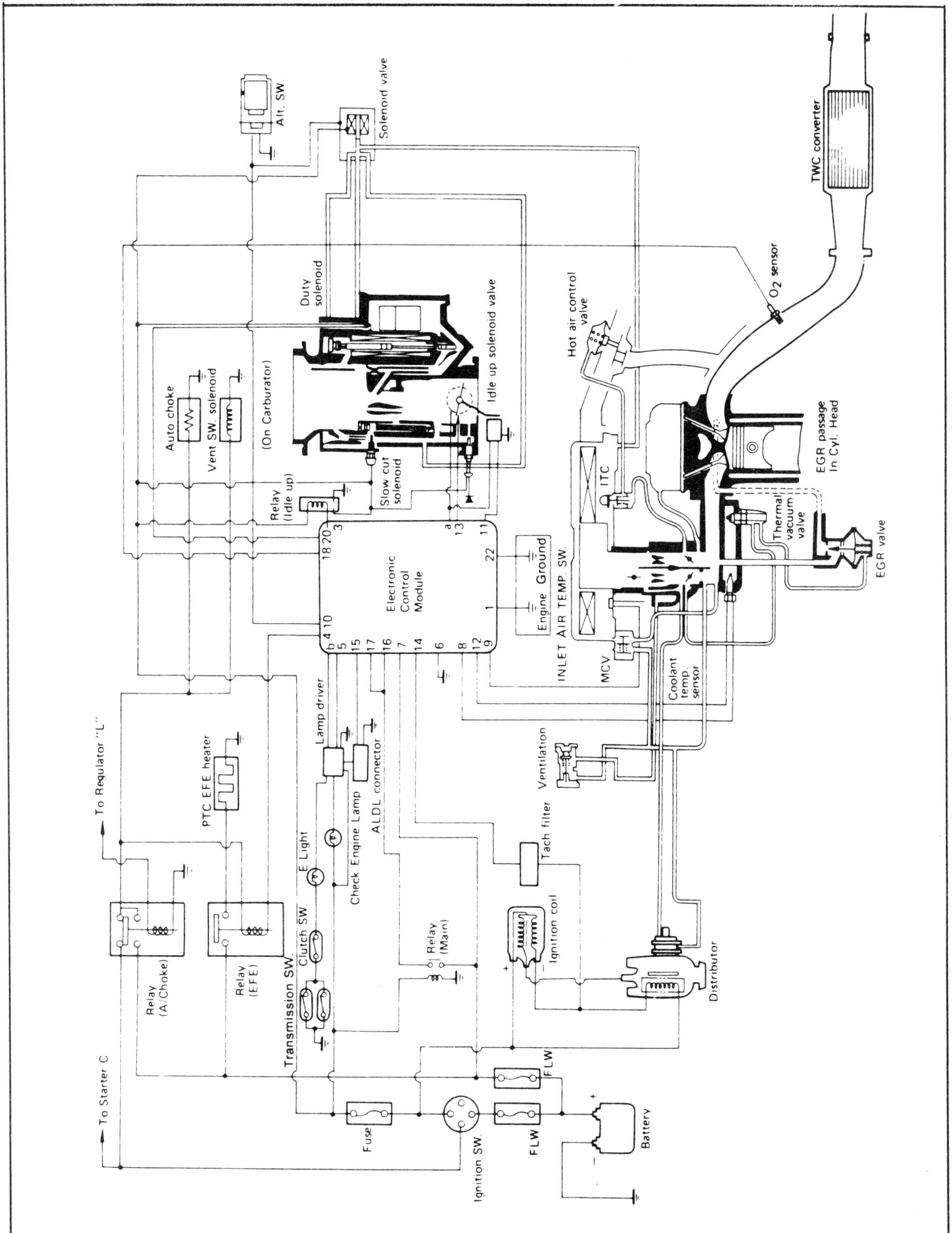

Engine Emission Control System

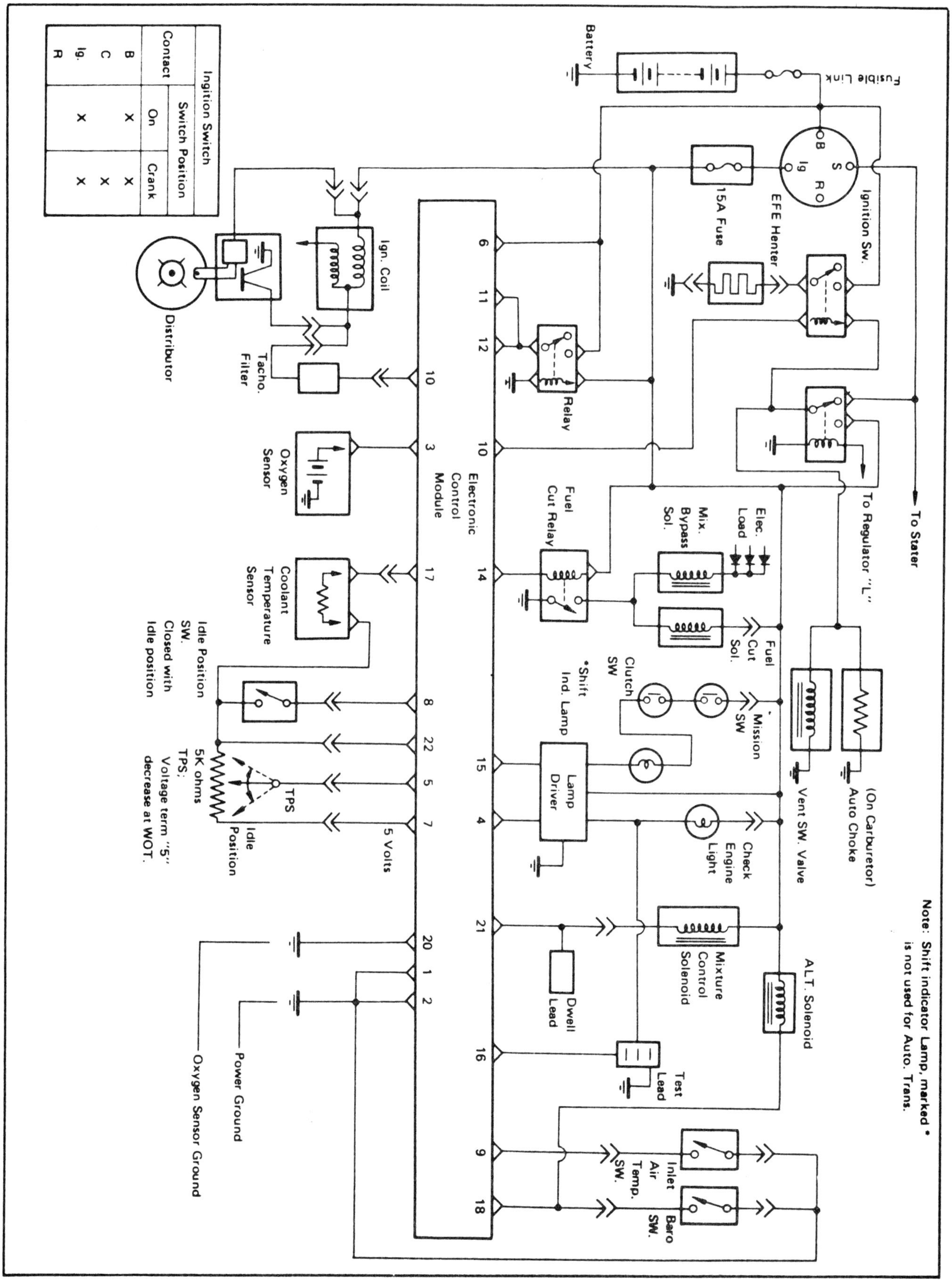

Engine Emission Electrical Circuit

Sprint (Chevrolet)

INDEX

VEHICLE IDENTIFICATION NUMBER AND ENGINE IDENTIFICATION

Vehicle Identification Number (VIN)

The vehicle identification number (VIN) is stamped on a metal tab that is mounted on the top left side of the instrument panel. The metal tab is mounted so that the VIN number can be seen through the windshield from outside the vehicle. The eighth digit of the VIN number is the engine identification code and the tenth digit represents the model year. ''F''represents 1985 and ''G'' represents 1986.

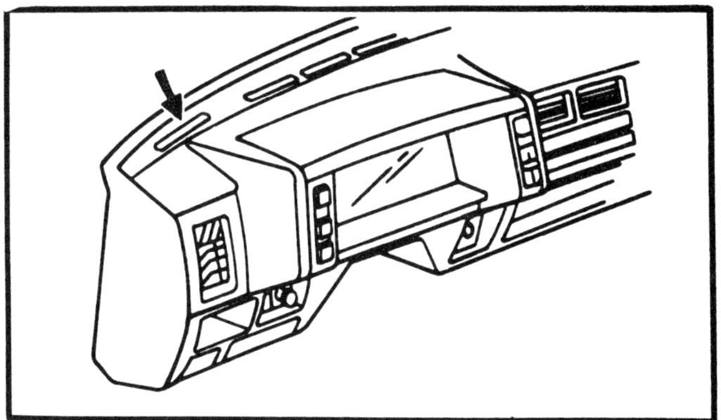

Vehicle identification number location

Engine Identification

The engine serial number is stamped into the left rear side of the engine casing. The first three digits indicate engine type and displacement. The fourth number shows models year with the remaining numbers referring to production sequence.

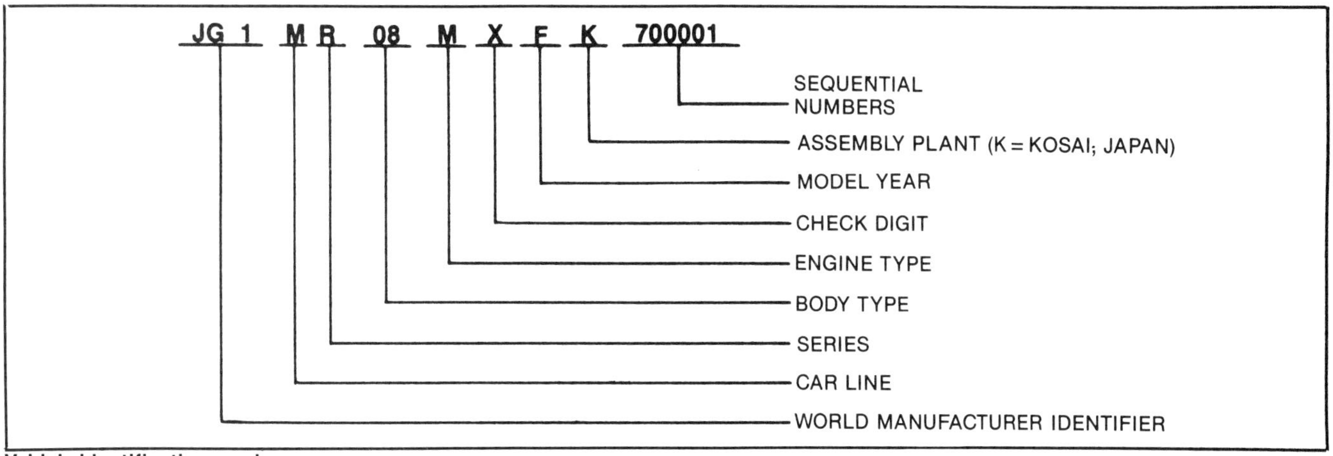

Vehicle identification number

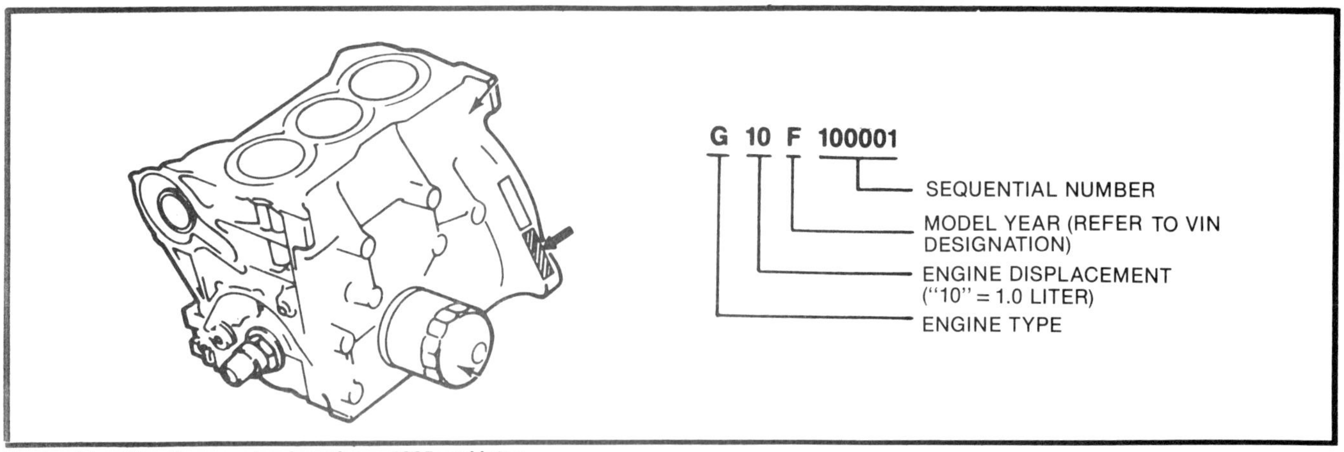

Engine identification number location—1985 and later

TUNE-UP SPECIFICATIONS

(When analyzing compression test results, look for uniformity among cylinders rather than specific pressures.)

Year	Eng. V.I.N. Code	Engine No. Cyl. Displace-ment (cu. in.)	Eng. Mfg.	hp.	Spark Plugs Orig. Type	Spark Plugs Gap (in.)	Ignition Timing (deg)▲ ● Man. Trans.	Ignition Timing (deg)▲ ● Auto. Trans.	Valves Intake Opens (deg)■	Fuel Pump Pressure (psi)	Idle Speed (rpm)▲ ● Man. Trans.	Idle Speed (rpm)▲ ● Auto. Trans.
'85–'86	M	61	Suzuki	48	①	③	③	③	N.A.	3.5 ②	③	③

① NGK:BPRGES-11 or Nippondenso WIGEXR-U11
② @ 5,000 rpm
③ Refer to underhood specifications sticker

FIRING ORDER

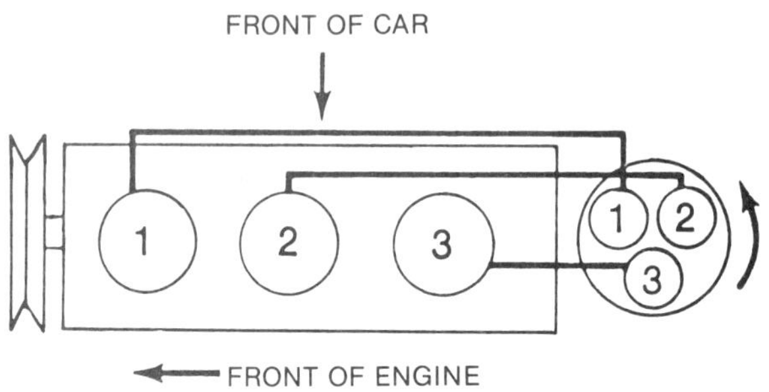

GM (Suzuki) 63-3 (1.0L) Engine firing order: 1-3-2 Distributor rotation: counterclockwise

EMISSION EQUIPMENT USED

1985–86

Positive crankcase ventilation (PCV)
Charcoal Canister
Computerized emission control system
Thermostatically controlled air cleaner (THERMAC)
Hot idle compensator (HIC)
Pulse air control system
Idle up system
Bowl ventilation system
Catalytic converter
Exhaust gas recirculation (EGR)

MAINTENANCE SCHEDULE
Severe Driving Conditions

TO BE SERVICED	WHEN TO PERFORM Miles (Kilometers) or Months, Whichever Occurs First	The services shown in this schedule up to 48,000 miles (80 000 km) are to be performed after 48,000 miles at the same intervals															
	MILES (000)	3	6	9	12	15	18	21	24	27	30	33	36	39	42	45	48
	KILOMETERS (000)	5	10	15	20	25	30	35	40	45	50	55	60	65	70	75	80
Engine Oil & Oil Filter Change*	Every 3,000 Miles (5 000 km) or 3 Months	•	•	•	•	•	•	•	•	•	•	•	•	•	•	•	•
Chassis Lubrication	Every other oil change		•		•		•		•		•		•		•		•
Carburetor Choke Inspection*	At 6,000 Miles (10 000 km), then every 30,000 Miles (50 000 km)		•								•†						
Engine Idle Speed Inspection*	At 6,000 Miles (10 000 km), then every 15,000 Miles (25 000 km)		•			•					•					•	
Valve Lash Inspection*	Every 15,000 Miles (25 000 km)					•					•					•	
Water Pump Belt Inspection*	Every 30,000 Miles (50 000 km) or 24 Months										•†						
Fuel Cut System Inspection*	Every 30,000 Miles (50 000 km), or 24 Months										•						
Transaxle Service	See Explanation in Section B																
Fuel Filter Replacement	Every 30,000 Miles (50 000 km)										•						
Spark Plug Replacement*	Every 30,000 Miles (50 000 km)										•†						
PCV System Inspection*	Every 30,000 Miles (50 000 km) or 24 Months										•						
Air Cleaner Element Replacement*	Every 30,000 Miles (50 000 km) or 24 Months										•						
Oxygen Sensor Inspection*	Every 30,000 Miles (50 000 km)										•						
Pulse Air Control System Inspection*	Every 30,000 Miles (50 000 km) or 24 Months										•						
Spark Plug Wires & Distributor Inspection*	Every 30,000 Miles (50 000 km) or 24 Months										•						

MAINTENANCE SCHEDULE
Severe Driving Conditions

TO BE SERVICED	WHEN TO PERFORM Miles (Kilometers) or Months, Whichever Occurs First	The services shown in this schedule up to 48,000 miles (80 000 km) are to be performed after 48,000 miles at the same intervals															
	MILES (000)	3	6	9	12	15	18	21	24	27	30	33	36	39	42	45	48
	KILOMETERS (000)	5	10	15	20	25	30	35	40	45	50	55	60	65	70	75	80
Fuel Tank, Cap & Lines—Inspection	Every 15,000 Miles (25 000 km) or 12 Months					•					•					•	
Thermostatically Controlled Air Cleaner*	Every 30,000 Miles (50 000 km) or 12 Months										•						

*An Emission Control Service.
†In California, these are the minimum Emission Control Maintenance Services an owner must perform according to the California Air Resources Board. General Motors, however, urges that all Emission Control Maintenance Services shown be performed.

MAINTENANCE SCHEDULE
Normal Driving Conditions

TO BE SERVICED	WHEN TO PERFORM Miles (Kilometers) or Months, Whichever Occurs First	The services shown in this schedule up to 45,000 miles (75 000 km) are to be performed after 45,000 miles at the same intervals					
	MILES (000)	7.5	15	22.5	30	37.5	45
	KILOMETERS (000)	12.5	25	37.5	50	62.5	75
Engine Oil & Oil Filter Change*	Every 7,500 Miles (12 500 km) or 12 Months	•	•	•	•	•	•
Chassis Lubrication	Every 7,500 Miles (12 500 km) or 12 Months	•	•	•	•	•	•
Carburetor Choke Inspection*	Every 30,000 Miles (50 000 km) or 24 Months				•†		
Engine Idle Speed Inspection*	At 7,500 Miles (12 500 km), then at 15,000 Miles (25 000 km) Intervals	•	•		•		•
Valve Lash Inspection*	Every 15,000 Miles (25 000 km)		•		•		•
Water Pump Belt Inspection*	Every 30,000 Miles (50 000 km) or 24 Months				•†		
Fuel Cut System Inspection*	Every 30,000 Miles (50 000 km), or 24 Months				•		
Transaxle Service	See Explanation in Section B						

MAINTENANCE SCHEDULE
Normal Driving Conditions

TO BE SERVICED	WHEN TO PERFORM Miles (Kilometers) or Months, Whichever Occurs First	The services shown in this schedule up to 45,000 miles (75 000 km) are to be performed after 45,000 miles at the same intervals					
	MILES (000)	7.5	15	22.5	30	37.5	45
	KILOMETERS (000)	12.5	25	37.5	50	62.5	75
Fuel Filter Replacement	Every 30,000 Miles (50 000 km)				•		
Spark Plug Replacement*	Every 30,000 Miles (50 000 km)				•†		
PCV System Inspection*	Every 30,000 Miles (50 000 km) or 24 Months				•		
Air Cleaner Element Replacement*	Every 30,000 Miles (50 000 km) or 24 Months				•		
Oxygen Sensor Inspection*	Every 30,000 Miles (50 000 km)				•		
Pulse Air Control System Inspection*	Every 30,000 Miles (50 000 km) or 24 Months				•		
Spark Plug Wires & Distributor Inspection*	Every 30,000 Miles (50 000 km) or 24 Months				•		
Fuel Tank, Cap & Lines—Inspection	Every 15,000 Miles (25 000 km) or 12 Months		•		•		•
Thermostatically Controlled Air Cleaner*	Every 30,000 Miles (50 000 km) or 12 Months				•		

*An Emission Control Service.
†In California, these are the minimum Emission Control Maintenance Services an owner must perform according to the California Air Resources Board. General Motors, however, urges that all Emission Control Maintenance Services shown be performed.

REQUIRED EMISSION CONTROL MAINTENANCE

NOTE: The following information is from the latest information available at the time of publication. If this information differs from any data on the underhood emission label or in the owners manual, use the information in the owners manual or on the emissions label, since that information reflects the latest changes or modifications to the vehicle.

EXHAUST GAS RECIRCULATION

Every 30,000 miles (48,000 km), the EGR flow passages should be checked for deposits. If the EGR passages in the inlet manifold exhibit signs of excessive build-up, the passages should be cleaned. All loose particles should be removed in order to prevent the valve from clogging. All mounting surfaces of the intake manifold and valve assembly should be cleaned using a wire wheel to remove the exhaust deposits. Use a new gasket when installing a new EGR valve. Do not use any kind of solvent or cleaning fluids. They will damage the valve diaphragm.

SPARK PLUGS

The spark plugs must fire properly to assure proper engine performance, economy and emissions control. New spark plugs should operate satisfactorily in normal service for 30,000 miles (50,000 km).

UNDERHOOD RUBBER AND PLASTIC COMPONENTS (EMISSION HOSES)

Inspect the hose surfaces for evidence of heat and/or mechanical damage. Hard or brittle rubber, cracking, checking, tears, cuts, abrasions and excessive swelling indicate deterioration of the rubber. Inspect all hose connections, couplings and clamps for leakage. Replace as required.

AUTOMATIC CHOKE SYSTEM

With the engine off, partially open the throttle and check the entire choke system for freedom of operation for its full travel. Check the vacuum kick and fast idle cam position settings.

CARBURETOR AIR FILTER

The filter in the carburetor air filter should be replaced every 30,000 miles (48,000 km). Replace the filter more often if the vehicle is driven in dusty or sandy conditions.

CRANKCASE INLET AIR CLEANER

The crankcase inlet air filter must be kept clean and lubricated. The inlet air cleaner should be washed in a cleaning solution or similar solvent. Lubricate the filter by inverting it and filling it with SAE 30 engine oil. Position the air cleaner to allow excess oil to drain through the vent nipple located in the top of the air cleaner. More frequent service may be required on vehicles that are operated extensively on short run, stop and go or extended idle service.

CHARCOAL CANISTER

Replace the filter element in the base of the canister with a new element every 30,000 miles (48,000 km). The filter should be replaced more often if the vehicle is driven under dusty or sandy conditions.

MANIFOLD HEAT CONTROL VALVE

Certain vehicles are equipped with a manifold heat control valve. The manifold heat control valve must work freely for proper engine warm-up and acceleration. Every 30,000 miles (48,000 km), the heat control valve should be checked for free operation and have manifold heat control valve solvent applied.

POSITIVE CRANKCASE VENTILATION (PCV) VALVE

Every 30,000 miles (48,000 km), the valve should be checked. If the valve is found to be clogged, replace it with a new valve. Never attempt to clean an old PCV valve.

IDLE SPEED AND MIXTURE ADJUSTMENT

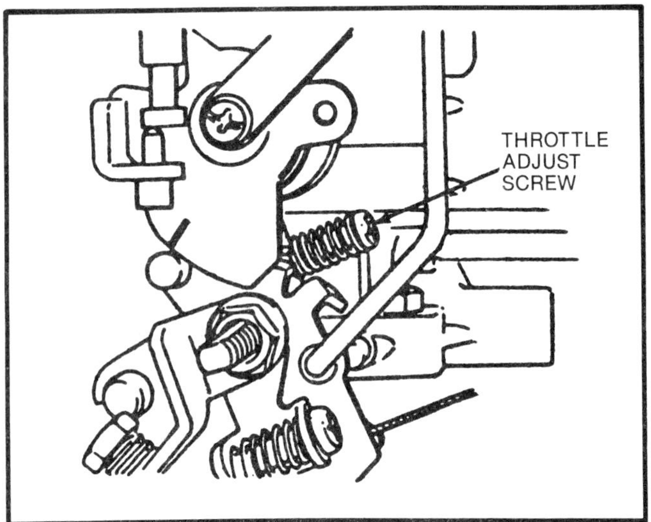

Throttle adjusting screw

Mixture adjusting screw pin location

Carburetor

Chevrolet Sprint uses a Model MR08, 2 barrel, downdraft type carburetor. It is equipped with primary and secondary systems. The primary system operates under normal conditions and the secondary system operates under high load or high speed driving conditions. A choke valve is provided on the primary system.

Idle Speed

Adjustment

NOTE: Disconnect the engine cooling fan before making the following adjustments. Make sure that emission control hoses are connected and the accelerator cable has a small amount of play. The ignition timing must be within 9–11 degrees BTDC at 750 rpm for vehicles equipped with manual transaxles and 5–7 degrees BTDC at 850 rpm for automatic transaxles.

1. Position the transaxle gear shift lever in the neutral position. Block the drive wheels and set the parking brake securely.
2. Start the engine and allow it to reach normal operating temperature.
3. With all of the accessories off, check to ensure that the idle speed is set according to the underhood emission control label.
4. Adjust the idle speed by turning the throttle adjusting screw until the idle is within the specified range.

Idle Mixture

Adjustment

NOTE: Disconnect the engine cooling fan before making the following adjustments. Make sure that the emission hoses are connected and that the accelerator cable has a small amount of play. The ignition timing must be within 9–11 degrees BTDC at 750 rpm for vehicles equipped with manual tranaxles and 5–7 degrees BTDC at 850 rpm for automatic transaxles.

1. Remove the carburetor from the intake manifold in order to gain access to the mixture adjusting screw pin which covers the mixture adjusting screw.

Mixture adjusting screw

2. Using a 4–4.5mm drill bit, drill a hole in the pin. After drilling, remove the pin.

3. Reinstall the carburetor and air cleaner assembly on the vehicle.

4. Place the gear shift selector lever in the neutral position. Set the parking brake and block the drive wheels securely.

5. Start the engine and allow it to reach normal operating temperature.

6. Remove the coupler on the duty check connector located in the engine compartment. Connect a dwell meter to the connector.

7. Connect the positive terminal of the dwell meter to the blue/red wire and the negative terminal to the black/green wire.

8. Set the dwell meter to the six cylinder position. Check to see that the indicator moves.

9. Check the idle speed and adjust if necessary.

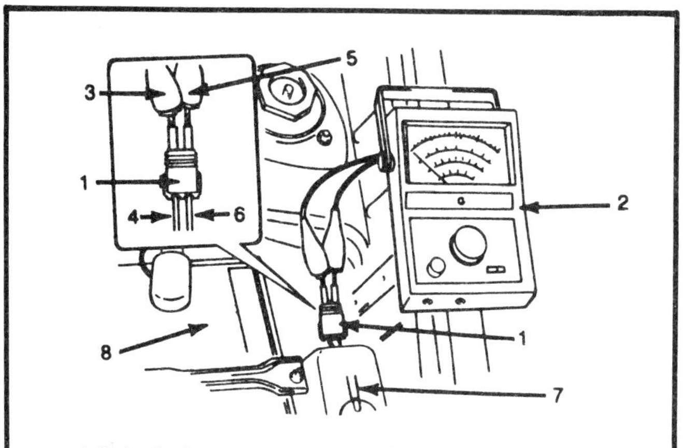

1. Duty check connector
2. Dwell meter
3. Positive (+) terminal
4. "Blue/Red" wire
5. Negative (−) terminal
6. "Black/Green" wire
7. Water reservoir tank
8. Battery

Dwell meter connections

10. Run the engine at idle speed and adjust the mixture adjusting screw slowly until a dwell of 21—27 degrees is obtained. If the dwell is too high, turn the screw in. If the dwell is too low, back the screw out.

11. After adjusting the dwell, check the idle speed and readjust if necessary.

12. Install the new mixture adjusting screw pin in the throttle chamber.

IGNITION TIMING AND IGNITION SYSTEM

IGNITION SYSTEM

The Chevrolet Sprint 3 cylinder engine uses a solid state ignition system. It is comprised of the spark plugs, ignition coil and distributor. The distributor has a rotor, a module, a pole piece, a vacuum advance and a centrifugal advance.

The signal generator is used to generate the ignition signal and consists of a pole piece, a magnet and a pick-up coil. The pole piece is attached to the distributor shaft. The magnet and pick-up coil are attached to the pick-up coil base plate.

When the distributor shaft rotates, the magnetic flux passing through the pick-up coil varies due to the change in the air gap between the

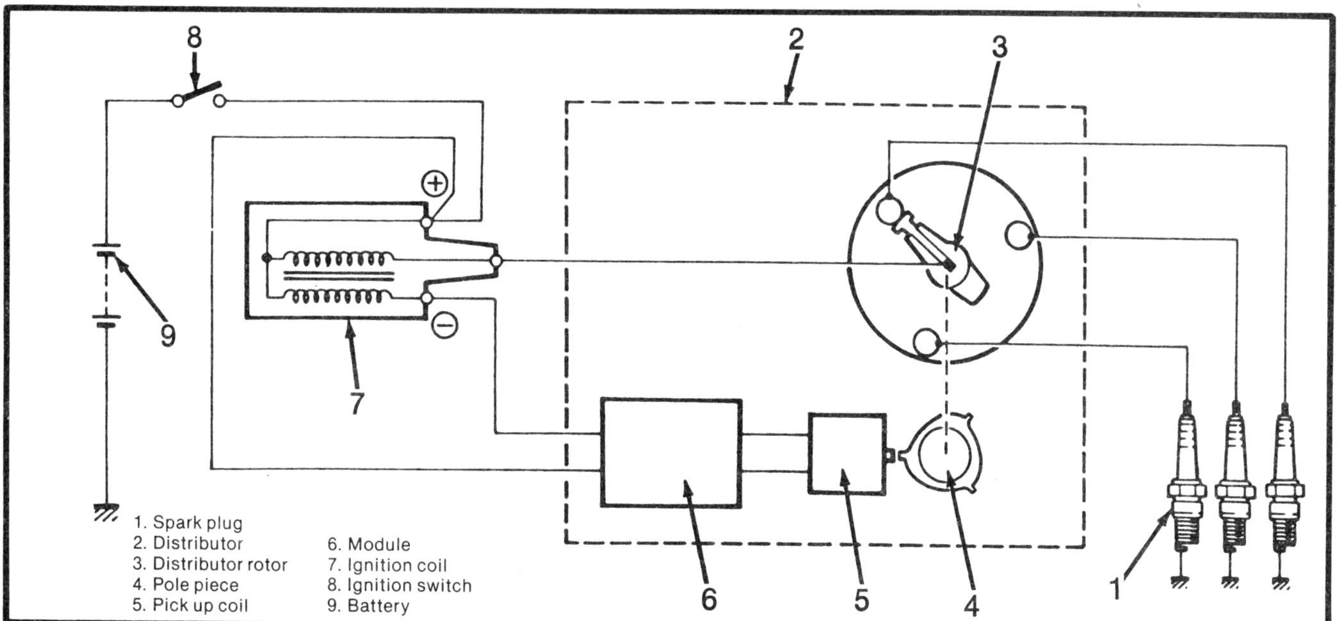

1. Spark plug
2. Distributor
3. Distributor rotor
4. Pole piece
5. Pick up coil
6. Module
7. Ignition coil
8. Ignition switch
9. Battery

Ignition circuit schematic

pick-up coil and the pole piece. Because of this, the alternating current voltage is induced in the pick-up coil. The induced voltage turns the module on and off which switches off the ignition coil primary current. The high voltage is induced in the secondary winding of the ignition coil and ignition sparks are generated at the spark plugs.

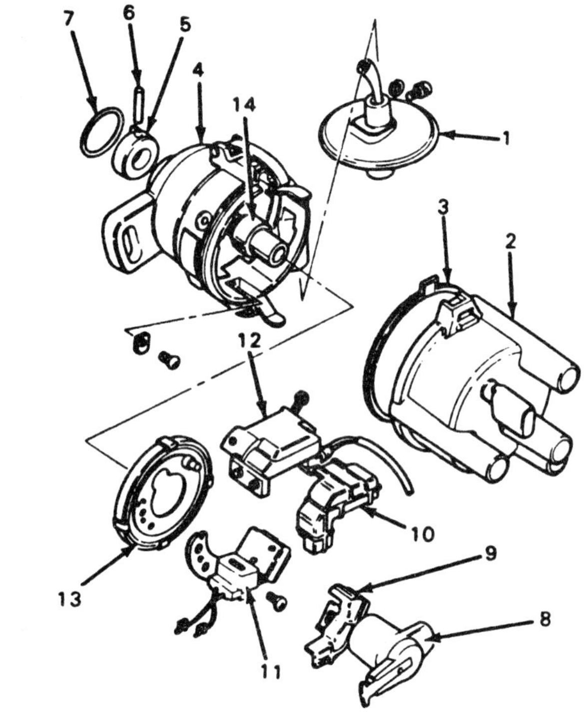

1. Vacuum advance unit
2. Distributor cap
3. Seal
4. Distributor housing
5. Distributor coupling
6. Pin
7. Seal
8. Rotor
9. Pick up coil dust cover
10. Module dust cover
11. Pick up coil
12. Module
13. Pick up coil base plate
14. Pole piece

Distributor-exploded view

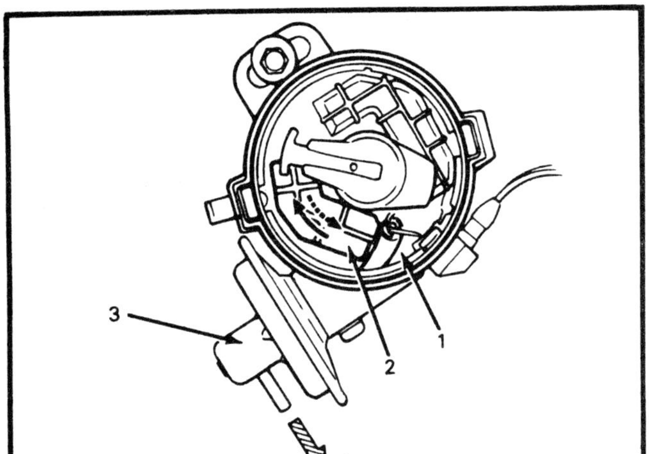

1. Pick up coil base plate
2. Pick up coil
3. Vacuum advance unit
4. Vacuum

Vacuum advance unit testing

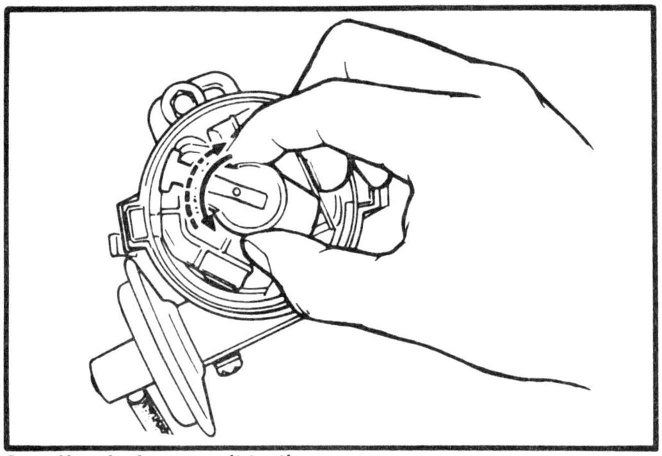

Centrifugal advance unit-testing

CENTRIFUGAL ADVANCE

Testing

1. Disconnect the negative battery cable.
2. .Remove the distributor cap.
3. Turn the rotor counterclockwise and release it.
4. The rotor should return smoothly by spring force. If not, replace the defective components as required.

VACUUM ADVANCE

Testing

1. Disconnect the negative battery cable. Remove the distributor cap.
2. Disconnect the vacuum hose from the vacuum advance unit.
3. Connect a vacuum tester to the advance unit.

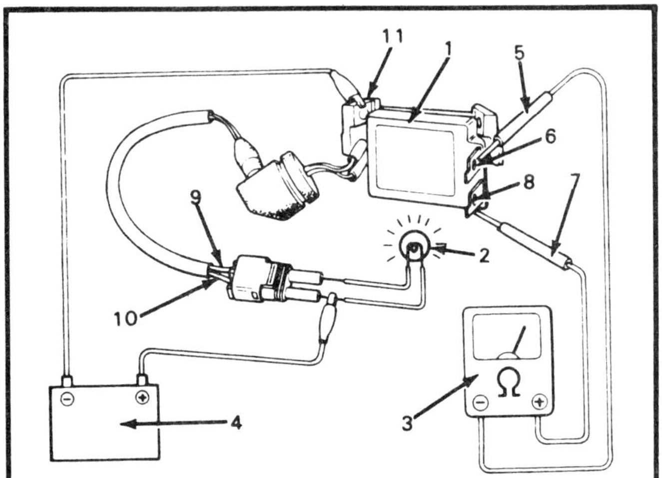

1. Module
2. Bulb
3. Ohm meter
4. Battery (12 volts)
5. Negative prod
6. Red wire's terminal
7. Positive prod
8. White wire's terminal
9. Brown wire
10. White/black wire
11. Ground

Module testing wiring

4. Apply approximately 16 in. Hg. of vacuum and release it. Check to see that the pick-up base plate moves smoothly.
5. If the plate does not move smoothly, inspect the plate and vacuum advance unit. Repair or replace as required.

SPARK PLUGS

Year	Eng. V.I.N. Code	Engine No. Cyl. Disp. (cu.in.)	Liters	Spark Plugs orig. type	Gap (in.)
'85–'86	M	61	1.0	NGK; PBRGES-11	0.039–0.043
				Denso; WIGEXR-U11	0.039–0.043
				AC; R43CXLS	0.039–0.043

MODULE

Testing

1. Remove the dust cover from the module. Tag and disconnect the white and red wires from the module.
2. Connect a light bulb, an ohmmeter and a 12 volt battery to the module.
3. Set the ohmmeter to the 1–10 ohm range. Allow the ohmmeter negative probe to touch the red wire terminal of the module and the positive lead to touch the white wire terminal.
4. If the light bulb begins to light, the module is good. If not, replace the module. Repair or replace as required.

NOTE: Failure to connect the ohmmeter in the described manner can result in damage to the ohmmeter and/or module.

IGNITION TIMING

Ignition timing for the Chevrolet Sprint should be set at 6° BTDC with the engine running between 800–900 rpm. Before adjusting the ignition timing, make sure that the head lights, heater fan, rear defogger (if equipped), air conditioner (if equipped) and the engine cooling fan are off. If any of the above components are on, the idle up system will operate and the engine idle speed will be higher than necessary.

Inspection

1. Start the engine and allow it to reach normal operating temperature.
2. Connect a tachometer to the engine according to the manufacturers instructions. Check to see that the idle speed is between 800–900 rpm. Adjust as necessary.
3. Connect a timing light to the number one cylinder spark plug wire.
4. While the engine is running at the specified idle speed, aim the timing light at the crankshaft pulley.
5. If the timing mark on the timing tab appears to be aligned with the timing notch on the crankshaft pulley, the ignition is properly timed. Adjust as necessary as outlined below.

Adjustment

1. Loosen the distributor flange retaining bolt.
2. Turn the distributor case clockwise to advance the timing and counterclockwise to retard the timing.
3. Tighten the flange bolt. Recheck the timing.

EMISSION CONTROL SYSTEMS

CRANKCASE EMISSION CONTROL SYSTEM

Positive Crankcase Ventilation System

The Positive Crankcase Ventilation System (PCV) is used on all vehicles to provide a more complete evacuation of the crankcase vapors. Outside vehicle air is routed through the air cleaner to the crankcase. It is then mixed with blow- by gases and is passed through the PCV valve. It is then routed into the intake manifold. The PCV valve meters the air flow rate which varies under engine operation depending on manifold vacuum. In order to maintain idle quality, the PCV valve limits the air flow when the intake manifold vacuum is high. If abnormal operating conditions occur, the system will allow excessive blow-by gases to back flow through the crankcase vent tube into the air cleaner. These blow-by gases will then be burned by normal combustion.

Trouble Diagnosis

When rough engine idling is encountered, check for a clogged PCV valve, a pinched or plugged hose. Replace or repaired as required. The following procedure is recommended.

1. Remove the PCV valve from the rocker arm cover.
2. Start the engine and allow it to reach normal operating temperature. Allow the engine to run at idle speed.

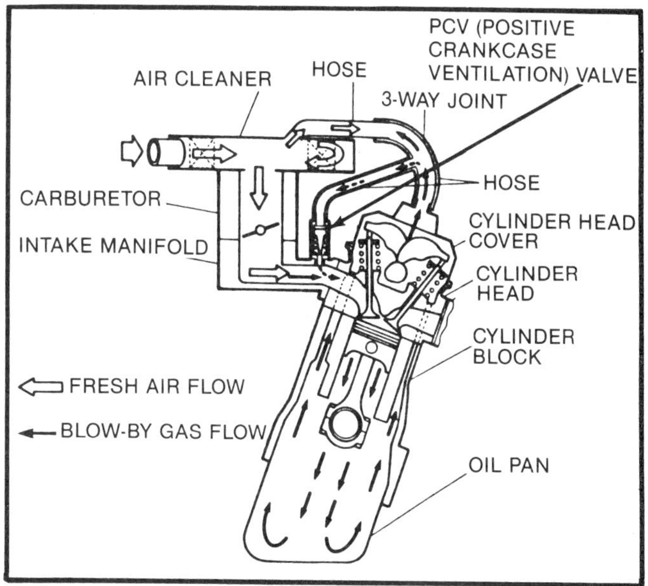

PCV system-Sprint 3 cylinder

3. Check for vacuum at the end of the valve. If there is no vacuum at the end of the valve, check for defective hoses or a clogged manifold port.

4. Turn off the engine and remove the PCV valve. Shake the valve and listen for the rattle of the check needle inside the valve. If no rattle is present, replace the valve.

5. Check the systems hoses and clamps and replace any that show signs of deterioration.

EVAPORATIVE EMISSION CONTROL SYSTEM

The Evaporative Emission Control System (EECS) is designed to trap fuel vapors emitted from the fuel tank and carburetor during normal engine operation. This will prevent gasoline vapor discharge into the atmosphere. Gasoline vapors are absorbed through the use of a fuel vapor charcoal canister. The charcoal canister absorbs the gasoline vapors and stores them until they can be removed and burned in the engine. The vapors pass through a 2-way valve and enters the charcoal canister. While the engine is running, the fuel vapor stored in the canister, is drawn into the intake manifold together with fresh air to be burned. From the charcoal canister, the vapors are routed to the PCV system. This allows the vapors to be burned during engine operation.

CHARCOAL CANISTER

Testing

1. Disconnect the rubber hoses from the charcoal canister.

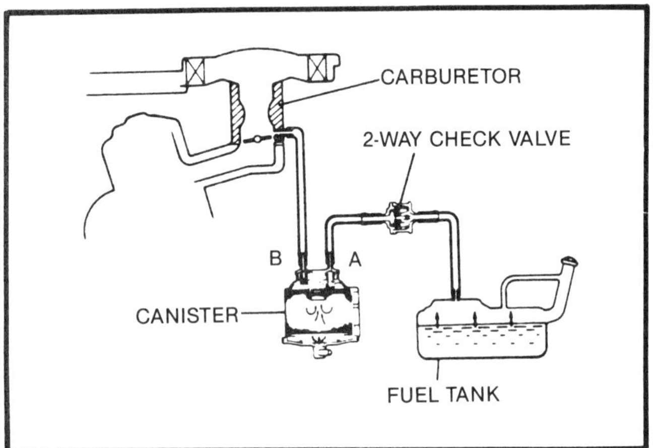

Charcoal canister schematic

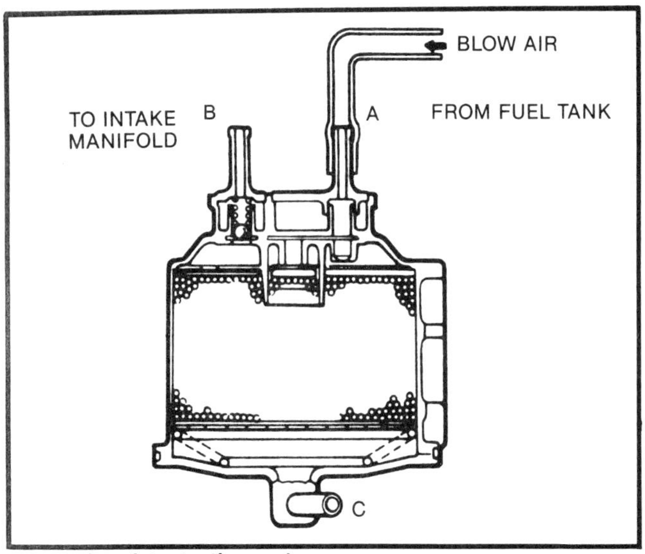

Charcoal canister-testing ports

2. Attempt to blow air through pipe A. There should be no restriction through pipes B and C (see illustration).

3. When air is blown into pipe B, air should not pass through either pipe A or C.

4. If operation differs from the above procedure, replace the canister. The canister is cleaned by blowing 40 psi of compressed air into pipe A while sealing pipe B.

TWO-WAY CHECK VALVE

Testing

1. Disconnect the rubber hoses from the two-way check valve. Remove the check valve.

2. Air should be able to be blown through the valve from the fuel tank side (black side) to the orange side.

3. From the orange side, air should come out of the black side.

4. If air does not pass through the valve on either attempt, replace the check valve with a new one.

HOT IDLE COMPENSATOR (HIC)

The hot idle compensator is attached to the carburetor assembly. It provides the best air/fuel mixture during hot idle which helps to maintain a stable idle speed. The HIC has a bimetal strip which warps as the transferred heat temperature rises above 120 degrees F. At this point,

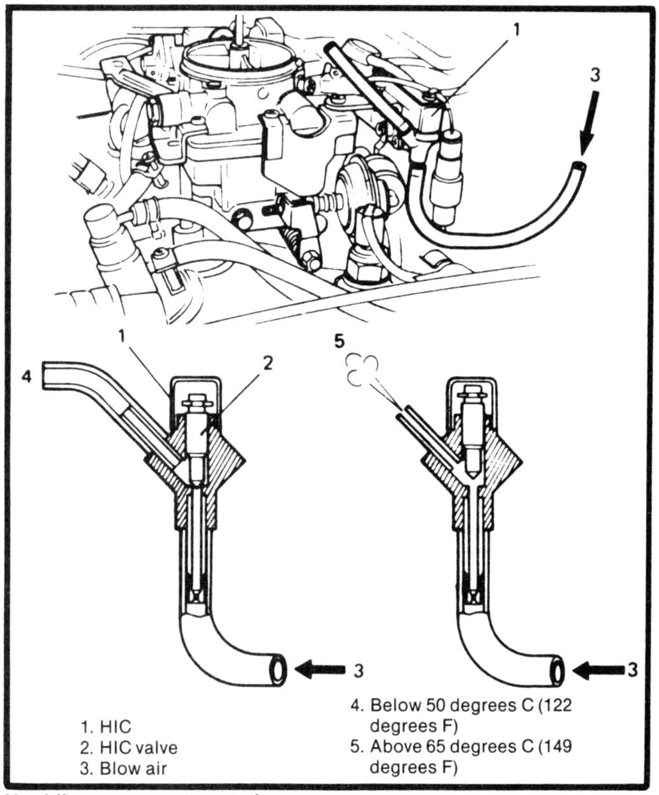

Hot idle compensator testing

the HIC valve begins to open. As the HIC valve opens, the air from the air cleaner side of the carburetor is drawn through the HIC valve into the intake manifold in order to prevent the air/fuel mixture from getting too rich during hot idle.

Testing

1. Remove the air cleaner assembly.
2. Tag and disconnect the vacuum hoses from the carburetor.
3. Using a thermometer or equivalent, check the temperature around the HIC. If the temperature is below 122 degrees F., air should not come out of the HIC nozzle. If the temperature is above 149 degrees F., air should come out of the HIC nozzle.
4. Replace the HIC if defective.

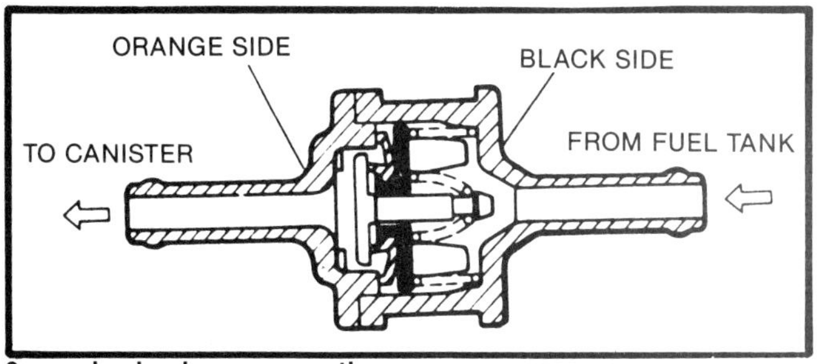

2-way check valve-cross section

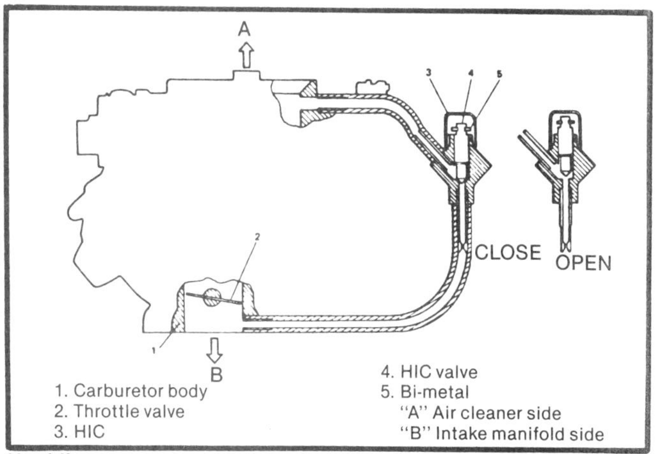

Hot idle compensator

EXHAUST EMISSION CONTROL SYSTEM

Thermostatically Controlled Air Cleaner

The Thermostatically Controlled Air Cleaner (TAC) system helps to improve fuel vaporization by controlling the temperature of the intake air. The intake air is almost always at a constant level, regardless of outside temperature and driving conditions. This helps to distribute the mixture evenly to each cylinder and to stabilize the air/fuel mixture ratio.

The system consists of the Air Control Actuator (ACA) and the thermo sensor. The thermo sensor located in the air cleaner case, senses the intake air temperature and controls the vacuum line by opening and closing the passage to the ACA. According to this opening and closing

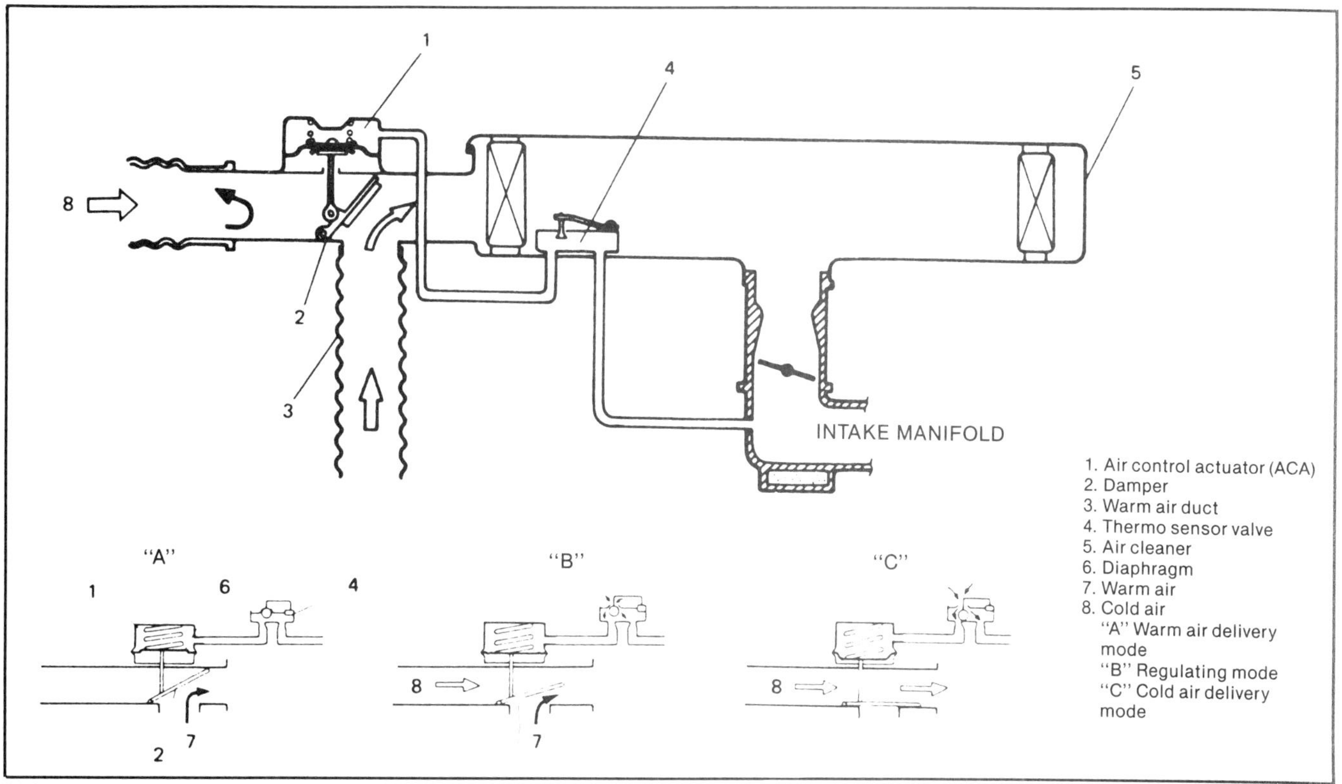

TAC operation and components

operation, the vacuum in the intake manifold acts upon the damper through the diaphragm in the ACA. For warm air, the air is warmed up in the exhaust manifold cover. For cold air, the outside air is drawn through the fresh air passage where both enter the air cleaner.

During cold weather engine starting, the thermo valve is closed because the intake air temperature in the air cleaner is low. Therefore, the vacuum is transmitted to the ACA diaphragm, which then pulls up the damper linked to the diaphragm to open the warm air duct fully. As the engine runs and warms up, the temperature of the intake air rises and the thermo valve starts opening. At this point, the vacuum sent to the ACA diaphragm decreases, and the damper pushed down by the spring force, lessens the warm air duct opening.

When the engine is operating under a high load or high rpm, the temperature of the air coming from the warm air duct rises, which causes the thermo valve opening to become larger and the damper opening smaller.

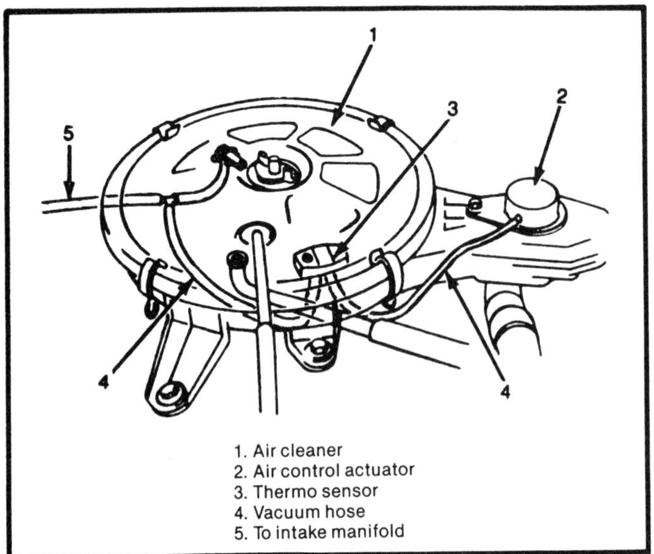

Thermostatically controlled air cleaner system

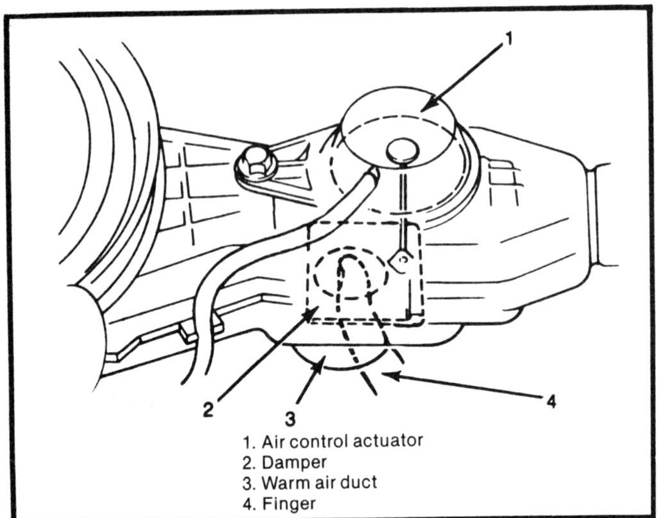

Damper door in the closed position

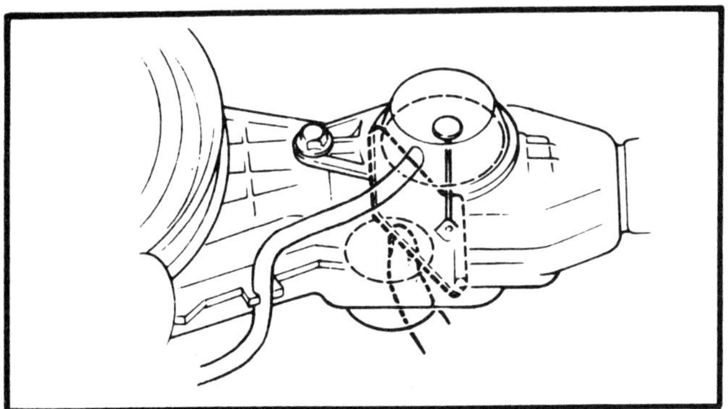
Damper door in the open position

THERMOSTATICALLY CONTROLLED AIR CLEANER

Inspection

1. Check that all vacuum hoses are connected. Inspect the hoses for deterioration or cracking. Replace as required.
2. Remove the air hose from the warm air duct, when the air cleaner is cool.
3. Make sure that the damper closes the warm air duct with the engine stopped.
4. Start the engine and run at idle. The damper should fully open the warm air duct immediately.
5. If the above circumstances do not happen, replace or repair defective parts as required.

AIR CONTROL ACTUATOR (ACA)

Inspection

1. Disconnect the vacuum hose from the air control actuator.
2. Check to see that the damper opens fully when more than 7.87 in.Hg. of vacuum is applied.
3. Repair or replace defective parts as required.

THERMO SENSOR

Inspection

1. Remove the air cleaner assembly.
2. Tag and disconnect the vacuum hoses (2) from the thermo sensor.
3. Measure the temperature around the thermo sensor.
4. Close the long nozzle (intake manifold side nozzle) and blow compressed air into the short nozzle (ACA side nozzle). If the measured temperature is above 104 degrees F., air should come out of the thermo sensor valve (valve is open).
5. If the temperature is below 77 degrees F., air should not come out (valve is closed).
6. Repair or replace defective components as required.
7. Reinstall the air cleaner assembly. Connect the vacuum hoses.

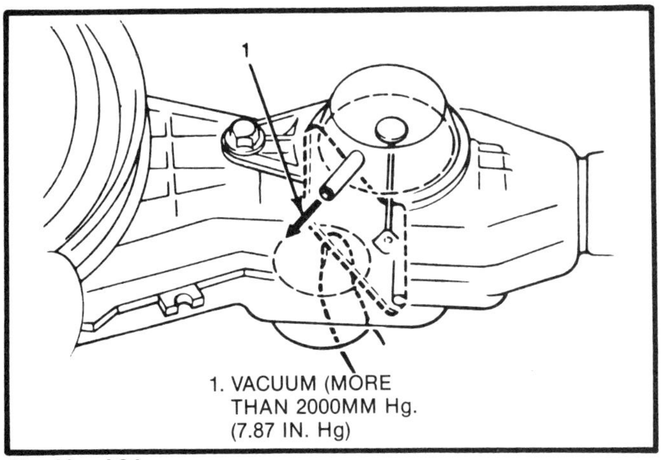

Checking ACA

Exhaust Gas Recirculation (EGR) System

The EGR system controls the formation of oxides of nitrogen (NOx) by recirculating the exhaust gas into the combustion chamber, through the intake manifold.

The diaphragm, which is mounted in the EGR modulator, is operated by the back pressure of the exhaust gas which will open and close the EGR valve. The EGR modulator controls vacuum routed to the EGR valve.

Under low speed driving, the exhaust pressure is low. When exhaust pressure is low, the diaphragm in the EGR modulator is pushed down by the spring force and the modulator valve opens to allow air into the vacuum passage. As a result, vacuum transmitted to the EGR valve becomes less and so does the opening of the EGR valve.

Under high speed driving or other high load conditions, the exhaust pressure is high. During high exhaust pressure, the diaphragm in the modulator is pushed up and closes the valve. Since air does not enter the vacuum passage, the vacuum sent to the EGR valve becomes larger and so does the EGR valve opening. Because of this, a larger amount of exhaust gas is recirculated to the intake manifold.

When coolant temperature is low, the vacuum passage of the EGR valve is opened to the air through the bi-metal Vacuum Switching Valve (BVSV) on the intake manifold. In this state, because vacuum is not transmitted to the EGR valve, it remains closed.

When the coolant temperature is normal, the BVSV is closed. So the EGR valve opens and closes according to the EGR modulator operation.

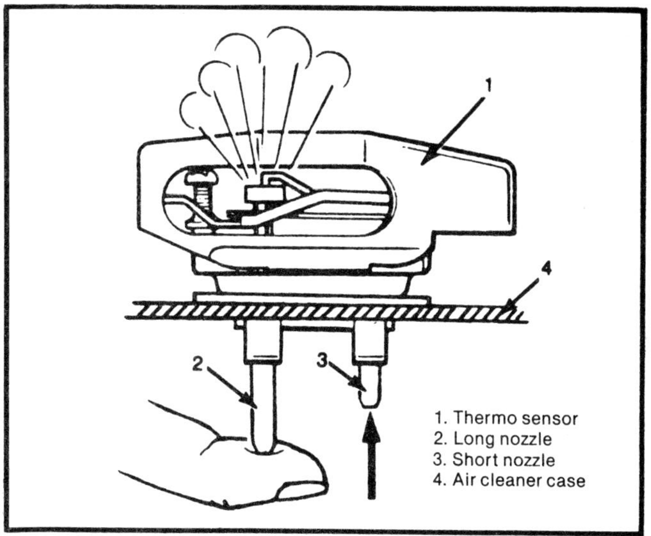

Thermo sensor inspection

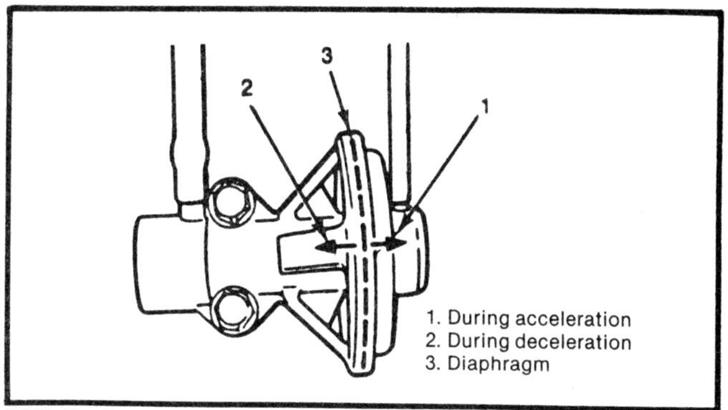

EGR diaphragm movement

EGR SYSTEM

Testing

1. Start the engine when cool and check that the EGR valve is not operating by placing a finger on the diaphragm.
2. Allow the engine to reach normal operating temperature. Movement of the EGR diaphragm should occur.

EGR DIAGNOSIS

Condition	Cause	Correction
Engine idles abnormally rough and/or stalls.	1) EGR valve vacuum hoses misrouted.	a) Check EGR valve vacuum hose routing.
	b) Leaking EGR valve.	b) Check EGR valve for correct operation.
	c) EGR valve gasket failed or loose EGR attaching bolts.	c) Check EGR attaching bolts for tightness. If not loose, remove EGR valve and inspect gasket.
	d) Improper vacuum to EGR valve at idle.	d) Check vacuum from carburetor EGR port with engine at stabilized operating temperature and at curb idle speed.
Engine runs rought on light throttle acceleration and has poor part load performance.	a) EGR valve vacuum hose misrouted.	a) Check EGR valve vacuum hose routing.
	b) Check for loose valve. Sticky or binding EGR valve.	b) Torque valve. Same as listing in "Engine Idles Abnormally Rough and/or Stalls" condition. Clean EGR passage deposits. Perform EGR System Check.
	c) Wrong or no EGR gasket.	c) Install new gasket, torque attaching parts.

EGR DIAGNOSIS

Condition	Cause	Correction
Engine stalls on decelerations.	a) EGR modulator valve blocked or air flow restricted.	a) Check EGR modulator valve operation.
	b) Restriction in EGR vacuum line.	b) Check EGR vacuum lines for kinks, bends, etc. Remove or replace hoses as required.
	c) Sticking or binding EGR valve.	c) Check EGR valve for excessive deposits causing sticky or binding operation.
Part throttle engine detonation. **NOTE:** Non-functioning EGR valve could contribute to part throttle detonation. Detonation can be caused by several other engine variables. Perform ignition and carburetor related diagnosis.	a) EGR modulator valve blocked or air flow restricted.	a) Check internal control valve operation.
	b) Insufficient exhaust gas recirculation flow during part throttle accelerations.	b) Check EGR valve hose routing. Check EGR valve operation. Repair or replace as required. Check EGR passages and valve for excessive deposit.
Engine Starts but immediately stalls when cold. **NOTE:** Stalls after start can also be caused by carburetor problems.	a) EGR valve hoses misrouted.	a) Check EGR valve hose routings.
	b) BVSV is out of order.	b) Check BVSV. Replace as necessary.

BI-METAL VACUUM SWITCHING VALVE (BVSV)

Testing

1. Drain the engine cooling system.
2. Remove the BVSV from the intake manifold.
3. While keeping the BVSV cool (below 113 degrees F.), blow compressed air into the two nozzles. Air should come out of the filter.
4. While keeping the BVSV warm (above 140 degrees F.), blow compressed air into the two nozzles. Air should not come out of the filter.
5. Replace defective part, if necessary.
6. Install the BVSV on the intake manifold. Connect the vacuum hoses. Fill the cooling system.

EGR MODULATOR

Testing

1. Inspect the EGR filter for any signs of damage or excessive wear. Using compressed air, clean the filter.
2. With the engine stopped, tag and disconnect the vacuum hoses from the EGR modulator. Plug the nozzle. Blow air into the other nozzle and check that air passes through the air filter side freely.
3. Start the engine and allow it to reach normal operating temperature.
4. Run the engine to 5000 rpm and plug one of the EGR nozzles. Attempt to blow air into the other nozzle. There should be a strong resistance to air flow.
5. Replace any defective parts as required. Reinstall the EGR modulator.

EGR VALVE

Testing

1. Remove the EGR valve from the vehicle.
2. Inspect the valve for sticking or exhaust desposits.
3. If found to be good, reinstall the EGR valve, using a new gasket.
4. Connect the vacuum hoses.

Catalytic Converter

Inspection

1. Raise the vehicle and support safely.
2. Inspect all exhaust pipe rubber mountings for deterioration, cracks or other signs of damage.
3. Inspect all components for leaks, loose connections or dents.
4. Inspect nearby body areas for any structural damage such as open seams which could allow exhaust fumes into the passenger compartment.
5. Repair or replace damaged components as required.

VACUUM CIRCUITS

VEHICLE EMISSION CONTROL INFORMATION LABEL

The Vehicle Emission Control Information Label is located under hood. The label contains important emission specifications and setting procedures, as well as a vacuum hose schematic with emission components identified.

When servicing the engine or emission systems, the Vehicle Emission Control Information Label should always be checked for up-to-date information.

VEHICLE EMISSION CONTROL INFORMATION
SUZUKI MOTOR CO., LTD.

ENGINE DISPLACEMENT:	1.0 LITER
ENGINE FAMILY IDENTIFICATION:	FSK1.0V2FFC4
EVAPORATIVE FAMILY IDENTIFICATION:	EV1

ENGINE TUNE-UP SPECIFICATIONS AND ADJUSTMENTS:
ALL ADJUSTMENTS ARE TO BE PERFORMED WITH THE TRANSMISSION IN NEUTRAL.

IDLE SPEED:		750 RPM
IGNITION TIMING:		10° BTDC/750 RPM
VALVE LASH(COLD):	INTAKE	0.006 IN.
	EXHAUST	0.008 IN.
(HOT):	INTAKE	0.010 IN.
	EXHAUST	0.012 IN.

TURN IDLE-ADJUSTING-SCREW TO OBTAIN THE CORRECT IDLE SPEED WHILE ENGINE IS HOT.

THIS VEHICLE CONFORMS TO U.S. EPA AND CALIFORNIA REGULATIONS APPLICABLE TO 1985 MODEL YEAR NEW PASSENGER CARS.

CATALYST TWC/CL/AIV/EGR/EGS 11955-82040

EMISSION HOSE ROUTING

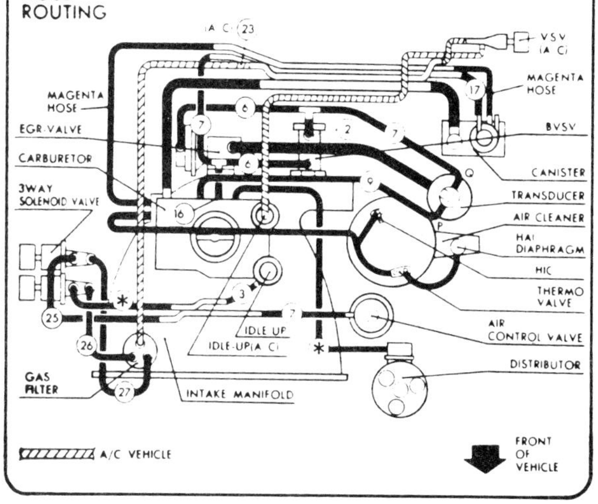

Vehicle Emission Control Information Label (Sample)

VACUUM CIRCUITS

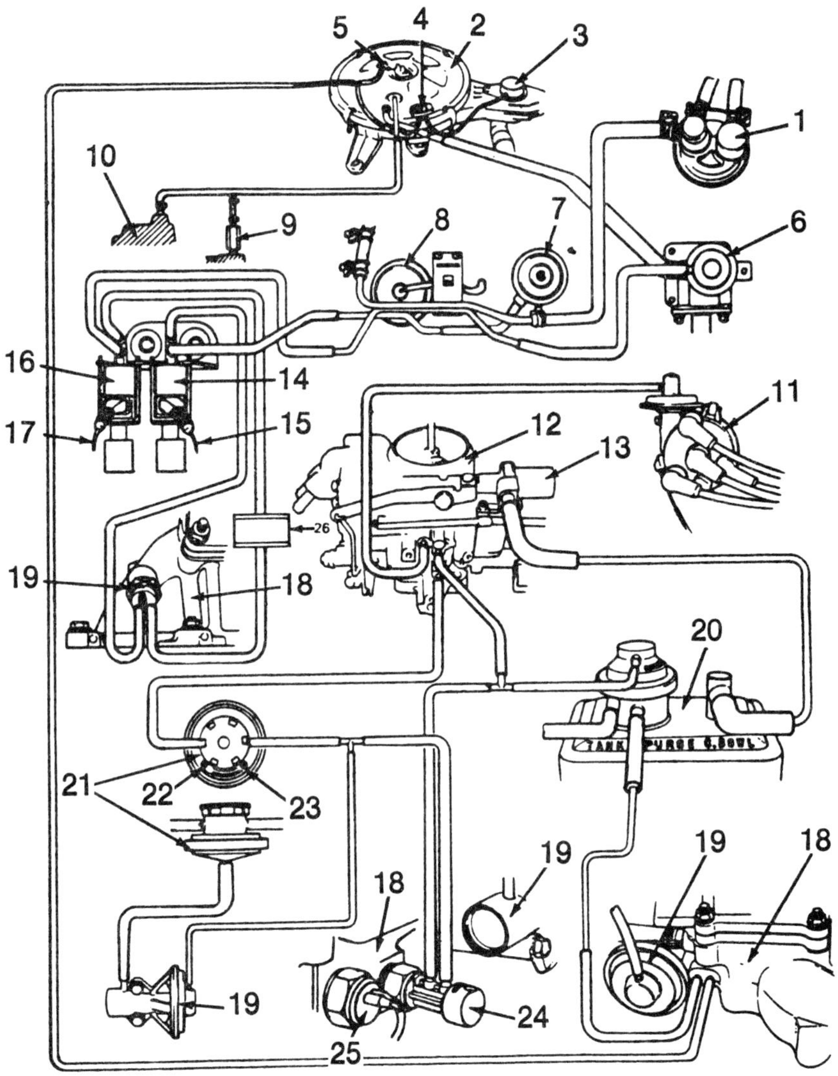

1. Fuel pump
2. Air cleaner
3. Air control actuator
4. Thermo sensor
5. Hot idle compensator
6. Second air valve
7. Idle-up actuator
8. Secondary diaphragm
9. PCV valve
10. Cylinder head cover
11. Distributor
12. Carburetor
13. Switch vent solenoid
14. Three way solenoid valve (black)
15. Black wire
16. Three way solenoid valve (blue)
17. Blue wire
18. Intake manifold
19. EGR valve
20. Canister
21. EGR modulator
22. P
23. Q
24. Bi-metal vacuum switching valve
25. Thermal switch
26. VTV (vacuum transmitting valve) A/T only

Vacuum hose routing for emission control system

VACUUM CIRCUITS

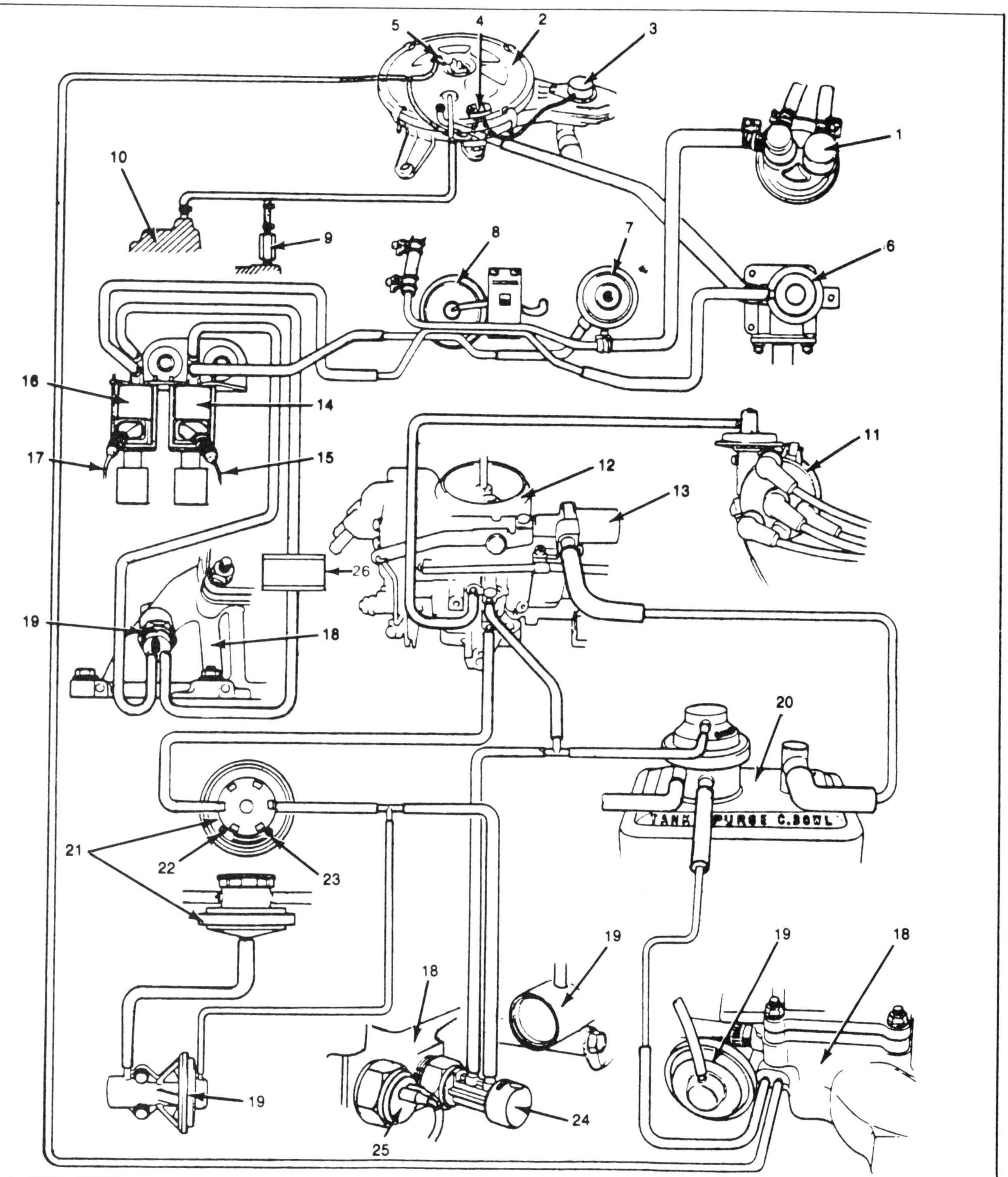

1. FUEL PUMP
2. AIR CLEANER
3. AIR CONTROL ACTUATOR
4. THERMO SENSOR
5. HOT IDLE COMPENSATOR
6. SECOND AIR VALVE
7. IDLE-UP ACTUATOR
8. SECONDARY DIAPHRAGM
9. PCV VALVE
10. CYLINDER HEAD COVER
11. DISTRIBUTOR
12. CARBURETOR
13. SWITCH VENT SOLENOID
14. THREE WAY SOLENOID VALVE (BLACK)
15. BLACK WIRE
16. THREE WAY SOLENOID VALVE (BLUE)
17. BLUE WIRE
18. INTAKE MANIFOLD
19. EGR VALVE
20. CANISTER
21. EGR MODULATOR
22. P
23. Q
24. BI-METAL VACUUM SWITCHING VALVE
25. THERMAL SWITCH
26. VTV (VACUUM TRANSMITTING VALVE) A/T ONLY

Engine Emission Control Systems Schematic

VACUUM CIRCUITS

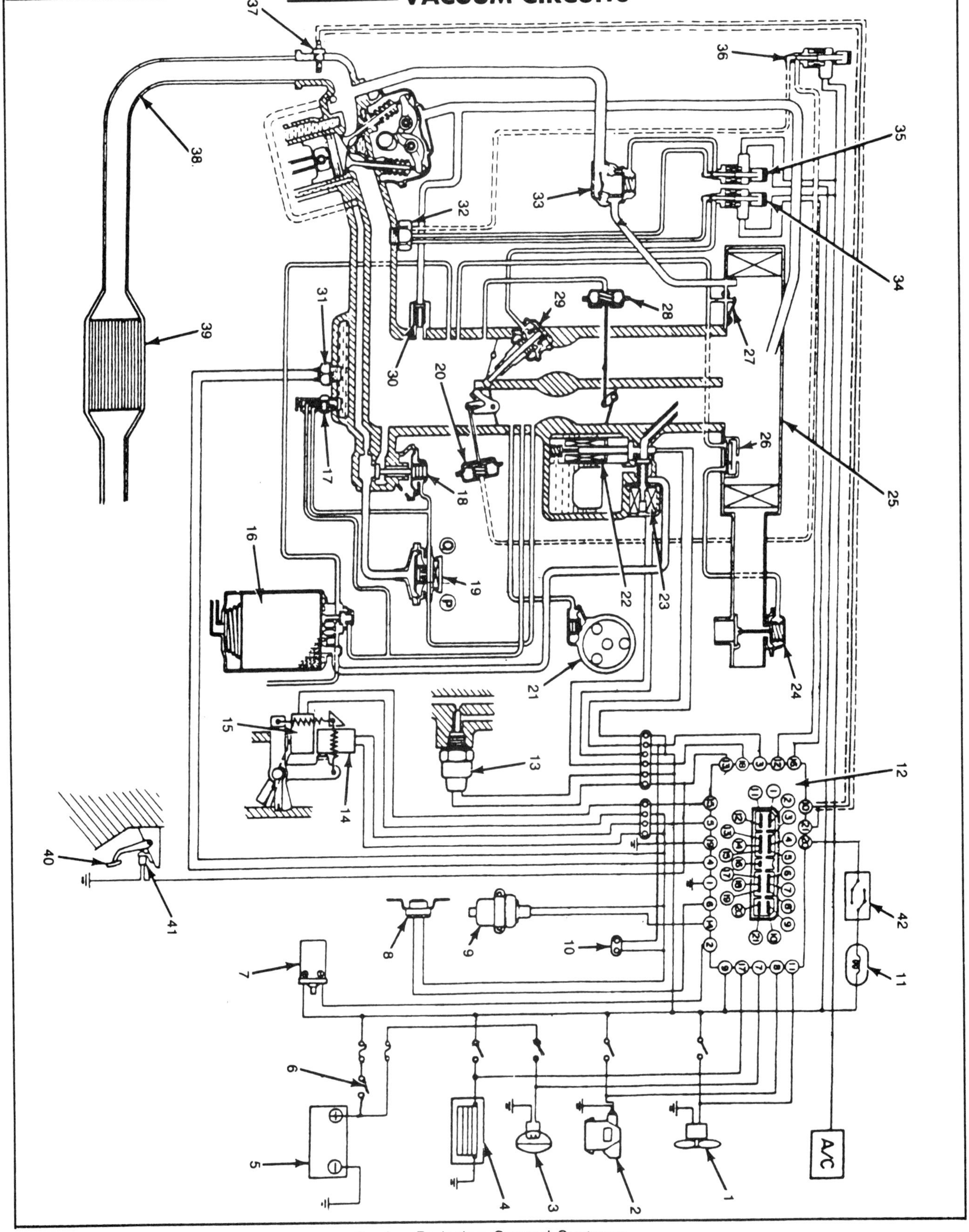

Engine Emission Control Systems

VACUUM CIRCUITS

1. ENGINE COOLING FAN
2. HEATER FAN
3. SMALL LIGHT, TAIL LIGHT, SIDE MARKER LIGHT & LICENSE LIGHT
4. REAR DEFOGGER
5. BATTERY
6. IGNITION SWITCH
7. IGNITION COIL
8. THERMAL ENGINE ROOM SWITCH
9. HIGH ALTITUDE COMPENSATOR
10. DUTY CHECK CONNECTOR
11. "SENSOR" LIGHT
12. ELECTRONIC CONTROL MODULE (ECM)
13. FUEL CUT SOLENOID VALVE
14. WIDE OPEN MICRO SWITCH
15. IDLE MICRO SWITCH
16. CANISTER
17. BI-METAL VACUUM SWITCHING VALVE (BVSV)
18. EGR VALVE
19. EGR MODULATOR
20. CAR AIR CON IDLE-UP
21. DISTRIBUTOR
22. MIXTURE CONTROL SOLENOID (MCS) VALVE
23. SWITCH VENT SOLENOID
24. AIR CONTROL ACTUATOR
25. AIR CLEANER
26. THERMO SENSOR
27. HOT IDLE COMPENSATOR
28. CHOKE PISTON
29. IDLE-UP ACTUATOR
30. POSITIVE CRANKCASE VENTILATION (PCV) VALVE
31. THERMAL SWITCH
32. GAS FILTER
33. SECOND AIR VALVE
34. THREE WAY SOLENOID VALVE (WITH BLACK LEAD WIRE)
35. THREE WAY SOLENOID VALVE (WITH BLUE LEAD WIRE)
36. IDLE-UP VACUUM SWITCHING VALVE (FOR A/C)
37. OXYGEN SENSOR
38. EXHAUST NO. 1 PIPE
39. THREE WAY CATALYST
40. CLUTCH PEDAL
41. CLUTCH SWITCH
42. CANCEL SWITCH

Engine Emission Control Systems

VACUUM CIRCUITS

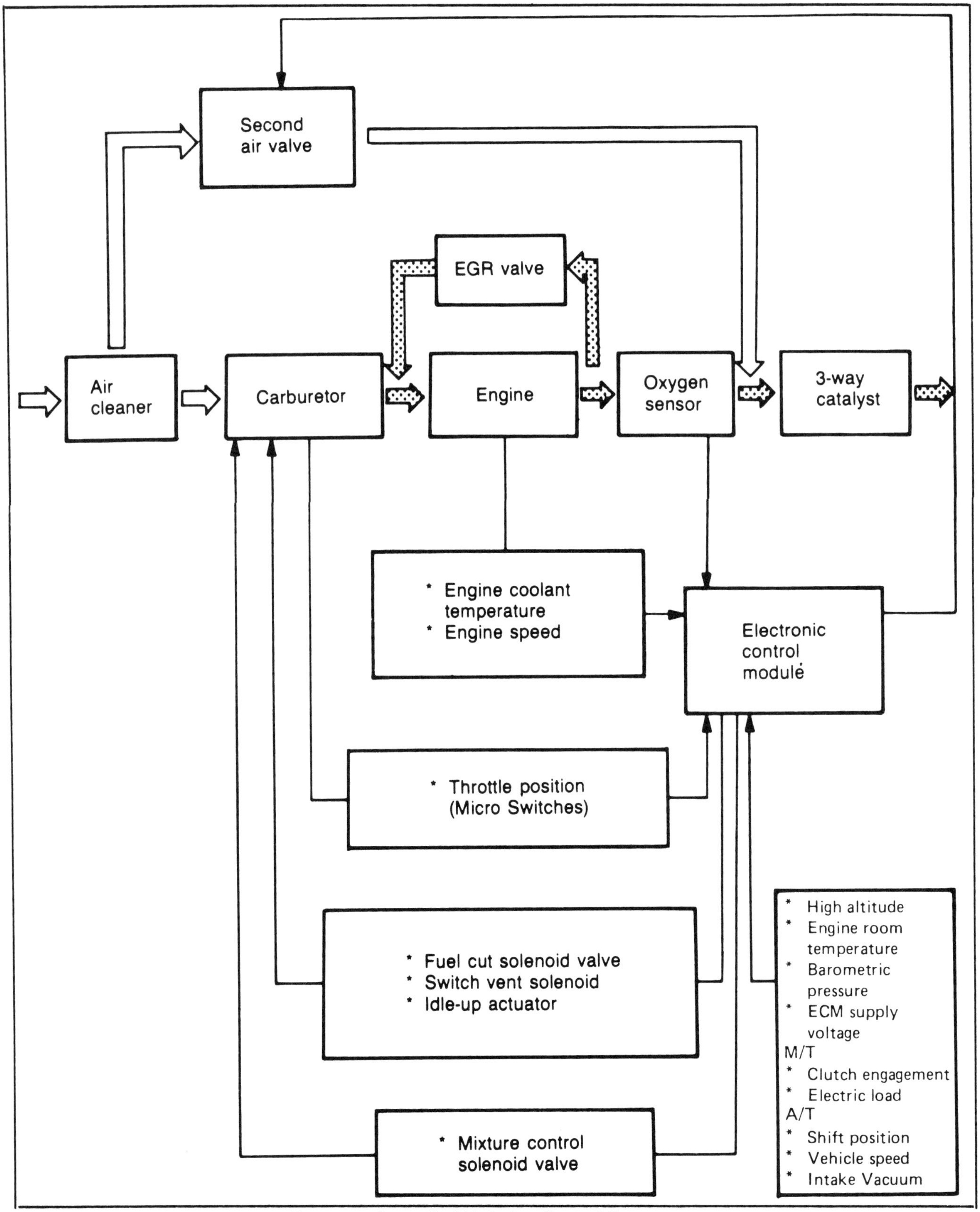

Computer Controlled Emission Control System

Subaru

INDEX

VEHICLE IDENTIFICATION NUMBER AND ENGINE IDENTIFICATION

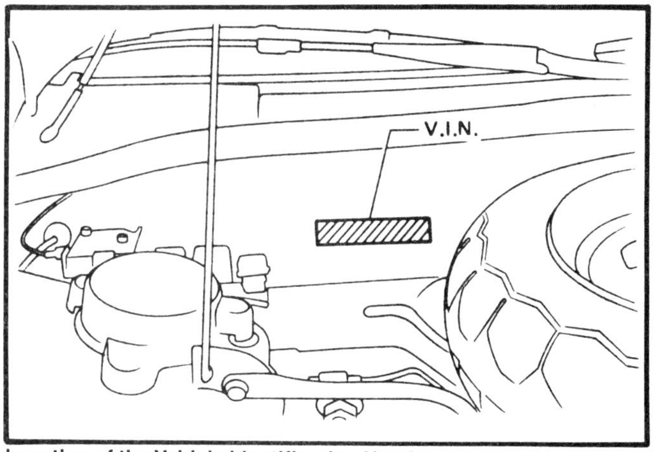

Location of the Vehicle Identification Number
(© Subaru of America, Inc.)

Vehicle Identification Number (VIN)

The vehicle identification number (VIN) is stamped on a medal tab, located on the firewall panel in the engine compartment. The VIN is a 17 digit number, with the fifth digit representing the engine series/type and the ninth digit representing the model year.

Engine Serial Number

The engine serial number is located on a metal tab, attached to the right side of the engine block at the front of the engine.

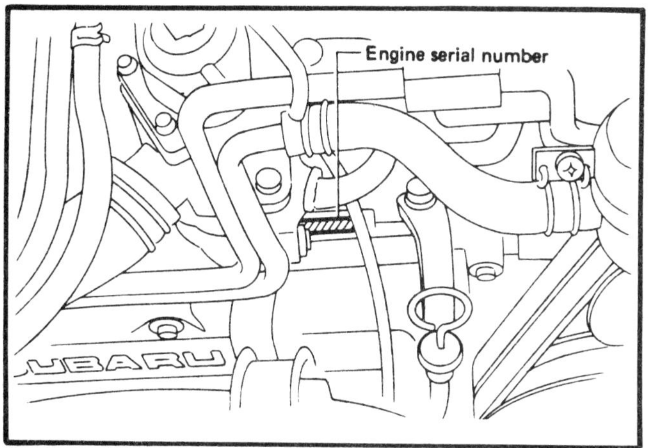

Engine serial number location (© Subaru of America, Inc.)

TUNE-UP SPECIFICATIONS

(When analyzing compression test results, look for uniformity among cylinders, rather than specific pressures)

Year	Model	Engine Displacement (cu. in.)	Spark Plugs ⑤ Type	Spark Plugs Gap (in.)	Distributor Point Dwell (deg)	Distributor Point Gap (in.)	Ignition Timing (deg)	Intake Valve Opens (deg)	Fuel Pump Pressure (psi)	Idle Speed (rpm)	Valve Clearance (in.) In	Valve Clearance (in.) Ex
'85–'86	DL-45 2-WD	1800	BPR6ES-11	0.039 0.043	Electronic		6B	20B	1.3–2.0	650	.010 ①	.014
'85–'86	DL-55 2-WD	1800	BPR6ES-11	0.039 0.043	Electronic		8B	20B	1.3–2.0	700	.010	.014
'85–'86	GL-55 2-WD	1800	BPR6ES-11	0.039 0.043	Electronic		8B	20B	1.3–2.0	700	.010	.014
'85–'86	GL-AT 2-WD	1800	BPR6ES-11	0.039 0.043	Electronic		8B	20B	1.3–2.0	800	0	0
'85–'86	GL-10 MPFI-55	1800	BPR6ES-11	0.039 0.043	Electronic		6B	20B	1.3–2.0	700	.010	.014
'85–'86	GL-10 MPFI-A/T	1800	BPR6ES-11	0.039 0.043	Electronic		6B	20B	1.3–2.0	800	0	0
'85–'86 BRAT	GL-A/T 4-WD	1800	BPR6ES-11	0.039 0.043	Electronic		8B	20B	1.3–2.0	800	0	0
'85–'86	Turbo 55	1800	BPR6ES-11	0.039 0.043	Electronic		15B	16B	434	700	.010	.014
'85–'86	XT-Coupe 55	1800	BPR6ES-11	0.039 0.043	Electronic		25B	18B	40	700	.010	.014
'85–'86	XT-Coupe A/T	1800	BPR6ES-11	0.039 0.043	Electronic		25B	18B	40	800	0	0

NOTE: The underhood specifications sticker often reflects tune-up specification changes made in production. Sticker figures must be used if they disagree with those in this chart.

4S Four speed transmission
5S Five speed transmission
A/T Automatic transmission
MPFI Multi-Point Fuel Injection
B Before top dead center
O Automatic transmission models are equipped with hydraulic lifters

FIRING ORDER

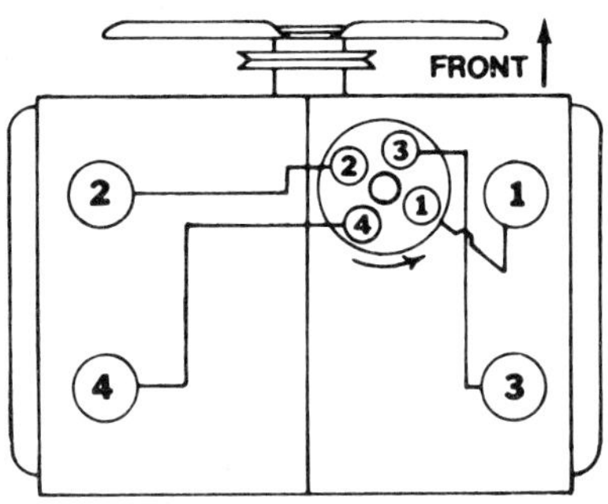

Firing order is 1-3-2-4

1800 engine firing order (© Subaru of America, Inc.)

EMISSION EQUIPMENT USED

ALL NON-TURBOCHARGED MODELS

Crankcase Emission Control
- Closed system
- Positive crankcase vent valve

Evaporative Emission Control
- Canister (Single)
- Bowl vent vacuum valve
- Purge control valve
- Fuel filler cap with relief valve
- Overfill limiter two way valve
- Fuel check valve
- Thermo-valve

Exhaust Emission Control
- Catalytic converter (dual three way type with Electronically controlled carburetor, Non 4WD models)
- Oxidation catalyst
- Air injection system
- Exhaust gas recirculation (EGR) system
- Heated air intake system
- Altitude compensator (Non 4WD models only)
- Anti-afterburning system (4WD models only)
- Shift up control system (Non 4WD models only)
- High-altitude kit (4WD models only)
- Ignition control system

ALL TURBOCHARGED MODELS

Crankcase Emission Control
- Closed system
- Positive crankcase vent valve

Evaporative Emission Control
- Canister (Single)
- Bowl vent vacuum valve
- Purge control valve
- Fuel filler cap with relief valve
- Overfill limiter two way valve
- Fuel check valve
- Thermo-valve

Exhaust Emission Control
- Catalytic converter (dual three way type with Electronically Controlled Gasoline Injection, 4WD models)
- Exhaust gas recirculation (EGR) system
- Ignition control system

REQUIRED EMISSION CONTROL MAINTENANCE

SCHEDULE OF INSPECTION AND MAINTENANCE SERVICES

		Maintenance Item	Maintenance Interval [Number of months or km (miles) whichever occurs first]									
			Months		7.5	15	22.5	30	37.5	45	52.5	60
			x 1000 miles	1	7.5	15	22.5	30	37.5	45	52.5	60
			x 1000 km	1.6	12	24	36	48	60	72	84	96
E	1	Intake and exhaust valve clearance (Except hydraulic valve lifter equipped engines)		1		1		1		1		1
E	2	Drive belt		A		(I)		(I)		(I)		(I)
E	3	Retighten cylinder head nuts and manifold bolts		P								
E	4	Engine oil		I	R	R	R	R	R	R	R	R
E	5	Engine oil filter			R	R	R	R	R	R	R	R
E	6	Engine coolant				(I)		(R)		(I)		(R)
E	7	Cooling system, hoses and connections						(I)				(I)
E	8	Engine idle speed		I								
E	9	Choke mechanism lubrication						P				P
E	10	Fuel filter				(R)		(R)		(R)		(R)
E	11	Fuel and evaporative system hoses and connections						(I)				(I)
E	12	Air filter elements (air cleaner, PCV air filter)						R				R
E	13	Spark plugs						R				R

NOTE: When operating your vehicles under any of following conditions, change the engine oil every 6000 km (3750 miles) or 3 months whichever occurs first.
- Operation in extremely cold weather
- Repeated short trips
- Driving on dusty roads

Symbols used
- A : Adjust
- R : Replace
- I : Inspect, correct or replace if necessary
- P : Perform
- E : Item for emission control system
- (I) or (R) : Recommended service for safe vehicle operation

IDLE SPEED AND MIXTURE ADJUSTMENTS

NOTE: The following specifications are published from the latest information available. If the published information differs from the information on the vehicle emission label, use the data on the emission label.

Idle Speed Specifications

IDLE SPEED

Adjustment

ALL MODELS

NOTE: Before making any idle speed adjustment, be sure that the ignition timing and valve lash adjustments are within specifications. Check all vacuum hoses for breaks and correct routing.

1. Start the engine and let it run until it reaches normal operating temperature. Increase the engine speed to 2500 rpm for a least a minute

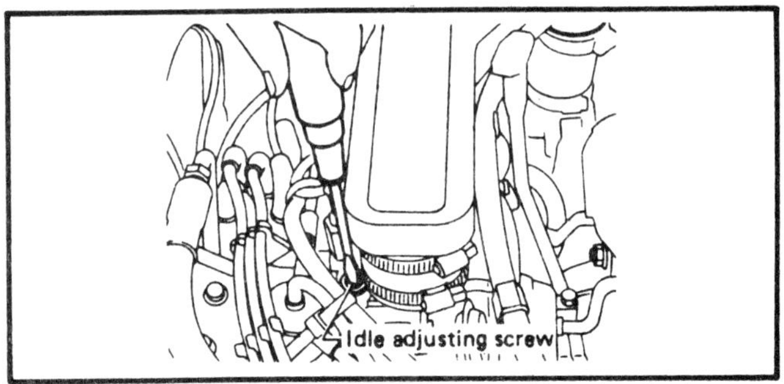

Location of idle adjusting screw (© Subaru of America, Inc.)

ENGINE IDLE SPEED ADJUSTMENT CARBURETED MODELS

Engine Model	Idle RPM
1800 with M/T 4-speed	650 + 100
1800 with M/T 5-speed	700 + 100
1800 with A/T	800 + 100

M/T Manual Transmission
A/T Automatic Transmission

ENGINE IDLE SPEED AND CO ADJUSTMENT FUEL INJECTION MODELS

Engine Model	Idle RPM	CO Level
1800 with M/T	700 + 100	1.5 + 1.0
1800 with A/T	800 + 100	1.5 + 1.0

NOTE: The CO level is without secondary air. The CO level with secondary air is 0–0.4%.
M/T Manual Transaxle
A/T Automatic Transaxle

to ensure that the oxygen sensor has also been sufficiently warmed up.

2. Remove and plug the purge hose to the intake manifold and be sure that the auxiliary air valve is completely closed.

3. Adjust the idle speed by using the idle adjusting screw located on the throttle body. After adjusting the idle speed, check both idle CO and HC contents in the exhaust gas.

NOTE: The CO content adjustment screw of the air flow meter does not have to be adjusted as the air-fuel ratio is feedback controlled.

4. If the CO and HC contents are not within specifications, check the following components:

   a. Check the EGI system by connecting a jumper wire to the system to see if the ECS lamp flickers with the engine at idle speed. If it does flicker on and off, the EGI system is operating properly.

   b. Check the fuel pressure and the fuel injectors. Remove the fuel injector and apply direct air pressure into the injector at approximately 28 psi., to see if air leaks at the nozzle tip. If air leaks, the injector is faulty and must be replaced.

Idle Mixture Adjustment

ALL MODELS EXCEPT NON 4WD

NOTE: The adjustment method is with the use of a dwell meter attached to the mixture main and slow duty solenoid valves feed wire, with a reading taken from the needle indication as the adjustment is performed. The meter range switch must be replaced on the " Four cylinder, Four cycle"position. To determined the proper specifications for both the Hitachi and Carter/Weber carburetors, follow the procedure as outlined.

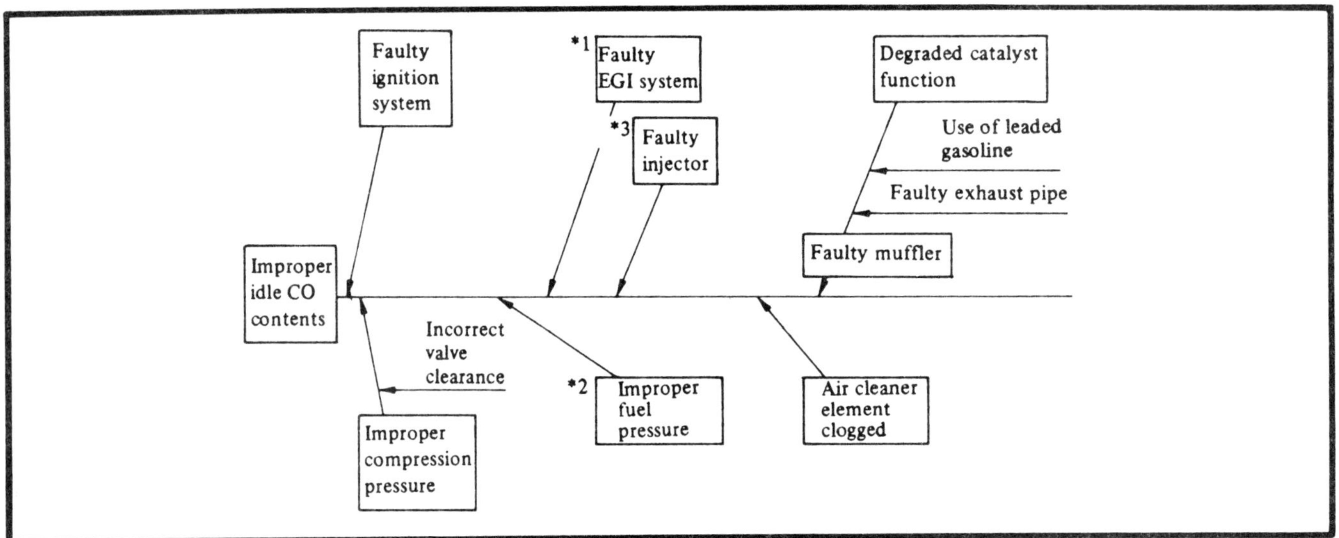

Improper idle CO troubleshooting schematic (© Subaru of America, Inc.)

CALCULATION FORMULA

Dwell (degree) = $^{90}/_{100}$ x duty (%)

HITACHI CARBURETOR

Duty ratio = Z (Z is determined according to altitude)
Example: 64 ft. (50m) = 35% (31.5 degrees)
5,250 ft. (1600 M) = 55% (49.5 degrees)

CARTER/WEBER CARBURETOR

Duty ratio = Z (Z = 35% or 31.5 degrees)

NOTE: Z is constant, regardless of altitude.

Example of Calculation Dwell (degree) = 90/100 x Duty (%)
Example #1 Dwell (degree) = 90/100 x 35%
(164 ft.)
Dwell = .9 x 35% Dwell = 31.5 degrees
Example #2 Dwell = .9 x 45%
(2543 ft.) Dwell = 40.5 degrees

Adjustment

ALL NON TURBOCHARGED 4WD MODELS

1. Operate the engine until it reaches normal operating temperature. Inspect the idle speed and CO percentage in the exhaust gas with the secondary air system connected.
2. Disconnect the air suction hose between the air suction valve and the secondary air cleaner.
3. Plug the air suction hose, inspect the engine idle speed and CO percentage in the exhaust gas with the secondary air system disconnected.

NOTE: If the vehicle is not equipped with a idle mixture screw, drill the concealment plug out and install a new idle mixture adjustment screw. When the adjustment is complete, be sure to seal the adjustment screw again.

4. Adjust both throttle adjusting screw and idle mixture adjusting screw to obtain the specified idle speed and CO percentage without the secondary air system connected.
5. Remove the plug from the air suction hose and connect the hose to the air cleaner.
6. With secondary air system connected, recheck the idle speed and CO percentage to see if they meet specifications.

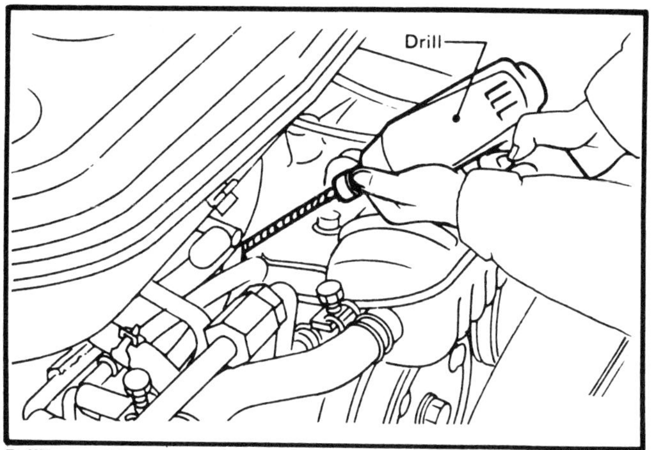

Drilling out the concealment plug (© Subaru of America, Inc.)

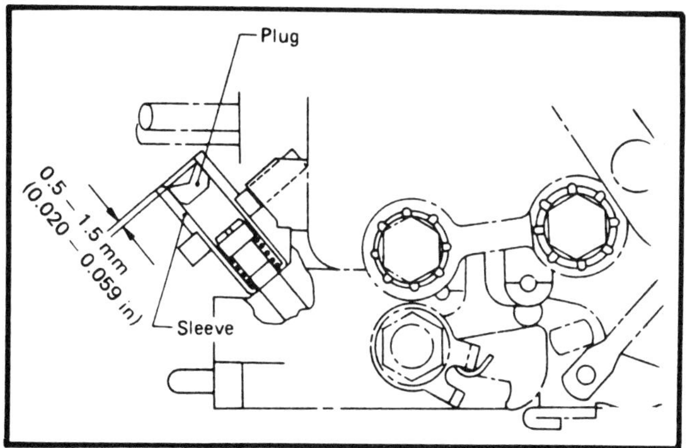

Location of the concealment plug (© Subaru of America, Inc.)

NOTE: This adjustment is not recommended on Turbocharged vehicles.

* N = 700 ± 100 rpm (MT); 800 ± 100 rpm (AT)

Z =

		Vehicle with HITACHI carburetor	Vehicle with C-W carburetor
Standard value		Duty ratio = Z 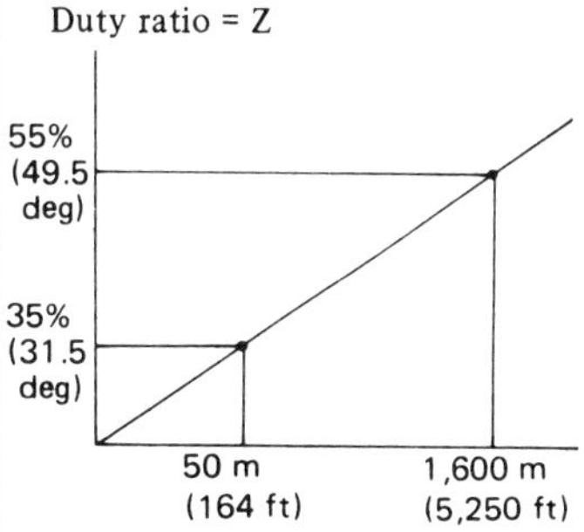 (Z is determined according to the altitude.)	Z = 35% (31.5 deg) (Z is constant regardless of the altitude.)
Allowance	At adjusting	±5% (±4.5 deg)	
	At checking	±20% (±18 deg)	

** Adjust idle adjusting screw so that duty meter indicates 35 ± 10%.

Diagram for finding carburetor duty cycle (© Subaru of America, Inc.)

IGNITION TIMING

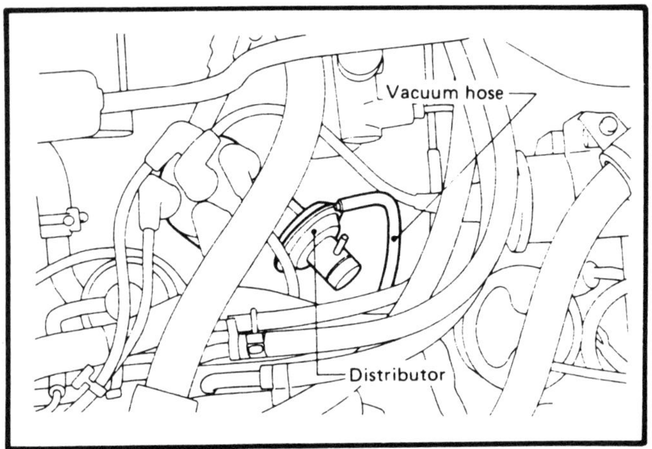

Disconnecting the vacuum hose to the advancer unit
(© Subaru of America, Inc.)

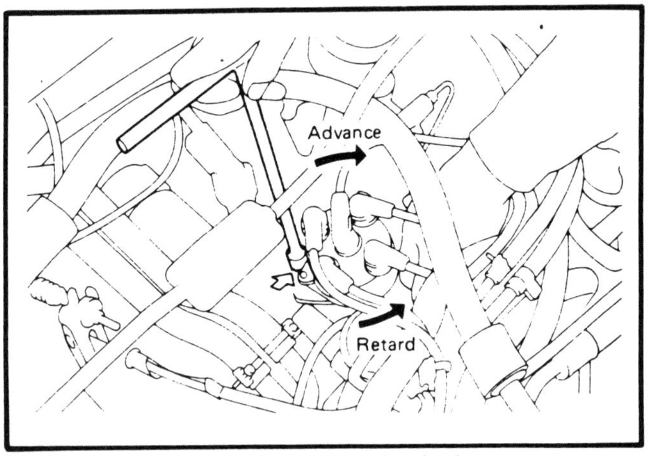

Adjusting the timing (© Subaru of America, Inc.)

Basic Adjustment

1. Clean the crankshaft damper and pointer on the water pump housing with a solvent-soaked rag so that the marks can be seen. Connect a timing light to the number one spark plug and the negative battery terminal. Scribe a mark on the crankshaft damper and on the marker with chalk or luminescent (day-glo) paint to highlight the correct timing mark.

2. Disconnect and plug the vacuum line to the advancer unit on the distributor. On the turbocharged models, disconnect the black connector (2-pole connector) between the distributor and the knock control unit.

3. Start the engine and let it run until it reaches normal operating temperature. Loosen the 6mm bolt on the mounting plate of the distributor.

4. Aim the timing light at the timing marks and turn the distributor clockwise to advance the timing and counterclockwise to retard the timing. Turn the distributor as necessary to align the proper timing marks.

5. Tighten the 6mm distributor hold-down bolt. Recheck the timing and adjust the idle speed to specifications. Remove all test equipment.

VACUUM SIGNAL (ADVANCE SIGNAL ONLY)

ALL NON TURBOCHARGED MODELS

Between 59–95 degrees F. of the intake manifold wall temperature, the vacuum passage is opened to atmospheric pressure through the thermo vacuum valve (I). At temperature other than the above-mentioned, the vacuum passage is connected to above throttle ported vacuum.

ALL TURBOCHARGED MODELS

Between 59–122 degrees F. of the coolant temperature, the pressure passage is opened to atmospheric pressure through the thermo vacuum valve. At temperatures other than the above-mentioned, the pressure passage is connected to above throttle ported pressure.

TIMING SPECIFICATIONS

Model	Degrees
DL—4 speed	6 @ BTDC
DL—5 speed	8 @ BTDC
GL—5 speed	8 @ BTDC
GL—A/T	8 @ BTDC
GL—10—5 speed	6 @ BTDC
GL—10—A/T	6 @ BTDC
Brat—M/T and A/T	8 @ BTDC
Turbo Models M/T and A/T	25 @ BTDC

NOTE: The underhood sticker often reflects tune-up specification changes made in production. Underhood sticker figures must be used if they disagree with those in the chart above.

M/T Manual Transaxle
A/T Automatic Transaxle

	Item	Designation					
Distributor	Type	Breakerless type with control unit, centrifugal governor and vacuum diaphragm advancers					
	Model	Carburetor vehicle				TURBO (with MPFI) vehicle	MPFI vehicle
		*1	*2	*3	*4		
		100291-0890	100291-0900	100291-0920	D4R84-16	D4R84-20	100291-0880
	Manufacturer	NIPPONDENSO			HITACHI		NIPPONDENSO
	Firing order	1-3-2-4					
	Rotating direction	Counterclockwise					
	Air gap mm (in)	0.2 – 0.4 (0.008 – 0.016)			0.3 – 0.5 (0.012 – 0.020)		0.2 – 0.4 (0.008 – 0.016)
	Cap insulation resistance	More than 50 MΩ					
	Rotor head insulation resistance	More than 50 MΩ					
Ignition coil	Type	K-31			CIT-116	CIT-117	K-31
	Manufacturer	NIPPONDENSO			HITACHI		NIPPONDENSO
	Primary coil resistance Ω	1.13 – 1.38			1.04 – 1.27	0.93 – 1.02	1.13 – 1.38
	Secondary coil resistance Ω	10795 – 14605			7360 – 11040	8000 – 12000	10795 – 14605
	Insulation resistance between primary terminal and case	More than 10 MΩ					

Distributor specifications (© Subaru of America, Inc.)

	Item	Designation	
Distributor	Type	Breakerless type with control unit, centrifugal governor and vacuum diaphragm advancers	
	Model	[Non-TURBO] 100291-0880	[TURBO] D4R84-20
	Manufacturer	NIPPONDENSO	HITACHI
	Firing order	1-3-2-4	
	Rotating direction	Counterclockwise	
	Air gap mm (in)	0.2 – 0.4 (0.008 – 0.016)	0.3 – 0.5 (0.012 – 0.020)
	Cap insulation resistance	More than 50 MΩ	
	Rotor head insulation resistance	More than 50 MΩ	
Ignition coil	Type	K-31	CIT-117
	Manufacturer	NIPPONDENSO	HITACHI
	Primary coil resistance Ω	1.13 – 1.38	0.93 – 1.02
	Secondary coil resistance kΩ	10.8 – 14.6	8.0 – 12.0
	Insulation resistance between primary terminal and case	More than 10 MΩ	

Distributor specifications for the XT models (© Subaru of America, Inc.)

SPARK PLUGS

ORIGINAL EQUIPMENT

ALL MODELS

NGK
- BPR6ESUI–11
- BPR5ESUI–11
- BPR7ESUI–11

NIPPONDENSO
- W20EPR–U11
- W16EPR–U11
- W22EPR–U11

CANADA MODELS

CHAMPION
- RN11YC–4
- RN9YC–4

GAP—ALL PLUGS
0.039–0.043 in. (1.0–1.1mm)

CARBURETOR SPECIFICATIONS

Vehicle			49-state 2WD and California		49-state 4WD and Canada	
			MT	AT	MT	AT
Carburetor			DCZ328-502	DCZ328-503	DCZ328-504	DCZ328-505
Air horn dia. (Inner dia. × Outer dia.)			59 × 63 mm (2.32 x 2.48 in)			
Throttle bore (P-S)			28—32 mm (1.10—1.26 in)			
Inner dia. of needle valve			1.7 mm (0.067 in)			
Main system	Main jet (P-S)		#115—#160		#112—#156	#114—156
	Main air bleed (P-S)		#60—#100		#65—#100	
	Main nozzle dia. (Inner dia. × Outer dia.)	P	2.3 × 3.8 mm (0.091 × 0.150 in)			
		S	2.8 × 4.0 mm (0.110 × 0.157 in)			
	Main nozzle and surface angle	P	5°			
		S	5°			
	Large venturi dia.	P	23 mm (0.91 in)			
		S	29 mm (1.14 in)			
	Small venturi dia.	P	8 mm/15 mm (0.31 in/0.59 in)			
		S	11 mm/15 mm (0.43 in/0.59 in)			
Slow system	Slow jet (P-S)		#45—#135		#46—#100	
	Slow air bleed (P-S)		#150—#100		#160—#100	
	Slow economizer		3 mm/1.8 mm (0.118 in/0.071 in)			
	Economizer bleed (P)		#90			
Power jet—vacuum required			————		#40—-20.0 kPa (-150 mmHg, -5.91 inHg)	
Accelerating pump system	Accelerating pump nozzle dia.		0.5 mm (0.020 in)			
	Pump delivery/stroke		0.6 cm³ (0.6 cc, 0.037 cu in)	0.7 cm³ (0.7 cc, 0.043 cu in)		
	Weight of accelerating pump injector		3.0 g (0.106 oz)		4.0 g (0.14 oz)	
Starting system	Choke system		Automatic choke			
	Choke valve angle when fully closed		16°			
Inner dia. of auxiliary passage (at nipple)		P.M.	#300			
		S.M.	#300 (Safety orifice: #110)		#300 (Safety orifice: #110)	
		P.S.	#300			

EMISSION CONTROL SYSTEMS

CRANKCASE EMISSION CONTROL SYSTEM

Positive Crankcase Ventilation System

The positive crankcase ventilations (PCV) system is employed to prevent air pollution which will be caused by blow-by gas being emitted from the crankcase. The system consists of a sealed oil filler cap, rocker covers with an emission outlet and fresh air inlet, connecting hoses, PCV valve and air cleaner.

At part throttle, the blow-by gas in the crankcase flows into the intake manifold through the connecting hose of the rocker cover on the #2–#4 cylinder side and PCV valve by the strong vacuum of the intake manifold. Under this condition, the fresh air is introduced into the crankcase through the connecting hose of the rocker cover on the #1–#3 cylinder side and drawn to the intake manifold, through the PCV valve, together with the blow-by gas.

At wide open throttle, a part of the blow-by gas flows into the air cleaner through the connecting hose of rocker cover on the #1–#3 cylinder side and is drawn to the carburetor. Under this condition, the intake manifold vacuum is not so strong as to introduce all blow-by gases, increasing with engine speed, directly through the PCV valve.

Under the special operating conditions, such as steep right turn driving. The engine oil sometimes blows up into connecting hose of the rocker cover on #2–# 4 cylinder side and flows into the intake manifold by the force of the vacuum. However, in this case, the connecting hose between the air cleaner case and the connecting hose of rocker cover on the #2–#4 cylinder side, reduces the vacuum to prevent this.

Note: For the turbo vehicles, the operational principle is almost the same as that for the Non-turbo vehicle.

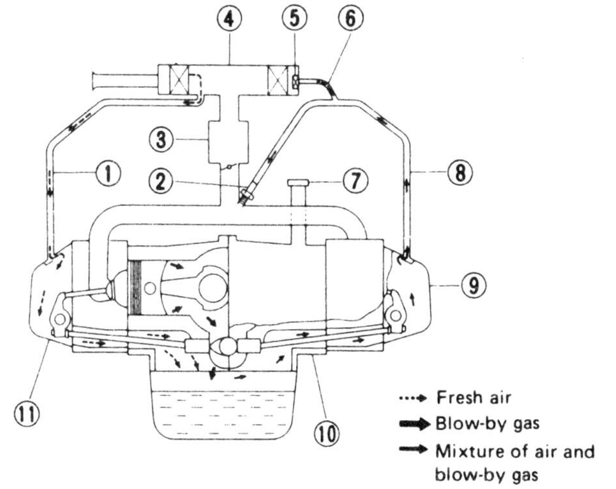

1 Connecting hose
2 P.C.V. valve
3 Carburetor
4 Air cleaner
5 Air filter
6 Connecting hose
7 Oil filler cap (Sealed)
8 Connecting hose
9 Rocker cover on #2 – #4 side
10 Crankcase
11 Rocker cover on #1 – #3 side

Crankcase emission control system schematic—typical
(© Subaru of America, Inc.)

System Service

NOTE: The PCV valve should be checked every 15,000 miles and replaced as required. If the valve is found to be defective, do not attempt to clean it, replace it.

1. With the engine running, remove the valve from its mounting. A hissing sound should be heard and vacuum should be felt from the inlet side of the valve.

2. Reinstall the valve. Remove the crankcase inlet air cleaner. Loosely hold a piece of stiff paper over the opening in the rocker cover. Allow one minute for the crankcase pressure to reduce itself. The paper should then suck itself against the rocker cover with noticeable force. Replace the inlet air cleaner in the rocker cover.

3. With the engine stopped. Remove the PCV valve and shake the valve. A clicking sound should be heard indicating that the valve is not stuck.

4. If the valve fails any of the above tests, it should be replaced.

FUEL EVAPORATIVE EMISSION CONTROL SYSTEM

The function of the evaporation control system is to prevent the emission of gas vapors from the fuel tank and carburetor from being expelled into the atmosphere. When fuel evaporates in the gas tank or the carburetor float chamber, the vapors pass through the vent hoses to the charcoal canister where they are stored until they can be drawn into the intake manifold, when the engine is running. All vehicles are equipped with the charcoal canister for the storage of fuel vapors. The fuel bowls of all carburetors are vented internally and on some applications, do not require venting the canister. If this is the case, the bowl vent port on the canister will be capped. Most carburetors are also externally vented to the charcoal canister.

System Service

The charcoal canister itself is a non serviceable component. The only service required for the system is to replace the filter pad on the bottom of the canister. The filter requires replacement every 12 months or 12,000 miles. If the vehicle is driven under severe conditions, the filter should be replaced more often.

BOWL VENT VACUUM VALVE

The bowl vent valve is connected to the carburetor fuel bowl, the charcoal canister and the air induction discharge. When the engine is running, there is no air pressure applied and there is a direct connection

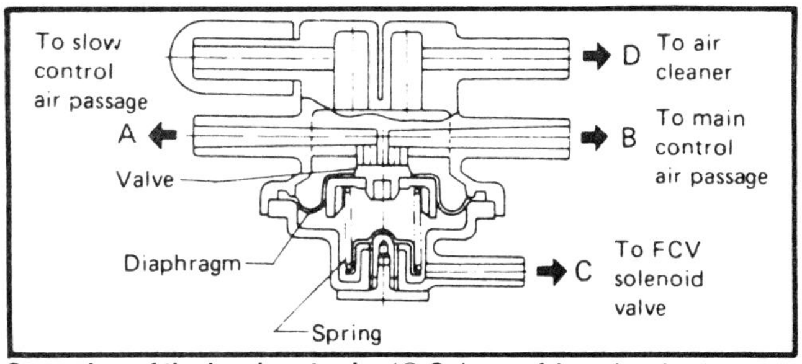

Crossview of the bowl vent valve (© Subaru of America, Inc.)

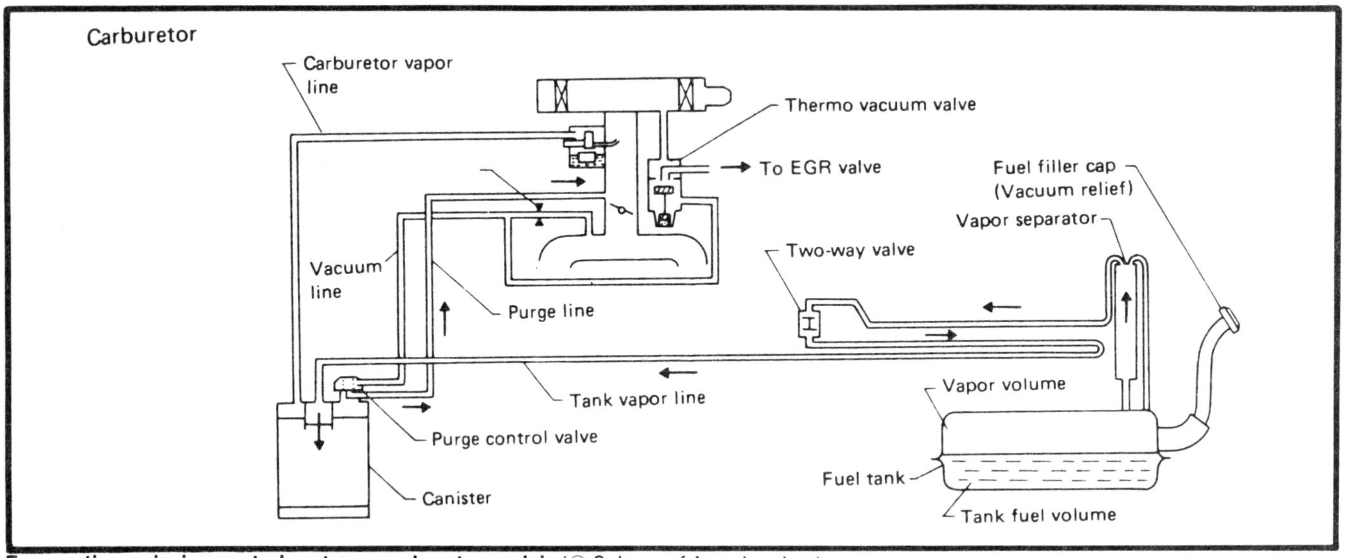

Evaporative emission control system—carburetor models (© Subaru of America, Inc.)

between the fuel bowl and the canister. When the engine is shut down, the valve air pressure bleeds down and the fuel bowl is allowed to vent into the canister.

OVERFILL LIMITER (TWO-WAY VALVE)

The overfill limiter consists of a pressure valve and a vacuum valve. The pressure valve is designed to open when the fuel tank internal pressure has increased over the normal pressure and the vacuum valve opens when a vacuum has been produced in the tank.

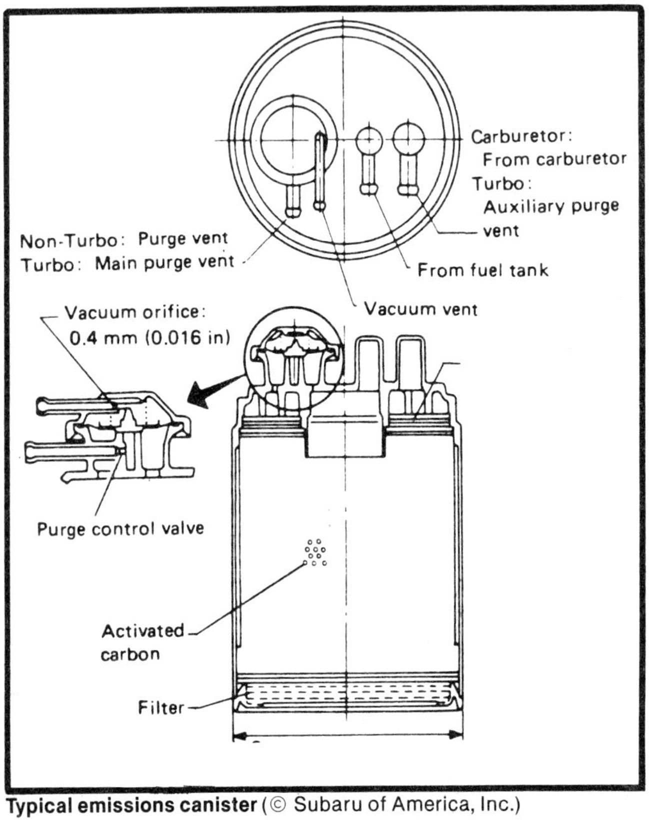

Typical emissions canister (© Subaru of America, Inc.)

Purge Control Valve

CARBURETED MODELS

The purge control valve is kept closed during idling in order to prevent vaporized fuel from entering into the intake manifold for positive control of of high idle CO emission, which is a particular problem under high ambient temperatures. When the carburetor vacuum working on the diaphragm of the valve exceeds the pre-set value, the purge control valve is opened.

ECI AND TURBOCHARGED EQUIPPED ENGINES

The purge control valve is closed at idling to prevent vaporized fuel from entering into the air intake hose for positive control of high idle CO emissions, which is particular problem under high ambient temperature. Once the pressure difference between the turbocharger and the throttle ported pressures exceeds the pre-set value, the purge control valve is opened.

THERMO VACUUM VALVE

A thermo valve is incorporated in the EGR system for sensing the coolant temperature at the intake manifold. It closes the purge control valve when the coolant temperature is lower than a pre-set value, so as to reduce CO and HC emissions under warm-up conditions and opens the purge control valve when the coolant temperature reaches the preset temperature level.

Testing

1. Connect vinyl tubes to the output ports of the air cleaner and the EGR valve. Plug the output port to the charcoal canister.

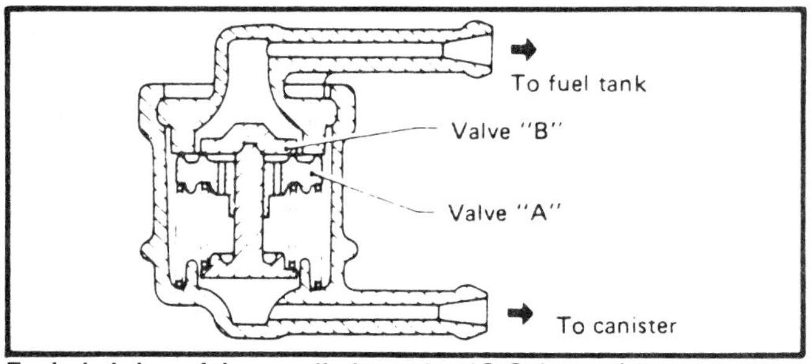

Exploded view of the over-limiter valve (© Subaru of America, Inc.)

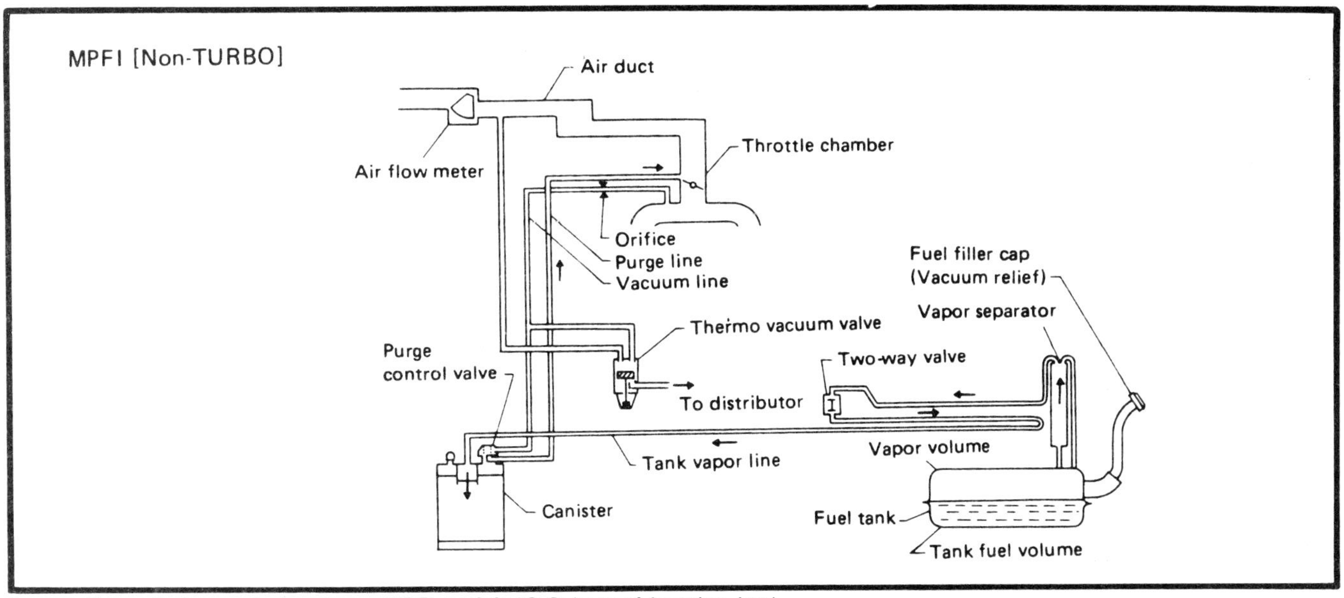

Evaporative emission control system—MPFI models (© Subaru of America, Inc.)

2. Blow air into the valve and confirm that the valve opens or closes in response to the intake manifold wall temperature (with the valve submerged in a heated container), specified below.

a. The thermo valve should be completely open below 86 degrees F. and completely closed above 104 degrees F.

3. If the valve does not operate properly within these specifications, remove and replace the valve.

NOTE: Perform the same test, only connect the vinyl tubes to the output ports to the EGR valve and the charcoal canister. Plug the output port to the air cleaner. On the models equipped with MPFI, the readings are 113 degrees F. open and 131 degrees F. closed.

FUEL CHECK VALVE

The fuel check valve is used to prevent fuel leaks should the vehicle suddenly roll over. The valve is connected in the fuel vapor line (between canister and overfill limiter) and is installed on the firewall. The fuel check valve contains two balls and under normal conditions, the gasoline vapor passage in the valve is opened. Should the vehicle roll over, one of the balls closes the fuel passage, thus preventing fuel leaks.

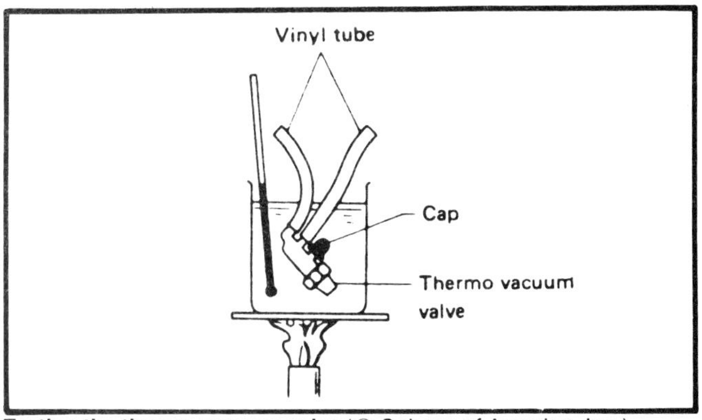

Testing the thermo-vacuum valve (© Subaru of America, Inc.)

FUEL FILLER CAP

The fuel filler cap is equipped with relief valve to prevent the escape of fuel vapor into the atmosphere.

FUEL SEPARATOR

The fuel separator is used to prevent liquid fuel from flowing into the canister in case of abrupt cornering, etc.

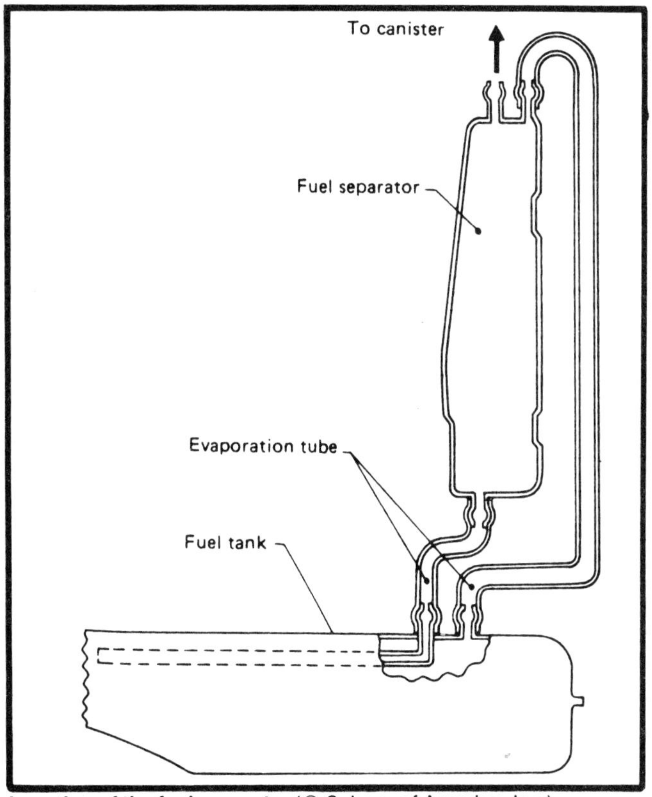

Location of the fuel separator (© Subaru of America, Inc.)

EXHAUST EMISSION CONTROL SYSTEM

Three-Way Catalyst With ECC System

The basic materials of the three-way catalyst is palladium (PD), platinum (Pt), rhodium (Rh). A thin film of their mixture is applied onto a honeycomb or porous ceramics of an oval shape carrier. To avoid damaging the catalyst, only unleaded gasoline should be used. The catalyst is used to reduce HC, CO and NOx in the exhaust gases and permits simultaneous oxidation and reduction. To obtain an excellent purification efficeincy on all the components HC, CO, and NOx, a balance should be kept among the concentrations of the components. These concentrations vary with the air/fuel ratio.

The air/fuel ratio needs to be controlled to a value within the very narrow range covering around the theoretical (stoichiometeric) air/fuel ratio to purify the components efficiently. Electronically Controlled Carburetor (ECC) system is employed with three-way catalyst for this purpose. ECC system is mainly made up of the following component parts.

a. Oxygen (O_2) Sensor
b. Electronic Control Module (ECM)
c. Duty solenoid
d. Carburetor

They compose a feedback system to control the air/fuel ratio during operation by supplying a measured amount of air into air bleeders of the carburetor. To avoid application of feedback during certain driving condition, vacuum switches, a thermosensor and an engine speed sensing circuit are also provided in this system.

Component Parts of the EEC System

OXYGEN SENSOR

The oxygen sensor is a kind of concentration cell that generates electromotive force according to the ratio of the oxygen sensor concentration in air to that in exhaust gases. It also has a characteristic that the electromotive force is changed drastically with respect to the stoichiometric air/fuel ratio. The force is larger on the rich side (smaller air/fuel ratio) and smaller on the lean side (larger air/fuel ratio) of the mixture. The oxygen sensor is installed in the exhaust manifold.

ELECTRONIC CONTROL MODULE (ECM)

According to the electronic signal from the oxygen sensor, the ECM judges whether the exhaust gas at the manifold is composed of a rich or lean mixture against the stoichiometeric ratio and issues signals to duty solenoid valves. The current air/fuel ratio is judged to be a rich mixture when the electromotive force of the oxygen sensor is higher than the specified level. As the result, the ECM issues signals to the duty solenoid valves to let more air into the carburetor.

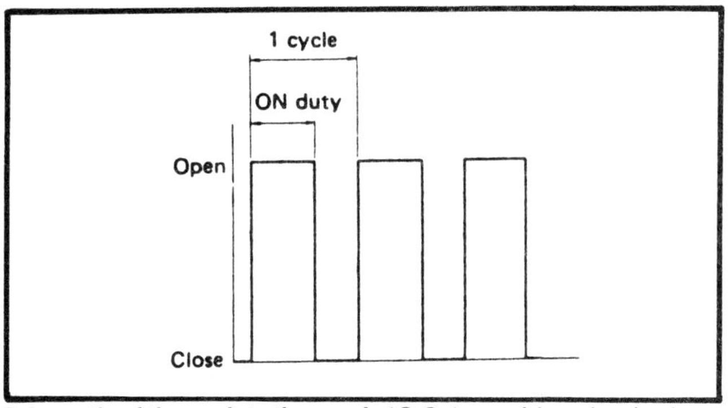

Schematic of the on duty time cycle (© Subaru of America, Inc.)

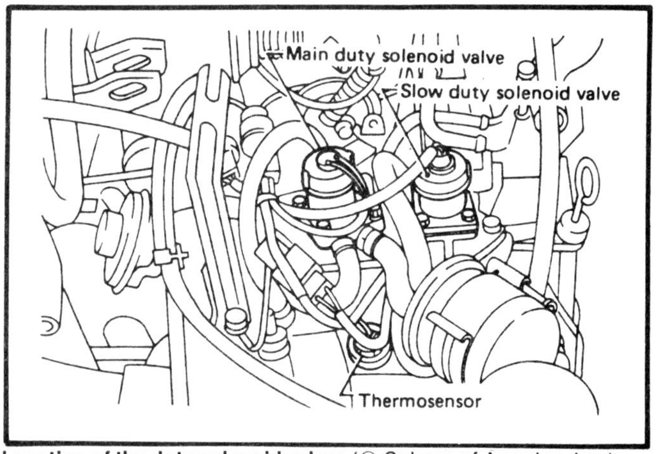

Location of the duty solenoid valves (© Subaru of America, Inc.)

On the other hand, the air/fuel ratio is judged to be a lean mixture when the electromotive force is lower. The ECM issues signals to the duty solenoid valves to let less air into the carburetor, so that the the air/fuel ratio will be changed toward a rich mixture. According to these feedback operations, the air/fuel ratio maintains the stoichiometric value.

DUTY SOLENOID VALVE

On Hitachi carburetors, the two duty solenoid valves are installed on the intake manifold. While on the Carter–Weber carburetor, one duty solenoid valve is unified in the carburetor. The light weight valve repeats opening/closing at a short cycle according to the signal put out from the ECM. The average flow rate of air passing through this valve to the carburetor changes, depending upon the period of time during which the valve is opened, Thereby varying the air/fuel ratio. The status that a voltage is applied to solenoid, that is, the valve is lifted, is called "On-duty", and duty ratio is given by the following equations. Duty ratio = On-duty time ÷ cycle x 100%.

The On-duty duration varies at a fixed acceleration, with the Rich-Lean mixture as measured by the oxygen sensor. In steady state operation (where the vacuum switch is on), the acceleration rate beyond a certain engine speed is changed in steps to obtain greater acceleration at higher rpm. This allows the valve to sensitively respond to the Rich-Lean mixture as measured by the oxygen sensor.

VACUUM SWITCH (II)

The acceleration (increasing/decreasing rate of the On-duty duration) increases for several seconds immediately after this switch is turned off, allowing the valve to sensitively respond to the Rich-Lean mixture as measured by the oxygen sensor. After a laspe of 5 seconds, the acceleration returns to the steady mode operating condition.

PRESSURE SENSOR AND SOLENOID VALVE

The sensor provides accurate, linear and continuous sensing of changes in the intake manifold vacuum pressure. A solenoid valve is placed in the pressure line between the intake manifold and pressure sensor. This valve senses changes in the atmospheric pressure and corrects the air/fuel ratio in the high altitude.

THERMOSENSOR AND ENGINE SPEED SENSING CIRCUIT

These sensors are additionally provided in the ECC system so that better driveability is assured by avoiding the application of feedback.

ALTITUDE COMPENSATOR

The altitude compensator is used for compensation of the air/fuel ratio in response to the elevation, by means of supplying additional air to the control air passages (at the primary main and secondary main system). Below specified elevation, the bellows in the altitude compensation system do not operate because the air passages in the altitude compensator is automatically closed by the needle valve of the altitude compensator. Above specified elevation, the bellows in the altitude compensator are expanded, moving the lever, which pushes the needle valve to allow additional air passages to open the control air passages. The air coming into the carburetor through the altitude compensator is metered and supplied to the control air passages, thereby compensating the overall air/fuel ratio of the carburetor.

OXIDATION CATALYST

The basic material of the three-way catalyst is Palladium (Pd), Rhodium (Rh) and a thin film of their mixture. Which is applied onto a honeycomb or porous ceramics of an ovel shape (carrier). To avoid damaging the catalyst, only unleaded gasoline should be used. The catalyst is used to reduce HC, CO and NOx in the exhaust gases and permits simultaneous oxidation and reduction.

EGR System

The function of the exhaust gas recirculation (EGR) system is the reduction of NOx by reducing the top combustion temperature through recirculating a part of the exhaust gas into the cylinders. The EGR valve opens in response to the engine driving conditions and a part of the exhaust gas flows into cylinders through the intake manifold. The vacuum signal to control the EGR valve is picked up from the port near the slightly up stream portion of the throttle valve

EGR VALVE AND FLOW PASSAGES

Testing

1. Viewing through an opening in the EGR valve body, check if the valve shaft moves when the engine speed reaches 3000–3500 rpm under no-load conditions, after the engine reaches normal operating temperature.

NOTE: On 49 state non 4WD models; after a lapse of 8 minutes or more from the engine start.

2. Check if the EGR valve moves when vacuum (7.87 in. Hg.) is applied to the EGR valve by a hand held vacuum pump or equivalent.
3. Either rough idling or engine stall should occur when the vacuum (7.87 in. Hg.) is applied to the EGR valve by a hand held vacuum pump or equivalent.

NOTE: On non 4WD of non California models; after a lapse of 8 minutes or more from the engine start.

4. If none of the above operate properly, remove the EGR valve. Clean the passages, both vacuum and exhaust and/or replace the EGR valve.

Three Way Catalyst With EGI System

The EGI system is a system that supplies the optimum air/fuel mixture to the engine for all the various operating conditions through the use of the latest electronic technology. With this system, fuel, which is pressurized at a constant pressure, is injected into the intake air passage of the cylinder head. The injected quantity of fuel is controlled by an intermittent injection system, where the electro-magnetic injection valve (fuel-injector) opens only for a short period of time, depending on the quantity of air required for one cycle operation. In actual operation, the injection quantity is determined by the duration of an electric pulse applied to the fuel injector and this permits a simple, yet highly precise metering of the fuel. Furthermore, all the operating conditions of the engine are converted into electric signals and this results in additional features of the system, such as large improved adaptability, easier addition of compensating element etc. The EGI system also has the following features:

1. Reduced emission of harmful exhaust gases.
2. Reduced in fuel consumption.
3. Increased engine output.
4. Superior acceleration and deceleration.
5. Good matching with turbocharger.
6. Superior start-ability and warm-up performance in cold weather since compensation is made for coolant and intake air temperature.

Component Parts of the EGI System

FUEL PUMP

A pin roller type magnet pump is used as the fuel pump.

FUEL DAMPER

The fuel damper is provided to reduce change in the fuel pressure generated in the fuel pump. Fuel pressure pulses entering the fuel damper are absorbed by the vibration of the diaphragm and almost constant pressure is maintained at the outlet of the fuel damper. This assures a stable fuel supply.

FUEL FILTER

A metal casing that can withstand the fuel pressure is adopted for the fuel filter, which uses a paper filter element.

PRESSURE REGULATOR

This regulator is located at the end of the injection manifold. It is used to maintain the fuel pressure in the injection manifold at a constant 36.3 psi. above the intake manifold pressure (which the regulator is connected to). All of the excess fuel is drain off through a fuel return line to the fuel tank.

FUEL INJECTOR

The fuel injector injects fuel according to the valve open signal received from the EGI control unit. The nozzle is attached on the top of the fuel injector. The needle valve is lifted by the solenoid coil through the plunger on arrival of the valve open signal. Since the injection opening, the lifted level of the needle valve and the regulator-controlled fuel pressure are kept constant. The amount of fuel to be injected can be controlled only by the valve open signal from the EGI control unit. At the fuel inlet of the injector, the filter is mounted to prevent dust from entering.

AIR CLEANER

In the EGI turbo system, the newly designed air cleaner is equipped on the body side. The outside air introducing method is adopted to enhance the aspiration efficiency, in which the air introduction part is connected with the upper frame. The air cleaner element is of a flat plate type and the air flow meter is mounted on the upper cover of the air cleaner.

AIR FLOW METER

The air flow meter measures the amount of air aspirated and provides signals to the EGI control unit accordingly. All air to be taken in the combustion chamber is passed through this meter mounted on the air cleaner cover. The aspirated air applies force to the flap so that its opening angle is balanced with the potentiometer. Then the voltage

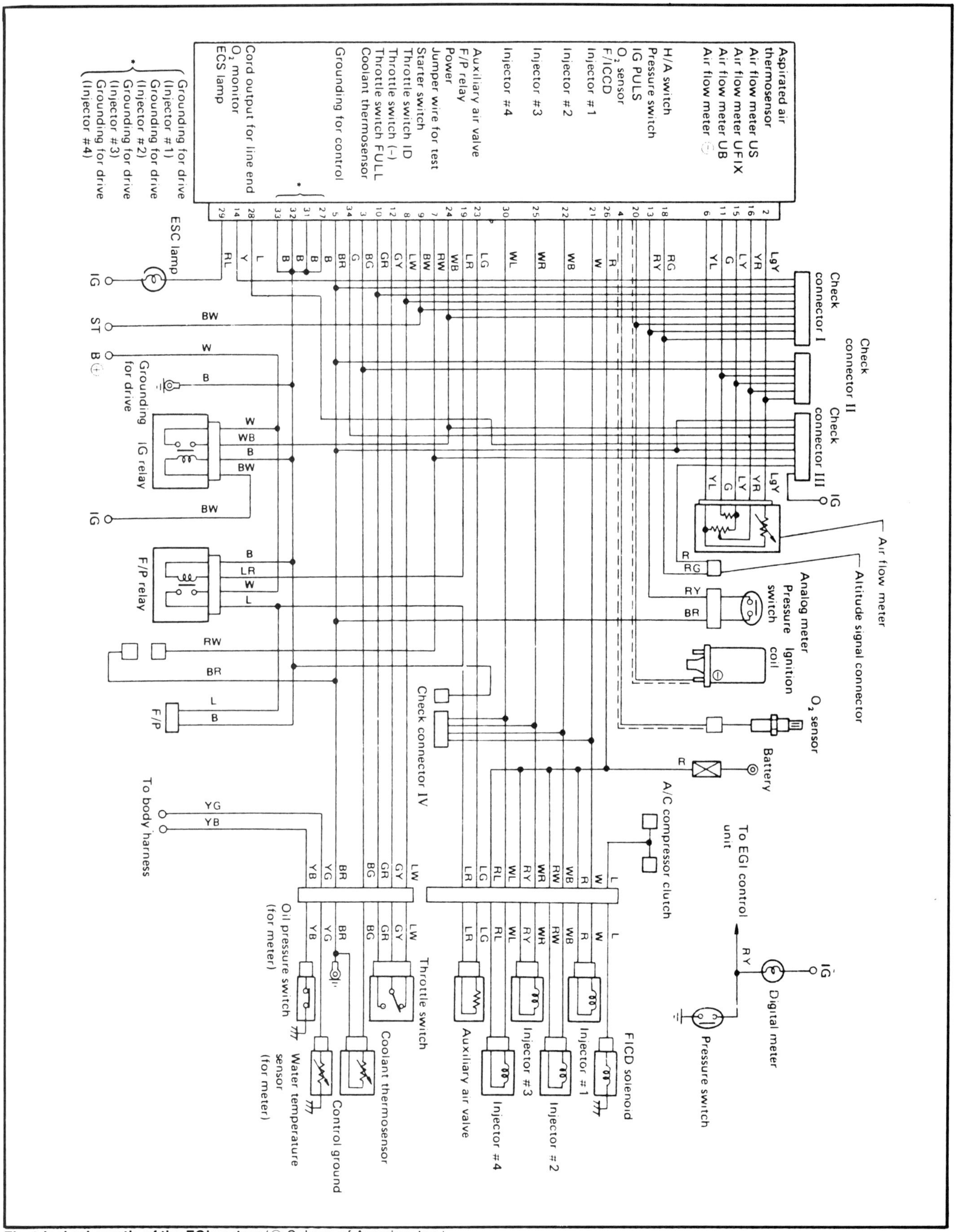

Electrical schematic of the EGI system (© Subaru of America, Inc.)

signal, corresponding to its angle, is sent to the EGI control unit. Also the air flow meter is provided with the aspirated air thermosensor.

THROTTLE BODY

In response to the depressing stroke of the throttle pedal, the throttle body opens/closes its valve to regulate the air volume to be taken into the combustion chamber. Negative pressure (positive pressure at the supercharging) generated according to the opening of the throttle valve, is applied to the pressure ports for ignition timing, EGR control and canister purge control. This pressure is used for controlling the distributor, EGR valve and purge control valve. During idling, the throttle valve is almost fully closed and the air flow through the body is less than that passing through the carburetor. More than half of the air necessary for idling is supplied to the intake manifold via the idle bypass passage. Turning the idle adjust screw on the idle bypass passage, can change the air flow to adjust the number of revolutions in idling. Furthermore, to prevent the number of revolutions from decreasing while the air conditioner is turned on, a fast idle bypass passage is provided, which has the valve operated by the fast idle solenoid.

THROTTLE SWITCH

The throttle switch, mounted on the throttle body, interworks with the throttle shaft. Its idle contact turns on in an idle state and its full contact turns on in a full throttle state to give signals to the EGI control unit. The throttle switch has three contacts; idle contact, full contact and moving contact. The idle contact and full contact are fixed , while the moving contact travels, depending on the open angle of the throttle valve. This moving contact is actuated by the groove of the guide cam, which is driven by the throttle valve shaft. When the idle contact is in the ON position, the EGI control unit operates in the throttle fully closed position (that is, idling state). If the full contact is on, the EGI contract is on and the EGI control unit operates in the throttle fully open position . Accordingly, the full-enrichment compensation of fuel occurs when the full contact is in the ON position.

OXYGEN SENSOR

The oxygen sensor is mounted on the front exhaust pipe between the engine and the turbocharger. It is used to sense oxygen concentration in the exhaust gas. If the fuel/ratio is leaner than the stoichiometric ratio in the mixture. (i.e. excessive amount of air), the exhaust gas contains more oxygen. On the other hand, if the fuel/ratio is richer than the stoichiometric ratio, the exhaust gas hardly contains any oxygen.

NOTE: A quick way to check the oxygen sensor is to connect a dry cell battery (size D, 1.5 volts) to the wiring connector of the oxygen sensor. Connect the positive post of the dry cell battery to the the oxygen sensor wiring connector and ground the negative side of the dry cell battery to the engine block. This set up will check the oxygen sensor and the sensor wiring for defects.

AUXILIARY AIR VALVE

The auxiliary air valve is used to increase the air flow when the engine is started up at a low temperature and allowedf to warm up. The auxiliary air valve consists of a coiled bimetal spring, the bimetal spring operated shutter valve and the electric heater element for the bimetal spring. The passing air flow (at start up) is increased as the temperature becomes lower. After the start up of the engine, the heating is performed by the heater to which current is supplied from the fuel relay circuit. Therefore, the shutter valve turns gradually to decrease the air flow. After a certain elapsed time, the shutter valve is closed.

COOLANT THERMOSENSOR

The coolant thermosensor is equipped on the thermostat casing of the intake manifold. Its thermister changes the resistance with respect to temperature. The thermosensor sends the coolant temperature signal to the EGI control unit, which is decisive for the fuel volume to be injected.

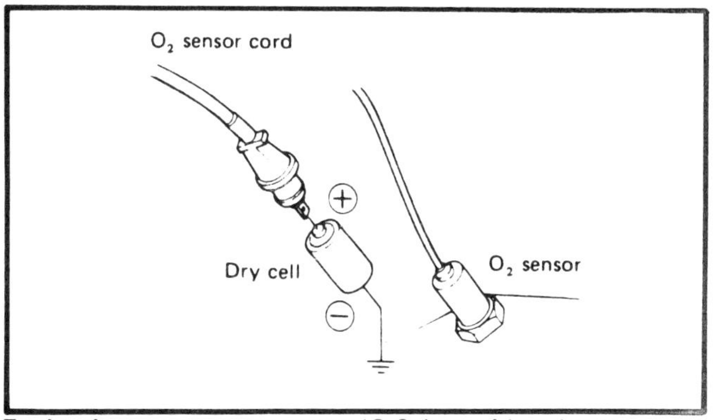

Testing the oxygen sensor system (© Subaru of America, Inc.)

ASPIRATED AIR THERMOSENSOR

The aspirated air thermosensor is provided in the intake air passage of the air flow meter. It uses a thermistor, whose resistance value varies with temperature. This sensor senses a change in temperature of intake air as a change in the resistance value of the thermistor, converting it into voltage and inputs this voltage signal to the EGI control unit, which enriches the fuel according to the intake air temperature.

PRESSURE SWITCH

The positive pressure switch consists of pressure proof diaphragm and microswitch. It lights the turbo indicator light (located on the instrument panel), showing the supercharge zone and also sends the heavy load signal to the EGI Control unit.

AIR INJECTION (AI) SYSTEM

The AI system is a major emission control system, whose purpose is to promote oxidation of hydrocarbons (HC) and carbon monoxide (CO) while the engine is cold.

CALIFORNIA MODELS

1. Component parts
   a. Air suction valve with cut-off valve
   b. Silencer
   c. Air suction pipe
   d. Air introduction hoses
   e. Air cleaner
   f. Thermosensor
   g. Air suction valve solenoid valve
   h. Air suction valve check valve
   i. Electronic control module

Secondary air is supplied to the exhaust port on the #2–#4 cylinder side by the air suction valve.

OPERATIONAL PRINCIPLE

The exhaust gas pulsation is transmitted to the air suction valve (a reed check valve), through the air suction pipe. When the negative pressure of pulsation is transferred to the air suction valve, the reeds of this valve are opened and simultaneously fresh air from the air cleaner is sucked into the exhaust passage.

When, on the other hand, the positive pressure reaches the air suction valve, the reeds are closed to prevent adverse flow of exhaust gas. The air injection system is controlled both by coolant temperature sensed by the thermosensor and time governed by the ECM. When the coolant temperature is below 95 degrees F., the ECM gives a command to open the intake manifold vacuum passage of the solenoid valve 1 for the period of 123 seconds, whereby the cut-off valve is opened by the diaphragm vacuum actuator to operate the air suction valve.

Under such driving condition as wide open throttle acceleration, the check valve closes the vacuum circuit to confine the vacuum, so as to hold the cut-off valve open for that period of time. After that period,

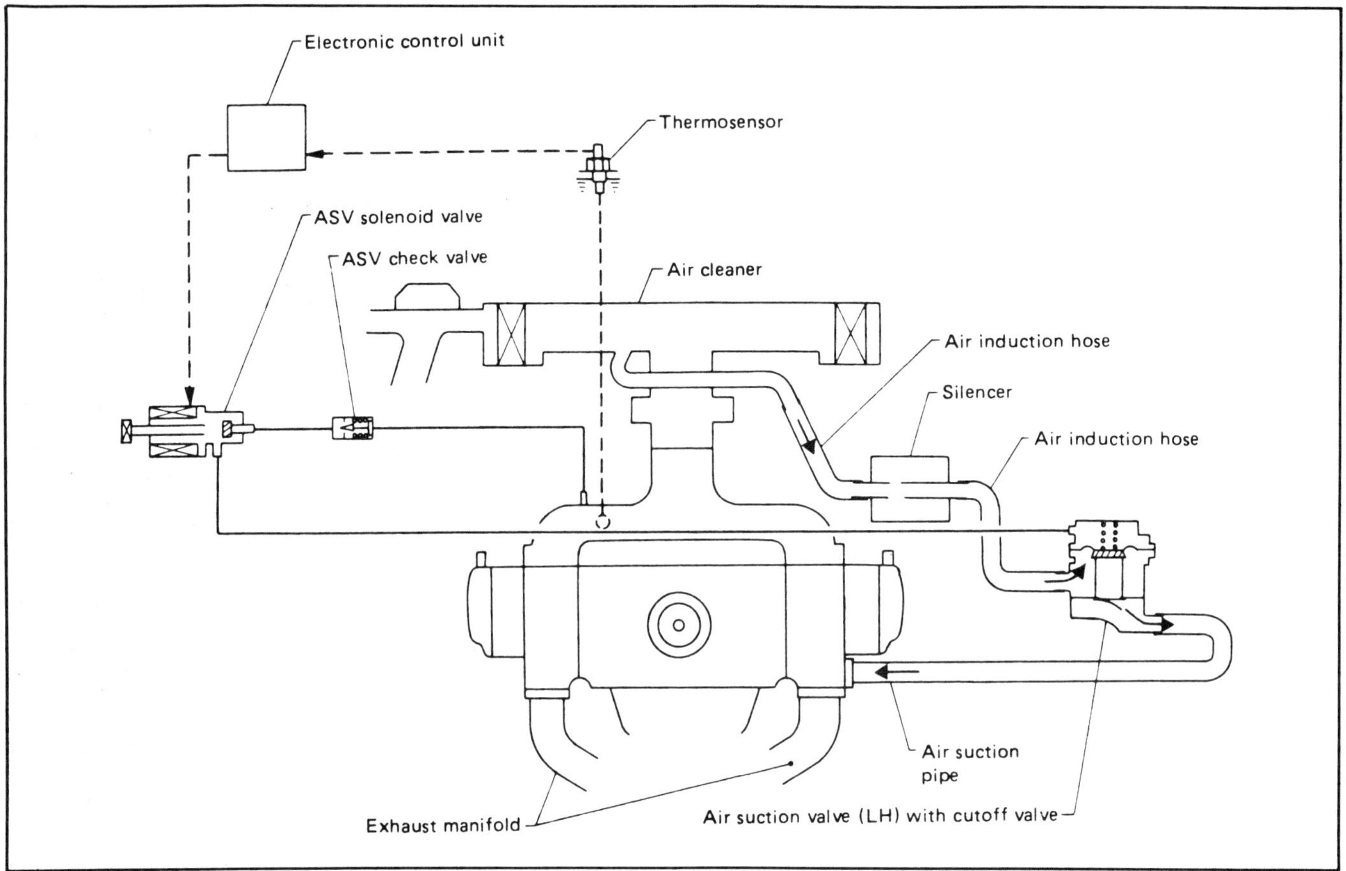

Air Injection system schematic—California models (© Subaru of America, Inc.)

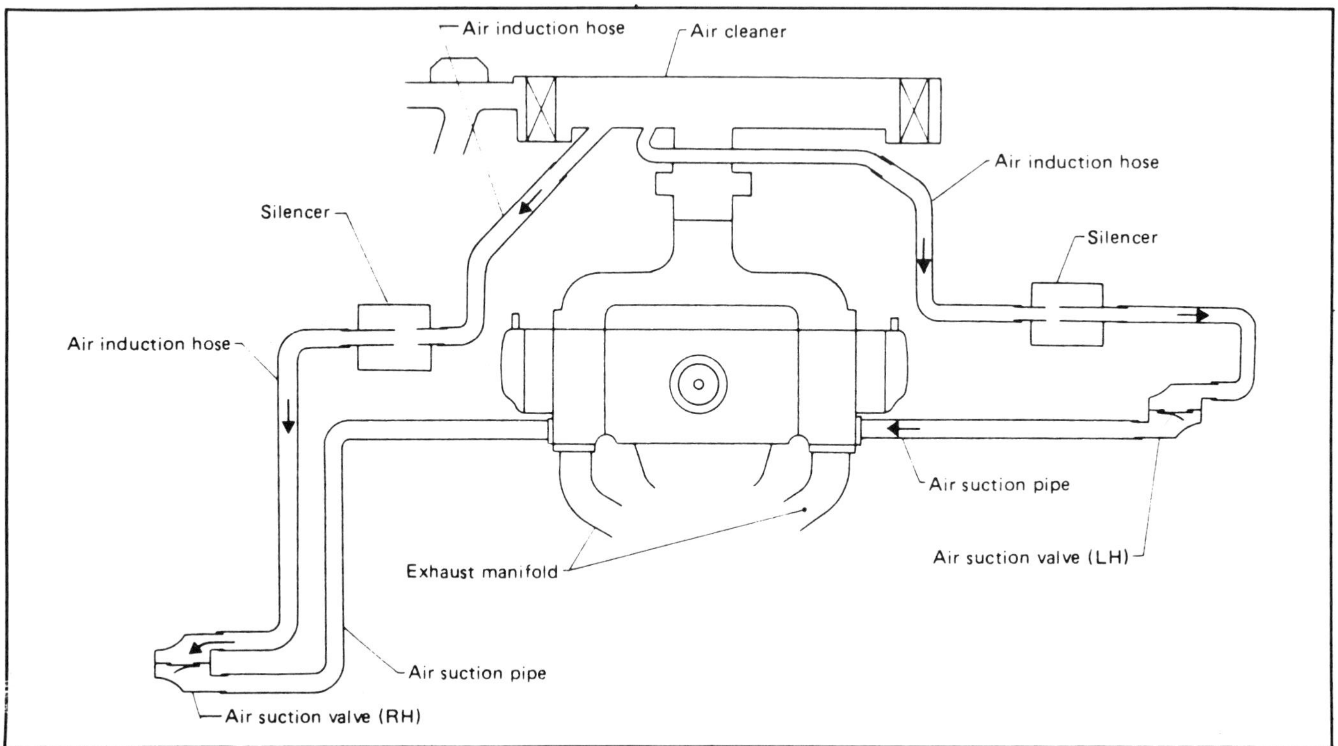

Air injection system schematic—49 state 4-WD and Canada (© Subaru of America, Inc.)

the ECM gives a command to close the vacuum passage of the solenoid valve and open the atmospheric pressure passage. Then the vacuum in the vacuum chamber of the cut-off valve goes out through the air suction valve solenoid valve, whereby the cut-off valve is closed to deactivate the air suction valve.

49 STATE 4WD AND CANADA

1. Component parts
   a. Air suction valve right hand side
   b. Air suction valve left hand side
   c. Silencer
   d. Air suction pipes
   e. Air introduction hoses
   f. Air cleaner

Secondary air is supplied to the exhaust ports on #1–#3 and #2–#4 cylinder group sides by the air suction valve.

OPERATIONAL PRINCIPLE

The exhaust gas pulsation is transmitted to the air suction valve (a reed check valve), through the air suction pipe, where the negative pressure of pulsation is transferred to the air suction valve. The reeds of this valve are opened and simultaneously fresh air from the air cleaner is sucked into the exhaust passage.

When, on the other hand, the positive pressure reaches the air suction valve, the reeds are closed to prevent adverse flow of exhaust gas.

49 STATE 2WD MODELS

1. Component parts
   a. Air suction valve right hand side with cut-off valve
   b. Air suction valve left hand side with cut-off valve
   c. Air suction valve (rear) with out cut-off valve
   d. Silencer
   e. Air suction pipe
   f. Air introduction hoses
   g. Air cleaner
   h. Thermosensor
   i. Air suction valve solenoid valve
   j. Air suction valve check valve
   k. Electronic control module

Secondary air is supplied to the exhaust port on the #2–#4 cylinders side by the air suction valve.

OPERATIONAL PRINCIPLE

The exhaust gas pulsation is transmitted to the air suction valve (a reed check valve), through the air suction pipe, where the negative pressure of pulsation is transferred to the air suction valve. The reeds of the valve are opened and simultaneously fresh air from the air cleaner is sucked into the exhaust passage.

When, on the other hand, the positive pressure reaches the air suction valve, the reeds are closed to prevent adverse flow of exhaust gas. An air suction valve (rear) without a cut-off valve, always operates regardless of the coolant temperature and the ECM signal. On the other hand, the air suction valve with cut-off valves, are controlled by both coolant and temperature, sensed by the thermosensor and the time governed by the ECM. When the coolant temperature is below 95 degrees F., the ECM gives a command to open the intake manifold vacuum passage of the solenoid valve 1 for the period of 123 seconds, whereby the cut-off valve is opened by the diaphragm vacuum actuator to operate the air suction valve.

Under such driving condition as wide open throttle acceleration, the check valve closes the vacuum circuit to confine the vacuum so as to hold the cut-off valve open for that period of time. After that period, the ECM gives a command to close the vacuum passage of the solenoid valve and open the atmospheric pressure passage. The vacuum in the vacuum chamber of the cut-off valve, goes out through the air suction valve solenoid valve, whereby the cut-off valve is closed to deactivate the air suction valve.

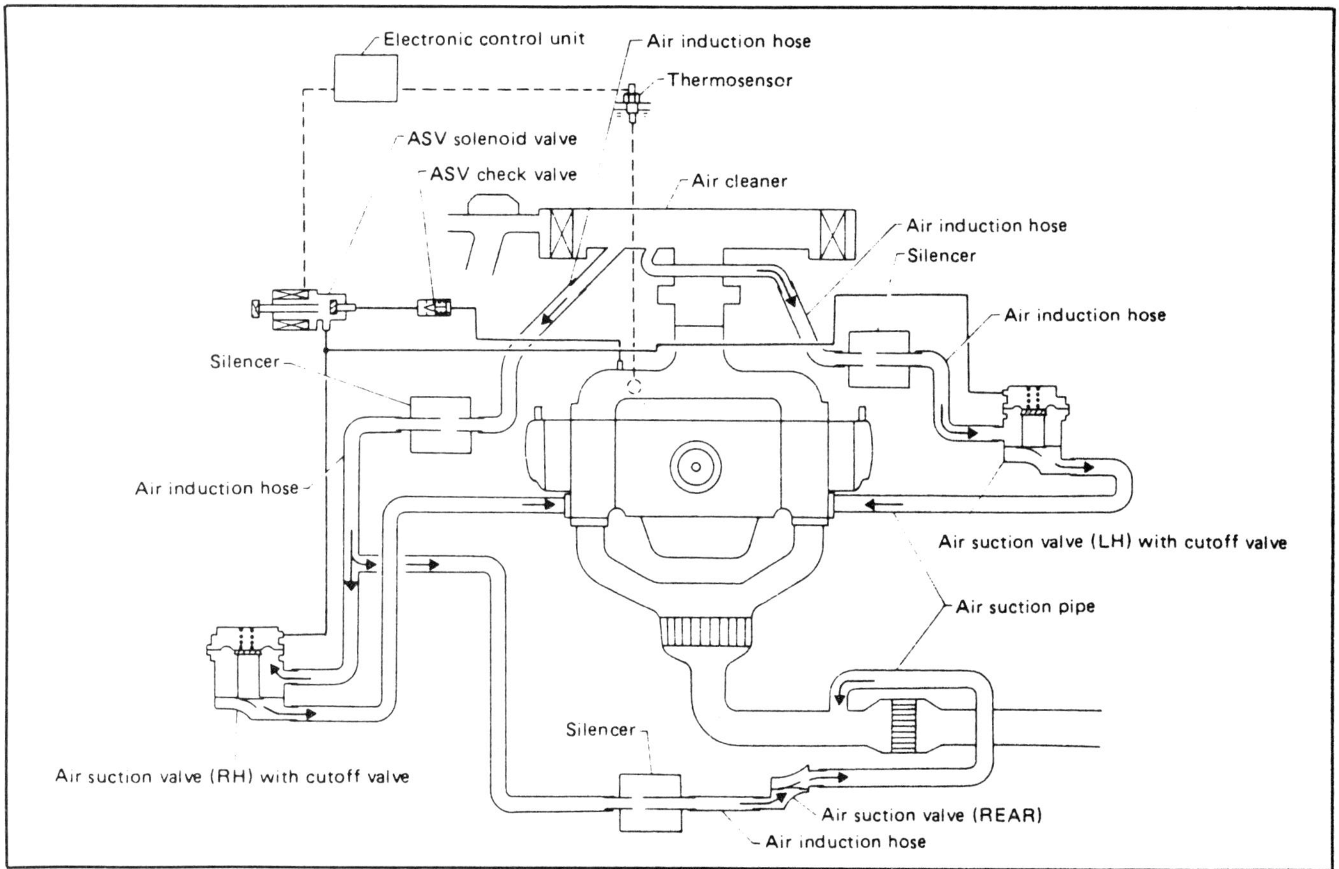

Air injection system schematic—49 state 2-WD (© Subaru of America, Inc.)

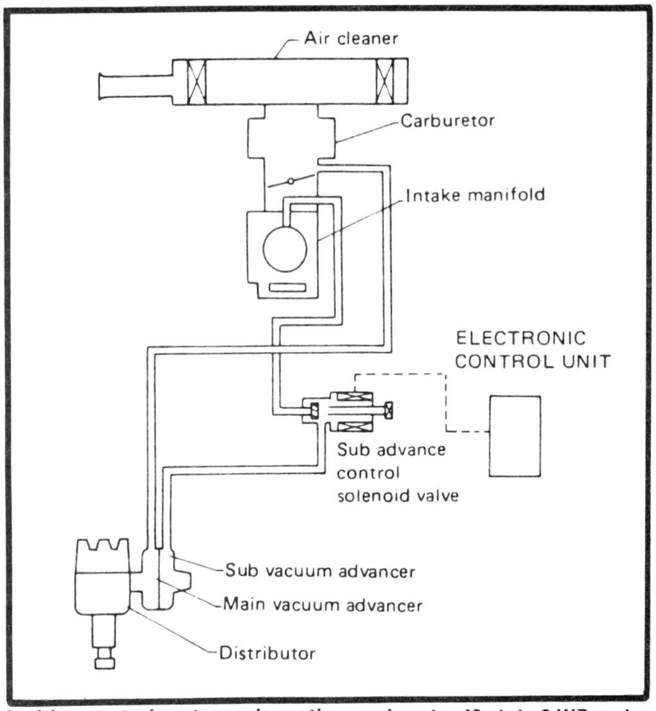

Ignition control system schematic—carburetor 49 state 2-WD and California (© Subaru of America, Inc.)

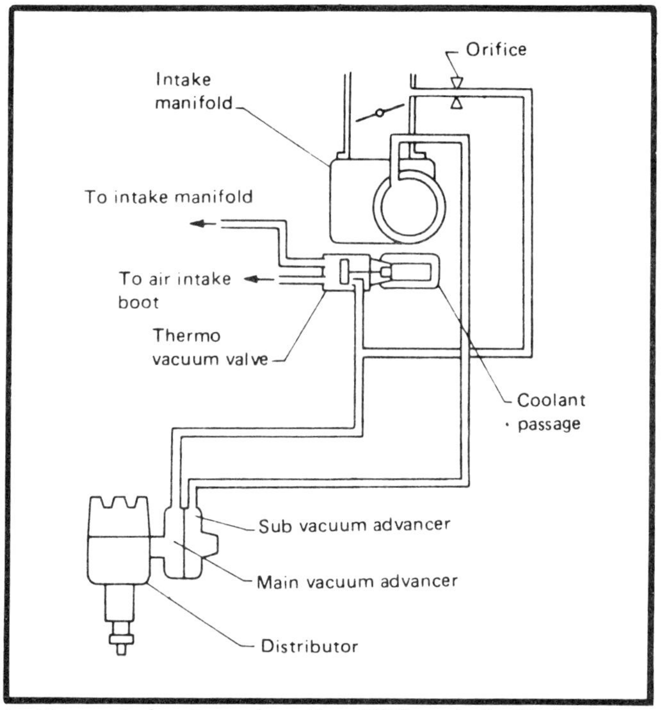

Ignition control system schematic—MPFI non-turbo (49 state 2-WD and California) (© Subaru of America, Inc.)

NOTE: Air suction valve left hand side with cut-off valve is the same as that for the California models.

Ignition Control System

The ignition control system is aimed to reduce HC, CO and NOx emission through all of the operating conditions. Actual ignition timing is controlled by the combination of a centrifugal advance and a vacuum controlled distributor. A double action (main and sub advancer) type vacuum controller is used in 49 state 2WD and California models, equipped with carburetor and the Multi-Point Fuel Injection non-turbo models. A single action, advance type vacuum control is used on the 49 state 4WD and Canada models, equipped with carburetor and Multi-Point Fuel Injection turbo models.

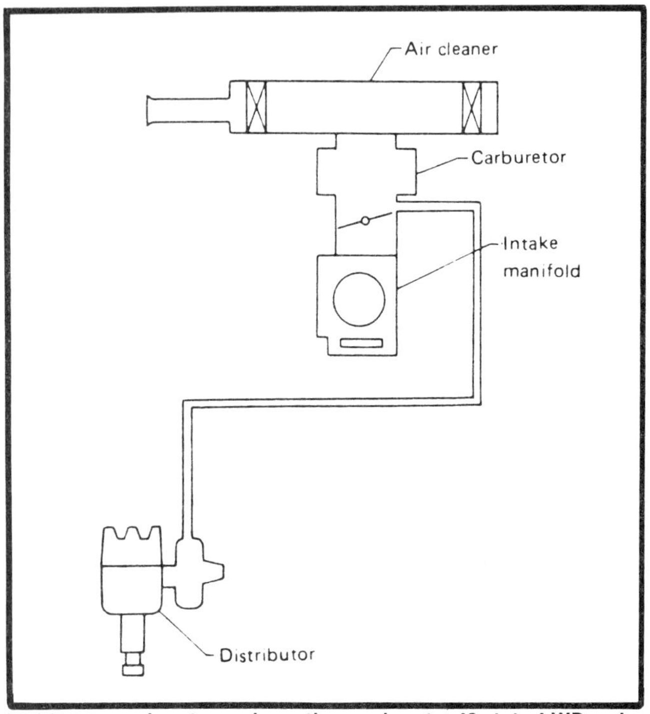

Ignition control system schematic—carburetor 49 state 4-WD and Canada (© Subaru of America, Inc.)

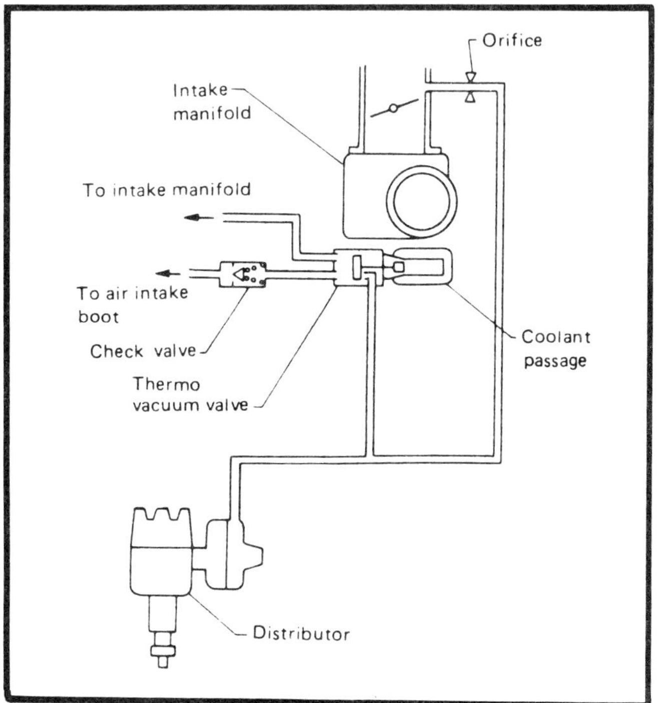

Ignition control system schematic—MPFI non-turbo (49 state 4-WD and Canada) (© Subaru of America, Inc.)

SUB-ADVANCE CONTROL (SAC) SOLENOID VALVE

Testing

1. Using an ohmmeter, check the resistance between the positive and negative terminals of the solenoid valve. Resistance should be 32.7–39.9 ohms.
2. Check the resistance between the positive or negative terminal and the body. The resistance should be one thousand ohms or more.
3. Check the vacuum passages for opening and closing operation, while applying electric current to both positive and negative terminals.
4. If the SAC solenoid valve does not pass any one of these test. Remove and replace the solenoid.

ANTI-AFTERBURNING SYSTEM

During rapid deceleration the air/fuel mixture becomes heavily concentrated temporarily, as the vacuum pressure in the intake manifold increases, causing fuel residue on the inside wall of the manifold to vaporize and to enter the combustion chamber. The anti-afterburning system prevents this heavy concentration from occurring, by introducing air into the intake manifold at this time. This prevents afterburning in the exhaust system. During constant-speed running, the valve is closed because chamber A and B areas have the same pressure. In deceleration (when the throttle is closed), the vacuum pressure in the intake manifold increases and the pressure in chamber B decreases. The diaphragm is forced down to open the valve. This permits outer air to be drawn into the intake manifold through the filter. After that, the pressure in chamber A gradually becomes equal to that of chamber B as the air flows through the orifice. The diaphragm is then pushed up by the spring, which in turn closes the valve to stop the suction of air.

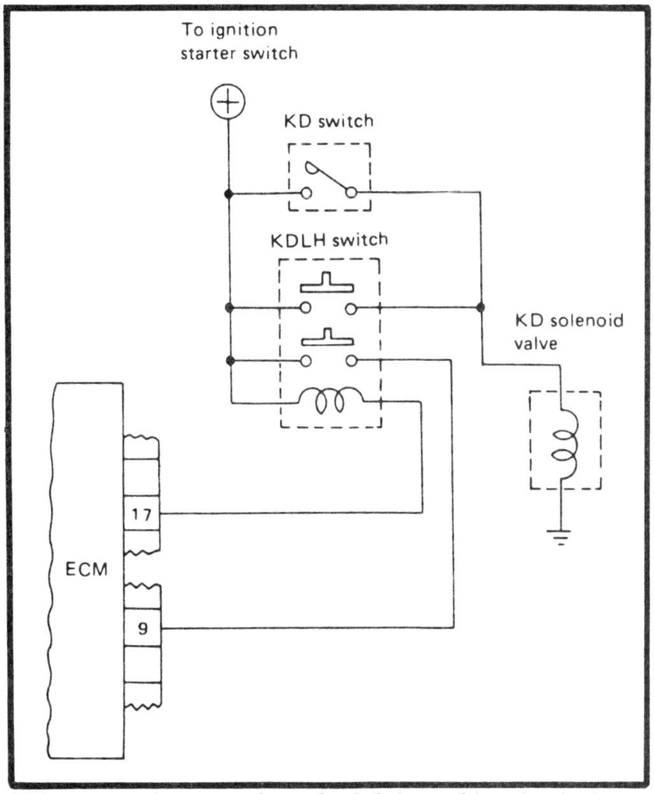

Shift-up control system schematic (© Subaru of America, Inc.)

ANTI–AFTERBURNING SYSTEM

Testing

1. Disconnect the vacuum hose from the Anti-Afterburning Valve (AAV) and operate the engine.
2. Check to see if air is being sucked into the vacuum hose. If there is no air, the vacuum line is defective.
3. Hold a piece of paper under the AAV and run the engine up to 3000 rpm. Close the throttle valve quickly.
4. The paper should be sucked up at the moment the throttle valve is released. If it is sucked up, the AAV is defective and should be replaced.

SHIFT-UP CONTROL SYSTEM

Shift-up control system is used on vehicles equipped with automatic transmission only. The function of this system is to reduce CO emissions due to a rapid warm-up. If the engine is started while the coolant is in the preset temperature range, the ECM keeps the automatic transmission in a kick-down state for a certain period of time.

HIGH-ALTITUDE KIT

When operating the vehicle over 4000 ft. (1200 m), the emission control system requires minor modifications to meet the emission standards at high altitude. A modification kit is available, through an authorized dealer and contains modification instructions that must be followed when modifications are required.

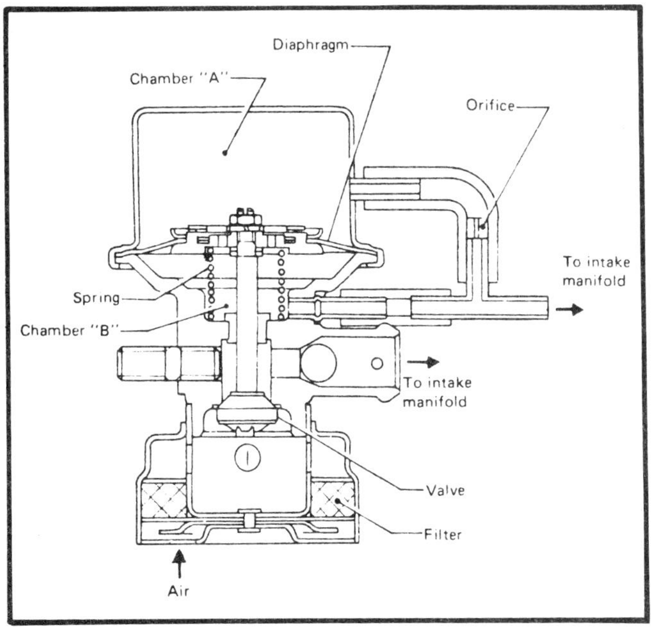

Anti afterburning valve (© Subaru of America, Inc.)

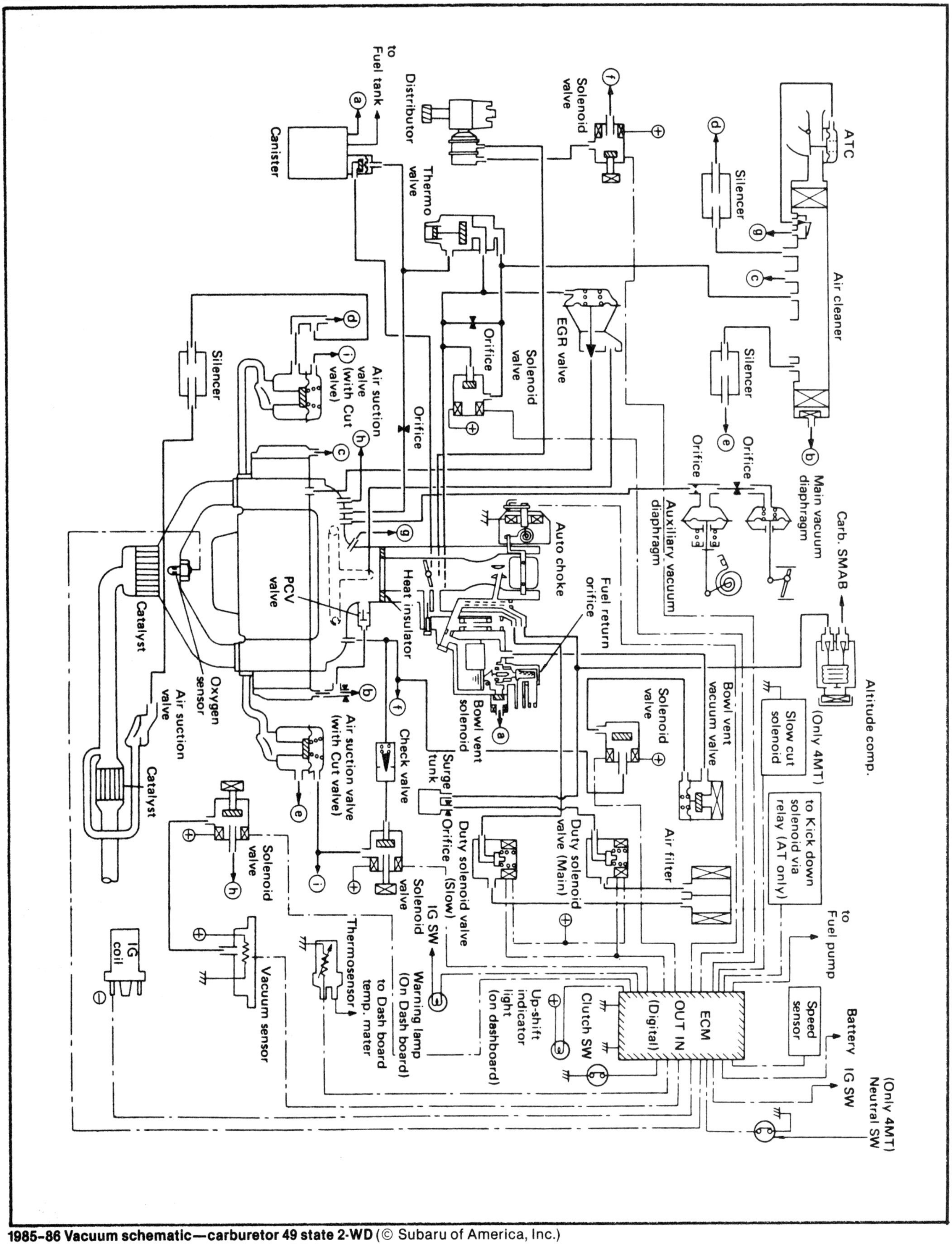

1985–86 Vacuum schematic—carburetor 49 state 2-WD (© Subaru of America, Inc.)

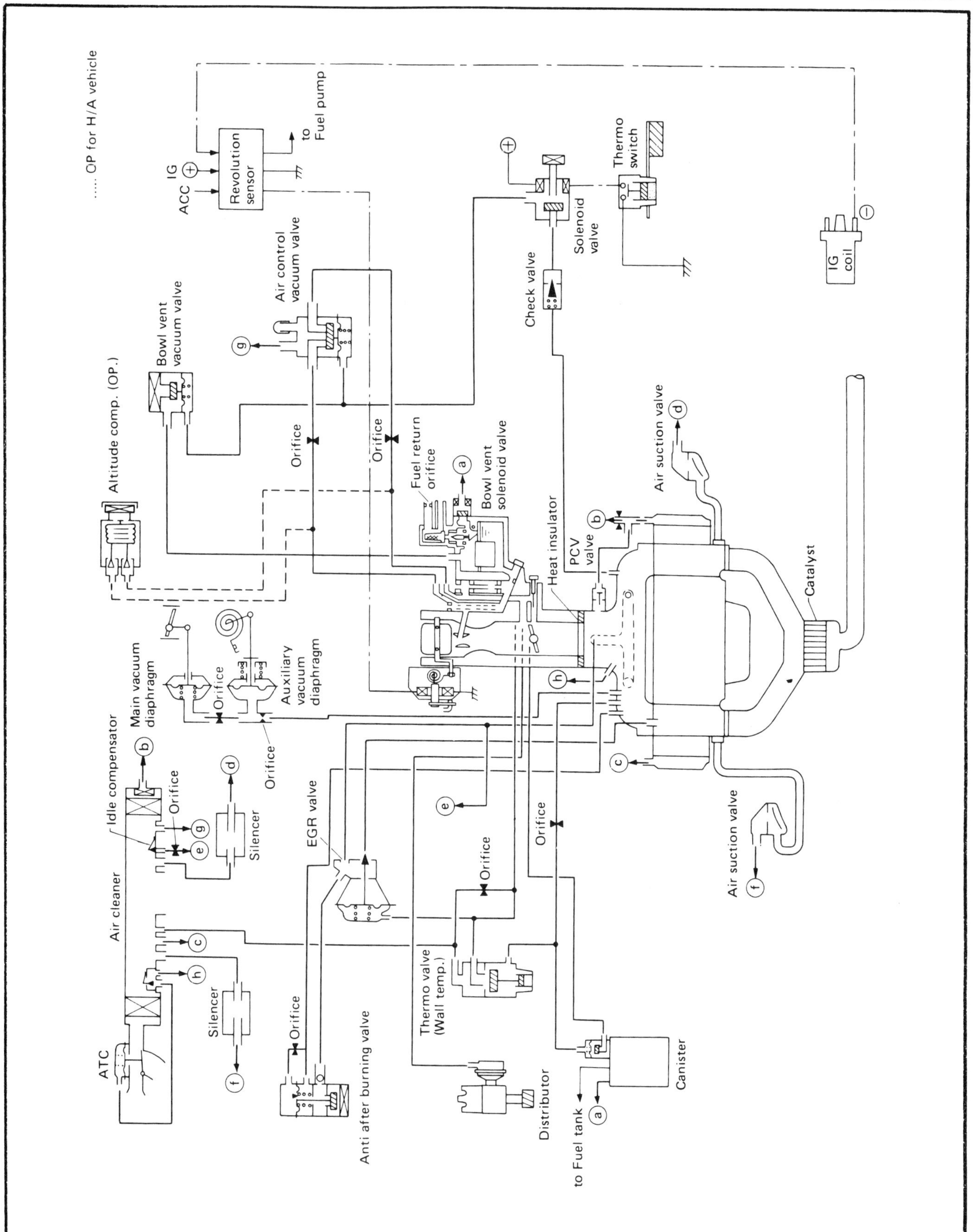

1985–86 Vacuum schematic—carburetor 49 state 4-WD and Canada (© Subaru of America, Inc.)

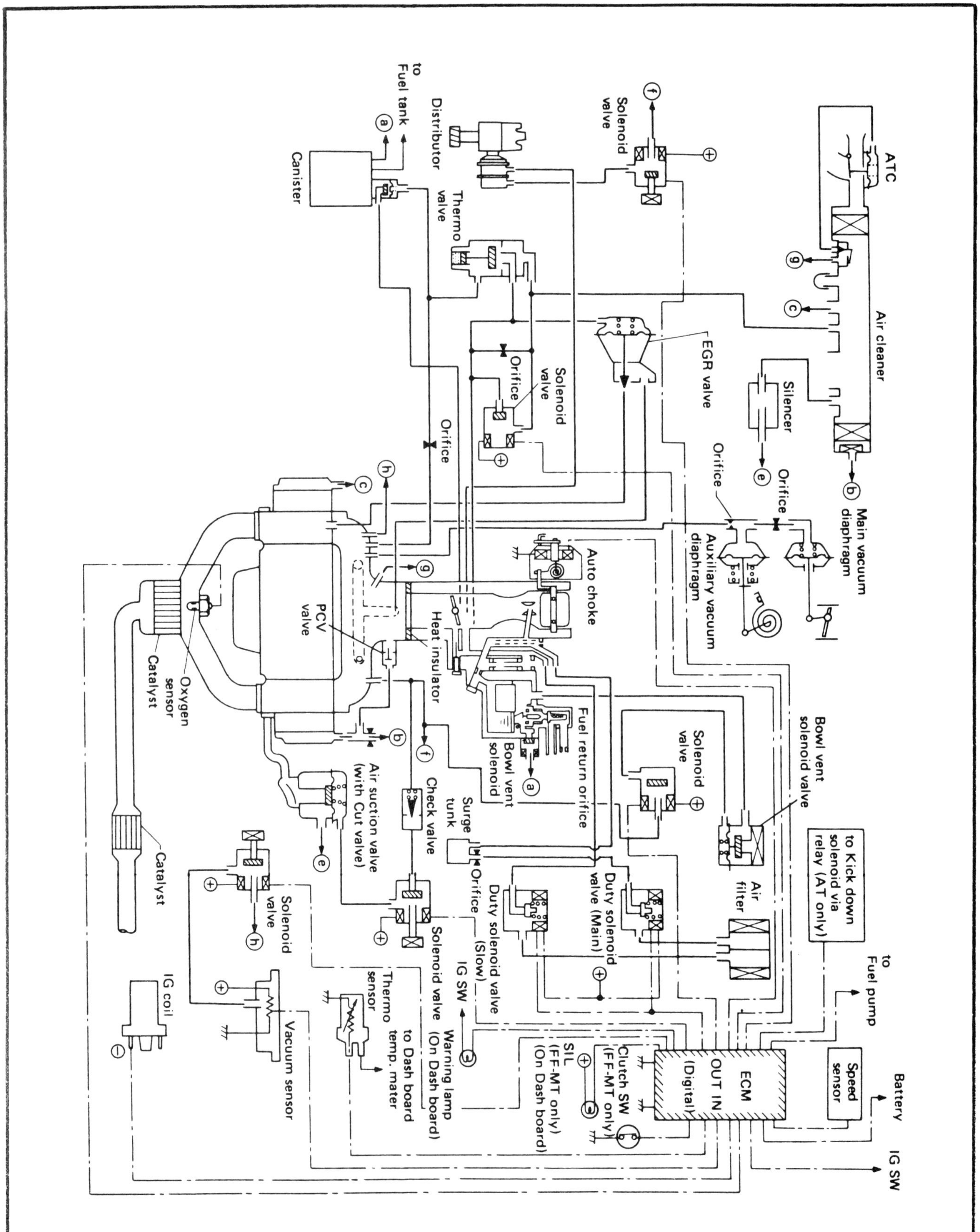

1985–86 Vacuum schematic—carburetor California (© Subaru of America, Inc.)

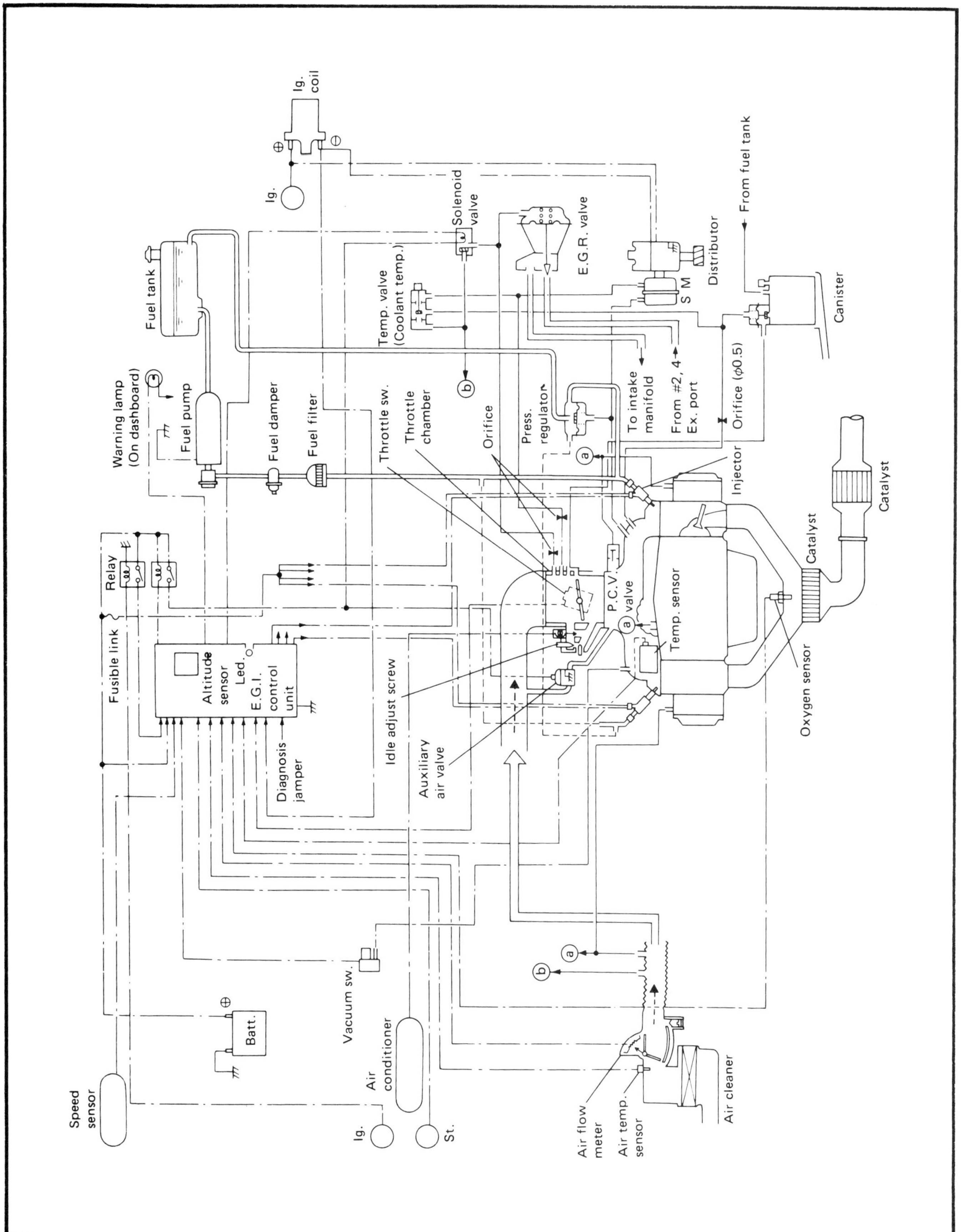

1985–86 Vacuum schematic—MPFI non-turbo (© Subaru of America, Inc.)

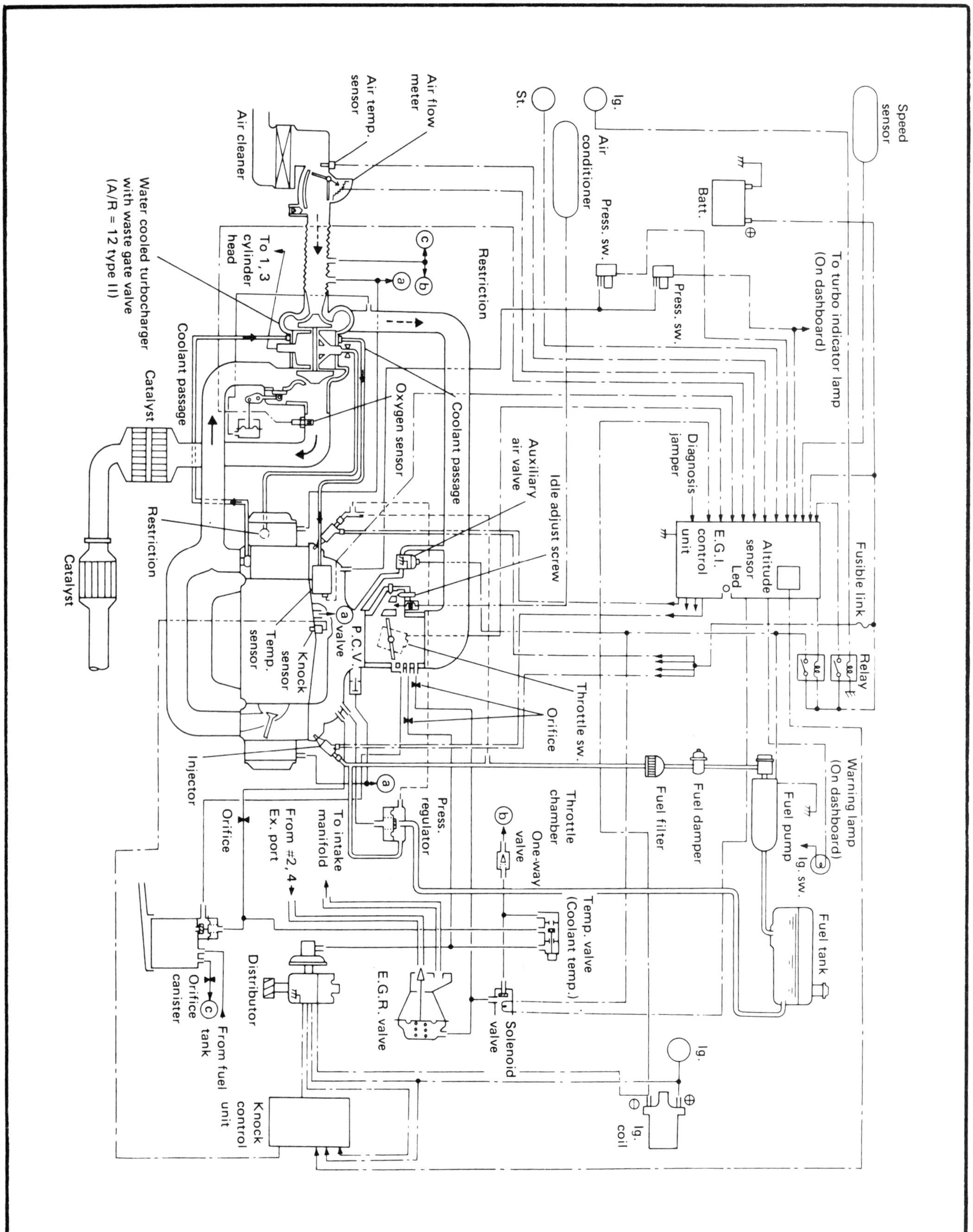

1985-86 Vacuum schematic—MPFI turbo (© Subaru of America, Inc.)

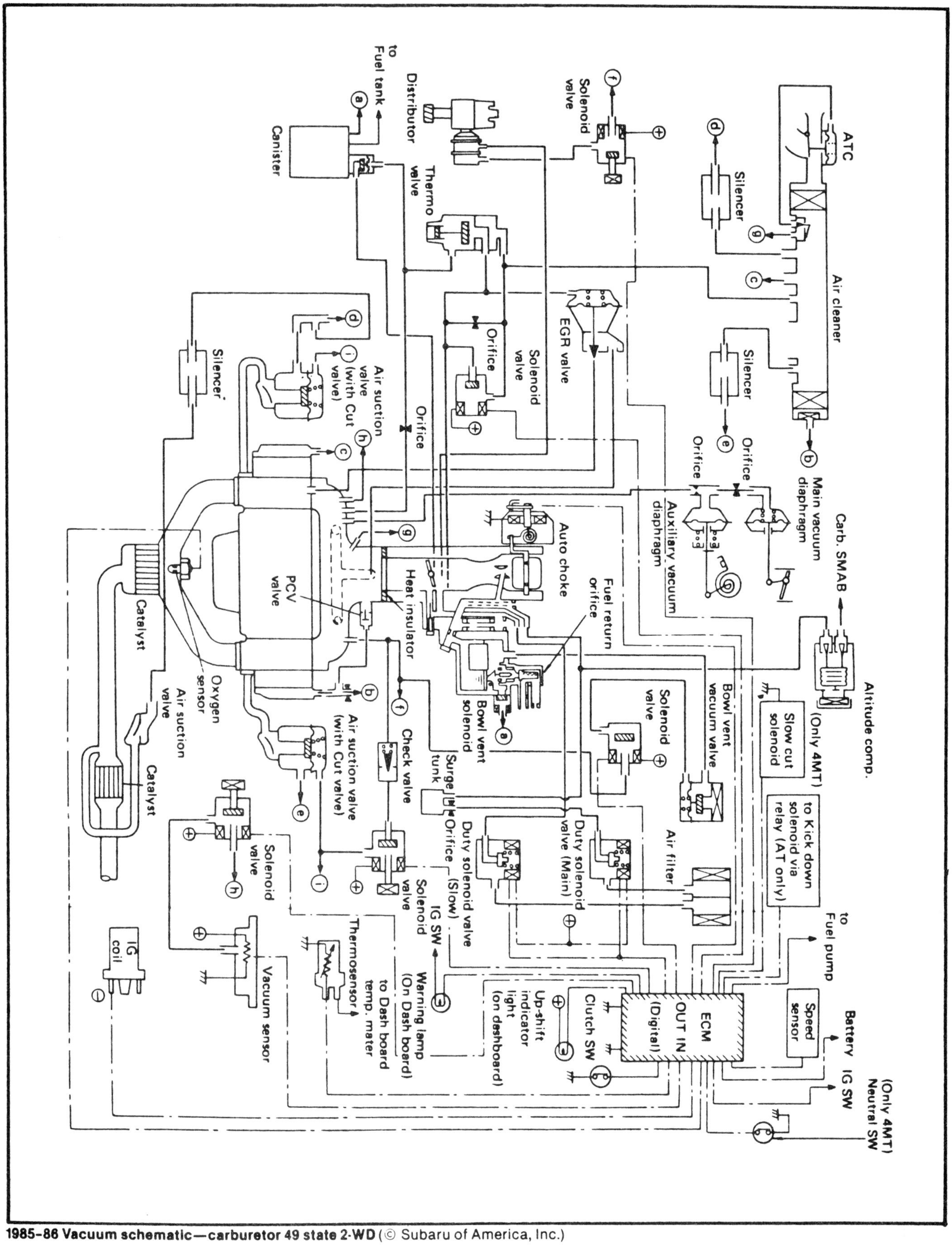

1985-86 Vacuum schematic—carburetor 49 state 2-WD (© Subaru of America, Inc.)

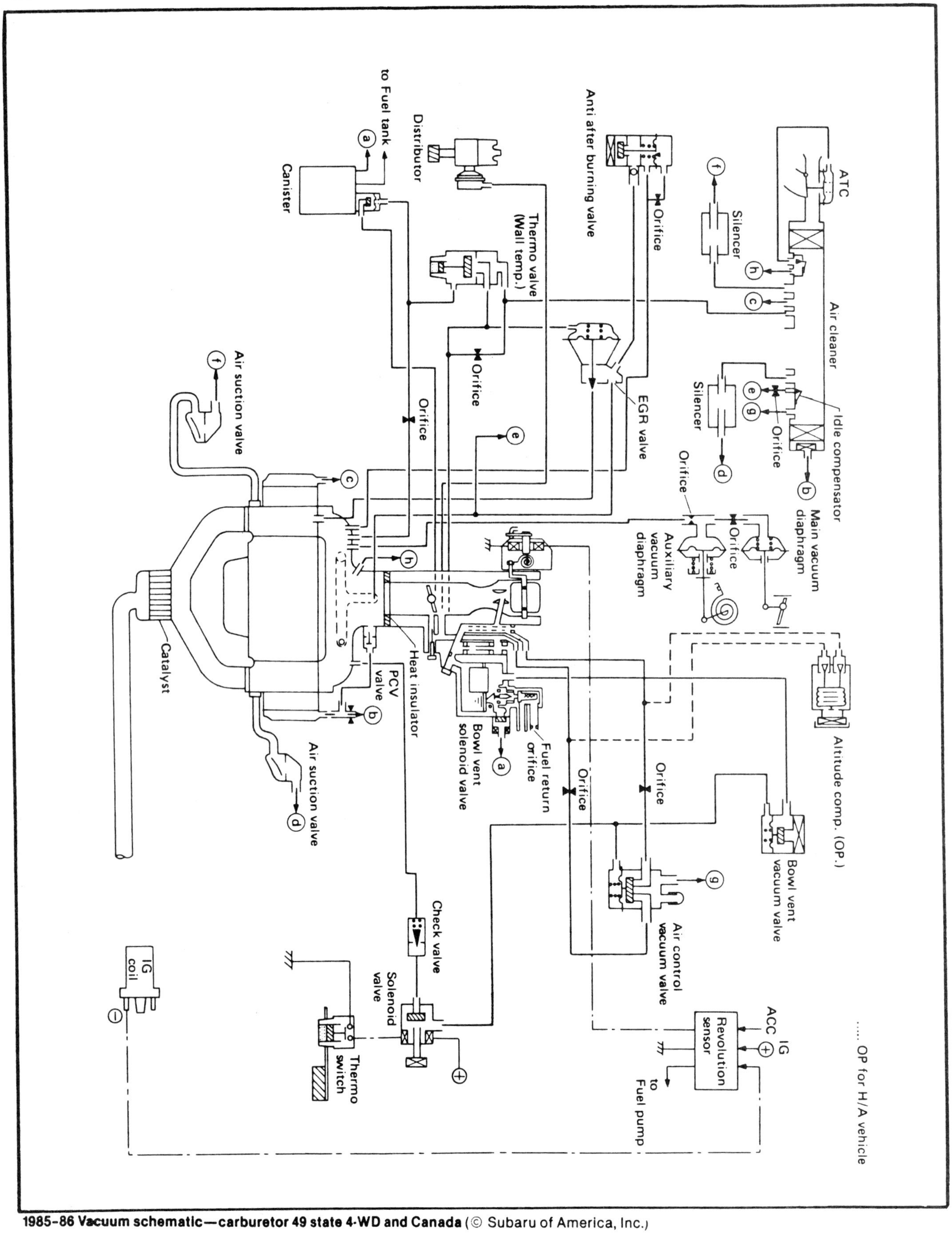

1985–86 Vacuum schematic—carburetor 49 state 4-WD and Canada (© Subaru of America, Inc.)

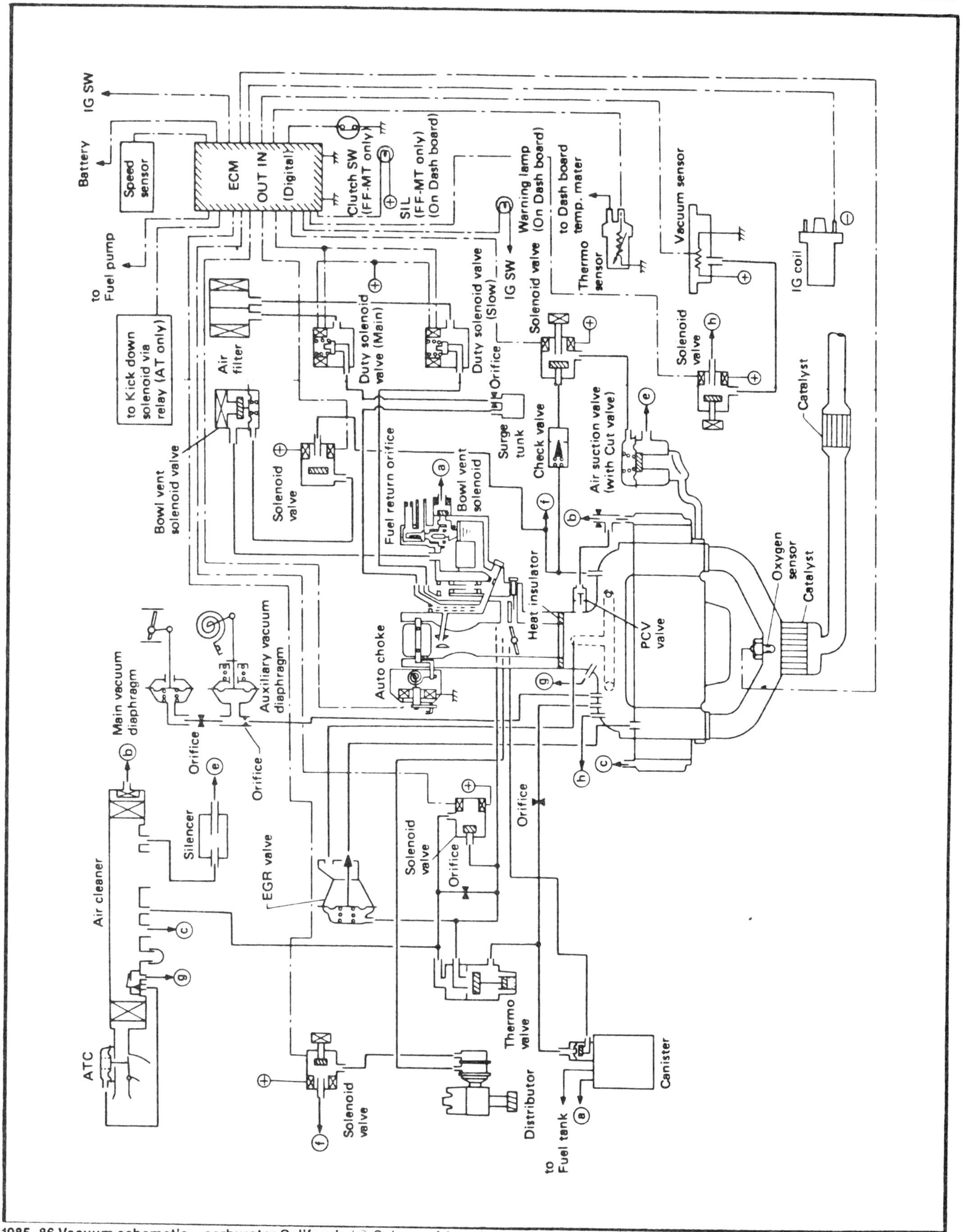

1985-86 Vacuum schematic—carburetor California (© Subaru of America, Inc.)

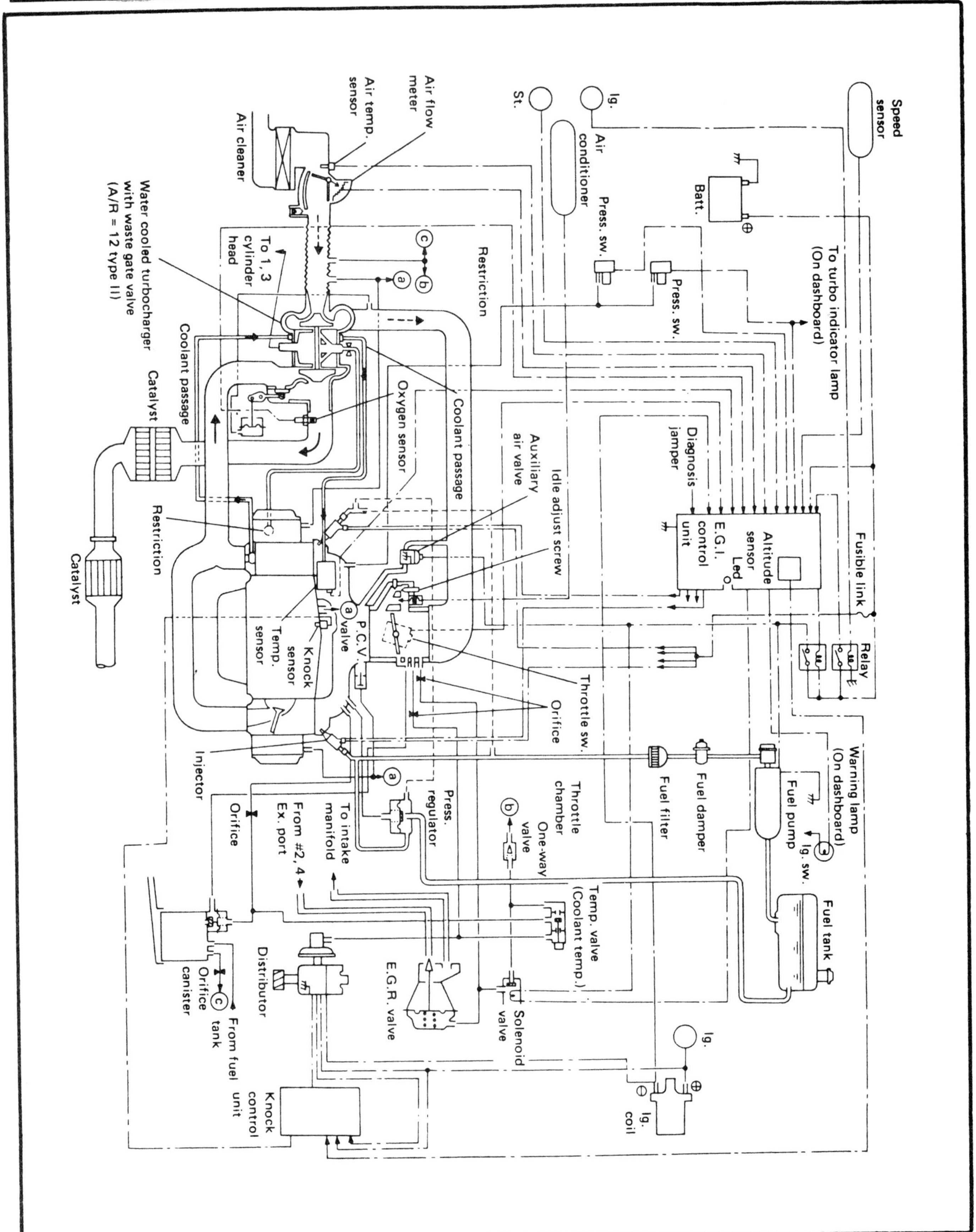

1985-86 Vacuum schematic—MPFI turbo (© Subaru of America, Inc.)

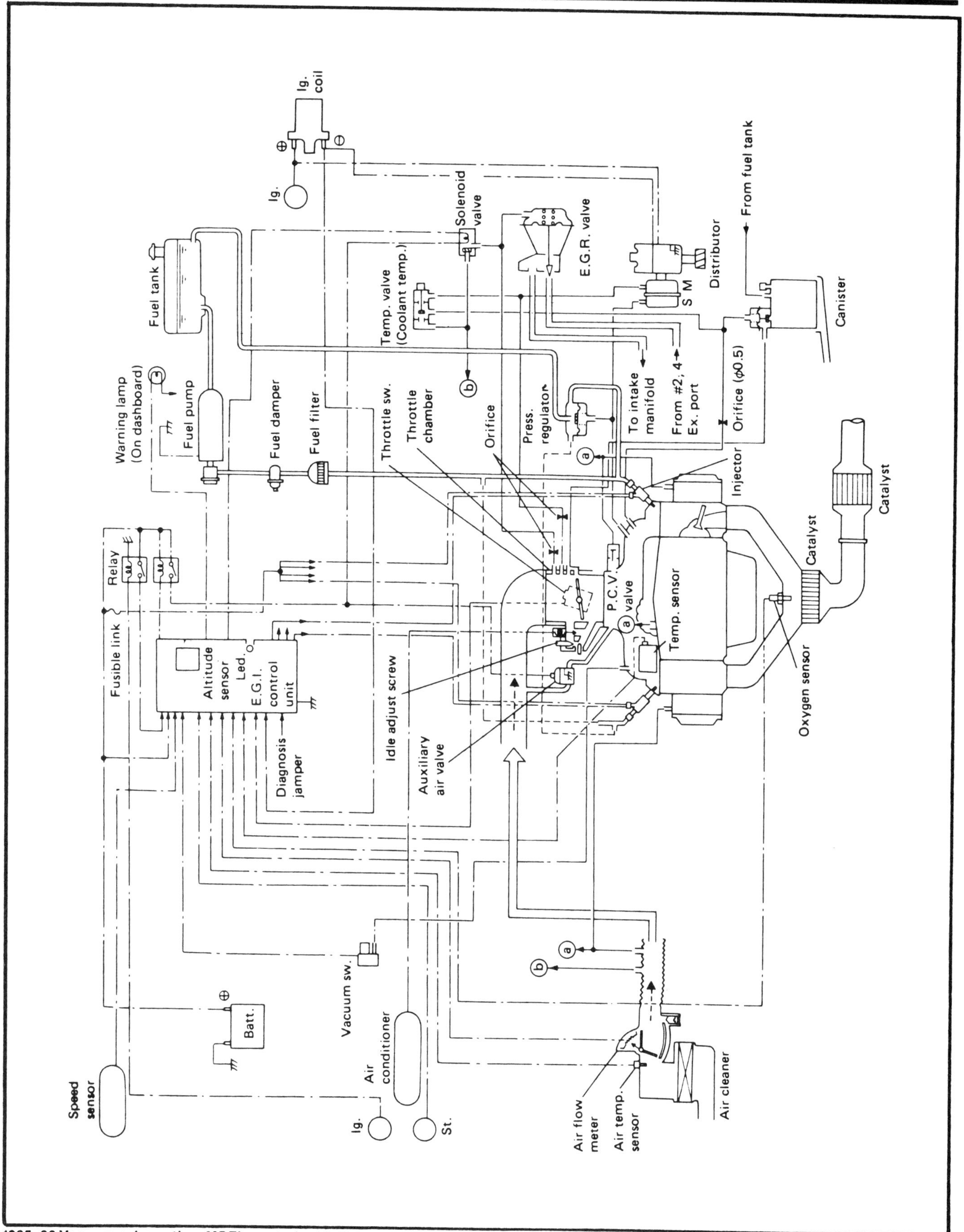

1985–86 Vacuum schematic—MPFI non-turbo (© Subaru of America, Inc.)

IGNITION CONTROL SYSTEM SCHEMATIC DRAWING

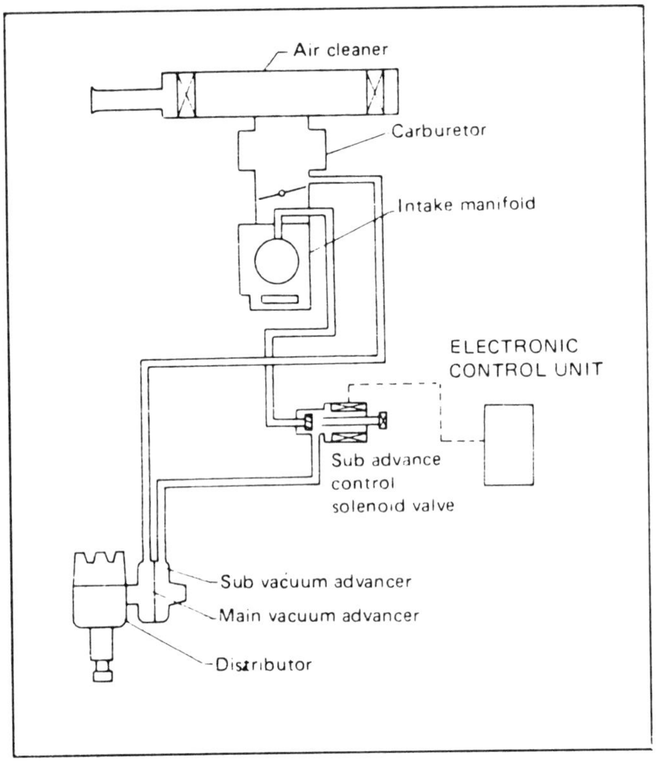

Carburetor [49-state 2WD and California]

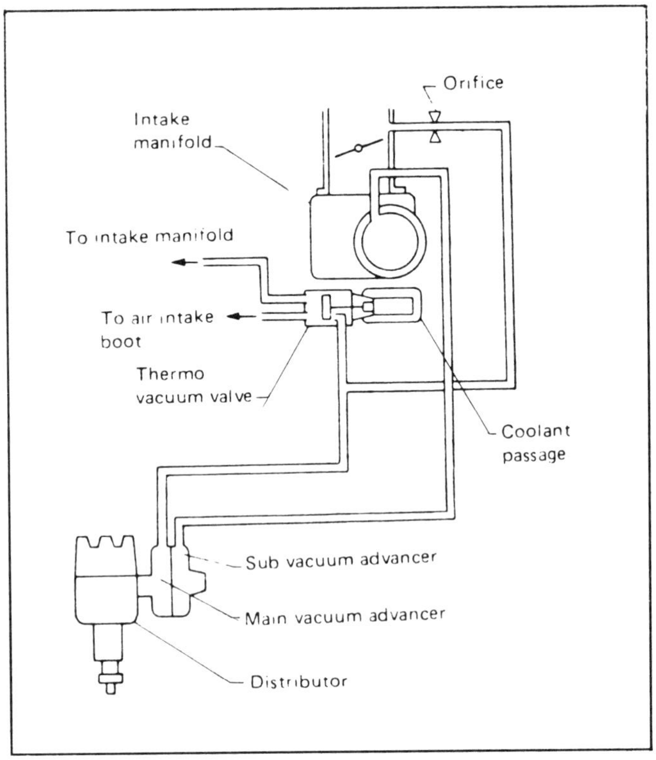

MPFI [Non-TURBO]

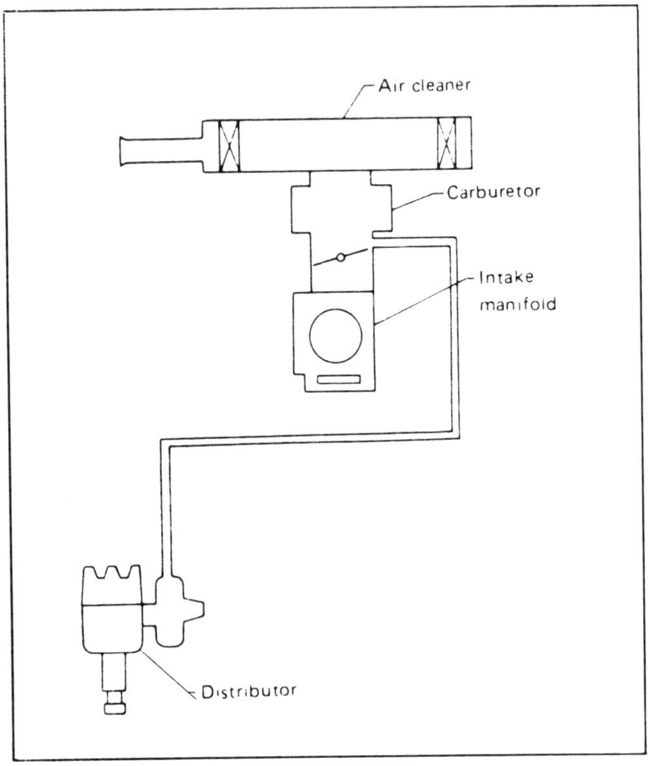

Carburetor [49-state 4WD and Canada]

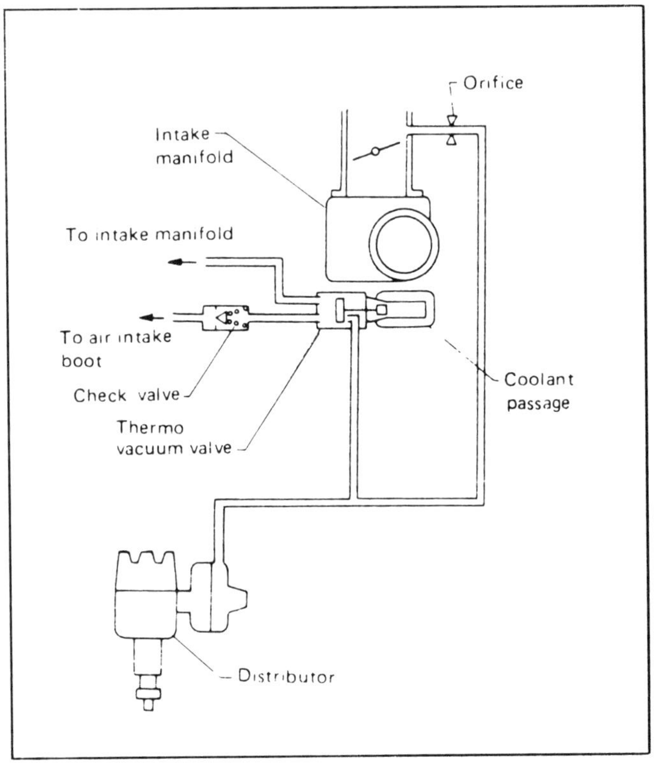

MPFI [Non-TURBO]

IGNITION CONTROL SYSTEM SCHEMATIC DRAWING

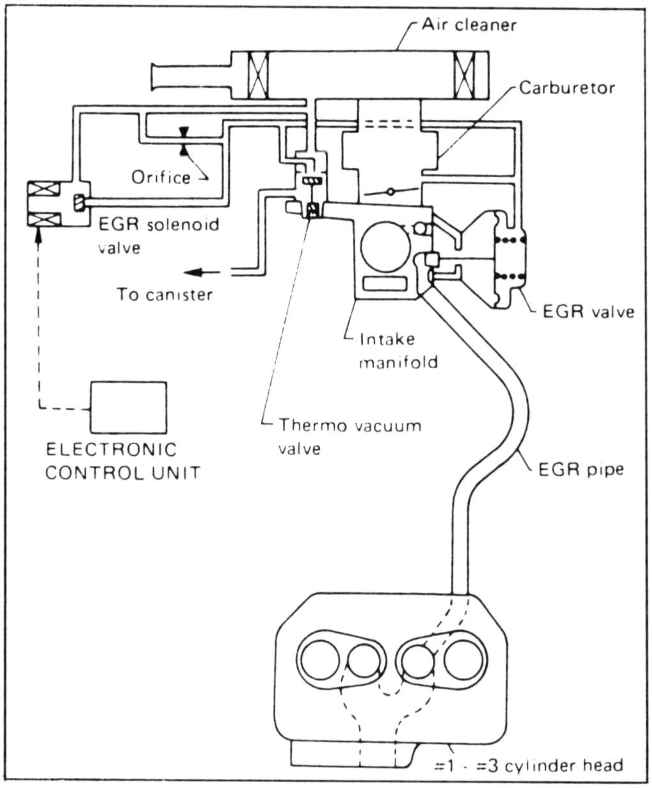

Carburetor [49-state 2WD MT and California]

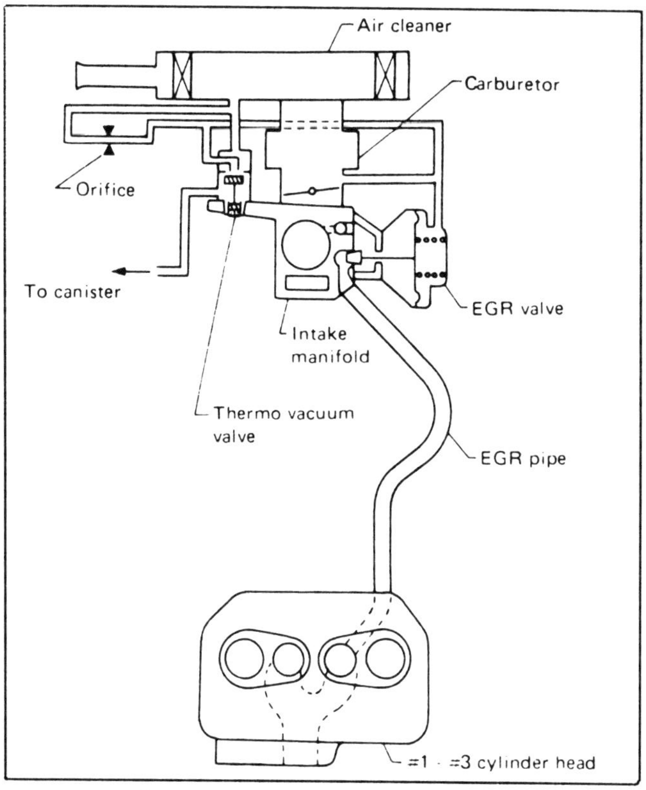

Carburetor [49-state 4WD MT and Canada MT]

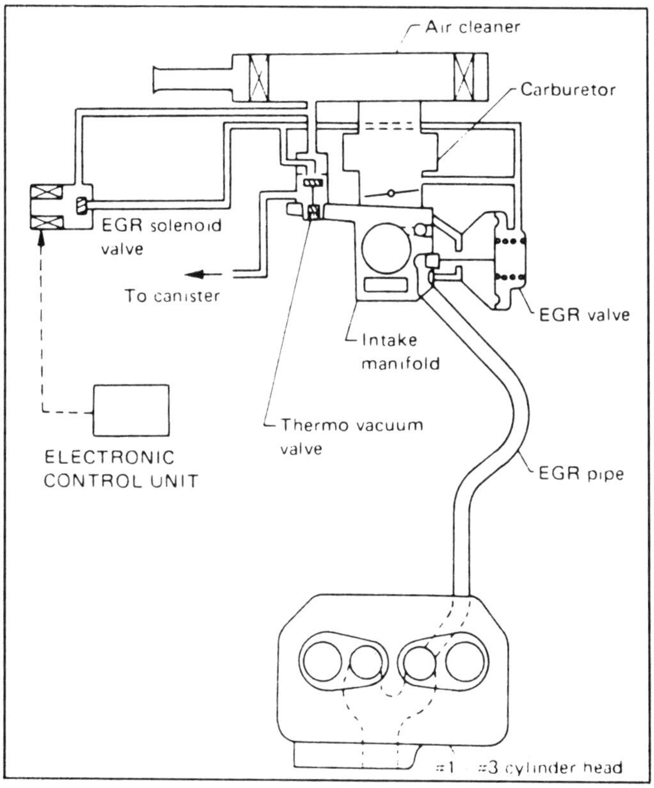

Carburetor [49-state 2WD AT]

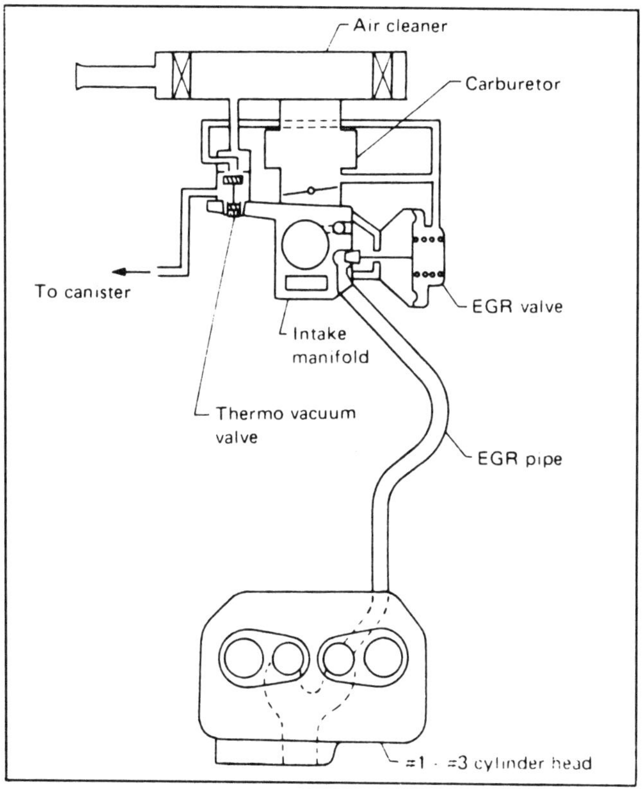

Carburetor [49-state 4WD AT and Canada AT]

IGNITION CONTROL SYSTEM SCHEMATIC DRAWING

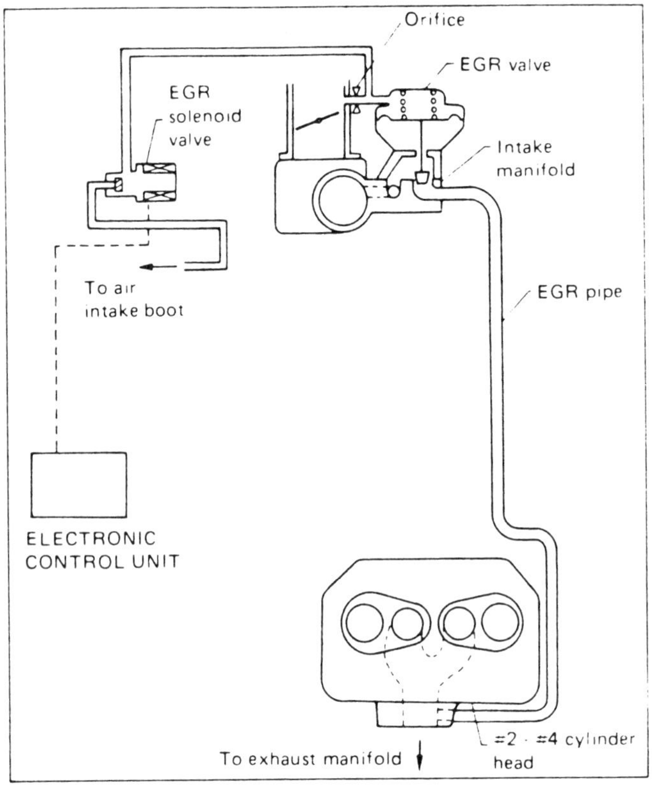

MPFI [Non-TURBO]

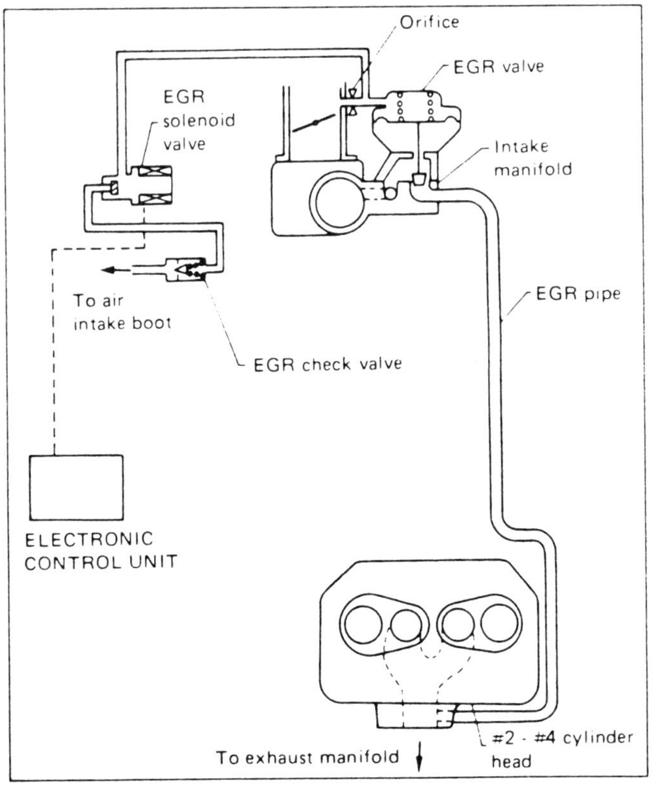

MPFI [TURBO]

Toyota

INDEX

VEHICLE IDENTIFICATION NUMBER AND ENGINE IDENTIFICATION

Vehicle Identification Number (VIN)

All models have the vehicle identification number (VIN) stamped on a plate which is attached to the left side of the instrument panel. This plate is visible through the windshield. The serial number consists of a series identification number followed by a six-digit production number.

Engine Identification

Basically, 1985–86 Toyota vehicles use five types of engines: the "A" series (1A-C, 3A, 3A-C, 4A-C, 4A-LC and 4A-GE); "M" series (5M-GE); "R" series (22R and 22R-E); "S" series (2S-E) and "C" series diesel (1C-TL, 1C-L and 1C-LC). Engines within each series are similar as the cylinder block designs are the same. Variations within each series may be due to displacement (bore x stroke), cylinder head design (single or double overhead camshafts) and fuel system type (fuel injection or carburetor). The engine serial number will be found in the following locations:

"A" series engines—stamped vertically on the left side rear of the engine block. "M" series engines—stamped horizontally on the passenger side of the engine block, behind the alternator.

"R" series engines—stamped horizontally on the driver's side of the engine block, behind the alternator. "S" and "C" series engines—stamped horizontally on the front side of the block.

ENGINE IDENTIFICATION

Model	Year	Engine Displacement Cu. In. (cc)	Engine Series Identification	No. of Cylinders	Engine Type
Camry	'85–'86	121.7 (1995)	2S-E	4	SOHC
	'85–'86	112.2 (1839)	1C-TL (Diesel)	4	SOHC, TURBO
Celica	'85–'86	144.4 (2367)	22R, 22R-E	4	SOHC
Supra	'85–'86	168.4 (2759)	5M-GE	6	DOHC
Corolla	'85–'86	97 (1587)	4A-C, 4A-LC	4	SOHC
	'85–'86	97 (1587)	4A-GE	4	DOHC
	'85–'86	112.2 (1839)	1C-L, 1C-LC (Diesel)	4	SOHC

ENGINE IDENTIFICATION

Model	Year	Engine Displacement Cu. In. (cc)	Engine Series Identification	No. of Cylinders	Engine Type
Tercel	'85–'86	88.6 (1452)	1A-C, 3A, 3A-C	4	SOHC
Cressida	'85–'86	168.4 (2759)	5M-GE	6	DOHC
MR2	'85–'86	97 (1587)	4A-GE	4	DOHC

DOHC—Double-overhead camshaft
SOHC—Single-overhead camshaft

GASOLINE ENGINE TUNE-UP SPECIFICATIONS

Year	Engine Type	Spark Plugs		Distributor		Ignition Timing (deg) ①		Com-pres-sion Press.	Fuel Pump Press.	Idle Speed (rpm)		Valve Clearance (in.) (hot)	
		Type	Gap (in.)	Point Dwell (deg)	Point Gap (in.)	MT	AT			MT	AT	Intake	Exhaust
'85–'86	2S-E	BPR5EA-L11	0.043	Electronic ②		5B	5B	171	28-36	700 ③	750 ③	Hyd.	Hyd.
	22R-E	BPR5EY	0.031	Electronic ②		5B ⑧	5B ⑧	171	35-38	750	750	0.008	0.012
	3A	BPR5EY	0.031	Electronic ②		5B	5B	178	2.6-3.5	④	④	0.008	0.012
	3A-C	BPR5EY-11 ⑨	0.043	Electronic ②		5B	5B	178	2.6-3.5	⑤	⑤	0.008	0.012
	3Y-EC	BPR5EP-11	0.043 ⑩	Electronic ②		8B	8B	171	33-38	700	750	Hyd.	Hyd.
	4A-C, 4A-CL	BPR5EY-11 ⑪	0.043	Electronic ②		5B ⑥	5B ⑥	178	2.6-3.5	⑦	⑦	0.008	0.012
	4A-GE	BCPR5EP-11	0.043	Electronic		10B ⑧	10B ⑧	179	33-39	800	—	0.008 ⑫	0.010 ⑫
	5M-GE	BPR5EP-11	0.043	Electronic		10B ⑧	10B ⑧	164	35-38	650	650	Hyd.	Hyd.

NOTE: If the information given in this chart disagrees with the information on the emission control specification decal, use the specifications on the decal.

MT Manual Transmission
AT Automatic Transmission
TDC Top dead center
B Before top dead center
Hyd. Hydraulic valve lash adjusters
① With vacuum advance disconnected
② Air gap 0.008–0.016 inch
③ With cooling fan OFF; trans. in Neutral
④ W/PS: MT—800
AT—900
⑤ W/PS: MT—800
AT—900
W/O PS: 4 spd—550
5 spd—650
AT—800
⑥ Figure given is with vacuum advance OFF; if ON: 13B
⑦ W/PS: MT—800
AT—900
W/O PS: MT—700
AT—800
⑧ Check connector shorted, see text
⑨ Canada wagon w/MT: BRP5EY; 0.031 in. gap
⑩ Maximum: 0.055 in.
⑪ Canada: BPR5EY; 0.031 in. gap
⑫ Cold clearance

DIESEL ENGINE TUNE-UP SPECIFICATIONS

Year	Engine	Valve Clearance (cold)		Intake Valve Opens (deg)	Injection Pump Setting (deg)	Injection Nozzle Pressure (psi)		Idle Speed (rpm)	Cranking Compression Pressure (psi)
		Intake (In.)	Exhaust (In.)			New	Used		
'85–'86	1C-TL	0.008–0.012	0.010–0.014	11B	25-30B ①	2062–2205	1920–2205	750	427

B—Before Top Dead Center
① See text for injection timing procedure

FIRING ORDERS

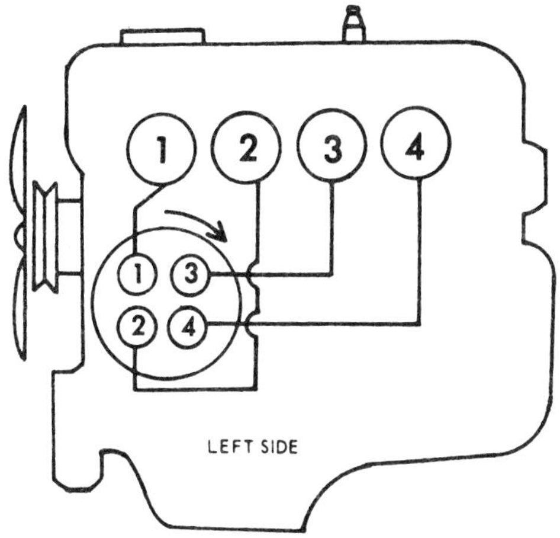

R series engines

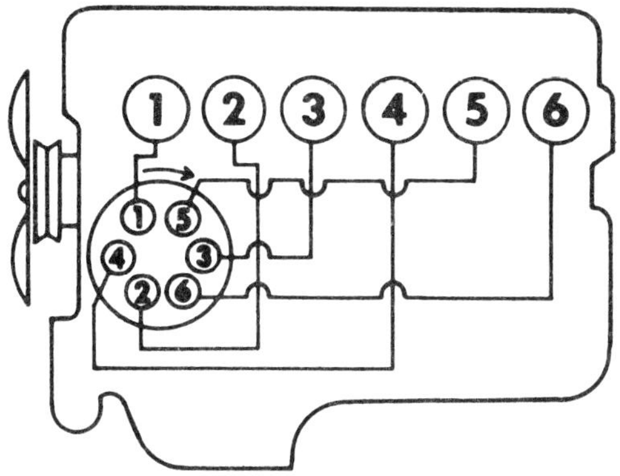

5M-GE engines

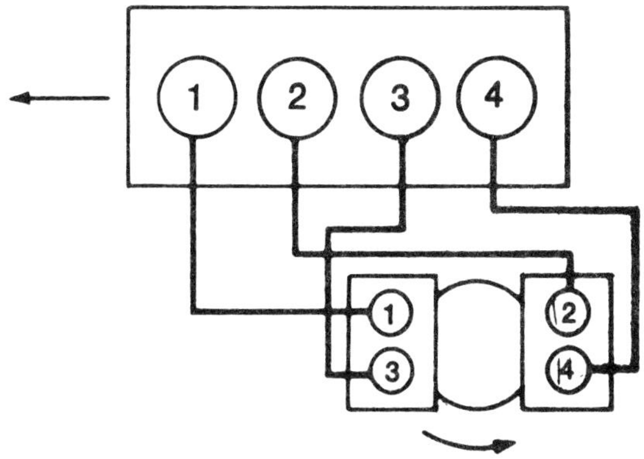

A series engines

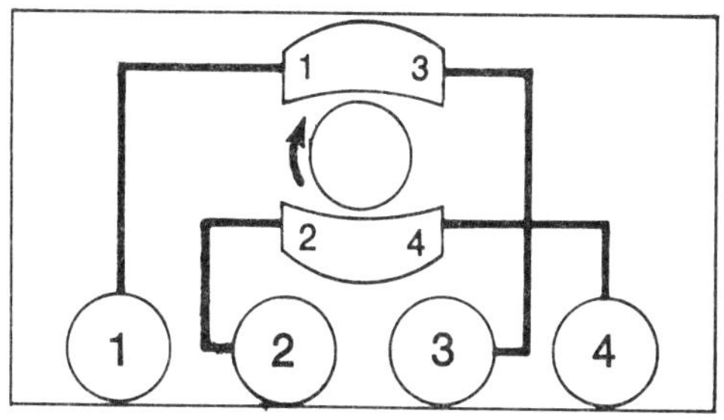

2-SE engine

EMISSION EQUIPMENT USED

NOTE: Canadian Tercel with 3A-C/4 spd and wagon with automatic transmission, uses the same equipment as the California system

1985–86 TERCEL, COROLLA (FWD)

Positive Crankcase Ventilation (PCV)
Fuel Evaporation Emission Control (EVAP)
Throttle Positioner (TP)
Mixture Control (MC)—Canada only
Feedback Carburetor (FBC)—USA only
Exhaust Gas Recirculation (EGR)
Air Suction (AS)
Three-Way and Sweeper Catalyst (TWO-OC)—Federal models
Three-Way Catalyst (TWC)—California Models
Oxidation Catalyst (OC)—Canada models
High Altitude Compensation (HAC)—Federal models
Auxiliary Acceleration Pump (AAP)
Automatic Hot Air Intake (HAI)
Hot Idle Compensator (HIC)
Automatic Choke
Choke Breaker (CB)
Choke Opener
Deceleration Fuel Cut
Heat Control Valve
Oxygen Sensor (OXS)—USA only
Cold Mixture Heater (CMH)—USA only

1985–86 MR2, CELICA, COROLLA (RWD)

Positive Crankcase Ventilation (PCV)
Fuel Evaporation Emission Control (EVAP)
Dash Pot (DP)
Exhaust Gas Recirculation (EGR)
Three-Way Catalyst (TWC)
Electronic Fuel Injection (EFI)
Oxygen Sensor (OXS)

1985–86 SUPRA, CRESSIDA

Positive Crankcase Ventilation (PCV)
Fuel Evaporation Emission Control (EVAP)
Dash Pot (DP)
Exhaust Gas Recirculation (EGR)
Three-Way Catalyst (TWC)
Electronic Fuel Injection (EFI)
Oxygen Sensor (OXS)

1985–86 CAMRY

Positive Crankcase Ventilation (PCV)
Fuel Evaporation Emission Control (EVAP)

Three-Way Catalyst (TWC)—USA only
Oxidation Catalyst (OC)—Canada only
Electronic Fuel Injection (EFI)
Oxygen Sensor (OXS)

REQUIRED EMISSION CONTROL MAINTENANCE

NOTE: In addition to the services listed below, Toyota recommends an oil and filter change every 10,000 miles and an air filter replacement every 30,000 miles. Cut the mileage intervals in half if the vehicle is driven daily in heavy traffic, under dusty conditions or is used to tow a trailer. Valve and Idle Speed/Mixture adjustments are also considered part of emission maintenance.

PCV SYSTEM

Check operation and replace the PCV valve (carbureted engine), or clean any gum deposits from the orifices (fuel injected engine) every 30,000 miles. Inspect all hoses for leaks or deterioration and replace as ecessary.

FUEL EVAPORATION EMISSION CONTROL (EVAP) SYSTEM

Inspect the system for proper operation every 30,000 miles. Check all hose connections and hoses for leaks or deterioration and replace as necessary. Check the charcoal canister for raw gas contamination or damage and replace if necessary.

EXHAUST GAS RECIRCULATION (EGR) SYSTEM

Inspect the system for proper operation every 30,000 miles. Check the EGR valve operation and inspect all hoses and connections for leaks or deterioration and replace as necessary.

OXYGEN SENSOR (OXS) SYSTEM

Although no replacement interval is specified by the manufacturer, the oxygen sensor should be checked for proper operation every 30,000 miles and replaced every 60,000 miles. Any oxygen sensor found to be malfunctioning should be replaced.

IDLE SPEED AND MIXTURE ADJUSTMENT

Carbureted engines

To conform with Federal regulations, the mixture adjusting screw is plugged at the factory to prevent tampering with the adjustment. Normally, this plug should not be removed. When troubleshooting a rough idle, check all other possible causes before removing the plug and adjusting the idle mixture.

MIXTURE ADJUSTING SCREW PLUG

Removal

1. Remove the carburetor from the engine.
2. Plug each carburetor vacuum port to prevent entry of shavings when drilling.
3. Mark the center of the plug with a punch.
4. Drill a 0.256 in. hole in the center of the plug. Drill carefully and slowly to avoid drilling into the screw, since there is only 0.04 in. clearance between the plug and the screw. The drill may force the plug off.
5. Lightly seat the mixture screw by inserting a screwdriver into the drilled hole and turning the screw clockwise. Be careful not to tighten the screw or damage to the needle tip may result.
6. If the plug is still in place, use a 0.295 in. drill bit to force the plug out of the hole.
7. Remove the mixture adjusting screw and inspect it for damage. If the tapered needle portion is damaged or scored, replace the mixture adjusting screw.

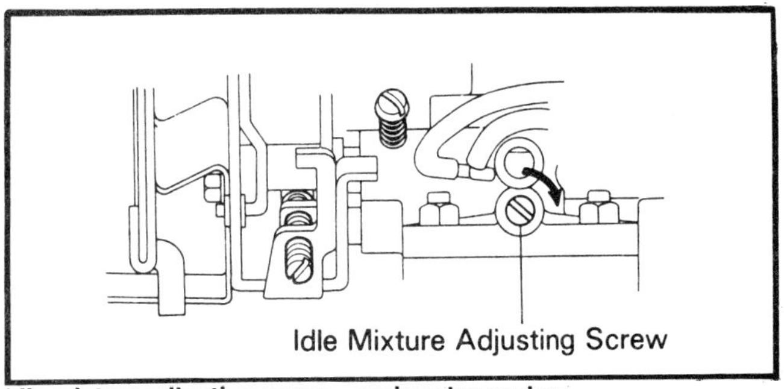

Idle mixture adjusting screw—carburetor engines

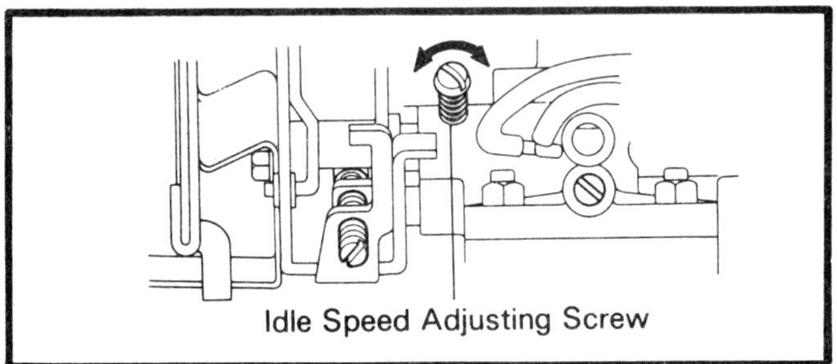

Idle speed adjusting screw—carburetor engines

8. Fully seat the mixture adjusting screw lightly once again, **then** back it out the following number of turns:
   a. Tercel (except Canada 4WD wagon)—3 $^5/_8$ turns counterclockwise
   b. Tercel Canada 4WD wagon—2 $^1/_2$ turns counterclockwise
   c. Corolla (USA models)—3 $^1/_4$ turns counterclockwise
   d. Corolla (Canada models)—2 $^1/_2$ turns counterclockwise
9. Install the carburetor and continue the idle speed and mixture adjustments as outlined below.

IDLE SPEED AND MIXTURE

Adjustment

1. All adjustments should be made with the engine at normal operating temperature under the following conditions:
   a. Air cleaner installed
   b. Choke fully open
   c. All accessories switched off
   d. All vacuum lines connected
   e. Ignition timing set to specifications
   f. Transmission in Neutral
2. Start the engine and allow it to reach normal operating temperature.
3. Turn the idle mixture screw to obtain the maximum idle speed, then use the idle speed screw to adjust the idle to 700 rpm. Continue going back and forth until the idle speed doesn't rise when the mixture screw is adjusted.

NOTE: The cooling fan should be off for all adjustments.

4. Adjust the idle speed down to 650 rpm by turning in the idle mixture screw (lean drop method of adjustment).

5. Adjust the idle speed to specifications by turning the idle speed adjusting screw.

6. Once the idle speed is adjusted, install a new tamper-proof plug over the mixture adjusting screw.

Fuel Injected engines

IDLE SPEED

Adjustment

CAMRY AND CELICA

1. This adjustment should be made with the engine at normal operating temperature under the following conditions:
   a. Air cleaner installed
   b. All hoses and wires connected
   c. All accessories switched off
   d. Transmission in Neutral
   e. Engine cooling fan off
2. Connect the tachometer positive (+) terminal to the ignition coil negative (–) terminal.

CAUTION

Never allow the tachometer terminal to touch ground as it could cause damage to the igniter or ignition coil. Always check with the instructions supplied with the tachometer manufacturer as some tachometers are not compatible with this ignition system.

3. Raise the engine speed to about 2500 rpm for approximately two minutes.
4. Allow the engine speed to stabilize and set the idle speed to specifications by turning the idle speed adjusting screw. The cooling fan should be off for all adjustments.

SUPRA AND CRESSIDA

NOTE: To perform this procedure, a voltmeter and EFI idle adjusting wire harness (No. 0984-42-14010) is necessary.

1. Behind the battery on the left front fender apron is a service connector. Remove the rubber caps from the connector and connect the EFI idle adjusting wire harness.
2. Connect the positive lead of the voltmeter to the red wire of the wiring harness and the negative lead to the black wire.
3. Connect a tachometer according to the manufacturer's instructions. Make sure the tachometer used is compatible with the ignition system.
4. Start and raise the engine speed to 2500 rpm for approximately two minutes to warm up the oxygen sensor. The needle of the voltmeter should be fluctuating at this time. If not, turn the idle mixture adjusting screw until the voltmeter fluctuates eight times in ten seconds. If the voltmeter cannot be made to fluctuate, replace the oxygen sensor and start the adjustment procedure again.

NOTE: The idle speed should be set immediately after warm-up while the needle of the voltmeter is fluctuating.

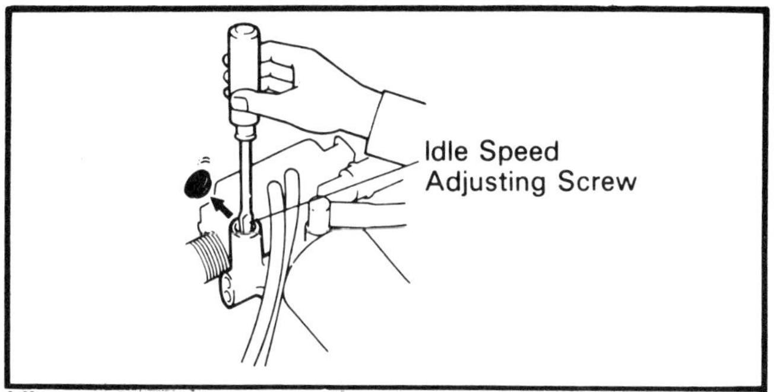

Idle speed adjusting screw—Celica with EFI (typical)

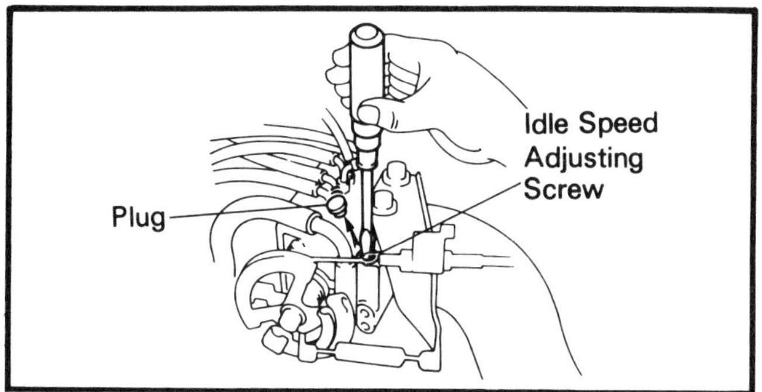

Idle speed adjusting screw—Camry with gas engine

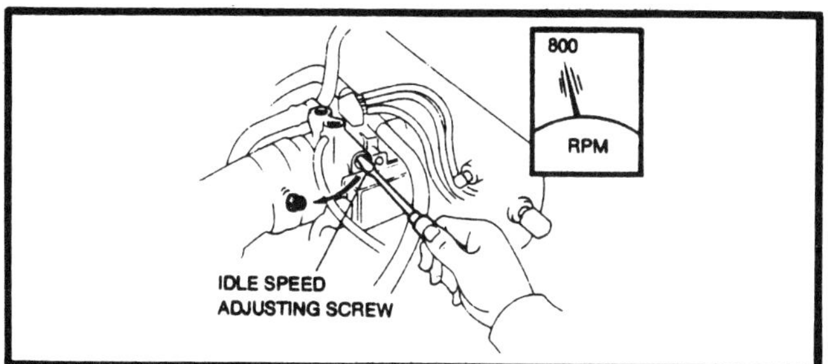

Idle speed adjusting screw—Celica Supra and Cressida with EFI (typical)

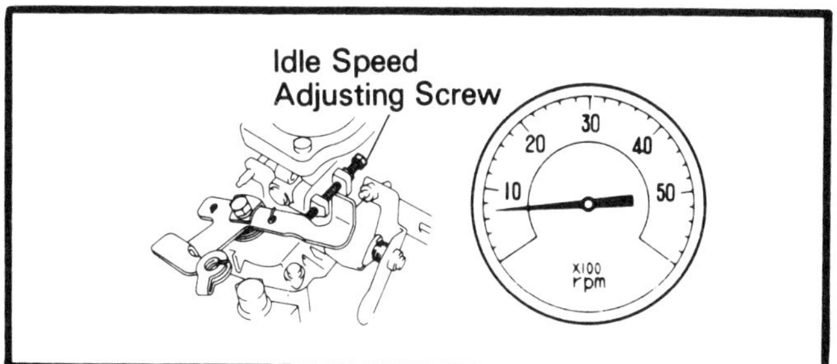

Idle speed adjusting screw Corolla and Camry with diesel engine

Diesel Engine

IDLE SPEED

Adjustment

COROLLA AND CAMRY

1. This adjustment should be made with the engine at normal operating temperature under the following conditions:
   a. Air cleaner installed
   b. All accessories switched off
   c. Transmission in Neutral
2. Connect a suitable diesel tachometer to the engine according to the manufacturer's instructions.
3. Start the engine and allow it to reach normal operating temperature. The cooling fan should be off for all adjustments.
4. Check the idle speed with the tachometer. It should be 700 rpm. If not, adjust the idle speed by loosening the locknut and turning the idle speed adjusting screw located on the injection pump.

IDLE SPEED SPECIFICATIONS

Year	Engine	Idle Speed (rpm) MT	Idle Speed (rpm) AT
'85–'86	2S-E	700 ①	750 ①
	22R-E	750	750
	3A	650 ②	800 ②
	3A-C	800 ③	900 ③
	3Y-EC	700	750
	4A-C, 4A-CL	700 ④	800 ④
	4A-GE	800	—
	5M-GE	650	650

NOTE: If the information in this chart disagrees with the specifications on the underhood emission decal, use the decal figures.

MT Manual Transmission
AT Automatic Transmission
① Cooling fan off and transmission in Neutral
② With power steering: MT—800 rpm
AT—900 rpm
③ Without power steering: 4 spd—550 rpm
5 spd—650 rpm
AT—800 rpm
④ With power steering: MT—800 rpm
AT—900 rpm

INITIAL TIMING SETTINGS

Gasoline Engines

NOTE: Refer to the Tune-Up Specifications Chart for dynamic timing settings. If the information on the underhood emission decal disagrees with the specifications listed in the chart, use the decal figures.

Adjustment

1. Start the engine and allow it to reach normal operating temperature. Connect a tachometer and check that the idle speed is within specifications. If not, adjust the idle speed and turn off the ignition. On models with electronic ignition, attach the tachometer to the negative (–) side of the ignition coil, not to the distributor primary lead. Damage to the ignition control unit will result from improper connections.

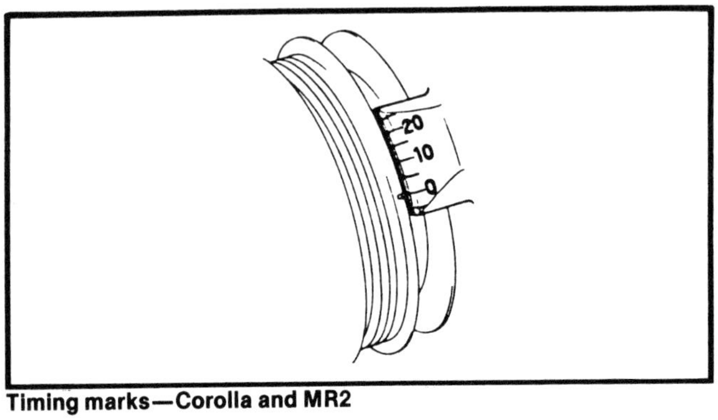

Timing marks—Corolla and MR2

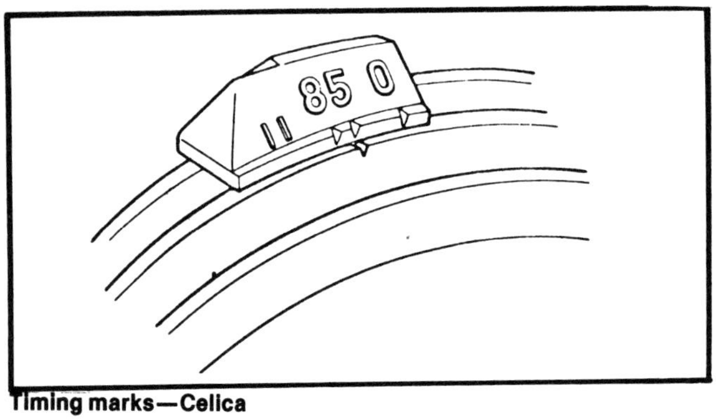

Timing marks—Celica

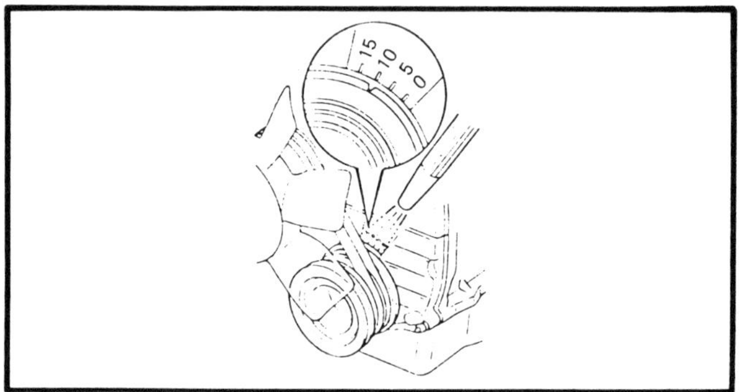

Timing marks—Celica Supra and Cressida

NOTE: A-series engines require a special type of tachometer which hooks up to the service connector wire coming out of the distributor. As many tachometers are not compatible with this hookup, consult with the tachometer manufacturer's instructions to make sure the unit will work on this type of system.

2. Connect a timing light to the engine according to the manufacturer's instructions. If the timing marks are difficult to see, use chalk or a dab of paint to make them more visible.
3. Disconnect the vacuum line from the distributor vacuum advance and plug the line. If a vacuum advance/retard unit is used, disconnect and plug both hoses.
4. Start and run the engine at idle with the transmission in Neutral on manual models, or Drive on automatic models. Make sure the parking brake is securely set and the wheels are chocked before allowing the car to idle in gear.
5. Point the timing light at the timing marks. With the engine at idle, timing should be set at the specifications given in the tune-up chart at the beginning of this section, or according to the values listed on the underhood emission control sticker. If not, loosen the pinch bolt at the base of the distributor, then rotate the distributor to advance or retard the timing as required.

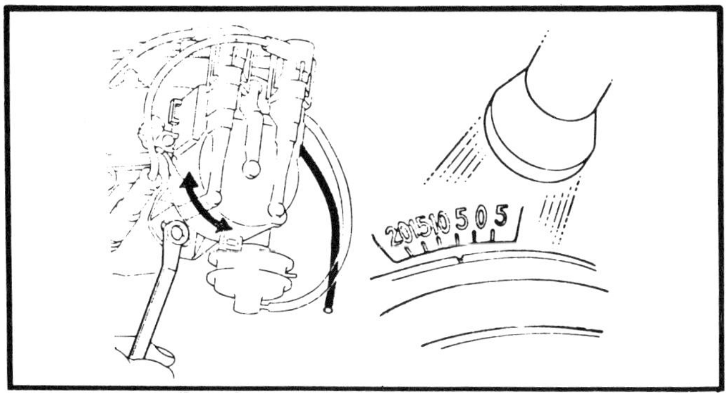

Timing marks—Tercel

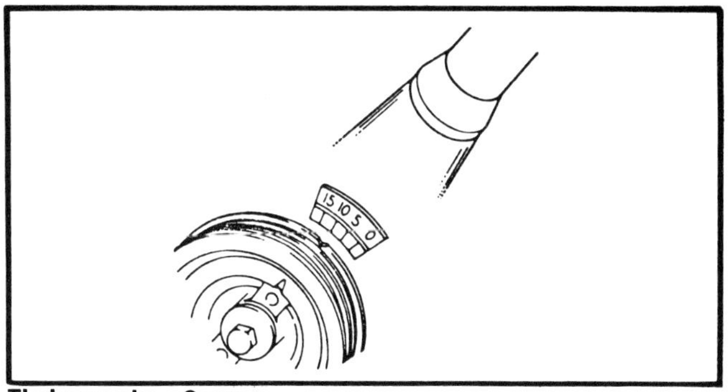

Timing marks—Camry

6. Stop the engine and tighten the pinch bolt. Start the engine and recheck the timing.

7. Stop the engine and disconnect the timing light and the tachometer. Reconnect the vacuum lines to the distributor advance unit.

Diesel Injection Timing

1C-L, 1C-LC and 1C-TL ENGINES

NOTE: This procedure requires the use of a plunger stroke measuring tool and dial indicator.

1. Remove the injection pump head bolt and install stroke measuring tool 09275-54010 or equivalent, along with the dial indicator.
2. Rotate the engine in the normal direction of rotation to set No. 1 cylinder to approximately 25–30°BTDC on the compression stroke.
3. Use a screwdriver to turn the cold start lever 20° counterclockwise, then place a metal plate 0.335–0.394 in. (8.5–10mm) thick between the cold start lever and thermo wax plunger.
4. Zero the dial indicator, then check to make sure the indicator remains at zero while rotating the crankshaft pulley slightly to the left and right.
5. Slowly rotate the crankshaft pulley until the No. 1 cylinder comes to TDC/compression, then measure the plunger stroke. It should be 0.0315 in. (0.80mm), as read on the dial indicator.
6. To adjust the injection timing, loosen the four injection lines and the union bolt of the fuel inlet line. Loosen the injection pump mounting bolts and nuts.
7. Adjust the plunger stroke by slightly tilting the injection pump body. If the stroke is less than specifications, tilt the pump toward the engine. If greater than specifications, tilt the pump away from the engine.
8. Once the pump stroke is within specifications (as described in Step 5), tighten the injection pump mounting bolts and nuts. Torque the bolts to 34 ft. lbs. (47 Nm) and the nuts to 13 ft. lbs. (18 Nm). Torque all union nuts and bolts to 22 ft. lbs. (29 Nm).
9. Remove the metal plate from the cold start lever and the pump stroke measuring tool from the injection pump. Install the distributor head bolt and torque to 12 ft. lbs. (17 Nm). Replace the head bolt washer when installing. Bleed any air from the injection pump by cranking the starter motor, then start the engine and check for leaks.

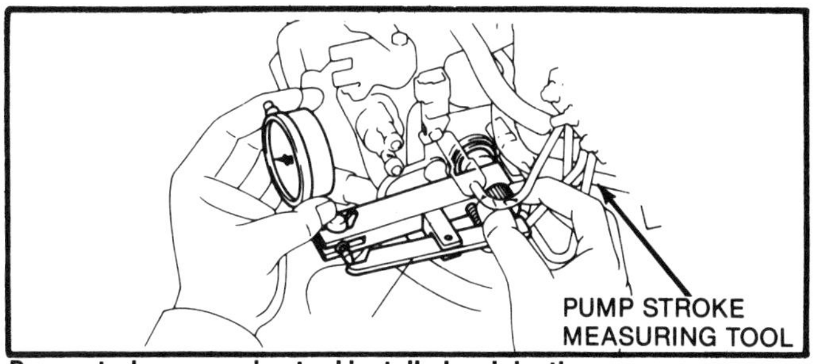

Pump stroke measuring tool installed on injection pump

Timing mark alignment at 25-30°BTDC

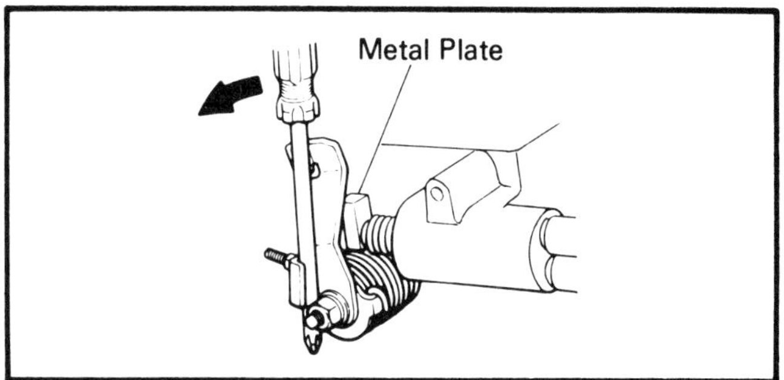

Install plate on cold start lever as shown

SPARK PLUGS

Engine	Type	Gap
2S-E	BPR5EA-L11	.043
22R-E	BPR5EY	.031
3A	BPR5EY	.031
3A-C	BPR5EY-11 ①	.043
3Y-EC	BPR5EP-11	.043 ②

SPARK PLUGS

Engine	Type	Gap
4A-C, 4A-CL	BPR5EY-11 ③	.043
4A-GE	BCPR5EP-11	.043
5M-GE	BPR5EP-11	.043

① Canada wagon w/MT: BPR5EY—.031 in. gap
② .055 in. maximum
③ Canada: BPR5EY—.031 in. gap

EMISSION CONTROL SYSTEMS

CRANKCASE EMISSION CONTROLS

Positive Crankcase Ventilation (PCV) System

A positive crankcase ventilation system is used on all Toyota engines sold in the United States. Blow-by gases are routed from the crankcase to the carburetor or intake manifold (fuel injection models), where they are combined with the air/fuel charge and burned in the combustion process.

On carbureted engines, a PCV valve is used in the line to prevent the gases in the crankcase from being ignited in case of a backfire, and to regulate the amount of blow-by gases entering the fuel mixture. On fuel injected engines, a pair of fixed orifices in the PCV line regulates the blow-by flow.

Maintenance and Testing

Every 30,000 miles, the PCV system should be checked for tight hose connections, leaks, cracks or deterioraton. Any hoses found to be leaking should be replaced. To check the PCV valve operation, remove the valve from the valve cover and blow through both ends. When blowing from the intake manifold side, very little air should pass through the valve. When blowing from the crankcase (valve cover) side, air should pass through freely. If the valve fails either of these tests, it should be replaced. Do not attempt to clean the PCV valve. On fuel injected engines, simply clean the calibrated orifices with solvent and blow dry with compressed air.

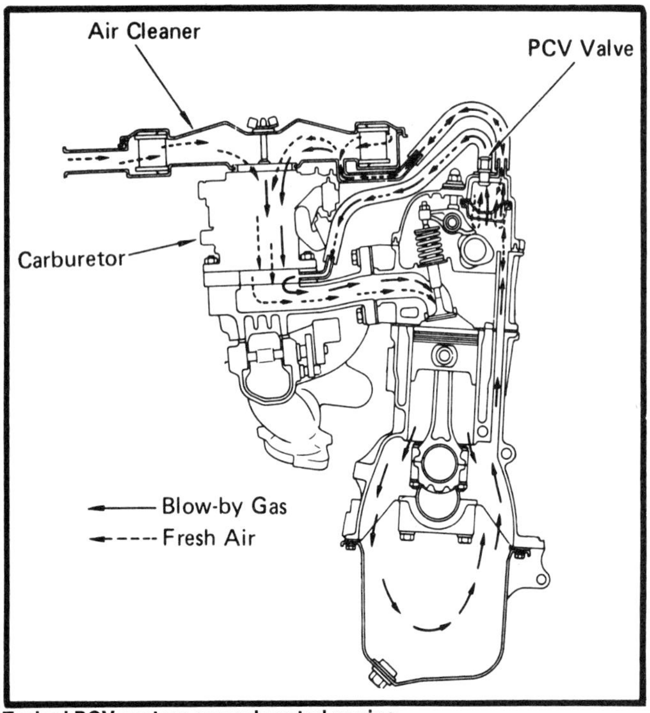

Typical PCV system on carbureted engine

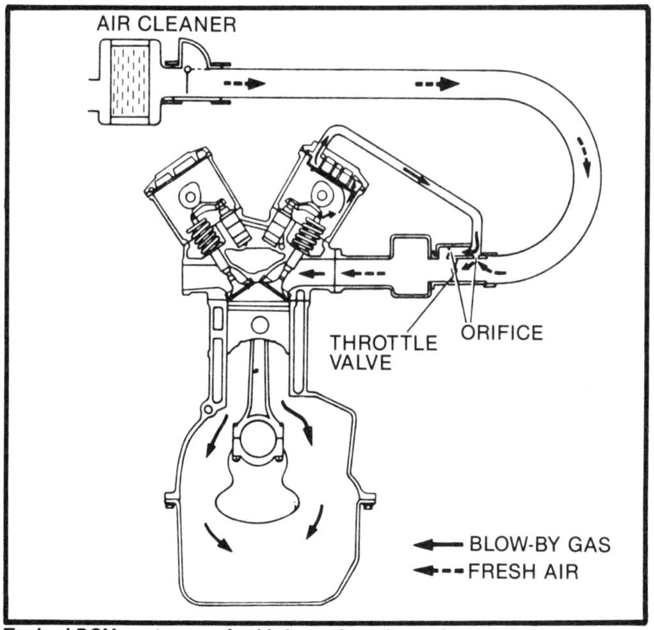

Typical PCV system on fuel injected engine—Supra shown

FUEL EVAPORATIVE EMISSION CONTROL SYSTEM

To reduce hydrocarbon (HC) emissions, evaporated fuel from the fuel tank and float chamber (on carbureted models) is routed through the charcoal canister to the intake manifold for combustion in the engine. A fuel filler cap with a check valve allows air to enter the fuel tank as the fuel is used to prevent a vacuum build-up and equalize the pressure, but does not allow fuel vapors to escape to the atmosphere. When the engine coolant temperature reaches 129° F., the vacuum switching valve opens to allow vapors trapped in the charcoal canister to enter the intake

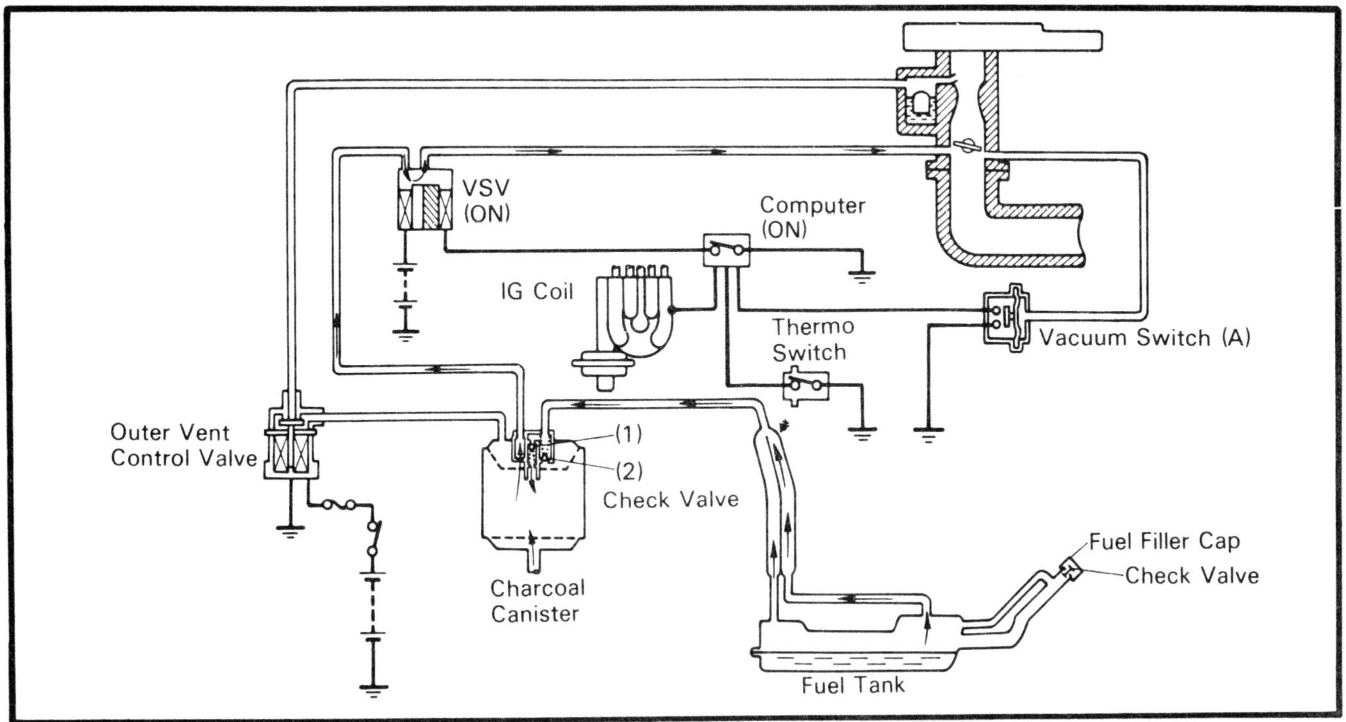

Typical EVAP system on carbureted engine—Tercel shown

manifold, where they are drawn into the combustion chambers and burned with the fuel charge.

Maintenance and Testing

The EVAP system should be checked for proper operation every 60,000 miles. Check all hoses and connections for leaks or deterioration and remove the charcoal canister to inspect for cracks or damage. Check for a clogged filter or frozen check valve by blowing low pressure compressed air into the tank and check that air flows freely to the canister port. Blow into the charcoal canister purge pipe and make sure no air flows from the tank connection or from the fresh air inlet at the bottom of the canister. If it does, replace the canister. Clean the canister filter by blowing 43 psi of compressed air into the tank pipe while holding the purge connection closed. Do not attempt to wash the canister in any way.

Check the BVSV operation by allowing the engine to cool completely and blowing through the valve. When cold, the valve should be closed (no air should pass). Start and warm up the engine, then blow through the valve again. Air should now pass freely (switch open). Any results other than these, replace the BVSV. On carbureted engines, an electric vacuum switching valve (VSV) is used. Check that the VSV is open (air passes through) when battery voltage is applied to the terminals, and closed when voltage is removed. If not, replace the VSV.

The outer vent control valve (carbureted engines only) may be checked by disconnecting the hoses and making sure air passes through the valve when the ignition key is OFF and doesn't pass through when the ignition is switched ON. If the valve does not operate, check the fuse and wiring connections.

The thermo switch is tested by allowing the engine to cool completely and checking for continuity across the thermo switch terminals. Start and warm up the engine and check that the thermo switch is open (no continuity) at normal operating temperature. If not, replace the thermo switch.

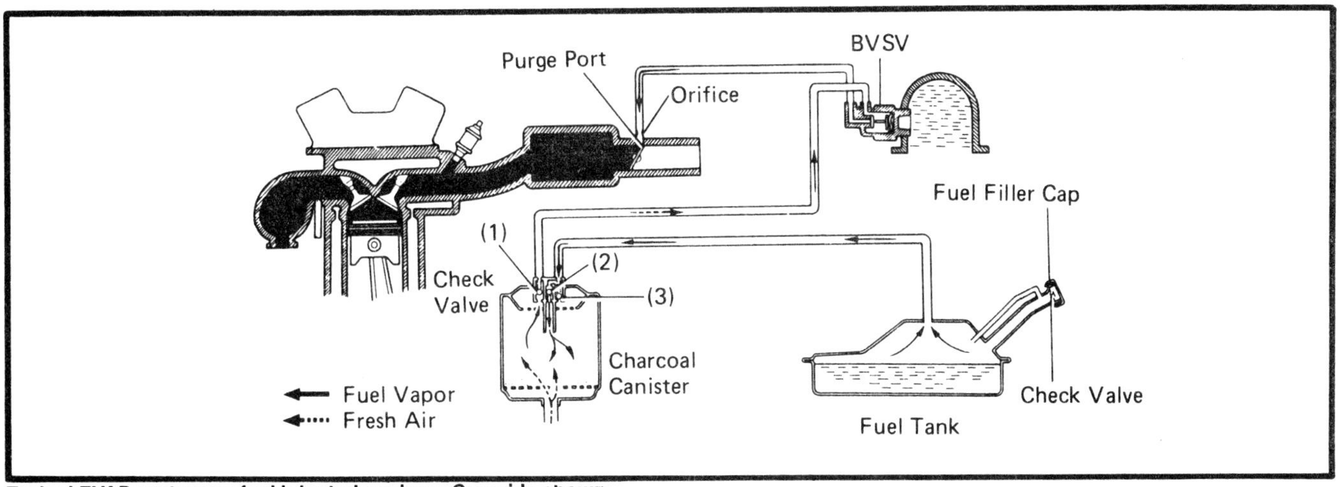

Typical EVAP system on fuel injected engine—Cressida shown

EXHAUST EMISSION CONTROLS

Air Suction System

On carbureted engines, fresh air is drawn through the filter and air suction valve, then into the catalytic converter (Federal models) or No. 3 branch of the exhaust manifold (California models) to burn excess HC and CO emissions. On Canadian vehicles, the fresh air is directed to the exhaust manifold. The system consists of an air suction valve mounted on the air cleaner, with a vacuum-operated shut-off valve fitted to California and Canada models. An in-line vacuum transducer valve (VTV) is also used on models equipped with a check valve.

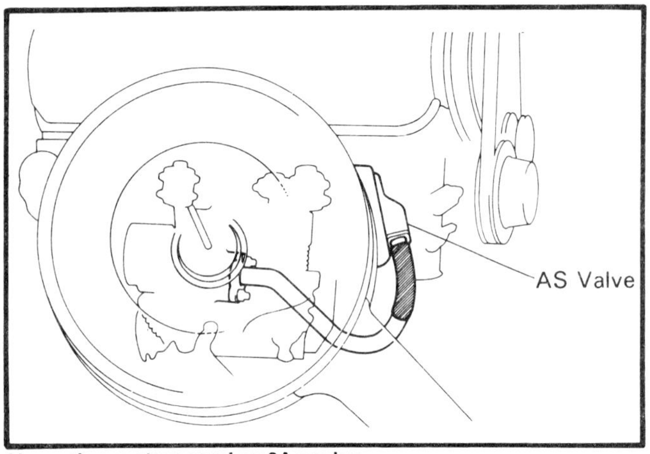

Air suction system used on 3A engine

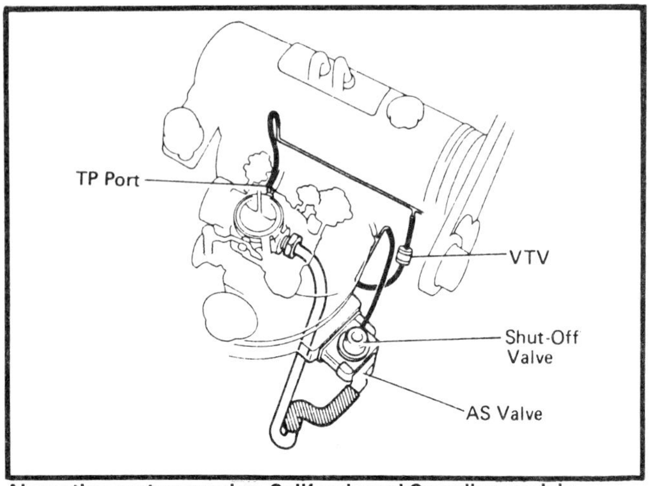

Air suction system used on California and Canadian models

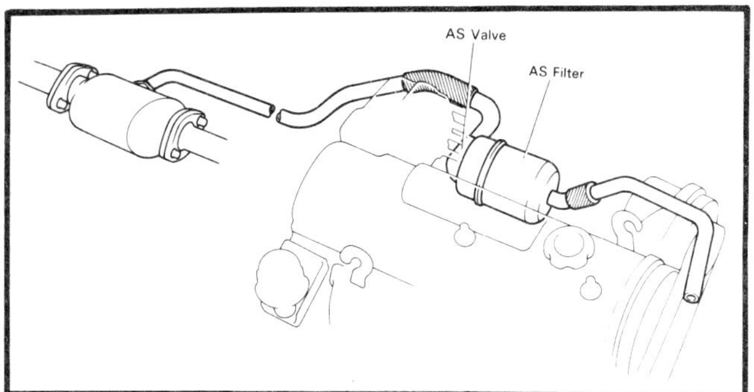

Air suction system used on Corolla RWD models

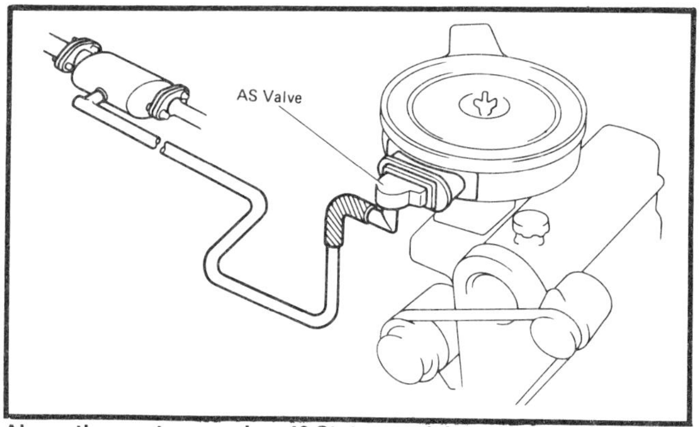

Air suction system used on 49 States models

Testing

The air suction system can be checked for proper operation by simply removing the air cleaner cover and listening for a bubbling sound with the engine at idle. On models with a check valve, disconnect the vacuum line and start the engine. Reconnect the vacuum line and listen for the bubbling sound within 2–6 seconds. If the noise is audible, the air suction system is working properly.

Throttle Positioner System

The throttle positioner opens the throttle valve slightly more than at idle when decelerating to reduce HC and CO emissions from the temporarily rich fuel mixture. The system consists of a dual-diaphragm throttle positioner, vacuum switching valve and vacuum transducer (delay) valve. The switching valve is electrically operated and performs an "idle-up" function to raise the engine rpm when energized.

Testing

1. Start the engine and allow it to reach normal operating temperature.
2. Check and adjust the idle speed, if necessary.
3. Disconnect the vacuum hose from the "M" port of the thermal vacuum switching valve (TVSV) and plug it. This shuts off the choke opener and EGR system.

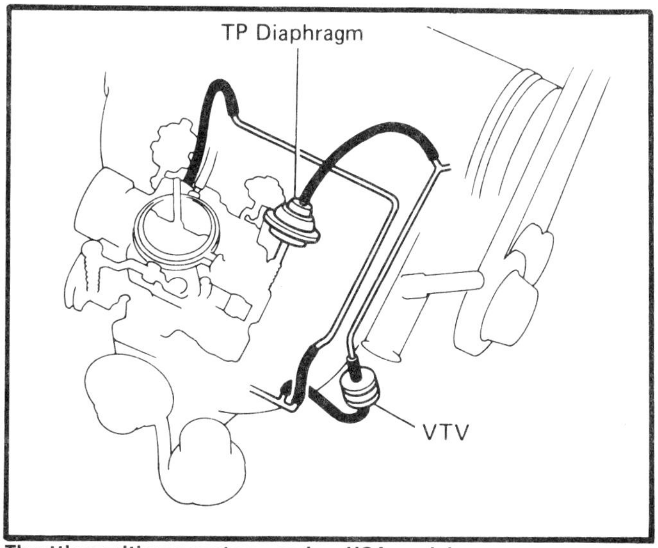

Throttle positioner system used on USA models

4. Disconnect the top vacuum hose from the throttle positioner diaphragm and make sure the throttle positioner is set at the first step (idle-up) position. The engine should idle at 800 rpm (900 rpm with automatic trans). If not, adjust the idle by turning the adjusing screw.

NOTE: Make all adjustments with the cooling fan off.

5. Disconnect the lower vacuum hose from the throttle positioner diaphragm and plug the hose end. Make sure the positioner is on the second step. Engine rpm should rise to 1100–1500 rpm (1200–1600 rpm with automatic trans).

6. Reconnect the lower vacuum hose and check that the engine returns to the first step setting within 2–6 seconds. If so, reconnect all vacuum hoses.

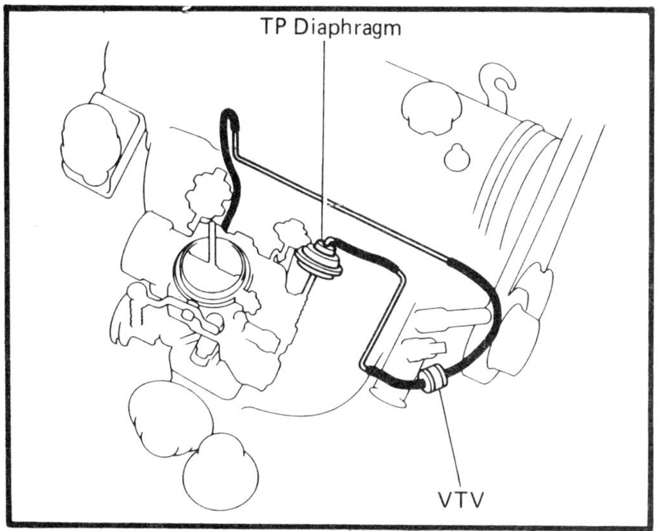

Throttle positioner system used on Canadian models

Dash Pot (DP) System

On fuel injected models, a dash pot system performs the same function as the throttle positioner described above. When decelerating, the dash pot opens the throttle valve slightly to lean out the temporarily rich fuel mixture and reduce the amount of HC and CO emissions. The system consists of the dash pot and a vacuum transducer valve with an air filter.

Testing

1. Start the engine and allow it to reach normal operating temperature.
2. Check the idle speed and adjust, if necessary.
3. Raise and hold the engine speed at 3000 rpm, then pinch the vacuum hose to the dash pot. Release the throttle valve and check that the dash pot is set at 2000 rpm.
4. If not, adjust using the dash pot adjusting bolt.
5. Release the pinched hose and check that the engine returns to idle speed in about one second. If it does, the system is working properly and no further testing is required.
6. Clean the filter on the VTV by using compressed air.

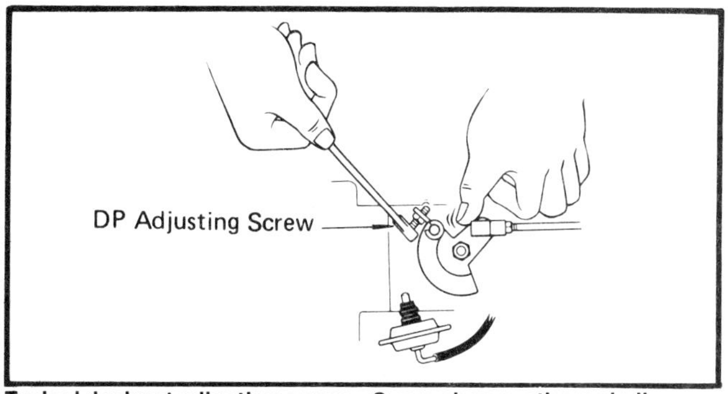

Typical dashpot adjusting screw—Supra shown, others similar

Exhaust Gas Recirculation (EGR) System

The EGR system recirculates part of the exhaust gas to lower the combustion chamber temperatures and reduce NOx emissions. The EGR system should be checked for proper operation every 60,000 miles. Check the filters on the vacuum modulator for contamination or damage and replace as necessary. Clean the filters with compressed air.

Testing

CARBURETED ENGINES

1. Using a three-way-connector, tee a vacuum gauge into the hose between the EGR valve and vacuum pipe. Make sure the engine starts and runs at idle.

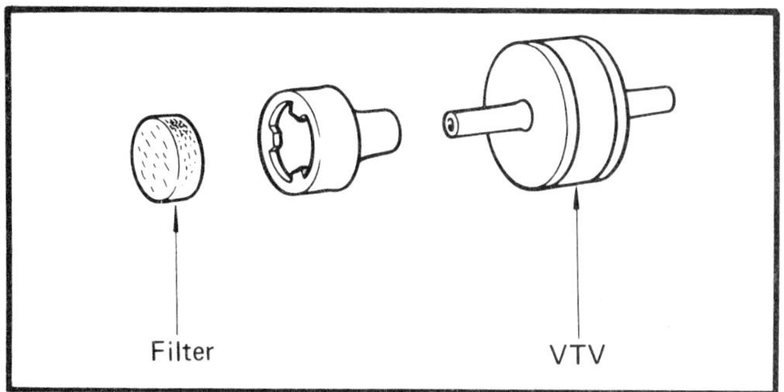

VTV filter

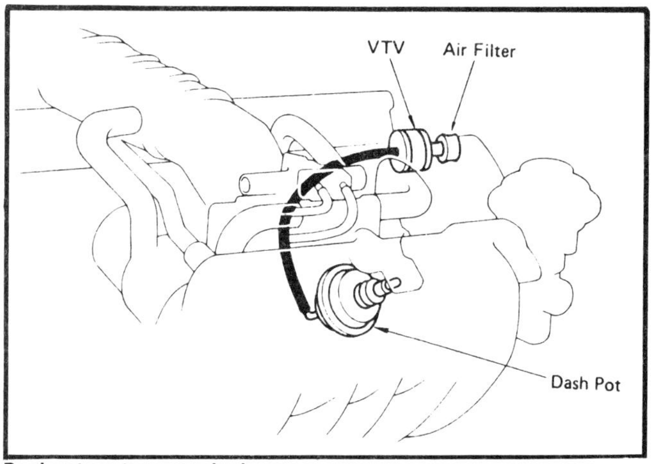

Dashpot system—typical

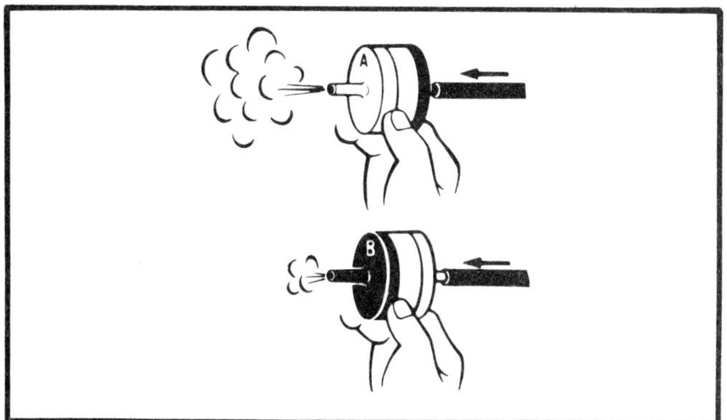
VTV testing

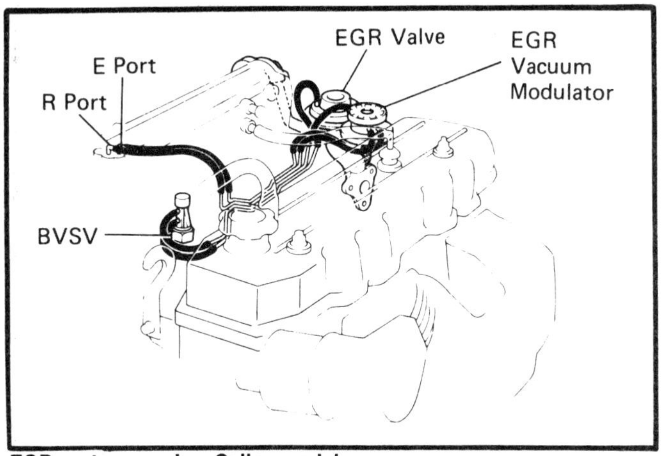

EGR system used on Celica models

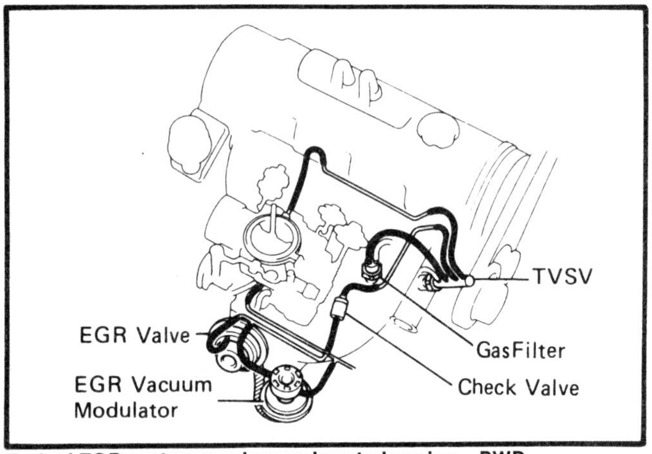

Typical EGR system used on carbureted engine—RWD Corolla shown

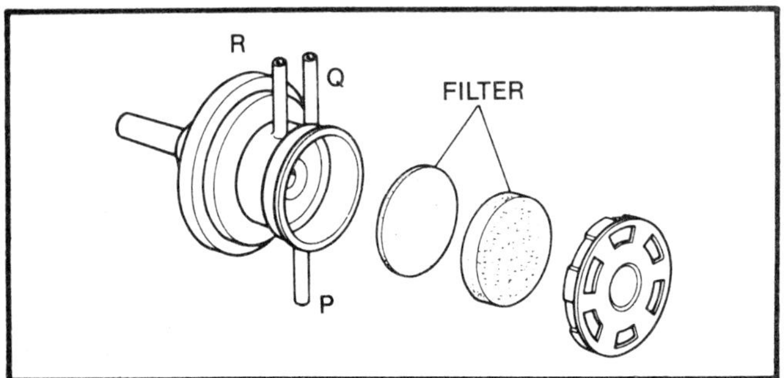

Typical EGR vacuum modulator showing port locations and filter

2. With the coolant temperature below 122°F (50°C), the gauge should read zero vacuum at 2000 rpm. If not, replace the thermal vacuum switching valve (TVSV).

3. Allow the engine to reach normal operating temperature, then check that the vacuum gauge indicates low vacuum at 2000 rpm.

4. Disconnect the vacuum hose from the "R" port of the EGR vacuum modulator and connect the "R" port directly to the intake manifold with another hose. The vacuum gauge should now indicate high vacuum at 2000 rpm.

NOTE: The engine should misfire slightly as the large amount of EGR gas enters the engine.

5. Disconnect the vacuum gauge and reconnect the vacuum hoses to their proper connections.

6. Apply vacuum directly to the EGR valve (using a hand vacuum pump) with the engine idling. It should stall. If not, remove the EGR valve from the engine and check the valve for sticking and heavy carbon deposits. Clean any light deposits, replace the EGR valve and gasket and retest. If heavy deposits are noted, replace the EGR valve and gasket.

7. Disconnect the vacuum hoses from ports P, Q and R of the EGR vacuum modulator. Plug ports Q and R, then blow air into port P and check that air passes through to the air filter side freely.

8. Start the engine and hold the engine speed at 2000 rpm. Repeat the above test and check that there is now strong resistance to air flow. If so, reconnect the vacuum hoses. If not, replace the EGR vacuum modulator and retest.

9. Inspect the check valve operation by blowing through it. Air should flow from the orange to black side, but not from black to orange. If not, replace the check valve.

FUEL INJECTED ENGINES

1. Disconnect the vacuum hose from the EGR valve and tee a vacuum gauge in-line.

2. Start the engine and make sure it idles smoothly. With the engine cold (below 135°F), the vacuum gauge should read zero at 2500 rpm on 5M-GE engines (3500 rpm on 4A-GE engines).

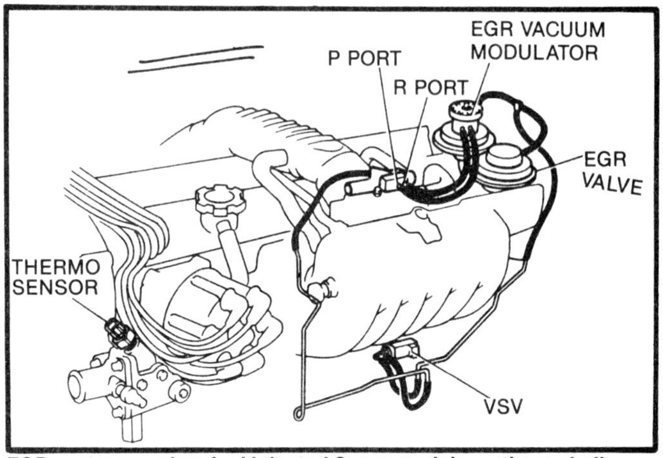

EGR system used on fuel injected Supra models—others similar

3. Allow the engine to reach normal operating temperature and check that the vacuum gauge now indicates about 3 in.Hg. at 2500 rpm (3500 rpm on 4A-GE engines). On 4A-GE engines, the vacuum gauge should read zero at 5000 rpm.

4. Disconnect the vacuum hose from the "R" port of the vacuum modulator, then connect the port directly to the intake manifold using another hose. The vacuum gauge should read high vacuum at 2500 rpm (3500 rpm on 4A-GE engines).

NOTE: The engine should misfire slightly as a large amount of EGR gas enters the engine.

5. If so, the EGR system is working properly. Disconnect the vacuum gauge and restore all disconnected vacuum hoses to their proper connections.

6. Check the EGR vacuum modulator operation by disconnecting the hoses from ports P, Q and R. Plug ports P and R, then blow air into port Q and check that air passes freely. Start the engine and hold the engine speed at 2500 rpm. Repeat the test and check that there is now a strong resistance to air flow. If not, replace the EGR vacuum modulator and retest. If so, restore the vacuum hoses to their proper connections.

7. Check the vacuum switching valve (VSV) operation by applying battery voltage to the terminals and blowing through the valve. On 5M-GE engines, the VSV should be open (air passes through) when energized and closed (no air passage) when power is removed. On 4A-GE engines, air should pass between the two vacuum ports when the VSV is energized, and from the back port to the air filter when power is removed. Any results other than these, replace the VSV and retest.

Carburetor Feedback System

This system uses the voltage signal from an oxygen sensor to control carburetor primary side main air bleed and slow air bleed volume, thereby maintaining optimum air/fuel mixture in relation to driving conditions. The oxygen sensor should be checked for proper operation every 30,000 miles and replaced if necessary.

TVSV WITH COLD ENGINE

Testing

1. Make sure the engine is cold (coolant below 45°F).
2. Disconnect the vacuum hose from vacuum switch B.
3. Start the engine and check that no vacuum is left in the disconnected vacuum hose.
4. Reconnect the vacuum hose.

EBCV WITH HOT ENGINE

Testing

1. Start the engine and allow it to reach normal operating temperature.
2. Disconnect the EBCV connector.
3. Hold the engine speed at 2500 rpm and reconnect the connector. The engine speed should drop 300 rpm momentarily.
4. Allow the engine to idle and once again disconnect and reconnect the EBCV connector. The engine speed should not change.
5. Disconnect the vacuum hose from vacuum switch B.
6. Repeat Steps 2 and 3 and make sure the engine speed does not change.
7. If the system does not check out as described, perform the component tests outlined below.

EBCV

Testing

Using an ohmmeter, check that there is no continuity between the positive (+) terminal and the EBCV body. If there is continuity, replace the EBCV. Check for an open circuit by measuring the resistance between the positive (+) and negative (–) terminals; it should be 11–13 ohms at 68°F (20°C). If the resistance does not fall within the specified range, replace the EBCV.

VACUUM SWITCH

Testing

Using an ohmmeter, check for continuity between the switch terminal and the switch body. There should be continuity with the engine cold on vacuum switch A and no continuity on vacuum switch B. Start and warm up the engine, then check again for continuity. Switch B should now show continuity and switch A should show none. Any results other than these, replace the vacuum switch and retest.

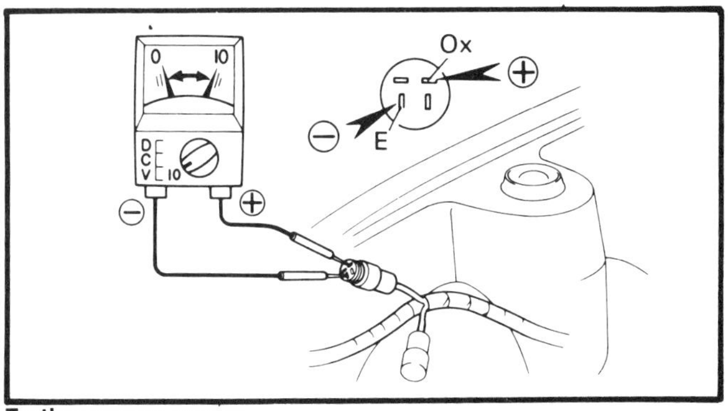

Testing oxygen sensor

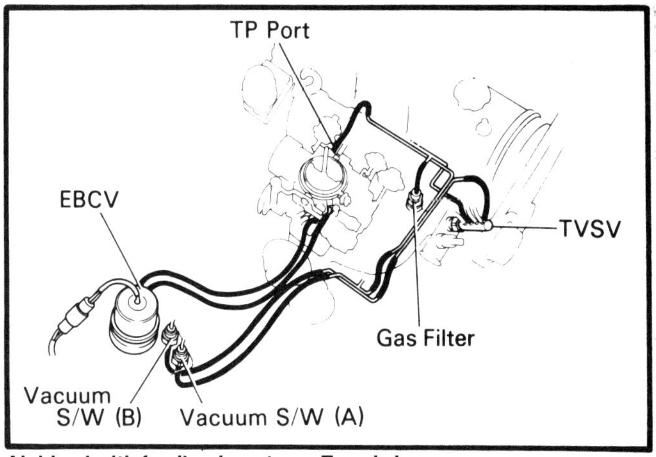

Air bleed with feedback system—Tercel shown

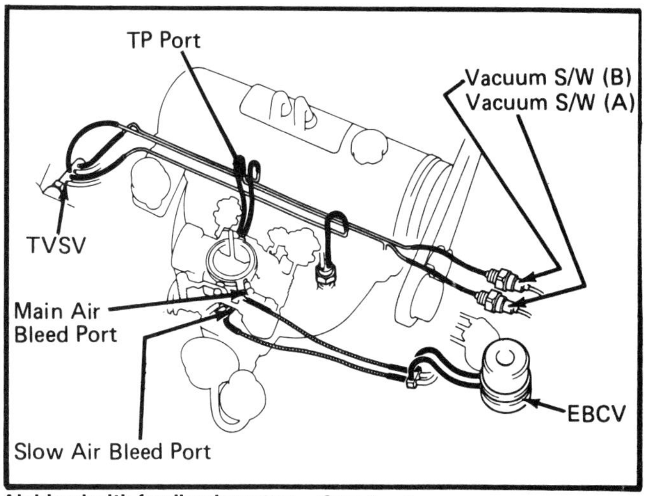

Air bleed with feedback system—Corolla shown

OXYGEN SENSOR

Testing

1. With the engine at normal operating temperature, connect a voltmeter to the service connector located on the right fender apron below the wiper motor or EBCV. Connect the positive (+) probe to the sensor terminal and the negative (–) probe to terminal E.
2. Raise the engine speed to 2500 rpm for about 90 seconds.
3. Maintain the engine speed at 2500 rpm and check that the needle of the voltmeter fluctuates 8 times or more in 10 seconds within 0–7 volts.

NOTE: If this test is positive, the oxygen sensor is functioning properly. If not, inspect the other parts, hose connections and wiring of the air bleed system. If no problem is found, replace the oxygen sensor.

High Altitude Compensation (HAC) System

This system is used on carbureted 49 States models only to compensate for the rich air/fuel mixture at high altitudes. In operation, it supplies additional air to the primary low and high speed circuit of the carburetor and advances the ignition timing to improve driveability at elevations above 3,930 ft. The air cleaner in the HAC valve should be visually checked and cleaned periodically and replaced if excessively contaminated with dust and dirt.

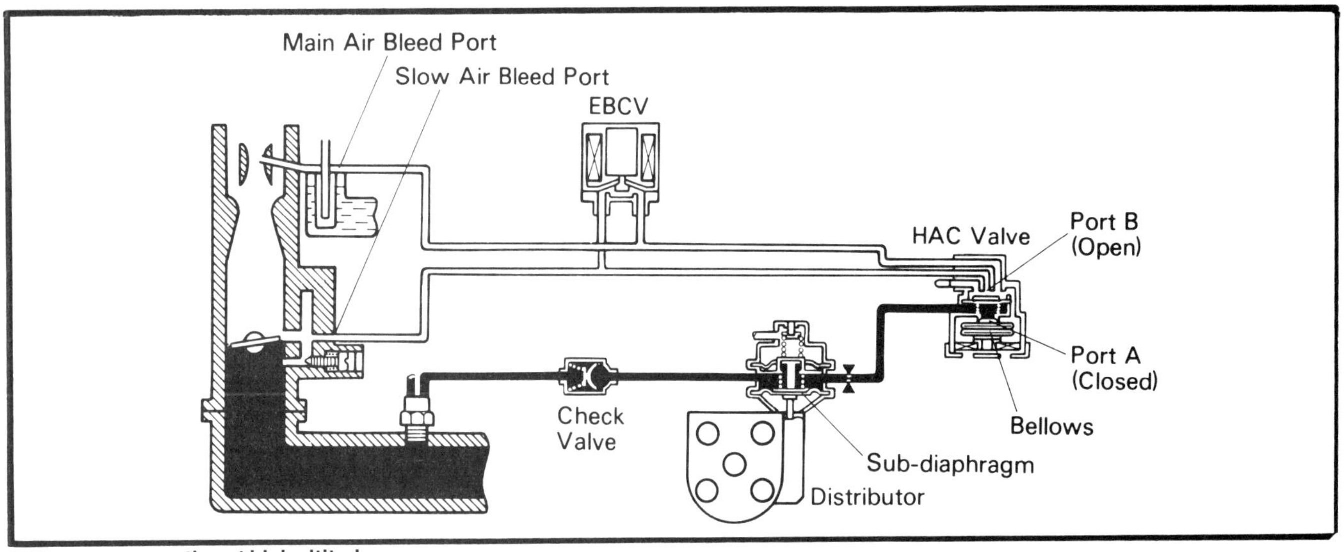

HAC system operation at high altitude

Testing

1. If located in a high altitude area (elevations above 3,930 ft.), blow into either of the two ports on top of the HAC valve with the engine idling and make sure the HAC valve is open. If the local elevation is below 2,570 ft., the valve should be closed when air is blown through either port.

2. Disconnect the vacuum hose from the distributor sub-diaphragm and plug the hose end. Check the ignition timing; it should be 5°BTDC

950 rpm. Reconnect the vacuum hose to the distributor and again check the ignition timing; it should now be approximately 13°BTDC
950 rpm.

3. Disconnect the vacuum hose from the check valve at the black side and plug the hose end. Check that the ignition timing remains stationary for more than one minute. If so, stop the engine and reconnect the vacuum hoses to their proper locations.

4. Disconnect the two hoses on top of the HAC valve. Blow air into each hose and check that air flows freely into the carburetor. If not, check the hoses for blockage and replace as necessary. If so, reconnect the hoses.

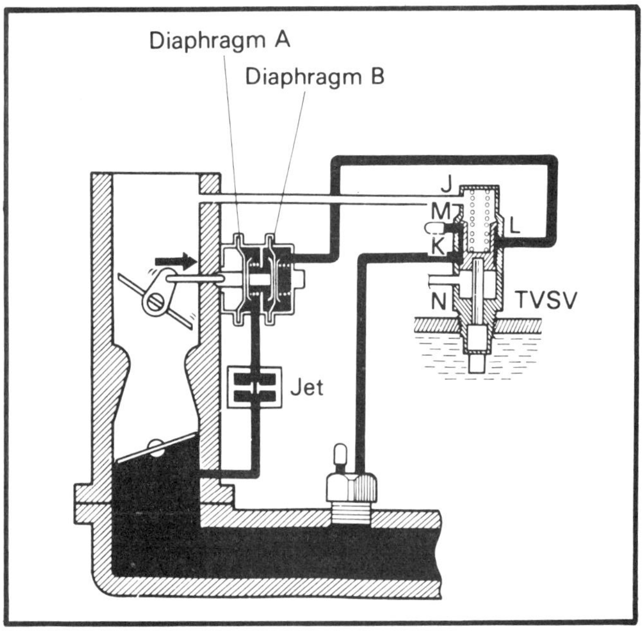

Choke breaker operation with hot engine—typical

Hot Idle Compensation (HIC) System

This system allows the air controlled by the HIC valve to enter the intake manifold to maintain proper air-fuel mixture during high temperatures at idle.

Testing

Check that air flows from the HAI diaphragm side to the carburetor side while closing the atmospheric port and that air does not flow from the carburetor side to the HAI diaphragm side. Below 72°F (22°C), make sure that air does not flow from the HAI diaphragm side to the atmospheric port while closing the intake manifold side. Start and warm up the engine to heat the HIC valve to above 84°F (29°C) and make sure that air flows from the HAI diaphragm side to the atmospheric port while closing the intake manifold side.

Choke Breaker (CB) System

When the choke is closed, this system opens the choke valve slightly to prevent too rich a mixture and forcible opens the choke when the engine warms up.

Testing

1. Start the engine and, with the coolant temperature below 45°F (7°C), disconnect the vacuum hose from choke breaker diaphragm B and make sure the choke linkage does not move. Reconnect the vacuum hose.

2. Disconnect the vacuum hose from choke breaker diaphragm A and make sure the linkage moves. Reconnect the vacuum hose and verify that the choke linkage moves within 1–5 seconds.

3. After the engine warms up, disconnect the vacuum hose from diaphragm B and make sure the choke linkage returns. Reconnect the vacuum hose. If no problem is found, the system is operating properly.

Choke Opener System

After warm-up, this system forcibly holds the choke valve open to prevent an over rich mixture and releases the fast idle to the third step to lower engine rpm.

Testing

TVSV WITH COLD ENGINE

1. Disconnect the vacuum hose from the choke opener diaphragm.

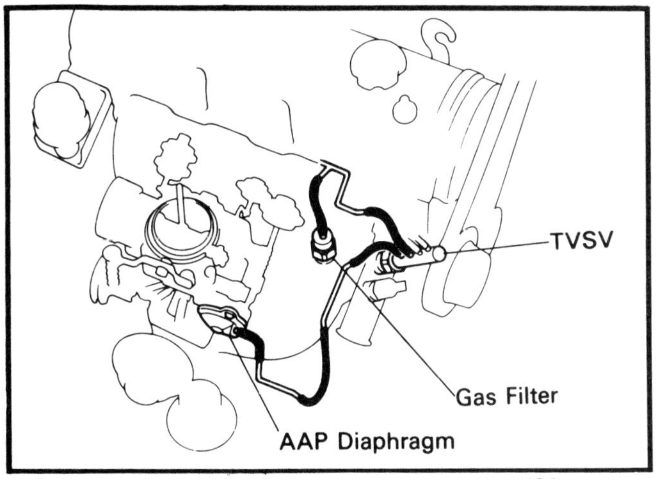

Auxiliary Acceleration Pump System (AAP)—Corolla, USA

2. With the coolant temperature below 122°F, press the accelerator down, then release it.

3. Start the engine, then reconnect the vacuum hose and check that the choke linkage does not move.

TVSV WITH WARM ENGINE

1. Warm the engine to normal operating temperature, then stop the engine.

2. Disconnect the vacuum hose from the choke opener diaphragm.

3. Set the fast idle cam by holding the throttle slightly open, pushing the choke valve closed and releasing the throttle valve.

4. Start the engine without touching the accelerator pedal.

5. Reconnect the vacuum hose and check that the choke linkage moves and that the fast idle cam is released to the third step.

6. If no problem is found, the system is operating properly. Any results other than these, check that the choke linkage moves in accordance with applied vacuum at the diaphragm and the TVSV operation as outlined under "EGR System Tests."

Auxiliary Accelerator Pump (AAP) System

When accelerating with a cold engine, the main accelerator pump capacity is insufficient to provide good acceleration, since the air/fuel mixture is very lean. The AAP system compensates for this by forcing more fuel into the accelerator nozzle to obtain better cold engine performance.

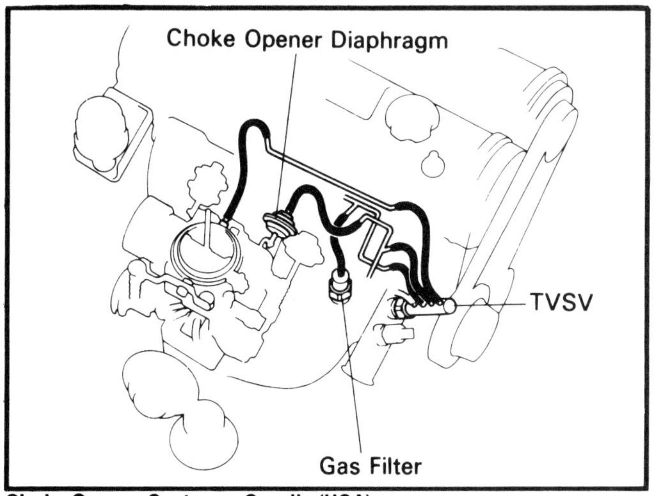

Choke Opener System—Corolla (USA)

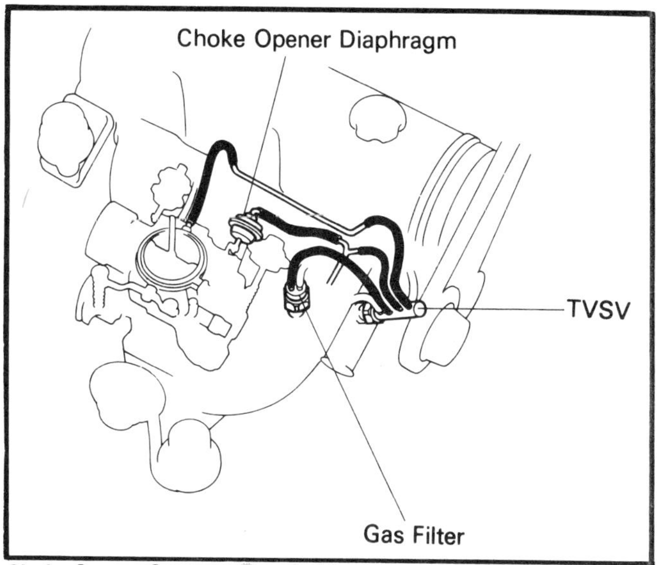

Choke Opener System—Tercel with 4-speed

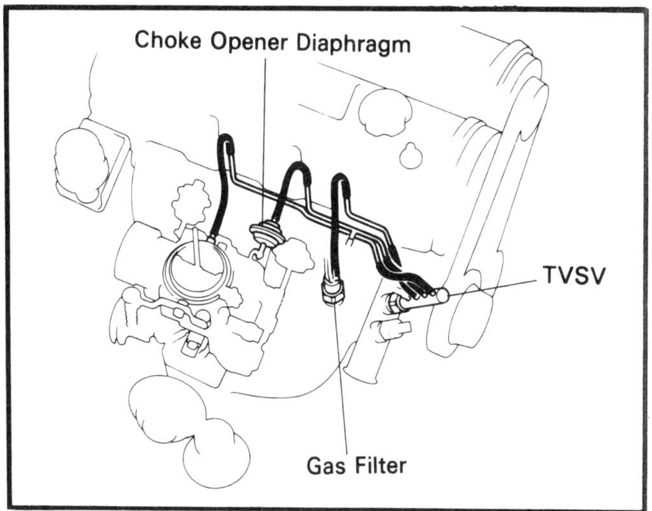

Choke Opener System—Corolla and Tercel without 4-speed (Canada)

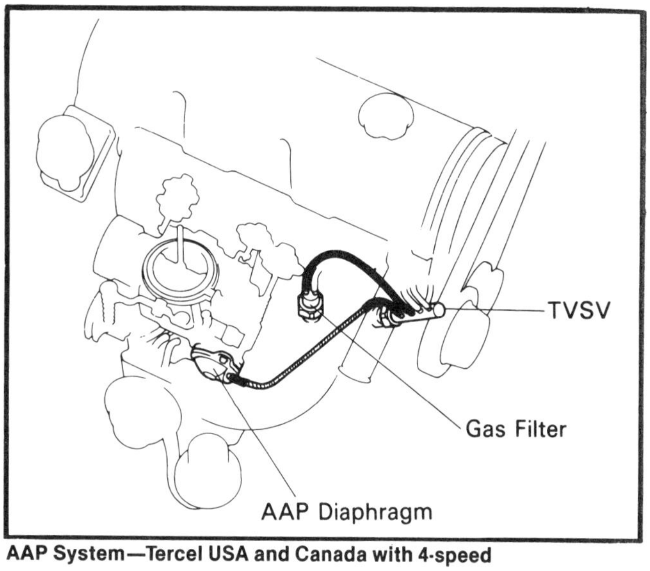

AAP System—Tercel USA and Canada with 4-speed

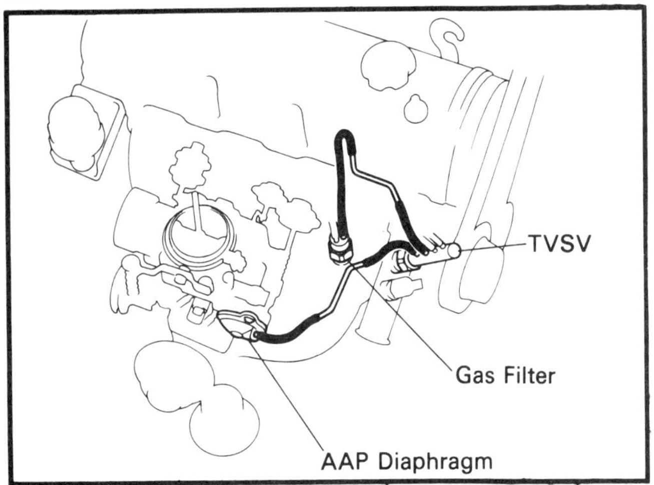

AAP System—Canada except Tercel without 4-speed

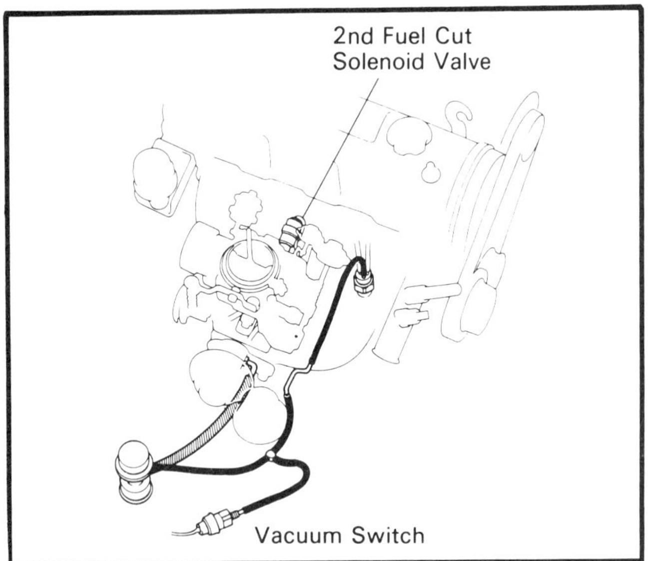

Deceleration Fuel Cut System—Tercel wagon (Canada)

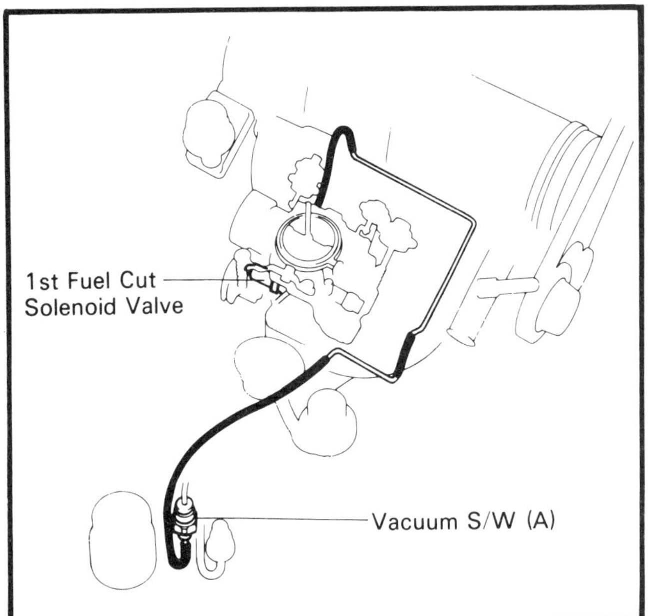

Deceleration Fuel Cut System—Tercel, 3A-C engine with 4-speed

Testing

COLD ENGINE

1. Make sure the coolant temperature is below 122°F.
2. Remove the air cleaner cover and start the engine.
3. Pinch the AAP hose and stop the engine.
4. With the engine off, release the hose and check that gasoline spurts from the accelerator pump nozzle.

WARM ENGINE

1. With the engine at normal operating temperature, repeat Steps 2 and 3 of the cold engine test.
2. With the engine off, release the hose and make sure gasoline does not spurt from the accelerator nozzle. If so, the test is complete. If not, continue testing.
3. Start the engine and disconnect the hose from the AAP diaphragm.
4. Apply and release the vacuum directly to the AAP diaphragm at idle and check that the engine rpm changes when vacuum is released. If so, reconnect the AAP hose. If not, replace the diaphragm and retest.
5. Test the TVSV operation as outlined under "EGR System Testing."

Deceleration Fuel Cut System

This system cuts off part of the fuel in the second slow circuit of the carburetor to prevent overheating and backfiring in the exhaust system.

Testing

USA MODELS

CAUTION

Perform this test quickly to avoid overheating of the catalytic converter.

1. Connect a tachometer to the engine.
2. Start the engine and make sure it runs normally.
3. Disconnect the vacuum switch connector.
4. Gradually increase engine speed to 2300 rpm and make sure engine speed fluctuates.
5. Reconnect the vacuum switch connector and again gradually increase engine speed to 2300 rpm; engine operation should return to normal. If no problem is found, test is complete. Any results other than these, continue testing.
6. Remove the solenoid valve and apply battery voltage to the terminals. You should feel a "click" from the solenoid valve as the voltage is applied and removed. If so, inspect the O-ring for damage and reinstall the valve. If not, replace the solenoid valve and retest.
7. Check vacuum switch A operation as outlined under "Carburetor Feedback System."

CANADIAN MODELS

1. With the engine cold (coolant below 45° F.), start the engine and disconnect the vacuum hose from the vacuum switch and plug the hose end.
2. Check that no "click" is felt from the second fuel cut solenoid valve when the vacuum hose is connected and disconnected at idle.

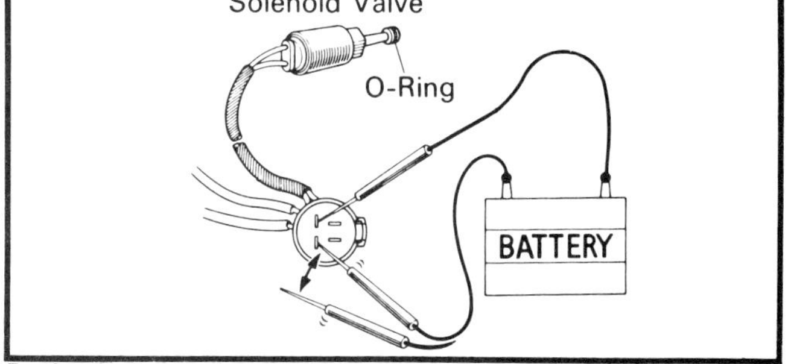

Testing fuel cut solenoid valve

3. Allow the engine to reach normal operating temperature.

4. Disconnect the vacuum hose from the vacuum switch and plug the hose end. Check that you can feel a "click" from the second fuel cut solenoid valve when the vacuum hose is connected and disconnected at idle.

5. If no problem is found, the test is complete. Any results other than these, check the solenoid valve and vacuum switch A as described in Steps 6 and 7 above. Check TVSV operation as described under "EGR System Testing."

Cold Mixture Heater (CMH) System

This system is used to reduce engine emissions and improve driveability by heating the intake manifold during cold engine operation and accelerate vaporization of the fuel.

Testing

1. Start the engine and use a voltmeter to check that there is voltage between the CMH positive (+) terminal (wire color WL) and the intake manifold with the coolant below 109°F (43°C). Insert the voltmeter probe from the rear side of the connector.

2. Allow the engine to reach normal operating temperature (coolant above 131° F.) and check that there is no voltage between the CMH positive (+) terminal and the intake manifold.

3. If test results are as described, the CMH system is operating properly and no further testing is required. Any results other than these, continue testing.

4. Check the CMH resistance by using an ohmmeter connected between the terminals of the CMH connector. Normal resistance is 0.5–2.0 ohms.

5. Check the CMH relay located on the left fender apron by verifying that continuity exists between terminals 1 and 2, but not between 3 and 4. Apply battery voltage between terminals 1 and 2, then verify that continuity now exists between terminals 3 and 4. Any results other than these, replace the CMH relay.

6. Check the thermo switch by allowing the engine to cool completely and checking for continuity across the thermo switch terminals. Start and warm up the engine and check that the thermo switch is open (no continuity) at normal operating temperature. If not, replace the thermo switch.

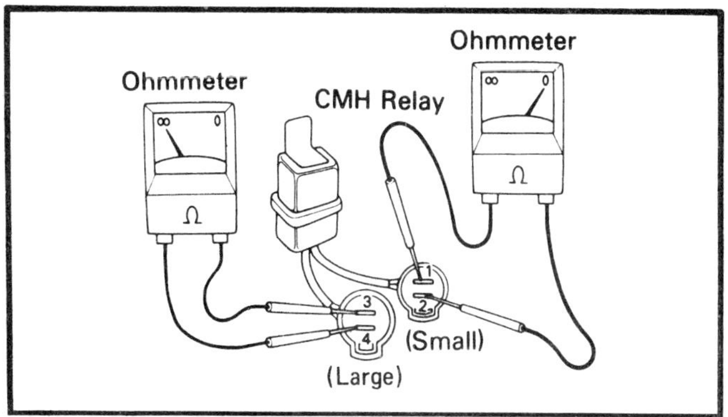

Testing CMH relay

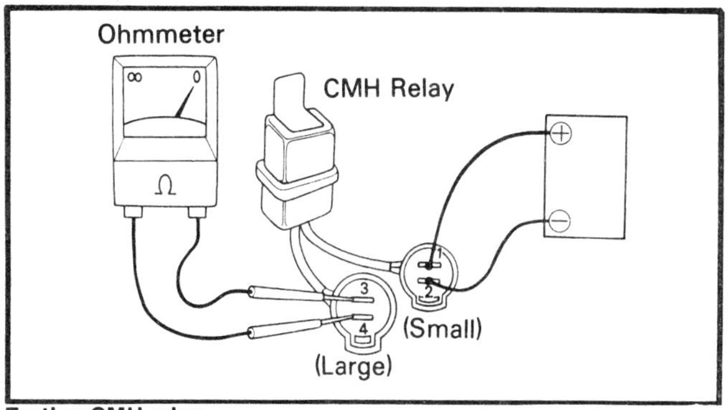

Testing CMH relay

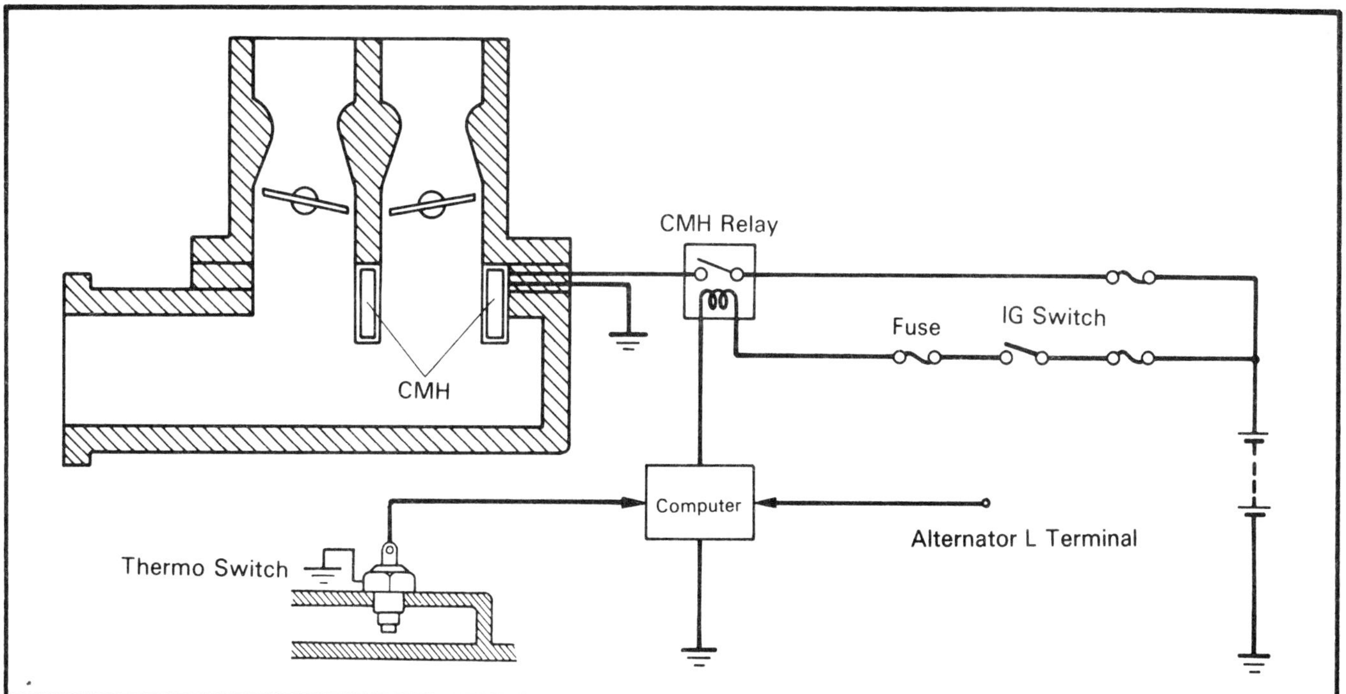

Cold Mixture Heater System (CMH)

VACUUM CIRCUITS

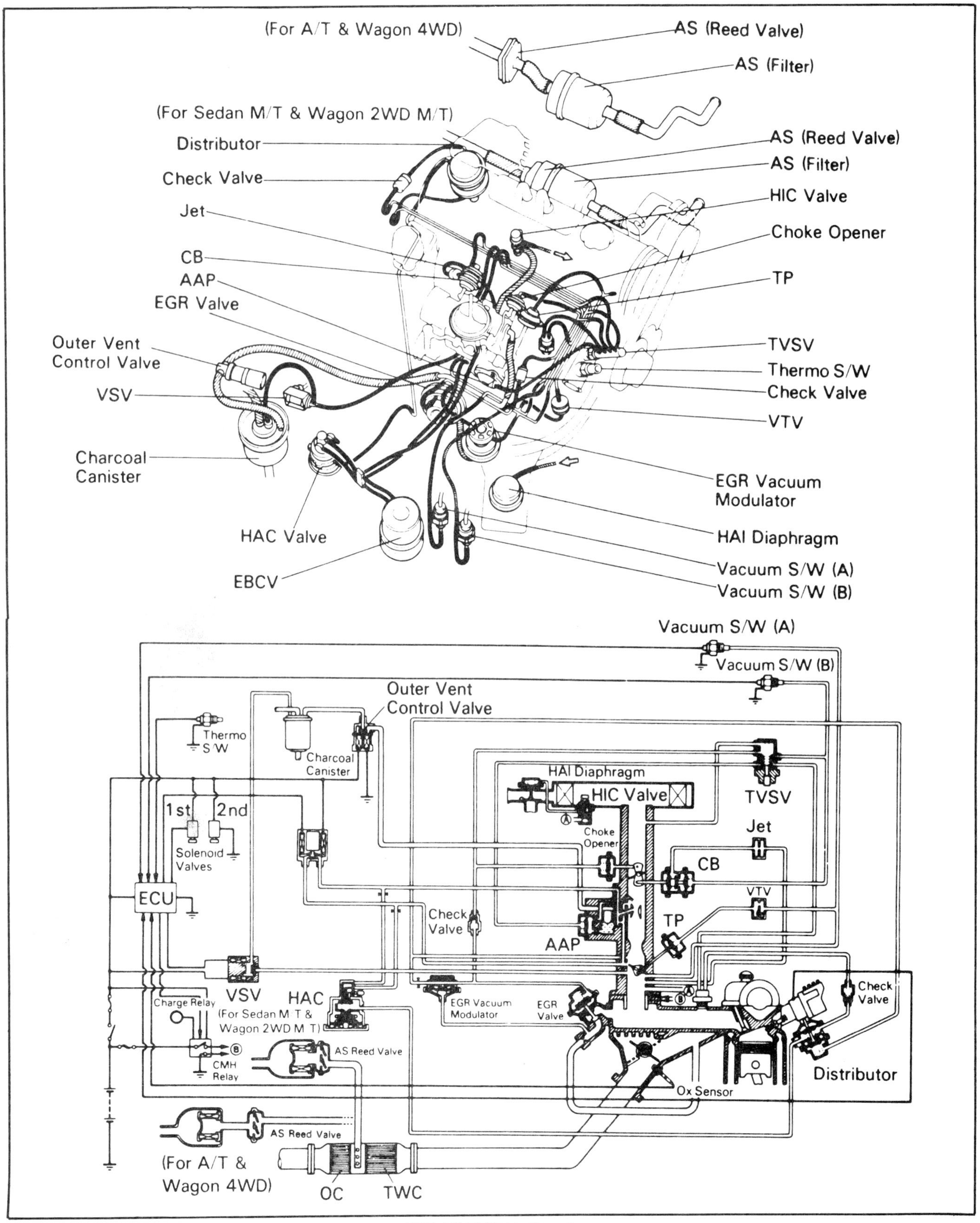

1985–86 49 States Tercel

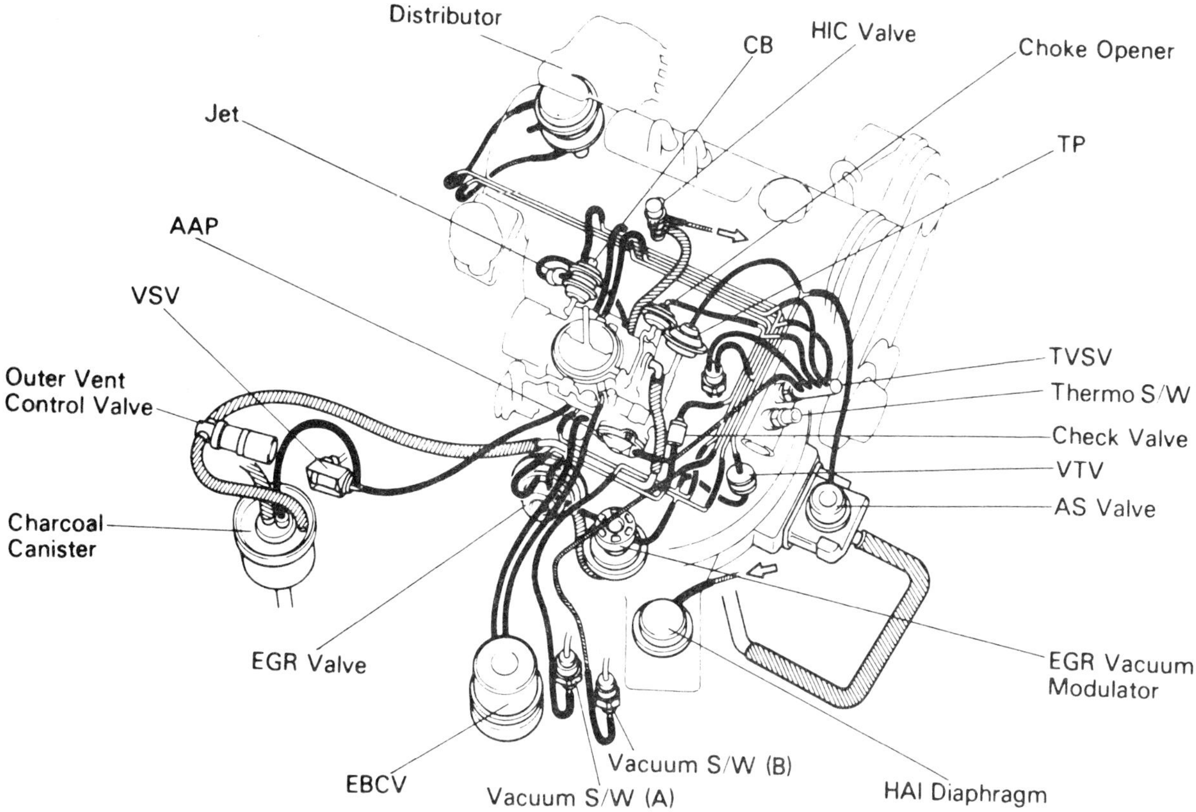

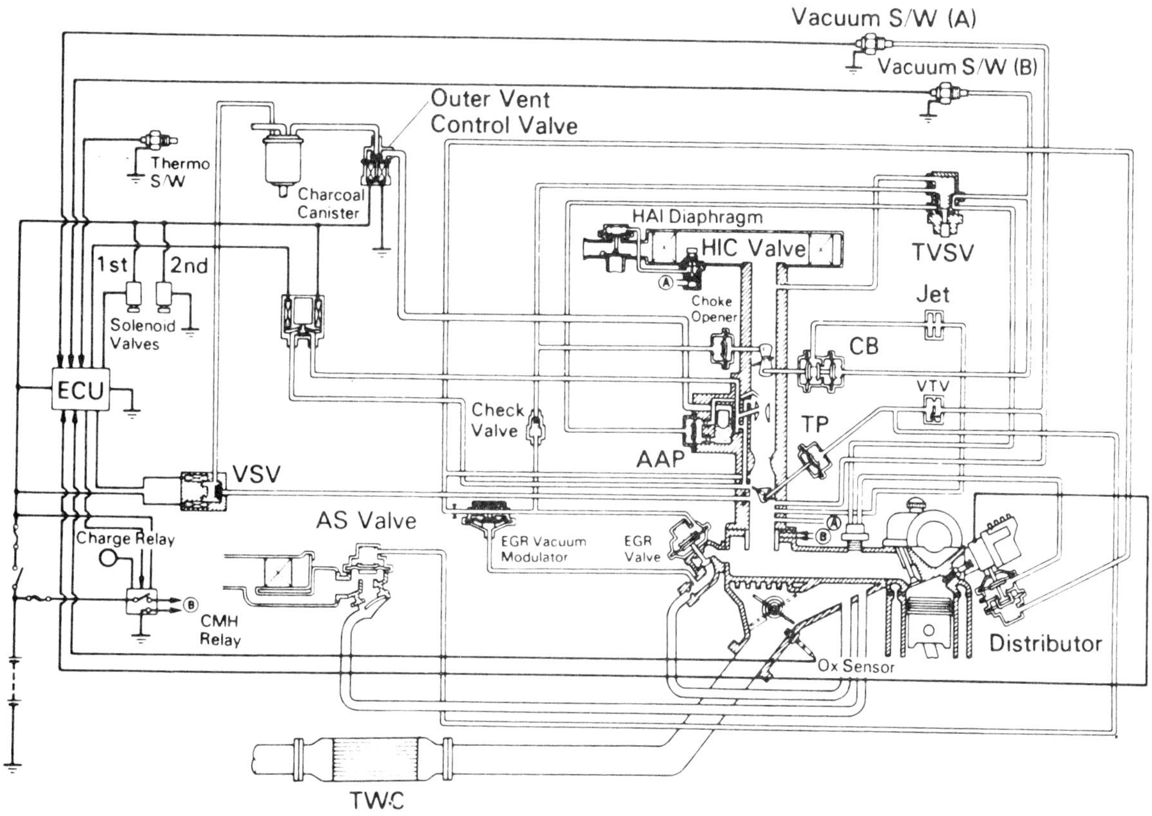

1985–86 California and Canadian Tercel

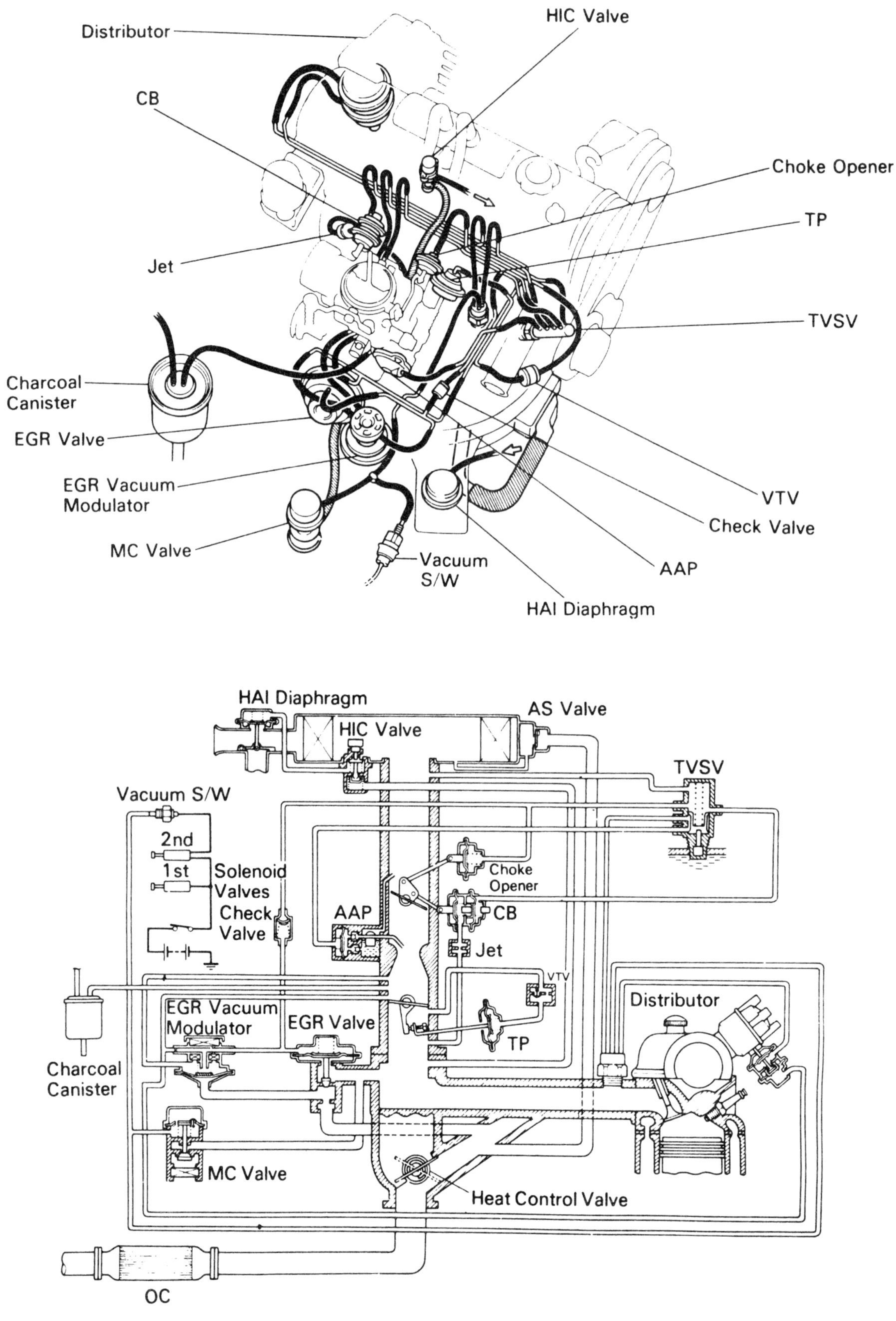

1985–86 Canadian Tercel wagon with M/T

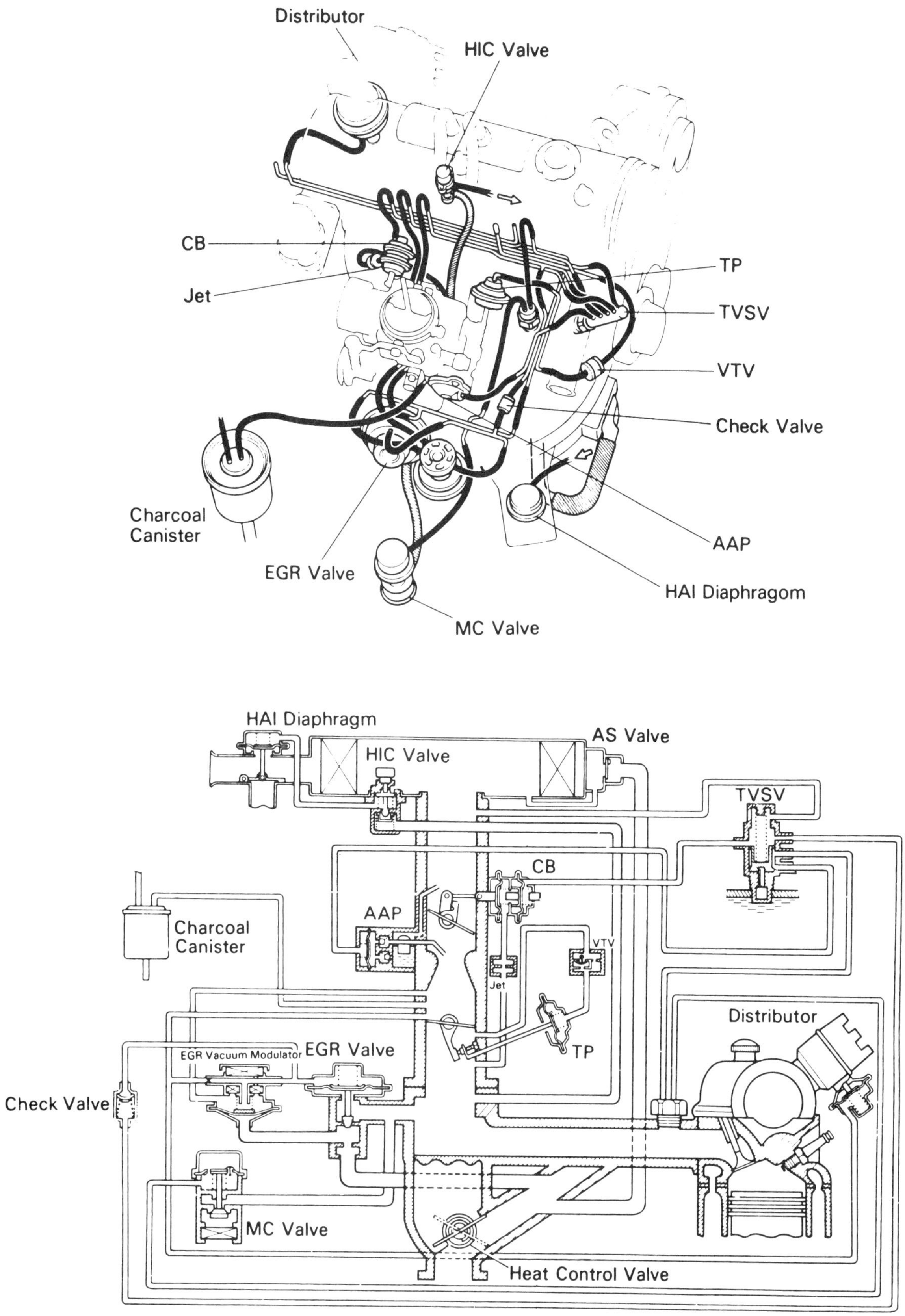

1985–86 Canadian Tercel with 3A engine

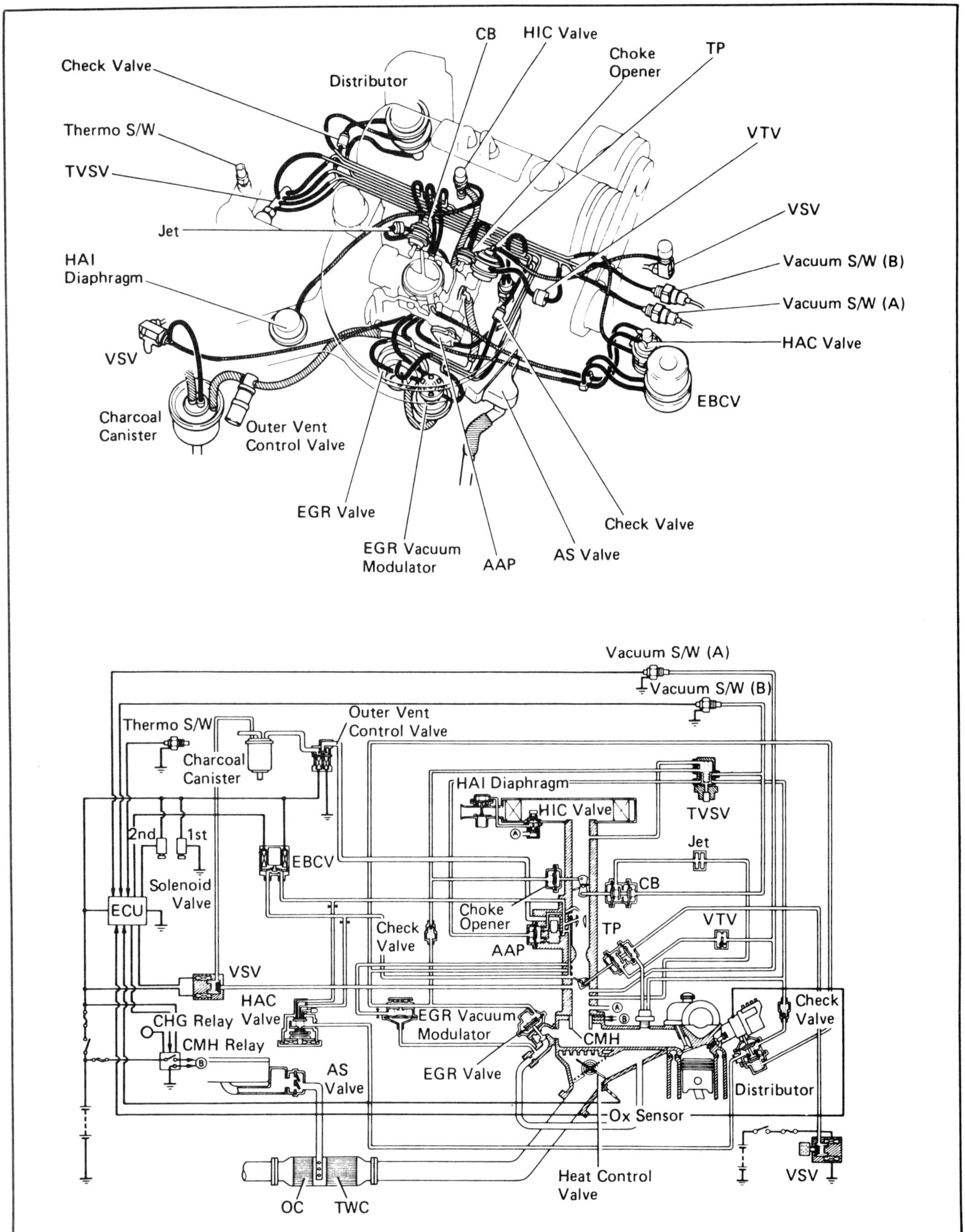

1985-86 49 States Corolla (FWD)

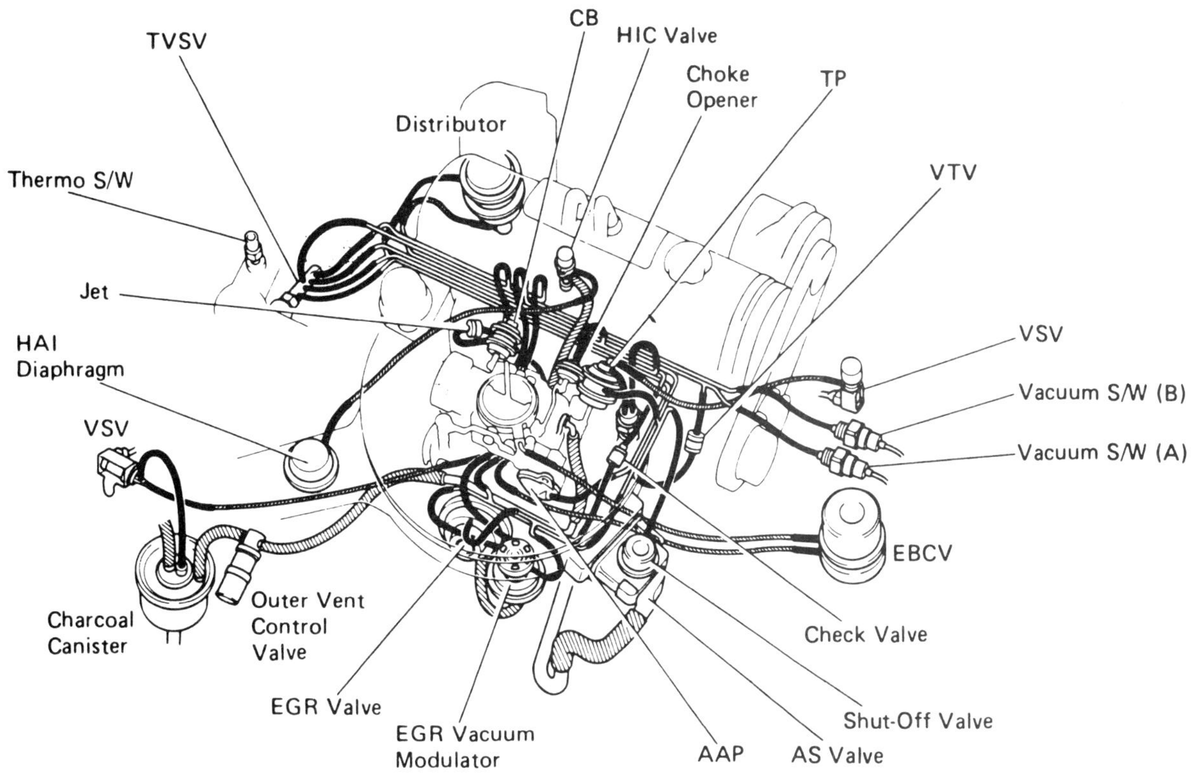

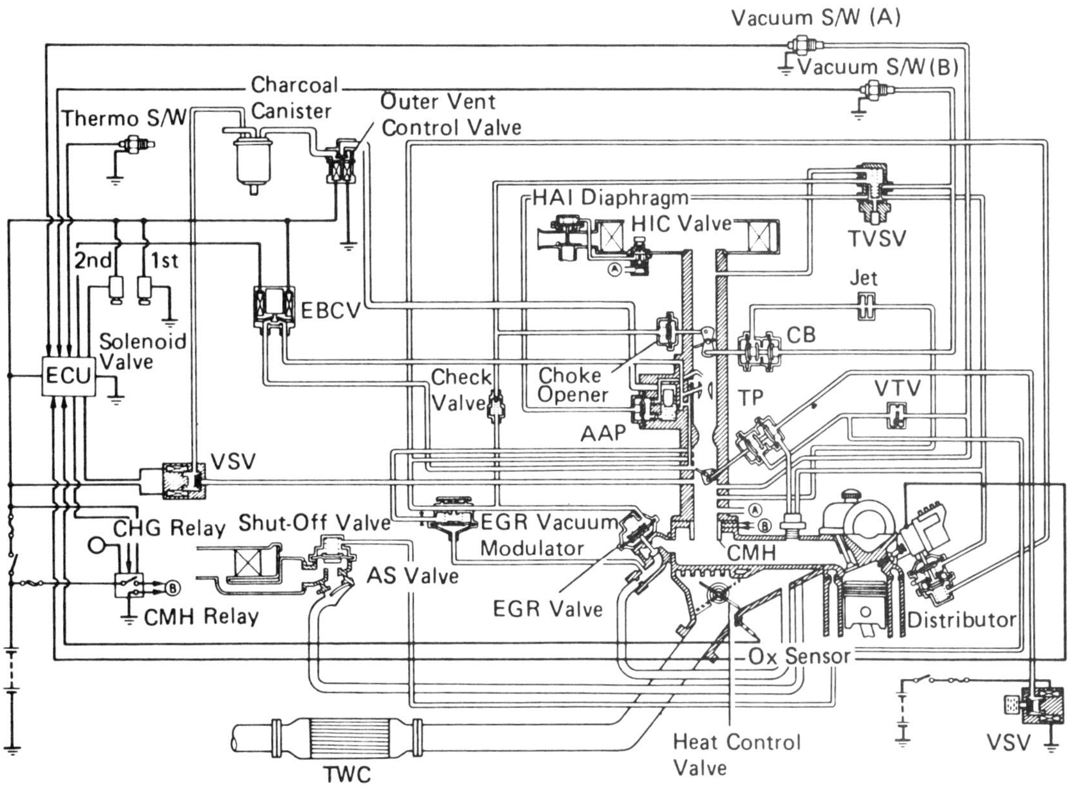

1985–86 California Corolla (FWD)

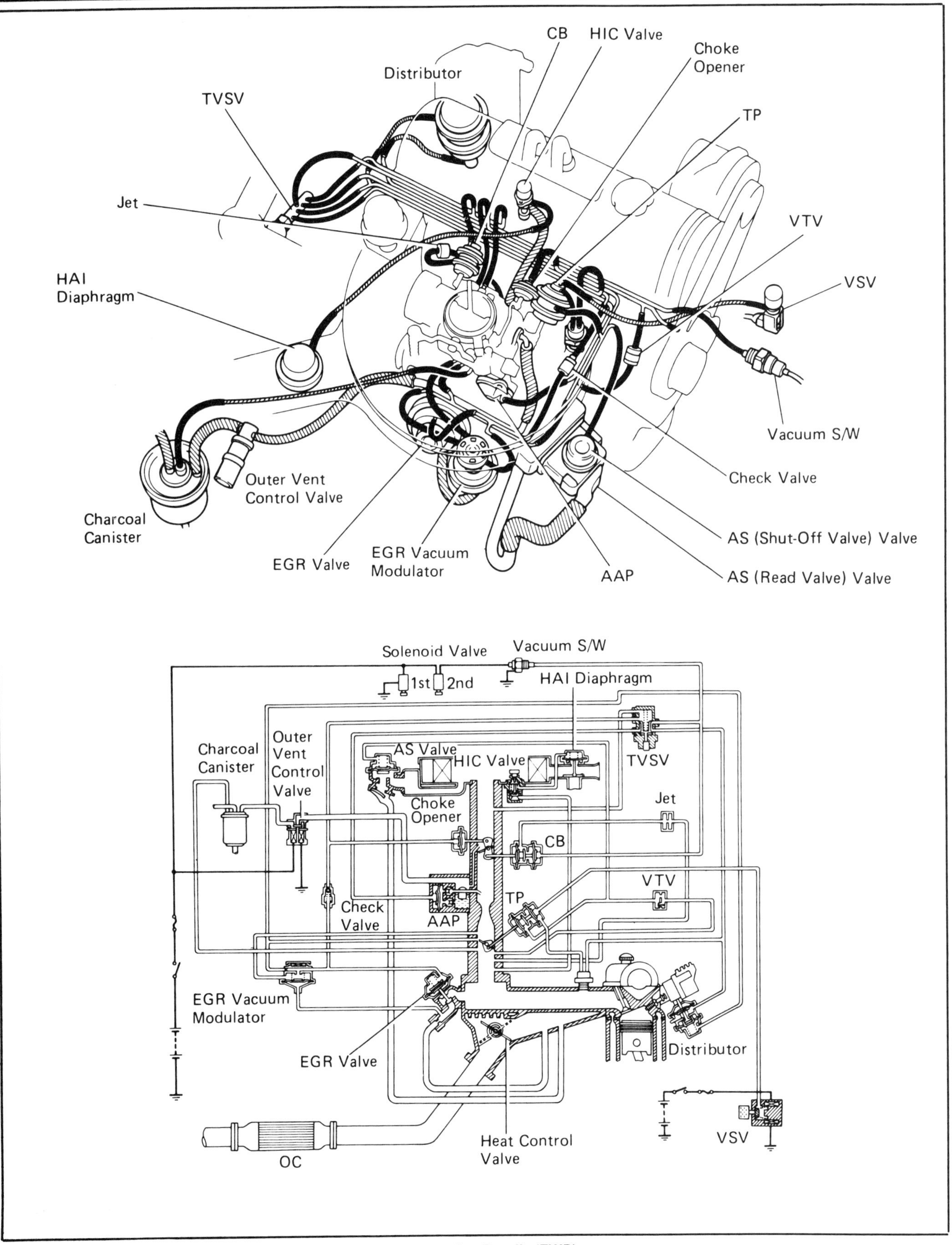

1985-86 Canadian Corolla (FWD)

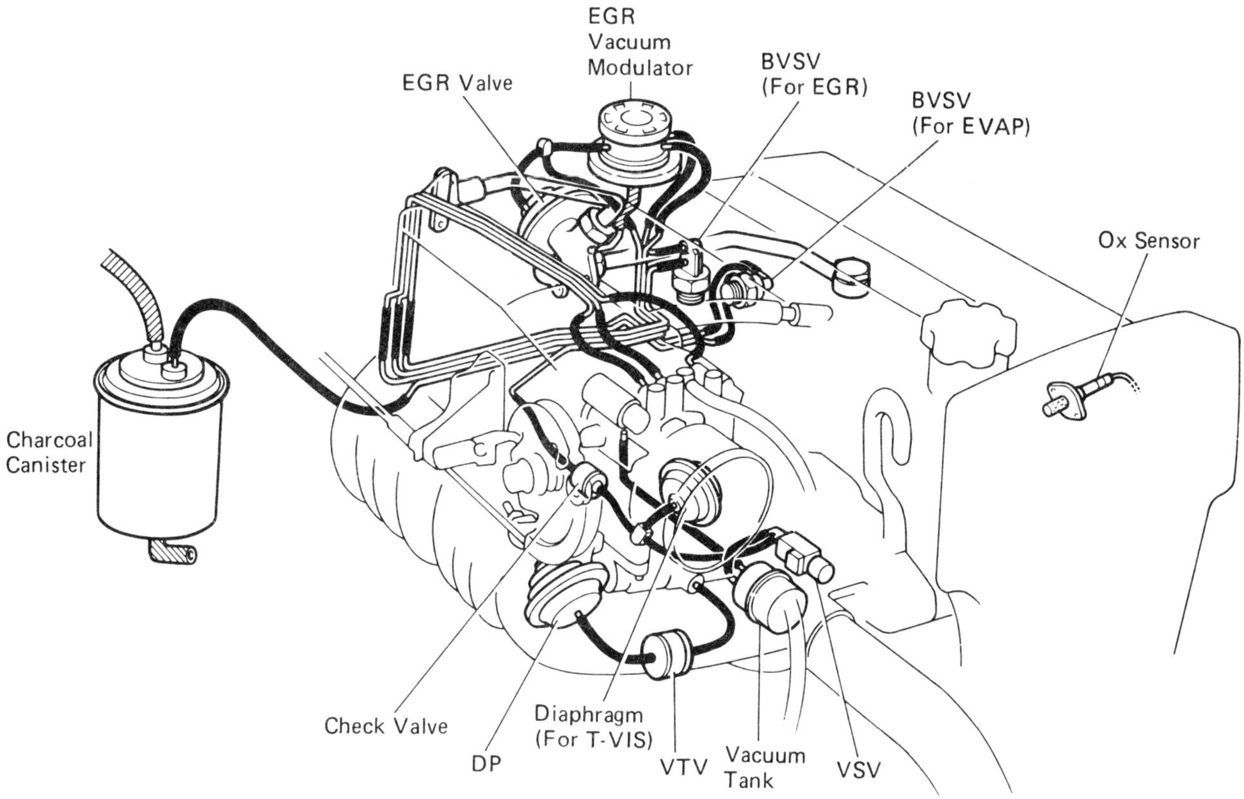

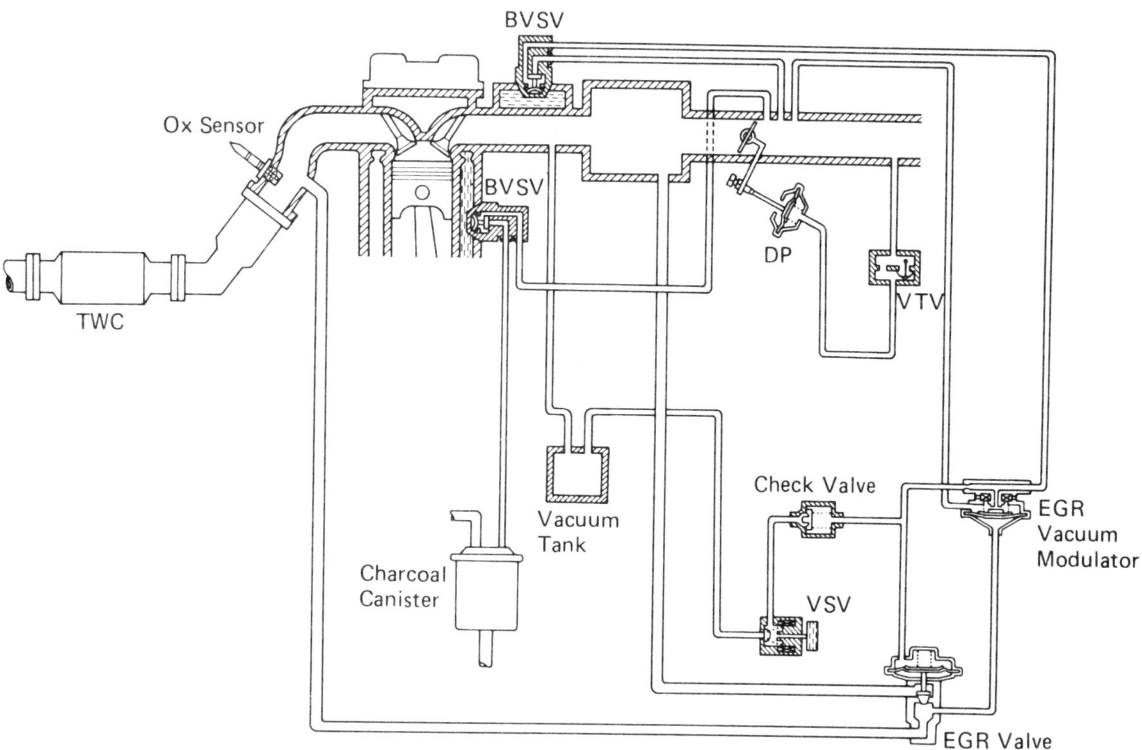

1985–86 Corolla (RWD)

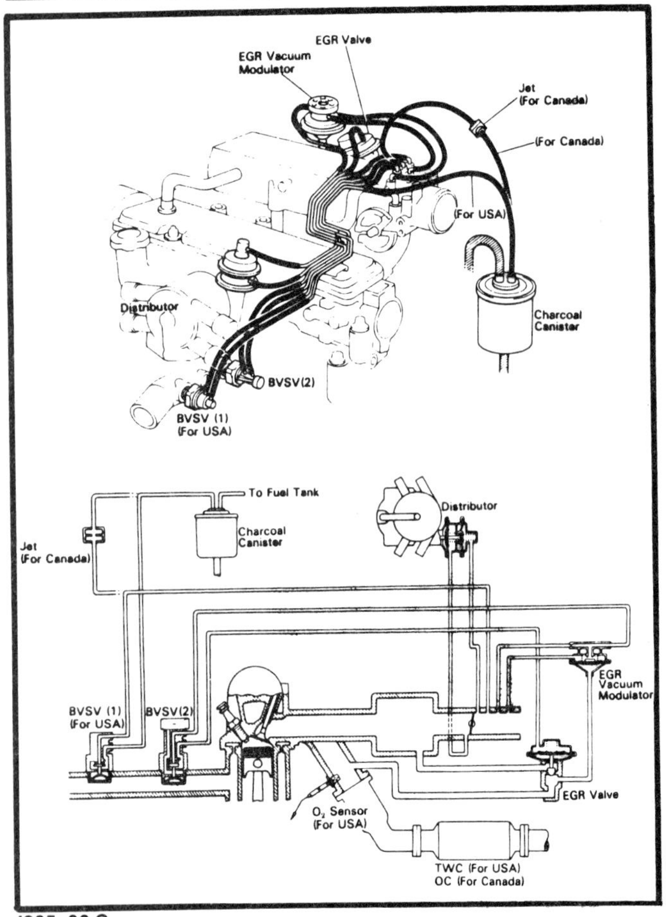

1985-86 Camry

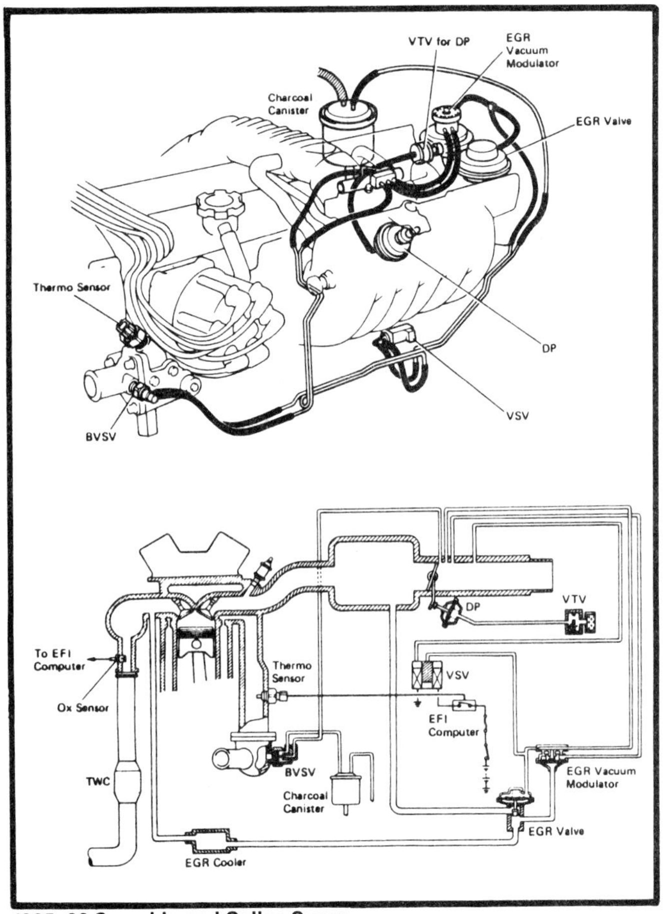

1985-86 Cressida and Celica Supra

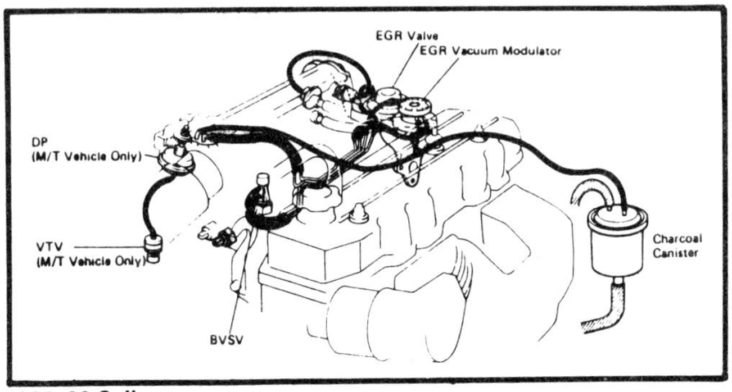

1985-86 Celica

Toyota Truck Specifications

VEHICLE IDENTIFICATION NUMBER AND ENGINE IDENTIFICATION

Vehicle Identification Number (VIN)

Vehicle Identification Number (VIN)All models have the vehicle identification number (VIN) stamped on a plate which is attached to the left side of the instrument panel. This plate is visible through the windshield. The serial number consists of a series identification number followed by a six-digit production number.

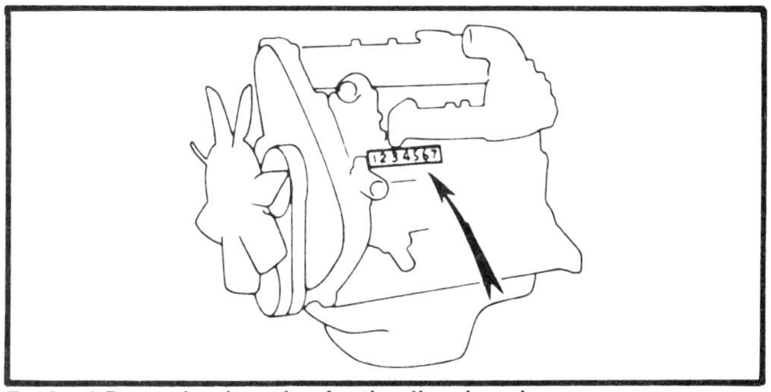

Engine I.D. number location for the diesel engine

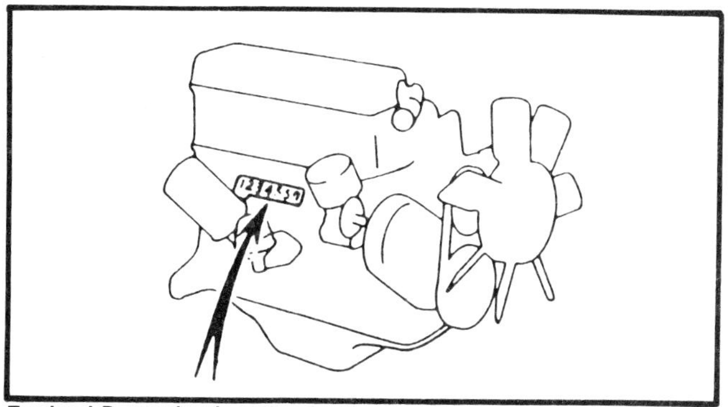

Engine I.D. number location for the gasoline engine

Engine Identification Number

The engine identification number for all the gasoline engines is stamped on the right side of the engine block. On the diesel engine the number is stamped on the left side of the engine block, under the exhaust manifold.

GASOLINE ENGINE TUNE-UP SPECIFICATIONS

(When analyzing compression test results, look for uniformity among cylinders, rather than specific pressures)

Year	Engine Type	Spark Plugs Type	Spark Plugs Gap (in.)	Distributor Air Gap (in.)	Ignition Timing (deg) MT	Ignition Timing (deg) AT	Fuel Pump Pressure (psi)	Idle Speed (rpm) MT	Idle Speed (rpm) AT	Valve Clearance (in.) In	Valve Clearance (in.) Ex
'85–'86	22R	W16EXR-U	0.031	0.008–0.016	0	0	2.1–4.3	700	750	0.008	0.012
	22R-E	W16EXR-U	0.031	0.008–0.016	5B T	5B T	33–38	750	750	0.008	0.012
	22R-TE	W16EXR-U	0.031	0.008–0.016	5B T	5B T	33–38	750	750	0.008	0.012
	2F	W14EXR-U	0.031	0.008–0.016	7B	7B	3.4–4.7	650	650	0.008	0.014
	3Y-EC	P16R	0.043	0.008–0.016	8B	8B	33–38	700	750	Hyd.	Hyd.

NOTE: If the information given in this chart disagrees with the information on the emission control specification decal, use the specifications on the decal.

MT Manual transmission
AT Automatic transmission
T Terminal shorted
B Before top dead center
Hyd. Hydraulic valve lash adjusters

DIESEL ENGINE TUNE-UP SPECIFICATIONS

Year Model	Engine Displacement cu. in. (cc)	Warm Valve Clearance (in.)		Intake Valve Opens (deg)	Injection Pump Setting (deg)	Injection Nozzle Pressure (psi)		Idle Speed (rpm)	Compression Pressure (psi)
		In	Ex			New	Used		
'85–'86	2L-153 (2466)	0.010	0.014	14B	0TDC ①	2,276– 2,389	2,062 2,389	700	284–455
	2LT-153 (2466)	0.010 0.010	0.014 0.014	14B	0TDC ②	2,276– 2,389	2,062 2,389	700	284-455

TDC Top dead center
B Before top dead center
① A plunger lift of 0.045 inches at 0 degrees top dead center
② A plunger lift of 0.039 inches at 0 degrees top dead center (non-California models)
A plunger lift of 0.035 inches at 0 degrees top dead center (California models)

FIRING ORDERS

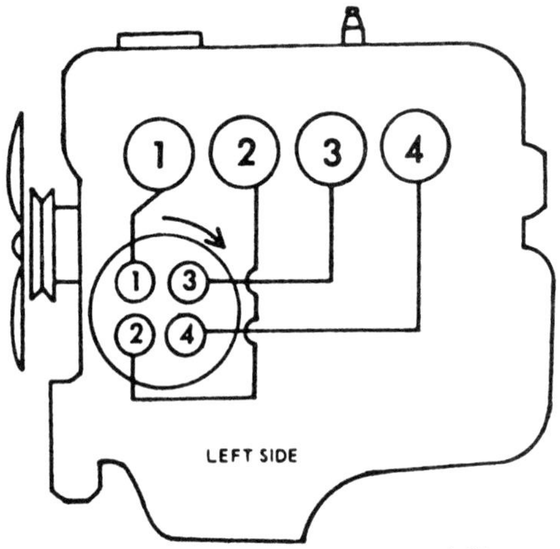

Firing order for the 22R engines—Pick up and 4Runner

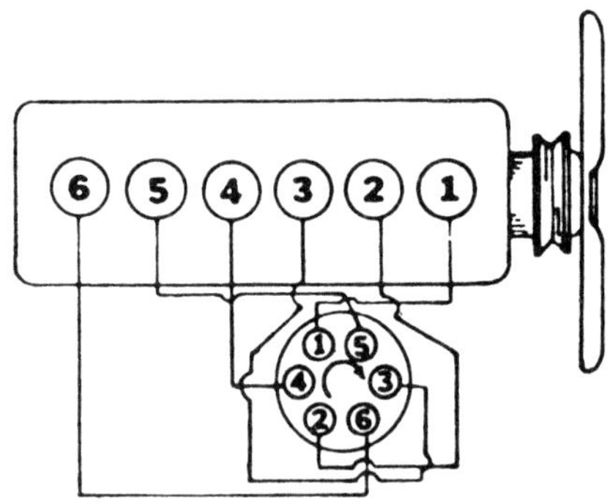

Firing order for the 2F engine—Land Cruiser

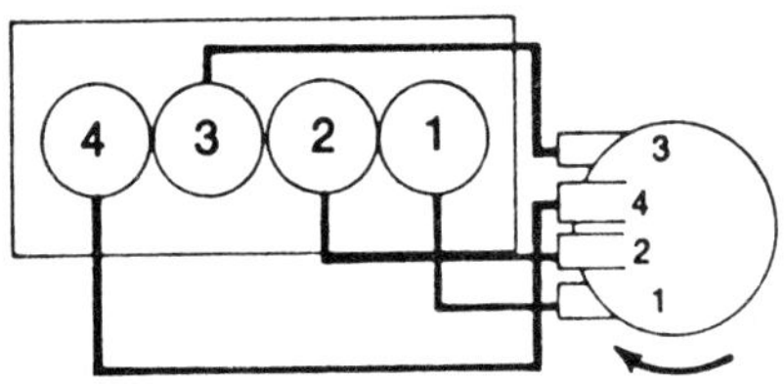

Firing order for the 3Y-EC engine—Van

Firing order for the 2L diesel engine—1-3-4-2

EMISSION EQUIPMENT USED

PICK UP AND 4 RUNNER CARBURETED ENGINES

Positive Crankcase Ventilation
Evaporation Emission Control
Mixture Control
Exhaust Gas Recirculation (EGR)
Air Suction
Air injection system
Catalytic converter
Early fuel evaporation
Electronic controlled carburetor
Fuel shut off system
Spark control system
Thermostatic air cleaner
Three way catalyst
Dashpot
Oxygen Sensor

PICK UP AND 4 RUNNER FUEL INJECTED ENGINES

Positive Crankcase Ventilation
Evaporation Emission Control
Air flow controlled fuel injection
Exhaust Gas Recirculation (EGR)
Electronic control unit
Catalytic converter
Oxygen Sensor
Dashpot

LAND CRUISER (OPTIONAL FOR 1985-86)

Positive Crankcase Ventilation
Evaporation Emission Control
Exhaust Gas Recirculation (EGR)
Air injection system
Catalytic converter
Spark control system
Thermostatic air cleaner
Dashpot
Oxygen Sensor
Deceleration control system
High altitude compensator
Idle control system

VAN

Positive Crankcase Ventilation
Evaporation Emission Control
Air flow controlled fuel injection
Exhaust Gas Recirculation (EGR)
Electronic control unit
Catalytic converter
Oxygen Sensor
Dashpot

DIESEL ENGINES

Positive Crankcase Ventilation
Exhaust Gas Recirculation (EGR)

REQUIRED EMISSION CONTROL MAINTENANCE

NOTE: In addition to the services listed below, Toyota recommends an oil and filter change every 10,000 miles and an air filter replacement every 30,000 miles. Cut the mileage intervals in half if the vehicle is driven daily in heavy traffic, under dusty conditions or is used to tow a trailer. Valve and Idle Speed/Mixture adjustments are also considered part of emission maintenance.

PCV SYSTEM

Check operation and replace the PCV valve (carbureted engine), or clean any gum deposits from the orifices (fuel injected engine) every 30,000 miles. Inspect all hoses for leaks or deterioration and replace as necessary.

FUEL EVAPORATION EMISSION CONTROL (EVAP) SYSTEM

Inspect the system for proper operation every 30,000 miles. Check all hose connections and hoses for leaks or deterioration and replace as necessary. Check the charcoal canister for raw gas contamination or damage and replace if necessary.

EXHAUST GAS RECIRCULATION (EGR) SYSTEM

Inspect the system for proper operation every 30,000 miles. Check the EGR valve operation and inspect all hoses and connections for leaks or deterioration and replace as necessary.

OXYGEN SENSOR (OXS) SYSTEM

Although no replacement interval is specified by the manufacturer, the oxygen sensor should be checked for proper operation every 30,000 miles and replaced every 60,000 miles. Any oxygen sensor found to be malfunctioning should be replaced.

IDLE SPEED AND MIXTURE ADJUSTMENTS

IDLE SPEED SPECIFICATIONS

Engine	Transaxle	Curb Idle Speed (RPM)
22R	M/T	700 [± −] 100
22R	A/T	750 [± −] 100
22R-E	M/T & A/T	750 [± −] 100
22R-TE	M/T & A/T	750 [± −] 100
2F	M/T	650 [± −] 100
2F	A/T	600 [± −] 100

IDLE SPEED SPECIFICATIONS

Engine	Transaxle	Curb Idle Speed (RPM)
3Y-EC	M/T	700 [± −] 100
3Y-EC	A/T	750 [± −] 100
2L (Diesel)	M/T & A/T	700 [± −] 100
2LT (Diesel)	M/T & A/T	700 [± −] 100

M/T Manual Transmission
A/T Automatic Transmission
E Electronic fuel injection
TE Turbocharged electronic fuel injection
T Turbocharged

NOTE: The following information is being published from the latest information available at the time of publication. If the information differs from the information on the vehicle emissions label, use the information on the emissions label.

IDLE SPEED

Adjustment

2F AND 22R ENGINES

1. Before starting this adjustment make sure that the following items are checked first.
   a. The air cleaner is installed and the choke valve is fully opened.
   b. All accessories are switched off and all vacuum lines are connected.
   c. The transmission is in neutral, with the drive wheels blocked. The engine is idling at normal operating temperature.
2. Connect the positive lead of a tachometer to the service connector of the ignition igniter.

NOTE: Never allow the positive tachometer terminal to touch a ground, because it could lead to damage of the igniter and/or the igniter coil. Some tachometers are not compatible with this igniter system. If this is the case, refer to the manufacturer for the proper tachometer.

3. With the tachometer now installed, set the idle speed to specifications, by turning the idle adjustment screw, located next to the choke linkage on the carburetor.

22R-E,22R-TE AND 3Y-EC ENGINES

1. Before starting this adjustment, make sure that the following items are checked first.
   a. The air cleaner is installed and the choke valve is fully opened.
   b. All accessories are switched off. All vacuum lines and air pipes are connected.
   c. The EFI wiring connectors are fully engaged.
   d. The transmission is in neutral with the drive wheels blocked. The engine is idling at normal operating temperature.
2. Connect the positive lead of a tachometer to the service connector of the ignition coil negative terminal.

NOTE: Never allow the positive tachometer terminal to touch a ground, because it could lead to damage to the igniter and/or the ignition coil. Some tachometers are not compatible with this ignition system. If this is the case, refer to the manufacturer for the proper tachometer.

3. With the tachometer now installed, race the engine at 2500 rpm for about 2 minutes and then return it to idle speed.
4. Set the idle speed to specifications by turning the idle adjustment screw, located next to the throttle lever linkage.

2L AND 2LT DIESEL ENGINES

1. Before starting this adjustment, make sure that the following items are checked first.
   a. The air cleaner is installed and all accessories are switched off.b.
   The transmission is in neutral with the drive wheels blocked. The engine is idling at normal operating temperature.
2. Connect a diesel tachometer to the engine.
3. With the tachometer now installed, set the idle speed to specifications (700 rpm M/T & A/T), by turning the idle adjustment screw, located next to the throttle lever linkage.

Fast Idle Speed Adjustment

2F AND 22R ENGINES

1. After adjusting the idle speed at previously outlined, leave the tachometer installed and shut off the engine.
2. Disconnect and plug the hoses for the heated air injection system and the mixture control system, to prevent rough idling.
3. Disconnect and plug the hose from the choke opener diaphragm, so the choke opener system will be in-operative.
3. Disconnect and plug the hose to the EGR valve to bypass the EGR system.
4. While holding the throttle valve slightly opened, push the choke valve closed and hold it closed while releasing the throttle valve.
5. Start the engine, but do not touch the accelerator pedal. Set the fast idle speed to specifications (2600 rpm), by turning the fast idle speed adjusting screw, located on the lower end of the throttle linkage of the carburetor.
6. Reconnect all hoses, remove the tachometer and reinstall the air cleaner.

IDLE MIXTURE

Adjustment

2F AND 22R ENGINES

NOTE: The idle mixture adjustments are factory set and sealed. No adjustments are required. The mixture control adjustment screw opening is plugged with a steel plug to prevent any adjustments. Mixture control adjustments should be done when the mixture control unit or carburetor is replaced or when the vehicle fails the emissions test.

1. Tag all vacuum lines going in and out of the carburetor. Then disconnect and plug all of these vacuum lines.
2. Remove the carburetor from the engine and cover the intake manifold with a clean rag, preventing any foreign objects from entering the engine.
3. Plug off the carburetor vacuum ports to prevent the entry of the steel particles from drilling. Mark the center of the steel concealment plug with a punch.
4. Drill through the concealment plug using a ¼ in. (6.5mm) drill bit. There is only 0.04 in. (1mm) of clearance between the concealment plug and the idle mixture screw. Drill slowly and stop when you feel the bit break through the concealment plug.
5. After removing the concealment plug, screw the idle mixture screw all the way in towards the base of the carburetor. Now back out the idle mixture screw about four and a half turns.
6. Reinstall the carburetor on the engine and reconnect all the vacuum lines.
7. Before starting the idle mixture adjustment, make sure that the following items are checked first.
   a. The air cleaner is installed and the choke valve is fully opened.
   b. All accessories are switched off and all vacuum lines are connected.
   c. The transmission is in neutral with the drive wheels blocked. The engine is idling at normal operating temperature.
   d. The fuel level should be about even with the correct level in the sight glass.
8. Install a tachometer to the engine and start the engine.
9. After the engine has reached normal operating temperature, set the carburetor to maximum speed by turning the idle mixture adjusting screw.
10. Set the mixture speed by turning the idle speed adjusting screw. Mixture speed specifications;
    a. 22R—740 rpm on manual transmissions
    b. 22R—790 rpm on automatic transmissions
    c. 2F—690 rpm on manual transmissions
    d. 2F—640 rpm on automatic transmissions
11. Before going on to the next step, continue to adjust the idle mixture and the idle speed until the maximum speed will not rise any further. No matter how much the idle mixture adjustment screw is adjusted.
12. Set the idle speed to specifications by turning in the idle mixture adjustment screw. This is the lean drop method of setting the idle speed and idle mixture.
13. Remove the air cleaner and tap in a new concealment plug until it is even with the carburetor surface.
14. Remove all test equipment and reinstall the air cleaner.

22R-E,22R-TE AND 3Y-EC ENGINES

Idle mixture adjustment cannot be accomplished with Electronically Controlled Fuel Injection system. The fuel delivery is controlled by the Electronic Control Unit (ECU), reacting to varied operating conditions.

Volkswagen

INDEX

VEHICLE IDENTIFICATION NUMBER AND ENGINE IDENTIFICATION

Vehicle Identification Number (VIN)

On the Golf, Jetta, Scirocco and Quantum, the vehicle identification plate is on the top of the body crossmember above the grille. The same plate on the Dasher is attached to the inner right fender. The date of manufacture and the chassis number are stamped on the plate.

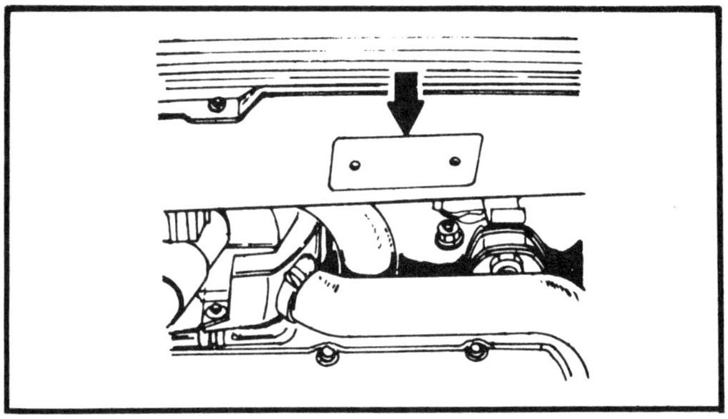

Vehicle identification plate (© Volkswagen of America)

Chassis Vehicle Identification Number

The chassis number plate is located on the drivers' side windshield pillar on the Dasher and Scirocco, and on the left front corner of the dashboard on the Jetta, Golf and Quantum. These numbers are visible through the windshield. Certain Dasher and Quantum chassis numbers are also stamped on the firewall over the windshield washer reservoir. The Jetta and Scirocco chassis number is also found on top of the right front suspension pillar. It also appears on the vehicle identification plate. 1985 and later models use a seventeen digit code. The fifth position

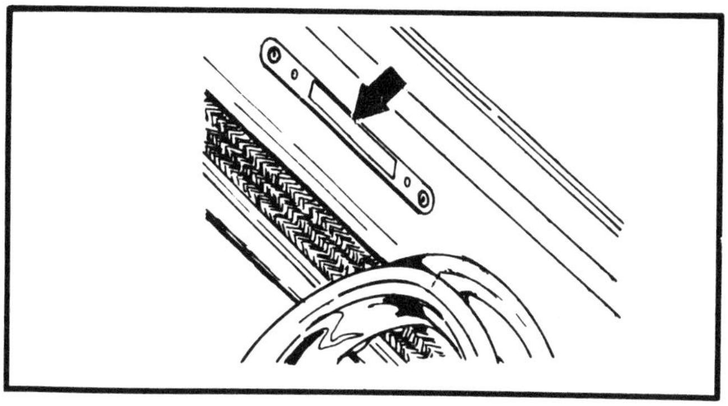

Scirocco chassis plate (© Volkswagen of America)

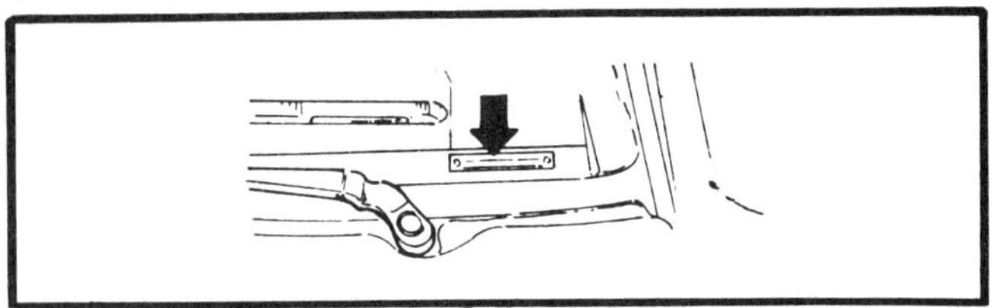

Golf, Jetta, Quantum chassis plate
(© Volkswagen of America)

Jetta chassis number (© Volkswagen of America)

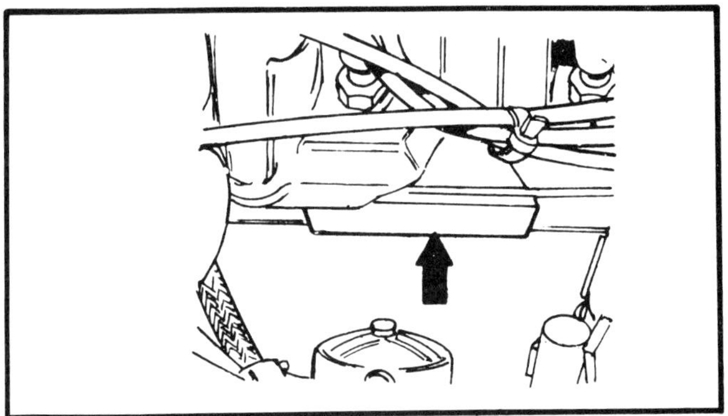

Engine number location (© Volkswagen of America)

indicates engine and the tenth the year. The year code will be a letter. "F"- 1985; "G"- 1986; etc.

Engine Number

The engine number is stamped on a flat boss on the left side (front on the Golf, Jetta, Scirocco and Quantum) of the engine block, just below the cylinder head between the fuel pump and the distributor.

GASOLINE ENGINE TUNE-UP SPECIFICATIONS
Volkswagen Front Wheel Drive

Year Model	Engine Displacement cu. in (cc)	Spark Plugs Type	Spark Plugs Gap (in.)	Distributor Point Dwell (deg)	Distributor Point Gap (in.)	Ignition Timing (deg)	Intake Valve Opens (deg)	Compression Pressure (psi)	Idle Speed (rpm)	Valve Clearance (in.) In ②	Valve Clearance (in.) Ex ②
'85 Golf, GTI, Jetta, Scirocco, Quantum, Cabriolet	109.0 (1780)	WR7DS N8GY	.024–.032	Electronic		6 BTDC @ Idle	6 BTDC	123–174	850–1000	0.008–0.012	0.016–0.020
'85 Quantum	136.0 (2226)	W7DG N9YC ③	.028–.023	Electronic		①	6 BTDC	123–174	850–1000	0.008–0.012	0.016–0.020
'86	All	See Underhood Specifications Sticker									

NOTE: The underhood specifications sticker often reflects tune-up specification changes made in production. Sticker figures must be used if they disagree with those in this chart.
① M/T: 6° BTDC; A/T: 3° ATDC
② Valve clearance need not be adjusted unless it varies more than 0.002 in. from specifications.
③ California: WR7DS: N8GY

GASOLINE ENGINE TUNE-UP SPECIFICATIONS
Volkswagen Rear Wheel Drive

Year	Code	Type	Common Designation	Spark Plugs Type ①	Spark Plugs Gap (in.)	Distributor Point Dwell (deg)	Distributor Point Gap (in.)	Ignition Timing (deg) MT	Ignition Timing (deg) AT	Fuel Pump Pressure (psi) @ 4000 rpm	Compression Pressure (psi)	Idle Speed (rpm) MT	Idle Speed (rpm) AT	Valve Clearance (in. cold) In	Valve Clearance (in. cold) Ex
'85	DH	2	Water-boxer (1900)	W7CO N288 Beru 14L-7C	.028	②	②	5A	5A	④	116–189	800–900 ③	800–900 ③	Hyd.	Hyd. ⑤
'86	See Underhood Specifications Sticker														

NOTE: The underhood specifications sticker often reflects tune-up specification changes made in production. Sticker figures must be used if they disagree with those in this chart.
A After Top Dead Center
B Before Top Dead Center
MT Manual trans.
AT Automatic trans.
① Recommended by manufacturer
② '85–'86 electronic ignition; point gap and dwell preset and non-adjustable
③ With vacuum hoses connected
④ 29 psi @ idle speed @ approx. 2.0 bar with vacuum hose connected
⑤ Valves must still be adjusted when cylinder heads have been removed; see "Valve Lash" in text.

DIESEL ENGINE TUNE-UP SPECIFICATIONS
Volkswagen Front and Rear Wheel Drive

Model	Valve Clearance (cold) ① Intake (in.)	Valve Clearance (cold) ① Exhaust (in.)	Intake Valve Opens (deg)	Injection Pump Setting (deg)	Injection Nozzle Pressure (psi) New	Injection Nozzle Pressure (psi) Used	Idle Speed (rpm)	Cranking Compression Pressure (psi)
Diesel (All models)	0.008–0.012	0.016–0.020	N.A.	Align marks	1885 ③	1706 ③	800-850 ② ④	406 minimum

N.A. Not Available
① Warm clearance given—
Cold clearance:
Intake 0.006-0.010
Exhaust 0.014-0.018
② Volkswagen has lowered the idle speed on early models to this specification.
Valve clearance need not be adjusted unless it varies more than 0.002 in. from specification.
③ Turbo diesel: New—2306; Used—2139
④ Turbo diesel: 900-1000

FIRING ORDER

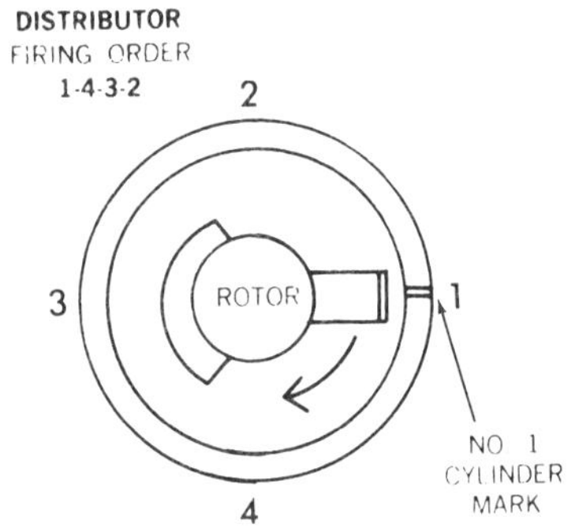

4 cylinder diesel engine firing order
(© Volkswagen of America)

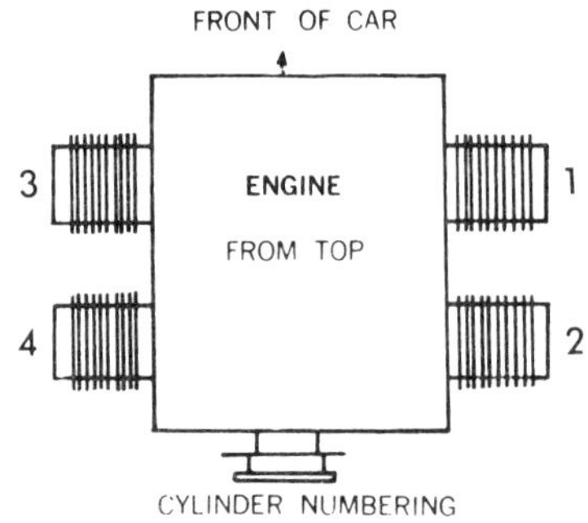

Air cooled engine firing order
(© Volkswagen of America)

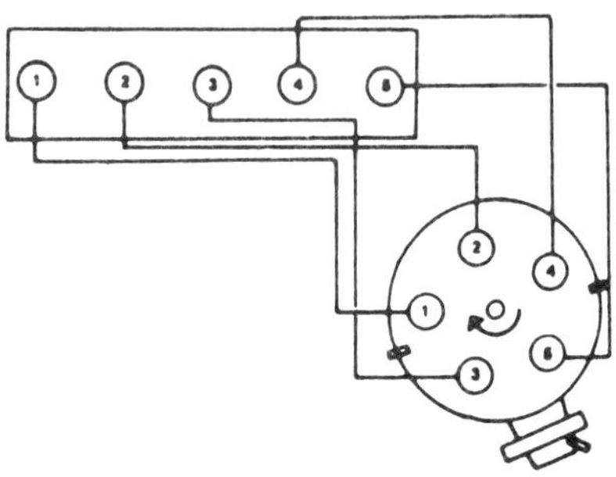

Firing order: 5 cylinder gas engine: 1-2-4-5-3
(© Volkswagen of America)

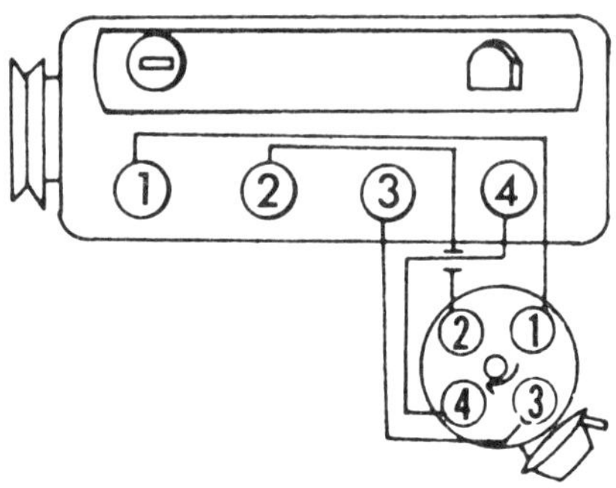

Firing order: 4 cylinder gas engine: 1-3-4-2
(© Volkswagen of America)

EMISSION EQUIPMENT USED

GOLF, JETTA, SCIROCCO AND QUANTUM

1985–86

Positive crankcase ventilation
Evaporative emission control
Dual diaphragm distributor
Exhaust gas recirculation
Air injection (Carbureted Rabbit only)
Catalytic converter
Oxygen sensor

VANAGON

1985–86

Positive crankcase ventilation
Evaporative emission control
Catalytic converter
AFC injection

DIESEL MODELS

1985–86

Positive crankcase ventilation

REQUIRED EMISSION CONTROL MAINTENANCE

NOTE: The following information is being published from the latest information available at the time of publication. If the information contained herein differs from that which is listed on the vehicles emission label, use the specifications given on the label.

POSITIVE CRANKCASE VENTILATION (PCV)

Every 15.000 miles, check the PCV valve. Every 30,000 miles replace the PCV valve. Inspect and clean all hoses and connections. Replace any that show signs of deterioration.

EVAPORATIVE EMISSION CONTROL SYSTEM

Make a visual check of all system hoses and filters every 10,000 miles. Replace the charcoal filter every 50,000 miles.

DUAL DIAPHRAGM DISTRIBUTOR

Check the condition of the vacuum lines every 10,000 miles.

EXHAUST GAS RECIRCULATION SYSTEM

Inspect and check the hoses regularly. Reset the mileage switch if equipped, as necessary. Replace the EGR filter every 30,000 miles.

AIR INJECTION

Visually inspect the pump, control valve and hoses every 10,000 miles. Clean the pump filter every 10,000 miles. Replace the pump filter every 20,000 miles or 2 years.

CATALYTIC CONVERTER

Check for damage and tight connections every 30,000 miles. Reset the indicator light as necessary.

AIR CLEANER

Replace the air cleaner element every 15,000 miles.

IDLE SPEED AND MIXTURE ADJUSTMENT

GASOLINE ENGINE FUEL INJECTION

Adjustment

NOTE: Certain 1.8 Liter engines are equipped with a manual pre-heat valve located on the air cleaner housing. The valve is marked "S" (summer) and "W" (winter). When servicing, position the valve to S (unless work area is below freezing). After servicing, return the valve to the position that matches climate conditions.

1. The engine must be at normal operating temperature.
2. Leave the crankcase breather hose open with the exception of the 5 cylinder models. Plug the hose on the 5 cylinder models.
3. Disconnect the two plugs on the idle stabilizer at the control unit and plug them together.
4. Make sure that all the electrical accessories are "OFF".
5. Connect a tachometer and timing light according to manufacturers instructions. Check the timing and adjust in the conventional manner, if necessary.
6. Check the idle speed. Adjust the idle speed by turning the idle adjustment screw located on the throttle chamber.

NOTE: Only adjust the idle when the radiator fan is off.

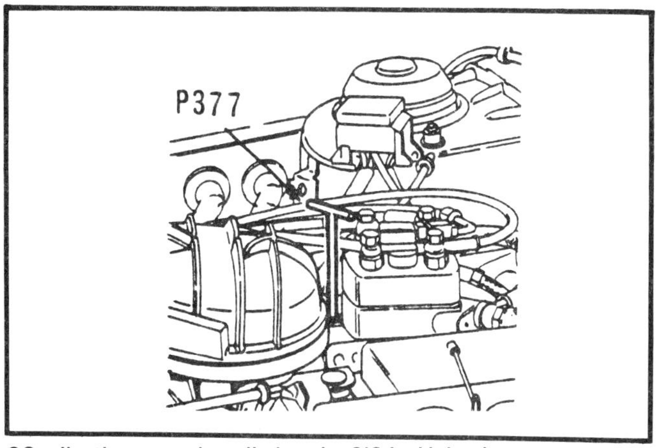

CO adjusting screw installed on the CIS fuel injection system (© Volkswagen of America)

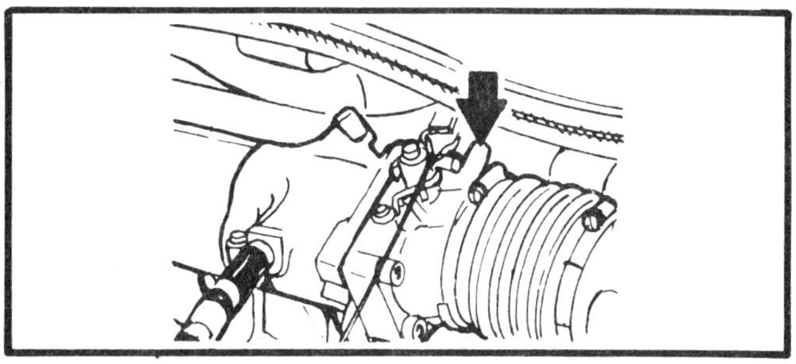
Fuel injected Jetta, Scirocco idle speed adjustment screw (© Volkswagen of America)

DIESEL ENGINE FUEL INJECTION

Idle Speed Maximum Speed Adjustment

Volkswagen diesel engines have both an idle speed and a maximum speed adjustment. The maximum engine speed adjustment prevents the engine from over-speeding and self-destructing. The adjusters are located side by side on top of the injection pump. The screw closest to the engine is the maximum speed adjuster.

The idle speed and maximum speed must be adjusted with the engine at normal operating temperature. Because the diesel engine has no conventional ignition, you will need a special adapter (VW 1324 or equivalent) to connect a tachometer. The tachometer in the instrument panel may be substituted, if equipped. Adjust the engine to the specified idle speed.

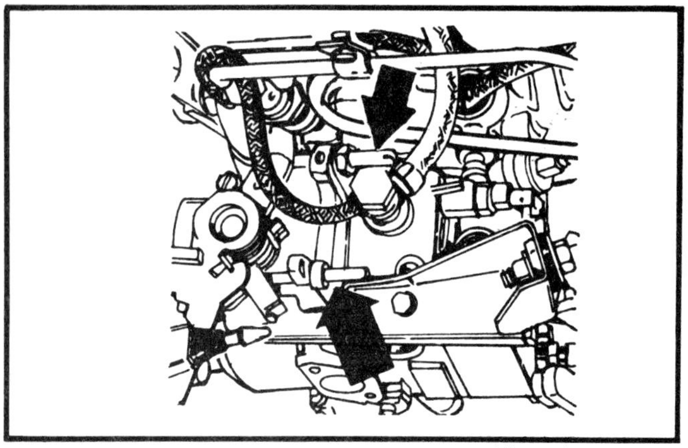

Diesel engine speed (upper) and maximum speed (lower) adjusting screws (© Volkswagen of America)

When adjustment is correct, lock the locknut on the screw and apply a commercial brand of thread sealer in order to prevent the screw from vibrating loose.

The maximum speed for all engines is 5300–5400 rpm or 5050–5150 rpm for turbo models. If it is not in this range, loosen the screw and correct the speed (turning the screw clockwise decreases rpm). Lock the nut on the adjusting screw and apply thread sealer.

Special adapter (VW 1324 or equivalent) is necessary to use an external tachometer on the diesel (© Volkswagen of America)

CAUTION

Do not attempt to squeeze more power out of the engine by raising the maximum speed (rpm). If this is attempted, serious engine may result.

IGNITION TIMING

ELECTRONIC IGNITION PRECAUTIONS

When performing work on the ignition, observe the following precautions in order to prevent damage to the ignition system.

1. Connect and disconnect test equipment when the ignition switch is off.
2. Do not crank the engine with the starter for compression tests until the high tension coil wire is grounded.
3. Do not install any kind of condenser to the coil terminal.
4. Do not use a battery booster for longer than one minute.

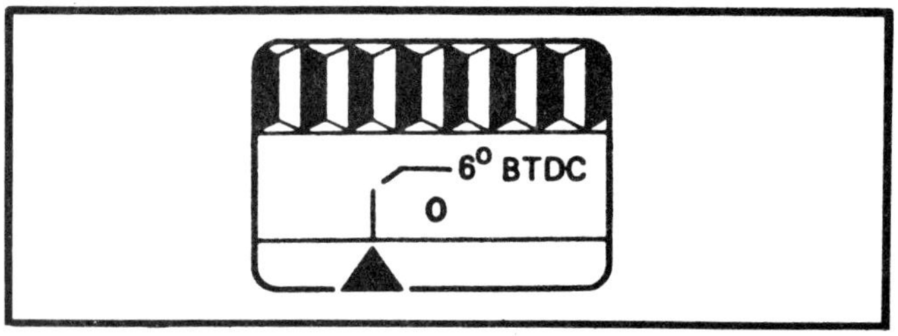

Timing mark on Quantum with M/T (© Volkswagen of America)

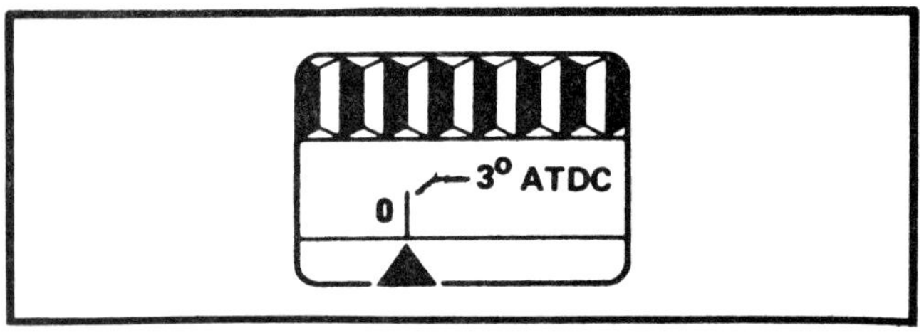

Timing mark on Quantum with A/T (© Volkswagen of America)

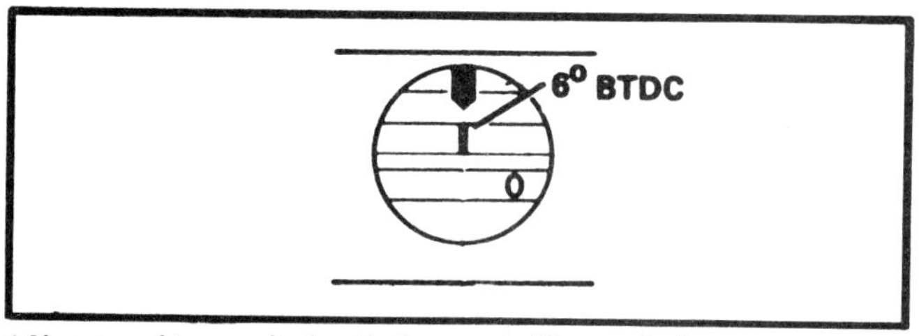

1.8L manual transmission timing mark. The mark shown represents 7½ degrees (© Volkswagen of America)

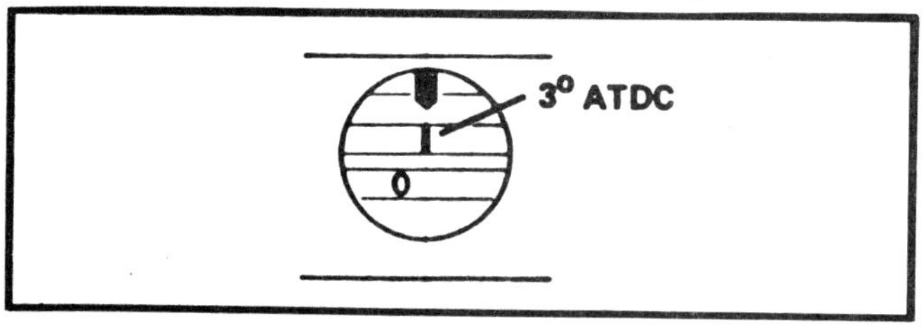

Timing mark on the Jetta and Scirocco models w/CIS except the 1.8L carbureted engines (© Volkswagen of America)

5. Do not tow cars with defective ignition systems without disconnecting the plugs on the idle stabilizer at the ignition control unit.

Timing Procedure

1. Start the engine and allow it to reach normal operating temperature. Connect a tachometer. See "Electronic Ignition Precautions" above.
2. Stop the engine and disconnect the plugs on the idle stabilizer at the control unit and plug them together.
3. Check the idle speed. It should be between 800–1000 rpm.

4. With the timing light attached according to the manufacturers instructions, shine the light on the timing hole. The pointer in the hole must line up with the notch in the flywheel. To adjust the timing, loosen the distributor at its base and turn it until the timing marks line up. Tighten the hold down bolts. after adjustment is complete.

5. Stop the engine and reconnect the plugs at the control unit.

NOTE: On fuel injected models, the flywheel must be marked at 6° BTDC to allow the use of a standard (strobe) timing light. The 3° ATDC timing mark does not apply to these models.

SPARK PLUGS

Year	Model		Type	Gap (in.)
'85–'86	Jetta, Scirocco, Quantum, Golf, exc Calif.		W175 T30, N8Y	.024–.032
		California	WR7DS, N8GY	.024–.032
	GTI, GLI, Scirocco 1.8L, Convertible 1.8L		WR7DS, N8GY	.024–.032
	Quantum 5 cyl. exc Calif.		W7D, N8Y	.024–.032
		California	WR7DS, N8GY	.024–.032
	Vanagon		W145M2, N288	①

NOTE: The underhood specification sticker often reflects changes or improvements made to the vehicle line while in production. Sticker figures must be used if they disagree with those in this chart.
① See Underhood Specification Sticker

EMISSION CONTROL SYSTEMS

CRANKCASE EMISSION CONTROL SYSTEM

Positive Crankcase Ventilation (PCV)

The Positive Crankcase Ventilation (PCV) system keeps harmful vapor by-products of combustion from escaping into the atmosphere and prevents the build up of crankcase pressure which can lead to oil leakage. Crankcase vapors are recirculated from the camshaft cover through a hose to the air cleaner. Here they are mixed with the air/fuel mixture and burned in the combustion chamber.

Trouble Diagnosis

When rough engine idling is encountered, check for a clogged PCV valve or a pinched or plugged hose. Inspect the PCV valve assembly for any unusual conditions or damage. Repair or replace as required.

Service

The only maintenance required on the crankcase ventilation system is a periodic inspection. At every tune-up, examine the hoses for clogging or deterioration. Clean or replace the hoses as necessary.

EVAPORATIVE EMISSION CONTROL SYSTEM

The Evaporative Emission Control System prevents the escape of raw fuel vapors (unburned hydrocarbons) into the atmosphere. The system consists of an unvented fuel tank filler cap, fuel tank expansion chamber, an activated charcoal filter canister and the connector hoses. Fuel vapors which reach the filter desposit hydrocarbons on the surface of the charcoal filter element. Fresh air enters the filter when the engine is running and forces the hydrocarbons to the air cleaner where they join the air/fuel mixture and are burned.

Charcoal Canister

This fuel vapor canister is used to absorb and store fuel vapors emitted from the fuel tank. While the engine is idling, a small amount of fuel vapor is drawn from the canister and routed to the engine for combustion. This allows purge of the canister through the PCV hose while the engine is idling. Normal servicing requires that the filter element be replaced periodically.

Trouble Diagnosis

Incorrect operation of the above system will result in poor engine idling, poor driveability or engine stalling.

Service

Inspect the system for damaged charcoal canister, split or dry rotted hoses or hoses not connected. Repair or replace as necessary. Periodically inspect the condition of the various connecting lines and the charcoal filter. The charcoal filter should be replaced at 48,000 mile intervals.

EXHAUST EMISSION CONTROL SYSTEM

To reduce oxides of nitrogen (NOx) emissions, metered amounts of exhaust gases are added to the air/fuel mixture. The recirculated exhaust gas lowers the peak flame temperature during combustion. Exhaust gas from the manifold passes through a filter where it is cleaned. The vacuum operated EGR valve controls the volume of this exhaust gas which is allowed into the intake manifold. Since the exhaust gas contains little or no oxygen, it cannot react with or influence the air/fuel mixture. However, the exhaust gas does (by volume) take up space in the combustion chambers (space that would otherwise be occupied by a heat producing, explosive air/fuel mixture), and does serve to lower peak combustion chamber temperature. The amount of exhaust gas routed to the combustion chambers is regulated by means of a vacuum operated EGR valve. There is no EGR at idle, partial at slight throttle and full EGR at mid throttle.

The EGR valve on fuel infected models is controlled by a temperature valve and a vacuum amplifier. The valve is located at the front of the intake manifold.

EGR Valve

Testing

Be sure that the vacuum lines are not leaking. Replace any that are leaking or cracked.

1. Start the engine and allow it to reach normal operating temperature.
2. Run the engine at idle.
3. Remove the vacuum hose from the EGR valve.
4. Connect the line from the brake booster to the EGR valve (this can be done by installing a "T" in the vacuum line to the retard side of the distributor diaphragm and running a separate hose from there to the EGR valve).
5. If the engine speed does not change, the EGR valve is clogged or damaged. Repair or replace as required.

EGR VALVE

Removal and Installation

1. Disconnect the vacuum hose from the EGR valve.
2. Unbolt the EGR fitting on the opposite side of the valve.
3. Remove the remaining bolts and lift the EGR valve from the intake manifold.
4. Installation is the reverse of the removal procedure. Use a new gasket at the intake manifold.

EGR TEMPERATURE VALVE

Testing

Warm the engine to normal operating temperature.

1. With the engine at idle, attach a vacuum gauge between the EGR temperature control valve and the EGR valve. The valve should be replaced if the gauge shows less than 2 in.Hg.

EGR DECELERATION VALVE

Testing

1. Remove the hose from the deceleration valve. Plug the hose.
2. Run the engine for a few seconds at 3000 rpm.
3. Snap the throttle valve closed.
4. Check for suction at the hose connection.
5. Remove the hose from the connector.
6. Run the engine at 3000 rpm for a few seconds. No suction should be felt. If no suction is felt, replace the EGR valve and check for a clogged or pinched hose.

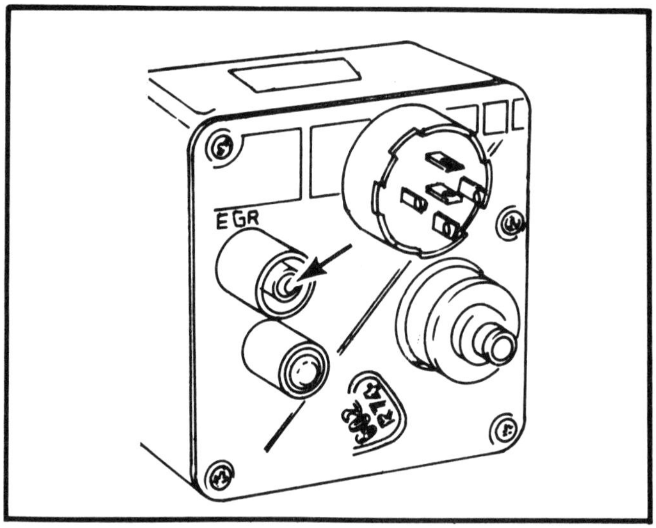

Resetting the EGR service indicator light
(© Volkswagen of America)

EGR VACUUM AMPLIFIER

Testing

1. Run the engine at idle.
2. Connect a vacuum gauge between the vacuum amplifier and the throttle valve port.
3. The gauge should read 0.2–0.3 in. Hg. If not, check the throttle plate for correct position and inspect the port for obstruction.
4. Connect a vacuum gauge between the vacuum amplifier and the temperature valve.
5. Replace the vacuum amplifier if the gauge reads less than 2 in. Hg.

Milage Reminder Light/Switch

Reset

The EGR reminder light in the speedometer should light up every 15,000 miles as a reminder for maintenance. To reset the light switch, press the white button located behind the speedometer assembly. The reminder light should go out.

Catalytic Converter

All models are equipped with a catalytic converter. The converter is installed in the exhaust system, upstream and adjacent to the muffler.

Catalytic converters change noxious emissions of hydrocarbons (HC) and carbon monoxide (CO) into harmless carbon dioxide and water vapor. The reaction takes place inside the converter using platinum and palladium metals as the catalyst. If the engine is operated on lead free fuel, the converter is designed to last 50,000 miles before replacement.

Maintenance

Required maintenance on the catalytic converter involves checking the condition of the insert every 30,000 miles. Inspect the catalytic converter housing and hanger brackets for any signs of damage or deterioration. Repair or replace as required. As the interval is reached, an indicator light on the dash will glow. Once the service to the converter

Manual transmission – U.S.A. (except Calif.)
Auto. transmission – U.S.A. (except Calif.)

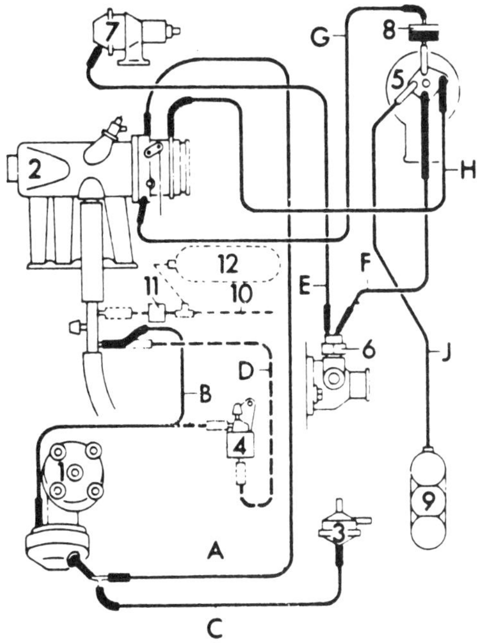

1. Ignition distributor
2. Intake manifold
3. Charcoal filter valve
4. Two-way valve (A/C cars only)
5. Vacuum booster
6. EGR temperature valve
7. EGR valve
8. Deceleration valve (49 states autom. transm. only)

8. Deceleration valve (49 states autom.transm.only)

9. Vacuum tank

10. Hose to air conditioner (A/C cars only)
11. Check valve (A/C cars only)

A. Black
B. White
C. Violet
D. Pink
E. Yellow
F. Light blue
G. Gray
H. Red
I. Light green

1. Ignition distributor
2. Intake manifold

Manual transmission – California only
Auto. transmission – California only

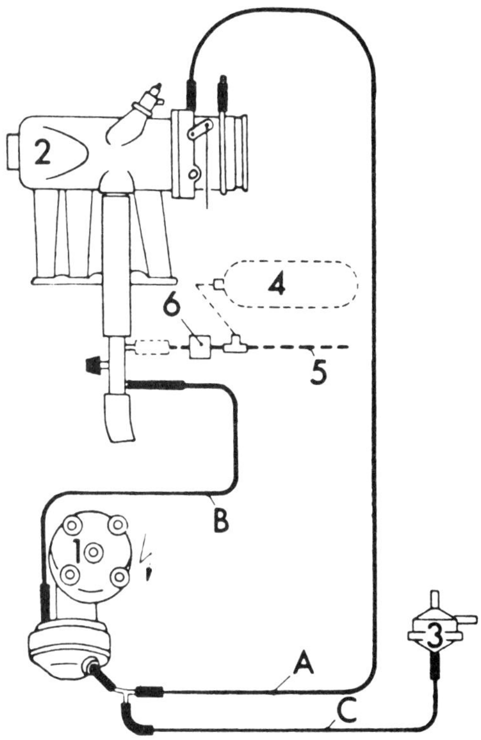

3. Charcoal filter valve
4. Air conditioner vacuum tank (A/C cars only)
5. Hose to air conditioner (A/C cars only)
6. Check valve (A/C cars only)

A. Black
B. White
C. Violet

Volkswagen-Jetta, Golf and Scirocco vacuum circuits, others similar (© Volkswagen of America)

is completed, the indicator light must be reset. Check or replace the converter as follows:

1. Raise and support the vehicle safely.
2. Disconnect the temperature sensor located at forward section of the converter.
3. Loosen and remove the bolts holding the converter to the exhaust system and the chassis.
4. Remove the converter from the vehicle.
5. Hold the converter up to a strong light and look through both ends, checking for blockages. If the converter is blocked, replace it.
6. Install the converter in the reverse order of removal.
7. Reset the indicator light as outlined.

Oxygen Sensor

All models are equipped with an oxygen sensor system which lowers toxic exhaust emissions while increasing fuel economy. The sensor system monitors the oxygen content in the exhaust system and makes adjustments to the air/fuel mixture in order to achieve maximum fuel efficiency over a wide range of operating conditions. The system incorporates a warning light marked OXS and is located in the instrument panel. The oxygen sensor must be replaced every 30,000 miles.

Resetting the oxygen catalytic converter service indicator light (© Volkswagen of America)

Emission Indicator Lights

Reset

After service is performed to the catalytic converter and/or the oxygen sensor, the indicator lights, located on the instrument panel, will require resetting. This can be accomplished by pushing the white button on the front of the switch until an audible click is heard. The switch is located on the firewall.

VACUUM CIRCUITS

Manual transmission — U.S.A. (except Calif.)
Auto. transmission — U.S.A. (except Calif.)

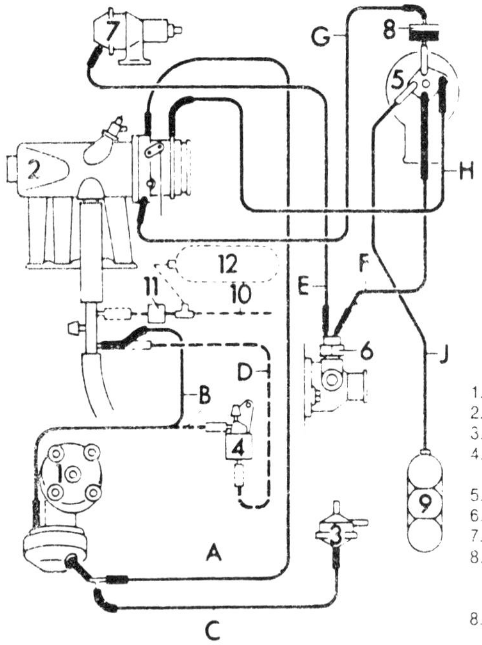

1. Ignition distributor
2. Intake manifold
3. Charcoal filter valve
4. Two-way valve (A/C cars only)
5. Vacuum booster
6. EGR temperature valve
7. EGR valve
8. Deceleration valve (49 states autom. transm. only)
8. Deceleration valve (49 states autom.transm.only)
9. Vacuum tank
10. Hose to air conditioner (A/C cars only)
11. Check valve (A/C cars only)

A. Black
B. White
C. Violet
D. Pink
E. Yellow
F. Light blue
G. Gray
H. Red
I. Light green

Manual transmission — California only
Auto. transmission — California only

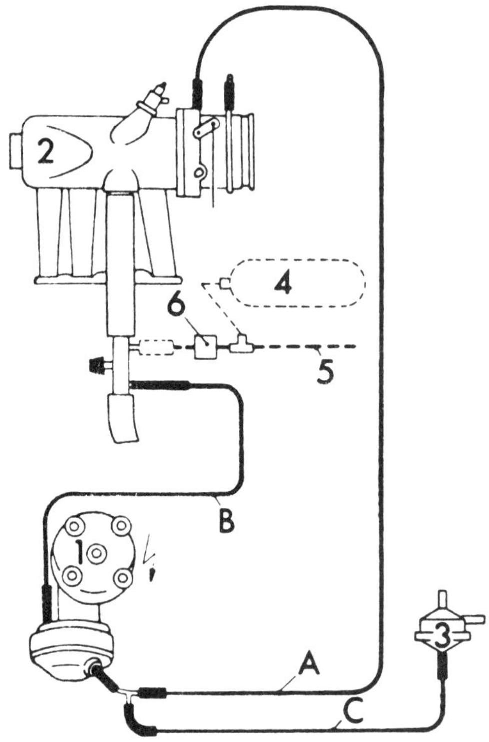

1. Ignition distributor
2. Intake manifold
3. Charcoal filter valve
4. Air conditioner vacuum tank (A/C cars only)
5. Hose to air conditioner (A/C cars only)
6. Check valve (A/C cars only)

A. Black
B. White
C. Violet

Golf Model—Typical (© Volkswagen of America)

Volvo

INDEX

VEHICLE IDENTIFICATION NUMBER AND ENGINE IDENTIFICATION

Vehicle Identification Number (VIN)

The vehicle identification number consists of seventeen digits. This number is embossed on a gray colored plate which is located on the upper left side of the engine compartment, either on the shock tower or the dash panel. The seventh digit of the number represents the engine code and the tenth digit of the number represents the year of the vehicle, 'F' equals 1985 'G' equals 1986. All engines carry an identification number.The VIN is also stamped on the right side door pillar. The is also a emissions label, located on the left front wheel housing (black print on a white background).

Engine Identification Number

The engine identification number on the B-230F, B-230F turbo and the B-28F engines is located on a decal on the timing belt cover and stamped on the block in the lower left front corner above the the front cover of the oil pan. The number on the B-21F turbo and B-23F engines is stamped on the left side of the engine block near the distributor. On the D-24 and D-24 turbo diesel engines, the engine identification numbers are stamped on the block on the left side of the engine under the vacuum pump.

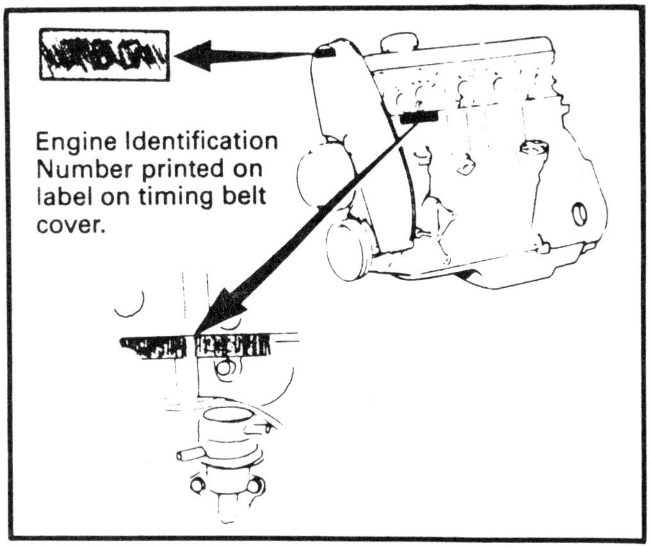

B-230F and B-28F engine number location (© AB Volvo Car Corp.)

Vehicle identification number

Y V1 D X 69 4 2 C 1 001234

YV1	Manufacturer's Code	
D	Series and safety system. D = 760 GLE with 3 point seat belts	
X	Vacant	
69	Engine type	69 = B 28 F 76 = D 24 T
4	Body version. 4 = 4-door with standard wheelbase	
2	Transmission	2 = M 46 7 = AW 71 8 = BW 55
C	Model year designation	C = 1982 model year D = 1983 model year
1	Manufacturing plant	0 = Sweden 1 = Sweden 2 = Belgium 3 = Canada
001234	Chassis number	

Breakdown of a typical vehicle identification number (© AB Volvo Car Corp.)

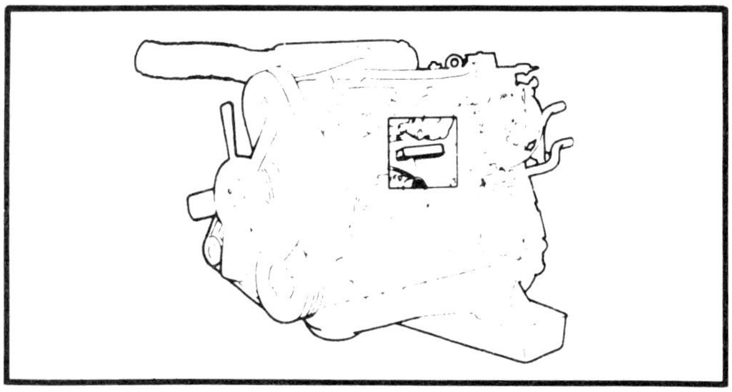

D-24T diesel engine number location (© AB Volvo Car Corp.)

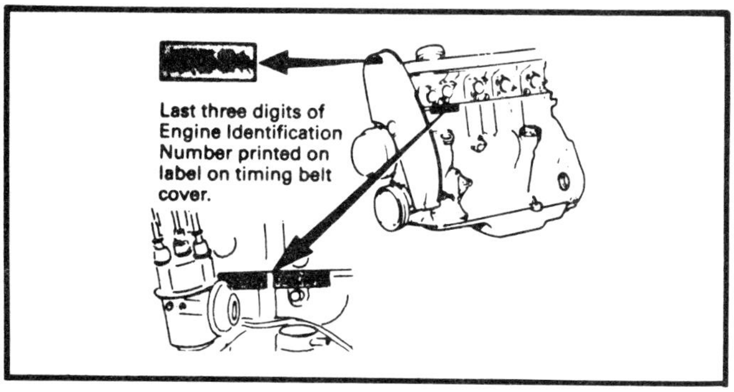

B-21F and B-23F engine number locations (© AB Volvo Car Corp.)

GASOLINE ENGINE TUNE-UP SPECIFICATIONS

(When analyzing compression test results, look for uniformity among cylinders, rather than specific pressures)

Year	Engine Model and Displacement cu. in.	Spark Plugs Type	Spark Plugs Gap (in.)	Distributor Point Dwell (deg)	Distributor Point Gap (in.)	Ignition Timing (deg) MT	Ignition Timing (deg) AT	Intake Valve Opens (deg)	Fuel Pump Pressure (psi)	Idle Speed (rpm) MT	Idle Speed (rpm) AT	Valve Clearance (cold) (in.) In	Valve Clearance (cold) (in.) Ex
'85–'86	B-21F 130	WR7DC	0.028–0.032	Electronic		12B	12B	—	64–75	900	900	0.014–0.016	0.014–0.016
'85–'86	B-21FT	WR7DC	0.028–0.032	Electronic		12B	12B	—	64–75	900	900	0.014–0.016	0.014–0.016
'85–'86	B-23F	WR7DS	0.028–0.032	Electronic		12B	12B	—	64–75	750	750	0.014–0.016	0.014–0.016
'85–'86	B-28F	HR6DS	0.024–0.028	Electronic		23B	23B	—	64–75	900	900	0.004–0.006	0.010–0.012
'85–'86	B-230F	WR7DC	0.028–0.032	Electronic		12B	12B	—	64–75	750	750	0.012–0.016	0.012–0.016
'85–'86	B-230FT	WR7DC	0.028–0.032	Electronic		12B	12B	—	64–75	750	750	0.012–0.016	0.012–0.016

DIESEL ENGINE TUNE-UP SPECIFICATIONS

Year	Model	Valve Clearance ① Intake (in.)	Valve Clearance ① Exhaust (in.)	Injection Pump Setting ② (in.)	Injector Nozzle Opening Pressure (psi)	Idle Speed (rpm)	Compression Pressure (psi)
'85–'86	D-24	0.006–0.010	0.014–0.018	0.0283–0.0315	1700–1850	750	340 (min)–455 (max) ③
'85–'86	D-24T	0.006–0.010	0.014–0.018	0.0323–0.0355	2062–2318	830	313 (min)–455 (max) ③

① Cold
② Plunger stroke
③ Maximum difference between cylinders 115 lbs. psi.

FIRING ORDERS

1-6-3-5-2-4

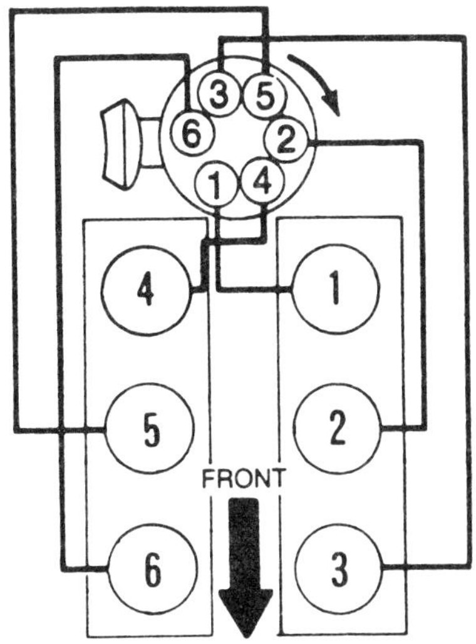

B-28F engine firing order (© AB Volvo Car Corp.)

FIRING ORDERS

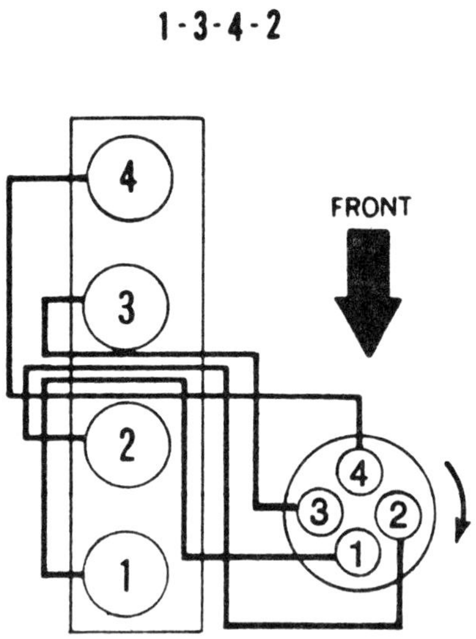

B-21F and B-23F engine firing order (© AB Volvo Car Corp.)

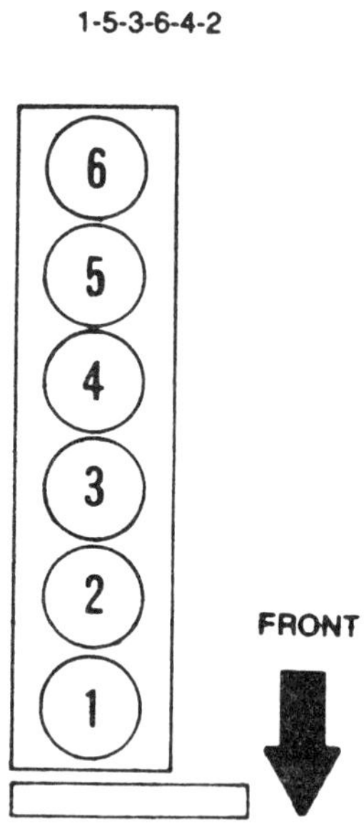

D-24T diesel engine firing order (© AB Volvo Car Corp.)

EMISSION EQUIPMENT USED

VOLVO DL, B-21F TURBO (130 CID)
- K-Jetronic fuel injection
- Constant injection system (CIS)
- Evaporative emission control system (EEC)
- Oxygen sensor (OS)
- Positive crankcase ventilation (PCV)
- Three way catalytic converter
- Idle speed compensator (ISC)
- Constant idle speed system
- Air flow sensor

VOLVO DL, GL, B-23F (142 CID)
- LH-Jetronic 11 fuel injection
- Evaporative emission control system (EEC)
- Knock sensor (KS)
- Oxygen sensor (OS)
- Idle speed control (ISC)
- Constant idle speed system
- Three way catalytic converter
- Positive crankcase ventilation
- Air flow sensor

VOLVO DL, GL, D-24 DIESEL (145 CID)
- Oil separator valve (OSV)
- Positive crankcase ventilation (PCV)

VOLVO 760, GLE, D-24T TURBO DIESEL (145 CID)
- Oil separator valve (OSV)
- Positive crankcase ventilation (PCV)

VOLVO 740 GL, B-230F (141 CID)
- LH-Jetronic 2.2 fuel injection
- Evaporative emission control system (EEC)
- Knock sensor (KS)
- Oxygen sensor (OS)
- Idle speed control (ISC)
- Constant idle speed system
- Three way catalytic converter
- Positive crankcase ventilation
- Air flow sensor
- Computer controlled ignition
- Computer model number—Bosch EZ–117K

VOLVO 740 GLE, B-230F (141 CID)
- LH-Jetronic 2.2 fuel injection
- Evaporative emission control system (EEC)
- Knock sensor (KS)
- Oxygen sensor (OS)
- Idle speed control (ISC)
- Constant idle speed system
- Three way catalytic converter
- Positive crankcase ventilation
- Air flow sensor
- Computer controlled ignition
- Computer model number—Bosch EZ–117K

VOLVO 740 TURBO B-230F (141 CID)
- LH-Jetronic 2.2 fuel injection
- Evaporative emission control system (EEC)
- Knock sensor (KS)
- Oxygen sensor (OS)
- Idle speed control (ISC)
- Constant idle speed system
- Three way catalytic converter
- Positive crankcase ventilation
- Air flow sensor
- Computer controlled ignition
- Computer model number—Bosch EZ–117K

VOLO 760 GLE B-28F (174 CID)

Bosch constant injection system (CIS)
Evaporative emission control system (EEC)
Knock sensor (KS)
Oxygen sensor (OS)
Idle speed control (ISC)
Constant idle speed system
Three way catalytic converter
Positive crankcase ventilation
Air flow sensor
Computer controlled ignition
Computer model number—Bosch EZ–117K

Volvo 760 Turbo B-230F (174 CID)

LH-Jetronic 2.2 fuel injection
Evaporative emission control system (EEC)
Knock sensor (KS)
Oxygen sensor (OS)
Idle speed control (ISC)
Constant idle speed system
Three way catalytic converter
Positive crankcase ventilation
Air flow sensor
Computer controlled ignition
Computer model number—Bosch EZ–117K

REQUIRED EMISSION CONTROL MAINTENANCE

Emission Maintenance Intervals

Procedure	Mileage in Thousands
Air filter change	30,000
Idle speed adjustment	15,000
Oxygen sensor replacement	30,000
PCV system check	15,000
Evaporative canister replacement	15,000
Ignition timing check	45,000
Fuel filter replacement	15,000
Fuel tank pick-up screen cleaning	15,000
Spark plug replacement	15,000
Valve lash adjustment	15,000
Fuel filter draining—diesel	8,000
Fuel filter replacement—diesel	15,000

IDLE SPEED AND MIXTURE ADJUSTMENT

IDLE MIXTURE

Adjustment

Note: The following specifications are published from the latest information available. if the published information differs from the information on the vehicle emission label, use the data on the emission label. The idle mixture adjustments are factory set and sealed and no adjustments are required. The mixture control adjustment screw opening is plugged to prevent any adjustment. Mixture control adjustments should be done when the mixture control unit is replaced or when the vehicle fails the emissions test.

FOUR CYLINDER TURBO MODELS

1. Remove the mixture control unit from the vehicle, separate the control unit and drive out the mixture plug with a punch.
2. With the mixture control plug removed, reassemble the control unit and install it onto the engine.
3. Disconnect the oxygen sensor electrical connector and remove the plug in the exhaust pipe in front of the catalytic converter, then insert the carbon monoxide (CO) probe.
4. If the CO reading is out of specifications, insert a long hex head wrench or equivalent into the adjustment hole and adjust the CO reading to specifications.
5. Reconnect the oxygen sensor electrical connector and check the CO reading. If the CO reading is still out of specifications, then repeat the adjustment procedure.
6. After every adjustment, remove the adjustment wrench and cover the adjustment hole. By covering the hole it will prevent a lean mixture while checking the CO level.
7. When the CO adjustment has been completed, remove the test equipment, seal the mixture adjustment hole and insert exhaust pipe plug.

FOUR CYLINDER NON-TURBO MODELS

1. Ground the blue and white wire at the test point and adjust the idle speed with the idle screw to 720 rpm. Remove the ground from the test point; now the engine speed should increase to 730–770 rpm.
2. Disconnect the oxygen sensor and insert the CO probe to check

IDLE SPEED AND CO LEVEL

Engine Model	Idle rpm	CO Level
B-21F	900	0.4–0.8
B-21F Turbo	900	0.7–1.3
B-23F	750	0.4–0.8
B-28F	900	0.7–1.3
B-230F	750	0.4–0.8
B-230F Turbo	750	0.4–0.8

NOTE: The CO level is with the oxygen sensor disconnected. The level will be less 1.0% with the oxygen sensor connected.

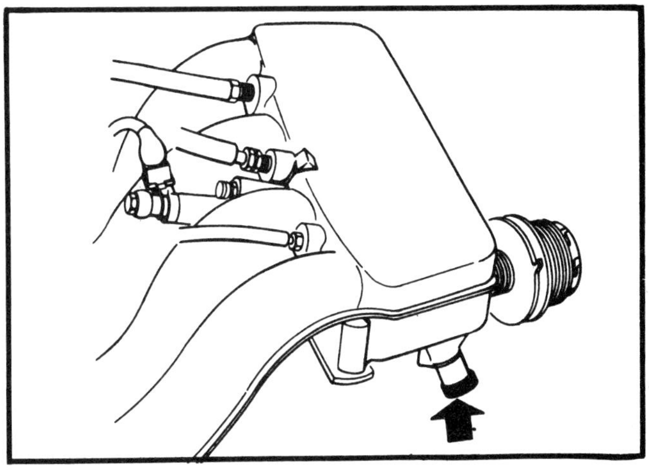

B-21F turbo idle mixture (CO) adjustment (© AB Volvo Car Corp.)

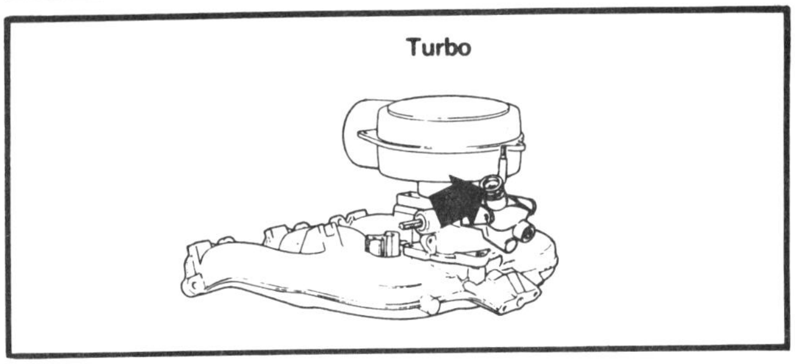

B-21F turbo idle adjustment screw (© AB Volvo Car Corp.)

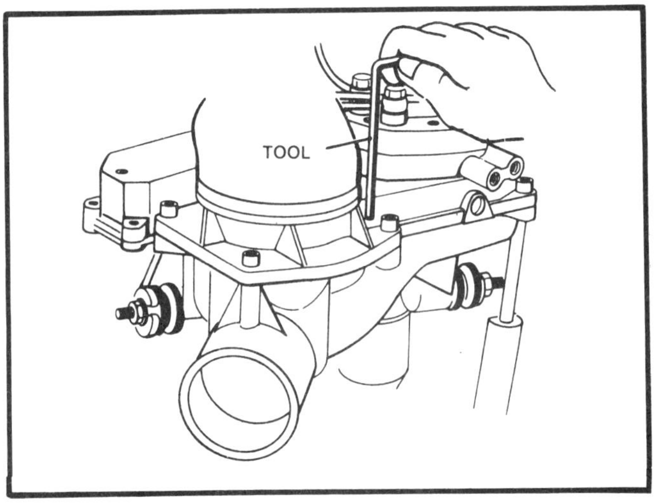

B-21F idle speed adjustment screw (© AB Volvo Car Corp.)

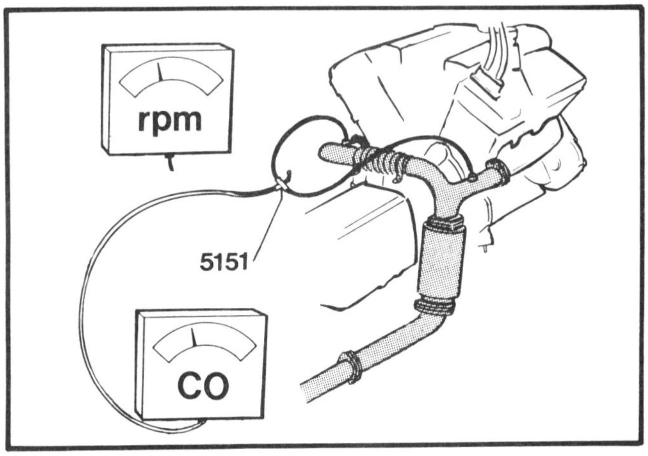

Connecting the CO meter probes into the exhaust pipes (© AB Volvo Car Corp.)

the CO level. If the CO level is out of specifications, turn off the engine and drill two 5/16 in. holes in the adjustment seal and pull out the mixture plug.

3. Adjust the CO level by turning the adjustment screw counter-clockwise to decrease the CO level and clockwise to increase the CO level.

4. Reconnect the oxygen sensor, recheck the CO level and seal the adjustment with a new plug.

V-6 MODELS

1. Remove the mixture control unit from the vehicle, separate the control unit and the drive out the mixture plug with a punch. Reinstall the mixture control unit on the vehicle.

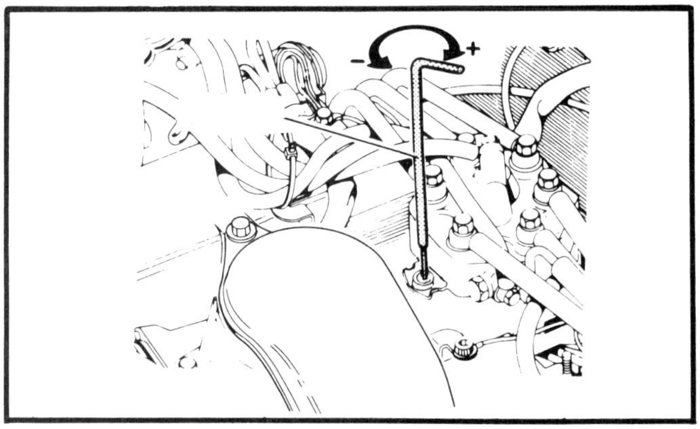

Adjusting the CO level (© AB Volvo Car Corp.)

2. Disconnect the oxygen sensor electrical connector. Remove the exhaust plugs from the header pipes and connect a CO probe to each pipe.

3. Turn the dual probe adapter to the center position. With the adapter in this position the exhaust gases are admitted from both cylinder banks for a total CO reading.

4. With the CO meter and tachometer still installed, install the air cleaner and connect the hoses. To adjust the CO level, insert a long hex head wrench or equivalent into the adjusting hole and adjust the CO level to specifications.

5. After each adjustment, remove the adjusting wrench and cover the adjustment hole. By doing this it will prevent a lean mixture while checking the CO level.

6. Turn the dual probe adapter toward the left cylinder bank and check the CO reading. This will also check the CO balance between the left and right cylinder banks.

7. If the left bank CO level is not within specifications, correct it by removing the balance screw plug and then adjusting the balance screw number two.

NOTE: When turning the balance adjusting screw, the left screw goes toward the right side manifold and the right balance screw goes toward the left side manifold.

8. Turn the dual probe adapter toward the right cylinder bank and check the CO reading. The CO reading should be equal for both banks and correct for a total system.

9. If the right bank CO level is not within specifications, correct it by removing the balance screw plug and then adjusting the balance screw number one. With a dual probe adapter in center position recheck the CO level.

10. When all CO adjustments are completed, reconnect the oxygen sensor electrical connector and remove all test equipment. Insert the exhaust pipe plugs and seal the mixture adjustment hole.

IDLE SPEED

Adjustment

B-21F AND B-23F ENGINES WITH FUEL INJECTION

1. Run the engine to normal operating temperature. Disconnect the throttle control rod at the lever, be sure the cable and pulley run smoothly and do not bind in any position. Stop the engine.

2. Remove the ECU cover panel, deactivate the ECU by grounding the #10 terminal of the blue connector (with the ECU connector in place), inserting a copper wire along the terminal wire.

3. Connect a tachometer to the engine and connect a test light across the battery positive terminal and the terminal on the throttle micro switch with the yellow wire connected.

4. Start the engine, the test light should not light up. If it does, adjust the micro switch position by loosening the switch retaining screws.

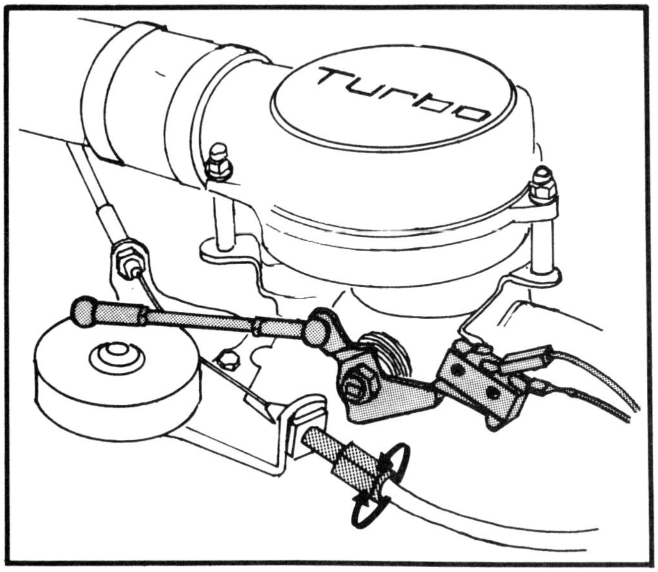

Adjusting the throttle cable (© AB Volvo Car Corp.)

Move the switch down until the light goes out, then tighten the switch retaining screws.

NOTE: This adjustment is temporary, the final adjustment will follow later in this procedure.

5. The idle speed should be 700 rpm with the ECU deactivated and 750 with the ECU activated. If the idle speed is out of specification, adjust the idle speed by proceeding to the next step if the idle speed is within specifications proceed to Steps 7 and 8.

6. Use the throttle position adjustment screw to increase or decrease the idle speed. The test light should not light up while adjusting the idle speed, if it does, readjust the micro switch position and activate the ECU by removing the ground wire from terminal # 10. With the ECU activated the idle speed should be 750 rpm, stop the engine.

7. Reconnect the throttle control rod at the lever, being sure the cable pulley is completely retracted. If the control rod length must be adjusted, disconnect the throttle cable kickdown cable (if so equipped). Loosen the lock-nuts on either end of the rod and adjust the rod as necessary by turning it, then tighten the lock-nuts.

8. Attach the throttle cable and adjust if necessary by turning the nut on the end of the cable. Automatic transmission kickdown cable length should be checked at closed and open throttle with the engine off. Open throttle cable measurement should be checked with the throttle pedal in the vehicle depressed. not by actuating the linkage by hand. The cable should be pulled out 1.9 in. (50mm).

9. Adjust the micro switch by moving the switch up, with the engine off and the throttle closed. Loosen the switch retaining screws, and move the switch up until the test light lights up. Set the switch position by moving the switch down 0.08–0.10 in. (2–5.5mm). If the test light does light up again, the the adjustment procedure must be performed again.

10. Remove all test equipment and reinstall the ECU panel.

B-21F TURBOCHARGED ENGINES

1. Run the engine to normal operating temperature. Disconnect the throttle control rod at the lever, be sure the cable and pulley run smoothly and do not bind in any position. Stop the engine.

2. Remove the ECU cover panel, deactivate the ECU by grounding the number ten terminal (with the ECU connector in place), inserting a copper wire along the terminal wire.

3. Connect a test light across the battery positive terminal and the terminal on the throttle micro switch with the orange wire connected.

4. The test light should not light up. If it does, adjust the micro switch position by loosening the switch retaining screws. Move the switch down until the light goes out, then tighten the switch retaining screws.

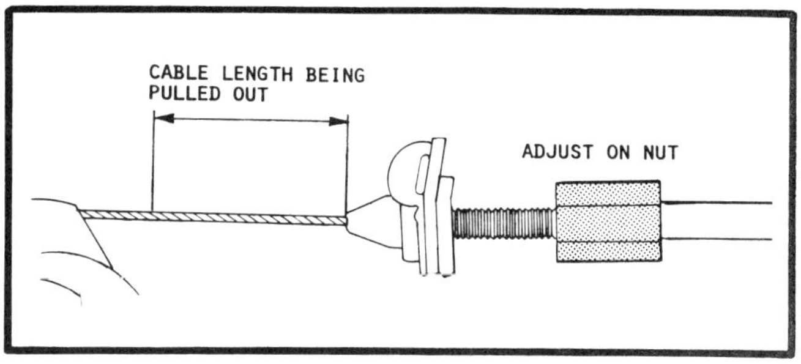

Adjusting the kickdown cable (© AB Volvo Car Corp.)

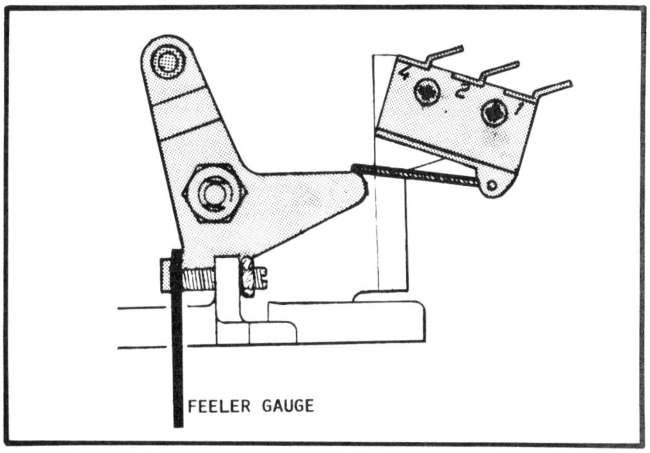

Adjusting throttle switch—B-21F Turbo insert 0.3mm feeler gauge between the throttle adjustment screw and the throttle control lever (© AB Volvo Car Corp.)

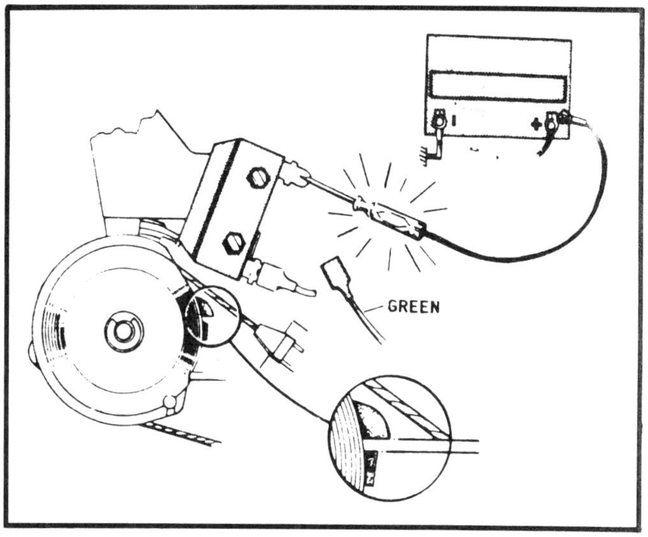

Adjusting the full throttle enrichment switch (© AB Volvo Car Corp.)

NOTE: This adjustment is temporary, the final adjustment will follow later in this procedure.

5. Connect a tachometer to the engine and start the engine. Check the idle speed, the idle speed should be 850 rpm. If the idle speed is outside these specifications, follow the procedure in the next step.

6. Use the throttle position adjustment screw to increase or decrease the idle speed. The test light should not light up while adjusting the idle speed, if it does, readjust the micro switch position and activate

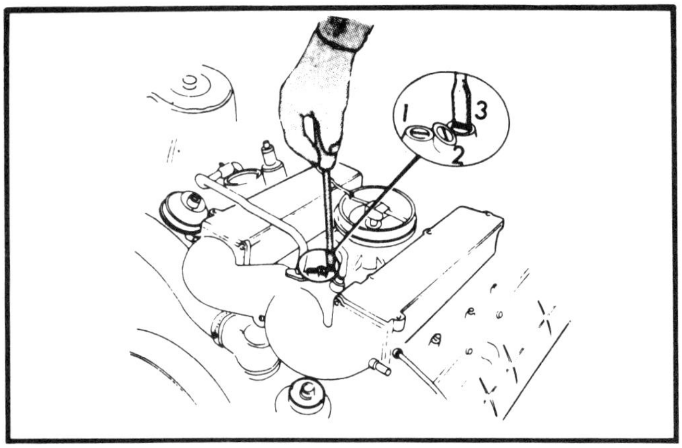

B-28F idle balance (1 and 2) and air adjusting screw #3
(© AB Volvo Car Corp.)

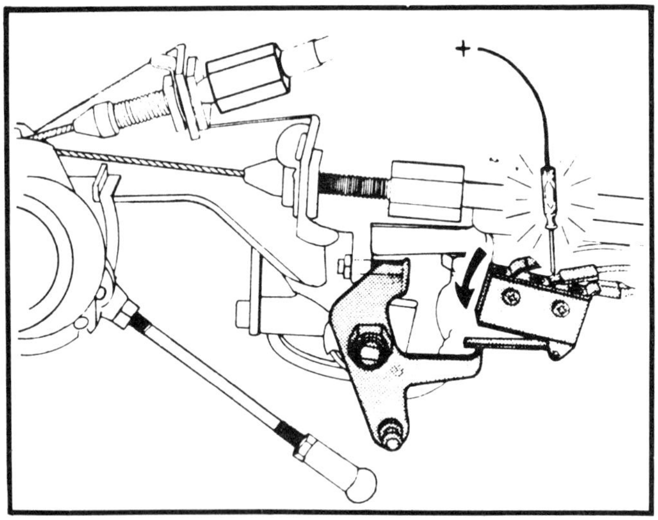

Adjusting throttle micro switch—B-21F with K-Jetronic injection. Adjust idle if test light lights up at idle speed
(© AB Volvo Car Corp.)

the ECU by removing the ground wire from terminal # 10 of the blue connector. With the ECU activated the idle speed should be 900 rpm, stop the engine and reinstall the ECU panel.

7. Reconnect the throttle control rod at the lever, being sure the cable pulley is completely retracted. If the control rod length must be adjusted, disconnect the throttle cable kickdown cable (if so equipped). Loosen the lock-nuts on either end of the rod and adjust the rod as necessary by turning it, then tighten the lock-nuts.

8. Attach the throttle cable and adjust if necessary by turning the nut on the end of the cable. Automatic transmission kickdown cable length should be checked at closed and open throttle with the engine off. Open throttle cable measurement should be checked with the throttle pedal in the vehicle depressed. not by actuating the linkage by hand. The cable should be pulled out 1.9 in. (50mm).

9. Adjust the throttle switch on this model, by inserting a 0.3mm feeler gauge between the throttle adjustment screw and the throttle control lever. Move the switch up until the test light lights up and set the switch by moving it down until the test light goes out. Remove all test equipment and reinstall the ECU panel, if it has not be done already.

B-28F ENGINE

1. Disconnect the throttle control rod at the lever and the cable pulley, be sure the cable and pulley run smoothly and do not bind in any position. Check to see that the throttle shaft and plate do not bind during operation.

2. Screw the idle speed adjustment screw in all the way until it just seats.

NOTE: This screw is used to adjust the idle speed on engines without the CIS system.

3. Remove the ECU cover panel, deactivate the ECU by grounding the number ten terminal (with the ECU connector in place), inserting a copper wire along the terminal wire.

4. Connect a test light across the battery positive terminal and the terminal on the throttle micro switch with the orange wire connected.

5. The test light should not light up. If it does, adjust the micro switch position by loosening the switch retaining screws. Move the switch down until the light goes out, then tighten the switch retaining screws.

NOTE: This adjustment is temporary, the final adjustment will follow later in this procedure.

6. Connect a tachometer to the engine and start the engine, let it run until it reaches normal operating temperatures. Check the idle speed, the idle speed should be 700 rpm.

7. Use the throttle position adjustment screw to increase or decrease the idle speed. Do not adjust the idle speed screw, for it should already be screwed in on its seat. Activate the ECU by removing the ground wire from terminal #10, when the ECU is activated the idle speed should read at 750 rpm.

8. Reconnect the throttle control rod at the cable pulley, being sure the cable pulley is completely retracted. Disconnect the throttle cable kickdown cable (if so equipped). Loosen the lock-nuts on either end of the rod and adjust the rod as necessary by turning it, then tighten the lock-nuts.

9. Attach the throttle cable and adjust if necessary by turning the nut on the end of the cable. Automatic transmission kickdown cable length should be checked at closed and open throttle with the engine off. Open throttle cable measurement should be checked with the throttle pedal in the vehicle depressed. not by actuating the linkage by hand. The cable should be pulled out 1.9 in. (50mm).

10. Adjust the throttle micro switch on this model, by moving the switch up until the test light lights up and set the switch by moving it down until the test light goes out. Remove all test equipment and reinstall the ECU panel, if it has not be done already.

B-230F ENGINE

The CO adjustment on this model is done by turning an adjustment screw on the side of the air mass meter. The screw is connected by a spring to a linear potentiometer. The potentiometer is located on the circuit board inside the air mass meter. It has a range of 15 turns and lacks distinct end stops. The CO setting is adjusted at the factory and a steel seal plug must be drilled out prior to making any later adjustment. The amount of fuel injected can be altered by as much as 40%.

FULL THROTTLE ENRICHMENT SWITCH

B-28F V-6 ENGINE

NOTE: This engine is equipped with the two micro switches actuated by throttle control. This second micro switch closes a Lambda-sond (the oxygen sensor) circuit at full throttle to provide richer air/ fuel mixture at maximum acceleration. Vehicles sold in high-altitude areas have this switch disconnected.

1. To adjust this switch disconnect the yellow terminal at the micro switch and connect a test light between the micro switch terminal and the positive battery terminal.

2. Turn the throttle cable pulley slowly to the full throttle stop. The test light should light up $^{5}/_{32}$-$^{1}/_{32}$ in. (4.0-1.0mm) before the pulley touches the stop.

3. If the test light does not light up at that point. Loosen the switch retaining screws and adjust the switch so it will light the test light at the specified distance as stated above.

4. Remove all test equipment and reconnect the yellow terminal on the micro switch.

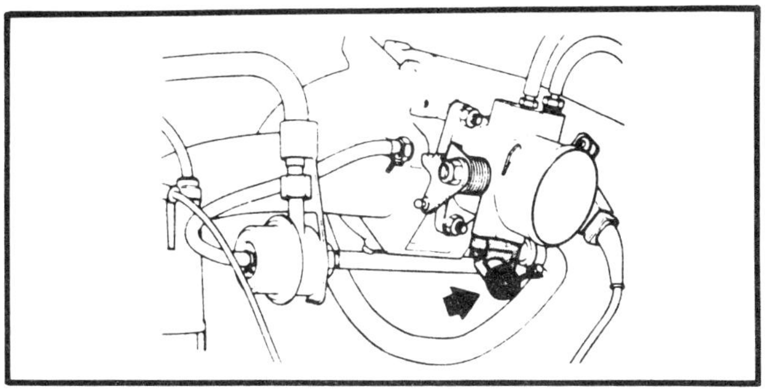

LH-Jetronic idle adjustment location (© AB Volvo Car Corp.)

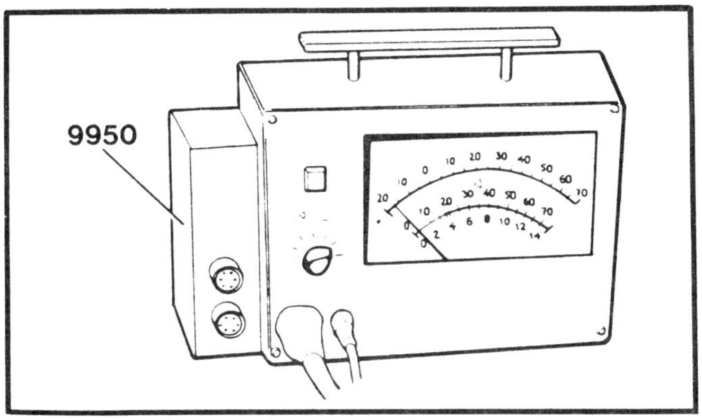

Volvo Mono-tester #9950 (© AB Volvo Car Corp.)

JETRONIC FUEL INJECTION

B-23F AND B-230F ENGINES

1. Seat the throttle butterfly valve by loosening the stop nut on the adjuster screw. Unscrew the adjuster a couple of turns, set the adjuster screw by screwing it in until it just touches the lever, then screw it in an additional ¼ turn. Tighten the lock nut.
2. Disconnect the controlled idle speed connector on the firewall. This is the connector that is directly behind the engine.
3. Connect a test light across the battery positive terminal and the orange wire terminal in the connector. Start the engine and the test light should light up at idle speed. If it does not, readjust the adjuster screw for the throttle butterfly valve position.
4. Open the throttle slightly by hand at the throttle control lever, with the engine running. The test light should go out, if it does not run through this procedure again and try a new throttle switch.

DIESEL ENGINE

NOTE: To correctly set the idle speed you will need either the Volvo Monotester and adapter 9950 or a suitable photo-electric tachometer, since a gasoline engine tachometer by itself can not be used on a diesel engine, because a diesel engine does not have an electric ignition system.

1. Connect the Volvo tachometer or a suitable photo- electric tachometer to the engine and run the engine until it reaches normal operating temperature.
2. Normal idle speed on this engine is 830 rpm, if the idle speed

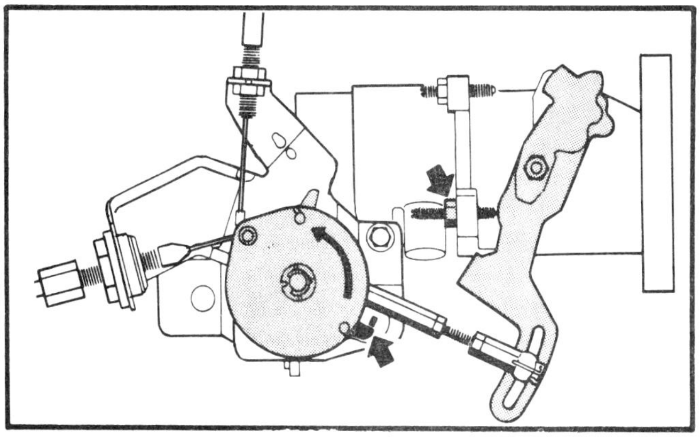

Diesel engine-maximum speed stop. Turn the pulley to bring the engine to maximum speed. (© AB Volvo Car Corp.)

is out of specifications, adjust the idle speed by loosening the lock-nut and turning the idle speed screw on the fuel injection pump.
3. Tighten the lock-nut and apply a dab of paint or thread sealer to the adjusting screw to prevent it from vibrating loose.
4. After adjusting idle speed and maximum engine speed, adjust the engine throttle linkage.

SETTING THE MAXIMUM ENGINE SPEED

The diesel engine is governed by the fuel injection pump so that the engine rpms will not exceed 5400 rpm. Because of the extremely high compression ratio (23.0:1) and the great stored energy diesel oil contains, the diesel engine cannot be run at the high rpm levels of modern gasoline engines, as it would place a tremendous strain on the pistons, wrist pins connecting rods and bearings of the engine. To adjust the maximum idle speed you will need a special tachometer which will work on the diesel engine. See the 'Note' under 'Setting The Idle Speed' on the diesel engine.

1. Connect the special tachometer and run the engine to normal operating temperature. Turn the throttle cable pulley counterclockwise to run the engine at maximum speed.

CAUTION

Do not race the engine longer than absolutely necessary.

2. Maximum speed should be 5400 rpm. If it is not, loosen the lock-nut and adjust, using the maximum speed adjusting screw.
3. Tighten the locknut and apply a dab of paint or thread sealer to the adjusting screw to keep from vibrating loose.

CAUTION

Do not attempt to squeeze more power out of your diesel by extending the maximum speed, it could cause internal engine damage.

4. After adjusting the maximum idle speed, adjust the engine throttle linkage.

IDLE SPEED RPM

Engine Model	Idle rpm	Maximum rpm
D-24	750	5200
D-24 Turbo	830	5400

INITIAL TIMING

Adjustment

ALL MODELS EXCEPT DIESEL ENGINE

1. Clean the crankshaft damper and pointer on the water pump housing with a solvent-soaked rag so that the marks can be seen. Connect a timing light to the number one spark plug and the negative battery terminal. Scribe a mark on the crankshaft damper and on the marker with chalk or luminescent (day-glo) paint to highlight the correct timing mark.
2. Disconnect and plug the distributor vacuum line (except non turbo four cylinder models) disconnect the hose between the air cleaner and the inlet duct at the duct, also place the A/C in the off position if so equipped.
3. Disconnect and plug the hose to the EGR valve and attach a tachometer to the engine, set the idle speed to specifications.
4. With the engine running at normal operating temperature, aim the timing light at the timing mark on the damper. If the marks do not coincide, stop the engine and loosen the the distributor hold down nut.
5. While observing the timing light flashes, slowly rotate the distributor left or right until the timing marks are aligned and stop the engine.
6. Tighten the distributor hold down, be careful not to disturb the distributor. Start the engine and recheck the timing.

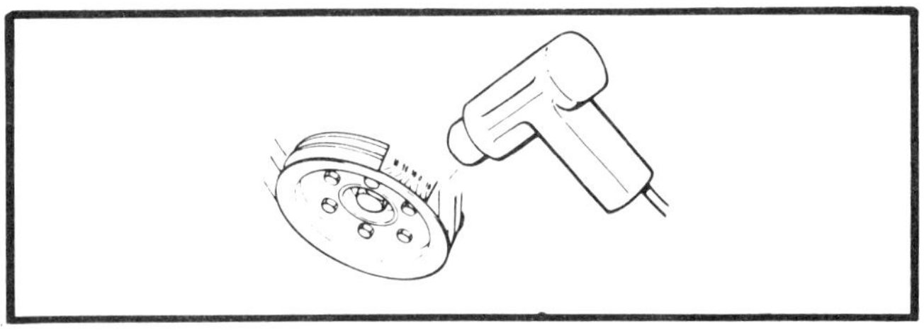

Setting the timing (© AB Volvo Car Corp.)

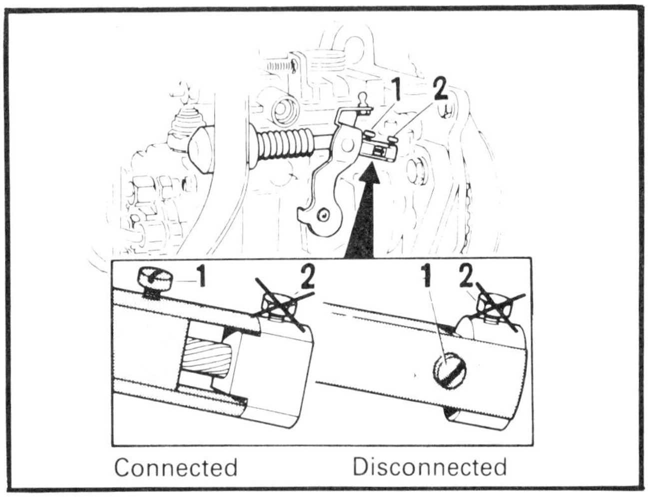

Disconnecting the cold start device (© AB Volvo Car Corp.)

7. If the timing is within the proper specifications stop the engine and reconnect all the vacuum lines, disconnect the timing light. If the timing is out of specification, adjust as necessary.

DIESEL ENGINE, D-24 AND D-24 TURBO

1. Remove the rear timing gear cover and disconnect the cold-start device. Loosen the forward screw on the cold-start device control lever, press the lever back toward the stop.

NOTE: Do not loosen the screw closest to the timing belt.

TIMING SPECIFICATIONS

Engine Model	Degrees
B-21F	12 @ BTDC
B-21F Turbo	12 @ BTDC
B-23F	12 @ BTDC
B-28F	23 @ BTDC
B-230F	12 @ BTDC
B-230F Turbo	12 @ BTDC

NOTE: The underhood sticker often reflects tune-up specifications changes made in production. Underhood sticker figures must be used if they disagree with those in the chart above.

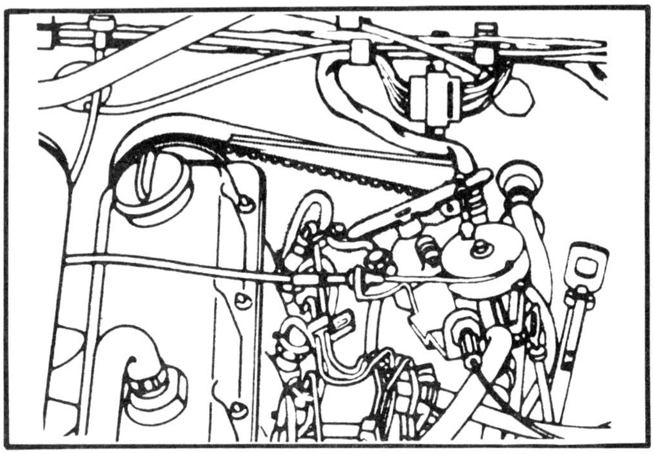

Removing the rear timing cover (© AB Volvo Car Corp.)

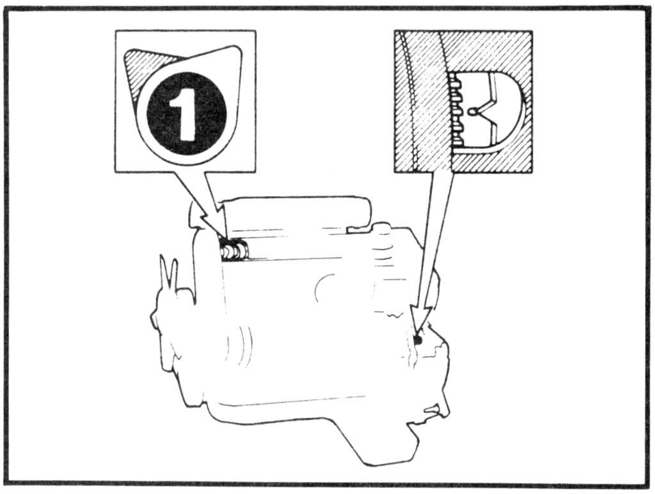

Setting the No. one cylinder to TDC and injection position (© AB Volvo Car Corp.)

2. Rotate the engine to align the mark on the injection gear with the mark on the pump bracket. The '0' mark on the flywheel should be centered at the timing mark window and with the number one cylinder at top dead center.
3. Remove the plug from the rear of the pump and install the Volvo

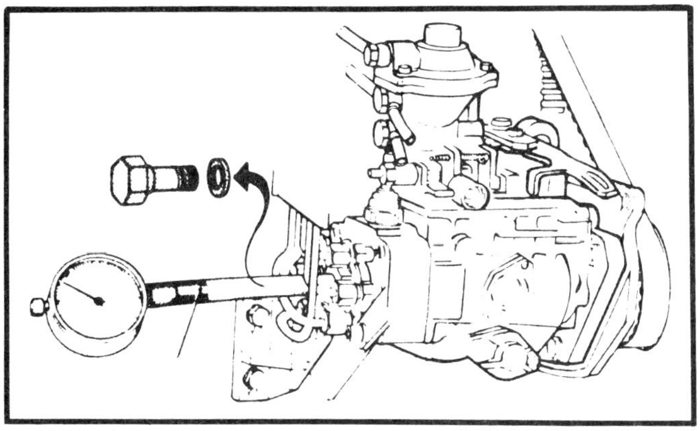

Installing the dial indicator (© AB Volvo Car Corp.)

DIESEL INJECTION TIMING

All Models	Dial Indicator Reading In. (mm)
Checking Specifications	0.0283–0.0315 (0.72–0.80)
Adjusting Specifications	0.0295 (0.75)

dial indicator adapter #5194 or equivalent, with a measuring range of 0–0.1 in. (0–3mm). Set the indicator gauge at approximately 0.08 in. (2mm) .

4. Rotate the engine slowly counterclockwise until the lowest reading on the dial indicator is observed, then re-set the dial indicator to zero.

5. Rotate the engine slowly in the clockwise direction until the '0' mark on the flywheel is centered at the timing mark window. The dial indicator should read within the specified range.

NOTE: When the engine is turned past the timing mark, turn the engine back a ¼ of a turn. Rotate the engine in the clockwise direction until the '0' mark on the flywheel is centered in the timing mark window.

6. If the dial indicator setting is outside the specified range, loosen

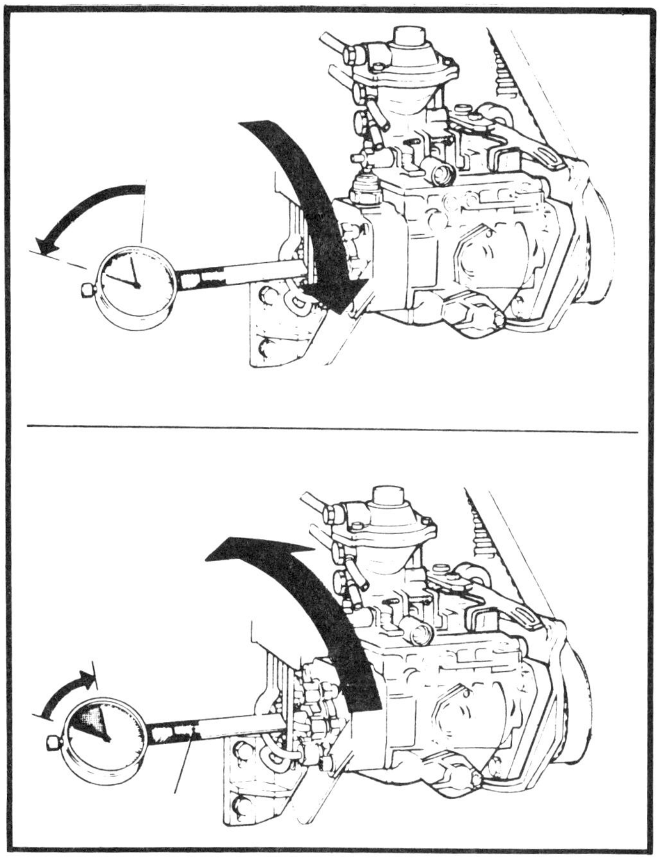

Setting the injection pump (© AB Volvo Car Corp.)

the pump bolts and turn the pump until the dial indicator shows the correct setting. Then re-tighten the pump bolts.

7. Crank the engine over by hand two revolutions and re-check the injection pump timing. If it is still out of specifications readjust as necessary.

8. Remove the dial indicator and adapter, reinstall the rear plug and rear timing gear cover, then reconnect the cold start device.

SPARK PLUGS

Engine Model	Spark Plug (Bosch)	Gap
B-21F	WR7DC	0.028–0.032 in.
B-21F Turbo	WR7DC	0.028–0.032 in.
B-23F	WR7DS	0.028–0.032 in.

Engine Model	Spark Plug (Bosch)	Gap
B-28F	HR6DS	0.024–0.028 in.
B-230F	WR7DC	0.028–0.032 in.
B-230F Turbo	WR7DC	0.028–0.032 in.

EMISSION CONTROL SYSTEMS

CRANKCASE EMISSON CONTROLS

Positive Crankcase Ventilation

All Volvo vehicles are equipped with a closed crankcase ventilation system. The system consists of a PCV valve mounted on the cylinder head cover, with a hose extending from the valve to the base of the carburetor. A closed engine oil inlet air cleaner with a hose connecting it to the carburetor air cleaner housing provides a source of air to the system. The ventilation system operates by manifold vacuum, which is air drawn from the carburetor air cleaner through the crankcase air cleaner and into the engine crankcase. It is then circulated through the engine and drawn out through the PVC valve hose and into a passage at the base of the carburetor where it becomes part of the air/fuel mixture. It is then drawn into the combustion chamber where it is burned and expelled with the exhaust gases.

This system is essentially the same as the system used on vehicles equipped with electronic fuel injection. If the vehicle is equipped with

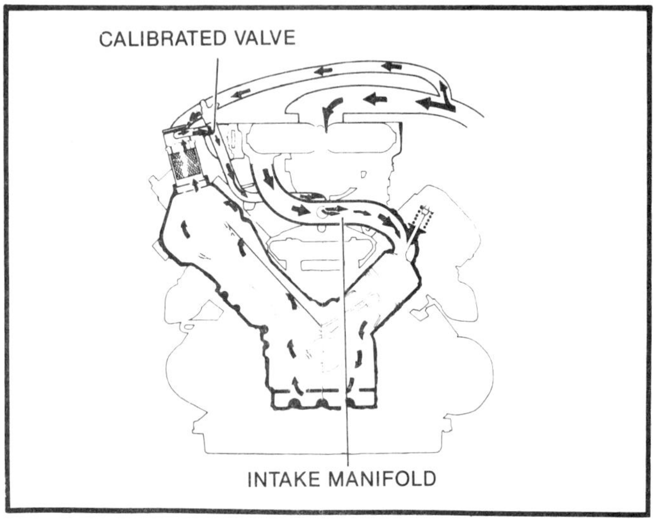

PCV system—B-28F engine (© AB Volvo Car Corp.)

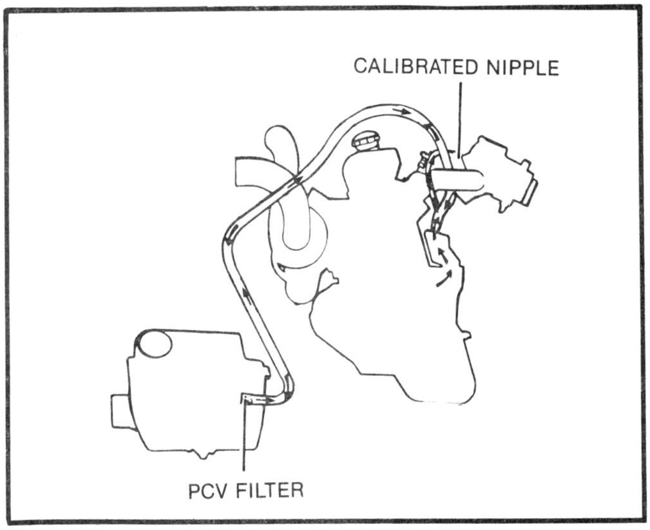

PCV system—B-230F Turbo engine (© AB Volvo Car Corp.)

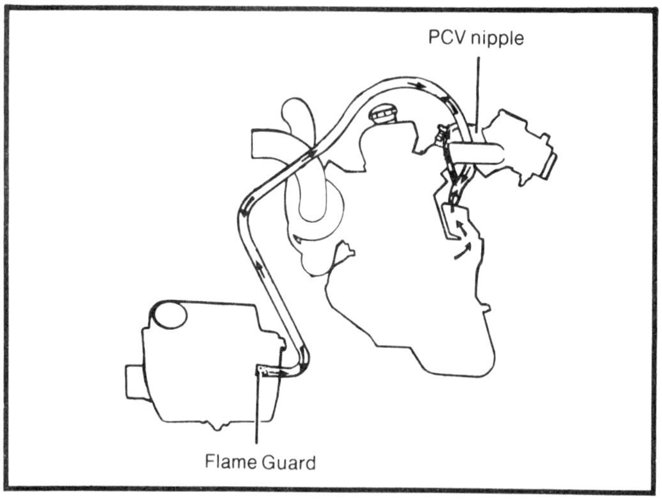

PCV system—B-230F engine (© AB Volvo Car Corp.)

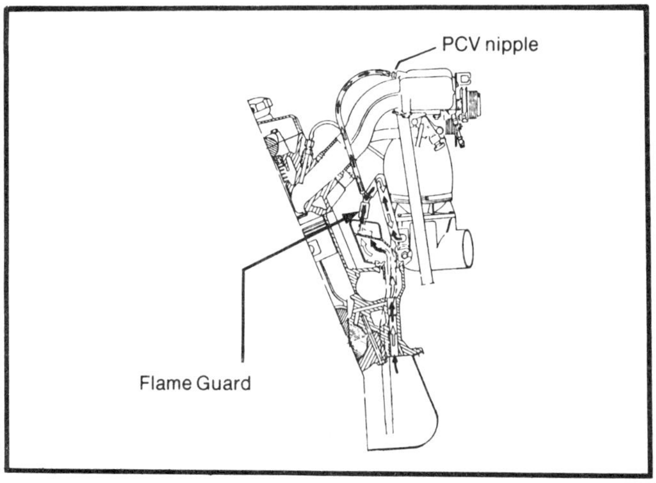

PCV system—B-21F Turbo engine (© AB Volvo Car Corp.)

electronic fuel injection, the engine, when operating, generates crankcase pressure the is used to purge crankcase vapors through the PCV valve to a port in the throttle body. When the engine is shut down the cooling gases in the crankcase causes a partial vacuum which is relieved by a breather hose that draws filtered air from the air cleaner assembly to the valve cover. However, all air entering the intake manifold must be measured by the air flow meter to insure proper air/fuel ratio. It is very important that the PCV system on a vehicle equipped with electronic fuel injection remain a closed system and be free from any disconnects or breaks in the hoses which may allow un-metered air to enter the intake system. Such conditions would cause poor driveability, fuel economy and performance.

System Service

Note: The PCV valve should be checked every 15,000 miles and replaced as required. If the valve is found to be defective, do not attempt to clean it. Replace it.

1. With the engine running, remove the valve from its mounting.

PCV SYSTEM DIAGNOSIS

Symptom	Possible Cause	Repair
Rough idle, hard starting	Clogged PCV valve, vacuum leak in system	Replace PCV valve, repair vacuum leak
No PCV flow	Defective or clogged PCV valve, blocked line	Clean or replace line and PCV valve
Engine stalls at idle	Excessive PCV flow, dirty air filter element	Check PCV valve and air filter
Surging at road speed, hesitation	Improper PCV flow	Check PCV system, clean all lines
Excessive oil consumption	PCV line blocked	Clear line

A hissing sound should be heard and vacuum should be felt from the inlet side of the valve.

2. Reinstall the valve. Remove the crankcase inlet air cleaner. Loosely hold a piece of stiff paper over the opening in the rocker cover. Allow one minute for the crankcase pressure to reduce itself. The paper should then suck itself against the rocker cover with noticeable force. Replace the inlet air cleaner in the rocker cover.

3. With the engine stopped, remove the PCV valve. Shake the valve, a clicking sound should be heard indicating that the valve is not stuck.

4. If the valve fails any of the above tests, it should be replaced.

With the engine stopped, remove the PCV valve. Shake the valve, a clicking sound should be heard indicating that the valve is not stuck. 4. If the valve fails any of the above tests, it should be replaced.

OIL SEPARATOR VALVE

D-24 DIESEL TURBO

The oil separator valve is located on the top of the valve cover and incorporates a spring loaded diaphragm, which is activated by the turbocharger compressor inlet pipe vacuum. The oil separator valve is normally in the open position which permits the crankcase gases to be drawn into the turbocharger inlet. A clogged engine air filter, high engine rpm or a blocked air inlet will close the oil separator valve. This will prevent excessive crankcase pressure from damaging the crankcase seals.

MAINTENANCE REMINDER INDICATOR

Reset Procedure

1. The oxygen sensor system on the Volvo models is equipped with a warning indicator light. This light is located in the dash panel and will light up every 30,000 miles as a reminder that the oxygen sensor has to be replaced.

Resetting the oxygen sensor maintenance reminder light
(© AB Volvo Car Corp.)

2. Remove the old oxygen sensor from the exhaust manifold, apply some form of anti-sieze compound on the threaded part of the new oxygen sensor.

3. Install the new sensor into the exhaust manifold and be sure not to get any of the anti-sieze compound on the slotted part of the oxygen sensor.

4. Locate the mileage counter in-line with the speedometer cable. Remove the retaining screw and the switch cover, then push the reset button. Check to make sure that the reminder light in the dash is out, then replace the switch cover.

NOTE: The DL and GL models are equipped with an electrically heated sensor that does not require periodic servicing.

EVAPORATIVE EMISSION CONTROL SYSTEM

Evaporation Control System

The function of the evaporation control system is to prevent the emission of gas vapors from the fuel tank and carburetor to be expelled into the atmosphere. When fuel evaporates in the gas tank or the carburetor float chamber the vapors pass through the vent hoses to the charcoal canister where they are stored until they can be drawn into the intake manifold when the engine is running. All vehicles are equipped with the charcoal canister for the storage of fuel vapors. The fuel bowls of all carburetors are vented internally and on some applications do not require venting the canister. If this is the case the bowl vent port on the canister will be capped. Most carburetors are also externally vented to the charcoal canister.

System Service

The charcoal canister itself is a non-serviceable component. The only

FUEL EVAPORATION EMISSION CONTROLS

FUEL EVAPORATIVE SYSTEM DIAGNOSIS

Symptom	Possible Cause	Repair
Noticeable fuel odor or leaks	Damaged or loose lines, saturated canister, broken or leaking two-way valve or liquid/vapor separator	Repair lines, replace canister or two-way valve. Check liquid/vapor separator
Fuel tank deformed	Canister clogged, tank cap defective, two-way valve not relieving pressure or vacuum	Replace defective canister or cap, clear vapor lines
Insufficient fuel delivery or vapor lock	Blocked or collapsed vapor lines, two-way valve or canister. Blocked fuel feed line	Repair as necessary
Tank won't take fuel	Clogged or defective two-way valve	Replace two-way valve
Charcoal canister saturated with fuel	Defective thermosensor, purge control/unloader solenoid or diaphragm valve. Blocked or disconnected vacuum or vapor lines	Check operation of all FES components. Check all lines and connections. Replace charcoal canister

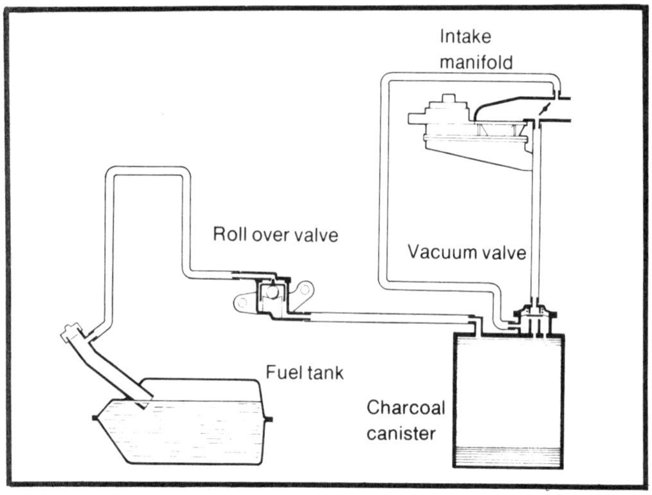

Typical fuel evaporative control system (© AB Volvo Car Corp.)

service required for the system is to replace the filter pad on the bottom of the canister. The filter requires replacement every 12 months or 12,000 miles. If the vehicle is driven under severe conditions, the filter should be replaced more often.

FUEL ROLLOVER VALVE

The fuelroll over valve is used to prevent fuel leaks should the vehicle suddenly roll over. The valve is connected in the fuel vapor line between canister and fuel tank. The fuel roll over valve contains a check ball and under normal conditions, the gasoline vapor passage in the valve is opened, but if a vehicle roll-over occurs the ball closes the fuel passage, thus preventing fuel leaks.

FUEL FILLER CAP

The fuel filler cap is equipped with relief valve to prevent the escape of fuel vapor into the atmosphere.

EXHAUST EMISSION CONTROL SYSTEMS

Exhaust Gas Recirculation System

EGR VALVE

D-24 TURBO DIESEL ONLY

This valve is used to reduce the amount of nitrous oxides (NOx) in the exhaust. Recirculating a certain amount of the exhaust gases reduces the combustion temperature which will reduce the emission of nitrous gases. The EGR system operates in proportion to the opening of the throttle lever. The controlling vacuum is taken from the engine vacuum pump, which also used for the brake operation.

THERMO VACUUM VALVE

This valve opens the controlling vacuum line when the engine coolant temperature has reached approximately 113 degrees F. This means that the EGR valve is inoperative on a cold engine. When the engine is cold, it will produce only small amounts of nitrous oxides and the amount of hydrocarbons and exhaust gas smoke would be increased if the EGR system operated.

IDLE SWITCH

This switch is controlled by a cam on the injection pump lever. Above the idle speed, the idle switch signals the EGR relay to cut out the

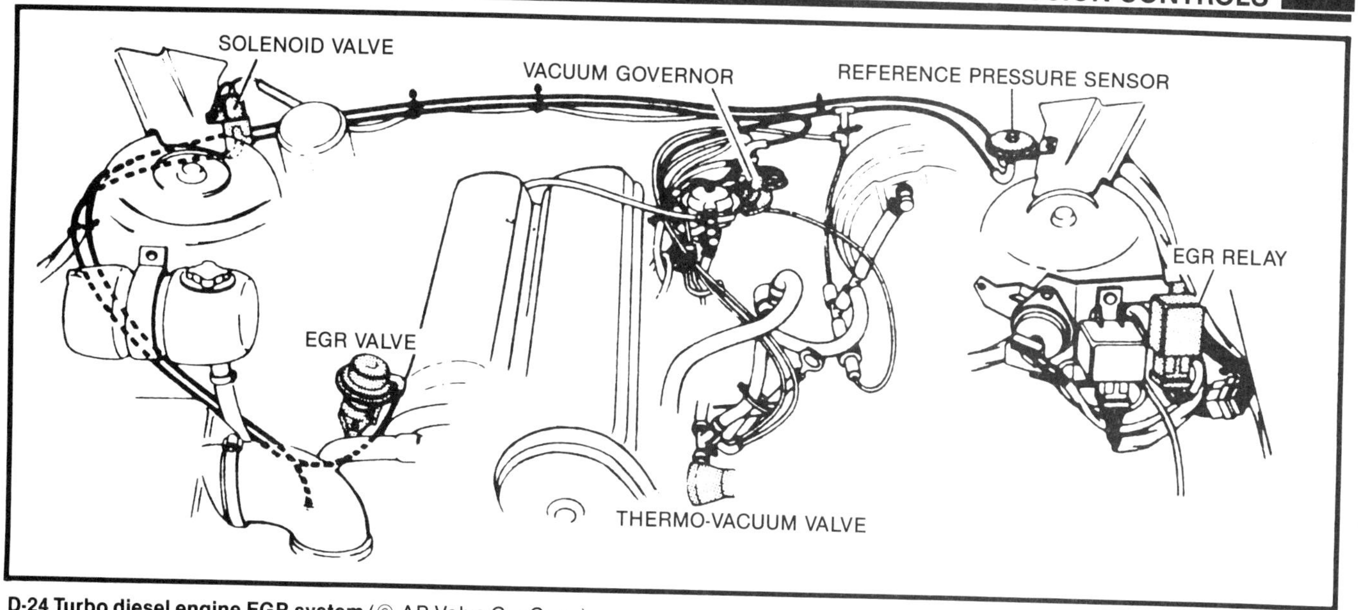

D-24 Turbo diesel engine EGR system (© AB Volvo Car Corp.)

control voltage to the solenoid valve for approximately 5 seconds. This will decrease smoke and hydrocarbon emissions when accelerating from idle.

VACUUM GOVERNOR

The vacuum governor is controlled by a cam on the injection pump lever and its function is to control the amount of vacuum going to the EGR valve. When the throttle opening is increased the vacuum is decreased. The EGR valve opening, and corresponding return of exhaust gases, is therefore reduced. The circuit through the switch incorporated in the vacuum governor is cut out when the control vacuum reaches approximately 2.8 psi. The solenoid valve voltage disappears and the EGR system is disengaged.

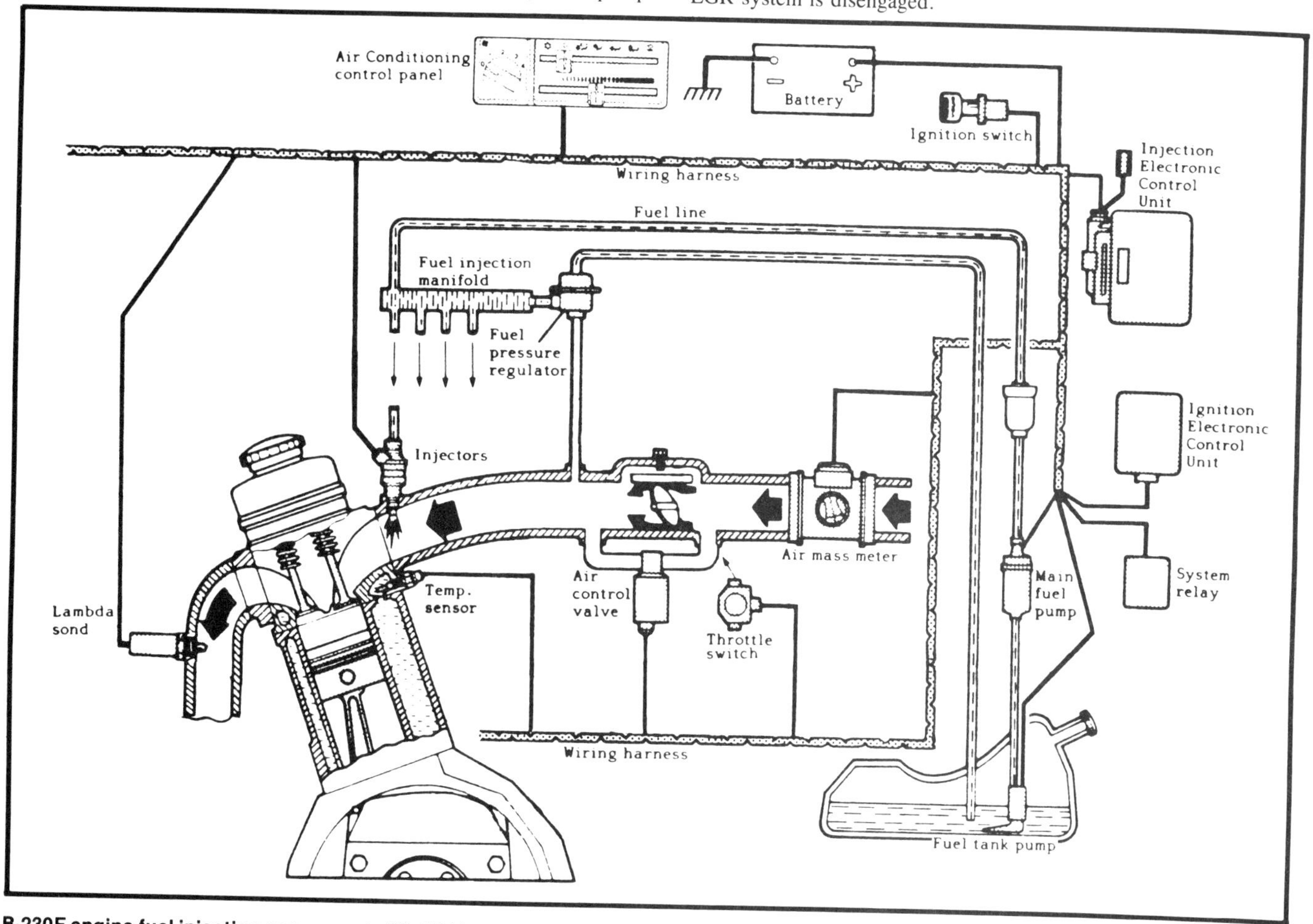

B-230F engine fuel injection components (© AB Volvo Car Corp.)

EGR SYSTEM DIAGNOSIS

Symptom	Possible Cause	Repair
EGR valve does not move during test	Leaking or clogged vacuum line or hose connection. Defective EGR valve	Repair vacuum leak, replace EGR valve
Engine won't idle or stalls on return to idle	EGR control system blocked or inoperative. Defective speed sensor	Check all vacuum connections and control valves. Replace defective control valves or speed sensor
EGR valve leaks in closed position. Poor wide-open throttle performance	EGR valve defective or blocked with carbon. Defective EGR control valve	Clean or replace EGR valve replace control valve
No vacuum to EGR valve	Thermosensor defective, leaking vacuum line	Replace thermosensor or repair vacuum hose connection

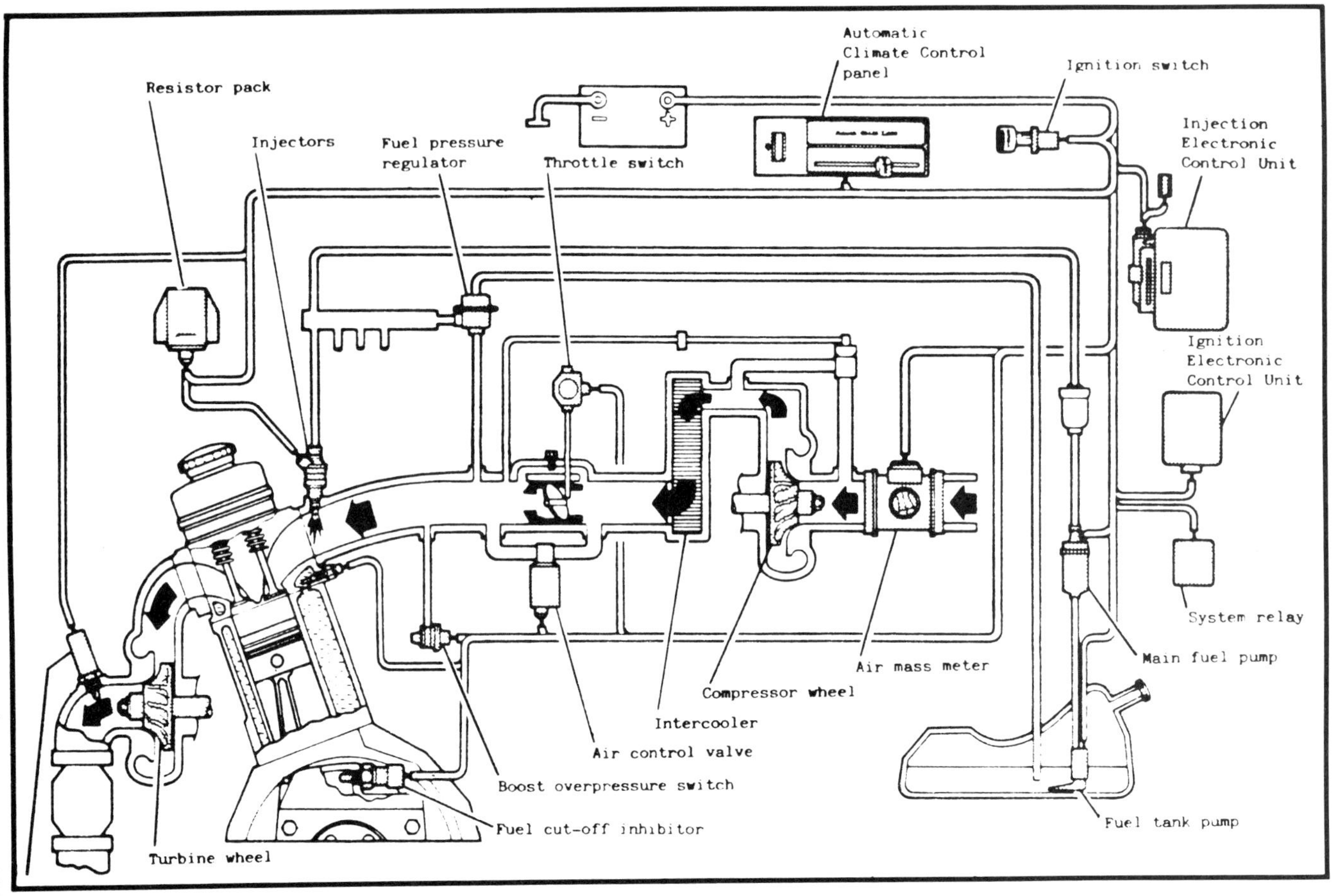

B-230F Turbo engine fuel injection components (© AB Volvo Car Corp.)

OVERDRIVE RELAY

The function of this relay is to cut out the voltage to the solenoid valve when the overdrive is engaged. The EGR system is then disengaged, because when driving in overdrive, the engine load is moderate and relatively constant. The emission of nitrous oxides is low and there is no need for the EGR valve.

Catalytic Converter (Gas Engines Only)

Catalytic converter require the use of unleaded fuel only. Leaded fuel will destroy the effectiveness of the catalysts as an emission control device. Under normal operating conditions, the converter will not re-

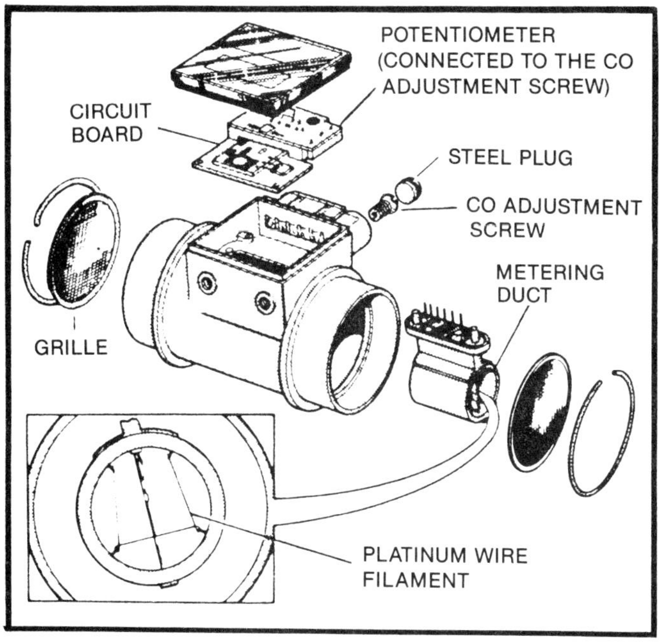

Typical air mass meter (© AB Volvo Car Corp.)

quire any maintenance. However, it is important to keep the engine properly tuned. If the engine is not kept properly tuned engine misfiring may cause overheating of the catalysts. This may cause heat damage to the converter or vehicle components. This situation can also occur during diagnostic testing if any spark plug cables are removed and the engine is allowed to idle for a long period of time.

Fuel Injection System

FUEL INJECTION COMPONENTS

B-230F AND B-230F TURBO ENGINES

FUEL INJECTION ELECTRONIC CONTROL UNIT (ECU)

This unit is located inside the passenger compartment for its own protection and is positioned in the front of the front right-side door. It contains a micro-processor which receives input signals from various sensors. This electronic control unit controls the engagement and disengagement of the system relay which supplies voltage to the control unit, injectors, and the other various sensors. It also controls fuel injection by varying the injector spray duration, and it controls the idle speed by varying the air control valve opening.

A main ignition ECU provides the injection ECU with essential information on the engine speed. At the extreme, the engine speed is used to cut-out injection at excessive speeds of 6200 rpm. The injection ECU provides the ignition ECU with information on the air mass meter position, the ignition ECU then takes this information and compares it with the engine speed to determined the engine load.

AIR MASS METER

The job of this meter is to measure the air flow mass and report it to the injection ECU. The basic sensing information is provided by a heated platinum wire filament. The air mass meter housing is constructed of plastic. The injection ECU on the B-230F turbo engine is different then the one on the B-230F engine.

Fuel Pump

There are two fuel pumps with this system, the first pump is a vane type pump mounted on the fuel tank gauge sender unit assembly. The job of this pump is to maintain pressure between the fuel tank and the main fuel pump. The main fuel pump is a roller type, pump which is mounted together with the fuel filter and mounted underneath the floor of the drivers seat. The job of this pump is to maintain pressure up to the fuel injection manifold.

INJECTION MANIFOLD

The purpose of this manifold is to distribute fuel to the fuel injectors, the fuel injectors are attached directly to the injection manifold.

PRESSURE REGULATOR

This regulator is located at the end of the injection manifold and is used to maintain the fuel pressure in the injection manifold at a constant 43 psi. above the intake manifold pressure (which the regulator is connected to). All of the excess fuel is drain off and back through a fuel return line to the fuel tank.

FUEL INJECTORS

The fuel injectors atomize and inject furl into the intake manifold. The injectors are controlled by electrical impulses from the injection ECU. The fuel nozzle is opened when the coil is energized. The duration of the injector opening determines the amount of fuel being injected. The injectors use O-rings as seals at the injection manifold and the engine intake manifold. The injectors are held in place on the injection manifold by a snap clamp around the base of the injector.

THROTTLE SWITCH

This switch is located on the throttle housing and provides indications of throttle position at idle to the injection ECU. The switch is a microswitch and reacts very precisely. Always adjust the throttle switch if the throttle plate has been adjusted. The throttle switch should click immediately after the throttle valve starts to open.

AIR CONTROL VALVE

Regulates the idle speed by increasing or decreasing the amount of air bypassing the throttle valve. A small electric motor rotates clockwise or counterclockwise, depending on signal from the injection ECU.

COOLANT TEMPERATURE SENSOR

This sensor is attached to a coolant channel in the engine head. It senses the coolant temperature and sends as appropriate signal to terminal 2 of the injection ECU.

KNOCK SENSOR

This sensor is located below the engine intake manifold and is really a piezoelectric transducer. The sensor has a sensing range of 13–15 kHz, but the ignition ECU only registers frequencies of approximately 7.5 kHz, which appear when the knock appears. When the sensor picks up an engine knock, the timing is retarded by 2.8 degrees and if the knocking still occurs the ECU retards the timing another 2.8 degrees until the knocking stops or the maximum retardation of 10–15 degrees is reached. After the knock has been eliminated, the timing is advanced in increments of 0.35 degrees until the knock starts again and then the process is repeated.

LIMP IN MODE

If anything should go wrong with the system a limp in mode retards the timing by 1–15 degrees. It is very important that the knock sensor

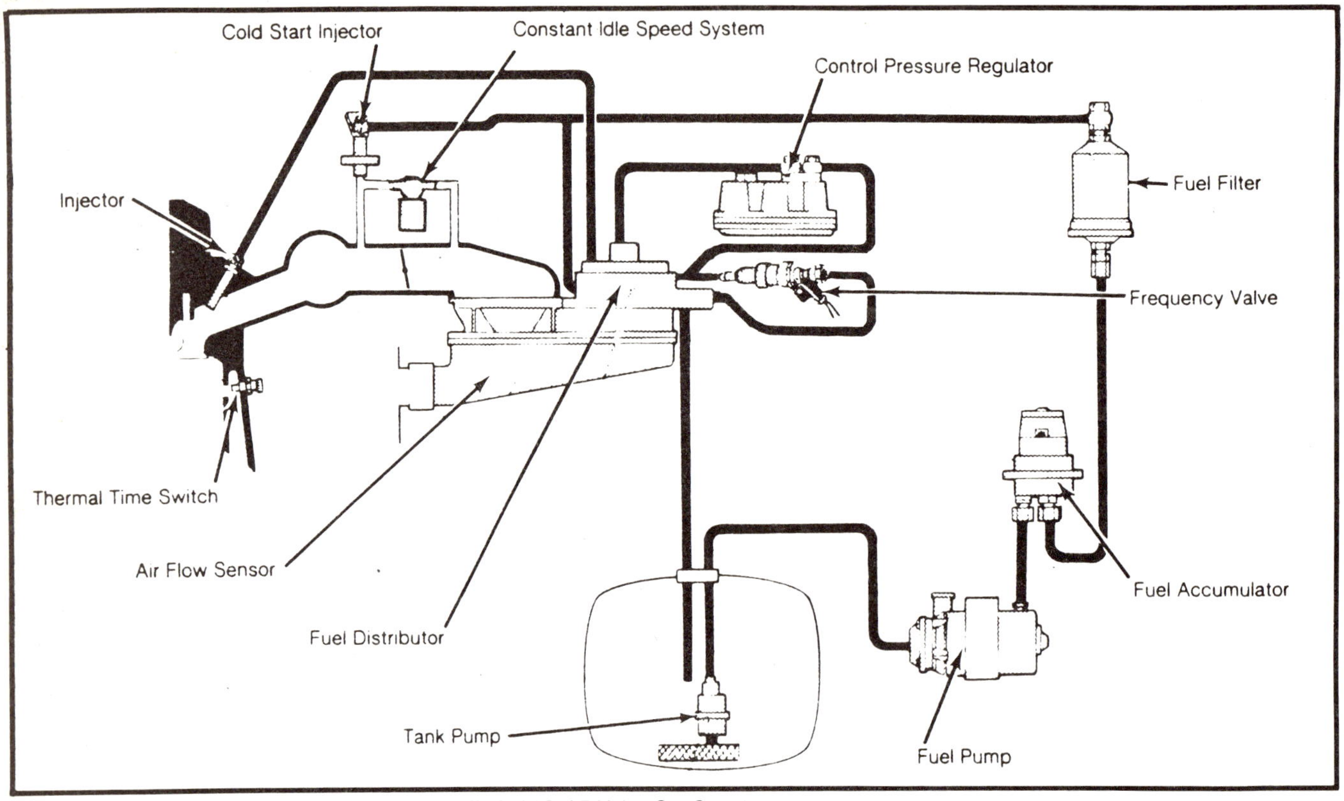

Volvo—CIS fuel system vacuum schematic (4 cylinder) (© AB Volvo Car Corp.)

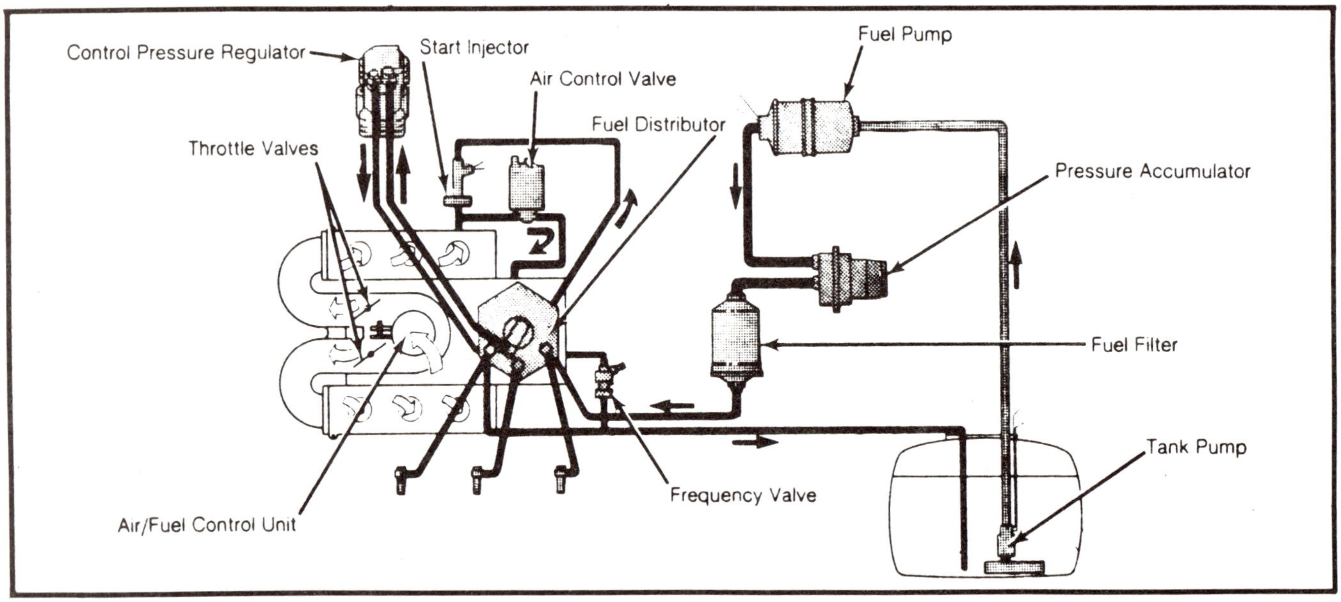

Volvo—B-28F vacuum schematic (© AB Volve Car Corp.)

does not work its way loose. Always use the correct bolt in the knock sensor and torque it down to specifications (use a thread sealer on the bolt).

OVERPRESSURE SWITCH

B-230F TURBO ENGINE ONLY

This switch is located on the firewall above the pedal carrier, it is a safety device, which deactivates the fuel pumps in case the turbocharger pressure should become too high. This could happen if the wastegate valve does not operate correctly. The engine stops momentarily but the overpressure switch resets when the pressure drops and then the vehicle can be driven, cautiously to be checked.

BALLAST RESISTOR PACK

Limits current supply to the injectors and protects the output control section of the injection ECU. The injectors are of high capacity with short duration opening and use copper wires with the low resistance in the injector solenoids. A pack of four ballast resistors protect the injectors from excessive current.